DNA Damage and Repair

Contemporary Cancer Research

DNA Damage and Repair, Volume 1: DNA Repair in Prokaryotes and Lower Eukaryotes, edited by **Jac A. Nickoloff** and **Merl F. Hoekstra,** 1998

DNA Damage and Repair, Volume 2: DNA Repair in Higher Eukaryotes, edited by **Jac A. Nickoloff** and **Merl F. Hoekstra,** 1998

DNA Damage and Repair

Volume 2: DNA Repair in Higher Eukaryotes

Edited by

Jac A. Nickoloff

University of New Mexico, Albuquerque, NM

and

Merl F. Hoekstra

Signal Pharmaceuticals, San Diego, CA

Humana Press Totowa, New Jersey

999 Riverview Drive, Suite 208
Totowa, New Jersey 07512

This publication is printed on acid-free paper. ∞
ANSI Z39.48-1984 (American Standards Institute) Permanence of Paper for Printed Library Materials.

Cover design by Patricia F. Cleary.

For additional copies, pricing for bulk purchases, and/or information about other Humana titles, contact Humana at the above address or at any of the following numbers: Tel: 973-256-1699; Fax: 973-256-8341; E-mail: humana@humanapr.com, or visit our Website: http://humanapress.com

Printed in the United States of America. 10 9 8 7 6 5 4 3 2 1

Library of Congress Cataloging in Publication Data

DNA damage and repair/edited by Jac A. Nickoloff and Merl F. Hoekstra
p. cm.—(Contemporary cancer research)
Includes index.
Contents: v. 1. DNA repair in prokaryotes and lower eukaryotes—v. 2 DNA repair in higher eukaryotes.
ISBN 0-89603-356-2 (v.1: alk. paper.—ISBN-89603-500-X (v.2: alk. paper)
1. DNA repair. I. Hoekstra, Merl F. II. Series.
[DNLM: 1. DNA Repair. 2. DNA Damage. 3. DNA—physiology. 4. Prokaryotic Cells. 5. Eukaryotic Cells. QH 467 D629 1998]
QH467.D15 1998
572.8'6—dc21
DNLM/DLC
for Library of Congress 97-28562
CIP

Preface

DNA Damage and Repair grew from a conversation between the editors at the 1994 Radiation Research Meeting in 1994. At that time, Errol Friedberg, Graham Walker, and Wolfram Seide were just releasing the second edition of the outstanding textbook *DNA Repair and Mutagenesis,* and many of the human genes involved in nucleotide excision repair were being described and their connection to xeroderma pigmentosum and trichothiodystrophy was being outlined. Many of these mammalian genes had homologs in model systems and investigators were making connections between NER and transcription. This was an exciting and confusing time. Nomenclature for radiation repair genes lacked consistency and new genes (and even pathways) were being described in many organisms.

The result of our conversation was the decision ask experts working with various organisms and in subdisciplines of DNA repair whether a volume reviewing these major advances might be helpful. The resounding answer was yes, quickly followed by the question of why anyone would want to take on this task. We took on the challenge with the goal of collecting peer-reviewed chapters on a variety of repair topics, from a variety of organisms, written by experts in the field. It has been quite a learning experience about the challenges of editorship and the breadth of DNA repair.

We thank all of the authors for their timely submissions. We also owe a great debt to the many individuals who made suggestions about our initial list of chapter titles and/ or reviewed and corrected chapters, including Brenda Andrews, David Ballie, James Cleaver, Pricilla Cooper, Elena Hildago, Lawrence Grossman, Philip Hanawalt, John Hays, Etta Kafer, Robert Lahue, Richard Kolodner, David Lilley, Michael Liskay, Lawrence Loeb, Kenneth Minton, Rodney Rothstein, Roy Rowley, Leona Samson, Alice Schroeder, Gerry Smith, Michel Sicard, Wolfram Siede, and Graham Walker. A comparison of this list with the table of contents shows that many of these individuals did double-duty, both as a chapter contributor and as reviewer. We owe special thanks to Phil Hanawalt for providing an excellent overview of the field and for his continuing encouragement. We are in debt to these individuals for their time and assistance, without which this volume would not have been possible.

During the formation of the book, a variety of exciting breakthroughs occurred and we are pleased that a number of authors were able to make last-minute changes to include these discoveries. In particular, the fields of mismatch repair, cell-cycle checkpoints, and DNA helicases have grown significantly by the discovery of disease genes, such as *MLH, MSH,* and *ATM,* and the genes involved in Bloom's and Werner's syndromes. Unfortunately we were not able to cover all topics and regret that we could not dedicate individual chapters for such organisms as bacteriophage and *Drosophila* and on such processes as X-ray crystallography/NMR and the impact this has on DNA repair. Nevertheless, we are grateful to all that have participated in this effort: It started modestly at 25–30 chapters, but rapidly grew to two volumes, a testament to the explosive growth that the field has enjoyed in recent years.

Finally, we thank our families for their patience and understanding: Denise, Jake, Ben, Courtney, Debra, Brad, Lauren and Brielle.

Jac A. Nickoloff
Merl F. Hoekstra

Contents

Contents for the Companion Volume: DNA Damage and Repair, Volume 1: DNA Repair in Prokaryotes and Lower Eukaryotes

Contributors

CARL W. ANDERSEN • *Biology Department, Brookhaven National Laboratory, Upton, NY*
SEAN BAKER • *Department of Molecular and Medical Genetics, Oregon Health Sciences University, Portland, OR*
NATHAN A. BERGER • *Department of Medicine and Cancer Research Center, Case Western Reserve University, Cleveland, OH*
JEFF BESTERMAN • *Department of Cell Biology, Glaxo Research Institute, Research Triangle Park, NC*
PAUL C. BILLINGS • *Department of Pathology, School of Dental Medicine, University of Pennsylvania, Philadelphia, PA*
DAVID A. BOOTHMAN • *Department of Human Oncology, University of Wisconsin, Madison, WI*
SATADAL CHATTERJEE • *Department of Medicine and Cancer Research Center, Case Western Reserve University, Cleveland, OH*
JINGWEN CHEN • *Department of Cell Biology, Glaxo Research Institute, Research Triangle Park, NC*
GILBERT CHU • *Departments of Medicine and Biochemistry, Stanford University Medical Center, Stanford, CA*
PATRICK CONCANNON • *Virginia Mason Research Center, Seattle, WA*
MICHAEL N. CORNFORTH • *Department of Radiation Therapy, The University of Texas Medical Branch, Galveston, TX*
ALAN D'ANDREA • *Dana-Farber Cancer Research Institute, Boston, MA*
THOMAS W. DAVIS • *Department of Human Oncology, University of Wisconsin, Madison, WI*
ANTHONY DIPPLE • *Chemistry of Carcinogenesis Laboratory, ABL-Basic Research Program, NCI-Frederick Cancer Research and Development Center, Frederick, MD*
SCOTT K. DURUM • *Laboratory of Molecular Immunoregulation, National Cancer Institute, Germantown, MD*
JEAN-MARC EGLY • *Institut de Génétique et de Biologie Moléculaire et Cellulaire, Illkirch, France*
ANDRÉ P. M. EKER • *Department of Cell Biology and Genetics, MGC-Erasmus University, Rotterdam, The Netherlands*
PAOLA GALLINARI • *IRBM, Pomezia, Italy*
RICHARD GATTI • *UCLA School of Medicine, Los Angeles, CA*

MYRIAM GOROSPE • *Gene Expression and Aging Section, National Institute on Aging, Baltimore, MD*
PHILIP C. HANAWALT • *Department of Biological Sciences, Stanford University, Stanford, CA*
HIROSHI HAYAKAWA • *Department of Biochemistry, School of Medicine, Kyushu University, Fukuoka, Japan*
MERL F. HOEKSTRA • *Signal Pharmaceuticals, San Diego, CA*
NIKKI J. HOLBROOK • *Gene Expression and Aging Section, National Institute on Aging, Baltimore, MD*
EDWARD N. HUGHES • *Department of Radiation Oncology, Kettering Medical Center, Kettering, OH*
INTISAR HUSAIN • *Department of Cell Biology, Glaxo Research Institute, Research Triangle Park, NC*
P. A. JEGGO • *MRC Cell Mutation Unit, University of Sussex, Brighton, UK*
JOSEF JIRICNY • *IRBM, Pomezia, Italy*
KUMKUM KHANNA • *Queensland Institute of Medical Research, Herston, Australia*
TIMOTHY J. KINSELLA • *Department of Human Oncology, University of Wisconsin, Madison, WI*
MARTIN LAVIN • *Queensland Institute of Medical Research, Herston, Australia*
LEONORA J. LIPINSKI • *Chemistry of Carcinogenesis Laboratory, ABL-Basic Research Program, NCI-Frederick Cancer Research and Development Center, Frederick, MD*
R. MICHAEL LISKAY • *Department of Molecular and Medical Genetics, Oregon Health Sciences University, Portland, OR*
YUSEN LIU • *Gene Expression and Aging Section, National Institute on Aging, Baltimore, MD*
A-LIEN LU • *Department of Biochemistry and Molecular Biology, University of Maryland, Baltimore, MD*
DAVID B. MANSUR • *Department of Radiation and Cellular Oncology, University of Chicago Hospitals, Chicago, IL*
MARK MEYERS • *Department of Human Oncology, University of Wisconsin, Madison, WI*
VINCENT MONCOLLIN • *Institut de Génétique et de Biologie Moléculaire et Cellulaire, Illkirch, France*
PETRA NEDDERMANN • *IRBM, Pomezia, Italy*
RUTH NETA • *Office of International Health Programs, Department of Energy, and Radiation Oncology Branch, National Cancer Institute, Germantown, MD*
JOHN L. NITISS • *Molecular Pharmacology Department, St. Jude Children's Research Hospital, Memphis, TN*
PEGGY L. OLIVE • *Medical Biophysics Department, British Columbia Cancer Research Centre, Vancouver, BC, Canada*
RUSSELL O. PIEPER • *Department of Medicine and Pharmacology, Loyola University Medical Center, Maywood, IL*

Tomas A. Prolla • *Department of Molecular and Human Genetics and Howard Hughes Medical Institute, Baylor College of Medicine, Houston, TX*
W. Kimryn Rathmell • *Departments of Medicine and Biochemistry, Stanford University Medical Center, Stanford, CA*
Roy Rowley • *Division of Radiation Oncology, Department of Radiation Oncology, University of Utah Medical Center, Salt Lake City, UT*
Navneet Sharda • *Department of Human Oncology, University of Wisconsin, Madison, WI*
Mutsuo Sekiguchi • *Department of Biology, Fukuoka Dental College, Fukuoka, Japan*
Rakesh K. Singhal • *Sealy Center for Molecular Science, University of Texas Medical Branch at Galveston, TX*
Michael J. Smerdon • *Department of Biochemistry and Biophysics, Washington State University, Pullman , WA*
Fritz Thoma • *Institut für Zellbiologie, Eidgenössische Technische Hochschule, Zürich, Switzerland*
Larry H. Thompson • *Biology and Biotechnology Research Program, Lawrence Livermore National Laboratory, Livermore, CA*
Alan E. Tomkinson • *Institute of Biotechnology, Center for Molecular Medicine, The University of Texas Health Science Center at San Antonio, TX*
Paul Vichin • *Institut de Génétique et de Biologie Moléculaire et Cellulaire, Illkirch, France*
John F. Ward • *Department of Radiology, University of California at San Diego, La Jolla, CA*
Corry M. R. Weemaes • *University of Neimegen, The Netherlands*
Ralph R. Weichselbaum • *Department of Radiation and Cellular Oncology, University of Chicago Hospitals, Chicago, IL*
Samuel H. Wilson • *Laboratory of Structural Biology, National Institute of Environmental Health Sciences, Research Triangle Park, NC*
Carmell Wilson-Van Patten • *Department of Human Oncology, University of Wisconsin, Madison, WI*
Chin-Rang Yang • *Department of Human Oncology, University of Wisconsin, Madison, WI*
Akira Yasui • *Department of Neurochemistry and Molecular Biology, Institute of Development, Aging, and Cancer, Tohoku University, Sendai, Japan*

1
Overview

Philip C. Hanawalt

In recent years, the field of DNA repair has attained widespread recognition and interest appropriate to its fundamental importance in genomic maintenance. An essential set of repair pathways must be operative in all living systems to maintain genomic stability in the face of the natural endogenous threats to DNA as well as those resulting from cellular exposures to radiation and chemicals in the external environment. The intrinsic chemical lability of the DNA molecule poses a formidable threat to its persistence and even the polymerases that replicate DNA occasionally make mistakes that must be rectified. Fortunately, the inherent redundancy of the genetic message assured by the two complementary strands of the duplex DNA molecule facilitates the recovery of information through excision repair when one of the strands is damaged, or when incorrect base pairings or small loops of unpaired bases are present. The fundamental research that led to the discovery of excision repair in *Escherichia coli* in the early 1960s has now developed to the point that we realize that DNA repair is ubiquitous and essential for life. Furthermore, we have learned that DNA repair interfaces in some manner with each of the other cellular DNA transactions, including replication, transcription, recombination, and regulation of the cell cycle. We are also finding remarkable sequence and functional homologies among the essential repair enzymes as different species are compared. This has spurred the rate of discovery, most remarkably through the parallel analyses in yeast and mammalian systems in which the complex multicomponent systems for excision repair are virtually identical.

The collection of chapters in this remarkable two-volume treatise attests to the explosive rate at which the DNA repair field is evolving. Recently published texts in this field are in need of updating as soon as they hit the bookshelves. Fortunately, the cutting edge researchers who have contributed chapters in the immediate area of their expertise are in a good position to anticipate significant upcoming developments in their respective subfields. The unique perspectives of researchers attacking similar problems using diverse approaches and different biological systems constitute another important strength of the present collection of chapters. For general background and an introduction to the field of DNA repair and mutagenesis, the reader is referred to the comprehensive treatment by Friedberg et al. *(5)*. An excellent set of current reviews covering excision repair *(15,18)*, mismatch repair *(13)*, and relationships between repair and transcription *(4)* can be found in the 1996 volume of *Annual Reviews of Biochemistry*.

From: DNA Damage and Repair, Vol. 2: DNA Repair in Higher Eukaryotes
Edited by: J. A. Nickoloff and M. F. Hoekstra © Humana Press Inc., Totowa, NJ

In this short overview, I would like to step back and peruse the broad field of DNA repair to ask: What is left to learn? What are some of the important new directions in which the field is moving—or should be moving? What is the significance of this field for human health? I will attempt to address just a few of these questions to provoke thought and debate as the reader approaches the many facets of the subject covered in the following comprehensive chapters.

To set the stage, we should begin with a cursory review of the sorts of deleterious alterations that can occur in DNA and the various classes of mechanisms by which those alterations can be repaired or accommodated in the cell. Nearly twenty years ago, my colleagues and I published a comprehensive review on "DNA repair in bacteria and mammalian cells" for *Annual Reviews of Biochemistry (7)*. At that time most of the cellular pathways for repairing or tolerating DNA damage were already known but the detailed enzymology had not been elucidated. We now know essentially all of the proteins required for each of the DNA repair pathways and for most of them we can also assign them unique roles *(5,13,15,18)*. However, there is still much to be learned about how the damaged DNA substrate is recognized.

We define damage as "any modification of DNA that alters its coding properties or its normal function in replication or transcription." Thus, some minor base modifications or base replacements might simply alter the coding sequence, whereas other types of damage, such as bulky chemical adducts or UV-induced cyclobutane pyrimidine dimers, may distort the DNA structure and interfere with the translocation of polymerases. To be repairable, the damage must be recognized by a protein that can initiate a sequence of biochemical reactions leading to its elimination and the restoration of the intact DNA structure. It is a formidable challenge to the lesion-recognition enzymes to detect the damaged sites in the context of the variable and dynamic structure of the normal DNA molecule. The normal DNA structure includes nucleotide-sequence-dependent bends and kinks, as well as unique secondary structures, such as Z-DNA. The DNA in cells is also decorated with a variety of ligands, including tightly bound proteins that may alter its intrinsic structural features and that would be expected to encumber any simple damage recognition system. The repair machinery must rigorously avoid mistaking a normal DNA–protein complex for damaged DNA. The detection machinery must also be able to distinguish a bona fide lesion from a normal variant of the DNA structure. An additional yet related problem is raised by the requirement that the first step in the excision-repair sequence involves cutting the DNA. It could be potentially disastrous if there were many individual nucleases diffusing freely about the cell in search of generalized structural distortions in DNA. The nucleotide excision-repair mechanism employs multienzyme complexes rather than single proteins to recognize broad classes of structure-distorting lesions. Thus, in the *E. coli* system the individual polypeptides encoded by the *uvrA*, *uvrB*, and *uvrC* genes are not independently operating nucleases. Lesion recognition requires a complex interaction between UvrA and UvrB. UvrB is loaded onto the DNA at the lesion site, and only then does the UvrC protein (the limiting element with only a dozen copies per cell) join the UvrB-DNA complex to unleash cryptic nuclease activities to produce dual incisions bracketing the lesion to initiate the excision repair process.

In the case of base excision repair, the initiation steps are simpler and evidently more specific. The general mechanisms appear to require the swinging out of the altered or incorrect base from the DNA helix so that its detailed dimensions and charge structure can be assessed in a "pocket" of the relevant enzyme. Such a recognition scheme also operates in the direct repair of cyclobutane pyrimidine dimers by photolyase. In that case the catalytic cofactor that splits the dimer is found in the pocket and is activated by light once the dimer has been bound. It is likely that the photon-catalyzed reversal of pyrimidine dimers *in situ* was one of the earliest repair mechanisms to emerge as life evolved on the primordial earth. The intense flux of ultraviolet light from the sun, unattenuated by an ozone layer, must have been the predominant threat to the survival of primitive life forms.

Some types of damage to DNA may not be recognized as such and may have no deleterious consequences. Thus, phosphotriesters in mammalian DNA do not appear to be subject to repair and do not appear to have any adverse effects for the cell. On the other hand, in bacteria the phosphotriesters are repaired and are also utilized to warn the cell of the presence of DNA alkylating agents that cause them so that an adaptive response to upregulate repair of alkylated DNA is triggered. A number of the enzymatic pathways for dealing with environmental threats to cellular DNA appear to be similarly inducible whereas, not surprisingly, a constitutive level is maintained of those enzymes needed to repair endogenous damage to DNA.

The principal endogenous or so-called spontaneous DNA lesions are caused by deamination of cytosine to uracil (or methyl cytosine to thymine), loss of purines to yield abasic sites, and reactive oxygen species that produce strand breaks directly but also at least 20 different sorts of base damage *(10,12)*. Notable are thymine glycols and 8-oxo-dG, which are repaired by the base excision repair pathway. However, it is important to remember that the other sorts of base damage as well as the strand breaks may also contribute to the biological consequences. Some potential problems are dealt with at the nucleotide level, even before incorporation into DNA might occur from precursor pools. Thus, the dUTP levels are modulated by dUTPases, and a major product of endogenous reactive oxygen species, 8-oxo-dGTP, is converted to the benign 8-oxo-dGMP by a dedicated phosphatase. The mere existence of the latter pathway attests to the need to control the level of damage inflicted by endogenous reactive oxygen species.

One of the most important excision-repair schemes is mismatch repair, which reduces the error rate in cellular DNA replication by three orders of magnitude. Destabilization of tracts of simple repetitive DNA occurs in bacteria defective in mismatch repair genes. There also may be important interactions between mismatch repair and other repair pathways, notably nucleotide excision repair. Deficiency in mismatch repair has been shown to attenuate transcription-coupled repair in human cells as well as in bacteria. The importance of these connections is highlighted by the implication of the human MutS and MutL mismatch-repair homologs in hereditary nonpolyposis colorectal cancer. The enhanced tumorigenesis is likely caused by the resultant microsatellite instability when mismatch repair is defective *(13)*.

Among the more impressive successes in recent years have been the development of cell-free systems from bacteria, yeast, and mammalian cells that carry out all of the essential steps in excision repair; indeed, such soluble DNA repair factories have now been reconstituted from purified proteins *(15,18)*. These successes have spawned models in which the requisite proteins are assembled into large, unwieldy complexes termed "repairasomes" that presumably diffuse through the cell in search of defective sites in the genomic DNA. This would be analogous to taking the lumber company into the woods in search of trees to cut. Models are usually drawn to show repair enzymes diffusing up to (or along) the DNA duplex. It is notable, however, that these soluble in vitro biochemically reconstituted repair reactions are remarkably inefficient, as are many other such reconstituted systems, including those for transcription. It is likely, indeed certain, that in the intact cell these transactions are carried out by spatially organized arrays of enzymes on surfaces (e.g., the nuclear matrix) and in subcellular compartments. These features could enhance the efficiency and certainty of repair by reducing a three-dimensional diffusion search for lesions to two dimensions, and/or by concentrating particular repair events to those genomic domains in which the repair process is essential to normal cellular functions. Recent studies have hinted that on recognition of lesions, the damaged DNA may be recruited to the nuclear matrix, where the repair factories may reside *(8)*. The damaged DNA once repaired would then be released from these sites (Fig. 1). An important emerging area for research is the elucidation of the intracellular localization of repair events.

We have learned a lot in recent years about the complex relationships between DNA damage and biological endpoints, such as mutagenesis, cell death, and tumorigenesis. We have learned that mutagenesis is under genetic control—thus, for example, ultraviolet (UV) light is not mutagenic in *E. coli* deficient in the *umuC* gene, evidently because almost no translesion DNA synthesis occurs. We have also been surprised to learn that cells carrying damaged genomes may choose to commit suicide, a process technically termed apoptosis, although severely damaged cells may be nonviable in any case. Apoptosis embodies a concept that is difficult to understand in the context of freely living cells and seemingly at odds with the idea of cellular competition for survival. Why should a single free-living cell exhibit such seemingly altruistic behavior? The answer is probably that apoptosis is a process that has evolved to promote development and maintenance of tissues in multicellular organisms; and that in fact it has no relevance to cells in culture. When a tissue is damaged, such as skin from intense solar exposure, the tissue must be remodeled. The dead and damaged cells must be removed so that a new epidermal layer can be assembled. The pathways to apoptosis are complex and under genetic control in damaged cells just as in the programmed cell death that occurs during normal embryonic development.

Susceptibility to cancer can be the consequence of a complex interplay between intrinsic hereditary factors and persisting damage to DNA. We need to learn the nature of these interactions as well as the genetic defects that confer enhanced risk. In xeroderma pigmentosum (XP) the increased risk of skin cancer correlates with a defect in nucleotide excision repair (NER). In Cockayne syndrome (CS) a specific defect in the subpathway of transcription-coupled DNA repair (TCR) does not predispose the patients to the sunlight-induced skin cancer characteristic of XP. The TCR pathway is targeted to lesions in the transcribed strand of expressed genes that arrest the translocation of RNA polymerase. The demonstration of TCR in rodent cells, which are generally deficient in the global genomic DNA-repair pathway, indicates that UV resistance correlates with repair of cyclobutane pyrimidine dimers in expressed genes rather than with global NER. The other major lesion produced by short wavelength UV light, the pyrimidine (6-4) pyrimidone photoproduct, is more efficiently recognized by excision

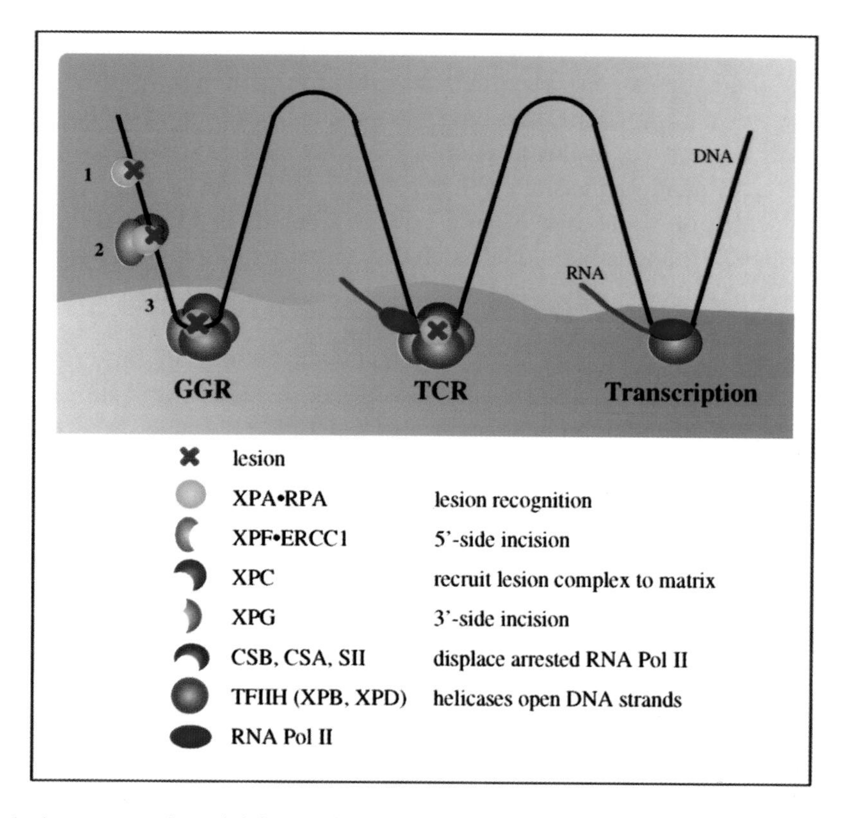

Fig. 1. A conceptual model for the localization of nucleotide excision repair in the nuclear matrix. Transcription occurs in association with the nuclear matrix, as shown, and requires a number of factors for its initiation, including TFIH, which is also essential for excision repair. In the model we assume that TFIIH is localized in the matrix to carry out its dual responsibilities in transcription and repair. When transcription is arrested at a lesion, various factors are required to transiently displace the polymerase and facilitate access of repair enzymes, including those that recognize the lesion and those that produce incisions in the damaged strand to initiate repair. As shown, the XPC protein is not required for transcription-coupled repair (TCR) but it is essential for global genomic repair (GGR). For GGR, we postulate that lesion recognition takes place in regions not associated with the matrix (Step 1), but that additional proteins are then employed (Step 2) to recruit the damaged DNA segments to "repair factory" sites in the nuclear matrix (Step 3) to complete the repair process. Once repaired, these genomic regions would then be released from the matrix. (Figure prepared by Graciela Spivak).

repair enzymes than is the cyclobutane pyrimidine dimer, such that the TCR of this lesion is usually masked by efficient global repair. It is important to consider genomic locations of DNA damage for a meaningful assessment of the biological importance of particular DNA lesions *(6)*.

Mutations in the p53 tumor suppressor genes occur in most human tumors. In the cancer-prone Li-Fraumeni syndrome, fibroblasts expressing only mutant p53 exhibit less apoptosis and are more UV resistant than are normal human cells. The p53-defective cells are deficient in global NER but have retained TCR. The loss of p53 function may lead to genomic instability by reducing the efficiency of global NER, whereas cellular survival may be assured through the operation of TCR and the elimination of apoptosis *(3)*. A specific role for the wild-type p53 gene product in the induction of global NER and apoptosis has been demonstrated in p53 homozygous mutant fibroblasts into which an exogenous wild-type p53 gene has been introduced under control of a tetracycline (Tet)-repressible promoter. Withdrawal of Tet results in the induction of wild-type p53 gene expression, transcriptional activation of p21, and cell-cycle checkpoint activation. Induction of wild-type p53 results in the recovery of normal levels of global NER and enhanced UV-induced apoptosis, compared to the NER-deficient p53 mutant cells *(3a)*.

The arrest of transcription at unrepaired lesions may be a signal for p53 stabilization and apoptosis *(11,19)*. In the case of CS, deficient in the removal of transcription blocking lesions, apoptosis is induced by very low UV doses, thus eliminating many cells that might otherwise undergo oncogenic transformation. Of course, dead cells do not form tumors. Those CS cells that survive are fully proficient in global genome repair. This may explain the lack of sunlight-induced tumors. In contrast, the cells from xeroderma pigmentosum, complementation group C, are proficient in TCR, so survival is high; unfortunately, these cells are defective in global genome repair so a high level of mutation may be induced. Cells from CS patients have also been shown to be deficient in TCR of certain lesions, including thymine glycol, produced by ionizing radiation *(9)*. Ionizing radiation produces lesions largely through oxidation by free radicals that attack DNA. The same types of free radicals are generated endogenously as byproducts of oxidative metabolism in nonirradiated cells. The characteristic developmental problems in CS may be the consequence of inappropriate apoptosis initiated by endogenous DNA damage in expressed genes *(1,6)*.

The importance of the TCR pathway to normal human development is highlighted by the clinical features of Cockayne syndrome patients *(14)*. On the other hand, the clinical consequences of xeroderma pigmentosum, except for the cases in which CS is also present, are not as severe; with the notable exception of the adverse response to sunlight. The lack of a major predisposition to internal cancers in XP patients assures us that there is not a substantial amount of carcinogenic DNA damage from endogenous sources that requires the intervention of nucleotide excision repair for its amelioration. However, the revelation that some rare lesions resulting from reactive oxygen species are subject to nucleotide excision repair has led to the suggestion that such lesions, if unrepaired, may cause the neurological degeneration that typically accompanies later development in the most severe XP patients *(16)*.

It is important that we continue to analyze DNA damage processing in so-called simple bacterial model systems. There have been some recent surprises, such as the discoveries that the genes defective in Bloom's syndrome and in Werner's syndrome in humans are both homologs of the *E. coli recQ* gene that was originally isolated as a gene that, when mutated, confers resistance to the lethality caused by thymidine starvation (i.e., thymineless death). RecQ has been shown to be a helicase and, in fact, there are three homologues of the *recQ* gene in humans. Adding to the intrigue is the fact that in *E. coli* RecQ is implicated in the very minor *recF* recombination pathway. Recent studies suggest that an important—if not the most important—role of *recF* is in enabling the resumption of DNA replication at arrested replication forks, such as at lesions and following their repair *(2)*. The situation of a blocked DNA polymerase complex at a lesion may be quite analogous to that of a blocked transcription complex. The polymerase encumbers access of repair enzymes to the lesion until it is displaced; furthermore, the DNA strands must be reannealed at the lesion site before excision repair can take place. It is likely that basic studies of the process of DNA polymerase displacement, repair, and replication restart in *E. coli* will help us to understand how the process might operate in mammalian cells.

New genes implicated in DNA repair continue to surface as genomes are sequenced and cancer susceptibility genes are characterized. Thus, the BRCA1 and BRCA2 genes that predispose to breast cancer, if either one is inherited in mutated form, have been shown to interact with RAD51, so they both may be involved in the same DNA-repair pathway *(17)*. Mouse embryos lacking BRCA2 are extremely sensitive to ionizing radiation. Once again, a possible implication is that endogenous DNA damage caused by reactive oxygen species may be the culprit in the mutation cascade from normal cells to tumors. In the realm of risk assessment it is becoming increasingly important to determine the relative contributions to DNA damage from endogenous vs environmental factors. Such understanding should lead to new approaches to cancer prevention.

REFERENCES

1. Cooper, P. K., T. Nouspikel, S. G. Clarkson, and S. A. Leadon. 1997. Defective transcription-coupled repair of oxidative base damage in Cockayne syndrome patients from XP group G. *Science* **275,** 990–993.
2. Courcelle, J., C. Carswell-Crumpton, and P. C. Hanawalt. 1997. *recF* and *recR* are required for the resumption of replication at DNA replication forks in *Escherichia coli*. *Proc. Natl. Acad. Sci. USA* **94,** 3714–3719.
3. Ford, J. M. and P. C. Hanawalt. 1995. Li-Fraumeni syndrome fibroblasts homozygous for p53 mutations are deficient in global DNA repair but exhibit normal transcription-coupled repair and enhanced UV-resistance. *Proc. Natl. Acad. Sci. USA* **92,** 8876–8880.

3a. Ford, J. M. and P. C. Hanawalt. 1997. Expression of wild-type p53 is required for efficent global genomic nucleotide excision repair in UV-irradiated human fibroblasts. *J. Biol. Chem.*, in press.

4. Friedberg, E. C. 1996. Relationships between DNA repair and transcription. *Annu. Rev. Biochem.* **65,** 13–42.
5. Friedberg, E. C., G. Walker, and W. Siede. 1995. DNA repair and mutagenesis. ASM, Washington, DC.
6. Hanawalt, P. C. 1994. Transcription-coupled repair and human disease. *Science* **266,** 1957–1958.
7. Hanawalt P. C., P. K. Cooper, A. K. Ganesan, and C. A. Smith. 1979. DNA repair in bacteria and mammalian cells. *Ann. Rev. Biochem.* **48,** 783–836.
8. Koehler, D. R. and P. C. Hanawalt. 1996. Recruitment of damaged DNA to the nuclear matrix in hamster cells following ultraviolet irradiation. *Nucleic Acids Res.* **24,** 2877–2884.
9. Leadon, S. A. and P. Cooper. 1993. Preferential repair of ionizing radiation-induced damage in the transcribed strand of an active human gene is defective in Cockayne syndrome. *Proc. Natl. Acad. Sci. USA* **90,** 10,499–10,503.
10. Lindahl, T. 1993. Instability and decay of the primary structure of DNA. *Nature* **362,** 709–715.
11. Ljungman, M., and F. Zhang. 1996. Blockage of RNA polymerase as a possible trigger for U. V. light-induced apoptosis. *Oncogene* **13,** 823–831.
12. Marnett, L. J. and P. C. Burcham. 1993. Endogenous DNA adducts: potential and paradox. *Chem. Res. Tox.* **6**:771–785.
13. Modrich, P. and R. Lahue. 1996. Mismatch repair in replication fidelity, genetic recombination, and cancer biology. *Ann. Rev. Biochem.* **65,** 101–33.

13a. Mullenders, L. H. F., A. C. V. van Leeuwen, A. A. van Zeeland, and A. T. Natarajan. 1988. Nuclear matrix associated DNA is preferentially repaired in normal human fibroblasts. *Nucleic Acids Res.* **16,** 10,607–10,622.
14. Nouspikel, T., P. Lalle, S. A. Leadon, P. K. Cooper, and S. G. Clarkson. 1997. A common mutational pattern in Cockayne syndrome patients from xeroderma pigmentosum group G: implications for a second XPG function. *Proc. Natl. Acad. Sci. USA* **94,** 3116–3121.
15. Sancar, A. 1996. DNA excision repair. *Ann. Rev. Biochem.* **65,** 43–81.
16. Satoh, M. S., C. J. Jones, R. D. Wood, and T. Lindahl. 1993. DNA excision-repair defect of xeroderma pigmentosum prevents removal of a class of oxygen free radical-induced base lesions. *Proc. Natl. Acad. Sci. USA* **90,** 6335–6339.
17. Sharon S. K., M. Morimatsu, U. Albrecht, D.-S. Lim, et al. 1997. Embryonic lethality and radiation hypersensitivity mediated by Rad51 in mice lacking *Brca2*. *Nature* **386,** 804–810.
18. Wood, R. D. 1996. DNA repair in eukaryotes. *Ann. Rev. Biochem.* **65,** 135–167.
19. Yamaizumi, M. and T. Sugano. 1994. UV-induced nuclear accumulation of p53 is evoked through DNA damage of actively transcribed genes independent of the cell cycle. *Oncogene* **9,** 2775–2784.

2
DNA Photolyases

Akira Yasui and André P. M. Eker

1. INTRODUCTION

Photoreactivation was discovered by chance in 1949 during a study of UV-induced mutagenesis when unexpected results were obtained that could be attributed to a greatly enhanced survival of UV-irradiated *Streptomyces griseus* conidia after illumination with visible light *(38)*. A similar phenomenon was found for the survival of UV-irradiated bacteriophages in *Escherichia coli (11)*.

Photoreactivation is a very potent repair mechanism: a 10^4 to 10^5-fold increase in cell survival has been reported *(18)*. Photoreactivation has been demonstrated in a wide variety of organisms, ranging from simple bacteria to multicellular eukaryotes. It is not a purely photophysical reaction, but is mediated by enzymes, termed photoreactivating enzymes or photolyases (EC 4.1.99.3). Numerous attempts have been made to isolate photolyase from natural sources. Only in a few instances a pure photolyase was obtained, mainly because of the very low content in cells. At present, photolyases are obtained mainly by recombinant techniques.

The substrate of photolyase is UV-irradiated DNA. UV irradiation induces the formation of cyclobutane pyrimidine dimers (CPDs) from two neighboring pyrimidine bases (Fig. 1). Treatment with photolyase and visible light removes dimers *(105)* by monomerization into the constituent pyrimidines *(9)*, i.e., a reversion of the dimerization reaction. Photoreactivation follows a simple Michaelis-Menten reaction scheme (Fig. 2) *(76)*. The enzyme first binds (in the dark) to UV-damaged DNA to form an enzyme–substrate complex. Then, in a light-dependent step, the dimer is split and photolyase dissociates from the repaired DNA.

In contrast to other repair pathways like nucleotide excision repair (NER), which are complex multistep processes and require multiple enzymes, photoreactivation is mediated by a single enzyme. Moreover, no energy-rich factors, like ATP or nucleotide triphosphates, are used. In nature, the energy for the reaction is provided by the same source (the sun) that introduced the lesions in DNA.

Recently, important advances have been made in photolyase research. Several photolyases have been purified and characterized. The crystal structure of *E. coli* photolyase was determined *(71)*, and the mechanism of photoreactivation was elucidated in great detail *(28)*. Photolyases of several animals have been isolated, and the

From: DNA Damage and Repair, Vol. 2: DNA Repair in Higher Eukaryotes
Edited by: J. A. Nickoloff and M. F. Hoekstra © Humana Press Inc., Totowa, NJ

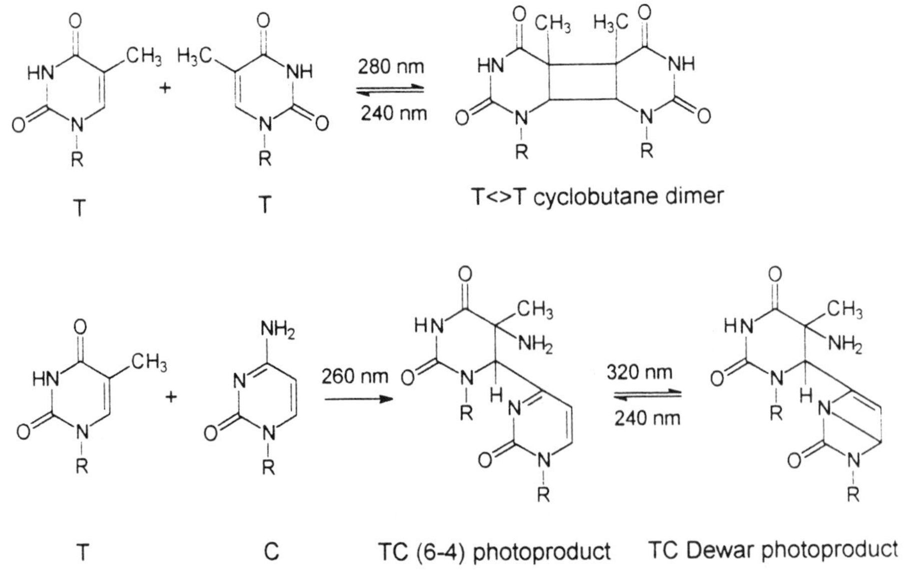

Fig. 1. UV-induced lesions in DNA. Two main lesions are induced by UV-irradiation: CPDs and (6-4)PDs, both formed from two neighboring pyrimidine bases. The most abundant lesions, T<>T CPD and TC (6-4)PD, are depicted. CPDs can be reversed (monomerized) by short wavelength irradiation, giving the same net result as enzymatic photoreactivation. Irradiation with 320 nm light converts (6-4)PDs into the Dewar isomer, which is reversed to (6-4) PD by short wavelength irradiation.

$$E + T{<>}T \underset{K_a}{\rightleftharpoons} E \cdot T{<>}T \xrightarrow{h\nu} E + TT$$

complex formation — photolysis

Fig. 2. Michaelis-Menten reaction scheme for photoreactivation. The photolyase–substrate complex is formed in the dark. In the subsequent photolysis step, T<>T CPD is split on illumination with visible or near-UV light.

presence of a possible human homolog analyzed *(91,100)*. Also, a new type of photolyase specific for pyrimidine(6-4)pyrimidone photodimers [(6-4)PDs], another major UV-induced lesion in DNA, has been reported *(98,100)*. This chapter describes recent developments in the field, emphasizing photoreactivation in higher eukaryotes.

2. BACTERIAL AND LOWER EUKARYOTE CPD PHOTOLYASES

2.1 Chromophores in Photolyases

The properties of some bacterial and lower eukaryote photolyases are shown in Table 1. All are single-chain proteins with molecular weights of 50–65 kDa. A common characteristic is the presence of two different chromophoric cofactors. The first, found in all known photolyases, is reduced FAD. For the second chromophore, two different structures are known: 5,10-methenyl tetrahydrofolate (MTHF) and 8-hydroxy-

Table 1
Properties of Bacterial and Lower Eukaryote Photolyases[a]

Organism	Mol wt kDa	Absorption band λ_{max}, nm	Absorption band ε	Chromophore ratio	Redox state FAD	Ref.
MTHF-type photolyases						
E. coli	49	382	27,000	0.6–0.79	Semiquinone	*20,26,59*
S. typhimurium	54	384	29,500	n.d.	Semiquinone	*54*
B. firmus	50	410	20,000	n.d.	Semiquinone	*61*
S. cerevisiae	57	381	27,800	1.06	Reduced	*20,84*
N. crassa	66	391	34,800	1.02	oxidized	*20*
8-HDF-type photolyases						
S. griseus	49	443	44,000	1.05	n.d.	*15*
A. nidulans	53	437	53,000	0.97	Semiquinone	*17*
S. acutus	56	437	52,200	0.94	n.d.	*16*

[a]The table comprises properties of highly purified photolyases. The molecular weights were determined by SDS-PAGE. Chromophore ratio: mol second chromophore (MTHF or 8-HDF)/mol FAD. n.d.: not determined.

5-deazaflavin (8-HDF) (Fig. 3). The photolyase chromophores have different functions and share a functional homology with the photosynthetic apparatus. MTHF and 8-HDF, which have high molar extinction coefficients (ε), efficiently harvest light energy (photoantenna), and pass this to the reduced FAD chromophore (low molar extinction coefficient), which acts as the reaction center in dimer splitting. The spectral properties of photolyases are mainly determined by the second chromophore. 8-HDF type photolyases have maximum absorption in the 435–445 nm region (ε = 45,000–55,000), whereas MTHF photolyases absorb at 380–410 nm (ε = 25,000–30,000) (Fig. 4). Photoreactivation action spectra are very similar to the absorption spectra. Comparison of both types indicates that 8-HDF is more efficient than MTHF photolyase, since both the molar extinction coefficient and the quantum yield of photoreactivation are higher (*see* Section 2.3.2.), whereas the absorption spectrum has a better overlap with the emission spectrum of the sun (Fig. 4).

Most photolyases contain an equimolar amount of both chromophores (Table 1). Sometimes however, the second chromophore is spontaneously lost. *E. coli* photolyase becomes partly depleted of MTHF during purification, yielding molar ratios of 0.3–0.79 MTHF/FAD *(20,26,59)*. A similar problem is encountered with *S. griseus* photolyase, which gradually releases its 8-HDF chromophore *(12)*. Moreover, the MTHF chromophore in *E. coli* photolyase is destroyed on illumination with visible/near-UV light *(27)*. Other photolyases, e.g., from *Anacystis nidulans*, *Saccharomyces cerevisiae*, and *Neurospora crassa*, are apparently more stable.

Totally or partly depleted photolyases can be reconstituted with their chromophores. In vitro incubation of depleted *E. coli* photolyase with synthetic MTHF restores the correct chromophore ratio *(103)*. Overproduction of 8-HDF photolyases in *E. coli* yields photolyases completely devoid of 8-HDF chromophore *(94)*. Since *E. coli* cells do not synthesize 8-OH-5-deazaflavins MTHF photolyases do not suffer from this problem. Subsequent incubation of, for example, depleted *A. nidulans* photolyase with synthetic 8-HDF chromophore (F0; *see* Fig. 3) gives efficient and complete reconstitution *(19,60,63)*. The binding site of the second chromophore is specific: MTHF-type

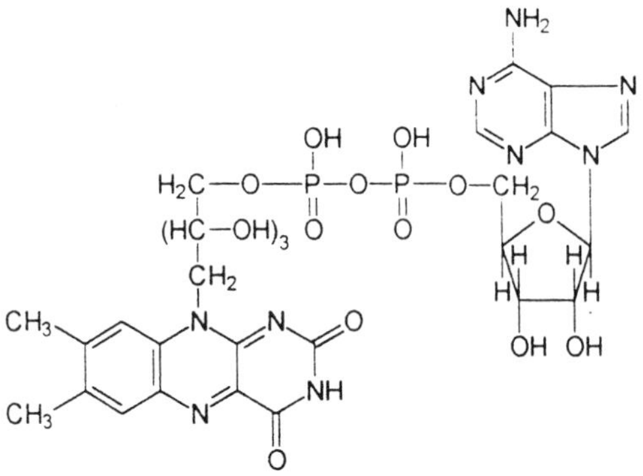

FAD

1st chromophore

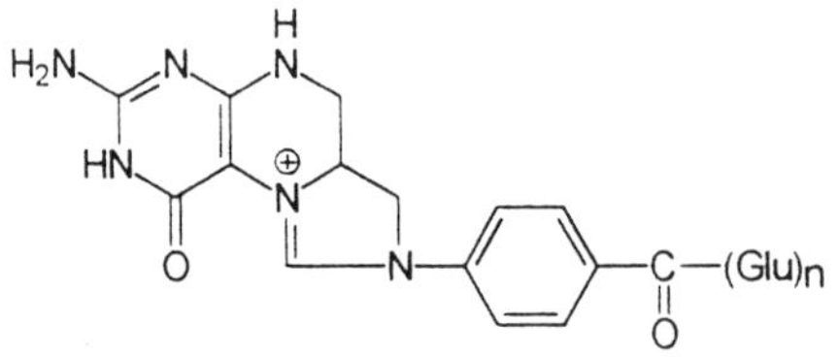

5,10-methenyl-tetrahydrofolate (MTHF)

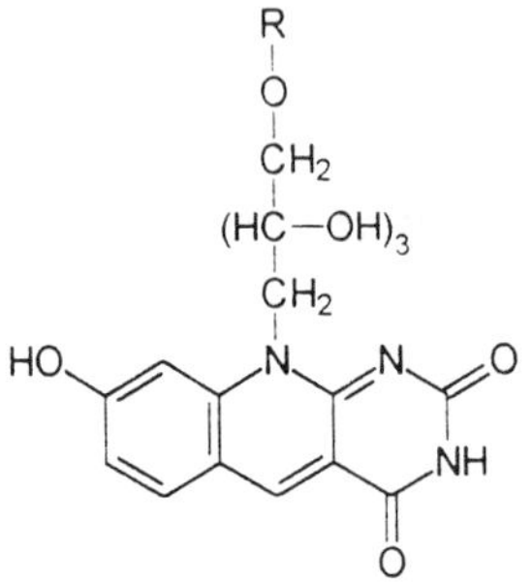

8-OH-5-deazaflavin (8-HDF)

FO : R = H

2nd chromophore

Fig. 3. Structure of photolyase chromophores. DNA photolyases contain two chromophores. The first is invariably FAD (in various redox states). The second chromophore is either a reduced folate (MTHF) or a 8-OH-5-deazaflavin (8-HDF) *(13)*. MTHF chromophores with a variable poly-glutamate tail (n = 3–6) are present in *E. coli* and yeast photolyases *(34,102)*. *A. nidulans* and *Scenedesmus acutus* photolyases contain 8-OH-5-deazariboflavin (F0, R = H) *(16,17)*.

photolyase cannot be reconstituted with 8-HDF and vice versa *(19,36,73)*. Depletion of both 8-HDF/MTHF and FAD chromophores yields an inactive enzyme incapable of substrate binding, but reconstitution with chromophore pairs is possible *(36,60)*.

2.2. Substrate Specificity

Photolyases catalyze monomerization of CPDs in DNA. Early work *(89)* indicated that T<>T CPDs are split more rapidly than T<>C or C<>T dimers, which in turn are split more efficiently than C<>C dimers. Using defined substrates, these differences can be analyzed more precisely in terms of affinity (equilibrium association constant, K_a) and photochemical efficiency (quantum yield, ϕ).

Both 8-HDF and MTHF photolyases bind tightly to UV-irradiated DNA ($K_a \sim 3 \times 10^8 M^{-1}$) compared to unirradiated DNA ($K_a \sim 10^3$–$10^4 M^{-1}$) (*see* Table 2) *(14,54,60,61)*.

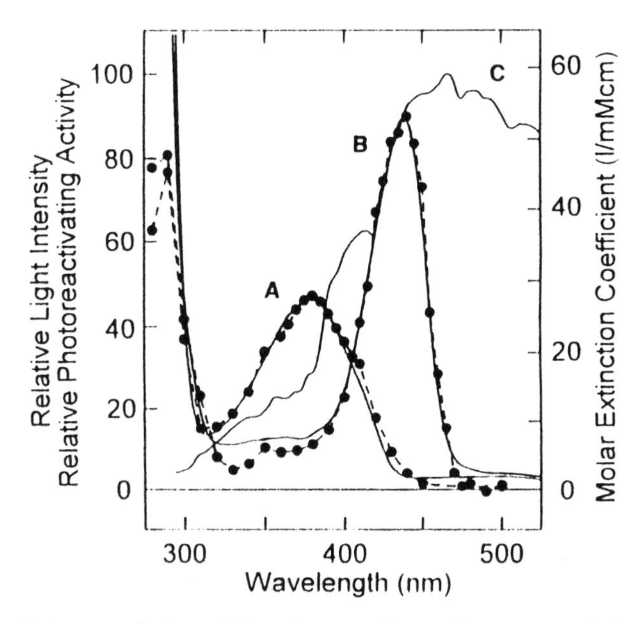

Fig. 4. Spectral characteristics of photolyases. Absorption spectra (—) of highly purified *S. cerevisiae* (MTHF type, curve A) *(20)* and *A. nidulans* (8-HDF type, curve B) *(17)* show a high degree of overlap with the action spectra of photoreactivation (• - - •). The action spectrum is corrected for inactivation below 310 nm. For comparison, the emission spectrum of the sun at earth surface is also shown (curve C).

Table 2
Complex Constant K_a and Overall Quantum Yield ϕ for Pyrimidine Dimer Splitting in Various Photolyase–Substrate Combinations

Organism	Type of photolyase	Substrate	K_a, M^{-1}	ϕ	Ref.
S. griseus	8-HDF	UV DNA	4×10^8	1	*14,15*
		Unirradiated DNA	6×10^4		*14*
A. nidulans	8-HDF	ssDNA with c,s T<>T	2×10^8	1	*60*
S. typhimurium	MTHF	dsDNA with c,s T<>T	6×10^8	0.5	*54*
B. firmus	MTHF	dsDNA with c,s T<>T	2×10^9	0.75, 0.9[a]	*61*
E. coli	MTHF	dsDNA with c,s T<>T	3×10^8	0.59, 0.88[a]	*31,41,74*
		dsDNA without dimer	3×10^3		*31*
		dsDNA with t,s T<>T	$\sim 10^4$	+	*46*
		UV-$(dT)_{200}$	5×10^7	0.91[a]	*41*
		UV-$(dT)_{12-18}$	5×10^7	0.92[a]	*41*
		UV-$(dC)_{12-18}$	6×10^6	0.05[a]	*41*
		UV-$(dU)_{200}$	2×10^7	0.60[a]	*41*
		UV-$(rU)_{200}$	8×10^2	0.54[a]	*41*
		dT<>dT	2×10^4	0.89[a]	*41*
		T<>T	$<10^3$	+	*41*
		cis-platinum treated DNA	2×10^7	–	*68*

[a]For photolyase depleted of MTHF chromophore: since the energy transfer MTHF → FADH2 is omitted, a higher value is found compared to nondepleted photolyase. +: dimer splitting detected, –: no dimer splitting

No significant difference is found between double- and single-stranded UV-treated DNA. Photolyase is highly specific for *cis/syn* dimers: the binding to *trans/syn* T<>T containing DNA is similar to undamaged DNA, although dimer splitting is detectable *(46)*. DNA chain length is not critical: a lower activity was found only for short oligonucleotides ($n < 8$) *(35)*, whereas the efficiency of the photochemical reaction is not affected *(41)*. The presence of the sugar-phosphate chain is not an absolute requirement, since *E. coli* photolyase is able to split single T<>T dimers. Binding to C<>C containing DNA is rather strong, but the quantum yield is greatly reduced *(41)*.

Recently, it was found that photolyase also binds to non-CPD lesions, such as inter- and intrastrand crosslinks in DNA *(23)*. For the *cis*-platinum GG diadduct, the K_a was only 10-fold lower than for a *cis/syn* T<>T dimer. Illumination with photoreactivating light had no influence *(68)*. The binding to non-CPD lesions is probably important for a second function of photolyases as an auxiliary factor in NER (*see* Section 5.).

Although there are reports of photoreactivation of RNA in plants *(30)* and insects *(33)*, no RNA-specific photolyase has yet been isolated. *E. coli* DNA photolyase, however, is able to split U<>U dimers in RNA with slightly reduced quantum yield although binding is much weaker compared to DNA *(41)*.

In addition to CPDs, UV irradiation introduces other lesions in DNA, of which (6-4)PDs (*see* Fig. 1) are most important. Previously photolyases were considered to be specific for CPDs *(5)*. Recently, however, photoreactivation of (6-4)PDs has been described, mediated by specific (6-4)photolyases as discussed in Section 4.

2.3. Reaction Mechanism

2.3.1. Photoactivation of Photolyase

In purified photolyases, three different redox states of FAD have been found. Most photolyases contain the half-reduced form (blue semiquinone radical FADH·), but oxidized (in *N. crassa*) and fully reduced (in *S. cerevisiae*) forms are also found (Table 1). The half-reduced and oxidized forms are probably artifacts formed by slow oxidation during purification. In vivo the FAD chromophore is most likely in the fully reduced form *(72)*, which is the only redox state of FAD that is active in dimer splitting.

Illumination with visible light activates photolyase by converting bound FAD or FADH· into $FADH_2$ (or rather its anion $FADH^-$) before or during the photoreactivation process. Absorption of 400–600 nm light generates an excited state of reduced FAD, which abstracts an electron from a neighboring tryptophan residue in the photolyase protein (W306 in *E. coli* photolyase; *45,55)* to form fully reduced $FADH^-$ (Fig. 5). The presence of an exogenous reducing agent, such as DTT or 2-mercaptoethanol, is crucial in this process. The intrinsic property of photolyases to restore the active, fully reduced form of FAD by photoreduction is essential for in vitro photoreactivation experiments.

2.3.2. Photochemical Mechanism of Dimer Splitting

Experiments with photosensitizers as a model for enzymatic photoreactivation indicate that dimers can be split by either abstraction or donation of an electron. Examples are the efficient splitting of T<>T by anthraquinone-β-sulfonate (by electron abstraction) *(52)* and the less efficient monomerization by indole (tryptophan) derivatives (by electron donation) *(29)*. Because of the high overall quantum yield (0.5–1), an electron abstraction (rather than donation) mechanism was suggested for enzymatic

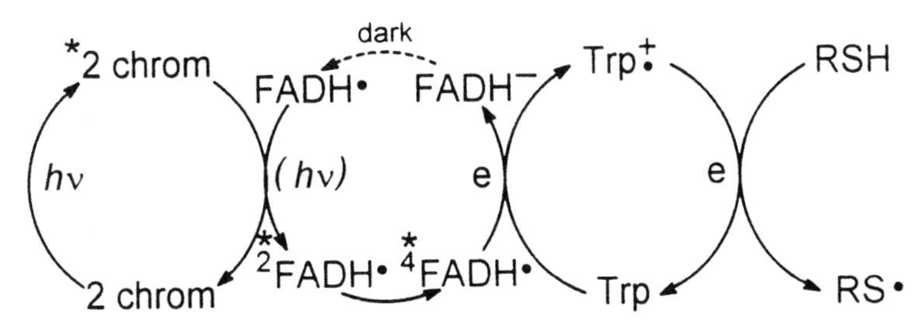

Fig. 5. Reaction scheme for photoactivation of photolyase *(28)*. Most photolyases contain FADH·, the half-reduced (blue semiquinone radical) form of FAD, which is inactive in dimer splitting. Illumination with 400–600 nm light produces the $*^2$FADH· excited doublet state, which is rapidly converted into the $*^4$FADH· lowest excited quartet state, which in turn abstracts an electron from a neighboring tryptophan in the photolyase protein yielding fully reduced $FADH^-$. Formation of excited FADH· is also possible through energy transfer from the excited second chromophore (*2 chrom: either MTHF or 8-HDF). The tryptophan radical cation $Trp^{+\cdot}$ is reduced by an exogenous reductor RSH in the solvent. When no such reductor is present, activation is unsuccessful because an electron is rapidly transferred back with formation of FADH· and Trp. In vitro, the fully reduced form $FADH^-$ is gradually oxidized in the presence of oxygen.

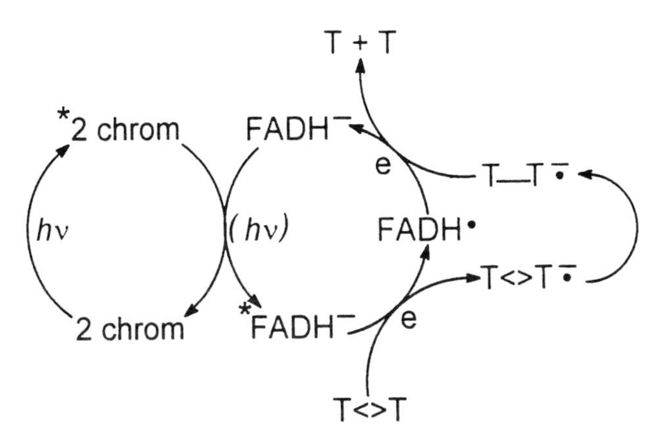

Fig. 6. Photochemical reaction mechanism of dimer splitting *(28,81)*. Absorption of a photon (300–500 nm) generates the excited singlet state of the second chromophore (*2 chrom: either MTHF or 8-HDF), which in turn generates by energy transfer the excited state of fully reduced FAD ($*FADH^-$). This species donates an electron to the bound T<>T CPD, yielding the radical anion $T<>T^{-\cdot}$. This species is unstable and one of the cyclobutane bonds is disrupted yielding $T—T^{-\cdot}$. Next, an electron is abstracted by the semiquinone radical FADH· with concomitant breakage of the remaining cyclobutane bond yielding two pyrimidines and fully reduced $FADH^-$.

photoreactivation. However, covalent linking of dimer substrate to indole sensitizers, together with the use of an apolar solvent, greatly increased the quantum yield of dimer splitting by electron donation in model systems ($\phi = 0.4$) *(40)*.

Ample evidence indicates that dimer splitting in enzymatic photoreactivation occurs through electron donation (for reviews, *see 28,81*). Absorption of visible or near-UV light by photolyase generates the excited state of the second chromophore (Fig. 6), which in turn generates by energy transfer the excited state of the fully reduced anion $FADH^-$. This species donates an electron to a pyrimidine dimer bound to photolyase. Disruption of one of the bonds of the cyclobutane ring of the dimer is followed by back-donation of an electron from dimer to FAD chromophore with breakage of the remaining bond of the cyclobutane ring, yielding two pyrimidines and $FADH^-$. Direct excitation of reduced FAD chromophore is also effective, since 8-HDF- or MTHF-depleted photolyase is active in dimer splitting.

Photoreactivation is very efficient. For 8-HDF-type photolyases an overall quantum yield of approx 1 has been found (Table 2) *(15,60)*, indicating that every photon absorbed by a photolyase/dimer complex leads to dimer splitting. Slightly lower quantum yields of 0.5–0.75 are found for MTHF-type photolyases *(41,54,61)*. The overall quantum yield is determined by the efficiency of energy transfer from excited second chromophore to reduced FAD, which is approx 0.6 for MTHF *(36,42)* compared to approx 1 for 8-HDF *(44)*, and the efficiency of electron transfer (0.9–1) from excited reduced FAD to the dimer. In MTHF-depleted photolyase, only the second step occurs and a higher overall quantum yield of 0.8–0.9 is found compared to complete photolyase (Table 2).

Action spectra of photoreactivation (Fig. 4) indicate a high rate of photoreactivation around 280 nm, which cannot be attributed to the activity of chromophoric cofactors only *(15)*. This suggests that the protein part of photolyase could be photochemically active. Indeed it was found that a conserved tryptophan (W277 in *E. coli* photolyase), which also is involved in substrate binding *(53)*, can act as a chromophore for efficient dimer splitting by electron transfer *(43)*.

2.3.3. *Structure of* E. coli *Photolyase*

Photolyases of *E. coli* and *A. nidulans* have been crystallized *(63,70,96)*, and recently the three-dimensional crystallographic structure of *E. coli* photolyase was determined *(71)*. Given this structure, a number of reasonable explanations for previously obtained biochemical data are possible. The 471 amino acid polypeptide chain of *E. coli* photolyase is folded into two domains: an amino-terminal α/β domain and a carboxyl-terminal domain, which are interconnected by a 72 aa loop. The MTHF chromophore resides in a cleft between the two domains. The catalytic and essential FAD chromophore is bound to the protein in a U-shaped conformation in the center of the helical domain. Light energy absorbed by MTHF is transferred to FAD over a distance of 16.8 Å. The putative substrate binding site is on the flat surface of the helical domain along a trace of positive electrostatic potential running across the surface. Near the middle of the surface, there is a hole in which the pyrimidine dimer substrate is presumably caught and approaches FAD to allow van der Waals contact.

This model, in which the pyrimidine dimer is flipped out from the DNA helix, is similar to the structures obtained for *Hha*I DNA cytosine-5-methyltransferase *(49)* and uracil-DNA glycosylase *(87)*. The proposed binding to the indicated surface is supported by the effect of substitution of conserved amino acids in *S. cerevisiae* photolyase *(4)*. Further details of substrate binding and the repair process must await the structural analysis of enzyme–substrate cocrystals.

3. CPD PHOTOLYASES FROM HIGHER EUKARYOTES

3.1. Photolyase Genes from Higher Eukaryotes

Photolyase activity has been demonstrated in cell extracts of a large number of higher eukaryotes *(77)*. The first photolyase gene from a multicellular organism was isolated from cultured goldfish (*Carassius auratus*) cells by complementing the photolyase deficiency of *phr*⁻ *E. coli* host cells with a goldfish cDNA library *(110)*. The cloned gene encodes a protein that has only limited amino acid sequence similarity to previously isolated microbial photolyases. Using this functional complementation method, photolyase genes were also independently isolated from *Drosophila melanogaster (99,115)* and another fish species, *Oryzias latipes (115)*. Overexpression of this photolyase gene after transfection provided *O. latipes* cells with increased photoreactivating activity, but microinjected embryos failed to show enhanced photoreactivation *(25)*. The deduced aa sequences of these photolyase genes are quite similar to each other, but differ from those of microorganisms. The sequence conservation among these genes made it possible to isolate homologous genes from two established cell lines of aplacental mammals, the South American opossum *Monodelphis domestica* and the rat kangaroo *Potorous tridactylis (37,115)*. Both marsupial genes provided *E. coli* mutant cells with photolyase activity.

There are contradicting reports on the photolyase activity in human cells. Recently the presence of photoreactivating activity in human white blood cells was reported *(92)*, but another report provided evidence that HeLa cells and human white blood cells do not possess any photolyase activity, thus casting doubt on the existence of a human photolyase gene *(56)*. A cDNA probe of the cloned rat kangaroo photolyase gene does not hybridize to human genomic DNA, whereas the probe does hybridize to genomic DNA of fish *(115)*. Furthermore, a cDNA probe derived from the opossum photolyase gene does not hybridize to genomic DNA or polyA$^+$ mRNA from various placental mammals *(37)*. Although the data suggest that, in contrast to aplacental mammals, the placental mammals, including humans, may not possess photolyase activity owing to the absence of a photolyase gene, a human photolyase-like gene has recently been isolated *(91,100)* (*see* Section 4.).

3.2. Comparison of Higher Eukaryote and Microbial Photolyase Genes

Table 3 lists the presently known photolyase genes. Based on differences and similarities in the deduced amino acid sequences, these genes fall into two classes *(115)*. Class I comprises microbial and lower eukaryote photolyases, whereas insect, fish, and marsupial photolyases fall into class II, suggesting this class is specific for higher eukaryotes. However, a photolyase gene isolated from the archaebacterium *Methanobacterium thermoautotrophicum* has a deduced amino acid sequence that is quite different from microbial photolyases, but very similar to those of higher eukaryotes *(115)*. Furthermore, a gene was isolated from the Gram-negative bacterium *Myxococcus xanthus*, which is also highly similar to photolyases of higher eukaryotes and provides *E. coli* host cells with photolyase activity *(67)*. Both genes clearly belong to class II, indicating that class II photolyases are distributed not only in animals, but also in archae and eubacteria, suggesting an ancient origin (*see* Section 8.).

From the alignment of deduced sequences shown in Fig. 7, it is obvious that all photolyases share a core of approx 500 amino acids. Eukaryote photolyases

Table 3
Cloned Photolyase Genes

Organism		Length aa[a]	Ref.
Class I MTHF-type CPD photolyases			
E. coli	Bacterium	472	*82,111*
S. typhimurium	Bacterium	473	*54*
B. firmus	Bacterium	485	*61*
S. cerevisiae	Yeast	565[b]	*111,83*
N. crassa	Fungus	615/640[b]	*106*
Class I 8-HDF-type CPD photolyases			
H. halobium	Archaeon	481/489	*93*
S. griseus	Bacterium	455	*50*
A. nidulans	Cyanobacterium	484	*112*
Class I (6-4) photolyase and its related gene			
D. melanogaster	Insect	540	*100*
H. sapiens	Mammal	586[c]	*91,100*
Class I photolyase-like blue-light receptors			
C. reinhardtii	Green alga	867[c]	*90*
A. thaliana	Plant	681[c]	*1*
S. alba	Plant	501	*3*
Class II CPD photolyases			
M. thermoautotrophicum	Archaeon	444	*115*
M. xanthus	Bacterium	446	*67*
D. melanogaster	Insect	536/640[b]	*99,115*
C. auratus	Fish	556[b]	*110*
O. latipes	Fish	504[b]	*115*
P. tridactylis	Marsupial	532[b]	*115*
M. domesticum	Marsupial	470	*37*

[a]Deduced length including N-terminal Met: two values indicate possible translational starts.
[b]Contains N-terminal extension with putative localization signals.
[c]Contains C-terminal extension.

have an N-terminal extension of variable length with putative nuclear and mitochondrial localization signals (*see* Section 7.). In Fig. 7, amino acids conserved in almost all photolyases, and those found almost exclusively in class I or class II photolyases are indicated. In the most conserved carboxyl-terminal region, there are a number of residues specific to each class. However, trp residues, which are common and highly conserved within each class, are also fairly well conserved between the classes. From the two trp residues that are involved in electron transfer reactions (*see* Section 2.3.), one that is involved in photoactivation (W306 in *E. coli* photolyase), seems to be conserved in both classes. The other trp (W277 in *E. coli* photolyase), involved in pyrimidine dimer splitting by short-wavelength light, is apparently absent in class II. In the amino-terminal region, distal to the eukaryote-specific extension and including the first trp residue, sequences are well conserved in photolyases of both classes. Furthermore, some of the catalytic residues identified from the

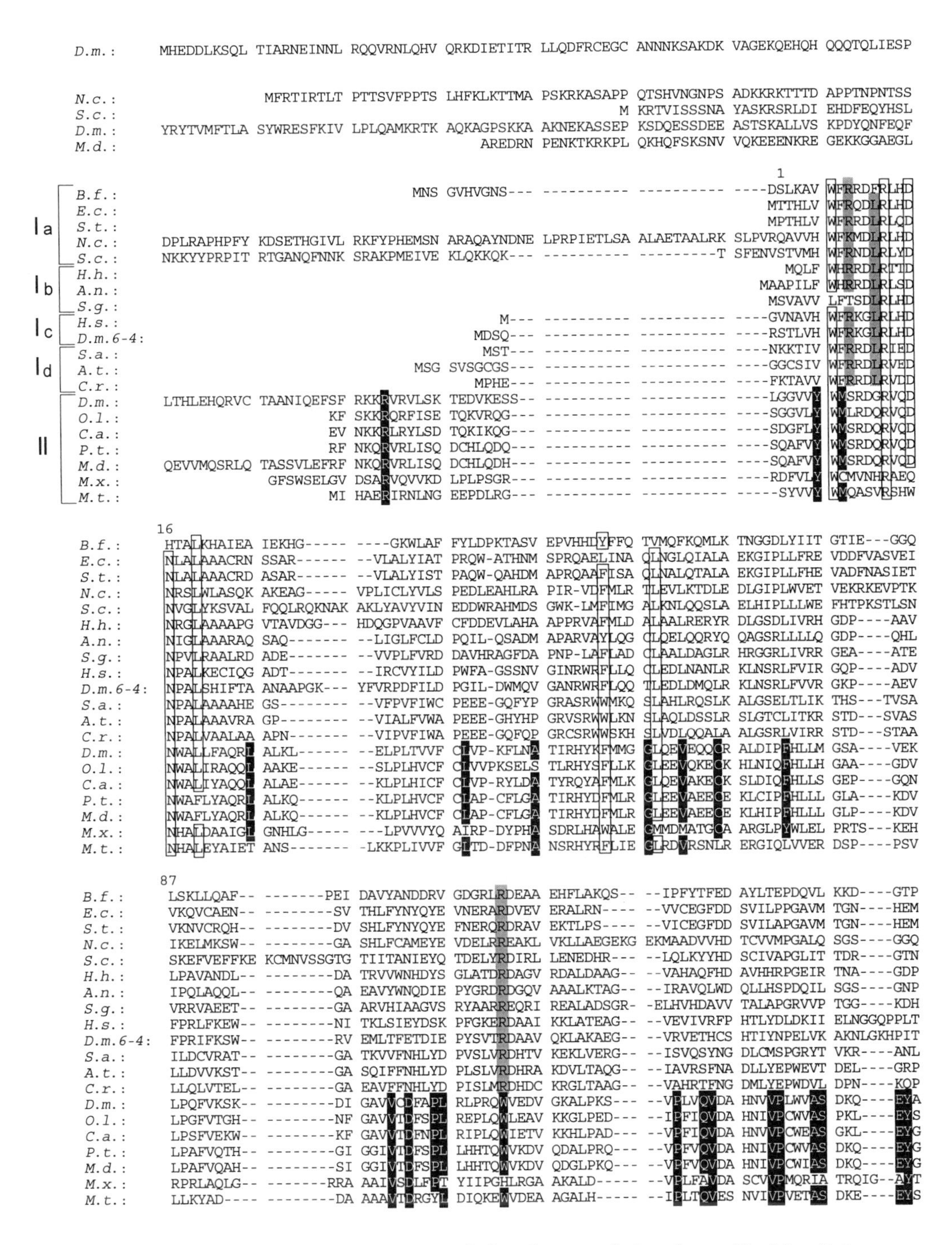

Fig. 7. Amino acid sequence alignment of photolyase and photolyase-like blue-light receptor genes obtained with Clustal W *(97)* and manual alignment. Abbreviations: Ia (class I MTHF-type photolyases), *B.f.*: *Bacillus firmus*, *E.c.*: *E. coli*, *S.t.*: *Salmonella typhimurium*, *N.c.*: *N. crassa*, *S.c.*: *S. cerevisiae*; Ib (class I 8-HDF-type photolyases), *H.h.*: *Halobacterium halobium*, *A.n.*: *A. nidulans*, *S.g.*: *S. griseus*; Ic (class I [6-4] photolyase family), *H.s.*: *Homo*

(Figure 7 caption continued on next page)

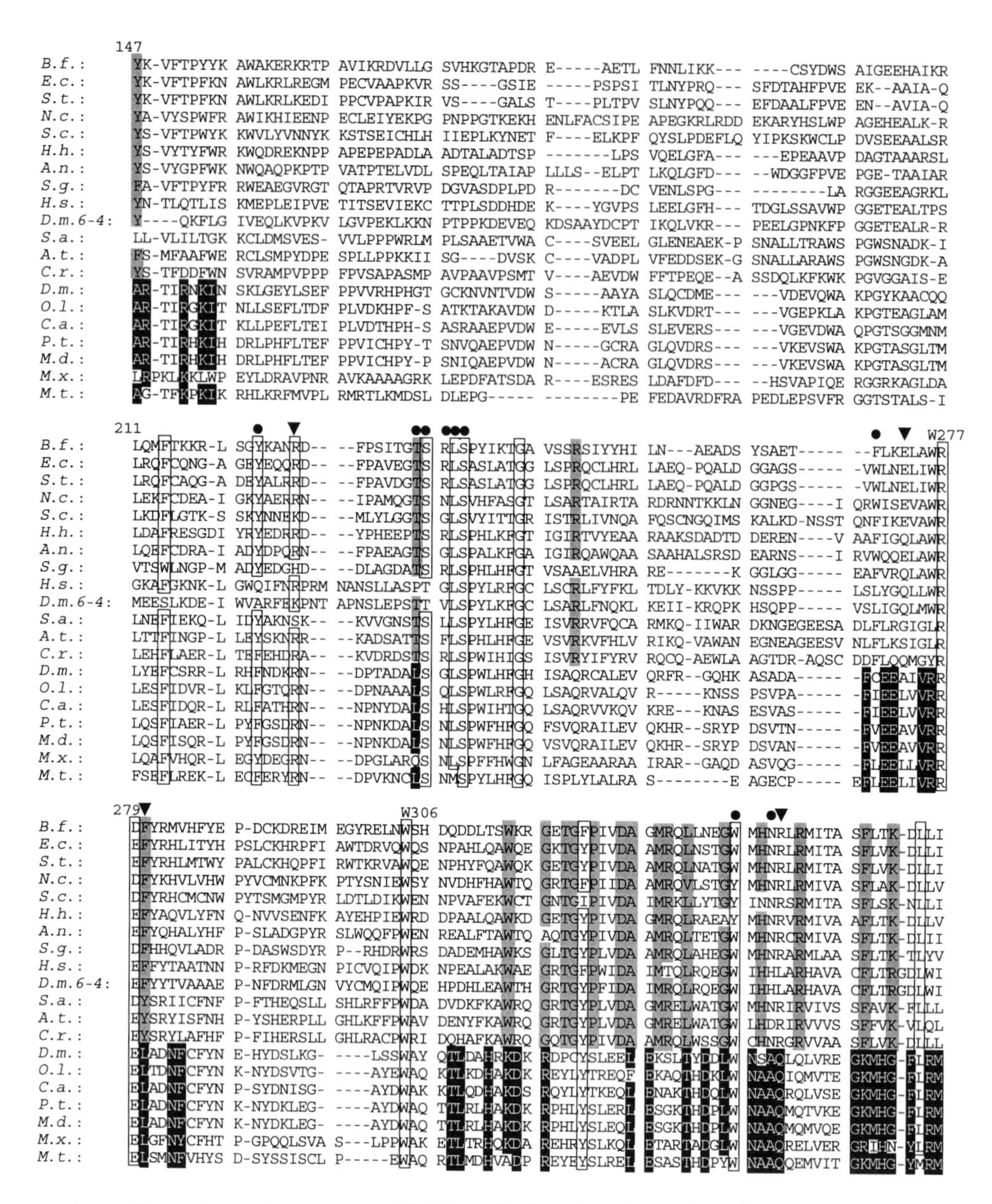

sapiens, *D.m.*: *D. melanogaster*; Id (class I photolyaselike blue-light receptor family), *A.t.*: *A. thaliana*, *C.r.*: *C. reinhardtii*, *S.a.*: *S. alba*; II (class II photolyases), *D.m.*: *D. melanogaster*; *0.l.*: *O. latipes*, *C.a.*: *C. auratus*, *P.t.*: *P. tridactylis*, *M.d.*: *M. domesticum*, *M.x.*: *M. xanthus*, *M.t.*: *M. thermoautotrophicum*. The amino acid residue numbers for *E. coli* photolyase are given; amino acid residues conserved in all species are boxed, those conserved only in class I photolyases are on a shadowed background, and those conserved only in class II photolyases are on a black background. Different amino acid residues with similar characteristics, namely W, Y, and F as well as R and K, are considered conserved. Amino acids involved in FAD binding (●) and positively charged amino acids located on the putative DNA binding surface of *E. coli* photolyase (▼) are indicated.

(Figure 7 caption continued from previous page)

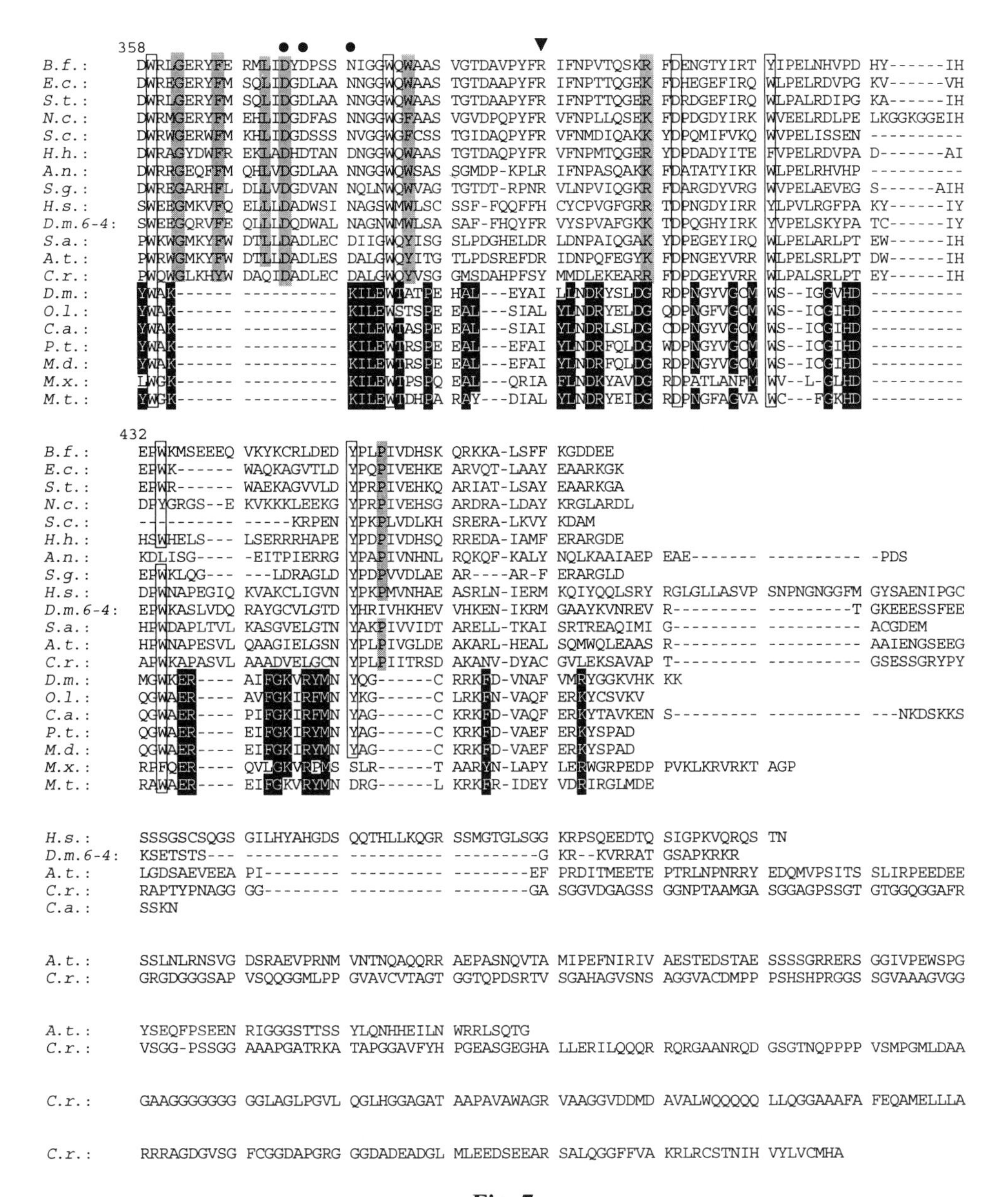

Fig. 7.

structural analysis of *E. coli* photolyase are also conserved in photolyases of higher eukaryotes. In Fig. 7, residues directly involved in the binding of FAD as well as those found along the positively charged trace on the putative DNA binding surface in *E. coli* photolyase are indicated. Some amino acid residues in the central region of *E. coli* photolyase have polar interactions with FAD, and three residues that may be involved in substrate binding are also conserved in class II photolyases. This suggests that the two photolyase classes may have a similar three-dimensional structure (*see also* ref. *67*).

For the binding site of the second cofactor, residues have been proposed that are exclusively conserved in either MTHF or 8-HDF photolyases *(61,106)*. However, structural analysis of *E. coli* photolyase revealed that residues forming hydrogen bonds to MTHF are not well conserved among the MTHF photolyases *(71)*. Therefore, prediction of 8-HDF or MTHF binding sites from sequence alone is difficult for class II photolyases. Determination of the structure of an 8-HDF photolyase would yield more information on residues specifically interacting with the second chromophore.

3.3. Characterization of Class II Photolyases

A 62 kDa photolyase was partially purified from *D. melanogaster* embryos with an action spectrum having maximal activity around 440 nm *(99)*. The cloned photolyase gene of *P. tridactylis* was fused with glutathione *S*-transferase and purified as fusion protein from *E. coli* cells *(115)*. This fusion protein is biologically active and contains FAD as a cofactor, indicating that FAD is the common chromophore for the two classes of evolutionarily diverged photolyases. Neither MTHF nor 8-HDF could be detected in purified recombinant *P. tridactylis* photolyase. The previously reported action spectra for photoreactivation in marsupials with a maximum around 360 nm *(8,32,78)* are in reasonable agreement with the presence of a single reduced FAD chromophore. Photolyase purified from *M. thermoautotrophicum* cells contains 8-HDF in addition to FAD *(39)*, suggesting the presence of photolyases with and without second cofactors in this class.

3.4. Photolyase and Photolyase-Like Blue-Light Receptors in Plants

The occurrence of photoreactivation in algae and plants is well documented, using both biological end points and light-dependent removal of CPDs. Biochemical studies indicated the presence of photolyase in *Phaseolus vulgaris (79)* and *Arabidopsis thaliana (69)*. Attempts to isolate a functional photolyase gene from plants are in progress.

A. thaliana plants harboring a mutation at the *HY4* locus fail to show inhibition of hypocotyl elongation in response to blue light. The gene responsible for this response was isolated and found to encode a 75-kDa protein (cryptochrome, CRY1) with sequence similarity to microbial photolyases *(1)* (*see* Fig. 7). It contains a C-terminal extension of 181 amino acids, with some similarity to tropomyosin A (but not to photolyases), which is essential for the blue-light response *(2)*. This protein seems to act as a blue-light receptor and not as a photolyase, because expression of the gene does not complement *E. coli phr*$^-$ mutants *(62)*, but causes a hypersensitive light response in transgenic plants *(58)*. Recombinant CRY1 fusion proteins purified from insect cells or *E. coli* (as maltose binding protein–fused protein) did not show enzymatic photolyase activity in vitro, although the purified proteins contain FAD as cofactor *(57,62)*. A similar situation was found for a gene from the plant *Sinapis alba*, which encodes a protein highly similar to microbial photolyases, lacking an extended C-terminus *(3)*. However, no photolyase activity was found for protein overexpressed in *E. coli*, and therefore, the gene is thought to encode a blue-light receptor *(62)*. The presence of MTHF chromophore was reported in *A. thaliana* and *S. alba* blue-light receptors overexpressed in *E. coli (62)*. Another possible blue-light receptor gene, encoding a protein of 867 amino acids, was isolated from the unicellular green alga *Chlamydomonas reinhardtii (90)*. The first 500 amino acids show a high similarity to microbial photolyases, whereas the C-terminal 367 amino acids show no homology to the *Arabidopsis* protein (Fig. 7). The locus of the photolyase gene is not linked to the putative blue-light receptor gene, suggesting the existence of separate photolyase and blue-light receptor genes in *C. reinhardtii*.

4. (6-4)PD PHOTOLYASES

The photolyases described in the previous sections are specific for CPDs. A different photoreactivating activity ([6-4] photolyase) has been found in *D. melanogaster (98)*, which removes (6-4)PDs, but apparently not their Dewar isomers (Fig. 1) *(47)*. Biochemical evidence indicates that (6-4)PDs are split into the constituent pyrimidines, possibly through an oxetane intermediate, and not into dihydrothymine *(47)*. This is remarkable, because in contrast to CPDs, (6-4)PDs cannot be converted photochemically into the constituent pyrimidines: instead the Dewar isomer is formed. The action spectrum of photoreactivation has a maximum at 400 nm, but it is likely an inefficient reaction (the photolytic cross-section is very low, $\varepsilon\phi \sim 200$). The *D. melanogaster* (6-4) photolyase (*64phr*) gene has been isolated and it provides *E. coli* cells with additional photoreactivating activity *(100)*. It contains a 540 amino acid open reading frame with homology (21% identical) to class I photolyases. The homology to class II photolyases, including *D. melanogaster* CPD photolyase, is lower (13%). Within class I photolyases, highest homology was found with blue-light receptors (*see* Section 3.4.). The chromosomal localization of the *64phr* gene in *D. melanogaster* is different from that of the *CPDphr* gene *(99)*. A similar (6-4) photolyase activity has been detected in cultured *Xenopus laevis* (South African clawed toad) and *Crotalus atrox* (rattlesnake) cells *(48)*. *X. laevis* photolyase showed optimal activity at 430 nm. No intrinsic chromophores have been identified yet. A comparable (6-4) photolyase is probably present in the plant *A. thaliana (7)*. At present, the occurrence of (6-4) photolyases seems to be restricted to higher eukaryotes.

Interestingly, a human gene has been found *(91,100)* that shows homology (48%) to the *D. melanogaster 64phr* gene. However, it shares a C-terminal extension with the blue-light receptor genes, which is not present in the *64phr* gene of *D. melanogaster*. The human gene is located on chromosome 12q24, and it is highly expressed in testis, skeletal muscle and other organs *(91)*. The biological function of this gene is uncertain. A second human photolyase homolog with 73% identity to the first one was also isolated *(29a)*. Recombinant proteins of both human photolyase homologs were found to be flavoproteins, but did not show photolyase activity on CPD as well as 64PD.

5. PHOTOLYASE AS A MODIFYING FACTOR FOR LIGHT-INDEPENDENT NUCLEOTIDE EXCISION REPAIR

In addition to photoreactivation activity, *E. coli* photolyase enhances NER in *E. coli* cells *(80,108)*. It is proposed that the structural deformation of DNA at a pyrimidine dimer site is increased by the binding of photolyase, increasing the accessibility for the NER machinery. The NER supporting activity of photolyase is apparently a subtle interaction between photolyase and the NER system, because, in contrast to *E. coli* photolyase, expression of *A. nidulans*, yeast or goldfish photolyase genes in *E. coli* considerably disturbs the NER activity of host cells *(51,85,110)*. Photolyase of *S. cerevisiae* binds to pyrimidine dimers, resulting in an enhancement of NER in yeast cells *(85)*. It also binds to DNA damage induced by *cis*-platinum or MNNG, resulting in a disturbance instead of enhancement of NER of these lesions *(23)*. This may be because of the difference in photolyase binding modes for various lesions, as suggested by the substrate binding model for *E. coli* photolyase *(71)*. *E. coli* photolyase also binds to *cis*-platinum lesions but, in contrast to yeast, it slightly stimulates NER both in vitro and in vivo *(68)*.

6. REGULATION OF PHOTOLYASE GENE EXPRESSION IN EUKARYOTES

The photolyase gene of the yeast *S. cerevisiae* is inducible by UV irradiation as well as by treatment with DNA-damaging agents, such as 4-NQO, MMS, or MNNG *(88)*. In experiments with yeast cells with multicopy plasmids harboring a photolyase–*lacZ* fusion, an up to 10-fold induction was found relative to nontreated cells. A transcription element was identified in the upstream region of the gene where a DNA binding factor (photolyase regulatory protein) binds and represses transcription of the photolyase gene in response to DNA damage. Other promoter elements of the yeast photolyase gene have been studied in detail *(86)*.

In higher eukaryotes, UV-inducibility of photolyase activity was first reported in an established frog cell line *(6)*. More recently, photolyase activity in fish cells was found to be inducible by visible light *(65,101,109)*. Northern blot analysis using cloned fish (*C. auratus*) cDNA indicated that gene expression increased >10-fold 4–8 h after exposure to visible light *(110)*. The time-course for inducibility agrees with the increase of enzymatic activity in cultured fish cells. Treatment of fish cells with H_2O_2 also increases expression of the photolyase gene *(64)*. Since visible light is a well-known inducer of activated oxygen, this suggests that at least part of the induction in fish cells is initiated by oxygen stress. It is interesting that, in contrast to yeast, the fish photolyase gene is only weakly induced by UV or other DNA-damaging agents, whereas the yeast photolyase gene is not induced by visible light *(88)*. The visible light inducibility of fish photolyase may have profound effects on the repair of UV-damaged DNA and, therefore, on the survival of fish, because most fish species live near the water surface where abundant visible light and a significant amount of UV-B are present.

Another interesting mode of regulation of photolyase gene expression was reported for *D. melanogaster (99)*. A large number of photolyase molecules is found in embryos and adult ovaries, whereas the photolyase gene is highly expressed only in ovaries. Apparently, photolyase molecules are abundantly transferred from ovary to eggs in order to protect eggs from harmful UV present in sunlight. The silkworm *Bombyx mori* may have a similar protection mechanism for eggs *(66)*.

7. PHOTOREACTIVATION IN MITOCHONDRIA AND CHLOROPLASTS

Photoreactivation is found to act not only in the nucleus, but also in mitochondria and chloroplasts. In the yeast *S. cerevisiae*, photoreactivation of mitochondrial DNA

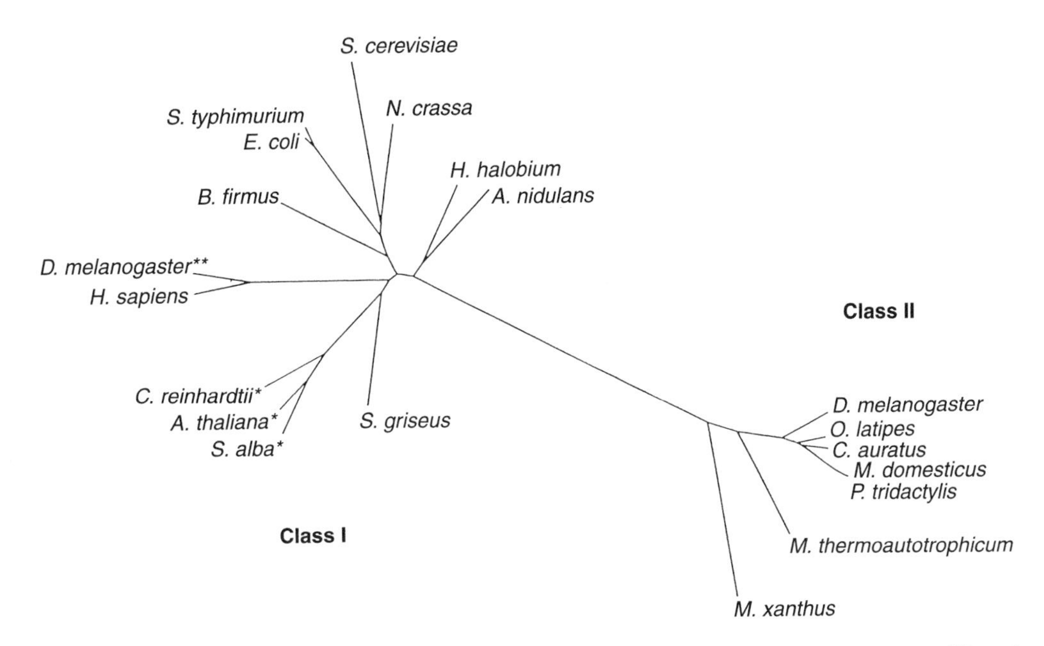

Fig. 8. Unrooted phylogenetic tree of photolyase, photolyase-like blue-light receptor (*) and (6-4) photolyase (**) genes as specified in Table 3. The tree was obtained with the neighbor-joining method (PHYLIP; *21*) using a pairwise distance matrix calculated from the aligned amino acid sequences shown in Fig. 7.

was identified based on a decrease of petite mutants as well as a decrease in the number of pyrimidine dimers after exposure of UV-irradiated cells to visible light *(75,104)*. The yeast photolyase gene has an extended amino terminal sequence (*see* Fig. 7), which contains intracellular transport signals for both mitochondria and nucleus. A *lacZ*-fused photolyase expressed in yeast cells was found in mitochondria and the nucleus with a ratio of 1:5. *E. coli* photolyase carrying the extended amino-terminal sequence of yeast photolyase is able to repair mitochondrial DNA in yeast cells *(114)*.

Photoreactivation of chloroplast DNA in the green alga *C. reinhardtii* is, in contrast to yeast, independent of photoreactivation of nuclear DNA. Although the Phr^- yeast mutant cannot photorepair its mitochondrial DNA, an almost normal photoreactivation of chloroplast DNA was found in a Phr^- green alga mutant *(10)*, suggesting the presence of two distinct photolyase genes. This difference between yeast and green alga may reflect the different symbiotic origin of organelles.

8. EVOLUTION OF PHOTOLYASE GENES

A phylogenetic tree derived from amino acid sequences clearly shows the evolutionary distance between the two classes of photolyases (Fig. 8). Three putative blue-light receptors, which are also incorporated in the tree, form a distinct cluster within class I photolyases. Another cluster containing *D. melanogaster* (6-4) photolyase and the human photolyase homolog is related to the photoreceptors (*see* Section 4.). As far as analyzed, class I comprises three functions, namely, photoreactivation (both for CPDs and [6-4]PDs), NER-supporting activity, and blue-light receptor activity, whereas for class II, only photoreactivating activity has been documented. It might be that the distribution of both classes in various organisms results from their functional differences.

In view of the simplicity and effectiveness of photoreactivation repair of a major UV-induced DNA damage, it is curious that there are a number of organisms, for example, the fission yeast *Schizosaccharomyces pombe*, the Gram-positive bacterium *Bacillus subtilis*, the nematode *Caenorhabditis elegans*, and various placental cell lines, which do not show any photoreactivation in cells or in cell extracts. Expression of the *S. cerevisiae* photolyase gene provided *S. pombe* cells with photoreactivating activity *(113)*, indicating that the lack of photoreactivating activity is the result of the absence of a functional photolyase gene in *S. pombe* and not a lack of chromophoric cofactors. The absence of a photolyase gene in some organisms was recently confirmed when the complete genomic sequences of *Haemophilus influenzae (22)* and *Mycoplasma genitalium (24)* were determined. However, most organisms have photolyase activity, some of which have an inducible regulatory mechanism for increased production of photolyase molecules. *D. melanogaster* and other higher organisms have, in addition to NER, not only a photolyase for CPDs, but also a (6-4)PD photolyase *(98)*. *S. pombe* and *B. subtilis* possess a UV endonuclease that repairs both CPDs and (6-4)PDs *(95,107)*. The different extent to which organisms utilize photorepair most likely reflects the environment of the organisms as well as the presence of other DNA repair mechanisms for UV damage. Analysis of UV-repair mechanisms, therefore, could clarify the history and future of these organisms in environments with different levels of UV radiation.

ACKNOWLEDGMENT

We are grateful to S. Yasuhira for assistance with alignment of photolyase sequences.

REFERENCES

1. Ahmad, M., and A. R. Cashmore. 1993. *HY4* gene of *A. thaliana* encodes a protein with characteristics of a blue-light photoreceptor. *Nature* **366:** 162–166.
2. Ahmad, M., C. Lin, and A. R. Cashmore. 1995. Mutations throughout an *Arabidopsis* blue-light photoreceptor impair bluelight-responsive anthocyanin accumulation and inhibition of hypocotyl elongation. *Plant. J.* **8:** 653–658.
3. Batschauer, A. 1993. A plant gene for photolyase: an enzyme catalyzing the repair of UV-light-induced DNA damage. *Plant J.* **4:** 705–709.
4. Baer, M. E., and G. B. Sancar. 1993. The role of conserved amino acids in substrate binding and discrimination by photolyase. *J. Biol. Chem.* **268:** 16,717–16,724.
5. Brash, D. E., W. A. Franklin, G. B. Sancar, A. Sancar, and W. A. Haseltine. 1985. *Escherichia coli* DNA photolyase reverses cyclobutane pyrimidine dimers but not pyrimidine-pyrimidone (6-4) photoproducts. *J. Biol. Chem.* **260:** 11,438–11,441.
6. Chao, C. C.-K., and S. Lin-Chao. 1987. Regulation of photorepair in growing and arrested frog cells in response to ultraviolet light. *Mutat. Res.* **192:** 211–216.
7. Chen, J.-J., D. L. Mitchell, and A. B. Britt. 1994. A light-dependent pathway for the elimination of UV-induced pyrimidine (6-4) pyrimidinone photoproducts in *Arabidopsis*. *Plant Cell* **6:** 1311–1317.
8. Chiang, T., and C. S. Rupert 1979. Action spectrum for photoreactivation of ultraviolet-irradiated marsupial cells in tissue culture. *Photochem. Photobiol.* **30:** 525–528.
9. Cook, J. S. 1967. Direct demonstration of the monomerization of thymine-containing dimers in U.V.-irradiated DNA by yeast photoreactivating enzyme and light. *Photochem. Photobiol.* **6:** 97–101.

10. Cox, J. L., and G. D. Small. 1985. Isolation of a photoreactivation-deficient mutant of Chlamydomonas. *Mutat. Res.* **146:** 249–255.
11. Dulbecco, R. 1949. Reactivation of ultra-violet-inactivated bacteriophage by visible light. *Nature* **163:** 949,950.
12. Eker, A. P. M. 1980. Photoreactivating enzyme from *Streptomyces griseus* III. Evidence for the presence of an intrinsic chromophore. *Photochem. Photobiol.* **32:** 593–600.
13. Eker, A. P. M., R. H. Dekkor, and W. Berends. 1981. Photoreactivating enzyme from *Streptomyces griseus* IV. On the nature of the chromophoric cofactor in *Streptomyces griseus* photoreactivating enzyme. *Photochem. Photobiol.* **33:** 65–72.
14. Eker, A. P. M. 1985. Evidence for the presence of an essential arginine residue in photoreactivating enzyme from *Streptomyces griseus. Biochem. J.* **229:** 469–476.
15. Eker, A. P. M., J. K. C. Hessels, and R. H. Dekker. 1986. Photoreactivating enzyme from *Streptomyces griseus*-VI. Action spectrum and kinetics of photoreactivation. *Photochem. Photobiol.* **44:** 197–205.
16. Eker, A. P. M., J. K. C. Hessels, and J. van de Velde. 1988. Photoreactivating enzyme from the green alga *Scenedesmus acutus*. Evidence for the presence of two different flavin chromophores. *Biochemistry* **27:** 1758–1765.
17. Eker, A. P. M., P. Kooiman, J. K. C. Hessels, and A. Yasui. 1990. DNA photoreactivating enzyme from the cyanobacterium *Anacystis nidulans. J. Biol. Chem.* **265:** 8009–8015.
18. Eker, A. P. M., L. Formenoy, and L. E. A. de Wit. 1991. Photoreactivation in the extreme halophilic archaebacterium *Halobacterium cutirubrum. Photochem. Photobiol.* **53:** 643–651.
19. Eker, A. P. M., and A. Yasui. 1991. Probing the chromophore binding site of 8-hydroxy-5-deazaflavin type photolyase. *Photochem. Photobiol.* **53:** 17S,18S.
20. Eker, A. P. M., H. Yajima, and A. Yasui. 1994. DNA photolyase from the fungus *Neurospora crassa*. Purification, characterization and comparison with other photolyases. *Photochem. Photobiol.* **60:** 125–133.
21. Felsenatein, J. 1993. *PHYLIP*. Department of Genetics, University of Washington, Seattle, WA, version 3.5c.
22. Fleischmann, R. D., et al. 1995. Whole-genome random sequencing and assembly of *Haemophilus influenzae* Rd. *Science* **269:** 496–512.
23. Fox, M. E., B. J. Feldman, and G. Chu. 1994. A novel role for DNA photolyase: binding to DNA damaged by drugs is associated with enhanced cytotoxicity in *Saccharomyces cerevisiae. Mol. Cell. Biol.* **14:** 8071–8077.
24. Fraser, C. M., et al. 1995. The minimal gene complement of *Mycoplasma genitalium. Science* **270:** 397–403.
25. Funayama, T., H. Mitani, and A. Shima. 1996. Overexpression of medaka (*Oryzias latipes*) photolyase gene in medaka cultured cells and early embryos. *Photochem. Photobiol.* **63:** 633–638.
26. Hamm-Alvarez, S., A. Sancar, and K. V. Rajagopalan. 1989. Role of enzyme-bound 5,10-methenyltetrahydropteroylpolyglutamate in catalysis by *Escherichia coli* DNA photolyase. *J. Biol. Chem.* **264:** 9649–9656.
27. Heelis, P. F., G. Payne, and A. Sancar. 1987. Photochemical properties of *Escherichia coli* DNA photolyase: selective photodecomposition of the second chromophore. *Biochemistry* **26:** 4634–4640.
28. Heelis, P. F., S.-T. Kim, T. Okamura, and A. Sancar. 1993. The photorepair of pyrimidine dimers by DNA photolyase and model systems. *J. Photochem. Photobiol. B: Biol.* **17:** 219–228.
29. Hélène, C., and M. Charlier. 1977. Photosensitized splitting of pyrimidine dimers by indole derivatives and by tryptophan-containing oligopeptides and proteins. *Photochem. Photobiol.* **25:** 429–434.
29a. Hsu, D. S., X. D. Zhao, S. Y. Zhao, A. Kazantsev, R. P. Wang, T. Todo, Y. F. Wei, and A. Sancar. 1996. Putative human blue-light photoreceptors hCRY1 and hCRY2 are flavoproteins. *Biochemistry* **35:** 13,871–13,877.

30. Hurter, J., M. P. Gordon, J. P. Kirwan, and A. D. McLaren. 1974. *In vitro* photoreactivation of ultraviolet-inactivated ribonucleic acid from tobacco mosaic virus. *Photochem. Photobiol.* **19:** 185–190.
31. Husain, I., and A. Sancar. 1987. Binding of *E. coli* DNA photolyase to a defined substrate containing a single T<>T dimer. *Nucleic Acids Res.* **15:** 1109–1120.
32. Ishizaki, K., and H. Takebe. 1985. Comparative studies on photoreactivation of ultraviolet light-induced T4 endonuclease susceptible sites and sister-chromatid exchanges in Potorous cells. *Mutat. Res.* **150:** 91–97.
33. Jaeckle, H., and K. Kalthoff. 1978. Photoreactivation of RNA in UV-irradiated insect eggs (*Smittia* sp., *Chironomidae, Diptera*) I. Photosensitized production and light-dependent disappearance of pyrimidine dimers. *Photochem. Photobiol.* **27:** 309–315.
34. Johnson, J. L., S. Hamm-Alvarez, G. Payne, G. B. Sancar, K. V. Rajagopalan, and A. Sancar. 1988. Identification of the second chromophore in *Escherichia coli* and yeast DNA photolyases as 5,10-methenyltetrahydrofolate. *Proc. Natl. Acad. Sci. USA* **85:** 2046–2050.
35. Jorns, M. S., G. B. Sancar, and A. Sancar. 1985. Identification of oligothymidylates as new simple substrates for *Escherichia coli* DNA photolyase and their use in a rapid spectrophotometric enzyme assay. *Biochemistry* **24:** 1856–1861.
36. Jorns, M. S., B. Wang, S. P. Jordan, and L. P. Chanderkar. 1990. Chromophore function and interaction in *Escherichia coli* DNA photolyase: reconstitution of the apoenzyme with pterin and/or flavin derivatives. *Biochemistry* **29:** 552–561.
37. Kato, T., T. Todo, H. Ayaki, K. Ishizaki, T. Morita, S. Mitra, and M. Ikenaga. 1994. Cloning of a marsupial DNA photolyase gene and the lack of related nucleotide sequences in placental mammals. 1994. *Nucleic Acids Res.* **22:** 4119–4124.
38. Kelner, A. 1949. Effect of visible light on the recovery of *Streptomyces griseus* conidia from ultra-violet irradiation injury. *Proc. Natl. Acad. Sci. USA* **35:** 73–79.
39. Kiener, A., I. Hussain, A. Sancar, and C. Walsh. 1989. Purification and properties of *Methanobacterium thermoautotrophicum* DNA photolyase. *J. Biol. Chem.* **264:** 13,880–13,887.
40. Kim, S.-T., R. F. Hartman, and S. D. Rose. 1990. Solvent dependence of pyrimidine dimer splitting in a covalently linked dimer-indole system. *Photochem. Photobiol.* **52:** 789–794.
41. Kim, S.-T., and A. Sancar. 1991. Effect of base, pentose, and phosphodiester backbone structures on binding and repair of pyrimidine dimers by *Escherichia coli* DNA photolyase. *Biochemistry* **30:** 8623–8630.
42. Kim, S.-T., P. F. Heelis, T. Okamura, Y. Hirata, N. Mataga, and A. Sancar. 1991. Determination of rates and yields of interchromophore (folate → flavin) energy transfer and intermolecular (flavin → DNA) electron transfer in *Escherichia coli* photolyase by time-resolved fluorescence and absorption spectroscopy. *Biochemistry* **30:** 11,262–11,270.
43. Kim, S.-T., Y. F. Li, and A. Sancar. 1992. The third chromophore of DNA photolyase: Trp-277 of *Escherichia coli* DNA photolyase repairs thymine dimers by direct electron transfer. *Proc. Natl. Acad. Sci. USA* **89:** 900–904.
44. Kim, S.-T., P. F. Heelis, and A. Sancar. 1992. Energy transfer (deazaflavin → $FADH_2$) and electron transfer ($FADH_2$ → T<>T) kinetics in *Anacystis nidulans* photolyase. *Biochemistry* **31:** 11,244–11,248.
45. Kim, S.-T., A. Sancar, C. Essenmacher, and G. T. Babcock. 1993. Time-resolved EPR studies with DNA photolyase: excited-state FADH· abstracts an electron from Trp-306 to generate $FADH^-$, the catalytically active form of the cofactor. *Proc. Natl. Acad. Sci. USA* **90:** 8023–8027.
46. Kim, S.-T., K. Malhotra, C. A. Smith, J.-S. Taylor, and A. Sancar. 1993. DNA photolyase repairs the *trans-syn* cyclobutane thymine dimer. *Biochemistry* **32:** 7065–7068.
47. Kim, S.-T., K. Malhotra, C. A. Smith, J.-S. Taylor, and A. Sancar. 1994. Characterization of (6-4) photoproduct DNA photolyase. *J. Biol. Chem.* **269:** 8535–8540.

48. Kim, S.-T., K. Malhotra, J.-S. Taylor, and A. Sancar. 1996. Purification and partial characterization of (6-4)photoproduct DNA photolyase from *Xenopus laevis*. *Photochem. Photobiol.* **63:** 292–295.
49. Klimasauskas, S., S. Kumar, R. J. Roberts, and X. Cheng. 1994. HhaI methyltransferase flips its target base out of the DNA helix. *Cell* **76:** 357–369.
50. Kobayashi, T., M. Takao, A. Oikawa, and A. Yasui. 1989. Molecular characterization of a gene encoding a photolyase from *Streptomyces griseus*. *Nucleic Acids Res.* **17:** 4731–4744.
51. Kobayashi, T., M. Takao, A. Oikawa, and A. Yasui. 1990. Increased UV sensitivity of *Escherichia coli* cells after introduction of foreign photolyase genes. *Mutat. Res.* **236:** 27–34.
52. Lamola, A. A. 1972. Photosensitization in biological systems and the mechanism of photoreactivation. *Mol. Photochem.* **4:** 107–133.
53. Li, Y. F., and A. Sancar. 1990. Active site of *Escherichia coli* DNA photolyase: mutations at Trp277 alter the selectivity of the enzyme without affecting the quantum yield of photorepair. *Biochemistry* **29:** 5698–5706.
54. Li, Y. F., and A. Sancar. 1991. Cloning, sequencing, expression and characterization of DNA photolyase from *Salmonella typhimurium*. *Nucleic Acids Res.* **19:** 4885–4890.
55. Li, Y. F., P. F. Heelis, and A. Sancar. 1991. Active site of DNA photolyase: Tryptophan-306 is the intrinsic hydrogen atom donor essential for flavin radical photoreduction and DNA repair in vitro. *Biochemistry* **30:** 6322–6329.
56. Li, Y. F., S.-T. Kim, and A. Sancar. 1993. Evidence for lack of DNA photoreactivating enzyme in humans. *Proc. Natl. Acad. Sci. USA* **90:** 4389–4393.
57. Lin, C., D. E. Robertson, M. Ahmad, A. A. Raibekas, M. S. Jorns, P. L. Button, and A. R. Cashmore. 1995. Association of flavin adenine dinucleotide with the *Arabidopsis* blue light receptor CRY1. *Science* **269:** 968–970.
58. Lin, C., N. Ahmad, D. Gordon, and A. R. Cashmore. 1995. Expression of an *Arabidopsis* cryptochrome gene in transgenic tobacco plants results in hypersensitivity to blue, UV-A, and green light. *Proc. Natl. Acad. Sci. USA* **92:** 8423–8427.
59. Lipman, R. S. A., and M. S. Jorns. 1992. Direct evidence for singlet-singlet energy transfer in *Escherichia coli* DNA photolyase. *Biochemistry* **31:** 786–791.
60. Malhotra, K., S.-T. Kim, C. Walsh, and A. Sancar. 1992. Roles of FAD and 8-hydroxy-5-deazaflavin chromophores in photoreactivation by *Anacystis nidulans* DNA photolyase. *J. Biol. Chem.* **267:** 15,406–15,411.
61. Malhotra, K., S.-T. Kim, and A. Sancar. 1994. Characterization of a medium wavelength type DNA photolyase: purification and properties of photolyase from *Bacillus firmus*. *Biochemistry* **33:** 8712–8718.
62. Malhotra, K., S.-T. Kim, A. Batschauer, L. Dawut, and A. Sancar. 1995. Putative blue-light photoreceptors from *Arabidopsis thaliana* and *Sinapis alba* with a high degree of sequence homology to DNA photolyase contain the two photolyase cofactors but lack DNA repair activity. *Biochemistry* **34:** 6892–6899.
63. Miki, K., T. Tamada, H. Nishida, K. Inaka, A. Yasui, P. E. de Ruiter, and A. P. M. Eker. 1993. Crystallization and preliminary X-ray diffraction studies of photolyase (photoreactivating enzyme) from the cyanobacterium *Anacystis nidulans*. *J. Mol. Biol.* **233:** 167–169.
64. Mitani, H., and A. Shima. 1995. Induction of cyclobutane pyrimidine dimer photolyase in cultured fish cells by fluorescent light and oxygen stress. *Photochem. Photobiol.* **61:** 373–377.
65. Mitchell, D. L., J. T. Scoggins, and D. C. Morizot. 1993. DNA repair in the variable platyfish (*Xiphophorus variatus*) irradiated in vivo with ultraviolet B light. *Photochem. Photobiol.* **58:** 455–459.
66. Muraoka, N., A. Okuda, and M. Ikenaga. 1980. DNA photoreactivating enzyme from silkworm. *Photochem. Photobiol.* **32:** 193–197.

67. O'Connnor, K. A., M. J. McBride, M. West, H. Yu, L. Trinh, K. Yuan, T. Lee, and D. R. Zusman. 1996. Photolyase of *Myxococcus xanthus*, a gram-negative eubacterium, is more similar to photolyases found in archae and "higher" eukaryotes than to photolyases of other eubacteria. *J. Biol. Chem.* **271:** 6252–6259.
68. Oezer, Z., J. T. Reardon, D. S. Hsu, K. Malhotra, and A. Sancar. 1995. The other function of DNA photolyase: stimulation of excision repair of chemical damage to DNA. *Biochemistry* **34:** 15,886–15,889.
69. Pang, Q., and J. B. Hays. 1991. UV-B-inducible and temperature-sensitive photoreactivation of cyclobutane pyrimidine dimers in *Arabidopsis thaliana*. *Plant Physiol.* **95:** 536–543.
70. Park, H.-W., A. Sancar, and J. Deisenhofer. 1993. Crystallization and preliminary crystallographic analysis of *Escherichia coli* photolyase. *J. Mol. Biol.* **231:** 1122–1125.
71. Park, H.-W., S.-T. Kim., A. Sancar, and J. Deisenhofer. 1995. Crystal structure of DNA photolyase from *Escherichia coli*. *Science* **268:** 1866–1872.
72. Payne, G., P. F. Heelis, B. R. Rohrs, and A. Sancar. 1987. The active form of *Escherichia coli* DNA photolyase contains a fully reduced flavin and not a flavin radical, both in vivo and in vitro. *Biochemistry* **26:** 7121–7127.
73. Payne, G., M. Wills, C. Walsh, and A. Sancar. 1990 Reconstitution of *Escherichia coli* photolyase with flavins and flavin analogues. *Biochemistry* **29:** 5706–5711.
74. Payne, G., and A. Sancar. 1990. Absolute action spectrum of E-FADH$_2$ and E-FADH$_2$-MTHF forms of *Escherichia coli* DNA photolyase. *Biochemistry* **29:** 7715–7727.
75. Prakash, L. 1975. Repair of pyrimidine dimers in nuclear and mitochondrial DNA of yeast irradiated with low doses of ultraviolet light. *J. Mol. Biol.* **98:** 781–795.
76. Rupert, C. S. 1962. Photoenzymatic repair of ultraviolet damage in DNA. *J. Gen. Physiol.* **45:** 703–724.
77. Rupert, C. S. 1975. Enzymatic photoreactivation: overview, in *Molecular Mechanisms for Repair of DNA*, part A (Hanawalt, P. C. and Setlow, R. B., eds.), Plenum, New York, pp. 73–87.
78. Sabourin, C. L. K., and R. D. Ley. 1988. Isolation and characterization of a marsupial DNA photolyase. *Photochem. Photobiol.* **47:** 719–723.
79. Saito, N., and H. Werbin. 1969. Action spectrum for a DNA-photoreactivating enzyme isolated from higher plants. *Photochem. Photobiol.* **9:** 421–424.
80. Sancar, A., K. A. Franklin, and G. B. Sancar. 1984. *Escherichia coli* DNA photolyase stimulates uvrABC excision nuclease *in vitro*. *Proc. Natl. Acad. Sci. USA* **81:** 7379–7401.
81. Sancar, A. 1994. Structure and function of DNA photolyase. *Biochemistry* **33:** 2–9.
82. Sancar, G. B., F. W. Smith, M. C. Lorence, C. S. Rupert, and A. Sancar. 1984. Sequences of the *Escherichia coli* photolyase gene and protein. *J. Biol. Chem.* **259:** 6033–6038.
83. Sancar, G. B. 1985. Sequence of the *Saccharomyces cerevisiae PHR1* gene and homology of the *PHR1* photolyase to *E. coli* photolyase. *Nucleic Acids Res.* **13:** 8231–8246.
84. Sancar, G. B., F. W. Smith, and P. F. Heelis. 1987. Purification of the yeast *PHR1* photolyase from an *Escherichia coli* overproducing strain and characterization of the intrinsic chromophores of the enzyme. *J. Biol. Chem.* **262:** 15,457–15,465.
85. Sancar, G. B., and F. W. Smith. 1989. Interactions between yeast photolyase and nucleotide excision repair proteins in *Saccharomyces cerevisiae* and *Escherichia coli*. *Mol. Cell. Biol.* **9:** 4767–4776.
86. Sancar, G. B., R. Ferris, F. W. Smith, and B. Vandeberg. 1995. Promoter elements of the *PHR1* gene of *Saccharomyces cerevisiae* and their roles in the response to DNA damage. *Nucleic Acids Res.* **23:** 4320–4328.
87. Savva, R., K. McAuley-Hecht, T. Brown, and L. Pearl. 1995. The structural basis of specific base-excision repair by uracil-DNA glycosylase. *Nature* **373:** 487–493.
88. Sebastian, J., B. Kraus, and G. B. Sancar. 1990. Expression of the yeast *PHR1* gene is induced by DNA-damaging agents. *Mol. Cell. Biol.* **10:** 4630–4637.
89. Setlow, R. B. and W. L. Carrier. 1966. Pyrimidine dimers in ultraviolet-irradiated DNA's. *J. Mol. Biol.* **17:** 237–254.

90. Small, G. D., B. Min, and P. A. Lefebvre. 1995. Characterization of a *Chlamydomonas reinhardtii* gene encoding a protein of the DNA photolyase/blue light receptor family. *Plant Mol. Biol.* **28:** 443–454.
91. Spek, P. J. van der, K. Kobayashi, D. Bootsma, N. Takao, A. P. M. Eker, and A. Yasui. 1996. Cloning, tissue expression, and mapping of a human photolyase homolog with similarity to plant blue-light receptors. *Genomics* **37:** 177–182.
92. Sutherland, B. M., and P. V. Bennett. 1995. Human white blood cells contain cyclobutyl pyrimidine dimer photolyase. *Proc. Natl. Acad. Sci. USA* **92:** 9732–9736.
93. Takao, M., T. Kobayashi, A. Oikawa, and A. Yasui. 1989. Tandem arrangement of photolyase and superoxide dismutase genes in *Halobacterium halobium*. *J. Bacteriol.* **171:** 6323–6329.
94. Takao, M., A. Oikawa, A. P. M. Eker, and A. Yasui. 1989. Expression of an *Anacystis nidulans* photolyase gene in *Escherichia coli*; functional complementation and modified action spectrum of photoreactivation. *Photochem. Photobiol.* **50:** 633–637.
95. Takao, M., R. Yonemasu, K. Yamamoto, and A. Yasui. 1996. Characterization of a UV endonuclease gene from the fission yeast *Schizosaccharomyces pombe* and its bacterial homolog. *Nucleic Acids Res.* **24:** 1267–1271.
96. Tamada, T., H. Nishida, K. Inaka, A. Yasui, P. E. de Ruiter, A. P. M. Eker, and K. Miki. 1995. A new crystal form of photolyase (photoreactivating enzyme) from the cyanobacterium *Anacystis nidulans*. *J. Struct. Biol.* **115:** 37–40.
97. Thompson, J. D., D. G. Higgins, and T. J. Gibson. 1994. Clustal W: improving the sensivity of progressive multiple sequence alignment through sequence weighting, positions specific gap penalties and weight matrix choice. *Nucleic Acids Res.* **22:** 4673–4680.
98. Todo, T., H. Takemori, H. Ryo, M. Ihara, T. Matsunaga, O. Nikaido, K. Sato, and T. Nomura. 1993. A new photoreactivating enzyme that specifically repairs ultraviolet light-induced (6-4)photoproducts. *Nature* **361:** 371–374.
99. Todo, T., H. Ryo, H. Takemori, H. Toh, T. Nomura, and S. Kondo. 1994. High-level expression of the photorepair gene in *Drosophila* ovary and its evolutionary implications. *Mutat. Res.* **315:** 213–228.
100. Todo, T., H. Ryo, K. Yamamoto, H. Toh, T. Inui, H. Ayaki, T. Nomura, and M. Ikenaga. 1996. Similarity among the *Drosophila* (6-4)photolyase, a human photolyase homolog, and the DNA photolyase-blue-light photoreceptor family. *Science* **272:** 109–112.
101. Uchida, N., H. Mitani, and A. Shima. 1995. Multiple effects of fluorescent light on repair of ultraviolet-induced DNA lesions in cultured goldfish cells. *Photochem. Photobiol.* **61:** 79–83.
102. Wang, B., S. P. Jordan, and M. S. Jorns. 1988. Identification of a pterin derivative in *Escherichia coli* DNA photolyase. *Biochemistry* **27:** 4222–4226.
103. Wang, B. and M. S. Jorns. 1989. Reconstitution of *Escherichia coli* DNA photolyase with various folate derivatives. *Biochemistry* **28:** 1148–1152.
104. Waters, R. and E. Moustacchi. 1974. The fate of ultraviolet-induced pyrimidine dimers in the mitochondrial DNA of *Saccharomyces cerevisiae* following various post-irradiation cell treatments. *Biochim. Biophys. Acta* **366:** 241–250.
105. Wulff, D. L. and C. S. Rupert. 1962. Disappearance of thymine photodimer in ultraviolet irradiated DNA upon treatment with a photoreactivating enzyme from baker's yeast. *Biochem. Biophys. Res. Commun.* **7:** 237–240.
106. Yajima, H., H. Inoue, A. Oikawa, and A. Yasui. 1991. Cloning and functional characterization of a eucaryotic DNA photolyase gene from *Neurospora crassa*. *Nucleic Acids Res.* **19:** 5359–5362.
107. Yajima, H., M. Takao, S. Yasuhira, J. H. Zhao, C. Ishii, H. Inoue, and A. Yasui. 1995. A eukaryotic gene encoding an endonuclease that specifically repairs DNA damaged by ultraviolet light. *EMBO J.* **14:** 2393–2399.

108. Yamamoto, K., M. Satake, H. Shinagawa, and Y. Fujiwara. 1983. Amelioration of ultraviolet sensitivity of an *Escherichia coli recA* mutant in the dark by photoreactivating enzyme. *Mol. Gen. Genet.* **190:** 511–515.
109. Yasuhira, S., H. Mitani, and A. Shima. 1991. Enhancement of photorepair of ultraviolet-damage by preillumination with fluorescent light in cultured fish cells. *Photochem. Photobiol.* **53:** 211–215.
110. Yasuhira, S. and A. Yasui. 1992. Visible light-inducible photolyase gene from the goldfish *Carassius auratus*. *J. Biol. Chem.* **267:** 25,644–25,647.
111. Yasui, A. and S. A. Langeveld. 1985. Homology between the photoreactivation genes of *Saccharomyces cerevisiae* and *Escherichia coli*. *Gene* **26:** 349–355.
112. Yasui, A., M. Takao, A. Oikawa, A. Kiener, C. T. Walsh, and A. P. M. Eker. 1988. Cloning and characterization of a photolyase gene from the cyanobacterium *Anacystis nidulans*. *Nucleic Acids Res.* **16:** 4447–4463.
113. Yasui. A., A. P. M. Eker, and M. Koken. 1989. Existence and expression of photoreactivation repair genes in various yeast species. *Mutat. Res.* **217:** 3–10.
114. Yasui, A., H. Yajima, T. Kobayashi, A. P. M. Eker, and A. Oikawa. 1992. Mitochondrial DNA repair by photolyase. *Mutat. Res.* **273:** 231–236.
115. Yasui, A., A. P. M. Eker, S. Yasuhira, H. Yajima, T. Kobayashi, M. Takao, and A. Oikawa. 1994. A new class of DNA photolyases present in various organisms including aplacental mammals. *EMBO J.* **13:** 6143–6151.

3

Cellular Responses to Methylation Damage

Russell O. Pieper

1. INTRODUCTION

The study of DNA methylation damage and the cellular response to methylation damage has contributed significantly to our understanding of cell function. The mutagenicity of DNA methylating agents, such as methyl methane sulfonate (MMS) and 1-methyl-1-nitrosourea (MNU), is widely appreciated, and is useful in creating mutants in bacterial and human cells *(42)*. These mutants, and the identification of mutant genes leading to specific measurable biological end points, have been key in understanding a variety of cell functions. Although the study of methylation damage has made significant contributions to understanding cell function, it has also been key to understanding the processes of DNA repair and carcinogenesis. Methylating agents have long been recognized as cancer-causing agents, and it was with these agents that the link between carcinogenicity and DNA damage was most clearly demonstrated *(66,79)*. The subsequent study of the repair of DNA methylation damage has led to the identification of base lesions, to the description of specific mutations induced by these lesions, and to the isolation of repair proteins that deal with these lesions. This knowledge has, in turn, led to the identification of cancer syndromes and sporadic cancers characterized by defective DNA repair, as well as novel cancer chemotherapeutic strategies designed to manipulate DNA repair. DNA methylation damage is widespread among living organisms because methylating agents are present in the environment as byproducts of industrial processes *(38,70)*. Endogenous methylation also occurs by a nonenzymatic mechanism involving *S*-adenosyl-L-methionine *(4,68)*. As such, all organisms undergo, and respond to, methylation damage. Although much is known about the response of lower organisms (*see* Chapter 3 in vol. 1), this chapter will focus on recent advances in the understanding of methylation damage and repair in higher eukaryotes, and will highlight how this information may facilitate detection and treatment of human disease.

2. METHYLATION DAMAGE AND CELLULAR CONSEQUENCES

Most studies of DNA methylation damage have been performed with agents that fall into one of four broad categories: methyl sulfates, methylsulfonates, methyl nitrosamines, and methyl nitrosamides. Methyl sulfates, such as dimethyl sulfate (DMS),

From: DNA Damage and Repair, Vol. 2: DNA Repair in Higher Eukaryotes
Edited by: J. A. Nickoloff and M. F. Hoekstra © *Humana Press Inc., Totowa, NJ*

and methylsulfonates, such as MMS, are direct alkylating agents, while most methyl nitrosamines (*N*-dimethylnitrosoamine, DMN) and methylnitrosamides (*N*-methyl-*N*'-nitro-*N*-nitrosoguanidine, MNNG) act indirectly, requiring p450-mediated oxidative demethylation to form the ultimate reactive methyldiazonium ion *(12,38,66)*. The methyl nitrosoureas (e.g., streptozotocin, STZ), however, are a subcategory of methyl nitrosamides that spontaneously decompose at physiologic pH to yield a reactive methyldiazonium ion, and do not require metabolic activation *(38)*. Regardless of the activation process, the end result is a DNA-reactive, electrophilic, methylating species.

The type of DNA damage caused by various methylating agents, and the extent of the damage are dependent on a variety of factors, including the electrophilicity of the reactive species and the substituents present in the parent compound. Electrophilic methylating species interact with nucleophilic sites in the DNA. The more electrophilic the reagent, the greater its chances of interacting with a broad range of targets including even weakly nucleophilic DNA sites *(38)*. Thus, weakly electrophilic compounds (such as MMS) will react predominantly with the most nucleophilic position (N^7-guanine), whereas compounds, such as MNNG, will react with many sites in DNA. Compounds that are extremely electrophilic react with very weak nucleophiles, such as water, and never reach the target DNA *(38)*. As such, there appears to be an optimal degree of electrophilicity for DNA methylating agents. Compounds with very low reactivity (such as MMS) or very high reactivity (such as *N*-ethyl-*N*-nitrosourea) tend to induce less DNA damage than compounds with an intermediate degree of reactivity (*N*-methyl-*N*-nitrosourea). A second influence on the extent and distribution of DNA damage is the substituent present on the reactive species. In general, methyl-substituted compounds are 10- to 20-fold more reactive than those containing ethyl substituents *(32)*. Compounds containing ethyl substituents are in turn significantly more reactive than compounds containing higher-order substituents. The substituents on the alkylating agent also significantly influence the distribution of lesions. Relative to ethylating agents, methylating agents are much less likely to react with oxygen atoms and much more likely to react with nitrogen atoms *(79)*. Nonetheless, highly electrophilic methylating agents create significant damage at several sites in DNA, including oxygen *(5)*. Therefore, damage induced by methylating agents involves a wide range of DNA adducts.

Although the type and extent of DNA damage depend on characteristics of the methylating agent, they also depend on characteristics of the target DNA. In theory, all ring oxygen and nitrogen molecules in DNA can be damaged, with the ability to verify such damage dependent on the sensitivity of the detection system and the stability of the adducts. In practice, however, methylation damage is not distributed uniformly among all possible sites, but rather is clustered at several sites as detailed in the following sections.

2.1. Purines

2.1.1. N^7-*Methylguanine (N7MG)*

The N^7 position of guanine (Fig. 1) is the most nucleophilic site in DNA, and this, along with its accessible location in the major groove, makes N7MG the principal product of weakly electrophilic methylating agents *(38,39,70)*. For compounds, such as DMS and MMS, N7MG accounts for >80% of the total adducts (Table 1), whereas for

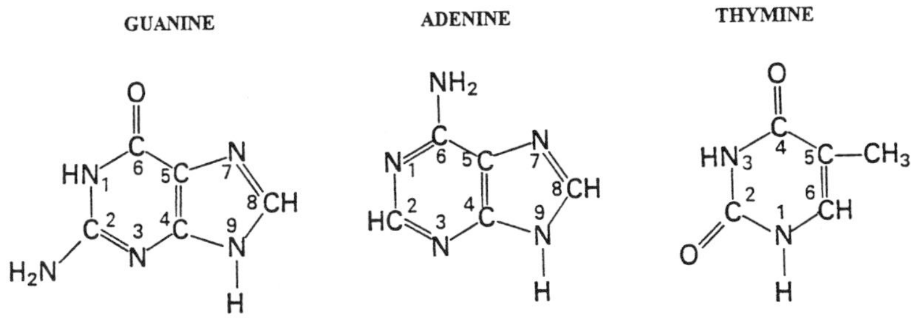

Fig. 1. Chemical structures of guanine, adenine, and thymine.

Table 1
Relative Reactivities of Sites in DNA Toward Methylating Agents

	Category of methylating agent			
Site in DNA	Methyl sulfate DMS[a]	Methyl sulfonate MMS[b]	Methyl nitrosamine DMN[c]	Methyl nitrosamide MNU[a]
Adenine				
N^1	1.9	1.1	0.8	0.9
N^3	11.3	9.8	5.0	8.4
N^7	1.8	0.3	1.7	2.0
Guanine				
N^1	0[d]	na[e]	nd	0
N^2	0.6	nd	nd	0
N^3	0.3	0.7	0.7	0.6
O^6	0.2	0.3	7.8	6.0
N^7	81.4	82.0	74.0	66.4
Cytosine				
O^2	0	nd	0.1	0
N^3	0.7	nd	0.5	0.5
Thymine				
O^2	0	0	0.1	0.1
N^3	0.08	0.06	0.1	0
O^4	0	0	0.05	0.7
Phosphodiester	0.82	1.1	9.6	12.1

Values are expressed as percentage of total alkylation products
[a]From ref. *5*.
[b]From ref. *41*.
[c]From refs. *17,79*.
[d]nd = not detected.
[e]na = not analyzed.

N-ethyl-*N*-nitrosourea, N^7-ethylguanine accounts for only 11% of the total adducts *(5)*. The biologic consequences of N7MG are not clear. Early studies with DMS and MMS showed that N7MG was the most abundant adduct, leading to the proposal that it was the most biologically significant adduct *(80)*. Subsequent work suggested that other adducts, and in particular those at the O^6 position of guanine, were more biologically important *(82)*. Further studies demonstrated that N7MG incorporated into RNA or DNA did not interfere with transcription *(48)* or DNA replication *(64)*, and did not cause mispairing during these processes *(48)*. Additionally, compounds that produced predominantly N7MG adducts were among the least mutagenic, least carcinogenic, and least cytotoxic among the methylating agents, indicating that N7MG is of little biologic significance *(75)*. It has, however, been suggested that the N7MG is not itself detrimental, but that rather the unstable glycosidic bond linking N7MG to the DNA backbone *(50)* may give rise to large numbers of biologically significant apurinic (AP) sites in the genome. Although the importance of N7MG remains unclear, it is apparent that cells of higher eukaryotes, as well as other organisms, have a mechanism to repair N7MG *(57,58)*. Identification of the N7MG repair protein as *N*-methylpurine DNA glycosylase (MPG) (Section 3.1.) has renewed interest in the biologic relevance of N7MG.

2.1.2. O^6-*Methylguanine (O6MG)*

Despite the fact that the O^6 position of guanine is a weak nucleophile and reacts with methylating agents at a much lower frequency than the N^7 position, O6MG is an important mutagenic and carcinogenic lesion. Because the O^6 position of guanine is a weak nucleophile, it is methylated primarily by strong electrophilic agents, such as the methyl nitrosamines and the methyl nitrosamides (Table 1). The biologic relevance of O^6 guanine methylation was made clear by studies that compared the distribution of adducts and the carcinogenic potential of two methylating agents, MNNG and DMS. MNNG was much more carcinogenic, and was much more efficient at methylating the O^6 position of guanine than was DMS *(40)*. These studies suggested that with regard to mutagenic and carcinogenic potential, the ratio of O6MG to N7MG formed by a particular agent was far more important than the total amount of methylation. Hence, MNU (O6MG:N7MG ratio of 0.1) is considerably more carcinogenic than MMS (O6MG:N7MG ratio of 0.004). It was demonstrated that O6MG mispairs with thymine during DNA synthesis *(1,47)* and with uridine during RNA synthesis *(25)*, suggesting that the mutagenic and carcinogenic potential of methylating agents is associated with mispairing of unrepaired O6MG following DNA replication.

Although O6MG is considered to be important in the mutagenicity and carcinogenicity of most methylating agents, its cytotoxic effects are only beginning to be understood. The cytotoxicity of methylating agents was initially associated with their mutagenic potential, the cytotoxicity being a result of an accumulation of mutagenic lesions. Studies with lymphoblastoid cells resistant to the cytotoxic potential of MNNG clearly suggest, however, that mutagenicity and cytotoxicity are related, but separate issues. In these studies, resistant cells accumulated MNNG-induced mutations at a rate similar to sensitive cells *(36)*. The resistant cells, however, were deficient in DNA mismatch repair and could not remove thymine from the mismatched base pair O6MG-thymine. The defect could be complemented in vitro with nuclear fractions from repair-proficient cells. These results suggest that although methylating agents are mutagenic and carcinogenic because O6MG (and other adducts) mispairs during replication, they are cytotoxic because O6MG-thymine mismatches trigger futile mismatch repair. The mismatch repair system does not remove O6MG, but rather removes thymine from the nondamaged strand, allowing additional rounds of mispairing by O6MG and reinitiation of mismatch repair. The mechanism by which futile repair cycling leads to cell death is unknown. It also remains possible that other mechanisms may contribute to methylating agent-induced cytotoxicity, since in yeast, the linkage among mismatch repair, O6MG repair, and cytotoxicity is considerably less clear that in human lymphoblastoid cells *(88)*. Nonetheless, studies of methylating agent-induced cytotoxicity have had an important impact on the understanding of the relationships between methylation damage, mutation, mismatch repair, and cell death.

2.1.3. N^3-*Methyladenine (N3MA) and Other Lesions*

The N^3 position of adenine is a strong nucleophilic site in DNA, and as such represents a target for both strong and weak electrophilic methylating agents (Table 1). Although lesions of the N^3 position of adenine are for most agents second in frequency only to the N^7 position of guanine, the biologic impact of N3MA remains uncertain. When incorporated into DNA, N3MA does not mispair, although DNA replication is inhibited *(59)*. As such, N3MA is a toxic, though not particularly mutagenic adduct.

A variety of other methylpurines are formed in low amounts on exposure of cells to methylating agents, including N1MA, N7MA, N2MG, and N3MG. Few detailed studies have examined the biologic function of these lesions in mammalian systems. Studies with the 1-methylpurines suggest that these adducts are strongly inhibitory to DNA synthesis *(38)*. Since N1MG is not reliably detected in vivo, however, these results are of questionable biologic significance.

2.2. *Pyrimidines*

2.2.1. O^4-*Methylthymine (O4MT)*

O4MT, along with O6MG, represents the major premutational lesion induced by methylating agents. The weakly nucleophilic nature of the O^4 position of thymine suggests that only strong electrophiles, such as methyl nitrosamines and methyl nitrosamides, would form significant amounts of O4MT in cells. This supposition is supported by a number of studies. O4MT lesions have been demonstrated in vitro and in vivo to mispair with guanine in DNA replication. This mispairing leads to TA $\rightarrow$ GC transversions *(65,76)*, consistent with the mutagenic and carcinogenic nature of compounds that induce O4MT lesions. O4MT does not appear to interfere with DNA synthesis, although this is to some degree dependent on the system and polymerase used *(52)*. The in vivo significance of O4MT lesions was suggested by studies that noted that in hepatocellular tumors induced in rats by continuous administration of DMN, O6MG was rapidly lost from DNA, whereas O4MT accumulated to much higher levels *(20,67,83)*. The persistence of O4MT is also consistent with the relative stability of O4MT in DNA ($t_{1/2}$ = 20 h)*(67)* and the apparent lack of repair of this lesion *(77)*. Thus, although formed relatively infrequently, mutagenic O4MT lesions likely accumulate over time to biologically significant levels.

2.2.2. O^2-*Methylthymine (O2MT)*

The O^2 position of thymine, like the O^4 position, reacts preferentially with those strongly electrophilic methylating agents capable of reacting with weak nucleophiles (Table 1). Little is known about O2MT, although studies of O^2-ethylthymine suggest that this lesion inhibits DNA synthesis when it pairs appropriately with adenine, but not when it mispairs with thymine *(27)*. As such, O^2 ethylthymine, and perhaps O2MT, may account for rare AT → TA transversions following treatment with methylating or ethylating agents.

2.3. *Phosphotriesters*

Most methylating agents, and particularly strongly electrophilic agents, produce significant amounts of methylation of the DNA phosphodiester backbone. Sixteen different phosphotriesters can be produced by methylation of the DNA backbone, and as such, phosphotriester levels are generally reported as one overall value. These adducts are extremely stable in vivo with a reported half-life of 1 mo. The biological significance of phosphotriesters remains unclear, since they do not cause misincorporation or inhibition of DNA synthesis *(84)*.

3. REPAIR OF METHYLATION DAMAGE

As noted, the various DNA lesions produced by methylating agents have different biologic half-lives. Some of this difference relates to the instability of the adducts. For example, N7MG disappears from cells over relatively long periods of time in a manner consistent with spontaneous depurination. The less stable N3A adduct disappears more rapidly than N7MG. Spontaneous depurination cannot, however, entirely account for the short biologic half-life of 3MA, nor can it account for the short biologic half-life of the very stable O6MG adduct. It is this rapid loss of adducts that initially stimulated interest in methylation adduct repair. In theory, methylation repair systems could evolve in multiple ways. Separate repair systems could evolve for each adduct. Alternatively, a single broadly specific repair system could recognize and repair all forms of methylation damage. A hierarchy of repair could also be superimposed on these systems, such that not all lesions would be repaired with equal speed and efficiency. In cells, various aspects of all these theoretical systems have been uncovered. In mammalian systems, a relatively specific repair protein, O^6-methylguanine DNA methyltransferase (MGMT), repairs only highly mutagenic lesions at the O^6 position of guanine. There also exists, however, a broader specificity repair protein, MPG. MPG initiates repair of mutagenic N^3-adenine lesions, but also repairs, at lower efficiency, a wide spectrum of other methyl adducts.

3.1. N-*Methylpurine-DNA Glycosylase (MPG)*

3.1.1. *Substrates and Enzyme Characteristics*

Prior to the purification of a repair activity for 3MA from mammalian cells, it was shown that bacteria possessed two distinct 3MA repair activities. One activity was mediated by the constitutively expressed Tag protein and removed 3MA exclusively *(7)*, whereas a second activity was mediated by the inducible AlkA protein. The AlkA protein was shown to function in the removal of a variety of substrates, including 3MA, N3MG, N7MG, N3MA, N7MA, O2MT, and O2MC *(45)*. Both Tag and AlkA were shown to have glycosylase activity. Subsequent to the cloning of the *tag* and *alkA* genes, a glycosylase that removes 3MA, and to a lesser extent N3MG and N7MG was purified from human cells *(31,78)*. The isolation of the cDNA encoding this glycosylase *(8,57,73)* allowed not only for verification of the glycosylase activity in human cells, but also for a more detailed analysis of the protein. The 32-kDa human protein, designated MPG, contains 293 amino acids *(8)*. The protein readily repairs 3MA (K_m = 130 n*M*), and to a lesser degree N7MG (K_m = 860 n*M*) and 3MG *(58,61)*. In vitro, and in high amounts, MPG also appears to repair 8-hydroxyguanine *(6)*, although the relevance of this repair in vivo is uncertain. In terms of substrate specificity, MPG most closely resembles the bacterial AlkA protein. MPG is a low abundance, constitutively expressed protein (1000–2000 copies/cell)*(58)*, which suggests that its primary purpose is the repair of 3MA lesions induced by nonenzymatic methylation by *S*-adenosyl methionine. This supposition has been questioned recently, however, by the observation that MPG repairs a variety of cyclic adducts formed by chloroacetaldehyde, and that MPG repairs 1,N^6-ethenoadenine 10 times more readily than 3MA *(18,19)*. Therefore, although much is now known about the substrate specificity of MPG, its primary substrate remains unclear.

3.1.2. MPG Repair Mechanism

The wide range of MPG substrates have focused attention on the mechanism of substrate recognition and base removal by MPG. The mechanisms by which glycosylases function are poorly defined, although it is possible that the 3MA glycosylases from various organisms might share invariant amino acids in regions responsible for common functions, such as DNA binding and *N*-glycosyl bond cleavage. Comparisons of MPG to Tag and AlkA have proven to be of little use, since there is little sequence conservation between these proteins. There is some suggestion that amino acids 82 and 97 (glutamine and arginine, respectively), which are invariant among known 3MA glycosylases, may be responsible for recognition of *N*-methyladenine and *N*-methylguanine *(8)*, although this has not been tested directly. Advances in the understanding of other glycosylases, and in some cases crystal structures of these glycosylases, will contribute to our understanding of MPG function.

3.2. Methylguanine DNA Methyltransferase (MGMT)

3.2.1. Substrates and Protein Characteristics

All organisms examined possess one or more related proteins whose function is to remove mutagenic methyl and alkyl lesions from the O^6 position of guanine. Human cells contain a single such protein designated MGMT. MGMT is a 25-kDa, 207 amino acid protein that appears to be localized to the nucleus *(3,44)*. The substrate range of the human MGMT is extremely limited. Although alkyltransferases from bacteria repair a variety of lesions, including O4MT and phosphomethyltriesters, MGMT removes lesions exclusively from the O^6 position of guanine. No convincing evidence exists that the protein removes methyl lesions from other base positions. There exists, however, a degree of substrate specificity for repair at the O^6 position of guanine such that methyl groups are repaired more efficiently than ethyl and larger substituents *(61)*.

3.2.2. MGMT Repair Mechanism

Repair of O6MG is brought about by the stoichiometric transfer of the methyl lesion to a cysteine residue in the MGMT protein. Following transfer, the methylated MGMT protein is degraded. The protein is frequently stated to work by a "suicide" mechanism, and is thus most accurately called a DNA repair protein rather than a repair enzyme.

The isolation of the MGMT cDNA *(85)*, as well as cDNAs encoding alkyltransferase proteins from a variety of other organisms has facilitated structure/function studies of the protein. Truncations of MGMT result for the most part in loss of methylation repair capacity *(86)*. The carboxy-terminal 35 amino acids of the protein, however, can be removed without affecting MGMT activity *(21,54)*. The function of this "tail" is uncertain, although its absence in bacterial alkyltransferases suggests that it may function in repair-independent protein–protein interactions in human cells. In contrast, the amino-terminus of MGMT is well documented to be responsible for O6MG repair. MGMT transfers the O6MG group in DNA to cysteine-145. In all alkyltransferases cloned to date, this cysteine is part of a conserved proline-cysteine-histidine-arginine motif *(59,60)*.

The transfer mechanism is believed to involve several steps. MGMT initially forms hydrogen bonds with the O^6 position, the N^1 position, and the exocyclic amino group of O6MG *(81)*. On binding, MGMT undergoes a conformational change *(9)*. This change has been suggested to break a bond between histidine-147 and glutamine-173, and to generate a nucleophilic thiolate anion from cysteine-145 *(53)*. The thiolate anion is then thought to displace the methyl group from the O^6 position of the substrate. Donation of a proton to the now demethylated oxygen at the 6 position of guanine subsequently neutralizes the charge *(81)*. This mechanism is consistent with the observations that cysteine-145, histidine-147, and glutamine-173 are conserved among known alkyltransferases and that all are required for MGMT stability *(15,34,46)*. The action of MGMT also involves DNA binding because O6MG in DNA serves as a better substrate for the protein than either O6MG in single-stranded DNA or the free base O6MG. Additionally, repair of the adducted base O^6-benzylguanine (BG) is stimulated by the presence of undamaged DNA *(26)*. Recent studies suggest that at least two amino acids, arginine-128 and tyrosine-114, play a role in the DNA binding of MGMT, since mutation of these amino acids to alanine results in a 10- to 1000-fold reduction in the ability of the mutant proteins to repair damaged DNA vs damaged free base *(35)*. It has recently been speculated that the methyltransferase reaction involves movement of the methylated base out of the double helix (possibly through a single-stranded intermediate) to facilitate interaction between the methyl group and the cysteine acceptor of the protein *(35)*. These suggestions will undoubtably stimulate more research into how the protein is targeted to damaged DNA.

3.2.3. *Regulation of MGMT Expression*

The MGMT gene contains four coding regions, which are spread over >150 kbp on chromosome 10q26 *(56,69)*. MGMT is expressed in every human tissue, consistent with the finding that the 5'-regulatory region of the MGMT gene is contained within an approx 1000-bp CpG island *(29)*. CpG islands are CG-rich stretches of DNA, which are frequently found at the 5'-end of housekeeping genes, and are devoid of cytosine methylation *(24)*. MGMT expression appears to be driven by binding of the ubiquitous transcription factor Sp1 at six consensus Sp1 binding sites flanking the transcription start site in the MGMT promoter/CpG island *(13)*. Although the MGMT gene is constitutively expressed, the amount of MGMT protein varies dramatically by tissue, and within a given tissue in the same individual *(71)*. MGMT expression is highest in liver, whereas brain and lung have 10–100 times less MGMT *(11,74)*. The basis for this difference is at present unclear, although in cultured cells from these tissues, the difference in expression appears to be mediated at the transcriptional level *(62)*. The chromatin structure of the MGMT promoter/CpG island, and the accessibility of this region to Sp1, may play a role in MGMT tissue-specific expression *(14)*, as may tissue-specific transcription-regulating sequences outside the MGMT promoter. A 60-bp enhancer region in the nontranslated exon 1 may influence MGMT tissue-specific expression *(30)*, although this possibility has not been fully examined.

A small percentage of human tumors (between 2 and 20%) do not express MGMT *(11)*. In $MGMT^-$ cultured tumor cell lines, the loss of MGMT expression appears to be associated with extensive cytosine methylation of the 5' CpG island/promoter of the gene, and with transcriptional inactivation *(14,87)*. In $MGMT^+$ cells, the island remains devoid of cytosine methylation *(14)*. In connection with the methylation noted in $MGMT^-$ cells, the MGMT promoter lacks interactions with transcription factors (most notably Sp1), and appears to be in a condensed, inaccessible chromatin state *(13)*. Consistent with this finding is the demonstration that $MGMT^-$ cells contain all the machinery necessary to express a transiently transfected reporter gene linked to an MGMT promoter devoid of higher-order chromatin structure *(28)*. Taken together, these studies suggest that an open chromatin structure associated with an unmethylated MGMT promoter allows for binding of Sp1 and subsequent MGMT expression in most normal cells, whereas in $MGMT^-$ tumor lines, a condensed chromatin structure excludes transcription factors and prevents MGMT expression.

MPG and MGMT play key roles in the repair of methylation damage, although other repair proteins may also contribute. Recent studies have suggested that the human nucleotide excision repair system may repair some forms of methylation damage. In these studies, cell-free extracts containing mammalian excinuclease were found to repair O6MG and a variety of potentially methylation-induced base mismatches, although the rate of repair of O6MG was quite low *(33)*. Thus, the repair of methylation damage probably occurs at two different levels. Rapid and relatively specific repair is accomplished by MPG and MGMT. In the absence of MPG and MGMT, cells may rely on an inefficient, broad specificity backup system. This potential redundancy in repair capacity highlights the importance of repair of methylation damage and the complexity of response to methylation damage.

4. CELLULAR RESPONSES TO METHYLATION DAMAGE

The cellular response to DNA methylation damage depends to a large degree on the type of damage produced and the repair mechanisms available to the cell to deal with the damage. N7MG and phosphotriesters are the primary type of damage produced by the majority of methylating agents. These lesions do not, however, appear to be directly harmful to cells, and do not elicit large cellular responses. Although it is unclear how phosphotriesters are removed, or if they are removed, the disappearance of N7MG is primarily dependent on the inherent lability of the base. Spontaneous depurination of N7MG results in AP sites, which are known to block DNA replication in vitro *(72)*. AP sites are repaired by the sequential action of AP endonucleases (types I and II, which cleave the DNA on either side of the AP site), DNA polymerase, and DNA ligase (*see* Chapter 6). The repair of AP sites occurs continuously and does not appear to require delay in the cell cycle. MPG may also contribute to the generation of

AP sites from N7MG, although its contribution in vivo remains uncertain. The most common lesions produced by methylating agents, therefore, appear to elicit little discernible effect in damaged cells.

In contrast to the abundant N7MG lesions, the rarer, more cytotoxic, and more mutagenic lesions, such as 3MA and O6MG, appear in large part to be responsible for the well-described cellular responses to methylation damage. 3MA lesions block DNA synthesis in vitro and likely requires repair prior to S phase *(37)*. Repair of these adducts is accomplished initially by the removal of the base by MPG to create an AP site, followed by repair of the AP site as described for N7MG adducts. The replication-terminating effects of 3MA might be expected to result in S phase arrest following exposure of cells to methylating agents. This is not typically seen, however, probably because 3MA is rare and repaired efficiently. Rather, what is noted is cell-cycle arrest in the second S phase following damage *(89)*. This effect has been ascribed to the way in which cells deal with O6MG.

MGMT-proficient cells repair O6MG lesions throughout the cell cycle in a manner that prevents the mutagenic O6MG lesion from mispairing during replication or transcription. Because of the suicide nature of the MGMT repair mechanism, and because cells have limited amounts of MGMT, the amount of O6MG lesions can exceed the repair capacity of the cell. If this occurs, O6MG lesions can escape repair and can be replicated, resulting in O6MG-thymine mispairs. Following replication, the mismatch repair system continuously removes the mispaired thymine, only to have it replaced during repair synthesis. It has been suggested that at the second S phase, the mismatch repair-generated gaps in the nonmethylated DNA strand signal termination of DNA synthesis in both damaged and undamaged DNA, leading to cell-cycle arrest *(89)*.

Cells that lack both MGMT and mismatch repair thus appear to be tolerant to the cytotoxic effects of methylating agents, although their DNA becomes highly mutated. Cells lacking MGMT and mismatch repair may exist in cancer-prone individuals and individuals with mismatch repair-deficient colon tumors *(43)*. Cells that lack MGMT, but retain mismatch repair capacity are thought to die as a result of either S-phase arrest and/or DNA breakage as a result of futile mismatch repair. $MGMT^+$ cells with normal mismatch repair capacity could also, in this scenario, suffer low levels of mutation if O6MG were replicated before repair, and if the levels of O6MG were too low to cause cell cycle arrest and/or large degrees of DNA breakage. It seems likely that O4MT lesions also contribute to the cellular response to methylating agents, since these lesions similarly allow for DNA synthesis, yet mispair during replication. Given the low frequency of this adduct, its slow disappearance, and the lack of an identifiable repair protein in human cells, the contribution of O4MT to the cellular response to methylating agents remains unclear. The predominant cellular effect of methylating agents appears to be most closely associated with the O6MG lesion, and it is the ability of the cell to carry out an appropriate response that determines whether the methylation damage suffered is innocuous, mutagenic, or cytotoxic.

5. THERAPEUTIC IMPLICATIONS OF DNA DAMAGE AND REPAIR

Although the vast majority of DNA methylating agents are not used therapeutically, the study of methylating agent-induced damage, and the repair of this damage may have direct and significant clinical impact. Chloroethylnitrosoureas (CENUs), including the well-studied compound, BCNU, are cancer chemotherapeutic agents that spontaneously decompose under physiologic conditions to form carbamoylating and chloroethylating species *(12)*. The chloroethylating species attacks guanine at the O^6 position. The initial reaction is followed by an intramolecular rearrangement and attack of the paired cytosine on the opposite DNA strand to yield a N^1-guanine-N^3-cytosine DNA interstrand crosslink. The relationship of CENU-induced damage to methylation repair was suggested by studies demonstrating that few CENU-induced DNA interstrand crosslinks were formed in cells that were able to remove O6MG from their DNA *(16,22)*. Subsequent to these studies, it was recognized that MGMT repaired O6MG, and that the initial chloroethyl adduct formed at the O^6 position of guanine by BCNU was a substrate for MGMT. This work as a whole suggested that variation in MGMT activity could influence therapeutic outcome, and that modulation of MGMT activity may be a means of modulating CENU cytotoxicity.

Several different strategies involving modulation of MGMT activity have been employed to enhance CENU efficacy. In theory, depletion of cellular MGMT should render a cell incapable of repairing the initial O^6-guanine adduct produced by CENU. Increased amounts of this initial adduct should in turn result in greater DNA interstrand crosslinking and greater toxicity to tumor cells. Nucleic acid-based approaches designed to inhibit MGMT activity have been attempted in the form of antisense oligonucleotides *(10)* and ribozyme RNAs *(63)* targeted to the MGMT mRNA. These approaches have shown some promise, although they are in the early stages of development. Considerably more effort has gone into the use of methylating agents to deplete methylation repair capacity. Initial studies suggested that pretreatment of $MGMT^+$ tumor cells with MNU, MMS, or STZ could deplete cells of MGMT to a degree that would sensitize cells to the cytotoxic effects of CENU (reviewed in ref. *23*). These studies have subsequently been pursued with STZ, which is already approved for clinical use. The strategy of MGMT depletion has also been pursued from the standpoint of nonmutagenic, direct inhibitors of the protein. A number of presumably nontoxic compounds, including O6MG, and most promisingly BG, have been developed, which directly interact with, and inactivate MGMT. In vitro and in vivo, these MGMT inhibitors sensitize cells to CENU-induced cytotoxicity (reviewed in ref. *60*). There are also efforts to combine therapeutic strategies such that the rapid depletion of MGMT by BG can be supplemented by a more long-lasting depletion caused by STZ-induced O6MG adducts *(49)*. Because the major side effect of CENU therapy, particularly following MGMT depletion, is myelosuppression, efforts are under way to protect hematopoietic progenitor cells by introduction of MGMT cDNA in retroviral vectors *(2,51,55)*.

Since many chemotherapeutic agents also directly or indirectly create N^7-guanine damage and/or apurinic sites in DNA, the possibility exists that inhibition of MPG or apurinic endonucleases could also be of therapeutic benefit. Conversely, protection of normal cells could also likely be afforded by introduction of MPG and/or apurinic endonucleases into stem cells. Such protective therapies could be of particular use in combination with high-dose chemotherapy/bone marrow transplantation regimens. Further work directly addressing methylation damage and repair should help turn therapeutic manipulation of enzymes involved in methylation repair from a concept to reality.

6. CONCLUSION

In the last 20 years, the study of DNA methylation damage and repair has gone from an appreciation of the mutagenic and carcinogenic effects of the compounds to a fairly detailed molecular understanding of these effects. Along the way, an understanding of the repair of methylation damage, an identification of the cellular responses to methylation damage, and an appreciation of how aberrant processing of methylation damage can contribute to cancer has been gained. This knowledge has also allowed for the initial steps in manipulating methylation damage and repair for therapeutic benefit. There are, however, many questions that remain unanswered with regard to methylation damage and repair. It remains unclear what triggers S-phase arrest in methylating agent-damaged cells, and how this signal is transmitted. It also remains unclear how the methylation repair proteins MGMT and MPG recognize and remove lesions, and to what extent this information could be used to develop inhibitors of the proteins. Studies addressing these questions are in progress, whereas studies addressing other questions, such as how DNA topology and chromatin structure influence methylation damage and repair, remain for the future.

ACKNOWLEDGMENTS

I would like to thank Mark Kelley and Sankar Mitra for their numerous helpful comments in the preparation of this chapter. Also thanks are given to Dawn Graunke, Sonal Patel, Tanja Dubravcic, Shelby Ting, and Len Erickson for comments on the work.

REFERENCES

1. Abbott, P. J., and R. Saffhill. 1979. DNA synthesis with methylated poly(dC-dG) templates. Evidence for a competitive nature to miscoding by *O*6-methylguanine. *Biochim. Biophys. Acta* **562:** 51–61.
2. Allay, J. A., L. L. Dumenco, O. N. Koc, L. Liu, and S. L. Gerson. 1995. Retroviral transduction and expression of the human alkyltransferase cDNA provides nitrosourea resistance to hematopoietic cells. *Blood* **85:** 3342–3351.
3. Ayi, T. C., K. C. Loh, R. B. Ali, and B. F. L. Li 1992. Intracellular localization of human DNA repair enzyme *O*6-methylguanine-DNA methyltransferase by antibodies and its importance. *Cancer Res.* **52:** 6423–6430.
4. Barrows, L. W., and P. N. Magee. 1982. Nonenzymatic methylation of DNA by *S*-adenosylmethionine in vitro. *Carcinogenesis* **3:** 349–351.
5. Beranek, D. T., C. C. Weis, and D. H. Swenson. 1980. A comprehensive quantitative analysis of methylated and ethylated DNA using high pressure liquid chromatography. *Carcinogensis* **1:** 595–606.
6. Bessho, T., R. Roy, K. Yamamoto, H. Kasai, S. Nishimura, K. Tano, and S. Mitra. 1993. Repair of 8-hydroxyguanine in DNA by mammalian *N*-methylpurine-DNA glycosylase. *Proc. Natl. Acad. Sci. USA* **90:** 8901–8904.
7. Bjelland, S., and E. Seeberg. 1987. Purification and characterization of 3-methyladenine DNA glycosylase I from *Escherichia coli*. *Nucleic Acids Res.* **15:** 2787–2801.
8. Chakravarti, D., G. C. Ibeanu, K. Tano, and S. Mitra. 1991. Cloning and expression in *Escherichia coli* of a human cDNA encoding the DNA repair protein *N*-methylpurine-DNA glycosylase. *J. Biol. Chem.* **266:** 15,710–15,715.
9. Chan, C., Z. Wu, T. Ciardelli, A. Eastman, and E. Bresnick. 1993. Kinetic and DNA-binding properties of recombinant human *O*6-methylguanine-DNA methyltransferase. *Arch. Biochemistry Biophys.* **300:** 193–200.

10. Citti, L., L. Boldrini, and G. Rainaldi. 1994. The genotoxicity of the chloroethylating agent mitozolomide is enhanced in CHO mex+ cells by the administration of antimessenger oligonucleotide targeted against methylguanine-DNA methyltransferase gene (MGMT). *Anticancer Res.* **14:** 2667–2672.
11. Citron, M., R. Decker, S. Chen, S. Schneider, M. Graver, L. Kleynerman, L. B. Kahn, A. White, M. Schoenhaus, and D. Yarosh. 1991. *O*6-Methylguanine-DNA methyltransferase in human normal and tumor tissue from brain, lung, and ovary. *Cancer Res.* **51:** 4131–4134.
12. Colvin, M. 1982. The alkylating agents, in *Pharmacologic Principles of Cancer Treatment* (Chabner, B., ed.), W. B. Saunders, Philadelphia, PA, pp. 276–308.
13. Costello, J. F., B. W. Futscher, R. A. Kroes, and R. O. Pieper. 1994. Methylation-related chromatin structure is associated with exclusion of transcription factors from and suppressed expression of the *O*-6-methylguanine DNA methyltransferase gene in human glioma cell lines. *Mol. Cell. Biol.* **14:** 6515–6521.
14. Costello, J. F., B. W. Futscher, K. Tano, D. M. Graunke, and R. O. Pieper. 1994. Graded methylation in the promoter and body of the *O*6-methylguanine DNA methyltransferase (MGMT) gene correlates with MGMT expression in human glioma cells. *J. Biol. Chem.* **269:** 17,228–17,237.
15. Crone, T. M., K. Goodtzova, S. Edara, and A. E. Pegg. 1994. Mutations in human *O*6-alkylguanine-DNA alkyltransferase imparting resistance to *O*6-benzylguanine. *Cancer Res.* **54:** 6221–6227.
16. Day, R. S. III, C. H. J. Ziolkowski, D. A. Scudiero, S. A. Meyer, A. S. Lubiniecki, A. J. Girardi, S. M. Galloway, and G. D. Bynum. 1980. Defective repair of alkylated DNA by human tumour and SV-40-transformed human cell strains. *Nature* **288:** 724–727.
17. den Engelse, L. D., G. J. Menkveld, R. J. DeBrij, and A. D. Tates. 1986. Formation and stability of alkylated pyrimidines and purines (including imidazole ring-opened 7-alkylguanine) and alkylphosphotriesters in liver DNA of adult rats treated with ethylnitosourea or dimethylnitrosamine. *Carcinogenesis* **7:** 393–403.
18. Dosanjh, M. K., A. Chenna, A. Kim, H. Fraenkel-Conrat, L. Samson, and B. Singer. 1994. All four known cyclic adducts formed in DNA by the vinyl chloride metabolite chloroacetaldehyde are released by a human DNA glycosylase. *Proc. Natl. Acad. Sci. USA* **91:** 1024–1028.
19. Dosanjh, M. K., R. Roy, S. Mitra, and B. Singer. 1994. 1,*N*6-ethenoadenine is preferred over 3-methyladenine as substrate by a cloned human *N*-methylpurine-DNA glycosylase (3-methyladenine-DNA glycosylase). *Biochemistry* **33:** 1624–1628.
20. Dyroff, M. C., F. C. Richardson, J. A. Popp, M. A. Bedell, and J. A. Swenberg. 1986. Correlation of *O*4-ethyldeoxythymidine accumulation, hepatic initiation and hepatocellular carcinoma induction in rats continuously administered diethylnitrosamine. *Carcinogenesis* **7:** 241–246.
21. Elder, R. H., J. Tumelty, K. T. Douglas, G. P. Margison, and J. A. Rafferty. 1992. C-Terminally truncated human *O*6-alkylguanine-DNA alkyltransferase retains activity. *Biochem. J.* **285:** 707–709.
22. Erickson, L. C., G. Laurent, N. A. Sharkey, and K. W. Kohn. 1980. DNA cross-linking and monoadduct repair in nitrosourea-treated human tumour cells. *Nature* **288:** 727–729.
23. Erickson, L. C. 1991. The role of *O*-6 methylguanine DNA methyltransferase (MGMT) in drug resistance and strategies for its inhibition. *Semin. Cancer Biol.* **2:** 257–265.
24. Gardiner-Garden, M., and M. Frommer. 1987. CpG islands in vertebrate genomes. *J. Mol. Biol.* **196:** 261–282.
25. Gerchman, L. L., and D. Ludlum. 1973. The properties of *O*6-methylguanine in templates for RNA polymerase. *Biochim. Biophys. Acta* **308:** 310–316.
26. Goodtzova, K., T. Crone, and A. E. Pegg. 1994. Activation of human *O*6-alkylguanine-DNA alkyltransferase by DNA. *Biochemistry* **33:** 8385–8390.

27. Grevatt, P. C., J. J. Solomon, and Bhanot, O. S. 1992. In vitro mispairing specificity of *O*2-ethylthymidine. *Biochemistry* **31:** 4181–4188.
28. Harris, L. C., P. M. Potter, J. S. Remack, and T. P. Brent. 1992. A comparison of human O6-methylguanine DNA methyltransferase promoter activity in Mer+ and Mer– cells. *Cancer Res.* **52:** 6404–6406.
29. Harris, L. C., P. M. Potter, K. Tano, S. Shiota, S. Mitra, and T. P. Brent. 1991. Characterization of the promoter region of the human *O*6-methylguanine-DNA methyltransferase gene. *Nucleic Acids Res.* **19:** 1663–1667.
30. Harris, L. C., J. S. Remack, and T. P. Brent. 1994. Identification of a 59 bp enhancer located at the first exon/intron boundary of the human *O*6-methylguanine DNA methyltransferase gene. *Nucleic Acids Res.* **22:** 4614–4619.
31. Helland, D. E., R. Male, B. L. Haukanes, L. Olsen, I. Haugen, and K. Kleppe. 1987. Properties and mechanism of action of eukaryotic 3-methyladenine-DNA glycosylases. *J. Cell Sci.* **Suppl. 6:** 139–146.
32. Hemminki, K., K. Falck, and H. Vainio. 1980. Comparisons of alkylation rates and mutagenicity of directly acting industrial and laboratory chemicals. *Arch. Toxicol.* **46:** 277–285.
33. Huang, J.-C., D. S. Hsu, A. Kazantsev, and A. Sancar. 1994. Substrate spectrum of human excinuclease: repair of abasic sites, methylated bases, mismatches, and bulky adducts. *Proc. Natl. Acad. Sci. USA* **91:** 12,213–12,217.
34. Ihara, K., H. Kawate, L. L. Chueh, H. Hayakawa, and M. Sekiguchi. 1994. Requirement of the pro-cys-his-arg sequence for *O*6-methylguanine-DNA methyltransferase activity revealed by saturation mutagenesis with negative and positive screening. *Mol. Gen. Genet.* **243:** 379–389.
35. Kanugula, S., K. Goodtzova, S. Edara, and A. E. Pegg. 1995. Alteration of arginine-128 to alanine abolishes the ability of human *O*6-alkylguanine-DNA alkyltransferase to repair methylated DNA but has no effect on its reaction with *O*6-benzylguanine. *Biochemistry* **34:** 7113–7119.
36. Kat, A., W. G. Thilly, W.-H. Fang, M. J. Longley, G.-M. Li, and P. Modrich. 1993. An alkylation-tolerant, mutator human cell line is deficient in strand-specific mismatch repair. *Proc. Natl. Acad. Sci. USA* **90:** 6424–6428.
37. Larson, K. J., R. Sahm, R. Shenkar, and B. Strauss. 1985. Methylation-induced blocks to in vitro DNA replication. *Mutat. Res.* **15:** 77–84.
38. Lawley, P. D. 1984. Carcinogenesis by alkylating agents, in *Chemical Carcinogens*, vol. 1 (Searle, C. E., eds.), ACS Monograph 182, American Chemical Society, Washington, DC, pp. 326–484.
39. Lawley, P. D., and D. J. Orr. 1970. Specific excision of methylation products from DNA of *Escherichia coli* treated with *N*-methyl-*N*'-nitro-*N*-nitrosoguanidine. *Chem. Biol. Interact.* **2:** 154–157.
40. Lawley, P. D., and C. J. Thatcher. 1970. Methylation of deoxyribonucleic acid in cultured mammalian cells by *N*-methyl-*N*'-nitro-*N*-nitrosoguanidine. *Biochem. J.* **116:** 693–707.
41. Lawley, P. D., D. J. Orr, and M. Jarman. 1975. Isolation of products from alkylation of nucleic acids: ethyl- and isopropyl-purines. *Biochem. J.* **145:** 73–84.
42. Lawrence, C. W. 1991. Classical mutagenesis techniques. *Methods Enzymol.* **194:** 273–281.
43. Leach, F. S., N. C. Nicolaides, et al. 1993. Mutations of a mut S homolog in hereditary nonpolyposis colorectal cancer. *Cell* **75:** 1215–1225.
44. Lee, S. M., J. A. Rafferty, R. H. Elder, C.-Y. Fan, M. Bromley, M. Harris, N. Thatcher, P. M. Potter, H. J. Altermatt, T. Perinat-Frey, T. Cerny, P. J. O'Connor, and G. P. Margison. 1992. Immunohistological examination of the inter-and intracellular distribution of *O*6-alkylguanine DNA-alkyltransferase in human liver and melanoma. *Br. J. Cancer* **66:** 355–360.
45. Lindahl, T., B. Sedgwick, M. Sekiguchi, and Y. Nakabeppu. 1988. Regulation and expression of the adaptive response to alkylating agents. *Annu. Rev. Biochemistry* **57:** 133–157.

46. Ling-Ling, C., T. Nakamura, Y. Nakatsu, K. Sakumi, H. Hayakawa, and M. Sekiguchi. 1992. *Carcinogenesis* **13:** 837–843.
47. Loechler, E. L., C. L. Green, and J. M. Essigmann. 1984. In vivo mutagenesis by *O*6-methylguanine built into a unique site in a viral genome. *Proc. Natl. Acad. Sci. USA* **81:** 6271–6275.
48. Ludlum, D. B. 1970. The properties of 7-methylguanine-containing templates for ribonucleic acid polymerase. *J. Biol. Chem.* **245:** 477–482.
49. Marathi, U. K., M. E. Dolan, and L. C. Erickson. 1994. Extended depletion of *O*6-methylguanine-DNA methyltransferase activity following *O*6-benzyl-2'-deoxyguanosine or *O*6-benzylguanine combined with streptozotocin treatment enhances 1,3-bis(2-chloroethyl)-1-nitrosourea cytotoxicity. *Cancer Res.* **54:** 4371–4375.
50. Margison, G. P., M. J. Capps, and O'Connor, P. J. 1973. Loss of 7-methylguanine from rat liver DNA after methylation in vivo with methyl methanesulfonate or dimethylnitrosamine. *Chem. Biol. Interact.* **6:** 119–124.
51. Maze, R., Carney, J. P., Kelley, M. R., Glassner, B. J., Williams, D. A., and Samson, L. 1996. Increasing DNA repair methyltransferse levels via bone marrow stem cell transduction rescues mice from the toxic effects of 1,3-bis(2-chloroethyl)-1-nitrosourea, a chemotherapeutic alkylating agent. *Proc. Natl. Acad. Sci. USA* **93:** 206–210.
52. Menichini, P., M. M. Mroczkowska, and B. Singer. 1994. Enzyme-dependent pausing during in vitro replication of *O*4-methylthymine in a defined oligonucleotide sequence. *Mutat. Res.* **307:** 53–59.
53. Moore, M. H., J. M. Gulbis, E. J. Dodson, B. Demple, and P. C. E. Moody. 1994. Crystal structure of a suicidal DNA repair protein: the Ada *O*6-methylguanine-DNA methyltransferase from *E coli. EMBO J.* **13:** 1495–1501.
54. Morgan, S. E., M. R. Kelley, and R. O. Pieper. 1993. The role of the carboxy-terminal tail in human *O*6-methylguanine DNA methyltransferase substrate specificity and temperature sensitivity. *J. Biol. Chem.* **268:** 19,802–19,809.
55. Moritz, T., W. Mackay, B. J. Glassner, D. A. Williams, and L. Samson. Retrovirus-mediated expression of a DNA repair protein in bone marrow protects hematopoietic cells from nitrosourea-induced toxicity in vitro and in vivo. *Cancer Res.* **55:** 2608–2614.
56. Nakatsu, Y., K. Hattori, H. Hayakawa, K. Shimizu, and M. Sekiguchi. 1993. Organization and expression of the human gene for *O*6-methylguanine-DNA methyltransferase. *Mutat. Res.* **293:** 119–132.
57. O'Connor, T. R., and J. Laval. 1991. Human cDNA expressing a functional DNA glycosylase excising 3-methyladenine and 7-methylguanine. *Biochem. Biophys. Res. Commun.* **176:** 1170–1177.
58. O'Connor, T. R. 1993. Purification and characterization of human 3-methyl-adenine-DNA glycoslyase. *Nucleic Acids Res.* **21:** 5561–5569.
59. O'Connor, T. R., S. Boiteux, and J. Laval. 1988. Ring-opened 7-methylguanine residues in DNA are a block to in vitro DNA synthesis. *Nucleic Acids Res.* **16:** 5879–5894.
60. Pegg, A. E., and T. L. Byers. 1992. Repair of DNA containing *O*6-alkylguanine. *FASEB J.* **6:** 2302–2310.
61. Pegg, A. E. 1990. Mammalian *O*6-alkylguanine-DNA alkyltransferase: regulation and importance in response to alkylating carcinogenic and therapeutic agents. *Cancer Res.* **50:** 6119–6129.
62. Pieper, R. O., B. W. Futscher, Q. Dong, T. M. Ellis, and L. C. Erickson. 1990. Comparison of *O*6-methylguanine DNA methyltransferase (MGMT) mRNA levels in Mer^+ and Mer^- human tumor cell lines containing the MGMT gene by the polymerase chain reaction technique. *Cancer Commun.* **2:** 13–20.
63. Potter, P. M., Harris, L. C., Remack, J. S., Edwards, C. C., and T. P. Brent. 1993. Ribozyme-mediated modulation of human *O*6-methylguanine-DNA methyltransferase expression. *Cancer Res.* **53:** 1731–1734.

64. Prakash, L., and B. Strauss. 1970. Repair of alkylation damage: stability of methyl groups in *Bacillus subtilis* treated with methyl methanesulfonate. *J. Bacteriol.* **102:** 760–766.
65. Preston, B. D., B. Singer, and L. A. Loeb. 1986. Mutagenic potential of *O*4-methylthymine in vivo determined by an enzymatic approach to site-specific mutagenesis. *Proc. Natl. Acad. Sci. USA* **83:** 8501–8505.
66. Preussmann, R., and B. W. Stewart. 1984. *N*-Nitroso carcinogens, in *Chemical Carcinogens* (Searle, C. E., ed.), vol. 2. ACS Monograph 182, American Chemical Society, Washington, DC, pp. 643–828.
67. Richardson, F. C., M. C. Dyroff, J. A. Boucheron, and J. A. Swenberg. 1985. Differential repair of *O*4-alkylthymine following exposure to methylating and ethylating hepatocarcinogens. *Carcinogenesis* **6:** 625–629.
68. Rydberg, B., and T. Lindahl. 1982. Nonenzymatic methylation of DNA by the intracellular methyl group donor *S*-adenosyl-L-methionine is a potentially mutagenic reaction. *EMBO J.* **1:** 211–216.
69. Rydberg, B., N. Spurr, and P. Karran. 1990. cDNA cloning and chromosomal assignment of the human *O*6-methylguanine-DNA methyltransferase. *J. Biol. Chem.* **265:** 9563–9569.
70. Saffhill, R., G. P. Margison, and P. J. O'Connor. 1985. Mechanisms of carcinogenesis induced by alkylating agents. *Biochim. Biophys. Acta.* **823:** 111–145.
71. Sagher, D., T. Karrison, J. L. Schwartz, R. Larson, P. Meier, and B. Strauss. 1988. Low *O*6-alkylguanine DNA alkyltransferase activity in the peripheral blood lymphocytes of patients with therapy-related acute nonlymphocytic leukemia. *Cancer Res.* **48:** 3084–3089.
72. Sagher, D., and B. Strauss. 1983. Insertion of nucleotides opposite AP/apyrimidinic sites in deoxyribonucleic acids during in vitro synthesis: uniqueness of adenine nucleotides. *Biochemistry* **22:** 4518–4526.
73. Samson, L., B. Derfler, M. Boosalis, and K. Call. 1991. Cloning and characterization of a 3-methyladenine DNA glycosylase cDNA from human cells whose gene maps to chromosome 16. *Proc. Natl. Acad. Sci. USA* **88:** 9127–9131.
74. Silber, J. R., B. A, Mueller, T. G. Ewers, and M. S. Berger. 1993. Comparison of *O*6-methylguanine-DNA methyltransferase activity in brain tumors and adjacent normal brain. *Cancer Res.* **53:** 3416–3420.
75. Singer, B. 1975. The chemical effects of nucleic acid alkylation and their relationship to mutagenesis and carcinogenesis. *Prog. Nucleic Acids Res. Mol. Biol.* **15:** 219–284.
76. Singer, B., H. Fraenkel-Conrat, and J. T. Kusmierek. 1978. Preparation and template activities of polynucleotides containing *O*2- and *O*4-alkyluridine. *Proc. Natl. Acad. Sci. USA* **75:** 1722–1726.
77. Singer, B. 1986. *O*-Alkyl pyrimidines in mutagenesis and carcinogenesis: Occurrence and significance. *Cancer Res.* **46:** 4879–4885.
78. Singer, B., and T. P. Brent. 1981. Human lymphoblasts contain DNA glycosylase activity excising *N*-3 and *N*-7 methyl and ethyl purines but not *O*6-alkylguanines or 1-alkyladenines. *Proc. Natl. Acad. Sci. USA* **78:** 856–860.
79. Singer, B. 1985. In vivo formation and persistence of modified nucleosides resulting from alkylating agents. *Env. Health Perspect.* **62:** 41–48.
80. Singer, B. 1979. *N*-Nitroso alkylating agents: formation and persistence of alkyl derivatives in mammalian nucleic acids as contributing factors in carcinogenesis. *J. Natl. Cancer Inst.* **62:** 1329–1339.
81. Spratt, T. E., and H. de los Santos. 1992. Reaction of *O*6-alkylguanine–DNA alkyltransferase with *O*6-methylguanine analogues: evidence that the oxygen of *O*6-methylguanine is protonated by the protein to effect methyl transfer. *Biochemistry.* **31:** 3688–3694.
82. Swann, P. F., and P. N. Magee. 1971. Nitrosamine-induced carcinogensis. The alkylation of *N*-7 guanine of nucleic acids of the rat by diethylnitrosamine, *N*-ethyl-*N*-nitrosourea and ethyl methylmethanesulfonate. *Biochem. J.* **125:** 841–847.

83. Swenberg, J. A., M. C. Dyroff, M. A. Bedell, J. A. Popp, N. Huh, V. Kirstein, and M. F. Rajeewsky. 1984. *O*4-Ethyldeoxythymidine, but not *O*6-ethyldeoxyguanosine, accumulates in hepatocyte DNA of rats exposed continuously to diethylnitrosamine. *Proc. Natl. Acad. Sci. USA* **81:** 1692–1695.
84. Swenson, D. H., and P. D. Lawley. 1978. Alkylation of deoxyribonucleic acid by carcinogens dimethyl sulphate, ethyl methanesulphonate, *N*-ethyl-*N*-nitrosourea and *N*-methyl-*N*-nitrosourea. Relative reactivity of the phosphodiester site thymidylyl (3'-5') thymidine. *Biochem. J.* **171:** 575–587.
85. Tano, K., S. Shiota, J. Collier, R. S. Foote, and S. Mitra. 1990. Isolation and structural characterization of a cDNA clone encoding the human DNA repair protein for O6-alkylguanine. *Proc. Natl. Acad. Sci. USA* **87:** 686–690.
86. Unpublished results.
87. von Wronski, M. A., L. C. Harris, K. Tano, S. Mitra, D. D. Bigner, and T. P. Brent. 1992. Cytosine methylation and suppression of *O*6-methylguanine-DNA methyltransferase in human rhabdomyosarcoma cell lines and xenografts. *Oncol. Res.* **4:** 167–174.
88. Xiao, W., Rathgeber, L., Fontanie, T., and Bawa, S. 1995. DNA mismatch repair mutants do not increase *N*-methyl-*N'*-nitro-*N*-nitrosoguanidine tolerance in O6-methylguanine DNA methyltransferase-deficient yeast cells. *Carcinogenesis* **16:** 1933–1939.
89. Zhukovskaya, N., P. Branch, G, Aquilina, and P. Karran. 1994. DNA replication arrest and tolerance to DNA methylation damage. *Carcinogenesis* **15:** 2189–2194.

4

Exogenous Carcinogen-DNA Adducts and Their Repair in Mammalian Cells

Anthony Dipple and Leonora J. Lipinski

1. CHEMICAL CARCINOGENS AND DNA ADDUCT FORMATION

A very wide range of chemical structures have been found to be carcinogenic in experimental animals, and according to the criteria used by the International Agency for Research on Cancer, 60 or more of these chemicals are probably carcinogenic in humans *(28)*. Despite recent exciting discoveries of inherited susceptibilities to cancer, such as the association of certain colon cancers with a deficiency in mismatch repair (*23*; *see* Chapter 20), Doll has argued that exogenous agents and lifestyle continue to play a major role in determining human cancer incidences *(20)*. For this reason, the mechanism of action of chemical carcinogens remains an important area of research. Although it has been argued that chemicals can induce carcinogenesis in the rodent bioassay system through either a mitogenic or mutagenic mechanism *(1)*, most of the known potent carcinogens are strong mutagens, and it is these genotoxic carcinogens that are the subject of this chapter.

Many correlations over the years have suggested that DNA is the principal target for chemical carcinogens, and the most compelling evidence in favor of this view is the demonstration that the phenotype of chemically transformed cells could be transmitted to other cells through transfection of the transformed cell DNA *(59)*. This finding is consistent with the view that the structural features in potent chemical carcinogens that are responsible for carcinogenic properties are those that endow carcinogens with reactivity toward DNA (and other cellular macromolecules) or that allow DNA-reactive metabolites to be formed through metabolic activation *(18,49)*.

Chemical carcinogen activation occurs when a normal metabolic reaction, intended to introduce polarity into foreign chemicals to facilitate their excretion, is subverted by some structural feature of the carcinogen into generating a chemically reactive metabolite that is not readily inactivated by further metabolism. Most carcinogen-activating reactions are oxidations, but reduction reactions are sometimes involved. Reactive metabolites can also result from the action of conjugation reactions on oxidation or reduction products, or on the carcinogen itself *(14,15)*. It would be unwise to assume that all possible carcinogen-activating reactions are currently known, because the

From: DNA Damage and Repair, Vol. 2: DNA Repair in Higher Eukaryotes
Edited by: J. A. Nickoloff and M. F. Hoekstra © Humana Press Inc., Totowa, NJ

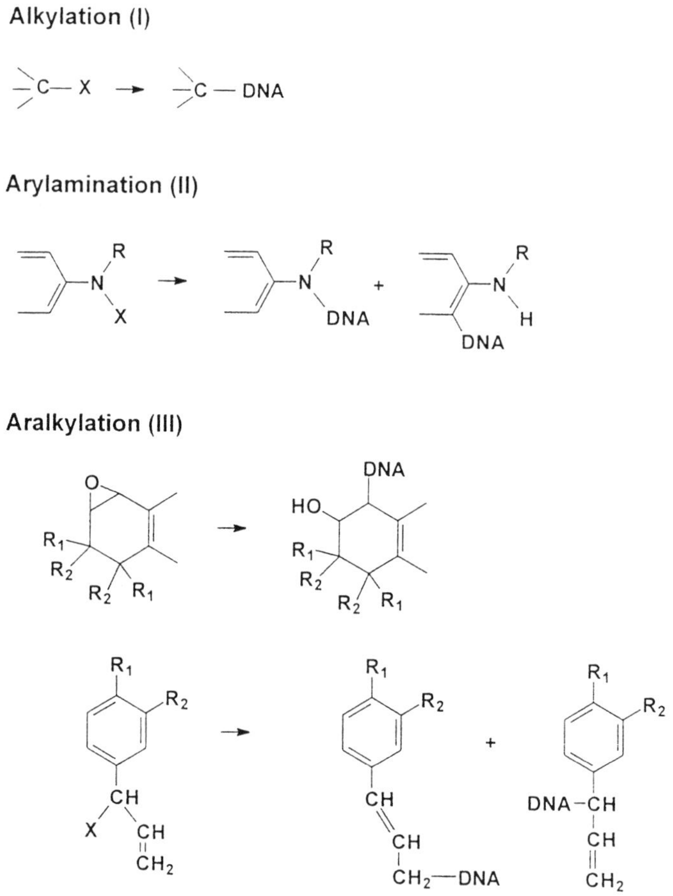

Fig. 1. Summary of typical sites of attachment of DNA bases to alkylating, arylaminating, and aralkylating agents.

metabolites of many chemical structures, both natural and synthetic, have yet to be explored.

As pointed out previously *(14,15)*, most chemical carcinogens transfer to DNA either an alkyl residue (I), an arylamino residue (II), or an aralkyl residue (III). All of these residues can be quite complex chemically. Alkylating agents (I) are derived from nitrosamines, aliphatic epoxides, aflatoxins, lactones, nitrosoureas, mustards, haloalkanes, alkyl triazenes, and sultones, but their binding to DNA is always through a saturated carbon atom that is not conjugated to an aromatic system (Fig. 1). Arylaminating agents (II) arise from aminoazo dyes, aromatic amines, nitroaromatics, and heterocyclic aromatic amines, and these residues attach to DNA through the amino nitrogen itself or through an aromatic carbon conjugated with this nitrogen (Fig. 1). The agents that transfer aralkyl residues (III) to DNA are alkenyl benzenes, pyrrolizidine alkaloids, nitroaromatics activated through dihydrodiol epoxide mechanisms, aralkyl halides, and the polycyclic aromatic hydrocarbons, and these agents react with DNA through a carbon atom that is conjugated with an aromatic system (Fig. 1).

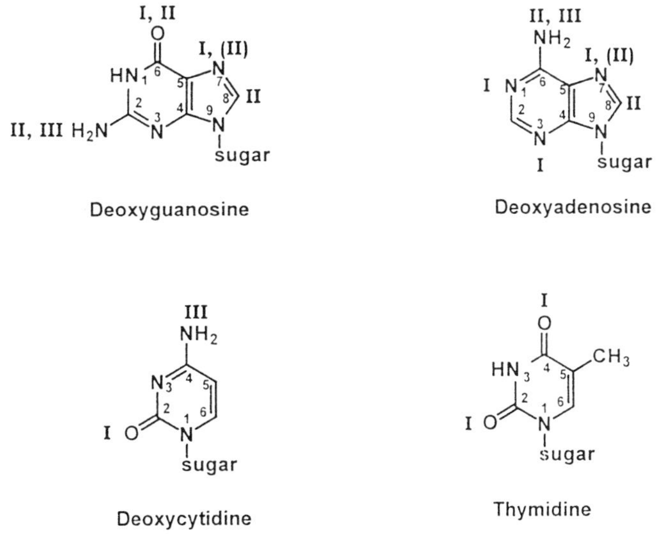

Deoxyguanosine

Deoxyadenosine

Deoxycytidine

Thymidine

Fig. 2. Sites of substitution of DNA bases by genotoxic carcinogens. Sites modified by alkylating agents are marked by the numeral I, those modified by arylaminating agents by a II, and those modified by polycyclic aralkylating agents by a III. Since it has been suggested that C^8 substituted arylamino adducts may have arisen from N^7-substituted precursors *(27)*, the arylaminating agents are listed parenthetically at the 7-position of the purines. Figure taken from Dipple *(14)*.

The above division into three broad categories is based not only on the nature of the residue that is transferred to DNA, but also on the sites in DNA to which these residues bind (Fig. 2) *(14)*. Thus, the sites on DNA bases most frequently substituted by the alkylating agents (I) are the ring nitrogens and exocyclic oxygen atoms *(36)*. In contrast, the arylaminating agents (II) primarily substitute C^8 of guanine residues (occasionally C^8 of adenine residues) and the amino group of guanine residues (occasionally the amino group of adenine residues) *(30)*. The polycyclic aralkylating agents (III) also react primarily with the amino groups of adenine or guanine residues in DNA, but unlike the arylaminating agents, they do not target the C^8 atoms of purines.

Detailed discussions of the chemistry of DNA modification by carcinogens *(29)* and of the basis for organizing carcinogens into the three broad categories above have been presented earlier *(14,15)*. The present chapter focuses on at least one example from each of these three categories and summarizes the DNA repair responses to these different classes of DNA damage.

2. METHODS FOR MONITORING ADDUCT EXCISION

The repair of DNA adducts of exogenous chemicals was originally studied over the entire cellular genome, but more recently, it has been possible to measure repair at subgenomic levels, i.e., in mitochondrial DNA, in specific genes, in the transcribed or nontranscribed strand of specific genes, and even at the level of individual nucleotides within a given gene. Two general approaches have been followed to monitor the repair of chemical damage to DNA. Thus, after exposure to a DNA-reactive chemical, both the incorporation of a radiolabeled DNA precursor during the repair synthesis step of excision repair and the time-dependent disappearance of chemical damage from cellular DNA have been extensively used to monitor DNA repair responses to chemical damage. As discussed below, a wide variety of techniques have been used in concert with these two basic methodologies (*see also* Chapter 24).

In the classical technique of measuring unscheduled DNA synthesis, cells are grown in the presence of ^{3}H-thymidine following treatment with a DNA-reactive chemical *(11)*. During the resynthesis step of excision repair, the radiolabel is incorporated into DNA, and labeled nuclei can be visualized by autoradiography. Cells in S-phase during this time also incorporate the radiolabel, and their nuclei are heavily labeled. This latter precludes measurement of repair in replicating cells using this technique, because replicative synthesis has to be blocked by addition of a drug, such as hydroxyurea.

More complex versions of the repair synthesis approach eliminate the problems arising from DNA replication during the assay by physical separation of this DNA from that originally present. Removal of this newly synthesized DNA is achieved by using 5-bromo-2'-deoxyuridine to density label the new DNA. In one version *(62)*, cellular DNA is prelabeled with ^{14}C-thymidine, and then following treatment with the chemical of interest, ^{3}H-bromodeoxyuridine is added and DNA is isolated at various times. DNA replicated during the experimental period contains bromodeoxyuridine in one strand, and is more dense than unreplicated DNA and can be separated from unreplicated DNA by equilibrium density gradient centrifugation. The ratio of $^{3}H/^{14}C$ in the unreplicated DNA is taken as a measure of DNA repair.

These repair synthesis-based assays have mostly been applied to investigations of repair in the genome overall. However, if chromatin fractionation methods are included in the protocol, these methods can be applied to repair in subgenomic fractions *(42,67)*.

The earlier methods that monitored the disappearance of DNA adducts use radioactive carcinogens *(41)*, and in replicating cells, correction for the decrease in specific radioactivity of cellular DNA owing to dilution from newly synthesized DNA is necessary. This is achieved either through prelabeling of the DNA with a different isotope *(22)* or through the physical removal of this DNA after density-labeling with 5-bromodeoxyuridine. Using this radioactive carcinogen technique, the chemical nature of the repaired DNA lesions can also be determined by digestion of the DNA to nucleosides and chromatographic separation of the products *(19)*.

These same types of repair studies can be undertaken using the more sensitive adduct postlabeling approaches *(5,24)*, although the quantitative aspects of this methodology are somewhat less clear than those associated with radioactively labeled carcinogen methods. As is the case for the repair synthesis approaches, subgenomic studies utilizing the radiolabeled adduct approach can be undertaken using chromatin fractionation approaches *(53)* or some other physical separation of the genome into different fractions.

Recent major advances in DNA repair measurement have been based on marking the presence of adducts in DNA, not by labeling the adducts, but by converting these adducts into strand breaks. Strand breaks may be created using enzymes *(70)*, photochemical cleavage *(2,40)*, or through the intrinsic chemical instability of a carcinogen-DNA adduct *(65)*. Enzymic cleavage at UV-induced dimers by T4 endonuclease V was used by Bohr et al. *(3)* in their initial studies of gene-specific repair in which Southern blotting was used to detect a restriction fragment of the gene of interest. With time, T4 endonuclease V cleavage of the restriction fragment diminished as an indicator of repair (*see* Chapter 24). A similar assay for study of chemical damage to DNA employs the *Escherichia coli* UvrABC excinuclease, which cleaves DNA containing a range of bulky adducts *(70)*, to incise DNA near the location of carcinogen adducts. A decrease in excinuclease cleavage is observed as adducts are repaired *(66)*. Repair in transcribed and nontranscribed strands of a particular gene can be studied with an adaptation of the Southern blotting technique that utilizes strand-specific probes *(48)*. This general technique has also been adapted for studying repair within mitochondria by probing for mitochondrial genes *(39)*.

Very recently, methods for measuring DNA repair at the level of individual nucleotides within a gene were established, initially for UV-induced damage *(68)*, and subsequently for bulky chemical damage to DNA *(72)*. The approach again depends on the enzymic conversion of DNA adducts into strand breaks. A gene-specific primer is then extended by a DNA polymerase to the sites of enzyme cleavage creating blunt-ended, double-stranded fragments. A linker of known sequence is ligated to these fragments, and PCR is then used to amplify them. Repair of a specific lesion prevents enzymic cleavage at that site, and therefore, no ligation-mediated PCR product is generated. The disappearance of a ligation-mediated PCR fragment serves to monitor repair at the corresponding specific nucleotide.

The techniques summarized above detect adducts in DNA by strand cleavage. Related approaches have been developed based on the ability of bulky DNA adducts to block progress of polymerase along a modified template. For example, as lesions that block replication are removed from DNA, repair within a specific gene can be studied by measuring the increase in PCR product generated by amplification with gene-specific primers *(32)*. Through the use of a linear amplification (i.e., using only one gene-specific primer), repair at individual nucleotides can be monitored *(8)*.

3. REPAIR OF SPECIFIC TYPES OF CARCINOGEN-DNA DAMAGE

The primary mechanism for removal of bulky adducts from DNA is the nucleotide excision repair (NER) pathway in which a complex of several proteins recognizes DNA damage, excises the damage along with surrounding nucleotides, synthesizes new DNA to replace the removed segment, and ligates the newly synthesized DNA to produce a continuous repaired structure (*see* Chapters 2 and 15 in vol. 1 and Chapter 18 in this vol.). However, DNA adducts produced by simple alkylating agents, such as ethyl methanesulfonate, methyl methanesulfonate (MMS) and *N*-ethyl- and *N*-methyl-*N*-nitrosourea (NMU), are also repaired by base excision repair mechanisms, in which repair is initiated by removal of a damaged base by a DNA glycosylase, as well as by direct removal of an alkyl residue by an alkyltransferase. Although detailed discussions of base excision repair mechanisms are provided in Chapter 3 in vol. 1 and Chapter 3 in this vol., the influence of adduct structure on these various repair responses is summarized below, with a focus on the repair of DNA damage from one or two examples of each of the three types of carcinogen discussed in Section 1. These types, based on the sites of reaction with DNA (Fig. 2), were alkylating agents (I), arylaminating agents (II), and polycyclic aralkylating agents (III).

3.1. Alkylating Agents (I)

Although the literature is limited for the more complex carcinogens whose reactions with DNA were classified in Section 1. as alkylations, the repair of damage from simple alkylating agents in mammalian cells has been studied extensively and was reviewed recently by Mitra and Kaina *(51)*. The major sites of modification of DNA bases by alkylating agents are the ring nitrogen atoms and exocyclic oxygens (Fig. 2) and different repair mechanisms respond to damage at these different sites. The chemical nature of the substituent is also a key factor in determining repair responses.

For example, it is believed that all bulky base adducts are substrates for NER, but special repair mechanisms exist for *O*-alkylated bases and *N*-alkylpurines generated by the simple alkylating agents *(51)*. Thus, O^6-alkylguanine residues in DNA are restored to guanine residues through the action of a repair protein that transfers the alkyl group to an active cysteine residue within the protein *(54)*. Ethyl and butyl residues are also removed from the corresponding O^6-alkylguanines, but at much slower rates than for a methyl residue *(52)*. The mammalian alkyltransferase protein will also remove a methyl group from O^4-methylthymine in DNA in vitro.

The mechanism of repair of ring nitrogen-substituted adducts derived from the simple alkylating agents differs from both the repair of the *O*-alkyl derivatives and from NER of bulky adducts. Mammalian cells encode an alkylpurine DNA glycosylase that has a broad substrate specificity and acts on 7-alkylguanines, 3-alkylguanines, and 3-alkyladenines to generate apurinic sites. The repair process is then continued by apurinic/apyrimidinic endonucleases that cleave the DNA strand adjacent to the apurinic site prior to the action of exonucleases, polymerase, and ligase that complete the repair process *(51)*.

7-Alkylguanine residues in DNA are unstable and slowly undergo an imidazole ring-opening reaction to yield a formamidopyrimidine derivative *(35)*. These lesions can be removed by a formamidopyrimidine-DNA glycosylase that has an associated apurinic/apyrimidinic lyase activity *(51)*. Most research has concerned the *E. coli* enzyme (*34*; *see* Chapter 3 in vol. 1), but an analogous activity has been reported in extracts from mammalian systems *(25,46)*. Despite the known existence of the mammalian activity, imidazole ring-opened 7-substituted guanines resulting from methylating agents *(31)* or bulky agents, such as aflatoxin *(12)*, are known to be persistent lesions in vivo.

Since an alkyltransferase repairs simple *O*-alkylguanine or *O*-alkylthymine damage, and different DNA glycosylases remove *N*-alkylpurine damage and formamidopyrimidine lesions in DNA, it is not surprising that these different lesions can be repaired at different rates. For example, there are alkyltransferase-deficient (Mer^-) cells *(61)* in which O^6-methylguanine residues are not repaired *(51)*, although other alkylation lesions are presumably repaired. As noted above, formamidopyrimidine derivatives have been found to persist much longer than their 7-alkylguanine precursors in vivo, and there is heterogeneity in the repair of different *N*-alkylpurines. Thus, the half-life for 7-alkylguanine has been found to be much longer than the half-life of 3-alkyladenine or 3-alkylguanine in rat liver cells *(71)*, showing that the repair responses vary for the same chemical alkylation at different sites on DNA bases. These differing susceptibilities to repair can be of importance, of course, in determining the ultimate biological response elicited by different adducts.

Interesting variations in the gene-specific repair of alkylation damage to DNA have been reported. For MMS and dimethylsulfate, repair was similar in the DHFR gene and its 3'-downstream nontranscribed region *(56,58)*. However, for cells treated with NMU, the removal of adducts in active genes was greater than in inactive genes *(37,38)*. Since NMU is able to methylate sites that are relatively resistant to MMS and dimethyl sulfate, it is conceivable that the chemical cleavage method used, i.e., neutral thermal depurination followed by alkaline treatment, may detect adducts other than *N*-methylpurines in the methylnitrosourea experiments.

3.2. *Arylaminating Agents (II)*

The principal sites of adduct formation for the arylaminating agents are the C^8 and the amino groups of the purine bases in DNA *(30)*. Since all these adducts are fairly bulky, it might be reasonable to anticipate that they would all be repaired through the same NER pathway. However, numerous studies indicate that there is considerable variation in the persistence of different adducts in DNA, possibly indicative of selective repair.

In studies of primary rat hepatocyte cultures following treatment with radiolabled *N*-hydroxy-2-acetylaminofluorene *(26)*, Howard et al. found three major DNA adducts identified as *N*-(deoxyguanosin-8-yl)-2-acetylaminofluorene, *N*-(deoxyguanosin-8-yl)-2-aminofluorene, and 3-(deoxyguanosin-N^2-yl)-2-acetylaminofluorene. With time, the acetylated C^8-substituted deoxyguanosine adduct was lost far more rapidly than the other adducts, which were removed slowly, if at all. Although it can be diffficult to distinguish completely between repair of an acetylated adduct and deacetylation in the biological milieu, it has been shown that deacetylation does not occur in human cells *(45)*. Many studies that measure carcinogen-DNA adducts at various times in cells in culture and in animal models do not make corrections for *de novo* DNA synthesis. In these cases, relative rates of disappearance of individual adducts are clearly established, but there is not always a clear distinction between repair and dilution of DNA adducts owing to DNA synthesis.

Under alkaline conditions in vitro, the persistent nonacetylated C^8 substituted deoxyguanosine adduct, above, can undergo imidazole ring opening at the 7,8-bond. Although this ring-opened adduct is a substrate for the *E. coli* formamidopyrimidine DNA glycosylase in vitro *(4)*, it has not been found in cellular DNA *(33)* and may not, therefore, play a role in repair in mammalian cells.

Studies with other arylaminating carcinogens showed that, in all cases, the N^2-substituted deoxyguanosine adducts were generally persistent (slowly repaired), whereas the persistence of the C^8-substituted deoxyguanosine adducts varied, depending on the tissue and species under investigation *(69)*. In a direct comparison of N^2-substituted deoxyguanosine adducts and N^6-substituted deoxyadenosine adducts formed from *N*-methyl-4-aminoazobenzene in rats, it was noted that the minor deoxyadenosine adduct was short-lived, whereas the deoxyguanosine adduct was persistent in hepatic DNA. It was also found that the C^8-substituted deoxyguanosine derivative underwent facile opening of the 8,9-bond of the imidazole ring, and it was suggested this might occur in vivo providing a potential substrate for the formamidopyrimidine DNA-glycosylase-mediated repair pathway *(69)*. As was the case for the alkylating agents, the site of substitution of DNA bases influences rates of adduct removal, and the nature of the substituent also has profound effects on the fate of the adduct.

Sophisticated studies of the repair of the *N*-(deoxyguanosin-8-yl)-2-aminofluorene adduct within the DHFR gene of CHO cells showed that adduct removal was similar in both coding and noncoding sequences of the gene *(64)*. This is in contrast with the removal of other bulky lesions such as those induced by UV irradiation, which are repaired preferentially in coding regions *(3)*.

3.3. *Polycyclic Aralkylating Agents (III)*

The polycyclic aralkylating agents are typified by the polycyclic aromatic hydrocarbon carcinogens that react with DNA through metabolically formed dihydrodiol epoxides. These metabolites have been found to react with DNA by substitution of the amino groups of the DNA bases *(13,55)* in the same general fashion described for the synthetic bromomethylbenz[*a*]anthracenes *(16,57)*. In studies utilizing radioactive 7-bromomethylbenz[*a*]anthracene to measure adduct disappearance and bromodeoxyuridine to correct for new DNA synthesis, excision of the minor N^6-substituted deoxyadenosine adduct was faster than excision of the major N^2-substituted deoxyguanosine adduct *(19,47)*, similar to findings for *N*-methyl-4-aminoazobenzene adducts described in Section 3.2. However, N^6-substituted deoxyadenosine adducts constitute over 50% of the total adducts formed from dihydrodiol epoxide metabolites of 7,12-dimethylbenz[*a*]anthracene *(50)*, and excision of these adducts was very slow in cells that effectively removed the 7-bromomethylbenz[*a*]anthracene-DNA adducts *(17)*.

Most studies of repair of hydrocarbon carcinogen-DNA adducts have utilized benzo[*a*]pyrene derivatives to introduce DNA damage. The overall rate of removal of radioactive benzo[*a*]pyrene-DNA adducts has been reported to be fairly slow in some cell cultures *(22,43,60,73)*. However, by using partially purified UvrABC proteins to detect adducts by incision, normal human cells were shown to remove 60% of the enzyme sensitive sites in a 12-h period *(70)*.

Benzo[*a*]pyrene dihydrodiol epoxide places >90% of the DNA substitutions on the N^2-amino group of deoxyguanosine residues and, therefore, does not present an ideal means of testing the relative rates of excision of adenine vs guanine adducts. Nevertheless, a faster rate for excision of deoxyadenosine adducts compared with deoxyguanosine adducts has been reported for human lung carcinoma cells *(22)*, further emphasizing the different rates of adduct removal associated with different sites of substitution on the DNA bases.

Rates of DNA repair are also affected by the nature of the substituent in the adduct as evidenced by the finding that radioactive adducts generated by the nontumorigenic *syn* diastereomer of benzo[*a*]pyrene dihydrodiol epoxide are removed from CHO cell DNA considerably faster than adducts from the tumorigenic *anti* diastereomer *(44)*. Studies in human lymphocytes indicated different rates of removal of radioactive adducts for these two diastereomeric dihydrodiol epoxides, but in this case, the adducts derived from the *anti* dihydrodiol epoxide were the most rapidly removed *(7)*.

Studies of repair of aralkylating agent-induced damage in specific genes have largely utilized UvrABC excinuclease nicking adjacent to DNA adduct sites to monitor repair. Several such studies of repair of benzo[*a*]pyrene dihydrodiol epoxide damage in transcribed genes have produced results that present some contrasts. Preferential repair was reported for the HPRT gene in human fibroblasts, and repair was more rapid for the transcribed strand than for the nontranscribed strand *(9)*. These data correspond well with mutational data from the same system showing fewer mutations in the transcribed stand than in the nontranscribed stand *(10)*. However, in the DHFR and the adenine phosphoribosyltransferase (APRT) genes in CHO cell lines, repair rates were similar to those found in inactive regions of genome, and correspondingly, no strand-specific repair was noted *(63)*. In the same CHO system, the lack of gene- and strand-specific repair in response to benzo[*a*]pyrene dihydrodiol epoxide exposure was confirmed using a laser to create strand breaks at the sites of adduct formation *(2)*. However, with a different aralkylating agent, a benzo[*c*]phenanthrene dihydrodiol epoxide that generates comparable amounts of N^2-substituted deoxyguanosine and N^6-substituted deoxyadenosine adducts, and with a different CHO cell line, preferential adduct repair in both the expressed DHFR gene and within the transcribed strand of this gene was detected *(6)*.

Different rates of repair at different nucleotides within the HPRT gene of human fibroblasts, determined using ligation-mediated PCR, were reported recently for benzo[*a*]pyrene dihydrodiol epoxide *(72)*. Rates of repair were slow at individual nucleotides that had previously been shown to be hotspots of mutation *(9,10)*. This indicates that, as shown earlier for UV-induced mutations *(68)*, rates of repair at individual nucleotides may be strong determinants of susceptibility to mutation.

4. CONCLUSIONS

This brief summary of some of the literature concerning the repair of DNA damage induced by exogenous chemicals illustrates the somewhat intimidating complexity of this field. The variety of chemical damage that can be introduced into cellular DNA is very extensive, and these chemical residues may be placed at several available receptor sites on the DNA bases depending on the chemistry of the agent. In this chapter, three main types of chemical damage, alkylation, arylamination, and aralkylation, have been discussed. However, there are types of DNA damage that do not fit within these categories, e.g., etheno adducts formed by chloroacetaldehyde, and novel DNA glycosylases that act on these lesions have been reported recently *(21)*. In addition to the complexity of the chemical damage, it seems clear that the repair capacity or competence of different cells or organs can differ substantially and that some of the enzymic activities detected in crude cell extracts do not necessarily make realistic contributions to repair in the living cell. The combined effect of cellular and chemical variables adds to the complexity.

Although a great deal remains to be learned about the repair of carcinogen adducts in mammalian cells, some generalizations can be made. The repair of simple alkylating agent damage involves single repair proteins with DNA glycosylase or, in one case, alkyltransferase activity. At present, the findings for repair of simple alkyl lesions in actively transcribed genes depend on the alkylating agent used, but no evidence to suggest preferential repair of the transcribed strand has been reported *(58)*. The gene-specific repair of the simple alkyl lesions may reflect the accessibility of the transcribed genes to the simple repair proteins that act on these lesions.

The more bulky adducts are repaired through the NER pathway, and though these were divided into arylaminations and aralkylations herein, there are interesting overlaps. There are examples of preferential repair of N^6-substituted deoxyadenosine adducts over N^2-substituted deoxyguanosine adducts for both aromatic amines and polycyclic aralkylating agents, suggesting that the site of substitution might be a more important determinant of relative excision rates than the nature of the substituent itself.

Findings with regard to preferential repair in expressed genes is somewhat confusing at present. The arylaminating agents do not seem to be subject to preferential repair in expressed genes, whereas aralkylated adducts from benzo[*c*]phenanthrene dihydrodiol epoxide are subject to preferential repair; both positive and negative findings have been reported for benzo[*a*]pyrene dihydrodiol epoxide adducts. The apparent discrepancy in the benzo[*a*]pyrene findings may be owing to inherent differences in the activity of the UvrABC excinuclease used, in sources of the chemical carcinogen, or because different cell types were studied. Any real differences between the benzo[*a*]pyrene and benzo[*c*]phenanthrene dihydrodiol epoxide findings might be attributable to different susceptibilities of adenine and guanine residues in DNA to these two carcinogens, but further studies will be necessary to determine this.

The advent of new techniques to measure repair of exogenous carcinogen-DNA adducts and further developments in the understanding of the enzymology and function of the repair machinery itself will ultimately allow the repair fate of the wide array of potential chemical damage to DNA to be more predictable.

ACKNOWLEDGMENTS

We thank W. M. Baird, R. C. Moschel, and G. T. Pauly for helpful comments on this manuscript. Research was supported by the National Cancer Institute, DHHS, under contract with ABL. The contents of this chapter do not necessarily reflect the views or policies of the Department of Health and Human Services, nor does mention of trace names, commercial products, or organizations imply endorsement by the US government.

REFERENCES

1. Ames, B. N., M. K. Shigenaga, and L. S. Gold. 1993. DNA lesions, inducible DNA repair, and cell division: Three key factors in mutagenesis and carcinogenesis. *Environ. Health Perspect.* **101(Suppl. 5):** 35–44.
2. Baird, W. M., C. A. Smith, G. Spivak, R. J. Mauthe, and P. C. Hanawalt. 1994. Analysis of the fine structure of the repair of *anti*-benzo[a]pyrene-7,8-diol-9,10-epoxide-DNA adducts in mammalian cells by laser-induced strand cleavage. *Polycyclic Aromatic Compounds* **6:** 169–176.
3. Bohr, V. A., C. A. Smith, D. S. Okumoto, and P. C. Hanawalt. 1985. DNA repair in an active gene: removal of pyrimidine dimers from the DHFR gene of CHO cells is much more efficient than the genome overall. *Cell* **40:** 359–369.
4. Boiteux, S., M. Bichara, R. P. P. Fuchs, and J. Laval. 1989. Excision of the imidazole ring-opened form of *N*-2-aminofluorene-C(8)-guanine adduct in poly(dG-dC) by *Escherichia coli* formamidopyrimidine—DNA glycosylase. *Carcinogenesis* **10:** 1095–1909.
5. Carothers, A. M., W. Yuan, B. E. Hingerty, S. Broyde, D. Grunberger, and E. G. Snyderwine. 1994. Mutation and repair induced by the carcinogen 2-(hydroxyamino)-1-methyl-6-phenylimidazo[4,5-*b*]pyridine (*N*-OH-PhIP) in the dihydrofolate reductase gene of Chinese hamster ovary cells and conformational modeling of the dG-C8-PhIP adduct in DNA. *Chem. Res. Toxicol.* **7:** 209–218.
6. Carothers, A. M., W. Zhen, J. Mucha, Y.-J. Zhang, R. M. Santella, D. Grunberger, and V. A. Bohr. 1992. DNA strand-specific repair of (±)-3α,4β–dihydroxy-1α,2α-epoxy-1,2,3,4-

tetrahydrobenzo[*c*]phenanthrene adducts in the hamster dihydrofolate reductase gene. *Proc. Natl. Acad. Sci. USA* **89:** 11,925–11,929.

7. Celotti, L., P. Ferraro, D. Furlan, N. Zanesi, and S. Pavanello. 1993. DNA repair in human lymphocytes treated in vitro with (±)-*anti*- and (±)-*syn*-benzo[*a*]pyrene diolepoxide. *Mutat. Res.* **294:** 117–126.
8. Chandrasekhar, D. and B. Van Houten. 1994. High resolution mapping of UV-induced photoproducts in the *Escherichia coli lacI* gene: Inefficient repair of the non-transcribed strand correlates with high mutation frequency. *J. Mol. Biol.* **238:** 319–332.
9. Chen, R.-H., V. M. Maher, J. Brouwer, P. Van de Putte, and J. J. McCormick. 1992. Preferential repair and strand-specific repair of benzo[*a*]pyrene diol epoxide adducts in the *HPRT* gene of diploid human fibroblasts. *Proc. Natl. Acad. Sci. USA* **89:** 5413–5417.
10. Chen, R.-H., V. M. Maher, and J. J. McCormick. 1990. Effect of excision repair by diploid human fibroblasts on the kinds and locations of mutations induced by (±)-7β,8α-dihydroxy-9α,10α-epoxy-7,8,9,10-tetrahydrobenzo[*a*]pyrene in the coding region of the *HPRT* gene. *Proc. Natl. Acad. Sci. USA* **87:** 8680–8684.
11. Cleaver, J. E. and G. H. Thomas. 1981. Measurement of unscheduled synthesis by autoradiography, in *DNA Repair—A Laboratory Manual of Research Procedures*, (Freidberg, E. C. and P. C. Hanawalt, eds.), Dekker, New York, pp. 277–287.
12. Croy, R. G. and G. N. Wogan. 1981. Temporal patterns of covalent DNA adducts in rat liver after single and multiple doses of aflatoxin B1. *Cancer Res.* **41:** 197–203.
13. Dipple, A. 1994. Reactions of polycyclic aromatic hydrocarbons with DNA, in *DNA Adducts: Identification and Biological Significance* (Hemminki, K., A. Dipple, D. E. G. Shuker, F. F. Kadlubar, D. Segerbäck, and H. Bartsch, eds.), IARC, Lyon, pp. 107–129.
14. Dipple, A. 1995. DNA adducts of chemical carcinogens. *Carcinogenesis* **16:** 437–441.
15. Dipple, A. 1997. DNA-reactive agents. *Comp. Toxicol.* (in press).
16. Dipple, A., P. Brookes, D. S. Mackintosh, and M. P. Rayman. 1971. Reaction of 7-bromomethylbenz[*a*]anthracene with nucleic acids, polynucleotides, and nucleosides. *Biochemistry* **10:** 4323–4330.
17. Dipple, A. and M. E. Hayes. 1979. Differential excision of carcinogenic hydrocarbon-DNA adducts in mouse embryo cell cultures. *Biochem. Biophys. Res. Commun.* **91:** 1225–1231.
18. Dipple, A., P. D. Lawley, and P. Brookes. 1968. Theory of tumour initiation by chemical carcinogens: dependence of activity on structure of ultimate carcinogen. *Eur. J. Cancer* **4:** 493–506.
19. Dipple, A. and J. J. Roberts. 1977. Excision of 7-bromomethylbenz[*a*]anthracene-DNA adducts in replicating mammalian cells. *Biochemistry* **16:** 1499–1503.
20. Doll, R. 1996. Nature and nurture: possibilities for cancer control. *Carcinogenesis* **17:** 177–184.
21. Dosanjh, M. K., A. Chenna, E. Kim, H. Fraenkel-Conrat, L. Samson, and B. Singer. 1994. All four known cyclic adducts formed in DNA by the vinyl chloride metabolite chloroacetaldehyde are released by a human DNA glycosylase. *Proc. Natl. Acad. Sci. USA* **91:** 1024–1028.
22. Feldman, G., J. Remsen, T. V. Wang, and P. Cerutti. 1980. Formation and excision of covalent deoxyribonucleic acid adducts of benzo[*a*]pyrenediol epoxide I in human lung cells A549. *Biochemistry* **19:** 1095–1101.
23. Fishel, R., M. K. Lescoe, M. R. S. Rao, N. G. Copeland, N. A. Jenkins, J. Garber, M. Kane, and R. Kolodner. 1993. The human mutator gene homolog MSH2 and its association with hereditary nonpolyposis colon cancer. *Cell* **75:** 1027–1038.
24. Gupta, R. C., M. V. Reddy, and K. Randerath. 1982. ^{32}P-Postlabeling analysis of non-radioactive aromatic carcinogen-DNA adducts. *Carcinogenesis* **3:** 1081–1092.
25. Hall, J., H. Brésil, F. Donato, C. P. Wild, N. A. Loktionova, O. I. Kazanova, I. P. Komyakov, V. G. Lemekhov, A. J. Likhachev, and R. Montesano. 1993. Alkylation and

oxidative-DNA damage repair activity in blood leukocytes of smokers and non-smokers. *Int. J. Cancer* **54:** 728–733.

26. Howard, P. C., D. A. Casciano, F. A. Beland, and J. G. Shaddock, Jr. 1981. The binding of N-hydroxy-2-acetylaminofluorene to DNA and repair of the abducts in primary rat hepatocyte cultures. *Carcinogenesis* **2:** 97–102.
27. Humphreys, W. G., F. F. Kadlubar, and F. P. Guengerich. 1992. Mechanism of C^8 alkylation of guanine residues by activated arylamines: Evidence for initial adduct formation at the N^7 position. *Proc. Natl. Acad. Sci. USA* **89:** 8278–8282.
28. IARC. 1987. Overall evaluations of carcinogenicity: an updating of IARC monographs 1–42, IARC, Lyon.
29. IARC. 1994. DNA adducts: identification and biological significance, IARC, Lyon.
30. Kadlubar, F. F. 1994. DNA adducts of carcinogenic aromatic amines, in *DNA Adducts: Identification and Biological Significance* (Hemminki, K., A. Dipple, D. E. G. Shuker, F. F. Kadlubar, D. Segerbäck, and H. Bartsch, eds.), IARC, Lyon, pp. 199–216.
31. Kadlubar, F. F., D. T. Beranek, C. C. Weis, F. E. Evans, R. Cox, and C. C. Irving. 1984. Characterization of the purine ring-opened 7-methylguanine and its persistence in rat bladder epithelial DNA after treatment with the carcinogen *N*-methyinitrosourea. *Carcinogenesis* **5:** 587–592.
32. Kalinowski, D. P., S. Illenye, and B. Van Houten. 1992. Analysis of DNA damage and repair in murine leukemia L1210 cells using a quantitative polymerase chain reaction assay. *Nucleic Acids Res.* **20:** 3485–3494.
33. Kriek, E., L. den Engelse, E. Scherer, and J. G. Westra. 1984. Formation of DNA modifications by chemical carcinogens identification, localization and quantification. *Biochim. Biophys. Acta* **738:** 181–201.
34. Laval, J., F. Lopes, J. C. Madelmont, D. Godeneche, G. Meyniel, Y. Habraken, T. R. O'Connor, and S. Boiteux. 1991. Excision of imidazole ring-opened *N*7-hydroxyethylguanine from chloroethylnitrosourea-treated DNA by *Escherichia coli* formamidopyrimidine-DNA glycosylase, in *Relevance to Human Cancer of N-Nitroso Compounds, Tobacco Smoke and Mycotoxins* (O'Neill, I. K., J. Chen, and H. Bartsch, eds.), IARC, Lyon, pp. 412–416.
35. Lawley, P. D. 1966. Effects of some chemical mutagens and carcinogens on nucleic acids. *Prog. Nucleic Acid Res. Mol. Biol.* **5:** 89–131.
36. Lawley, P. D. 1994. From fluorescence spectra to mutational spectra, a historical overview of DNA-reactive compounds, in *DNA Adducts: Identification and Biological Significance* (Hemminki, K., A. Dipple, D. E. G. Shuker, F. F. Kadlubar, D. Segerbäck, and H. Bartsch, eds.), IARC, Lyon, pp. 3–22.
37. LeDoux, S. P., N. J. Patton, J. W. Nelson, V. A. Bohr, and G. L. Wilson. 1990. Preferential DNA repair of alkali-labile sites within the active insulin gene. *J. Biol. Chem.* **265:** 14,875–14,880.
38. LeDoux, S. P., M. Thangada, V. A. Bohr, and G. L. Wilson. 1991. Heterogeneous repair of methylnitrosourea-induced alkali-labile sites in different DNA sequences. *Cancer Res.* **51:** 775–779.
39. LeDoux, S. P., G. L. Wilson, E. J. Beecham, T. Steusner, K. Wassermann, and V. A. Bohr. 1992. Repair of mitochondrial DNA after various types of damage in Chinese hamster ovary cells. *Carcinogenesis* **13:** 1967–1973.
40. Li, B., B. Mao, T.-M. Liu, J. Xu, A. Dourandin, S. Amin, and N. E. Geacintov. 1995. Laser pulse-induced photochemical strand cleavage of site-specifically and covalently modified (+)-*anti*-benzo[*a*]pyrene diol epoxide-oligonucleotide adducts. *Chem. Res. Toxicol.* **8:** 396–402.
41. Lieberman, M. W. and A. Dipple. 1972. Removal of bound carcinogen during DNA repair in nondividing human lymphocytes. *Cancer Res.* **32:** 1855–1860.

42. Lieberman, M. W. and M. C. Poirier. 1974. Intragenomal distribution of DNA repair synthesis: repair in satellite and mainband DNA in cultured mouse cells. *Proc. Natl. Acad. Sci. USA* **71:** 2461–2465.
43. Lo, K.-Y. and T. Kakunaga. 1982. Similarities in the formation and removal of covalent DNA adducts in benzo[*a*]pyrene-treated BALB/3T3 variant cells with different induced transformation frequencies. *Cancer Res.* **42:** 2644–2650.
44. MacLeod, M. C., A. Daylong, G. Adair, and R. M. Humphrey. 1991. Differences in the rate of DNA adduct removal and the efficiency of mutagenesis for two benzo[*a*]pyrene diol epoxides in CHO cells. *Mutat. Res. Genet. Toxicol. Testing* **261:** 267–279.
45. Mah, M. C.-M., J. Boldt, S. J. Culp, V. M. Maher, and J. J. McCormick. 1991. Replication of acetylaminofluorene-adducted plasmids in human cells: Spectrum of base substitutions and evidence of excision repair. *Proc. Natl. Acad. Sci. USA* **88:** 10,193–10,197.
46. Margison, G. P. and A. E. Pegg. 1981. Enzymatic release of 7-methylguanine from methylated DNA by rodent liver extracts. *Proc. Natl. Acad. Sci. USA* **78:** 861–865.
47. McCaw, B. A., A. Dipple, S. Young, and J. J. Roberts. 1978. Excision of hydrocarbon-DNA adducts and consequent cell survival in normal and repair defective human cells. *Chem. Biol. Interact.* **22:** 139–151.
48. Mellon, I., G. Spivak, and P. C. Hanawalt. 1987. Selective removal of transcription-blocking DNA damage from the transcribed strand of the mammalian DHFR gene. *Cell* **51:** 241–249.
49. Miller, J. A. and E. C. Miller. 1971. Chemical carcinogenesis: mechanism and approaches to its control. *J. Natl. Cancer Inst.* **47:** v–xiv.
50. Milner, J. A., M. A. Pigott, and A. Dipple. 1985. Selective effects of selenium on 7,12-dimethylbenz[*a*]anthracene-DNA binding in fetal mouse cell cultures. *Cancer Res.* **45:** 6347–6354.
51. Mitra, S. and B. Kaina. 1993. Regulation of repair of alkylation damage in mammalian genomes. *Prog. Nucleic Acid Res. Mol. Biol.* **44:** 109–142.
52. Morimoto, K., M. E. Dolan, D. Scicchitano, and A. E. Pegg. 1985. Repair of O^6-propylguanine and O^6-butylguanine in DNA by O^6-alkylguanine-DNA alkyltransferases from rat liver and *E. coli*. *Carcinogenesis* **6:** 1027–1031.
53. Oleson, F. B., B. L. Mitchell, A. Dipple, and M. W. Lieberman. 1979. Distribution of DNA damage in chromatin and its relation to repair in human cells treated with 7-bromomethylbenz(a)anthracene. *Nucleic Acids Res.* **7:** 1343–1361.
54. Pegg, A. E., M. E. Dolan, and R. C. Moschel. 1995. Structure, function, and inhibition of O^6-alkylguanine-DNA alkyltransferase. *Prog. Nucleic Acid Res. Mol. Biol.* **51:** 167–223.
55. Peltonen, K. and A. Dipple. 1995. Polycyclic aromatic hydrocarbons: chemistry of DNA adduct formation. *J. Occup. Env. Med.* **37:** 52–58.
56. Pirsel, M. and V. A. Bohr. 1993. Methyl methanesulfonate adduct formation and repair in the DHFR gene and in mitochondrial DNA in hamster cells. *Carcinogenesis* **14:** 2105–2108.
57. Rayman, M. P. and A. Dipple. 1973. Structure and activity in chemical carcinogenesis. Comparison of the reactions of 7-bromomethylbenz[*a*]anthracene and 7-bromomethyl-12-methylbenz[*a*]anthracene with deoxyribonucleic acid in vitro. *Biochemistry* **12:** 1202–1207.
58. Scicchitano, D. A. and P. C. Hanawalt. 1989. Repair of *N*-methylpurines in specific DNA sequences in Chinese hamster ovary cells: absence of strand specificity in the dihydrofolate reductase gene. *Proc. Natl. Acad. Sci. USA* **86:** 3050–3054.
59. Shih, C., B. Shilo, M. P. Goldfarb, A. Dannenberg, and R. A. Weinberg. 1979. Passage of phenotypes of chemically transformed cells via transfection of DNA and chromatin. *Proc. Natl. Acad. Sci. USA* **76:** 5714–5718.
60. Shinohara, K. and P. A. Cerutti. 1977. Excision repair of benzo[*a*]pyrene-deoxyguanosine adducts in baby hamster kidney 21/C13 cells and in secondary mouse embryo fiboblasts C57BL/6J. *Proc. Natl. Acad. Sci. USA* **74:** 979–983.

61. Sklar, R. and B. Strauss. 1981. Removal of O^6-methylguanine from DNA of normal and xeroderma pigmentosum-derived lymphoblastoid lines. *Nature* **289:** 417–420.
62. Smith, C. A., P. K. Cooper, and P. C. Hanawalt. 1981. Measurement of repair replication by equilibrium sedimentation, in *DNA Repair—A Laboratory Manual of Research Procedures* (Freidberg, E. C. and P. C. Hanawalt, eds.), Dekker, New York, pp. 289–305.
63. Tang, M., A. Pao, and X. Zhang. 1994. Repair of benzo(a)pyrene diol epoxide- and UV-induced DNA damage in dihydrofolate reductase and adenine phosphoribosyltransferase genes of CHO cells. *J. Biol. Chem.* **269:** 12,749–12,754.
64. Tang, M.-S., V. A. Bohr, X.-S. Zhang, J. Pierce, and P. C. Hanawalt. 1989. Quantification of aminofluorene adduct formation and repair in defined DNA sequences in mammalian cells using the UVRABC nuclease. *J. Biol. Chem.* **264:** 14,455–14,462.
65. Teebor, G. W. and T. P. Brent. 1981. Measurement of alkali-labile sites, in *DNA Repair—A Laboratory Manual of Research Procedures* (Freidberg, E. C. and P. C. Hanawalt, eds.), Dekker, New York, pp. 203–212.
66. Thomas, D. C., A. G. Morton, V. A. Bohr, and A. Sancar. 1988. General method for quantifying base adducts in specific mammalian genes. *Proc. Natl. Acad. Sci. USA* **85:** 3723–3727.
67. Tlsty, T. D. and M. W. Lieberman. 1978. The distribution of DNA repair synthesis in chromatin and its rearrangement following damage with *N*-acetoxy-2-acetylaminofluorene. *Nucleic Acids Res.* **5:** 3261–3273.
68. Tornaletti, S. and G. P. Pfeifer. 1994. Slow repair of pyrimidine dimers at p53 mutation hotspots in skin cancer. *Science* **263:** 1436–1440.
69. Tullis, D. L., K. L. Dooley, D. W. Miller, K. P. Baetcke, and F. F. Kadlubar. 1987. Characterization and properties of the DNA adducts formed from *N*-methyl-4-aminoazobenzene in rats during a carcinogenic treatment regimen. *Carcinogenesis* **8:** 577–583.
70. Van Houten, B., W. E. Maskerj, W. L. Carrier, and J. D. Regan. 1986. Quantitation of carcinogen-induced DNA damage and repair in human cells with the UVR ABC excision nuclease from *Escherichia coli*. *Carcinogenesis* **7:** 83–87.
71. Vogel, E. W. and A. T. Natarajan. 1995. DNA damage and repair in somatic and germ cells in vivo. *Mutat. Res.* **330:** 183–208.
72. Wei, D., V. M. Maher, and J. J. McCormick. 1995. Site-specific rates of excision repair of benzo[*a*]pyrene diol epoxide adducts in the hypoxanthine phosphoribosyltransferase gene of human fibroblasts: Correlation with mutation spectra. *Proc. Natl. Acad. Sci. USA* **92:** 2204–2208.
73. Yang, L. L., V. M. Maher, and J. J. McCormick. 1980. Error-free excision of the cytotoxic, mutagenic N^2-deoxyguanosine DNA adduct formed in human fibroblasts by (±)-7β,8α-dihydroxy-9α,10α-epoxy-7,8,9,10-tetrahydrobenzo[α]pyrene. *Proc. Natl. Acad. Sci. USA* **77:** 5933–5937.

5

Nature of Lesions Formed by Ionizing Radiation

John F. Ward

1. INTRODUCTION

A major goal of those studying ionizing radiation-induced damage in DNA is the identification of the alterations that (when produced within a cell) are the source of higher-level damage (killing, mutation, transformation). Of course, the damage that is produced initially by radiation is subject to enzymatic repair, and hence, to assess the ability of a cell to repair the change, it is important to know the identities of the types of damage. Product identification in the DNA of irradiated cells is not straightforward. Yields of specific base damages have been measured after high doses, but damage to the deoxyribose moieties is less studied, except as the general consequence of such damage, i.e., strand breakage. The majority of studies in which products have been identified and characterized have been carried out in model systems. In these systems, a variety of routes to damage production are known, and an understanding of these routes in the context of the intracellular environment of the DNA makes possible valid extrapolation to the types of damage produced in the cell and, hence, to the identification of the important lesions.

Ionizing radiation produces a broad spectrum of damage types, and the mechanisms by which the various damages are produced should be described to the extent that the effects of radiation modifiers (protectors, sensitizers, oxygen, radiation quality) can also be understood. The intricate details of the mechanisms of damage production are not covered here, and the reader is referred to recent reviews on this topic *(3,53)*. Rather, this chapter concentrates on the damage produced in DNA irradiated within mammalian cells in oxygenated conditions as well as on factors that affect this damage production. Thus, the contributions from the various modes of damage production to intracellular DNA are extrapolated from the data from model systems and are compared with what has been measured in mammalian cells.

2. MECHANISMS OF DAMAGE INDUCTION

To extrapolate information obtained in model systems to cells, it is necessary to consider the mechanisms by which damage is induced in both. Ionizing radiation deposits its energy in all matter in amounts almost directly proportional to the mass of each molecular species present (for a discussion of this topic, *see* ref. *35*). The radiation

From: DNA Damage and Repair, Vol. 2: DNA Repair in Higher Eukaryotes
Edited by: J. A. Nickoloff and M. F. Hoekstra © Humana Press Inc., Totowa, NJ

energy is deposited in the form of ionization and excitations, but there is no evidence that the excitations play a role in causing biologically significant damage. All molecules within a cell are ionized by radiation to form cation radicals:

$$RH \xrightarrow{\text{Ionization}} H^+ + \text{electron} \quad (1)$$

The electron produced by the ionization can attach to another molecule or it can become solvated prior to reacting. Note that electrons resulting from ionization of all species are identical. On the other hand, the cation radical represents an alteration in the molecule ionized. This alteration can be transferred to another molecule:

$$RH^+ + X \longrightarrow RH + X^+ \quad (2)$$

Cation radicals can also react, losing a proton and becoming a neutral radical, as is the case of the initial cation radical produced by ionization of water: H_2O^+ reacts immediately (10^{-14} s) with a neighboring water molecule to produce the hydroxyl radical (·OH):

$$H_2O^+ + H_2O \longrightarrow H_3O^+ + \cdot OH \quad (3)$$

Thus, damage to cellular DNA can be caused by:

1. Direct ionization of the DNA;
2. Reaction with electrons or solvated electrons;
3. Reactions of ·OH or H_2O^+; or
4. Reactions with other radicals.

2.1. Free Radicals

Two types of studies of the radiation chemistry of DNA have been carried out: irradiating the molecule itself or irradiating it in dilùte aqueous solution, e.g., simulating what were called the "direct" and "indirect" effects, respectively. Thus, the direct effect was considered to be the deposition of energy directly in the DNA, but the indirect effect resulted from reactions of radicals resulting from ionizations of the solvent molecules. Thus, such studies of pure systems have provided a wealth of information about mechanisms of formation of products that are of importance in extrapolating to the cell. The initial reactions of ·OH radicals with DNA constituents are established. Both the sites of attack and rates of reaction have been determined: ·OH adds to the unsaturated bonds of the bases at almost diffusion controlled rates—there appears to be little preference for one base over another *(53,69)*. Subsequent reactions of the radicals so formed, with oxygen for instance, lead to the final products. ·OH radicals abstract hydrogen atoms from the deoxyribose moiety, probably from all sites at a rate about 25% of the diffusion controlled rate *(69)*. Again subsequent reactions with oxygen and other molecules or other radicals lead to products. Direct ionization of the DNA produces cation radicals, many of which have been studied and described by electron spin resonance investigations. There has been much discussion of the identity of the cation radical in DNA; the current view is that it resides on guanine *(3)*. Of course, the initial ionizations do not occur only on the guanine, but radical transfer to guanine occurs following the initial ionization. The work identifying the guanine as the preferred site was carried out in conditions where reactions of the cation radicals with other species could not occur. Thus, it is possible that in the cell, the initial cation radicals formed (evenly distributed among all DNA moieties) might react with other species (oxygen, water, and so forth) prior to transferring the radical to guanine.

Recently, Becker and Sevilla *(3)* have proposed a modification of the classification of the mechanisms by which damage is produced in DNA—they suggested that three effects be considered: direct, indirect, and quasi-direct. The former two terms have the same meaning as previously, and the latter refers to an ionization that occurs initially in a molecule adjacent to DNA, but that transfers its radical site to the DNA. One reason for considering the new route to damage production is that ionizations transferred from water close to DNA may have a different preference of site for reaction than that which results from direct ionization of the DNA. Hence, they should be distinguished as quasi-direct. However, it should be pointed out that if transfer of the sites of these radical cations occurs, then the same radicals as those produced by direct ionization would result.

Although the products of reaction of ·OH radical reactions with DNA are well established, there are few data on those resulting from direct ionization. There is a problem in devising a system to study such damage, since to simulate intracellular reactions, the ion radicals must carry out their subsequent reactions in an aqueous environment. Thus, irradiating dried material and analyzing the products are inadequate; the radicals formed cannot react with an aqueous environment and, in addition, problems arise when the samples are dissolved prior to product analysis—during dissolution, radicals in the solid react in an uncontrolled manner with water, oxygen, other molecules, or other radicals. Some studies have been done in which DNA was oxidized by reaction with cation radicals *(23)* or via biphotonic excitation *(54)*. However, reactions initiated by these methods may give rise to different radical cations in the DNA than those formed by the direct or quasi-direct effect.

The system being developed by Swarts et al. *(66,67)* should be a valid means of measuring damage from direct ionization. In their system, the relative contributions of cation radicals and ·OH radicals is controlled by limiting the amount of water associated with the DNA. This is an advantage over controlling these contributions using radical scavengers (*see* Section 2.4.). Swarts et al. *(66)* conclude from their data that two distinct compartments of water are involved in causing strand breakage: the hydration layer and the more loosely bound water molecules. Their data have been examined in the context of the contributions to damage of energy deposited in water and in DNA in mammalian cells *(75)*. It was found that the absolute yields of nonscavengable strand break type reactions measured by Swarts et al. *(66)* agree well with those measured in mammalian cells. The data also suggest that strand breaks in a cell are caused by direct ionization of the DNA and the ·OH radicals from 120 water molecules/deoxynucleotide. Thus, the consequences of direct ionization of the DNA can be monitored as the cation radicals react subsequently in an aqueous environment. This system also measures the consequences of the quasi-direct effect. Therefore, together with the descriptions of damage produced by ·OH reactions, this system should permit a complete extrapolation of the description of the singly damaged sites produced by irradiation of DNA intracellularly. Thus far, results from the hydrated DNA system are limited to the measurement of the damage produced in anoxia.

2.2. Radiation Modifiers

The biological effectiveness of ionizing radiation is markedly dependent on whether oxygen is present during radiolysis. Oxygen *per se* is a biradical species and, as such, reacts readily with many free radicals. For example, its rate of addition to several of the base-hydroxyl adduct radicals is rapid *(83)*. In addition, oxygen is an avid scavenger of reducing radicals, such as the hydrated electron and the hydrogen atom. The manner in which oxygen sensitizes cells is related to its reactivity with DNA radicals "fixing the damage." The fate of DNA radicals within the cell in the absence of oxygen is not entirely understood. It is generally accepted that reducing species, such as glutathione, can react with radicals chemically repairing the molecule (in the case of those formed by hydrogen atom abstraction). However, reactions of thiols with radical sites does not necessarily reconstitute the parent molecule. The addition of a hydrogen atom to a base hydroxyl adduct radical, for instance, forms a hydrate and not the original base. In the case of cytosine, the hydrate can break down to form uracil *(72)*. Raleigh *(58)* pointed out that hydrogen atom donation to a deoxyribose site from which a hydrogen atom has been abstracted can lead to the formation of an isomer of the sugar, and Raleigh also pointed out the potential challenge of such a site to cellular repair processes.

2.3. Cellular Studies of Damage Induction

Paralleling radiation chemistry studies, work in cellular systems probed the source of the radiation damage causing cell death. The history of these explorations is traced by Alper *(1)*. The first mechanistic investigation was that of Johansen *(31)*. He reasoned that if radicals produced from water are important in killing cells, it should be possible to reduce the extent of such reactions by adding compounds that would react with these radicals prior to their reaction with the intracellular target (i.e., scavenge). His studies of radioprotection of *Escherichia coli* confirmed this hypothesis and allowed him to show that approx 65% of the cell killing was caused by ·OH radicals. However, no radioprotection was afforded when scavengers of electrons (reducing species) were used. Later, Roots and Okada *(61)*, and Chapman et al. *(11)* carried out similar experiments in mammalian cells and came to a similar conclusion: that 65% of the cell killing was caused by ·OH radicals. These studies have been extended by others, and evidence now exists for low linear energy transfer (LET) radiation, such as X-rays or γ-rays, that high concentrations of ·OH radical scavengers can, as well as protecting against cell killing, also reduce by a factor of three the yield of single-strand breaks (SSBs) *(61)*, double-strand breaks (DSBs) *(18,62)*, mutations *(62)*, and chromosome aberrations *(41)*.

Recent discussions of literature data have suggested a pragmatic classification of DNA damage origins in cells *(19)*. It was pointed out that the reduction of radiosensitivity achieved using radical scavengers in cells does not reflect the fraction of damage that originates from energy deposited in cellular water. This fraction originates from those radicals that can be scavenged before they reacted with the DNA. Any damage that is a consequence of radical transfer from ionized water molecules next to DNA is not affected. Thus, it was suggested that damage be classified as scavengable and nonscavengable.

If we are to extrapolate from model systems to cellular DNA, it is important that we know the contributions of the various damage mechanisms. It is clear that the 65% of the damage that is scavengable is indeed caused by the indirect effect (·OH radicals) and that the majority of the unscavengable damage is caused by direct + quasi-direct effects. However, what is not yet clear is what fraction of the unscavengable damage is caused by unscavengable ·OH radicals.

However, the question of the products arising from the different original radicals (cation- or ·OH-induced) may to some extent be moot, since they may be the same from both routes *(53,70)*. Cation radicals formed on the base moieties could react by picking up a hydroxyl ion from the water to form an ·OH adduct, or could deprotonate to give the same radical as that formed after hydrogen atom abstraction by an ·OH radical. Examination of the base damage products formed in DNA irradiated intracellularly *(49–51)* indicates that the range of products is similar to that produced by ·OH radical reactions. However, the spacing of the individual products on the macromolecule would be different from the two mechanisms. The production of a direct ionization on the DNA necessitates that the site be close to the position at which the initial energy deposition event occurred—in the absence of charge migration *(53)*. Hence, ionized sites on the DNA are more likely to be participants in locally multiply damaged site* (MDS) production (*see* Section 5.).

2.4. Other Damage Induction Mechanisms

Two other sources of damaged bases in cellular DNA consequent on exposure to ionizing radiation are possible. Damage of the same types as those produced by UV radiation may result from the Cerenkov radiation consequent on electrons produced by the radiation having velocities greater than the speed of light in the medium surrounding the DNA *(59)*. However, at the ionizing radiation doses typically considered for mammalian cells, this contribution to the total damage is minor.

It is possible that lesions are produced by mechanisms other than those involving direct ionization of the DNA and ·OH radicals. In studies of a model system *(47)*, we have shown that methyl peroxyl radicals ($CH_3O_2^\bullet$) react with DNA to cause base damage. The base damages were measured using base damage-specific enzymes. However, the $CH_3O_2^\bullet$ radicals do not react to produced strand breaks *per se*. An additional characteristic of the peroxyl radical induced damages is that they will not be contributors to MDS, since their rates of reaction are low; hence, their life-time is long, and they will diffuse relatively long distances before reacting. It is possible that peroxyl radicals formed elsewhere in the cell (by direct ionization or radical attack) can diffuse to and react with the DNA, causing base alterations.

One possible corollary of base damage introduced by peroxyl radicals is the double-damaged neighboring bases, which have been discovered and described by Box et al. *(8)*. These products are formed when oligodeoxynucleotides are irradiated in dilute aqueous solution. They are produced linearly with dose, indicating that they are initiated by reaction of a single ·OH radical. The fact that peroxyl radicals can initiate base damage suggests that a peroxyl radical on a base moiety can react further, causing damage to a neighboring base. To date there has been no consideration of the means by which these kinds of adjacent damages are excised from the DNA of irradiated cells.

*E. M. Fielden has pointed out that the original term "locally multiply damaged sites" contains a tautology, now "multiply damaged sites" (MDS) is used.

3. DNA BASE DAMAGE

3.1. Base Damage in Solution

There had been many studies of the mechanisms of production and yields of base damage prior to the availability of newer techniques *(70)*. The new approaches to base damage measurement are described and compared by Cadet and Weinfeld *(10)*. The assays are classified into those that require hydrolysis of the DNA into monomers for separation and analysis of products, and those in which the damage can be assayed in the intact DNA. The methods that employ DNA hydrolysis cannot provide information about the original positions of the damage in the DNA—they do not detect doubly damaged lesions *(8)*, nor do they provide information about MDS (*see* Section 5.). Hydrolysis approaches do, however, permit the handling of large DNA molecules and can also be extrapolated to use in cellular systems. The prime example of this type of study is that developed by Dizdaroglu and coworkers *(4,22,49–51)*. Six base-damaged products from DNA irradiated in solution in either single- or double-stranded state in the presence of oxygen were identified and quantitated *(22)*. These yields are shown in Table 1, where they are quoted as percentages of the number of ·OH radicals reacting with the DNA. It can be seen that the major products measured are thymine glycol, cytosine glycol, and 8-hydroxyguanine. The total yield of these products does not equal 100% of the ·OH radicals reacting. There are several reasons for this:

1. Some ·OH radicals react with deoxyribose moieties causing strand breaks (*see* Section 4.).
2. Some base damage products may not be measured in the protocol, e.g., the formyl product identified by Box et al. *(8)*.
3. Products that are volatile *(70)* would not be detected.
4. Other scavengers may have been present in the irradiated solution reducing the number of ·OH radicals reacting with the DNA.
5. Reaction of an ·OH radical with a base moiety does not necessarily lead to an altered base.

All of the base damage products measured by gas chromotography-mass spectrometry (GCMS) are consistent with the known mechanisms for damage production *(53,69)* following ·OH radical attack.

The products formed by irradiation in solution can be compared with the those formed as a result of ·OH radicals produced by a different route. Blakely et al. *(4)* treated an aqueous solution of DNA with H_2O_2 in the presence of adventitious metal ions and measured the products formed by GCMS. The yields of these products formed in double-stranded DNA are included in Table 1, where for comparison to the radiation yields, they are normalized to equal thymine glycol yields. The spectrum of products is similar for the two cases, but with differences in the relative yields. There are several possible reasons for these differences, the most probable being the location of the adventitious metal ion with which the hydrogen peroxide reacts. In addition, differences may occur in the subsequent reactions of the base radicals formed, e.g., some could back react with the oxidized metal ion.

There have been few studies of the radiation products resulting from reactions of ion radicals produced in DNA. Electron paramagnetic resonance (EPR) studies of solid DNA give precise information about the structures of the radicals induced in the various DNA moieties *(3)*. However, to determine the subsequent reactions of these radicals leading to product formation, it is necessary to simulate the cellular environment

Table 1
Relative Yields of DNA Base Damage Caused by ·OH Radical Attack

	Ionizing radiation		Hydrogen peroxide[d]
Product	Single-strand DNA	Double-strand DNA	Double-strand DNA
Thymine glycol	28[a]	15[a]	15[c]
Cytosine glycol	15[a]	8[a]	2.5[c]
4,6-Diamino-5-formamidopyrimidine (FAPY-adenine)	0.2[a]	1.4.[a]	23.6[c]
2-Amino-6,8-dihydroxypurine (8-Hydroxyadenine)	4.4[a]	4[a]	1.5[c]
2,6-Diamino-4-hydroxy-5-formamidopyrimidine (FAPY-guanine)	0.8[a]	3.4[a]	5.1[c]
2-amino-6,8-dihydroxypurine 8-hydroxyguanine	24[a]	24[a]	16.4[c]
SSBs		11[b]	

[a]Ref. *22*.
[b]Ref. *43*.
[c]Ref. *4*.
[d]For comparison with radiation in double-stranded DNA, yields from hydrogen peroxide treatment are normalized by scaling to equal yields of thymine glycol.

around the radicals. Most importantly, the DNA cation radicals should react within an aqueous environment. At first glance, it would seem that the use of scavengers to reduce the amount of ·OH radical reactions would be a valid method to modulate damage induction mechanisms. However, in some cases the radicals produced from the scavenger after its reaction with ·OH radicals react with the DNA *(45,47)*. The hydrated DNA model being developed by Swarts et al. *(66,67)* (described above) should be a route to the determination of damaged bases induced by direct and quasi-direct mechanisms.

3.2. *Base Damage Produced in Cells*

Earlier calculations, which assumed contributions from scavengable and unscavengable reactions and extrapolated relative yields from studies in model systems, arrived at the conclusion that the total yield of base damage in mammalian cells should be on the order of 2.5–3 times the yield of SSBs. With the yield of SSBs taken as 1000/cell/Gy, the yield of base damage would be 2500–3000/cell/Gy. Measurements of base damage in irradiated cells have now been made by several groups.

The most comprehensive studies have been carried out by Dizdaroglu's group, who measured the yields of a variety of base damages in several cellular systems *(49–51)*. They measured more than 10 altered bases, and a comparison of their identities with the altered bases produced in model systems indicates that these yields constitute the majority of the altered bases produced. The yields measured after irradiation in animals may be lower than the total of base damage induced, since enzymatic removal of damaged bases could have occurred prior to DNA isolation. However, yields in cells irradiated at 0°C should be the actual yields produced. The yields of the damaged bases produced in these systems ("total" yield) have been summed and plotted against dose *(76)*. For the data that constitute the closest to the initial yield (irradiation at 0°C) the amounts of "total" base damage/cell/Gy are 1–2 × 10^4 base damages/cell, e.g., about 10 times the expected yield. The previous consideration of these data *(76)* had no explanation of the major difference of the yield between this and the calculated yield. However, the recent finding of the induction of base damages by peroxyl radicals *(47)* indicates that base damage measured in irradiated cells may arise from reactions of these species, in addition to ·OH radicals and direct ionization.

Others have measured the yields of base damage in irradiated cells *(17,38)*. In these instances, the full spectrum of base damage was not monitored so that the total yield of base damage can only be calculated. Such a calculation can be made assuming that the yields of the specific base damage measured represents the same fraction of total base damage as that in the measurements of Fuciarelli et al. *(22)*. From this assumption, the total yields are closer to the yields previously calculated *(73)*. The difference in these yields from those measured by GCMS could be owing to variability among the cell lines of the reactions of the peroxyl radicals—perhaps the detoxification of the peroxyl radicals is more efficient in some cells than others. Mori and Dizdaroglu *(49)* did find that the glutathione level was lower in cells in which higher yields of base damage were measured. They point out that glutathione is excluded from the vicinity of the DNA, and this would preclude its scavenging of ·OH radicals that migrate short distances. However, it would not preclude scavenging of peroxyl radicals which are longer-lived and would therefore travel further.

Ames and his coworkers have shown that base damages of the same types are produced by normal oxidative metabolism in mammalian cells in relatively large yields *(60)*. Contributing to the production of these damaged bases may be peroxyl radicals (*see* Section 2.4.). Such radicals can diffuse long distances from their sites of origin. The base damages are excised from cellular DNA in significant amounts per day *(55)*. Thus, it is somewhat expected that enzymatic mechanisms for the repair of these base damages exist: Lindahl has emphasized this in his discussion of the removal of base damages from cellular DNA *(40)*. With this reasoning as background, it was suggested that the yields of oxidized bases induced in cellular DNA by biologically significant radiation doses are unimportant as causes of such biological effects as mutagenesis *(76)*.

4. DNA STRAND BREAKS

The earliest measurements of damage in the DNA of irradiated cells were of the induction of SSBs *(15,42)*. Determinations were readily made from changes in the average length of single strands determined by velocity gradient sedimentation through alkaline sucrose gradients. Later, more sensitive techniques for the measurement of SSBs, such as alkaline unwinding and alkaline elution, were used *(21)*. These techniques measuring DNA size, although they are a means of accurately determining the number of breaks, provide little information about the nature of the damage induced or the mechanisms of its formation. It rested with studies in model systems to provide information about mechanisms and to identify the end groups produced.

4.1. Mechanisms of Strand Break Formation

It seems clear that in order to cause a strand break, the moiety that must be damaged is the deoxyribose. However, whether this damage originates from initial radical attack on the sugar moiety and/or radical attack on the base is the subject of some debate. Studies of dilute solutions of single-stranded polyuridylic acid (poly U) as a model showed that the yield of SSBs is much greater than the amount of ·OH radical reaction with the sugar moiety *(39)*. Thus, in this work and subsequent work with poly U, the major fraction of the strand breakage yield is a consequence of base radicals attacking the ribose moiety *(63)*. Further work with other single-stranded polyribosides also suggested the involvement of base radicals in causing strand breaks. In the deoxyribose series, there is no conclusive evidence that base radicals are precursors of strand breaks even in single-stranded DNA *(53)*. In fact, the yield of SSBs in model DNA systems is close to the yield of ·OH radical reaction with the deoxyribose. In model studies *(44)*, the yield of SSBs is ~11% of the yield of ·OH radicals reacting with the DNA. This yield is consistent with many studies showing that the fraction of the ·OH radicals reacting with the bases is ~80%. The reasons for the low fraction of ·OH radicals reacting with the deoxyribose moieties are several, and include the lower relative rate constant for reaction with saturated vs unsaturated compounds, and the accessibility of the participant atoms to ·OH radical attack within the double-helical structure *(73)*. Here, in the absence of data to the contrary, we will consider that only initial radical reactions with the deoxyribose moieties lead to SSBs.

The mechanisms by which ·OH radical attack on deoxyribose can lead to a strand breaks have been well described by van Sonntag *(69)*. He points out which reactions might be expected to produce immediate breaks and which would lead to alkali-labile sites (ALS), such as those measured by others (e.g., *33*). ALS are an important consideration, because measurements of SSBs in cells are necessarily carried out after alkali treatment and thus measure SSBs that do not occur as such in cells. The yields of SSBs determined in aqueous supercoiled plasmid systems cannot be directly extrapolated to those measured in cells because of the alkaline conditions used in the latter assays. Ionizing radiation has long been known to induce damage in DNA, which on treatment with alkali is converted to an SSB. These ALS have been shown not to be abasic sites, since the pH at which they break is different from that of the abasic sites *(37)*.

Thus, the abstraction of a hydrogen atom from the deoxyribose moiety by an ·OH radical is the source of the strand break caused by the indirect mechanism. The question remains regarding which hydrogen atom. Von Sonntag *(69)* describes mechanisms whereby strand breakage can arise from abstraction of any of the deoxyribose hydrogens. It has been pointed out, however *(82)*, that abstraction from the 2'-position of deoxyribose is expected to be energetically less favorable than abstraction from the other sites. Since at least three products result from deoxyribose damage (*see* Section 4.2.), it is probable that attack on more than one site is involved, although the so-called 4'-mechanism is favored in the literature.

Mechanisms of strand breakage from the direct effect/quasi-direct effect might also be presumed to originate in deoxyribose damage. However, EPR studies of irradiated DNA have failed to detect radicals on the deoxyribose site *(3,14)*. Swarts et al. *(67)* present a schematic of the reactions that follow irradiation of DNA in the presence of carefully controlled amounts of water, and conclude that in their system, the radical precursors of strand breaks are deoxyribose radicals and that these radicals are formed from deoxyribose radical cations by proton loss. They find no evidence of a pathway to a strand break that involves transfer of radical sites from base cation radicals, nor is the radical site on the deoxyribose transferred to a base site.

4.2. *Structure of SSB Ends*

Early studies of SSB termini were made in DNA irradiated in solution after relatively high radiation doses *(7,33)*. Recent investigations of these termini that remain at the site of a strand break showed that all of the 5'-termini are monophosphate groups, whereas 70% of the 3'-termini are monophosphate and 30% are phosphoglycolate termini *(27)*. Since both termini of the strand break contain phosphate *(26,27)*, it is implicit that a base moiety (possibly attached to a sugar fragment) is released at the site of each SSB. Several earlier measurements *(68,79)* of radiation-induced base release showed that the yields were approximately equal, except guanine release was lower by a factor of two. However, the sum of these yields was approximately equal to the yield of SSBs. It is of interest that all of the products isolated after strand cleavage (apart from a small yield of released 8-hydroxyadenine; *25*) are free intact bases without any sugar fragment attached. This finding is in contrast to the products released from the DNA after its treatment with bleomycin *(84)*, which is presumed to damage DNA by an ·OH radical mechanism. Although the strand break termini are the same for treatment with this agent as for radiation, the bases are released as propenals, indicating that the chemistry involved in the strand breakage is not identical. The differences probably lie in the subsequent reactions of the radicals on the deoxyribose.

During assays of SSB termini structures, there is necessarily extensive handling of the DNA. Since it has been established that radiation induces labile sites in DNA, the SSB terrnini analyzed may not be those present immediately after irradiation, e.g., the termini processed by cellular repair systems. However, the assay used in the measurement of base release (precipitation of the macromolecular DNA with ethanol at 0°C immediately after irradiation) would not be expected to break labile bonds. Henle et al. *(25)* measured base release in supercoiled plasmid DNA using controlled dialysis to separate the released bases from the macromolecular DNA, and compared directly the yield of SSBs to the yield of bases released in the same system. The yields of SSBs were measured by electrophoresis at ice temperature to control hydrolysis of labile bonds. They found that at time zero after irradiation, the yield of strand breaks was ~30% higher than the total yield of released bases. However, on incubation of the irradiated DNA at 37°C, both yields increased and the yields became approximately the same after 24 h. Their conclusion is that immediately after irradiation, damaged sites occur in the sugar, one type causing an immediate break with release of free base, a second type exists as a break, but does not release the base immediately, whereas a third type slowly causes a strand break and releases the base. One complicating issue here is that the protocols used to measure the two end points are not identical; the measurement of strand breaks requires time for electrophoresis, which even though it is carried out at 0°C may allow labile bonds to hydrolyze. Thus, it is possible that only two types of sites are involved, one being an immediate strand break releasing a base and the other a slow hydrolysis, both releasing a base and causing a strand break.

4.3. Other Deoxyribose Damage

An additional consequence of damage to the deoxyribose moiety has been described *(29,33,70)*: After irradiation in solution, DNA contains moieties produced by oxidation of the sugar, which react with thiobarbituric acid in a manner indicative of the presence of malondialdehyde (MDA). It was shown that this moiety remains bound to the macromolecular DNA. Since end-group structure determination does not reflect the presence of a three-carbon sugar fragment and the strand break yield is equal to the yield of released bases, it must be concluded that the damage that is the MDA precursor does not represent a strand break.

Thus, there are at least three consequences of the reaction of ·OH radicals with the deoxyribose of DNA:

1. Strand break leaving 5'- and 3'-phosphate termini;
2. Strand break leaving 5'-phosphate and 3'-phosphoglycolate termini; and
3. Production of an MDA precursor.

5. MULTIPLY DAMAGED SITES (MDS)

Although the majority of studies of radiation damage in DNA have been carried out on individually damaged sites, there has been a growing realization that such lesions when produced in mammalian cells are biologically unimportant at the levels induced by biologically relevant radiation doses. It was shown, for instance, that SSBs introduced into the DNA of mammalian cells by treatment with hydrogen peroxide at 0°C do not cause cell killing *(77)*. One argument against this conclusion could be that the SSBs introduced are not the same as those caused by ionizing radiation. However, the production of these breaks was inhibited by the presence of an ·OH radical scavenger, indicating that the intermediate reactive species causing the breaks is indeed an ·OH radical whose reactions should be independent of its mode of initiation.

Subsequent to this work, the numbers of lesions per cell present after a lethal dose of a variety of agents were shown to separate into two classes *(78)*: Agents that caused singly damaged sites and those that produce MDS (such as DSBs). At a lethal dose of the former class of agents, the numbers of lesions present per cell are on the order of 10^5–10^6, whereas for the latter, the numbers were lower than 100. Thus, it was concluded that MDS in DNA are much more biologically effective than singly damaged sites.

The possibility that singly damaged bases are a cause of radiation-induced mutations was also examined *(76)*, and it was again concluded that singly damaged sites in the form of base damages are insignificant as precursors of radiation-induced mutations. Thus, in addition to cell killing, the cause of radiation induced mutations was argued to be MDS.

5.1. Mechanism of Production

It has been known for some time that ionizing radiation energy is not deposited homogeneously, nor is the amount deposited quantized. It is deposited in events that range in energy up to hundreds of eV (even for low LET radiation) with the average amount being 60 eV. (This amount of energy can be compared to the 4.8 eV energy of a quantum of 260 nm light.) Since the energy to form an ion pair (H_2O^+ and an electron) from water is ~20 eV *(13)*, the average energy deposited per event is sufficient to produce ~3 ion pairs. Of course, the number of radicals produced in an event is proportional to the energy deposited in that event. The deposition of the energy in small volumes of nanometer dimensions means that clusters of reactive species (radicals) are produced at high local concentrations.

The distance that ·OH radicals move prior to reacting intracellularly can be determined from scavenging data. The concentration of dimethyl sulfoxide necessary to reduce the amount of ·OH radical-induced damage in mammalian cells by a factor of two has been determined to be 0.14*M* (e.g., *70*). From these data, it can be calculated that the average half-life of ·OH radicals in a mammalian cell is 0.85×10^{-9} s. During this time, an ·OH radical in free solution diffuses 3.1 nm, which is therefore the average distance that a scavengable ·OH radical moves before it reacts with the DNA. If the ·OH radical arose from an event in which several ion pairs were produced, it is likely—since the migration distance is so short—that other radicals are present in the vicinity and may react within the same region of the DNA. From the average distance that a radical migrates and the size of the energy deposition event of similar dimensions *(73)*, it can be seen that the spacings of the positions at which two radicals react will vary up to about 10 nm.

If the mass of water from which ·OH radicals migrate 3 nm *(61)* to react with cellular DNA is compared to the mass of the DNA involved, it is clear that direct + quasi-direct ionizations of DNA will also contribute to the reactions involved in producing MDS. A reasonable ratio to assume is 65% ·OH radical-induced damage and 35% direct + quasi-direct *(73)*. Of course, MDS will have origins from mixtures of these sources, and the individual damage types within the site are the same as those produced in singly damaged sites (as described in Sections 3.1. and 4.1.).

Another variable of this kind of damage is the relative contributions of base damage and strand breaks. As discussed in Section 3.2., base damage is more prevalent, and hence, the majority of the individual lesions that make up these MDS *(73)* are base damages.

To summarize, MDS produced in cellular DNA have three variables:

1. The number of damaged entities within the site;
2. The distance over which these damages are distributed; and
3. The relative contributions of base damage and strand breaks.

5.2. *Evidence for MDS*

MDS damage is produced in DNA even when it is irradiated in dilute aqueous solution. However, irradiation under these conditions allows ·OH radicals to migrate large distances before reacting with the DNA. Therefore, damage arising from the latter reactions, singly damaged sites, is present in much larger amounts than MDS in dilute solution. If the distance that the ·OH radicals move before reacting with the DNA *(36,46)* is restricted by the use of scavengers or by limiting the amount of water present *(66)*, then MDS become a significant fraction of the total damage produced in dilute solution.

It was previously assumed that the distance ·OH radicals move before reacting with the DNA was controlled by the presence of low-molecular compounds and histones. Now it has been hypothesized *(52)* that the major factor controlling this distance is the presence of spermine. This polyamine causes compaction and aggregation (PICA) of the DNA by neutralizing the macromolecule's negative charge. In the presence of spermine at physiological concentrations the yield of SSBs in DNA in SV40 minichromosomes in dilute solution was reduced to the yield of SSBs in cellular DNA. Thus, studies of DNA in simulated cellular environment will be facilitated by the use of spermine; it will no longer be necessary to use high concentrations of compounds that scavenge ·OH radicals in bulk solution.

Of course two strand breaks produced by the MDS mechanism, if they occur on the opposite strands of the DNA, constitute the well-known DNA DSB. Other mechanisms for the production of DSBs have been suggested *(2,6,64)*, but the preponderance of the present evidence is that the MDS mechanism is that which operates in mammalian cells. When the idea of MDS was first introduced in 1981 *(71)*, there was little direct evidence for their production. However, several lines of evidence have appeared supporting the MDS mechanism. Hagen et al. *(24)* summarized earlier pioneering work where damage to DNA irradiated *in situ* was compared to that after irradiation in vitro. The differences in damage are interpreted in terms of clusters of damaged bases being produced in the former case. Additional evidence for MDS was reported by Krisch et al. *(36)*, who measured the ratio of the yields of SSBs to DSBs as a function of ·OH radical scavenger concentration and found that this ratio decreased from ~100 at low scavenger concentration to ~25 at high concentration. This change was interpreted in terms of the MDS mechanism. Folkard et al. *(20)* determined the dependence of the yields of SSBs and DSBs in plasmids on the energy of the irradiating electrons—they found that SSBs had a threshold of ~25 eV, but that the DSB threshold is between 25 and 50 eV. This is consistent with two radical requirement for DSB production. However, in later work using photon irradiation, this differential dependence on energy was not apparent *(43)*. Other work investigating the reactions of strand break precursors showed that SSBs were produced from a single radical sites and DSBs from a double radical site *(57)*.

5.3. *Structures of MDS*

Experimental information about the characteristics of MDS is beginning to emerge. As mentioned in Section 5.2., a DSB from low LET radiation has a two-radical precursor *(57)*; this has been confirmed using a steady state radiolysis technique *(46)*. For higher LETs, the precursors of DSBs have an average of more than two radicals, i.e., sites have a multiplicity of >2. It has been hypothesized that the greater radiobiological effectiveness of high LET radiation is owing to the greater complexity of the damaged sites produced *(5,74)*, which makes them more difficult to repair enzymatically. It is necessary to invoke such a hypothesis, since the yields of the biologically significant DSBs do not increase with LET *(9)*. Indeed the repair of DSBs after high LET radiation has been shown to be slower than and less complete than that after low LET radiation *(5,30)*. (Of course, these measurements of DSB repair provide no information about the fidelity of the repair.)

Within the complexity of the damage, there are not only strand breaks, but base-damaged sites. The latter lesions were measured in the DNA of irradiated cells many years ago *(65)*. Now with the availability of high-activity purified base damage-specific enzymes, a more detailed study of the production of base damage in irradiated DNA is possible. Formamidopyrimidine-glycosylase (FPG) and endonuclease III have been used to investigate base damage produced in DNA irradiated under simulated cell scavenging conditions. Base-damaged sites are recognized by these enzymes and converted to strand breaks. FPG recognizes 8-hydroxyguanine and the formamido-pyrimidine residues of adenine and guanine, which comprises 90% of the damaged purines detected by GCMS analysis of irradiated DNA *(22)*. Endonuclease III causes strand breaks at the site of cytosine and thymine glycols, the major pyrimidine damaged products. Therefore, these two enzymes together convert the majority (95%; *22*) of the base damage sites to strand breaks. Treatment of irradiated DNA with FPG increases the SSB yield by a factor of ~3, whereas endonuclease III increases it by ~2.5. The two enzymes together increase the yield of SSBs by a factor of ~4.5, indicating that ~3.5 times as many base damage sites are produced as strand breaks *(47)*. When the increase in DSBs produced by treatment with these enzymes was measured, it was found to be the square of the increase of the SSB yield, i.e., treatment of the DNA with both enzymes increases the DSB yield by a factor of ~13 *(48)*. Such an increase had been predicted in the earlier discussion of the mechanism of MDS production *(73)*. This indicates that the DSBs measured in cells are <10% of the total MDS produced.

The increase in DSBs caused by treatment with base damage-specific enzymes should also be considered against the findings of Chaudhry and Wernfeld *(12)*, who examined the ability of endonuclease III to cleave base-damaged sites (dihydrothymine) opposite other lesions. They found that the enzyme does not cleave if the two damaged sites are within 4 bp of each other, and that the activity is still diminished if the damaged sites are 6 bp apart. This result would suggest that MDS would not be susceptible to endonuclease III cutting if the individual damaged sites are within 4–6 bp. There are several ways in which these data can be reconciled to those described in the previous paragraph:

1. The damages produced in the irradiated DNA strands could have a greater separation distance than the 6 bp examined by Chaudhry and Weinfeld *(12)*;
2. Milligan et al. *(48)* required a higher concentration of enzyme to maximize the DSB yield than that needed to maximize the SSB yield;
3. The major fraction of the base damage is of the purines—it is possible that FPG does not have the same restriction to activity by neighboring damage as the endonuclease III; and
4. Dihydrothymine is not formed when DNA is irradiated under oxygenated conditions.

The data of Milligan et al. suggest that the MDS sites contain base damages in the same ratio to strand breaks as for singly damaged sites.

Preliminary data indicate that the size of the region over which an MDS is produced after γ-radiation is up to ~15 bases *(81)*. In that work, it was pointed out that two strand breaks in a MDS have equal probability of being on the same strand as of being on opposite strands and the separations of the breaks would be the same in both instances. The occurrence of the two breaks on the same strand should release an oligonucleotide. Hence, the single-stranded oligonucleotides produced after irradiation of plasmid DNA in simulated cell scavenging conditions were denatured from and separated from the macromolecule, end labeled with ^{32}P, and sized on a sequencing gel. A spread of sizes of oligonucleotides was observed with the maximum length ~15 deoxynucleotides. This suggests that the DSBs that are measured after irradiation of DNA within cells using various techniques, such as pulse-field gel electrophoresis, neutral elution, and neutral sucrose gradient sedimentation, contain DSBs with their constituent SSBs separated on opposite strands by different distances up to 15 bp. It is probable that the cell can repair these widely separated strand breaks as SSBs, and hence, part of the DSB repair observed (probably that repairing faster) represents damage that never forms a DSB within the cell. This type of quasi-DSB is probably not important biologically, since repair of its constituent lesions occurs readily and the repair patch involved does not require synthesis on a damaged template. This is in contrast to damaged sites that are close together on opposite strands, where repair synthesis involves a damaged template or requires an undamaged homolog as a template for recombinational repair—the former can be expected to be the significant biological damage.

Currently, the biological importance of DSBs is clear, but it is not known whether MDS containing damaged bases (but no DSB) are significant. As argued above, problems associated with the repair of base damage when closely opposed to other damage suggests that all such MDS have the potential to be biologically significant. An argument can be made for the importance of non-DSB MDS: Of the mutations produced by ionizing radiation, small-scale mutations (those mutants showing no change in RFLP pattern) are presumably caused by sequence errors incorporated during repair of an MDS. In the case of *HPRT* mutation, the yield of these small scale mutations is approximately equal to the yield of DSBs produced within the exons of the gene in the population of cells irradiated *(80)*. However, as discussed above, a significant fraction of the measured DSB are probably repaired accurately. Thus, some other MDS lesions must contribute to the formation of small-scale mutations. An additional argument for this contention is that the yield of small-scale mutations is greater by a factor of ~12 if α-particle radiation is used *(32)*. Since the yield of DSBs is the same for α-particles as for γ-radiation, again some other precursor of these mutations must be invoked to explain the increased yield of mutations—MDS that are not DSBs are this possible precursor.

6. CONCLUDING REMARKS

The singly damaged sites produced in cellular DNA by ionizing radiation are readily and accurately repaired after biologically significant radiation doses. This might be expected since they are the same lesions as those produced by endogenous oxidation. The more significant radiation products are the multiply damaged sites that have several degrees of complexity and that can be predicted to be problematic for the cell to repair. The manner in which these MDS are produced is in line with what is known about ionizing radiation energy deposition and the subsequent reactions of the free radicals produced.

It is questionable whether it is fruitful to attempt to determine the yields of all possible MDS. The number of different doubly damaged sites can be calculated. The individual lesions participating include the six major base-damaged products listed in Table 1 and two strand break structures: one with a 3'-phosphate and the other with a 3'-phophoglycolate end group, both having a 5'-phosphate end group. Considering this range of eight radiation products, there are 64 possible combinations of products in doubly damaged sites (only four of these combinations are DSBs). An additional variable is the distance that these lesions are separated on opposite strands—preliminary estimates of this distance (Section 5.3.) indicate that it can vary up to 15 bases. Of course this separation can be either 3' or 5'. This indicates that there are ~1800 different combination lesions of the doubly damaged type. The number of possible products will increase markedly with increasing numbers of damages per MDS.

Of more importance than attempting to catalog these structures would be the determination of the ability of cellular systems to repair them. Such investigations could proceed along the lines initiated by Hodgkins et al. *(28)*, who examined the ability of human cell extracts to repair SSBs in plasmid DNA. Such extracts have not been able to rejoin DSBs (P. O'Neill, personal communication; P. Pfeiffer, personal communication). Hence, a more effective system for probing the "repairability" of these lesions might be the *Xenopus* extracts used successfully by Pfeiffer et al. to rejoin DSBs introduced into plasmids by restriction enzymes *(56)*.

Again determining the abilities of damage-specific enzymes to repair each of the large number of possible doubly damaged lesions would be counterproductive. It has been reported that there are MDS structures from which the bacterial base damage-specific enzyme endonuclease III cannot excise the damaged base *(12)*. It has been suggested (M. Weinfeld, personal communication) that correlation of the abilities of base damage-specific repair enzymes to excise specific damage from defined MDS structures with the structure of the enzyme binding site may permit some general understanding of the factors controlling activity.

Some consideration might be given to the significance of MDS produced by other agents that are lethal only at 10^5–10^6 DNA lesions/cell. What is the probability of the larger patch size involved in many instances encountering a damaged site on the opposite strand at these levels of damage?

ACKNOWLEDGMENT

Work in the author's laboratory is supported by grant CA46295.

REFERENCES

1. Alper, T. 1979. *Cellular Radiobiology*. Cambridge University Press, Cambridge, UK.
2. Baverstock, K. F. 1985. Abnormal distribution of double strand breaks in DNA after direct action of ionizing energy. *Int. J. Radiat. Biol.* **47:** 369–374.
3. Becker, D., and M. D. Sevilla. 1993. The chemical consequences of radiation damage to DNA. *Adv. Radiat. Biol.* **17:** 121–180.
4. Blakely, W. F., A. F. Fuciarelli, B. J. Wegher, and M. Dizdaroglu. 1990. Hydrogen peroxide induced base damage in deoxyribonucleic acid. *Radiat. Res.* **121:** 338–343.
5. Blöcher, D. 1988. DNA double strand break repair determines the RBE of α-particles. *Int. J. Radiat. Biol.* **54:** 761–771.
6. Boon, P. J., P. M. Cullis, M. C. R. Symons, and B. W. Wren. 1984. Effects of ionizing radiation on deoxyribosenucleic acid and related systems. Part 1. The role of oxygen. *J. Chem. Soc. Perkin Trans.* **II:** 1393–1399.
7. Bopp, A. and U. Hagen. 1970. End group determination in gamma-irradiated DNA. *Biochim. Biophys. Acta.* **209:** 320–326.
8. Box, H. C., H. G. Freund, E. E. Budzinski, J. C. Wallace, and A. E. McCubbin. 1995. Free radical-induced double base lesions. *Radiat. Res.* **141:** 91–94.
9. Brenner, D. J. and J. F. Ward. 1992. Constraints on energy deposition and target size of multiply-damaged sites associated with DNA double-strand breaks. *Int. J. Radiat. Biol.* **61:** 737–748.

10. Cadet, J. and M. Weinfeld. 1993. Detecting DNA damage. *Analyt. Chem.* **65:** 675A–682A.
11. Chapman, E. D., A. Reuvers, J. Borsa, and C. L. Greenstock. 1973. Chemical radioprotection and radiosensitization of mammalian cells growing in vitro. *Radiat. Res.* **56:** 291–306.
12. Chaudhry, M. A. and M. Weinfeld. 1995. The action of *Escherichia coli* endonuclease III on multiply damaged sites in DNA. *J. Mol. Biol.* **249:** 914–922.
13. Hagen, U., D. Harder, H. Jung, and C. Streffer, eds. 1995. *Proceedings of the 10th International Congress of Radiation Research*, vol. 1, in Radiation Research 1895–1995, Würzburg, Germany, Aug. 27–Sept. 1.
14. Close, D. M. 1995. Where are the sugar radicals in irradiated DNA? *Xth International Congress of Radiation Research*, Würzburg, Germany, Abstract P18-18.
15. Dean, C. J., P. Feldschreiber, and J. T. Lett. 1966. Repair of X-ray damage to the DNA in Micrococcus radiodurans. *Nature*, **209:** 49.
16. deLara, C. M., T. J. Jenner, K. M. S. Stewart, S. J. Marsden, and P. O'Neill. 1995. The effect of dimethyl sulfoxide on the induction of DNA double strand breaks in V79-4 mammalian cells by alpha particles. *Radiat. Res.* **144:** 43–49.
17. Epe, B., M. Pflaum, M. Häring, J. Hegler, and H. Rüdinger. 1993. Use of repair endonucleases to characterize DNA damage induced by reactive oxygen species in cellular and cell-free systems. *Toxicol. Lett.* **67:** 57–72.
18. Evans, J. W., J. F. Ward, and C. L. Limoli. 1985. Effects of dimethyl sulfoxide a radioprotector on intracellular DNA. Abstract Ehl2, 33rd Radiat. Res. Soc. Mtg., Los Angeles, CA.
19. Fielden, E. M. and P. O'Neill. 1991. *The Early Effects of Radiation on DNA*. Springer Verlag, Berlin.
20. Folkard, M., K. M. Prise, B. Vojnovic, S. Davies, M. J. Roper, and B. D. Michael. 1993. The measurement of DNA damage by electrons with energies between 25eV and 4000eV. *Int. J. Radiat. Biol.* **64:** 651–658.
21. Friedberg, E. and P. C. Hanawalt. 1981. *DNA Repair*, vol. 1B. Dekker, New York, pp. 363–418.
22. Fuciarelli, A. F., B. J. Wegher, W. F. Blakely, and M. Dizdaroglu. 1990. Yields of radiation induced base products in DNA: effects of DNA conformation and gassing conditions. *Int. J. Radiat. Biol.* **58:** 397–415.
23. Görner, H., C. Stradowski, and D. Schulte-Frohlinde. 1988. Photoreactions of tris(2,2'-bipyridyl)-ruthenium (II) with peroxydisulphate in deoxygenated aqueous solution in the presence of nucleic acid components, polynucleotides and DNA. *Photochem. Photobiol.* **47:** 15–29.
24. Hagen, U., H. Bertram, E.-M. Geigl, E. Kohfeldt, and S. Wendel. 1989. Radiation induced clustered damage in DNA. *Free Radical Res. Commun.* **6:** 177,178.
25. Henle, E. S., R. Roots, W. R. Holley, and A. Chatterjee. 1995. DNA strand breakage is correlated with unaltered base release after gamma irradiation. *Radiat. Res.* **143:** 144–150.
26. Henner, W. D., L. O. Rodriguez, S. M. Hecht, and W. A. Haseltine. 1983. γ-Ray induced deoxyribonucleic acid strand breaks. *J. Biol. Chem.* **258:** 711–713.
27. Henner, W. D., S. M. Grunberg, and W. A. Haseltine. 1982. Sites and structure of γ-radiation-induced DNA strand breaks. *J. Biol. Chem.* **257:** 11,750–11,754.
28. Hodgkins, P. S., M. P. Fairman, and P. O'Neill. 1996. Rejoining of gamma-radiation-induced single-strand breaks in plasmid DNA by human cell extracts: Dependence on the concentration of the hydroxyl radical scavenger, Tris. *Radiat. Res.* **145:** 24–30.
29. Janicek, M. F., W. A. Haseltine, and W. D. Henner. 1985. Malonaldehyde precursors in gamma-irradiated DNA deoxynucleotides and deoxynucleosides. *Nucleic Acids Res.* **13:** 9011–9029.
30. Jenner, T. J., C. M. deLara, P. O'Neill, and D. L Stevens. 1993. Induction and rejoining of DNA double strand breaks in V79-4 cells following γ- and α-irradiation. *Int. J. Radiat. Biol.* **64:** 265–274.

31. Johansen, L. 1965. The contribution of water-free radicals to the X-ray inactivation of bacteria, in *Cellular Radiation Biology*, Williams and Wilkins, Baltimore, pp. 103–106.
32. Jostes, R. F., E. W. Fleck, T. L. Morgan, G. L. Stiegler, and F. T. Cross. 1994. Southern blot and polymerase chain reaction exon analyses of $HPRT^-$ mutations by radon and radon progeny. *Radiat. Res.* **137:** 371–379.
33. Kapp, D. S. and K. C. Smith 1970. Chemical nature of chain breaks produced in DNA by X-irradiation in vitro. *Radiat. Res.* **42:** 34–49.
34. Katcher, H. and S. S. Wallace. 1978. The production of alkali-labile lesions in X-irradiated PM2 DNA. *Int. J. Radiat. Biol.* **34:** 497–500.
35. Kraft, G., and M. Krämer. 1993. Linear energy transfer and track structure. *Adv. Radiat. Biol.* **17:** 1–52.
36. Krisch, R. E., M. B. Flick, and C. N. Trumbore. 1991. Radiation chemical mechanisms of single and double-strand break formation in irradiated SV40 DNA. *Radiat. Res.* **126:** 251–259.
37. Lafleur, M. V. M., J. Woldhuis, and H. Loman. 1991. Alkali-labile sites in biologically active DNA: comparison of radiation induced potential breaks and apurinic sites. *Int. J. Radiat. Biol.* **39:** 113–118.
38. Leadon, S. A. 1990. Production and repair of DNA damage in mammalian cells. *Health Phys.* **59:** 15–22.
39. Lemaire, D. G. E., E. Bothe, and D. Schulte-Frohlinde. 1984. Yields of radiation induced main chain scission of poly U in aqueous solution: strand break formation via base radicals. *Int. J. Radiat. Biol.* **45:** 351–358
40. Lindahl, T. 1990. Repair of intrinsic DNA lesions. *Mutat. Res.* **238:** 305–311.
41. Littlefield, L. G., E. E. Joiner, S. P. Colyer, A. M. Sayer, and E. L. Frome. 1988. Modulation of radiation induced chromosomal aberration by DMSO, an OH radical scavenger. I. Dose response studies in human lymphocytes exposed to 220 kV X-rays. *Int. J. Radiat. Biol.* **53:** 875–890.
42. McGrath, R. A. and R. W. Williams. 1966. Reconstruction in vivo of irradiated *Escherichia coli* deoxyribonucleic acid; the rejoining of the broken pieces. *Nature* **212:** 534–537.
43. Michael, B. D., K. M. Prise, M. Folkard, B. Vojnovic, B. Brocklehurst, I. H. Munro, and A. Hopkirk. 1995. Critical energies for SSB and DSB induction in plasmid DNA: studies using synchrotron irradiation, in *Radiation Damage to DNA: Structure Function Relationships at Earlier Times* (Fuciarelli, A. F. and J. D. Zimbrick, eds.), Battelle, Columbus, OH, pp. 251–258.
44. Milligan, J. R., J. A. Aguilera, and J. F. Ward. 1993. Variation of single strand break yield with scavenger concentration for plasmid DNA irradiated in aqueous solution. *Radiat. Res.* **133:** 151–157.
45. Milligan, J. R. and J. F. Ward. 1994. Yield of strand breaks due to attack of scavenger derived radicals. *Radiat. Res.* **137:** 295–299.
46. Milligan, J. R., J. Y.-Y. Ng, C. C. L. Wu, J. A. Aguilera, and J. F. Ward. 1995. DNA repair by thiols in air shows two radicals make a double strand break. *Radiat. Res.* **143:** 273–280.
47. Milligan, J. R., J. Y.-Y. Ng, C. C. L. Wu, J. A. Aguilera, J. F. Ward, Y. W. Kow, S. S. Wallace, and R. P. Cunningham. 1996. Methyperoxyl radicals as intermediates in DNA damage by ionizing radiation. *Radiat. Res.* **146:** 36–444.
48. Milligan, J. R., J. Y.-Y. Ng, J. A. Aguilera, J. F. Ward, Y. W. Kow, and S. S. Wallace. 1996, unpublished.
49. Mori, T. and M. Dizdaroglu. 1994. Ionizing radiation causes greater DNA base damage in radiation-sensitive mutant M10 cells than in parent mouse lymphoma L5178Y cells. *Radiat. Res.* **140:** 85–90.
50. Mori, T., Y. Hori, and M. Dizdaroglu. 1993. DNA base damage generated *in vivo* in hepatic chromatin of mice upon whole body γ-irradiation. *Int. J. Radiat. Biol.* **64:** 645–650.

51. Nackerdien, Z., R. Olinski, and M. Dizdaroglu. 1992. DNA base damage in chromatin of γ-irradiated cells. *Free Radical Res. Commun.* **16:** 259–273.
52. Newton, G. L., J. A. Aguilera, J. F. Ward, and R. C. Fahey. 1996. Polyamine-induced compaction and aggregation of DNA—a major factor in radioprotection of chromatin under physiologic conditions. *Radiat. Res.* **145:** 776–780.
53. O'Neill, P., and E. M. Fielden. 1993. Primary free radical processes in DNA. *Adv. Radiat. Biol.* **17:** 53–120.
54. Opitz, J., and D. Schulte-Frohlinde. 1987. Laser-induced photoionization and single strand break formation for polynucleotide and single strand DNA in aqueous solution: Model studies for the direct effect of high energy radiation on DNA. *J. Photochem.* **39:** 145–163.
55. Park, E.-M., M. K. Shigenaga, P. Degan, T. S. Korn, J. W. Kitzler, C. M. Wehr, P. Kolachana, and B. N. Ames. 1992. Assay of excised oxidative DNA lesions: Isolation of 8-oxoguanine and its nucleoside derivatives from biological fluids with a monoclonal antibody column. *Proc. Natl. Acad. Sci. USA* **89:** 3375–3379.
56. Pfeiffer, P., S. Thode, S. Hancke, P. Keohavong, and W. G. Thilly. 1994. Resolution and conservation of mismatches in DNA end joining. *Mutagenesis* **9:** 527–535.
57. Prise, K. M., M. Folkard, and B. D. Michael. 1995. A comparison of the fast kinetics of radiation-induced DNA lesions from low and high LET radiations: Clustered damage and dependence on OER, in *Radiation Damage to DNA: Structure Function Relationships at Earlier Times* (Fuciarelli, A. F. and J. D. Zimbrick, eds.) Battelle, Columbus, OH, pp. 185–190.
58. Raleigh, J. A. 1989. Secondary reaction in irradiated nucleotides: possible significance for chemical repair mechanisms. *Free Radical Res. Comm.* **6:** 141–143.
59. Redpath, J. L, E. Zabilansky, T. Morgan, and J. F. Ward. 1981. Cerenkov light and the production of photoreactivatable damage in X-irradiated *E. coli*. *Int. J. Radiat. Biol.* **39:** 569–575.
60. Richter, C., J.-W. Park, and B. N. Ames. 1988. Normal oxidative damage to mitochondrial and nuclear DNA is extensive. *Proc. Natl. Acad. Sci. USA* **85:** 6465–6467.
61. Roots, R., and S. Okada. 1972. Protection of DNA molecules of cultured mammalian cells from radiation-induced single-strand scissions by various alcohols and SH compounds. *Int. J. Radiat. Biol.* **21:** 329–342.
62. Sapora, O., F. Barone, M. Belli, A. Maggi, M. Quintilliani, and M. A. Tabocchini. 1991. Relationships between cell killing, mutation induction and DNA damage in X-irradiated V79 cells: The influence of oxygen and DMSO. *Int. J. Radiat. Biol.* **60:** 467–482.
63. Schulte-Frohlinde, D. 1986. Mechanism of radiation induced strand break formation in DNA and polynucleotide. *Adv. Space Res.* **6:** 89–96.
64. Siddiqi, M. A. and E. Bothe. 1987. Single- and double-strand break formation in DNA irradiated in aqueous solution: Dependence on dose and OH radical scavenger concentration. *Radiat. Res.* **112:** 449–463.
65. Skov, KA., B. Palcic, and L. D. Skarsgard. 1979. Radiosensitization of mammalian cells by misonidazole and oxygen: DNA damage exposed my *Micrococcus luteus* enzymes. *Radiat Res.* **79:** 591–600.
66. Swarts, S. G., M. D. Sevilla, D. Becker, C. K. Tokar, and K. T. Wheeler. 1993. Radiation induced DNA damage as a function of hydration. 1. Release of unaltered bases. *Radiat. Res.* **129:** 333–344.
67. Swarts, S. G., D. Becker, M. D. Sevilla, and K. T. Wheeler. 1996. Radiation-induced DNA damage as a function of hydration II. Oxidative base damage. *Radiat. Res.* **145:** 304–314.
68. Ullrich, M., and U. Hagen. 1971. Base liberation and concomitant reaction in irradiated DNA solutions. *Int. J. Radiat. Biol.* **19:** 507–517.
69. von Sonntag, C. 1987. *The Chemical Basis of Radiation Biology*. Taylor and Francis, London.

70. Ward, J. F. 1975 Molecular mechanisms of radiation induced damage to nucleic acids. *Adv. Radiat. Biol.* **5:** 181–239.
71. Ward, J. F. 1981. Some biochemical consequences of the spatial distribution of ionizing radiation produced free radicals. *Radiat. Res.* **86:** 185–195.
72. Ward, J. F. 1983. Chemical aspects of DNA radioprotectors, in *Radioprotectors and Anticarcinogens* (Nygaard, O. F. and M. G. Simic, eds.), Academic, New York, pp. 73–85.
73. Ward, J. F. 1985. Biochemistry of DNA lesions. *Radiat. Res.* **104:** S103–S111.
74. Ward, J. F. 1992. The intracellular molecular damage which is dependent on radiation energy deposition patterns at the nanometer level. *Genes, Cancer and Radiation Protection*, N. C. R. P. Proceedings No. 13, pp. 38–48.
75. Ward, J. F. 1994. The complexity of DNA damage—relevance to biological consequences. *Int. J. Radiat. Biol.* **66:** 427–432.
76. Ward, J. F. 1995. Radiation mutagenesis: The initial DNA lesions responsible. *Radiat. Res.* **142:** 362–368.
77. Ward, J. F., W. F. Blakely, and E. I. Joner. 1985. Mammalian cells are not killed by DNA single strand breaks caused by hydroxyl radicals from hydrogen peroxide. *Radiat. Res.* **104:** 383–393.
78. Ward, J. F., J. W. Evans, C. L. Limoli, and P. M. Calabro-Jones. 1987. Radiation and hydrogen peroxide induced free radical damage to DNA. *Br. J. Cancer* **55:** Suppl. VIII 105–112.
79. Ward, J. F. and I. Kuo. 1976. Strand breaks, base release and post-irradiation changes in DNA γ-irradiated in dilute O_2-saturated aqueous solution. *Radiat. Res.* **66:** 485–498.
80. Ward, J. F., J. R. Milligan, and G. D. D. Jones. 1994. Biological consequences of non-homogeneous energy deposition by ionizing radiation. *Radiat. Protect. Dosimetry* **52:** 271–276.
81. Hagen, U., D. Harder, H. Jung, and C. Streffer, eds. 1995. *Proceedings of the 10th International Congress of Radiation Research*, vol. 2, in Radiation Research 1895–1995, Würzburg, Germany, Aug. 27–Sept. 1.
82. Washino, K., O. Denk, and W. Schnabel. 1982. OH radical-induced main-chain scission of poly(ribonucleic acids) under anoxic conditions. *Z. Naturforsch.* **38c:** 100–106.
83. Willson, R. L. 1970. The reaction of oxygen with radiation-induced free radicals in DNA and related compounds. *Int. J. Radiat. Biol.* **17:** 349–358.
84. Wu, J. C., J. W. Kozarich, and J. Stubbe. 1983. The mechanism of free base formation from DNA by bleomycin. A proposal based on site specific tritium release from Poly (dA. dU). *J. Biol. Chem.* **258:** 4694–4697.

6

Mammalian Enzymes for Preventing Mutations Caused by Oxidation of Guanine Nucleotides

Mutsuo Sekiguchi and Hiroshi Hayakawa

1. INTRODUCTION

Oxygen radicals are produced through normal cellular metabolism, and formation of such radicals is further enhanced by ionizing radiation and by various chemicals *(3)*. The oxygen radicals attack nucleic acids and generate various modified bases in DNA *(9,16)*. Among them, 8-oxo-7,8-dihydroguanine (8-oxoG) is the most abundant, and appears to play critical roles in carcinogenesis and in aging *(6,22)*. 8-OxoG can pair with both cytosine and adenine during DNA synthesis, and as a result, G-C to T-A transversions are induced *(28,31)*. Oxidation of guanine also occurs in the cellular nucleotide pool. 8-Oxo-dGTP thus formed is a potent mutagenic substrate for DNA synthesis, since it can be incorporated opposite adenine as well as cytosine in DNA, at almost equal efficiencies *(24)*. In this case, both types of transversions, A-T to C-G and G-C to T-A, would be induced *(13)*.

Studies with *Escherichia coli* mutator mutants revealed that cells possess elaborate mechanisms that prevent mutations caused by oxidation of the guanine base, in both DNA and free nucleotide forms (*see* Chapters 4 and 6 in vol. 1). 8-OxoG residues in DNA can be removed by an enzyme that is coded by the *mutM* gene of *E. coli (6,14,25)*. *mutM* mutant cells deficient in the enzyme activity show a 10-fold higher frequency of G-C to T-A transversion, compared with the wild-type strain *(11)*. Another mutator gene, named *mutY*, also suppresses specifically G-C to T-A transversion *(4,29)*. MutY protein removes adenine from an adenine:8-oxoG mismatch *(5,26)*. Thus, two proteins, MutM and MutY, act consecutively at the site of oxidized guanine residue in the DNA to prevent occurrence of mutations in *E. coli (26,33)*. On the other hand, mutations owing to misincorporation of 8-oxo-dGTP can be prevented by the *mutT* gene product, which hydrolyzes 8-oxo-dGTP to 8-oxo-dGMP *(24)*. The *mutT* mutant induces specifically A-T to C-G transversion *(35)*, and this mutational specificity occurs through the concerted actions of the MutM and MutY proteins (*32*; *see* Chapter 6 in vol. 1).

8-OxoG-related mutagenesis may account for a considerable number of spontaneous mutations in mammalian cells. A significant amount of 8-oxoG is formed in mammalian DNA, and most of the modified bases are excised from the DNA *(3)*. Enzyme activities that cleave DNA at the site of 8-oxoG were detected in extracts of mamma-

From: DNA Damage and Repair, Vol. 2: DNA Repair in Higher Eukaryotes
Edited by: J. A. Nickoloff and M. F. Hoekstra © Humana Press Inc., Totowa, NJ

lian cells *(7,8,36)*, and hence, some are likely to be mammalian counterparts of the *E. coli* MutM and MutY proteins. An enzyme similar to the MutT protein found in human cells *(27)* apparently has the capability to suppress the mutator phenotype of *E. coli mutT* mutant cells *(30)*. These results suggest that mechanisms similar to those found in *E. coli* appear to function in mammalian cells to prevent the occurrence of mutations.

2. MAMMALIAN ENZYMES FOR OXYGEN-DAMAGED NUCLEOTIDES

2.1. *Purification and Characterization of Human 8-Oxo-dGTPase*

In search of an activity similar to *E. coli* MutT in mammalian cells, it was found that Jurkat cells (a human T-cell leukemia cell line) contained a high level of such activity. Taking advantage of this high level of activity, the enzyme was purified from a crude extract of Jurkat cells. The extract was processed through ammonium sulfate fractionation and four cycles of column chromatography. Purification as followed by sodium dodecyl sulfate-polyacrylamide gel electrophoresis (SDS-PAGE) as well as enzyme assays, and an approx 800-fold purification was achieved *(27)*.

The size of native human 8-oxo-dGTPase was determined by gel-filtration column chromatography. From the elution volume, the Stokes radius of the human protein was estimated to be 1.87 nm, corresponding to a molecular mass of 18 kDa as a globular protein. Thus, the human enzyme is slightly larger than MutT.

Using purified preparations, several properties of human 8-oxo-dGTPase were examined. The pH optimum for the activity was at pH 8.0. Mg^{2+} was essential for the activity, and the maximal activity was obtained with 2–6 m*M* $MgCl_2$. The addition of 40 m*M* NaCl to the assay stimulated the activity 1.6-fold; however, higher NaCl concentrations inhibited the activity, and the reaction with 160 m*M* NaCl showed only 20% of the maximum activity.

Substrate specificity of the enzyme was examined with α-^{32}P-labeled dNTPs. Although dGTP and dATP were also hydrolyzed to the corresponding nucleoside monophosphates, product yields were about 5% of that with 8-oxo-dGTP. Neither TTP nor dCTP was hydrolyzed by the enzyme. The apparent K_m for hydrolysis of 8-oxo-dGTP was 70 times lower than that for the degradation of dGTP, whereas the maximal reaction rates observed with both substrates were similar.

8-Oxo-dGMP can be placed onto a poly(dA) template on incubation of 8-oxo-dGTP with the α-subunit of *E. coli* DNA polymerase III, and this misincorporation is prevented by the *E. coli* MutT protein with a distinct 8-oxo-dGTPase activity *(2)*. The purified preparation of human 8-oxo-dGTPase protein possessed such a potential. With the reconstituted system, the activity to prevent 8-oxo-dGMP misincorporation copurified with the 8-oxo-dGTPase.

2.2. *Cloning and Sequence of cDNA for 8-Oxo-dGTPase*

The human 8-oxo-dGTPase protein was purified to physical homogeneity and its partial amino acid sequence was determined. Based on this, a cDNA for human 8-oxo-dGTPase was cloned, and its nucleotide sequence was determined *(30)*. The molecular weight of the protein, calculated from the predicted amino acid sequence, was 17.9 kDa, a value close to that estimated from analysis by SDS-PAGE *(27)*. When the amino acid sequences of the peptide fragments, derived from the purified 8-oxo-dGTPase protein, were aligned with the deduced amino acid sequence, there was a complete match.

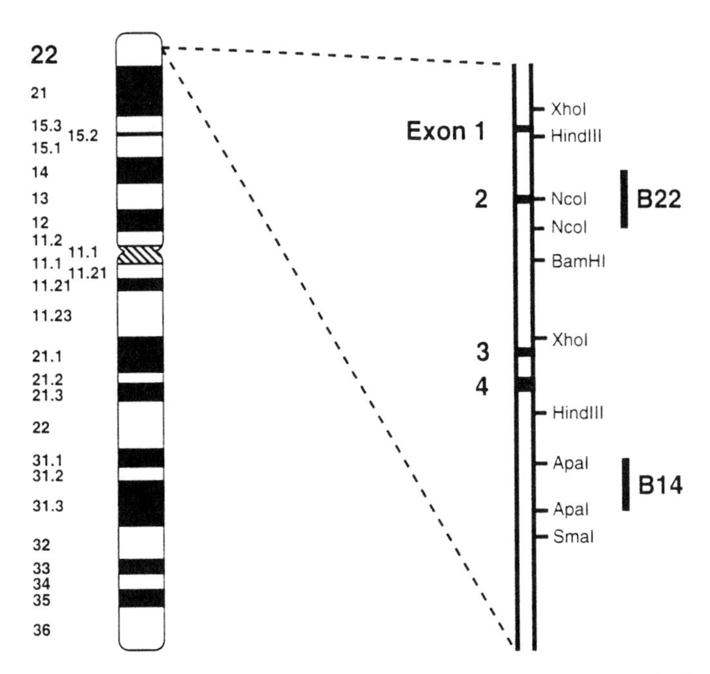

Fig. 1. Organization and chromosomal location of the *MTH1* gene. **(Left)** Ideogram of human chromosome 7 showing position of the *MTH1* gene on 7p22. **(Right)** The structure of the gene, together with representative restriction enzyme sites, is shown. Filled boxes represent the exons. Thick lines represent DNA fragments (B14 and B22) used for fluorescence *in situ* hybridization analysis. Data were taken from ref. *(15)*.

To confirm that the human cDNA isolated indeed codes for 8-oxo-dGTPase, the cDNA was expressed in *E. coli mutT*⁻ cells and the resulting protein was examined for the presence of 8-oxo-dGTPase activity. By placing the cDNA under control of the *lac* promoter, a high level of 8-oxo-dGTPase activity was induced in the presence of isopropyl-1-thio-β-D-galactopyranoside. When the extract was resolved by DEAE column chromatography, a distinct peak for 8-oxo-dGTPase was detected in addition to a small peak for nucleoside triphosphatase, which has a broader substrate specificity; the latter is also present in cells carrying a vector lacking the cDNA insert.

The effects of cDNA expression on the frequency of spontaneous mutation toward rifampicin resistance in *E. coli mutT*⁻ cells was also examined. It was found that the mutation frequency of *mutT*⁻ cells carrying the cDNA was considerably lower than that of *mutT*⁻ cells, which show about 400-fold higher mutation frequency than wild-type. Thus, the human 8-oxo-dGTPase functions in *E. coli* cells to prevent mutations caused by accumulation of 8-oxo-dGTP in the nucleotide pool.

2.3. Genomic Structure and Chromosome Location of the MTH1 Gene

The human gene for 8-oxo-dGTPase was named *MTH1* for *mutT* homolog. To elucidate the structure of gene, human genomic DNA libraries were screened by using the cDNA as a probe *(15)*. The sequences isolated were aligned according to patterns of restriction enzyme digestion, and the exon regions were identified by Southern blot hybridization. As shown in Fig. 1, the gene spans approx 10 kbp and contains at least four exons. The nucleotide sequences of all the exons and their flanking regions were determined. The coding sequence resides on exons 2, 3, and 4. There is a possibility that one or more exons exist in the upstream region.

Chromosomal assignment of the gene was made with the use of fragments B22 and B14 as probes (*see* Fig. 1). Both fragments consistently hybridized to a single locus on the short arm of chromosome 7 at p22. Whether or not 7p22 is related to a certain type of inherited disease or to familial cancer is unknown, but it is possible that a defect in this locus could lead to a high incidence of cancer as well as to a high frequency of spontaneous mutation.

2.4. Intracellular Localization of 8-Oxo-dGTPase

In eukaryotic cells, a pool of dNTP for nuclear DNA replication is present mainly in the cytosol. Mitochondria preserve a pool of dNTP for mitochondrial DNA synthesis, consisting of more than 10% of the total intracellular dNTP. The mitochondrial respiratory chain located on inner membranes is a major site for the initiation of lipid peroxidation, which can lead to oxidation of the guanine to 8-oxo-G. In addition, the mitochondrial respiratory chain produces superoxide, which can be converted to hydroxyl radical via hydrogen peroxide. The hydroxyl radical is the main species of active oxygen that attacks the guanine base. Thus, DNA and dNTP in the mitochondrial pool may be exposed to a greater oxidative stress than that in the nucleus. To investigate this, the intracellular location of 8-oxo-dGTPase was determined *(21)*. In human Jurkat cells, most of the enzyme activity was found in the cytosolic and the mitochondrial fractions. The specific activity of 8-oxo-dGTPase in the mitochondrial fraction was about 17% of that in the cytosolic fraction; almost no enzyme activity was found in the nuclear and the microsomal fractions. Electron microscopic immunocytochemistry, using a specific antibody against MTH1 protein, showed MTH1 protein localized to the mitochondrial matrix. An identical molecular form of MTH1 protein appears to be present both in cytosol and mitochondria.

2.5. Metabolism of Oxidized Guanine-Containing Nucleotides

Reduction of ribonucleotides to deoxyribonucleotides occurs at the nucleoside diphosphate level by the action of nucleoside diphosphate reductase. Human cells contain nucleoside diphosphate kinase, an enzyme activity that phosphorylates various nucleoside diphosphates to the corresponding nucleoside triphosphates. This enzyme can convert 8-oxo-dGDP to 8-oxo-dGTP, although the rate of phosphorylation of 8-oxo-dGDP was one-third that of dGDP *(17)*. Thus, in addition to direct oxidation of dGTP, 8-oxo-dGTP may be generated by phosphorylation of 8-oxo-dGDP in the cellular nucleotide pool.

8-Oxo-dGMP, produced by the action of 8-oxo-dGTPase, cannot be rephosphorylated by cellular enzymes. Human guanylate kinase, which phosphorylates both GMP and dGMP to the corresponding nucleoside diphosphates, is totally inactive on 8-oxo-dGMP *(17)*, probably reflecting the importance of excluding this mutagenic substrate from the DNA precursor pool. Instead, 8-oxo-dGMP is dephosphorylated to yield the corresponding nucleoside, 8-oxo-deoxyguanosine. Nucleosides are readily transported through the cell membrane, and extracellular nucleosides can be excreted in the

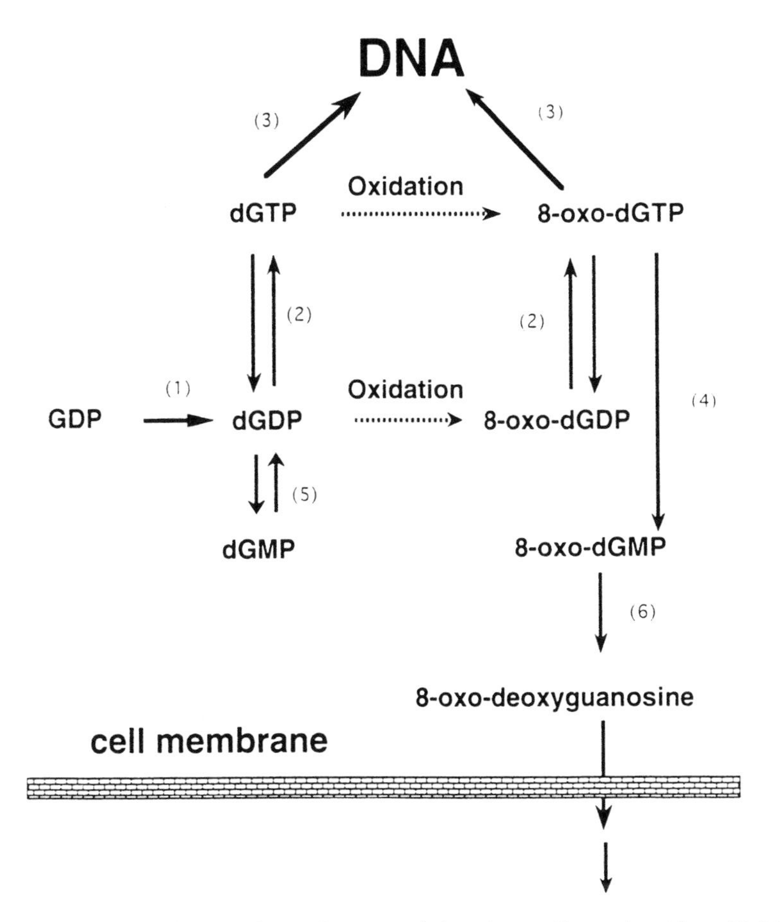

Fig. 2. A metabolic pathway of guanine-containing deoxyribonucleotides. (1) Nucleoside diphosphate reductase. (2) Nucleoside diphosphate kinase. (3) DNA polymerase. (4) 8-Oxo-dGTPase. (5) Guanylate kinase. (6) 8-Oxo-dGMPase.

urine. Dephosphorylation of 8-oxo-dGMP may be an essential step for excretion of 8-oxoguanine-containing materials. The enzyme that catalyzes this reaction, 8-oxo-dGMPase, was partially purified from an extract of human Jurkat cells, and the mode of action was elucidated. 8-Oxo-dGMP is the preferred substrate of the enzyme; other nucleoside monophosphates are cleaved at significantly lower rates *(17)*. Figure 2 illustrates the metabolic pathways for normal and oxidized forms of guanine-containing nucleotides.

3. BIOLOGICAL SIGNIFICANCE OF 8-OXO-dGTPASE

A certain degree of sequence homology has been noted in the *E. coli* MutT and the human MTH1 protein *(1,30)*. Genes for analogous functions were isolated from *Proteus vulgaris* and *Streptococcus pneumoniae*, bacteria distantly related to *E. coli* *(10,20)*. The products of the latter two genes carry enzyme activity degrading specifically dGTP to dGMP and are functionally related to the *E. coli* MutT protein. More recently, cDNAs for mouse and rat 8-oxo-dGTPase were isolated, and their structures

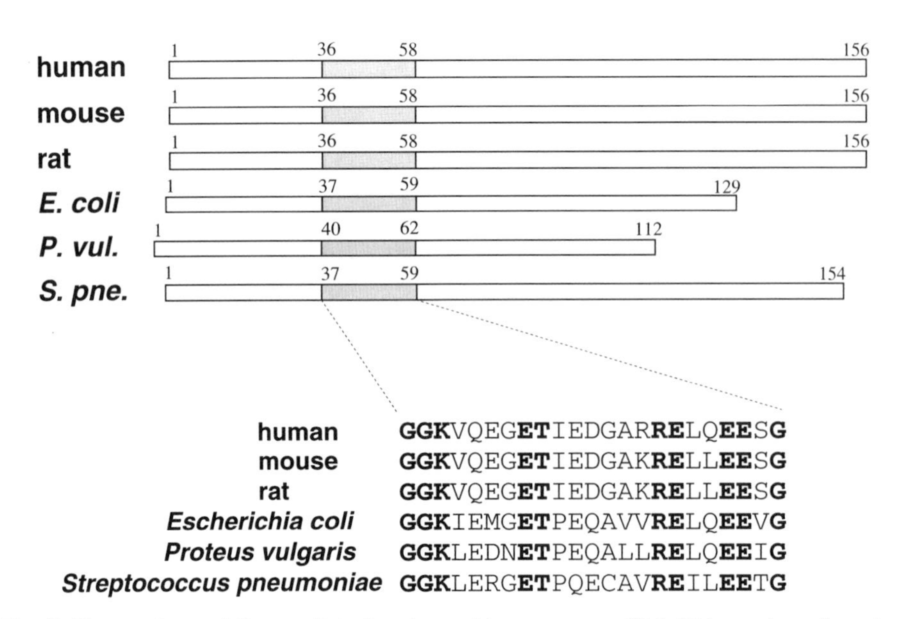

Fig. 3. Comparison of the predicted amino acid sequences of MutT homologs from human, mouse, rat, *S. pneumoniae*, *E. coli*, and *P. vulgaris* are shown. Gaps were inserted to maximize homology. Amino acid residues in the putative active site region and conserved through these six species are shown as boldface letters.

were elucidated *(12,19)*. Both proteins are comprised of 156 amino acid residues, as is human MTH1 protein.

These six proteins are similar in size, and alignment of the sequences shows that all carry a highly conserved sequence in nearly the same region (Fig. 3), corresponding to amino acids 36–58 for human MTH1. Ten of 23 amino acid residues in this region are identical, and it is likely that this constitutes an active center for the enzyme. This view is supported by the secondary structure of the MutT protein, as elucidated by NMR analysis *(34)*.

The biological significance of 8-oxo-dGTPase has been established on the basis of *E. coli* mutants defective in the *mutT* gene *(24,32,35)*. Lack of the gene causes an increased level of A-T to C-G transversion 1000-fold over the wild type cells. The elevated level of spontaneous mutation frequency of *mutT*⁻ cells reverted to normal when cDNAs for mouse and rat 8-oxo-dGTPase were expressed in such cells *(12,19)*. Specific suppression of A-T to C-G transversion by expression of the human cDNA was also demonstrated *(15)*. It seems likely that mammalian 8-oxo-dGTPase proteins have the same antimutagenic capacity as the *E. coli* MutT protein.

When examining the distribution of 8-oxo-dGTPase protein in mouse, all organs except the small intestine contained substantial amounts as detected by Western blot analysis. The highest levels were in the liver when comparisons were made on the basis of either protein or DNA content. An even higher level of the enzyme activity was detected in embryonic stem cell line CCE, which has great proliferation capacity. High oxygen consumption may correlate with high levels of oxidative damage, and the level of expression of the *MTH1* gene might be regulated in this context.

4. CONCLUDING REMARKS

An important notion arising from *E. coli* mutator studies is that organisms are equipped with elaborate mechanisms to reduce the occurrence of spontaneous mutations. Isolation of the cDNA and the gene for a human protein with antimutator activity will facilitate studies on mammalian mutation avoidance. By isolating the corresponding gene from mouse, it will be possible to prepare knockouts to examine roles of the enzyme in the control of spontaneous mutagenesis and carcinogenesis.

Human genes homologous to the *E. coli mutS* and *mutL*, both of which are involved in the postreplicational mismatch correction, were isolated *(23)*. These loci are related to causative mutations for a major group of hereditary nonpolyposis colon cancer (HNPCC), which is characteristic of the high incidence of alterations in microsatellite sequence (RER^+ phenotype) (*18*; *see* Chapter 20). Since the *MTH1* gene product functions in the metabolism of DNA substrates, its defect would not be expected to cause alterations in repeated sequences. *MTH1* mutations might be associated with a minor group of HNPCC with RER^- phenotype. It is also possible that defects in the *MTH1* gene may be related to the cause of other types of cancers.

ACKNOWLEDGMENTS

The research from this laboratory was supported by Grant-in-Aid from Ministry of Education, Science and Culture of Japan and by Human Frontier Science Program.

REFERENCES

1. Akiyama, M., T. Horiuchi, and M. Sekiguchi. 1987. Molecular cloning and nucleotide sequence of the *mutT* mutator of *Escherichia coli* that causes A:T to C:G transversion. *Mol. Gen. Genet.* **206:** 9–16.
2. Akiyama, M., H. Maki, M. Sekiguchi, and T. Horiuchi. 1989. A specific role of MutT protein: To prevent dG·dA mispairing in DNA replication. *Proc. Natl. Acad. Sci. USA* **86:** 3949–3952.
3. Ames, B. N. and L. S. Gold. 1991. Endogenous mutagens and the causes of aging and cancer. *Mutat. Res.* **250:** 3–16.
4. Au, K. G., M. Cabrera, J. H. Miller, and P. Modrich. 1988. *Escherichia coli mutY* gene product is required for specific A·G → C·G mismatch correction. *Proc. Natl. Acad. Sci. USA* **85:** 9163–9166.
5. Au, K. G., S. Clark, J. H. Miller, and P. Modrich. 1989. *Escherichia coli mutY* gene encodes an adenine glycosylase active on G-A mispairs. *Proc. Natl. Acad. Sci. USA* **86:** 8877–8881.
6. Bessho, T., K. Tano, H. Kasai, and S. Nishimura. 1992. Deficiency of 8-hydroxyguanine DNA endonuclease activity and accumulation of the 8-hydroxyguanine in mutator mutant (*mutM*) of *Escherichia coli*. *Biochem. Biophys. Res. Commun.* **188:** 372–378.
7. Bessho, T., K. Tano, H. Kasai, E. Ohtsuka, and S. Nishimura. 1993. Evidence for two DNA repair enzymes for 8-hydroxyguanine(7,8-dihydro-8-oxoguanine) in human cells. *J. Biol. Chem.* **268:** 19,416–19,421.
8. Bessho, T., R. Roy, K. Yamamoto, H. Kasai, S. Nishimura, K. Tano, and S. Mitra. 1993. Repair of 8-hydroxyguanine in DNA by mammalian *N*-methylpurine-DNA glycosylase. *Proc. Natl. Acad. Sci. USA* **90:** 8901–8904.
9. Boiteux, S., E. Gajewski, J. Laval, and M. Dizdaroglu. 1992. Substrate specificity of the *Escherichia coli* Fpg protein (formamidopyrimidine-DNA glycosylase): excision of purine lesions in DNA produced by ionizing radiation or photosensitization. *Biochemistry* **31:** 106–110.

10. Bullions, L. C., V. Méjean, J.-P. Claverys, and M. J. Bessman. 1994. Purification of the MutX protein of *Streptococcus pneumoniae*, a homologue of *Escherichia coli* MutT. *J. Biol. Chem.* **269:** 12,339–12,344.
11. Cabrera, M., Y. Nghiem, and J. H. Miller. 1988. *mutM*, a second mutator locus in *Escherichia coli* that generates G·C → T·A transversions. *J. Bacteriol.* **170:** 5405–5407.
12. Cai, J.-P., T. Kakuma, T. Tsuzuki, and M. Sekiguchi. 1995. cDNA and genomic sequences for rat 8-oxo-dGTPase that prevents occurrence of spontaneous mutations due to oxidation of guanine nucleotides. *Carcinogenesis* **16:** 2343–2350.
13. Cheng, K. C., D. S. Cahill, H. Kasai, S. Nishimura, and L. A. Loeb. 1992. 8-Hydroxyguanine, an abundant form of oxidative DNA damage, causes G → T and A → C substitutions. *J. Biol. Chem.* **267:** 166–172.
14. Chung, M. H., H. Kasai, D. S. Jones, H. Inoue, H. Ishikawa, E. Ohtsuka, and S. Nishimura. 1991. An endonuclease activity of *Escherichia coli* that specifically removes 8-hydroxyguanine residues from DNA. *Mutat. Res.* **254:** 1–12.
15. Furuichi, M., M. C. Yoshida, H. Oda, T. Tajiri, Y. Nakabeppu, T. Tsuzuki, and M. Sekiguchi. 1994. Genomic structure and chromosome location of the human *mutT* homologue gene *MTH1* encoding 8-oxo-dGTPase for prevention of A:T to C:G transversion. *Genomics* **24:** 485–490.
16. Gajewski, E., G. Rao, Z. Nackerdien, and M. Dizdaroglu. 1990. Modification of DNA bases in mammalian chromatin by radiation-generated free radicals. *Biochemistry* **29:** 7876–7882.
17. Hayakawa, H., A. Taketomi, K. Sakumi, M. Kuwano, and M. Sekiguchi. 1995. Generation and elimination of 8-oxo-7,8-dihydro-2'-deoxyguanosine 5'-triphosphate, a mutagenic substrate for DNA synthesis, in human cells. *Biochemistry* **34:** 89–95.
18. Ionov, Y., M. A. Peinado, S. Malkhosyan, D. Shibata, and M. Perucho. 1993. Ubiquitous somatic mutations in simple repeated sequences reveal a new mechanism for colonic carcinogenesis. *Nature* **363:** 558–561.
19. Kakuma, T., J. Nishida, T. Tsuzuki, and M. Sekiguchi. 1995. Mouse MTH1 protein with 8-oxo-7,8-dihydro-2'-deoxyguanosine 5'-triphosphatase activity that prevents transversion mutation. *J. Biol. Chem.* **270:** 25,942–25,948.
20. Kamath, A. V. and C. Yanofsky. 1993. Sequence and characterization of *mutT* from *Proteus vulgaris*. *Gene* **134:** 99–102.
21. Kang, D., J. Nishida, A. Iyama, Y. Nakabeppu, M. Furuichi, T. Fujiwara, M. Sekiguchi, and K. Takeshige. 1995. Intracellular localization of 8-oxo-dGTPase in human cells, with special reference to the role of the enzyme in mitochondria. *J. Biol. Chem.* **270:** 14,659–14,665.
22. Kasai, H., P. F. Crain, Y. Kuchino, S. Nishimura, A. Ootsuyama, and H. Tanooka. 1986. Formation of 8-hydroxyguanine moiety in cellular DNA by agents producing oxygen radicals and evidence for its repair. *Carcinogenesis* **7:** 1849–1851.
23. Loeb, L. A. 1994. Microsatellite instability: marker of a mutator phenotype in cancer. *Cancer Res.* **54:** 5059–5063.
24. Maki, H. and M. Sekiguchi. 1992. MutT protein specifically hydrolyses a potent mutagenic substrate for DNA synthesis. *Nature* **355:** 273–275.
25. Michaels, M. L., L. Pham, C. Cruz, and J. H. Miller. 1991. MutM, a protein that prevents G·C → T·A transversions, is formamidopyrimidine-DNA glycosylase. *Nucleic Acids Res.* **19:** 3629–3632.
26. Michaels, M. L., C. Cruz, A. P. Grollman, and J. H. Miller. 1992. Evidence that MutY and MutM combine to prevent mutations by an oxidatively damaged form of guanine in DNA. *Proc. Natl. Acad. Sci. USA* **89:** 7022–7025.
27. Mo, J.-Y., H. Maki, and M. Sekiguchi. 1992. Hydrolytic elimination of a mutagenic nucleotide, 8-oxodGTP, by human 18-kilodalton protein: sanitization of nucleotide pool. *Proc. Natl. Acad. Sci. USA* **89:** 11,021–11,025.

28. Moriya, M. 1993. Single-stranded shuttle phagemid for mutagenesis studies in mammalian cells: 8-Oxoguanine in DNA induces targeted G·C → T·A transversions in simian kidney cells. *Proc. Natl. Acad. Sci. USA* **90:** 1122–1126.
29. Nghiem, Y., M. Cabrera, C. G. Cupples, and J. H. Miller. 1988. The *mutY* gene: A mutator locus in *Escherichia coli* that generates G·C → T·A transversions. *Proc. Natl. Acad. Sci. USA* **85:** 2709–2713.
30. Sakumi, K., M. Furuichi, T. Tsuzuki, T. Kakuma, S. Kawabata, H. Maki, and M. Sekiguchi. 1993. Cloning and expression of cDNA for a human enzyme that hydrolyzes 8-oxo-dGTP, a mutagenic substate for DNA synthesis. *J. Biol. Chem.* **268:** 23,524–23,530.
31. Shibutani, S., M. Takeshita, and A. P. Grollman. 1991. Insertion of specific bases during DNA synthesis past the oxidation-damaged base 8-oxodG. *Nature* **349:** 431–434.
32. Tajiri, T., H. Maki, and M. Sekiguchi. 1995. Functional cooperation of MutT, MutM and MutY proteins in preventing mutations caused by spontaneous oxidation of guanine nucleotide in *Escherichia coli. Mutat. Res.* **336:** 257–267.
33. Tchou, J. and A. P. Grollman. 1993. Repair of DNA containing the oxidatively-damaged base, 8-oxoguanine. *Mutat. Res.* **299:** 277–287.
34. Weber, D. J., C. Abeygunawardana, M. J. Bessman, and A. S. Mildvan. 1993. Secondary structure of the MutT enzyme as determined by NMR. *Biochemistry* **32:** 13,081–13,088.
35. Yanofsky, C., E. C. Cox, and V. Horn. 1966. The unusual mutagenic specificity of an *E. coli* mutator gene. *Proc. Natl. Acad. Sci. USA* **55:** 274–281.
36. Yeh, Y.-C., D.-Y. Chang, J. Masin, and A.-L. Lu. 1991. Two nicking enzyme systems specific for mismatch-containing DNA in nuclear extracts from human cells. *J. Biol. Chem.* **266:** 6480–6484.

7

Biochemistry of Mammalian DNA Mismatch Repair

A-Lien Lu

1. INTRODUCTION

Multiple mismatch repair pathways with different mispair specificities and different size repair tracts are utilized by all organisms to reduce replicative errors and to protect their DNA from various types of damage *(31,97)*. Mismatch repair systems enhance the fidelity of DNA replication by correcting replicative errors skipped by the DNA polymerase during editing. Mutants defective in mismatch repair enzymes are marked by genome instability and/or mutator phenotypes *(31)*. During homologous genetic recombination, mismatch repair of heteroduplex DNA causes gene conversion and modulates the degree of genetic diversification *(45)*. Mismatch repair is best understood in *Escherichia coli (16,97,121)* (*see* Chapter 11 in vol. 1). *E. coli* mutants defective in repair have been isolated and the repair enzymes have been purified and characterized in vitro. A considerable amount of information is also available about mismatch repair in the yeast *Saccharomyces cerevisiae (31,57)* (*see* Chapter 19 in vol. 1).

Similar repair systems in higher eukaryotes have been reported *(29,31,57,66,97,99)* and are currently the subject of intense study. Remarkably, the basic processes of mismatch repair are highly conserved among diverse organisms; they share high homology in mismatch specificities, repair enzymes, and repair mechanisms. There are two major repair pathways that differ in the length of the repair patch. Short-patch repair systems have repair tracts shorter than 10 nucleotides and are dictated by the nature of the mismatches. Long-patch repair systems involve a long-patch excision and resynthesis (up to several hundred nucleotides) and direct repair to a particular DNA strand.

Two short-patch repair pathways have been characterized in mammalian cells. Thymine glycosylase removes thymines from T-G and other mismatches *(13,100,101,143,144)*, and is functionally equivalent to the T-G-specific very short-patch (VSP) pathway found in *E. coli (74,75,122)*, which repairs mispairs arising from the deamination of 5-methylcytosine. Mammalian MYH (*M*ut*Y h*omolog) glycosylase specifically removes mispaired adenines from A-G, A-7,8-dihydro-8-oxo-deoxyguanine (8-oxoG), and A-C mismatches *(92,145)*, and is believed to be responsible for protection against oxidative damage and for increased replicative fidelity. MYH shares about 40% identity to *E. coli* MutY protein *(86)* and crossreacts with antibodies raised against the *E. coli* MutY protein *(92)*. The high homology of MutY

From: DNA Damage and Repair, Vol. 2: DNA Repair in Higher Eukaryotes
Edited by: J. A. Nickoloff and M. F. Hoekstra © Humana Press Inc., Totowa, NJ

homologs among different organisms underscores the importance of these proteins in cellular functions. Refer to Chapter 11 in vol. 1 and Chapter 8 in this vol. for information on these short-patch pathways.

The long-patch repair system repairs the majority of base–base mismatches and insertion/deletion loops (IDLs) with unpaired bases. This pathway is dependent on the MutS and MutL homologs, and is believed to be involved in the avoidance of DNA replicative errors, alkylation tolerance, and the processing of recombination intermediates. Recent discoveries have illustrated that inactivation of the human long-patch mismatch repair pathway is linked with certain types of hereditary and sporadic cancers *(51,57,98)* (*see* Chapter 20). This chapter focuses on the biochemistry of mammalian long-patch mismatch repair.

2. LONG-PATCH MISMATCH REPAIR

This repair pathway is directed to the daughter strand, which bears replicative errors. MutS and MutL homologs are involved in this pathway, which repairs all eight base–base mismatches and insertion/deletion loops. Based on the similarity of specificities, repair enzymes, and repair mechanisms, this pathway is functionally equivalent to the *E. coli* methyl-directed MutHLS pathway (*see* Chapter 11 in vol. 1).

2.1. *In Vitro Mismatch Repair Systems*

In vitro assays with mammalian cell extracts have facilitated the purification and characterization of the repair enzymes involved in the long-patch mismatch repair pathway *(46,134)*. Extracts of human HeLa cells efficiently catalyze mismatch repair reactions. Two systems have been described: one uses nuclear extracts *(46)* and the other uses cytoplasmic extracts *(134)*. Presumably, the repair enzymes leak to the cytoplasm during cell lysis in the second method *(134)*. Circular heteroduplex DNA substrates used in these experiments contain a nick placed several hundred nucleotides away from a mismatch site. A repair reaction requires added Mg^{2+}, ATP, and four dNTPs in HeLa nuclear extracts *(46)*. Heteroduplexes developed in Modrich's laboratory utilize base–base mismatches within overlapping sequences for two restriction endonucleases within f1 phage DNA *(132)*. This allows scoring of strand-specific repair by cleavage of the resulting substrates with either of the restriction endonucleases. A strand break can be placed either on the complementary (–) strand by cleavage with a restriction enzyme or on the viral (+) strand by cleavage with phage fd gpII *(22,46)*.

The method developed by Kunkel and his coworkers *(134)* employs M13 DNA containing a covalently closed (+) strand and a (–) strand with a nick located several hundred base pairs away from a mismatch or unpaired base in the *lacZ* α-complementation sequence. The two DNA strands encode different plaque colors (e.g., blue vs colorless) on indicator plates. When unrepaired heteroduplex DNA is transformed into a repair-deficient *E. coli* strain, mixed plaques are observed because the phenotypes of both strands are expressed. If correction is directed to the nicked (–) DNA strand during the incubation with cell extracts, the percentage of mixed plaques is reduced and the ratio of the (+) strand phenotype to the (–) strand phenotype is increased.

In vitro repair of the heteroduplex DNA is strongly biased in favor of nicked strands *(22,23,46,134)*. Different mismatches are not repaired with the same efficiencies (Table 1). C-C mispairs, which are refractory to repair in *E. coli* extracts *(97,132)*, are repaired

Table 1
Mismatch Repair Specificities of *E. coli* and Humans

	Relative rate of repair		% of repair
Mismatch	*E. coli* extract[a]	HeLa extract[b]	Monkey CV1 cells[c]
G-T	100	100	96 (92 G/C, 4 A/T)
A-C	54	68	78 (37 A/T, 41 G/C)
A-A	74	66	64
G-G	66	110	92
A-G	61	38	39 (12 A/T, 27 C/G)
T-C	62	70	72 (12 T/A, 60 G/C)
T-T	68	27	39
C-C	≤5	13	66

[a]Repair in *E. coli* extracts is from the data of Su et al. *(132)*. Rates shown are normalized to that observed with a G-T substrate (8.5 fmol/min/mg).

[b]Repair in human HeLa extracts is from the data of Fang et al. *(22)*. Rate shown are normalized to that observed with a G-T substrate (5.6 fmol/min/mg).

[c]Repair frequencies were measured by transfection of SV40 heteroduplexes in monkey kidney CV1 cells *(14)*. Numbers in parenthesis represent % of plaques types.

to some extent in the human cell extracts *(20,22,46,134)*. Heteroduplexes containing an IDL are also repaired *(20,113,135)*. Mismatch repair can be provoked by a single-strand break up to several hundred base pairs away from the mismatch site *(23)*. Bidirectional repair is accomplished by DNA synthesis localized to the region between the mismatch and the nicked site, and is sensitive to aphidicolin (an inhibitor of DNA polymerases α, δ, and ε) *(22,23)*. Moreover, as discussed in Section 2.4., extracts derived from human cells with mutations in *hMLH1* (*M*ut*L h*omolog) *(73,113)*, *hMSH2* (*M*ut*S h*omolog) *(136)*, *hGTBP* (*G/T-b*inding *p*rotein, a MutS homolog) *(20)*, and *hPMS2* (*p*ost*m*eiotic *s*egregation, a MutL homolog) *(125)* genes are defective in base–base mismatch repair. These results strongly suggest that this mammalian repair pathway involving multiple MutS and MutL homologs is analogous to the *E. coli* methyl-directed pathway.

2.2. *Mismatch Specificities*

The repair of different base–base mismatches has been tested in vivo by transfection of monkey kidney CV1 cells *(14,40,41)*, and in vitro using extracts of human HeLa cells *(46,134)*. All eight possible base–base mismatches can be repaired in vivo *(14)* and in vitro *(22,46)*, although with variable efficiencies and specificities *(22,23,46,134)* (Table 1). For example, G-T and G-G are repaired more efficiently than other mispairs. C-C mispairs are repaired to some extent in CV1 cells in vivo *(14)* and in human cell extracts *(20,22,46,134)*. Because different DNA heteroduplexes with different neighboring sequences are used in these experiments, a direct comparison of the specificities between the two systems is difficult. In addition, mammalian cell lines deficient in either a MutS or MutL homolog have not been used in the transfection assays, so in vivo mismatch repair may be controlled by more than one pathway. For example, closed circular SV40 DNA with a single T-G mismatch is repaired in CV1 cells to yield 92% of C-G, 4% of A-T, and 4% of mixed type. This strong bias in vivo is consistent with the property of thymine glycosylase, which removes T from T-G mismatches *(13,100,101,143,144)*. Similarly, A-G mismatches can be corrected in favor of C-G by the MYH short-patch pathway *(92,145)*. Therefore, both the long-patch and short-patch pathways (*see* Section 1. and Chapters 8 and 20) may repair T-G and A-G mismatches in vivo.

The similarities in base specificities of mismatch repair in *E. coli*, *Streptococcus pneumoniae*, *S. cerevisiae*, *Drosophila melanogaster*, and human HeLa cells *(16,97,121)* suggest that some characteristics of mismatch repair are shared among diverse organisms. Mammalian cells exhibit greater specificity of long-patch mismatch repair over bacteria and yeast by their ability to repair C-C mispairs more efficiently *(8,16,60,97)*. Coversely, A-G and T-T mispairs are less efficiently repaired by the mammalian cells than by *E. coli* (Table 1).

Heteroduplexes containing IDLs can also be repaired in human cell extracts *(20,113,135)*. Repair of heteroduplexes containing unpaired loops of different lengths seems to involve different enzymes. Repair of DNA with a loop of five nucleotides is dependent on hMSH2, but not hMLH1 *(135)*, and purified recombinant hMSH2 can bind to DNA containing loops of up to 14 nucleotides *(27)*. Moreover, extracts of MT1 and HCT15 cell lines, which are deficient in hGTBP, are defective in the repair of single nucleotide IDLs, but are proficient in correcting 2-, 3-, and 4-nucleotide loops *(20,55)*. These results suggest that larger IDLs may be repaired by a pathway that is distinct from that responsible for base–base mismatch repair. Recently, a distinct repair activity specific for IDLs with 4–9 extra bases has been identified in *S. cerevisiae (95)*. The ability of human cells to repair larger IDLs is also different from that of *E. coli*, which cannot repair loops of more than five unpaired bases *(15,19,70,112)*. *E. coli* has essentially no repetitive DNA, and therefore, recombination between repetitive sequences or polymerase slippage is unlikely to produce large IDLs. Therefore, there is no need for such repair activity in *E. coli*. Defects in loop repair may be relevant to the microsatellite instability observed in some cancers and other hereditary diseases *(1,7,50,62,133)*.

Lee et al. *(71)* have used electron microscopy and gel retardation assays to show that the p53 tumor suppressor protein can form highly stable complexes with DNA substrates containing one or three-nucleotide IDLs, but not with DNA containing a G-T mismatch. p53 protein is involved in the control of the cell-cycle checkpoint after DNA damage. These results raise an interesting question. Since both hMSH2 and p53 can specifically recognize IDLs, what, if any, is the functional relationship between these proteins? One possibility is that binding of p53 to a lesion site may provide a scaffold for other repair enzymes, such as hMSH2, to initiate DNA repair *(71)*.

2.3. Strand Specificity

A major unanswered question in eukaryotic mismatch repair is what dictates strand specificity. Hare and Taylor *(40,41)* have reported that cytosine methylation can play a role in directing repair to unmethylated daughter strands in mammalian cells in a manner similar to the role of adenine methylation in *E. coli (16,97,121)*. However, these results have not been further supported, because an endonuclease activity similar to *E. coli* MutH has not yet been identified and attempts to isolate the *mutH* gene homolog in eukaryotes have been unsuccessful. If mismatch repair in mammalian cells is directed by cytosine methylation, one would expect that an endonuclease exists to provide the nick on the unmethylated DNA strand.

Although in vitro mismatch repair assays use nicked heteroduplexes, it is unclear whether a strand break is an intermediate or a true signal. Although covalently closed heteroduplexes are repaired in vitro less efficiently and without bias *(46,134)*, the mechanism for this type of repair is poorly understood. In *E. coli*, strand breaks are sufficient to direct mismatch repair, bypassing the requirement for MutH and hemimethylated sites *(67,68)*. However, primary strand discrimination is provided by methylation at GATC sites *(83,97)*. Because *S. pneumoniae*, *S. cerevisiae*, and *D. melanogaster* do not contain any methylated bases, strand breaks may suffice to direct repair to newly synthesized DNA strands in these organisms *(16,117,137)*. In this case, strand discontinuities in the daughter strands may direct the repair to these strands. Alternately, a direct interaction between repair components and the replication machinery may provide a structural difference between daughter and parental strands.

2.4. Cloning Human Mismatch Repair Genes

Several homologs of bacterial MutS and MutL have been identified in mammals *(6,12,30,32,47,48,69,77,80,103,108,110,138)*. Table 2 lists the known human MutS and MutL homologs.

2.4.1. MutS Homologs

The first potential mammalian MutS homolog was identified on the basis of DNA sequence similarity to bacterial genes. The human *DUG* and mouse *Rep-3* genes *(32,77,80)*, which are located upstream of the dihydrofolate reductase gene (*DHFR*), contain open reading frames with significant homologies to the MutS protein of *Salmonella typhimurium (38)* and the HexA protein of *S. pneumoniae (116)*. The role of the human DUG protein in mismatch repair has been demonstrated by formation of a heterodimer with hMSH2 that specifically binds to IDL larger than 1 nucleotide *(2a,108a)*. These properties of hDUG (hMSH3) are similar to that of the yeast MSH3 protein *(90a)*. The nucleotide sequences of *DUG/Rep-3* have facilitated the cloning of several yeast *MutS* homologs *(102,124)* and the human (*hMSH2*) gene *(30)* by PCR. Independently, the *hMSH2* gene was isolated by position cloning of a gene associated with hereditary nonpolyposis colon cancer (HNPCC) *(69)*. The hMSH2 protein sequence is about 40% identical with yeast Msh2p *(30,69,124)*. Expression of hMSH2 in *E. coli* causes a dominant mutator phenotype, suggesting that hMSH2 can bind to heteroduplex DNA, but fails to interact with other repair enzymes in *E. coli (30)*. Jiricny and colleagues have identified a mixture of 100- and 160-kDa proteins that bind to G-T and other mismatches *(48,51,52)*. The sequences of tryptic peptides of this 100 kDa protein can be found within the amino acid sequence of hMSH2 *(108)* identifying the 100 kDa protein as hMSH2. The relationship of hMSH2 to another 100 kDa protein that binds to A-C-, T-C-, and T-T-containing DNA in human cells *(129)* has not been established. The *hMSH2* gene maps to chromosome 2p21-22, and mutations of *hMSH2* have been associated with half of all HNPCC cases *(30,69)*.

hGTBP was isolated from HeLa cell extracts by its preferential binding to G-T-containing DNA *(52)*. Purification of this binding activity yielded a mixture of a 160-kDa protein and the 100-kDa hMSH2 protein *(48,107)*. Independently, the 160-kDa hGTBP

Table 2
Human MutS and MutL Homologs

Protein	*E. coli* homolog	Chromosome	Function	Repair type	Prevalence in HNPCC[a], %	Defective cell lines	Reference
hMSH2	MutS	2p21–22	MMR	Base, IDL	50	LoVo, RB, HEC59, EA1, 2774	*20,30,69 108,135,136*
hGTBP/hMSH6	MutS	~2p15–16	MMR	Base, IDL (1 nucl.)	<5	MT1, HCT15/DLD1, 543X	*2a,20,48,107, 109,136*
hDUG/hMSH3	MutS	5q11–12	MMR	IDL	<5	?	*2a,77,80,108a*
hMLH1	MutL	3p21	MMR, meiosis	Base, IDL	30	H6/HCT116, 595X, SW-48, AN3CA, SK-OV-3, DU145	*12, 73,110, 113,125 135,136*
hPMS1	MutL	2q31–32	MMR	Base, IDL	5	?	*103,110*
hPMS2	MutL	7p22	MMR, meiosis	Base, IDL	5	GC, HEC-1-A	*6,103 110,125*
hPMS3–11	MutL	7q11.23, 7q22, ?	?	?	?	?	*47,103*

[a]Abbreviations: HNPCC, hereditary nonpolyposis colon cancer; MMR, mismatch repair; Base, base–base mismatch; IDL, insertion/deletion loop.

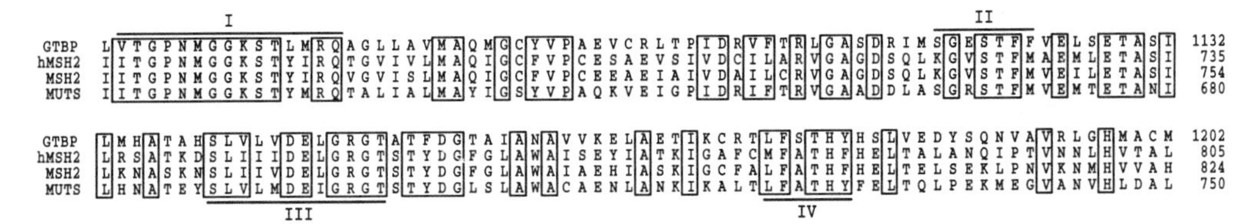

Fig. 1. Alignment of the amino acid sequences of the conserved C-terminal regions of the mismatch-binding proteins human hGTBP (GTBP), human hMSH2, *S. cerevisiae* Msh2p, and *E. coli* MutS. Conserved amino acids are boxed. The ATP binding site consensus sequences are indicated by numerals I–IV. Reprinted with permission from Palombo et al. (1995) *Science* **268:** 1912–1914 *(107)*. Copyright 1995 American Association for the Advancement of Science.

and the 100-kDa hMSH2 heterodimer (called hMutSα) was isolated from HeLa extracts by its ability to restore mismatch repair activity to an extract from an *hMSH2*-deficient cell line *(20)*. Crosslinking experiments have demonstrated that the 160-kDa protein preferentially binds to G-T-containing DNA *(48)*. DNA probes derived from peptide sequences were used to screen a HeLa cDNA library and thereby isolate the *hGTBP* (or hMSH6) gene *(107)*. The full-length hMSH6 cDNA and genomic clones were recently isolated and characterized *(2a)*. Computer analysis indicates that hGTBP belongs to the MutS family. As in human cells, Msh6p of *S. cerevisiae*, a homolog to human hGTBP *(90)*, forms a heterodimer with Msh2p protein *(49)*. The *hGTBP* gene localizes to within 1 Mb of the *hMSH2* gene on chromosome 2p15–16 *(109)*. It is suspected that these two genes may have arisen by gene duplication.

The MutS family of proteins shares a highly conserved carboxyl-terminal segment of about 100–150 amino acids (Fig. 1) *(30,69,102,107,124)* containing nucleotide binding sites. Like *E. coli* MutS protein, hMSH2/hGTBP complex has an ATPase activity and a weak helicase activity *(48)*. Human cell lines containing a mutation in either *hMSH2* or *hGTBP* are defective in mismatch repair *(10,20,136)*, indicating both gene products are involved in the repair process. However, repair of IDLs is different in *hMSH2*- and *hGTBP*-defective cell extracts. Unlike *hMSH2*, which is involved in the repair of 1–5 nucleotide IDLs *(20,135,136)*, cells with *hGTBP* mutations only show limited instability in mononucleotide tracts *(109)*. Extracts from hGTBP-deficient cells are defective in the repair of single-nucleotide IDLs, but are proficient in the repair of IDLs with two to five unpaired bases *(20)*. Interestingly, recent findings suggest hDUG (hMSH3) has reciprocal specificity to hGTBP (hMSH6) *(2a,108a)*. hMSH2 can form a heterodimer with hMSH3 or hMSH6, similar to protein complexes demonstrated by studies of the *S. cerevisiae* Msh2p, Msh3p, and Msh6p *(90)*. Complex of hMSH2/hMSH3 specifically binds to IDLs larger than 1 nucleotide *(2a,108a)*. The yeast *msh3* mutant has a higher mutation rate for dinucleotide repeats, but not for base substitutions *(130)*.

2.4.2. MutL Homologs

Human MutL homologs have been identified by two approaches. In one approach, the homology among yeast Pms1p *(61)*, bacterial MutL *(87)*, and *S. pneumoniae* HexB proteins *(120)* permitted the design of PCR primers used to clone the *hMLH1* gene and another gene similar to the yeast *PMS1* gene *(12)*. Another approach using a computer search of expressed sequence tag (EST) cDNA libraries also identified three human *MutL* homologs, *hMLH1*, *hPMS1*, and *hPMS2 (110)*. Human *PMS2* is more homologous to yeast *PMS1* than is *hPMS1 (110)*. Thus, human cells contain several *mutL* homologs: *hMLH1* on chromosome 3p21 *(12,110)*, *hPMS1* on chromosome 2q31-32 *(110)*, and *hPMS2* on chromosome 7p22 *(103,110)* (Table 2). In addition, at least two other *hPMS2*-related genes are located on chromosome 7 *(103)*. At least 11 PMS members are included in the human MutL homolog superfamily *(47)*.

Given the large number of *hPMS* genes, questions remain concerning which are functional in mismatch repair. Mutations have been found in HNPCC families in *hMLH1*, *hPMS1*, and *hPMS2 (12,103,110,125)*, implicating roles in mismatch repair. Further, extracts of colon tumor cell lines with mutations in *hMLH1* or *hPMS2* are defective in mismatch repair *(10,73,113,125)*. However, the involvement of other *hPMS* members in mismatch repair remains to be established. Li and Modrich *(73)* have purified hMutLα by its ability to complement mismatch repair activity in extracts from *hmlh1* mutant tumor cells *(73)*. Analysis of hMutLα shows that it is composed of equimolar amounts of hMLH1 and a second protein that is likely to be hPMS2, but not hPMS1. Therefore, MutL homologs function as a heterodimer of hMLH1 and hPMS2 or a closely related hPMS2 homolog.

2.5. Mechanism of Mismatch Repair

In order to study the biochemical mechanism of mismatch repair, DNA repair enzymes have been purified by using DNA substrate binding assays or complementation assays of in vitro repair systems. The mechanism governing mammalian mismatch repair is less well understood than that of the bacterial systems because the proteins involved in the process have not all been identified. Nevertheless, as is the case in bacteria and yeast, the first step of mismatch repair in mammalian cells is the recognition of the mismatch site by MutS homologs. After the specific binding of these MutS homologs, MutL homologs and probably other unidentified factors may search for a nick or a hemimethylated site to trigger the repair process. Figure 2 illustrates possible mechanisms of mammalian long-patch mismatch repair.

2.5.1. hMSH Binds to the Mismatch Site

Fishel and his coworkers have demonstrated that recombinant hMSH2 expressed in *E. coli* specifically interacts with T-G mismatches and IDLs *(27,28)*. The hMSH2 protein appears to have a higher affinity for IDLs than for single base–base mismatches *(27)*. Electron microscopic analysis indicates hMSH2 complexes with 8–14 nucleotide IDL-containing DNA in multiple oligomeric forms *(27)*. The presence of ATP appears to increase the binding affinity of hMSH2 for heteroduplex DNA *(27,28)*. However, Palombo et al. *(107)* using an in vitro translation system to express hMSH2 and hGTBP have found that separately neither hMSH2 nor hGTBP exhibits mismatch binding activity, but a mixture of the two proteins is active for mismatch binding.

A heterodimer consisting of hMSH2 and hGTBP has been isolated from HeLa cells by virtue of its ability to bind G-T-containing DNA and restore mismatch repair activity to nuclear extracts of hMSH-deficient cells *(20,107)*. The heterodimer of hMSH2 and hGTBP, called hMutSα, binds with high specificity to G-T mispairs and to 1–3 nucleotide IDLs *(20,107)*. Surprisingly, the effect of ATP on mismatch binding by

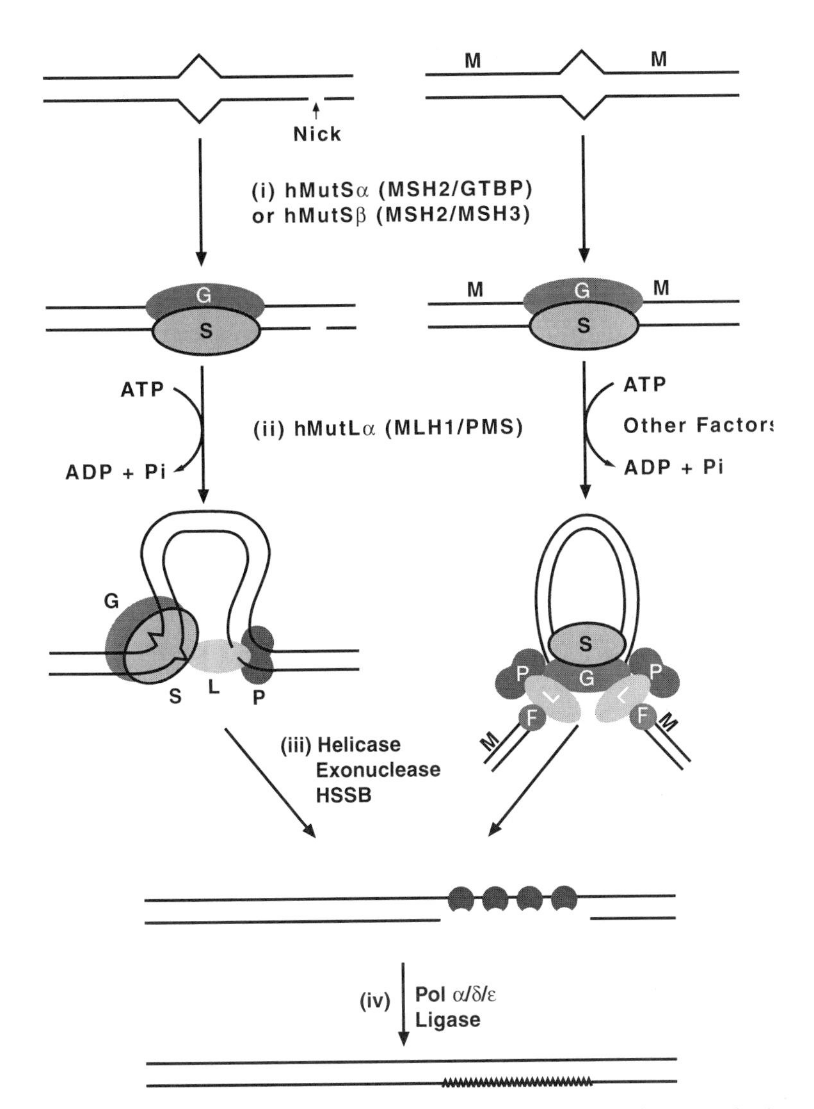

Fig. 2. Model for mammalian long-patch mismatch repair. The first step is the recognition of the mismatch site (represented by a bubble) by MutS homologs. The heterodimer of hMSH2 and hGTBP, called hMutSα, binds with high specificity to base–base mispairs and to 1–3 nucleotide IDLs. Alternatively, the heterodimer of hMSH2 and hMSH3, called hMutSβ, may bind to larger TDLs. When the hMSH2/hGTBP (S/G) heterodimer recognizes a mismatch site, the two subunits may interact with the site differently, since hGTBP can be crosslinked to T-G-containing DNA, but is not required for repair of large IDLs. In the second step, MutL homologs (hMutLα composed of hMLH1 [L] and one member of PMS [P]) are predicted to bind to hMutSα (or hMutSβ)-DNA complexes and enhance strand discrimination. If heteroduplex substrates contain a nick (shown on the left), a quaternary complex of hMSH2, hGTBP, hMLH1, and hPMS may bring the nick to the proximity of the mismatch site. If hemimethylated DNA is the substrate (shown on the right), other unidentified factors (F) may be involved to search for and cleave a hemimethylated site. In the third step, DNA helicase, exonuclease, and single-stranded DNA binding protein are required to generate intermediates with gaps that span the shorter path between the strand break and the mismatch, irrespective of the polarity of the strand breaks. Finally, DNA polymerase (polα, ε, or δ) and DNA ligase are predicted to fill in the gap and seal the nick, respectively.

hMutSα is different from that observed with recombinant hMSH2 expressed in *E. coli* *(20)*. The affinity of hMutSα for mispairs is greatly reduced in the presence of ATP *(20)*, an effect similar to that observed with bacterial MutS protein *(36)*. Studies by Modrich and colleagues indicate that hMutSα is the major functional mismatch repair complex in human HeLa cells *(20,99)*.

When the hMSH2/hGTBP heterodimer recognizes a mismatch site, the two subunits may interact with the site differently because hGTBP can be crosslinked to T-G-containing DNA *(48)*, but is not required for repair of large IDLs *(20,55)*. One additional intriguing finding is that germline mutations of *hMSH2* gene account for about 50% of the all HNPCC cases *(30,69)*, yet *hGTBP* mutations are very rare in HNPCC *(109)*. In fact, *hGTBP* defects have only been identified in sporadic colorectal cancer, but not in HNPCC kindreds *(20,107–109)*. Could *hGTBP* mutation cause lethality or no phenotype at all? Based on the partial redundancy of *S. cerevisiae* of Msh3p (human DUG homolog) and Msh6p (human hGTBP homolog) in mismatch repair *(90)*, *hGTBP* mutation in humans may not have a deleterious effect even though the MT1 cell line with an *hGTBP* mutation shows a moderate mutator phenotype *(33)*. It remains to be determined whether hGTBP plays a role in the maintenance of genome stability. The redundancy of hMSH3 and hMSH6 (hGTBP) in mismatch recognition suggests that in the absence of hGTBP (hMSH6), hMSH2 can form a homodimer or a heterodimer with hMSH3. Consequently, it has been suggested that hGTBP may participate in processes other than mismatch repair, or it may be differentially expressed in various tissues *(99)*.

Exactly how mammalian MutS homologs recognize the subtle structural changes in heteroduplexes is not clear. Although a potential helix-turn-helix DNA binding domain has been identified at the C-termini of MutS homologs *(124)*, its role in substrate recognition has not been demonstrated.

2.5.2. Interaction of hMutSα and Human MutL Homologs

The large number of human MutL homologs (*see* Section 2.4.2. and Table 2) hints at the complexity of human mismatch repair. Multiple functions and/or tissue-specific expression has been suggested for MutL homologs *(99)*. Thus far, biochemical studies have been limited to hMLH1 and hPMS2 proteins. Although mutations have been found in *hPMS1* in HNPCC families, implicating its role in mismatch repair, cell lines with *hPMS1* mutations have not been available for biochemical analysis. Little information is available about the role of other hPMS members in mismatch repair.

A heterodimer of hMLH1 and hPMS2 (or its homolog), named hMutLα, has been isolated as an activity that complements extracts of *hmlh1* mutant tumor cells *(73)*. No specific biochemical activities for hMutLα have been found. Similar results have also been observed with yeast MutL homologs. The yeast Mlh1p and Pms1p proteins have been shown to function as a heterodimer that can bind a complex consisting of yMsh2 and mismatch-containing DNA *(119)*. Both yeast Mlh1p and Pms1p are required for mismatch repair in *S. cerevisiae (118)*. Because human hPMS2 is more homologous to yeast Pms1p than is hPms1 *(110)*, hMutLα may be functionally equivalent to the yeast Mlh1p/Pms1p complex. As in bacteria and yeast, hMutLα is predicted to bind to DNA complexes and enhance strand discrimination (Fig. 2), but further experiments are required to confirm this. Thus, it appears that mammalian mismatch repair may involve a quaternary complex of hMSH2, hGTBP (hMSH3), hMLH1, and hPMS2 (or its homolog) (Fig. 2, left panel).

As mentioned above (Section 2.3.), cytosine methylation may provide strand discrimination in mammalian mismatch repair *(40,41)*. If this is the case, other factors including an endonuclease similar to *E. coli* MutH protein may be involved to search for and cleave a hemimethylated site (Fig. 2, right panel). A "repairosome" composed of multiple components may bend the DNA substrate and bring the hemimethylated or nicked site to the proximity of the mismatch site, as found in *E. coli*.

2.5.3. Excision and Resynthesis

Unfortunately, biochemical information regarding the later steps of mismatch repair in mammals is very limited. Fang et al. *(22,23)* have examined the excision tracts generated in vitro by repair of heteroduplexes with HeLa nuclear extracts in the absence of dNTPs or in the presence of aphidicolin. Intermediates were found with gaps that span the shorter path between the strand break and the mismatch, irrespective of the polarity of the strand breaks. Excision of nucleotides proceeds from the nick to around 100 nucleotides past the mismatch. This type of repair tract is similar to that found in *E. coli* mismatch repair reactions *(17)*. Based on the similarity of bacterial and eukaryotic systems, enzymes including DNA helicase, exonuclease, single-stranded DNA binding protein, DNA polymerase, and DNA ligase are predicted to be involved in these steps *(31,57,67)*. Some of these enzymes have been studied and purified, but their roles in mismatch repair have not been made clear. For example, the exact DNA polymerase responsible for repair resynthesis in the human mismatch repair pathway has not been identified; it may be polα, ε, or δ, because mismatch repair is sensitive to aphidicolin *(23,46,134)*.

3. MISMATCH REPAIR AND REPLICATION FIDELITY

The major biological function of mismatch repair is mutation avoidance, increasing the fidelity of DNA replication by three orders of magnitude. Human tumor cell lines defective in mismatch repair show a mutator phenotype and microsatellite instability similar to that observed in bacteria and yeast *(1,30,50,69,113,133)*. Mismatch repair enzymes act after DNA replication and direct repair to the newly synthesized daughter strands. Transition-type mismatches (A-C and G-T) are frequently generated by DNA polymerase *(64)*. Interestingly, these mismatch types are favorably repaired by the long-patch repair system. C-C mispairs, which are poorly repaired in bacteria *(16,97)*, are repaired efficiently by mammalian cell lines *(14,20,22,46,134)*. G-G mispairs are also more efficiently repaired by mammalian cells than by *E. coli* (Table 1). These may account for a lower mutation rate for C-G to G-C transversions in mammalian cells.

The long-patch mismatch repair pathway is not the only pathway to increase replication fidelity. A-G mismatches can be repaired by two pathways differing in strand specificities and repair tracts. The long-patch repair pathway can repair A-G mismatches to either C-G or A-T depending on the state of methylation or strand breaks. The short-patch MYH pathway acts on A-G to restore C-G base pairs exclusively *(92,145)* (*see* Chapter 8). In order for the MYH pathway to increase replicative fidelity, the frequency of A-G mismatches with A on the daughter strand has to be higher than that of A-G mismatches with G on the daughter strand. A mammalian MutT homolog may act during DNA replication to prevent the incorporation of guanines or 7,8-dihydro-8-oxo-deoxyguanines to template adenines *(96,126)* (*see* Chapter 6). Therefore, MYH would rarely encounter A-G mismatches with G on the daughter strands in normal cells.

Insertion or deletion mutations can arise during copying of repetitive sequences by DNA polymerase as a consequence of slippage *(63,65)*. This type of replicative error results in a loop structure that is efficiently removed by mismatch repair enzymes *(20,27,113,135)*. Human cells containing mutations in mismatch repair genes have a replicative error (RER^+) phenotype with mutability of $(CA)_n$ repeats at least two orders of magnitude higher than wild-type cells *(113)*. There is evidence that larger IDLs may be repaired by a pathway distinct from that responsible for base–base mismatch repair in mammalian cells *(135)* and yeast *(95)* (*see* Section 2.2.). Because higher organisms contain a large proportion of repetitive sequences ranging from 1–6 nucleotides *(7)*, it may be necessary to have a special pathway to repair larger IDLs. Defects in loop repair may be relevant to the microsatellite instability observed in cancers and other hereditary diseases (examples: Huntington's disease, Fragile X syndrome) that have been linked to the expansion of trinucleotide repeats *(1,7,50,62,133)*.

4. MISMATCH REPAIR AND DNA LESION REPAIR

Agents such as *N*-methyl-*N'*-nitro-*N*-nitrosoguanidine (MNNG) and *N*-methyl-*N*-nitrosourea (MNU) are mutagenic because they induce the formation of O^6-methylguanine (O^6meG) *(31)*. The cellular enzyme, methylguanine-DNA methyltransferase, is responsible for the repair of O^6meG. Cells that are defective in this repair protein are very sensitive to killing and mutagenesis by alkylating agents *(53,127)*. Alkylation-tolerant cells are more resistant to killing, but remain sensitive to mutagenesis by alkylating agents and have been shown to contain a second site mutation *(33,54)*. Alkylation-tolerant cells have been shown to be cross-tolerant to 6-thioguanine *(3,5,34)*.

Karran and colleagues have isolated alkylation-tolerant CHO and Raji Mex^- cell lines (Mex^- cells do not express active O^6meG DNA methyltransferase), which have a weak spontaneous mutator phenotype *(11)*. Extracts of these cell lines are deficient in binding to G-T mismatches and IDLs *(4,11)*, but mutations in these cells have not been identified.

A methylation-tolerant lymphoblastic cell line, MT1, exhibits a defective G2 cell-cycle checkpoint after treatment with high doses of MNNG and has a higher spontaneous mutation rate than the parental TK6 lymphoblastoid cells *(33)*. MT1 cells have been shown to contain mutations in both alleles of *hGTBP (109)*. The spontaneous mutations in MT1 cells are mainly A $\rightarrow$ G transitions and 1-nucleotide frameshifts *(55,109)*. Consistent with this mutation specificity, extracts of MT1 cells are defective in repair of base–base mismatches and 1-nucleotide IDLs, but are proficient in the repair of larger IDLs *(20,55)*.

Human HCH116 colorectal tumor cells exhibit microsatellite instability *(56)*, are deficient in mismatch repair *(113,136)*, are defective in *hMLH1 (110)*, and are tolerant to killing by MNNG *(43,56)*. The transfer of a normal copy of *hMLH1* on chromosome 3 into this cell line restored mismatch repair activity, increased sensitivity to MNNG, and restored G2 arrest following treatment with 6-thioguanine *(43,56)*. Hawn et al.

(43) have isolated an MNNG-resistant revertant (M2) clone from the HCH116 + chromosome 3 cell line that, like HCH116 tumor cells, is defective in mismatch repair activity, exhibits loss of hMLH1 function, is tolerant to 6-thioguanine, and shows no G2 arrest following treatment with 6-thioguanine *(43)*. These results strongly suggest that tolerance to alkylating agents and 6-thioguanine is the result of deficiency in mismatch repair. In addition, mouse *MSH2*-deficient cells have also lost mismatch binding, and as a consequence show increased microsatellite instability and tolerance to methylating agents *(18)*. Thus far, however, it has not been determined if mutations in *PMS* can also cause alkylation tolerance.

It has been suggested that O^6meG-C or O^6meG-T mispairs are recognized by the mismatch repair system *(33,35)* and it has been shown recently that this is the case *(20a)*. Incorporation of C or T opposite to O^6meG during DNA replication may trigger mismatch repair. In mismatch repair-proficient cells, futile attempts to correct mismatches generated during DNA replication of the alkylated bases may result in single-stranded regions that will lead to double-strand breaks and lethality *(53,54)*. It is also possible that repair may occur on both DNA strands to generate double-strand breaks and lead to lethality. The loss of mismatch repair ability would prevent this process, thus conferring alkylation tolerance.

Mismatch repair enzymes can also recognize other types of DNA damage. The *E. coli* methylation-directed pathway responds to UV-induced lesions *(25,26)*. Recently, MutS and MutL and their homologs were found to be necessary for transcription-coupled repair of UV-induced lesions in both *E. coli* and human cells, respectively *(93,94)*. The ability of hMutSα and hMSH2 to recognize the cisplatin but not transplatin DNA adducts *(20a,26a,92a)* indicates a direct participation of mismatch-repair enzymes in the cytotoxic effects of these adducts and the correlation between mismatch-repair deficiency and resistance to these agents.

5. MISMATCH REPAIR AND CANCER

Cancers are the result of multiple, progressive somatic mutations within critical genes. Alterations in the expression or function of genes that control cell growth or differentiation are considered to be the main causes of cancer *(9,24,128,139)*. It has also been suggested that genetic instability may facilitate cancer development *(81,82,105)*. In this "mutator hypothesis," an early spontaneous mutation occurs in one of many genes that maintain the genetic information. This leads to a mutator phenotype, reflecting either the accumulation of replicative errors or gross chromosome rearrangement in genes that regulate growth, invasion, or metastasis. Recently, this hypothesis has been supported by the demonstration that HNPCC is linked to a deficiency in mismatch repair (reviewed in refs. *29,51,57,98*). HNPCC (also known as Lynch syndromes I and II and Muir-Torre syndrome) is a common cancer predisposition syndrome characterized by an autosomal-dominant mode of transmission *(85)*. HNPCC families are susceptible to colon cancer (Lynch I) as well as epithelial tumors, such as endometrium, ovary, stomach, small intestine, kidney, and uterine tumors (Lynch II) *(141)*. HNPCC patients comprise 3–15% of the sporadic colon cancers in the US population with a frequency of about 1 in 200 *(110)*. Genetic linkage studies have identified several HNPCC loci. One locus mapped to chromosome 2p-16 (or 2p22-

21 by fine mapping) *(30,106,115)*, which accounts for about 50% of HNPCC. A second locus mapped to 3p21 accounts for an additional 30% of HNPCC *(76,106)*. The remaining loci contribute to about 10–20% of HNPCC.

One important finding that points to the linkage between tumor development and increased mutation rate or replicative error (RER^+) is the observation that sporadic tumors and tumors from HNPCC patients contain frequent mutations in microsatellite DNA *(1,21,50,133)*. Because defects in mismatch repair in both *E. coli* and yeast are known to cause high mutation rates of simple repeats *(72,131)*, it has been proposed that HNPCC could be caused by a defect in mismatch repair. Therefore, the coincidental location of the *hMSH2* gene and the HNPCC locus on chromosome 2 indicates that loss of hMSH2 function is linked to HNPCC *(30,69)*.

Thus far, four mismatch repair genes have been implicated in HNPCC: *hMSH2 (30,39,59,69,79,91)*, *hMLH1 (12,44,58,84,110)*, *hPMS1*, and *hPMS2 (103)*. Analysis of chromosome 2-linked HNPCC families has demonstrated that *hmsh2* mutations are present in the affected patients *(30,59,69,79,91)*. Extracts of cell lines derived from the *hMSH2* defective tumor of HNPCC are defective in mismatch repair activity *(10,20,136)*. The *hMLH1* gene was mapped to the same region of HNPCC 3p21 *(12)*. Similarly, the *hmlh1* mutation cosegregates with HNPCC in chromosome 3p-linked kindreds *(12)*, and cells with *hmlh1* mutations are defective in mismatch repair *(10,73,113)*. Mutations at two additional *mutL* homolog genes, *hPMS1* and *hPMS2*, have also been linked to HNPCC *(103,125)*. However, mismatch repair has not been shown to be defective in *hpms1* mutant cell extracts. Mutations in *hGTBP* have been found in several sporadic tumor cell lines, but not in HNPCC *(20,107–109)*.

The majority of HNPCC kindreds examined to date harbor mutations in either *hMSH2* or *hMLH1*. Mutations in *hPMS1* and *hPMS2* only occur in a small fraction of HNPCC cases *(12,30,69,103,109,110)*. This raises a particularly interesting question. As mentioned above (Section 2.5.), hMSH2/hGTBP, hMSH2/hMSH3, and hMLH1/hPMS2 (or its homolog) are heterodimers, but it is unclear why mutations at *hGTBP*, *hPMS1*, and *hPMS2* genes are less prevalent in cancer cells than mutations at *hMSH2* and *hMLH1*. These different degrees of prevalence in cancer may suggest an asymmetry of function within the heterodimers or a redundancy of repair enzymes. As in the case of hMutSα, different mutation specificities were observed between *hmsh2* and *hgtbp* mutants (*see* Sections 2.4.1. and 2.5.1.) *(20,109)*. The relative stability of repetitive sequences in cells with *hgtbp* mutations argues against the relationship of microsatellite instability and mismatch repair, and suggests the use of microsatellite instability as a molecular marker of tumor cells.

Mice harboring a knockout mutation of a mismatch repair gene, *PMS2*, *MLH1*, or *MSH2*, display microsatellite instability *(6,6a,18,20a)*, and some are prone to sarcomas and lymphomas. Colon cancer apparently is not observed in these knockout mice. This represents a major tissue difference in cancer development between mice and humans, since colon cancer is the predominant cancer found in HNPCC patients.

Exactly how a defect in mismatch repair gives rise to an increased incidence of cancer is unclear. Most tumor cell lines have a higher spontaneous mutation rate, but some are stable in microsatellite sequences. HNPCC patients are heterozygous for mutations in mismatch repair genes, but the tumors from these patients carry mutations in both alleles *(69,78,103,114)*. Thus, the inactivation of both alleles of repair genes is required for HNPCC tumor development. The reduced mismatch repair activity in heterozygous HNPCC patients may be caused by a dominant negative effect and may increase the frequency of mutation of the wild-type allele. Once both alleles are mutated, greatly increased mutation rates accelerate the accumulation of other mutations. It is likely that mutations in proto-oncogenes and/or tumor suppressor genes then lead to cancer development. One of the likely targets for mismatch repair is the human Type II transforming growth factor-β (TGF-β) receptor gene *(88)*. An A_{10} repeat and a $(GT)_3$ repeat within this gene are frequently altered in human RER^+ tumor cells *(88)*. Since cells with mutations in the TGF-β receptor gene fail to exhibit growth regulation by TGF-β and show tumorigenicity *(89,111)*, genetic instability will predispose these cells for tumor development. The amount of repetitive sequences in these downstream target genes is proportional to the risk of mutation. This may explain cancer predisposition in different tissues in different animals that are defective in mismatch repair.

6. MISMATCH REPAIR AND DNA RECOMBINATION

The involvement of mismatch repair in genetic recombination has been studied in detail in bacteria and yeast. This information can be found in a number of recent reviews *(2,37,104,123,142)* and in Chapters 16 and 19 in vol. 1. In general, the mismatch repair activities of the long-patch pathway suppress recombination between similar DNA sequences with some degree of diversity (so-called homeologous recombination). Thus, mismatch repair can safeguard the genome from rearrangement by controlling genetic recombination. During homologous genetic recombination, mismatch repair of heteroduplex DNA causes gene conversion and modulates the degree of genetic diversification *(45)*.

The relationship between mismatch repair and DNA recombination in mammalian cells is just emerging. It has been shown that sequence divergence decreases the frequency of recombination in mouse cells *(42,123)*. Intrachromosomal gene conversion between sequences diverged by 19% are reduced 100-fold compared to homologous controls *(140)*. Recently, it was shown that mismatch repair has antirecombination effects in mouse cells. Target integration of plasmid DNA at the *Rb* locus in mouse ES cells is increased 50-fold in *MSH2* mutant cells *(18)*. As in *E. coli*, heteroduplexes formed between two similar DNA sequences may be recognized by MSH2, and could prevent elongation of strand exchange and abort recombination. Mammalian genomes contain many repetitive, but diverged sequences. Long-patch mismatch repair reduces the recombination between these sequences and maintains genomic stability.

Mice harboring knockouts of *MSH2*, *MLH1*, or *PMS2* have distinctly different meiotic phenotypes *(6,6a,18,20a)*. The homozygous *msh2* mice have normal fertility and *pms* mice exhibit male infertility with normal female fertility, whereas male and female *mlh1* mice are sterile. Examination of meiotic chromosomes has suggested that aberrant synapsis is responsible for the spermatogenesis defect in *PMS2* knockout mice *(6)*. *MLH1*-deficient spermatocytes exhibit high levels of prematurely separated chromosomes and arrest during the first division of meiosis. In contrast, mutation of *pms1* in yeast (homologous to *hPMS2*) had no effect on recombination *(130)*. It remains to be determined if unidentified MutS homologs or any other PMS homologs are involved in mammalian meiotic recombination.

7. CONCLUSIONS

Mismatch repair systems increase genomic stability by enhancing the fidelity of DNA replication and homologous genetic recombination. Mismatch repair defects have been shown to be correlated with a significant fraction of certain types of hereditary and sporadic cancers and the cause of alkylation tolerance. The basic processes of mammalian mismatch repair are highly similar to *E. coli* and yeast. They share high homologies in mismatch specificities, repair enzymes, and repair mechanisms. However, the mammalian long-patch repair system differs from *E. coli* methyl-directed pathway in three aspects:

1. C-C mismatches are efficiently repaired;
2. Larger insertion/deletion loops may be repaired by a special pathway; and
3. Multiple MutS and MutL homologs are involved in the initiation steps.

There are several questions that remain with regard to mismatch repair mechanisms in mammalian cells. First, methylation and strand breaks have been suggested to dictate strand specificity. What is the signal to direct the repair to newly synthesized DNA strands? Second, hGTBP and hMLH1 may not be required for the larger IDL repair. What other proteins are involved in larger IDL repair? Could that be hMSH3 or some other proteins? Third, hMSH2, hGTBP, hMLH1, hPMS1, and hPMS2 have been shown to be involved in mismatch repair, but there are many MutS and MutL homologs without assigned functions. Why do human cells need multiple MutS and MutL homologs? Studies of yeast suggest that some of these proteins may be involved in genetic recombination. Finally, based on the similarity of bacterial and eukaryotic systems, activities including DNA helicase, exonuclease, single-stranded DNA binding protein, DNA polymerase, and DNA ligase are predicted to be involved in the later steps of excision and repair synthesis. Which enzymes are involved in catalyzing these steps?

Several questions remain about the relationship of mismatch repair and cancer. First, at least two MutS homologs and three MutL homologs are involved in mismatch repair, but why do mutations in these genes have different degrees of prevalence in cancer? Second, are individuals defective in genes for other components of the long-patch pathway cancer-prone? Third, do defects in genes involved in other mismatch repair pathways cause cancer susceptibility? Fourth, what is the mechanism for tissue specificity in cancer predisposition among different mismatch repair defective animals? It is likely that transgenic animals will be suitable models for further investigation of the relationship between mismatch repair and carcinogenesis.

ACKNOWLEDGMENTS

Work in the author's laboratory was supported by Public Service grant GM35132 from the National Institute of General Medical Science. The author thanks William Fawcett and Jason Cillo for comments on the manuscript.

REFERENCES

1. Aaltonen, L. A., P. Peltmaki, F. S. Leach, P. Sistonen, L. Rylkkanen, J. P. Mecklin, H. Jarvinen, S. M. Powell, J. Jen, S. R. Hamilton, G. M. Petersen, K. W. Kinzler, B. Vogelstein, and A. D. Chapelle. 1993. Clues to the pathogenesis of familial colorectal cancer. *Science* **260:** 812–816.

2. Alani, E., R. A. Reenan, and R. D. Kolodner. 1994. Interaction between mismatch repair and genetic recombination in *Saccharomyces cerevisiae*. *Genetics* **137:** 19–39.
2a. Acharya, S., T. Wilson, S. Gradia, M. F. Kane, S. Guerrette, G. T. Marsischky, R. Kolodner, and R. Fishel. 1996. hMSH2 forms specific mispair-binding complexes with hMSH3 and hMSH6. *Proc. Natl. Acad. Sci. USA* **93:** 13,629–13,634.
3. Aquilina, G., A. M. Giammarioli, A. Zijno, A. DiMuccio, E. Dogliotti, and M. Bignami. 1990. Tolerance to O^6-methylguanine and 6-thioguanine cytotoxic effects: a cross-resistant phenotype in *N*-methylnitrosourea-resistant Chinese hamster overy cells. *Cancer Res.* **50:** 4248–4253.
4. Aquilina, G., P. Hess, P. Branch, C. MacGeoch, I. Casciano, P. Karran, and M. Bignami. 1994. A mismatch recognition defect in colon carcinoma confers DNA microsatellite instability and a mutator phenotype. *Proc. Natl. Acad. Sci. USA* **91:** 8905–8909.
5. Aquilina, G., A. Zijno, N. Moscufo, E. Dogliotti, and M. Bignami. 1989. Tolerance to methylnitrosourea-induced DNA damage is associated with 6-thioguanine resistance in CHO cells. *Carcinogenesis* **10:** 1219–1223.
6. Baker, S. M., C. E. Bronner, L. Zhang, A. W. Plug, M. Robatzek, G. Warren, E. A. Elliott, J. Yu, T. Ashley, N. Arnheim, and R. M. Liskay. 1995. Male mice defective in the DNA mismatch repair gene PMS2 exhibit abnormal chromosome synapsis in meiosis. *Cell* **82:** 309–319.
6a. Baker, S. M., A. W. Plug, T. A. Prolla, C. E. Bronner, A. C. Harris, X. Yao, D. M. Christie, C. Monell, N. Arnheim, A. Bradley, T. Ashley, and R. M. Liskay. 1996. Involvement of mouse MLH1 in DNA mismatch repair and meiotic crossing over. *Nature Genet.* **13:** 336–342.
7. Beckman, J. S., and J. L. Weber. 1992. Survey of human and rat microsatellites. *Genomics* **12:** 627–631.
8. Bishop, D. K., J. Andersen, and R. D. Kolodner. 1989. Specificity of mismatch repair following transformation of *Saccharomyces cerevisiae* with heteroduplex plasmid DNA. *Proc. Natl. Acad. Sci. USA* **86:** 3713–3717.
9. Bishop, J. M. 1991. Molecular themes in oncogenesis. *Cell* **64:** 235–248.
10. Boyer, J. C., A. Umar, J. I. Risinger, R. Lipford, M. Kane, J. C. Barett, R. D. Kolodner, and T. A. Kunkel. 1995. Microsatellite instability, mismatch repair deficiency and genetic defects in human cancer cell lines. *Cancer Res.* **55:** 6063–6070.
11. Branch, P., G. Aquilina, M. Bignami, and P. Karran. 1993. Defective mismatch binding and a mutator phenotype in cells tolerant to DNA damage. *Nature* **362:** 652–654.
12. Bronner, C. E., S. M. Baker, P. T. Morrison, G. Warren, L. G. Smith, M. K. Lescoe, M. Kane, C. Earabino, J. Lipford, A. Lindblom, P. Tannergard, R. J. Bollag, A. R. Godwin, D. C. Ward, R. Nordenskjoid, R. Fishel, R. Kolodner, and R. M. Liskay. 1994. Mutation in the DNA mismatch repair gene homologue *hMLH1* is associated with hereditary nonpolyposis colon cancer. *Nature* **368:** 258–261.
13. Brown, T. C. and J. Jiricny. 1987. A specific mismatch repair event protects mammalian cells from loss of 5-methylcytosine. *Cell* **50:** 945–950.
14. Brown, T. C. and J. Jiricny. 1988. Different base/base mispairs are corrected with different efficiencies and specificities in monkey kidney cells. *Cell* **54:** 705–711.
15. Carraway, M. and M. G. Marinus. 1993. Repair of heteroduplex DNA molecules with multibase loops in *Escherichia coli*. *J. Bacteriol.* **175:** 3972–3980.
16. Claverys, J.-P. and S. A. Lacks. 1986. Heteroduplex deoxyribonucleic acid base mismatch repair in bacteria. *Microbiol. Rev.* **50:** 133–165.
17. Cooper, D. L., R. S. Lahue, and P. Modrich. 1993. Methyl-directed mismatch repair is bidirectional. *J. Biol. Chem.* **268:** 11,823–11,829.
18. de Wind, N., M. Dekker, A. Berns, M. Radman, and H. te Riele. 1995. Inactivation of the mouse Msh2 gene results in mismatch repair deficiency, methylation tolerance, hyperrecombination, and predisposition to cancer. *Cell* **82:** 321–330.

19. Dohet, C., R. Wagner, and M. Radman. 1986. Methyl-directed repair of frameshift mutations in heteroduplex DNA. *Proc. Natl. Acad. Sci. USA* **83:** 3395–3397.
20. Drummond, J. T., G. M. Li, M. J. Longley, and P. Modrich. 1995. Isolation of an hMSH2-p160 heterodimer that restores DNA mismatch repair to tumor cells. *Science* **268:** 1909–1912.
20a. Duckett, D. R., J. T. Drummond, A. I. H. Murchie, J. Reardon, A. Sancar, D. M. J. Lilley, and P. Modrich. 1996. Human MutSα recognize damaged DNA base pairs containing O^6-methylguanine, O^4-methylthymine, or the cisplatin-d(GpG) adduct. *Proc. Natl. Acad. Sci. USA* **93:** 6443–6447.
20b. Edlmann, W., P. E. Cohen, M. Kane, K. Lau, B. Morrow, S. Bennett, A. Umar, T. Kunkel, G. Cattoretti, R. Chaganti, J. W. Pollard, R. D. Kolodner, and R. Kucherlapati. 1996. Meiotic pachytene arrest in MLH1-deficient mice. *Cell* **85:** 1125–1134.
21. Eshleman, J. R. and S. D. Markowitz. 1995. Microsatellite instability in inherited and sporadic neoplasms. *Curr. Opinion Oncol.* **7:** 83–89.
22. Fang, W. H., G. M. Li, M. Longley, J. Holmes, W. Thilly, and P. Modrich. 1993. Mismatch repair and genetic stability in human cells. *Cold Spring Harb. Symp. Quant. Biol.* **58:** 597–603.
23. Fang, W. H. and P. Modrich. 1993. Human strand-specific mismatch repair occurs by a bidirectional mechanism similar to that of the bacterial reaction. *J. Biol. Chem.* **268:** 11,838–11,844.
24. Fearon, E. R. and B. Vogelstein. 1990. A genetic model for colorectal tumorigenesis. *Cell* **61:** 759–767.
25. Feng, W. Y. and J. B. Hays. 1995. DNA structures generated during recombination initiated by mismatch repair of UV-irradiated nonreplicating phage DNA in *Escherichia coli*—requirements for helicase, exonucleases, RecF and RecBCD functions. *Genetics* **140:** 1175–1186.
26. Feng, W. Y., E. H. Lee, and J. B. Hays. 1991. Recombinagenic processing of UV-light photoproducts in nonreplicating phage DNA by the *Escherichia coli* methyl-directed mismatch repair system. *Genetics* **129:** 1007–1020.
26a. Fink, D., S. Nebel, S. Aebi, H. Zheng, B. Cenni, A. Nehme, R. D. Christen, and S. B. Howell. 1996. The role of DNA mismatch repair in platinum drug resistance. *Cancer Res.* **56:** 4881–4886.
27. Fishel, R., A. Ewel, S. Lee, M. K. Lescoe, and J. Griffith. 1994. Binding of mismatched microsatellite DNA sequences by the human MSH2 protein. *Science* **266:** 1403–1405.
28. Fishel, R., A. Ewel, and M. K. Lescoe. 1994. Purified human MSH2 protein binds to DNA containing mismatched nucleotides. *Cancer Res.* **54:** 5539–5542.
29. Fishel, R. and R. D. Kolodner. 1995. Identification of mismatch repair genes and their role in the development of cancer. *Curr. Opinion Genet. Dev.* **5:** 382–395.
30. Fishel, R., M. K. Lescoe, M. R. S. Rao, N. G. Copeland, N. A. Jenkins, J. Garber, M. Kane, and R. Kolodner. 1993. The human mutator gene homolog *MSH2* and its association with hereditary nonpolyposis colon cancer. *Cell* **75:** 1027–1038.
31. Friedberg, E. C., G. C. Walker, and W. Siede. 1995. *DNA Repair and Mutagenesis*. ASM, Washington, DC.
32. Fujii, H. and T. Shimada. 1989. Isolation and characterization of cDNA clones derived from the divergently transcribed gene in the region uptream from the human dihydrofolate reductase gene. *J. Biol. Chem.* **264:** 10,057–10,064.
33. Goldmacher, V. S., R. A. Cuzick, and W. G. Thilly. 1986. Isolation and partial characterization of human cell mutants differing in sensitivity to killing and mutation by methylnitrosourea and *N*-methyl-*N'*-nitro-*N*-nitrosoguanidine. *J. Biol. Chem.* **261:** 12,462–12,471.
34. Green, M. H. L., J. E. Lowe, C. Petit-Frere, P. Karran, J. Hall, and H. Kataoka. 1989. Properties of *N*-methyl-*N*-nitrosourea-resistant, Mex$^-$ derivatives of an SV40-immortalized human fibroblast cell line. *Carcinogenesis* **10:** 893–898.

35. Griffin, S., P. Branch, Y. Z. Xu, and P. Karran. 1994. DNA mismatch binding and incision at modified guanine bases by extracts of mammalian cells: implications for tolerance to DNA methylation damage. *Biochemistry* **33:** 4787–4793.
36. Grilley, M., K. M. Welsh, S.-S. Su, and P. Modrich. 1989. Isolation and characterization of the *Escherichia coli mutL* gene product. *J. Biol. Chem.* **264:** 1000–1004.
37. Haber, J. E. 1992. Exploring the pathways of homologous recombination. *Curr. Opinion Cell Biol.* **4:** 401–412.
38. Haber, L. T., P. P. Pang, D. I. Sobell, J. A. Mankovich, and G. C. Walker. 1988. Nucleotide sequence of *Salmonella typhimurium mutS* gene required for mismatch repair: homology of MutS and HexA of *Streptococcus pneomoniae. J. Bacteriol.* **170:** 197–202.
39. Hall, N. R., G. R. Taylor, P. J. Finan, R. D. Kolodner, W. F. Bodmer, S. E. Cottrell, I. Frayling, and D. T. Bishop. 1994. Intron splice acceptor site sequence variation in the hereditary non-polyposis colorectal cancer gene hMSH2. *Eur. J. Cancer* **30A:** 1550–1552.
40. Hare, J. T. and J. H. Taylor. 1988. Hemi-methylation dictates strand selection in repair of G/T and A/C mismatches in SV40. *Gene* **74:** 158–161.
41. Hare, J. T., and J. H. Taylor. 1985. One role for DNA methylation in vertebrate cells is strand discrimination in mismatch repair. *Proc. Natl. Acad. Sci. USA* **82:** 7350–7354.
42. Hasty, P., J. Rivera-Perez, and A. Bradley. 1991. The length of homology required for gene targeting in embryonic stem cells. *Mol. Cell* Biol. **11:** 5586–5591.
43. Hawn, M. T., A. Umar, J. M. Carethers, G. Marra, T. A. Kunkel, C. R. Boland, and M. Koi. 1995. Evidence for a connection between the mismatch repair system and the G2 cell cycle checkpoint. *Cancer Res.* **55:** 3721–3725.
44. Hemminki, A., P. Peltomaki, J. P. Mecklin, H. Jarvinen, R. Salovaara, M. Nystrom-Lahti, A. de la Chapelle, and L. A. Aaltonen. 1994. Loss of the wild type MLH1 gene is a feature of hereditary nonpolyposis colorectal cancer. *Nature Genet.* **8:** 405–410.
45. Holliday, R. A. 1964. A mechanism for gene conversion in fungi. *Genet. Res.* **5:** 283–304.
46. Holmes, J., Jr., S. Clark, and P. Modrich. 1990. Strand-specific mismatch correction in nuclear extracts of human and *Drosophila melanogaster* cell lines. *Proc. Natl. Acad. Sci. USA* **87:** 5837–5841.
47. Horii, A., H. J. Han, S. Sasaki, M. Shimada, and Y. Nakamura. 1994. Cloning, characterization and chromosomal assignment of the human genes homologous to yeast PMS1, a member of mismatch repair genes. *Biochem. Biophys. Res. Commun.* **204:** 1257–1264.
48. Hughes, M. J. and J. Jiricny. 1992. The purification of a human mismatch-binding protein and identification of its associated ATPase and helicase activities. *J. Biol. Chem.* **267:** 23,876–23,882.
49. Iaccarino, I., F. Palombo, J. Drummond, N. F. Totty, J. J. Hsuan, P. Modrich, and J. Jiricny. 1996. MSH6, a *Saccharomyces cerevisiae* protein that binds to mismatches as a heterodimer with MSH2. *Curr. Biol.* **6:** 484–486.
50. Ionov, Y., M. A. Peinado, S. Malkbosyan, D. Shibata, and M. Perucho. 1993. Ubiquitous somatic mutations in simple repeated sequences reveal a new mechanism for colonic carcinogenesis. *Nature* **363:** 558–561.
51. Jiricny, J. 1994. Colon cancer and DNA repair: have mismatches met their match? *Trends Genet.* **10:** 164–168.
52. Jiricny, J., M. Hughes, N. Corman, and B. B. Rudkin. 1988. A human 200-kDa protein binds selectively to DNA fragment containing G/T mismatches. *Proc. Natl. Acad. Sci. USA* **85:** 8860–8864.
53. Karran, P. and M. Bignami. 1992. Self-destruction and tolerance in resistance of mammalian cells to alkylation damage. *Nucleic Acids Res.* **20:** 2933–2940.
54. Karran, P. and M. Bignami. 1994. DNA damage tolerance, mismatch repair and genome instability. *BioEssays* **16:** 833–839.

55. Kat, A., W. G. Thilly, W. H. Fang, M. J. Longley, G. M. Li, and P. Modrich. 1993. An alkylation-tolerant, mutator human cell line is deficient in strand-specific mismatch repair. *Proc. Natl. Acad. Sci. USA* **90:** 6424–6428.
56. Koi, M., A. Umar, D. P. Chauhan, S. P. Cherian, J. M. Carethers, T. A. Kunkel, and C. R. Boland. 1994. Human chromosome 3 corrects mismatch repair deficiency and microsatellite instability and reduces *N*-methyl-*N*'-nitro-*N*-nitrosoguanidine tolerance in colon tumor cells with homozygous hMLH1 mutation. *Cancer Res.* **54:** 4308–4312.
57. Kolodner, R. D. and E. Alani. 1994. Mismatch repair and cancer susceptibility. *Curr. Opinion Biotech.* **5:** 585–594.
58. Kolodner, R. D., N. R. Hall, J. Lipford, M. F. Kane, P. T. Morrison, P. J. Finan, J. Burn, P. Chapman, C. Earabino, and E. Merchant. 1995. Structure of the human MLH1 locus and analysis of a large hereditary nonpolyposis colorectal carcinoma kindred for *mlh1* mutations. *Cancer Res.* **55:** 242–248.
59. Kolodner, R. D., N. R. Hall, J. Lipford, M. F. Kane, M. R. Rao, P. Morrison, L. Wirth, P. J. Finan, J. Burn, and P. Chapman. 1994. Structure of the human MSH2 locus and analysis of two Muir-Torre kindreds for msh2 mutations. *Genomics* **24:** 516–526.
60. Kramer, B., W. Kramer, M. S. Williamson, and S. Fogel. 1989. Heteroduplex DNA correction in *Saccharomyces cerevisiae* is mismatch specific and requires functional PMS genes. *Mol. Cell. Biol.* **9:** 4432–4440.
61. Kramer, W., B. Kramer, M. S. Williamson, and S. Fogel. 1989. Cloning and nucleotide sequence of DNA mismatch repair gene *PMS1* from *Saccharomyces cerevisiae*: homology of PMS1 to procaryotic MutL and HexB. *J. Baterol.* **171:** 5339–5346.
62. Kuhl, D. P. and C. T. Caskey. 1993. Trinucleotide repeats and genome variation. *Curr. Opinion Genet. Develop.* **3:** 404–407.
63. Kunkel, T. A. 1990. Misalignment-mediated DNA synthesis errors. *Biochemistry* **29:** 8003–8011.
64. Kunkel, T. A. 1992. DNA replication fidelity. *J. Biol. Chem.* **267:** 18,251–18,254.
65. Kunkel, T. A. 1993. Nucleotide repeats. Slippery DNA and diseases. *Nature* **365:** 207,208.
66. Kunkel, T. A. 1995. The intricacies of eukaryotic spell-checking. *Curr. Biol.* **5:** 1091–1094.
67. Lahue, R. S., K. G. Au, and P. Modrich. 1989. DNA mismatch correction in a defined system. *Science* **245:** 160–164.
68. Langle-Rouault, F., M. G. Maehaut, and M. Radman. 1987. GATC sequences, DNA nicks and the MutH function in *Escherichia coli* mismatch repair. *EMBO J.* **6:** 1121–1127.
69. Leach, F. S., N. C. Nicolaides, N. Papadopoulos, B. Liu, H. Hen, R. Parsons, P. Peltomaki, P. Sistonen, L. A. Aaltonen, M. Nystrom-Lahti, X. Y. Guan, J. Zhang, P. S. Meltzer, J.-W. Yu, F.-T. Kao, D. J. Chen, K. M. Cerosaletti, R. E. K. Fournier, S. Todd, T. Lewis, R. J. Leach, S. L. Naylor, J. Weisenbach, J.-P. Mecklin, H. Jarvinen, G. M. Petersen, S. R. Hamilton, J. Green, J. Jas, P. Watson, H. T. Lynch, J. M. Trent, A. de la Chapelle, K. W. Kinzler, and B. Vogelstein. 1993. Mutations of a *mutS* homolog in hereditary nonpolypopsis colorectal cancer. *Cell* **75:** 1215–1225.
70. Learn, B. A. and R. H. Grafstrom. 1989. Methyl-directed repair of frameshift heteroduplexes in cell extracts from *Escherichia coli*. *J. Bacteriol.* **171:** 6473–6481.
71. Lee, S., B. Elenbaas, A. Levine, and J. Griffith. 1995. p53 and its 14 kDa C-terminal domain recognize primary DNA damage in the form of insertion/deletion mismatches. *Cell* **81:** 1013–1020.
72. Levinson, G. and G. A. Gutman. 1987. High frequencies of short frameshifts in poly-CA/TG tandem repeats born by bacteriophage M13 in *Escherichia coli* K-12. *Nucleic Acids Res.* **15:** 5323–5338.
73. Li, G. M. and P. Modrich. 1995. Restoration of mismatch repair to nuclear extracts of H6 colorectal tumor cells by a heterodimer of human MutL homologs. *Proc. Natl. Acad. Sci. USA* **92:** 1950–1954.

74. Lieb, M. 1983. Specific mismatch correction on bacteriophage lambda by very short patch repair. *Mol. Gen. Genet.* **181:** 118–125.
75. Lieb, M., E. Allen, and D. Read. 1986. Very short patch repair in phage lambda: repair sites and length of repair tracts. *Genetics* **114:** 1041–1060.
76. Lindblom, A., P. Tannergard, B. Werelius, and M. Nordenskjold. 1993. Genetic mapping of a second locus predisposing to hereditary non-polyposis colon cancer. *Nature Genet.* **5:** 279–282.
77. Linton, J. P., J.-Y. J. Yen, E. Selby, Z. Chen, J. M. Chinsky, K. Liu, R. E. Kellems, and G. F. Crouse. 1989. Dual bidirectional promoters at the mouse *dhfr* locus: cloning and characterization of two mRNA classes of the divergently transcribed *Rep-1* gene. *Mol. Cell. Biol.* **9:** 3058–3072.
78. Liu, B., N. C. Nicolaides, S. Markowitz, J. K. Willson, R. E. Parsons, J. Jen, N. Papadopolous, P. Peltomaki, A. de la Chapelle, and S. R. Hamilton. 1995. Mismatch repair gene defects in sporadic colorectal cancers with microsatellite instability. *Nature Genet* **9:** 48–55.
79. Liu, B., R. E. Parsons, S. R. Hamilton, G. M. Petersen, H. T. Lynch, P. Watson, S. Markowitz, J. K. Willson, J. Green, and A. de la Chapelle. 1994. hMSH2 mutations in hereditary nonpolyposis colorectal cancer kindreds. *Cancer Res.* **54:** 4590–4594.
80. Liu, K., L. Niu, J. P. Linton, and G. F. Crouse. 1994. Characterization of the mouse Rep-3 gene: sequence similarities to bacterial and yeast mismatch-repair proteins. *Gene* **147:** 169–177.
81. Loeb, L. A. 1991. Mutator phenotype may be required for multistage carcinogenesis. *Cancer Res.* **51:** 3075–3079.
82. Loeb, L. A., C. F. Springgate, and N. Battula. 1974. Errors in DNA replication as a basis of malignant changes. *Cancer Res.* **34:** 2311–2321.
83. Lu, A.-L., S. Clark, and P. Modrich. 1983. Methyl-directed repair of DNA base-pair mismatches in vitro. *Proc. Natl. Acad. Sci. USA* **80:** 4639–4643.
84. Lynch, H. T., T. Drouhard, S. Lanspa, T. Smyrk, P. Lynch, J. Lynch, B. Vogelstein, M. Nystrom-Lahti, P. Sistonen, and P. Peltomaki. 1994. Mutation of an mutL homologue in a Navajo family with hereditary nonpolyposis colorectal cancer. *J. Natl. Cancer Inst.* **86:** 1417–1419.
85. Lynch, H. T., T. C. Smyrk, P. Watson, S. J. Lanspa, J. F. Lynch, P. M. Lynch, J. Caralieri, and C. R. Boland. 1993. Genetic, natural history, tumor spectrum, and pathology of hereditary nonpolyposis colorectal cancer: an updated review. *Gastroenterology* **104:** 1535–1549.
86. Malgorzata, M. S., C. Baikalov, W. M. Luther, J.-H. Chiang, Y.-F. Wei, and J. H. Miller. 1996. Cloning and sequencing a human homolog (*hMYH*) of the *Escherichia coli* mutY gene whose function is required for the repair of oxidative DNA damage. *J. Bacteriol.* **178:** 3885–3892.
87. Mankovich, J. A., C. A. McIntyre, and G. C. Walker. 1989. Nucleotide segence of the *Salmonella typhimuruim mutL* gene required for mismatch repair: homology of *mutL* to *hexB* of *Streptococcus pneumoniae* and to *PMS1* of the yeast *Saccharomyces cerevisiae*. *J. Bacteriol.* **171:** 5325–5331.
88. Markowitz, S., J. Wang, L. Myeroff, R. Parsons, L. Sun, J. Lutterbaugh, R. S. Fan, E. Zborowska, K. W. Kinzler, and B. Vogelstein. 1995. Inactivation of the type II TGF-beta receptor in colon cancer cells with microsatellite instability. *Science* **268:** 1336–1338.
89. Markowitz, S. D., L. Myeroff, M. J. Cooper, J. Traicoff, M. Kochera, J. Lutterbaugh, M. Swiriduk, and J. K. Willson. 1994. A benign cultured colon adenoma bears three genetically altered colon cancer oncogenes, but progresses to tumorigenicity and transforming growth factor-beta independence without inactivating the p53 tumor suppressor gene. *J. Clin. Invest.* **93:** 1005–1013.
90. Marsischky, G. T., N. Filosi, M. F. Kane, and R. Kolodner. 1996. Redundancy of *Saccharomyces cerevisiae* MSH3 and MSH6 in MSH-dependent mismatch repair. *Genes Dev.* **10:** 407–420.

91. Mary, J. L., T. Bishop, R. Kolodner, J. R. Lipford, M. Kane, W. Weber, J. Torhorst, H. Muller, M. Spycher, and R. J. Scott. 1994. Mutational analysis of the hMSH2 gene reveals a three base pair deletion in a family predisposed to colorectal cancer development. *Hum. Mol. Genet.* **3:** 2067–2069.
92. McGoldrick, J. P., Y.-C. Yeh, M. Solomon, J. M. Essigmann, and A.-L. Lu. 1995. Characterization of a mammalian homolog of the *Escherichia coli* MutY mismatch repair protein. *Mol. Cell. Biol.* **15:** 989–996.
92a. Mello, J. A., S. Acharya, R. Fishel, and J. M. Essigmann. 1996. The mismatch-repair protein hMSH2 binds selectively to DNA adducts of the anticancer drug cispatin. *Chem. Biol.* **3:** 579–589.
93. Mellon, I. and G. N. Champe. 1995. Products of DNA mismatch repair genes mutS and mutL are required for transcription-coupled nucleotide excision repair of the lactose operon in *Escherichia coli. Proc. Natl. Acad. Sci. USA* **93:** 1292–1297.
94. Mellon, I., D. K. Rajpal, M. Koi, C. R. Boland, and G. N. Champe. 1996. Transcription-coupled repair deficiency and mutations in human mismatch repair genes. *Science* **272:** 557–560.
95. Miret, J. J., B. O. Parker, and R. S. Lahue. 1996. Recognition of DNA insertion/deletion mismatches by an activity in *Saccharomyces cerevisiae. Nucleic Acids Res.* **24:** 721–729.
96. Mo, J.-Y., H. Maki, and M. Sekiguchi. 1992. Hydrolytic elimination of a mutagenic nucleotide,8-oxodGTP, by human 18-kilodalton protein: sanitization of nucleotide pool. *Proc. Natl. Acad. Sci. USA* **89:** 11,021–11,025.
97. Modrich, P. 1991. Mechanisms and biological effects of mismatch repair. *Annu. Rev. Genet.* **25:** 229–253.
98. Modrich, P. 1994. Mismatch repair, genetic stability, and cancer. *Science* **266:** 1959–1960.
99. Modrich, P. and R. S. Lahue. 1996. Mismatch repair in replication fidelty, genetic recombination and cancer biology. *Ann. Rev. Biochem.* **65:** 101–133.
100. Neddermann, P. and J. Jiricny. 1993. The purification of a mismatch-specific thymine-DNA glycosylase from HeLa cells. *J. Biol. Chem.* **268:** 21,218–21,224.
101. Neddermann, P. and J. Jiricny. 1994. Efficient removal of uracil from G · U mispairs by the mismatch-specific thymine DNA glycosylase from HeLa cells. *Proc. Natl. Acad. Sci. USA* **91:** 1642–1646.
102. New, L., K. Liu, and G. F. Crouse. 1993. The yeast gene MSH3 defines a new class of eukaryotic MutS homologues. *Mol. Gen. Genet.* **239:** 97–108.
103. Nicolaides, N. C., N. Papadopoulos, B. Liu, Y. F. Wei, K. C. Carter, S. M. Ruben, C. A. Rosen, W. A. Haseltine, R. D. Fleischmann, C. M. Fraser, M. D. Adams, J. C. Venter, M. G. Dunlop, S. R. Hamilton, G. M. Petersen, A. de la Chapelle, B. Vogelstein, and K. W. Kinzler. 1994. Mutations of two PMS homologues in hereditary nonpolyposis colon cancer. *Nature* **371:** 75–80.
104. Nicolas, A. and T. D. Petes. 1994. Polarity of meiotic gene conversion in fungi: contrasting views. *Experientia* **50:** 242–252.
105. Nowell, P. C. 1976. The clonal evolution of tumor cell populations. *Science* **194:** 23–28.
106. Nystrom-Lahti, M., R. Parsons, P. Sistonen, L. Pylkkanen, L. A. Aaltonen, F. S. Leach, S. R. Hamilton, P. Watson, E. Bronson, and R. Fusaro. 1994. Mismatch repair genes on chromosomes 2p and 3p account for a major share of hereditary nonpolyposis colorectal cancer families evaluable by linkage. *Am. J. Hum. Genet.* **55:** 659–665.
107. Palombo, F., P. Gallinari, I. Iaccarino, T. Lettieri, M. Hughes, A. D'Arrigo, O. Truong, J. J. Hsuan, and J. Jiricny. 1995. GTBP, a 160-kilodalton protein essential for mismatch-binding activity in human cells. *Science* **268:** 1912–1914.
108. Palombo, F., M. Hughes, J. Jiricny, O. Truong, and J. Hsuan. 1994. Mismatch repair and cancer. *Nature* **367:** 417.72.
108a. Palombo, F., I. Iaccarino, E. Nakajima, M. Ikejima, T. Shimada, and J. Jiricny. 1996. hMSHβ, a heterodimer of hMSH2 and hMSH3, binds to insertion/deletion loops in DNA. *Curr. Biol.* **6:** 1181–1184.

109. Papadopoulos, N., N. C. Nicolaides, B. Liu, R. Parsons, C. Lengauer, F. Palombo, A. D'Arrigo, S. Markowitz, J. K. Willson, K. W. Kinzler, J. Jiricny, and B. Vogelstein. 1995. Mutations of GTBP in genetically unstable cells. *Science* **268:** 1915–1917.
110. Papadopoulos, N., N. C. Nicolaides, Y.-F. Wei, S. M. Ruben, K. C. Carter, C. A. Rosen, W. A. Haseltine, R. K. Fleischmann, C. M. Graser, M. D. Adams, J. C. Venter, S. R. Hamilton, G. M. Petersen, P. Watson, H. T. Lynch, P. Peltomaki, J.-P. Mecklin, A. de la Chapelle, K. W. Kinzler, and B. Vogelstein. 1994. Mutation of a mutL homolog in hereditary colon cancer. *Science* **263:** 1625–1629.
111. Park, K., S. J. Kim, Y. J. Bang, J. G. Park, N. K. Kim, A. B. Roberts, and M. B. Sporn. 1994. Genetic changes in the transforming growth factor beta (TGF-beta) type II receptor gene in human gastric cancer cells: correlation with sensitivity to growth inhibition by TGF-beta. *Proc. Natl. Acad. Sci. USA* **91:** 8772–8776.
112. Parker, B. O. and M. G. Marinus. 1992. Repair of DNA heteroduplexes containing small heterologous sequences in *Escherichia coli*. *Proc. Natl. Acad. Sci. USA* **89:** 1730–1734.
113. Parsons, R., G.-M. Li, M. J. Longley, W. Fang, N. Papadopoulos, J. Jen, A. de la Chapelle, K. W. Kinzler, B. Vogelstein, and P. Modrich. 1993. Hypermutability and mismatch repair deficiency in RER^+ tumor cells. *Cell* **75:** 1227–1236.
114. Parsons, R., G. M. Li, M. Longley, P. Modrich, B. Liu, T. Berk, S. R. Hamilton, K. W. Kinzler, and B. Vogelstein. 1995. Mismatch repair deficiency in phenotypically normal human cells. *Science* **268:** 738–740.
115. Peltomaki, P., L. A. Aaltonen, P. Sistonen, L. Pylkkanen, J. P. Mecklin, H. Jarvinen, J. S. Green, J. R. Jass, J. L. Weber, and F. S. Leach. 1993. Genetic mapping of a locus predisposing to human colorectal cancer. *Science* **260:** 810–812.
116. Priebe, S. D., S. M. Hadi, B. Greenberg, and S. A. Lacks. 1988. Nucleotide sequence of the *hexA* gene for DNA mismatch repair in *Streptococcus pneumoniae* and homology of *hexA* to *mutS* of *Escherichia coli* and *Salmonella typhimurium*. *J. Bacteriol.* **170:** 190–196.
117. Proffitt, J. H., J. R. Davie, D. Swinton, and S. Hattman. 1984. 5-Methylcytosine is not detectable in *Saccharomyces cerevisiae* DNA. *Mol. Cell. Biol.* **4:** 985–988.
118. Prolla, L. A., D.-M. Christie, and R. M. Liskay. 1994. Dual requirement in yeast DNA mismatch repair for *MLH1* and *PMS1*, two homologs of the bacterial *mutL* gene. *Mol. Cell. Biol.* **14:** 407–415.
119. Prolla, T. A., Q. Pang, E. Alani, R. D. Kolodner, and R. M. Liskay. 1994. MLH1, PMS1, and MSH2 interactions during the initiation of DNA mismatch repair in yeast. *Science* **265:** 1091–1093.
120. Prudhomme, M., B. Martin, V. Mejean, and J.-P. Claverys. 1989. Nucleotide sequence of *Streptococcus pneumoniae hexB* mismatch repair gene: homology of *hexB* to *mutL* of *Salmonella typhimurium* and to *PMS1* of *Saccharomyces cerevisiae*. *J. Bacteriol.* **171:** 5332–5338.
121. Radman, M. and R. Wagner. 1986. Mismatch repair in *Escherichia coli*. *Annu. Rev. Genet.* **20:** 523–528.
122. Raposa, S. and M. S. Fox. 1987. Some features of base pair mismatch and heterology repair in *Escherichia coli*. *Genetics* **117:** 381–390.
123. Rattray, A. J. and L. S. Symington. 1995. Multiple pathways for homologous recombination in *Saccharomyces cerevisiae*. *Genetics* **139:** 45–56.
124. Reenan, R. A. and R. D. Kolodner. 1992. Isolation and characterization of two *Saccharomyces cerevisiae* genes encoding homologs of the baterial HexA and MutS mismatch repair proteins. *Genetics* **132:** 963–9733.
125. Risinger, J. I., A. Umar, J. C. Barrett, and T. A. Kunkel. 1995. A hPMS2 mutant cell line is defective in strand-specific mismatch repair. *J. Biol. Chem.* **270:** 18,183–18,186.
126. Sakumi, K., M. Furuichi, T. Tsuzuki, T. Kakuma, S.-I. Kawabata, H. Maki, and M. Sekiguchi. 1993. Cloning and expression of cDNA for a human enzyme that hydrolyzes 8-oxo-dGTP, a mutagenic substrate for DNA synthesis. *J. Biol. Chem.* **268:** 23,524–23,530.

127. Sibghat, U. and R. S. Day. 1992. Incision at O^6-methylguanine: thymine mispairs in DNA by extracts of human cells. *Biochemistry* **31:** 7998–8008.
128. Stanbridge, E. J. 1990. Human tumor suppressor genes. *Annu. Rev. Genet.* **24:** 615–657.
129. Stepheson, C. and P. Karran. 1989. Selective binding to DNA base pair mismatches by proteins from human cells. *J. Biol. Chem.* **264:** 21,177–21,182.
130. Strand, M., M. C. Earley, G. F. Crouse, and T. D. Petes. 1995. Mutations in the *MSH3* gene preferentially lead to deletions within tracts of simple repetitive DNA. *Proc. Natl. Acad. Sci. USA* **92:** 10,418–10,421.
131. Strand, M., T. A. Prolla, R. M. Liskay, and T. Petes. 1993. Destablization of tracts of simple repetitive DNA in yeast by mutations affecting DNA mismatch repair. *Nature* **365:** 274–276.
132. Su, S.-S., R. S. Lahue, K. G. Au, and P. Modrich. 1988. Mispair specificity of methyl-directed DNA mismatch correction *in vitro*. *J. Biol. Chem.* **263:** 6829–6835.
133. Thibodeau, S. N., G. Bren, and D. Schaid. 1993. Microsatellite instability in cancer of the proximal colon. *Science* **260:** 816–819.
134. Thomas, D. C., J. D. Roberts, and T. A. Kunkel. 1991. Heteroduplex repair in extracts of human HeLa cells. *J. Biol. Chem.* **266:** 3744–3751.
135. Umar, A., J. C. Boyer, and T. A. Kunkel. 1994. DNA loop repair by human cell extracts. *Science* **266:** 814–816.
136. Umar, A., J. C. Boyer, D. C. Thomas, D. C. Nguyen, J. I. Risinger, J. Boyd, Y. Ionov, M. Perucho, and T. A. Kunkel. 1994. Defective mismatch repair in extracts of colorectal and endometrial cancer cell lines exhibiting microsatellite instability. *J. Biol. Chem.* **269:** 14,367–14,370.
137. Urieli-Shoval, S., Y. Gruenbaum, J. Sedat, and A. Razin. 1982. The absence of detectable methylated bases in *Drosophila melanogaster* DNA. *FEBS Lett.* **146:** 148–152.
138. Varlet, I., C. Pallard, M. Radman, J. Moreau, and N. de Wind. 1994. Cloning and expression of the *Xenopus* and mouse Msh2 DNA mismatch repair genes. *Nucleic Acids Res.* **22:** 5723–5728.
139. Vogelstein, B. and K. W. Kinzler. 1993. The multistep nature of cancer. *Trends Genet.* **9:** 138–141.
140. Waldman, A. S. and R. M. Liskay. 1987. Differential effects of base-pair mismatch on intrachromosomal versus extrachromosomal recombination in mouse cells. *Proc. Natl. Acad. Sci. USA* **84:** 5340–5344.
141. Watson, P. and H. T. Lynch. 1993. Extracolonic cancer in hereditary nonpolyposis colorectal cancer. *Cancer* **71:** 677–685.
142. West, S. C. 1994. The processing of recombination intermediates: mechanistic insights from studies of bacterial proteins. *Cell* **76:** 9–15.
143. Wiebauer, K., and J. Jiricny. 1989. In vitro correction of G · T mispairs to G · C pairs in nuclear extracts from human cells. *Nature* **339:** 234–236.
144. Wiebauer, K., and J. Jiricny. 1990. Mismatch-specific thymine DNA glycosylase and DNA polymerase β mediate the correction of G · T mispairs in nuclear extracts from human cells. *Proc. Natl. Acad. Sci. USA* **87:** 5842–5845.
145. Yeh, Y.-C., D.-Y. Chang, J. Masin, and A.-L. Lu. 1991. Two nicking enzymes systems specific for mismatch-containing DNA in nuclear extracts from human cells. *J. Biol. Chem.* **266:** 6480–6484.

8

Short Patch Mismatch Repair in Mammalian Cells

Paola Gallinari, Petra Neddermann, and Josef Jiricny

1. INTRODUCTION

Until several years ago, our knowledge of mismatch correction mechanisms was based almost entirely on data obtained with bacterial systems. Thus, *Escherichia coli*, *Salmonella typhimurium*, and *Streptococcus pneumoniae* were shown to possess a postreplicative mismatch correction system whose function involved the resynthesis of long tracts of DNA, up to several kilobases in length *(7,36,44)*. In addition, *E. coli* was shown by Lieb *(24,25)* to possess an additional, very short-patch (VSP) mismatch repair pathway, which is dedicated to the correction of G-T mispairs arising by hydrolytic deamination of 5-methylcytosine residues in the sequence CCA/TGG *(17)* modification of the inner cytosine by the DCM methylase *(30)*. In the late 1980s, following with mammalian systems indicated that vertebrate cells also possess two distinct mismatch correction pathways, which appeared to be functionally analogous to those of bacteria. The serendipitous discovery of human (*DUG1/hMSH3*; *15*) and murine (*Rep-3*; *27*) homologs of the *E. coli mutS* gene suggested that the long-patch mismatch repair system might be conserved not only mechanistically, but also at the structural level of the constituent proteins. The association of mismatch repair deficiency with microsatellite instability in yeast *(50)* helped establish the link between microsatellite instability and hereditary nonpolyposis colon cancer (HNPCC) *(19)*, resulting in the identification of the first human genes involved in mismatch correction *(14)*. To date, five proteins, hMSH2, GTBP, hMLH1, hPMS1, and hPMS2, have been implicated in the long-patch mismatch repair pathway, and mutations in their respective genes have been associated with human cancer (for reviews, *see* refs. *13,20*).

In recent years, several reports have appeared in the literature that describe mismatch correction processes involving short repair tracts. The study of the genetics and biochemistry of these short-patch mismatch repair processes, both in bacteria and in higher organisms, has received considerable attention, and already there is some evidence that at least some short-patch pathways may also turn out to be evolutionarily conserved. In general, short-patch mismatch repair involves mismatch-specific DNA glycosylases, which initiate a base-excision process at the site of the mispair. These activities represented a novel class of DNA glycosylases, since they apparently recognize and remove normal bases from DNA. Indeed, all glycosylase-like enzymes

From: DNA Damage and Repair, Vol. 2: DNA Repair in Higher Eukaryotes
Edited by: J. A. Nickoloff and M. F. Hoekstra © Humana Press Inc., Totowa, NJ

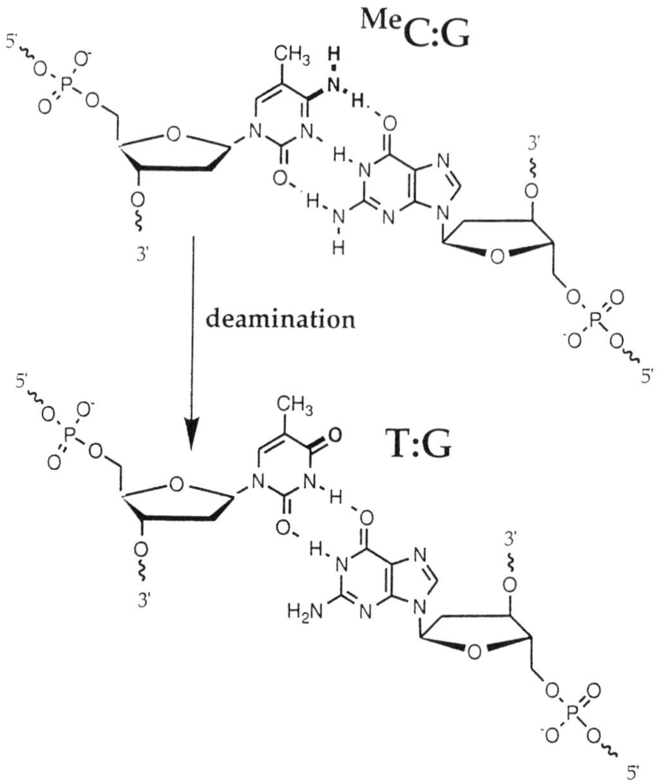

Fig. 1. Formation of a G-T mismatch in double-stranded DNA by hydrolytic deamination of 5-methylcytosine.

described earlier had one thing in common: their substrates were damaged or modified DNA bases. Moreover, since the antimutator function of some of these activities was not patently obvious, the question had been asked whether base–base mismatches are the true physiological substrates of these enzymes. This chapter will attempt to review the recent data, with the view to answering the latter query, as well as to put in perspective the relative importance of these highly specialized repair pathways in DNA metabolism.

2. G-T MISMATCH REPAIR

The sites of cytosine methylation in DNA are mutagenic hotspots, both in prokaryotes and in eukaryotes. Correspondingly, the genomes of organisms that methylate cytosine are often depleted of the methylatable sequence motifs, such as the CpG dinucleotide in the DNA of vertebrates *(18)*. The mechanism underlying this mutagenic process is known: 5-methylcytosine deaminates to thymine, giving thus rise to a C to T transition mutation if unrepaired (Fig. 1). However, cytosine and adenine, which are also subject to hydrolytic deamination (to uracil and hypoxanthine, respectively) and would thus also be expected to give rise to transition mutations, do not do so, most likely because their deamination is efficiently countered by specific repair processes. These findings led Miller and colleagues to postulate that G-T mispairs arising by deamination of 5-methylcytosine are not repaired *(10)*. However, a very approximate calculation, taking into account the rate of hydrolysis of 5-methylcytosine *(12)*, indicated that if deaminations at CpGs were not repaired, the human genome would by now contain no CpG dinucleotides at all. Since this is clearly not the case, it was anticipated that G-T mismatches associated with 5-methylcytosine deamination must be repaired.

To substantiate this hypothesis experimentally, a search for a G-T-specific repair process in mammalian cells was initiated. A synthetic oligonucleotide duplex, containing a single mismatch in a defined orientation, was ligated into the intron of the large T antigen gene of SV40, and these well-characterized heteroduplexes, lacking potential strand-differentiating signals that might have made them substrates for postreplicative mismatch correction, were transfected into CV-1 (African green monkey kidney) cells. Analysis of progeny viral DNA revealed that the heteroduplexes containing the G-T mispair were repaired extremely efficiently, and that the repair was biased almost exclusively in favor of G-C *(3)*. Since none of the other mismatch-containing heteroduplexes were corrected with a similar efficiency or directionality *(4)*, these results were taken as evidence for the existence of a G-T-specific mismatch repair process in CV-1 cells, which most likely evolved to counteract the loss of 5-methylcytosine residues through deamination. The G-T mismatch was corrected with similar efficiency and directionality in cell lines of human origin, including normal human fibroblasts, xeroderma pigmentosum type A, and Bloom's syndrome cells *(6)*.

By incubating synthetic 90-bp oligonucleotide duplexes containing single G-T or A-C mismatches with extracts from HeLa cells, it was shown that only the G-T substrate was processed under conditions where the long-patch repair was inactive *(52)*. In contrast to the *E. coli* VSP repair, where the Vsr endonuclease initiates the G-T repair process by cleaving the sugar-phosphate backbone 5' to the mispaired thymine *(17)*, the mammalian extracts catalyzed the removal of the mispaired thymine by a G-T mismatch-specific DNA glycosylase to generate an apyrimidinic site opposite the guanine (Fig. 2, step 1). Mammalian cells thus repair G-T mispairs differently from bacteria, although both organisms employ a mechanism that ensures an absolute strand-specificity and does not require secondary signals to dictate the directionality of the repair event.

The subsequent steps of the G-T mismatch repair process have not been strictly defined. However, in common with other base-excision repair processes (*see* Chapters 3 and 6), they most likely involve the cleavage of the sugar-phosphate backbone of the DNA at the 5'-side of the AP site by a class II AP-endonuclease (Fig. 2, step 2), such as HAP-1 (also called APE; *8,45*), followed by the removal of the abasic sugar (Fig. 2, step 3) either by the action of dRPase *(42)* or by the lyase activity of the 8-kDa subunit of β-polymerase *(31)*. The resulting single-nucleotide gap was shown to be filled in by DNA polymerase-β (*53*; Fig. 2, step 5), and the remaining nick is most likely sealed by DNA ligase III (*51*; Fig. 2, step 6).

2.1. Mismatch-Specific Thymine DNA Glycosylase

The G-T-specific thymine glycosylase was purified from HeLa cells to apparent homogeneity *(38)*. The 55-kDa polypeptide appeared to possess no lyase activity, such that it removed the thymine from the mispair without concurrent cleavage of the sugar-phosphate backbone of the DNA. The purified enzyme was shown to recognize and remove thymine from T-T and T-C mismatches, although with an efficiency consider-

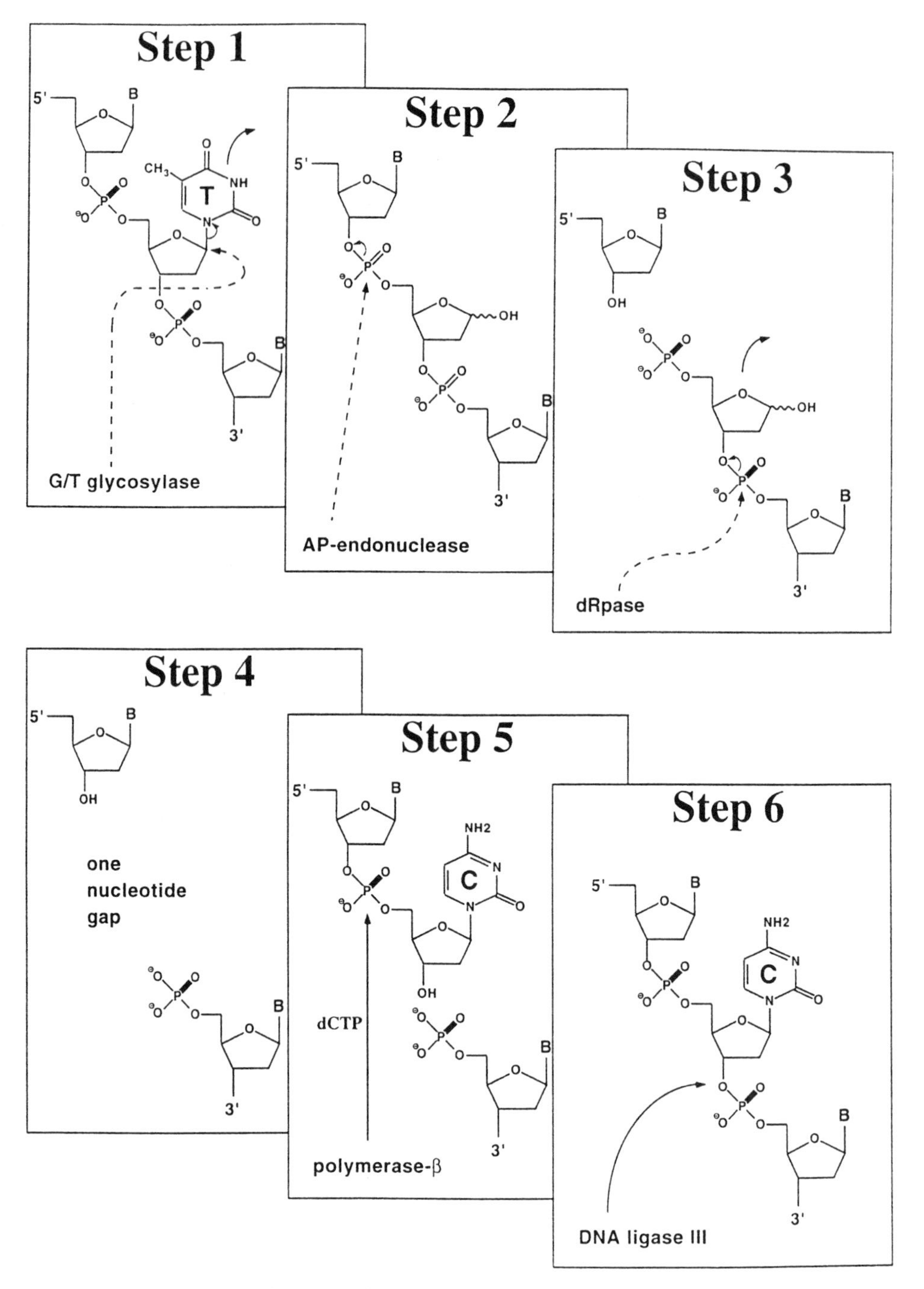

Fig. 2. G-T mismatch repair by the mismatch-specific thymine DNA glycosylase. Step 1: Removal of the mispaired thymine by the glycosylase generates an apyrimidinic site; step 2: cleavage of the apyrimidinic site by an AP-endonuclease generates 3'-OH and 5'-phosphate termini; step 3: removal of the abasic sugar phosphate can be mediated either by a dRPase as shown or, alternatively, by the 8-kDa N-terminal domain of polymerase-β. In either case, the one nucleotide gap so generated will have a 3'-OH and 5'-phosphate termini as shown in step 4; step 5: incorporation of dCMP into the one nucleotide gap by polymerase-β generates a nicked substrate, which is sealed by DNA ligase in step 6. The long dashed arrows symbolize bond-breaking, hydrolytic reactions, whereas the long solid arrows indicate bond-forming steps. For simplicity, only the strand undergoing processing is shown; however, all steps take place in duplex DNA, where a guanine base is positioned opposite the processed site.

ably lower than that observed with G-T mismatches. Moreover, it removed uracil and 5-bromouracil from mispairs with guanine *(39)*. Two cDNA clones encoding the human G-T-specific thymine glycosylase were recently identified *(40)*. Both cDNAs contain an ORF encoding a polypeptide of 410 amino acids, with a calculated molecular mass of 46 kDa. The protein expressed from this ORF, both in *E. coli* and in HeLa cells, migrates through denaturing polyacrylamide gels with an apparent molecular size of 60 kDa. The previously reported 55-kDa species *(38)* thus most likely corresponds to a proteolytic degradation product of the full-length glycosylase.

2.2. Sequence and Substrate Specificity of Recombinant G-T Glycosylase

The G-T glycosylase expressed in *E. coli* demonstrated a substrate specificity identical to the purified human protein *(39)*. Thus, the G-U mispair remains the best substrate for the enzyme, followed by G-U^{Br} and G-T. Moreover, the enzyme also recognizes the thymine residues in C-T and T-T mispairs, although with an efficiency at least an order of magnitude lower than that for guanine/pyrimidine pairs. Although it would seem likely that the G-U^{Br}, C-T, and T-T mispairs are not biologically relevant substrates of the enzyme, it is still unclear whether the G-U mispair is processed by this enzyme in vivo. In one report, the uracil glycosylase was reported to be less efficient in the excision of uracil residues from DNA with high G + C content *(11)*. Although this report was initially accepted with some skepticism, the recent elucidation of the mechanism of action of this enzyme *(37,47)* puts the data in a different light. The crystal structure studies of uracil glycosylase have revealed that, in order to cleave the glycosidic bond, the enzyme has to flip the uracil out of the helix and into its active site. Since this action requires the local melting of at least one neighboring base pair on either side of the uracil, the energy needed to perform this operation in a G + C-rich context would be higher than in a less stable duplex. It might therefore be anticipated that the uracil glycosylase would indeed be less proficient in the removal of uracil from G + C-rich DNA, increasing the likelihood that the G-T glycosylase participates in the repair of G-U mispairs.

The enzyme also recognizes and removes the thymine residue from mispairs with O^6-methylguanine and 2-amino-6-methylaminopurine, although with an efficiency lower than that observed for the G-T mispair. The purified enzyme does not act on the thymine-2,6-diaminopurine pair *(49)*, although these were processed in HeLa extracts *(48)*. Thus, this latter base-pair homolog may be a substrate for an as yet unidentified enzyme.

In terms of sequence preference, the purified G-T glycosylase appears to favor G-T mispairs in a CpG context *(40,49)* and is an order of magnitude less efficient in processing G-Ts in the context of GpG dinucleotides. However, this flanking sequence effect may not be strong enough to prevent the enzyme from acting on G-T mispairs in other sequence contexts. Indeed, in vivo studies showed that a G-T mispair in an ApG context was repaired with an efficiency and directionality equal to that of a G-T in a CpG dinucleotide *(3)*. In these SV40 transfection experiments, a time limit is imposed by the onset of replication of the transfected viral DNA, which gives the G-T repair system several hours to complete the correction process. This time period was clearly sufficient for efficient repair to have taken place. Thus, although the kinetics of repair for some G-T mispairs may be slower, the enzyme will most likely address all G-T mispairs—given enough time. In any case, the relevance of these results may be purely academic, since in vertebrate DNA, all deamination-associated G-T mispairs will be found in the context of methylated CpG dinucleotides. Since the enzyme does indeed prefer the G-T substrate in a CpG context, this is further evidence that the G-T glycosylase has evolved to counter the mutagenic effect of 5-methylcytosine deamination in CpGs.

2.3. Substrate Recognition Requirements

In double-stranded DNA, the deamination of adenine in an A-T base pair results in the formation of a hypoxanthine-thymine (H-T) mispair. The repair process is initiated by the removal of the hypoxanthine base by 3-methyladenine DNA glycosylase *(46)*. Similarly, a U-G mispair will form at sites of cytosine deamination, and the uracil will be removed most efficiently by uracil DNA glycosylase *(9)*. Despite the fact that these structures bring about a significant distortion of duplex DNA, neither uracil DNA glycosylase nor hypoxanthine DNA glycosylase would appear to utilize these distortions for damage recognition, since both enzymes can remove U or H, respectively, from single-stranded as well as from double-stranded DNA *(26)*. The recently elucidated crystal structure of uracil glycosylase *(37,47)* suggests that the enzyme removes the base by first flipping it out of the helix into the deep active site cleft, which facilitates the hydrolysis of the glycosidic bond. The flipping-out process presumably requires little energy in the case of single-stranded substrates, whereas the lower melting temperature of the H-T or G-U mispairs, as compared to Watson-Crick base pairs, might similarly lower the energy requirements of the flipping-out process in double-stranded substrates.

In the case of the hydrolytic deamination of 5-methylcytosine, the G-T mispair arises in the DNA in a way analogous to the abovementioned H-T and U-G mispairs. However, any processing of the G-T mismatch would be dependent solely on its recognition in double-stranded DNA, because in single-stranded DNA, a thymine arising from 5-methylcytosine deamination cannot be distinguished from any other thymine. This mode of recognition imposes an important constraint on the repair system, since the G-T mispair is also one of the most frequent polymerase errors arising in DNA during replication. Thus, a G-T arising from the deamination of 5-methylcytosine must in all cases be corrected to a G-C, presumably via the G-T glycosylase pathway. In contrast, the other type of a G-T mispair, structurally indistinguishable from the former, must be corrected by the postreplicative mismatch correction pathway either to a G-C or to an A-T, depending on whether the polymerase misincorporated a G opposite a T or a T opposite a G. Thus, for misincorporated bases, the repair directionality must always favor the parent template strand.

Can the cell restrict the action of these two distinct systems solely to their respective substrates? Transfection experiments with mismatch-containing SV40 heteroduplexes *(3)* provided circumstantial evidence for the existence of interference between the long-patch and the G-T-specific mismatch repair pathways. Transfection of the G-T heteroduplex resulted in 96% correction, with 92% of plaques having been repaired to G-C and 4% to A-T. It is now known that the G-T-specific repair is mediated by a glycosylase that removes the mispaired thymine, i.e., by a pathway that cannot catalyze the repair of G-T to A-T. However, the A-T-containing plaques were the result of repair, since they were found in pure bursts. (Had they resulted from unrepaired heteroduplexes segregated during replication, the plaques would have been mixed 50% G-C and 50% A-T.) The result was also not an artifact of the assay system, since the transfection of a G-U heteroduplex, which is repaired by the abundant uracil glycosylase, yielded 100% G-C *(5)*. This implies that the G-T heteroduplex was processed by a repair pathway distinct from the G-T-specific glycosylase, presumably the long-patch mismatch repair system. Indeed, it is likely that a small fraction of the covalently closed G-T-containing heteroduplexes acquired random nicks in either strand during the tranfection process and that these nicks were used as signals by the strand-discrimination function of the long-patch repair system for mismatch correction. Thus, 8% of the transfected G-T heteroduplex was processed by the long-patch repair system, since correction of 4% of the G-T heteroduplex to A-T stipulates that an equal fraction of the G-T mispairs was corrected by the same mechanism to G-C.

These data suggest that the G-T-specific repair process is not as efficient as uracil repair and, thus, that competition between the mismatch repair pathways in this transfection system was possible. However, this does not mean that the two pathways compete for G-T mispairs in vivo. It is probable that the long-patch system, which should function primarily in postreplicative mismatch correction, is cell-cycle regulated, whereas the glycosylase should be expressed in a constitutive fashion. Moreover, the long-patch system requires, in addition to mispairs, strand-discrimination signals that may not be present in the vicinity of deaminated 5-methylcytosines. It is thus highly unlikely that the long-patch system interferes with the deamination-associated G-T repair. In contrast, the inverse is possible, i.e., that the G-T glycosylase also processes G-T mispairs that arise by misincorporation. However, in such a scenario, it would be assumed that the enzyme present in the cell has a higher affinity for the replication-associated G-T mispair than the postreplicative mismatch repair machinery. Indeed, the results of the SV40 transfection experiments *(3)*, as well as some new data obtained in our laboratory *(23)* argue that the G-T glycosylase does not interfere with postreplicative mismatch correction.

It is interesting to note that the G-T glycosylase appears to differ from related enzymes in its slow dissociation from the substrate. The enzyme complex with a G-T mismatch-containing oligonucleotide is sufficiently stable to be separated from the unbound oligonucleotide by native polyacrylamide gel electrophoresis. Interestingly, the protein remains attached to the substrate even after cleaving the glycosidic bond, as demonstrated by the fact that the bound oligonucleotide recovered from the gel is subject to cleavage at the site of the mismatch with NaOH *(41)*. Indeed, this property of the enzyme allows purification by affinity chomatography. It is worthy of note that although no such stable complex was observed with uracil glycosylase *(41)*, the *E. coli* MutY protein, as well as its mammalian homolog, MYH, also produce stable complexes (*32*; *see* Section 3.).

2.4. 5-Methylcytosine Deamination and Cancer

The inefficient correction of deamination-associated G-T mispairs would lead to an increase in the number of C to T transitions in CpG dinucleotides that might result in the activation of an oncogene or, the inactivation of a tumor suppressor gene. Significantly, a recent analysis of p53 mutations in a large number of tumors showed that in colon cancers, nearly 50% are C to T transitions in CpGs *(16)*. Although highly suggestive, the link between these mutations and the possible malfunction of the G-T glycosylase has yet to be established.

Aside from bringing about transition mutations, hydrolytic deamination of 5-methylcytosine may have yet other consequences. In terminally differentiated cells, genes whose products are not needed for the maintenance of a particular phenotype are often transcriptionally silenced by DNA methylation. In many instances, such an inactivation involves the modification of most or all CpGs in a long stretch of DNA, including the body of the gene as well as several kilobases upstream and downstream. In such cases, the loss of a single CpG would most likely be without effect. However, some genes appear to be inactivated by the methylation of only one or a few CpGs *(18)*. In such instances, the loss of a critical CpG may lead to an erroneous reactivation and, thus, to the expression of a protein that is not tolerated in the particular differentiated cell. For example, were this protein involved in cell proliferation, its expression may result in uncontrolled cell division and thus, conceivably, may lead to cancer.

It would appear that despite the existence of an active DNA repair pathway that corrects G-T mismatches arising through the spontaneous hydrolytic deamination of 5-methylcytosine to G-C, the sites of cytosine methylation remain significantly mutagenic. Unfortunately, in the absence of a cell line deficient in this pathway, the biological importance of G-T-specific DNA repair cannot be assessed. The development of mice null for this activity should establish whether the deamination of 5-methylcytosine does indeed present a serious threat to the organism if unrepaired.

3. A-G MISMATCH REPAIR

During the study of the repair efficiencies of different kinds of mismatches, it was observed that in certain sequence contexts, the G-A mispair was not a good substrate for postreplicative mismatch repair in *E. coli*, both in vitro *(29)* as well as in vivo *(36)*. Later data from Fox, Lu and Modrich *(1,28,43)* revealed that the G-A mispair was processed by a novel short-patch repair pathway, where the principal enzyme was an adenine-specific DNA glycosylase encoded by the *mutY* gene *(2)*. It was not until several years later that it was recognized that the MutY protein was one of the components of a sophisticated system of oxidative damage repair, whereby the true substrate of the enzyme appears to be adenine mispaired with 8-oxoguanine *(34)*. This so-called GO system (*35*; *see* Chapter 6 in vol. 1) appears to be evolutionarily conserved, since a cDNA species highly homologous to the *mutY* sequence has been identified as part of the human genome sequencing project. Available biochemical data, obtained using the protein purified from HeLa cells *(54)* and calf thymus *(32)*, suggest that the enzyme, termed MYH (MutY Homolog) is a glycosylase/lyase *(32,54)*, which can remove adenine from mispairs with 8-oxoguanine, as well as with guanine and cytosine. The MYH enzyme is likely a member of the family of iron-sulfur cluster proteins, since the activity is sensitive to oxidation with potassium ferricyanide and since the protein sequence of the bacterial homolog contains the characteristic four-cysteine motif that is conserved between MutY and endonuclease III *(33)* and is thought to participate in the formation of a $[4Fe\text{-}4S]^{2+}$ cluster.

Interestingly, the mammalian MYH enzyme appears to be uncharacteristically large for a DNA glycosylase, with an apparent molecular mass of 65 kDa, nearly twice the

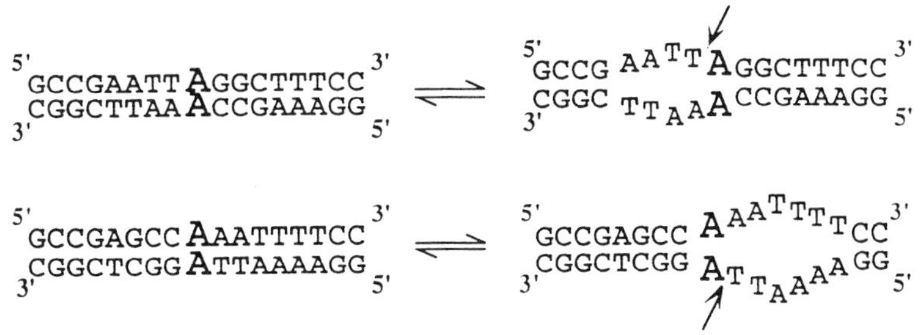

5' GCCGAATTAGGCTTTCC 3' ⇌ 5' GCCG AATTAGGCTTTCC 3'
3' CGGCTTAAACCGAAAGG 5' ⇌ 3' CGGC TTAAACCGAAAGG 5'

5' GCCGAGCCAAATTTTCC 3' ⇌ 5' GCCGAGCC AAATTTTCC 3'
3' CGGCTCGGATTAAAAGG 5' ⇌ 3' CGGCTCGG ATTAAAAGG 5'

Fig. 3. Schematic diagram of the mismatched duplex substrates for ATE/topoisomerase I. The sites of incision are indicated by arrows. The drawing indicates an equilibrium between an annealed and partially denatured duplex, whereby the enzyme would recognize and cut the denatured structure and remain covalently bound to the 3'-terminus of the single-stranded region.

size of the 39-kDa bacterial homolog MutY *(32)*. The biological role of the extra amino acids is puzzling at the moment, especially considering that the two enzymes are highly conserved even at the tertiary structure level, since MYH crossreacts with MutY antisera and they fulfill identical roles. However, it is interesting to note that in this context, the mammalian G-T glycosylase is also unusually large *(38)*.

4. ALL-TYPE ENDONUCLEASE (ATE)/TOPOISOMERASE I

Lu and colleagues identified a mismatch-specific endonuclease in HeLa cells *(54)* and in calf thymus *(55)*, which cleaves one strand of the heteroduplex immediately 5' from the mismatch and remained covalently attached to the 3'-end of the cleaved strand. Since the enzyme recognized all mispairs, they named it ATE. Active ATE was purified more than 1500-fold from calf thymus *(55)*, and its substrate and strand preferences were tested. The enzyme cleaved the mismatches in the order C-C > A-A = C-A = C-T > A-G > G-G > T-T > T-G. The strand preference was shown to be sequence-dependent. Although the precise criteria for strand selection are not known, the limited data available indicate that the enzyme preferentially cleaves denatured DNA. Thus, in the case of the A-A mispair studied by Yeh and colleagues, the mispair was flanked on one side by an A + T-rich region and by a G + C-rich sequence on the other. Nicking was observed on the strand in which the mispaired adenine was 3' of the A + T-rich stretch (Fig. 3).

The purified enzyme was shown to remain covalently bound to the 3'-terminus of the nicked strand. This is a common property of DNA topoisomerases, and indeed, protein sequencing revealed that the calf thymus ATE was topoisomerase I. The biological significance of the mismatch-specific nicking activity of ATE is not yet clear. However, the data provide valuable insight into the mechanism of action of DNA topoisomerases. The principal biological role of these enzymes is thought to lie in the relaxation of supercoiled DNA. Since this process introduces a considerable amount of stress into the helical structure, it is probable that local regions of strand separation will arise as a result of excessive supercoiling. The results of Lu and colleagues *(55)* indicate that the topoisomerase may recognize such partially melted structures, and proceed to introduce a nick into one of the strands at the site bordering the denatured and the annealed structure. Owing to the presence of a mismatch at the site of incision, the religation step would most likely be hindered, and the nick may therefore persist. During the following round of replication, a double-strand break would appear at this site. These findings imply that topoisomerase I could be one of the factors that contribute to the chromosomal instability of mismatch repair-deficient cells.

5. SUMMARY

Although three distinct enzymatic activities capable of cleaving duplex DNA at the sites of mismatches have been identified to date, the true biological substrates of two of these enzymes, MYH and topoisomerase I, are modified structures, such as the 8-oxoguanine-A mispair and a "bubble" in the DNA, respectively. The G-T-specific thymine DNA glycosylase remains the only known enzyme that can be said to have evolved specifically to correct mismatches. This statement is only true, however, if we define "mismatch" as a structure in duplex DNA composed of two unmodified nucleotides in a non-Watson-Crick base pair. The caveat of this definition lies in the fact that the thymine residue in the G-T mispair generated by hydrolytic deamination is not an unmodified base in a strict sense of the word: it is in fact a deaminated 5-methylcytosine. Thus, according to this definition, the only process that can be said to have evolved specifically for mismatch repair is postreplicative mismatch repair, the substrates for which arise during replication or recombination. All short-patch processes can now be regarded as having evolved to remove damaged DNA bases, the G-T being no longer the odd one out. However, the G-T glycosylase and the MYH enzymatic activities discussed above constitute a new class of DNA glycosylases, which recognize and therefore process their respective substrates solely in duplex DNA.

ACKNOWLEDGMENTS

The authors wish to express their gratitude to their collaborators, past and present, who have contributed to this work over the years.

REFERENCES

1. Au, K. G., M. Cabrera, J. H. Miller, and P. Modrich. 1988. *Escherichia coli mutY* gene product is required for specific A•G to C•G mismatch correction. *Proc. Natl. Acad. Sci. USA* **85:** 9163–9166.
2. Au, K. G., S. Clark, J. H. Miller, and P. Modrich. 1989. *Escherichia coli mutY* gene encodes an adenine glycosylase active on G/A mispairs. *Proc. Natl. Acad. Sci. USA* **86:** 8877–8881.
3. Brown, T. C. and J. Jiricny. 1987. A specific mismatch repair event protects mammalian cells from loss of 5-methylcytosine. *Cell* **50:** 945–950.
4. Brown, T. C. and J. Jiricny. 1988. Different base/base mispairs are corrected with different efficiencies and specificities in monkey kidney cells. *Cell* **54:** 705–711.
5. Brown, T. C. and M. L. Brown-Luedi. 1989. G/U lesions are efficiently corrected to G/C in SV40 DNA. *Mutat. Res.* **227:** 233–236.
6. Brown, T. C., I. Zbinden, P. A. Cerutti, and J. Jiricny. 1989. Modified SV40 for analysis of mismatch repair in simian and human cells. *Mutat. Res.* **220:** 115–123.
7. Claverys, J.-P. and S. A. Lacks. 1986. Heteroduplex deoxyribonucleic acid base mismatch repair in bacteria. *Microbiol. Rev.* **50:** 133–165.

8. Demple, B., T. Herman, and D. Chen. 1991. Cloning and expression of APE—a cDNA encoding the major human AP endonuclease. Definition of a family of DNA repair enzymes. *Proc. Natl. Acad. Sci. USA* **88:** 11,450–11,454.
9. Dianov, G. and T. Lindahl. 1994. Reconstitution of the DNA base excision-repair pathway. *Curr. Biol.* **4:** 1069–1076.
10. Duncan, B. K. and J. H. Miller. 1980. Mutagenic deamination of cytosine residues in DNA. *Nature* **287:** 560–563.
11. Eftedal, I., P. H. Guddal, G. Slupphaug, G. Volden, and H. E. Krokan. 1993. Consensus sequences for good and poor removal of uracil from double stranded DNA by uracil-DNA glycosylase. *Nucleic Acids Res.* **21:** 2095–2101.
12. Ehrlich, M., K. F. Norris, R. Y.-H. Wang, K. C. Kuo, and C. W. Gehrke. 1986. DNA cytosine methylation and heat-induced deamination. *Biosci. Rep.* **6:** 387–393.
13. Fishel, R. A. and R. D. Kolodner. 1995. Identification of mismatch repair genes and their role in the development of cancer. *Curr. Opinion Gen. Dev.* **5:** 382–395.
14. Fishel, R. A., M. K. Lescoe, M. R. S. Rao, N. Copeland, N. Jenkins, J. Garber, M. Kane, and R. Kolodner. 1993. The human mutator gene homolog MSH2 and its association with hereditary nonpolyposis colon cancer. *Cell* **75:** 1027–1038.
15. Fujii, H. and T. Shimada. 1989. Isolation and characterization of cDNA clones derived from the divergently transcribed gene in the region upstream from the human dihydrofolate reductase gene. *J. Biol. Chem.* **264:** 10,057–10,064.
16. Greenblatt, M. S., W. P. Bennett, M. Hollstein, and C. C. Harris. 1994. Mutations in the p53 tumor suppressor gene: clues to cancer etiology and molecular pathogenesis. *Cancer Res.* **54:** 4855–4878.
17. Hennecke, F., H. Kolmar, K. Brundl, and H.-J. Fritz. 1991 The vsr gene product of *E. coli* K-12 is a strand- and sequence-specific DNA mismatch endonuclease. *Nature* **353:** 776–778.
18. Hergersberg, M. 1991. Biological aspects of cytosine methylation in eukaryotic cells. *Experientia* **47:** 1171–1185.
19. Ionov, Y., M. A. Peinado, S. Malkbosyan, D. Shibata, and M. Perucho. 1993. Ubiquitous somatic mutations in simple repeated sequences reveal a new mechanism for colonic carcinogenesis. *Nature* **363:** 558–561.
20. Jiricny, J. 1994. Colon cancer and DNA repair: have mismatches met their match? *Trends Genet.* **10:** 164–168.
21. Jones, M., R. Wagner, and M. Radman. 1987. Mismatch repair of deaminated 5-methylcytosine. *J. Mol. Biol.* **194:** 155–159.
22. Leach, F. S., N. C. Nicolaides, N. Papadopoulos, B. Liu, J. Jen, R. Parsons, P. Peltomäki, P. Sistonen, L. A. Aaltonen, M. Nyström-Lahti, Y.-Y. Guan, J. Zhang, P. S. Meltzer, J.-W. Yu, F.-T. Kao, D. J. Chen, K. M. Cerosaletti, R. E. K. Fournier, S. Todd, T. Lewis, R. J. Leach, S. L. Naylor, J. Weissenbach, J.-P. Mecklin, H. Järvinen, G. M. Petersen, S. R. Hamilton, J. Green, J. Jass, P. Watson, H. T. Lynch, J. M. Trent, A. de la Chapelle, K. W. Kinzler, and B. Vogelstein. 1993. Mutations of a *mutS* homolog in hereditary nonpolyposis colorectal cancer. *Cell* **75:** 1215–1225.
23. Lettieri, T., P. Gallinari, and J. Jiricny. Unpublished results.
24. Lieb, M. 1983. Specific mismatch correction in bacteriophage lambda crosses by very short patch repair. *Mol. Gen. Genet.* **191:** 118–125.
25. Lieb, M. 1985. Recombination in the lambda repressor gene: evidence that very short patch (VSP) mismatch correction restores a specific sequence. *Mol. Gen. Genet.* **199:** 465–470.
26. Lindahl, T. 1982. DNA repair enzymes. *Ann. Rev. Biochemistry* **5:** 61–87.
27. Linton, J. P., J. J. Yen, E. Selby, Z. Chen, J. M. Chinsky, K. Liu, R. E. Kellems, and G. F Crouse. 1989. Dual bidirectional promoters at the mouse dhfr locus: Cloning and characterization of two mRNA classes of the divergently-transcribed Rep-1 gene. *Mol. Cell Biol.* **9:** 3058–3072.

28. Lu, A.-L. and D.-Y. Chang. 1988. A novel nucleotide excision repair for the conversion of an A/G mismatch to C/G base pair in *E. coli*. *Cell* **54:** 805–812.
29. Lu, A.-L. and D.-Y. Chang. 1988. Repair of single base-pair transversion mismatches of *Escherichia coli in vitro*: correction of certain A/G mismatches is independent of *dam* methylation and host *mutHLS* gene functions. *Genetics* **118:** 593–600.
30. Marinus, M. G. 1984. Methylation of prokaryotic DNA, in *DNA Methylation, Biochemistry and Biological Significance* (Razin, A., H. Cedar, and A. D. Riggs, eds.), Springer-Verlag, New York, pp. 81–109.
31. Matsumoto, Y. and K. Kim. 1995. Excision of deoxyribose phosphate residues by DNA polymerase β during DNA repair. *Science* **269:** 699–702.
32. McGoldrick, J. P., Y.-C. Yeh, M. Solomon, J. M. Essigmann, and A.-L. Lu, 1995. Characterization of a mammalian homolog of the *Escherichia coli* MutY mismatch repair protein. *Mol. Cell. Biol.* **15:** 989–996.
33. Michaels, M. L., L. Pham, Y. Nghiem, C. Cruz, and J. H. Miller. 1990. MutY, an adenine glycosylase active on G-A mispairs, has homology to endonuclease III. *Nucleic Acids Res.* **18:** 3841–3845.
34. Michaels, M. L., C. Cruz, A. P. Grollman, and J. H. Miller. 1992. Evidence that MutM and MutY combine to prevent mutations by an oxidatively damaged form of guanine in DNA. *Proc. Natl. Acad. Sci. USA* **89:** 7022–7025.
35. Michaels, M. L. and J. H. Miller. 1992. The GO repair system protects organisms from the mutagenesis effect of 8-hydroxyguanine (7,8-dihydro-8-oxo-guanine). *J. Bacteriol.* **174:** 6321–6325.
36. Modrich, P. 1989. Methyl-directed DNA mismatch correction. *J. Biol. Chem.* **264:** 6597–6600.
37. Mol, C. D., A. S. Arvai, G. Slupphaug, B. Kavli, I. Alseth, H. E. Krokan, and J. A. Tainer. 1995. Crystal structure and mutational analysis of human uracil-DNA glycosylase: structural basis for specificity and catalysis. *Cell* **80:** 869–878.
38. Neddermann, P. and J. Jiricny. 1993. The purification of a mismatch-specific thymine-DNA glycosylase from HeLa cells. *J. Biol. Chem.* **268:** 21,218–21,224.
39. Neddermann, P. and J. Jiricny. 1994. Efficient removal of uracil from G-U mispairs by the mismatch-specific thymine DNA glycosylase from HeLa cells. *Proc. Natl. Acad. Sci. USA* **91:** 1642–1646.
40. Neddermann, P., P. Gallinari, T. Lettieri, D. Schmid, O. Truong, J. J. Hsuan, K. Wiebauer, and J. Jiricny. 1996. Cloning and expression of human G/T mismatch-specific thymine-DNA glycosylase. *J. Biol. Chem.* **271:** 12,767–12,774.
41. Neddermann, P. and J. Jiricny. Unpublished results.
42. Price, A. and T. Lindahl. 1991. Enzymatic release of 5'-terminal deoxyribose phosphate residues from damaged DNA in human cells. *Biochemistry* **30:** 8631–8637.
43. Radicella, J. P., E. A. Clark, and M. S. Fox. 1988. Some mismatch repair activities in *Escherichia coli*. *Proc. Natl. Acad. Sci. USA* **85:** 9674–9678.
44. Radman, M. and R. Wagner. 1986. Mismatch repair in *Escherichia coli*. *Ann. Rev. Genet.* **20:** 523–538.
45. Robson, C., A. Milne, D. Pappin, and I. Hickson. 1991. Isolation of cDNA clones encoding an enzyme from bovine cells that repairs oxidative DNA damage in vitro: homology with bacterial repair enzymes. *Nucleic Acids Res.* **19:** 1087–1092.
46. Saparbaev, M. and J. Laval. 1994. Excision of hypoxanthine from DNA containing dIMP residues by the *Escherichia coli*, yeast, rat, and human alkylpurine DNA glycosylases. *Proc. Natl. Acad. Sci. USA* **91:** 5873–5877.
47. Savva, R., K. McAuley-Hecht, T. Brown, and L. Pearl. 1995. The structural basis of specific base-excision repair by uracil-DNA glycosylase. *Nature* **373:** 487–493.
48. Sibghat-Ullah, Y.-Z. Xu, and R. S. Day III. 1995. Incision at diaminopurine: thymine base pairs but not at O^4-methylthymine: guanine base pairs in DNA by extracts of human cells. *Biochemistry* **34:** 7438–7442.

49. Sibghat-Ullah, P. Gallinari, Y.-Z. Xu, M. F. Goodman, L. B. Bloom, J. Jiricny, and R. S. Day III. 1996. Base analog and neighboring base effects on substrate specificity of recombinant human G: T mismatch-specific thymine DNA-glycosylase. *Biochemistry* **35:** 12,926–12,932.
50. Strand, A., T. A. Prolla, R. M. Liskay, and T. D. Petes. 1993. Destabilization of tracts of simple repetitive DNA in yeast by mutations affecting DNA mismatch repair. *Nature* **365:** 274–276.
51. Wei, Y.-F., P. Robin, K. Carter, K. Caldecott, D. J. C. Pappin, G.-L. Yu, R.-P. Wang, B. K. Shell, R. A. Nash, P. Schär, D. E. Barnes, W. A. Haseltine, and T. Lindahl. 1995. Molecular cloning and expression of human cDNAs encoding a novel DNA ligase IV and DNA ligase III, an enzyme active in DNA repair and recombination. *Mol. Cell. Biol.* **15:** 3206–3216.
52. Wiebauer, K., and J. Jiricny. 1989. In vitro correction of G/T mispairs to G/C pairs in nuclear extracts from human cells. *Nature* **339:** 234–236.
53. Wiebauer, K., and J. Jiricny. 1990. Mismatch-specific thymine DNA glycosylase and DNA polymerase β mediate the correction of G/T mispairs in nuclear extracts from human cells. *Proc. Natl. Acad. Sci. USA* **87:** 5842–5845.
54. Yeh, Y.-C., D.-Y. Chang, J. Masin, and A.-L. Lu. 1991. Two nicking enzyme systems specific for mismatch-containing DNA in nuclear extracts from human cells. *J. Biol. Chem.* **266:** 6480–6484.
55. Yeh, Y.-C., H.-F. Liu, C. A. Ellis, and A.-L. Lu. 1994. Mammalian topoisomerase I has base mismatch nicking activity. *J. Biol. Chem.* **269:** 15,498–15,504.

9

Role of HMG and Other Proteins in Recognition of Cisplatin DNA Damage

Paul C. Billings and Edward N. Hughes

1. INTRODUCTION

Numerous biochemical processes are mobilized in cells in response to DNA damage. A key early event in this process requires detection, or recognition, of DNA damage. Presumably, damage recognition is mediated by specific damage-recognition proteins (DRPs), which recognize particular types of DNA lesions and/or specific DNA conformations resulting from adduct formation, such as that following binding of *cis*-diaminedichloroplatinum (II) (cisplatin or CDDP) to DNA. Once bound to DNA, DRPs are likely to have a major influence on the subsequent processing of specific types of DNA damage. For example, DRPs may influence the kinetics of repair of particular DNA abducts and, hence, determine the physiological response of cells to specific DNA-damaging agents.

In the last few years, a number of DRPs have been described. However, the functional roles that these proteins play in directing the cellular response to DNA damage are just beginning to be elucidated. This chapter describes contemporary approaches used to detect factors that recognize specific and well-characterized types of DNA damage, and speculates on the functional significance of these proteins.

2. EXPERIMENTAL APPROACHES TO DETECT DRPS

Several strategies are available to identify DRPs. These include using defined oligonucleotides as substrates for DRPs, as well as damaged DNA affinity precipitation techniques.

2.1. *Oligonucleotides*

Oligonucleotides are the most widely used reagents for the detection of DRPs. Following modification by CDDP, UV light, or other agents, oligonucleotides are subsequently end labeled with the Klenow fragment of *Escherichia coli* DNA polymerase I or T4 polynucleotide kinase. The modified DNAs are used:

1. In gel shift or electrophoretic mobility shift (EMSA) assays *(12,38)*;
2. As probes for Southwestern blots *(12,38)*; and
3. To screen cDNA libraries *(8)*.

From: DNA Damage and Repair, Vol. 2: DNA Repair in Higher Eukaryotes
Edited by: J. A. Nickoloff and M. F. Hoekstra © Humana Press Inc., Totowa, NJ

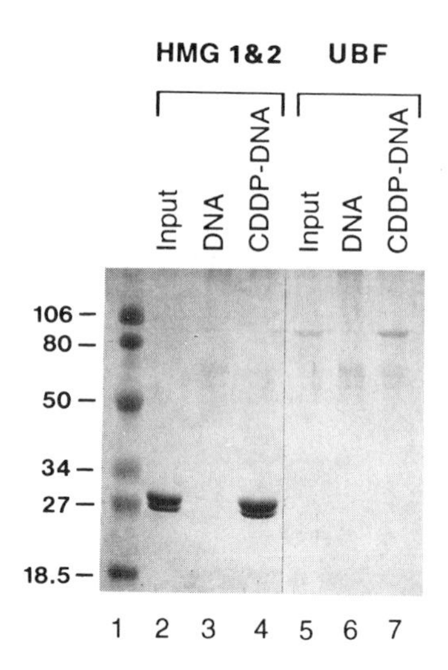

Fig. 1. Binding of purified proteins to untreated and CDDP-DNA. Proteins were incubated with untreated DNA (lanes 3 and 6) or CDDP-DNA (lanes 4 and 7), and bound protein was extracted and analyzed on an SDS-polyacrylamide gel *(3,25)*. For visualization, the gel was stained with Coomassie blue. Lane 1, M_r standards; lanes 2–4, HMG1 and HMG2; lanes 5–7, UBF. Input protein is shown in lanes 2 and 5. Ten times the amount of protein shown in each input lane was used in binding assays. DNA was modified with CDDP at an R_f of 0.03. Numbers on left, M_r in kDa.

Lippard and colleagues have further refined this strategy, using defined oligonucleotides with specific platination and restriction sites engineered into the probes. This approach makes it possible to position precisely cis-Pt-adduct formation in the DNA *(35)*.

2.2. *Affinity Precipitation*

A damaged DNA affinity precipitation assay has also been developed that allows the direct isolation of proteins that specifically bind to damaged DNA *(25)*. In this technique, DNA cellulose is damaged by a specific agent and mixed with protein samples (purlfied proteins, or extracts from cells or tissues). The protein–DNA cellulose complexes are then washed, and bound proteins are eluted and analyzed on SDS-polyacrylamide gels (Fig. 1). The damaged DNA affinity precipitation assay employs relatively high-mol-wt DNA of random sequence rather than oligonucleotides of defined sequence. This system provides a direct assessment of protein binding to damaged DNA and also allows the quantitation of DRPs present in different cell types and tissues. This strategy was used to isolate CDDP-DNA binding proteins *(25)*. The sensitivity of the assay can be enhanced by use of bound proteins that are radioiodinated following binding to damaged DNA *(5)*. He et al. *(23)* have utilized a similar approach, where a defined biotinylated DNA sequence is modified with *N*-acetoxy-2-acetylaminofluorene (AAAF). Modified DNA is immobilized on streptavidin beads and used as substrate in protein binding reactions.

3. DAMAGE RECOGNITION PROTEINS

Several eukaryotic proteins have been identified that recognize damaged DNA (Table 1; reviewed in ref. *37*) and some of these proteins are discussed below.

Table 1
Proteins That Bind Damaged DNA

Protein[a]	Damaged DNA	Reference
HMG1 (28) and HMG2 (26.5)	CDDP	*3,25,37*
UBF (97)	CDDP	*44*
RPA (3 subunits: 70, 32, 14)	CDDP,UV light	*18,23*
XPA (40)	CDDP, UV light	*35*
XPE (125)	CDDP, UV light, alkylating agents	*1,26,43*

[a]Numbers in parentheses, mass of protein in kDa.

High Mobility Group Protein 2

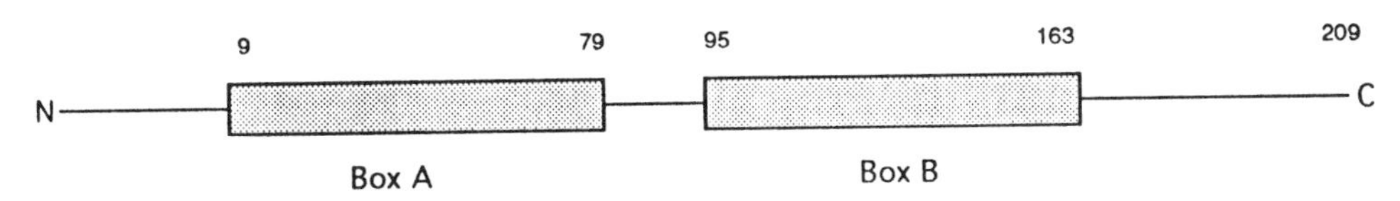

Fig. 2. Schematic diagram of HMG2. The two HMG boxes (Box A and Box B) are indicated by the shaded regions. The C-terminal region (residues 186–209) of HMG2 is highly acidic; 23 out of 24 residues are Asp or Glu. Numbers on top of the figure indicate amino acid residues from the N-terminal (N) initiation Met *(31)*.

3.1. CDDP-Damaged DNA Binding Proteins

CDDP has proven to be an extremely valuable reagent for isolating DRPs. CDDP is a widely used cancer chemotherapy drug. It produces specific and well-characterized intrastrand DNA crosslinks resulting from the covalent binding to the N^7 positions of adjacent purines *(16)*. The resultant cis-GpG and cis-ApG adducts bend DNA 32–34° toward the major groove *(2)*. The crystal structure of CDDP-DNA has recently been determined *(40)*.

Most CDDP-DNA binding factors are members of the high mobility group (HMG) box family of proteins. HMG boxes are basic domains (~80 amino acids) containing three α-helical regions and well-conserved hydrophobic amino acids within their sequences (*43*; reviewed in refs. *30,39*). HMG boxes have been identified in a large number of proteins involved in chromatin structure/function and gene expression *(30,39)*. A characteristic feature of HMG box proteins is their ability to recognize "bent" DNA structures *(39)*. They bind to the minor groove of DNA and once bound, induce additional bending *(39,44)*. The binding properties of these proteins observed in vitro likely have functional significance in vivo.

The interactions of two HMG box proteins, HMG1 and HMG2, with CDDP-DNA have been extensively examined. HMG proteins are a family of abundant, low-mol-wt chromatin proteins that are widespread and highly conserved in eukaryotic organisms *(10)*. HMG1 and HMG2 contain two HMG boxes in the amino terminal region (Fig. 2), and a carboxy-terminal region characterized by a high composition of aspartic and glutamic acid residues *(10)*. HMG1 and HMG2 bind DNA modified with CDDP and carboplatin with high affinity ($k_d \sim 10^{-10}$), with an estimated minimum lesion density of 8 Pt adduct/10^3 nucleotides *(3,25,36)*. In contrast, these proteins do not recognize DNA damaged with the therapeutically inactive trans-isomer of CDDP, *trans*-diamine-dichloroplatinum (II) (TDDP) (which forms DNA interstrand crosslinks vs the intrastrand crosslinks formed by CDDP), or with UV light *(25,36)*. Hence, the recognition specificity of HMG1 and HMG2 is highly selective and probably results from CDDP-induced structural changes in DNA resulting from cis-Pt-DNA adduct formation. The recognition site for damaged DNA has been localized to the N-terminal region of HMG2, containing both HMG boxes *(31)*. Protein constructs of HMG2 containing both HMG boxes bind CDDP-DNA with roughly the same affinity as the wild-type protein *(31)*, whereas fragments containing a single HMG box bind CDDP-DNA with reduced affinity *(14,31)*. The HMG box docks into the widened minor groove, resulting from cis-Pt-DNA adduct formation, and increases bending from 32 to 70° *(15,33,39)*.

3.2. Protein Binding and Tumor Cell Sensitivity to CDDP

CDDP is widely used as a cancer chemotherapeutic agent. Unfortunately, the clinical benefit of platinum-based therapy regimens may be compromised by the development of drug-resistant tumor cells. The role that HMG box proteins play in this process is unclear, since conflicting results have been obtained depending on the system utilized.

Using indirect immunofluorescence, changes in the intracellular distribution of HMG1, HMG2, and upstream binding factor (UBF) were observed in cells treated with CDDP, but not in cells treated with TDDP *(13)*. These proteins do not recognize DNA modified with TDDP in vitro (*see* Section 3.1.), suggesting that these proteins are functionally active DRPs in vivo. HMG1 blocks excision of cis-Pt-DNA adducts in an in vitro assay *(24)*. One interpretation of these findings is that HMG1 blocks excision repair by shielding the adducts from repair enzymes *(21,24)*. Inactivation of yeast Ixr1p (an HMG box protein that binds CDDP-DNA) resulted in a modest (approximately twofold) increase in CDDP resistance *(7)*. From this study, it was suggested that Ixr1p shields intrastrand crosslinks from DNA repair enzymes, sensitizing cells to CDDP *(7,21)*. UBF binds to rRNA promoter sequences and CDDP-DNA with comparable affinity *(42)*. Treiber et al. have proposed that cis-Pt-DNA adducts selectively compete with the rRNA promoter for UBF binding, resulting in "cisplatin-mediated transcription factor hijacking," which reduces rRNA synthesis in growing cells *(42)*. The results of these studies suggest that HMG box protein binding to cis-Pt-DNA adducts alters normal cellular function, i.e., by blocking repair of these lesions by HMG1 and HMG2 or reducing UBF-dependent transcription.

Experiments in other systems suggest that increased HMG protein expression is correlated with CDDP resistance. Human tumor cell lines resistant to CDDP contain the same or elevated levels of HMG1 and/or HMG2 relative to the sensitive parental cells *(4)* and are efficient at removing cis-Pt-DNA adducts *(27)*. Further, CDDP-damaged supercoiled plasmid DNA is efficiently repaired by extracts obtained from human cells and tissues; these extracts contain HMG1 and HMG2 *(22)*. Hence, results from these studies suggest that adduct shielding by HMG1 and/or HMG2 may not be operative in vivo. The efficiency of cis-Pt-DNA adduct repair may be influenced by HMG1/HMG2 binding as well as the ratio of HMG1 and HMG2 to other repair proteins. Addition of exogenous HMG1 to in vitro repair assays will alter its normal stoichiometric relationship with other DNA repair enzymes, and may result in slower repair kinetics. It should be stressed that information regarding the effect of HMG1 on the excision repair of CDDP-DNA has been obtained from in vitro assays, and the physiological significance of these findings is unclear.

3.3. *Replication Protein A*

Replication protein A (RPA) is a multisubunit DNA binding protein that has been found to bind damaged DNA preferentially. This factor is required for DNA replication in eukaryotes, DNA repair and recombination, and SV40 DNA replication *(6,9,28,45)*. RPA has much higher affinity for single-stranded DNA than for double-stranded DNA or RNA *(28)*. The holoenzyme is a heterotrimer consisting of 70-, 32-, and 14-kDa subunits *(6)*. The 70-kDa subunit possesses single-stranded DNA binding activity; no known biochemical function has been assigned to the smaller subunits *(6)*. The 32-kDa subunit is phosphorylated in a cell cycle-dependent manner *(19)* by the high-mol-wt DNA-activated protein kinase *(9)*. Interestingly, phosphorylation of the 32 kDa subunit also occurs in cells exposed to UV and ionizing radiation *(11,32)*. The significance of RPA phosphorylation is unknown, but it may serve a signaling function, indicating that specific DNA structures are present, such as partially duplex DNA structures which form during replication and repair *(9)*. RPA binds to CDDP-DNA *(18)* and also binds as a complex with XP-A (the protein defective in patients with xeroderma pigmentosum [XP] group A) to DNA damaged with AAAF *(23)*.

3.4. *Xeroderma Pigmentosum*

It has been known for many years that XP patients are prone to UV-induced skin cancer (*see* Chapter 18). Cells derived from these patients have specific defects in excision repair, and exhibit increased sensitivity to UV light and CDDP *(35)*. Both XP-A and the protein defective in XP group E patients (XP-E) have been shown to bind to damaged DNA *(34)*. XP-E has been purified to homogeneity from human and monkey cells, has a mass of 125 kDa, and exists in solution as a homodimer *(1,26)*. The human and monkey-derived proteins show 99% homology *(21,41)*. XP-E has a relatively broad spectrum of binding activity, and recognizes DNA damaged with UV, CDDP, and alkylating agents *(21,26)*. Surprisingly, XP-E shows much higher affinity for UV-induced pyrimidine(6-4)pyrimidone photoproducts [(6-4)PDs] than cyclobutane pyrimidine dimers (CPDs) *(35)*. XP-A also binds to UV-damaged DNA with a K_d ~ 1000-fold lower than XP-E *(35)*.

Since (6-4)PDs are good substrates for excision repair *(34)*, these results suggest that XP-E functions in excision repair as a DRP. However, patients in the XP-E complementation group have only a 50% reduction in excision repair activity, indicating that other components are involved *(34)*. Increased levels (three- to fivefold) of XP-E are observed in cells made resistant to CDDP by stepwise selection *(17)*. Further, an ~130-kDa protein is induced in cells following treatment with CDDP; this protein binds both CDDP and UV-modified DNA. Although the identity of this protein was not determined in this study, its mass and binding characteristics are the same as those for XP-E

(12). These results suggest that XP-E may be operative in tumor cell resistance to CDDP and other platinum analogs *(16,26,41)*. It has been proposed that XP-E is involved in the recognition step of nucleotide excision repair and may also play a role in drug resistance *(21)*.

4. EVOLUTION OF DRPS

What are the evolutionary or selective forces that would result in cells expressing DRPs? It is unlikely that cells would express proteins whose sole function is to recognize DNA damaged with CDDP or other agents. One hypothesis supported by several studies *(18,23,39,42)* is that DRPs recognize specific DNA structures, arising in cells under normal physiological conditions (including DNA replication and repair, transcription, and other processes). DNA damage resulting from the covalent binding of specific agents, such as CDDP and other alkylating agents, or CPD and (6-4)PD formation by UV light may induce DNA structures that mimic those that ordinarily arise in dividing cells. As a group, DRPs are likely to be structure-specific recognition proteins and, as such, they may recognize specific types of DNA damage because of structural similarities to naturally occurring DNA conformations.

For example, HMG1 and HMG2 have high affinity for bent DNA structures *(39)*. CDDP forms specific adducts that induce a 32° bend in DNA *(16)*. These proteins recognize CDDP-DNA because of the bent DNA structure, resulting from cis-Pt-DNA adduct formation. UBF binds CDDP-DNA and RNA polymerase I promoter sequences with the same affinity *(42)*. It may well be that this promoter takes on a conformation that is similar to that of CDDP-DNA. XP-E binds to DNA damaged with UV light and has an ~1000-fold higher affinity for (6-4)PDs than CPDs *(34)*. One explanation for this difference is that XP-E prefers the structure of (6-4)PD adducts vs the structure of CPD adducts *(29)*. Most, if not all, DRPs identified to date (such as those listed in Table 1) are constitutively expressed in cells, suggesting they have a normal physiological function, i.e., they may well bind to particular DNA conformations. Consequently, in addition to yielding insights into the interaction of proteins with specific types of damaged DNA, studies of DRPs will likely provide information regarding the interaction of these proteins with naturally occurring DNA conformations as well.

5. FUTURE DIRECTIONS

Although a number of DRPs have been identified over the last decade, the functional roles that these proteins play in DNA adduct recognition and repair are not well understood. Future work will need to focus on establishing the precise function of these proteins in living cells. For example, what are the physiological consequences of proteins binding to damaged DNA? It appears likely that DRPs play a central role in DNA repair, since damage recognition is an early, if not the first step in this process. Further, these proteins could potentially serve a signaling function and recruit other proteins to the site of damage *(23)*, thereby influencing the response of cells to specific DNA-damaging agents. DRP binding could be involved in cell-cycle arrest induced by DNA damage, altering the kinetics of DNA repair or, alternatively, directing cells to undergo apoptosis *(16,20)*. Several DRPs have been cloned, and high-affinity antibodies have been generated against some of these proteins. These reagents should greatly facilitate this work. In addition to providing information on protein function, these studies will undoubtedly yield important insights regarding protein–DNA interactions. In addition, they have potential clinical significance as well. For example, it may be possible to alter the levels of specific DRPs using pharmacological agents and, thereby, enhance the ability of cancer chemotherapy drugs, such as CDDP, to kill tumor cells. This information will be extremely useful to clinical oncologists treating patients with chemotherapeutic agents that damage DNA.

ACKNOWLEDGMENTS

We would like to thank Beatrice Engelsberg, Jonathan Cryer, and David Lawrence for their contributions to this work. The work in the authors' laboratory was funded in part by grants from the National Cancer Institute.

REFERENCES

1. Abramic, M., A. S. Levine, and M. Protic. 1991. Purification of an ultraviolet-inducible, damage-specific DNA-binding protein from primate cells. *J. Biol. Chem.* **266:** 22,493–22,500.
2. Bellon, S. F. and S. J. Lippard. 1990. Bending studies of DNA site-specifically modified by cisplatin, trans-diamminedichloroplatinum (II) and cis-$[Pt(NH_3)_2(N3\text{-}cytosine)CI)]^+$. *Biophys. Chem.* **35:** 179–188.
3. Billings, P. C., R. J. Davis, B. N. Engelsberg, K. A. Skov, and E. N. Hughes. 1992. Characterization of high mobility group protein binding to cisplatin-damaged DNA. *Biochem. Biophys. Res. Comm.* **188:** 1286–1294.
4. Billings, P. C., B. N. Engelsberg, and E. N. Hughes. 1994. Proteins binding to cisplatin-damaged DNA in human cell lines. *Cancer Invest.* **12:** 597–604.
5. Billings, P. C., L. Moy, J. E. Cryer, and B. N. Engelsberg. 1995. A post-labeling technique for the iodination of DNA damage recognition proteins. *Cancer Biochem. Biophys.* **14:** 223–230.
6. Blackwell, L. J. and J. A. Boroweic. 1994. Human replication protein A binds single stranded DNA in two distinct complexes. *Mol. Cell Biol.* **14:** 3993–4001.
7. Brown, S. J., P. J. Kellett, and S. J. Lippard. 1993. Ixrl, a yeast protein that binds to platinated DNA and confers sensitivity to cisplatin. *Science* **261:** 603–605.
8. Bruhn, S. L., P. M. Pil, J. M. Essigmann, D. E. Housman, and S. J. Lippard. 1992. Isolation and characterization of human cDNA clones encoding a high mobility group box protein that recognizes structural distortions to DNA caused by binding of the anticancer agent cisplatin. *Proc. Natl. Acad. Sci. USA* **89:** 2307–2311.
9. Brush, G. S., C. W. Anderson, and T. J. Kelly. 1994. The DNA-activated protein kinase is required for the phosphorylation of replication protein A during simian virus 40 DNA replication. *Proc. Natl. Acad. Sci. USA* **91:** 12,520–12,524.
10. Bustin, M., D. A. Lehn, and D. Landsman. 1990. Structural features of the HMG chromosomal proteins and their genes. *Biochem. Biophys. Acta.* **1049:** 231–243.
11. Carty, M. P., M. Zernik-Kobak, S. McGrath, and K. Dixon. 1994. UV light-induced DNA synthesis arrest in Hela cells is associated with changes in phosphorylation of human single stranded DNA binding protein. *EMBO J.* **13:** 2114–2123.
12. Chao, C. C. K., S. L. Huang, L. Y. Lee, and S. Lin-Chao. 1991. Identification of inducible damage-recognition proteins that are overexpressed in HeLa cells resistant to cis-diamminedichloroplatinum (II). *Biochem. J.* **277:** 875–878.
13. Chao, J. C., X. S. Wan, B. N. Engelsberg, L. I. Rothblum, and P. C. Billings. 1996. Intracellular distribution of HMG1, HMG2 and UBF change following treatment with cisplatin. *Biochim. Biophys. Acta* **1307:** 213–219.

14. Chow, C. S., C. M. Barnes, and S. J. Lippard. 1995. A single HMG domain in high mobility group 1 protein binds to DNAs as small as 20 base pairs containing the major cisplatin adduct. *Biochemistry* **34:** 2956–2964.
15. Chow, C. S., J. P. Whitehead, and S. J. Lippard. 1994. HMG domain proteins induce sharp bends in cisplatin-modified DNA. *Biochemistry* **33:** 15,124–15,130.
16. Chu, G. 1994. Cellular responses to cisplatin. The roles of DNA-binding proteins and DNA repair. *J. Biol. Chem.* **269:** 787–790.
17. Chu, G. and E. Chang. 1990. Cisplatin-resistant cells express increased levels of a factor that recognizes damaged DNA. *Proc. Natl. Acad. Sci. USA* **87:** 3324–3327.
18. Clugston, C. K., K. McLaughlin, M. K. Kenny, and R. Brown. 1992. Binding of human single-stranded DNA binding protein to DNA damaged by the anticancer drug cis-diaminedichloroplatinum (II). *Cancer Res.* **52:** 6375–6379.
19. Din, S., S. J. Brill, M. P. Fairman, and B. Stillman. 1990. Cell-cycle-regulated phosphorylation of DNA replication factor A from human and yeast cells. *Genes Dev.* **4:** 968–977.
20. Enoch, T. and C. Norbury. 1995. Cellular responses to DNA damage: cell-cycle check points, apoptosis and the roles of p53 and ATM. *TIBS* **20:** 426–430.
21. Fox, M. E., B. J. Feldman, and G. Chu. 1994. A novel role for DNA photolyase: binding to DNA damaged by drugs is associated with enhanced cytotoxicity in *Saccharomyces cerevisiae*. *Mol. Cell. Biol.* **14:** 8071–8077.
22. Hansson, J. and R. D. Wood. 1989. Repair synthesis by human cell extracts in DNA damaged by cis- and trans-diamminedichloroplatinum (II). *Nucleic Acids Res.* **17:** 8073–8091.
23. He, Z., L. E. Henricksen, M. S. Wold, and C. J. Ingles. 1995. RPA involvement in the damage-recognition and incision steps of nucleotide excision repair. *Nature* **374:** 566–569.
24. Huang, J. C., D. B. Zamble, J. T. Reardon, S. J. Lippard, and A. Sancar. 1994. HMG domain proteins specifically inhibit the repair of the major DNA adduct of the anticancer drug cisplatin by human excision nuclease. *Proc. Natl. Acad. Sci. USA* **91:** 10,394–10,398.
25. Hughes, E. N., B. N. Engelsberg, and P. C. Billings. 1992. Purification of nuclear proteins that bind to cisplatin-damaged DNA: identity with high mobility group proteins 1 and 2. *J. Biol. Chem.* **267:** 13,520–13,527.
26. Hwang, B. J. and G. Chu. 1993. Purification and characterization of a human protein that binds to damaged DNA. *Biochemistry* **32:** 1657–1666.
27. Johnson, S. W., R. P. Perez, A. K. Godwin, A. T. Yeung, L. M. Handel, R. F. Ozols, and T. C. Hamilton. 1994. Role of platinum-DNA adduct formation and removal in cisplatin resistance in human ovarian cancer cell lines. *Biochem. Pharm.* **47:** 689–697.
28. Kim, C., R. O. Snyder, and M. S. Wold. 1992. Binding properties of replication protein A from human and yeast cells. *Mol. Cell Biol.* **12:** 3050–3059.
29. Kim, J.-K., D. Patel, and B.-S. Choi. 1995. Contrasting structural impacts induced by cis-syn cyclobutane dimer and (6-4) adduct in DNA duplex decamers. Implication in mutagenesis and repair activity. *Photochem. Photobiol.* **62:** 44–50.
30. Landsman, D. and M. Bustin. 1993. A signature for the HMG Box DNA-binding proteins. *BioEssays* **15:** 539–546.
31. Lawrence, D. L., B. N. Engelsberg, R. S. Farid, E. N. Hughes, and P. C. Billings. 1993. Localization of the binding region of high mobility group protein 2 to cisplatin-damaged DNA. *J. Biol. Chem.* **268:** 23,940–23,945.
32. Liu, V. F. and D. T. Weaver. 1993. The ionizing radiation-induced replication protein A phosphorylation response differs between ataxia telangiectasia and normal human cells. *Mol. Cell. Biol.* **13:** 7222–7231.
33. Locker, D., M. Decoville, J. C. Maurizot, M. E. Bianchi, and M. Leng. 1995. Interaction between cisplatin-modified DNA and the HMG boxes of HMG1: Dnase 1 footprinting and circular dichroism. *J. Mol. Biol.* **246:** 243–247.
34. Naegeli, H. 1995. Mechanisms of DNA damage recognition in mammalian nucleotide excision repair. *FASEB J.* **9:** 1043–1050.

35. Pil, P. M., C. S. Chow, and S. J. Lippard. 1993. High mobility group 1 protein mediates DNA bending as determined by ring closure. *Proc. Natl. Acad. Sci. USA* **90:** 9465–9469.
36. Pil, P. M. and S. J. Lippard. 1992. Specific binding of chromosomal protein HMG1 to DNA damaged by the anticancer drug cisplatin. *Science* **256:** 234–237.
37. Protic, M. 1994. Eukaryotic damaged DNA-binding proteins. *Ann. NY Acad. Sci.* **726:** 333–335.
38. Protic, M. and A. S. Levine. 1993. Detection of DNA damage-recognition proteins using the band shift assay and Southwestern blots. *Electrophoresis* **14:** 682–692.
39. Read, C. M., P. D. Cary, C. Crane-Robinson, P. C. Driscoll, M. O. M. Carrillo, and D. G. Norman. 1995. The structure of the HMG box and its interaction with DNA, in *Nucleic Acids and Molecular Biology*, vol. 9 (Eckstein, F. and D. M. J. Lilley, eds.), Springer-Verlag, Berlin, pp. 222–250.
40. Takahara, P. M., A. C. Rosenzweig, C. A. Frederick, and S. J. Lippard. 1995. Crystal structure of double stranded DNA containing the major adduct of the anticancer drug cisplatin. *Nature* **377:** 649–652.
41. Takao, M., M. Abramic, M. Moos, V. R. Otrin, J. C. Wootton, M. Mclenigan, A. S. Levine, and M. Protic. 1993. A 127 kDa component of a UV-damaged DNA binding complex which is defective in some xeroderma-pigmentosum group-E patients is homologous to a slime mold protein. *Nucleic Acids Res.* **21:** 4111–4118.
42. Treiber, D. K., X. Zhai, H. M. Jantzen, and J. M. Essigmann. 1994. Cisplatin-DNA adducts are molecular decoys for the ribosomal RNA transcription factor hUBF (human upstream binding factor). *Proc. Natl. Acad. Sci. USA* **91:** 5672–5676.
43. Weir, H. M., P. J. Kraulis, C. S. Hill, A. R. C. Raine, E. D. Laue, and J. O. Thomas. 1993. Structure of the HMG box motif in the B-domain of HMG1. *EMBO J.* **12:** 1311–1319.
44. Werner, M. H., J. R. Huth, A. M. Gronenborn, and G. M. Clore. 1995. Molecular basis of human 46X, Y sex reversal revealed from the three-dimensional solution strucutre of the human SRY-DNA complex. *Cell* **81:** 705–714.
45. Wold, M. S., D. H. Weinberg, D. M. Virshup, J. J. Li, and T. J. Kelly. 1989. Identification of cellular proteins required for simian virus 40 DNA replication. *J. Biol. Chem.* **264:** 2801–2809.

10

TFIIH

A Transcription Factor Involved in DNA Repair and Cell-Cycle Regulation

Vincent Moncollin, Paul Vichi, and Jean-Marc Egly

1. INTRODUCTION

TFIIH is a multisubunit protein complex that was originally identified as a basal transcription factor of class II genes (protein coding genes). The discovery that one of its subunits was XP-B (ERCC3), a protein known to be involved in nucleotide excision repair (NER; *see* Chapter 18), put it at the focus of both transcription and DNA repair fields. Further studies in humans and yeast have shown that additional subunits of TFIIH, including XP-D, p62, and p44, may also have an essential role in DNA repair. Indeed, it may turn out that all subunits that comprise the core of the complex are required for both functions. This chapter summarizes recent data, which with the identification of the cyclin-dependent kinase (Cdk7) as another integral component of TFIIH, extends the possible roles of the complex to cell-cycle regulation and point to a crucial role for the complex in the overall function of the cell.

2. COMPONENTS OF TRANSCRIPTION INITIATION

2.1. *Basal Transcription Factors*

The class II transcription machinery is now well characterized. In addition to RNA polymerase II (pol II), at least six factors (TFIIA, B, D, E, F and H) are required to initiate transcription at the minimal promoter, which includes a TATA box and an initiation site (reviewed in ref. *11*). The assembly of an active transcription complex (Fig. 1) is initiated by sequence-specific binding of TFIID to the TATA box element. TFIID is a multisubunit protein that contains the TATA box binding protein (TBP) and several TBP-associated factors (TAFs) required in various combinations to mediate the effect of activators. The TFIID–promoter complex is stabilized by the binding of transcription factors TFIIA and TFIIB forming the TFIID-A-B (DAB) complex. Pol II, in association with TFIIF, binds to the DAB complex *(46)*. The final association of TFIIE and TFIIH allows the enzyme to enter the elongation process in the presence of nucleotides *(76)*. This view of a progressive formation of the preinitiation complex involving a stepwise assembly of the various components has been recently challenged

From: DNA Damage and Repair, Vol. 2: DNA Repair in Higher Eukaryotes
Edited by: J. A. Nickoloff and M. F. Hoekstra © Humana Press Inc., Totowa, NJ

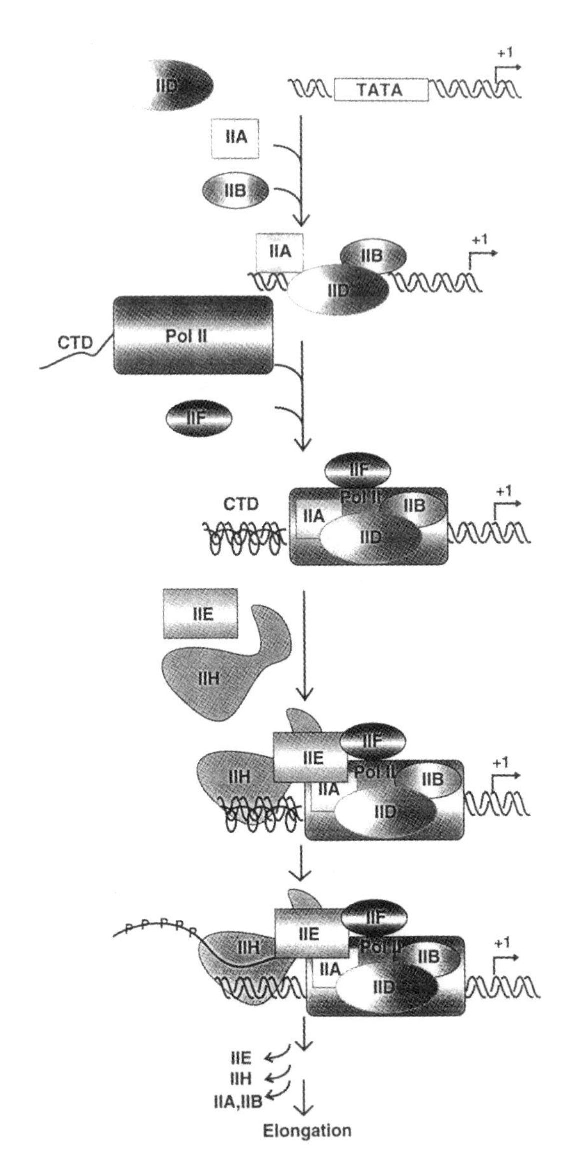

Fig. 1. Formation of the transcription initiation complex on a TATA box containing class II promoter. TFIID (IID), TFIIA (IIA), and TFIIB (IIB) form the initial DAB complex. Pol II is integrated as a complex with TFIIF (IIF). The pol II CTD (bold line) is in its unphosphorylated form when entering the complex. In an ATP-dependent step, TFIIE (IIE) in conjunction with the helicase activity of TFIIH, creates a minimal open region around the start site in the DNA duplex. After formation of the first phosphodiester bond and movement of the initiation bubble, the TFIIH (IIH)-associated kinase phosphorylates the CTD and stimulates promoter clearance. This leads to the dissociation of TFIIA, TFIIB, TFIIE, and finally TFIIH. TFIID is thought to remain on the TATA box sequence for subsequent rounds of initiation. TFIIH is known to enter the complex late in the process, but its association with the elongation complex is not clearly defined. TFIIH may thus leave the transcription complex before elongation and then be involved, independently of transcription, in DNA repair. Alternatively, TFIIH could remain in the transcription complex or repeatedly associate with the complex to promote the elogation of stalled complexes and also to facilitate the removal of lesions in transcribed strands of DNA.

Table 1
Subunits of Human and Yeast TFIIH Complexes

Subunits, kDa, ref.	Name	Properties, ref.		Yeast homolog, ref.	
89 *(74)*	XP-B/ERCC3[a]	3'-5' Helicase *(20,32,74,77)*		Rad25p/Ssl2p[a] *(23,32)*	
80 *(75)*	XP-D/ERCC2[a]	5'-3' Helicase *(20,75,77,82)*		Rad3p[a] *(23,31,85)*	
62 *(25)*	p62[a]	?		Tfb1p[a] *(85)*	
52 *(53)*	p52	?		Tfb2p *(7)*	
44 *(28,39)*	p44[a]	Zinc finger *(39)* DNA binding		Ssl1p[a] *(85,92)*	
34 *(28,39)*	p34[a]	Zinc finger *(39)*		?	
40 *(71,80,81)*	Cdk7	CTD kinase	CAK activity	Kin28p *(24,84,85)*	TFIIK
34 *(2,27,51,87)*	Cyclin H	Regulation of Cdk7	CAK activity	Cclp *(24,83)*	TFIIK
32 *(2,18,26,99)*	MAT1	Ring finger kinase stimulation	CAK activity	Tfb3p *(7,85)*	TFIIK

[a]These subunits are implicated in NER in human or yeast.

by reports that demonstrate the presence of a holoenzyme containing, in addition to pol II, all the components required for promoter-specific transcription initiation, both in yeast *(47,48)* and in mammals *(64)*. Existence of the holoenzyme demonstrates that pol II can associate with various factors in the absence of the promoter DNA to form the "ready to work" initiation complex. These findings emphasize the numerous interactions that occur between pol II and other factors involved in the transcription process.

2.2. Activators of Transcription

In addition to the "basal" transcription factors, which permit a low, but specific level of transcription, activating factors can regulate transcription by binding to specific DNA sequences usually located upstream of the TATA box. These regulators have been shown in most cases to interact with the basal transcription complex either directly, through one of the basal factors, or via additional proteins called adapters (reviewed in ref. *19*). As discussed below, in addition to TFIID and TFIIB, TFIIH is probably one of the major targets of these activators.

3. THE TFIIH COMPLEX

3.1. Subunit Composition

TFIIH, initially identified as a basal transcription factor, is a multisubunit protein complex containing nine polypeptides ranging from 32–89 kDa (*20,25,28*; *see* Table 1 and Fig. 2). Characterization of the various subunits of TFIIH has been essential in understanding the reactions driven by this protein complex. The two largest subunits, p89 and p80, are XP-B/ERCC3 and XP-D/ERCC2, two helicases of opposite polarity with DNA-dependent ATPase activities *(20,23,32,74,75,77,82)*. Three other polypeptides of TFIIH, Cdk7 (also named MO15), cyclin H, and MAT1, form an active kinase *(2,26,71)*, capable of phosphorylating the carboxy-terminal domain (CTD) of the largest subunit of pol II *(50)*, the cyclin-dependent kinases cdc2 and cdk2 *(7,51, 80,81,87,99)*, and other components of the transcriptional machinery, such as TBP,

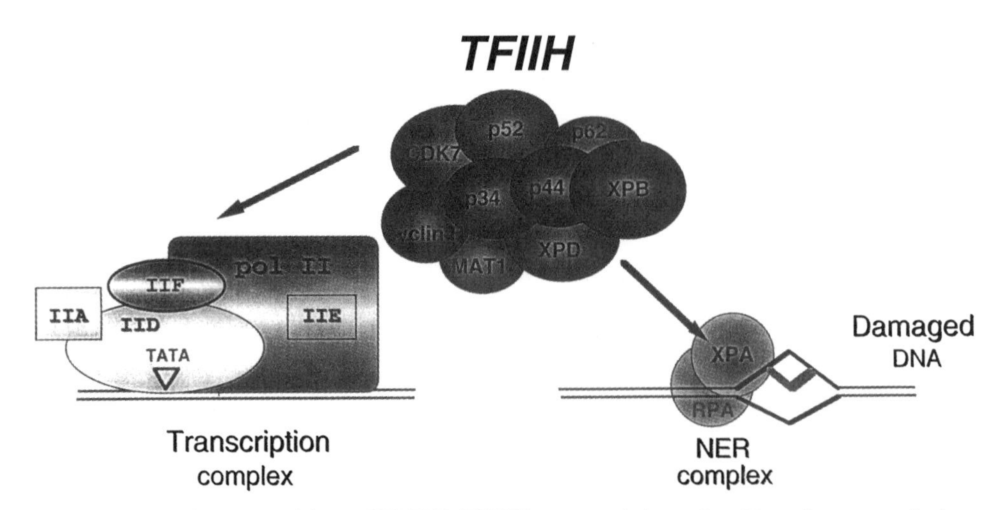

Fig. 2. Subunit composition of TFIIH. TFIIH can participate in either the transcription or DNA repair complexes. The involvment in DNA repair has been definitively shown for several of the subunits (XP-B, XP-D, p44, p62) and may also involve other subunits that comprise the core of the complex.

TFIIE, and TFIIF *(62)*. In addition to the helicases, three other subunits may confer some DNA binding properties to TFIIH: MAT1, which possesses a conserved ring finger domain in its aminc-terminus *(18,26,87)*, or p44 and p34, which contain TFIIA-like or CX_2C zinc finger motifs *(39)*. With respect to p44 and p34, these domains may function to stabilize the complex around the DNA and thus facilitate the positioning of the helicase. Two additional polypeptides of mol wt 62 *(25,29)* and 52 kDa *(53)*, which have no identified motifs or enzymatic functions, are also part of TFIIH.

It is not known whether TFIIH exists as a unique and homogeneous complex. Indeed, Western-blot analyses of TFIIH immunoprecipitated either from crude fractions, such as whole-cell extract, or from highly purified fractions reveal differences in stoichiometry among the various polypeptides contained in the protein complex *(28)*. Furthermore, depending on the methods used to isolate the TFIIH complex, there is an observed variability with regard to subunit composition and/or associated polypeptides. For example, the kinase complex (called TFIIK in yeast) also exists in free form in the cell *(24,80,83)*. In addition, the p80 subunit, as well as the kinase complex, can be dissociated from the five other subunits (p89, p62, p52, p44, and p34), which may constitute the core TFIIH *(2,20,28,75)*. These preliminary observations strongly suggest that several TFIIH complexes coexist in the cell and may be involved in different tasks.

Various forms of TFIIH were also observed in yeast. The core TFIIH (previously named factor b) contains Rad25, Rad3, Ssl1, Tfb1, Tfb2, and Tfb3, the homologs of XP-B, XP-D, p44, p62, p52, and MAT1, respectively (Table 1) *(7)*. The kinase moiety, TFIIK, contains Kin28p, the *Saccharomyces cerevisiae* homolog of Cdk7, and one of two additional proteins encoded by *CCL1*, the homolog of cyclin H *(83)*. The two polypeptides, Kin28p and Ccl1p, constitute TFIIK, which, when associated with the core TFIIH, yields a transcriptionally active TFIIH complex *(24,85)*.

3.2. ATPase and Helicase Activity

TFIIH is the only transcription factor known to possess several enzymatic activities. Although none of these activities has been definitively associated with a precise role in one of the various mechanisms in which TFIIH is thought to function, a picture can be tentatively drawn based on studies that reveal a requirement for the ATP-dependent helicase activites during formation of an open complex *(38)* and a participation of the CTD kinase activity of TFIIH in the stimulation of promoter clearance *(3)*.

ATP is required by pol II as a substrate for several different reactions during RNA synthesis *(10,72)*. These include the use of ATP as a phosphate donor for phosphorylation of the pol II CTD or any of the other basal or activated transcription factors, and for the formation of an open (active) transcription complex *(42,43)*. Indeed, AMP-PNP, a nonhydrolyzable β-γ analog of ATP, cannot substitute for ATP in pol II transcription from specific promoters in vitro, whereas addition of dATP with AMP-PNP restores accurate transcription *(72)*. This demonstrates that ATP plays a dual role in transcription: first as a pol II substrate and second as an energy source. In consideration of such demands, it was not surprising to find an ATPase activity among one of the various factors involved in pol II transcription *(13)*. TFIIH fills the expected requirement perfectly for such an activity, since two of its subunits, XP-B and XP-D, contain seven helicase motifs, ATP and DNA binding sites, and exhibit ATP-dependent helicase activity in vitro *(31,74,75,77,82)*. The strict requirement for ATP during initiation appears to be a function of the topology of the promoter and suggests a requirement for the helicase activities supplied by TFIIH. This is complicated by the fact that the K_M value for ATP during promoter opening is much less than the level needed to support helicase activity *(42)*. Nonetheless, negatively supercoiled, but not linear DNA templates circumvent the requirement of ATP and TFIIH for the formation of open promoter complexes *(37,38)*. AMP-PNP supports transcription from supercoiled templates but not from linear templates and cannot be used as a cofactor for TFIIH helicase activity *(72)*. Thus, ATP may be required at several stages during transcription, including the initial opening of the DNA helix and promoter clearance *(30,43,56)*. An intervening step during transcription, the forward movement of the initiation bubble, does not require ATP *(43)*. The activities of TFIIH fit quite well with this paradigm and allow the complex to serve as an important site for regulation via interactions with other transcription factors and activator proteins.

The helicase activity of TFIIH is thought to be associated with precise functions in both transcription and DNA repair. In transcription, the XPB and XPD helicases have been suggested to be responsible for the opening of the DNA duplex at a preinitiation step (open complex formation) to allow pol II to read the template strand. Binding of TFIIH to the preinitiation complex leads to the formation of a single-stranded region from –9 to +1 relative to the transcription start site, which is extended to +8 with formation of the first phosphodiester bond *(38)*. In the NER process, the helicases could be involved in the formation of a single-stranded region to allow the excision step. The 3'–5' helicase activity (Rad25p/XP-B) appears to be more crucial for transcription as well as DNA repair than the 5'–3' helicase activity (Rad3p/XP-D). Mutations in the ATP binding site of Rad25p/XP-B result in defects in both transcription and NER *(32)*. In contrast, mutations in the ATP binding site of *RAD3p* affect only NER, but leave transcription intact *(23,31)*. These results suggest that the ATP binding domain/helicase activity of XP-D is dispensable for transcription, but do not discount the role of other domains that function in DNA binding or protein–protein interactions as having an important contribution toward transcription. In support, a temperature-sensitive mutation of Rad3p has been shown to lead to deficient transcription *(31)*. Moreover, patients with mutations in *XP-D* fall into a hybrid classification of xeroderma pigmentosa (XP)—Cockayne's syndrome (CS) (XP-CS) or the classical trichothiodystrophy (TTD) *(86)*. The phenotypes of these disorders cannot be explained simply by deficiencies in NER, but suggest disruptions in TFIIH-mediated transcription as well.

3.3. CTD Kinase of TFIIH

Pol II exists in the cell in a highly phosphorylated (pol II0) or a dephosphorylated form (pol IIA). Phosphorylation occurs on the largest subunit of pol II, which possesses at the CTD an imperfect repeat of the consensus heptapeptide sequence YSPTSPS (repeated 52 times in humans), which is used as a substrate by protein kinases *(15)*. This CTD tail is particularly sensitive to degradation, and pol II is purified in part as a CTD-deleted form (pol IIB). Phosphorylation of the CTD is believed to play a role in the transition from initiation to elongation *(50,68)* and in the response to activators *(14,21,73)*. In support of this, pol IIA preferentially associates with the preinitiation complex (and is also found in the holoenzyme; *64*), whereas pol II0 is found in the elongation complex *(44)*. The CTD kinase of TFIIH is able to phosphorylate pol II as well as other transcription factors in the initiation complex. However, the activity directed toward the CTD is thought to be critical for transition of the complex from initiation to elongation (Fig. 1). In this case, CTD phosphorylation is thought to induce the release of the initiation complex from TFIID, which remains bound to the promoter *(14)*. Although several CTD kinases have been isolated *(69)*, the kinase associated with TFIIH appears to be an ideal candidate to perform this function owing to the proximity of the enzyme with the CTD in the initiation complex. The TFIIH kinase resides in Cdk7, a cyclin-dependent kinase involved in cell-cycle regulation *(24,71,80,81,87)*. Cdk7 is associated with two other polypeptides, cyclin H and MAT1 *(2,26,99)*, to constitute the Cdk-activating kinase (CAK), which phosphorylates cdc2 and cdk2, two cyclin-dependent kinases of the cell-cycle (for a review of cyclin-dependent kinases, *see 60*). The addition of MAT1 appears to optimize the activity of CTD-phosphorylation by Cdk7 *(2,26)*. The kinase complex can be resolved from TFIIH under high-salt conditions and exists as a free form in the cell *(2,26,81)*. Although the substrate specificity remains similar, the phosphorylation of pol II by TFIIH, but not by the free complex, is strongly stimulated by the basal transcription machinery and promoter elements *(50)*. TFIIH containing a mutated form of Cdk7 is unable to phosphorylate pol II even in the presence of basal factors and template, indicating that Cdk7 is most likely the enzyme responsible for the CTD kinase activity of TFIIH *(52)*. Nonetheless, the Cdk7 mutant was as active as the wild-type protein in supporting TFIIH-dependent basal transcription, which demonstrates that the CTD kinase is not essential for basal transcription from all promoters in vitro. This observation is supported by the fact that transcription from several class II promoters, like the Adenovirus 2 major late promoter (AdMLP), can take place in vitro using a CTD-less form of pol II *(3)*. However, other promoters, such as the TATA-less *DHFR* promoter, require the presence of pol

IIA (CTD-containing pol II) and phoshorylation of its CTD for efficient transcription *(3,45)*. Although these data suggest that the CTD and the CTD kinase are dispensable for transcription initiation in vitro *(78)* both are absolutely required in vivo *(16,61)*. Thus, the Cdk7 kinase appears to be critical for pol II phosphorylation in vivo. If the CTD is required for initiation, its phosphorylation appears to be necessary at a subsequent step, probably promoter clearance *(3)*.

Interestingly, TFIIH also phosphorylates TBP the TATA-box-binding protein and the large subunits of TFIIE and TFIIF in vitro *(62)*. This phosphorylation might be involved as well in the efficient release of pol II by providing a high concentration of phosphorylated proteins in the initiation complex.

In *S. cerevisiae*, Kin28p and Ccl1p are subunits of TFIIK, the protein kinase complex of the holo-TFIIH. Thermosensitive (ts) mutants have been used to demonstrate that Kin28p encodes a CTD kinase, but lacks CAK activity *(12,90)*. Thus, Kin28p appears to be involved in transcription by pol II, but not in cell-cycle regulation. This is complicated by recent data from *Schizosaccaromyces pombe* documenting the isolation of mop1p/crk1p, a homolog of Cdk7 *(9,17)*. mop1p can associate with a cyclin H-related protein, mcs2p, and phoshorylate a CTD substrate as well as human Cdk2, indicative of CAK activity. It is not known, however, if in *S. pombe* the mop1-Mcs2 complex is part of the transcription complex homologous with TFIIH.

4. DUAL FUNCTION OF TFIIH: TRANSCRIPTION–REPAIR COUPLING

Previous studies suggested a connection between transcription and DNA repair. Following UV irradiation, the recovery of transcription was defective in patients suffering from XP or CS *(57)*. Furthermore, Mellon et al. *(58)* showed that repair of the transcribed strand of specific genes was faster than the nontranscribed strand. The finding that the basal transcription factor TFIIH is involved in NER is in agreement with these observations *(20,74)*. Several lines of evidence have demonstrated an involvement of the p89, p80, p62, and p44 subunits of TFIIH in both reactions: immunodepletion or inhibition of TFIIH reduced NER and transcription in vitro *(20,75)* and in vivo *(91)*; DNA repair activity of XP-B- or XP-D-deficient cell extracts was restored in vitro and in vivo by addition of highly purified TFIIH *(91)*; mutations in TFIIH subunits decreased DNA repair in yeast *(55)*. The p34 subunit, which exhibits homology with the Ssl1p and p44 proteins *(39)* shown to be required for both transcription and NER *(92)*, may also be involved in DNA repair. Thus, TFIIH can be considered as a DNA repair/transcription factor. Work performed with yeast strongly supports this hypothesis. As mentioned in Section 3.2., a mutation in the nucleotide binding site of Rad25p is lethal *(32)*; this is likely owing to a loss of part, if not all of its transcription and repair functions. Although similar mutations in the ATP binding site of Rad3p disrupt NER, but do not affect trancription *(23)*, other mutations demonstrate an essential function for Rad3p in transcription *(31)* and emphasize the important role that TFIIH has in both functions.

The fact that TFIIH participates not only in transcription, but NER as well, suggests a role in the transcription-repair coupling mechanism. Two distinct genes, CSA and CSB, which are defective in CS, are required in strand-specific repair of transcriptionally active genes, but not in global genome repair *(50,51)*. The recent discovery that

CSA interacts directly with TFIIH (through the p44 and XP-B subunits) and with CSB *(36)* may explain why CS symptoms are sometimes associated with XP symptoms. In these cases, a mutation in p44 or XPB may affect the normal function of TFIIH in NER as well as disrupt the interaction of TFIIH with CSA, resulting in a deficiency of both total genome repair and the transcription–repair coupling mechanism involving CSA and CSB proteins. In view of the relationship between CSA and CSB, it has been suggested that these factors may have additionally unidentified roles directly in pol II transcription, the inhibition of which could lead to the various phenotypes indicative of CS and TTD.

5. FUNCTIONS OF TFIIH IN TRANSCRIPTION

5.1. *Interactions with Basal Transcription Factors*

In addition to TBP and pol II, TFIIH also interacts strongly with TFIIE (through XP-B and TFIIEα; *20,28,57,62,63*). TFIIH and TFIIE appear to function cooperatively in transcription, and as with the requirement for ATP during initiation, their role might depend on the DNA template topology *(66,67)*. For example, the transcription of negatively supercoiled templates of the AdMLP does not require TFIIH or TFIIE, although TFIIE is stimulatory *(37,38)*. In contrast, transcription of relaxed or linear templates requires both TFIIE and TFIIH *(37)*.

TFIIE appears to have multiple functions, which can modulate the activities of TFIIH. For example, TFIIE can participate in limited melting of the DNA promoter independent of TFIIH *(38)*, a function that would facilitate loading of TFIIH. Indeed, most helicases usually require some single-stranded DNA to begin unwinding the template. In addition, TFIIE, which was shown to bind to the promoter complex a step earlier, may facilitate the recruitment and the positioning of TFIIH close to the CTD *(42,56)*. The consequence of this may be the phosphorylation of the CTD by TFIIH. In support, TFIIE stimulates the kinase and the ATPase activity of TFIIH *(62,63,79)* and negatively regulates its helicase activity *(20)*. Since TFIIE is released from the preinitiation complex before TFIIH, the following scenario can be envisaged. In a first step, TFIIE binds TFIIH and recruits it to the initiation complex. In an ATP-dependent process, the helicase activities of TFIIH generate a small single-stranded region around the promoter. Interactions with TFIIE stimulate the CTD kinase of TFIIH (the phosphorylation of the CTD may lead to disruption of the TFIID–pol II interaction) and at the same time repress its helicase function. The dissociation of TFIIE (which may be a consequence of the phosphorylation of the CTD or of its own phosphorylation by TFIIH) allows the TFIIH helicases to promote extension of the bubble region generated by the original complex. This may induce the release of all the factors with the exception of TFIID, which remains bound to the promoter for subsequent rounds of initiation. Finally, pol II (in association with TFIIF) pursues elongation. The finding that the CTD kinase activity of TFIIH is not required for production of abortive initiation products, but was necessary for the detection of run-off transcripts *(3)* suggests that TFIIH may also have a direct role in elongation. Analysis of proteins contained in the elongation complex reveal that TFIIH dissociates after approx +30 nucleotides *(100)*. These results suggest that TFIIH is only associated in the beginning stages of transition to elongation. However, additional data suggest a more extensive role. Using an in vitro kinase assay, the elongation inhibitor 5,6-dichloro-1-β-D-ribofuranosyl-benzimidazole (DRB) inhibited the CTD kinase activity of TFIIH *(97)*. Furthermore, microinjection of antibodies to TFIIH into *Xenopus* oocytes, led to an inhibition of elongation *(98)*. This was not observed when antibodies to the initiation-specific protein TFIIB were used. It is uncertain if TFIIH is the direct substrate of DRB, since other evidence indicates that this compound can also inhibit p-TEFb, a CTD kinase present in *Drosophila* and HeLa cells that is essential for the production of long run-off transcripts *(4,54)*. Additional support for a role of TFIIH in the stimulation of elongation, however, is provided by a correlation between the ability of gene-specific activators to stimulate elongation and their affinity to TFIIH *(8)*.

5.2. *Interactions of TFIIH with Transcription Activators*

TFIIH, as a central protein required for essential processes and with several important enzymatic functions, is a particularly appropriate target for regulatory proteins. Indeed, activators or repressors may be capable of targeting, either directly or indirectly, some of the components of TFIIH with the consequence of modulating either DNA melting or kinase activity. Several transcription regulators, such as p53 and VP16, can interact with TFIIH *(49,93,95)*.

Acidic activators have been suggested to stimulate a slow step involving TFIID and TFIIA or to lead to a recruitment of TFIIB in the preinitiation complex *(70)*. They may also stimulate subsequent steps of the preinitiation complex assembly. The Epstein-Barr virus nuclear antigen 2 (EBNA2), an acidic transactivator essential for viral transformation of B-lymphocytes, interacts both in vitro and in vivo with the p62 and XP-D subunits of TFIIH. A mutation in the acidic activation domain of EBNA2, which disrupts transactivation, also compromises binding to TFIIH *(88)*. Two other acidic transactivators, herpes simplex virus VP16 and human p53, directly interact with TFIIH (via p62, XP-B, or XP-D) or with its yeast homolog, factor b (via Tfb1) *(49,93,95)*. Binding of TFIIH to wild-type p53 negatively modulates both XP-B and XP-D helicase activities in vitro, whereas mutants of p53, such as Arg 273 (found in numerous cancer cells) and others, have a much weaker effect *(49,93)*. In similar fashion to EBNA2, a point mutation that reduces the transactivation activity of VP16 decreases its binding to human and yeast TFIIH *(95)*. Such binding of acidic activators to TFIIH could stimulate melting of the DNA at the promoter, lead to more processive transcription by RNA polymerase II, or stabilize a complex containing TFIIH and other components of the preinitiation complex.

6. FUNCTIONS OF TFIIH IN DNA REPAIR

TFIIH has been shown to be required for NER both in vivo and in vitro *(1,59)*. Experiments performed in yeast have provided evidence for a large number of interactions between TFIIH subunits and other proteins involved in NER. Gel-shift assays suggest that TFIIH can be recruited to the excision repair complex by the XP-A protein *(65)*, which recognizes the DNA lesion in combination with replication protein A (RPA) *(35)*, another NER factor. Since Rad3 interacts with Ssl2p and Ssl1p in vitro *(5)*, the zinc finger region of Ssl1p may anchor both helicases to the DNA allowing TFIIH to function in transcription and/or NER. Ssl1p, as a damage recognition polypeptide, may thus modulate the processivity and/or allow damage-recognition by the Rad3p helicase.

Furthermore, the efficient recognition of the DNA lesion could be facilitated by the interaction of TFIIH with Rad23p and Rad14p *(33)*, two proteins also required for NER. In these circumstances, Rad23p appears to be an intermediary factor that promotes association of TFIIH with Rad14p.

Several other polypeptides involved in DNA repair also interact with TFIIH in vitro. Rad2p and Rad4p in yeast, and their mammalian homologs XP-G and XP-C have been shown to bind to several subunits of TFIIH *(5,6,20,40,59)*. In turn, XP-G has also been shown to bind to the CSB protein *(40)*. In a similiar fashion to the CSA and CSB proteins described in Section 4., these observations may partially explain why XP groups B, D, and G have been found to be associated with CS, XP group D, and TTD clinical features, as well as why some CSA and CSB cells show a reduced level of pol II transcription *(36,89)*. Thus, the three proteins (XP-B, -D, and -G) implicated in combined XP/CS or XP/TTD DNA repair disorders may also be involved in specific protein–protein interactions important for transcriptional functions.

The function of TFIIH in NER was also analyzed using in vitro reconstituted systems *(1,59)*. These studies have defined the minimal set of proteins required for the incision stage: RPA, XP-A, XP-G, XP-C, TFIIH, ERCC1/XP-F in addition to a factor named IF7. RPA, a heterotrimer that binds single-stranded DNA *(35,94)* and is involved in the recognition of DNA damage *(35)*, could participate with XP-C/HHR23B in stabilizing the open structure generated by the helicase activity of TFIIH in order to allow the formation of the active incision complex (*see also* ref. *22*). In yeast, a similar reconstitution assay defines Rad14p, Rad4p/Rad23p, Rad1p/Rad10p, Rad2p, RPA, and TFIIH as the components required for the incision step of NER *(34)*.

7. RELATIONSHIPS AMONG TRANSCRIPTION, NER, AND THE CELL CYCLE

DNA damage induces cell-cycle arrest, and this is thought to provide the DNA repair system time to eliminate lesions in DNA that would interfere with transcription and replication (*see* Chapter 17 in vol. 1 and Chapter 21 in this vol.). Thus, the Cdk7 kinase and its cyclin regulatory subunit associated with TFIIH could be the checkpoint between transcription blocked by a DNA lesion and cell-cycle control. In this view, Cdk7, either through the transcription apparatus or directly, may initiate the signaling cascade that leads to cell-cycle arrest in G1, S, or G2 when transcription is hampered. Alternatively, the protein kinase activity may be required for recruiting TFIIH from the transcription mode into the repair mode or vice versa.

As discussed in Section 3.1., Cdk7 and its two partners cyclin H and MAT1 can be found in two forms, one associated with TFIIH, probably through XP-D, and one free. It is unknown if these two complexes work independently or are in equilibrium. In the latter case, a change in the ratio between these two forms could assign the Cdk7 kinase to play a role either in transcription, NER, or cell-cycle regulation. Neither the overall subunit composition nor the kinase activity of TFIIH (or of the free complex) is modified during a normal cell-cycle. However, the protein kinase activity associated with TFIIH, but not with the free kinase complex, decreases after induction of DNA damage by UV treatment in human cells *(2)*. Kin28p interacts with Cdc37p, the cell division cycle gene, and Ccl1p, the analog of the mammalian cyclin H, which is also essential for cell proliferation *(90)*. Moreover, a decrease in the phosphorylation of the CTD is observed when a Kin28p thermosensitive mutant is grown at the restrictive temperature, which could explain the decrease in pol II transcription. Although Kin28p acts as a positive regulator of pol II transcription in vivo and possesses CTD kinase activity in vitro, it neither regulates the yeast cell-cycle proteins Cdk and Cdc28p nor possesses CDK-activating kinase activity in vitro *(12)*. Thus, the implication of the TFIIH-associated kinase in the cell-cycle remains uncertain.

A possible mechanism of cell-cycle control through TFIIH after UV-irradiation could involve p53 *(44)*. Damage-induced G1/S cell-cycle arrest can be mediated by posttranslational modification of p53. It has also been demonstrated that p53 interacts with TFIIH through its p62 and XP-B subunits *(49,88,93)*. Furthermore, the accumulation of p53 after UV irradiation depends on the presence of damage in the actively transcribed genes, rather than in the overall genome *(96)*. TFIIH may participate in the transcription initiation of class II promoters and then be subsequently released, or it may remain with elongation complex to provide a further role. In the latter case, when TFIIH is still associated with pol II, it might aide in the detection of a DNA lesion (probably in association with CSB) and help to recruit p53. This would stimulate cell-cycle arrest and allow DNA repair to proceed before replication. p53 binding to TFIIH may also induce a novel conformation of the preformed complex, which would be more appropriate to bind other DNA repair proteins leading to an activation of NER.

Another possibility is that the recruitment of a protein to one process (DNA repair) may prevent its participation in another one (transcription or cell-cycle regulation). In that case, these processes may have different priorities, with DNA repair being favored over replication and transcription. The eukaryotic DNA repair system has evolved to repair transcribed regions of the genome more efficiently. In order to be efficient in this task, the DNA repair complex (which could be a repairosome) should be able to detect transcribed regions. This may be signaled by a stalled pol II. However, since natural pause sites occur during elongation of transcription, the NER system should distinguish between paused complexes and complexes stalled at lesions. If not, nonspecific repair of undamaged DNA would lead to the introduction of mutations owing to the higher error frequency of NER. The recruitment of the repair factors to the lesion could be due to the interaction of CSB (involved in transcription coupled repair) with TFIIH. One can imagine that pause sites are initially recognized by TFIIS, which is known to enhance elongation and nascent transcript cleavage activities of pol II in a stalled elongation complex *(41)*. If elongation does not continue, the formation of a more complex structure (the repairosome) will have time to proceed.

8. CONCLUDING REMARKS

Although characterization of the TFIIH subunits is almost completed, our knowledge concerning their in vivo function(s) and their regulation is still somewhat fragmentary. Continued work is required to define clearly the role of TFIIH and its various enzymatic activities in transcription as well as in DNA repair and cell-cycle regulation. The discovery that mutations in at least two of its subunits, XP-B and XP-D, give rise to disorders that are now explained not only on the basis of a deficiency in NER, but also in transcription, adds a new clinical dimension to the intricacies of TFIIH. Future studies will determine whether or not transcription of specific genes is affected as a function of TFIIH mutations and, thereby, provide further insight into the role of transcription factors in hereditary diseases.

ACKNOWLEDGMENTS

We are thankful to all members of our group for fruitful discussions. This work was supported by grants from the Institut National de la Santé et de la Recherche Médicale (INSERM), Center National de la Recherche Scientifique (CNRS) and Association pour la Recherche contre le Cancer (ARC).

REFERENCES

1. Aboussekhra, A., M. Biggerstaff, M. K. K. Shivji, J. A. Vilpo, V. Moncollin, V. N. Podust, M. Protic, U. Hübscher, J.-M. Egly, and R. D. Wood. 1995. Mammalian DNA nucleotide excision repair reconstituted with purified protein components. *Cell* **80:** 859–868.
2. Adamczewski, J. P., M. Rossignol, J.-P. Tassan, E. A. Nigg, V. Moncollin, and J.-M. Egly. 1996. MAT1, MO15 and cyclin H form a kinase complex which is UV light sensitive upon association with TFIIH. *EMBO J.* **15:** 1877–1884.
3. Akoulitchev, S., T. P. Mäkelä, R. A. Weinberg, and D. Reinberg. 1995. Requirement for TFIIH kinase activity in transcription by RNA polymerase II. *Nature* **377:** 557–560.
4. Amendt, B. A., Z. Xie, and D. H. Price. 1996. Elongation by human RNA polymerase II is partially controlled through the action of P-TEF. Proc. Transcription Mechanisms, Keystone Symp. Taos, NM, p. 68.
5. Bardwell L., A. J. Bardwell, W. J. Feaver, J. Q. Svejstrup, R. D. Kornberg, and E. C. Friedberg. 1994. Yeast RAD3 protein binds directly to both SSL2 and SSL1 proteins: implications for the structure and function of transcription factor b. *Proc. Natl. Acad. Sci. USA* **91:** 3926–3930.
6. Bardwell A. J., L. Bardwell, N. Iyer, J. Q. Svejstrup, W. J. Feaver, R. D. Kornberg, and E. C. Friedberg. 1994. Yeast nucleotide excision repair proteins Rad2 and Rad4 interact with RNA polymerase II basal transcription factor b (TFIIH). *Mol. Cell. Biol.* **14:** 3569–3576.
7. Bhatia, P. K., Z. Wang, and E. C. Friedberg. 1996. DNA repair and transcription. *Curr. Opinion Genet. Dev.* **6:** 146–150.
8. Blau, J., H. Xiao, S. McCraken, P. O'Hare, J. Greenblatt, and D. Bentley. 1996. Three functional classes of transcriptional activation domains. *Mol. Cell. Biol.* **16:** 2044–2055.
9. Buck, V., P. Russel, and J. B. A. Millar. 1995. Identification of a cdk-activating kinase in fission yeast. *EMBO J.* **14:** 6173–6183.
10. Bunick, D., R. Zandomeni, S. Ackerman, and R. Zeinmann. 1982. Mechanism of RNA polymerase II-specific initiation of transcription in vitro: ATP requirement and uncapped runoff transcripts. *Cell* **29:** 877–886.
11. Chalut, C., V. Moncollin, and J.-M. Egly. 1994. Transcription by RNA polymerase II: a process linked to DNA repair. *BioEssays* **16:** 651–655.
12. Cismowski, M. J., G. M. Laff, M. J. Solomon, and S. I. Reed. 1995. KIN28 encodes a C-terminal domain kinase that controls mRNA transcription in *Saccharomyces cerevisiae* but lacks cyclin-dependent kinase-activating kinase (CAK) activity. *Mol. Cell. Biol.* **15:** 2983–2992.
13. Conaway, R. C. and J. W. Conaway. 1989. An RNA polymerase II transcription factor has an associated DNA-dependent ATPase (dATPase) activity strongly stimulated by the TATA region of promoters. *Proc. Natl. Acad. Sci. USA* **86:** 7356–7360.
14. Conaway, R. C., J. N. Bradsher, and J. W. Conaway. 1992. Mechanism of assembly of the RNA polymerase II preinitiation complex. Evidence for a functional interaction between

the carboxyl-terminal domain of the largest subunit of RNA polymerase II and a high molecular mass form of the TATA factor. *J. Biol. Chem.* **267:** 8464–8467.

15. Corden, J. L., D. L. Cadena, J. M. Ahearn, Jr., and M. E. Dahmus. 1985. A unique structure at the carboxyl terminus of the largest subunit of eukaryotic RNA polymerase II. *Proc. Natl. Acad. Sci. USA* **82:** 7934–7938.
16. Dahmus, M. E. 1995. Phosphorylation of the C-terminal domain of RNA polymerase II. *Biochim. Biophys. Acta* **1261:** 171–182.
17. Damagnez, V., T. P. Makela, and G. Cottarel. 1995. *Schizosaccharomyces pombe* Mop1-Mcs2 is related to mammalian CAK. *EMBO J.* **14:** 6164–6172.
18. Devault, A., A.-M. Martinez, D. Fesquet, J-C. Labbè, N. Morin, J.-P. Tassan, E. A. Nigg, J.-C. Cavadore, and M. Dorèe. 1995. MAT1 ('menage à trois') a new RING finger protein subunit stabilizing cyclin H-cdk7 complexes in starfish and *Xenopus* CAK. *EMBO J.* **14:** 5027–5036.
19. Drapkin, R., A. Merino, and D. Reinberg. 1993. Regulation of RNA polymerase II transcription. *Curr. Opinion Cell Biol.* **5:** 469–476.
20. Drapkin, R., J. T. Reardon, A. Ansari, J. C. Huang, L. Zawel, K. Ahn, A. Sancar, and D. Reinberg. 1994. Dual role of TFIIH in DNA excision repair and in transcription by RNA polymerase II. *Nature* **368:** 769–772.
21. Emili, A. and C. J. Ingles. 1995. The RNA polymerase II carboxy-terminal domain: link to a bigger and better "holoenzyme"? *Curr. Opinion Genet. Dev.* **5:** 204–209.
22. Fairman, M. P. and B. Stillman. 1988. Cellular factors required for multiple stages of SV40 DNA replication *in vitro*. *EMBO J.* **7:** 1211–1218.
23. Feaver W. J., J. Q. Svejstrup, L. Bardwell, A. J. Bardwell, S. Buratowski, K. D. Gulyas, T. F. Donahue, E. C. Friedberg, and R. D. Kornberg. 1993. Dual roles of a multiprotein complex from *S. cerevisiae* in transcription and DNA repair. *Cell* **75:** 1379–1387.
24. Feaver W. J., J. Q. Svejstrup, N. L. Henry, and R. D. Kornberg. 1994. Relationship of CDK-activating kinase and RNA polymerase II CTD kinase TFIIH/TFIIK. *Cell* **79:** 1103–1109.
25. Fischer, L., M. Gérard, C. Chalut, Y. Lutz, S. Humbert, M. Kanno, P. Chambon, and J.-M. Egly. 1992. Cloning of the 62-kilodalton component of basic transcription factor BTF2. *Science* **257:** 1392–1395.
26. Fisher, R. F., P. Jin, H. M. Chamberlin, and D. O. Morgan. 1995. Alternative mechanisms of CAK assembly require an assembly factor or an activating kinase. *Cell* **83:** 47–57.
27. Fisher, R. P. and D. O. Morgan. 1994. A novel cyclin associates with MO15/CDK7 to form the CDK-activating kinase. *Cell* **78:** 713–724.
28. Gérard, M., L. Fischer, V. Moncollin, J.-M. Chipoulet, P. Chambon, and J.-M. Egly. 1991. Purification and interaction properties of the human RNA polymerase B(II) general transcription factor BTF2. *J. Biol. Chem.* **266:** 20,940–20,945.
29. Gileadi, O., W. J. Feaver, and R. D. Kornberg. 1992. Cloning of a subunit of yeast RNA polymerase II transcription factor b and CTD kinase. *Science* **257:** 1389–1392.
30. Goodrich J. A. and R. Tjian. 1994. Transcription factors IIE and IIH and ATP hydrolysis direct promoter clearance by RNA polymerase II. *Cell* **77:** 145–156.
31. Guzder, S. N., H. Piu, C. H. Sommers, P. Sung, L. Prakash, and S. Prakash. 1994. DNA repair gene RAD3 of *S. cerevisiae* is essential for transcription by RNA polymerase II. *Nature* **367:** 91–94.
32. Guzder, S. N., P. Sung, V. Bailly, L. Prakash, and S. Prakash. 1994. RAD25 is a DNA helicase required for DNA repair and RNA polymerase II transcription. *Nature* **369:** 578–581.
33. Guzder, S. N., V. Bailly, P. Sung, L. Prakash, and S. Prakash. 1995. Yeast DNA repair protein RAD23 promotes complex formation between transcription factor TFIIH and DNA damage recognition factor RAD14. *J. Biol. Chem.* **270:** 8385–8388.
34. Guzder, S. N., Y. Habraken, P. Sung, L. Prakash, and S. Prakash. 1995. Reconstitution of yeast nucleotide excision repair with purified Rad proteins, replication protein A, and transcription factor TFIIH. *J. Biol. Chem.* **270:** 12,973–12,976.

35. He Z., L. A. Henricksen, M. S. Wold, and C. J. Ingles. 1995. RPA involvement in the damage-recognition and incision steps of nucleotide excision repair. *Nature* **374:** 566–569.
36. Henning, K. A, L. Li, N. Iyer, L. D. McDaniel, M. S. Reagan, R. Legerski, R. A. Schultz, M. Stefanini, A. R. Lehmann, L. V. Mayne, and E. C. Friedberg. 1995. The Cockayne syndrome group A gene encodes a WD repeat protein that interacts with CSB protein and a subunit of RNA polymerase II TFIIH. *Cell* **82:** 555–564.
37. Holstege, F. C. P., D. Tantin, M. Carey, P. C. van der Vliet, and H. TH. M. Timmers. 1995. The requirement for the basal transcription factor IIE is determined by the helical stability of promoter DNA. *EMBO J.* **14:** 810–829.
38. Holstege, F. C. P., P. C. van der Vliet, and F. TH. M. Timmers. 1996. Opening of an RNA polymerase II promoter occurs in two distinct steps and requires the basal transcription factors IIE an IIH. *EMBO J.* **15:** 1666–1677.
39. Humbert, S., H. van Vuuren, Y. Lutz, J. H. J. Hoeijmakers, J.-M. Egly, and V. Moncollin. 1994. p44 and p34 subunits of the BTF2/TFIIH transcription factor have homologies with SSL1, a yeast protein involved in DNA repair. *EMBO J.* **13:** 2393–2398.
40. Iyer, N., M. S. Reagan, K.-J. Wu, B. Canagarajah, and E. C. Friedberg. 1996. Interactions involving the human RNA polymerase II transcription: nucleotide excision repair complex TFIIH, the nucleotide excision repair protein XPG, and Cockayne syndrome group B (CSB) protein. *Biochemistry* **35:** 2157–2167.
41. Jeon, C., H. Joon, and K. Agarwal. 1994. The transcription factor TFIIS zinc ribbon dipeptide Asp-Glu is critical for stimulation of elongation and RNA cleavage by RNA polymerase II. *Proc. Natl. Acad. Sci. USA* **91:** 9106–9110.
42. Jiang, Y. and J. D. Gralla. 1995. Nucleotide requirements for activated RNA polymerase II open complex formation in vitro. *J. Biol. Chem.* **270:** 1277–1281.
43. Jiang, Y., M. Yan, and J. D. Gralla. 1996. A three-step pathway of transcription initiation leading to promoter clearance at an activated RNA ploymerase II promoter. *Mol. Cell. Biol.* **16:** 1614–1621.
44. Jones, J. and D. Wynford-Thomas. 1995. Is TFIIH an activator of the p53-mediated G1/S checkpoint. TIG **11:** 165,166.
45. Kang, M. E. and M. E. Dahmus. 1993. RNA polymerase IIA and II0 have distinct roles during transcription from the TATA-less murine dihydrofolate reductase promoter. *J. Biol. Chem.* **268:** 25,033–25,040.
46. Kang, M. E. and M. E. Dahmus. 1995. The photoactivated cross-linking of recombinant C-terminal domain to proteins in a HeLa cell transcription extract that comigrate with transcription factors IIE and IIF. *J. Biol. Chem.* **270:** 23,390–23,397.
47. Kim, Y., S. Björklund, Y. Li, H. M. Sayre, and R. D. Kornberg. 1994. A multiprotein mediator of transcriptional activation and its interaction with the C-terminal repeat domain of RNA polymerase II. *Cell* **77:** 599–608.
48. Koleske, A. J. and R. A. Young. 1994. An RNA polymerase II holoenzyme responsive to activators. *Nature* **368:** 466–469.
49. Léveillard, T., L. Andera, N. Bissonnette, L. Schaeffer, L. Bracco, J.-M. Egly, and B. Wasylyk. 1996. Functional interactions between p53 and the TFIIH complex are affected by tumor associated mutations. *EMBO J.* **15:** 1615–1624.
50. Lu, H., L. Zawel, L. Fischer, J.-M. Egly, and D. Reinberg. 1992. Human general transcription factor IIH phosphorylates the C-terminal domain of RNA polymerase II. *Nature* **358:** 641–645.
51. Makela, T. P., J.-P. Tassan, E. A. Nigg, S. Frutiger, G. J. Huges, and R. A. Weinberg. 1994. A cyclin associated with the CDK-activating kinase MO15. *Nature* **371:** 254–257.
52. Mäkelä, T. P., J. D. Parvin, J. Kim, L. J. Huber, P. A. Sharp, and R. A. Weinberg. 1995. A kinase-deficient transcription factor TFIIH is functional in basal and activated transcription. *Proc. Natl. Acad. Sci. USA* **92:** 5174–5178.

53. Marinoni, J. C., R. Roy, W. Vermeulen, P. Miniou, Y. Lutz, G. Weeda, T. Seroz, G. Molina, J. H. J. Hoeijmakers, and J.-M. Egly. 1996. Cloning and characterization of p52, the fifth subunit of the core of the transcription/DNA repair factor TFIIH. *EMBO J.*, in press.
54. Marshall, N. F. and D. H. Price. 1995. Purification of P-TEFb, a transcription factor required for the transition into productive elongation. *J. Biol. Chem.* **270:** 12,335–12,338.
55. Matsui, P., J. De Paulo, and S. Buratowski. 1995. An interaction between the Tfb1 and Ssl1 subunits of yeast TFIIH correlates with DNA repair activity. *Nucleic Acids Res.* **23:** 767–772.
56. Maxon M. E., J. A. Goodrich, and R. Tjian. 1994. Transcription factor IIE binds preferentially to RNA polymerase IIa and recruits TFIIH: a model for promoter clearance. *Genes Dev.* **8:** 515–524.
57. Mayne, L. V. and A. R. Lehmann. 1982. Failure of RNA synthesis to recover after UV irradiation: an early defect in cells from individuals with Cockayne's syndrome and xeroderma pigmentosum. *Cancer Res.* **42:** 1473–1478.
58. Mellon, I., G. Spivak, and P. C. Hanawalt. 1987. Selective removal of transcription-blocking DNA damage from the transcribed strand of the mammalian DHFR gene. *Cell* **51:** 241–249.
59. Mu, D., C.-H. Park, T. Matsunaga, D. S. Hsu, J. T. Reardon, and A. Sancar. 1995. Reconstitution of human DNA repair excision nuclease in a highly defined system. *J. Biol. Chem.* **10:** 2415–2418.
60. Nigg, E. A. 1995. Cyclin-dependent protein kinases: key regulators of the eukaryotic cell-cycle. *BioEssays* **17:** 471–480.
61. Nonet, M., D. Sweetser, and R. A. Young. 1987. Functional redundancy and structural polymorphism in the large subunit of RNA polymerase II. *Cell* **50:** 909–915.
62. Ohkuma, Y. and R. G. Roeder. 1994. Regulation of TFIIH ATPase and kinase activities by TFIIE during active initiation complex formation. *Nature* **368:** 160–163.
63. Ohkuma, Y., S. Hashimoto, C. K. Wang, M. Horikoshi, and R. G. Roeder. 1995. Analysis of the role of TFIIE in basal transcription and TFIIH-mediated carboxy-terminal domain phosphorylation through structure-function studies of TFIIE-a. *Mol. Cell. Biol.* **15:** 4856–4866.
64. Ossipow, V., J.-P. Tassan, E. A. Nigg, and U. Schibler. 1995. A mammalian RNA polymerase II holoenzyme containing all components required for promoter-specific transcription initiation. *Cell* **83:** 137–146.
65. Park, C.-H., D. Mu, J. T. Reardon, and A. Sancar. 1995. The general transcription-repair factor TFIIH is recruited to the excision repair complex by the XPA protein independent of the TFIIE transcription factor. *J. Biol. Chem.* **270:** 4896–4902.
66. Parvin, J. D., H. T. Timmers, and P. A. Sharp. 1992. Promoter specificity of basal transcription factors. *Cell* **68:** 1135–1144.
67. Parvin, J. D. and P. A. Sharp. 1993. DNA topology and a minimal set of basal factors for transcription by RNA polymerase II. *Cell* **73:** 533–540.
68. Payne, J. M., P. J. Laybourn, and M. E. Dahmus. 1989. The transition of RNA polymerase II from initiation to elongation is associated with phosphorylation of the carboxyl-terminal domain of subunit IIa. *J. Biol. Chem.* **264:** 19,621–19,629.
69. Poon, R. Y. C. and T. Hunter. 1995. Innocent bystanders or chosen collaborators? *Curr. Biol.* **5:** 1243–1247.
70. Roberts, S. G. E. and M. R. Green. 1994. Activator-induced conformational change in general transcription factor TFIIB. *Nature* **371:** 717–720.
71. Roy, R., J. P. Adamczewski, T. Seroz, W. Vermeulen, J.-P. Tassan, L. Schaeffer, E. A. Nigg, J. H. J. Hoeijmakers, and J.-M. Egly. 1994. The MO15 cell-cycle kinase is associated with the TFIIH transcription-DNA repair factor. *Cell* **79:** 1093–1101.
72. Sawadogo, M. and R. G. Roeder. 1984. Energy requirement for specific transcription initiation by the human RNA polymerase II system. *J. Biol. Chem.* **259:** 5321–5326.

73. Scafe, C., D. Chao, J. Lopes, J. P. Hirsch, S. Henry, and R. A. Young. 1990. RNA polymerase II C-terminal repeat influences response to transcriptional enhancer signals. *Nature* **347:** 491–494.
74. Schaeffer, L., R. Roy, S. Humbert, V. Moncollin, W. Vermeulen, J. H. J. Hoeijmakers, P. Chambon, and J.-M. Egly. 1993. DNA repair helicase: a component of BTF2 (TFIIH) basic transcription factor. *Science* **260:** 58–63.
75. Schaeffer, L., V. Moncollin, R. Roy, A. Staub, M. Mezzina, A. Sarasin, G. Weeda, J. H. J. Hoeijmakers, and J.-M. Egly. 1994. The ERCC2/DNA repair protein is associated with the class II BTF2/TFIIH transcription factor. *EMBO J.* **13:** 2388–2392.
76. Serizawa, H., R. C. Conaway, and J. W. Conaway. 1992. A carboxyl-terminal-domain kinase associated with RNA polymerase II transcription factor delta from rat liver. *Proc. Natl. Acad. Sci. USA* **89:** 7476–7480.
77. Serizawa, H., R. C. Conaway, and J. W. Conaway. 1993. Multifunctional RNA polymerase II initiation factor delta from rat liver. Relationship between carboxyl-terminal domain, ATPase, and DNA helicase activities. *J. Biol. Chem.* **268:** 17,300–17,308.
78. Serizawa, H., J. W. Conaway, and R. C. Conaway. 1993. Phosphorylation of C-terminal domain of RNA polymerase II is not required in basal transcription. *Nature* **363:** 371–374.
79. Serizawa, H., J. W. Conaway, and R. C. Conaway. 1994. An oligomeric form of the large subunit of transcription factor (TF) IIE activates phosphorylation of the RNA polymerase II carboxy-terminal domain by TFIIH. *J. Biol. Chem.* **269:** 20,750–20,756.
80. Serizawa, H., T. P. Makela, J. W. Conaway, R. C. Conaway, R. A. Weinberg, and R. A. Young. 1995. Association of Cdk-activating kinase subunits with transcription factor TFIIH. *Nature* **374:** 280–282.
81. Shiekhattar, R., F. Mermelstein, R. P. Fisher, R. Drapkin, B. Dynlacht, H. C. Wessling, D. O. Morgan, and D. Reinberg. 1995. Cdk-activating kinase complex is a component of human transcription factor TFIIH. *Nature* **374:** 283–287.
82. Sung, P., V. Bailly, C. Weber, L. H. Thompson, L. Prakash, and S. Prakash. 1993. Human xeroderma pigmentosum group D gene encodes a DNA helicase. *Nature* **365:** 852–855.
83. Svejstrup, J. Q., W. J. Feaver, and R. D. Kornberg. 1996. Subunits of yeast RNA polymerase II transcription factor TFIIH encoded by the CCL1 gene. *J. Biol. Chem.* **271:** 643–645.
84. Svejstrup, J. Q., Z. Wang, W. J. Feaver, X. Wu, D. A. Bushnell, T. F. Donahue, E. C. Friedberg, and R. D. Kornberg. 1995. Different forms of TFIIH for transcription and DNA repair: holo-TFIIH and a nucleotide excision repairosome. *Cell* **80:** 21–28.
85. Svejstrup, J. Q., W. J. Feaver, J. Lapointe, and R. D. Kornberg. 1994. RNA polymerase transcription factor IIH holoenzyme from yeast. *J. Biol. Chem.* **269:** 28,044–28,048.
86. Tanaka, K., K. Y. Kawai, Y. Kumahara, M. Ikenaga. 1981. Genetic complementation groups in Cockayne syndrome. *Somatic Cell Genet.* **7:** 445–456.
87. Tassan, J.-P., S. J. Schultz, J. Bartek, and E. A. Nigg. 1994. Cell cycle analysis of the activity, subcellular localization, and subunit composition of human CAK (CDK-activating kinase). *J. Cell. Biol.* **127:** 467–478.
88. Tong, X., R. Drapkin, D. Reinberg, and E. Kieff. 1995. The 62- and 80-kDa subunits of transcription factor IIH mediate the interaction with Epstein-Barr virus nuclear protein 2. *Proc. Natl. Acad. Sci. USA* **92:** 3259–3263.
89. Troelstra, C., A. van Gool, J. de Wit, W. Vermeulen, D. Bootsma, and J. H. J. Hoeijmakers. 1992. ERCC6, a member of a subfamily of putative helicase, is involved in Cockayne's syndrome and preferential repair of active genes. *Cell* **71:** 939–953.
90. Valay, J. G., M. Simon, M. F. Dubois, O. Bensaude, C. Facca, and G. Faye. 1995. The KIN28 gene is required both for RNA polymerase II mediated transcription and phosphorylation of the Rpb1p CTD. *J. Mol. Biol.* **249:** 535–544.
91. van Vuuren A. J., W. Vermeulen, L. Ma, G. Weeda, E. Appeldoorn, N. G. J. Jaspers, A. J. van der Eb, D. Bootsma, J. H. J. Hoeijmakers, S. Humbert, L. Schaeffer, and J.-M. Egly.

1994. Correction of xeroderma pigmentosum repair defect by basal transcription factor BTF2 (TFIIH). *EMBO J.* **13:** 1645–1653.
92. Wang, Z., S. Buratowski, J. Q. Svejstrup, W. J. Feaver, X. Wu, R. D. Kornberg, T. F. Donahue, and E. C. Friedberg. 1995. Yeast TFB1and SSL1 genes, which encode subunits of transcription factor IIH (TFIIH), are required for nucleotide excision repair and RNA polymerase II transcription. *Mol. Cell. Biol.* **15:** 2288–2293.
93. Wang, X. W., H. Yeh, L. Schaeffer, R. Roy, V. Moncollin, J.-M. Egly, Z. Wang, E. C. Friedberg, M. K. Evans, B. G. Taffe, V. A. Bohr, G. Weeda, J. H. J. Hoeijmakers, K. Forrester, and C. C. Harris. 1995. p53 modulation of TFIIH-associated nucleotide excision repair activity. *Nature Genet.* **10:** 188–195.
94. Wold, M. S. and T. J. Kelly. 1988. Purification and characterization of replication protein A, a cellular protein required for *in vitro* replication of simian virus 40 DNA. *Proc. Natl. Acad. Sci. USA* **85:** 2523–2527.
95. Xiao, H., A. Pearson, B. Coulombe, R. Truant, S. Zhang, J. L. Regier, S. J. Triezenberg, D. Reinberg, O. Flores, C. J. Ingles, and J. Greenblatt. 1994. Binding of basal transcription factor TFIIH to the acidic activation domains of VP16 and p53. *Mol. Cell. Biol.* **14:** 7013–7024.
96. Yamaizumi, M. and T. Sugano. 1994. UV-induced nuclear accumulation of p53 is evoked through DNA damage of actively transcribed genes independent of the cell-cycle. *Oncogene* **9:** 2775–2784.
97. Yankulov, K., K. Yamashita, R. Roy, J.-M. Egly, and D. L. Bentley. 1995. The transcriptional elongation inhibitor 5,6-dichloro-1-β-D-ribofuranosylbenzimidazole inhibits transcription factor IIH-associated protein kinase. *J. Biol. Chem.* **270:** 23,922–23,925.
98. Yankulov, K. Y., M. Pandes, S. McCracken, D. Bouchard, and D. Bentley. 1996. TFIIH functions in regulating transcriptional elongation by RNA polymerase II in *Xenopus* oocytes. *Mol. Cell. Biol.* **16:** 3291–3299.
99. Yee, A., M. A. Nichols, L. Wu, F. L. Hall; R. K. Kobayashi, and Y. Xiong. 1995. Molecular cloning of CDK7-associated human MAT1, a cyclin-dependent kinase-activating kinase (CAK) assembly factor. *Cancer Res.* **55:** 6058–6062.
100. Zawel, L., K. P. Kumar, and D. Reinberg. 1995. Recycling of the general transcription factors during RNA polymerase II transcription. *Genes Dev.* **9:** 1479–1490.

11

Mammalian DNA Repair and the Cellular DNA Polymerases

Samuel H. Wilson and Rakesh K. Singhal

1. INTRODUCTION

Genomic DNA is damaged by various physical and chemical agents during the life of an organism. For faithful reproduction and preservation of genomic DNA, damaged DNA sites must be repaired, and organisms have several DNA repair pathways that are vital in maintaining genome stability. Each individual DNA repair pathway (i.e., error-free repair, base excision repair, nucleotide excision repair, and so forth) is generally similar throughout nature, and this concept of conservation has been invaluable in rapidly advancing the mammalian DNA repair field. Several examples of DNA polymerase-independent "error-free repair" are well known, including photolyase reversal of UV damage and methyltransferase reversal of alkylation damage. In the various excision repair pathways, all of which involve some form of DNA polymerase-mediated gap-filling, the damaged site in DNA is first recognized and excised, the excision gap is tailored to allow gap-filling DNA synthesis, the nucleotide sequence is restored through DNA synthesis, and finally the phosphodiester backbone is ligated. Excision repair of damaged DNA is, therefore, a sequential multistep process. The specific enzyme(s) involved in an individual step may depend on the type of DNA lesion being repaired and the DNA structure/sequence context surrounding the lesion. Since resynthesis of a DNA sequence after excision of damaged DNA is catalyzed by a DNA polymerase, these enzymes clearly play a central role in DNA repair and hence in genomic stability.

Detailed information about the field of DNA polymerase biochemistry has been comprehensibly addressed in several recent reviews *(19,50,69,78,80,81,124,142,150)*. In this chapter, we discuss the involvement of DNA polymerases in mammalian DNA repair and their specific role(s) in individual DNA repair pathway(s). Our main emphasis will be on studies in the past several years.

Five distinct DNA polymerases have been recognized thus far in mammalian cells. Four of these DNA polymerases, designated α, β, δ, and ε, are considered nuclear in function, whereas the fifth enzyme, designated DNA polymerase γ, is localized mainly in mitochondria and is considered responsible for mitochondrial DNA replication *(69)*. Cloning of full-length cDNAs for each of these enzymes has been accomplished, and it

From: DNA Damage and Repair, Vol. 2: DNA Repair in Higher Eukaryotes
Edited by: J. A. Nickoloff and M. F. Hoekstra © Humana Press Inc., Totowa, NJ

is clear that they are products of distinct nuclear genes. From inhibitor studies and other approaches, roles of the DNA polymerases in DNA replication have been tentatively assigned: Pol α synthesizes primers for both leading and the lagging strand replication; a primase activity is uniquely associated with pol α. Pol δ is responsible for leading strand replication. There is no definite consensus on the role of pol ε. Yet, it is suggested to play a role in lagging strand synthesis; some studies also point to a pol ε role in leading-strand synthesis, whereas other studies point to a role in DNA repair. Pol β is considered to be a base excision repair enzyme. In addition, pol β may also play a role in DNA replication by gap-filling between Okazaki fragments. Intense research has not yet been devoted to the question of the biological role of pol γ, because it is assumed to conduct both replicative DNA synthesis and repair synthesis in mitochondria *(80)*.

The rate of repair of damaged DNA in most biological systems appears to be biphasic, where an initial rapid phase is followed by a much slower phase *(15)*. Several hypotheses have been suggested regarding the reason for the biphasic nature of DNA repair, including: the presence of more than one DNA repair pathway *(112)*, differential access to damaged sites by the DNA repair enzymes as a result, for example, of chromatin structure *(72)*, and preferential repair of certain DNA sites as, for example, in actively transcribing genes *(88)*. Historically, two excision repair pathways, called "long" patch repair (more than 20 nt incorporations during gap-filling DNA synthesis to fill the excision patch) and "short" patch repair (<10 nt incorporations) have been recognized, and recently these excision pathways have become well understood. The excision repair pathways are now classified in three general types termed base excision repair (BER), nucleotide excision repair (NER), and mismatch repair (MMR). Since the individual repair pathways are dealt with in detail in other sections of this book (*see* Chapters 7, 8, 18, and 20), our focus here will be on studies of the roles of the DNA polymerases in the repair pathways.

2. MAMMALIAN NUCLEAR DNA POLYMERASES

2.1. *DNA Polymerase α*

The first mammalian DNA polymerase described and characterized in detail *(69)*, DNA polymerase α (pol α) is localized in the nucleus *(11)* and involved in semiconservative DNA replication. Pol α activity is most abundant in rapidly growing cells, with much less expression in quiescent cells, and its expression is regulated in a tissue-specific fashion. In addition to dNMP incorporation, this polymerase catalyzes both pyrophosphorolysis and pyrophosphate exchange, but does not have intrinsic exonuclease activities. Pol α is a multisubunit enzyme: Although extensively studied, there is still some uncertainty about the precise composition of pol α, since the enzyme is highly susceptible to proteolysis *(78)* and is associated with a high molecular mass complex. Nevertheless, it seems clear that in cultured mammalian cell systems the core pol α holoenzyme has molecular mass > 250 kDa and is composed of at least four subunits of ~180, ~70, ~60, and ~50 kDa, respectively *(4,143)*. The largest subunit ranges in size from 140 to 185 kDa depending on proteolysis *(78)* and carries the DNA polymerase activity, whereas the 60- and 50-kDa subunits carry the DNA primase activity *(52,143)*.

Although there is no exonuclease activity associated with mammalian pol α *(44)*, a weak 3' → 5' exonuclease activity is revealed on dissociation of the 182- and 73-kDa subunits of *Drosophila* pol α *(32,78)*. Therefore, nuclease deficiency is not a general feature of eukaryotic DNA polymerase α. Interestingly, mammalian pol α can coordinate with a heterologous proofreading exonuclease, suggesting that replication by this polymerase in the cell can coordinate with proofreading exonucleases without direct physical association between the polymerase and exonuclease activities. Thus, addition of the *Escherichia coli* DNA polymerase III, 3' → 5' exonuclease-proficient subunit ε to calf thymus pol α resulted in enhanced replication fidelity by pol α, even though no physical association between the two proteins was observed *(106)*.

The primase activity of pol α, which synthesizes short ribonucleotide primers, is inherent in the 50- and 60-kDa primase subunits, independent of their association with the other two pol α subunits *(143)*. The length of oligoribonucleotide primers synthesized by the pol α-primase complex is normally limited to approx 10 bases or less. After synthesis of a molecule of this size range, the enzyme complex switches from a primer synthesis to a DNA chain elongation mode, through an unknown mechanism. The ideal substrate for pol α is duplex DNA with gaps of 20–70 single-stranded nucleotides; the enzyme is relatively inactive on nicked DNA *(144)*. Various higher molecular mass DNA polymerase α complexes containing accessory biochemical activities, such as DNA-dependent ATPase, RNase H, 3' → 5' exonuclease, C_1 and C_2 primer recognition, proteins, and diadenosine 5'-tetraphosphate (Ap_4A) binding protein *(81,104,122)* have been reported to function in DNA replication. However, it is not clear if any of these complexes or associated activities are involved in DNA repair.

Pol α is considered to be cell cycle regulated in proliferating cells. During the activation of quiescent (G_0 phase) cells, mRNA levels, protein content, and enzyme activity increase just prior to the peak of DNA synthesis *(140)*. On the other hand, studies in human fibroblast have demonstrated that a large increase in pol α activity on growth stimulation results from increased phosphorylation of the enzyme rather than new protein synthesis *(33)*. A cDNA of the polymerase subunit of the human enzyme has been isolated *(152)* and the human gene mapped to the X chromosome at the region p21.3–22.1 *(145)*.

2.2. DNA Polymerase δ

Historically DNA polymerase δ (pol δ) was considered as a "pol α-like enzyme" that could be distinguished from pol α by physicochemical differences, template–primer use differences, and the presence of an associated 3' → 5' exonuclease activity *(21–23,51,73–76)*. Furthermore, pol α activity in vitro is strongly stimulated by PCNA (proliferating cell nuclear antigen), in terms of increased processivity when copying long single-stranded DNA stretches *(132)*, whereas pol α is not similarly stimulated. Pol δ has been purified from rabbit bone marrow, calf thymus, human placenta, HeLa cells, and *Drosophila* embryos *(4,21,23,73–75,129)*. For example, pol δ from calf thymus consists of two subunits of 125- and 48-kDa *(75)* and the catalytic subunit was identified as the 125 kDa moiety *(6)*. Although pol δ was purified in association with DNA primase in one case *(34)*, this latter enzyme activity has not generally been found associated with pol δ. This fact, along with the ability of pol δ to conduct highly processive DNA replication on single-strand DNA templates, led to the idea that pol δ functions in leading strand synthesis during semiconservative DNA replication. Indeed, during DNA synthesis in vitro from the SV40 replication origin, pol δ is known to conduct replication of the leading-strand template *(137)*, and is generally considered to perform this role during genomic DNA replication.

The human pol δ cDNA encodes a polypeptide of ~124 kDa, and the nuclear gene has been mapped to chromosome 19 *(30)*. Recently, the catalytic polypeptide of pol δ was overexpressed in bacteria. The purified recombinant protein was fully active and was found to exhibit enzymatic properties similar to those of the native protein, including 3' → 5' exonuclease activity and stimulation by PCNA *(156,157)*. In addition, the PCNA binding site of pol δ was localized to the N-terminal region of the 125 kDa catalytic subunit *(156,157)*. Transcriptional studies suggest that pol δ gene expression is regulated during the cell cycle at the G1/S boundary; pulse labeling of cells metabolically arrested in G1, S, and G2/M showed that pol δ is a phosphoprotein, most abundantly phosphorylated during S phase *(154,155)*.

2.3. *DNA Polymerase ε*

DNA polymerase ε (pol ε), another of the "pol α-like enzymes," historically was first established as clearly distinct from the pol α in *Saccharomyces cerevisiae* *(20,93,151)*. Mammalian pol ε consists of subunits of 215 and 55 kDa *(29,62)*, contains 5' → 3' exonuclease activity *(118)*, as well as 3' → 5' exonuclease activity *(102)*. Similarity in the sizes of the catalytic subunits of pol ε and pol α *(19,63,76)* and in their responses to PCNA initially caused some confusion in the literature. Pol ε was first isolated and purified to homogeneity from mammalian tissue as a unique polymerase with both 3' → 5' and 5' → 3' exonuclease activities *(28,29,62,102,118)*. Pol ε is a highly processive enzyme that does not require PCNA *(29)*, although pol ε was shown to be PCNA-responsive *(82)* at physiological salt concentrations *(20,77)*. Pol ε is involved in long-patch DNA repair *(103,117)* as well as DNA replication *(82,93)*. The cDNA encoding the catalytic polypeptide of human pol ε has been cloned *(64)* and the nuclear gene localized to human chromosome 12q 24·3 (131). Pol ε is transcriptionally regulated during cell proliferation at the G1/S boundary *(138)* in a manner typical of a replicative DNA polymerase, and links the DNA replication machinery to the S phase checkpoint *(99)*.

2.4. *DNA Polymerase β*

DNA polymerase β (pol β), which is the smallest naturally occurring polymerase *(7,26,148)* (~39 kDa), consists of an N-terminal 8-kDa domain and 31-kDa C-terminal domain linked by a protease-sensitive hinge region *(150)*. The enzyme is devoid of DNA primase and of 3' → 5' or 5' → 3' exonuclease activities *(7,26,148)*. Pol β was first recognized in the early 1970s and eventually distinguished from proteolytic fragments of the larger pol α-like enzymes by tryptic peptide mapping *(2,108)*. Historically, however, pol β was not included in the "pol α-like family" (α-δ-ε) of DNA polymerases *(26)*, since pol β is present in only low levels, is constitutively expressed, and is not regulated in a cell-cycle-specific manner; pol β had distinct physicochemical properties, such as its size, isoelectric point, and sulfhydryl reagent sensitivity, from those of pol α and the two pol α-like enzymes (pol δ and pol ε) *(83,148)*. The amino acid sequence of pol β shows very little similarity with sequences of the other DNA polymerases *(3)*. Yet, X-ray crystallography has shown that pol β has global structural features, such as a cleft-like groove, that are reminiscent of the other DNA polymerase

(113). Recently, deoxyribosephosphodiesterase (dRpase) activity was shown to be associated with the 8-kDa N-terminal domain of pol β *(85,96,97,107)* and DNA polymerase activity associated with the 31 kDa domain *(71)*. Pol β has both distributive and processive DNA polymerase properties, depending on the gap size in the DNA substrate: The enzyme requires the presence of a 5' phosphate group on the downstream DNA to fill a short gap (of 6 nt or less) in a processive manner, but is distributive when filling longer gaps or replicating single-stranded templates *(120)*. DNA synthesis by pol β is relatively error-prone in vitro, compared with the other cellular DNA polymerases *(10,105)*. Since pol β prefers DNA with short gaps as in vitro substrates *(120,144)*, pol β has generally been assigned a role in short-patch DNA repair synthesis *(94,95,150)*, and the constitutive expression pattern of pol β is consistent with this hypothesis. In addition, there are reports indicating a role for pol β in DNA replication in some systems *(61,128)*.

The genes for pol β from both human and rodent sources have been cloned *(1,116,158,159)*. The human pol β gene has 14 exons and is single-copy with location at chromosome 8p11-p12 *(24,87)*. Gene expression of mammalian pol β genes studied to date is upregulated by monofunctional DNA alkylating agent treatment of cells *(98)* and by oxidative stress-inducing agents *(27)*.

2.5. *DNA Polymerase Accessory Factors*

Efforts to identify the mammalian cellular DNA polymerases and their function(s) in vivo led to the discovery of several important DNA polymerase auxiliary proteins. PCNA, for example, is a 36-kDa nuclear protein required for SV40 DNA replication in vitro. PCNA is cell-cycle regulated and is a necessary cofactor for processive DNA synthesis by pol δ *(132)*. It has been suggested that PCNA binds to DNA, by encasing it through forming a donut-shaped structure, in the presence of pol δ and/or replication factor C *(9,135)*.

Replication factor C (RF-C) is a multisubunit factor made up of a 50-kDa subunit that binds primer-template, a 41-kDa subunit that binds ATP, and a 37-kDa subunit. RF-C has a DNA-dependent ATPase activity and increases primer recognition by pol α in an in vitro replication system *(135,147)*. PCNA and RF-C are considered to be functional equivalents of the products of the T4 genes 45 and 44/62, respectively, which together with T4 DNA polymerase conduct DNA replication in vitro *(84,137)*.

Replication factor A (RF-A), also known as human single-stranded DNA-binding protein (HSSB), acts as an auxiliary protein for both pol α and δ for improved processivity and efficiency *(136,147)*. RF-A contains subunits of 70, 34, and 10-kDa that are analogous to yeast RF-A subunits of 69, 39, and 13-kDa, respectively *(16)*. Topoisomerases and DNA helicases also stimulate pols α, δ, and ε and are required for in vitro replication systems. The only known stimulatory factor for pol β is DNase V, which helps mediate a nick translation reaction *(111)*.

2.6. *Inhibition of DNA Polymerases*

In prokaryotes, convincing results have been obtained by creating mutations and/or gene deletions to ascertain cellular function(s) of enzymes. However, such studies are cumbersome in mammalian cells and the more challenging approach of using "specific" inhibitory reagents has been widely adopted. Use of such inhibitors in complex biological systems can result in problems with interpretation, especially when multiple enzymes of overlapping functions are involved. Despite such concerns about inhibitor studies, they have provided important insights about the roles of various mammalian DNA polymerases in DNA replication and repair *(50,57,62,69,123,142,153)*.

The simplest approach to the use of inhibitors for deducing function of an enzyme in vivo is to first demonstrate specific and complete inhibition of enzyme catalytic activity, both in vitro and in vivo. To infer the role of a particular DNA polymerase, inhibitor studies are done initially with purified enzymes, and then followed by studies of either whole-cell extracts or permeable cells using various concentrations of the inhibitor. Aphidicolin (an antibiotic isolated from the fungus *Cephalosporium aphidicola*), *N*-ethylmaleimide (NEM), and arabiofuranosyul nucleotide analogs (e.g., ara-CTP) effectively inhibit pol α, δ, and ε but inhibit pol β only weakly *(5,51,57,58,126,142)*. In addition, deoxynucleoside-5'-triphosphate (dNTP) analogs of usual DNA polymerase substrates have been employed to distinguish between different DNA polymerases. For example, butylphenol-dGTP and butylanilino-dATP inhibit pol α and pol δ at ~1 and 100 μ*M* concentrations, respectively *(22,66,123,141)* and do not inhibit pol δ. Pol β is preferentially inhibited by the 2',3'-dideoxynucleoside-5'-triphosphates (ddNTPs).

Recently, several new inhibitors have been identified for DNA polymerase studies. Carbonyldiphosphate acts as a competitive inhibitor of pol δ and ε, with four- to six-fold selectivity compared to pol α *(153)*. Two 2'-deoxynucleoside 5'-α-methyleno-phosphanyl-β, γ-diphosphates are incorporated into the DNA chain by pol α, but did not act as substrates for pol δ and ε from human placenta *(60)*. Acyclovir, ganciclovir, and penciclovir triphosphates preferentially inhibit pol δ (especially ganciclovir), although they also affect some pol α and δ enzymes *(59)*. Pol γ differs from pols α, β, δ, and ε in its ability to effectively incorporate a dTTP analog fialuridine 5'-monophosphate *(79)*. Finally, an important approach to specific inhibition of DNA polymerases has been use of enzyme-specific neutralizing and nonneutralizing antibodies. Such enzyme-specific antibodies are available for pol α, δ, and β *(76,121,127,133)*.

3. DNA REPAIR AND THE DNA POLYMERASES

3.1. Base Excision Repair

The base excision repair (BER) pathways are found throughout nature, presumably to preserve individual bases and sugars in genomic DNA. Various endogeneous cellular processes resulting in DNA lesions that are repaired by BER include the following: spontaneous base loss forming the AP site, deamination of cytosine, 5-methylcytosine, and adenine forming uracil, thymine, and hypoxanthine, respectively, strand breaks from oxidative stress, and reactions leading to oxidation or alkylation of bases and sugars. Cellular exposure to DNA damaging agents, especially monofunctional alkylating agents, is also a source of BER repaired lesions. If uncorrected, these DNA lesions lead to genomic instability and ultimately to deleterious effects for the cell.

Recent studies indicate the presence of two general pathways for BER in mammalian cells, characterized by the DNA polymerase, involving: single-nucleotide and short-gap (2–6 nt) BER, both of which are dependent on pol β, and short-gap (<10 nt) BER that is pol δ/PCNA-dependent.

3.2. *Multiple Pathways for Mammalian Base Excision Repair*

Single-nucleotide BER is a multistep pathway initiated by the AP site, created through spontaneous base loss or by enzymatic removal of a base. Cells contain several DNA glycosylases dedicated for different types of base bearing lesions, inferring lesion specificity to imitation of the BER pathway. Once an apurinic/apyrimidic (AP) site is created in double-stranded DNA, AP-endonuclease cleaves one strand of DNA immediately strand 5' to the AP site. The remaining deoxyribose phosphate residue is removed by the dRpase activity of pol β creating a single-nucleotide gap, which in turn is filled by pol β. Finally, the strand is sealed by DNA ligase I *(109)*. Based on reconstitution studies with purified mammalian enzymes, the combination of XRCC1 and either DNA ligase III *(67,70)* or DNA ligase I *(109)* can seal the intermediate nicked DNA in single-nucleotide BER, and also in BER with a gap size of ~2–6 nt *(67)*. This latter gap size may correspond to repair of 3'-OH-blocked gaps (e.g., 3'-phosphate) or 5'-blocked gaps (i.e., oxidized or reduced AP site sugar) *(67)*. Processing of such 2–6 nt gaps depends on the activity of "flap endonuclease" (i.e., FEN-1), and as for single-nucleotide gaps, the gap-filling DNA synthesis activity of DNA polymerase β *(67)*; the 2–6 nt gap size corresponds to the in vitro substrate specificity for pol β *(120)*.

The pol δ/PCNA BER pathway was first recognized in *Xenopus laevis* oocyte extracts and has been demonstrated to occur in mammalian cell as well *(13,45,49,100)*. This BER pathway appears to be a backup pathway to the single-nucleotide BER pathway *(13,45)*. However, the relative contribution of the two BER pathways may depend on the gene expression pattern of the cell, sequence context of the lesion, and the global context of the DNA segment *(13)*. The pol δ/PCNA-dependent pathway involves excision gaps of approx 10 nt or less *(48)*.

3.3. *Historical Perspective and Complexities*

Initial studies to establish a role of the DNA polymerases in DNA repair synthesis were conducted using inhibitors. These studies gave insight into the roles of the various polymerases in DNA repair synthesis, and conclusions drawn from the earlier inhibitor studies have been reevaluated and updated periodically *(62)*. As noted earlier, studies had revealed that aphidicolin, arabinofuranosyl nucleotide analogs, and NEM selectively inhibited mammalian pols α, δ, and ε, whereas ddNTPs selectively inhibited pol β *(50)*. DNA repair synthesis in human fibroblasts responding to bleomycin and certain other DNA-damaging agents was inhibited by ddTTP, whereas inhibitors of pols α, δ, and ε were relatively ineffective *(25,43,89,90)*, suggesting a role of pol β in DNA repair synthesis. These results also supported the view that more than one DNA repair pathway may function with any single type of DNA-damaging agent, since bleomycin, for example, produces DNA-strand breaks in addition to base lesions that leave a DNA strand with deoxyribose at 3'-terminus *(18)*; these termini cannot serve as primers for DNA synthesis in BER. However, pol β, in combination with an exonuclease to generate appropriate primers, can carry out repair synthesis in vitro *(94,115)*.

Involvement of various polymerases in a specific type of damage repair may also depend on the dose of DNA-damaging agent. Permeable human fibroblasts at low doses of methyl nitrosourea (MNU), bleomycin, *N*-2-acetyl-2-aminofluorene (AAF), or UV were more resistant to aphidicolin than were fibroblasts exposed to high doses *(42)*.

Similar observations were made using UV-damaged cells *(123)*. These results suggested that pol β is a key DNA polymerase for repair synthesis at lower doses of genotoxic agent, whereas an aphidicolin-sensitive polymerase(s) (pol α and pol α-like enzymes) may participate at higher doses of the same agent. However, there also is a report *(65)* indicating that even at low doses of UV, repair synthesis that was resistant to aphidicolin was also resistant to ddTTP, arguing against involvement of pol β. Some of these conflicting observations can possibly be explained by kinetic arguments (concerning substrate DNA template and dNTPs), as well as the idea that available amounts of soluble polymerase(s) may affect the degree of inhibition. This point on conflicting observations is illustrated by studies of ddTP inhibition of UV-repair synthesis in permeable cells: In two separate studies, 80% and 20% inhibition, respectively, were observed with ddTTP *(31,42)*. This discrepancy may be because of different dTTP concentrations used in the assays, as well as the possibility that DNA polymerases have different kinetic parameters for DNA synthesis during replication and repair *(38,43)*. Also, other factors, such as metal ions, phyrophosphate, and accessory proteins, may influence interpretations. Supported by a variety of inhibitor studies *(37,39–42,89,123)* and by in vitro gap-filling studies *(50,95,119,144)*, mammalian pol β has been suggested to play a role in short patch repair synthesis.

3.4. Recent Advances

Progress has been made in the past several years in further identifying the polymerase(s) involved in the DNA synthesis step in BER. Pol β was implicated in BER synthesis of G/T mispairs in human cell extracts; this lesion results from spontaneous deamination of 5'- methylcytosine *(149)*. In addition to inhibitor studies, specific antibody against pol β was used to show inhibition of BER synthesis in vitro *(149)*. A role for pol β in repair synthesis was also suggested from inhibitor studies using mammalian cell extracts and a defined substrate containing a uracil residue opposite G *(149)*. Although these studies suggested a role for pol β in BER, direct evidence was still lacking.

To identify DNA polymerase(s) required at the DNA synthesis step of BER, Singhal et al. *(121)* conducted a series of in vitro studies using bovine testis nuclear extract. Although many earlier studies (as noted in Section 3.2.) had pointed to pol β involvement during repair synthesis in BER, results showing pol β to be a processive enzyme on gaps of ≤6 nt *(120)* prompted further experiments to define the role of this enzyme in short gap-filling repair synthesis. Using a defined 51-bp substrate with a single uracil opposite G at position 22 from the 5'-end, a robust in vitro repair reaction was demonstrated for removal of U and incorporation of radiolabeled C using the bovine testis nuclear extract. Complete inhibition of this repair reaction by ddTTP and by neutralizing polyclonal antibodies against pol β were observed *(121)*. In addition, the AP-endonuclease-gap tailoring activity could be physically separated from the polymerase and ligase activities, and using a fraction devoid of all polymerase activity, Singhal et al. *(121)* demonstrated that, among the cellular DNA polymerases tested, only pol β could reconstitute the single-nucleotide gap-filling DNA synthesis step.

Although these in vitro studies seemed to establish a role for pol β in single-nucleotide BER, the implications of deregulation of pol β are being further examined by genetic studies. Embryonic mouse fibroblast cell lines homozygous for a deletion in the pol β gene have been established *(125)*. The pol β-deleted cell lines have normal viability. Extracts from these cell lines are deficient in uracil-initiated single-nucleotide BER. The cells exhibit mild hypersensitivity to monofunctional DNA-alkylating agents (e.g., *N*-methyl-*N*'-nitro-*N*-nitrosoguanidine [MNNG] and methyl methanesulfonate [MMS]), but not to several other DNA-damaging agents (e.g., UV). Both the deficiency in BER in vitro and the hypersensitivity to DNA-alkylating agents were rescued by stable transfection with a wild-type pol β expression vector *(125)*. These studies, therefore, suggest that cellular sensitivity to DNA-alkylating agents is linked to single-nucleotide BER in vivo.

Pol β is not the only DNA polymerase functioning in gap-filling synthesis in BER in eukaryotes. As noted above, the repair of abasic sites in *X. laevis* oocytes has been shown to occur by two distinct pathways, one dependent on pol β and the other dependent on pol δ/PCNA *(86)*. Pol β is functional only on natural AP-site repair, whereas the pol δ/PCNA pathway can act on a reduced sugar AP site, as well as on natural AP-sites *(67)*. A similar picture exists in mammalian cells *(13)*. The picture in yeast (*S. cerevisiae*), however, is even more complicated. The presence of pol β in yeast has been recently established *(110)*. Yet, an earlier in vitro study, using nuclear extracts from pol α, δ, and ε temperature-sensitive mutants, demonstrated that repair synthesis in yeast is catalyzed by pol ε and/or pol δ (pol ε/δ) *(146)*. Later, studies with *S. cerevisiae* nuclear extracts from wild-type cells and strains carrying a pol β gene deletion were conducted to examine the role of yeast pol β in BER *(121)*. Inhibitor and antibody studies, as well as reconstitution and genetic studies, confirmed the earlier finding *(146)* and supported a role of pol ε/δ. In another study, pol δ appeared to be the BER gap-filling DNA polymerase enzyme in *S. cerevisiae*, as judged from analysis of *cdc2* mutants that are alkylation-sensitive when transiently incubated at the non-permissible temperature *(14)*.

4. NUCLEOTIDE EXCISION REPAIR PATHWAY

Mammalian cells have two types of very closely related nucleotide excision repair (NER) pathways. One pathway is termed transcription-coupled NER and the other is termed global NER. The NER pathways fall into the "long" patch repair categories since the DNA damage site is removed as part of an oligonucleotide fragment. Interestingly, the length of the excised oligonucleotide fragment differs somewhat in prokaryotes (~12–13 nt) and eukaryotes (~28–30 nt). NER involves the products of many genes in mammalian cells and understanding the enzymology of NER represents an interesting biochemical challenge. NER has broad DNA lesion specificity, recognizing a wide variety of DNA lesions that are thought to distort the structure of DNA. In humans, NER deficiencies are associated with xeroderma pigmentosium and other disorders, such as trichothisdystrophy and Cockayne syndrome *(46,47)*.

Briefly, the initial step in the mammalian NER pathway is recognition of DNA damage by the XPA protein in association with RPA, which in turn recruits transcription factor TFIIH. Next, ERCC1-XPF and XPG cut the DNA on either side of the damage. This is followed by release or excision of the damaged oligonucleotide fragment containing the lesion, and then by DNA synthesis to fill the excision gap. Finally, joining of the nick by DNA ligase completes the pathway. NER has been an intensively studied pathway, and it is discussed in Chapters 10 and 18. We will concentrate here on discussion of recent progress concerning involvement of DNA polymerases at the DNA synthesis step of the pathway.

Exposure of cells to UV-irradiation results in two major products, T-T and C-T dimers *(8)*, and these photoproducts are thought to be repaired predominately via NER *(54)*. Repair synthesis in UV-irradiated HeLa cells, under conditions that minimized replicative synthesis, is inhibited by aphidicolin *(53)*. Aphihdicolin also inhibits repair synthesis in response to *N*-acetoxy-2-acetylaminofluorene (AAAF) *(12,31,42,53,90)*, with 80–90% inhibition of repair at saturating levels of inhibitor. In contrast, an inhibitor of pol β was required for inhibition of BER-mediated MNNG-induced repair synthesis *(89,90)*. These results point to a role of aphidicolin-sensitive DNA polymerase(s) in DNA repair synthesis in NER, and initially this sensitivity was thought to reflect inhibition of pol α , since pol α was the only known replicative polymerase at the time. With the discovery of pol δ and pol ε and of their sensitivity of aphidicolin, pol δ has been implicated in UV-damage repair *(38)*. In addition, studies using butylphenyl dGTP are consistent with a role of either pol δ or pol ε as the major UV-repair polymerase. Finally, data from permeable cell complementation studies *(103,130)* provided direct evidence for involvement of pol ε, and not pols α or β, in UV-repair synthesis. As discussed earlier (*see* Section 3.3.), other factors, such as damage dose, type of DNA damage, cofactors, and different kinetic parameters of various polymerases in replication and repairs, have confounded assignment of a role of the different DNA polymerases in NER.

Recently, attempts have been made to identify the polymerase responsible for DNA synthesis in the UV-irradiated DNA damage repair using antibody and complementation studies. Antibodies specific to pol δ or PCNA markedly inhibit the capacity of HeLa nuclear extracts to effect repair of UV-damaged plasmid DNA *(155)*. In addition, mRNA levels of both pol δ and PCNA were significantly stimulated subsequent to UV-irradiation in the same system; this is in keeping with a role of pol δ in this type of DNA repair synthesis. On the other hand, pol ε alone could fill the repair patch when UV-damaged DNA was preincubated with PCNA-depleted human cell extracts to create repair incision *(117)*. Upon addition of RPA, synthesis by Pol ε became dependent on both PCNA and RPC. Pol δ could perform very low repair synthesis in the presence of PCNA and RPC, but did not require RPA. These results taken together suggest that pol δ and pol ε, in combination with PCNA, RPC, and RPA, are responsible for NER synthesis in mammalian cells.

In contrast to BER, a similar picture is emerging in yeast and human cells with respect to involvement of polymerases in NER. Based on mutational studies, both pol ε and δ seem to be major polymerases responsible for repair synthesis in NER *(17)*. In *E. coli*, Pol I remains the major repair synthesis enzyme, both in BER and NER *(114)*.

5. MISMATCH REPAIR PATHWAY

The mismatch repair (MMR) pathway corrects base mispairs arising in genomic DNA during replication, genetic recombination, and as a result of damage to DNA. In addition, MMR appears to play a role in regulating recombination and parts of the cell-cycle check point system. MMR has been well studied in *E. coli (91)* and involves the proteins known as MutS, MutL, and MutH. The fundamental steps of MMR appear to be similar in prokaryotes and eukaryotes *(56)*. Here we will discuss only recent progress in identifying roles of the different polymerase(s) in MMR repair synthesis.

The methyl-directed "very long" patch MMR pathway *(92)* is uniquely suited to repair of DNA replication errors. In *E. coli*, DNA is methylated at GATC sites by the Dam methylase. However, after DNA replication the daughter strand is transiently unmethylated. MMR specifically repairs the unmethylated daughter strand and, hence, improves overall DNA replication fidelity. In *E. coli*, MMR is initiated by binding of MutS protein to a mismatch followed by binding of MutL, which is thought to increase the stability of the MutS-DNA complex and is required to activate MutH. MutH then nicks the unmethylated strand of DNA at heminethylated GATC sites, which can be located more than 1 Kb from the MutS binding site. Hemimethylated sites either 5' or 3' of the base mispair can be utilized by this repair system. Excision requires UvrD (helicase II) and one of the single-stranded DNA exonulceases—Exo I (3' exo), Exo VII (3' and 5' exo), or Rec J (5' exo)—depending on whether the nicked unmethylated site is 5' or 3' to the mispair. Once excision has occurred, resynthesis is mediated by DNA pol III holoenzyme, SSB, and DNA ligase *(68)*.

In general, MMR in human cell extracts shares many features with the *E. coli* system *(91)*. All possible mismatched base pairs are repaired in a strand-specific manner, although the precise signaling mechanism for strand-discrimination is unknown. In vitro strand discrimination is achieved artificially by nicking of the substrate DNA. Human MMR can be bidirectional, and excision and resynthesis occur between the nick and the mismatch. Despite these similarities, more than one gene with homology to both MutS and MutL have been identified in eukaryotes. There are only a few studies in the literature related to identifying the polymerase involvement in MMR. However, inhibitor studies using aphidicolin have consistently shown the involvement of aphidicolin-sensitive or pol α class enzyme (pol α, δ, and ε) in MMR *(56,134)*.

Recently, physical interaction between mismatch repair proteins and PCNA has been reported *(139)*, pointing to a role of pol δ or pol ε in mismatch and post-replication repair. Finally, mutations in the ExoIII box of human pol δ *(35)* suggest that replication errors resulting from the nonfunctional proofreading activity of pol δ can lead to cancer *(36)*.

6. CONCLUDING REMARKS

It has been a challenge to delineate roles of the five known mammalian cellular DNA polymerases in the various DNA repair pathways. Judging from information available to date, the situation within any given repair pathway is emerging to be much more complex than previously thought. Although various approaches have been used to identify roles of DNA polymerases in repair, results from in vivo studies with mutant DNA polymerases are required. Results of such genetic studies in *E. coli* and *S. cerevisiae* suggest some degree of overlap of polymerase function in DNA repair synthesis. In mammalian systems, the situation has been complicated by difficulties in conducting genetic studies. Yet, overlap in polymerase function most likely exists. Clearly, biologically relevant in vitro repair systems with inhibitors, neutralizing antibodies, and biochemical complementation, in combination with results of in vivo genetic studies will provide the best evidence. Studies assigning pol β in BER are an example of combining such in vitro studies *(121)* and in vivo genetic studies *(125)* in a mammalian system.

Finally, it is known that a number of human hereditary diseases result from defects in DNA repair and much progress has been made through studies of cell lines from affected individuals. Cells from some individuals with genetic susceptibility for cancer may contain altered DNA polymerases *(36)* or have altered polymerase function through modifications in accessory proteins. The roles of the mammalian DNA polymerases in repair will remain an active area of research in the fast moving field of genome DNA integrity.

REFERENCES

1. Abbotts, J., D. N. SenGupta, B. Zmudzka, S. G. Widen, B. Notario, and S. H. Wilson. 1988. Expression of human DNA polymerase beta in *Escherichia coli* and characterization of the recombinant enzyme. *Biochemistry* **27:** 901–909.
2. Albert, W., F. Grummt, U. Hubscher, and S. H. Wilson. 1982. Structural homology among calf thymus α-polymerase polypeptides. *Nucleic Acids Res.* **10:** 935–946.
3. Anderson, R. S., C. B. Lawrence, S. H. Wilson, and K. L. Beattie. 1987. Genetic relatedness of DNA polymerase beta and terminal deoxynucleotidyltransferase. *Gene* **60:** 163–173.
4. Aoyagi, N., S. Matsuoka, A. Furunobu, A. Matsukage, and K. Sakaguchi. 1994. Drosophila DNA polymerase delta. Purification and characterization. *J. Biol. Chem.* **269:** 6045–6050.
5. Bambara, R. A. and C. B. Jessee. 1991. Properties of DNA polymerases δ and ε, and their roles in eukaryotic DNA replication. *Biochim. Biophys. Acta* **1088:** 11–24.
6. Bambara, R. A., T. W. Myers, and R. D. Sabatino. 1990. DNA polymerase δ, in *The Eukaryotic Nucleus: Molecular Biochemistry and Macromolecular Assemblies* (Strauss, P. and S. Wilson, eds.), Telford Press, Caldwell, NJ, pp. 69–94.
7. Baril, E. F., O. E. Brown, M. D. Jenkins, and J. Laszlo. 1971. Deoxyribonucleic acid polymerase with rat liver ribosomes and smooth membranes, Purification and properties of the enzymes. *Biochemistry* **10:** 1981–1992.
8. Basu, A. K. and J. M. Essigmann. 1988. Site-specifically modified oligodeoxynucleotides as probes for the structural and biological effects of DNA-damaging agents. *Chem. Res. Toxicol.* **1:** 1–18.
9. Bauer, G. A. and P. M. Burgers. 1988. Protein-protein interactions of yeast DNA polymerase III with mammalian and yeast proliferating cell nuclear antigen (PCNA)/cyclin. *Biochim. Biophys. Acta* **951:** 274–279.
10. Beard, W. A., W. P. Osheroff, R. Prasad, M. Jaju, M. R. Sawaya, T. G. Wood, J. Kraut, T. A. Kunkel, and S. H. Wilson. 1996. Enzyme-DNA interactions required for efficient nucleotide incorporation and discrimination in human DNA polymerase β. *J. Biol. Chem.* **271:** 12,141–12,144.
11. Bensch, K. G., S. Tanaka, S. Z. Hu, T. S. Wang, and D. Korn. 1982. Intracellular localization of human DNA polymerase α with monoclonal antibodies. *J. Biol. Chem.* **257:** 8391–8396.
12. Berger, N. A., K. K. Kurohara, S. J. Petzoid, and G. W. Sikorski. 1979. Aphidicolin inhibits eukaryotic DNA replication and repair—implications for involvement of DNA polymerase α in both processes. *Biochem. Biophys. Res. Commun.* **89:** 218–225.
13. Biade, S., R. W. Sobol, S. H. Wilson, and Y. Matsumoto. Impairment of proliferating cell nuclear antigen (PCNA)-dependent base excision repair on linear DNA. Submitted.
14. Blank, A., B. Kim, and L. A. Loeb. 1994. DNA polymerase δ is required for base excision repair of DNA methylation damage in *Saccharomyces cerevisiae. Proc. Natl. Acad. Sci. USA* **91:** 9047–9051.
15. Bohr, V. A., D. H. Phillips, and P. C. Hanawalt. 1987. Heterogeneous DNA damage and repair in the mammalian genome. *Cancer Res.* **47:** 6426–6436.

16. Brill, S. J. and B. Stillman. 1989. Yeast replication factor-A functions in the unwinding of the SV40 origin of DNA replication. *Nature* **342:** 92–95.
17. Budd, M. E. and J. L. Campbell. 1995. DNA polymerases required for repair of UV-induced damage in *Saccharomyces cerevisiae. Mol. Cell. Biol.* **15:** 2173–2179.
18. Burger, R. M., A. R. Berkowitz, J. Peisach, and S. B. Horowitz. 1980. Origin of malondialdehyde from DNA degraded by Fe(II)-bleomycin. *J. Biol. Chem.* **255:** 11,832–11,838.
19. Burgers, P. M. 1989. Eukaryotic DNA polymerases α and δ: conserved properties and interactions, from yeast to mammalian cells. *Prog. Nucleic Acid Res. Mol. Biol.* **37:** 235–280.
20. Burgers, P. M. 1991. *Saccharomyces cerevisiae* replication factor C. II. Formation and activity of complexes with the proliferating cell nuclear antigen and with DNA polymerases δ and ε. *J. Biol. Chem.* **266:** 22,698–22,706.
21. Byrnes, J. J. 1984. Structural and functional properties of DNA polymerase δ from rabbit bone marrow. *Mol. Cell. Biochem.* **62:** 13–24.
22. Byrnes, J. J. 1985. Differential inhibitors of DNA polymerases α and δ. *Biochem. Biophys. Res. Commun.* **132:** 628–634.
23. Byrnes, J. J., K. M. Downey, B. L. Black, and A. G. So. 1976. A new mammalian DNA polymerase with 3' to 5' exonuclease activity: DNA polymerase δ. *Biochemistry* **15:** 2817–2823.
24. Cannizzaro, L. A., F. J. Bollum, K. Huebner, C. M. Croce, L. C. Cheung, X. Xu, B. K. Hecht, F. Hecht, and L. M. S. Chang. 1988. Chromosome sublocalization of a cDNA for human DNA polymerase β to 8p11–p12. *Cytogenet. Cell. Genet.* **47:** 121–124.
25. Castellot, J. J., Jr., M. R. Miller, D. M. Lehtomaki, and A. B. Pardee. 1979. Comparison of DNA replication and repair enzymology using permeabilized baby hamster kidney cells. *J. Biol. Chem.* **254:** 6904–6908.
26. Chang, L. M. S. and F. J. Bollum. 1971. Low molecular weight deoxyribonucleic acid polymerase in mammalian cells. *J. Biol. Chem.* **246:** 5835–5837.
27. Chen, K.-H., F. M. Yakes, D. K. Srivastava, R. K. Singhal, R. W. Sobol, J. K. Horton, B. Van Houten, and S. H. Wilson. Oxidative stress inducing agents up-regulate base excision repair in mouse cell lines. *Nucleic Acids Res.*, in press.
28. Chen, Y.-C., E. W. Bohn, S. R. Planck, and S. H. Wilson. 1979. Mouse DNA polymerase alpha: subunit structure and identification of a species with associated exonuclease. *J. Biol. Chem.* **254:** 11,678–11,687.
29. Chui, G. and S. Linn. 1995. Further characterization of HeLa DNA polymerase epsilon. *J. Biol. Chem.* **270:** 7799–7808.
30. Chung, D. W., J. Zhang, C. K. Tan, E. W. Davie, A. G. So, and K. M. Downey. 1991. Primary structure of the catalytic subunit of human DNA polymerase δ and chromosomal location of the gene. *Proc. Natl. Acad. Sci. USA* **88:** 11,197–11,201.
31. Ciarrocchi, G., J. G. Jose, and S. Linn. 1979. Further characterization of a cell-free system for measuring replicative and repair DNA synthesis with cultured human fibroblasts and evidence for the involvement of DNA polymerase α in DNA repair. *Nucleic Acids Res.* **7:** 1205–1219.
32. Cotterill, S. M., M. E. Reyland, L. A. Loeb, and I. R. Lehman. 1987. A cryptic proofreading 3' → 5' exonuclease associated with the polymerase subunit of the DNA polymerase-primase from *Drosophila melanogaster. Proc. Natl. Acad. Sci. USA* **84:** 5635–5639.
33. Cripps-Wolfman, J., E. C. Henshaw, and R. A. Bambara. 1989. Alterations in the phosphorylation and activity of DNA polymerase α correlate with the change in replicative DNA synthesis as quiescent cells re-enter the cell cycle. *J. Biol. Chem.* **264:** 19,478–19,486.
34. Crute, J. J., A. F. Wahl, and R. A. Bambara. 1986. Purification and characterization of two new high molecular weight forms of DNA polymerase δ. *Biochemistry* **25:** 26–36.

35. Cullman, G., R. Hindges, M. W. Berchtold, and U. Hübscher. 1993. Cloning of a mouse cDNA encoding DNA polymerase delta: refinement of the homology boxes. *Gene* **134:** 191–200.
36. Da Costa, L. T., B. Liu, W. El-Deiry, S. R. Hamilton, K. W. Kinzler, B. Vogelstein, S. Markowitz, J. K. Willson, A. de la Chapelle, K. M. Downey, and A. G. So. 1995. Polymerase delta variants in RER colorectal tumors [letter]. *Nat. Genet.* **9:** 10,11.
37. Di Giuseppe, J. A. and S. L. Dresler. 1989. Bleomycin-induced DNA repair synthesis in permeable human fibroblasts: mediation of long-patch and short-patch repair by distinct DNA polymerases. *Biochemistry* **28:** 9515–9520.
38. Dresler, S. L. 1984. Comparative enzymology of ultraviolet-induced DNA repair synthesis and semiconservative DNA replication in permeable diploid human fibroblasts. *J. Biol. Chem.* **259:** 13,947–13,952.
39. Dresler, S. L. and M. G. Frattini. 1986. DNA replication and UV-induced DNA repair synthesis in human fibroblasts are much less sensitive than DNA polymerase α to inhibition by butylphenyl-deoxyguanosine triphosphate. *Nucleic Acids Res.* **14:** 7093–7102.
40. Dresler, S. L. and M. G. Frattini. 1988. Analysis of butylphenyl-guanine, butylphenyl-deoxyguanosine, and butylphenyl-deoxyguanosine triphosphate inhibition of DNA replication and ultraviolet-induced DNA repair synthesis using permeable human fibroblasts. *Biochem. Pharmacol.* **37:** 1033–1037.
41. Dresler, S. L. and K. S. Kimbro. 1987. 2', 3'-Dideoxythymidine 5'-triphosphate inhibition of DNA replication and ultraviolet-induced DNA repair synthesis in human cells: evidence for involvement of DNA polymerase δ. *Biochemistry* **26:** 2664–2668.
42. Dresler, S. L. and M. W. Lieberman. 1983. Identification of DNA polymerases involved in DNA excision repair in diploid human fibroblasts. *J. Biol. Chem.* **258:** 9990–9994.
43. Dresler, S. L., M. G. Frattini, and R. M. Robinson-Hill. 1988. In situ enzymology of DNA replication and ultraviolet-induced DNA repair synthesis in permeable human cells. *Biochemistry* **27:** 7247–7254.
44. Fisher, P. A., T. S. Wang, and D. Korn. 1979. Enzymological characterization of DNA polymerase α. Basic catalytic properties processivity, and gap utilization of the homogeneous enzyme from human KB cells. *J. Biol. Chem.* **254:** 6128–6137.
45. Fortini, P., B. Pascucci, R. W. Sobol, S. H. Wilson, and E. Dogliotti. Different DNA polymerases are involved in the short- and long-patch base excision repair in mammalian cells. *Biochemistry* **37:** 3575–3580.
46. Friedberg, E. C. 1992. Xeroderma pigmentosum, Cockayne's syndrome, helicases, and DNA repair: what's the relationship? *Cell* **71:** 887–889.
47. Friedberg, E. C., G. C. Walker, and W. Siede. 1995. In *DNA Repair and Mutagenesis,* American Society for Microbiology, Washington, DC, pp. 317–365.
48. Frosina, G., P. Fortini, O. Rossi, F. Carrozzino, A. Abbondandolo, and E. Dogliotti. 1994. Repair of abasic sites by mammalian cell extracts. *Biochem. J.* **304:** 699–705.
49. Frosina, G., P. Fortini, O. Rossi, F. Carrozzino, G. Raspaglio, L. S. Cox, D. P. Lane, A. Abbondandolo, and E. Dogliotti. 1996. Two pathways for base excision repair in mammalian cells. *J. Biol. Chem.* **271:** 9573–9578.
50. Fry, M. and L. A. Loeb. 1986. *Animal Cell DNA Polymerases.* CRC, Boca Raton, FL.
51. Goscin, L. P. and J. J. Byrnes. 1982. DNA polymerase δ: one peptide, two activities. *Biochemistry* **21:** 2513–2518.
52. Goulian, M. and C. J. Heard. 1989. Intact DNA polymerase α/primase from mouse cells: purification and structure. *J. Biol. Chem.* **264:** 19,407–19,415.
53. Hanaoka, F., H. Kato, S. Ikegami, M. Ohashi, and M. Yamada. 1979. Aphidicolin does inhibit repair replication in HeLa cells. *Biochem. Biophys. Res. Commun.* **87:** 575–580.
54. Hanawalt, P. C., P. K. Cooper, A. K. Ganesan, and C. A. Smith. 1979. DNA repair in bacteria and mammalian cells. *Ann. Rev. Biochem.* **48:** 783–836.
55. Hindges, R. and U. Hübscher. 1997. DNA polymerase δ, an essential enzyme for DNA transactions. *Biol. Chem.* **378:** 345–362.

56. Holmes, J., S. J. Clark, and P. Modrich. 1990. Strand-specific mismatch correction in nuclear extracts of human and Drosophila melanogaster cell lines. *Proc. Natl. Acad. Sci. USA* **87:** 5837–5841.
57. Huberman, J. A. 1981. New views of the biochemistry of eukaryotic DNA replication revealed by aphidicolin, an unusual inhibitor of DNA polymerase α. *Cell* **23:** 647,648.
58. Ikegami, S., T. Taguchi, M. Ohashi, M. Oguro, H. Nagano, and Y. Mano. 1978. Aphidicolin prevents mitotic cell division by interfering with the activity of DNA polymerase-α. *Nature (Lond.)* **275:** 458–460.
59. Ilsley, D. D., S. H. Lee, W. H. Miller, and R. D. Kuchta. 1995. Acyclic guanosine analogs inhibit DNA polymerases alpha, delta, and epsilon with very different potencies and have unique mechanisms of action. *Biochemistry* **34:** 2504–2510.
60. Jasko, M. V., D. G. Semizarov, L. S. Victorova, Mozzherin, A. A. Krayevsky, and M. K. Kukhanova. 1995. New modified substrates for discriminating between human DNA polymerases alpha and epsilon. *FEBS Lett.* **357:** 23–26.
61. Jenkins, T. M., J. K Saxena, A. Kumar, S. H. Wilson, and E. J. Ackerman. 1992. DNA polymerase β and DNA synthesis in Xenopus oocytes and in a nuclear extract. *Science* **258:** 475–478.
62. Keeney, S. and S. Linn. 1990. A critical review of permeabilized cell systems for studying mammalian DNA repair. *Mutat. Res.* **236:** 239–252.
63. Kesti, T. and J. E. Syvaoja. 1991. Identification and tryptic cleavage of the catalytic core of HeLa and calf thymus DNA polymerase epsilon. *J. Biol. Chem.* **266:** 6336–6341.
64. Kesti, T., F. Hannele, and J. E. Syväoja. 1993. Molecular cloning of the catalytic subunit of human DNA polymerase epsilon. *J. Biol. Chem.* **268:** 10,238–10,245.
65. Keyse, S. M. and R. M. Tyrrell. 1985. Excision repair in permeable arrested human skin fibroblasts damaged by UV (254 nm) radiation: evidence that α- and β-polymerases act sequentially at the repolymerization step. *Muta. Res.* **146:** 109–119.
66. Khan, N. N., G. E. Wright, L. W. Dudycz, and N. C. Brown. 1984. Butyllphenyl dGTP: a selective and potent inhibitor of mammalian DNA polymerase α. *Nucleic Acids Res.* **12:** 3695–3706.
67. Klungland, A. and T. Lindahl. 1997. Second pathway for completion of human DNA base excision-repair: reconstitution with purified proteins and requirement for DNase IV (FEN1). *EMBO J.* **16:** 3341–3348.
68. Kolodner, R. D. 1995. Mismatch repair: mechanisms and relationship to cancer susceptibility. *Trends. Biochem. Sci.* **20:** 391–397.
69. Kornberg, A. and T. Baker. 1991. *DNA Replication*. Freeman, New York.
70. Kubota, Y., R. A. Nash, A. Klungland, P. Schär, D. Barnes, and T. Lindahl. 1996. Reconstitution of DNA base excision-repair with purified human proteins: interaction between DNA polymerase β and the XRCC1 protein. *EMBO J.* **15:** 6662–6670.
71. Kumar, A., J. Abbotts, E. Karawya, and S. H. Wilson. 1990. Identification and properties of the catalytic domain of mammalian DNA polymerase beta. *Biochemistry* **29:** 7156–7159.
72. Lan, S. Y. and M. J. Smerdon. 1985. A nonuniform distribution of excision repair synthesis in nucleosome core DNA. *Biochemistry* **24:** 7771–7783.
73. Lee, M. Y. W. T., and N. L. Toomey. 1987. Human placental DNA polymerase δ: identification of a 170-kilodalton polypeptide by activity staining and immunoblotting. *Biochemistry* **26:** 1076–1085.
74. Lee, M. Y. W. T., C.-K. Tan, A. G. So, and K. M. Downey. 1980. Purification of deoxyribonucleic acid polymerase δ from calf thymus: partial characterization of physical properties. *Biochemistry* **19:** 2096–2101.
75. Lee, M. Y. W. T., C.-K. Tan, K. M. Downey and A. G. So. 1984. Further studies on calf thymus DNA polymerase δ purified to homogeneity by a new procedure. *Biochemistry* **23:** 1906–1913.

76. Lee, M. Y., Y. Q. Jiang, S. J. Zhang, and N. L. Toomey. 1991. Characterization of human DNA polymerase delta and its immunochemical relationships with DNA polymerase alpha and epsilon. *J. Biol. Chem.* **266:** 2423–2429.
77. Lee, S. H., Z. Q. Pan, A. D. Kwong, P. M. Burgers, and J. Hurwitz. 1991. Synthesis of DNA by DNA polymerase ε in vitro. *J. Biol. Chem.* **266:** 22,707–22,717.
78. Lehman, I. R. and L. S. Kaguni. 1989. DNA polymerase α. *J. Biol. Chem.* **264:** 4265–4268.
79. Lewis, W., R. R. Meyer, J. F. Simpson, J. M. Colacino, and F. W. Perrino. 1994. Mammalian DNA polymerases alpha, beta, gamma, delta, and epsilon incorporate fialuridine (FIAU) monophosphate into DNA and are inhibited competitively by FIAU Triphosphate. *Biochemistry* **33:** 14,620–14,624.
80. Linn, S. 1991. How many pols does it take to replicate nuclear DNA? *Cell* **66:** 185–187.
81. Loeb, L. A., P. K. Liu, and M. Fry. 1986. DNA polymerase-alpha: enzymology, function, fidelity, and mutagenesis. *Prog. Nucleic Acid Res. Mol. Biol.* **33:** 57–110.
82. Maga, G. and U. Hübscher. 1995. DNA polymerase epsilon interacts with proliferating cell nuclear antigen in primer recognition and elongation. *Biochemistry* **34:** 891–901.
83. Matsukage, A., E. W. Bohn, and S. H. Wilson. 1974. Multiple forms of DNA polymerase in mouse myeloma. *Proc. Natl. Acad. Sci. USA* **71:** 578–582.
84. Matsumoto, T., T. Eki., and J. Hurwitz. 1990. Studies on the initiation and elongation reactions in the simian virus 40 DNA replication system. *Proc. Natl. Acad. Sci. USA* **87:** 9712–9716.
85. Matsumoto, Y. and K. Kim. 1995. Excision of deoxyribose phosphate residues by DNA polymerase β during DNA repair. *Science* **269:** 699–702.
86. Matsumoto, Y., K. Kim, and D. F. Bogenhagen. 1994. Proliferating cell nuclear antigen-dependent abasic site repair in *Xenopus laevis* oocytes: an alternating pathway of base excision repair. *Mol. Cell. Biol.* **14:** 6187–6197.
87. McBride, O. W., C. A. Kozak, and S. H. Wilson. 1990. Mapping of the gene for DNA polymerase β to mouse chromosome 8. *Cytogenet. Cell Genet.* **53:** 108–111.
88. Mellon, I., V. A. Bohr, C. A. Smith, and P. C. Hanawalt. 1986. Preferential DNA repair of an active gene in human cells. *Proc. Natl. Acad. Sci. USA* **83:** 8878–8882.
89. Miller, M. R. and D. N. Chinault. 1982. Evidence that DNA polymerases α and β participate differentially in DNA repair synthesis induced by different agents. *J. Biol. Chem.* **257:** 46–49.
90. Miller, M. R. and D. N. Chinault. 1982. The roles of DNA polymerases α, β, γ and in DNA repair synthesis induced in hamster and human cells by different DNA damaging agents. *J. Biol. Chem.* **257:** 10,204–10,209.
91. Modrich, P. 1994. Mismatch repair, genetic stability, and cancer. *Science* **266:** 1959,1960.
92. Modrich, P. 1991. Mechanisms and biological effects of mismatch repair. *Ann. Rev. Genet.* **25:** 229–253.
93. Morrison, A., H. Araki, A. B. Clark, R. K. Hamatake, and A. Sugino. 1990. A third essential DNA polymerase in *S. Cerevisiae. Cell* **62:** 1143–1151.
94. Mosbaugh, D. W. and S. Linn. 1983. Excision repair and DNA synthesis with a combination of HeLa DNA polymerase β and DNase V. *J. Biol. Chem.* **258:** 108–118.
95. Mosbaugh, D. W. and S. Linn. 1984. Gap-filling DNA synthesis by HeLa DNA polymerase alpha in an in vitro base excision DNA repair scheme. *J. Biol. Chem.* **259:** 10,247–10,251.
96. Mullen, G. P. and S. H. Wilson. 1997. Repair activity in DNA polymerases: a structurally conserved helix-hairpin-helix motif in base excision repair enzymes and in DNA polymerase β. *Biochemistry* **36:** 4713–4717.
97. Mullen, G. P., Antuch, W., Maciejewski, M. W., Prasad, R., and S. H. Wilson. 1997. Insights into the mechanism of the β-elimination catalyzed by the N-terminal domain of DNA polymerase β. *Tetrahedron* **53(35):** 12,057–12,066.

98. Narayan, S., F. He, and S. H. Wilson. 1996. Activation of the human DNA polymerase β promoter by a DNA-alkylating agent through induced phosphorylation of CREB-1. *J. Biol. Chem.* **271:** 18,508–18,513.
99. Navas, T. A., Z. Zhou, and S. J. Elledge. 1995. DNA polymerase epsilon links the DNA replication machinery to the S phase checkpoint. *Cell* **80:** 29–39.
100. Nealon, K., I. D. Nicholl, and M. K. Kenny. 1996. Characterization of the DNA polymerase requirement of human base excision repair. *Nucleic Acids Res.* **24:** 3763–3770.
101. Nicholl, I. D., K. Nealon, and M. K. Kenny. 1997. Reconstitution of human base excision repair with purified proteins. *Biochemistry* **36:** 7557–7566.
102. Niranjanakumari, S. and K. P. Gopinathan. 1993. Isolation and characterization of DNA polymerase epsilon from the silk glands of Bombyx mori. *J. Biol. Chem.* **268:** 15,557–15,564.
103. Nishida, C., P. Reinhard, and S. Linn. 1988. DNA repair synthesis in human fibroblasts requires DNA polymerase delta. *J. Biol. Chem.* **263:** 501–510.
104. Ottiger, H.-P. and U. Hübscher. 1984. Mammalian DNA polymerase α holoenzymes with possible functions at the leading and lagging strand of the replication fork. *Proc. Natl. Acad. Sci. USA* **81:** 3993–3997.
105. Pelletier, H., M. R. Sawaya, W. Wolfle, S. H. Wilson, and J. Kraut. 1996. A structural basis for metal ion mutagenicity and nucleotide selectivity in human DNA polymerase. *Biochemistry* **35:** 12,762–12,777.
106. Perrino, F. W. and L. A. Loeb. 1989. Proofreading by the ε subunit of *Escherichia coli* DNA polymerase III increases the fidelity of calf thymus DNA polymerase α. *Proc. Natl. Acad. Sci. USA* **86:** 3085–3088.
107. Piersen, C. E., R. Prasad, S. H. Wilson, and R. S. Lloyd. 1996. Evidence for an imino intermediate in the DNA polymerase β deoxyribose phosphate excision reaction. *J. Biol. Chem.* **271:** 17,811–17,815.
108. Planck, S. R., K. Tanabe, and S. H. Wilson. 1980. Distinction between mouse DNA polymerases α and β by tryptic peptide mapping. *Nucleic Acids Res.* **8:** 2771–2782.
109. Prasad, R., R. K. Singhal, D. K. Srivastava, A. E. Tomkinson, and S. H. Wilson. 1996. Specific interaction of DNA polymerase β and DNA ligase I in a multiprotein base excision repair complex from bovine testis. *J. Biol. Chem.* **271:** 16,000–16,007.
110. Prasad, R., S. G. Widen, R. K. Singhal, J. Watkins, L. Prakash, and S. H. Wilson. 1993. Yeast open reading frame YCR14C encodes a DNA polymerase β-like enzyme. *Nucleic Acids Res.* **21:** 5301–5307.
111. Randahl, H., G. C. Elliott, and S. Linn. 1988. DNA-repair reactions by purified HeLa DNA polymerases and exonucleases. *J. Biol. Chem.* **263:** 12,228–12,234.
112. Regan, J. D. and R. B. Setlow. 1974. Two forms of repair in the DNA of human cells damaged by chemical carcinogens and mutagens. *Cancer Res.* **34:** 3318–3325.
113. Sawaya, M. R., H. Pelletier, A. Kumar, S. H. Wilson, and J. Kraut. 1994. Crystal structure of rat DNA polymerase β reveals a conserved polymerase catalytic site. *Science* **264:** 2930–2935.
114. Seeberg, E., L. Eide, and M. Bjørås. 1995. The base excision repair pathway. *Trends Biochem. Sci.* **20:** 391–397.
115. Seki, S. and T. Oda. 1988. An exonuclease possibly involved in the initiation of repair of bleomycin-damaged DNA in mouse ascites sarcoma cells. *Carcinogenesis* **9:** 2239–2244.
116. SenGupta, D. N., B. Z. Zmudzka, P. Kumar, F. Cobianchi, J. Skowronski, and S. H. Wilson. 1986. Sequence of human DNA polymerase β mRNA obtained through cDNA cloning. *Biochem. Biophys. Res. Commun.* **136:** 341–347.
117. Shivji, M. K., V. N. Podust, U. Hübscher, and R. D. Wood. 1995. Nucleotide excision repair DNA synthesis by DNA polymerase epsilon in the presence of PCNA, RFC, and RPA. *Biochemistry* **34:** 5011–5017.

118. Siegal, G., J. J. Turchi, T. W. Myers, and R. A. Bambara. 1992. A 5' to 3' exonuclease functionally interacts with calf DNA polymerase ε. *Proc. Natl. Acad. Sci. USA* **89:** 9377–9381.
119. Siedlecki, J. A., J. Szyszko, I Pietrzykowska, and B. Zmudzka. 1980. Evidence implying DNA polymerase beta function in excision repair. *Nucleic Acids Res.* **8:** 361–375.
120. Singhal, R. K. and S. H. Wilson. 1993. Short gap-filling synthesis by DNA polymerase β is processive. *J. Biol. Chem.* **268:** 15,906–15,911.
121. Singhal, R. K., R. Prasad, and S. H. Wilson. 1995. DNA polymerase β conducts the gap-filling step in uracil-initiated base excision repair in a bovine testis nuclear extract. *J. Biol. Chem.* **270:** 949–957.
122. Skarnes, W., P. Bonin, and E. Baril. 1986. Exonuclease activity associated with a multiprotein form of HeLa cell DNA polymerase α. Purification and properties of the exonuclease. *J. Biol. Chem.* **261:** 6629–6636.
123. Smith, C. A. and D. S. Okumoto. 1984. Nature of DNA repair synthesis resistant to inhibitors of polymerase α in human cells. *Biochemistry* **23:** 1383–1391.
124. So, A. G. and K. M. Downey. 1988. Mammalian DNA polymerases α and δ: current status in DNA replication. *Biochemistry* **27:** 4591–4595.
125. Sobol, R. W., J. K. Horton, R. Kühn, H. Gu, R. K. Singhal, R. Prasad, K. Rajewsky, and S. H. Wilson. 1996. Requirement of mammalian DNA polymerase-β in base-excision repair. *Nature* **379:** 183–186.
126. Spadari, S., F. Sala, and G. Pedrali-Noy. 1984. Aphidicolin and eukaryotic DNA synthesis. *Adv. Exp. Med. Biol.* **179:** 169–181.
127. Srivastava, D. K., T. Y. Rawson, S. D. Showalter, and S. H. Wilson. 1995. Phorbol ester abrogates up-regulation of DNA polymerase β by DNA alkylating agents in Chinese hamster ovary cells. *J. Biol. Chem.* **270:** 16,402–16,408.
128. Sweasy, J. B., M. Chen, and L. A. Loeb. 1995. DNA polymerase beta can substitute for DNA polymerase I in the initiation of plasmid DNA replication. *J. Bacteriol.* **177:** 2923–2925.
129. Syvaoja, J. and S. Linn. 1989. Characterization of a large form of DNA polymerase δ from HeLa cells that is insensitive to proliferating cell nuclear antigen. *J. Biol. Chem.* **264:** 2489–2497.
130. Syvaoja, J., S. Suomensaari, C. Nishida, J. S. Goldsmith, and S. Linn. 1990. DNA polymerases α, δ, and ε: three distinct enzymes from HeLa cells. *Proc. Natl. Acad. Sci. USA* **87:** 6664–6668.
131. Szpirer, J., F. Pedeutour, T. Kesti, M. Riviere, J. E. Syvaoja, C. Turc-Carel, and C. Szpirer. 1994. Localization of the gene for DNA polymerase epsilon (POLE) to human chromosome 12q24.3 and rat chromosome 12 by somatic cell hybrid panels and fluorescence in situ hybridization. *Genomics* **20:** 223–226.
132. Tan, C.-K., C. Castillo, A. G. So, and K. M. Downey. 1986. An auxiliary protein for DNA polymerase δ from fetal calf thymus. *J. Biol. Chem.* **261:** 12,310–12,316.
133. Tanaka, S., S. Z. Hu, T. S.-F. Wang, and D. Korn. 1982. Preparation and preliminary characterization of monoclonal antibodies against human DNA polymerase α. *J. Biol. Chem.* **257:** 8386–8390.
134. Thomas, D. C., J. D. Roberts, and T. A. Kunkel. 1991. Heteroduplex repair in extracts of human HeLa cells. *J. Biol. Chem.* **266:** 3744–3751.
135. Tsurimoto, T. and B. Stillman. 1991. Replication factors required for SV40 DNA replication *in vitro*. II. DNA structure specific recognition primer template junction by eukaryotic DNA polymerases and their accessory factors. *J. Biol. Chem.* **266:** 1950–1960.
136. Tsurimoto, T. and B. Stillman. 1991. Replication factors required for SV40 DNA replication in vitro. II. Switching of DNA polymerase alpha and delta during initiation of leading and lagging strand synthesis. *J. Biol. Chem.* **266:** 1961–1968.

137. Tsurimoto, T., T. Melendy, and B. Stillman. 1990. Sequential initiation of lagging and leading strand synthesis by two different polymerase complexes at the SV40 DNA replication origin. *Nature (Lond.)* **346:** 534–539.
138. Tuusa, J., L. Uitto, and J. E. Syvaoja. 1995. Human DNA polymerase epsilon is expressed during cell proliferation in a manner characteristic of replicative DNA polymerases. *Nucleic Acids Res.* **23:** 2178–2183.
139. Umar, A., A. B. Buermeyer, J. A. Simon, D. C. Thomas, A. B. Clark, R. M. Liskay, and T. A. Kunkel. 1996. Requirement for PCNA in DNA mismatch repair at a step preceding DNA resynthesis. *Cell* **87:** 65–73.
140. Wahl, A. F., A. M. Geis, B. H. Spain, S. W. Wong, D. K. Korn, and T. S.-F. Wang. 1988. Gene expression of human DNA polymerase α during cell proliferation and the cell cycle. *Mol. Cell. Biol.* **8:** 5016–5025.
141. Wahl, A. F., J. J. Crute, R. D. Sabatino, J. B. Bodner, R. L. Marraccino, L. W. Harwell, E. M. Lord, and R. A. Bambara. 1986. Properties of two forms of DNA polymerase δ from calf thymus. *Biochemistry* **25:** 7821–7827.
142. Wang, T. S.-F. 1991. Eukaryotic DNA polymerases. *Ann. Rev. Biochem.* **60:** 513–552.
143. Wang, T. S.-F., S. Z. Hu, and D. Korn. 1984. DNA primase from KB cells. Characterization of a primase activity tightly associated with immunoaffinity-purified DNA polymerase-α. *J. Biol. Chem.* **259:** 1854–1865.
144. Wang, T. S.-F. and D. Korn. 1980. Reactivity of KB cell deoxyribonucleic acid polymerases α and β with nicked and gapped deoxyribonucleic acid. *Biochemistry* **19:** 1782–1790.
145. Wang, T. S.-F., B. E. Pearson, H. A. Suomalainen, T. Mohandas, L. J. Shapiro, J. Schroder, and D. Korn. 1985. Assignment of the gene for human DNA polymerase α to the X chromosome. *Proc. Natl. Acad. Sci. USA* **82:** 5270–5274.
146. Wang, Z., X. Wu, and E. C. Friedberg. 1993. DNA repair synthesis during base excision repair *in vitro* is catalyzed by DNA polymerase ε and is influenced by DNA polymerase α and δ in *Saccharomyces cerevisiae. Mol. Cell. Biol.* **13:** 1051–1058.
147. Weinberg, D. H., K. L. Collins, P. Simancek, A. Russo, M. S. Wold, D. M. Virshup, and T. J. Kelly. 1990. Reconstitution of simian virus 40 DNA replication with purified proteins. *Proc. Natl. Acad. Sci. USA* **87:** 8692–8696.
148. Weissbach, A., A. Schlabach, B. Fridlender, and A. Bolden. 1971. A DNA polymerase from human cells. *Nature (Lond.), New Biol.* **231:** 167–170.
149. Wiebauer, K. and J. Jiricny. 1990. Mismatch-specific thymine DNA glycosylase and DNA polymerase beta mediate the correction of G. T mispairs in nuclear extracts from human cells. *Proc. Natl. Acad. Sci. USA* **87:** 5842–5845.
150. Wilson, S., J. Abbotts, and S. Widen. 1988. Progress toward molecular biology of DNA polymerase β. *Biochim. Biophys. Acta* **949:** 149–157.
151. Wintersberger, E. 1974. Deoxyribonucleic acid polymerases from yeast; further purification and characterization of DNA-dependent DNA polymerases A and B. *Eur. J. Biochem.* **50:** 41–47.
152. Wong, S. W., A. F. Wahl, P.-M. Yuan, N. Arai, B. E. Pearson, K.-I. Arai, D. Korn, M. W. Hunkapiller, and T. S.-F. Wang. 1988. Human DNA polymerase α gene expression is cell proliferation dependent and its primary structure is similar to both prokaryotic and eukaryotic replicative DNA polymerases. *EMBO J.* **7:** 37–47.
153. Wright, G. E., U. Hübscher, N. N. Khan, F. Focher and A. Verri. 1994. Inhibitor analysis of calf thymus DNA polymerases alpha, delta and epsilon. *FEBS Lett.* **3411:** 128–130.
154. Zeng, X. R., H. Hao, Y. Jiang, and M. Y. Lee. 1994. Regulation of human DNA polymerase delta during the cell cycle. *J. Biol. Chem.* **269:** 24,027–24,033.
155. Zeng, X. R., Y. Jiang, S. J. Zhang, H. Hao, and M. Y. Lee. 1994. DNA polymerase delta is involved in the cellular response to UV damage in human cells. *J. Biol. Chem.* **269:** 13,748–13,751.

156. Zhang, P., I. Frugulhetti, Y. Jiang, G. L. Holt, R. C. Condit, and M. Y. Lee. 1995. Expression of the catalytic subunit of human DNA polymerase delta in mammalian cells using a vaccinia virus vector system. *J. Biol. Chem.* **270:** 7993–7998.
157. Zhang, S. J., X. R. Zhang, P. Zhang, N. L. Toomey, R. Y. Chuang, L. S. Chang, and M. Y. Lee. 1995. A conserved region in the amino terminus of DNA polymerase delta is involved in proliferating cell nuclear antigen binding. *J. Biol. Chem.* **270:** 7988–7992.
158. Zmudzka, B. Z., A. Fornace, Jr., J. Collins, and S. H. Wilson. 1988. Characterization of DNA polymerase β mRNA: cell-cycle and growth response in cultured human cells. *Nucleic Acids Res.* **16:** 9587–9596.
159. Zmudzka, B. Z., D. SenGupta, A. Matsukage, F. Cobianchi, P. Kumar, and S. H. Wilson. 1986. Structure of rat DNA polymerase β revealed by partial amino acid sequencing and cDNA cloning. *Proc. Natl. Acad. Sci. USA* **83:** 5106–5110.

12

Cellular Functions of Mammalian DNA Ligases

Alan E. Tomkinson, Jingwen Chen, Jeff Besterman, and Intisar Husain

1. INTRODUCTION

DNA strand breaks, in particular double-strand breaks, are potentially cytotoxic lesions. These breaks may be introduced directly by a DNA-damaging agent, such as ionizing radiation, or as a consequence of DNA repair proteins recognizing and excising DNA damage. Thus, the majority of DNA repair mechanisms, including recombinational repair pathways, share a common essential step, phosphodiester bond formation, that restores the integrity of the DNA substrate molecule(s). In addition, DNA-joining events are required to link together the Okazaki fragments generated during lagging strand DNA synthesis.

All the DNA-joining requirements of a prokaryotic cell are fulfilled by a single species of DNA ligase. Consequently, prokaryotic DNA ligase mutants exhibit a pleiotropic phenotype that includes conditional lethality, hyperrecombination, increased sensitivity to DNA-damaging agents, and an increased mutation frequency *(21,34)*. In contrast to prokaryotes, higher eukaryotes possess more than one species of DNA ligase *(36)*. Two and three biochemically distinct DNA ligase activities have been purified from extracts of *Drosophila melanogaster* embryos *(61)* and mammalian tissues *(68)*, respectively. Although these enzymes have different biochemical properties, including DNA substrate specificity, they all act by the same basic reaction mechanism (Fig. 1). In the first step of the DNA-joining reaction, eukaryotic DNA ligases interact with ATP to form a covalent enzyme–adenylate complex releasing pyrophosphate. In this complex, the AMP moiety is linked to a specific lysine residue within the polypeptide via a phosphoramidite bond. When the adenylylated enzyme encounters a DNA molecule containing a nick with 3'-hydroxyl and 5'-phosphate termini, it transfers the adenylate group to the 5'-phosphate terminus, producing a covalent DNA-adenylate reaction intermediate. In the final step of the reaction, the enzyme catalyzes phosphodiester bond formation between the 3'-hydroxyl terminus and the adenylylated 5'-phosphate terminus, releasing AMP.

Biochemical studies on mammalian DNA ligases resulted in the identification of a lysine-containing active site motif that is required for formation of the enzyme–adenylate intermediate (Fig. 2A). Sequences homologous to this motif have been found in all

From: DNA Damage and Repair, Vol. 2: DNA Repair in Higher Eukaryotes
Edited by: J. A. Nickoloff and M. F. Hoekstra © Humana Press Inc., Totowa, NJ

(1) Enz + ATP → Enz-AMP + PPi

(2) 3'OH 5'P + Enz-AMP → AMP 3'OH 5'P + Enz

(3) AMP 3'OH 5'P + Enz → + Enz + AMP

Fig. 1. DNA joining reaction catalyzed by ATP-dependent DNA ligases. DNA ligase interacts with ATP to form a covalent enzyme–AMP complex. The AMP moiety is then transferred from the polypeptide to the 5'-phosphate terminus at a nick in duplex DNA. Finally, the nonadenylated enzyme catalyzes phosphodiester bond formation, releasing AMP.

Fig. 2. (A) Conserved amino sequences in eukaryotic DNA ligases. Active site motif for enzyme–adenylate formation. The amino acid sequences of adenylated peptides from bovine DNA ligase I *(70)* and bovine DNA ligase II *(73)* have been aligned with homologous sequences in the open reading frames encoded by human DNA ligases I, III, and IV cDNAs, *S. cerevisiae CDC9*, and the vaccinia DNA ligase gene *(4,15,55,70,73,75)*. Conserved residues are indicated in bold face, and the active site lysine residue is underlined. **(B)** Conserved peptide. A conserved 16 amino acid sequence was identified in a comparison of vaccinia and yeast DNA ligases *(55)*. These sequences are aligned with homologous regions encoded by human DNA ligases I, III, and IV cDNAs *(4,15,75)*. Conserved residues are indicated in bold face.

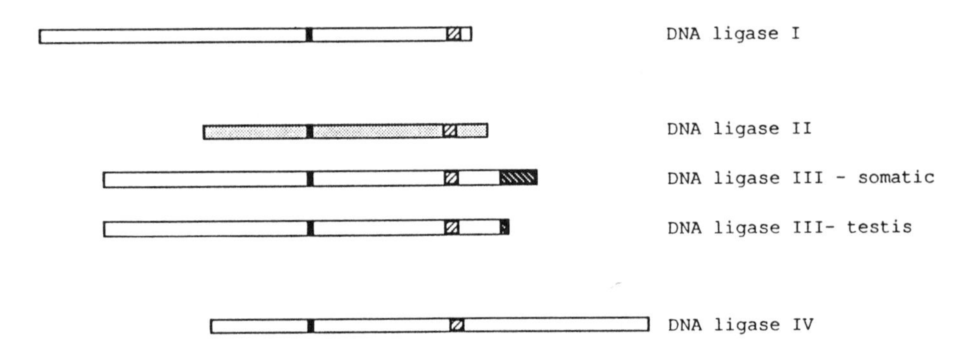

Fig. 3. Comparison of mammalian DNA ligases aligned at the active site motif. The polypeptides encoded by human DNA ligases I, III, and IV cDNAs *(4,15,75)* have been aligned at the active site motif for enzyme–adenylate formation (Fig. 2A), which is indicated by black boxes. The positions of the conserved peptide sequence (Fig. 2B) are indicated by unshaded boxes filled with diagonal lines. The different C-terminal sequences of DNA ligase III apparently encoded by alternatively spliced forms of DNA ligase III mRNA are indicated by shaded boxes with diagonal lines. Peptide sequences from bovine DNA ligase II can be aligned with homologus sequences encoded by human DNA ligase III cDNA *(15,29,73)*. The region encompassed by DNA ligase II peptides when aligned with homologous sequences in the human DNA ligase III coding sequence is indicated by the shaded box.

DNA ligases *(70)*, in RNA ligases *(26,79)*, and in GTP-dependent mRNA capping enzymes *(17,22)*. A second conserved 16 amino acid sequence (referred to as the conserved peptide) that also appears to be essential for catalytic function has been identified in eukaryotic DNA ligases (Fig. 2B). The high proportion of positively charged residues (30%) present at conserved positions within this sequence suggest that this region may be involved in the recognition of and interaction with the DNA substrate.

The active site motif and the conserved peptide appear to define a minimal catalytic domain, since the number of residues separating these highly conserved regions is relatively constant in eukaryotic DNA ligases (Fig. 3) and the intervening sequences also exhibit homology with each other. It seems reasonable to assume that, although the different species of mammalian DNA ligase catalyze the same reaction, these enzymes are not functionally redundant. Presumably, a DNA ligase is recruited to a particular DNA metabolic pathway by specific protein–protein interactions that involve regions of the DNA ligase other than the catalytic domain. This chapter focuses on recent progress that has been made in the isolation of mammalian DNA ligase genes and in defining the cellular roles of mammalian DNA ligases.

2. MAMMALIAN DNA LIGASE PROTEINS AND GENES

2.1. DNA Ligase I

DNA ligase I is the most extensively studied of the mammalian DNA ligases. The first evidence linking this enzyme with DNA replication was provided by a study showing that DNA ligase I enzyme levels were significantly higher in regenerating liver compared with normal liver *(57)*. This link has been further substantiated in subsequent biochemical studies that have demonstrated the colocalization of DNA ligase I with other DNA replication proteins *(76)*, the copurification of DNA ligase I with a

Table 1
Mammalian DNA Ligases

	I	II	III	IV
Molecular mass: by denaturing gel electrophoresis	125 kDa	72 kDa	100 kDa	100 kDa
Tissue source	Thymus	Liver	Testis	?
Ligation of:				
Oligo (dT)·poly (dA)	Yes	Yes	Yes	?
Oligo (dT)·poly (rA)	No	Yes	Yes	?
Oligo (rA)·poly (dT)	Yes	No	Yes	?
Associated proteins	Replication complex (21S) DNA pol β	?	Recombination complex ? Xrcc1	?
Predicted molecular weight from cDNA sequence	102	?	103/96	100
Gene	*LIG1*	?	*LIG3*	*LIG4*
Chromosomal localization	19q13.2-13.3	?	17q11-12	13q33-34
Cellular functions	Replication Repair	?	Repair Recombination	?

21S replication complex *(35)*, and the specific requirement for DNA ligase I to reconstitute lagging strand DNA synthesis in vitro *(71)*.

The highest levels of DNA ligase I activity are found in the thymus glands of young mammals, and this facilitated the purification of bovine DNA ligase I to homogeneity *(64,67)*. Studies on the physical properties of purified native DNA ligase I demonstrated that it is a monomeric enzyme with an elongated shape *(67)*. When measured by denaturing gel electrophoresis, the homogeneous enzyme has a molecular mass of 125 kDa (Table 1). Analysis of the enzyme by proteolysis identified an 85-kDa C-terminal catalytic domain *(67)*. The amino-terminal domain that is dispensable for catalytic activity is phosphorylated by casein kinase II, resulting in increased enzyme activity *(48)*. This phosphorylation does not appear to be essential for enzyme activity, as was originally suggested, since active nonphosphorylated DNA ligase I has been purified from *Escherichia coli (62)*. The biological significance of DNA ligase I phosphorylation remains to be elucidated. In addition to being the site of posttranslational modifications, the amino-terminal region may be involved in protein–protein interactions. At the present time, it is not known which replication enzymes interact with DNA ligase I in the 21S replication complex *(35)*. A polypeptide that specifically interacts with and inhibits DNA ligase I has been described *(80)*. The identity of this polypeptide and its biological role are not known. Recently DNA ligase I has been shown to interact directly with DNA polymerase β within a multiprotein complex that catalyzes DNA base excision repair (*47*; Section 3.1. and *see* Chapter 11).

The reactive lysine residue that forms a phosphoramidite bond with the AMP moiety in the first step of the DNA ligation reaction was initially identified in bovine DNA ligase I *(70)*. After incubation with [^{3}H]ATP, adenylylated bovine DNA ligase I was digested with trypsin. As expected, the sequence of the purified adenylylated peptide contained an internal lysine residue that was protected from tryptic digestion by the AMP moiety and was identical with a region of the open reading frame encoded by human DNA ligase I cDNA (Fig. 2A). The essential role of this lysine residue in the formation of the DNA ligase I–adenylate intermediate has been confirmed by site-directed mutagenesis *(32)*. Based on homology with the bovine DNA ligase I peptide, it was possible to identify both the active site lysine residue in all DNA ligases *(70)* as well as an active site motif, KXDGXR, required for the adenylylation reaction (Fig. 2A).

As mentioned previously, there is another region of amino acid sequence, known as the conserved peptide, that is present in all eukaryotic DNA ligases (Fig. 2B). This sequence was initially identified as a highly conserved region near to the C-termini of vaccinia DNA ligase and the DNA ligases encoded by the *cdc9* and *cdc17* genes of *Saccharomyces cerevisiae* and *Schizosaccharomyces pombe*, respectively *(55)*. Mammalian DNA ligase I was recognized by a rabbit polyclonal antibody raised against this peptide. Furthermore, treatment of DNA ligase I with carboxypeptidase rapidly removed the epitope recognized by the peptide antibody, suggesting that this sequence was also situated close to the C-terminus *(67)*. These observations were confirmed by the subsequent isolation and sequencing of human DNA ligase I cDNA *(4)*. The conservation of this peptide sequence in all eukaryotic DNA ligases examined thus far (Fig. 2B) implies that it plays an important role in the DNA ligation reaction, but this role remains unknown.

Human DNA ligase I cDNA was initially isolated by screening a human cDNA library with degenerate oligonucleotides that were designed using bovine DNA ligase I peptide sequences. In an independent approach, human DNA ligase I cDNAs were isolated based on their ability to complement the conditional lethal phenotype of an *S. cerevisiae cdc9* DNA ligase mutant *(4)*. Within their catalytic domains, human DNA ligase I and *S. cerevisiae* Cdc9p DNA ligase share 50% amino acid identity. Not surprisingly, mammalian DNA ligase I and *S. cerevisiae* Cdc9p DNA ligase have similar biochemical properties *(67–69)*. These results indicate that mammalian DNA ligase I and Cdc9p DNA ligase are functionally homologous, and therefore, suggest that mammalian DNA ligase I may play an essential role in DNA replication.

Human DNA ligase I cDNA fragments encoding only the catalytic C-terminal domain of the enzyme complement the conditional lethal phenotype of a *S. cerevisiae cdc9* strain *(4)*, suggesting that the amino-terminus of the enzyme is not required for biological function. The question of the essential role of DNA ligase I in mammalian cells has been addressed by an adaptation of the gene-targeting strategies developed in murine embryonic stem cells. In this approach, homozygous null mutants of an essential gene are allowed to survive by the ectopic expression of a functional cDNA. DNA ligase I homozygous null cell lines were obtained in the presence of a full-length DNA ligase I cDNA, but not with a DNA ligase I cDNA fragment encoding only the catalytic domain *(46)*. These studies demonstrate that DNA ligase I, including the noncatalytic amino-terminal domain, performs an essential function in mammalian cells and implies that the other DNA ligases cannot substitute for DNA ligase I in this function. In a surprising recent development, it has been reported that mouse embryos lacking DNA ligase I develop normally to midterm and then die because of a defect in liver erythropoiesis *(7)*. It is possible that sufficient DNA ligase I is stored in the heterozygous oocyte to permit partial embryonic development. Alternatively, one of the other DNA ligases may be able to substitute for DNA ligase I in DNA replication, at least during early embryogenesis. The establishment of cell lines, if it is possible, from the embryos lacking DNA ligase I should prove informative.

DNA ligase I is ubiquitously expressed in mammalian tissues and cells *(15,41)*. The highest steady-state levels of DNA ligase I mRNA were detected in the thymus with elevated levels also detected in the testis. These observations are in agreement with previous measurements of DNA ligase I activity in different tissues *(29,58)*. The steady-state levels of DNA ligase I mRNA increase when quiescent cells are stimulated to proliferate, consistent with the involvement of DNA ligase I in DNA replication *(41,45)*. However, the presence of DNA ligase I mRNA in terminally differentiated cells may indicate that DNA ligase I is involved in some forms of DNA repair (*41*; *see* Sections 3.1. and 3.2.). Expression of mammalian DNA ligase I is increased after UV irradiation, although the kinetics of this response indicate that DNA ligase I functions in a repair pathway distinct from nucleotide excision repair *(42)*. Furthermore, DNA ligase I appears to be involved in repair of DNA strand breaks resulting from DNA damage by alkylating agents and ionizing radiation (*47*; *see* Sections 3.1. and 3.2.).

2.2. DNA Ligase II

DNA ligase II was initially described as a minor ATP-dependent DNA-joining activity in extracts from calf thymus glands that was not recognized by a polyclonal antiserum raised against DNA ligase I *(36)*. Subsequently, it was shown that this 70-kDa DNA ligase has different catalytic properties than DNA ligase I, in particular, the ability to join oligo (dT) molecules hybridized to a poly(rA) template *(3)* (Table 1). Furthermore, a comparison of labeled peptides generated by proteolytic digestion of the adenylylated forms of DNA ligases I and II revealed significant differences in the active sites of these enzymes *(51)*.

DNA ligase II has been purified to apparent homogeneity from bovine thymus glands *(63,68)* and, more recently, from bovine liver nuclei, where it is the predominant DNA-joining activity *(73)*. Attempts to obtain a partial amino acid sequence from the amino-terminus of DNA ligase II were unsuccessful *(73)*, suggesting that the amino-terminal residue was resistant to Edman degradation because of modification. Since the amino-terminal residues of most primary translation products are *N*-acetylated *(9)* and, consequently, resistant to Edman degradation, the blocked amino-terminus of DNA ligase II suggests that it may be a primary translation product.

The amino acid sequence of a labeled, adenylylated bovine DNA ligase II peptide was obtained using essentially the same strategy employed for DNA ligase I *(70,73)*. The DNA ligase II active site peptide contained an internal lysine residue within a region that exhibited homology with the previously defined DNA ligase active site motif *(73)*. As predicted by previous peptide mapping studies *(51)*, the sequences immediately adjacent to the active site motif were significantly different from those of DNA ligase I and, in fact, shared more homology with the putative active site region of the DNA ligase encoded by vaccinia virus (Fig. 1A). Further amino acid sequencing of DNA ligase II peptides revealed a peptide that exhibited homology with the conserved peptide sequence, in agreement with the recognition of purified DNA ligase II by the conserved peptide antiserum in immunoblotting experiments. Other DNA ligase II peptides also exhibited homology with vaccinia DNA ligase (about 60% identity), but only 30% identity with DNA ligase I in similar sequence comparisons *(73)*.

Since the role of vaccinia DNA ligase in viral DNA metabolism is not well understood, the homology between vaccinia DNA ligase and DNA ligase II has not provided clues regarding the cellular function(s) of DNA ligase II. Vaccinia and other poxviruses, which replicate in the cytoplasm of infected cells, encode some of their own replicative enzymes. Thus, the vaccinia DNA ligase may function in viral DNA replication, and this idea is supported by the complementation of the conditional lethal phenotype of an *S. cerevisiae cdc9* DNA ligase mutant by expression of vaccinia DNA ligase *(31)*. Surprisingly, deletion of the vaccinia DNA ligase gene has no apparent effect on viral DNA replication and recombination *(16)*. However, the mutant virus is less virulent and more sensitive to treatment with DNA-damaging agents *(31)*. This latter observation suggests a role for the viral enzyme in DNA repair, although it is not immediately obvious how DNA repair occurs in the cytoplasm. A role in DNA repair has been proposed for DNA ligase II based on increases in DNA ligase II activity following treatment with DNA-damaging agents *(12,18)*. The results of these studies should be interpreted cautiously given the still ill-defined relationship between DNA ligases II and III (Section 2.3.) and the recent identification of DNA ligase IV (Section 2.4.). At the present time, it appears that DNA ligase II is generated from DNA ligase III by a specific proteolytic processing mechanism.

2.3. *DNA Ligase III*

A 100-kDa DNA ligase with a similar substrate specificity to DNA ligase II was first purified from rat liver extracts and designated as a high-mol wt form of DNA ligase II *(19)*. In a later study, DNA ligase I, DNA ligase II, and a 100-kDa DNA ligase activity were purified from the same calf thymus extract. The 100-kDa DNA ligase was clearly distinct from DNA ligase I, and although it was more similar to DNA ligase II, there were some differences in substrate specificity (Table 1). In addition, it was not possible to convert the 100-kDa enzyme into an active 70-kDa fragment by proteolysis. Thus, the 100-kDa enzyme was designated DNA ligase III *(68)*. Subsequent peptide mapping studies on enzyme–AMP intermediates revealed similarities between the 100-kDa DNA ligase from bovine thymus and bovine DNA ligase II *(51)*. A 100-kDa DNA ligase, presumed to be DNA ligase III, has been extensively purified from human 293 cells. In contrast to DNA ligase I, this enzyme did not join Okazaki fragments in a reconstituted SV40 DNA replication assay *(71)*. In this study, a second 100-kDa DNA ligase activity with different biochemical properties was detected, raising the possibility that mammalian cells contain two different DNA ligases, which have similar electrophoretic mobility in denaturing polyacrylamide gels (*see below* and Section 2.4.).

A 100-kDa DNA ligase activity, presumed to be DNA ligase III, has been purified to homogeneity from bovine testis nuclear extracts *(29)*. These extracts do not contain significant amounts of DNA ligase II, whereas similarly prepared extracts from bovine liver contain high levels of DNA ligase II, but no significant amounts of DNA ligase III *(29,73)*. A comparison of DNA ligase III peptide sequences with those obtained from DNA ligase II identified 10 pairs of peptides that contained identical amino acid sequences *(29)*. The absence of differences in amino acid sequence suggests that DNA ligases II and III are encoded by the same gene. Although the simplest explanation is that these enzymes are related by artifactual proteolysis, the failure to produce DNA ligase II by limited proteolytic digestion of DNA ligase III *(29,68)* and the apparent modification of the amino-terminus of DNA ligase II *(73)* suggest that this is not the case.

Preliminary information about the cellular roles of the 100-kDa DNA ligase(s) has been provided by studies revealing the specific association of a 100-kDa DNA ligase with proteins involved in DNA repair and/or genetic recombination. Overexpression of a histidine-tagged version of the human DNA strand-break repair protein, Xrcc1 in the mutant CHO cell line, EM9 resulted in the specific binding of a 100-kDa DNA ligase activity to an immobilized metal affinity matrix *(10)*. This association between the 100-kDa DNA ligase and Xrcc1 implicates the DNA ligase in the repair of DNA single-strand breaks *(10,11,66)*. In addition, a 100-kDa DNA ligase with similar biochemical properties to those described for DNA ligase III copurifies with a high-mol wt complex that repairs double-strand breaks by a mechanism involving homologous recombination *(30)*.

Human cDNAs encoding 100 kDa DNA ligases have recently been isolated by two different approaches. In one approach, three different human cDNAs were identified by searching an expressed sequence tagged (EST) data base with the conserved peptide sequence described earlier (Fig. 2B) *(75)*. One of these sequences was identical with DNA ligase I, whereas the full-length versions of the other cDNAs isolated from HeLa and prostate cDNA libraries encoded distinct DNA ligase polypeptides, both with calculated molecular weights of about 100 kDa. The predicted amino acid sequence of the polypeptide encoded by the HeLa cDNA (calculated molecular weight of 103 kDa) contained regions that were strikingly homologous with the peptide sequences from bovine DNA ligase II and the 100-kDa DNA ligase purified from bovine testes *(29,73,75)*. The gene encoding this polypeptide has been designated *LIG3*, since the translated product of the cDNA forms a specific complex with Xrcc1 *(75)*. The other DNA ligase, encoded by the *LIG4* gene, is described in Section 2.4.

In an independent approach, human cDNA fragments were amplified by PCR using degenerate primers deduced from putative DNA ligase III peptide sequences *(15,29)*. Using these PCR fragments as probes, partial cDNA fragments were isolated from a human testis cDNA library, and a full-length cDNA was constructed from these clones. The open reading frame encoded by this cDNA is identical with that encoded by the cDNA isolated from the HeLa cDNA library *(75)*, except for the C-terminal region, and it encodes a polypeptide with a calculated molecular weight of 96 kDa *(15)*. The 77 C-terminal residues of the 103-kDa DNA ligase III encoded by the HeLa cDNA are replaced by 17 different residues in the version of DNA ligase III encoded by the testis cDNA (Fig. 3). A comparison of the nucleotide sequences reveals that these sequences diverge at the same position as the amino acid sequences and then appear to be unrelated. The DNA sequences at the point of divergence are homologous with consensus splice donor and acceptor sequences, suggesting that these cDNAs are alternatively spliced mRNAs from the same gene. The 103-kDa form of DNA ligase III is present in the testis, because two peptides isolated from bovine testis DNA ligase III can be aligned within the unique C-terminal region of the larger polypeptide *(29,75)*.

Interestingly, the amino-terminal region of Xrcc1 exhibits weak homology with the C-terminus of 103-kDa DNA ligase III, but not with the C-terminus of 96 kDa DNA ligase III *(15,75)*. The availability of cDNAs encoding DNA ligase III and Xrcc1 should facilitate the identification of the regions of these proteins that interact. The amino-terminal region of DNA ligase III contains a putative zinc finger that is homologous with one of the two zinc fingers of poly(ADP-ribose)polymerase *(75)*. Since it is situated outside of the conserved DNA ligase catalytic domain, it is unlikely that the DNA ligase III zinc finger is involved in the DNA-joining reaction. However, the identification of homologous sequences in three proteins that are involved in DNA strand break repair, DNA ligase III, Xrcc1, and poly(ADP-ribose)polymerase, may reflect the function of these proteins in the same pathway, possibly as a multiprotein complex.

All of the peptide sequences obtained from bovine DNA ligase II can be aligned within the human DNA ligase III open reading frame prior to the divergence in sequence at the C-terminus *(15)*. Assuming that the intervening regions are also present in DNA ligase II, these peptides encompass a region of 601 amino acids whose calculated molecular weight (~66 kDa) is similar to the observed molecular mass (70 kDa) of DNA ligase II (Fig. 3). Although the possibility that DNA ligase II is encoded by a different gene cannot be excluded, this observation together with the absence of changes in amino acid sequence between homologous peptides from DNA ligases II and III suggests that these enzymes are encoded by the same gene. DNA ligase II does not appear to be derived from the *LIG3* gene by alternative splicing, since the liver, where DNA ligase II is the major DNA-joining activity, does not contain an mRNA species different in size from the one detected in other tissues and cells *(15)*. Thus, DNA ligase II is probably generated from DNA ligase III by a specific proteolytic mechanism that involves removal of amino- and carboxy-terminal sequences (Fig. 3).

DNA ligase III appears to be ubiquitously expressed in mammalian tissues and cells *(15)*. Consistent with measurements of DNA ligase III activity in different tissues, the highest steady-state levels of DNA ligase III mRNA were detected in the testis *(15,29)*. In addition to the 3.6 kb mRNA detected in somatic tissues and cells, a 3.4-kb mRNA, is the major species in adult testis *(15)*. It appears that the 3.4- and 3.6-kb mRNAs correspond to the alternatively spliced products described above and encode the 96-kDa and 103-kDa forms of DNA ligase III, respectively. Based on its interaction with Xrcc1, DNA ligase III appears to function in the repair of DNA single-strand breaks generated as a consequence of DNA damage by alkylating agents and ionizing radiation (*10*; Section 3.1.). Since both *LIG3* and *XRCC1* are highly expressed in the testis *(15,72)*, these DNA strand-break repair proteins may also be involved in meiotic recombination during germ cell development (Section 3.2.).

2.4. DNA Ligase IV

As mentioned previously, a cDNA distinct from the ones encoding DNA ligases I and III was identified in the screen of an EST data base with the conserved peptide sequence *(75)*. A full-length cDNA, which was isolated from a prostate cDNA library, encodes a 96-kDa polypeptide that has DNA ligase activity *(75)*. Compared with DNA ligases I and III, this DNA ligase, designated DNA ligase IV, has a significantly shorter amino-terminal region prior to the putative site of adenylylation, and a significantly longer C-terminal sequence after the conserved peptide (Fig. 2B). The 323 amino acid C-terminal extension, which presumably mediates protein–protein interactions, is not homologous with any protein sequence in the public data bases. The *LIG4* gene, which has been mapped to human chromosome 13q33-34 (Table 1), is highly expressed in the thymus and testis, suggesting that this enzyme may be involved in recombination *(75)*.

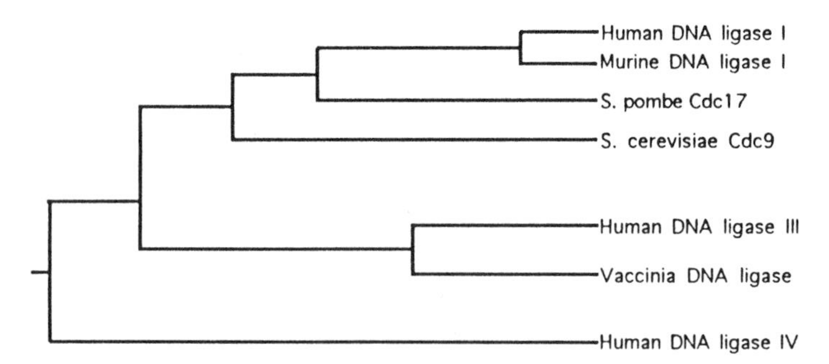

Fig. 4. Phylogenetic tree of eukaryotic DNA ligases. The amino acid sequences of the DNA ligases shown were obtained from public data bases. A phylogenetic tree was constructed with amino acid sequences from the active site motif to the conserved peptide using the MegAlign program (DNA Star).

2.5. Putative Evolutionary Relationship Between Eukaryotic DNA Ligases

A comparison of the catalytic domains of representative eukaryotic DNA ligases suggests that there may be three distinct families of eukaryotic DNA ligases (Fig. 4). The degree of amino acid identity shared between the conserved catalytic domains of the mammalian DNA ligases from different families ranges from 24–31%, whereas the catalytic domains of DNA ligases within the same family share more than 50% identity. The grouping of the replicative DNA ligases, mammalian DNA ligase I, Cdc9p DNA ligase, and Cdc17 DNA ligase, indicates that the conservation of amino acid sequence reflects functional homology. It is likely that future studies will identify functional homologs of mammalian DNA ligases III and IV in other eukaryotes and that different cellular roles will be assigned to the three DNA ligase families.

Presumably the differences in amino acid sequence within the catalytic domains of these enzymes account for the differences in DNA substrate specificity that were initially used to characterize DNA ligases I, II, and III (Table 1). For example, DNA ligase I and Cdc9p DNA ligase, which share 50% amino acid identity within their catalytic domains, cannot join oligo(dT) molecules that are hybridized to poly(rA), whereas DNA ligases II and III can catalyze joining of this substrate *(68,69)*. Reactivity with the oligo(dT)/poly(rA) substrate has been used to characterize DNA ligase activities as being distinct from the replicative DNA ligase I-type enzymes in extracts from *Xenopus laevis (2)* and *D. melanogaster (61)*. Until the reactivity of DNA ligase IV with the different homopolymer substrates has been examined, it cannot be assumed that oligo(dT)/poly(rA) joining activity is a unique feature of DNA ligase III-type enzymes.

Although assays with the homopolymer substrates (Table 1) have been useful for the biochemical characterization of DNA ligases, it seems unlikely that the observed differences in substrate specificity have any biological relevance. In assays with DNA substrates that contain a single defined nick, and therefore more closely resemble the in vivo substrate, it has been shown that DNA ligase I and cdc9p DNA ligase are much less tolerant of mismatched termini than DNA ligases II and III *(29,69)*. The more stringent DNA-joining requirements of these replicative DNA ligases may function to prevent the joining of Okazaki fragments with inappropriate termini and, thus, prevent incorporation of mismatched base pairs into the newly synthesized DNA molecule. It is possible that the ability to join DNA molecules with mismatched termini is tolerated or even required in certain DNA repair pathways.

3. DNA LIGASE FUNCTION

3.1. *Defective DNA Joining in Mammalian Cell Lines*

Studies on mammalian DNA ligase were stimulated by the association of a DNA ligation defect with the inherited, cancer-prone human disease, Bloom syndrome (BS) *(24)*. This rare disorder, which is inherited in an autosomal-recessive manner, appears to be caused by a defect in a single gene *(40)*. Cell lines established from BS patients exhibit chromosomal instability, in particular a high spontaneous frequency of sister-chromatid exchange (SCE), an elevated rate of spontaneous mutation, sensitivity to DNA-damaging agents and an abnormal profile of DNA replication intermediates (*24,38*; *see also* Chapter 19). These alterations in DNA metabolism mimic to a certain extent those observed in DNA ligase mutants of prokaryotes and lower eukaryotes. Consistent with a putative DNA ligation defect, BS cells are defective in the joining of linear plasmid DNA molecules *(52)* and have been reported to contain abnormal DNA ligase activity, presumed to be DNA ligase I *(12,13,77,78)*. However, the sequencing of BS DNA ligase I alleles failed to reveal mutations and demonstrated that defects in this gene are not responsible for BS *(5,45)*. This was confirmed by the localization of the BS gene to chromosome 15, whereas the DNA ligase I gene is on chromosome 19 *(4,40)*.

A single female patient, whose symptoms appear to have been caused by DNA ligase I mutations, has been identified. This patient, who initially presented with recurrent chest and ear infections, was found to have low levels of circulating immunoglobulins and T cells that did not respond to mitogens. She failed to develop normally and died at age 19 because of complications from a chest infection. Molecular genetic analysis of skin biopsies from this patient revealed a different point mutation in each of her DNA ligase I alleles *(5)*. Cell lines established from this patient exhibit a defect in the joining of Okazaki fragments, and are sensitive to a wide range of DNA-damaging agents and 3-aminobenzamide, an inhibitor of poly(ADP-ribose)polymerase *(27,33,39)*. One of the mutant alleles, whose origin is unknown, encodes an inactive polypeptide, whereas the other allele encodes a defective enzyme that retains about 10% activity *(5,49)*. Presumably this residual DNA ligase I activity is sufficient for DNA replication, but not for other DNA metabolic functions involving DNA ligase I. A similar apparent sequestration of DNA ligase activity for DNA replication has been observed in studies of conditional lethal DNA ligase mutants of *E. coli (21,34)*. The sensitivity of the DNA ligase I mutant cell lines to DNA damage, which can be complemented by expression of wild-type DNA ligase I cDNA *(59)*, implicates DNA ligase I in the repair of DNA single-strand breaks generated either directly by DNA-damaging agents or indirectly by DNA repair enzymes. In agreement with these observations, DNA ligase I has been identified as a component of a multiprotein complex that catalyzes the repair of a uracil-containing DNA substrate by a base excision repair mechanism *(47)*. Within this complex, DNA ligase I interacts with DNA polymerase β, which performs the DNA synthesis step of base excision repair (*47,56*; *see* Chapter 11). The severe combined immunodeficiency of the patient suggests that DNA ligase I is required for development and function the immune system. However, DNA ligase I mutant cells do not exhibit a defect in V(D)J recombination *(28,44)*. At the present time, the molecular link between the DNA ligase I defect and immunodeficiency is not understood.

The Chinese hamster ovary cell line, EM9, is hypersensitive to alkylating agents and is also sensitive to ionizing radiation *(65)*. Biochemical analysis of DNA repair in EM9 cells suggests that the defect is in the closure of DNA single-strand breaks. The human *XRCC1* gene complements the phenotype of the EM9 cell line *(66)* and, as described previously (Section 2.3.), the product of the *XRCC1* gene interacts with DNA ligase III *(10,11)*, implicating DNA ligase III in the repair of DNA single-strand breaks. EM9 cells, which have reduced levels of DNA ligase III activity *(37)*, were recently shown to have reduced amounts of both Xrcc1 and DNA ligase III polypeptides, suggesting that DNA ligase III is stabilized by its interaction with Xrcc1 *(11)*.

Interestingly, the EM9 cell line is the only other example of a mammalian cell line apart from BS cell lines that exhibits a high spontaneous frequency of SCE. *XRCC1* was excluded as a candidate gene for BS, because cell hybrids between BS and EM9 cell lines had normal levels of SCE *(50)*. Subsequently, the *XRCC1* and BS genes have been mapped to chromosomes 19 and 15, respectively *(40,54)*. The genes encoding DNA ligase III and DNA ligase IV have also been eliminated as candidates for BS by virtue of their chromosomal location (Table 1). Recently, the BS gene has been identified, and it appears to encode a DNA helicase *(20)*. It is likely that further investigation of the BS gene product will provide insights into the DNA-joining abnormalities of BS cells.

3.2. DNA Ligases Involved in DNA Repair and Genetic Recombination

There is substantial evidence indicating that DNA ligase I is the enzyme responsible for the joining Okazaki fragments during DNA replication *(35,36,46,57,71)*. The DNA ligase requirements of the different pathways of DNA repair and genetic recombination are less well established, and there is evidence indicating functional redundancy. As described above, deficiencies in DNA ligases I and III both cause a defect in the repair of DNA single-strand breaks. This observation can be explained if one assumes that there are different types of DNA single-strand breaks. For example, DNA strand breaks introduced by the base excision repair pathway should have appropriate termini for DNA ligation, whereas those introduced by ionizing radiation will usually have damaged termini that require processing prior to ligation *(23)*. Thus, DNA strand breaks may be repaired by two repair pathways that have overlapping, but nonidentical DNA substrate specificities and which utilize different DNA ligases. In this model, the extreme sensitivity of the DNA ligase I mutant cell line to 3-aminobenzamide *(33)*, which is already defective in the DNA ligase I-mediated DNA strand-break repair pathway, is caused by the inactivation of the second DNA single-strand-break repair pathway that involves poly(ADP-ribose)polymerase and possibly DNA ligase III, resulting in a cell with essentially no capacity to repair DNA single-strand breaks.

Although the mammalian DNA nucleotide excision repair pathway has been extensively studied (*1*; *see also* Chapters 10 and 18), there is no evidence linking a particular DNA ligase to the joining event that occurs after removal of a DNA-damage-containing oligonucleotide and subsequent gap filling. This DNA synthesis is most likely performed by DNA polymerase ε. Because a 100-kDa DNA ligase (possibly DNA ligase III or DNA ligase IV) is associated with a multiprotein complex containing DNA polymerase ε *(30)*, DNA ligase III or DNA ligase IV may be involved in nucleotide excision repair. However, the UV sensitivity of the DNA ligase I-deficient cell line 46BR *(74)* suggests that DNA ligase I may also be involved in nucleotide excision repair.

Mammalian cells appear to possess at least two recombination pathways that can repair DNA double-strand breaks. One pathway involves the Ku proteins and the p350 DNA-dependent protein kinase (*see* Chapters 16 and 17), which also function in the site-specific V(D)J recombination pathway of the immune system *(8,60)*. DNA ligase III and/or DNA ligase IV probably provides the DNA-joining activity in this pathway, since DNA ligase I deficiency has no effect on V(D)J recombination *(28,44)*. A second recombination pathway appears to be mediated by mammalian homologs of yeast genes in the *RAD52* epistasis group. In yeast, these gene products are involved in a homologous recombination pathway that repairs DNA double-strand breaks in vegetative cells and are required for meiotic recombination during sporulation (*23*; *see* Chapter 16 in vol. 1). Studies of the mammalian homolog of yeast Rad51p indicate that this protein also functions in DNA repair and in meiosis during spermatogenesis *(25,43,53)*. It is not known which DNA ligase completes these repair and recombination processes, although there is some evidence, outlined below, suggesting that DNA ligase III plays a role in meiotic recombination.

Both DNA ligase I and DNA ligase III are highly expressed in adult mammalian testis *(15)*. When expression of these genes was examined as a function of testis development in mice, the levels of DNA ligase I expression correlated with the contribution of spermatogonia to the testis. Thus, the testes of young animals, in which 20% of the cells are spermatogonia, contain the highest levels of DNA ligase I *(15)*. With increasing age, the contribution of spermatogonia to the testis gradually declines to about 1% in the adult animal *(6)*, and there is a corresponding decline in the steady-state expression levels of DNA ligase I *(15)*. Since spermatogonia are proliferating cells, the high level of DNA ligase I expression in these cells presumably reflects the involvement of this enzyme in DNA replication. In contrast, high levels of DNA ligase III expression correlated with the contribution of pachytene spermatocytes to the testis, and this was confirmed by *in situ* hybridization *(15)*. Since pachytene spermatocytes are undergoing meiotic recombination prior to the first meiotic division, it is possible that DNA ligase III plays a specific role in meiotic recombination. The expression of *XRCC1* is also elevated in pachytene spermatocytes *(72)* suggesting that DNA ligase III and Xrcc1 function together as a complex to repair DNA single-strand breaks generated as a consequence of meiotic recombination in germ cells, in the same manner as they repair DNA damage-induced single-strand breaks in somatic cells.

4. CONCLUDING REMARKS

DNA joining is required for the completion of DNA replication, DNA repair, and genetic recombination. The presence of three biochemically distinct DNA ligases in mammalian cell extracts suggests that each of these enzymes may function in different aspects of DNA metabolism. Biochemical and genetic studies have demonstrated that DNA ligase I joins Okazaki fragments during DNA replication and is also involved in

DNA repair. Similarly, DNA ligase III has been shown to function in the repair of DNA single-strand breaks and possibly in the completion of meiotic recombination. Studies have shown that DNA ligases II and III are very closely related, but it is not yet known whether they are encoded by the same gene. In addition to this unresolved question, a human cDNA encoding a novel DNA ligase of unknown function, DNA ligase IV, has been isolated.

It appears that many DNA transactions are carried out by multiprotein complexes. Therefore, the identification of polypeptides that interact with the different DNA ligases should provide insights into the cellular functions of these enzymes. The availability of DNA ligase cDNAs will also permit the investigation of DNA ligase function by genetic approaches, such as the isolation of homologous genes from lower eukaryotes and the use of gene targeting strategies to examine the phenotypic effects caused by the absence of a particular DNA ligase in whole-animal models. Thus, the recent progress made in studies of mammalian DNA ligases should lead to rapid advances in our understanding of the functions that these enzymes perform in DNA metabolism and the molecular mechanisms of DNA-joining in eukaryotes.

ACKNOWLEDGMENT

Studies in A. E. T.'s laboratory were supported by United States Public Health Service Grant GM47251.

REFERENCES

1. Aboussekhra, A., M. Biggerstaff, M. K. K. Shivji, J. A. Vilpo, V. Moncollin, V. N. Podust, M. Protic, U. Hubscher, J.-M. Egly, and R. D. Wood. 1995. Mammalian DNA nucleotide excision repair reconstituted with purified protein components. *Cell* **80:** 859–868.
2. Aoufouchi, S., C. Prigent, N. Theze, M. Phillipe, and P. Thiebaud. 1992. Expression of DNA ligases I and II during oogenesis and early development. *Dev. Biol.* **152:** 190–202.
3. Arrand, J. E., A. E. Willis, I. Goldsmith, and T. Lindahl. 1986. Different substrate specificities of the two DNA ligases of mammalian cells. *J. Biol. Chem.* **261:** 9079–9082.
4. Barnes, D. E., L. H. Johnston, K. Kodama, A. E. Tomkinson, D. D. Lasko, and T. Lindahl. 1990. Human DNA ligase I cDNA: Cloning and functional expression in *Saccharomyces cerevisiae*. *Proc. Natl. Acad. Sci. USA* **87:** 6679–6683.
5. Barnes, D. E., A. E. Tomkinson, A. R. Lehmann, A. D. B. Webster, and T. Lindahl. 1992. Mutations in the DNA ligase I gene of an individual with immunodeficiencies and cellular hypersensitivity to DNA damaging agents. *Cell* **69:** 495–503.
6. Bellve, A. R., J. C. Cavicchia, C. F. Millette, D. A. O'Brien, Y. M. Bhatnagar, and M. Dym. 1977. Spermatogenic cells of the prepubertal mouse: isolation and morphological characterization. *J. Cell. Biol.* **74:** 68–85.
7. Bentley, D. J., J. Selfridge, J. K. Millar, K. Samuel, N. Hole, J. D. Ansell, and D. W. Melton. 1996. DNA ligase I is required for fetal liver erythropoiesis but is not essential for mammalian cell viability. *Nat. Genet.* **13:** 489–491.
8. Blunt, T., N. J. Finnie, G. E. Taccioli, G. C. M. Smith, J. Demengeot, T. M. Gottlieb, R. Mizuta, A. J. Varghese, F. W. Alt, P. A. Jeggo, and S. P. Jackson. 1995. Defective DNA-dependent protein kinase activity is linked to V(D)J recombination and DNA repair defects associated with the murine scid mutation. *Cell* **80:** 813–823.
9. Brown, J. L. and W. K. Roberts. 1976. Evidence that approximately eighty percent of soluble proteins from Ehrlich ascites cells are *N*-acetylated. *J. Biol. Chem.* **251:** 1009–1014.
10. Caldecott, K. W., C. K. McKeown, J. D. Tucker, S. Ljunquist, and L. H. Thompson. 1994. An interaction between the mammalian DNA repair protein Xrcc1 and DNA ligase III. *Mol. Cell. Biol.* **14:** 68–76.

11. Caldecott, K. W., C. K. McKeown, J. D. Tucker, L. Stanker, and L. H. Thompson. 1996. Characterization of the Xrcc1-DNA ligase III complex in vitro and its absence from mutant hamster cells. *Nucleic Acids Res.* **23:** 4836–4843.
12. Chan, J. J. H., L. H. Thompson, and F. F. Becker. 1984. DNA ligase activites appear normal in the CHO mutant EM9. *Mutat. Res.* **131:** 209–214.
13. Chan, J. Y., F. F. Becker, G. J., and J. H. Ray. 1987. Altered DNA ligase I activity in Bloom's syndrome cells. *Nature* **325:** 357–359.
14. Chan, J. Y.-H., and F. F. Becker. 1988. Defective DNA ligase I in Bloom's syndrome cells. *J. Biol. Chem.* **263:** 18,231–18,235.
15. Chen, J., A. E. Tomkinson, W. Ramos, Z. B. Mackey, S. Danehower, C. A. Walter, R. A. Schultz, J. M. Besterman, and I. Husain. 1995. Mammalian DNA ligase III: molecular cloning, chromosomal localization, and expression in spermatocytes undergoing meiotic recombination. *Mol. Cell. Biol.* **15:** 5412–5422.
16. Colinas, R. J., S. J. Goebel, S. W. Davis, G. P. Johnson, E. K. Norton, and E. Paoletti. 1990. A DNA ligase gene in the Copenhagen strain of vaccinia virus is non-essential for viral replication and recombination. *Virology* **179:** 267–275.
17. Cong, P. and S. Shuman. 1993. Covalent catalysis in nucleotidyl transfer. A KTDG motif essential for enzyme-GMP formation by mRNA capping enzyme is conserved at the active site of RNA and DNA ligases. *J. Biol. Chem.* **268:** 7256–7260.
18. Creissen, D. and S. Shall. 1982. Regulation of DNA ligase activity by poly(ADP-ribose). *Nature* **296:** 271–272.
19. Elder, R. H. and J.-M. Rossignol. 1990. DNA ligases from rat liver. Purification and partial characterization of two molecular forms. *Biochemistry* **29:** 6009–6017.
20. Ellis, N. A., J. Groden, T. Z. Ye, J. Straughen, D. Lennon, S. Ciocci, M. Protycheva, and J. German. 1996. The Bloom's syndrome gene product is homologous to RecQ helicases. *Cell* **83:** 655–666.
21. Engler, M. J. and C. C. Richardson. 1982. DNA ligases, in *The Enzymes*, vol. XV (Boyer, P. D., ed.), Academic, New York, pp. 3–29.
22. Fresco, L. D. and S. Buratowski. 1994. Active site of the mRNA-capping enzyme guanylyltransferase from *Saccharomyces cerevisiae*: similarity to the nucleotidyl attachment motif of DNA and RNA ligases. *Proc. Natl. Acad. Sci. USA* **91:** 6624–6628.
23. Friedberg, E. C., G. C. Walker, and W. Seide. 1995. *DNA Repair and Mutagenesis*. ASM, Washington, DC.
24. German, J. and E. Passarge. 1989. Bloom's syndrome. XII. Report from the Registry for 1987. *Clin. Genet.* **35:** 57–69.
25. Haaf, T., B. Golub, G. Reddy, C. M. Radding, and D. C. Ward. 1995. Nuclear foci of mammalian Rad51 protein in somatic cells after DNA damage and its localization in synaptonemal complexes. *Proc. Natl. Acad. Sci. USA* **92:** 2298–2302.
26. Heaphy, S., M. Singh, and M. J. Gait. 1987. Effects of single amino acid changes in the region of the adenylylation site of T4 RNA ligase. *Biochemistry* **26:** 1688–1696.
27. Henderson, L. M., C. F. Arlett, S. A. Harcourt, A. R. Lehmann, and B. C. Broughton. 1985. Cells from an immunodeficient patient (46BR) with a defect in DNA ligation are hypomutable but sensitive to the induction of sister chromatid exchanges. *Proc. Natl. Acad. Sci. USA* **82:** 2044–2048.
28. Hsieh, C. L., C. F. Arlett, and M. R. Lieber. 1993. V(D)J recombination in ataxia telangiectasia, Bloom syndrome and a DNA ligase I-associated immunodeficiency. *J. Biol. Chem.* **268:** 20,105–20,109.
29. Husain, I., A. E. Tomkinson, W. A. Burkhart, M. B. Moyer, W. Ramos, Z. B. Mackey, J. M. Besterman, and J. Chen. 1995. Purification and characterization of DNA ligase III from bovine testes. *J. Biol. Chem.* **270:** 9683–9690.
30. Jessberger, R., V. Podost, U. Hubscher, and P. Berg. 1993. A mammalian protein complex that repairs double-strand breaks and deletions by recombination. *J. Biol. Chem.* **268:** 15,070–15,079.

31. Kerr, S. M., L. H. Johnston, M. Odell, S. A. Duncan, K. M. Law, and G. L. Smith. 1991. Vaccinia virus complements Saccharomyces cerevisiae cdc9, localizes in cytoplasmic factories and affects virulence and virus sensitivity to DNA damaging agents. *EMBO J.* **10:** 4343–4350.
32. Kodama, K., D. E. Barnes, and T. Lindahl. 1991. In vitro mutagenesis and functional expression in *Escherichia coli* of a cDNA encoding the catalytic domain of human DNA ligase I. *Nucleic Acids Res.* **19:** 6093–6099.
33. Lehman, A. R., A. E. Willis, B. C. Broughton, M. R. James, H. Steingrimdottir, S. A. Harcourt, C. F. Arlett, and T. Lindahl. 1988. Relation between the human fibroblast strain 46BR and cell lines representative of Bloom's syndrome. *Cancer Res.* **48:** 6343–6347.
34. Lehman, I. R. 1974. DNA ligase: structure, mechanism and function. *Science* **186:** 790–797.
35. Li, C., J. Goodchild, and E. F. Baril. 1994. DNA ligase I is associated with the 21S complex of enzymes for DNA synthesis in HeLa cells. *Nucleic Acids Res.* **22:** 632–638.
36. Lindahl, T. and D. E. Barnes. 1992. Mammalian DNA ligases. *Ann. Rev. Biochem.* **61:** 251–281.
37. Ljungquist, S., K. Kenne, L. Olsson, and M. Sandstrom. 1994. Altered DNA ligase III activity in the CHO EM9 mutant. *Mutat. Res.* **314:** 177–186.
38. Lonn, U., Lonn, S., Nylen, U., Winblad, G., and German J. 1990. An abnormal profile of DNA replication intermediates in Bloom's syndrome. *Cancer Res.* **50:** 3141–3145.
39. Lonn, U., S. Lonn, U. Nylen, and G. Winblad. 1989. Altered formation of DNA replication intermediates in human 46BR fibroblast cells hypersensitve to 3-amino benzamide. *Carcinogenesis* **10:** 981–985.
40. McDaniel, L. D. and R. A. Schultz. 1992. Elevated sister chromatid exchange phenotype of Bloom's syndrome cells is complemented by human chromosome 15. *Proc. Natl. Acad. Sci. USA* **89:** 7968–7972.
41. Montecucco, A., G. Biamonti, E. Savini, F. Focher, S. Spadari, and G. Ciarrocchi. 1992. DNA ligase I gene expression during differentiation and cell proliferation. *Nucleic Acids Res.* **20:** 6209–6214.
42. Montecucco, A., E. Savini, G. Biamonti, M. Stefanini, F. Focher, and G. Ciarrocchi. 1995. Late induction of human DNA ligase I after UV-C irradiation. *Nucleic Acids Res.* **23:** 962–966.
43. Morita, T., Y. Yoshimura, A. Yamamoto, K. Murata, M. Mori, H. Yamamoto, and A. Matsushiro. 1993. A mouse homolog of the *Escherichia coli recA* and *Saccharomyces cerevisiae RAD51* genes. *Proc. Natl. Acad. Sci. USA* **90:** 6577–6580.
44. Petrini, J. H. J, J. W. Donovan, C. Dimare, and D. T. Weaver. 1994. Normal V(D)J coding joint formation in DNA ligase I deficiency. *J. Immunol.* **152:** 176–183.
45. Petrini, J. H. J., K. G. Huwiler, and D. T. Weaver. 1991. A wild type DNA ligase I gene is expressed in Bloom's syndrome cells. *Proc. Natl. Acad. Sci. USA* **88:** 7615–7619.
46. Petrini, J. H. J., Y. Xiao, and D. T. Weaver. 1995. DNA ligase I mediates essential functions in mammalian cells. *Mol. Cell. Biol.* **15:** 4303–4308.
47. Prasad, R., R. K. Singhal, D. K. Srivastava, J. T. Molina, A. E. Tomkinson, and S. H. Wilson. 1996. Specific interaction of DNA polymerase β and DNA ligase I in a multiprotein base excision repair complex. *J. Biol Chem.* **271:** 16,000–16,007.
48. Prigent, C., D. D. Lasko, K. Kodama, J. R. Woodgett, and T. Lindahl. 1994. Activation of mammalian DNA ligase I through phosphorylation by casein kinase II. *EMBO J.* **11:** 2925–2933.
49. Prigent, C., M. S. Satoh, G. Daly, D. E. Barnes, and T. Lindahl. 1994. Aberrant DNA Repair and DNA Replication due to an inherited defect in Human DNA ligase I. *Mol. Cell. Biol.* **14:** 310–317.
50. Ray, J. H., E. Louie, and J. German. 1987. Different mutations are responsible for the elevated sister-chromatid exchange frequencies characteristic of Bloom's syndrome and hamster EM9 cells. *Proc. Natl. Acad. Sci. USA* **84:** 2368–2371.

51. Roberts, E., R. A. Nash, P. Robins, and T. Lindahl. 1994. Different active sites of mammalian DNA ligases I and II. *J. Biol. Chem.* **269:** 3789–3792.
52. Runger, T. M. and K. H. Kraemer. 1989. Joining of linear plasmid DNA is reduced and error-prone in Bloom's syndrome cells. *EMBO J.* **8:** 1419–1425.
53. Shinohara, A., H. Ogawa, Y. Matsuda, N. Ushio, K. Ikeo, and T. Ogawa. 1993. Cloning of the human, mouse and fission yeast recombination genes homologous to *RAD51* and *recA*. *Nature Genet.* **4:** 239–243.
54. Siciliano, M. J., L. J. Carrano, and L. H. Thompson. 1986. Assignment of a human DNA repair gene associated with sister chromatid exchange to human chromosome 19. *Mutat. Res.* **174:** 303–308.
55. Smith, G. L., Chan, Y. S., and Kerr, S. M. 1989. Transcriptional mapping and nucleotide sequence of a vaccinia virus gene encoding a polypeptide with extensive homology to DNA ligases. *Nucleic Acids Res.* **17:** 9051–9062.
56. Sobol, R. W., J. K. Horton, R. Kuhn, H. Gu., R. K. Singhal, R. Prasad, K. Rajewsky, and S. H. Wilson. 1996. Requirement of mammalian DNA polymerase β in base excision repair. *Nature* **379:** 183–186.
57. Soderhall, S. 1976. DNA ligases during rat liver regeneration. *Nature* **260:** 640–642.
58. Soderhall, S. and T. Lindahl. 1975. Mammalian DNA ligases. Serological evidence for two separate enzymes. *J. Biol. Chem.* **250:** 8438–8444.
59. Somia, N. V., J. K. Jessop, and D. W. Melton. 1993. Phenotypic correction of a human cell line (46BR) with aberrant DNA ligase I. *Mutat. Res.* **294:** 51–58.
60. Taccioli, G. E., T. M. Gottlieb, T. Blunt, A. Priestley, J. Demengeot, R. Mizuta, A. R. Lehmann, F. W. Alt, S. P. Jackson, and P. A. Jeggo. 1994. Ku80: Product of the *XRCC5* gene and its role in DNA repair and V(D)J recombination. *Science* **265:** 1442–1445.
61. Takahashi, M., and M. Senshu. 1987. Two distinct DNA ligases from *Drosophila melanogaster* embryos. *FEBS Lett.* **213:** 345–352.
62. Teraoka, H., H. Minami, S. Iijima, K. Tsukada, O. Koiwa, and T. Date. 1993. Expression of active human DNA ligase I in *E. coli* cells that harbor a full-length DNA ligase I cDNA construct. *J. Biol. Chem.* **268:** 24,156–24,162.
63. Teraoka, H., T. Sumikawa, and K. Tsukada. 1986. Purification of DNA ligase II from calf thymus and preparation of rabbit antibody against calf thymus DNA ligase II. *J. Biol. Chem.* **261:** 6888–6892.
64. Teraoka, H. and K. Tsukada. 1982. Eukaryotic DNA ligase. Purification and properties of the enzyme from bovine thymus and immunochemical studies of the enzyme from animal tissues. *J. Biol. Chem.* **257:** 4758–4763.
65. Thompson, L. H., K. W. Brookman, L. E. Dillehay, A. V. Carrano, J. A. Mazrimas, C. L. Mooney, and J. L. Minkler. 1982. A CHO-cell strain having hypersensitivity to mutagens, a defect in strand break repair, and an extraordinary baseline frequency of sister chromatid exchange. *Mutat. Res.* **95:** 247–254.
66. Thompson, L. H., K. W. Brookman, N. J. Jones, S. A. Allen, and A. V. Carrano. 1990. Molecular cloning of the human *XRCC1* gene, which corrects defective DNA strand break repair and sister chromatid exchange. *Mol. Cell. Biol.* **10:** 6160–6171.
67. Tomkinson, A. E., D. D. Lasko, G. Daly, and T. Lindahl. 1990. Mammalian DNA ligases. Catalytic domain and size of DNA ligase I. *J. Biol. Chem.* **265:** 12,611–12,617.
68. Tomkinson, A. E., E. Roberts, G. Daly, N. F. Totty, and T. Lindahl. 1991. Three distinct DNA ligases in mammalian cells. *J. Biol. Chem.* **286:** 21,728–21,735.
69. Tomkinson, A. E., N. J. Tappe, and E. C. Friedberg. 1992. DNA ligase from *Saccharomyces cerevisiae*: physical and biochemical characterization of the *CDC9* gene product. *Biochemistry* **31:** 11,762–11,771.
70. Tomkinson, A. E., N. F. Totty, M. Ginsburg, and T. Lindahl. 1991. Location of the active site for enzyme-adenylate formation in DNA ligases. *Proc. Natl. Acad. Sci. USA* **88:** 400–404.

71. Waga, S., G. Bauer, and B. Stillman. 1994. Reconstitution of complete SV40 replication with purified replication factors. *J. Biol. Chem.* **269:** 10,923–10,934.
72. Walter, C. A., J. Lu, M. Bhakta, Z.-Q. Zhou, L. H. Thompson, and J. R. McCarrey. 1995. Testis and Somatic *XRCC1* DNA repair gene expression. *Somatic Cell Mol. Genet.* **20:** 451–461.
73. Wang, Y. C., W. A. Burkhart, Z. B. Mackey, M. B. Moyer, W. Ramos, I. Husain, J. Chen, J. M. Besterman, and A. E. Tomkinson. 1994. Mammalian DNA ligase II is highly homologous with vaccinia DNA ligase; identification of the DNA ligase II active site for enzyme-adenylate formation. *J. Biol. Chem.* **269:** 31,923–31,928.
74. Webster, D., C. F. Arlett, S. A. Harcourt, I. A. Teo, and L. Henderson. 1982. A new syndrome of immunodeficiency and increased sensitivity to DNA damaging agents, in *Ataxia telangiectasia—A Cellular and Molecular Link Between Cancer, Neuropathology and Immune Deficiency* (Bridges, B. A. and D. G. Harnden, eds.), Wiley, New York, pp. 379–386.
75. Wei, Y.-F., P. Robins, K. Carter, K. Caldecott, D. J. C. Pappin, G.-L. Yu, R.-P. Wang, B. K. Shell, R. A. Nash, P. Schar, D. E. Barnes, W. A. Haseltine, and T. Lindahl. 1995. Molecular cloning and expression of human cDNAs encoding a novel DNA ligase IV and DNA ligase III, an enzyme active in DNA repair and genetic recombination. *Mol. Cell. Biol.* **15:** 3206–3216.
76. Wilcock, D. and D. P. Lane. 1991. Localization of p53, retinoblastoma and host replication proteins at sites of viral replication in herpes-infected cells. *Nature* **349:** 429–431.
77. Willis, A. E. and T. Lindahl. 1987. DNA ligase I deficiency in Bloom's syndrome. *Nature* **325:** 355–357.
78. Willis, A. E., R. Weksberg, S. Tomlinson, and T. Lindahl. 1987. Structural alterations of DNA ligase I in Bloom's syndrome. *Proc. Natl. Acad. Sci. USA* **84:** 8016–8020.
79. Xu, Q., D. Teplow, T. D. Lee, and J. Abelson. 1990. Domain structure in Yeast tRNA ligase. *Biochemistry* **29:** 6132–6138.
80. Yang, S. W., F. F. Becker, and J. Y. H. Chan. 1992. Identification of a specific inhibitor for DNA ligase I in human cells. *Proc. Natl. Acad. Sci. USA* **89:** 2227–2231.

13

Modulations in Chromatin Structure During DNA Damage Formation and DNA Repair

Michael J. Smerdon and Fritz Thoma

1. INTRODUCTION

It has long been recognized that the "target" of DNA-damaging agents and the "substrate" of DNA repair enzymes in eukaryotes is the highly compact and dynamic structure of chromatin. Understanding the modulation of DNA damage and repair in chromatin, as well as the modulation of chromatin structure by DNA damage and its repair processing, is necessary for understanding the fate of potential mutagenic and carcinogenic lesions in DNA. The central idea to be discussed in this chapter is that DNA damage, DNA repair (as well as other DNA-processing mechanisms), and chromatin structure are intimately associated in the cell (Fig. 1).

Many reports have appeared that directly or indirectly relate to the accessibility of DNA damage to repair enzymes in chromatin. Our understanding of DNA repair at the nucleosome level (or primary level of DNA packaging) has been increased significantly by:

1. High-resolution mapping of DNA lesions in chromatin, including well-defined nucleosome substrates;
2. Structure/function studies on how proteins interact with nucleosomes and modulate chromatin structure;
3. Monitoring the maturation of nascent repair patches in chromatin; and
4. Monitoring DNA damage and repair heterogeneity in chromatin.

Preferential repair of transcriptionally active genes in chromatin has been reviewed extensively (*see* Chapters 10 and 18), and this topic will only be addressed here in the context that transcriptionally active regions of chromatin are known to have unique structural features, which may play a role in the preferential repair of active genes.

We start with a review of chromatin structure, focusing on features that may play a role in modulation of DNA damage and/or DNA repair processing. The rest of the text focuses on DNA damage (Section 3.) and DNA repair (Section 4.) in chromatin.

2. CHROMATIN STRUCTURE AND MODIFICATION

Chromatin is organized in several structural levels: nucleosomes, chromatin fibers, and higher-order structures that condense into recognizable chromosomes during mito-

From: DNA Damage and Repair, Vol. 2: DNA Repair in Higher Eukaryotes
Edited by: J. A. Nickoloff and M. F. Hoekstra © Humana Press Inc., Totowa, NJ

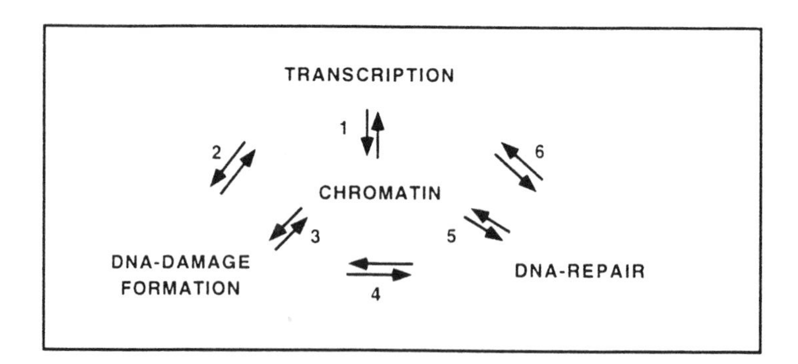

Fig. 1. Mutual effects between chromatin structure, transcription, DNA damage formation, and DNA repair. Chromatin is the substrate of all DNA-dependent processes. (1) Chromatin structures are altered when chromatin is prepared for transcription. Transcription generates loss of nucleosomes (or nucleosome rearrangement) and local supercoils. (2) DNA damage may block RNA polymerase or alter transcription factor binding. DNA damage formation may be affected by generation of transcription-dependent supercoils or nucleosome destabilization. (3) Folding DNA in chromatin affects the accessibility and reactivity of DNA; DNA lesions may affect folding in nucleosomes (*see* Fig. 2). (4) DNA damage induces DNA repair pathways; DNA repair removes DNA lesions. (5) Chromatin structure and dynamic properties modulate accessibility to damage recognition and repair proteins. DNA repair leads to opening of chromatin and rearrangement of chromatin structures. (6) Transcription can enhance DNA repair (transcription-coupled repair) and these processes may "share" proteins. Stalled RNA polymerase or tight-binding transcription factors may block access to damage-recognition proteins.

sis. Histone proteins serve a general role in packaging DNA into nucleosomes and chromatin fibers, and act as general repressors of transcription, but also as specific regulators of gene transcription (for reviews, *see* refs. *27,58,86,93,103*).

2.1. Histones

Each of the four types of core histones, H2A, H2B, H3, and H4, contain a region of four α-helices flanked at the N-terminus, and for H2A also at the C-terminus, by an apparently unstructured tail (refs. in *3,58,93,103*). Two copies of the lysine-rich H2A-H2B dimer and one of the H3-H4 tetramer assemble into a histone octamer. Histone H1 (or "linker histone") contains a short unstructured N-terminal tail, a globular central region, and a long unstructured C-terminal tail. The globular domain appears to be responsible for specific binding on the nucleosome. The unstructured terminal regions of the histones are not required for formation of the histone octamer nor the nucleosome core, but they contain the sites for posttranslational modifications. These include methylation, acetylation, phosphorylation, ubiquitination, and poly-ADP-ribosylation, and appear to be involved in the modulation of structural and functional properties of nucleosomes.

2.2. Nucleosomes

Histone proteins are associated with every 160–240 bp of genomic DNA to form arrays of nucleosomes, called "nucleosome filaments" *(90)*. The nucleosomes may be released and purified by mild digestion with micrococcal nuclease (refs. in *93,103*). These nucleosomes contain a full histone octamer and about one histone H1. The average DNA content varies between 160 and 240 bp. After more extensive digestion, a chromatosome particle can be purified that consists of 168 bp of DNA wrapped in two left-handed superhelical coils around the histone octamer and histone H1. Further digestion removes about 10 bp of DNA at each end of the chromatosome as well as histone H1, and produces the nucleosome core particle. Although the core histones, as well as the nucleosome core particle, are well conserved in evolution from yeast to humans, the DNA that connects core particles (here referred to as linker DNA) may vary between 0 and 90 bp, depending on the organism, cell type, and even within the same cell. Histone H1 may be absent in some organisms such as yeast *Saccharomyces cerevisiae* or at certain stages during development.

2.3. Nucleosome Core

Crystal structure data on the histone octamer *(3)* and the nucleosome core particle *(60)*, in combination with numerous biophysical and biochemical data, allowed development of an accurate picture of the nucleosome core particle organization *(3,58,61)*. The core of a nucleosome is a cylindrical particle, about 5.7 nm high and 11 nm in diameter. In nucleosome cores, about 145 bp of DNA are wound in 1.8 left-handed superhelical turns around the octamer. H3 and H4, as a tetramer $(H3\text{-}H4)_2$, tightly interact with and organize the central six turns of the DNA double helix, whereas each of the two H2A-H2B heterodimers bind less tightly on opposite faces of the tetramer and organize the next four turns of DNA. This makes the nucleosome core a pseudosymmetric particle with a dyad axis running through the minor groove about 72 bp from each end. One superhelical turn of nucleosomal DNA is shown in Fig. 2B,D).

The H3-H4 tetramer binds DNA more tightly than the H2A-H2B dimer. Thereby, the ends of the nucleosome core are easier to unfold and, in this respect, are more reactive. Binding of H1 stabilizes the nucleosome and prevents unfolding. The formation of nucleosome cores requires only the structured domains of core histones. The N-terminal tails are sensitive to proteolysis and, hence, appear to extend outside of the particle. They may serve a structural role by interacting with linker DNA or in stabilizing chromatin fibers. The exposed location of the N-terminal tails and the fact that they carry the protein modification sites support a role of N-terminal tail modification in remodeling of chromatin for participation in DNA-dependent processes.

Folding of DNA affects its structure and thereby its reactivity. Because the DNA is spooled around the outside of the histone octamer, it has an inner surface that is relatively inaccessible to bulky drugs and proteins, and an outer surface readily accessible from solution (Fig. 2B). The inner and outer surfaces define the orientation of the DNA on the nucleosome surface, the "rotational setting." The DNA in the nucleosome is not smoothly folded in the superhelix, but displays four points of tight bending. In addition, it is underwound and overwound in the central turns and toward the ends, respectively, and minor and major grooves are compressed when facing the histones and expanded when facing the solvent *(28,60)*.

2.4. Nucleosome Positioning

Owing to the limited accessibility of DNA in a nucleosome, the location of the histone octamer on the DNA sequence (a nucleosome position) can influence all processes that require DNA recognition by proteins *(86,92)*. The following terms need to

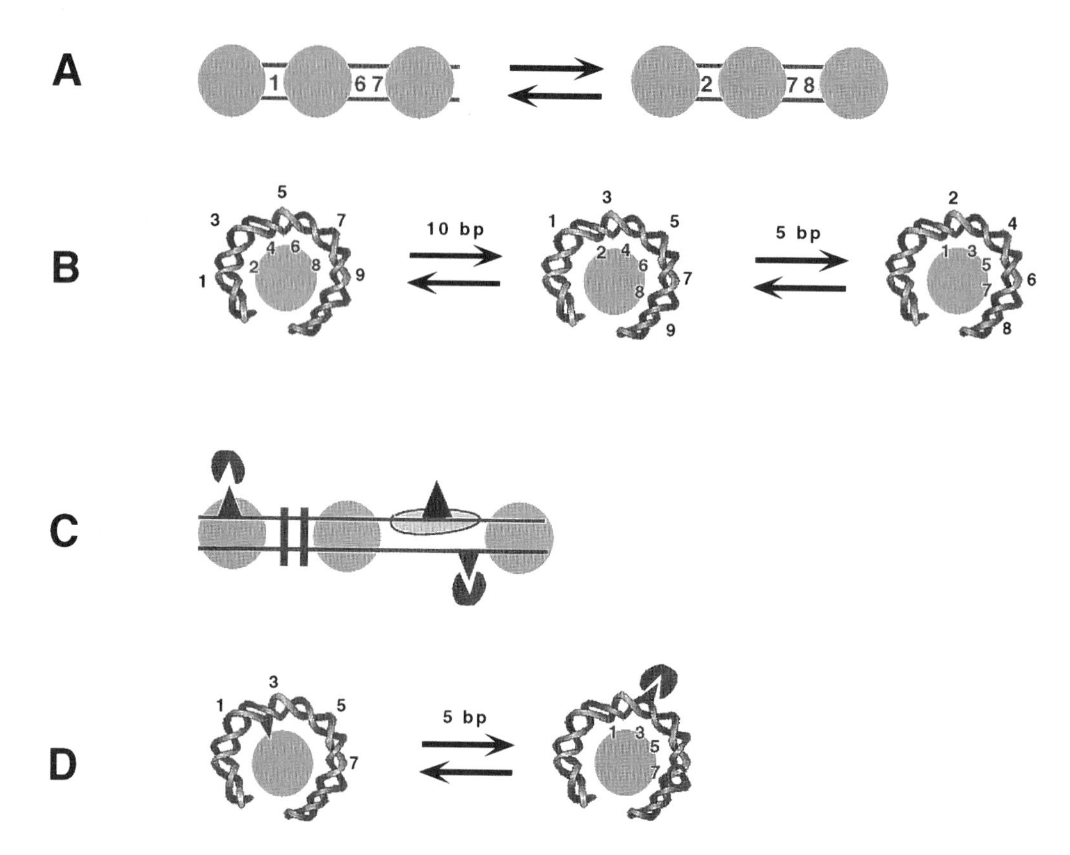

Fig. 2. (A) The location of a histone octamer (gray circles) on the DNA sequence defines a nucleosome position ("translational positioning"). DNA may be moved in and out of nucleosomes by DNA-processing mechanisms and dynamic properties that affect nucleosome positioning. **(B)** Schematic illustration of one superhelical turn of nucleosomal DNA on the histone octamer (adopted from ref. *60*). The DNA structure is severly distorted. The "inner surface" (even numbers) facing the histone octamer and the "outer surface" (odd numbers) define the "rotational setting." A translational shift of the octamer by 10 bp maintains their rotational setting, whereas a shift by 5 bp turns the inner surface outside. Different positions may exist in an equilibrium (arrows). **(C)** Linker DNA, nucleosomes "(black circles)" or factors "(open ovals)" bound to DNA affect the formation of DNA lesions (monoadducts, triangles; crosslinks, vertical bars), and lesion accessibility to repair enzymes (black circles). **(D)** Damage formation on the nucleosome surface can be modulated by local DNA distortions. Accommodation of DNA damage may require alterations in the rotational and translational setting, and affect the equilibrium. DNA damage recognition may depend on the "visibility" of lesions on the surface.

be distinguished: "rotational positioning" refers to the orientation of the DNA molecule on the nucleosome surface, "translational positioning" refers to the exact position of a histone octamer on the DNA sequence. A shift in translational positioning by 10 bp maintains the rotational setting, whereas a shift of 5 bp turns the "inner surface" outside (Fig. 2A,B).

Several parameters are known to affect nucleosome positioning (reviewed in refs. *86,92*).

1. Mechanical properties of DNA (bendability) that favor folding around histone octamers result in rotationally positioned nucleosomes. In particular, statistical analysis of nucleosome core DNA showed alternating $(A/T)_3$ and $(G/C)_3$ elements with a total period of about 10.2 bp, and suggested that such sequences are favored at sites of minor groove compression and major groove compression, respectively. In contrast, long poly dA·dT tracts, which show unusual rigid structure in solution, appear to be less favorable for incorporation in nucleosomes and in particular are not found in the center of nucleosomes.
2. Proteins other than histones can position nucleosomes by direct or protein-mediated contacts with histones.
3. Proteins, positioned nucleosomes, or DNA that does not fold into nucleosomes could affect nucleosome positioning in the flanking regions by a "boundary" effect *(34)*. Indeed, exchanging nuclease-sensitive regions in yeast chromatin showed that nuclease-sensitive regions could alter positions of nucleosomes in the flanking region.
4. Three-dimensional folding of chromatin was also shown to affect nucleosome positioning.

2.5. *Dynamic Properties of Nucleosomes*

Nucleosomes are dynamic structures that undergo various structural transitions, such as disassembly/reassembly, unfolding, or changing positions (reviewed in ref. *86*). They may partially or completely dissociate and reassemble. In vitro, nucleosomes dissociate in high salt (>1*M* NaCl), and histone octamers reassemble on DNA when the ionic strength is reduced. Reconstitution in vitro is efficient with short linear DNA (<200 bp) facilitating the study of nucleosome cores on defined sequences of damaged and undamaged DNA (*see* Section 3.). Cellular extracts, containing chromatin assembly factors, may provide a more natural spacing of nucleosomes on long DNA, but nucleosomes appear not to be positioned. Disassembly and reassembly in vitro mimic processes in vivo during replication and transcription. For example, during chromatin replication in vivo, nucleosomes dissociate in front of the replication fork and reassemble within a few hundred base pairs behind the replication fork (reviewed in ref. *78*). Chaperone proteins exist that can combine with histones while they are displaced from DNA and histone acetylation, which destabilizes histone–DNA contacts, and may facilitate dissociation (reviewed in ref. *103*).

Nucleosomes may unfold and refold. These structural transitions can be generated in vitro at very low ionic strength or in the presence of urea. Although these transitions could provide a basic mechanism to facilitate transcription, replication, or DNA repair through nucleosomes, evidence that unfolding occurs in vivo is limited (e.g., *see* ref. *12*; *see also* Section 4.).

The histone octamers may slide along DNA ("nucleosome mobility"), both in vitro and in vivo *(45)*. This sliding may be promoted on long chromatin by incubation at elevated temperature and/or by removal of histone H1. Nucleosome positions frequently differ by multiples of 10 bp and maintain the rotational setting on the *Xenopus* 5S rDNA in vitro and in vivo *(9,56)*. In the yeast *URA3* gene, multiple positions that do not maintain the rotational setting are observed close to positions that maintain the rotational setting, indicating that there is no tight coordination in positioning at that level *(84)*. Generally, there seems to be a preference to maintain the rotational setting, while translational positioning is less strongly enforced. Several states may exist in a dynamic equilibrium (Fig. 2A,B). These dynamic properties are essential to allow DNA-dependent reactions, such as transcription, replication, and repair, to occur within a chromatin template (reviewed in ref. *86*).

2.6. Nucleosome Structure and Positioning Affect Function

The functional consequences of packaging DNA in nucleosomes have been extensively studied with respect to transcription and transcriptional regulation, both in vitro and in vivo. In particular, molecular genetic work in yeast contributed significantly to our understanding of the complex roles of histone proteins and their domains in gene regulation *(27)*. Generally, transcriptional regulation requires accessibility of regulatory sequences to transactivating or repressing factors. The factors need to interact with proteins bound to promoter elements, and eventually they affect loading of the transcription complex on the DNA. Chromatin structures contribute to gene regulation in various ways (reviewed in ref. *86*).

1. In eukaryotes, regulatory sequences and promoter elements can be spread over long distances. Hence, a first role of nucleosomes and higher-order chromatin structure is to mediate loop formation required for efficient interaction of the transcription factors. One or two turns of the nucleosomal supercoiled DNA brings sites into close proximity that are ~80 and 160 bp apart, respectively, or contacts between nucleosomes along the chromatin fiber axis (with about 6–8 nucleosomes/turn) bridge over a thousand base pairs.
2. Since nucleosomes restrict the accessibility of DNA to proteins, nucleosomes located in the promoter regions are generally repressive for transcription initiation.
3. Nucleosomes located in regulatory regions can participate in the inhibition or facilitation of factor binding by the rotational setting of DNA on the nucleosome surface (Fig. 2A,B). Furthermore, distortions of DNA on the nucleosome surface might directly affect properties of binding sites.

Nucleosome assembly in vivo occurs immediately after replication *(78)*. Hence, proteins present at high concentrations or with high affinity for their recognition site may compete successfully for the binding site with histones, thereby preventing nucleosome formation. This mechanism is proposed to establish the open, nonnucleosomal regions in "preset promoters" *(96)*. For regulated promoters, nucleosomes appear to be present in the repressed state, but they are removed, dissolved, or structurally altered by competitive interaction with activator proteins *(96)*. This "remodeling" of chromatin structure from a nucleosomal to an accessible nuclease-sensitive state may require additional proteins (e.g., the "Swi/Snf" complex) *(102)*. Hence, the dissolution or "opening" of nucleosomes is a complex mechanism involving numerous protein partners.

2.7. Fate of Nucleosomes During Transcription

During transcription elongation, RNA polymerases face the problem of how to transcribe through nucleosomes *(85)*. Based on recent in vitro experiments, it seems possible that RNA polymerases enter nucleosome cores and histones are transferred to the back of the polymerase *(81)*. For many genes transcribed by RNA polymerase II (pol II) in vivo, it has been shown that nucleosomes are present immediately before and immediately after transcription, whereas a few other examples support a loss of nucleosomes or an altered nucleosome structure (*11,89*; reviewed in ref. *94*). In the ribosomal DNA cluster, only the nontranscribed genes are folded in nucleosomes, whereas transcription by RNA polymerase I (pol I) generates a loss of nucleosomes *(15,17,79)*. It is likely that the presence of nucleosomes on transcribed genes in vivo depends on many parameters, including the RNA polymerase, the rate of transcription, sequence-specific differences in histone binding, and the chromosomal locus *(85)*.

2.8. Higher-Order Chromatin Structures

The nucleosome filament, which at low ionic strength can be visualized in the electron microscope as a string of beads connected in a "zigzag" conformation, is condensed to about a 30-nm diameter fiber at physiological salt concentration *(90)*. This higher-order nucleosome structure or chromatin fiber is the dominant structure of chromatin in nuclei and the substrate on which all DNA-dependent reactions are initiated. In the simplest model, the higher-order structure is a low pitch, single start helix with 6–8 nucleosomes/10 nm of fiber length. The connectivity and location of the linker DNA in the 30 nm fiber is unknown, but is envisioned to occupy the central cylinder along the fiber axis. This model also places the H1 histones, which are required for fiber formation, in the center of the fiber. Additional stability may be provided by the core histone tails, which could contact adjacent nucleosomes and linker DNA. Hence, acetylation of core histone tails, or removal (or modification) of histone H1, may destabilize chromatin fibers and temper them for transcription or DNA repair. Although the structure of nucleosome core particles is highly conserved in evolution, the linker DNA varies between individual nucleosome pairs along the same filament. This variation may be a major reason for the observed irregularity of the fiber structure (reviewed in ref. *88*).

3. DNA DAMAGE IN CHROMATIN

3.1. Modulation of DNA Damage in Chromatin

3.1.1. Distribution of DNA Damage in Mixed-Sequence Nucleosomes

The location of DNA damage sites in chromatin may play an important role in their accessibility to repair enzymes. The distribution of a number of different DNA lesions in bulk (or "mixed-sequence") chromatin was reported during the late 1970s and the 1980s. As shown in Table 1, different classes of DNA lesions either form preferentially in linker DNA of nucleosomes or about equally (per unit DNA) between linker and core regions. As expected, DNA lesions caused by certain bulky chemicals (e.g., bleomycin strand breaks and trimethylpsoralen crosslinks) show a marked preference for linker DNA regions (e.g., *14,35*). It is surprising that some small alkylating agents (e.g., methylnitrosourea) also show this preference (Table 1). These studies formed a basis for understanding chromatin rearrangements during DNA repair (*see* Section 4.2.).

A number of laboratories have also reported on the distribution of UV photodamage in bulk chromatin. It was established early on that the major UV photoproduct in DNA (*cis-syn* cyclobutane pyrimidine dimer [or CPD]) forms almost randomly (per unit DNA) between linker and core regions of the genome in irradiated cells (*51*; Table 1). This observation has been confirmed using anti-CPD antibodies to compare the yield of these photoproducts in genomic DNA with either "nuclease-resistant" or isolated nucleosome core DNA *(47,83)*. In contrast the second most prevalent (stable) UV photoproduct in DNA (pyrimidine [6-4] pyrimidone dimers, [6-4] PDs), forms preferentially in linker regions in chromatin of irradiated cells (Table 1).

The distribution of UV photoproducts has also been mapped within nucleosome cores using a T4 polymerase-exonuclease blockage assay and isolated 145-bp core DNA *(23,25,26)*. Formation of CPDs in irradiated cells, isolated chromatin, or isolated nucleosome cores is significantly modulated in core DNA with an average periodicity

Table 1
Nucleosome Distribution of Different Classes of DNA Lesions

Class	Examples[a]	Adduct size and/or helix distortion[b]	Nucleosome preference[c]
UV radiation (254 nm)	CPD	++	Random
	(6-4)PD	+++	Linker
Bulky chemicals	NA-AAF, AFB1 BPDE	+++	Linker
Small alkylating agents	DMS, MMS	+	Random[d]
	MNU, DMN	+	Linker
Ionizing radiation	X-ray, γ-ray	+	Linker (?)
X-ray mimetic chemicals	Bleomycin	+[e]	Linker
Cross-linking agents	DDP, TMP	+++	Linker

[a]Abbreviations: CPD, *cis-syn* cyclobutane pyrimidine dimer, (6-4)PD, pyrimidine (6-4) pyrimidone; NA-AAF, *N*-acetoxy-2-acetylaminofluorene; AFB1, aflatoxin B_1; BPDE, benzo[a]pyrene-diol-epoxide; DMS, dimethylsulfate; MMS, methymethane sulfonate; MNU, methylnitrosourea; DMN, dimethylnitrosoamine; DDP, cis-dichloroplatinum (II); TMP, trimethylpsoralen crosslinks.

[b]Relative amount of perturbation associated with the major adducts in each case (e.g., *see* refs. *16,33,66*)

[c]Region of nucleosome (linker or core) most frequently damaged (on a unit DNA basis), (e.g., *see* refs. *14,35,37,48,51*).

[d]Assumes distribution of MMS damage is similar to DMS *(6,43,48)*.

[e]Refers to the DNA lesion induced by the bulky bleomycin compound.

of 10.3 bases *(23,26)*. Maxima in this pattern are at positions in the DNA helix farthest from the histone surface (e.g., *see* Fig. 2), and this pattern is not observed when core histones are removed prior to irradiation. Thus, although CPD formation makes only a small distinction between nucleosome core and linker regions in bulk chromatin, formation of these photoproducts is extremely sensitive to the histone binding surface of the DNA molecule.

Two different studies strongly suggest that this UV "photofootprint" arises from the bending of DNA around the histone octamer. In one study, the UV photoproduct pattern was examined in an artificial DNA sequence, consisting of two binding motifs for the tetrameric λ repressor separated by five helical turns of diverse (or mixed) sequence DNA *(53)*. An ~10-base periodic pattern of CPDs in the diverse sequence was observed only when λ repressor was bound prior to UV irradiation (i.e., where the diverse sequence is bent by binding to the λ repressor). In another study, the UV photofootprint was examined in isolated nucleosome cores exposed to very low ionic strengths, where core particles unfold to a much more extended conformation while the histones remain tightly bound *(8)*. Following the unfolding transition and irradiation of the extended nucleosome cores, the photofootprint changes to a much more random pattern resembling that of irradiated free DNA.

The modulation of CPDs in bent DNA may reflect the structural alterations in DNA at these sites. Recent NMR studies of a specific decamer sequence containing a CPD at a single site indicate that these photoproducts compress the major groove (and widen the minor groove), causing a 9° bend in the long axis of the DNA helix *(33)*. This result confirms an earlier estimate of the degree of bending at CPD sites in synthetic $dA_n{\cdot}dT_n$ tracts by gel retardation *(97)*. After photon absorption, the [2 + 2] cycloaddition that occurs during the excited state of the 5–6 double bond in a pyrimidine base *(10)* will be more probable when adjacent 5–6 double bonds are aligned more frequently. Therefore, in nucleosome core DNA, dipyrimidines with the minor groove facing away from the histone surface should be more favorable sites for CPD formation than dipyrimidines with the minor groove next to the histone surface (*see* Fig. 2).

The modulated pattern of CPDs is also observed when preirradiated mixed-sequence DNA is reconstituted into nucleosomes *(82)*. Therefore, pre-existing CPDs in the DNA molecule can influence its rotational setting during nucleosome assembly. Furthermore, after nucleosome formation, both the yield and the bias of CPDs toward the outer surface of the DNA helix is less pronounced in the central three helical turns encompassing the dyad axis of nucleosomes *(82)*, where the helix is underwound (*see* Section 2.). This observation indicates that CPDs influence the rotational setting of DNA on nucleosome assembly, and the "energy penalty" for locating such lesions next to the histone surface is greater than positioning histone octamers in many sequences of genomic (or mixed-sequence) DNA (*see also* Section 3.2.).

The photofootprint has also been used as a diagnostic probe for the path of DNA in chromatin *(54,55)*. Exonuclease mapping studies were performed on both nucleosome core DNA and "trimmed" dinucleosome DNA (containing two cores and the connecting linker) from nuclei or H1-stripped chromatin irradiated in the presence of a photosensitizer to enhance formation of cyclobutyl thymine dimers. The photofootprints indicate that histone H1 binding alters thymine dimer formation near the midpoint of the core particle *(54)*. Furthermore, in irradiated chromatin, thymine dimer formation is not modulated in nucleosome linker DNA, indicating that linker DNA is bent less (on average) than DNA in nucleosome cores in higher-order chromatin structures *(55)*. This observation indicates that linker DNA is not bent the same way as nucleosome DNA, and seems to argue against "continuous coil" models for higher-order folding of nucleosomes in chromatin (e.g., *44*). However, it needs to be considered that the photosensitizer used in this study *(55)* could have altered the path of the linker DNA in chromatin.

Although (6-4)PDs are less prevalent (on average) than CPDs in UV-irradiated genomic DNA, the contribution of (6-4)PDs to UV-induced mutagenesis can be significant (reviewed in *7,46*). The yield of (6-4)PDs at specific sites within chromatin can be much higher than their yield in free DNA *(57)*. Futhermore, these photoproducts are distributed much more randomly in nucleosome core DNA than CPDs *(25)*. Thus, the distribution of (6-4)PDs in chromatin differs markedly from that of CPDs. However, as with CPDs, these features may reflect the structure of (6-4)PDs in DNA. Recent NMR studies indicate that these lesions produce a major distortion in the DNA helix, inducing a bend of 44° in the helix axis *(33)*. The energy penalty of such distortions should be greater in the more constrained core DNA than in the more flexible linker DNA, and could explain the preferential formation of these photoproducts in linker regions (Table 1). It is less obvious why these lesions form almost randomly within nucleosome core DNA (although at a lower yield). This may reflect a "selection" for a subset of (6-4)PD structures that can be accommodated in nucleosome core regions and/or a selection for nucleosomes with more flexible core DNA.

3.1.2. Distribution of DNA Damage in Single-Sequence Nucleosomes

Reconstitution of defined sequences into nucleosomes allows one to address the following questions:

1. How is the formation of DNA damage affected by nucleosome structure?
2. How does formation of damage affect a pre-existing nucleosome structure?
3. How do pre-existing DNA lesions affect the formation and structure of nucleosomes during assembly?

In contrast to mixed-sequence experiments, the results on defined sequences will strongly depend on the sequence specificity of damage formation, the location of the target sequence on the nucleosome surface, the local distortions and flexibility of DNA, and the stability of the nucleosome structure and its mobility.

Effects of nucleosome formation on the distribution of benzo[a]pyrene diol epoxide (BPDE) adducts have been examined in *Xenopus laevis (75)* and in *Xenopus borealis (91)* 5S rDNA fragments. The overall level of adduction is reduced by 2- to 2.5-fold in the nucleosome core region, where the central 60–90 bp of DNA are the least reactive during early reaction times. Similarly, reconstitution of 5S rDNA in nucleosomes inhibited cleavage by bleomycin (fivefold) and neocarzinostatin (2.4-fold) in the central region of the nucleosome *(76)*. Alkylation by melphalan at adenine N-3 was inhibited about twofold throughout the nucleosome, whereas alkylation at guanine N-7 was either slightly inhibited or enhanced depending on sequence position. None of the latter three drugs showed a 10 bp periodicity characteristic of hydroxyl radical cleavage *(76)*.

In contrast to chemical damage in the 5S nucleosome, irradiation of a nucleosome containing the HISAT sequence of yeast with UV light (254 nm) shows a strong modulation of CPD formation in nucleosomal DNA *(63)*. The 134 bp HISAT-sequence contains a 40 bp polypyrimidine region with a T_6-, two T_5-, and a T_9-tract, and forms a nucleosome with a defined rotational setting *(39)*. This sequence allows one to monitor CPD formation over three helical turns of DNA in a defined nucleosome. The CPD yields and distribution in nucleosomal DNA are dramatically different from those in free DNA. However, the distribution of CPDs in nucleosomes only partially resembles those of mixed-sequence nucleosomes (*see* Section 3.2.). In particular, high CPD yields are found at two sites where the minor groove faces the histones (i.e., where the mixed-sequence results predict low yields). This may reflect a somewhat different accommodation of T-tracts in nucleosomes compared to mixed-sequence DNA.

3.2. Modulation of Chromatin Structure by DNA Damage

3.2.1. Modulation of Nucleosome Structure by DNA Damage

Since DNA lesions present various degrees of structural distortions in DNA, damage induction in chromatin might directly affect chromatin structure. For example, DNA lesions might affect nucleosome stability, alter the rotational setting of nucleosomal DNA, or alter the translational mobility of nucleosomes. To address this topic, chromatin analysis prior to and after DNA damage is required.

Using a DNA supercoil assay to estimate nucleosome density, it was reported that only about half the number of nucleosomes can be reconstituted onto circular pBR322 DNA following irradiation with up to 3000 J/m^2 UV light, yielding up to 3 photoproducts/nucleosome *(42)*. However, reduced yields in nucleosome assembly are not obvious when nucleosomes are reconstituted onto the linear HISAT sequence irradiated with up to 4000 J/m^2 when assayed by nucleoprotein gel electrophoresis *(63)*. On the other hand, competitive reconstitution analysis of the energy of formation (ΔG) of nucleosomes onto linear 5S rDNA fragments irradiated with 500 or 2500 J/m^2 is increased from that of undamaged DNA, indicating UV photo lesions can inhibit nucleosome formation onto linear DNA fragments *(40)*. (Surprisingly, nucleosome formation is *enhanced* [i.e., has a lower ΔG] when 5S rDNA is damaged with BPDE *[40]*.) Thus, UV lesions can reduce nucleosome stability in linear relaxed DNA, but the extent of this effect may be governed by the DNA sequence of these regions.

To date, the question of whether the rotational setting of nucleosome DNA can change as a consequence of DNA damage induction has been studied directly only in the reconstituted HISAT nucleosome *(63)*. In this case, no evidence for a change in rotational setting is obtained following UV irradiation up to 4000 J/m^2, illustrating that a particular rotationally positioned nucleosome can accommodate the DNA distortion of CPDs.

Two studies show that reassembly appears to be required in order to change the rotational setting of at least some DNA sequences *(63,82)*. As mentioned previously (Section 3.1.1.), the rotational setting of mixed-sequence DNA can change to optimize accommodation of CPDs during nucleosome reconstitution *(82)*. Furthermore, as mentioned above, irradiation of HISAT DNA in nucleosomes does not alter the rotational setting. However, when this DNA is extracted and reconstituted into nucleosomes, the rotational setting is clearly changed *(63)*. These results indicate that nucleosomes can tolerate some DNA distortion in the intact complex and override the energy penalty of having lesions at unfavorable positions. To what extent the lesions are tolerated will eventually depend on the strength of the rotational signal as well as on the parameters that affect nucleosome mobility in a chromosomal context.

3.2.2. Modulation of Higher-Order Chromatin Structure by DNA Damage

Studies on the modulation of higher-order chromatin structures by DNA lesions is not as straightforward, since these structures are heterogeneous and less well characterized (*see* Section 2.). Studies using premature chromatin condensation (or fusion of interphase and mitotic cells) indicate that large enough sections of chromatin are stably "decondensed" following DNA damage by UV radiation that they are visible in the light microscope (reviewed in ref. *29*). This decondensation appears to result from DNA repair processing rather than a direct physical distortion at chromatin damage sites. On the other hand, differential scanning calorimetry indicates that certain anticancer drugs directly alter the DNA melting profile of chromatin in intact nuclei *(1)*. Physical studies on polynucleosome folding in vitro indicate that even massive doses of trimethylpsoralen crosslinks or UV photoproducts (up to ~1 lesion/nucleosome) are "accommodated" during polynucleosome condensation at elevated salt concentrations

(24). Therefore, direct physical alterations by at least some DNA lesions in chromatin appear to be much more subtle compared to the chromatin processing response by repair of these lesions (*see* Section 4.2.).

4. DNA REPAIR IN CHROMATIN

4.1. Modulation of DNA Repair in Chromatin

4.1.1. Repair Heterogeneity in Nucleosomes

The distribution of repair synthesis in nucleosomes has been examined in cultured human fibroblasts. Excision repair of UV damage in these cells occurs in two distinct phases (reviewed in ref. *69*). Repair synthesis occurring during early times after irradiation (3–6 h) is rapid and nonuniform in nucleosomes, having a strong bias toward the 5'-ends (and somewhat less toward the 3'-ends) of core DNA *(36)*. On the other hand, repair synthesis occurring during late times after irradiation (e.g., >16 h in AG1518 cells) is much slower and much more randomly distributed in core DNA, with an apparent decrease in repair patch length *(32)*. Therefore, UV photoproducts are either more accessible to repair enzymes in the 5'-ends of nucleosome cores (preferential repair) or UV photoproducts form preferentially in the 5'-ends of core DNA (preferential damage). A direct test of these possibilities was performed using the T4 polymerase-exonuclease blockage assay (*see* Section 3.1.) to map the distribution of CPDs in core DNA of human cells during the early and late repair phases *(32)*. Little (or no) change in the periodic pattern was observed during the fast repair phase, indicating that this phase is not the result of preferential repair in the 5'-ends of nucleosome cores. The results demonstrated that CPDs are removed at nearly equal rates from the inner and outer faces of the DNA helix in nucleosome cores (*see* Fig. 2C,D). However, CPDs form preferentially in the 5'-ends of nucleosome core DNA, and this bias accounts for nearly all of the nonuniform distribution of repair patches observed during the early rapid repair phase *(32)*. Thus, preferential damage at the nucleosome level can account for most of the nonuniform distribution of repair synthesis within core DNA, and other factors must be responsible for the two repair phases in human cells.

The rapid repair observed in transcriptionally active chromatin accounts for some of the early rapid phase observed when monitoring repair of the total genome (*see* Chapters 10 and 18). However, since the early rapid phase of repair of UV photoproducts in human cells comprises at least 50% of the total CPDs in the genome (*see* ref. *69*), more than just active gene repair is needed to account for this repair phase. A second contributer to repair synthesis during this phase is the repair of (6-4)PDs. Although both CPDs and (6-4)PDs are removed by nucleotide excision repair, the overall repair of (6-4)PDs in genomic DNA is much more rapid than CPDs (reviewed in ref. *46*). From the chromatin distribution of these lesions (*see* Section 3.1.), a possible explanation for their rapid repair could be that they are more accessible to repair enzymes than CPDs owing to their preferential location in nucleosome linker regions. This possibility was recently examined in isolated nucleosome core DNA of UV-irradiated human fibroblasts *(83)*. Using radioimmunoassays for detection of the two different UV photoproducts, it was observed that (6-4)PDs are removed much faster (~75% in 2 h) from nucleosome cores than CPDs (10–15% in 2 h). Similar rates are observed for the removal of (6-4)PDs from the genome overall in these cells *(98)*. Thus, most of these photoproducts are removed during the early rapid repair phase in human cells and could account for up to half of the repair synthesis observed during this period. Therefore, the nonuniform distribution of repair synthesis observed in nucleosomes during this period must, in part, reflect repair at (6-4)PD sites. The preferential location of these lesions in linker DNA (Table 1) could lead to extension of repair patches into the end regions of nucleosome cores, and this bias would be superimposed on the repair synthesis at CPD sites, which are preferentially located near core DNA ends *(32)*.

4.1.2. Repair Heterogeneity in Transcribed Chromatin

Repair in active genes is reviewed in Chapters 10 and 18 in this vol. Therefore, this section focuses only on studies that relate to the unique properties of chromatin in these regions (*see* Section 2.). One series of investigations involves the use of well-characterized yeast minichromosomes as model substrates to study repair in chromatin (e.g., *49,70*). Advantages of these substrates include the following:

1. They can be constructed with defined sequences;
2. The locations of nucleosomes can be precisely mapped; and
3. The rates of repair at specific sites can be determined.

These are coupled with the obvious advantage of having numerous repair mutants available in yeast (*see* Chapter 15 in vol. 1). One of these minichromosomes (called TRURAP) contains a single selectable gene (*URA3*), an origin of replication (*ARS1*), and nucleosomes of known position and stability *(5,87)*. The overall repair rate of this plasmid reflects the repair rate of genomic chromatin in wild-type, *rad1*, and *rad7* cells *(49)*, but not in *rad23* cells *(50)*. Repair rates at over 40 different CPD sites in TRURAP are found to vary markedly, being highest in the transcribed strand of *URA3* and in both strands upstream of this gene, and lowest in the nontranscribed strand of *URA3* and in both strands of the *ARS1* region *(70)*. Four different (presumably nonsense) transcripts are made from TRURAP in addition to *URA3* mRNA, and these transcripts encompass all of the efficiently repaired regions outside of the *URA3* gene *(5)*. There is a good correlation between the transcription rate (relative to the *URA3* gene) and the repair rate of four of the five template regions of TRURAP, which includes the "zero transcription rate" of the nontranscribed strand of the *URA3* gene *(73)*. The fifth region, however, is only weakly transcribed, yet repaired very rapidly *(5,73)*. This template contains the only two nucleosomes in TRURAP that are less stable in the "liquid holding" conditions used for these experiments *(5,73)*, and the lack of correlation between repair and transcription rates may reflect a modulation of excision repair by nucleosome stability when the transcription rate is low.

The increased repair of CPDs (and other DNA lesions) in transcriptionally active chromatin is owing primarily to efficient removal of damage from the transcribed strand genes actively transcribed by pol II (*see* Chapter 18). Determining if the multicopy ribosomal genes (rDNA) transcribed by pol I are also preferentially or strand-specifically repaired is more complex, because only a fraction of these genes is transcriptionally active *(15)*. This fraction changes with cell type and, in the yeast *S. cerevisiae*, with growth conditions *(17)*.

Although several laboratories reported on repair of total rDNA sequences (*13,80,95*; *see* Chapter 18), it was not until recently that repair of CPDs was measured in the individual strands of transcriptionally active ribosomal chromatin *(22)*. These regions are free of canonical nucleosomes and can be seperated from inactive chromatin by selective psoralen crosslinking and/or restriction enzyme digestion *(15)*. Transcription coupled repair is not observed in active (or total) ribosomal chromatin, and repair of CPDs is very slow in each strand *(22)*, in agreement with the earlier reports *(13,80)*. Furthermore, chromatin rearrangements may not occur in ribosomal chromatin following UV irradiation *(22)*, indicating that a (presumed) early event in nucleotide excision repair may not occur in these regions (*see* Section 4.2.).

It is somewhat surprising that both strands of active ribosomal genes are repaired slowly. Stalled pol I complexes at CPD sites could block repair of the transcribed strand, since these complexes may lack the "polymerase dislodging" activity associated with pol II transcription (Chapter 10), but this does not account for the lack of repair of nontranscribed strands in rDNA. It is possible, for example, that nucleolar chromatin is less "permeable" to excision repair complexes than the rest of the chromatin in nuclei.

4.2. Modulation of Chromatin Structure During DNA Repair

4.2.1. Rearrangement of Nucleosome Structure During DNA Repair

Almost two decades have passed since it was first reported that rearrangement of nucleosome structure occurs following excision repair of UV-induced photoproducts in DNA of human cells *(71)*. Since that time, several laboratories have observed this phenomenon following repair synthesis induced by a variety of DNA lesions, including bulky chemical adducts and alkylated bases, which differ in distribution between linker and core DNA (Table 1). Furthermore, this phenomenon has been observed in several different mammalian cell types, including human xeroderma pigmentosum cells and monkey cells (reviewed in ref. *69*).

At the nucleosome level, there is an initial rapid change in nuclease sensitivity of repair patches, which may represent refolding or sliding of nucleosomes, and a prolonged slow change, which involves a slow repositioning of nucleosomes in repaired regions in intact cells (*see* Section 4.2.3.). The rapid nucleosome rearrangement phase occurs regardless of the time after UV damage that repair synthesis takes place *(67)*. Thus, following UV irradiation of human cells, whether repair synthesis occurs during the early rapid repair phase or the late slow repair phase, nucleosome rearrangement is observed. However, differences are apparent in both the rate and extent of nucleosome rearrangement following repair during these two phases, and these differences may reflect the repair of different structural regions in chromatin (*67*; *see* Fig. 3).

During the early phase of rearrangement, nascent repair patches do not yield an ~10-base "ladder" on denaturing gels following DNase I digestion *(67)*. This footprint is diagnostic for core histone binding to DNA in nucleosomes (reviewed in ref. *93*). This observation strongly supports the notion that during and immediately after repair synthesis, newly repaired DNA is not tightly bound to a surface of core histones, and an altered nucleosome structure during excision repair appeared to be the most likely explanation for these results *(38)*. However, the timing of repair patch ligation (relative to the association of these patches with nucleosome cores) is crucial for the interpretation of the DNase I results. Therefore, the relationship between ligation of nascent repair patches and nucleosome formation in these regions was examined *(31,68)*. It was observed that repair patch ligation preceded nucleosome formation (indicated by loss of nuclease sensitivity), even when ligation was delayed by inhibitors *(31,68)* or in

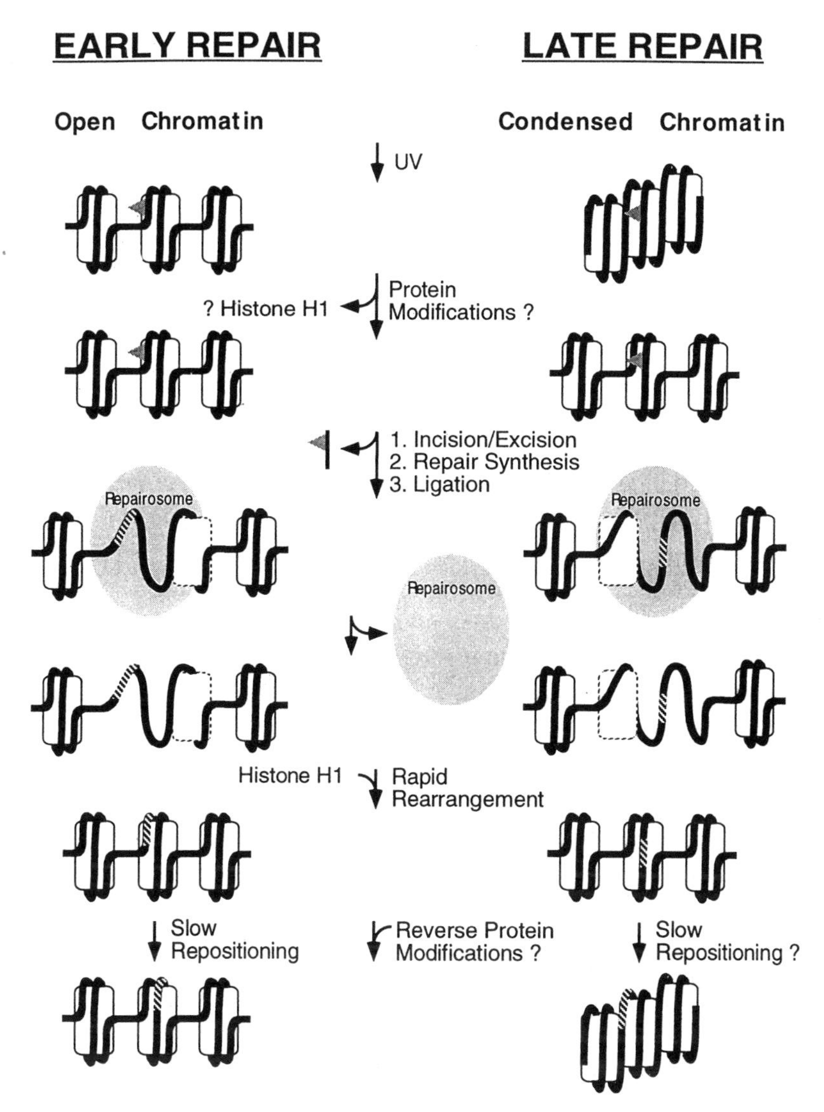

Fig. 3. Schematic model of possible transitions in chromatin structure during DNA repair processing. Model depicts possible differences between early and late repair in chromatin (DNA, thick solid lines; core histones, open rectangles). UV photoproducts and repair patches are denoted as stippled triangles and hatched bars, respectively. Dashed rectangles denote altered and/or displaced histone octamers during invasion of the "repairosome." Steps with question marks denote steps with little (or no) data obtained to date. (The structures shown are schematic and are not meant to imply actual structures in chromatin.)

partially ligase-deficient human cells *(68)*. Furthermore, it was shown that the majority of repair patches associated with isolated nucleosome core particles, shortly after rearrangement, contain a ligated 3'-end *(68)*. Thus, completion of excision repair during the rapid phase of rearrangement in nucleosomes appears to progress from an unligated, nonnucleosome structure to a ligated, nonnucleosome structure, and finally to a ligated, nucleosome structure (*see* Fig. 3).

The structure(s) of the nonnucleosome state(s) of nascent repair patches in chromatin is unknown. Nascent repair patches are much more accessible to exogenous nucleases and lack a "canonical" nucleosome structure (e.g., where core DNA is resistant to micrococcal nuclease digestion). An indication of the structure(s) comes from in vitro studies on the salt- and temperature-induced rearrangement of nascent repair patches in intact, or H1-depleted, human cell nuclei *(100)*. These studies demonstrated that nucleosome rearrangement in bulk chromatin in vitro occurs in two distinct phases, involving a reduction in the average nucleosome repeat (i.e., DNA in linker plus core) from 192 bp to either ~168 bp (phase 1) or ~145 bp (phase 2). However, following the first salt-induced transition, presumably because of histone octamer sliding, little additional nucleosome formation is observed in newly repaired DNA. This observation is consistent with a mechanism involving the transient formation of an altered nucleosome structure during excision repair *(101)*. The rapid phase of rearrangement is therefore envisioned as the refolding of these regions back into a native, or near-native, nucleosome conformation (Fig. 3).

The structural alterations observed for "long-patch" excision repair (20–30 bases) may differ for "short-patch" excision repair (<5 bases; *see* Chapter 11), particularly when repair is confined to open regions of chromatin and/or nucleosome linker DNA. Indeed, in contact-inhibited human fibroblasts, repair of low levels of bleomycin-induced damage is not associated with the rapid phase of nucleosome rearrangement *(65)*. Under these conditions, bleomycin-induced strand breaks are located almost exclusively in open regions of chromatin and nucleosome linker regions (Table 1). However, in permeabilized human cells, where higher levels of bleomycin damage are induced in nucleosome cores *(64)*, both phases of rearrangement are observed *(65)*. These differences may reflect a heterogeneity in repair pathways operating on bleomycin damage and/or differences in repair of open and compact regions of chromatin. Indeed, although repair of bleomycin damage consists of both long-patch and short-patch components *(19)*, initiation of the long-patch component may differ from that induced by UV radiation *(20)*.

4.2.2. Histone Modifications During DNA Repair

It appears that in a large fraction of mammalian cell chromatin, significant structural alterations are required for the multistep excision repair process to occur. One modification that correlates with an "open" chromatin structure is the reversible acetylation of lysines on core histones (Section 2.). An indication that this chromatin modification may play an active role in excision repair came from the observation that treatment of nonreplicating human cells with sodium butyrate, under conditions where the core histones are maximally acetylated, enhances excision repair immediately following UV irradiation *(74)*. Although this short-chained fatty acid can have a number of effects in addition to the reversible inhibition of histone deacetylase enzymes (*see* discussion in *74*), the simplest interpretation of these results was that either DNA lesions become more accessible to repair enzymes in "hyperacetylated" chromatin, or sodium butyrate stimulates the production of repair proteins. Importantly, this observation was repeated in permeable human cells *(21)*, where repair synthesis is dependent on exogeneous nucleotides, since pretreatment with sodium butyrate can result in changes in cellular nucleotide pools, which could confound the labeling results with intact cells (discussed in ref. *69*).

The increased rate of repair in butyrate-treated cells is short-lived, lasting only a few hours after UV irradiation, and involving no more than 20% of the total CPDs *(74)*. Since only about half of the total core histones are hyperacetylated *(69,74)*, it is possible that the enhanced repair in butyrate-treated cells occurs in regions of chromatin not containing hyperacetylated histones. However, it was found that most of the enhanced repair synthesis is associated with core regions of hyperacetylated nucleosomes in butyrate-treated human cells, and does not appear to reflect increased UV damage in these regions *(59)*. In addition, this association is transient, lasting several hours after repair incorporation, and may reflect a slow deacetylation of newly repaired chromatin *(69)*. Thus, it appears that increased acetylation of core histones in chromatin influences the rate of excision repair at a fraction of the CPDs in hyperacetylated domains. Furthermore, to date, it is not known if butyrate treatment enhances repair at (6-4)PD sites, although most of the repair synthesis yielding maximal butyrate enhancement (~10%) can be accounted for by the enhanced removal of CPDs during this period *(59)*.

Since excision repair is a multistep process, any one of these steps may be affected by butyrate treatment. The rate of incision at UV photoproducts appears to increase in butyrate-treated cells *(21,68,77)*. For example, the number of ligatable single-strand breaks in DNA 30 min after UV irradiation increases approximately twofold in human cells treated with sodium butyrate *(69)*. This is similar to the enhancement observed in repair synthesis (or CPD removal) following this same time period. Since butyrate treatment has no effect on the average size or composition of repair patches after UV irradiation *(21)*, this treatment must result in either increased levels of incision-mediating proteins or increased accessibility of some damaged sites to repair enzymes in a fraction of the total chromatin. It remains to be determined if this chromatin modulation plays an active role in excision repair.

Another chromatin modification that may play a role in DNA repair is poly(ADP-ribosyl)ation (*see* Chapter 22). Chromatin condensation in vitro is prevented by poly(ADP-ribosyl)ation, and this inhibition is reversed on digestion of the poly(ADP-ribose) chains *(30)*. Indeed, this modification of histone H1 yields an extended nucleofilament structure in electron micrographs, under conditions where unmodified nucleofilaments fold into the more compact 30-nm fiber *(18)*. It was also observed that depletion of poly(ADP-ribose) blocks chromatin redistribution of repair patches in rat hepatocytes treated with *N*-acetoxy-2-acetylaminofluorene *(41)*. This observation, along with a series of in vitro experiments, led to the proposal of a histone "shuttle" mechanism during repair of chromatin DNA in regions of DNA strand breaks *(2)*. The enzyme responsible for poly(ADP-ribosyl)ation of chromatin proteins, poly(ADP-ribose) polymerase, is a zinc finger protein that requires DNA strand breaks (or ends) for activity (*see* Chapter 22). Surprisingly, mice lacking this activity apparently develop normally, with the exception of increased susceptibility to certain skin diseases *(99)*. It remains to be determined if one of the roles of poly(ADP-ribosyl)ation in nuclei is to decondense chromatin during repair, and if this role is essential for efficent repair of certain chromatin domains. Thus, the possibility of an active role for chromatin protein modification during repair is still in question (Fig. 3).

4.2.3. Maturation of Newly Repaired Regions of Chromatin

As mentioned above, a consistent feature of nucleosome rearrangement in newly repaired regions has been the biphasic nature of this process. The early rapid phase, involving the most dramatic loss in nuclease sensitivity, has an associated half-life of about 20 min following repair at early times after UV *(71)*. The subsequent slow phase of rearrangement accounts for, at most, 20% of the total change in nuclease sensitivity and takes many hours before the nuclease digestion kinetics are the same as those of bulk chromatin *(52)*. Since histone H1 seems to be less prevalent in transcriptionally active regions of chromatin (reviewed in ref. *93*), and appears to be the last histone to associate with nascent nucleosomes following replication (reviewed in ref. *78*), it was possible that the slow phase of rearrangement reflected a slow reassociation of histone H1 with these regions. Since histone H1 leaves a "signature" on the nucleosome digestion pattern when it is bound in cells *(72)*, the "submonomer" pattern for newly repaired DNA was examined in nascent repair patches immediately after their association with nucleosome core structures (i.e., during the rapid phase of rearrangement) *(74a)*. The pattern for H1-containing nucleosomes in these regions is observed even when only 40% of the total repair patches had undergone the rapid phase of rearrangement and were selectively isolated. Therefore, histone H1 rapidly associates with the newly formed nucleosomes of nascent repair patches in chromatin, and this association cannot explain the slow phase of nucleosome rearrangement (Fig. 3).

Since the distribution of repair patches in nucleosomes is nonrandom during the early repair phase in human cells *(32,36)*, this distribution was measured in core DNA from confluent (nondividing) cells that were pulse-labeled for short times following UV irradiation, and the label chased for long times after the pulse period *(52)*. Over a 72-h period, while the cells are contact-inhibited, the distribution of patches becomes random in core DNA. The time-course of this randomization process is similar to that of the slow change observed in the nuclease digestion data *(52)*. Therefore, maturation of nascent repair patches inserted early after UV irradiation in chromatin must involve: (1) the "preservation," at least to some degree, of the nonuniform alignment of repair patches in nucleosomes after the rapid phase of rearrangement, and (2) a slow "repositioning" of core histones along the DNA in these regions long after repair takes place. This migration of core histones could be quite limited, however, since they would need to migrate only about 50 bp (on average) to completely randomize repair patches in these regions *(4)*.

A schematic model of these features is shown in Fig. 3, where the rapid rearrangement phase is depicted as a refolding of nucleosomes immediately after repair synthesis and ligation, and the slow rearrangement phase as limited nucleosome migration during the completion of repair processing of chromatin domains. Regardless of the details of the rapid and slow rearrangement phases, it is expected that rearrangement of the basic chromatin unit would also cause significant changes in higher-order chromatin folding. The transient resistance of large domains in chromatin to premature chromatin condensation *(29)* may indicate that these regions are temporarily "fixed" in a more open conformation that resists the normal chromatin compaction occuring during mitosis.

5. CONCLUDING REMARKS

In the last few years, considerable progress has been made in understanding structural and dynamic properties of nucleosomes in vivo and in vitro, and in particular, how proteins involved in transcription, transcriptional regulation, or replication interact with nucleosomes. A major step has been the realization that the structure is not static, but changes according to the varying requirements for processing genetic information. On the other hand, research in the DNA repair field achieved some dramatic breakthroughs in identification, cloning, purification, and characterization of numerous protein components of various repair mechanisms, and the structural parameters of several DNA lesions and protein/DNA complexes have been determined. Fusion of these research areas will lead to tests for DNA damage recognition in chromatin substrates and provide a molecular background on the role of chromatin structures in DNA damage formation and repair. Since the development of techniques based on ligation-mediated PCR and primer extension by *Taq* polymerase for measuring DNA damage and repair in vivo and in vitro that allow the mapping of chromatin structures and DNA lesions at nucleotide resolution, it seems reasonable to speculate that within the next decade major problems of DNA damage formation and repair can be solved at the nucleosome level. In contrast, new approaches are required to reveal structural and functional organization of DNA at higher-order chromatin levels, chromatin fiber dynamics, and the dynamic roles of chromatin modification during DNA processing events.

ACKNOWLEDGMENTS

The work referred to in this chapter from the authors' laboratories was supported by NIH grants ES02614, ES04106, ES03720 and NIH research career development award ES00110 (to MJS), and grants from the Swiss National Science Foundation and Krebsliga des Kantons Zürich (to FT).

REFERENCES

1. Almagor, M. and R. D. Cole. 1989. Differential scanning calorimetry of nuclei as a test for the effects of anticancer drugs on human chromatin. *Cancer Res.* **49:** 5561–5566.
2. Althaus, F. R., L. Hofferer, H. E. Kleczkowska, M. Malanga, H. Naegeli, P. Panzeter, and C. Realini. 1993. Histone shuttle driven by the automodification cycle of poly (ADP-ribose) polymerase. *Environ. Mol. Mutagen.* **22:** 278–282.
3. Arents, G. and E. N. Moudrianakis. 1993. Topography of the histone octamer surface—repeating structural motifs utilized in the docking of nucleosomal DNA. *Proc. Natl. Acad. Sci. USA* **90:** 10,489–10,493.
4. Arnold, G., A. K. Dunker, and M. J. Smerdon. 1987. Limited nucleosome migration can completely randomize DNA repair patches in intact human cells. *J. Mol. Biol.* **654:** 433–436.
5. Bedoyan, J., R. Gupta, F. Thoma, and M. J. Smerdon. 1992. Transcription, nucleosome stability and DNA repair in a yeast minichromosome. *J. Biol. Chem.* **267:** 5996–6005.
6. Berkowitz, E. M. L. and H. Silk. 1981. Methylation of chromosomal DNA by two alkylating agents differing in carcinogenic potential. *Cancer Lett.* **12:** 311–321.
7. Brash, D. E. 1988. UV mutagenic photoproducts in *Escherichia coli* and human cells: amolecular genetics perspective on human skin cancer. *Photochem. Photobiol.* **48:** 59–66.
8. Brown, D. W., L. J. Libertini, C. Suquet, E. W. Small, and M. J. Smerdon. 1993. Unfolding of nucleosome cores dramatically changes the distribution of UV photoproducts. *Biochemistry* **32:** 10,527–10,531.

9. Buttinelli, M., E. D. DiMauro, and R. Negri. 1993. Multiple nucleosome positioning with unique rotational setting for the Saccharomyces cerevisiae 5S rRNA gene *in vitro* and *in vivo*. *Proc. Natl. Acad. Sci. USA* **90:** 9315–9319.
10. Cadet, J., C. Anselmino, T. Douki, and L. Voituriez. 1992. Photochemistry of nucleic acids in cells. *J. Photochem. Photobiol. B.* **15:** 277–298.
11. Cavalli, G. and F. Thoma. 1993. Chromatin transitions during activation and repression of galactose-regulated genes in yeast. *EMBO J.* **12:** 4603–4613.
12. Chencleland, T. A., M. M. Smith, S. Y. Le, R. Sternglanz, and V. G. Allfrey. 1993. Nucleosome structural changes during derepression of silent mating-type loci in yeast. *J. Biol. Chem.* **268:** 1118–1124.
13. Christians, F. C. and P. C. Hanawalt. 1993. Lack of transcription-coupled repair in mammalian ribosomal RNA genes. *Biochemistry* **32:** 10,512–10,518.
14. Conconi, A., R. Losa, Th. Koller, and J. M. Sogo. 1984. Psoralen-crosslinking of soluble and of H1-depleted soluble rat liver chromatin. *J. Mol. Biol.* **178:** 920–928.
15. Conconi, A., R.-M. Widmer, T. Koller, and J. M. Sogo. 1989. Two different chromatin structures coexist in ribosomal RNA genes. *Cell* **57:** 753–761.
16. Cosman, M., C. de los Santos, R. Fiala, B. E. Hingerty, S. B. Singh, V. Ibanez, L. A. Margulis, D. Live, N. E. Geacintov, S. Broyde, et al. 1992. Solution conformation of the major adduct between the carcinogen (+)-anti-benzo[a]pyrene diol epoxide and DNA. *Proc. Natl. Acad. Sci. USA* **89:** 1914–1918.
17. Dammann, R., R. Lucchini, T. Koller, and J. M. Sogo. 1993. Chromatin structures and transcription of rDNA in yeast *S. cerevisiae*. *Nucleic Acids Res.* **21:** 2321–2338.
18. de Murcia, G., A. Huletsky, D. Lamarre, A. Gaudreau, J. Pouyet, M. Daune, and G. G. Poirier. 1986. Modulation of chromatin superstructure induced by poly (ADP-ribose) synthesis and degradation. *J. Biol. Chem.* **510:** 7011–7017.
19. Digiuseppe, J. A. and S. L. Dresler. 1989. Bleomycin-induced DNA repair synthesis in permeable human fibroblasts: Mediation of long-patch and short-patch repair by distinct DNA polymerases. *Biochemistry* **28:** 9515–9520.
20. Digiuseppe, J. A., D. J. Hunting, and S. L. Dresler. 1990. Aphidicolin-sensitive DNA repair synthesis in human fibroblasts damaged with bleomycin is distinct from UV-induced repair. *Carcinogenesis* **11:** 1021–1026.
21. Dresler, S. L. 1985. Stimulation of deoxyribonucleic acid excision repair in human fibroblasts pretreated with sodium butyrate. *Biochemistry* **287:** 6861–6869.
22. Fritz, L. K. and M. J. Smerdon. 1995. Repair of UV damage in actively transcribed ribosomal genes. *Biochemistry* **34:** 13,117–13,124.
23. Gale, J. M. and M. J. Smerdon. 1988. Photofootprint of nucleosome core DNA in intact chromatin having different structural states. *J. Mol. Biol.* **204:** 949–958.
24. Gale, J. M. and M. J. Smerdon. 1988. UV-induced pyrimidine dimers and trimethylpsoralen crosslinks do not alter chromatin folding *in vitro*. *Biochemistry* **27:** 7197–7205.
25. Gale, J. M. and M. J. Smerdon. 1990. UV induced (6-4) photoproducts are distributed differently than cyclobutane dimers in nucleosomes. *Photochem. Photobiol.* **51:** 411–417.
26. Gale, J. M., K. A. Nissen, and M. J. Smerdon. 1987. UV-induced formation of pyrimidine dimers in nucleosome core DNA is strongly modulated with a period of 10.3 bases. *Proc. Natl. Acad. Sci. USA* **84:** 6644–6648.
27. Grunstein, M. 1990. Histone function in transcription. *Ann. Rev. Cell Biol.* **6:** 643–678.
28. Hayes, J., T. D. Tullius, and A. P. Wolffe. 1990. The structure of DNA in a nucleosome. *Proc. Natl. Acad. Sci. USA* **87:** 7405–7409.
29. Hittelman, W. N. 1990. Direct measurement of chromosome repair by premature chromosome condensation. *Prog. Clin. Biol. Res.* **340B:** 337–346.
30. Huletsky, A., G. de Murcia, S. Muller, M. Hengartner, L. Menard, D. Lamarre, and G. G. Poirier. 1989. The effect of poly(ADP-ribosyl)ation on native and H1-depleted chroma-

tin. A role of poly(ADP-ribosyl)ation on core nucleosome structure. *J. Biol. Chem.* **264:** 8878–8886.
31. Hunting, D. J., S. L. Dresler, and M. W. Lieberman. 1985. Multiple conformational states of repair patches in chromatin during DNA excision repair. *Biochemistry* **24:** 3219–3226.
32. Jensen, K. A. and M. J. Smerdon. 1990. DNA repair within nucleosome cores of UV-irradiated human cells. *Biochemistry* **29:** 4773–4782.
33. Kim, J.-K., D. Patel, and B.-S. Choi. 1995. Contrasting structural impacts induced by *cis-syn* cyclobutane dimer and (6-4) adduct in DNA duplex decamers: implication in mutagenesis and repair activity. *Photochem. Photobiol.* **62(1):** 44–50.
34. Kornberg, R. 1981. The location of nucleosomes in chromatin: specific or statistical? *Nature* **292:** 579–580.
35. Kuo, M. T. and T. C. Hsu. 1978. Bleomycin causes release of nucleosomes from chromatin and chromosomes. *Nature* **271:** 83,84.
36. Lan, S. Y. and M. J. Smerdon. 1985. A nonuniform distribution of excision repair synthesis in nucleosome DNA. *Biochemistry* **24:** 7771–7783.
37. Lang, M. C., G. deMurcia, A. Mazen, R. P. P. Fuchs, M. Leng and M. Daune. 1982. Non-random binding of *N*-acetoxy-*N*-2-acetylaminoflourene to chromatin subunits as visualized by immunoelectron microscopy. *Chem.-Biol. Interactions* **41:** 83–93.
38. Lieberman, M. W., M. J. Smerdon, T. D. Tlsty, and F. B. Oleson. 1979. The role of chromatin structure in DNA repair in human cells damaged with chemical carcinogens and ultraviolet radiation, in *Environmental Carcinogenesis* (Emmelot, P. and E. Kriek, eds.), Elsevier/North Holland Biomedical Press, Amsterdam, pp. 345–363.
39. Losa, R., S. Omari, and F. Thoma. 1990. Poly(dA)'poly(dT) rich sequences are not sufficient to exclude nucleosome formation in a constitutive yeast promoter. *Nucleic Acids Res.* **18:** 3495–3502.
40. Mann, D. B., D. L. Springer, and M. J. Smerdon. 1997. DNA damage can alter the stability of nucleosomes: effects are dependent on damage type. *Proc. Natl. Acad. Sci. USA*, in press.
41. Mathis, G. and F. R. Althaus. 1990. Uncoupling of DNA excision repair and nucleosomal unfolding in poly(ADP-ribose)-depeleted mammalian cells. *Carcinogenesis* **11:** 1237–1239.
42. Matsumoto, H., A. Takakusu, and T. Ohnishi. 1994. The effects of ultraviolet C on *in vitro* nucleosome assembly and stability. *Photochem. Photobiol.* **60:** 134–138.
43. McGhee, J. D. and G. Felsenfeld. 1979. Reaction of nucleosome DNA with dimethylsulfate. *Proc. Natl. Acad. Sci. USA* **76:** 2133–2137.
44. McGhee, J. D., D. C. Rau, E. Charney, and G. Felsenfled. 1980. Orientation of the nucleosome within the higher order structure of chromatin. *Cell* **22:** 87–96.
45. Meersseman, G., S. Pennings, and E. M. Bradbury. 1992. Mobile nucleosomes—a general behavior. *EMBO J.* **11:** 2951–2959.
46. Mitchell, D. L. and R. S. Nairn. 1989. The biology of the (6-4) photoproduct. *Photochem. Photobiol.* **49:** 805–819.
47. Mitchell, D. L., T. D. Nguyen, and J. E. Cleaver. 1990. Nonrandom induction of pyrimidine-pyrimidone (6-4) photoproducts in ultraviolet-irradiated human chromatin. *J. Biol. Chem.* **265:** 5353–5356.
48. Moyer, R., K. Marien, K. van Holde, and G. Bailey. 1989. Site-specific aflatoxin B1 adduction of sequence-positioned nucleosome core particles. *J. Biol. Chem.* **264:** 12,226–12,231.
49. Mueller, J. P. and M. J. Smerdon. 1995. Repair of plasmid and genomic DNA in a *rad7Δ* mutant of yeast. *Nucleic Acids Res.* **23:** 3457–3464.
50. Mueller, J. P. and M. J. Smerdon. 1996. Rad23 is required for transcription coupled repair and efficient overall repair in yeast. *Mol. Cell. Biol.* **16:** 2361–2368.
51. Niggli, H. and P. Cerutti. 1982. Nucleosomal distribution of thymine photodimers following far- and near-ultraviolet irradiation. *Biochem. Biophys. Res. Commun.* **105:** 1215–1223.

52. Nissen, K. A., S. Y. Lan, and M. J. Smerdon. 1986. Stability of nucleosome placement in newly repaired regions of DNA. *J. Biol. Chem.* **261:** 8585–8588.
53. Pehrson, J. R. and L. H. Cohen. 1992. Effects of DNA looping on pyrimidine dimer formation. *Nucleic Acids Res.* **20:** 1321–1324.
54. Pehrson, J. R. 1989. Thymine dimer formation as a probe of the path of DNA in and between nucleosomes in intact chromatin. *Proc. Natl. Acad. Sci. USA* **86:** 9149–9153.
55. Pehrson, J. R. 1995. Probing the conformation of nucleosome linker DNA *in situ* with pyrimidine dimer formation. *J. Biol. Chem.* **270(38):** 22,440–22,444.
56. Pennings, S., G. Meersseman, and E. M. Bradbury. 1991. Mobility of positioned nucleosomes on 5-S-rDNA. *J. Mol. Biol.* **220:** 101–110.
57. Pfeifer, G. P., R. Drouin, A. D. Riggs, and G. P. Holmquist. 1992. Binding of transcription factors creates hot spots for UV photoproducts *in vivo. Mol. Cell. Biol.* **12:** 1798–1804.
58. Pruss, D., J. J. Hayes, and A. P. Wolffe. 1995. Nucleosomal anatomy—where are the histones? *Bioessays* **17:** 161–170.
59. Ramanathan, B. and M. J. Smerdon. 1989. Enhanced DNA repair synthesis in hyperacetylated nucleosomes. *J. Biol. Chem.* **264:** 11,026–11,034.
60. Richmond, T. J., J. T. Finch, B. Rushton, D. Rhodes, and A. Klug. 1984. Structure of the nucleosome core particle at 7 Å resolution. *Nature* **311:** 532–537.
61. Richmond, T. J., T. Rechsteiner, and K. Luger. 1993. Studies of nucleosome structure. *Cold Spring Harb. Symp. Quant. Biol.* **58:** 265–272.
62. Roth, S. Y., A. Dean, and R. T. Simpson. 1990. Yeast α2 repressor positions nucleosomes in TRP1/ARS1 chromatin. *Mol. Cell. Biol.* **10:** 2247–2260.
63. Schieferstein, U. and F. Thoma. 1996. Modulation of cyclobutane pyrimidine dimer formation in a positioned nucleosome containing polydA·dT tracts. *Biochemistry* **35:** 7705–7714.
64. Sidik, K. and M. J. Smerdon. 1990. Bleomycin-induced DNA damage and repair in human cells permeabilized with lysophosphatidylcholine. *Cancer Res.* **50:** 1613–1619.
65. Sidik, K., and M. J. Smerdon. 1990. Nucleosome rearrangement in human cells following short patch repair of DNA damaged by bleomycin. *Biochemistry* **29:** 7501–7511.
66. Singer, B. and J. T. Kusmierek. 1982. Chemical mutagenesis. *Ann. Rev. Biochemistry* **51:** 655–693.
67. Smerdon, M. J. and M. W. Lieberman. 1980. Distribution within chromatin of deoxyribonucleic acid repair synthesis occurring at different times after ultraviolet radiation. *Biochemistry* **19:** 2992–3000.
68. Smerdon, M. J. 1986. Completion of excision repair in human cells. Relationship between ligation and nucleosome formation. *J. Biol. Chem.* **261:** 244–252.
69. Smerdon, M. J. 1989. DNA excision repair at the nucleosome level of chromatin, in *DNA Repair Mechanisms and Their Biological Implications in Mammalian Cells* (Lambert, M. W., and J. Laval, eds.), Plenum, New York, pp. 271–294.
70. Smerdon, M. J. and F. Thoma. 1990. Site-specific DNA repair at the nucleosome level in a yeast minichromosome. *Cell* **61:** 675–684.
71. Smerdon, M. J. and M. W. Lieberman. 1978. Nucleosome rearrangement in human chromatin during UV-induced DNA repair synthesis. *Proc. Natl. Acad. Sci. USA* **75:** 4238–4241.
72. Smerdon, M. J. and M. W. Lieberman. 1981. Removal of histone H1 from intact nuclei alters the digestion of nucleosome core DNA by staphylococcal nuclease. *J. Biol. Chem.* **256:** 2480–2483.
73. Smerdon, M. J., R. Gupta, and A. O. Murad. 1993. DNA repair in transcriptionally active chromatin, in *DNA Repair Mechanisms* (Bohr, V. A., K. Wassermann, and K. H. Kraemer, eds.), Munksgaard, Copenhagen, pp. 258–270.
74. Smerdon, M. J., S. Y. Lan, R. E. Calza, and R. Reeves. 1982. Sodium butyrate stimulates DNA repair in UV-irradiated normal and xeroderma pigmentosum human fibroblasts. *J. Biol. Chem.* **257:** 13,441–13,447.

74a. Smerdon, M. J., J. F. Watkins, and M. W. Lieberman. 1982. Effect of histone H1 removal on the distribution of UV-induced DNA repair synthesis within chromatin. *Biochemistry* **21:** 3879–3885.
75. Smith, B. L. and M. C. Macleod. 1993. Covalent binding of the carcinogen benzo[a]pyrene diol epoxide to Xenopus laevis [5]S DNA reconstituted into nucleosomes. *J. Biol. Chem.* **268:** 20,620–20,629.
76. Smith, B. L., G. B. Bauer, and L. F. Povirk. 1994. DNA damage induced by bleomycin, neocarzinostatin, and melphalan in a precisely positioned nucleosome—asymmetry in protection at the periphery of nucleosome-bound DNA. *J. Biol. Chem.* **269:** 30,587–30,594.
77. Smith, P. J. 1986. *n*-Butyrate alters chromatin accessibility to DNA repair enzymes. *Carcinogenesis* **7:** 423–429.
78. Sogo, J. M. and R. A. Laskey. 1995. Chromatin replication and assembly, in *Chromatin Structure and Gene Expression* (Elgin, S., ed.), IRL at Oxford University Press, pp. 49–70.
79. Sogo, J. M., P. J. Ness, R. M. Widmer, R. W. Parish, and T. Koller. 1984. Psoralen crosslinking of DNA as a probe for the structure of active nucleolar chromatin. *J. Mol. Biol.* **178:** 897–928.
80. Stevnsner, T., A. May, L. N. Petersen, F. Larminat, M. Pirsel, and V. A. Bohr. 1993. Repair of ribosomal RNA genes in hamster cells after UV irradiation, or treatment with cisplatin or alkylating agents. *Carcinogenesis* **14:** 1591–1596.
81. Studitsky, V. M., D. J. Clark, and G. Felsenfeld, G. 1995. Overcoming a nucleosomal barrier to transcription. *Cell* **83:** 19–27.
82. Suquet, C. and M. J. Smerdon. 1993. UV damage to DNA strongly influences its rotational setting on the histone surface of reconstituted nucleosomes. *J. Biol. Chem.* **268:** 23,755–23,757.
83. Suquet, C., D. L. Mitchell, and M. J. Smerdon. 1995. Repair of UV induced (6-4) photoproducts in nucleosome core DNA. *J. Biol. Chem.* **270:** 16,507–16,509.
84. Tanaka, S., M. Livingstone, and F. Thoma. 1996. Chromatin structure of the yeast URA3 gene at high resolution provides insight into structure and positioning of nucleosomes in the chromosomal context. *J. Mol. Biol.* **257:** 919–934.
85. Thoma, F. 1991. Structural changes in nucleosomes during transcription—strip, split or flip. *Trends Genet.* **7:** 175–177.
86. Thoma, F. 1992. Nucleosome positioning. *Biochim. Biophys. Acta* **1130:** 1–19.
87. Thoma, F. 1986. Protein-DNA interactions and nuclease-sensitive regions determine nucleosome positions on yeast plasmid chromatin. *J. Mol. Biol.* **190:** 177–190.
88. Thoma, F. 1988. The role of histone H1 in nucleosomes and chromatin fibers, in *Architecture of Eukaryotic Genes* (Kahl, G., ed.), VCH, Germany, pp. 163–185.
89. Thoma, F. and J. M. Sogo. 1988. Structures of bulk and transcriptionally active chromatin revealed by electron microscopy, in *Chromosomes and Chromatin*, vol. I (K. W. Adolph, ed.), CRC, Boca Raton, FL, pp. 85–107.
90. Thoma, F., T. Koller, and A. Klug. 1979. Involvement of histone H1 in the organization of the nucleosome and of the salt dependent superstructures of chromatin. *J. Cell Biol.* **83:** 403–427.
91. Thrall, B. D., D. B. Mann, M. J. Smerdon, and D. L. Springer. 1994. Nucleosome structure modulates benzo[a]pyrenediol epoxide adduct formation. *Biochemistry* **33:** 2210–2216.
92. Travers, A. A. and A. Klug. 1987. The bending of DNA in nucleosomes and its wider implications. *Phil. Trans. R. Soc. Lond. B* **317:** 537–561.
93. Van Holde, K. E. 1989. *Chromatin*. Springer-Verlag, New York.
94. Van Holde, K. E., D. E. Lohr, and C. Robert. 1992. What happens to nucleosomes during transcription. *J. Biol. Chem.* **267:** 2837–2840.
95. Vos, J.-M. and E. L. Wauthier. 1991. Differential introduction of DNA damage and repair in mammalian genes transcribed by RNA polymerases I and II. *Mol. Cell. Biol.* **11:** 2245–2252.

96. Wallrath, L. L., Q. Lu, H. Granok, and S. C. R. Elgin. 1994. Architectural variations of inducible eukaryotic promoters—preset and remodeling chromatin structures. *Bioessays* **16:** 165–170.
97. Wang, C. I. and J. S. Taylor. 1991. Site-specific effect of thymine dimer formation on $dA_n \cdot dT_n$ tract bending and its biological implications. *Proc. Natl. Acad. Sci. USA* **88:** 9072–9076.
98. Wang, Y. C., V. M. Maher, D. L. Mitchell, and J. J. McCormick. 1993. Evidence from mutation spectra that the UV hypermutability of xeroderma pigmentosum variant cells reflects abnormal, error-prone replication on a template containing photoproducts. *Mol. Cell. Biol.* **13:** 4276–4283.
99. Wang, Z. Q., B. Auer, L. Stingl, H. Berghammer, D. Haidacher, M. Schweiger, and E. F. Wagner. 1995. Mice lacking ADPRT and poly(ADP-ribosyl)ation develop normally but are susceptible to skin disease. *Genes Dev.* **9:** 509–520.
100. Watkins, J. F. and M. J. Smerdon. 1985. Nucleosome rearrangement *in vitro*. 2. Formation of nucleosomes in newly repaired regions of DNA. *Biochemistry* **24:** 7288–7295.
101. Watkins, J. F. and M. J. Smerdon. 1985. Nucleosome rearrangement *in vitro*. 1. Two phases of salt-induced nucleosome migration in nuclei. *Biochemistry* **24:** 7279–7287.
102. Wolffe, A. P. 1994. Transcriptional activation: switched-on chromatin. *Curr. Biol.* **4:** 525–528.
103. Wolffe, A. P. 1995. *Chromatin Structure and Function*, Academic, New York.

14

Transcriptional Responses to Damage Created by Ionizing Radiation

Molecular Sensors

Thomas W. Davis, Mark Meyers, Carmell Wilson-Van Patten, Navneet Sharda, Chin-Rang Yang, Timothy J. Kinsella, and David A. Boothman

1. INTRODUCTION

Accurate and complete DNA repair after a genetic insult, such as ionizing radiation (IR) treatment, is essential for the integrity and survival of all living cells. IR causes a spectrum of DNA lesions *(63,133,321,336,337)*, including DNA single-strand breaks (SSBs), double-strand breaks (DSBs), DNA–protein crosslinks (intra- and interstrand), and apurinic/apyrimidinic sites (*see* Chapter 5). The extent of DSBs correlates well with lethality following IR *(138)*. IR also damages other cellular components, causing peroxidation of membrane lipids and formation of protein free radical intermediates *(17,25,43,81,95,118,232,255,266,281,326,336,337)*. The observed cellular responses to IR damage include cell-cycle arrest (*see* Chapter 17 in vol. 1 and Chapter 21 in this vol.) and/or cell death (via necrosis or apoptosis). In order to repair, recover, and survive, mammalian cells must first recognize this DNA or membrane damage. Otherwise, the consequences of DNA damage tolerance have been shown to lead to increased rates of mutation and ultimately carcinogenesis, as observed in DNA mismatch repair-deficient cells *(44,104,145,184,210,279)*.

Based on known biphasic kinetics of DNA repair and survival recovery following IR, it was hypothesized more than 10 years ago that the fast component of potentially lethal damage repair (PLDR) may be mediated by constitutively synthesized enzymes (e.g., DNA ligases or polymerases) *(29,36,49,227)*, which act immediately to repair X-ray-damaged DNA. Partial PLDR is observed in the presence of the protein synthesis inhibitor, cycloheximide (CHX) *(29)*. In contrast, the slow phase of PLDR in human cells may require the induction of specific genes and proteins involved in the repair of potentially lethal or carcinogenic DNA lesions, including more complex chromosomal damage, since CHX or actinomycin D partially blocked this late repair *(29)*. Once recognition processes have been activated, mammalian cells (as with prokaryotic and lower eukaryotic cells) respond by activating DNA repair processes. It is thought that

From: DNA Damage and Repair, Vol. 2: DNA Repair in Higher Eukaryotes
Edited by: J. A. Nickoloff and M. F. Hoekstra © Humana Press Inc., Totowa, NJ

the constitutively expressed proteins must include sensors of DNA damage that can trigger repair pathways.

Over the past five years, considerable evidence has accumulated to indicate that following cellular insult, mammalian cells alter certain key enzymes (e.g., cyclin-dependent and signal transduction protein kinases *[62,70,80,96,110,173,276,312]* and poly[ADP-ribosyl] polymerase) *(268,288)*; *see* Chapter 22) which initiate a cascade of poorly understood signal transduction pathways. These pathways eventually lead to the activation of certain key transcription factors. These transcription factors, in turn, act as "molecular response sensors," which induce new gene transcripts and result in the synthesis of new proteins. Thus, the simultaneous activation of key cytosolic and nuclear transcription factors act to amplify initial damage-detection sensors and, subsequently, regulate cell-cycle progression (at key checkpoints), trigger additional DNA repair processes, or induce apoptosis (under certain conditions) *(4,21,31,39,126, 287,297,300,355–357,364)*. This chapter reviews the known intracellular events occurring in mammalian cells that act to sense DNA or membrane damage following IR. Data are discussed that support the theory that initial DNA damage recognition and DNA repair responses are "coupled" to transcriptional events. Coupling processes probably evolved in mammalian cells in order to simultaneously activate constitutively expressed DNA repair processes and induce additional DNA repair processes if needed, DNA synthetic enzymes (for unscheduled DNA synthesis if the cell is not in S phase), checkpoint functions to allow time for repair, or apoptosis under adverse conditions.

2. IR-INDUCIBLE EXPRESSION IN MAMMALIAN CELLS

2.1. *IR-Inducible Proteins in Mammalian Cells*

The synthesis of new proteins in highly radioresistant, human malignant melanoma (U1-Mel) cells was examined immediately following IR treatment under conditions that allow PLDR *(29)*. Cells were grown to confluence arrest in which >90% of these cells were in G0/G1 (and remained in this state for 48 h). U1-Mel cells were chosen since they exhibit extremely high levels of PLDR *(29,32,36)*. Several X-ray-induced proteins (XIPs) were identified by two-dimensional gel electrophoreses in U1-Mel cells and in a variety of other human cells *(29,152)*. Eight polypeptides of 126–275 kDa (i.e., XIP126–XIP275) were coordinately induced 5- to 10-fold following IR. XIP induction was dose-dependent, but not evident until 4–5 h following IR. All proteins induced following IR were generally present at very low levels in untreated cells, and except for XIP269, were not visualized in silver-stained gels. Unirradiated, log-phase U1-Mel cells synthesized polypeptides with identical apparent molecular weights and isoelectric points as the XIPs described above. Treatment of these proliferating cells with IR led to an additional increase in their levels *(29)*. Therefore, XIPs were normally synthesized in proliferating U1-Mel cells, switched off in confluence-arrested cells, and resynthesized in response to X-ray treatment *(164,225,239)*.

XIP269 was observed in nearly all cycling human cells examined *(29,152)*. Induction levels of XIP269 correlated well with PLDR capacities of various human cells; the higher the induction of XIP269, the greater the capacity of the cell for repair. XIP269 was not observed in any untreated quiescent human cells tested. Caffeine, an inhibitor of PLDR following IR *(152)*, also reduced the level of XIP269 in IR-treated U1-Mel cells. Expression of the other XIPs remained unchanged following caffeine treatment. Therefore, XIP269 may be a regulatory protein required for repair of IR damage and may have a cell-cycle regulatory function *(152)*. However, purification and further study of this protein have proven difficult. Another study indicated that proliferating cell nuclear antigen (PCNA) levels in normal rat fibroblasts decreased after X-ray exposure, but not in radiosensitive, transformed fibroblasts *(192)*. It was speculated that the downregulation of PCNA leads to enhanced radioresistance. Further studies of inducible IR responses are needed to clarify the roles of these regulated proteins.

2.2. *IR-Inducible Genes in Mammalian Cells*

Over the past five years, multiple laboratories have reported altered gene expression or enzymatic activities in mammalian cells by agents that cause genotoxic stress, including the tumor promoter TPA (12-0-tetradeconyl phorbol-13-acetate), UV, and IR (summarized in Table 1). They can be classified into several groups, which include:

1. Proto-oncogenes, such as c-*fos*, c-*jun*, and c-*myc*;
2. Growth-related or cell-cycle regulatory genes, such as PCNA, TGF-α, thymidine kinase, protein kinase C, and *p53*;
3. RNA binding proteins such as spr-II, the functions of which are poorly understood; and
4. Other genes that may have multiple functions, including other *GADD* (growth-arrest and DNA-damage) transcripts, *XIP* transcripts, proteases, cell matrix proteins, histones, RNA polymerase β, TNF-α, metallothioneins, and DT diaphorase. For a complete listing of these proteins with references, *see* Table 1.

Considerable heterogeneity exists among these induced proteins/transcripts in the specific responses to various DNA-damaging agents. Additionally, responses of particular products to IR are variable from one cell type to another. Current research is focused on determining the functions of these induced proteins, and the intracellular signaling pathways (e.g., damage specificity, signal location and mechanism, and transcription factor activation) that result in new transcripts after IR treatment. Although many of these transcripts/proteins are modulated by IR, many are not altered at the level of transcription, but posttranslationally modified *(164,223)*, or increase or decrease owing to cell-cycle changes (e.g., blocks in G1 or G2). It is still poorly understood how mammalian cells respond rapidly to DNA damage by altering transcription of various genes, and how these early responses influence DNA repair, survival, apoptosis, and ultimately genetic alterations and carcinogenesis.

2.3. *Induction of Enzymes Involved in DNA Metabolism*

2.3.1. *Thymidine Kinase*

Three XIPs were identified by differential hybridization and found to be known proteins: thymidine kinase (TK), tissue-type plasminogen activator (t-PA), and DT diaphorase *(35)*. mRNA and enzyme levels of these were induced 6- to 80-fold in X-irradiated (6 Gy), confluence-arrested, U1-Mel cells, and to a lesser extent (two- to threefold) in comparably irradiated normal human cells *(30,33,35,37)*.

TK is an essential enzyme required for DNA synthesis and may be important in unscheduled DNA synthesis (short-patch nucleotide excision repair, NER) following IR *(315,334)*. *TK* expression is necessary for low-dose radiation resistance in rat glioma

Table 1
Damage-Inducible Transcripts/Proteins[a]

Gene or gene product	UV-C	IR[b]	Other[c]	UV-A and/or reactive O_2	References
c-*fos*	+	+	+	+	*46,66,150,297,300*
c-*jun*	+	+		+	*77,300*
jun-B	+	+		+	*77,300*
NFκB		+			*40*
EGR-1		+			*142*
c-*myc*	+	+	+	+	*66,283,317*
c-H-*ras*	+	+			*5,283*
c-*src*		+			*5*
C-5 (Trk-2 homolog)	+				*22*
PKC	–	+		+	*256,348*
CL100 (tyrosine phosphatase)	–	–	–	+	*168*
EGF receptor				+	*363*
α-Interferon		+			*346*
Interleukin-1	+	+			*117,186,347*
TNF-α	+				*141*
Basic fibroblast growth factor	+	+			*135,186*
TGF-α	+				*93*
p53	+	+	+		*114,164,213,214, 243,367*
MDM-2	+				*257,267*
WAF1/CIP1	+	+			*87*
MHC (class I)	+		+	+	*193*
MHC (class II)		+	+	+	*273*
uPA	+				*231*
tPA	+	+			*35,37*
Nmo-1 (=DT diaphorase)		+	+		*35,106,241*
Pgp (P-glycoprotein)		+			*249*
Collagenase	+	+	+		*187,310*
Stromelysin	+				*159*
α-Tubulin		+			*350*
α-Actin		+			*350*
Keratins	+		+		*160*
Metallothioneins	+		+		*10,107,204*
Heme oxygenase				+	*169*
Ornithine decarboxylase	+	+		+	*217,284*
Thymidine kinase	+	+			*35*
Ribosomal protein L7A	+		+		*22*
p-Glycoprotein		+			*148*
O^6-Methylguanine-DNA methyltransferase	+	+	+		*115,202,242*
N^3-Methyladenine-DNA glycosylase		+	+		*196*
DNA polymerase β	–	–	+	+	*108*
DNA ligase	+				*226*
DDI (class 1)	+	–			*105*
DDI (class 2) (=GADD)	+	–	+		*105,106,251*
spr-II (small proline-rich proteins)	+		+		*48,161*

[a]Table modified from Friedberg et al. *(113)*.
[b]Ionizing radiation.
[c]Other DNA-damaging agents.

cells *(2)*. During the cell-cycle, *TK* mRNA and enzyme levels rise sharply at the G_1/S border and remain elevated throughout S phase *(112,127,128,184)*. *TK* transcript levels increased in normal and U1-Mel neoplastic cells in response to 1–7 Gy IR, but are not increased when cells were treated with lower or higher doses of X-rays *(30,33,35)*. *TK* mRNA levels increase in U1-Mel cells within 5 min after 5 Gy IR exposure, peak at 4.5 h, and fall to basal levels by 9 h. TK enzyme levels increase concomitantly with its transcript, and actinomycin D (5 μg/mL) prevents *TK* induction at all X-ray doses tested *(33)*. *TK* mRNA and enzymatic levels in tumor cells were induced seven- to ninefold after radiation treatment compared to only two- to threefold induction in normal human and Chinese hamster embryo fibroblasts *(30)*. However, induction of this enzyme in human cells, as with many IR-induced proteins (Table 1), appears to be cell-type- and cell-cycle-dependent. In fact, since the initial report of *TK* induction, a number of cell types (e.g., many human colon cell lines) have been shown to exhibit only a modest increase in *TK* expression following IR treatment *(364)*.

Enhanced and altered binding of proteins in the TK promoter region has been observed in transformed compared to normal cells *(41,50,51,127,179)*, and comparative investigations of *TK* promoter regulation have led to a number of recent important discoveries. Pardee et al. *(84,85,237)* observed that during serum deprivation, transcription factor (TF) binding to the murine *TK* promoter was more stable in transformed than in normal cells. Later this group defined three regions by DNase I footprinting and gel-shift analyses to which proteins bound in a cell-cycle-dependent fashion. The complex was named *Yi (84,85)*. Although the proteins that bind to these regions remain unknown, *Yi* was reported to be affected by retinoblastoma protein (pRb) expression *(85)*. A prominent feature of the *TK* promoter is the 25-bp G1/S control element, named the cell-cycle regulatory unit (CCRU), located between bases –63 and –119 *(178,179)*. This element contains one *Yi* binding region, a CCAAT binding region, and an Sp1 consensus site. The CCRU is required for cell-cycle-dependent expression, and deregulated (altered) TF binding to this region was cited as an excellent marker of neoplasia, indicating it may play a role in tumorigenesis *(41,50–52,84,85,127,178,179,237)*. Protein binding to this region is enhanced following IR treatment *(30,288)*. Increases in *TK* transcripts also correlate with increased transcription factor binding in the CCRU of the *TK* promoter *(30)*. At least one component of these complexes has been identified as the transcription factor, Sp1 *(30)*. Thus, enhanced Sp1 binding to this promoter may play a role in increased TK transcription leading to increased thymidine triphosphate pools required for DNA repair. Alterations in Sp1 binding are discussed further in Section 3.4.2., and a model in which Sp1 binding may be modulated by a cascade of events initiated by the putative DSB sensor, DNA-PK, is outlined in Section 4.

2.3.2. Ribonucleotide Reductase

Another enzyme involved in DNA metabolism and suspected to play a role in DNA repair is ribonucleotide reductase (RR). RR is a cell-cycle-regulated enzyme that is principally responsible for the conversion of ribonucleotide diphosphates to the deoxyribonucleotide form utilized in either *de novo* DNA synthesis or DNA repair *(92,97,220,278,279,317)*. RR is responsible for maintaining balanced nucleotide pools necessary for optimal fidelity of DNA replication.

Since RR activity is closely correlated with DNA synthesis, cell-cycle regulation of RR has been examined in detail *(92,97,220,278,279,317)*. In addition to the well-characterized allosteric and cell-cycle regulation of the RR holoenzyme, RR synthesis is also tightly regulated at the level of transcription. The mRNA for both subunits of RR are cell-cycle-regulated and inducible by various types of DNA damage in all organisms so far studied, including *Escherichia coli*, *Saccharomyces cerevisiae*, and mice *(91,153,319)*. In Balb/c 3T3 cells, Hurta and Wright *(153)* reported a two- to fivefold increase in RR holoenzyme activity within 2–8 h following exposure to the alkylating agent, chlorambucil, using a concentration (200 μ*M*) known to cause DNA damage. Increased RR enzyme activity was associated with a 12-fold increase in mRNA levels of the R_2 regulatory subunit within 30 min of exposure to chlorambucil. Chlorambucil does not affect the rate of DNA synthesis. Mammalian RR activity is rapidly increased in the presence of TPA, a tumor promoter *(54–56)* apparently by increasing the half-lives of both R_1 an R_2 mRNA. *(3,56)*. It was hypothesized that DNA damage induction of RR proceeds indirectly through nucleotide depletion brought about by repair synthesis. DNA damage regulation is believed to modulate deoxyribonucleotide levels to facilitate DNA replication repair.

Surprisingly, there is no direct evidence supporting a role for RR in IR-associated DNA damage repair in mammalian cells. There is, however, an increasing laboratory and clinical data base correlating hydroxyurea (HU) exposure with enhanced radiosensitivity in mammalian cells *(11,221)*. HU is a well-characterized inhibitor of RR, which acts by destabilizing the Fe^{3+} center of the R_2 protein and destroying the tyrosyl free radical required for RR activity. Over 25 years ago, Sinclair reported enhanced radiosensitivity in vitro of Chinese hamster V79 cells when exposed to HU for 4 h before or after X-irradiation *(302)*. The observed radiosensitization was believed to result from a cell-cycle synchronization into a more radiosensitive (early S) phase with subsequent inhibition of DNA repair following IR damage. This drug–radiation interaction was also found in vivo in studies of spontaneous C3H/He mouse mammary carcinomas *(262)*. Importantly, clinical trials in several human solid tumors have suggested some benefit in tumor response and patient survival with the use of orally administered HU during conventional radiation therapy (*11,221*; *see* Chapter 27). The most convincing clinical data are seen in locally advanced cervix cancer, where a 15–25% improvement in tumor control and survival was found without any increase in acute and late normal tissue complications compared to radiation therapy alone *(263,309)*.

Based on the experimental data of the 1960–1970s and the more recent clinical data summarized above, laboratory studies attempting to understand better the role of RR in IR-induced DNA damage repair and its role in HU-induced radiosensitization have been initiated. In 647V human bladder cancer cells, in vitro exposure to a noncytotoxic dose of HU for one cell-cycle time resulted in significant radiosensitization without any change in cell-cycle distribution *(190)*. An increase (block) was found in early S with 12 h of exposure to HU but without a change in radiosensitivity. RR activity was significantly reduced after the 24 h of exposure. These in vitro observations were confirmed in two additional human squamous cell lines, Caski (human cervical cancer) and KB (human nasopharyngeal cancer) when HU exposure followed irradiation *(181,190)*. In addition to an HU-resistant (15-fold) KB clone with a threefold increase in R_2 mRNA and RR activity, there was a 1.5-fold increased resistance to radiation by clonogenic survival *(190)*.

3. TRANSCRIPTION FACTOR RESPONSES TO IR

Of the estimated 5×10^4 transcriptionally active genes in the human genome, relatively few have been reported to be induced by IR at the transcriptional level (Table 1). Some reported X-ray-inducible genes/proteins have yet to be confirmed as being transcriptionally activated following radiation. A number of questions remain unanswered about X-ray-induced transcription:

1. What are the common promoter elements (i.e., consensus sites) in X-ray-inducible genes which control transcription?
2. What TFs and TF complexes are responsible for X-ray-mediated transcription?
3. Once identified and isolated, how are these TFs activated by IR, and are these intracellular signaling pathways specific for IR or are they general stress responses, which are somehow modulated by other proteins to give damage-specific responses?
4. What other cell type- or cell-cycle-specific factors (i.e., pRB, p53, and so forth) act to regulate X-ray-responsive elements and TFs?

Transcriptional rate and mRNA level differences between human normal, cancer-prone, and neoplastic cells, and differences in inducibility/repression between cells in different stages of the cell-cycle have been well documented (*see* Table 1). In addition, comparison of X-ray-induced transcriptional regulation in normal vs neoplastic cells may yield information for antineoplastic and gene therapy treatments *(33,341,342)*. From the following information demonstrating increased TF binding after IR, a model is proposed wherein DNA break repair and the alteration of a series of retinoblastoma control proteins are "coupled" and act to control subsequent transcriptional responses in mammalian cells after IR treatment.

The specific activation of a number of TFs has been observed following IR exposure. Singh and Lavin *(303)* discovered a protein that bound to the SV40 promoter region and was overexpressed in Bloom's syndrome cells. The nuclear translocation and DNA binding of this 43-kDa protein was induced by IR treatment, but not by UV or heat shock. Datta et al. *(71,72)* described the activation (not requiring new synthesis) of a TF complex that bound to *CArG* elements within the *egr*-1 promoter, although very high doses of IR were required for its activation. Several other TFs (CREB, Sp1, and NFκB) were activated 30% to eightfold following low-dose IR of human U1-Mel cells *(288)*. Binding of the tumor suppressor and transcription factor, p53, was also induced within these cells, but its induction required considerably higher doses (in excess of 9 Gy) of IR in confluence-arrested cells *(364)*. Other cell lines demonstrated increased nuclear localization or DNA binding of p53 with more moderate doses of IR using log-phase cells *(120,123,286)*. X-ray treatment activated TFs that bound to the *t-PA* and *TK* promoters *(30,35)*, both of which were X-ray-inducible in certain human cells *(35)*. One common feature within these promoters was a combinational array of up to six Sp1/retinoblastoma control elements (RCE) and NFκB sites which lie within 400 bp of the 5'-end of the TATAA boxes of each X-ray-inducible gene promoter. The precise X-ray response element (XRE) sequence has not been elucidated to date. In the following sections are reviewed the known transcription factors that are "activated" (not requiring *de novo* protein synthesis) following IR exposure. However, it is important to note that not all of these TF changes are usually found in a given cell type under a given experimental setting. The factors that govern which TFs (or combinations of

TFs) are involved are complex and still poorly understood. The TFs that have been reported to be altered following IR exposure are p53, AP-1, NFκB, and members of the Sp1 family.

3.1. The p53 Transcription Factor

p53 is the most commonly mutated gene in human cancer *(151)*. Wild-type *p53* is a tumor suppressor gene *(90,101)*. In normal cells, p53 is an unstable nuclear phosphoprotein with a half-life of 20–35 min *(215)* that functions as a transcription factor *(99,100)*. Its consensus DNA binding site consists of two 10 bp motifs, 5'PuPuPuC(A/T)(T/A)GPyPyPy3', separated by 0–13 bp *(88)*. When DNA is damaged, p53 levels increase *(114,164,214,215)*.

p53 modulates the transcription of several genes and interacts with many proteins (reviewed in *65,82,163,206,296*). p53 activates the transcription of mouse muscle-specific creatine kinase, growth-arrest and DNA damage-inducible protein 45 *(GADD45)*, wild-type p53-activated fragment 1 *(WAF1)*, murine double-minute 2 protein *(MDM2)*, κ light-chain immunoglobulin, thrombospondin-1, Bcl-2-associated X protein (Bax), and the novel transcription factor Hic-1 (hypermethylated in cancer). p53 negatively regulates the promoters of c-*fos*, c-*jun*, β-actin, c-*myc*, heat-shock cognate 70 (hsc70), interleukin (IL)-6, retinoblastoma susceptibility gene *(Rb)*, multidrug resistance gene (*MDR1*), heat-shock protein 70 gene (*HSP*70), cyclin G, proliferating cell nuclear antigen (PCNA), b-*myb*, DNA polymerase-α, B-cell lymphoma gene 2 (*BCL*-2), and the promoters and enhancers of various viruses (SV40, RSV, CMV, HIV-1, HSV-1, and HTLV-I). Many mutant p53 proteins found in tumors exhibit partial or complete loss of sequence-specific DNA binding activity *(19,168,276)*. p53 interacts with the SV40 large T-antigen, the E6 protein from human papillomavirus, adenovirus E1b protein, MDM2, CCAAT binding factor (CBF), DNA nuclear excision repair protein ERCC3, TATA binding protein (TBP), Sp1, and DNA replication factor, RPA. Binding to MDM2 is also thought to repress p53 function *(248)*.

The abilities of p53 to block proliferation as well as induce apoptosis are widely reported *(65,163,206,296)*. The basic model of p53 function, which will be elaborated below, asserts that DNA damage leads to increased nuclear p53 levels *(114,164)*. This causes a transient delay in G1 (presumably to permit repair of damaged DNA before DNA replication in S phase) *(164,189)* and possibly in G2 (to prevent segregation of damaged chromosomes) *(241,314)*. The appearance of cell-cycle delays after treatment with DNA-damaging agents has long been reported (*209*; *see* Chapter 21). If repair fails, cell death by mechanisms involving apoptosis may occur *(320)* in order to prevent the propagation of cells that have sustained mutation. For this reason, p53 has been dubbed the "guardian of the genome" *(195)*.

The effect of p53 status on IR sensitivity has been studied extensively. Cells mutant in p53 were found to be more resistant to IR *(198,222,253,360)*. Normal human fibroblasts that express E6 (which binds to and causes the degradation of p53) were also more radioresistant *(325)*. However, one report showed that rat epithelial cells mutant in p53 were actually more radiosensitive than their normal counterparts *(24)*. p53-deficient mice also exhibit an increase in chromosomal abnormalities following IR exposure *(38)*.

Kastan et al. *(164)* found that IR treatment led to a three- to fivefold increase in total p53 protein levels within 1 h in human cells. Since there were no changes in p53 mRNA levels, posttranscriptional mechanisms apparently increased the half-life of the protein. In fact, a variety of DNA-damaging agents have now been shown to induce nuclear accumulation of wild-type p53 levels or stimulate p53-specific DNA binding activity; these include UV, mitomycin C, methylmethane sulfonate, cisplatin, hydrogen peroxide, actinomycin D, adriamycin, 5-fluorouracil, camptothecin, and etoposide *(114,324,369)*. One critical lesion appears to be DNA strand breaks, since electroporation of a restriction enzyme (*Alu*I) or DNase I into cells increases p53 levels *(244)*.

The increase in p53 levels correlates with G1 cell-cycle arrest *(164,189)*. Cells that lack endogenous p53 gain the ability to arrest in G1 after γ-ray treatment following transfection of wild-type p53, and cells with wild-type endogenous p53 loose the radiation-induced G1 arrest when transfected with mutant p53 *(189)*. Furthermore, primary fibroblasts from p53 null mice do not arrest in G1 following treatment with IR *(165)*. Human fibroblasts that express E6 show altered cell-cycle regulation *(345)*. The presence of wild-type p53, however, does not ensure that IR-induced G1 arrest will occur *(210,241,255)*, and the lack of this G1 checkpoint is not necessarily associated with an increase in IR sensitivity *(306)*. Kastan et al. *(164)* found that G2 arrest is unaffected by p53 status, whereas others have reported a possible role for p53 in regulating a G2/M cell-cycle arrest checkpoint *(67,129,210,314)*.

Ataxia telangectasia (AT) is a human autosomal-recessive disorder characterized by progressive cerebellar ataxia, hypersensitivity to IR, and a marked increase in cancer incidence *(223)*. Fibroblasts from AT patients lack a radiation-induced increase in p53 protein levels and correspondingly do not arrest in G1 *(165)*. Interestingly, AT cells also fail to upregulate *GADD45*, which is induced by DNA-damaging agents or other treatments eliciting growth arrest in many different mammalian cells *(106)*. The human *GADD45* gene has a conserved p53 binding site in its third intron. Therefore, the AT gene product may be part of a radiation-response pathway, which includes both p53 and GADD45 *(165)*. GADD45 has also been reported to be associated with PCNA and to participate in NER *(308)*, although this finding is somewhat controversial *(166)*.

How does p53 cause cell-cycle arrest? In a search for genes activated by p53, el-Deiry and coworkers *(89)* identified a 21-kDa protein they named WAF1 (wild-type p53 activated fragment), which had a p53 binding site in its distal upstream region. This gene was found to be identical to CIP1 (Cdk-interacting protein), isolated by its ability to bind the cyclin-dependent kinase (Cdk), Cdk2 *(143)*, as well as to the p21 protein isolated by Beach and others found in complexes with cyclins and Cdks *(362)*. $p21^{WAF1/CIP1}$ mRNA and protein levels increase following exposure to IR in a p53-dependent fashion *(87)*. Since it inhibits cyclin-dependent kinase complexes, $p21^{WAF1/CIP1}$ may cause G1 arrest by preventing the phosphorylation of pRb *(305)*. pRb is a substrate for, in addition to other Cdks, Cdk2 *(207)*. pRb in its hypophosphorylated form (present only in G0/G1) binds to and inhibits members of the E2F family of transcription factors *(245)*. When pRb is phosphorylated (normally in late G1), E2F dissociates from this inhibitory complex to direct the transcription of genes involved in S-phase entry *(344)*. p53 also interacts with the transcription factors $TAF_{II}40$ and $TAF_{II}60$ *(323)*, and also the TATAA binding protein (TBP), which is required for transcription from promoters possessing a TATA box *(58,273)*. Wild-type p53 also abrogates the expression of the proliferation-associated genes PCNA *(154)*, c-*fos* *(119)*, and cyclin G *(247)*. Other evidence suggests a role for p53 in S-phase arrest *(65,214)*.

It can promote reannealing of DNA strands *(246)*, inhibit nuclear DNA replication in a transcription-free system *(64)*, block large T-antigen DNA helicase activity *(42)*, and interact with the single-stranded DNA binding protein complex containing RPA *(86)*.

p53 has also been implicated in the process of radiation-induced programmed cell death (PCD) in many cell types in vitro and in animals *(59,60,213,225,300)*. PCD, or apoptosis, plays an important role in normal development *(94)*, and is characterized by nuclear condensation and cleavage of chromosomes into small DNA fragments *(356)*. The specific activating signal appears to be the presence of DNA strand breaks *(244)*. However, wild-type p53 is not absolutely required for all apoptotic responses; cells mutant in p53 can still undergo apoptosis *(60)* and p53-null mice exhibit relatively normal development *(83)*.

The molecular mechanism that mediates p53-induced apoptosis remains unclear. Wild-type p53 upregulates the expression of Bax, while simultaneously downregulating Bcl-2 expression *(235)*. Bax has been shown to block the apoptosis-inhibiting ability of Bcl-2 by heterodimerizing with it. The ratio between Bax and Bcl-2 is thought to determine cellular survival or death following an apoptotic stimulus *(249)*. Coexpression of wild-type p53 and E2F-1 in a murine cell line (having endogenous mutant p53) resulted in apoptosis, indicating that p53 and E2F-1 may cooperate to mediate apoptosis *(355)*.

It should also be noted that p53-mediated G1 arrest and apoptosis are not necessarily coupled. M1 leukemic cells (expressing a temperature-sensitive p53) underwent apoptosis, but not G1 arrest, when p53 was active at the permissive temperature *(367)*. Arrest only occurred in the presence of IL-6, which actually protected these cells from p53-mediated apoptosis. Thus, p53 may cooperate with IL-6-specific signaling to induce arrest and prevent apoptosis in these cells. It has been generally hypothesized that low levels of DNA damage lead to cell-cycle arrest, whereas high levels lead to apoptosis; however, it may be that this decision is cell-type- and condition-specific *(163,206,306)*.

How does p53 detect DNA damage? Four possible mechanisms have been suggested *(136)*. First, in addition to its sequence-specific binding, p53 binds nonspecifically to double- and single-stranded DNA *(16,167,312)*. p53 catalyzes the renaturation of short complementary single-stranded DNA fragments and promotes DNA strand transfer reactions *(16)*. p53 also induces RNA conformational changes, suggesting a possible role in RNA metabolic processes *(246)*. It was also shown that human p53 binds ATP and exhibits an intrinsic ATP-stimulated DNA strand reassociation activity *(42)*. Therefore, p53 may directly detect DNA damage; however, some signal must then lead from DNA damage detection to the increase in p53 levels observed in the cytoplasm. Second, p53 might respond to changes in nucleotide metabolism *(164)*. Caffeine was found to block p53 induction, suggesting the involvement of a cAMP- or cGMP-dependent process (possibly involving a cyclic nucleotide-dependent kinase). This may indicate a role for p53 in G2 arrest, since caffeine may abrogate the block observed in G2 after DNA damage *(196)*. Third, p53 may merely respond to the highly reactive oxygen species generated from the genotoxic stress created by IR exposure. Sulfhydryl oxidation was found to disrupt wild-type p53 structure and function, whereas reducing conditions acted to stabilize p53 and stimulate DNA binding *(137)*. Alternatively, genotoxic stress could affect p53 function through an enzymatic cascade involving the mitogen-activated protein (MAP) kinase, which can phosphorylate p53 *(229)*. Finally, DNA damage could be signaled by the double-stranded DNA-dependent protein kinase (DNA-PK), which has been shown to phosphorylate p53 at two different residues in vitro (Ser 15 and Ser 37) and which is activated by DNA damage *(122,202)*. This model is further elaborated in Section 4. It is clear, however, that activation of p53 and its downstream genes play key roles in the G1 (and possibly G2) cell-cycle arrest checkpoints, as well as in IR-induced apoptosis.

3.2. The AP-1 Transcription Factor

AP-1 is an ubiquitous nuclear transcription factor that is highly responsive to genotoxic stress. It was initially found to be required for metallothionein 2A expression *(199)*. Subsequently, AP-1 was noted to mediate gene induction following treatment with TPA. Thus, the specific DNA binding sequence, TGA(G/C)TCA, was named the TPA response element (TRE) *(8,199)*. Following exposure to a host of widely different stimuli, the DNA binding and transcriptional activity of AP-1 are markedly increased *(109,191)*, with resultant elevation in the synthesis of various downstream proteins involved in DNA repair, cell-cycle control, and tumor progression *(9,287)*.

AP-1 is a dimer made up of the gene products of the c-*fos* and c-*jun* proto-oncogenes. Three c-*jun* variants and four c-*fos* variants have been characterized *(9)*. c-Fos does not form homodimers, although c-Jun homodimerizes rapidly following cellular injury. Temporal analysis of nuclear extracts from log-phase cells exposed to a variety of mitogenic agents initially shows only c-Jun/c-Jun homodimers form within 15 min, with a gradual predominance of heterodimers occurring at later time-points *(9,353)*. ATF2 and c-Jun have also been noted to form heterodimers possessing transcriptional activity *(134)*. Normally, c-Jun is expressed at low levels, primarily in G1/S *(217)*. c-*fos* and c-*jun* expression increases throughout the cell-cycle in a tissue-specific manner following exposure of cells to IR, oxidative stress, UV, TPA, and a variety of other agents *(216,217)*.

Relatively high IR doses increase cellular levels of c-Fos and c-Jun, and enhance the binding activity of AP-1 *(140,301)*. AP-1 induction is temporally related to the appearance of DNA strand breaks, through separate, but functionally related signaling cascades. DNA damage activates poly(ADP-ribosyl) polymerase (PARP), an enzyme involved in DNA repair. Cotreatment of cells with irradiation and specific PARP enzyme inhibitors, such as 3-amino benzamide (3-AB), nicotinamide, and theophylline, abolish the induction of c-*jun* as well as the appearance of DNA breaks *(217)*. Similar experiments with mitomycin C as the DNA damaging agent resulted in both the induction of c-*jun* and DNA breaks, but cotreatment with 3-AB demonstrated no effect on either c-*jun* induction or DNA breaks. These data support the hypothesis that there are specific molecular sensors of DNA damage induced by IR *(217)*. AP-1 induction is clearly regulated via reactive oxygen intermediates (ROIs), which cause the majority of the DNA strand breaks following IR exposure. The inhibition of c-Jun synthesis has been noted following cotreatment of cells with IR and *N*-acetyl-L-cysteine, an antioxidant *(69)*. Other studies have shown that intracellular levels of glutathione affect binding of AP-1 to the glutathione *S*-transferase gene promoter and subsequent activation of the gene *(23)*. These findings support a relationship between generation of ROIs and AP-1-mediated gene expression *(261)*.

The redox state of the cell plays a vital role in regulating the binding activity of AP-1. The DNA binding ability of both c-Fos/c-Jun and c-Jun/c-Jun dimers requires these proteins to be in a reduced state *(1,109)*, although the mechanism of redox regulation of AP-1 is unclear. The redox regulation of AP-1 binding is linked to a highly conserved cysteine residue, which lies in the DNA binding domain of both proteins *(1)*. This cysteine residue is flanked by basic amino acids that may also play a role. This cysteine residue is protected from modification when AP-1 binds DNA, but is able to undergo modification when unbound, supporting its role in the DNA binding domain *(254)*. Replacement of cysteine 154 in c-Fos and cysteine 272 in c-Jun with serine residues results in increased DNA binding and a loss of redox regulation *(1)*. A nuclear factor called Ref-1/HAP-1 is a 37-kDa protein that markedly enhances the DNA binding of c-Fos, c-Jun, and AP-1 to the TRE in nuclear extracts *(358)*. Ref-1 reduces the conserved cysteine residue and is itself reduced for full activity by other cellular agents. In addition to its redox activity, Ref-1 is an apurinic/apyrimidinic endonuclease involved in DNA repair *(358)*, and serves as a coupling factor regulating DNA repair and transcription.

Multiple signaling cascades participate in the DNA binding and transcriptional activity of AP-1. These signaling cascades also participate in the activation of other nuclear transcription factors, such as NFκB and egr-1, and are thus thought to be general stress-response pathways. Three separate kinase families, grouped under the general heading of the MAP kinases, have roles in the molecular response to genotoxic stress *(160)*. Extracellular stimulus responsive kinases (ERKs) phosphorylate a ternary complex factor and Elk-1, which in turn causes an increase in the synthesis of c-Fos *(307)*. Fos-regulating kinases (FRKs) cause an increase in the transcriptional activity of c-*fos* *(76)*. Jun N-terminal kinases (JNKs) phosphorylate both c-Jun and ATF-2, leading to an enhancement of AP-1 transcriptional activity *(78,130)*.

Although these kinases have similar actions, they are regulated via different stimuli. Human tonsillar B cells exposed to 15 Gy have a marked increase in the binding activity of AP-1. Cotreatment with a protein kinase inhibitor results in inhibition of AP-1 binding as well as c-Fos/c-Jun expression *(346)*. Although all these kinase pathways appear to be stimulated via the c-H-Ras pathway, ERKs are also specifically activated by growth factors and phorbol esters. In contrast, FRKs are growth factor-responsive but not TPA-responsive *(76)*. JNKs respond to exposure to various agents, including UV, TPA, tumor necrosis factor (TNF), and heat shock *(146,231)*. More recently, another family of protein kinases termed the stress-activated protein (SAP) kinases were shown to be involved in the cellular response to IR and a variety of cellular stresses *(171,191)*. These are believed to be a subfamily of the JNKs *(173)*.

Although the molecular events that cause an increase in AP-1 binding and transcriptional activity are well understood, less is known about the regulation of kinase activity and the actual molecular sensors of DNA damage that activate the signaling cascade. It is known that MAP kinases need to be phosphorylated to stimulate their activity, but the specific protein kinases that perform these functions remain unclear *(354)*. A host of different protein kinases have been identified and characterized, and there is a complex interplay in response to different cellular insults *(354)*.

The c-Abl tyrosine kinase is selectively activated by agents that damage DNA. Treatment of c-Abl positive lymphocytes with 2 Gy IR causes a 300% increase in its tyrosine kinase activity. There is a concurrent increase in SAP activity that was not noted in c-Abl-negative cells *(171)*. Increases in c-Abl kinase activity were also noted following treatment of cells to other agents that damage DNA, but not by TNF, which induces AP-1 by a different mechanism *(171)*. Thus, c-Abl appears to be a general sensor of DNA damage. Activation of c-*jun* following IR is not necessarily dependent on DNA strand breaks, and the specific pattern of DNA damage (damage to deoxyribose groups vs oxidization of DNA bases) influences c-*jun* regulation *(217)*. In addition, not all cells demonstrate AP-1-induced DNA binding after IR *(288)*. Thus, considerable work remains to be done to characterize the role of AP-1 as a molecular sensor of IR-induced DNA damage. How it interacts with the various kinase pathways, and why most human epithelial cells do not demonstrate increased AP-1 DNA binding activity following IR (i.e., which alternative pathways are utilized) remains to be determined.

3.3. The NFκB Transcription Factor

NFκB was first characterized in lymphoid cells as a specific protein that bound to a decameric κB oligonucleotide (GGGACTTTCC) located in the J-C intron of the κ light-chain immunoglobulin gene *(297)*. NFκB has been shown to be ubiquitously expressed and, thus, is likely to be involved in the regulation of a large number of genes through their κB DNA binding sequences. Its specific DNA binding activity has been detected in a wide variety of lymphoid cell lines (CD4-positive T-cell line Jurkat, the pre-B-cell line 70Z/3, and the promonocytic lines HL-60 and U937), spleen cells, T lymphocytes, tracheal epithelial cells, and human melanoma cells *(13,157,288,297,340)*.

NFκB consists of two DNA binding subunits, p50 and RelA (formerly p65) *(14,234)*. Both can bind independently as homodimers, but they have greater affinity for DNA as a heterodimer. These subunits share a region of 300 amino acids that is homologous to sequence-specific DNA binding and dimerization *(14,26)*. Only the RelA subunit has a strong potential to activate transcription and nuclear localization signals *(234)*. In transient transfection assays using a κB-controlled chloramphenicol acetyl transferase reporter gene, high activation of transcription was only seen with the RelA-containing dimers *(291)*. p50 is known as a "helper" subunit that interacts with high affinity with the RelA subunit, allowing binding to low-symmetry κB sites *(334,368)*. Both subunits are transported as a complex into the nucleus owing to positively charged amino acid residues in their structure *(27)*. Both subunits are also known to bend DNA on binding *(294)*. It is unclear to what extent homodimers of p50 may control gene expression. Future work should lead to answers about the role(s) of p50 and RelA in activation of transcription of genes involved in cell proliferation, inflammation, and apoptosis.

The activity of NFκB and related factors is negatively controlled by members of the IκB family. IκBα is the most studied member of this family. Inducible activation of NFκB requires its dissociation from IκB. When NFκB is associated with IκB, it is sequestered in the cytoplasm and thus unable to interact with DNA *(13,15)*. Extracellular stimulation leads to release of IκB, and restores NFκB nuclear uptake and transcriptional activation of NFκB-responsive genes *(12)*. Therefore, IκB release seems to be the principal event required for translocation/activation of NFκB. The release of IκB is apparently driven by direct phosphoryl transfer from ATP to IκB, possibly by casein kinase II and subsequent degradation of IκB *(20,234)*.

Several extracellular conditions are known to activate NFκB. These include the inflammatory cytokines, such as TNF and IL-1, double-stranded RNA, bacterial lipopolysaccharide (LPS), phorbol esters, UV, protein synthesis inhibitors, viruses, asbestos, H_2O_2, and IR *(40,111,124,157,288,293,297,310,336,341)*. When human KG-1 myeloid leukemia cells were exposed to IR, binding of NFκB to DNA increased. This effect was first seen at a dose of 2 Gy and reached a maximum at 5–20 Gy. At the highest dose tested, the increased DNA binding activity was maximal at 2–4 h and decreased to basal levels by 12 h *(40)*. In contrast, IR doses of 0.25–2 Gy induce expression of NFκB in EBV-transformed 244B human lymphoblastoid cells, with maximal expression at a dose of 0.5 Gy after 8 h. Interestingly, a difference in p50 and RelA DNA binding and nuclear localization was noted. RelA was maximally present in nuclear extracts at 0.5 Gy, and remained at basal levels when exposed to 1 Gy or above. However, p50 is maximally expressed at 0.5–2 Gy *(266)*. These differences could be owing to the transcriptional activation of the p50 gene by NFκB; the RelA gene is not regulated by NFκB *(234)*. The effects of IR on NFκB DNA binding was explored using radioresistant, human melanoma cells (U1-Mel). After 4.5 Gy and 4 h, the DNA binding activity of NFκB to its consensus site increased 30-fold *(288)*, but the increased binding was transient (lasting 12 h) and decreased with higher doses of IR. It remains to be determined how increased NFκB affects cell functions, such as repair and apoptosis following IR.

To determine whether the activation of NFκB binding activity by IR was owing to the activation/translocation of pre-existing protein (by the dissociation of IκB from NFκB) or owing to new synthesis of p50 and/or RelA, a time-course study was performed on KG-1 cells treated with 20 Gy IR *(40)*. Northern analysis showed low levels of NFκB transcripts in the nonirradiated cells and a significant increase in expression at the 3 and 6 h following IR. By 15 h, expression returned to control levels. It was then reasoned that since there was not a detectable increase in the level of expression at 1 h, the binding of NFκB to DNA must be owing to the activation of pre-existing NFκB protein. This theory was confirmed by treatment of cells with 10 mg/mL of CHX. Increased DNA binding by NFκB in band-shift experiments was found after 2 h of drug treatment and 20 Gy IR exposure. These results indicated that the effect of IR on the expression of NFκB may occur at two levels: protein translocation/activation and transcription *(40)*. No clear relationship between increased NFκB binding and the induction of the X-ray-induced proteins in Table 1 has been found to date, although such a relationship has been suggested by our lab *(288)*.

The signal transduction pathways responsible for the activation of NFκB remain to be determined. A few signaling pathways have been implicated; the most convincing involve protein kinase C (PKC) activation *(328)* and/or the ROI cascade (probably involving PKC) *(236)*. EBV-transformed 244B human lymphoblastoid cells were exposed to a dose of 0.5–2.0 Gy, and then immediately treated with the antioxidant, *N*-acetyl-L-cysteine (NAC). DNA binding activity was maximally reduced to 50% at a dose of 0.5 Gy (with 30 m*M* NAC) after 8 h. It is thought that NAC blocks NFκB induction by scavenging free radicals (ROIs) created by the IR exposure. However, the lack of complete suppression by NAC may indicate that either the NAC is not completely effective or that there is another signaling pathway involved. Alternatively, the antioxidant (NAC) might be more effective if added prior to IR exposure. There are two ways in which the ROI pathway can lead to activation of NFκB. First, through an intranuclear signaling cascade, free radical formation from IR might lead to DNA damage, which eventually could result in a reverse signaling mechanism from the nucleus to the cytoplasm *(40)*. This pathway may involve PKC, which could represent a convergence of the two likely pathways. This reasoning is based on past studies that have indicated that IR can induce c-*jun* transcription, which is known to occur through the PKC pathway *(172)*. Second, a signal cascade initiated at the plasma membrane was found in HeLa S3 cells that were exposed to UV light *(79)*. These studies were based on the observation that activation of Ha-Ras and c-Raf-1 increase the phosphorylation/activation of c-Jun, again implicating PKC *(78)*. Band-shift assays directly demonstrated that tyrosine kinases and Ha-Ras are involved in the activation of NFκB binding following UV *(79)*.

There is also more direct evidence that the activation of NFκB binding activity is initiated through the PKC pathway. PKC is a key participant in the transduction of many growth-regulatory signals. Its activation is stimulated by the phosphatidylinositol-specific phospholipase C-γ (PLC-γ). As stated before, IR causes the formation of free radicals. These oxidants, such as hydrogen peroxide, can stimulate the activation of PLC-γ, which in turn leads to Ca^{2+} mobilization and, ultimately, PKC activation. These data led to studies that were performed on human B-lymphocyte cells in which protein-tyrosine kinases (PTKs) play an important role in the initiation of signal cascades that affect proliferation and survival. Tyrosine phosphorylation by receptor and nonreceptor-type PTKs regulates the catalytic activity of PLC-γ *(290,329)*. Cells treated overnight with the potent PTK inhibitor herbimycin A, and then irradiated with 10–40 Gy 2 h prior to nuclear extraction significantly decreased the DNA binding activity of NFκB. Although this does not rule out the possibility that other mechanisms are involved, it does give supportive evidence that tyrosine phosphorylation is important in NFκB activity *(290,328)*.

Therefore, the current data for NFκB activation supports the theory that IR-induced free radicals activate PLC-γ, leading to mobilization of Ca^{2+} and activation of PKC *(139,292)*. Activated PKC may then cause the release of IκB from the p50–RelA-IκB complex and the subsequent translocation of the p50–RelA complex into the nucleus *(208)*. All studies seem to indicate a relationship between both pathways in the activation of DNA binding activity of NFκB. The importance of IR-induced NFκB activation on survival, apoptosis and mutagenesis remains to be determined.

3.4. Transcription Factors Controlled by the Retinoblastoma Susceptibility Gene Product and Their Response to IR

pRb is a member of a family of proteins characterized by a protein binding "pocket" that includes p107 and p130 *(61,68,98,295)*. pRb regulates traversal of the cell-cycle and initiation of differentiation by associating with and/or controlling the activity of a number of transcription factors. This list includes the E2F family (E2F-1, E2F-2 and E2F-3) *(53,147,299,347,363)*, ATF-2 *(177)*, Elf-1 *(272,337)*, MDM2 *(361)*, and members of the Sp family (Sp1, Sp3, and Sp4) *(176,299)*, which are also known as retinoblastoma control proteins. These proteins are discussed in terms of the known pRb-regulated transcriptional changes observed following IR. Of the above-mentioned interactions, few have been examined following IR. A model is discussed that relates

DNA damage recognition and repair to transcriptional activation and induction of gene products important for cell-cycle arrest, DNA repair, and apoptosis.

3.4.1. The E2F-1 Transcription Factor

E2F-1 is a member of a family of transcription factors necessary for transition through G1 and S phases of the cell-cycle *(343)*. pRb associates with E2F-1 in early to mid-G1 prior to it becoming hyperphosphorylated *(245)*. The transcription of many growth-related genes is activated by E2F-1 binding; these include cyclin D1, RR, PCNA, B-*myb*, cyclins E and A, DNA polymerase-α, and thymidylate synthase *(75)*. Exogenous expression of pRb in SAOS2 (pRb-/pRb-) cells elicits a G1 block, which can be bypassed by coexpression of E2F-1 *(271)*. Thus, a likely mechanism of pRb-mediated G1 arrest is downregulation of E2F-1-mediated transcription, either by competitive binding or regulatory modification. It was also reported that infection of serum/confluence-arrested rat embryo fibroblasts (REF52) with an adenovirus-E2F1 vector induced DNA synthesis and the expression of a number of growth-related genes, such as PCNA, cyclin A, B-*myb*, and cyclin E *(75)*. Cyclin E is interesting in light of the fact that it, along with its associated kinase, cdk4, is one of the chief candidates for in vivo phosphorylation (inactivation) of pRb *(144,149,281)*. Thus, a feedback loop involving E2F-1, cyclin E/cdk4, and the phosphorylation status of pRb is suggested. Degregori et al. *(75)* also demonstrated that overexpression of E2F-1 could overcome a G1 block following exposure of REF52 cells to IR. Along with the reduced G1 block, an enhanced cyclin A and cyclin E response was observed *(75)*. These data indicate a pivotal role for pRb-mediated downregulation of E2F-1 for radiation-induced G1 arrest.

3.4.2. The RCE Is Controlled by the SP1 Family

Although hypophosphorylated pRb has little or no affinity for DNA *(283,315)*, it indirectly regulates the functions of many TFs. Since pRb binds at least 10 different cell-cycle regulatory proteins, it is possible that a diverse collection of TFs may be targets. A few investigators have examined the role of pRb in regulating the transcription of immediate-early G1 response genes and other genes expressed prior to S phase. Some of this control is thought to be regulated via a DNA consensus sequence termed the RCE, which includes the Sp1 consensus sequence *(57,331)*. To date, six pRb-controlled, cell-cycle-regulatory target genes have been identified, including c-*fos*, c-*myc*, c-*jun*, TGF-β, *TK*, and *egr*-1, all of which contain RCEs in their promoter regions. Robbins et al. *(283)* demonstrated that the c-*fos* promoter was negatively regulated by pRb and that an RCE was necessary for pRb-mediated transcriptional repression. RCEs are also essential for pRb-mediated transcriptional regulation of the c-*fos*, c-*myc*, and TGF-β1 promoters in a cell-cycle-dependent fashion *(175–177,331)*. Thus, the transcriptional regulation of genes by pRb seems to be cell-cycle-dependent. The RCE consensus site, (CC[G/C/A]CCC), is crucial for conferring pRb-mediated gene expression *(57,74,176,192,330,331)*.

pRb does not interact with DNA directly as a TF *(283,315)*; rather, it interacts with a number of known retinoblastoma control proteins (RCPs, such as Sp1, Sp3, and Sp4). RCPs may then form complexes and ultimately bind to Sp1/RCE sites in certain promoters to regulate transcription of target genes. Recently, Chen et al. *(57)* and Udvadia et al. *(330)* discovered that the interaction of pRb with RCPs enhanced binding to Sp1/RCE elements within immediate-early response genes; induction of these genes was essential for G0/G1 to S-phase transition, but their transcription was repressed in S-phase cells. pRb (probably hypophosphorylated) enhances Sp1-mediated DNA binding *(57,74,176,177,330)*. Increased DNA binding activity is thought to be initiated via sequestration or dissociation (initiated by pRB) of an Sp1 inhibitor (Sp1-I), a 20-kDa protein that inhibits Sp1 binding to RCE oligonucleotides in band-shift assays *(57)*. Dissociation of Sp1 from its inhibitor was apparently mediated by phosphorylation of Sp1, similarly to the activation of the cytoplasmic NFκB–IκB complex *(293)*. Murata et al. *(239)* has reported yet another putative inhibitory protein that binds to Sp1. This one a ubiquitously expressed 74-kDa protein that is able to bind to a region of Sp1 necessary for transcriptional trans-activation.

Enhanced protein binding to an Sp1 consensus site was observed following IR in confluence-arrested human melanoma cells *(288)*. This binding was dose- and time-dependent, with maximal binding observed 2–4 h after 4.5 Gy *(364)*. Supershift DNA binding assays demonstrated that Sp1 was present in at least one of the three to four bands observed, and the enhanced DNA binding correlated with increased levels of the phosphorylated form of Sp1 *(364)*. The band-shift pattern is similar to that commonly seen with Sp1 consensus probes. It will be interesting to monitor Sp-factor DNA binding changes given the recent flood of information on the identity and function of the various RCPs. Kingsley and Winoto *(180)* and Hagan et al. *(131,132)* have recently identified two of these components as Sp3 and Sp4. They share homology with the DNA binding domain of Sp1, but otherwise are novel. Sp1 (~90 kDa), Sp3 (115 kDa), and Sp4 (80 kDa) bind with equal affinity to Sp1/RCE consensus sites and so have been hypothesized to compete for RCE binding. Udvadia et al. *(332)* have demonstrated that cotransfection of cells with expression vectors of pRb and either Sp1 or Sp3 can "superactivate" the binding of the Sp factor and induce the transcription of an Sp-responsive reporter construct (either c-*myc* or c-*fos* promoter regions). The "superinduced" binding was equivalent between the two Sp factors, but transcriptional activation was much higher from Sp1 than Sp3 *(131,332)*. This led to the hypothesis that Sp3 competes with Sp1 for binding, and its lower transcriptional activity would effectively inhibit RCE-mediated transcriptional activation. Davis et al. *(73)* recently observed a downregulation of Sp3 protein levels following X-ray exposure within human HCT116 colon carcinoma cells. Such a reduction could functionally activate Sp1-mediated transcription following IR without altering Sp1 levels directly.

Thus, regulation of Sp1-mediated transcription by pRb could occur via sequestration of a 20-kDa Sp1-inhibitor, direct association with RCPs, or by a combination of the two. Jensen et al. *(158)* recently reported that incubation of nuclear extracts with an antibody against pRb ablates Sp1 binding to a consensus site. There are two interpretations of this observation; either there is a direct association between pRb and the Sp1 transcriptional complex, or pRb sequesters an Sp1 inhibitory factor that is released by pRb–antibody association. Jensen et al. *(158)* favor the latter explanation.

Many of these studies were done in disparate cell lines using varying conditions, making it difficult to integrate the data into a unified picture of the regulatory roles and cellular effects of these interactions. However, the observations that IR treatment leads to an pRb-mediated G1 block, activation of Sp1 binding *(364)*, decreased Sp3 levels *(73)*, and increased expression of a number of Sp1-mediated transcripts believed to be needed for cell-cycle arrest/DNA repair suggest that these events will be important in our understanding of the effects of IR on gene expression and the control of cell-cycle progression, survival, and cell death.

4. THE KU/DNA-PK COMPLEX AS A DIRECT SENSOR OF DNA DAMAGE

Since DSBs are the primary lethal lesions following IR exposure, various mechanisms implicated in the repair of this type of damage have been investigated as potential candidates for sensoring/signaling of DNA damage *(92,121,270,370)*. The best current candidate for linking the presence of DNA damage to a signaling cascade is DNA-dependent protein kinase (DNA-PK). A serine/threonine kinase activity has been purified from HeLa cells *(47)*, and a ~460 kDa polypeptide has been identified and cloned (*XRCC7* gene) as the catalytic subunit of DNA-PK (DNA-PKcs) *(304)*. Its activation requires free DNA ends *(122)*; circular or intact DNA cannot activate this enzyme. The human Ku autoantigen has been identified as the regulatory and DNA binding subunit of the DNA-PK holoenzyme *(122)*. Ku consists of two subunits, Ku70 and Ku86 polypeptides. Only Ku70 has DNA binding ability *(230)*. Ku86 is encoded by a unique gene *(XRCC5) (321)*, whereas the Ku70 is likely to be a gene family *(125)*. It has been hypothesized that Ku70-related proteins may have different affinities for different DNA structures *(219)*. Absence of either Ku subunit (Ku70 and Ku86) or the DNA-PKcs itself (~460 kDa) causes aberrant V(D)J joining, and sensitivity to agents that cause DNA breaks *(6,7,28,45,49,102,121,122,182,183,186,201,238,304,321)*. Data suggest that the Ku subunits act as sensors/binders of free DNA ends *(6,49,122,275)*. It has also been suggested that the Ku subunits, once bound to DNA ends, may stabilize the free ends and bring them into proximity *(275)*. Recently, Ku has been identified as human DNA helicase II *(326)*, which unwinds damaged DNA ends and makes them more accessible to repair enzymes. Also, Ku can recruit the DNA-PKcs, which is activated by association with Ku and free DNA ends. The activated DNA-PK holoenzyme has been demonstrated in vitro to phosphorylate a variety of substrates, including TFs, such as c-Jun *(18)*, c-Fos, c-Myc, Sp1 *(6,7,122,155,251)*, and p53 *(202)*; the C-terminal domain of RNA polymerase II *(259,260)*; RPA *(45)*; and T-antigen *(200)*. DNA-PK also autophosphorylates both DNA-PKcs and Ku subunits, which may downregulate holo-enzyme activity *(7)* by affecting complex stability and/or protein half-life *(6)*.

From the above observations, it is attractive to hypothesize that the DNA-PK holoenzyme may be a DNA-break sensor, which, by phosphorylation, activates enzymes and transcription factors that mobilize DNA repair complexes, both constitutive and inducible. A model by which DNA-PK may regulate the inducible component of DNA repair is presented below.

Due to lack of an in vivo assay for DNA-PK activity, the true function of DNA-PK remains unclear. In vitro assays using double-stranded DNA oligonucleotides as activating agents have shown that DNA-PK can phosphorylate a variety of substrates (as discussed above), including TFs, such as Sp1, p53, c-Jun, c-Fos, and c-Myc. In light of these results and the following two lines of evidence, it is possible to construct a model for DNA-PK function that couples DNA repair and transcriptional activation.

First, PLDR survival (from 10% survival with no repair to 70% with 8 h of repair) of radioresistant U1-Mel cells was partially prevented by CHX *(29,152)*. This indicates that cellular repair responses may require new transcripts/proteins (e.g., they are coupled), and a possible involvement for these radiation-response transcripts/proteins in DNA repair complexes.

Second, IR induces immediate-early genes (IEGs), including c-*fos*, c-*jun*, c-*myc*, and *egr*-1 *(265,364)*. The transcription of these genes has been shown to be regulated by pRb through RCEs in their promoters *(176)*. pRb does not directly bind to promoter regions containing RCE sites, but interacts with RCPs that are regulated by pRb *(331)*. One such RCP is Sp1, and Sp1-mediated transcription is stimulated by pRb *(330)*. Sp1/RCP DNA binding and phosphorylation were optimally activated within 2–4 h in confluence-arrested human U1-Mel cells after radiation *(30,33,288,364)*. Activation of Sp1/RCP DNA binding correlates well with X-ray-induced, endogenous expression of XIPs *(35)* and IEGs *(265)*. It has been proposed that Sp1 phosphorylation causes its dissociation from Sp1-I, a process facilitated by hypophosphorylated pRb *(57)*. Phosphorylated Sp1 may subsequently induce transcription of IEGs in confluence-arrested cells. NFκB DNA binding also increases continuously over a 24-h period after radiation. Induction of NFκB may be caused by cell membrane damage, which activates c-Raf-1 and results in phosphorylation of IκB (the NFκB inhibitor). Phosphorylated IκB then dissociates from NFκB, which then migrates to the nucleus. Sp1/RCP and NFκB DNA binding, however, decrease at higher doses of radiation (>6 Gy) *(364)*. It is unknown whether the possible enhancing activity of simultaneous Sp1 and NFκB binding as an X-ray-responsive element may control expression of some of the genes in Table 1.

Interestingly, p53 DNA binding and nuclear protein levels increased only after a relatively high dose (12 Gy) of radiation within confluence-arrested U1-Mel cells. Increased p53 DNA binding and subsequent transcriptional stimulation/repression may have cascade-like signals in the cell, possibly involving apoptosis *(212,300)*. Both p53 and Sp1 are substrates for DNA-PK in vitro *(212,300)*. Phosphorylation of p53 at Ser15 by DNA-PK may stabilize p53 and increase its half-life in the nucleus *(103,333)*. Phosphorylated Sp1 increases following IR exposure *(364)*; this phosphorylation may increase its transcriptional activity *(156)*. Thus, these considerations lead to the hypothesis that DNA-PK is a DNA damage sensor (Fig. 1). It may recognize and participate in the repair of SSBs and DSBs generated by radiation, and in response, phosphorylates and activates Sp1 and/or p53, depending on the state and type of the cell, and the levels of these two proteins in the nucleus. The entire Sp1/RCP DNA binding process may be dependent on hypophosphorylated pRb, but the mechanistic details remain unclear. Increased RCP binding to the promoter regions of certain genes, possibly in combination with the coordinate activation of NFκB DNA binding, might result in the transcriptional induction of unique, X-ray-inducible gene products, such as IEGs and XIPs *(30,34,35,116,152,265,351)*, which are presumably needed for unscheduled DNA synthesis and survival recovery (i.e., PLDR) *(29,364)*. This would result in a "coupling" of DNA repair to the control of transcription in order to regulate coordinately DNA repair, cell-cycle regulation, and possibly apoptosis.

Thus, a cellular decision to carry out DNA repair (Sp1-mediated) or apoptosis (p53-mediated) could be decided by the ratio or relative amounts of phosphorylated/activated Sp1/p53, respectively. It should be emphasized this model for DNA-PK as a

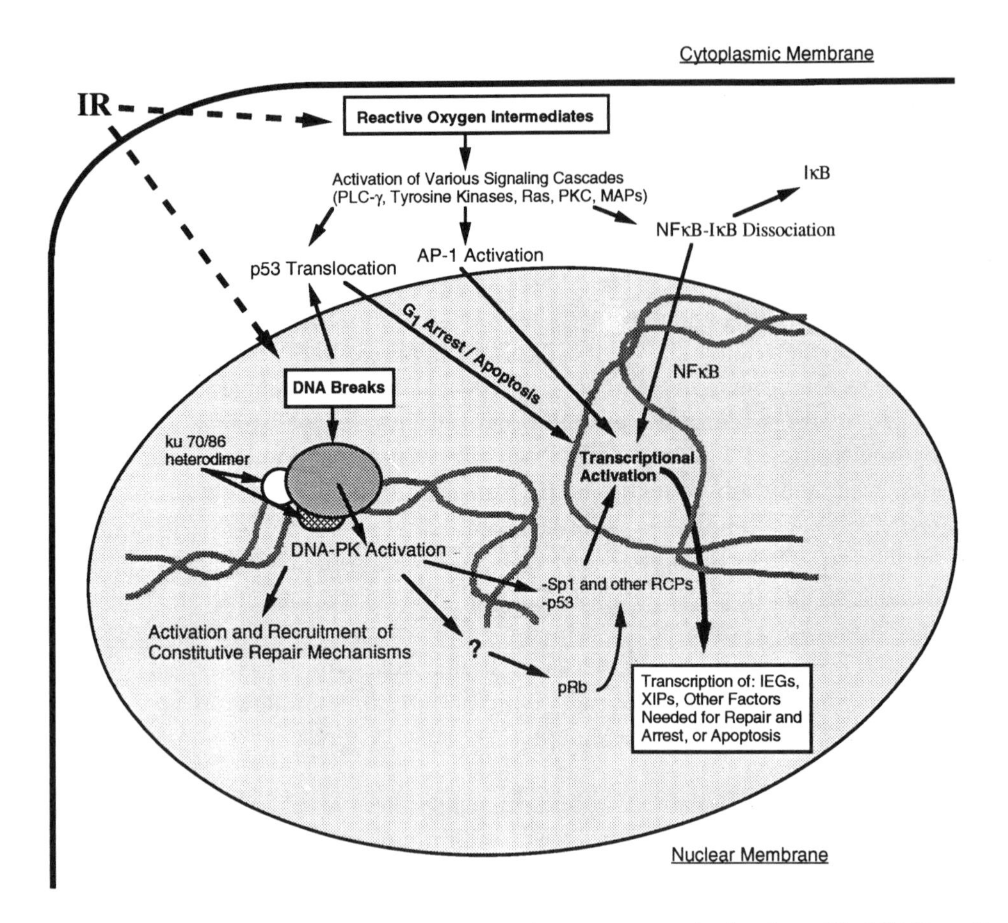

Fig. 1. Possible model for DNA-PK as a DNA damage sensor: Activation of Sp1/RCPs, p53, and NFκB after ionizing radiation. Sp1/RCP activation is "coupled" to and activated by DNA-PK, which is a direct sensor of DNA damage. Activation of DNA-PK may cause Sp1 phosphorylation and transcriptional activation, which associates with hypophosphorylated retinoblastoma (pRb). NFκB activation occurs via membrane damage and ROI generation, which leads to the subsequent phosphorylation and dissociation of its inhibitor, IκB. Activation of Sp1 and NFκB may result in the transcriptional induction of unique, X-ray-inducible genes (IEGs and XIPs), which may be involved in DNA repair and cell survival. p53 phosphorylation and nuclear protein levels also increase after DNA damage and, under some conditions, may lead the damaged cells to undergo apoptosis.

DNA damage sensor is highly speculative. Also, it is necessarily simplistic, since it does not take into account the complex interactions of other modified transcription factors, or other p53-independent mechanisms governing cellular decisions between repair and apoptosis.

ACKNOWLEDGMENTS

The authors would like to thank John Petrini, Scott Cuthill, Shigeki Miyamoto, and Ajit Verma for critical reading and suggestions in preparing this chapter.

REFERENCES

1. Abate, C., L. Patel, F. J. D. Rauscher, and T. Curran. 1990. Redox regulation of fos and jun DNA-binding activity in vitro. *Science* **249:** 1157–1161.
2. al-Nabulsi, I., Y. Takamiya, Y. Voloshin, A. Dritschilo, R. L. Martuza, and T. J. Jorgensen. 1994. Expression of thymidine kinase is essential to low dose radiation resistance of rat glioma cells. *Cancer Res.* **54:** 5614–5617.
3. Amara, F. M., F. Y. Chen, and J. A. Wright. 1995. Defining a novel cis element in the 3'-untranslated region of mammalian ribonucleotide reductase component R2 mRNA: role in transforming growth factor-beta 1 induced mRNA stabilization. *Nucleic Acids Res.* **23:** 1461–1467.
4. Amstad, P. A., G. Krupitza, and P. A. Cerutti. 1992. Mechanism of c-fos induction by active oxygen. *Cancer Res.* **52:** 3952–3960.
5. Anderson, A. and G. E. Woloschak. 1992. Cellular proto-oncogene expression following exposure of mice to gamma rays. *Radiat. Res.* **130:** 340–344.
6. Anderson, C. W. 1993. DNA damage and the DNA-activated protein kinase. *Trends Biochem. Sci.* **18:** 433–437.
7. Anderson, C. W. and S. P. Lees-Miller. 1992. The nuclear serine/threonine protein kinase DNA-PK. *Crit. Rev. Eukaryotic Gene Expression* **2:** 283–314.
8. Angel, P., M. Imagawa, R. Chiu, B. Stein, R. J. Imbra, H. J. Rahmsdorf, C. Jonat, P. Herrlich, and M. Karin. 1987. Phorbol ester-inducible genes contain a common cis element recognized by a TPA-modulated trans-acting factor. *Cell* **49:** 729–739.
9. Angel, P. and M. Karin. 1991. The role of Jun, Fos and the AP-1 complex in cell-proliferation and transformation. *Biochim. Biophys. Acta* **1072:** 129–157.
10. Angel, P., A. Poting, U. Mallick, H. J. Rahmsdorf, M. Schorpp, and P. Herrlich. 1986. Induction of metallothionein and other mRNA species by carcinogens and tumor promoters in primary human skin fibroblasts. *Mol. Cell. Biol.* **6:** 1760–1766.
11. Antoccia, A., F. Palitti, T. Raggi, C. Catena, and C. Tanzarella. 1994. Lack of effect of inhibitors of DNA synthesis/repair on the ionizing radiation-induced chromosomal damage in G2 stage of ataxia telangiectasia cells. *Int. J. Radiat. Biol.* **66:** 309–317.
12. Baeuerle, P. A. and D. Baltimore. 1988. Activation of DNA-binding activity in an apparently cytoplasmic precursor of the NF-kappa B transcription factor. *Cell* **53:** 211–217.
13. Baeuerle, P. A. and D. Baltimore. 1988. I kappa B: a specific inhibitor of the NF-kappa B transcription factor. *Science* **242:** 540–546.
14. Baeuerle, P. A. and T. Henkel. 1994. Function and activation of NF-kappa B in the immune system. *Ann. Rev. Immunol.* **12:** 141–179.
15. Baeuerle, P. A., M. Lenardo, J. W. Pierce, and D. Baltimore. 1988. Phorbol-ester-induced activation of the NF-kappa B transcription factor involves dissociation of an apparently cytoplasmic NF-kappa B/inhibitor complex. *Cold Spring Harbor Symp. Quant. Biol.* **53:** 789–798.
16. Bakalkin, G., T. Yakovleva, G. Selivanova, K. P. Magnusson, L. Szekely, E. Kiseleva, G. Klein, L. Terenius, and K. G. Wiman. 1994. p53 binds single-stranded DNA ends and catalyzes DNA renaturation and strand transfer. *Proc. Natl. Acad. Sci. USA* **91:** 413–417.
17. Bankson, D. D., M. Kestin, and N. Rifai. 1993. Role of free radicals in cancer and atherosclerosis. *Clin. Lab. Med.* **13:** 463–480.
18. Bannister, A. J., T. M. Gottlieb, T. Kouzarides, and S. P. Jackson. 1993. c-Jun is phosphorylated by the DNA-dependent protein kinase in vitro: definition of the minimal kinase recognition motif. *Nucleic Acids Res.* **21:** 1289–1295.
19. Bargonetti, J., P. N. Friedman, S. E. Kern, B. Vogelstein, and C. Prives. 1991. Wild-type but not mutant p53 immunopurified proteins bind to sequences adjacent to the SV40 origin of replication. *Cell* **65:** 1083–1091.

20. Barroga, C. F., J. K. Stevenson, E. M. Schwarz, and I. M. Verma. 1995. Constitutive phosphorylation of I kappa B alpha by casein kinase II. *Proc. Natl. Acad. Sci. USA* **92:** 7637–7641.
21. Belka, C., K. Wiegmann, D. Adam, R. Holland, M. Neuloh, F. Herrmann, M. Kronke, and M. A. Brach. 1995. Tumor necrosis factor (TNF)-alpha activates c-Raf-1 kinase via the p55 TNF receptor engaging neutral sphingomyelinase. *EMBO J.* **14:** 1156–1165.
22. Ben-Ishai, R., R. Scharf, R. Sharon, and I. Kapten. 1990. A human cellular sequence implicated in trk oncogene activation is DNA damage inducible. *Proc. Natl. Acad. Sci. USA* **87:** 6039–6043.
23. Bergelson, S., R. Pinkus, and V. Daniel. 1994. Intracellular glutathione levels regulate Fos/Jun induction and activation of glutathione S-transferase gene expression. *Cancer Res.* **54:** 36–40.
24. Biard, D. S. F., M. Martin, Y. L. Rhun, A. Duthu, J. L. Lefaix, E. May, and P. May. 1994. Concomitant p53 gene mutation and increased radiosensitivity in rat lung embryo epithelial cells during neoplastic development. *Cancer Res.* **54:** 3361–3364.
25. Billen, D. 1987. Free radical scavenging and the expression of potentially lethal damage in X-irradiated repair-deficient *Escherichia coli. Radiat. Res.* **111:** 354–360.
26. Blank, V., P. Kourilsky, and A. Israel. 1992. NF-kappa B and related proteins: Rel/dorsal homologies meet ankyrin-like repeats. *Trends Biochemistry Sci.* **17:** 135–140.
27. Blank, V., P. Kourilsky, and A. Israel. 1991. Cytoplasmic retention, DNA binding and processing of the NF-kappa B p50 precursor are controlled by a small region in its C-terminus. *EMBO J.* **10:** 4159–4167.
28. Blunt, T., N. J. Finnie, G. E. Taccioli, G. C. Smith, J. Demengeot, T. M. Gottlieb, et al. 1995. Defective DNA-dependent protein kinase activity is linked to V(D)J recombination and DNA repair defects associated with the murine scid mutation. *Cell* **80:** 813–823.
29. Boothman, D. A., I. Bouvard, and E. N. Hughes. 1989. Identification and characterization of X-ray-induced proteins in human cells. *Cancer Res.* **49:** 2871–2878.
30. Boothman, D. A., T. W. Davis, and W. M. Sahijdak. 1994. Enhanced expression of thymidine kinase in human cells following ionizing radiation. *Int. J. Radiat. Oncol. Biol. Phys.* **30:** 391–398.
31. Boothman, D. A., N. Fukunaga, and M. Wang. 1994. Down-regulation of topoisomerase I in mammalian cells following ionizing radiation. *Cancer Res.* **54:** 4618–4626.
32. Boothman, D. A., E. N. Hughes, and A. B. Pardee. 1990. The role of X-ray-induced DNA repair processes in mutagenesis and carcinogenesis. *Prog. Clin. Biol. Res.* **340E:** 319–327.
33. Boothman, D. A., I. W. Lee, and W. M. Sahijdak. 1994. Isolation of an X-ray-responsive element in the promoter region of tissue-type plasminogen activator: potential uses of X-ray-responsive elements for gene therapy. *Radiat. Res.* **138:** S68–S71.
34. Boothman, D. A., G. Majmudar, and T. Johnson. 1994. Immediate X-ray-inducible responses from mammalian cells. *Radiat. Res.* **138:** S44–S46.
35. Boothman, D. A., M. Meyers, N. Fukunaga, and S. W. Lee. 1993. Isolation of x-ray-inducible transcripts from radioresistant human melanoma cells. *Proc. Natl. Acad. Sci. USA* **90:** 7200–7204.
36. Boothman, D. A., D. K. Trask, and A. B. Pardee. 1989. Inhibition of potentially lethal DNA damage repair in human tumor cells by beta-lapachone, an activator of topoisomerase I. *Cancer Res.* **49:** 605–612.
37. Boothman, D. A., M. Wang, and S. W. Lee. 1991. Induction of tissue-type plasminogen activator by ionizing radiation in human malignant melanoma cells. *Cancer Res.* **51:** 5587–5595.
38. Bouffler, S. D., C. J. Kemp, A. Balmain, and R. Cox. 1995. Spontaneous and ionizing radiation-induced chromosomal abnormalities in p53-deficient mice. *Cancer Res.* **55:** 3883–3889.

39. Brach, M. A., H. J. Gruss, T. Kaisho, Y. Asano, T. Hirano, and F. Herrmann. Ionizing radiation induces expression of interleukin 6 by human fibroblasts involving activation of nuclear factor-kappa B. *J. Biol. Chem.* **268:** 8466–8472.
40. Brach, M. A., R. Hass, M. L. Sherman, H. Gunji, R. Weichselbaum, and D. Kufe. 1991. Ionizing radiation induces expression and binding activity of the nuclear factor kappa B. *J. Clin. Invest.* **88:** 691–695.
41. Bradley, D. W., Q. P. Dou, J. L. Fridovich-Keil, and A. B. Pardee. 1990. Transformed and nontransformed cells differ in stability and cell cycle regulation of a binding activity to the murine thymidine kinase promoter. *Proc. Natl. Acad. Sci. USA* **87:** 9310–9314.
42. Brain, R. and J. R. Jenkins. 1994. Human p53 directs DNA strand reassociation and is photolabelled by 8-azido ATP. *Oncogene* **9:** 1775–1780.
43. Breen, A. P. and J. A. Murphy. 1995. Reactions of oxyl radicals with DNA. *Free Rad. Biol. Med.* **18:** 1033–1077.
44. Bronner, C. E., S. M. Baker, P. T. Morrison, G. Warren, L. G. Smith, M. K. Lescoe, M. Kane, C. Earabino, J. Lipford, A. Lindblom, P. Tannergard, R. J. Bollag, A. R. Godwin, D. C. Ward, M. Nordenskjold, R. Fishel, R. Kolodner, and R. M. Liskay. 1994. Mutation in the DNA mismatch repair gene homologue *hMLH1* is associated with hereditary non-polyposis colon cancer. *Nature* **368:** 258–261.
45. Brush, G. S., C. W. Anderson, and T. J. Kelly. 1994. The DNA-activated protein kinase is required for the phosphorylation of replication protein A during simian virus 40 DNA replication. *Proc. Natl. Acad. Sci. USA* **91:** 12,520–12,524.
46. Buscher, M., H. J. Rahmsdorf, M. Litfin, M. Karin, and P. Herrlich. 1988. Activation of the c-fos gene by UV and phorbol ester: different signal transduction pathways converge to the same enhancer element. *Oncogene* **3:** 301–311.
47. Cao, Q. P., S. Pitt, J. Leszyk, and E. F. Baril. 1994. DNA-dependent ATPase from HeLa cells is related to human Ku autoantigen. *Biochemistry* **33:** 8548–8557.
48. Carrier, F., A. Gatignol, M. C. Hollander, K. T. Jeang, and A. J. Fornace, Jr. 1994. Induction of RNA-binding proteins in mammalian cells by DNA-damaging agents. *Proc. Natl. Acad. Sci. USA* **91:** 1554–1558.
49. Chang, C., K. A. Biedermann, M. Mezzina, and J. M. Brown. 1993. Characterization of the DNA double strand break repair defect in scid mice. *Cancer Res.* **53:** 1244–1248.
50. Chang, Z. F. and S. M. Cheng. 1993. Constitutive overexpression of DNA binding activity to the distal CCAAT box of human thymidine kinase promoter in human tumor cell lines. *Cancer Res.* **53:** 3253–3256.
51. Chang, Z. F. and S. M. Cheng. 1992. Methylation-sensitive protein-DNA interaction at the cell cycle regulatory domain of human thymidine kinase promoter. *Biochem. Biophys. Res. Commun.* **189:** 480–487.
52. Chang, Z. F. and C. J. Liu. 1994. Human thymidine kinase CCAAT-binding protein is NF-Y, whose A subunit expression is serum-dependent in human IMR-90 diploid fibroblasts. *J. Biol. Chem.* **269:** 17,893–17,898.
53. Chellappan, S. P., S. Hiebert, M. Mudryj, J. M. Horowitz, and J. R. Nevins. 1991. The E2F transcription factor is a cellular target for the RB protein. *Cell* **65:** 1053–1061.
54. Chen, F. Y., F. M. Amara, and J. A. Wright. 1994. Defining a novel ribonucleotide reductase r1 mRNA cis element that binds to an unique cytoplasmic trans-acting protein. *Nucleic Acids Res.* **22:** 4796,4797.
55. Chen, F. Y., F. M. Amara, and J. A. Wright. 1994. Regulation of mammalian ribonucleotide reductase R1 mRNA stability is mediated by a ribonucleotide reductase R1 mRNA 3'-untranslated region cis-trans interaction through a protein kinase C-controlled pathway. *Biochem. J.* **302:** 125–132.
56. Chen, F. Y., F. M. Amara, and J. A. Wright. 1994. Posttranscriptional regulation of ribonucleotide reductase R1 gene expression is linked to a protein kinase C pathway in mammalian cells. *Biochem. Cell Biol.* **72:** 251–256.

57. Chen, L. I., T. Nishinaka, K. Kwan, I. Kitabayashi, K. Yokoyama, Y. H. Fu, S. Grunwald, and R. Chiu. 1994. The retinoblastoma gene product RB stimulates Sp1-mediated transcription by liberating Sp1 from a negative regulator. *Mol. Cell. Biol.* **14:** 4380–4389.
58. Chen, X., G. Farmer, H. Zhu, R. Prywes, and C. Prives. 1993. Cooperative DNA binding of p53 with TFIID (TBP): a possible mechanism for transcriptional activation. *Genes Dev.* **7:** 1837–1849.
59. Clarke, A. R., S. Gledhill, M. L. Hooper, C. C. Bird, and A. H. Wyllie. 1994. p53 dependence of early apoptotic and proliferative responses within the mouse intestinal epithelium following gamma-irradiation. *Oncogene* **9:** 1767–1773.
60. Clarke, A. R., C. A. Purdie, D. J. Harrison, R. G. Morris, C. C. Bird, M. L. Hooper, and A. H. Wyllie. 1993. Thymocyte apoptosis induced by p53-dependent and independent pathways. *Nature* **362:** 849–852.
61. Claudio, P. P., C. M. Howard, A. Baldi, A. De Luca, Y. Fu, G. Condorelli, Y. Sun, N. Colburn, B. Calabretta, and A. Giordano. 1994. p130/pRb2 has growth suppressive properties similar to yet distinctive from those of retinoblastoma family members pRb and p107. *Cancer Res.* **54:** 5556–5560.
62. Coffer, P. J., B. M. Burgering, M. P. Peppelenbosch, J. L. Bos, and W. Kruijer. 1995. UV activation of receptor tyrosine kinase activity. *Oncogene* **11:** 561–569.
63. Collins, A. 1987. Cellular responses to ionizing radiation: effects of interrupting DNA repair with chemical agents. *Int. J. Radiat. Biol. Related Studies in Phys., Chem. Med.* **51:** 971–983.
64. Cox, L. S., T. Hupp, C. A. Midgley, and D. P. Lane. 1995. A direct effect of activated human p53 on nuclear DNA replication. *EMBO J.* **14:** 2099–2105.
65. Cox, L. S. and D. P. Lane. 1995. Tumour supressors, kinases and clamps: how p53 regulates the cell cycle in response to DNA damage. *Bioessays* **17:** 501–508.
66. Crawford, D., P. Zbinden, P. Amstad, and P. Cerutti. 1988. Oxidant stress induces the proto-oncogenes c-fos and c-myc in mouse epidermal cells. *Oncogene* **3:** 27–32.
67. Cross, S. M., C. A. Sanchez, C. A. Morgan, M. K. Schimke, S. Ramel, R. L. Idzerda, W. H. Raskind, and B. J. Reid. 1995. A p53-dependent mouse spindle checkpoint. *Science* **267:** 1353–1356.
68. Dagnino, L., L. Zhu, K. L. Skorecki, and H. L. Moses. 1995. E2F-independent transcriptional repression by p107, a member of the retinoblastoma family of proteins. *Cell Growth Diff.* **6:** 191–198.
69. Datta, R., D. E. Hallahan, S. M. Kharbanda, E. Rubin, M. L. Sherman, E. Huberman, R. R. Weichselbaum, and D. W. Kufe. 1992. Involvement of reactive oxygen intermediates in the induction of c-jun gene transcription by ionizing radiation [published erratum appears in *Biochemistry* 1992 **Oct 6;31(39):** 9512]. *Biochemistry* **31:** 8300–8306.
70. Datta, R., R. Hass, H. Gunji, R. Weichselbaum, and D. Kufe. 1992. Down-regulation of cell cycle control genes by ionizing radiation. *Cell Growth Diff.* **3:** 637–644.
71. Datta, R., E. Rubin, V. Sukhatme, S. Qureshi, D. Hallahan, R. R. Weichselbaum, and D. W. Kufe. 1992. Ionizing radiation activates transcription of the EGR1 gene via CArG elements. *Proc. Natl. Acad. Sci. USA* **89:** 10,149–10,153.
72. Datta, R., N. Taneja, V. P. Sukhatme, S. A. Qureshi, R. Weichselbaum, and D. W. Kufe. 1993. Reactive oxygen intermediates target $CC(A/T)_6GG$ sequences to mediate activation of the early growth response 1 transcription factor gene by ionizing radiation. *Proc. Natl. Acad. Sci. USA* **90:** 2419–2422.
73. Davis, T. W., and C. Wilson-Van Patten. unpublished data.
74. Defeo-Jones, D., P. S. Huang, R. E. Jones, K. M. Haskell, G. A. Vuocolo, M. G. Hanobik, H. E. Huber, and A. Oliff. 1991. Cloning of cDNAs for cellular proteins that bind to the retinoblastoma gene product. *Nature* **352:** 251–254.
75. Degregori, J., T. Kowalik, and J. R. Nevins. 1995. Cellular targets for activation by the E2f1 transcription factor include DNA synthesis- and G(1)/S-regulatory genes. *Mol. Cell. Biol.* **15:** 4215–4224.

76. Deng, T. and M. Karin. 1994. c-Fos transcriptional activity stimulated by H-Ras-activated protein kinase distinct from JNK and ERK. *Nature* **371:** 171–175.
77. Devary, Y., R. A. Gottlieb, L. F. Lau, and M. Karin. 1991. Rapid and preferential activation of the c-jun gene during the mammalian UV response. *Mol. Cell. Biol.* **11:** 2804–2811.
78. Devary, Y., R. A. Gottlieb, T. Smeal, and M. Karin. 1992. The mammalian ultraviolet response is triggered by activation of Src tyrosine kinases. *Cell* **71:** 1081–1091.
79. Devary, Y., C. Rosette, J. A. DiDonato, and M. Karin. 1993. NF-kappa B activation by ultraviolet light not dependent on a nuclear signal. *Science* **261:** 1442–1445.
80. Di Leonardo, A., S. P. Linke, K. Clarkin, and G. M. Wahl. 1994. DNA damage triggers a prolonged p53-dependent G1 arrest and long-term induction of Cip1 in normal human fibroblasts. *Genes Dev.* **8:** 2540–2551.
81. Dizdaroglu, M. 1992. Oxidative damage to DNA in mammalian chromatin. *Mutat. Res.* **275:** 331–342.
82. Donehower, L. A. and A. Bradley. 1993. The tumor suppressor p53. *Biochim. Biophys. Acta* **1155:** 181–205.
83. Donehower, L. A., M. Harvey, B. L. Slagle, M. J. McArthur, C. A. Montgomery, Jr., J. S. Butel, and A. Bradley. 1992. Mice deficient for p53 are developmentally normal but susceptible to spontaenous tumours. *Nature (Lond.)* **356:** 215–221.
84. Dou, Q. P., J. L. Fridovich-Keil, and A. B. Pardee. 1991. Inducible proteins binding to the murine thymidine kinase promoter in late G1/S phase. *Proc. Natl. Acad. Sci. USA* **88:** 1157–1161.
85. Dou, Q. P., P. J. Markell, and A. B. Pardee. 1992. Thymidine kinase transcription is regulated at G_1/S phase by a complex that contains retinoblastoma-like protein and a cdc2 kinase. *Proc. Natl. Acad. Sci. USA* **89:** 3256–3260.
86. Dutta, A., J. M. Ruppert, J. C. Aster, and E. Winchester. 1993. Inhibition of DNA replication factor RPA by p53. *Nature (Lond.)* **365:** 79–82.
87. el-Deiry, W. S., J. W. Harper, P. M. O'Connor, V. E. Velculescu, C. E. Canman, J. Jackman, J. A. Pietenpol, M. Burrell, D. E. Hill, Y. Wang, K. G. Wiman, W. E. Mercer, M. B. Kastan, K. W. Kohn, S. J. Elledge, K. W. Kinzler, and B. Vogelstein. 1994. WAF1/CIP1 is induced in p53-mediated G1 arrest and apoptosis. *Cancer Res.* **54:** 1169–1174.
88. el-Deiry, W. S., S. E. Kern, J. A. Pietenpol, K. W. Kinzler, and B. Vogelstein. 1992. Definition of a consensus binding site for p53. *Nature Genet.* **1:** 45–49.
89. el-Deiry, W. S., T. Tokino, V. E. Velculescu, D. B. Levy, R. Parsons, J. M. Trent, D. Lin, W. E. Mercer, K. W. Kinzler, and B. Vogelstein. 1993. WAF1, a potential mediator of p53 tumor suppression. *Cell* **75:** 817–825.
90. Eliyahu, D., D. Michalovitz, S. Eliyahu, O. Pinhasi-Kimhi, and M. Oren. 1989. Wild-type p53 can inhibit oncogene-mediated focus formation. *Proc. Natl. Acad. Sci. USA* **86:** 8763–8767.
91. Elledge, S. J. and R. W. Davis. 1987. Identification and isolation of the gene encoding the small subunit of ribonucleotide reductase from Saccharomyces cerevisiae: DNA damage-inducible gene required for mitotic viability. *Mol. Cell. Biol.* **7:** 2783–2793.
92. Elledge, S. J., Z. Zhou, J. B. Allen, and T. A. Navas. 1993. DNA damage and cell cycle regulation of ribonucleotide reductase. *Bioessays* **15:** 333–339.
93. Ellem, K. A., M. Cullinan, K. C. Baumann, and A. Dunstan. 1988. UVR induction of TGF alpha: a possible autocrine mechanism for the epidermal melanocytic response and for promotion of epidermal carcinogenesis. *Carcinogenesis* **9:** 797–801.
94. Ellis, R. E., J. Yuan, and H. R Horvitz. 1991. Mechanisms and functions of cell death. *Ann. Rev. Cell Biol.* **7:** 663–698.
95. Emerit, J., J. M. Klein, A. Coutellier, and F. Congy. 1991. Free radicals and lipid peroxidation in cell biology: physiopathologic prospects. *Pathologie Biologie* **39:** 316–327.
96. Engel, K., A. Ahlers, M. A. Brach, F. Herrmann, and M. Gaestel. 1995. MAPKAP kinase 2 is activated by heat shock and TNF-alpha: in vivo phosphorylation of small heat

shock protein results from stimulation of the MAP kinase cascade. *J. Cell. Biochem.* **57:** 321–330.

97. Engstrom, Y., S. Eriksson, I. Jildevik, S. Skog, L. Thelander, and B. Tribukait. 1985. Cell cycle-dependent expression of mammalian ribonucleotide reductase. Differential regulation of the two subunits. *J. Biol. Chem.* **260:** 9114–9116.
98. Ewen, M. E. 1994. The cell cycle and the retinoblastoma protein family. *Cancer Met. Rev.* **13:** 45–66.
99. Farmer, G., J. Bargonetti, H. Zhu, P. Friedman, R. Prywes, and C. Prives. 1992. Wild-type p53 activates transcription in vitro. *Nature (Lond.)* **358:** 83–86.
100. Fields, S. and S. K. Jang. 1990. Presence of a potent transcription activating sequence in the p53 protein. *Science* **249:** 1046–1049.
101. Finlay, C. A., P. W. Hinds, and A. J. Levine. 1989. The p53 proto-oncogene can act as a suppressor of transformation. *Cell* **57:** 1083–1093.
102. Finnie, N. J., T. M. Gottlieb, T. Blunt, P. A. Jeggo, and S. P. Jackson. 1995. DNA-dependent protein kinase activity is absent in xrs-6 cells: implications for site-specific recombination and DNA double-strand break repair. *Proc. Natl. Acad. Sci. USA* **92:** 320–324.
103. Fiscella, M., S. J. Ullrich, N. Zambrano, M. T. Shields, D. Lin, S. P. Lees-Miller, C. W. Anderson, W. E. Mercer, and E. Appella. 1993. Mutation of the serine 15 phosphorylation site of human p53 reduces the ability of p53 to inhibit cell cycle progression. *Oncogene* **8:** 1519–1528.
104. Fishel, R., M. K. Lescoe, M. R. Rao, N. G. Copeland, N. A. Jenkins, J. Garber, M. Kane, and R. Kolodner. 1993. The human mutator gene homolog MSH2 and its association with hereditary nonpolyposis colon cancer. *Cell* **75:** 1027–1038.
105. Fornace, A. J., I. Alamo, and M. C. Hollander. 1988. DNA damage-inducible transcripts in mammalian cells. *Proc. Natl. Acad. Sci. USA* **85:** 8800–8804.
106. Fornace, A. J., D. W. Nebert, M. C. Hollander, J. D. Luethy, M. Papathanasiou, J. Fargnoli, and N. J. Holbrook. 1989. Mammalian genes coordinately regulated by growth arrest signals and DNA-damaging agents. *Mol. Cell. Biol.* **9:** 4196–4203.
107. Fornace, A. J., Jr., H. Schalch, and I. Alamo, Jr. 1988. Coordinate induction of metallothioneins I and II in rodent cells by UV irradiation. *Mol. Cell. Biol.* **8:** 4716–4720.
108. Fornace, A. J., Jr., B. Zmudzka, M. C. Hollander, and S. H. Wilson. 1989. Induction of beta-polymerase mRNA by DNA-damaging agents in Chinese hamster ovary cells. *Mol. Cell. Biol.* **9:** 851–853.
109. Frame, M. C., N. M. Wilkie, A. J. Darling, A. Chudleigh, A. Pintzas, J. C. Lang, and D. A. Gillespie. 1991. Regulation of AP-1/DNA complex formation in vitro. *Oncogene* **6:** 205–209.
110. Franklin, C. C., T. Unlap, V. Adler, and A. S. Kraft. 1993. Multiple signal transduction pathways mediate c-Jun protein phosphorylation. *Cell Growth Diff.* **4:** 377–385.
111. Freimuth, W. W., J. M. Depper, and G. J. Nabel. 1989. Regulation of the IL-2 receptor alpha-gene. Interaction of a kappa B binding protein with cell-specific transcription factors. *J. Immunol.* **143:** 3064–3068.
112. Fridovich-Keil, J. L., J. M. Gudas, Q. P. Dou, I. Bouvard, and A. B. Pardee. 1991. Growth-responsive expression from the murine thymidine kinase promoter: genetic analysis of DNA sequences. *Cell Growth Diff.* **2:** 67–76.
113. Friedberg, E. C., G. C. Walker, and W. Siede. 1995. *DNA Repair and Mutagenesis*. ASM, Washington, DC.
114. Fritsche, M., C. Haessler, and G. Brandner. 1993. Induction of nuclear accumulation of the tumor-suppressor protein p53 by DNA-damaging agents. *Oncogene* **8:** 307–318.
115. Fritz, G., K. Tano, S. Mitra, and B. Kaina. 1991. Inducibility of the DNA repair gene encoding O^6-methylguanine-DNA methyltransferase in mammalian cells by DNA-damaging treatments. *Mol. Cell. Biol.* **11:** 4660–4668.

116. Fukunaga, N., H. L. Burrows, M. Meyers, R. A. Schea, and D. A. Boothman. 1992. Enhanced induction of tissue-type plasminogen activator in normal human cells compared to cancer-prone cells following ionizing radiation. *Int. J. Radiat. Oncol. Biol. Phys.* **24:** 949–957.
117. Gahring, L., M. Baltz, M. B. Pepys, and R. Daynes. 1984. Effect of ultraviolet radiation on production of epidermal cell thymocyte-activating factor/interleukin 1 in vivo and in vitro. *Proc. Natl. Acad. Sci. USA* **81:** 1198–1202.
118. Gantchev, T. G. 1995. Thymidine free radicals generated during metallo-phthalocyanine photosensitization: a comparison with gamma-radiation. *Int. J. Radiat. Biol.* **68:** 29–36.
119. Ginsberg, D., F. Mechta, M. Yaniv, and M. Oren. 1991. Wild-type p53 can down-modulate the activity of various promoters. *Proc. Natl. Acad. Sci. USA* **88:** 9979–9983.
120. Girinsky, T., C. Koumenis, T. G. Graeber, D. M. Peehl, and A. J. Giaccia. 1995. Attenuated response of p53 and p21 in primary cultures of human prostatic epithelial cells exposed to DNA-damaging agents. *Cancer Res.* **55:** 3726–3731.
121. Gottlieb, T. M. and S. P. Jackson. 1994. Protein kinases and DNA damage. *Trends Biochem. Sci.* **19:** 500–503.
122. Gottlieb, T. M. and S. P. Jackson. 1993. The DNA-dependent protein kinase: requirement for DNA ends and association with Ku antigen. *Cell* **72:** 131–142.
123. Graeber, T. G., J. F. Peterson, M. Tsai, K. Monica, A. J. Fornace, Jr., and A. J. Giaccia. 1994. Hypoxia induces accumulation of p53 protein, but activation of a G1-phase checkpoint by low-oxygen conditions is independent of p53 status. *Mol. Cell. Biol.* **14:** 6264–6277.
124. Griffin, G. E., K. Leung, T. M. Folks, S. Kunkel, and G. J. Nabel. 1989. Activation of HIV gene expression during monocyte differentiation by induction of NF-kappa B [*see* comments]. *Nature* **339:** 70–73.
125. Griffith, A. J., J. Craft, J. Evans, T. Mimori, and J. A. Hardin. 1992. Nucleotide sequence and genomic structure analyses of the p70 subunit of the human Ku autoantigen: evidence for a family of genes encoding Ku (p70)-related polypeptides. *Mol. Biol. Rep.* **16:** 91–97.
126. Gruss, H. J., M. A. Brach, and F. Herrmann. 1992. Involvement of nuclear factor-kappa B in induction of the interleukin-6 gene by leukemia inhibitory factor. *Blood* **80:** 2563–2570.
127. Gudas, J. M., G. B. Knight, and A. B. Pardee. 1988. Nuclear posttranscriptional processing of thymidine kinase mRNA at the onset of DNA synthesis. *Proc. Natl. Acad. Sci. USA* **85:** 4705–4709.
128. Gudas, J. M., G. B. Knight, and A. B. Pardee. 1990. Ordered splicing of thymidine kinase pre-mRNA during the S phase of the cell cycle. *Mol. Cell. Biol.* **10:** 5591–5595.
129. Guillouf, C., F. Rosselli, K. Krishnaraju, E. Moustacchi, B. Hoffman, and D. A. Liebermann. 1995. p53 involvement in control of G2 exit of the cell cycle: role in DNA damage-induced apoptosis. *Oncogene* **10:** 2263–2270.
130. Gupta, S., D. Campbell, B. Derijard, and R. J. Davis. 1995. Transcription factor ATF2 regulation by the JNK signal transduction pathway. *Science* **267:** 389–393.
131. Hagen, G., S. Muller, M. Beato, and G. Suske. 1994. Sp1-mediated transcriptional activation is repressed by Sp3. *EMBO J.* **13:** 3843–3851.
132. Hagen, G., S. Muller, M. Beato, and G. Suske. 1992. Cloning by recognition site screening of two novel GT box binding proteins: a family of Sp1 related genes. *Nucleic Acids Res.* **20:** 5519–5525.
133. Hagen, U. 1986. Current aspects on the radiation induced base damage in DNA. *Radiat. Environ. Biophys.* **25:** 261–271.
134. Hai, T. and T. Curran. 1991. Cross-family dimerization of transcription factors Fos/Jun and ATF/CREB alters DNA binding specificity. *Proc. Natl. Acad. Sci. USA* **88:** 3720–3724.
135. Haimovitz-Friedman, A., I. Vlodavsky, A. Chaudhuri, L. Witte, and Z. Fuks. 1991. Autocrine effects of fibroblast growth factor in repair of radiation damage in endothelial cells. *Cancer Res.* **51:** 2552–2558.

136. Hainaut, P. 1995. The tumor suppressor protein p53: a receptor to genotoxic stress that controls cell growth and survival. *Curr. Opinion Oncol.* **7:** 76–82.
137. Hainaut, P. and J. Milner. 1993. Redox modulation of p53 conformation and sequence-specific DNA binding in vitro. *Cancer Res.* **53:** 4469–4473.
138. Hall, E. J., M. Astor, J. Bedford, C. Borek, S. B. Curtis, M. Fry, et al. 1988. Basic radiobiology. *Am. J. Clin. Oncol.* **11:** 220–252.
139. Hallahan, D. E., D. Bleakman, S. Virudachalam, D. Lee, D. Grdina, D. W. Kufe, and R. R. Weichselbaum. 1994. The role of intracellular calcium in the cellular response to ionizing radiation. *Radiat. Res.* **138:** 392–400.
140. Hallahan, D. E., D. Gius, J. Kuchibhotla, V. Sukhatme, D. W. Kufe, and R. R. Weichselbaum. 1993. Radiation signaling mediated by Jun activation following dissociation from a cell type-specific repressor. *J. Biol. Chem.* **268:** 4903–4907.
141. Hallahan, D. E., D. R. Spriggs, M. A. Beckett, D. W. Kufe, and R. R. Weichselbaum. 1989. Increased tumor necrosis factor alpha mRNA after cellular exposure to ionizing radiation. *Proc. Natl. Acad. Sci. USA* **86:** 10,104–10,107.
142. Hallahan, D. E., V. P. Sukhatme, M. L. Sherman, S. Virudachalam, D. Kufe, and R. R. Weichselbaum. 1991. Protein kinase C mediates x-ray inducibility of nuclear signal transducers EGR1 and JUN. *Proc. Natl. Acad. Sci. USA* **88:** 2156–2160.
143. Harper, J. W., G. R. Adami, N. Wei, K. Keyomarsi, and S. J. Elledge. 1993. The p21 Cdk-interacting protein Cip1 is a potent inhibitor of G1 cyclin-dependent kinases. *Cell* **75:** 8015,8016.
144. Hatakeyama, M., J. A. Brill, G. R. Fink, and R. A. Weinberg. 1994. Collaboration of G1 cyclins in the functional inactivation of the retinoblastoma protein. *Genes Dev.* **8:** 1759–1771.
145. Hawn, M. T., A. Umar, J. M. Carethers, G. Marra, T. A. Kunkel, C. R. Boland, and M. Koi. 1995. Evidence for a connection between the mismatch repair system and the G2 cell cycle checkpoint. *Cancer Res.* **55:** 3721–3725.
146. Hibi, M., A. Lin, T. Smeal, A. Minden, and M. Karin. 1993. Identification of an oncoprotein- and UV-responsive protein kinase that binds and potentiates the c-Jun activation domain. *Genes Dev.* **7:** 2135–2148.
147. Hiebert, S. W., S. P. Chellappan, J. M. Horowitz, and J. R. Nevins. 1992. The interaction of RB with E2F coincides with an inhibition of the transcriptional activity of E2F. *Genes Dev.* **6:** 177–185.
148. Hill, B. T., K. Deuchars, L. K. Hosking, V. Ling, and R. D. Whelan. 1990. Overexpression of P-glycoprotein in mammalian tumor cell lines after fractionated X irradiation in vitro. *J. Natl. Cancer Inst.* **82:** 607–612.
149. Hinds, P. W., S. Mittnacht, V. Dulic, A. Arnold, S. I. Reed, and R. A. Weinberg. 1992. Regulation of retinoblastoma protein functions by ectopic expression of human cyclins. *Cell* **70:** 993–1006.
150. Hollander, M. C. and A. J. Fornace, Jr. 1989. Induction of fos RNA by DNA-damaging agents. *Cancer Res.* **49:** 1687–1692.
151. Hollstein, M., D. Sidransky, B. Vogelstein, and C. C. Harris. 1991. p53 mutations in human cancers. *Science* **253:** 49–53.
152. Hughes, E. N. and D. A. Boothman. 1991. Effect of caffeine on the expression of a major X-ray induced protein in human tumor cells. *Radiat. Res.* **125:** 313–317.
153. Hurta, R. A. and J. A. Wright. 1992. Alterations in the activity and regulation of mammalian ribonucleotide reductase by chlorambucil, a DNA damaging agent. *J. Biol. Chem.* **267:** 7066–7071.
154. Jackson, P., P. Ridgway, J. Rayner, J. Noble, and A. Braithwaite. 1994. Transcriptional regulation of the PCNA promoter by p53. *Biochem. Biophys. Res. Commun.* **203:** 133–140.

155. Jackson, S., T. Gottlieb, and K. Hartley. 1993. Phosphorylation of transcription factor Sp1 by the DNA-dependent protein kinase. *Adv. Second Messenger Phosphoprotein Res.* **28:** 279–286.
156. Jackson, S. P., J. J. MacDonald, S. Lees-Miller, and R. Tjian. 1990. GC box binding induces phosphorylation of Sp1 by a DNA-dependent protein kinase. *Cell* **63:** 155–165.
157. Janssen, Y. M., A. Barchowsky, M. Treadwell, K. E. Driscoll, and B. T. Mossman. 1995. Asbestos induces nuclear factor kappa B (NF-kappa B) DNA-binding activity and NF-kappa B-dependent gene expression in tracheal epithelial cells. *Proc. Natl. Acad. Sci. USA* **92:** 8458–8462.
158. Jensen, D. E., C. B. Rich, A. J. Terpstra, S. R. Farmer, and J. A. Foster. 1995. Transcriptional regulation of the elastin gene by insulin-like growth factor-I involves disruption of Sp1 binding. Evidence for the role of Rb in mediating Sp1 binding in aortic smooth muscle cells. *J. Biol. Chem.* **270:** 6555–6563.
159. Karin, M. 1995. The regulation of AP-1 activity by mitogen-activated protein kinases. *J. Biol. Chem.* **270:** 16,483–16,486.
160. Kartasova, T., B. J. Cornelissen, P. Belt, and P. van de Putte. 1987. Effects of UV, 4-NQO and TPA on gene expression in cultured human epidermal keratinocytes. *Nucleic Acids Res.* **15:** 5945–5962.
161. Kartasova, T., and P. van de Putte. 1988. Isolation, characterization, and UV-stimulated expression of two families of genes encoding polypeptides of related structure in human epidermal keratinocytes. *Mol. Cell. Biol.* **8:** 2195–2203.
162. Kastan, M. B., C. E. Canman, and C. J. Leonard. 1995. p53, cell cycle control and apoptosis: implications for cancer. *Cancer Met. Rev.* **14:** 3–15.
163. Kastan, M. B., O. Onyekwere, D. Sidransky, B. Vogelstein, and R. W. Craig. 1991. Participation of p53 protein in the cellular response to DNA damage. *Cancer Res.* **51:** 6304–6311.
164. Kastan, M. B., Q. Zhan, W. S. el-Deiry, F. Carrier, T. Jacks, W. V. Walsh, B. S. Plunkett, B. Vogelstein, and A. J. Fornace, Jr. 1992. A mammalian cell cycle checkpoint pathway utilizing p53 and GADD45 is defective in ataxia-telangiectasia. *Cell* **71:** 587–597.
165. Kazantsev, A., A. Sancar, J. M. Kearsey, M. K. K. Shivji, P. A. Hall, R. D. Wood, M. L. Smith, I.-T. Chen, and A. J. Fornace, Jr. 1995. Does the p53 up-regulated gadd45 protein have a role in excision repair? *Science* **270:** 1003–1006.
166. Kern, S. E., K. W. Kinzler, S. J. Baker, J. M. Nigro, V. Rotter, A. J. Levine, P. Friedman, C. Prives, and B. Vogelstein. 1991. Mutant p53 proteins bind DNA abnormally in vitro. *Oncogene* **6:** 131–136.
167. Kern, S. E., J. A. Pietenpol, S. Thiagalingam, A. Seymour, K. W. Kinzler, and B. Vogelstein. 1992. Oncogenic forms of p53 inhibit p53-regulated gene expression. *Science* **256:** 827–830.
168. Keyse, S. M., and E. A. Emslie. 1992. Oxidative stress and heat shock induce a human gene encoding a protein-tyrosine phosphatase. *Nature* **359:** 644–647.
169. Keyse, S. M., and R. M. Tyrrell. 1989. Heme oxygenase is the major 32-kDa stress protein induced in human skin fibroblasts by UVA radiation, hydrogen peroxide, and sodium arsenite. *Proc. Natl. Acad. Sci. USA* **86:** 99–103.
170. Kharbanda, S., R. Ren, P. Pandey, T. D. Shafman, S. M. Feller, R. R. Weichselbaum, and D. W. Kufe. 1995. Activation of the c-Abl tyrosine kinase in the stress response to DNA damaging agents. *Nature* **376:** 785–788.
171. Kharbanda, S., E. Rubin, H. Gunji, H. Hinz, B. Giovanella, P. Pantazis, and D. Kufe. 1991. Camptothecin and its derivatives induce expression of the c-jun protooncogene in human myeloid leukemia cells. *Cancer Res.* **51:** 6636–6642.
172. Kharbanda, S., A. Saleem, Y. Emoto, R. Stone, U. Rapp, and D. Kufe. 1994. Activation of Raf-1 and mitogen-activated protein kinases during monocytic differentiation of human myeloid leukemia cells. *J. Biol. Chem.* **269:** 872–878.

173. Kim, C. Y., A. J. Giaccia, B. Strulovici, and J. M. Brown. 1992. Differential expression of protein kinase C epsilon protein in lung cancer cell lines by ionising radiation. *Br. J. Cancer* **66:** 844–849.
174. Kim, S. J., H. D. Lee, P. D. Robbins, K. Busam, M. B. Sporn, and A. B. Roberts. 1991. Regulation of transforming growth factor beta 1 gene expression by the product of the retinoblastoma-susceptibility gene. *Proc. Natl. Acad. Sci. USA* **88:** 3052–3056.
175. Kim, S. J., U. S. Onwuta, Y. I. Lee, R. Li, M. R. Botchan, and P. D. Robbins. 1992. The retinoblastoma gene product regulates Sp1-mediated transcription. *Mol. Cell. Biol.* **12:** 2455–2463.
176. Kim, S. J., S. Wagner, F. Liu, M. A. O'Reilly, P. D. Robbins, and M. R. Green. 1992. Retinoblastoma gene product activates expression of the human TGF-beta 2 gene through transcription factor ATF-2. *Nature* **358:** 331–334.
177. Kim, Y. K., and A. S. Lee. 1992. Identification of a protein-binding site in the promoter of the human thymidine kinase gene required for the G1-S-regulated transcription. *J. Biol. Chem.* **267:** 2723–2727.
178. Kim, Y. K., and A. S. Lee. 1991. Identification of a 70–base-pair cell cycle regulatory unit within the promoter of the human thymidine kinase gene and its interaction with cellular factors. *Mol. Cell. Biol.* **11:** 2296–2302.
179. Kingsley, C., and A. Winoto. 1992. Cloning of GT box-binding proteins: a novel Sp1 multigene family regulating T-cell receptor gene expression. *Mol. Cell. Biol.* **12:** 4251–4261.
180. Kinsella, T. J., K. A. Kunugi, and Y. Yen. 1995. Hydroxyurea-mediated radiosensitization: an in vitro study in a human squamous cell cancer line (KB). Proceedings of the 9th International Conference on Chemical Modifiers of Cancer Treatment, pp. 181,182.
181. Kirchgessner, C. U., C. K. Patil, J. W. Evans, C. A. Cuomo, L. M. Fried, T. Carter, M. A. Oettinger, and J. M. Brown. 1995. DNA-dependent kinase (p350) as a candidate gene for the murine SCID defect. *Science* **267:** 1178–1183.
182. Kirchgessner, C. U., L. M. Tosto, K. A. Biedermann, M. Kovacs, D. Araujo, E. J. Stanbridge, and J. M. Brown. 1993. Complementation of the radiosensitive phenotype in severe combined immunodeficient mice by human chromosome 8. *Cancer Res.* **53:** 6011–6016.
183. Knight, G. B., J. M. Gudas, and A. B. Pardee. 1989. Coordinate control of S phase onset and thymidine kinase expression. *Jap. J. Cancer Res.* **80:** 493–498.
184. Koi, M., A. Umar, D. P. Chauhan, S. P. Cherian, J. M. Carethers, T. A. Kunkel, and C. R. Boland. 1994. Human chromosome 3 corrects mismatch repair deficiency and microsatellite instability and reduces N-methyl-N'-nitro-N-nitrosoguanidine tolerance in colon tumor cells with homozygous hMLH1 mutation [published erratum appears in *Cancer Res.* 1995 **Jan 1;55(1):** 201]. *Cancer Res.* **54:** 4308–4312.
185. Komatsu, K., N. Kubota, M. Gallo, Y. Okumura, and M. R. Lieber. 1995. The scid factor on human chromosome 8 restores V(D)J recombination in addition to double-strand break repair. *Cancer Res.* **55:** 1774–1779.
186. Kramer, M., C. Sachsenmaier, P. Herrlich, and H. J. Rahmsdorf. 1993. UV irradiation-induced interleukin-1 and basic fibroblast growth factor synthesis and release mediate part of the UV response. *J. Biol. Chem.* **268:** 6734–6741.
187. Kramer, M., B. Stein, S. Mai, E. Kunz, H. Konig, H. Loferer, H. H. Grunicke, H. Ponta, P. Herrlich, and H. J. Rahmsdorf. 1990. Radiation-induced activation of transcription factors in mammalian cells. *Radiat. Environ. Biophys.* **29:** 303–313.
188. Kuerbitz, S. J., B. S. Plunkett, W. V. Walsh, and M. B. Kastan. 1992. Wild-type p53 is a cell cycle checkpoint determinant following irradiation. *Proc. Natl. Acad. Sci. USA* **89:** 7491–7495.
189. Kuo, M.-L. and T. J. Kinsella. 1995. Radiosensitization by hydroxyurea in a cervical carcinoma cell line, Caski., *Proceedings of the 9th international Conference on Chemical Modifiers of Cancer Treatment*. London, pp. 183,184.

190. Kyriakis, J. M., P. Banerjee, E. Nikolakaki, T. Dai, E. A. Rubie, M. F. Ahmad, J. Avruch, and J. R. Woodgett. 1994. The stress-activated protein kinase subfamily of c-Jun kinases. *Nature* **369:** 156–160.
191. La Mantia, G., B. Majello, A. Di Cristofano, M. Strazzullo, G. Minchiotti, and L. Lania. 1992. Identification of regulatory elements within the minimal promoter region of the human endogenous ERV9 proviruses: accurate transcription initiation is controlled by an Inr-like element. *Nucleic Acids Res.* **20:** 4129–4136.
192. Lambert, M. and C. Borek. 1988. X-ray-induced changes in gene expression in normal and oncogene-transformed rat cell lines. *J. Natl. Cancer Inst.* **80:** 1492–1497.
193. Lambert, M. E., Z. A. Ronai, I. B. Weinstein, and J. I. Garrels. 1989. Enhancement of major histocompatibility class I protein synthesis by DNA damage in cultured human fibroblasts and keratinocytes. *Mol. Cell. Biol.* **9:** 847–850.
194. Lane, D. P. 1992. p53, guardian of the genome. *Nature (Lond.)* **358:** 15,16.
195. Lau, C. C. and A. B. Pardee. 1982. Mechanism by which caffeine potentiates lethality of nitrogen mustard. *Proc. Natl. Acad. Sci. USA* **79:** 2942–2946.
196. Laval, F. 1991. Increase of O6-methylguanine-DNA-methyltransferase and N3-methyladenine glycosylase RNA transcripts in rat hepatoma cells treated with DNA-damaging agents. *Biochem. Biophys. Res. Commun.* **176:** 1086–1092.
197. Lee, J. M. and A. Bernstein. 1993. p53 mutations increase resistance to ionizing radiation. *Proc. Natl. Acad. Sci. USA* **90:** 5742–5746.
198. Lee, W., P. Mitchell, and R. Tjian. 1987. Purified transcription factor AP-1 interacts with TPA-inducible enhancer elements. *Cell* **49:** 741–752.
199. Lees-Miller, S. P., Y. R. Chen, and C. W. Anderson. 1990. Human cells contain a DNA-activated protein kinase that phosphorylates simian virus 40 T antigen, mouse p53, and the human Ku autoantigen. *Mol. Cell. Biol.* **10:** 6472–6481.
200. Lees-Miller, S. P., R. Godbout, D. W. Chan, M. Weinfeld, R. S. R. Day, G. M. Barron, and J. Allalunis-Turner. 1995. Absence of p350 subunit of DNA-activated protein kinase from a radiosensitive human cell line. *Science* **267:** 1183–1185.
201. Lees-Miller, S. P., K. Sakaguchi, S. J. Ullrich, E. Appella, and C. W. Anderson. 1992. Human DNA-activated protein kinase phosphorylates serines 15 and 37 in the amino-terminal transactivation domain of human p53. *Mol. Cell. Biol.* **12:** 5041–5049.
202. Lefebvre, P. and F. Laval. 1986. Enhancement of O6-methylguanine-DNA-methyltransferase activity induced by various treatments in mammalian cells. *Cancer Res.* **46:** 5701–5705.
203. Li, C. J., C. Wang, and A. B. Pardee. 1995. Induction of apoptosis by beta-lapachone in human prostate cancer cells. *Cancer Res.* **55:** 3712–3715.
204. Lieberman, M. W., L. R. Beach, and R. D. Palmiter. 1983. Ultraviolet radiation-induced metallothionein-I gene activation is associated with extensive DNA demethylation. *Cell* **35:** 207–214.
205. Liebermann, D. A., B. Hoffman, and R. A. Steinman. 1995. Molecular controls of growth arrest and apoptosis: p53-dependent and independent pathways. *Oncogene* **11:** 199–210.
206. Lin, B. T. Y., S. Gruenwald, A. O. Morla, W. H. Lee, and J. Y. J. Wang. 1991. Retinoblastoma cancer suppressor gene product is a substrate of the cell cycle regulator cdc2 kinase. *EMBO J.* **10:** 857–864.
207. Link, E., L. D. Kerr, R. Schreck, U. Zabel, I. Verma, and P. A. Baeuerle. 1992. Purified I kappa B-beta is inactivated upon dephosphorylation. *J. Biol. Chem.* **267:** 239–246.
208. Little, J. B. 1968. Delayed initiation of DNA synthesis in irradiated human diploid cells. *Nature (Lond.)* **218:** 1064,1065.
209. Little, J. B., H. Nagasawa, P. C. Keng, Y. Yu, and C. Y. Li. 1995. Absence of radiation-induced G1 arrest in two closely related human lymphoblast cell lines that differ in p53 status. *J. Biol. Chem.* **270:** 11,033–11,036.

210. Loeb, L. A. 1994. Microsatellite instability: marker of a mutator phenotype in cancer. *Cancer Res.* **54:** 5059–5063.
211. Lowe, S. W., H. E. Ruley, T. Jacks, and D. E. Housman. 1993. p53-dependent apoptosis modulates the cytotoxicity of anticancer agents. *Cell* **74:** 957–967.
212. Lowe, S. W., E. M. Schmitt, S. W. Smith, B. A. Osborne, and T. Jacks. 1993. p53 is required for radiation-induced apoptosis in mouse thymocytes. *Nature* **362:** 847–849.
213. Lu, X., and D. P. Lane. 1993. Differential induction of transcriptionally active p53 following UV or ionizing radiation: defects in chromosome instability syndromes? *Cell* **75:** 765–778.
214. Maltzman, W. and L. Czyzyk. 1984. UV irradiation stimulates levels of p53 cellular tumor antigen in nontransformed mouse cells. *Mol. Cell. Biol.* **4:** 1689–1694.
215. Manome, Y., R. Datta, and H. A. Fine. 1993. Early response gene induction following DNA damage in astrocytoma cell lines. *Biochem. Pharmacol.* **45:** 1677–1684.
216. Manome, Y., R. Datta, N. Taneja, T. Shafman, E. Bump, R. Hass, R. Weichselbaum, and D. Kufe. 1993. Coinduction of c-jun gene expression and internucleosomal DNA fragmentation by ionizing radiation. *Biochemistry* **32:** 10,607–10,613.
217. Marsh, J. P. and B. T. Mossman. 1991. Role of asbestos and active oxygen species in activation and expression of ornithine decarboxylase in hamster tracheal epithelial cells. *Cancer Res.* **51:** 167–173.
218. May, G., C. Sutton, and H. Gould. 1991. Purification and characterization of Ku-2, an octamer–binding protein related to the autoantigen Ku. *J. Biol. Chem.* **266:** 3052–3059.
219. McClarty, G. A., A. K. Chan, Y. Engstrom, J. A. Wright, and L. Thelander. 1987. Elevated expression of M1 and M2 components and drug-induced posttranscriptional modulation of ribonucleotide reductase in a hydroxyurea-resistant mouse cell line. *Biochemistry* **26:** 8004–8011.
220. McGinn, C. J. and T. J. Kinsella. 1993. The clinical rationale for S-phase radiosensitization in human tumors. *Curr. Problems Cancer* **17:** 273–321.
221. McIlwrath, A. J., P. A. Vasey, G. M. Ross, and R. Brown. 1994. Cell cycle arrests and radiosensitivity of human tumor cell lines: dependence on wild-type p53 for radiosensitivity. *Cancer Res.* **54:** 3718–3722.
222. McKinnon, P. J. 1987. Ataxia-telangiectasia: an inherited disorder of ionizing-radiation sensitivity in man. *Hum. Genet.* **75:** 197–208.
223. McLaughlan, W. P., R. Shea, P. E. McKeever, and D. A. Boothman. 1993. Radiobiological effects and changes and in the central nervous system in response to ionizing radiation, in *Molecular Genetics of Nervous System Tumors* (Levine, A. J. and H. H. Schmidek, eds.), Wiley-Liss, New York, pp. 163–177.
224. Merritt, A. J., C. S. Potten, C. J. Kemp, J. A. Hickman, A. Balmain, D. P. Lane, and P. A. Hall. 1994. The role of p53 in spontaneous and radiation-induced apoptosis in the gastrointestinal tract of normal and p53-deficient mice. *Cancer Res.* **54:** 614–617.
225. Meyers, M., R. Shea, H. Seabury, A. Petrowski, W. P. McLaughlin, I. Lee, S. W. Lee, and D. A. Boothman. 1992. Role of X-ray-induced genes and proteins in adaptive survival responses, in *Low Dose Irradiation and Biological Defense Mechanisms* (T. Sugahara, L. A. Sagan, and T. Aoyama, eds.), Excerpta Medica, London, pp. 263–266.
226. Mezzina, M. and S. Nocentini. 1978. DNA ligase activity in UV-irradiated monkey kidney cells. *Nucleic Acids Res.* **5:** 4317–4328.
227. Mezzina, M., S. Nocentini, and A. Sarasin. 1982. DNA ligase activity in carcinogen-treated human fibroblasts. *Biochimie* **64:** 743–748.
228. Milne, D. M., D. G. Campbell, F. B. Caudwell, and D. W. Meek. 1994. Phosphorylation of the tumor suppressor protein p53 by mitogen-activated protein kinases. *J. Biol. Chem.* **269:** 9253–9260.
229. Mimori, T. and J. A. Hardin. 1986. Mechanism of interaction between Ku protein and DNA. *J. Biol. Chem.* **261:** 10,375–10,379.

230. Minden, A., A. Lin, T. Smeal, B. Derijard, M. Cobb, R. Davis, and M. Karin. 1994. c-Jun N-terminal phosphorylation correlates with activation of the JNK subgroup but not the ERK subgroup of mitogen-activated protein kinases. *Molec. Cell. Biol.* **14:** 6683–6688.
231. Miskin, R. and E. Reich. 1980. Plasminogen activator: induction of synthesis by DNA damage. *Cell* **19:** 217–224.
232. Mitchel, R. E. and D. P. Morrison. 1984. An oxygen effect for gamma-radiation induction of radiation resistance in yeast. *Radiat. Res.* **100:** 205–210.
233. Miyamoto, S. and I. M. Verma. 1995. Rel/NF-kappa B/I kappa B story. *Adv. Cancer Res.* **66:** 255–292.
234. Miyashita, T., S. Krajewski, M. Krajewska, H. G. Wang, H. K. Lin, D. A. Liebermann, B. Hoffman, and J. C. Reed. 1994. Tumor suppressor p53 is a regulator of bcl-2 and bax gene expression in vitro and in vivo. *Oncogene* **9:** 1799–1805.
235. Mohan, N. and M. L. Meltz. 1994. Induction of nuclear factor kappa B after low-dose ionizing radiation involves a reactive oxygen intermediate signaling pathway. *Radiat. Res.* **140:** 97–104.
236. Molnar, G., A. Crozat, and A. B. Pardee. 1994. The immediate-early gene Egr-1 regulates the activity of the thymidine kinase promoter at the G0-to-G1 transition of the cell cycle. *Mol. Cell. Biol.* **14:** 5242–5248.
237. Morozov, V. E., M. Falzon, C. W. Anderson, and E. L. Kuff. 1994. DNA-dependent protein kinase is activated by nicks and larger single-stranded gaps. *J. Biol. Chem.* **269:** 16,684–16,688.
238. Murata, Y., H. G. Kim, K. T. Rogers, A. J. Udvadia, and J. M. Horowitz. 1994. Negative regulation of Sp1 trans-activation is correlated with the binding of cellular proteins to the amino terminus of the Sp1 trans-activation domain. *J. Biol. Chem.* **269:** 20,674–20,681.
239. Muschel, R. J., H. B. Zhang, G. Iliakis, and W. G. McKenna. 1991. Cyclin B expression in HeLa cells during the G2 block induced by ionizing radiation. *Cancer Res.* **51:** 5113–5117.
240. Nagasawa, H., C. Y. Li, C. G. Maki, A. C. Imrich, and J. B. Little. 1995. Relationship between radiation-induced G1 phase arrest and p53 function in human tumor cells. *Cancer Res.* **55:** 1842–1846.
241. Nebert, D. W., D. D. Petersen, and A. J. Fornace, Jr. 1990. Cellular responses to oxidative stress: the [Ah] gene battery as a paradigm. *Environ. Health Perspect.* **88:** 13–25.
242. Nehls, P., D. van Beuningen, and M. Karwowski. 1991. After X-irradiation a transient arrest of L929 cells in G2-phase coincides with a rapid elevation of the level of O6-alkylguanine-DNA alkyltransferase. *Radiat. Environ. Biophys.* **30:** 21–31.
243. Nelson, W. G. and M. B. Kastan. 1994. DNA strand breaks: the DNA template alterations that trigger p53-dependent DNA damage response pathways. *Mol. Cell. Biol.* **14:** 1815–1823.
244. Nevins, J. R. 1992. E2F: a link between the Rb tumor suppressor protein and viral oncoproteins. *Science* **258:** 424–429.
245. Oberosler, P., P. Hloch, U. Ramsperger, and H. Stahl. 1993. p53-catalyzed annealing of complementary single-stranded nucleic acids. *EMBO J.* **12:** 2389–2396.
246. Okamoto, K. and D. Beach. 1994. Cyclin G is a transcriptional target of the p53 tumor suppressor protein. *EMBO J.* **13:** 4816–4822.
247. Oliner, J. D., K. W. Kinzler, P. S. Meltzer, D. L. George, and B. Vogelstein. 1992. Amplification of a gene encoding a p53-associated protein in human sarcomas. *Nature (London)* **358:** 80–83.
248. Oltvai, Z. N., C. L. Milliman, and S. J. Korsmeyer. 1993. Bcl-2 heterodimerizes in vivo with a conserved homolog, bax, that accelerates programed cell death. *Cell* **74:** 609–6019.
249. Osmak, M., S. Miljanic, and S. Kapitanovic. 1994. Low doses of gamma-rays can induce the expression of mdr gene. *Mutat. Res.* **324:** 35–41.

250. Pan, Z. Q., A. A. Amin, E. Gibbs, H. Niu, and J. Hurwitz. 1994. Phosphorylation of the p34 subunit of human single-stranded-DNA-binding protein in cyclin A-activated G1 extracts is catalyzed by cdk-cyclin A complex and DNA-dependent protein kinase. *Proc. Natl. Acad. Sci. USA* **91:** 8343–8347.
251. Papathanasiou, M. A., N. C. Kerr, J. H. Robbins, O. W. McBride, I. Alamo, Jr., S. F. Barrett, I. D. Hickson, and A. J. Fornace, Jr. 1991. Induction by ionizing radiation of the gadd45 gene in cultured human cells: lack of mediation by protein kinase C. *Mol. Cell. Biol.* **11:** 1009–1016.
252. Pardo, F. S., M. Su, C. Borek, F. Preffer, D. Dombkowski, L. Gerweck, and E. V. Schmidt. 1994. Transfection of rat embryo cells with mutant p53 increases the intrinsic radiation resistance. *Radiat. Res.* **140:** 180–185.
253. Patel, L., C. Abate, and T. Curran. 1990. Altered protein conformation on DNA binding by Fos and Jun [*see* comments]. *Nature* **347:** 572–575.
254. Peacock, J. W., S. Chung, R. G. Bristow, R. P. Hill, and S. Benchimol. 1995. The p53-mediated G1 checkpoint is retained in tumorigenic rat embryo fibroblast clones transformed by the human papillomavirus type 16 E7 gene and EJ-ras. *Mol. Cell. Biol.* **15:** 1446–1454.
255. Peak, J. G., T. Ito, F. T. Robb, and M. J. Peak. 1995. DNA damage produced by exposure of supercoiled plasmid DNA to high- and low-LET ionizing radiation: effects of hydroxyl radical quenchers. *Int. J. Radiat. Biol.* **67:** 1–6.
256. Peak, J. G., G. E. Woloschak, and M. J. Peak. 1991. Enhanced expression of protein kinase C gene caused by solar radiation. *Photochem. Photobiol.* **53:** 395–397.
257. Perry, M. E., J. Piette, J. A. Zawadzki, D. Harvey, and A. J. Levine. 1993. The mdm-2 gene is induced in response to UV light in a p53-dependent manner. *Proc. Natl. Acad. Sci. USA* **90:** 11,623–11,627.
258. Peterson, S. R., A. Dvir, C. W. Anderson, and W. S. Dynan. 1992. DNA binding provides a signal for phosphorylation of the RNA polymerase II heptapeptide repeats. *Genes Dev.* **6:** 426–438.
259. Peterson, S. R., S. A. Jesch, T. N. Chamberlin, A. Dvir, S. K. Rabindran, C. Wu, and W. S. Dynan. 1995. Stimulation of the DNA-dependent protein kinase by RNA polymerase II transcriptional activator proteins. *J. Biol. Chem.* **270:** 1449–1454.
260. Pinkus, R., L. M. Weiner, and V. Daniel. 1995. Role of quinone-mediated generation of hydroxyl radicals in the induction of glutathione S-transferase gene expression. *Biochemistry* **34:** 81–88.
261. Piver, M. S., A. D. Howes, H. D. Suit, and N. Marshall. 1973. Effects of hydroxyurea on the radiation response of C3H mouse mammary tumors. *Cancer* **29:** 407–412.
262. Piver, M. S., M. Khalil, and L. J. Emrich. 1989. Hydroxyurea plus pelvic irradiation versus placebo plus pelvis irradiation in nonsurgically staged stage IIIB cervical cancer. *J. Surg. Oncol.* **42:** 120–125.
263. Planchon, S. M., S. Wuerzberger, B. Frydman, D. T. Witiak, P. Hutson, D. R. Church, G. Wilding, and D. A. Boothman. 1995. Beta-lapachone-mediated apoptosis in human promyelocytic leukemia (HL-60) and human prostate cancer cells: a p53-independent response. *Cancer Res.* **55:** 3706–3711.
264. Prasad, A. V., N. Mohan, B. Chandrasekar, and M. L. Meltz. 1995. Induction of transcription of immediate early genes by low-dose ionizing radiation. *Radiat. Res.* **143:** 263–272.
265. Prasad, A. V., N. Mohan, B. Chandrasekar, and M. L. Meltz. 1994. Activation of nuclear factor kappa B in human lymphoblastoid cells by low-dose ionizing radiation. *Radiat. Res.* **138:** 367–372.
266. Pribush, A., G. Agam, T. Yermiahu, A. Dvilansky, D. Meyerstein, and N. Meyerstein. 1994. Radiation damage to the erythrocyte membrane: the effects of medium and cell concentrations. *Free Radical Res.* **21:** 135–146.

267. Price, B. D. and S. J. Park. 1994. DNA damage increases the levels of MDM2 messenger RNA in wtp53 human cells. *Cancer Res.* **54:** 896–899.
268. Prigent, C., D. D. Lasko, K. Kodama, J. R. Woodgett, and T. Lindahl. 1992. Activation of mammalian DNA ligase I through phosphorylation of casein kinase II. *EMBO J.* **11:** 2925–2933.
269. Protic, M. and A. S. Levine. 1993. Detection of DNA damage-recognition proteins using the band-shift assay and southwestern hybridization. *Electrophoresis* **14:** 682–692.
270. Qin, X. Q., D. M. Livingston, M. Ewen, W. R. Sellers, Z. Arany, and W. G. Kaelin, Jr. 1995. The transcription factor E2F-1 is a downstream target of RB action. *Mol. Cell. Biol.* **15:** 742–755.
271. Radulescu, R. T. 1995. The "LXCXE" hydropathic superfamily of ligands for retinoblastoma protein: a proposal. *Med. Hypotheses* **44:** 28–31.
272. Ragimov, N., A. Krauskopf, N. Navot, V. Rotter, M. Oren, and Y. Aloni. 1993. Wild-type but not mutant p53 can repress transcription initiation in vitro by interfering with the binding of basal transcription factors to the TATA motif. *Oncogene* **8:** 1183–1193.
273. Rahmsdorf, H. J., N. Harth, A. M. Eades, M. Litfin, M. Steinmetz, L. Forni, and P. Herrlich. 1986. Interferon-gamma, mitomycin C, and cycloheximide as regulatory agents of MHC class II–associated invariant chain expression. *J. Immunol.* **136:** 2293–2299.
274. Rathmell, W. K. and G. Chu. 1994. Involvement of the Ku autoantigen in the cellular response to DNA double-strand breaks. *Proc. Natl. Acad. Sci. USA* **91:** 7623–7627.
275. Raycroft, L., H. Wu, and G. Lozano. 1990. Transcriptional activation by wild-type but not transforming mutants of the p53 anti-oncogene. *Science* **249:** 1049–1051.
276. Reed, S. I., E. Bailly, V. Dulic, L. Hengst, D. Resnitzky, and J. Slingerland. 1994. G1 control in mammalian cells. *J. Cell Sci.* **Suppl. 18:** 69–73.
277. Reichard, P. 1993. From RNA to DNA, why so many ribonucleotide reductases? *Science* **260:** 1773–1777.
278. Reichard, P. 1985. Ribonucleotide reductase and deoxyribonucleotide pools. *Basic Life Sci.* **31:** 33–45.
279. Reid, T. M., M. Fry, and L. A. Loeb. 1991. Endogenous mutations and cancer. Princess Takamatsu Symposia **22:** 221–229.
280. Resnitzky, D., L. Hengst, and S. I. Reed. 1995. Cyclin a-associated kinase activity is rate limiting for entrance into S phase and is negatively regulated in G1 by p27(Kip1). *Mol. Cell. Biol.* **15:** 4347–4352.
281. Riley, P. A. 1994. Free radicals in biology: oxidative stress and the effects of ionizing radiation. *Int. J. Radiat. Biol.* **65:** 27–33.
282. Robbins, P. D., J. M. Horowitz, and R. C. Mulligan. 1990. Negative regulation of human c-fos expression by the retinoblastoma gene product [published erratum appears in *Nature* 1991 **May 30;351(6325):** 419]. *Nature* **346:** 668–671.
283. Ronai, Z. A., E. Okin, and I. B. Weinstein. 1988. Ultraviolet light induces the expression of oncogenes in rat fibroblast and human keratinocyte cells. *Oncogene* **2:** 201–204.
284. Rosen, C. F., D. Gajic, and D. J. Drucker. 1990. Ultraviolet radiation induction of ornithine decarboxylase in rat keratinocytes. *Cancer Res.* **50:** 2631–2635.
285. Rosselli, F., A. Ridet, T. Soussi, E. Duchaud, C. Alapetite, and E. Moustacchi. 1995. p53-dependent pathway of radio-induced apoptosis is altered in Fanconi anemia. *Oncogene* **10:** 9–17.
286. Saez, E., S. E. Rutberg, E. Mueller, H. Oppenheim, J. Smoluk, S. H. Yuspa, and B. M. Spiegelman. 1995. c-fos is required for malignant progression of skin tumors. *Cell* **82:** 721–732.
287. Sahijdak, W. M., C. R. Yang, J. S. Zuckerman, M. Meyers, and D. A. Boothman. 1994. Alterations in transcription factor binding in radioresistant human melanoma cells after ionizing radiation. *Radiat. Res.* **138:** S47–S51.

288. Satoh, M. S. and T. Lindahl. 1992. Role of poly(ADP-ribose) formation in DNA repair. *Nature* **356:** 356–358.
289. Schieven, G. L., J. M. Kirihara, D. E. Myers, J. A. Ledbetter, and F. M. Uckun. 1993. Reactive oxygen intermediates activate NF-kappa B in a tyrosine kinase-dependent mechanism and in combination with vanadate activate the $p56l^{ck}$ and $p59^{fyn}$ tyrosine kinases in human lymphocytes. *Blood* **82:** 1212–1220.
290. Schmitz, M. L. and P. A. Baeuerle. 1991. The p65 subunit is responsible for the strong transcription activating potential of NF-kappa B. *EMBO J.* **10:** 3805–3817.
291. Schreck, R., K. Albermann, and P. A. Baeuerle. 1992. Nuclear factor kappa B: an oxidative stress-responsive transcription factor of eukaryotic cells (a review). *Free Radical Res. Comm.* **17:** 221–237.
292. Schreck, R., P. Rieber, and P. A. Baeuerle. 1991. Reactive oxygen intermediates as apparently widely used messengers in the activation of the NF-kappa B transcription factor and HIV-1. *EMBO J.* **10:** 2247–2258.
293. Schreck, R., H. Zorbas, E. L. Winnacker, and P. A. Baeuerle. 1990. The NF-kappa B transcription factor induces DNA bending which is modulated by its 65-kD subunit. *Nucleic Acids Res.* **18:** 6497–6502.
294. Schwarz, J. K., S. H. Devoto, E. J. Smith, S. P. Chellappan, L. Jakoi, and J. R. Nevins. 1993. Interactions of the p107 and Rb proteins with E2F during the cell proliferation response. *EMBO J.* **12:** 1013–1020.
295. Selivanova, G. and K. G. Wiman. 1995. p53: a cell cycle regulator activated by DNA damage. *Adv. Cancer Res.* **66:** 143–180.
296. Sen, R., and D. Baltimore. 1986. Inducibility of kappa immunoglobulin enhancer-binding protein Nf-kappa B by a posttranslational mechanism. *Cell* **47:** 921–928.
297. Shah, G., R. Ghosh, P. A. Amstad, and P. A. Cerutti. 1993. Mechanism of induction of c-fos by ultraviolet B (290–320 nm) in mouse JB6 epidermal cells. *Cancer Res.* **53:** 38–45.
298. Shao, Z. and P. D. Robbins. 1995. Differential regulation of E2F and Sp1-mediated transcription by G1 cyclins. *Oncogene* **10:** 221–228.
299. Shaw, P., R. Bovey, S. Tardy, R. Sahli, B. Sordat, and J. Costa. 1992. Induction of apoptosis by wild-type p53 in a human colon tumor-derived cell line. *Proc. Natl. Acad. Sci. USA* **89:** 4495–4499.
300. Sherman, M. L., R. Datta, D. E. Hallahan, R. R. Weichselbaum, and D. W. Kufe. 1990. Ionizing radiation regulates expression of the c-jun protooncogene. *Proc. Natl. Acad. Sci. USA* **87:** 5663–5666.
301. Sinclair, W. K. 1968. The combined effects of hydroxyurea and x-rays on Chinese hamster cells in vivo. *Cancer Res.* **28:** 198–206.
302. Singh, S. P. and M. F. Lavin. 1990. DNA-binding protein activated by gamma radiation in human cells. *Mol. Cell. Biol.* **10:** 5279–5285.
303. Sipley, J. D., J. C. Menninger, K. O. Hartley, D. C. Ward, S. P. Jackson, and C. W. Anderson. 1995. Gene for the catalytic subunit of the human DNA-activated protein kinase maps to the site of the XRCC7 gene on chromosome 8. *Proc. Natl. Acad. Sci. USA* **92:** 7515–7519.
304. Slebos, R. J. C., M. H. Lee, B. S. Plunkett, T. D. Kessis, B. O. Williams, T. Jacks, L. Hedrick, M. B. Kastan, and K. R. Cho. 1994. p53-dependent G1 arrest involves pRB-related proteins and is disrupted by the human papillomavirus 16 E7 oncoprotein. *Proc. Natl. Acad. Sci. USA* **91:** 5320–5324.
305. Slichenmyer, W. J., W. G. Nelson, R. J. Slebos, and M. B. Kastan. 1993. Loss of a p53-associated G1 checkpoint does not decrease cell survival following DNA damage. *Cancer Res.* **53:** 4164–4168.
306. Smeal, T., P. Angel, J. Meek, and M. Karin. 1989. Different requirements for formation of Jun:Jun and Jun: Fos complexes. *Genes Dev.* **3:** 2091–2100.

307. Smith, M. L., I. T. Chen, Q. Zhan, I. Bae, C. Y. Chen, T. M. Gilmer, M. B. Kastan, P. M. O'Connor, and A. J. Fornace, Jr. 1994. Interaction of the p53-regulated protein Gadd45 with proliferating cell nuclear antigen. *Science* **266:** 1376–1380.
308. Stehman, F. B., B. N. Bundy, G. Thomas, H. M. Keys, G. d. d'Ablaing, W. C. Fowler, Jr., R. Mortel, and W. T. Creasman. 1993. Hydroxyurea versus misonidazole with radiation in cervical carcinoma: long-term follow-up of a Gynecologic Oncology Group trial. *J. Clin. Oncol.* **11:** 1523–1528.
309. Stein, B., M. Kramer, H. J. Rahmsdorf, H. Ponta, and P. Herrlich. 1989. UV-induced transcription from the human immunodeficiency virus type 1 (HIV-1) long terminal repeat and UV-induced secretion of an extracellular factor that induces HIV-1 transcription in nonirradiated cells. *J. Virol.* **63:** 4540–4544.
310. Stein, B., H. J. Rahmsdorf, A. Steffen, M. Litfin, and P. Herrlich. 1989. UV-induced DNA damage is an intermediate step in UV-induced expression of human immunodeficiency virus type 1, collagenase, c-fos, and metallothionein. *Mol. Cell. Biol.* **9:** 5169–5181.
311. Steinmeyer, K. and W. Deppert. 1988. DNA binding properties of murine p53. *Oncogene* **3:** 501–507.
312. Stevenson, M. A., S. S. Pollock, C. N. Coleman, and S. K. Calderwood. 1994. X-irradiation, phorbol esters, and H_2O_2 stimulate mitogen-activated protein kinase activity in NIH-3T3 cells through the formation of reactive oxygen intermediates. *Cancer Res.* **54:** 12–15.
313. Stewart, N., G. G. Hicks, F. Paraskevas, and M. Mowat. 1995. Evidence for a second cell cycle block at G2/M by p53. *Oncogene* **10:** 109–115.
314. Stirdivant, S. M., H. E. Huber, D. R. Patrick, D. Defeo-Jones, E. M. McAvoy, V. M. Garsky, A. Oliff, and D. C. Heimbrook. 1992. Human papillomavirus type 16 E7 protein inhibits DNA binding by the retinoblastoma gene product. *Mol. Cell. Biol.* **12:** 1905–1914.
315. Stuart, P., M. Ito, C. Stewart, and S. E. Conrad. 1985. Induction of cellular thymidine kinase occurs at the mRNA level. *Mol. Cell. Biol.* **5:** 1490–1497.
316. Stubbe, J. 1990. Ribonucleotide reductases: amazing and confusing. *J. Biol. Chem.* **256:** 5329–5332.
317. Sullivan, N. R. and A. E. Willis. 1989. Elevation of c-myc protein by DNA strand breakage. *Oncogene* **4:** 1497–1502.
318. Sum, L. and J. A. Fuchs. 1992. *Escherichia coli* ribonucleotide reductase expression is cell cycle regulated. *Mol. Cell. Biol.* **3:** 1095–1105.
319. Szumiel, I. 1994. Ionizing radiation-induced cell death. *Int. J. Radiat. Biol.* **66:** 329–341.
320. Taccioli, G. E., T. M. Gottlieb, T. Blunt, A. Priestley, J. Demengeot, R. Mizuta, A. R. Lehmann, F. W. Alt, S. P. Jackson, and P. A. Jeggo. 1994. Ku86: product of the XRCC5 gene and its role in DNA repair and V(D)J recombination. *Science* **265:** 1442–1445.
321. Teoule, R. 1987. Radiation-induced DNA damage and its repair. *Int. J. Radiat. Biol. Related Studies in Phys., Chem. Med.* **51:** 573–589.
322. Thut, C. J., J.-L. Chen, R. Klemm, and R. Tjian. 1995. p53 transcriptional activation mediated by coactivators TAFII40 and TAFII60. *Science* **267:** 100–104.
323. Tishler, R. B., S. K. Calderwood, C. N. Coleman, and B. D. Price. 1993. Increases in sequence specific DNA binding by p53 following treatment with chemotherapeutic and DNA damaging agents. *Cancer Res.* **53:** 2212–2216.
324. Tsang, N. M., H. Nagasawa, C. Li, and J. B. Little. 1995. Abrogation of p53 function by transfection of HPV16 E6 gene enhances the resistance of human diploid fibroblasts to ionizing radiation. *Oncogene* **10:** 2403–2408.
325. Tuteja, N., R. Tuteja, A. Ochem, P. Taneja, N. W. Huang, A. Simoncsits, et al. 1994. Human DNA helicase II: a novel DNA unwinding enzyme identified as the Ku autoantigen. *EMBO J.* **13:** 4991–5001.
326. Tyshkin, S. M., V. M. Taranenko, M. I. Rudniev, G. S. Voronkov, G. I. Pliushch, I. M. Isaiechkina, L. M. Popova, and V. V. Bratus. 1993. Activation of free-radical processes

as a factor of ionizing radiation-induced changes in contractile activity of a vascular wall. *Fiziologicheskii Zhurnal* **39:** 23–29.

327. Uckun, F., G. Schieven, L. Tuel-Ahlgren, I. Dibirdik, D. Myers, J. Ledbetter, and C. Song. 1993. Tyrosine phosphorylation is a mandatory proximal step in radiation-induced activation of the protein kinase C signaling pathway in human B-lymphocyte precursors. *Proc. Natl. Acad. Sci. USA* **90:** 252–256.
328. Uckun, F. M., I. Dibirdik, R. Smith, L. Tuel-Ahlgren, M. Chandan-Langlie, G. L. Schieven, K. G. Waddick, M. Hanson, and J. A. Ledbetter. 1991. Interleukin 7 receptor ligation stimulates tyrosine phosphorylation, inositol phospholipid turnover, and clonal proliferation of human B-cell precursors. *Proc. Natl. Acad. Sci. USA* **88:** 3589–3593.
329. Udvadia, A. J., K. T. Rogers, P. D. Higgins, Y. Murata, K. H. Martin, P. A. Humphrey, and J. M. Horowitz. 1993. Sp-1 binds promoter elements regulated by the RB protein and Sp-1-mediated transcription is stimulated by RB coexpression. *Proc. Natl. Acad. Sci. USA* **90:** 3265–3269.
330. Udvadia, A. J., K. T. Rogers, and J. M. Horowitz. 1992. A common set of nuclear factors bind to promoter elements regulated by the retinoblastoma protein. *Cell Growth Differ.* **3:** 597–608.
331. Udvadia, A. J., D. J. Templeton, and J. M. Horowitz. 1995. Functional interactions between the retinoblastoma (Rb) protein and Sp-family members: superactivation by Rb requires amino acids necessary for growth suppression. *Proc. Natl. Acad. Sci. USA* **92:** 3953–3957.
332. Ullrich, S. J., K. Sakaguchi, S. P. Lees-Miller, M. Fiscella, W. E. Mercer, C. W. Anderson, and E. Appella. 1993. Phosphorylation at Ser-15 and Ser-392 in mutant p53 molecules from human tumors is altered compared to wild-type p53. *Proc. Natl. Acad. Sci. USA* **90:** 5954–5958.
333. Urban, M. B. and P. A. Baeuerle. 1991. The role of the p50 and p65 subunits of NF-kappa B in the recognition of cognate sequences. *New Biologist* **3:** 279–288.
334. Visvanathan, K. V. and S. Goodbourn. 1989. Double-stranded RNA activates binding of NF-kappa B to an inducible element in the human beta-interferon promoter. *EMBO J.* **8:** 1129–1138.
335. Wang, C. Y., B. Petryniak, C. B. Thompson, W. G. Kaelin, and J. M. Leiden. 1993. Regulation of the Ets-related transcription factor Elf-1 by binding to the retinoblastoma protein. *Science* **260:** 1330–1335.
336. Ward, J. F. 1994. The complexity of DNA damage: relevance to biological consequences. *Int. J. Radiat. Biol.* **66:** 427–432.
337. Ward, J. F. 1986. Ionizing radiation induced DNA damage: identities and DNA repair. *Basic Life Sci.* **38:** 135–138.
338. Weichselbaum, R. R., D. Hallahan, Z. Fuks, and D. Kufe. 1994. Radiation induction of immediate early genes: effectors of the radiation-stress response. *Int. J. Radiat. Oncol. Biol. Phys.* **30:** 229–234.
339. Weichselbaum, R. R., D. E. Hallahan, M. A. Beckett, H. J. Mauceri, H. Lee, V. P. Sukhatme, and D. W. Kufe. 1994. Gene therapy targeted by radiation preferentially radiosensitizes tumor cells. *Cancer Res.* **54:** 4266–4269.
340. Weichselbaum, R. R., D. E. Hallahan, V. P. Sukhatme, and D. W. Kufe. 1992. Gene therapy targeted by ionizing radiation. *Int. J. Radiat. Oncol. Biol. Phys.* **24:** 565–567.
341. Weintraub, S. J. and D. C. Dean. 1992. Interaction of a common factor with ATF, Sp1, or TATAA promoter elements is required for these sequences to mediate transactivation by the adenoviral oncogene E1a. *Mol. Cell. Biol.* **12:** 512–517.
342. Weintraub, S. J., C. A. Prater, and D. C. Dean. 1992. Retinoblastoma protein switches the E2F site from positive to negative element. *Nature (Lond.)* **358:** 259–261.

343. White, A. E., E. M. Livanos, and T. D. Tlsty. 1994. Differential disruption of genomic integrity and cell cycle regulation in normal human fibroblasts by the HPV oncoproteins. *Genes Dev.* **8:** 666–677.
344. Wilson, R. E., S. L. Taylor, G. T. Atherton, D. Johnston, C. M. Waters, and J. D. Norton. 1993. Early response gene signalling cascades activated by ionising radiation in primary human B cells. *Oncogene* **8:** 3229–3237.
345. Wiman, K. G. 1993. The retinoblastoma gene: role in cell cycle control and cell differentiation. *FASEB J.* **7:** 841–845.
346. Woloschak, G. E., and C. M. Chang-Liu. 1990. Differential modulation of specific gene expression following high- and low-LET radiations. *Radiat. Res.* **124:** 183–187.
347. Woloschak, G. E., C. M. Chang-Liu, P. S. Jones, and C. A. Jones. 1990. Modulation of gene expression in Syrian hamster embryo cells following ionizing radiation. *Cancer Res.* **50:** 339–344.
348. Woloschak, G. E., C. M. Chang-Liu, and P. Shearin-Jones. 1990. Regulation of protein kinase C by ionizing radiation. *Cancer Res.* **50:** 3963–3967.
349. Woloschak, G. E., P. Felcher, and C. M. Changliu. 1995. Combined effects of ionizing radiation and cycloheximide on gene expression. *Mol. Carcinogenesis* **13:** 44–49.
350. Woloschak, G. E., P. Shearin-Jones, and C. M. Chang-Liu. 1990. Effects of ionizing radiation on expression of genes encoding cytoskeletal elements: kinetics and dose effects. *Mol. Carcinogenesis* **3:** 374–378.
351. Woodgett, J. R. 1990. Fos and jun: two into one will go. *Semin. Cancer Biol.* **1:** 389–397.
352. Woodgett, J. R., J. Avruch, and J. M. Kyriakis. 1995. Regulation of nuclear transcription factors by stress signals. *Clin. Exp. Pharm. Phys.* **22:** 281–283.
353. Wu, X. and A. J. Levine. 1994. p53 and E2F-1 cooperate to mediate apoptosis. *Proc. Natl. Acad. Sci. USA* **91:** 3602–3606.
354. Wyllie, A. H. 1980. Glucocorticoid-induced thymocyte apoptosis is associated with endogenous endonuclease activation. *Nature (Lond.)* **284:** 555,556.
355. Xanthoudakis, S. and T. Curran. 1992. Identification and characterization of Ref-1, a nuclear protein that facilitates AP-1 DNA-binding activity. *EMBO J.* **11:** 653–665.
356. Xanthoudakis, S., G. Miao, F. Wang, Y. C. Pan, and T. Curran. 1992. Redox activation of Fos-Jun DNA binding activity is mediated by a DNA repair enzyme. *EMBO J.* **11:** 3323–3335.
357. Xanthoudakis, S., G. G. Miao, and T. Curran. 1994. The redox and DNA-repair activities of Ref-1 are encoded by nonoverlapping domains. *Proc. Natl. Acad. Sci. USA* **91:** 23–27.
358. Xia, F., X. Wang, Y. H. Wang, N. M. Tsang, D. W. Yandell, K. T. Kelsey, and H. L. Liber. 1995. Altered p53 status correlates with differences in sensitivity to radiation-induced mutation and apoptosis in two closely related human lymphoblast lines. *Cancer Res.* **55:** 12–15.
359. Xiao, Z. X., J. D. Chen, A. J. Levine, N. Modjtahedi, J. Xing, W. R. Sellers, and D. M. Livingston. 1995. Interaction between the retinoblastoma protein and the oncoprotein Mdm2. *Nature (London)* **375:** 694–698.
360. Xiong, Y., G. J. Hannon, H. Zhang, D. Casso, R. Kobayashi, and D. Beach. 1993. p21 is a universal inhibitor of cyclin kinases. *Nature (Lond.)* **366:** 701–704.
361. Yamamoto, M., M. Yoshida, K. Ono, T. Fujita, N. Ohtani-Fujita, T. Sakai, and T. Nikaido. 1994. Effect of tumor suppressors on cell cycle-regulatory genes: RB suppresses p34cdc2 expression and normal p53 suppresses cyclin A expression. *Exp. Cell Res.* **210:** 94–101.
362. Yang, C.-R., E. J. Odegaard, and D. A. Boothman. unpublished data.
363. Yang, X. Y., Z. A. Ronai, R. M. Santella, and I. B. Weinstein. 1988. Effects of 8-methoxypsoralen and ultraviolet light A on EGF receptor (HER-1) expression. *Biochem. Biophys. Res. Commun.* **157:** 590–596.
364. Yao, K. S., S. Xanthoudakis, T. Curran, and P. J. O'Dwyer. 1994. Activation of AP-1 and of a nuclear redox factor, Ref-1, in the response of HT29 colon cancer cells to hypoxia. *Mol. Cell. Biol.* **14:** 5997–6003.

365. Yonish-Rouach, E., D. Grunwald, S. Wilder, A. Kimchi, E. May, J.-J. Lawrence, P. May, and M. Oren. 1993. p53-mediated cell death: relationship to cell cycle control. *Mol. Cell. Biol.* **13:** 1415–1423.
366. Zabel, U., R. Schreck, and P. A. Baeuerle. 1991. DNA binding of purified transcription factor NF-kappa B. Affinity, specificity, Zn^{2+} dependence, and differential half-site recognition. *J. Biol. Chem.* **266:** 252–260.
367. Zhan, Q., F. Carrier, and A. J. Fornace, Jr. 1993. Induction of cellular p53 activity by DNA-damaging agents and growth arrest. *Mol. Cell. Biol.* **13:** 4242–4250.
368. Zhou, Z. and S. J. Elledge. 1992. Isolation of crt mutants constitutive for transcription of the DNA damage inducible gene RNR3 in Saccharomyces cerevisiae. *Genetics* **131:** 851–866.

15

Posttranslational Mechanisms Leading to Mammalian Gene Activation in Response to Genotoxic Stress

Yusen Liu, Myriam Gorospe, Nikki J. Holbrook, and Carl W. Anderson

1. INTRODUCTION

Exposure of cells to treatments or agents that damage DNA evokes changes in gene expression. These effects were first documented in bacteria, but similar responses occur in unicellular (*see* Chapters 7 and 18 in vol. 1) and multicellular eukaryotes. In mammalian cells, changes in gene expression are initiated and signaled primarily through posttranslational modifications of key regulatory molecules. Almost immediately on exposing cells to genotoxic agents, increases can be detected in the activities of a number of protein kinases (reviewed in refs. *7,213,219*). These kinases form intracellular signaling cascades that result in the activation of transcription factors that, in turn, alter the pattern of gene expression. Although activation of transcription appears to account for most changes in gene expression, changes in mRNA *(112,156)* or in protein stability (e.g., p53) also can contribute to the changes in protein levels that occur in response to genotoxic stress.

An ability to damage DNA is the unifying feature of the so-called genotoxins, but most such agents also damage other cellular constituents, including proteins and lipids. Genetic responses that result directly from DNA damage appear to originate mainly from DNA strand breaks and are mediated largely through the induction and/or activation of the tumor suppressor protein p53 (Fig. 1). DNA strand breaks are created directly by genotoxic agents (e.g., by radioactive decay, ionizing radiation, or inhibition of enzymes), or as a secondary consequence of repair processes and effects on transcription and DNA replication. On the other hand, considerable evidence supports the view that an important part of the genotoxic response is initiated outside the cell nucleus and does not directly involve DNA damage. The major signaling pathways resulting from responses to effects on molecules other than DNA involve the activation of one or more members of the mitogen-activated protein kinase (MAPK) family (*see* Section 3.2.).

The ultimate fate of a cell following exposure to genotoxic agents is highly variable, and includes phenotypic alterations ranging from growth arrest to death, differentiation, and senescence *(24,124,185)*. The particular outcome depends on several factors,

From: DNA Damage and Repair, Vol. 2: DNA Repair in Higher Eukaryotes
Edited by: *J. A. Nickoloff and M. F. Hoekstra* © *Humana Press Inc., Totowa, NJ*

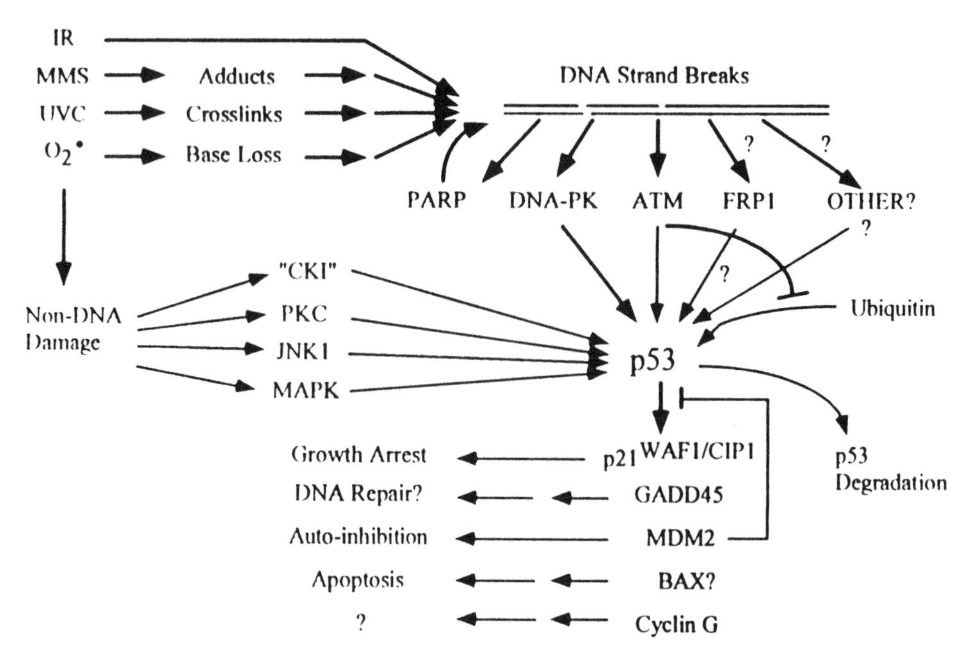

Fig. 1. Pathways of gene induction through p53 in response to genotoxic agents. Exposure to DNA-damaging agents produces a transient induction of p53 protein through posttranscriptional mechanisms that inhibit ubiquitin-mediated degradation; the accumulation of p53, in turn, activates transcription from several genes, including *p21*$^{WAF1/CIP1}$, *GADD45*, *MDM2*, cyclin G, and perhaps *BAX*. p53 accumulates in response to DNA strand breaks; responses to other agents may involve strand breaks created during repair or may be mediated by the effects of genotoxic agents on other cellular components via crosstalk mechanisms. Posttranslational modifications, including phosphorylations by several protein kinases, may activate p53 sequence-specific DNA binding or may affect p53 activity through its interactions with other proteins (e.g., MDM2, an inhibitor of transactivation; transcription components including TBP and TAFs). DNA-PK is activated by DNA strand breaks; ATM may be a protein kinase that responds to DNA strand breaks. FRP1, a homolog of the yeast checkpoint proteins *MEC1p* and *rad3p*, is another member of the DNA-PK/ATM kinase family. *See text* for details and references.

including the dose and specificity of the agent, and the type of cell and its state of differentiation. In general, the outcome appears to reflect the relative degree of activation of various signal transduction pathways and the particular effector genes that are induced. This chapter provides an overview of the genetic response to genotoxic stress in mammalian cells, and reviews common signaling events and transcription factors that mediate transcriptional activation. Particular consideration is given to the potential overlap and/or crosstalk between different signaling pathways and how this contributes to the overall cellular response.

2. MAMMALIAN GENES INDUCED IN RESPONSE TO GENOTOXIC STRESS

Table 1 lists genes whose mRNA expression increases in response to genotoxic stress; in most instances, this increase occurs as a consequence of an increased rate of

Table 1
Mammalian Genes Activated in Response to Genotoxic Stress

Inducible gene products[a]	p53-regulated[b]	References
Transcription factors		
c-Jun	–	*55,91,206,228*
c-Fos	–	*32,206,213,236*
Jun-B	–	*55*
c-Myc	–	*206*
Egr-1	–	*49,91*
C/EBPβ/NF-IL-6	–	*67*
GADD153	–	*71,72*
Proliferation/arrest		
Mdm-2	+	*16,118,193,199*
GADD45	+	*123,192,279*
Cyclin G	+	*188,276*
p21 *WAF1/CIP1*	+	*62,63*
BAX	+	*174,176,280*
Fas/APO-1	+	*191*
PCNA	+	*113,181*
IGFBP-3	+	*26*
MCK	+	*114,265*
MAP kinase phosphatase-1	–	*126,148*
RhoB	–	*77*
General stress		
Metallothionein-1	–	*8,236*
Heat-shock protein family	–	*179,180*
Heme oxygenase-1	–	*4,5,9*
MDR1	–	*248*
TNF-α	–	*90*
IL-1	–	*133,156*
IL-6	–	*156*
DNA metabolism/repair		
DNA ligase I	–	*178*
β-polymerase	–	*65*
MGMT	–	*76*
Thymidine kinase	–	*21*
Ribonucleotide reductase	–	*64*
Tissue remodeling		
Stromelysin	–	*119*
Collagenase	–	*119,236*
Urokinase-type plasminogen activator	–	*209*
Tissue-type plasminogen activator	–	*20*
Keratin	–	*233*
Other		
GADD7, GADD34	–	*71–73*
Ornithine decarboxylase	–	*207*
Spr2-1	–	*80*

[a]Abbreviations: C/EBP, CCAAT/enhancer binding protein; Egr-1, early growth response-1; NF-IL-6, nuclear factor for IL-6 expression, same as C/EBPβ; mdr1, multidrug resistance-1; GADD, growth arrest and DNA damage; Mdm-2, mouse double minute-2; IGFBP-3, insulin-like growth factor binding protein-3; PCNA, proliferating cell nuclear antigen; MCK, muscle-specific creatine kinase; Sgk, serum/glucocorticoid-inducible serine/threonine protein kinase; mdr, multidrug resistance; MGMT, O^6-methylguanine-DNA methyltransferase; Spr, small proline-rich protein.

[b](+) Genes whose expression is activated in part by p53; (–) genes whose induction in response to genotoxic stress have not been shown to depend on wild-type p53.

transcription. We have restricted the list to genes in which expression or function have been relatively well characterized. The list encompasses genes with diverse functions; most were identified through cloning approaches involving subtraction libraries, and included are the so-called DNA damage-inducible (DDI) transcripts described by Fornace et al. *(71)*, the small proline-rich (spr) genes of Kartasova and van de Putte *(121)*, and the X-ray-inducible transcripts identified by Boothman et al. (*22*; *see* Chapter 14). Although an attempt was made to organize the genes by function, it is clear that many could be placed in more than one category.

The first group includes several transcription factors. They are among the most rapidly induced genes, and are presumed to play a critical role in mediating the induction of other genes whose expression increases at later times. These transcription factors also play an important role in controlling gene activation in response to a variety of other external cues, most notably the proliferative response to mitogen stimulation. Indeed, such observations first led researchers to suspect that the induction of many of the DDI genes was initiated through insults other than DNA damage and that the MAPK signaling pathways were involved (*see* Section 3.2.).

A second set of genes is comprised of those directly associated with the cell-cycle machinery and regulation of growth. Several of these, particularly *p21*^{WAF1,CIP1}, *BAX*, and *GADD45*, are believed to play a major role in determining cellular fate following genotoxic stress (i.e., effecting growth arrest and/or cell death). Not surprisingly, the regulation of many of these depends on p53, at least in part.

The third group appears to function in general host defense mechanisms; these include the heat-shock proteins, metallothioneins, heme oxygenase, a multiple drug resistance gene, and three cytokines that are central to the regulation of acute-phase response proteins (*see* Chapter 26).

The fourth group includes several proteins that function in DNA metabolism and/or repair. It is worth noting, however, that, so far, none of these proteins involved in major DNA repair pathways have been found to be regulated by genotoxic agents. Their presence in all cells likely is a consequence of a continuous need for the capacity to repair DNA lesions and, hence, for a substantial constitutive activity.

A fifth group is comprised of gene products that appear to function in tissue remodeling; they likely respond to cell damage occurring outside the nucleus, which threatens the structure and integrity of tissues.

The transcriptional activation of most of the genes in Table 1 have been examined after treatment with genotoxic agents, and in some instances, critical *cis* regulatory elements necessary for transcriptional activation have been identified. Table 2 lists the major *cis* elements (and transcription factors which interact at these sites) known to regulate gene activation in response to genotoxic stress. Thus far, AP-1 has been the most common element identified. Many inducible genes contain several regulatory elements, with each element contributing in part to the magnitude of gene activation according to the response of the corresponding binding factor to particular agents. However, p53 appears to be unique in that it is absolutely required for the induction of many genes by ionizing radiation and for the associated G1 checkpoint that occurs in response to such DNA damage *(122,123)*.

It is important to emphasize that although the genes encoding the various transcription factors listed in the first group of Table 1 are themselves induced by genotoxic

Table 2
Promoter Elements Regulating Genotoxic Stress-Inducible Gene Expression

Cis-regulatory elements[a]	Binding sequence[b]	Transcription factors	Representative promoters[c]	References
AP-1 binding site	ATGACTCAT	c-Jun, Jun D, c-Fos, ATF-2	*JUN, FOS,* collagenase-1, *MKP-1, HMOX1, MT1, GADD153*	*8,27,88,135,236,252*
SRE	CCATATTAGG	SRF, TCF/Elk-1	*EGR1, FOS*	*49,81,245*
C/EBP binding site	GATTGCGCAATC	C/EBPβ/NF-IL-6	*GADD153*	*67,110*
HSE	GAANNTTCNNGAA	HSF1	*HSP70, HMOX1*	*179,180,189*
NFκB binding element	GGGRNNYYCC	NFκB	*HMOX1*	*4,254*
p53 binding site	RRRC(W)(W)GYYY	p53	*GADD45, MDM2, p21, BAX*	*16,61–63,123,174*

[a]Abbreviations: SRE, serum response element; AP-1, activating protein-1; HSE, heat shock element; NFκB, nuclear factor-κB; SRF, serum response factor; TCF, ternary complex factor; HSF1, heat-shock factor 1; ATF-2, activating transcription factor-2; C/EBP, CCAAT/enhancer binding protein; NF-IL-6, nuclear factor for IL-6 expression; *MKP-1*, MAP kinase phosphatase-1; *HMOX1*, heme oxygenase-1; *MT1*, metallothionein-1; *HSP*, heat-shock protein; *GADD*, growth arrest and DNA damage; *MDM2*, mouse double minute-2.

[b]Optimal or consensus transcription factors DNA binding sites are shown; however, in most cases variations of these sequences are efficiently recognized.

[c]List is partial and is intended simply to provide examples.

stress, most are present in the cell constitutively and are rapidly activated to a DNA binding or transcriptionally active form following cellular damage. Likewise, p53 protein levels and transactivating activity are elevated after DNA damage primarily through posttranslational alterations to the pre-existing p53 protein, which affects both DNA binding activity and stability. Thus, in most cases it is the modulation of existing transcription factor proteins that causes the rapid transcriptional activation of genes during genotoxic stress. Activation occurs largely through phosphorylation events, involving complex protein kinase cascades.

3. SIGNAL TRANSDUCTION PATHWAYS INVOLVED IN MEDIATING THE RESPONSE TO GENOTOXIC STRESS

3.1. p53-Mediated Events

The induction of gene expression mediated by the p53 tumor suppressor protein may be the best studied genetic response to direct DNA damage. Exposing normal cells to a wide variety of genotoxic agents, including ionizing radiation, UVC, and many anticancer drugs, arrests cell-cycle progression at one of several checkpoints (*see* Chapter 21). Arrest at the G1 checkpoint, but not in S or G2, requires the presence of a wild-type p53 gene *(28,29,122,134)*. The wild-type p53 protein has a short half-life (20 min to 2 h), and, in normal cells, is present at low levels (~1000 molecules/cell) *(203)*. Exposing cells to genotoxic agents causes it to accumulate in the nucleus *(75,122,157,183)*. Importantly, microinjecting restriction enzymes or other nucleases into cells also induces p53 accumulation, a fact that indicates that p53 induction can be the direct consequence of DNA strand breakage *(183)*.

3.1.1. Effects of p53 on Transcription

p53 is a sequence-specific DNA binding protein of about 390 amino acids (393 for human p53) that is present in vertebrates, but not in lower species. It functions primarily as a transcriptional enhancer that regulates the expression of a growing number of genes (reviewed in *11,45,223*). p53 binds the consensus recognition element RRRCWWGYYY$(N)_n$RRRCWWGYYY (where R is a purine [A or G], W is either an A or a T, and Y is a pyrimidine [T or C]) as a homo-tetramer *(61,79)*. The recognition element consists of a direct repeat of two inverted pentanucleotide elements (RRRCW); each p53 monomer in the tetramer interacts with one pentanucleotide element *(38,261)*. The two 10-bp half-sites may be separated by a spacer region ($[N]_n$, where n is 0 to ~13 bp), but both half-sites are required for efficient p53 binding. The p53-regulated genes so far identified contain one or two imperfect consensus sequences that often are located at a considerable distance upstream or downstream of the transcription start site. Thus, two p53 recognition elements were found in the upstream regulatory region of the human, mouse, and rat *p21*$^{WAF1/CIP1}$ genes (at –1.3 and –2.2 kbp from the transcription start site) *(63)*, whereas p53 elements were identified in the *GADD45* gene in intron 3 *(123)* and the *MDM2* gene in intron 1 *(16,118,268)*. Two p53 recognition elements were identified in the muscle creatine kinase gene; one in the far upstream region, at –3.1 kbp *(265)*, and the other approx 160 bp upstream of the transcription start site *(114)*. p53 elements were identified 1.5 kbp upstream of the transcription start site for the murine cyclin G gene *(188,277)*, and 217 nucleotides upstream of the human *PCNA* promoter *(181)*. Other p53-responsive genes have been reported; however, in some cases, binding elements were not identified. Certainly, additional p53-responsive genes will be discovered *(25,155,244)*. A search for human sequence elements that activate transcription in yeast identified 57 p53-responsive sequences, 46 of which had two copies of the half-site element. It was estimated that the human genome may contain 200–300 such sequences *(244)*.

Wild-type p53 can influence the promoter activity of some genes even in the absence of functional p53 binding sites *(83,154)*. p53 represses transcription from a number of cellular and viral genes, most of which have TATA-box elements in their promoters: these repressed genes include the genes for cytokine interleukin 6; c-Fos, a component of the AP-1 transcription factor; retinoblastoma protein (Rb), a regulator of the E2F transcription factor; Bcl-2, an inhibitor of apoptosis; and nitric oxide synthase (NOS-2), an inducible enzyme that synthesizes nitric oxide *(74,132,175,176,215,229)*. Depending on circumstances, p53 may either activate or inhibit transcription from the *PCNA* promoter *(113,163,181,271)*. p53 binding elements usually are not found in the promoters of repressible genes, and the mechanism of trans-repression in vivo and in vitro appears to be mediated by a more general mechanism involving factors that bind the DNA elements of the basal promoter. p53 forms complexes with several transcription modulators, including TBP, a component of TFIID, and certain TBP-associated factors (TAFs) *(35,101,147,152,159,225,243,246)*, as well as with other transcription factors, including Sp1 *(23)*, E2F *(187)*, and the cytomegalovirus immediate-early protein IE2 *(247)*. It is thought that p53 may reduce transcription by sequestering these factors or interfering with their role in forming initiation complexes.

p53 also can exert a positive influence on the transcriptional activity of genotoxic stress-inducible promoters in the absence of recognizable specific binding elements. For example, loss of p53 activity reduced transcription from the *GADD153* and *GADD45* promoter reporter constructs (the *GADD45* region examined does not contain a p53 binding element) after treatment with various genotoxic agents *(281)*. These effects could result from p53 interactions with either basal promoter elements or other stress-regulated transcription factors.

Surprisingly, a few genes, including the multiple drug resistance gene *(MDR1)* *(37,276)*, basic fibroblast growth factor *(bFGF)* *(249)*, and the vascular endothelial growth factor *(VEGF)* *(130)*, are transcriptionally activated by mutant p53, but not by wild-type p53. Not all types of mutant p53 activate transcription from these genes, and the mechanism of activation by mutant p53 has not been elucidated. Nevertheless, the gain-of-function ascribed to some p53 mutants may result, in part, from effects on transcription *(60)*.

3.1.2. Roles of p53 Effector Genes

Depending on the type of cell and the circumstances, the induction of p53 has one of two outcomes: the arrest of cell-cycle progression in G1 or the induction of apoptosis. Both outcomes protect organisms from the deleterious effects of genotoxic agents that lead to genome instability and to cancer, and the p53-regulated genes are believed to contribute to this protection. Although the exact role played by p53 in apoptosis is unclear, the arrest of cell-cycle progression in G1 is mediated largely through transcriptional induction of *p21*$^{WAF1/CIP1}$ *(52,62)*, a potent inhibitor of several cyclin-dependent kinase activities that are required for progression through the cell cycle

(93,270). MDM2 binds to the amino-terminal domain of p53 and inhibits its trans-activation; thus, MDM2 may function in a feedback loop to limit p53-mediated transcriptional responses *(33,177,190,195)*. The roles of other p53-regulated genes are less clear. PCNA is an essential DNA replication protein that also is required for DNA repair *(184,198,231)*. PCNA is found in complexes with cyclins, cyclin-dependent kinases, $p21^{WAF1/CIP1}$, and GADD45; the induction of $p21^{WAF1/CIP1}$ and *GADD45* in response to DNA damage may modulate the ability of PCNA to support replication and DNA repair *(144,234,256)*. Repression of Bcl-2 and the induction of Bax may sensitize cells to apoptosis, whereas the repression of NOS-2 may serve as a feedback mechanism to limit the formation of this genotoxic agent. In other cases (e.g., *MCK*, *Sgk [114,155,265]*), the purpose of gene induction has not been established.

3.1.3. *Regulation of p53 Expression and Activity*

Although transcriptional induction *(82,204,239)* and enhanced translation *(45,66,70)* may contribute to the accumulation of p53 in some circumstances, stabilization appears to be the primary mechanism responsible for its accumulation after genotoxic stress. p53 protein is degraded by the ubiquitin-mediated pathway *(39,41,205)*; thus, DNA damage either leads to inhibition of the system that ubiquitinates p53 or to modification of the p53 protein to diminish recognition by the ubiquitination machinery.

Recent studies suggest that the p53 protein is synthesized in a latent form that has a lesser capacity for sequence-specific DNA binding. In vitro, sequence-specific binding is enhanced by several conditions or treatments that affect the carboxy-terminus of the protein; these include removal or replacement of the 30 carboxy-terminal amino acids, phosphorylation by several protein kinases, the addition of proteins or peptides that interact with the carboxy-terminal region (e.g., DnaK and the monoclonal antibody [MAb] PAb421), and the addition of DNA fragments *(17,105–109,117,241,259)*. The carboxy-terminal domain of p53 has nonsequence-specific DNA and RNA binding *(142,202,260)* and strand-annealing activity *(15,186)*, but most importantly, the carboxy-terminus appears to act as a negative regulatory domain that can inhibit sequence-specific DNA binding. These observations suggest that in vivo, activation of p53-mediated transcription may require several steps, including;

1. p53 binding to sites of DNA damage through its carboxy-terminus;
2. Stabilization by a modification that inhibits ubiquitination; and
3. Activation of latent DNA binding by posttranslational modification(s) of p53.

This multistep model for transcriptional activation in response to genotoxins via p53 has considerable appeal. A requirement for activation would serve as a check on the checkpoint mechanism, and could prevent potentially catastrophic consequences of inadvertent or inappropriate p53 induction.

The most likely mechanisms for activating latent p53 in vivo are phosphorylation, glycosylation, and binding through the carboxy-terminus to damaged DNA. Nothing is known about posttranslational modifications to components of the ubiquitin system involved in p53 recognition, but the p53 protein is phosphorylated at several sites, and at least 11 charge isoforms have been observed (*164,250,258*; reviewed in *162*). A recent study also suggests that p53 is glycosylated *(227)*. Six phosphorylation sites, three in the amino-terminal transactivation segment (serines 9, 15, and 34) and three in the carboxy-terminal regulatory/tetramerization segment (serines 312, 378, and 392;

human p53 numbering unless otherwise specified) are conserved among mammalian p53 proteins *(235)*, an observation that argues for the importance of these sites in regulating p53 activity. The phosphorylation of serine 34 of mouse p53 is enhanced by exposing cells to UVC and, in vitro, this site is phosphorylated by JNK1, an MAPK family member that is activated in response to genotoxic agents (*see* Section 3.2.2.) *(169)*. Serine 15 of human and murine p53 is phosphorylated in vitro by DNA-PK, a kinase that is activated specifically by DNA strand breaks *(6,143,258)*. In the carboxy-terminal domain, serine 312 is phosphorylated by the $p34^{cdc2}$ kinase *(2,18)*, serine 378 by protein kinase C *(241)*, and serine 392 by casein kinase 2 *(161)*. There are several additional phosphorylation sites in the amino-terminal region, including serines 4 and 6, which can be phosphorylated by a casein kinase I-like enzyme that may be related to the DNA damage-induced *HRR* kinases of yeast *(58,166)*, and threonines 73 and 83, which in vitro are substrates for ERK *(168)*. Serine 4 of mouse p53 also is a potential DNA-PK site.

Interest in phosphorylation as a modulator of p53 activity was stimulated by a report indicating that mutating the casein kinase 2 site (serine 389) in murine p53 abolished its antiproliferative activity *(167)*; however, subsequent studies have not confirmed this effect *(69,78,160,196)*, and modifications that remove other sites only modestly affected p53 function. Changing several mouse p53 amino-terminal phosphorylation sites singly or simultaneously to alanine diminished, but did not abolish, transactivation, but changing the same sites to aspartic acid to mimic phosphorylation resulted in activity near the wild-type level *(78,160)*. Similarly, changing serine 15 of human p53 to alanine reduced p53-mediated growth arrest by about 50%, but markedly increased the half-life of the mutant protein *(68)*. Nevertheless, it is intriguing that several protein kinases that phosphorylate p53 are activated in response to genotoxic agents and that phosphorylation at the known carboxy-terminal sites activates sequence-specific DNA binding. Furthermore, there are at least two reasons why attempts to analyze mutant p53 proteins may have underestimated effects of phosphorylation on p53 function in vivo. First, p53 proteins with mutated phosphorylation sites inevitably have been expressed in amounts much above the levels of p53 in normal cells; under these conditions, regulatory effects may be lost or masked. Second, posttranslational activation may only be required in normal cells under certain circumstances, e.g., in resting (G0) cells; studies of mutant p53 proteins have employed rapidly growing, immortalized cell lines.

3.1.4. Role of DNA Damage in Initiating the Response

How DNA damage is detected and a signal transmitted to p53 is unknown (Fig. 1). A few DNA strand breaks per genome (perhaps just one) are sufficient to activate the G1 checkpoint *(59,103)*; thus, the strand-break-detection system must be extremely sensitive, yet resistant to inappropriate activation. Although p53 can bind to a DNA damage site and although this binding might contribute to p53 stabilization or activation, another damage detection and signaling mechanism clearly is required. In yeast, several protein kinases are known to be involved in detecting DNA damage and in transmitting the signals (*7,31,153*; *see* Chapters 17 and 18 in vol. 1). In mammalian cells, two enzymes are known to be activated by DNA strand breaks: DNA-PK and poly(ADP-ribose) polymerase (PARP) *(6)*. However, neither enzyme is essential for activating the G1 cell-cycle checkpoint. PARP is an abundant nuclear enzyme that tenaciously binds DNA nicks, and then modifies itself and other chromatin-bound proteins in an NAD-dependent reaction *(50,146; see* Chapter 22*)*. Mice lacking PARP are viable and have apparently normal cell-cycle checkpoints *(262)*. DNA-PK also is an abundant nuclear enzyme, and in vitro it phosphorylates several nuclear DNA binding proteins, including p53 *(7,143)*. Recently, DNA-PK was shown to be an essential component of a mechanism that repairs DNA double-strand breaks; it also is required for V(D)J recombination, which generates functional immunoglobulin and T-cell receptor genes *(115,208; see* Chapters 16 and 17*)*. Cells from the severe combined immunodeficient (SCID) mouse lack DNA-PK activity and cannot complete V(D)J recombination *(19,194,240)*. The gene for DNA-PK_{cs}, the 470-kDa catalytic subunit of DNA-PK, maps to the SCID locus *(19,131,194,232)*; thus, the SCID mutation probably resides in the gene for DNA-PK_{cs}. Although DNA-PK is a serine/threonine protein kinase, the sequence of DNA-PK_{cs} reveals that it belongs to a recently discovered family of cell-cycle and DNA replication/repair regulatory proteins with homology to phosphatidylinositol (PI) kinases *(94,104,125,274)*. Nevertheless, DNA-PK has no detectable PI kinase activity *(94)*. Cells from SCID mice are hypersensitive to ionizing radiation, but are normal with respect to activation of the G1/S and G2 cell-cycle checkpoints *(102)*. Thus, DNA-PK might contribute to the induction and activation of p53 by genotoxic agents in some circumstances, but it cannot be essential for this process in most normal cells.

Accumulation of p53 protein and induction of *p21*$^{WAF1/CIP1}$ and *GADD45* transcripts is diminished or delayed after exposing cells to ionizing radiation taken from patients with ataxia telangiectasia (AT) (*12,28,122,127,165*; Chapter 19). AT cells are hypersensitive to ionizing radiation, but are relatively normal with respect to the repair of DNA strand breaks. However, AT cells are unable to activate any cell-cycle progression checkpoint in response to ionizing radiation, including the G1 checkpoint *(211,275)*. In contrast, the induction of p53 by UVC light and other agents that do not directly make DNA strand breaks is normal in AT cells. These findings suggest that AT cells are defective in a component of a DNA damage-detection or signaling system that functions upstream of p53. Because p53 still responds to UVC irradiation, other mechanisms for detecting the effects of genotoxic agents must exist and remain intact in AT cells. The spectrum of genotoxic agents to which the p53 response is defective led Canman et al. *(28)* to propose that AT cells may be specifically defective in the ability to respond to single-stranded DNA breaks. The defective gene in AT cells, designated *ATM*, recently was identified, and the ATM protein, like DNA-PK, has a carboxy-terminal segment of ~400 residues that is homologous to the kinase domains of PI kinases *(216,217)*. Several budding yeast genes have been identified that belong to the ATM/DNA-PK_{cs} lipid kinase-like family; these include *TOR1*, *TOR2*, *MEC1*, and *TEL1*. *MEC1* and its fission yeast homolog, *rad3*, are yeast checkpoint genes; a human homologue of yeast *MEC1*, *ATR/FRP1*, was identified recently *(42,124a)*. *MEC1* mutants, like AT cells, are defective in all the radiation cell-cycle checkpoints (Chapter 17 in vol. 1). Although neither Mec1p/Frp1p nor the ATM protein have been shown to possess protein or lipid kinase activity, it is tempting to speculate from the comparison with DNA-PK that these proteins may have a similar function that is activated directly or indirectly in response to DNA strand breaks. With the genes in hand

and mutant cells lacking the activities available, rapid progress in characterizing the functions of these proteins should be possible. The availability of these genes and of the corresponding enzyme-defective cell lines also should aid in identifying and characterizing other signaling pathways involved in regulating gene expression in response to genotoxic agents.

3.2. Role of MAPK Signaling Pathways in the Genotoxic Response

At least three distinct MAPKs have been implicated in mediating gene activation in response to genotoxic insults. These are the extracellular signal-regulated kinases (ERKs), the stress-activated protein kinases or c-Jun N-terminal kinases (SAPK/JNK), and the p38/reactivating kinase (RK)/cytokine-suppressive anti-inflammatory drug binding protein (CSBP) kinases *(30,48,100,136,137)*. These proline-directed kinases play a key role in activating pre-existing transcription factors, as well as other regulatory proteins, by phosphorylating serine and threonine residues. Their activation occurs largely, if not exclusively, through nonnuclear signaling events, and requires phosphorylation of threonine and tyrosine residues in a homologous protein kinase subdomain VIII (Fig. 2). However, the particular tripeptide motif specifying phosphorylation differs for the three MAPKs (TEY, TPY, and TGY for ERK, JNK, and p38, respectively) *(30,48)*.

Table 3 lists a variety of transcription factors known to be regulated by MAPK. The references cited have additional information on sites of phosphorylation. It should be emphasized that not all of the transcription factors in Table 3 have been shown to be phosphorylated by MAPK in response to genotoxic stress. Furthermore, two transcription factors, HSF1 and NFκB, known to be activated in response to some genotoxic treatments, are not included in Table 3, since MAPKs have not been implicated in their regulation. Although HSF1 becomes hyperphosphorylated in heat-shocked cells *(179,180)*, the protein kinase responsible has not been identified. Phosphorylation also plays a critical role in activating NFκB (reviewed in ref. *254*). NFκB is sequestered in the cytoplasm as a complex with its inhibitor, IκB. In response to genotoxic stress, IκB is phosphorylated and then destroyed by the ubiquitin degradation system. Initial studies suggested phosphorylation of IκB in response to genotoxic stress is mediated by protein kinase C through a mechanism independent of MAPK, but a recent study suggests another protein kinase(s) may be responsible for phosphorylating IκB *(116)*.

3.2.1. Extracellular Signal-Regulated Protein Kinases

There is considerable overlap in the genetic responses to mitogenic stimulation and UVC irradiation, much of which is now attributed to the activation of one or more ERK isoforms in both responses. Mitogen-stimulated events leading to the activation of ERK1 and ERK2, the prototypes of this subfamily, have been extensively studied (reviewed in *13,222*). Here the critical steps involved in the pathway are summarized. As illustrated in Fig. 2, binding of growth factors to their receptors results in autophosphorylation of tyrosine residue(s) on the receptor tyrosine kinases and leads to the Grb2-mediated translocation of the GDP-exchange factor, mSos, from the cytoplasm to the cell membrane, thereby activating Ras. Ras activation, in turn, triggers a cascade of extranuclear phosphorylation events that involve sequential activation of Raf, MAPK kinase (MEK), and finally ERKs. Activated ERK then translocates to the nucleus to phosphorylate a variety of transcription factors, resulting in gene activation.

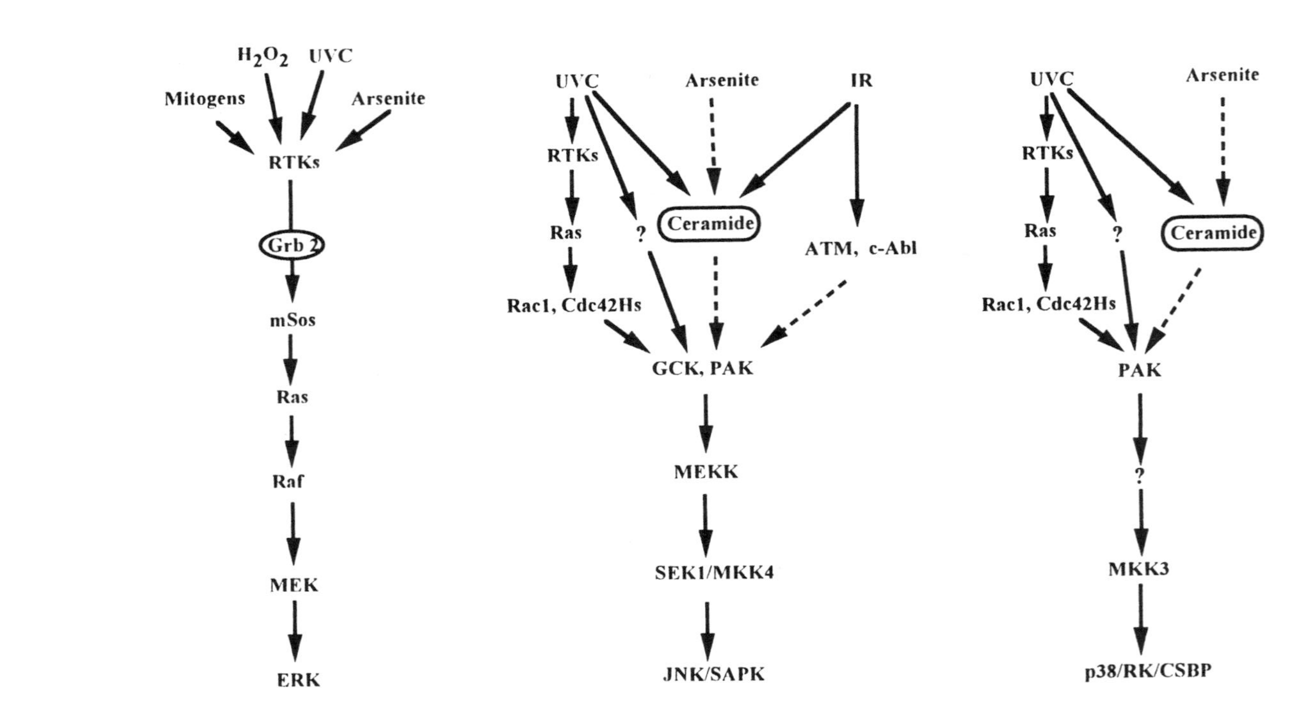

Fig. 2. MAPK signaling pathways involved in mediating the cellular response to genotoxic stress. Genotoxin-induced effects other than direct DNA damage are regulated largely through signaling pathways leading to activation of one or more MAPK families. Dotted arrows indicate links, although it is not clear whether the pathway is direct or involves other intermediates. The (?) indicates that the intermediate has not been identified. *See text* for details and references for the three pathways.

Table 3
Transcription Factors Activated by MAP Kinases

Transcription[a] factor	MAP kinase	Properties regulated by phosphorylation	References
c-Myc	ERK	Transactivation?	*47,224*
c-Jun	JNK	Transactivation; association with coactivator CBP	*10,53,97,120, 136*
c-Fos	FRK	Transactivation?	*51*
Elk-1/TCF	ERK, JNK, p38	Transactivation; ternary complex formation	*32,81,98,245, 267*
ATF-2	ERK, JNK, p38	Transactivation; DNA binding	*1,86,252*
GADD153/CHOP10	p38	Transactivation? Transrepression?	*257*
p53	ERK, JNK	Increased stability?	*168,169*
C/EBPβ/NF-IL-6	ERK	Transactivation	*182*
Stat[b]	ERK	Dimerization; nuclear localization	*46*
Tal-1[b]	ERK	Transactivation?	*36,255*
TTP[b]	ERK	Transactivation?	*242*

[a]Abbreviations: TCF, ternary complex factor; ATF-2, activating transcription factor-2; C/EBPβ, CCAAT/enhancer binding protein β; NF-IL-6, nuclear factor for IL-6 expression; GADD153, growth arrest and DNA damage-153 protein; CHOP10, C/EBP homologous protein 10; Stat, signal transducer and activator of transcription; TTP, tristetraprolin protein; Tal-1, T-cell acute leukemia-1 protein.

[b]MAPK-regulated transcription factors that have not been implicated in the response to genotoxic stress.

As shown in Table 3, a large number of transcription factors are activated by ERK; however, not all of these (e.g., Stat, TAL-1, and TTP) have been implicated in cellular responses to genotoxic stress.

Recent findings suggest that this same phosphorylation cascade is involved in activating gene expression in response to genotoxic stress. First, many genotoxic agents have been shown to activate ERK, although they are not as potent as mitogens *(100)*. Second, inhibition of ERK activity, by using dominant negative mutants of Raf, Ras, and ERK, overexpression of MAPK phosphatases, or specific inhibitors of MEK, attenuates the UVC-mediated activation of ERK-dependent transcription factors and ERK-dependent reporter gene expression *(87,150,200,213)*.

3.2.2. Stress-Activated c-Jun N-Terminal Kinases

The SAPK/JNK MAPK subfamily consists of at least 10 isoforms, with JNK1 and JNK2 being the most extensively studied *(48,53,85,136,137)*. The protein kinases of this subfamily are highly activated in response to stresses, including UVC irradiation, ionizing radiation, heat shock, and the DNA-alkylating agent methyl methanesulfonate (MMS). In contrast to ERKs, they show little activation in response to most growth factors *(48,53,137,148)*. The immediate upstream activator of JNK, SEK1/MKK4, shares significant homology with MEK (involved in the ERK activation) and was originally cloned on this basis *(54,145,214)*. It is phosphorylated by MAPK kinase kinase (MEKK) *(139,170,272)*. Although originally identified as an MEK regulator, MEKK does not appear to be involved in activating MEK under normal physiological conditions, but rather occupies a position in the JNK pathway analogous to Raf in the ERK pathway as depicted in Fig. 2. The role of Ras in activating JNK is not entirely clear, but current evidence indicates that there are both Ras-dependent pathways (as for UVC and growth factors) and Ras-independent pathways (as for arsenite and TNF-α) *(140,150,170)*. Rac1 and CDC42, small GTPases of the Rho family, have been implicated as intermediates in JNK activation in response to growth factors and inflammatory cytokines *(44,172)*. Both can bind and activate the protein kinase p65/PAK, which is a homolog of the yeast MEKK kinases Ste20p and Cla4p *(96,158)*. Thus, it was proposed that PAK, or a related kinase, links Rac1 and CDC42 to the MEKK-SEK-JNK module. Consistent with this view, overexpression of either PAK or a related protein kinase, GCK1, leads to the activation of JNK *(14,197)*. However, Rac1 and CDC42 are unlikely to be general mediators of the JNK cascade in response to stress, since dominant-negative mutants of Rac1 and CDC42 have no effect on JNK activation in response to the protein synthesis inhibitor anisomycin *(44,172)*.

JNK was identified by its ability to bind and phosphorylate the N-terminal portion of the c-Jun oncoprotein, leading to its enhanced transactivating capacity *(97)*. Recently, JNK was shown to phosphorylate ATF-2 and Elk-1/TCF *(1,32,86,252,267)*, illustrating its broader influence in activating stress-responsive genes, overlapping with that of ERK.

3.2.3. *p38/RK/CSBP Kinases*

The third group of MAPKs consists of the endotoxin and osmotic stress-activated protein kinase p38, MAPKAPK-2 reactivating kinase, two protein kinases involved in regulating inflammatory cytokine biosynthesis (CSBP1 and CSBP2), and the MAX-interacting protein, Mxi-2 *(92,141,210,278)*. In general, these protein kinases appear to be activated by the same conditions that activate JNKs, but little is known about the upstream regulators that control their activation. A dual-specificity protein kinase, designated RKK, that phosphorylates RK was purified from arsenite-treated PC12 cells *(210)*. Additionally, an MEK homolog, MKK3, has been isolated that can activate p38 in vitro *(54)*. Whether RKK and MKK3 are identical or related proteins is unclear. MKK3 shares significant homology with MKK4/SEK1 (the upstream activator of JNK), and therefore, some homology to MEK. In vitro studies demonstrated that SEK1/MKK4 phosphorylates and activates p38, but MKK3 cannot activate JNK *(54,145)*. Given the similarity in the conditions leading to the activation of JNK and p38, it is reasonable to believe that these MAPK subfamilies share some common upstream regulators. Indeed, PAK was recently shown to be an intermediate of the p38 pathway *(14,282)*.

The phosphorylation of the small heat-shock protein HSP25/27 was the first example of a protein regulated by p38 *(210)*; however, p38 was originally thought to function primarily in the extranuclear environment. Nevertheless, recent studies indicate that p38 also resides in the nucleus of stressed cells, raising the possibility that it also has nuclear functions *(201)*. Indeed, p38 can phosphorylate a number of transcription factors, including ATF-2, Elk-1/TCF, and GADD153/CHOP *(201,245,257)*. Phosphorylation of GADD153/CHOP enhances its transactivation activity as well as its ability to inhibit transactivation by other C/EBP family members. Given that GADD153/CHOP expression is induced by a variety of stresses, its modification by p38 is likely to influence a broad range of transcriptional responses.

3.2.4. Differential MAPK Activation by Different Genotoxins and Functional Consequences of the Response

Although all three MAPK subfamilies have been implicated in the cellular response to genotoxic stress, the magnitude and kinetics of activation for any particular MAPK can vary significantly for different genotoxic agents. For example, in most types of cells ERK, JNK, and p38 are all activated in response to UVC treatment. However, in comparative studies examining ERK and JNK activation with various genotoxic agents in HeLa cells, JNK was preferentially activated by MMS, whereas ERK was preferentially activated by hydrogen peroxide *(87,148)*. Further, although exposure to arsenite resulted in a significant activation of all three subfamilies, activation of ERK was transient, whereas that of JNK and p38 was sustained much longer *(150)*.

Activation of these protein kinase pathways in response to particular treatments also exhibits cell type specificity. For example, arsenite is a potent activator of ERK in rat pheochromocytoma PC12 cells, but activates it less strongly in HeLa cells *(150)*. On the other hand, PC12 cells are uniquely unresponsive to UVC irradiation, displaying less than a twofold increase in ERK, JNK, and p38 activities across a wide range of doses *(150)*. This effect is specific to UVC irradiation, since PC12 cells respond both to growth factor stimulation and a variety of genotoxic agents with high activation of one or more MAPKs *(149,150)*.

The immediate consequence of MAPK activation is the phosphorylation and altered activity of other proteins, some of which regulate gene expression. How these immediate effects influence the ultimate outcome of the response is less clear. As mentioned previously, the major phenotypic effects seen in cells treated with genotoxic agents are either growth arrest or death. Growth arrest may be transient or permanent, depending on the particular treatment, the dose, type of cells, and growth conditions. These same factors likewise affect the degree to which different MAPKs are activated. The physiologic consequences of each particular MAPK are undetermined; however, evidence is accumulating suggesting that ERK and JNK may act in opposition. ERK activation favors survival, whereas JNK activation is associated with cell death *(34,253,269)*. For example, Canman et al. *(29)* demonstrated that high levels of growth factor and overexpression of Raf oncoprotein, both of which elevate ERK activity, protect cells from death induced by ionizing radiation. Likewise, cell lines defective in ERK activity (through expression of dominant Ras mutants or use of other inhibitors of the pathway) are more sensitive to killing by hydrogen peroxide, but cells manipulated to express a constitutively active mutant of MEK (the immediate upstream activator of ERK) display enhanced resistance to killing relative to wild-type cells *(88)*. A role for JNK in programmed cell death is supported by the following observations:

1. Persistent JNK activation in response to ionizing radiation correlates with apoptotic cell death;
2. Activation of JNK via ectopic expression of a positive *MKK4* mutant results in apoptosis *(269)*; and
3. Inhibition of JNK activation by using a dominant negative *sek1* mutant blocks apoptosis *(34,253,269)*.

3.2.5. Role of Growth Factor Receptors, Ceramide, Oxidative Stress, and DNA Damage in Initiating MAPK Activation by Genotoxins

Much of the signal responsible for activating ERK in response to genotoxic stress is generated at the cell membrane through growth factor receptors or other nongrowth factor receptor-linked tyrosine protein kinases. Genotoxic stresses trigger tyrosine phosphorylation of growth factor-linked tyrosine protein kinases (e.g., epidermal growth factor) as well as nongrowth factor-linked ones (e.g., Src and Syc), and inhibitors of tyrosine protein kinases, such as herbamycin, genistein, tyrphostin, and lavendustin, block the response to genotoxic stress *(43,56,218,219,264)*. The growth factor receptor poison suramin (which blocks interactions between certain receptors and their ligands) largely inhibits the activation of ERK in response to a variety of genotoxic agents, including UVC irradiation, hydrogen peroxide, MMS, and sodium arsenite *(87,150,213)*. Finally, overexpression of a dominant negative EGF receptor mutant attenuates ERK activation after exposure to UVC *(43,150,213)*. Such treatments likewise inhibit ERK-dependent gene activation. The importance of growth factor receptors in mediating the UVC response is not limited to the EGF receptor, but also includes at least the receptors for fibroblast growth (FGF), interleukin-1α (IL-1α), insulin, and the T-cell receptor *(43,213,220)*. Prestimulation of cells with UVC attenuates the response to the relevant growth factors and vice versa. Therefore, the level of ERK activation in a given type of cell is likely to reflect the specificity and density of the growth factor receptors present.

The initiating signals responsible for activating the JNK and p38 protein kinase pathways are less well understood. Recent studies suggest that activation of growth factor receptors may contribute (via activation of Ras) to this pathway, since suramin partially inhibits JNK activation in response to some treatments (e.g., UVC irradiation) *(150)*. However, since mitogens, which potently activate the ERK pathway, are only weak activators of JNK and p38, growth factor receptors are unlikely to be the major mediators of these pathways.

The product of the sphingomyelin pathway, ceramide, was identified as a major intermediate in the JNK pathway *(253)*. Both UVC and ionizing radiation result in the generation of ceramide, which potently activates both JNK and p38 in cultured cells *(253,266)*. Thus, the Rac1- and CDC42-mediated signal from Ras and growth factor receptors, and the ceramide-mediated signal from the sphingomyelinase pathway may synergistically activate JNK and p38. Ceramide also is a potential activator of ERK, since it can activate Raf through ceramide-dependent protein kinase *(273)*.

Much evidence suggests that many genotoxic agents trigger their effects on MAPK pathways through a mechanism involving oxidative stress (reviewed in *89,219*). The free radical scavenger *N*-acetyl cysteine (NAC), which elevates cellular levels of glutathione, can block the activation of transcription factors (e.g., NFκB, AP-1), as well as gene induction in response to genotoxic agents, including UVC, hydrogen peroxide, MMS, and arsenite *(56,57,87,88,150,219,221)*. In addition, activation of either ERK or JNK or both in response to several genotoxic agents (UVC, sodium arsenite, and hydrogen peroxide) can be prevented by first treating cells with NAC *(87,150)*. In contrast, treating cells with buthionine sulfoximine depletes glutathione levels and potentiates gene activation in response to genotoxic stress *(88)*. How oxidative stress activates the response is unclear, but it may involve crosslinking of growth factor or cytokine receptors. Many of these receptors contain cysteine-rich regions, which play a role in the dimerization and activation of receptors following ligand binding *(95)*. Reactive oxygen species might react with these receptors and mimic the effects of ligands, resulting in their dimerization and activation.

Although the evidence summarized above indicates that the activation of MAPKs by genotoxic agents is initiated primarily through extranuclear signals, a role for direct DNA damage in contributing to the response cannot be excluded. Indeed, recent studies have demonstrated that ionizing radiation as well as other DNA damaging agents, including cis-platinum and mitomycin C, can potently activate JNK through a c-Abl-oncoprotein-dependent mechanism *(129)*. Furthermore, cells lacking c-Abl respond well to other treatments, such as TNF-α and UVC, suggesting that the dependency of JNK activation on c-Abl is specific to DNA damage. In addition, JNK activation by DNA-damaging agents, as well as by other stressors, is mediated by SEK1, since a *SEK1*-dominant negative mutant was found to inhibit c-Abl-mediated activation of JNK by the DNA-damaging-agent 1-β-D-arabinofuranosylcytosine *(128)*. The exact pathways mediating the detection of DNA damage are unknown, but may involve the AT gene product ATM; ATM appears to be required for JNK activation by ionizing radiation *(222)*. Another recent study supporting a role for DNA damage in the activation of JNK showed that adding damaged DNA from UVC-treated cells directly to cell lysates enhanced JNK activity *(3)*.

3.2.6. *Attenuation of the Response*

Since MAPKs play a central role in mediating the response to genotoxic stress, their inactivation is central to attenuating the response. Attenuation is accomplished largely by a growing family of tyrosine/threonine dual-specificity protein phosphatases that include MAPK phosphatase 1 (MKP-1, also known as 3CH134 and CL100), MKP-2, PAC-1 (a lymphocyte-specific protein phosphatase), and B23 *(111,126,173,237,263)*. These phosphatases are themselves regulated, at least in part, through MAPK-mediated gene activation, and they play a key role in regulating MAPK activity during mitogenesis *(135,237,238,263)*. Several observation suggest that at least one of these phosphatases, MKP-1, is important in the normal attenuation of MAPK activity during the cellular response to genotoxic stress. First, the time-course for inducing endogenous MKP-1 expression after exposure to genotoxic agents is such that maximum expression correlates with a decline in MAPK activity *(148)*. Second, overexpression of MKP-1 can block activation of ERK, JNK, and p38, and prevent induction of MAPK-dependent reporter genes in UVC and MMS-treated cells *(40,88,148,201)*. Third, treating cells with cycloheximide to prevent MKP-1 protein expression prolongs the activation of JNK in UVC-treated cells *(151)*. Thus, MKP-1 and other members of this phosphatase family probably function in a feedback loop mechanism to control MAPK-mediated gene activation during exposure to genotoxic stress. Recently, the CDK inhibitor p21$^{WAF1/CIP1}$ was shown to interact with and inhibit the activity of JNK. Since *p21*$^{WAF1/CIP1}$ is induced by several stresses, it also may function in a feedback circuit to attenuate JNK activity *(230)*.

3.3. Crosstalk and Networking Among the Response Pathways

As noted above, a given genotoxic agent can generate multiple signals activating a variety of pathways functioning in parallel to regulate gene expression. However, in reviewing these individual pathways, it is apparent that there are significant points for potential crosstalk between the various pathways at multiple levels. In this section, examples of such crosstalk at three different levels are discussed: the signaling pathways, the transcription factors modified, and the gene promoter regions with which transcriptional regulators interact (Fig. 3).

3.3.1. Signal Transduction Pathways

Unlike the situation with mitogenic stimulation, cells have no specific receptor to detect environmental stresses. Rather, they rely on multiple sensors to detect damage and transmit signals to the nucleus. These initiating signals can converge at a common point for further transmission to the nucleus, thereby streamlining the signaling process. One example of such convergence is the activation of multiple growth factor receptors (designated $TRK_{x,y,z}$ in Fig. 3A) by genotoxic agents, all of which result in the activation of Ras, and the subsequent activation of Raf (Fig. 3A). In addition, UVC treatment activates PKC, which can act synergistically with Ras to activate Raf *(138)*.

On the other hand, signaling through a single initiating event can diverge, allowing for transmission of that signal to several different pathways. For example, Ras is believed to contribute to the activation of both ERK, through Raf, and JNK, through Rac and CDC42. Another example of such divergence is the product of *ATM*, which appears to play a role in activating both p53 and JNK in response to DNA damage *(123,222)*.

3.3.2. Transcription Factors Activated

Transcription factors are important targets of the MAPK cascades activated by genotoxic stress; their conformation is altered in a fashion that either facilitates or inhibits their association with DNA or the basal transcription machinery. In many cases, transcription factors are phosphorylated by more than one kinase at either the same site or at different sites, exerting complementary or antagonistic effects. For example, ERK, JNK, and p38 all phosphorylate the Elk-1 transcription factor, thereby enhancing its ability to bind DNA and form ternary complexes with serum-response factor and the serum-response element (Fig. 3B) *(32,81,98,245,267)*. Although all three MAPKs appear to phosphorylate the same group of serine and threonine residues of Elk-1, subtle differences in the phosphorylation patterns are apparent. ERK phosphorylation results in equal amounts of phosphoserine and phosphothreonine, whereas JNK displays a preference for threonine residues. These subtle differences probably account for the difference in activation by JNK relative to ERK. ERK and JNK also both phosphorylate the c-Jun protein, but at different sites, resulting in antagonistic effects. JNK phosphorylation of c-Jun occurs at serine 63 and serine 73 to activate AP-1, but ERK phosphorylation occurs at the carboxy-terminal region of the protein and inhibits AP-1 activity *(171)*.

3.3.3. Gene Promoter Regulatory Elements

A third level at which the integration of different signaling pathways contributes to gene induction is determined by the promoter regions of the affected genes. Activation

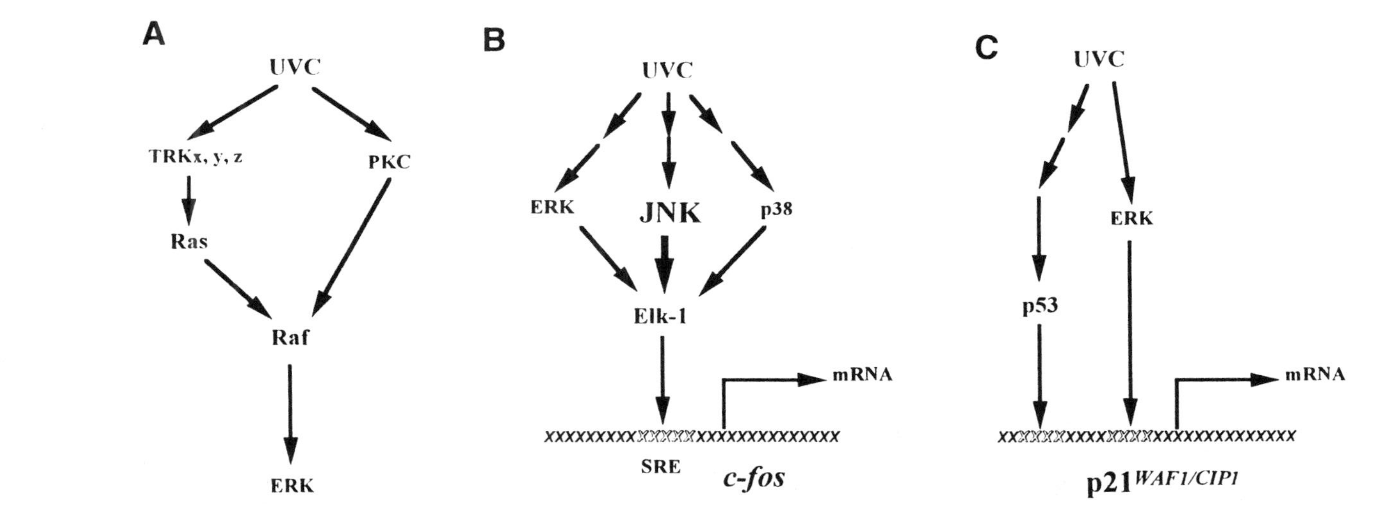

Fig. 3. Representative examples of crosstalk between the various pathways involved in controlling gene activation in response to genotoxic stress. **(A)** Multiple initiating signals converge to the same MAPK signaling pathway. Activation of Ras through multiple distinct tyrosine receptor kinases (designated x, y, and z) and activation of Raf by Ras and PKC, all lead to the activation of ERK. **(B)** ERK, JNK, and p38, all activated in response to the same genotoxin (UVC), can phosphorylate Elk-1, enhancing its DNA *(xxx)* binding activity and transactivation effects. The larger size of JNK denotes that it exerts the greatest influence on Elk activity. **(C)** p53-dependent and p53-independent response elements (interacting with different transcription factors) separately influence the activity of the *p21* promoter.

of many transcription factors requires the formation of heterodimers that bind a single *cis*-element; the AP-1 transcription factor is a typical example *(120)*. Although many genes have an AP-1 binding element that contributes to their regulation by genotoxic agents, in most cases, the particular players in genotoxic responses have not been identified. AP-1 binding complexes can contain any number of combinations of Fos and Jun protein isoforms, which are themselves differentially activated by different MAPK cascades (c-Fos mainly by ERK, c-Jun by JNK). Thus, the end result can be greatly affected by which members are activated by specific genotoxins in specific cell types, as well as by subtle differences in the particular AP-1 binding element or adjacent sequences *(99)*. A further complication is the fact that members of the AP-1 family can also interact with other leucine zipper proteins, such as ATF-2, allowing for further fine-tuning of the response in a particular cell type *(110,251)*.

Many genes have multiple regulatory elements that contribute to their overall responsiveness to genotoxic stress, and the transcription factors interacting at these sites may be regulated via different signaling pathways (*see* Table 2). Of particular interest are $p21^{WAF1/CIP1}$ and *GADD45*, which are regulated during DNA damage through both p53-dependent and p53-independent events (Fig. 3C) *(84,212)*. Although the particular p53-independent regulatory elements controlling transcriptional activation of either of these genes in response to genotoxic stress have not been identified, it is clear that significant induction and promoter activation of both genes by genotoxic agents (with the exception of ionizing radiation, which is highly dependent on p53) occur in the absence of functional p53. Indeed, there is strong evidence supporting the role of MAPK pathways in the transcriptional activation of $p21^{WAF1/CIP1}$ *(149)*.

4. CONCLUSIONS

Studies of mammalian cell responses to genotoxic insults are an outgrowth of, and have been heavily influenced by, early studies of the effects of DNA damage on gene expression in bacteria. Here, it was found that the mechanisms that repair DNA damage are inducible. In contrast, DNA repair mechanisms in mammalian cells, for the most part, are not inducible, but rather are expressed at appropriate steady-state levels. This difference likely reflects the fact that the environment, and therefore, the stresses and stimuli to which bacteria and higher eukaryotic organisms respond are qualitatively and quantitatively different.

Not surprisingly, the response of mammalian cells to genotoxic agents is complex and intimately linked with regulatory pathways governing cell growth and division. Major advances in our understanding of this stress response were the identification of p53 as a transcription factor that influences gene expression in response to DNA damage, and the demonstration that much of the induction seen with genotoxins is initiated through signals generated outside the nucleus that are attributable to activation of the MAPKs. However, much remains to be learned about the initiating signals responsible for triggering the activation of these pathways and the consequences of the response. Cells respond to genotoxic stress with either growth arrest or death (occurring largely through an apoptotic pathway). Clearly, p53 contributes to both responses, although the mechanisms associated with apoptosis are unclear. Likewise, current information indicates that MAPK-regulated responses contribute to influencing the path taken, with

ERK activation favoring survival and JNK activation favoring cell death. However, the particular effector genes involved in dictating these outcomes remain to be identified. Nonetheless, based on present understanding, the cellular response to genotoxic stress is likely to depend on the balance between levels of activation of the various signaling pathways involved and their potential interactions.

In closing, it is important to note that mammalian cells are rarely exposed to the genotoxic agents that are most conveniently studied in the laboratory. No UVC radiation from sunlight reaches the surface of the earth, and ionizing radiation is not a significant genotoxic component in the natural environment. However, genotoxic compounds are among the most effective agents available for treating cancer and other proliferative disorders. Increased knowledge of how these agents activate gene expression and influence other cellular processes affecting growth and survival could lead to the development of new approaches for cancer treatment through manipulation of these pathways.

REFERENCES

1. Abdel-Hafiz, H. A. M., L. E. Heasley, J. M. Kyriakis, J. Avruch, D. J. Kroll, G. L. Johnson, and J. P. Hoeffler. 1992. Activating transcription factor-2 DNA-binding activity is stimulated by phosphorylation catalyzed by p42 and p54 microtubule-associated protein kinases. *Mol. Endocrinol.* **16:** 2079–2089.
2. Addison, C., J. R Jenkins, and H.-W. Stürzbecher. 1990. The p53 nuclear-localization signal is structurally linked to a $p34^{cdc2}$ kinase motif. *Oncogene* **5:** 423–426.
3. Adler, V., S. Y. Fuchs, J. Kim, A. Kraft, M. P. King, J. Pelling, and Z. Ronai. 1995. Jun-NH_2-terminal kinase activation mediated by UV-induced DNA lesions in melanoma and fibroblast cells. *Cell Growth Diff.* **6:** 1437–1446.
4. Alam, J. 1994. Multiple elements within the 5' distal enhancer of the mouse heme oxygenase-1 gene mediate induction by heavy metals. *J. Biol. Chem.* **269:** 25,049–25,056.
5. Alam, J., S. Camhi, and A. M. Choi. 1995. Identification of a second region upstream of the mouse heme oxygenase-1 gene that functions as a basal level and inducer-dependent transcription factor enhancer. *J. Biol. Chem.* **270:** 11,977–11,984.
6. Anderson, C. W. 1993. DNA damage and the DNA-activated protein kinase. *Trends Biochem. Sci.* **18:** 433–437.
7. Anderson, C. W. 1994. Protein kinases and the response to DNA damage. *Semin. Cell. Biol.* **5:** 427–436.
8. Angel, P., A. Poting, U. Mallick, H. J. Rahmsdorf, M. Schorpp, and P. Herrlich. 1986. Induction of metallothionein and other mRNA species by carcinogens and tumor promoters in primary human skin fibroblasts. *Mol. Cell. Biol.* **6:** 1760–1766.
9. Applegate, L. A., P. Luscher, and R. M. Tyrrell. 1991. Induction of heme oxygenase: A general response to oxidant stress in cultured mammalian cells. *Cancer Res.* **51:** 974–978.
10. Arias, J., A. S. Alberts, P. Brindle, F. X. Claret, T. Smeal, M. Karin, J. Feramisco, and M. Montminy. 1994. Activation of cAMP and mitogen responsive genes relies on a common nuclear factor. *Nature* **370:** 226–229.
11. Arrowsmith, C. H. 1996. New insights into p53 function from structural studies. *Oncogene* **12:** 1379–1385.
12. Artuso, M., A. Esteve, H. Bresil, M. Vuillaume, and J. Hall. 1995. The role of the ataxia telangiectasia gene in the p53-, WAF1/CIP1(p21)- and GADD45-mediated response to DNA damage produced by ionizing radiation. *Oncogene* **11:** 1427–1435.
13. Avruch, J., X. F. Zhang, and J. M. Kyriakis. 1994. Raf meets Ras: completing the framework of a signal transduction pathway. *Trends Biochem. Sci.* **19:** 279–283.

14. Bagrodia, S., B. Dérijard, R. J. Davis, and R. A. Cerione. 1995. Cdc42 and PAK-mediated signaling leads to Jun kinase and p38 mitogen-activated protein kinase activation. *J. Biol. Chem.* **270:** 27,995–27,998.
15. Bakalkin, G., G. Selivanoa, T. Yakovleva, E. Kiseleva, E. Kashuba, K. P. Magnusson, L. Szekely, G. Klein, L. Terenius, and K. G. Wiman. 1995. p53 binds single-stranded DNA ends through the C-terminal domain and internal DNA segments via the middle domain. *Nucleic Acids Res.* **23:** 362–369.
16. Barak, Y., T. Juven, R Haffner, and M. Oren. 1993. mdm2 expression is induced by wild type p53 activity. *EMBO J.* **12:** 461–468.
17. Bayle, J. H., B. Ellenbaas, and A. J. Levine. 1995. The carboxy-terminal domain of the p53 protein regulates sequence-specific DNA binding through its nonspecific nucleic acid-binding activity. *Proc. Natl. Acad. Sci. USA* **92:** 5729–5733.
18. Bischoff, J. R, P. N. Friedman, D. R. Marshak, C. Prives, and D. Beach. 1990. Human p53 is phosphorylated by p60-cdc2 and cyclin B-cdc2. *Proc. Natl. Acad. Sci. USA* **87:** 4766–4770.
19. Blunt, T., N. J. Finnie, G. E. Taccioli, G. C. M. Smith, J. Demengeot, T. M. Gottlieb, R. Mizuta, A. J. Varghese, F. W. Alt, P. A. Jeggo, and S. P. Jackson. 1995. Defective DNA-dependent protein kinase activity is linked to V(D)J recombination and DNA repair defects associated with the murine *scid* mutation. *Cell* **80:** 813–823.
20. Boothman, D. A., and S. W. Lee. 1991. Induction of tissue type-specific plasminogen activator by ionizing radiation in human melanoma cells. *Proc. Am. Assoc. Cancer Res.* **32:** 74.
21. Boothman, D. A., T. W. Davis, and W. M. Sahijdak. 1993. Enhanced expression of thymidine kinase in human cells following ionizing radiation. *Int. J. Radiat. Oncol. Biol. Phys.* **30:** 391–398.
22. Boothman, D. A., M. Meyers, N. Fukunaga, and S. W. Lee. 1993. Isolation of X-ray-inducible transcripts from radioresistant human melanoma cells. *Proc. Natl. Acad. Sci. USA* **90:** 7200–7204.
23. Borellini, F., and R. I. Glazer. 1993. Induction of Sp1-p53 DNA-binding heterocomplexes during granulocyte/macrophage colony-stimulating factor-dependent proliferation in human erythroleukemia cell line TF-1. *J. Biol. Chem.* **268:** 7923–7928.
24. Bouvrette, M., B. Gowans, L. Yu, J. Stankiva, and D. Hunting. 1994. Cellular responses to genotoxic agents: DNA repair, activation of transcription, cell cycle arrest and apoptosis. *J. Chim. Phys.* **91:** 1005–1558.
25. Buckbinder, L., R. Talbott, B. R. Seizinger, and N. Kley. 1994. Gene regulation by temperature-sensitive p53 mutants: identification of p53 response genes. *Proc. Natl. Acad. Sci. USA* **91:** 10,640–10,644.
26. Buckbinder, L., R Talbott, S. Velasco-Miguel, I. Takenaka, B. Faha, B. R. Seizinger, and N. Kley. 1995. Induction of the growth inhibitor IGF-binding protein 3 by p53. *Nature* **377:** 646–649.
27. Camhi, S. L., J. Alam, L. Otterbein, S. L. Sylvester, and A. M. Choi. 1995. Induction of heme oxygenase-1 gene expression by lipopolysaccharide is mediated by AP-1 activation. *Am. J. Respir. Mol. Biol.* **13:** 387–398.
28. Canman, C. E., C.-Y. Chen, M.-H. Lee, and M. B. Kastan. 1994. DNA damage responses: p53 induction, cell cycle perturbations and apoptosis. *Cold Spring Harbor Symp. Quant. Biol.* **59:** 277–286.
29. Canman, C. E., T. M. Gilmer, S. B. Coutts, and M. B. Kastan. 1995. Growth factor modulation of p53-mediated growth arrest *versus* apoptosis. *Genes Dev.* **9:** 600–611.
30. Cano, E. and L. C. Mahadevan. 1995. Parallel signal processing among mammalian MAPKs. *Trends Biochem. Sci.* **20:** 117–122.
31. Carr, A. M. 1995. DNA structure checkpoints in fission yeast. *Semin. Cell Biol.* **6:** 65–72.

32. Cavigelli, M., F. Dolfi, F.-X. Claret, and M. Karin. 1995. Induction of c-fos expression through JNK-mediated TCF/Elk-1 phosphorylation. *EMBO J.* **14:** 5957–5964.
33. Chen, C. Y., J. D. Oliner, Q. Zhan, A. J. Fornace, Jr., B. Vogelstein, and M. B. Kastan. 1994. Interactions between p53 and MDM2 in a mammalian cell cycle checkpoint pathway. *Proc. Natl. Acad. Sci. USA* **91:** 2684–2688.
34. Chen, Y. R., C. F. Meyer, and T. H. Tan. 1996. Persistent activation of c-Jun terminal kinase 1 (JNK1) in gamma radiation-induced apoptosis. *J. Biol. Chem.* **271:** 631–634.
35. Chen, X., G. Farmer, H. Zhu, R. Prywes, and C. Prives. 1993. Cooperative DNA binding of p53 with TFIID (TBP): a possible mechanism for transcriptional activation. *Genes Dev.* **7:** 1837–1849.
36. Cheng, J. T., M. H. Cobb, and R. Baer. 1993. Phosphorylation of the TAL1 oncoprotein by the extracellular-signal-regulated protein kinase ERK1. *Mol. Cell. Biol.* **13:** 801–808.
37. Chin, K.-V., K. Ueda, I. Pastan, and M. M. Gottesman. 1992. Modulation of activity of the promoter of the human *MDR1* gene by ras and p53. *Science* **255:** 459–462.
38. Cho, Y., S. Gorina, P. D. Jeffrey, and N. P. Pavletich. 1994. Crystal structure of a p53 tumor suppressor-DNA complex: understanding tumorigenic mutations. *Science* **265:** 346–355.
39. Chowdary, D. R., J. J. Dermody, K. K. Jha, and H. L. Ozer. 1994. Accumulation of p53 in a mutant cell line defective in the ubiquitin pathway. *Mol. Cell. Biol.* **14:** 1997–2003.
40. Chu, Y., P. A. Solski, R. Khosravi-Far, C. J. Der, and K. Kelly. 1996. The mitogen-activated protein kinase phosphatases PAC1, MKP-1 and MKP-2 have unique substrate specificities and reduced activity *in vivo* toward the ERK2 *sevenmaker* mutation. *J. Biol. Chem.* **271:** 6497–6501.
41. Ciechanover, A., D. Shkedy, M. Oren, and B. Bercovich. 1994. Degradation of the tumor suppressor protein p53 by the ubiquitin-mediated proteolytic system requires a novel species of ubiquitin-carrier protein, E2. *J. Biol. Chem.* **269:** 9582–9589.
42. Cimprich, K. A., T. B. Shin, C. T. Keith, and S. L. Schreiber. 1996. cDNA cloning and gene mapping of a candidate human cell cycle checkpoint protein. *Proc. Natl. Acad. Sci. USA* **93:** 2850–2855.
43. Coffer, P. J., B. M. Burgering, M. P. Peppelenbosch, J. L. Bos, and W. Kruijer. 1995. UV activation of receptor tyrosine kinase activity. *Oncogene* **11:** 561–569.
44. Coso, O. A., M. Chiarello, J.-C. Yu, H. Teramoto, P. Crespo., N. Xu, T. Miki, and J. S. Gutkind. 1995. The small GTP-binding proteins Rac1 and Cdc42 regulate the activity of the JNK/SAPK signalling pathway. *Cell* **81:** 1137–1146.
45. Cox, L. S. and D. P. Lane. 1995. Tumour suppressor, kinases and clamps: how p53 regulates the cell cycle in response to DNA damage. *Bioessays* **17:** 501–508.
46. David, M., E. Petricoin, C. Benjamin, R. Pine, M. J. Weber, and A. C. Larner. 1995. Requirement for MAP kinase (ERK2) activity in interferon alpha- and interferon beta-stimulated gene expression through STAT proteins. *Science* **269:** 1721–1723.
47. Davis, R. J. 1993. The mitogen-activated protein kinase signal transduction pathway. *J. Biol. Chem.* **268:** 14,553–14,556.
48. Davis, R. J. 1994. MAPKs: New JNK expands the group. *Trends Biochem. Sci.* **19:** 470–473.
49. Datta, R., N. Taneja, V. P. Sukhatme, S. A. Qureshi, R. Weicheselbaum, and D. W. Kufe. 1993. Reactive oxygen intermediates target $CC(A/T)_6GG$ sequences to mediate activation of the early growth response 1 transcription factor gene by ionizing radiation. *Proc. Natl. Acad. Sci. USA* **90:** 2419–2422.
50. de Murcia, G. and J. M. de Murcia. 1994. Poly(ADP-ribose) polymerase: a molecular nick-sensor. *Trends Biochem. Sci.* **19:** 172–176.
51. Deng, T. and M. Karin. 1994. c-Fos transcriptional activity stimulated by H-Ras-activated protein kinase distinct from JNK and ERK. *Nature* **371:** 171–175.
52. Deng, C., P. Zhang, J. W. Harper, S. J. Elledge, and P. Leder. 1995. Mice lacking $p21^{CIP1/WAF1}$ undergo normal development, but are defective in G1 checkpoint control. *Cell* **82:** 675–684.

53. Dérijard, B., M. Hibi, I.-H. Wu, T. Barrett, B. Su, T. Deng, M. Karin, and R. J. Davis. 1994. JNK1: a protein kinase stimulated by UV light and Ha-Ras that binds and phosphorylates the c-Jun activation domain. *Cell* **76:** 1025–1037.
54. Dérijard, B., J. Raingeaud, T. Barrett, I.-H. Wu, J. Han, R. J. Ulevitch, and R. J. Davis. 1995. Independent human MAP kinase signal transduction pathways defined by MEK and MKK isoforms. *Science* **267:** 682–685.
55. Devary, Y., R. A. Gottlieb, L. F. Lau, and M. Karin. 1991. Rapid and preferential activation of the c-jun gene during the mammalian uv response. *Mol. Cell. Biol.* **11:** 2804–2811.
56. Devary, Y., R. A. Gottlieb, T. Smeal, and M. Karin. 1992. The mammalian ultraviolet response is triggered by activation of Src tyrosine kinases. *Cell* **71:** 1081–1091.
57. Devary, Y., C. Rossette, J. A. DiDonatto, and M. Karin. 1993. NF-κB by ultraviolet light not dependent on a nuclear signal. *Science* **261:** 1442–1445.
58. Dhillon, N. and M. F. Hoekstra. 1994. Characterization of two protein kinases from *Schizosaccharomyces pombe* involved in the regulation of DNA repair. *EMBO J.* **13:** 2777–2788.
59. Di Leonardo, A., S. P. Linke, K. Clarkin, and G. M. Wahl. 1994. DNA damage triggers a prolonged p53-dependent G1 arrest and long-term induction of Cip1 in normal human fibroblasts. *Genes Dev.* **8:** 2540–2551.
60. Dittmer, D., S. Pati, G. Zambetti, S. Chu, A. K. Teresky, M. Moore, C. Finlay, and A. J. Levine. 1992. Gain of function mutations in p53. *Nature Genet.* **4:** 42–46.
61. El-Deiry, W. S., S. E. Kern, J. A. Pietenpol, K. W. Kinzler, and B. Vogelstein. 1992. Definition of a consensus binding site for p53. *Nature Genet.* **1:** 45–49.
62. El-Deiry, W. S., T. Tokino, V. E. Velculescu, D. B. Levy, R. Parsons, J. M. Trent, D. Lin, W. E. Mercer, K. W. Kinzler, and B. Vogelstein. 1993. WAF-1, a potential mediator of p53 tumor suppression. *Cell* **75:** 817–825.
63. El-Deiry, W. S., T. Tokino, T. Waldman, J. D. Oliner, V. E. Velculescu, M. Burrell, D. E. Hill, E. Healy, J. L. Rees, S. R. Hamilton, K. W. Kinzler, and B. Vogelstein. 1995. Topological control of $p21^{WAF1/CIP1}$ expression in normal and neoplastic tissues. *Cancer Res.* **55:** 2910–2919.
64. Elledge, S. J., Z. Zhou, J. B. Allen, and T. A. Navas. 1993. DNA damage and cell cycle regulation of ribonucleotide reductase. *Bioessays* **15:** 333–339.
65. Englander, E. and S. H. Wilson. 1992. DNA damage response of cloned DNA beta polymerase promoter is blocked in mutant cell lines deficient in protein kinase A. *Nucleic Acids Res.* **20:** 5527–5531.
66. Ewen, M. E. and S. J. Miller. 1996. p53 and translational control. *Biochim. Biophys. Acta* **1242:** 181–184.
67. Fawcett, T. W., H. B. Eastman, J. L. Martindale, and N. J. Holbrook. 1996. Physical and functional association between GADD153 and C/EBPβ during cellular stress. *J. Biol. Chem.* **271:** 14,285–14,289.
68. Fiscella, M., S. J. Ullrich, N. Zambrano, M. T. Shields, D. Lin, S. P. Lees-Miller, C. W. Anderson, W. E. Mercer, and E. Appella. 1993. Mutation of the serine 15 phosphorylation site of human p53 reduces the ability of p53 to inhibit cell cycle progression. *Oncogene* **8:** 1519–1528.
69. Fiscella, M., N. Zambrano, S. J. Ullrich, T. Ungar, D. Lin, B. Cho, W. E. Mercer, C. W. Anderson, and E. Appella. 1994. The carboxy-terminal serine 392 phosphorylation site of human p53 is not required for wild-type activities. *Oncogene* **9:** 3249–3257.
70. Fontoura, B. M. A., E. A. Sorokina, E. David, and R. B. Carroll. 1992. p53 is covalently linked to 5.8S RNA. *Mol. Cell. Biol.* **12:** 5145–5151.
71. Fornace, A. J., Jr., I. Alamo, and C. Hollander. 1988. DNA damage-inducible transcripts in mammalian cells. *Proc. Natl. Acad. Sci. USA* **20:** 5527–5531.

72. Fornace, A. J., Jr., D. W. Nebert, M. C. Hollander, J. D. Luethy, M. Papathanasiou, J. Fargnoli, and N. J. Holbrook. 1989. Mammalian genes coordinately regulated by growth arrest signals and DNA-damage inducible agents. *Mol. Cell. Biol.* **9:** 4196–4203.
73. Fornace, A. J., Jr. 1992. Mammalian genes induced by radiation; activation of genes associated with growth control. *Annu. Rev. Genet.* **26:** 507–526.
74. Forrester, K., S. Ambs, S. E. Lupold, R. B. Kapust, E. A. Spillare, W. C. Weinberg, E. Felley-Bosco, X. W. Wang, D. A. Geller, E. Tzeng, T. R. Billar, and C. C. Harris. 1996. Nitric oxide-induced p53 accumulation and regulation of inducible nitric oxide synthase expression by wild-type p53. *Proc. Natl. Acad. Sci. USA* **93:** 2442–2447.
75. Fritsche, M., C. Haessler, and G. Brandner. 1993. Induction of nuclear accumulation of the tumor suppressor protein p53 by DNA-damaging agents. *Oncogene* **8:** 307–318.
76. Fritz, G., K. Tano, S. Mitra, and B. Kaina. 1991. Inducibility of the DNA repair gene encoding O^6-methylguanine DNA methyltransferase in mammalian cells by DNA-damaging treatments. *Mol. Cell. Biol.* **11:** 4660–4668.
77. Fritz, G., B. Kaina, and K. Aktories. 1995. The ras-related small GTP-binding protein RhoB is immediate-early inducible by DNA damaging treatments. *J. Biol. Chem.* **270:** 25,172–25,177.
78. Fuchs, B., D. O'Connor, L. Fallis, K. H. Scheidtmann, and X. Lu. 1995. p53 phosphorylation mutants retain transcription activity. *Oncogene* **10:** 789–793.
79. Funk, W. D., D. T. Pak, R. H. Karas, W. E. Wright, and J. W. Shay. 1992. A transcriptionally active DNA-binding site for human p53 protein complexes. *Mol. Cell. Biol.* **12:** 2866–2871.
80. Gibbs, S., F. Lohman, W. Teabel, P. van de Putte, and C. Backendorf. 1990. Characterization of the human spr2 promoter: induction after UV irradiation or TPA treatment and regulation during differentiation of cultured keratinocytes. *Nucleic Acids Res.* **18:** 4401–4407.
81. Gille, H., A. D. Sharrocks, and P. E. Shaw. 1992. Phosphorylation of transcription factor $p62^{TCF}$ by MAP kinase stimulates ternary complex formation at c-fos promoter. *Nature* **358:** 414–417.
82. Ginsberg, D., M. Oren, M. Yaniv, and J. Piette. 1990. Protein binding elements in the promoter region of the mouse p53 gene. *Oncogene* **5:** 1285–1290.
83. Ginsberg, D., F. Mechta, M. Yaniv, and M. Oren. 1991. Wild-type p53 can down-modulate the activity of various promoters. *Proc. Natl. Acad. Sci. USA* **88:** 9979–9983.
84. Gorospe, M., J. L. Martindale, M. S. Sheikh, A. J. Fornace, Jr., and N. J. Holbrook. 1996. Regulation of $p21^{Waf1/Cip1}$ expression by cellular stress: p53-dependent and p53-independent mechanisms. *Mol. Cell. Differ.* **4:** 47–65.
85. Gupta, S., T. Barrett, A. J. Whitmarsh, J. Cavanagh, H. K. Sluss, B. Derijard, and R. J. Davis. 1996. Selective interaction of JNK protein kinase isoforms with transcription factors. *EMBO J.* **15:** 2760–2770.
86. Gupta, S., D. Campbell, B. Dérijard, and R. J. Davis. 1995. Transcription factor ATF2 regulation by the JNK signal transduction pathway. *Science* **267:** 389–393.
87. Guyton, K. Z., Y. Liu, M. Gorospe, Q. Xu, and N. J. Holbrook. 1996. Activation of mitogen-activated protein kinase by H_2O_2: role in cell survival following oxidant injury. *J. Biol. Chem.* **271:** 4138–4142.
88. Guyton, K. Z., M. Gorospe, and N. J. Holbrook. 1996. Induction of the mammalian stress response gene GADD153 by oxidative stress: role of AP-1 element. *Biochem. J.* **314:** 547–554.
89. Guyton, K. Z., M. Gorospe, and N. J. Holbrook. 1996. Oxidative stress, gene expression and the aging process, in *The Molecular Biology of Free Radical Scavenging Systems* (Scandalios, J., ed.), Cold Spring Harbor Laboratory, Cold Spring Harbor, NY, pp. 247–272.

90. Hallahan, D. E., D. R. Spriggs, M. A. Beckett, D. W. Kufe, and R. R. Weichselbaum. 1989. Increased tumor necrosis factor α mRNA after cellular exposure to ionizing radiation. *Proc. Natl. Acad. Sci. USA* **86:** 10,104–10,107.
91. Hallahan, D. E., V. P. Sukhatme, M. L. Sherman, S. Virudachalam, D. W. Kufe, and R. R. Weichselbaum. 1991. Protein kinase C mediates x-ray inducibility of nuclear signal transducers EGR1 and JUN. *Proc. Natl. Acad. Sci. USA* **88:** 2156–2160.
92. Han, J., J.-D. Lee, L. Bibbs, and R. J. Ulevitch. 1994. A MAP kinase targeted by endotoxin and hyperosmolaritiy in mammalian cells. *Science* **265:** 808–811.
93. Harper, J. W., G. R. Adami, N. Wei, K. Keyomarsi, and S. J. Elledge. 1993. The p21 Cdk-interacting protein Cip1 is a potent inhibitor of G1 cyclin-dependent kinases. *Cell* **75:** 805–816.
94. Hartley, K. O., D. Gell, H. Zhang, G. C. M. Smith, N. Divecha, M. A. Connelly, A. Admon, S. P. Lees-Miller, C. W. Anderson, and S. P. Jackson. 1995. DNA-dependent protein kinase catalytic subunit: a relative of phosphatidylinositol 3-kinase and the ataxia telangiectasia gene product. *Cell* **82:** 849–856.
95. Heldin, C.-H. 1995. Dimerization of cell surface receptors in signal transduction. *Cell* **80:** 213–223.
96. Herskowitz, I. 1995. MAP kinase pathways in yeast: for mating and more. *Cell* **80:** 187–197.
97. Hibi, M., A. Lin, T. Smeal, A. Minden, and M. Karin. 1993. Identification of an oncoprotein- and UV-responsive protein kinase that bind and potentiate the c-Jun activation domain. *Genes Dev.* **7:** 2135–2148.
98. Hill, C. S., R. Marais, S. John, J. Wynne, S. Dalton, and R. Treisman. 1993. Functional analysis of a growth factor-responsive transcription factor complex. *Cell* **73:** 395–406.
99. Holbrook, N. J. and A. J. Fornace, Jr. 1991. Response to adversity: molecular control of gene activation following genotoxic stress. *New Biol.* **3:** 825–833.
100. Holbrook, N. J., Y. Liu, and A. J. Fornace, Jr. 1996. Signaling events controlling the molecular response to genotoxic stress, in *Stress-Induced Cellular Responses* (Feige, U., R. I. Morimoto, I. Yahara, and B. S. Polla, eds.), Birkhauser/Springer, Basel, Switzerland, pp. 273–288.
101. Horikoshi, N., A. Usheva, J. Chen, A. J. Levine, R. Weinmann, and T. Shenk. 1995. Two domains of p53 interact with the TATA-binding protein, and the adenovirus 13S E1A protein disrupts the association, relieving p53-mediated transcriptional repression. *Mol. Cell. Biol.* **15:** 227–234.
102. Huang, L.-C., K. C. Clarkin, and G. M. Wahl. 1996. p53-dependent cell-cycle arrests are preserved in DNA-activated protein kinase-deficient fibroblasts. *Cancer Res.* **56:** 2940–2944.
103. Huang, L.-C., K. C. Clarkin, and G. M. Wahl. 1996. Sensitivity and selectivity of the DNA damage sensor responsible for activating p53-dependent G1 arrest. *Proc. Natl. Acad. Sci. USA* **93:** 4827–4832.
104. Hunter, T. 1995. When is a lipid kinase not a lipid kinase? When it is a protein kinase. *Cell* **83:** 1–4.
105. Hupp, T. R., D. W. Meek, C. A. Midgley, and D. P. Lane. 1992. Regulation of the specific DNA binding function of p53. *Cell* **71:** 875–886.
106. Hupp, T. R. and D. P. Lane. 1994. Allosteric activation of latent p53 tetramers. *Curr. Biol.* **4:** 865–875.
107. Hupp, T. R. and D. P. Lane. 1994. Regulation of the cryptic sequence-specific DNA binding function of p53 by protein kinases. *Cold Spring Harbor Symp. Quant. Biol.* **59:** 195–206.
108. Hupp, T. R. and D. P. Lane. 1995. Two distinct signaling pathways activate the latent DNA binding function of p53 in a casein kinase II-independent manner. *J. Biol. Chem.* **270:** 18,165–18,174.
109. Hupp, T. R., A. Sparks, and D. P. Lane. 1995. Small peptides activate the latent sequence-specific DNA binding function of p53. *Cell* **83:** 237–245.

110. Hurst, H. 1994. Transcription factors 1: bZIP proteins. *Protein Profile* **1:** 123–152.
111. Ishibashi, T., D. P. Bottaro, P. Michieli, C. A. Kelley, and S. A. Aaronson. 1994. A novel dual specificity phosphatase induced by serum stimulation and heat shock. *J. Biol. Chem.* **47:** 29,897–29,902.
112. Jackman, J., I. Alamo, and A. J. Fornace, Jr. 1994. Genotoxic stress confers preferential and coordinate messenger RNA stability on five *gadd* genes. *Cancer Res.* **54:** 5656–5662.
113. Jackson, P., P. Ridgway, J. Rayner, J. Nobel, and A. Braithwaite. 1994. Transcriptional regulation of the PCNA promoter by p53. *Biochem. Biophys. Res. Commun.* **203:** 133–140.
114. Jackson, P., M. Shield, J. Buskin, S. Hawkes, M. Reed, K. Perrem, S. D. Hauschka, and A. Braithwaite. 1995. p53-dependent activation of the mouse MCK gene promoter: identification of a novel p53-responsive sequence and evidence for cooperation between distinct p53 binding sites. *Gene Express.* **5:** 19–33.
115. Jackson, S. P. 1996. The recognition of DNA damage. *Curr. Opinion Genet. Dev.* **6:** 19–25.
116. Janosch, P., M. Schellerer, T. Seitz, P. Reim, M. Eulitz, M. Brielmeier, W. Kolch, J. M. Sedivy, and H. Mischak. 1996. Characterization of IκB kinases. *J. Biol. Chem.* **271:** 13,868–13,874.
117. Jayaraman, L. and C. Prives. 1995. Activation of p53 sequence-specific DNA binding by short single strands of DNA requires the p53 C-terminus. *Cell* **81:** 1021–1029.
118. Juven, T., Y. Barak, A. Zauberman, D. L. George, and M. Oren. 1993. Wild type p53 can mediate sequence-specific transactivation of an internal promoter within the *mdm2* gene. *Oncogene* **8:** 3411–3416.
119. Kaina, B., B. Stein, A. Schonthal, H. J. Rahmasdorf, H. Ponta, and P. Herrlich. 1990. An update of the mammalian UV response, Gene regulation of a protective function, in *DNA Repair Mechanisms and Their Biological Implications in Mammalian Cells* (Lambert, M. W. and J. Laval, eds.), Plenum, New York, pp. 149–165.
120. Karin, M. 1995. The regulation of AP-1 activity by mitogen-activated protein kinases. *J. Biol. Chem.* **270:** 16,483–16,486.
121. Kartasova, T. and P. van de Putte. 1988. Isolation, characterization, and UV-stimulated expression of two families of genes encoding polypeptides of related structure in human epidermal keratinocytes. *Mol. Cell. Biol.* **8:** 2195–2203.
122. Kastan, M. B., O. Onyekwere, D. Sidransky, B. Vogelstein, and R. W. Craig. 1991. Participation of p53 protein in the cellular response to DNA damage. *Cancer Res.* **51:** 6304–6311.
123. Kastan, M. B., Q. Zhan, W. S. El-Deiry, F. Carrier, T. Jacks, W. V. Walsh, B. S. Plunkett, B. Vogelstein, and A. J. Fornace, Jr. 1992. A mammalian cell cycle checkpoint pathway utilizing p53 and GADD45 is defective in ataxia-telangiectasia. *Cell* **71:** 587–597.
124. Kastan, M. B., C. E. Canman, and C. J. Leonard. 1995. p53, cell cycle and apoptosis: implications for cancer. *Cancer and Metastasis Rev.* **14:** 3–15.
124a. Keegan, K. S., D. A. Holtzman, A. W. Plug, E. R. Christenson, E. E. Brainerd, G. Flaggs, N. J. Bently, E. M. Taylor, M. S. Meyn, S. B. Moss, A. M. Carr, T. Ashley, and M. F. Hoekstra. 1996. The Atr and Atm protein-kinases associate with different sites along meiotically pairing chromosomes. *Genes Dev.* **10:** 2423–2437.
125. Keith, C. T. and S. L. Schreiber. 1995. PIK-related kinases: DNA repair, recombination, and cell cycle checkpoints. *Science* **270:** 50,51.
126. Keyse, S. M. and E. A. Emslie. 1992. Oxidative stress and heat shock induce a human gene encoding a protein tyrosine phosphatase. *Nature* **359:** 644–647.
127. Khanna, K. K., H. Beamish, J. Yan, K. Hobson, R. Williams, I. Dunn, and M. F. Lavin. 1995. Nature of G1/S cell cycle checkpoint defect in ataxia-telangiectasia. *Oncogene* **11:** 609–618.
128. Kharbandra, S., P. Pandrey, R. Ren, B. Mayer, L. Zon, and D. Kufe. 1995. C-Abl activation regulates induction of the SEK1/stress-activated protein kinase pathway in the cellular response to 1-β-D-arabinoside. *J. Biol. Chem.* **270:** 30,278–30,281.

129. Kharbandra, S., R. Ren, P. Pandey, T. D. Shafman, S. M. Feller, R. R. Weichselbaum, and D. W. Kufe. 1995. Activation of the c-Abl tyrosine kinase in the stress response to DNA-damaging agents. *Nature* **376:** 785–788.
130. Kieser, A., H. A. Weich, G. Brandner, D. Marme, and W. Kolch. 1994. Mutant p53 potentiates protein kinase C induction of vascular endothelial growth factor expression. *Oncogene* **9:** 963–969.
131. Kirchgessner, C. U., C. K. Patil, J. W. Evans, C. A. Cuomo, L. M. Fried, T. Carter, M. A. Oettinger, and J. M. Brown. 1995. DNA-dependent kinase (p350) as a candidate gene for the murine *SCID* defect. *Science* **267:** 1178–1183.
132. Kley, N., R. Y. Chung, S. Fay, J. P. Loeffler, and B. R. Seizinger. 1992. Repression of the basal *c-fos* promoter by wild-type p53. *Nucleic Acids Res.* **20:** 4083–4087.
133. Kramer, M., C. Sachsenmaier, P. Herrlich, and H. J. Rahmsdorf. 1993. UV irradiation-induced interleukin-1 and basic fibroblast growth factor synthesis and release mediate part of the UV response. *J. Biol. Chem.* **268:** 6734–6741.
134. Kuerbitz, S. J., B. S. Plunkett, W. V. Walsh, and M. B. Kastan. 1992. Wild-type p53 is a cell cycle checkpoint determinant following irradiation. *Proc. Natl. Acad. Sci. USA* **89:** 7491–7495.
135. Kwak, S. P., D. J. Hakes, K. J. Martell, and J. E. Dixon. 1994. Isolation and characterization of a human dual specificity protein-tyrosine phosphatase gene. *J. Biol. Chem.* **269:** 3596–3604.
136. Kyriakis, J. M., P. Banerjee, E. Nikolakaki, T. Dai, E. A. Rubie, M. F. Ahmad, J. Avruch, and J. R. Woodgett. 1994. The stress-activated protein kinase subfamily of c-Jun kinases. *Nature* **369:** 156–160.
137. Kyriakis, J. M., J. R. Woodgett, and J. Avruch. 1994. The stress-activated protein kinases: A novel ERK subfamily response to cellular and inflammatory cytokines. *Ann. NY Acad. Sci.* **369:** 156–160.
138. Kolch, W., G. Heiddecker, G. Kochs, R. Hummel, H. Vahiddi, H. Mischak, G. Finkenzeller, D. Marmé, and U. R. Rapp. 1993. Protein kinase Cα activates RAF-1 by direct phosphorylation. *Nature* **364:** 249–252.
139. Lange-Carter, C. A., C. M. Pleiman, A. M. Gardner, K. J. Blumer, and G. L. Johnson. 1993. A divergence in the MAP kinase regulatory network defined by MEK kinase and Raf. *Science* **260:** 315–319.
140. Lange-Carter, C. A. and G. L. Johnson. 1994. Ras-dependent growth factor regulation of MEK kinase in PC12 cells. *Science* **265:** 1458–1461.
141. Lee, J. C., J. T. Laydon, P. C. McDonnell, T. F. Gallagher, S. Kumar, D. Green, D. McNulty, M. J. Blumenthal, J. R. Heys, S. W. Landvatter, J. E. Strickler, M. M. McLaughlin, I. R. Siemens, R. M. Fisher, G. P. Livi, J. R. White, J. L. Adams, and P. R. Young. 1994. A protein kinase involved in the regulation of inflammatory cytokine biosynthesis. *Nature* **372:** 739–800.
142. Lee, S., B. Elenbaas, A. Levine, and J. Griffith. 1995. p53 and its 14 kDa C-terminal domain recognize primary DNA damage in the form of insertion/deletion mismatches. *Cell* **81:** 1013–1020.
143. Lees-Miller, S. P., K. Sakaguchi, S. Ullrich, E. Appella, and C. W. Anderson. 1992. Human DNA-activated protein kinase phosphorylates serines 15 and 37 in the amino-terminal transactivation domain of human p53. *Mol. Cell. Biol.* **12:** 5041–5049.
144. Li, R., S. Waga, G. J. Hannon, D. Beach, and B. Stillman. 1994. Differential effects by the p21 CDK inhibitor on PCNA-dependent DNA replication and repair. *Nature* **371:** 534–537.
145. Lin, A., A. Minden, H. Martinetto, F. X. Claret, C. Lange-Carter, F. Mercurio, G. L. Johnson, and M. Karin. 1995. Identification of a dual specificity kinase that activates the Jun kinases and p38-Mpk2. *Science* **268:** 286–290.

146. Lindahl, T., M. S. Satoh, G. G. Poirer, and A. Klungland. 1995. Post-translational modification of poly(ADP-ribose) polymerase induced by DNA strand breaks. *Trends Biochem. Sci.* **20:** 405–411.
147. Liu, X., C. W. Miller, P. H. Koeffler, and A. J. Berk. 1993. The p53 activation domain binds the TATA box-binding polypeptide in holo-TFIID, and a neighboring p53 domain inhibits transcription. *Mol. Cell Biol.* **13:** 3291–3300.
148. Liu, Y., M. Gorospe, C. Yang, and N. J. Holbrook. 1995. Role of mitogen-activated protein kinase phosphatase during the cellular response to genotoxic stress. Inhibition of c-Jun N-terminal kinase activity and AP-1-dependent gene activation. *J. Biol. Chem.* **270:** 8377–8380.
149. Liu, Y., J. L. Martindale, M. Gorospe, and N. J. Holbrook. 1996. Regulation of p21/WAF1/CIP1 expression through mitogen-activated protein kinase signaling pathway. *Cancer Res.* **56:** 1–5.
150. Liu, Y., K. Z. Guyton, M. Gorospe, Q. Xu, and N. J. Holbrook. 1996. Differential activation of ERK, JNK/SAPK and p38/CSBP/RK MAP kinase family members during the cellular response to arsenite. *Free Radical Biol. Med.* **21:** 771–781.
151. Liu, Y. and N. J. Holbrook, unpublished observations.
152. Lu, H. and A. J. Levine. 1995. Human $TAF_{II}31$ protein is a transcriptional coactivator of the p53 protein. *Proc. Natl. Acad. Sci. USA* **92:** 5154–5158.
153. Lydall, D. and T. Weinert. 1996. From DNA damage to cell cycle arrest and suicide: a budding yeast perspective. *Curr. Opinion Genet. Dev.* **6:** 4–11.
154. Mack, D. H., J. Vartikar, J. M. Pipas, and L. A. Laimins. 1993. Specific repression of TATA-mediated but not initiator-mediated transcription by wild-type p53. *Nature* **363:** 281–283.
155. Maiyar, A. C., A. J. Huang, P. T. Phu, H. H. Cha, and G. L. Firestone. 1996. p53 stimulates promoter activity of the *sgk* serum/glucocorticoid-inducible serine/threonine-protein kinase gene in rodent mammary epithelial cells. *J. Biol. Chem.* **271:** 12,414–12,422.
156. Mallardo, M., V. Giordano, E. Dragonetti, G. Scala, and I. Quinto. 1994. DNA damaging agents increase the stability of interleukin-1α, interleukin-1β, and interleukin-6 transcripts and the production of the relative proteins. *J. Biol. Chem.* **269:** 14,899–14,902.
157. Maltzman, W. and L. Czyzyk. 1984. UV irradiation stimulates levels of p53 cellular tumor antigen in nontransformed mouse cells. *Mol. Cell. Biol.* **4:** 1689–1694.
158. Manser, E., T. Leung, H. Salihuddin, Z. Zhao, and L. Lim. 1994. A brain serine/threonine protein kinase activated by Cdc42 and Rac1. *Nature* **367:** 40–46.
159. Martin, D. W., R. M. Munoz, M. A. Subler, and S. Deb. 1993. p53 binds to the TATA-binding protein-TATA complex. *J. Biol. Chem.* **268:** 13,062–13,067.
160. Mayr, G. A., M. Reed, P. Wang, J. F. Schwedes, and P. Tegtmeyer. 1995. Serine phosphorylation in the NH_2 terminus of p53 facilitates transactivation. *Cancer Res.* **55:** 2410–2417.
161. Meek, D. W., S. Simon, U. Kikkawa, and W. Eckhart. 1990. The p53 tumor suppressor protein is phosphorylated at serine 389 by casein kinase II. *EMBO J.* **9:** 3253–3260.
162. Meek, D. 1994. Post-translational modification of p53. *Semin. Cancer Biol.* **5:** 203–210.
163. Mercer, W. E., M. T. Shields, D. Lin, E. Appella, and S. J. Ullrich. 1991. Growth suppression induced by wild-type p53 protein is accompanied by selective down-regulation of proliferating-cell nuclear antigen expression. *Proc. Natl. Acad Sci. USA* **88:** 1958–1962.
164. Merrick, B. A., P. M. Pence, C. He, R. M. Patterson, and J. K. Selkirk. 1995. Phosphor image analysis of human p53 protein isoforms. *Biotechniques* **18:** 292–299.
165. Meyn, M. S. 1995. Ataxia-telangiectasia and cellular responses to DNA damage. *Cancer Res.* **55:** 5991–6001.
166. Milne, D. M., R. H. Palmer, D. G. Campbell, and D. W. Meek. 1992. Phosphorylation of the p53 tumour-suppressor protein at three N-terminal sites by a novel casein kinase I-like enzyme. *Oncogene* **7:** 1361–1369.

167. Milne, D. M., R. H. Palmer, and D. W. Meek. 1992. Mutation of the casein kinase II phosphorylation site abolishes the anti-proliferative activity of p53. *Nucleic Acids. Res.* **20:** 5565–5570.
168. Milne, D., D. G. Campbell, F. B. Caudwell, and D. W. Meek. 1994. Phosphorylation of the tumor suppressor p53 by mitogen-activated protein kinases. *J. Biol. Chem.* **269:** 9253–9260.
169. Milne, D., L. E. Campbell, D. G. Campbell, and D. W. Meek. 1995. p53 is phosphorylated *in vitro* and *in vivo* by an ultraviolet radiation-induced protein kinase characteristic of the c-Jun kinase, JNK1. *J. Biol. Chem.* **270:** 5511–5518.
170. Minden, A., A. Lin, M. McMahon, C. Lange-Carter, B. Dérijard, R. J. Davis, G. L. Johnson, and M. Karin. 1994. Differential activation of ERK and JNK mitogen-activated protein kinases by Raf-1 and MEKK. *Science* **266:** 1719–1723.
171. Minden, A., A. Lin, T. Smeal, B. Dérijard, M. Cobb, R. J. Davis, and M. Karin. 1994. c-Jun N-terminal phosphorylation correlates with activation of the JNK subgroup but not the ERK subgroup of mitogen-activated protein kinases. *Mol. Cell. Biol.* **14:** 6683–6688.
172. Minden, A., A. Lin, F. X. Claret, A. Abo, and M. Karin. 1995. Selective activation of the JNK signaling cascade and c-Jun transcriptional activity by the small GTPases Rac and Cdc42Hs. *Cell* **81:** 1147–1157.
173. Misra-Press, A., C. S. Rin, H. Yao, M. S. Roberson, and P. J. S. Stork. 1995. A novel mitogen-activated protein kinase phosphatase. *J. Biol. Chem.* **270:** 14,587–14,596.
174. Miyashita, T. and J. C. Reed. 1995. Tumor suppressor p53 is a direct transcriptional activator of the human bax gene. *Cell* **80:** 293–299.
175. Miyashita, T., M. Harigai, M. Hanada, and J. C. Reed. 1994. Identification of a p53-dependent negative response element in the bcl-2 gene. *Cancer Res.* **54:** 3131–3135.
176. Miyashita, T., S. Krajewski, M. Krajewska, H. G. Wang, H. K. Lin, D. A. Liebermann, B. Hoffman, and J. C. Reed. 1994. Tumor suppressor p53 is a regulator of bcl-2 and bax gene expression *in vitro* and *in vivo*. *Oncogene* **9:** 1799–1805.
177. Momand, J., G. P. Zambetti, D. C. Olson, D. George, and A. J. Levine. 1992. The mdm-2 oncogene product forms a complex with the p53 protein and inhibits p53 mediated transactivation. *Cell* **69:** 1237–1245.
178. Montecucco, A., E. Savini, G. Biamonti, M. Stefanini, F. Focher, and G. Ciarrocchi. 1995. Late induction of human DNA ligase I after UV-C irradiation. *Nucleic Acids Res.* **23:** 962–966.
179. Morimoto, R. I. 1993. Cells in stress: transcriptional activation of the heat shock genes. *Science* **259:** 1409,1410.
180. Morimoto, R. I., K. D. Sarge, and K. Abravaya. 1992. Transcriptional regulation of heat shock genes. *J. Biol. Chem.* **267:** 21,987–21,990.
181. Morris, G. F., J. R. Bischoff, and M. B. Mathews. 1996. Transcriptional activation of the human proliferating-cell nuclear antigen promoter by p53. *Proc. Natl. Acad. Sci. USA* **93:** 895–899.
182. Nakajima, T., S. Kinoshita, T. Sasagawa, K. Sasaki, M. Naruto, T. Kishimoto, and S. Akira. 1993. Phosphorylation at threonine-235 by a *ras*-dependent mitogen-activated protein kinase cascade is essential for transcription factor NF-IL6. *Proc. Natl. Acad. Sci. USA* **90:** 2207–2211.
183. Nelson, W. G. and M. B. Kastan. 1994. DNA strand breaks: the DNA template alterations that trigger p53-dependent DNA damage response pathways. *Mol. Cell. Biol.* **14:** 1815–1823.
184. Nichols, A. F. and A. Sancar. 1992. Purification of PCNA as a nucleotide excision repair protein. *Nucleic Acids Res.* **20:** 2441–2446.
185. Niggli, H. J. 1993. Aphidicolin inhibits excision repair of UV-induced pyrimidine photodimers in low serum cultures of mitotic and mitomycin C-induced postmitotic human skin fibroblasts. *Mutat. Res.* **295:** 125–133.
186. Oberosler, P., P. Hloch, U. Ramsperger, and H. Stahl. 1993. p53-catalyzed annealing of complementary single-stranded nucleic acids. *EMBO J.* **12:** 2389–2396.

187. O'Connor, D. J., E. W. Lam, S. Griffin, S. Zhong, L. C. Leighton, S. A. Burbridge, and X. Lu. 1995. Physical and functional interactions between p53 and cell cycle co-operating E2F and DP1. *EMBO J.* **14:** 6184–6192.
188. Okamoto, K. and D. Beach. 1994. Cyclin G is a transcriptional target of the p53 tumor suppressor protein. *EMBO J.* **13:** 4816–4822.
189. Okinaga, S. and S. Shibahara. 1993. Identification of a nuclear protein that constitutively recognizes the sequence containing a heat-shock element. Its binding properties and possible function modulating heat-shock induction of the rat heme oxygenase gene. *Eur. J. Biochem.* **212:** 167–175.
190. Oliner, J. D., J. A. Pietenpol, S. Thiagalingam, J. Gyuris, K. W. Kinzler, and B. Vogelstein. 1993. Oncoprotein MDM2 conceals the activation domain of tumor suppressor p53. *Nature* **362:** 857–860.
191. Owen-Schaub, L. B., W. Zhang, J. C. Cusack, L. S. Angelo, S. M. Santee, T. Fujiwara, J. A. Roth, A. B. Deisseroth, W. W. Zhang, and E. Kruzel. 1995. Wild-type human p53 and a temperature-sensitive mutant induce Fas/APO-1 expression. *Mol. Cell. Biol.* **15:** 3032–3040.
192. Papathanasiou, M. A., N. C. Kerr, J. H. Robbins, O. W. McBride, I. Alamo, Jr., S. F. Barrett, I. D. Hickson, and A. J. Fornace, Jr. 1991. Induction by ionizing radiation of the *gadd45* gene in cultured human cells: lack of mediation by protein kinase C. *Mol. Cell. Biol.* **11:** 1009–1016.
193. Perry, M. E., J. Piette, J. A. Zawadzki, D. Harvey, and A. J. Levine. 1993. The *mdm-2* gene is induced in response to UV light in a p53-dependent manner. *Proc. Natl. Acad. Sci. USA* **90:** 11,623–11,627.
194. Peterson, S. R., A. Kurimasa, M. Oshimura, W. S. Dynan, E. M. Bradbury, and D. J. Chen. 1995. Loss of the catalytic subunit of the DNA-dependent protein kinase in DNA double-strand-break-repair mutant mammalian cells. *Proc. Natl. Acad. Sci. USA* **92:** 3171–3174.
195. Picksley, S. M. and D. P. Lane. 1993. The p53-mdm2 autoregulatory feedback loop: a paradigm for the regulation of growth control by p53. *Bioessays* **15:** 689,690.
196. Pietenpol, J. A., T. Tokino, S. Thiagalingam, W. S. El-Deiry, K. W. Kinzler, and B. Vogelstein. 1994. Sequence-specific transcriptional activation is essential for growth suppression by p53. *Proc. Natl. Acad. Sci. USA* **91:** 1998–2002.
197. Pombo, C. M., J. H. Kehrl, I. Sanchez, P. Katz, J. Avruch, L. Zon, J. R. Woodgett, T. Force, and J. M. Kyriakis. 1996. Activation of the SAPK pathway by the human STE20 homologue germinal centre kinase. *Nature* **377:** 750–754.
198. Prelich, G., M. Kostura, D. R. Marshak, M. B. Mathews, and B. Stillman. 1987. The cell-cycle regulated proliferating cell nuclear antigen is required for SV40 DNA replication *in vitro*. *Nature* **326:** 471–475.
199. Price, B. D. and S. J. Park. 1994. DNA damage increases the levels of MDM2 messenger RNA in wtp53 human cells. *Cancer Res.* **54:** 896–899.
200. Radler-Pohl, A., C. Sachsenmaier, S. Gebel, H.-P. Auer, J. T. Bruder, U. Rapp, P. Angel, H. J. Rahmsdorf, and P. Herrlich. 1993. UV-induced activation of AP-1 involves obligatory extranuclear steps including Raf-1 kinase. *EMBO J.* **12:** 1005–1012.
201. Raingeaud, J., S. Gupta, J. S. Rogers, M. Dickens, J. Han, R. J. Ulevitch, and R. J. Davis. 1995. Pro-inflammatory cytokines and environmental stress cause p38 mitogen-activated protein kinase activation by dual phosphorylation on tyrosine and threonine. *J. Biol. Chem.* **270:** 7420–7426.
202. Reed, M., B. Woelker, P. Wang, M. E. Anderson, and P. Tegtmeyer. 1995. The C-terminal domain of p53 recognizes DNA damage by ionizing radiation. *Proc. Natl. Acad. Sci. USA* **92:** 9455–9459.
203. Reich, N. C., M. Oren, and A. J. Levine. 1983. Two distinct mechanisms regulate the levels of a cellular tumor antigen, p53. *Mol. Cell. Biol.* **3:** 2143–2150.

204. Reich, N. C. and A. J. Levine. 1984. Growth regulation of a cellular tumour antigen, p53, in nontransformed cells. *Nature* **308:** 199–201.
205. Rolfe, M., P. Beer-Romero, S. Glass, J. Eckstein, I. Berdo, A. Theodoras, M. Pagano, and G. Draetta. 1995. Reconstitution of p53-ubiquitinylation reactions from purified components: the role of human ubiquitin-conjugating enzyme UBC4 and E6-associated protein (E6AP). *Proc. Natl. Acad. Sci. USA* **92:** 3264–3268.
206. Ronai, Z. A., E. Okin, and I. B. Weinstein. 1988. Ultraviolet light induces the expression of oncogenes in rat fibroblasts and human keratinocyte cells. *Oncogene* **2:** 201–204.
207. Rosen, C. F., D. Gajic, and D. J. Drucker. 1990. Ultraviolet radiation induction of ornithine decarboxylase in rat keratinocytes. *Cancer Res.* **50:** 2631–2635.
208. Roth, D. B., T. Lindahl, and M. Gellert. 1995. How to make ends meet. *Curr. Biol.* **5:** 496–499.
209. Rothem, N., J. H. Axelrod, and R. Miskin. 1987. Induction of urokinase-type plasminogen activator by UV light in human fetal fibroblasts is mediated through a UV-induced secreted protein. *Mol. Cell. Biol.* **7:** 622–631.
210. Rouse, J., P. Cohen, S. Trigon, M. Morange, A. Alonso-Liamazares, D. Zamanillo, T. Hunt, and A. R. Nebreda. 1994. A novel kinase cascade triggered by stress and heat shock that stimulates MAPKAP kinase-2 and phosphorylation of the small heat shock proteins. *Cell* **78:** 1027–1037.
211. Rudolph, N. S. and S. A. Latt. 1989. Flow cytometric analysis of X-ray sensitivity in ataxia telangiectasia. *Mutat. Res.* **211:** 31–41.
212. Russo, T., N. Zambrano, F. Esposito, R. Ammendola, F. Cimino, M. Fiscella, J. Jackman, P. M. O'Connor, C. W. Anderson, and E. Appella. 1995. A p53-independent pathway for activation of *WAF1/CIP1* expression following oxidative stress. *J. Biol. Chem.* **270:** 29,386–29,391.
213. Sachsenmaier, C., A. Radler-Pohl, R. Zinck, A. Nordheim, P. Herrlich, and H. J. Rahmsdorf. 1994. Involvement of growth factor receptors in the mammalian UVC response. *Cell* **78:** 963–972.
214. Sanchez, I., R. T. Hughes, B. J. Mayer, K. Yee, J. R. Woodgett, J. Avruch, J. M. Kyriakis, and L. I. Zon. 1994. Role of SAPK/ERK kinase-1 in the stress-activated pathway regulating transcription factor c-Jun. *Nature* **372:** 794–798.
215. Santhanam, U., A. Ray, and P. B. Sehgal. 1991. Repression of the interleukin 6 gene promoter by p53 and the retinoblastoma susceptibility gene product. *Proc. Natl. Acad. Sci. USA* **88:** 7605–7609.
216. Savitsky, K., A. Bar-Shira, S. Gilad, G. Rotman, Y. Ziv, M. Vanagaite, I. Pecker, M. Frydman, R. Harnik, S. R. Patanjali, A. Simmons, G. A. Clines, A. Sartiel, R. A. Gatti, L. Chessa, O. Sanal, M. F. Lavin, N. G. J. Jaspers, A. Malcom, R. Taylor, C. F. Arlett, T. Miki, S. M. Weissman, M. Lovett, F. S. Collins, and Y. Shiloh. 1995. A single ataxia telangiectasia gene with a product similar to PI-3 kinase. *Science* **268:** 1749–1753.
217. Savitsky, K., S. Sfez, D. Tagle, Y. Ziv, A. Sartiel, F. S. Collins, Y. Shiloh, and G. Rotman. 1995. The complete sequence of the coding region of the ATM gene reveals similarity to cell cycle regulators in different species. *Hum. Mol. Genet.* **4:** 2025–2032.
218. Schieven, G. L., J. M. Kirihara, L. K. Gilliland, F. M. Uckun, and J. A. Ledbetter. 1993. Ultraviolet radiation rapidly induces tyrosine phosphorylation and calcium signaling in lymphocytes. *Mol. Biol. Cell* **4:** 523–530.
219. Schieven, G. L., and J. A. Ledbetter. 1994. Activation of tyrosine kinase signal pathways by radiation and oxidative stress. *Trends Endocrinol. Metab.* **5:** 383–388.
220. Schieven, G. L., R. S. Mittler, S. G. Nadler, J. M. Kirihara, J. B. Bolen, S. B. Kanner, and J. A. Ledbetter. 1994. ZAP-70 tyrosine kinase, CD45, and T cell receptor involvement in UV- and H_2O_2-induced cell signal transduction. *J. Biol. Chem.* **269:** 20,718–20,726.

221. Schreck, R., P. Rieber, and P. A. Baeuerle. 1991. Reactive oxygen intermediates as apparently widely used messengers in the activation of the NF-κB transcription factor and HIV-1. *EMBO J.* **10:** 2247–2258.
222. Seger, R. and E. G. Krebs. 1995. The MAPK signaling cascade. *FASEB J.* **9:** 726–735.
223. Selivanova, G. and K. G. Wiman. 1995. p53: a cell cycle regulator activated by DNA damage. *Adv. Cancer Res.* **66:** 143–181.
224. Seth, A., E. Alvarez, S. Gupta, and R. J. Davis. 1991. A phosphorylation site located in the N-terminal domain of c-myc increases transactivation of gene expression. *J. Biol. Chem.* **266:** 23,521–23,524.
225. Seto, E., A. Usheva, G. P. Zambetti, J. Momand, N. Horikoshi, R. Weinmann, A. J. Levine, and T. Shenk. 1992. Wild-type p53 binds to the TATA-binding protein and represses transcription. *Proc. Natl. Acad. Sci. USA* **89:** 12,028–12,032.
226. Shafman, T. D., A. Saleem, J. Kyriakis, R. Weichselbaum, S. Kharbanda, and D. W. Kufe. 1995. Defective induction of stress-activated protein kinase activity in ataxia telangiectasia cells exposed to ionizing radiation. *Cancer Res.* **55:** 3242–3245.
227. Shaw, P., J. Freeman, R. Bovey, and R. Iggo. 1996. Regulation of specific DNA binding by p53: evidence for a role for *O*-glycosylation and charged residues at the carboxy-terminus. *Oncogene* **12:** 921–930.
228. Sherman, M. L., R. Datta, D. E. Hallahan, R. R. Weichselbaum, and D. W. Kufe. 1990. Ionizing radiation regulates expression of the c-jun protooncogene. *Proc. Natl. Acad. Sci. USA* **87:** 5663–5666.
229. Shiio, Y., T. Yamamoto, and N. Yamaguchi. 1992. Negative regulation of Rb expression by the p53 gene product. *Proc. Natl. Acad. Sci. USA* **89:** 5206–5210.
230. Shim, J., H. Lee, J. Park, H. Kim, and E.-J. Choi. 1996. A non-enzymatic p21 protein inhibitor of stress-activated protein kinases. *Nature* **381:** 805–807.
231. Shivji, K. K., M. K. Kenny, and R. D. Wood. 1992. Proliferating cell nuclear antigen is required for DNA excision repair. *Cell* **69:** 367–374.
232. Sipley, J. D., J. C. Menninger, K. O. Hartley, D. C. Ward, S. P. Jackson, and C. W. Anderson. 1995. Gene for the catalytic subunit of the human DNA-activated protein kinase maps to the site of the *XRCC7* gene on chromosome 8. *Proc. Natl. Acad. Sci. USA* **92:** 7515–7519.
233. Smith, M. D. and J. L. Rees. 1994. Wavelength-specific upregulation of keratin mRNA expression in response to ultraviolet radiation. *J. Invest. Dermatol.* **102:** 433–439.
234. Smith, M. L., I. Chen, Q. Zhan, I. Bae, C. Chen, T. Gilmer, M. B. Kastan, P. M. O'Connor, and A. J. Fornace, Jr. 1994. Interaction of the p53-regulated protein Gadd45 with proliferating cell nuclear antigen. *Science* **266:** 1376–1380.
235. Soussi, T., C. Caron de Fromentel, and P. May. 1990. Structural aspects of the p53 protein in relation to gene evolution. *Oncogene* **5:** 945–952.
236. Stein, B., H. J. Rahmsdorf, A. Steffen, M. Litfin, and P. Herrlich. 1989. UV-induced DNA damage is an intermediate step in UV-induced expression of human immunodeficiency virus type 1, collagenase, c-fos, and metallothionein. *Mol. Cell. Biol.* **9:** 5169–5181.
237. Sun, H., C. H. Charles, L. F. Lau, and N. K. Tonks. 1993. MKP-1 (3CH134), an immediate early gene product, is a dual specificity phosphatase that dephosphorylates MAP kinase *in vivo*. *Cell* **75:** 487–493.
238. Sun, H., N. K. Tonks, and D. Bar-Sagi. 1994. Inhibition of Ras-induced DNA synthesis by expression of the phosphatase MKP-1. *Science* **266:** 285–288.
239. Sun, X., H. Shimizu, and K.-I. Yamamoto. 1995. Identification of a novel p53 promoter element involved in genotoxic stress-inducible p53 gene expression. *Mol. Cell. Biol.* **15:** 4489–4496.
240. Taccioli, G. E. and F. W. Alt. 1995. Potential targets for autosomal SCID mutations. *Curr. Opinion Immunol.* **7:** 436–440.

241. Takenaka, I., F. Morin, B. S. Seizinger, and N. Kley. 1995. Regulation of the sequence-specific DNA binding function of p53 by protein kinase C and protein phosphatases. *J. Biol. Chem.* **270:** 5405–5411.
242. Taylor, G. A., M. J. Thompson, W. S. Lai, and P. J. Blackshear. 1995. Phosphorylation of tristetraprolin, a potential zinc finger transcription factor, by mitogen stimulation in intact cells and by mitogen-activated protein kinase *in vivo*. *J. Biol. Chem* **270:** 13,341–13,347.
243. Thut, C. J., J. L. Chen, R. Klemm, and R. Tjian. 1995. p53 transcriptional activation mediated by coactivators $TAF_{II}{}^{40}$ and $TAF_{II}{}^{60}$. *Science* **267:** 100–104.
244. Tokino, T., S. Thiagalingam, W. S. El-Deiry, T. Waldman, K. W. Kinzler, and B. Vogelstein. 1994. p53 tagged sites from human genomic DNA. *Human Mol. Genet.* **3:** 1537–1542.
245. Treisman, R. 1995. Journey to the surface of the cell: Fos regulation and the SRE. *EMBO J.* **14:** 4905–4913.
246. Truant, R., H. Xiao, C. J. Ingles, and J. Greenblatt. 1993. Direct interaction between the transcriptional activation domain of human p53 and the TATA box-binding protein. *J. Biol. Chem.* **268:** 2284–2287.
247. Tsai, H. L., G. H. Kou, S. C. Chen, C. W. Wu, and Y. S. Lin. 1996. Human cytomegalovirus immediate-early protein IE2 tethers a transcriptional repressor domain to p53. *J. Biol. Chem.* **271:** 3534–3540.
248. Uchiumi, T., K. Kohno, H. Tanimura, K.-I. Matwuo, S. Sato, Y. Uchida, and M. Kuwano. 1993. Enhanced expression of the human multidrug resistance 1 gene in response to UV light irradiation. *Cell Growth Differ.* **4:** 147–157.
249. Ueba, T., T. Nosaka, J. A. Takahashi, F. Shibata, R. Z. Florkiewicz, B. Vogelstein, Y. Oda, H. Kikuchi, and M. Hatanaka. 1994. Transcriptional regulation of basic fibroblast growth factor gene by p53 in human glioblastoma and hepatocellular carcinoma cells. *Proc. Natl. Acad. Sci. USA* **91:** 9009–9013.
250. Ullrich, S. J., K. Sakaguchi, S. P. Lees-Miller, M. Fiscella, W. E. Mercer, C. W. Anderson, and E. Appella. 1993. Phosphorylation at serine 15 and 392 in mutant p53s from human tumors is altered compared to wild-type p53. *Proc. Natl. Acad. Sci. USA* **90:** 5954–5958.
251. Van Dam, H., M. Duyndam, R. Rottier, A. Bosch, L. de Vries-Smits, P. Herrlich, A. Zantema, P. Angel, and A. J. Van der Eb. 1993. Heterodimer formation of c-Jun and ATF-2 is responsible for induction of c-jun by the 243 amino acid adenovirus E1A protein. *EMBO J.* **12:** 479–487.
252. Van Dam, H., D. Wilhelm, I. Herr, A. Steffen, P. Herrlich, and P. Angel. 1995. ATF-2 is preferentially activated by stress-activated protein kinases to mediate c-jun induction response to genotoxic stress. *EMBO J.* **14:** 1798–1811.
253. Verheij, M., R. Bose, X. H. Lin, B. Yao, W. D. Jarvis, S. Grant, M. J. Birrer, E. Szabo, L. I. Zon, J. M. Kyriakis, A. Haimovitz-Friedman, Z. Fuks, and R. N. Kolesnick. 1996. Requirement for ceramide-initiated SAPK/JNK signalling in stress-induced apoptosis. *Nature* **380:** 75–79.
254. Verma, I. M., J. K. Stevenson, E. M. Schwarz, D. Van Antwerp, and S. Miyamoto. 1995. Rel/NF-κB/IκB family: intimate tales of association and disassociation. *Genes Dev.* **9:** 2723–2735.
255. Wadman, I. A., H. L. Hsu, M. H. Cobb, and R. Baer. 1994. The MAP kinase phosphorylation site of TAL1 occurs within a transcriptional activation domain. *Oncogene* **9:** 3713–3716.
256. Waga, S., G. J. Hannon, D. Beach, and B. Stillman. 1994. The p21 inhibitor of cyclin-dependent kinases controls DNA replication by interaction with PCNA. *Nature* **369:** 574–578.
257. Wang, X. and D. Ron. 1996. Stress-induced phosphorylation and activation of the transcription factor CHOP (GADD153) by p38 MAP-kinase. *Science* **272:** 1347–1349.
258. Wang, Y. and W. Eckhart. 1992. Phosphorylation sites in the amino-terminal region of mouse p53. *Proc. Natl. Acad. Sci. USA* **89:** 4231–4235.

259. Wang, Y. and C. Prives. 1995. Increased and altered DNA binding of human p53 by S and G2/M but not G1 cyclin-dependent kinases. *Nature* **376:** 88–91.
260. Wang, Y., M. Reed, Y. Wang, G. Mayr, J. E. Stenger, M. E. Anderson, J. F. Schwedes, and P. Tegtmeyer. 1994. p53 domains: structure, oligomerization, and transformation. *Mol. Cell. Biol.* **14:** 5182–5191.
261. Wang, Y., J. F. Schwedes, D. Parks, K. Mann, and P. Tegtmeyer. 1995. Interaction of p53 with its consensus DNA-binding site. *Mol. Cell. Biol.* **15:** 2157–2165.
262. Wang, Z.-Q., B. Auer, L. Stingl, H. Berghammer, D. Haidacher, M. Schweiger, and E. F. Wagner. 1995. Mice lacking ADPRT and poly(ADP-ribosyl)ation develop normally but are susceptible to skin disease. *Genes Dev.* **9:** 509–520.
263. Ward, Y., S. Gupta, P. Jensen, M. Wartman, R. J. Davis, and K. Kelly. 1994. Control of MAP kinase activation by the mitogen-induced threonine/tyrosine phosphatase PAC1. *Nature* **367:** 651–654.
264. Warmuth, I., M. S. Harth, N. Matsui, N. Wang, and V. A. DeLeo. 1994. Ultraviolet radiation induces phosphorylation of the epidermal growth factor receptor. *Cancer Res.* **54:** 374–376.
265. Weintraub, H., S. Hauschka, and S. J. Tapscott. 1991. The MCK enhancer contains a p53 responsive element. *Proc. Natl. Acad. Sci. USA* **88:** 4570–4571.
266. Westwick, J. K., A. E. Bielawska, G. Dbaibo, Y. A. Hannun, and D. A. Brenner. 1995. Ceramide activates the stress-activated protein kinases. *J. Biol. Chem.* **270:** 22,689–22,692.
267. Whitmarsh, A. J., P. Shore, A. D. Sharrocks, and R. J. Davis. 1995. Integration of MAP kinase signal transduction pathways at the serum response element. *Science* **269:** 403–407.
268. Wu, X., J. H. Bayle, D. Olson, and A. J. Levine. 1993. The p53-mdm-2 autoregulation feedback loop. *Genes Dev.* **7:** 1126–1132.
269. Xia, Z., M. Dickens, J. Raingeaud, R. J. Davis, and M. E. Greenberg. 1995. Opposing effects of ERK and JNK on apoptosis. *Science* **270:** 1326–1331.
270. Xiong, Y., G. J. Hannon, H. Zhang, D. Casso, R. Kobayashi, and D. Beach. 1993. p21 is a universal inhibitor of cyclin kinases. *Nature* **366:** 701–704.
271. Yamaguchi, M., Y. Hayashi, S. Matsuoka, T. Takahashi, and A. Matsukage. 1994. Differential effect of p53 on the promoters of mouse DNA polymerase [beta] gene and proliferating-cell-nuclear antigen gene. *Eur. J. Biochem.* **221:** 227–237.
272. Yan, M., T. Dai, J. C. Deak, J. M. Kyriakis, L. I. Zon, J. R. Woodgett, and D. J. Templeton. 1994. Activation of stress-activated protein kinase by MEKK1 phosphorylation of its activator SEK1. *Nature* **372:** 798–800.
273. Yao, B., Y. Zhang, S. Delikat, S. Mathias, S. Basu, and R. Kolesnick. 1995. Phosphorylation of Raf by ceramide-activated protein kinases. *Nature* **378:** 307–310.
274. Zakian, V. A. 1995. *ATM*-related genes: what do they tell us about functions of the human gene? *Cell* **82:** 685–687.
275. Zampetti-Bosseler, F., and D. Scott. 1981. Cell death, chromosome damage and mitotic delay in normal human, ataxia telangiectasia and retinoblastoma fibroblasts after x-irradiation. *Int. J. Radiat. Biol. Relat. Stud. Phys. Chem. Med.* **39:** 547–558.
276. Zastawny, R. L., R. Salvino, J. Chen, S. Benchimol, and V. Ling. 1993. The core promoter region of the P-glycoprotein gene is sufficient to confer differential responsiveness to wild-type and mutant p53. *Oncogene* **8:** 1529–1535.
277. Zauberman, A., A. Lupo, and M. Oren. 1995. Identification of p53 target genes through immune selection of genomic DNA: the cyclin G gene contains two distinct p53 binding sites. *Oncogene* **10:** 2361–2366.
278. Zervos, A. S., L. Faccio, J. P. Gatto, J. M. Kyriakis, and R. Brent. 1995. Mxi2, a mitogen-activated protein kinase that recognizes and phosphorylates Max protein. *Proc. Natl. Acad. Sci. USA* **92:** 10,531–10,534.
279. Zhan, Q., I. Bae, M. B. Kastan, and A. J. Fornace, Jr. 1993. The p53-dependent gamma-ray response of GADD45. *Cancer Res.* **54:** 2755–2760.

280. Zhan, Q., S. Fan, I. Bae, C. Guillouf, D. A. Liebermann, P. M. O'Connor, and A. J. Fornace, Jr. 1994. Induction of bax by genotoxic stress in human cells correlates with normal p53 status and apoptosis. *Oncogene* **9:** 3743–3751.
281. Zhan, Q., S. Fan, M. L. Smith, I. Bae, K. Yu, I. Alamo, P. M. O'Connor, and A. J. Fornace, Jr. 1996. Abrogation of p53 function affects gadd gene responses to DNA-base damaging agents and starvation. *DNA and Cell Biol.* **15:** 805–815.
282. Zhang, S., J. Han, M. A. Sells, J. Chernoff, U. G. Knaus, R. J. Ulevitch, and G. M. Bokoch. 1995. Rho family GTPases regulate p38 mitogen-activated protein kinase through the downstream mediator Pak1. *J. Biol. Chem.* **270:** 23,934–23,936.

16

Mechanisms for DNA Double-Strand Break Repair in Eukaryotes

W. Kimryn Rathmell and Gilbert Chu

1. INTRODUCTION

Of all the forms of DNA damage, double-strand breaks (DSBs) are potentially the most problematic, since they may lead to broken or rearranged chromosomes, cell death, or cancer. Eukaryotic cells have evolved specific mechanisms for repairing DSBs by either homologous recombination (HR) or nonhomologous recombination (NHR) pathways. This chapter will summarize recent progress in understanding these pathways in eukaryotes. Of particular interest is the surprising discovery that the nonhomologous pathway shares at least four gene products with V(D)J recombination, a site-specific recombination pathway that generates immunological diversity.

2. SOURCES OF DSBs

DSBs are caused by a variety of exogenous agents. Ionizing radiation can produce DSBs directly. Alternatively, ionizing radiation can indirectly produce DSBs via oxidative free radicals, which can cause base damage or disruption of a phosphodiester bond or ribose ring leading to a DNA strand break (Chapter 5). A DSB then results from two closely spaced single-strand breaks. The antitumor drug bleomycin produces oxidative free radicals and causes a similar spectrum of DNA damage. The antitumor drug etoposide interferes with the activity of topoisomerase II to create protein-bridged DSBs. A DSB can also arise during the repair of other lesions, such as interstrand crosslinks.

DSBs are also caused by endogenous cellular processes. Oxidative metabolism generates oxidative free radicals, which cause strand breaks as described above. V(D)J recombination generates DSBs by site-specific cleavage of DNA during rearrangement of B-cell immunoglobulin and T-cell receptor genes.

3. REPAIR OF DSBs BY HOMOLOGOUS RECOMBINATION (HR)

3.1. HR in Yeast

HR refers to rearrangement or repair of DNA that is directed by homologous DNA sequences. This process will be discussed only briefly, since it is reviewed in Chapters

From: DNA Damage and Repair, Vol. 2: DNA Repair in Higher Eukaryotes
Edited by: J. A. Nickoloff and M. F. Hoekstra © Humana Press Inc., Totowa, NJ

16 and 26, vol. 1. HR has been studied extensively in the yeast *Saccharomyces cerevisiae* in the context of meiotic recombination, mitotic recombination, and mating type switching *(50)*. In yeast cells, HR occurs efficiently when homology in the recombining DNA molecules extends over several hundred base pairs. This process is stimulated to greater efficiency by the presence of DSBs. HR repairs DSBs conservatively, with no nucleotide loss at the break site.

The study of HR in yeast has been greatly facilitated by the identification of mutants. Screens for X-ray sensitivity have repeatedly yielded mutants defective in HR. These mutants fall into the *RAD52* epistasis group *(50)*, so named because *RAD52* is the gene that confers the greatest X-ray sensitivity when mutated or deleted. Mutations in any of the genes in the *RAD52* epistasis group lead to defects in the HR pathways of meiotic recombination, mitotic recombination, and mating type switching. Additionally, these mutations cause reduced HR in extra- and intrachromosomal assays and decreased double-strand break repair (DSBR). Thus, the repair of DSBs produced by ionizing radiation occurs predominantly by HR in yeast.

Pathways also exist in yeast to resolve DSBs by mechanisms not involving HR. However, these homology-independent pathways make only minor contributions to radioresistance in the presence of an intact *RAD52* pathway. These mechanisms have been studied in greater detail in a *rad52* mutant background *(35)*.

3.2. *HR in Amphibians*

Studies of HR in the frog *Xenopus laevis* have also yielded valuable insights into the process of DSBR (Chapter 26, vol. 1). Experiments in intact cells with injected DNA substrates have demonstrated that HR does occur and that HR activity is stimulated by the presence of DSBs *(11)*. However, some homologous substrates undergo NHR instead of HR *(31)*. Therefore, it appears that although the repair of DSBs can occur via HR, nonhomologous mechanisms may be important for the resolution of DSBs in vertebrate cells.

3.3. *HR in Mammals*

HR has also been studied in mammalian systems *(6)*. Experiments in intact cells have observed HR with a number of different substrate systems. HR occurs efficiently between extrachromosomal substrates, between intrachromosomal substrates, and between extra- and intrachromosomal substrates. As in the yeast and amphibian systems, HR is significantly stimulated by DSBs. HR between extra- and intrachromosomal DNA has been exploited in embryonic stem cells for gene targeting, which is used to generate whole animals containing deletions in the targeted genes.

Screens for X-ray sensitivity have identified several mutant cell lines defective for DSBR. The mammalian DSBR mutants xrs1 and xrs7 were not impaired in HR in an assay for intact cells *(26)*, whereas the mutant xrs5 showed no defect in extracts and a sixfold reduction in HR in intact cells *(46)*, a small effect compared to yeast *rad52* mutants. Thus, HR occurs in mammalian cells, and HR plays a role in DSBR. However, nonhomologous mechanisms for resolving strand breaks appear to be more important.

4. REPAIR OF DSBs BY NONHOMOLOGOUS RECOMBINATION (NHR)

4.1. NHR in Amphibians

NHR has been observed as a mode of DSBR in *Xenopus* oocytes (Chapter 26, vol. 1). Grzesiuk and Carroll injected linearized plasmid DNA substrates into *Xenopus* oocytes and observed a low level of plasmid recircularization *(24)*. Sequence analysis suggested that the end joining process was supported by homologous sequences of 1–10 bp, less than the homology required for bona fide HR. In addition, the homologous sequences had to be very close (within 20 bp) to the original DNA ends. Figure 1A,B shows examples of how end joining might utilize such short regions of microhomology.

Jeong-yu and Carroll also studied DSBR in *Xenopus* oocytes by introducing a DSB into one of two homologous tandem repeats *(31)*. Although this substrate could potentially be repaired by HR, the break was repaired predominantly by an NHR pathway. The repair process was not conservative, and nucleotide loss of up to 20 bp on either side of the break point was observed. Six repair joints were sequenced, and all of them appeared to utilize short segments (2–4 bp) of microhomology from the two sides of the break (Fig. 1A,B). These results demonstrate that DSBR in *Xenopus* cells may occur via NHR preferentially over HR.

Experiments in *Xenopus* extracts have yielded somewhat different results from those in intact cells. DSBR was analyzed by circularization of linear DNA substrates with different types of DNA ends *(54,55,71)*. In contrast to results in intact cells, no nucleotide loss was observed from either end beyond losses from protruding single strands (PSSs). Regions of microhomology were utilized at the repair joint, but only if the PSSs were of the same polarity (5'-PSS/5'-PSS or 3'-PSS/3'-PSS) and only if the homologous base pairs were available within the protruding single-stranded region (Fig. 1A). End joining could also occur without the utilization of microhomology. Blunt ends were aligned directly (Fig. 1C) as ligatable double strands. Substrates with differing polarities (blunt/5'-PSS, blunt/3'-PSS, 5'-PSS/3'-PSS), were joined by the alignment of single strands, gap filling by DNA polymerase, and ligation (Fig. 1D). Gap filling could occur before ligation, since PSSs containing nonligatable dideoxynucleotides were filled in by DNA polymerase activity *(71)*. This result also suggested the presence of an alignment factor to stabilize the DNA substrate for the polymerase.

4.2. NHR in Mammals

Roth et al. transfected linearized SV40 DNA into mammalian cells and recovered recircularized virion DNA to measure DSBR in mammalian cells *(62,63)*. End joining occurred independently of any homology (Fig. 1C,D), or utilized microhomology either from the protruding single strands or from the duplex DNA near the DNA ends (Fig. 1A,B). When noncohesive ends were studied, deletions of up to 20 bp usually occurred on either side of the repair joint and usually involved microhomology. Roth et al. proposed that these end joining events were directed by base pairing of homologous nucleotides from the two ends. Interestingly, addition of oligonucleotides was occasionally seen *(61)*. These end joining events could also have been directed by homology pairing with bases present in the added oligonucleotides. The added oligonucleotides appear to be derived from a number of sources, including other DNA molecules in the cell, copying of sequences near the end of the DSB by slipped

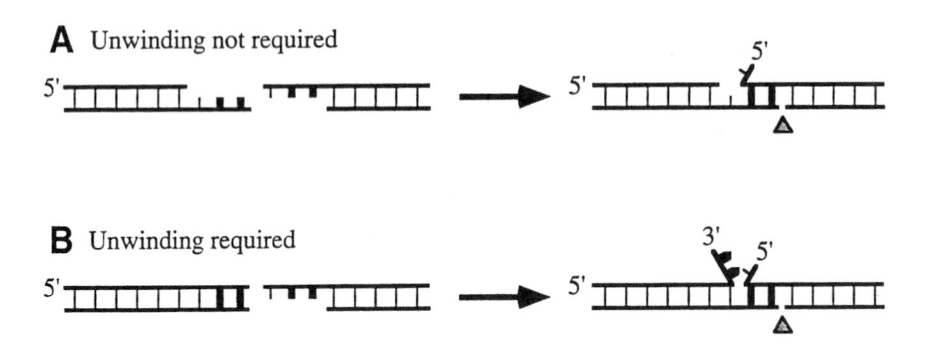

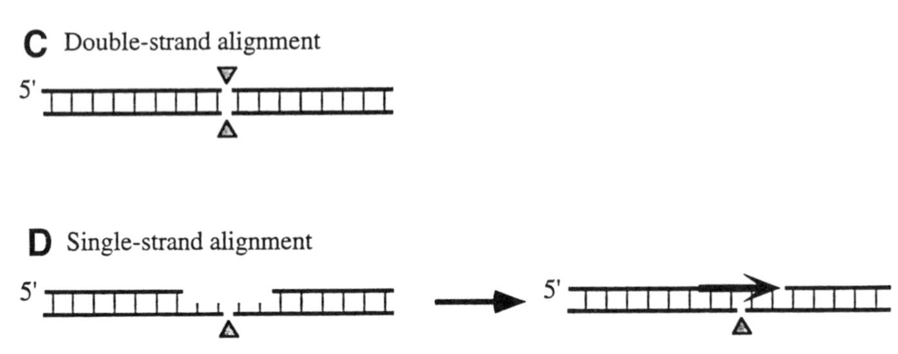

Fig. 1. End joining reactions during DSBR. End joining during DSBR can either be dependent or independent of regions of microhomology near the DNA ends. Shaded triangles denote sites of potential ligation. **(A)** End joining dependent on microhomology; unwinding not required. Microhomologies present within the protruding single strands are hypothesized to direct the alignment of the DNA ends by base pairing. The end joining involves processing of any unpaired ends (in this example, a 5'-overhang), filling in gaps, and ligation. The processing of unpaired ends could occur by exonuclease or flap endonuclease activity. The formation of the joint does not require unwinding of the duplex DNA, and only a few nucleotides are lost from the protruding single strands. **(B)** End joining dependent on microhomology; unwinding required. Microhomologies within one or both of the DNA duplexes are hypothesized to lead to base pairing after unwinding of duplex DNA. The formation of the joint must involve nucleotide loss and requires the processing of unpaired ends. **(C)** End joining independent of microhomology, with double-strand alignment. Blunt DNA ends are aligned on both strands, which can then be ligated directly. No homology is required for the reaction. Free DNA ends may be made into blunt ends by exonucleases and/or filling in by DNA polymerases. **(D)** End joining independent of microhomology, with single-strand alignment. DNA ends are hypothesized to be aligned along only one of the strands. The ends might then be joined either by a single-stranded ligase followed by gap filling, as postulated by Roth and Wilson *(63)*, or by gap filling across the aligned single strand followed by ligation of the resulting nicked DNA, as postulated by Pfeiffer and Vielmetter *(55)*.

mispairing and repair synthesis, and misincorporation of A residues during repair of a 5' single-stranded extension.

End joining has also been analyzed in extracts of mammalian cells. In one series of studies, pUC18 plasmid DNA was linearized by restriction cleavage in the polylinker region, disrupting the *lacZ* gene *(19,70)*. DSBR restored functional *lacZ* when the plasmid was recircularized by joining the cohesive restriction-digested ends (Fig. 1C). Misrejoined molecules with *lacZ* disruptions were observed at a very low frequency in wild-type cells. The misrejoined molecules usually contained deletions, but occasionally included additions. The mechanism for misrejoining involved the utilization of homology near the DNA ends (Fig. 1A,B).

Separate studies analyzed the products of DNA end joining using linearized pUC18 plasmid DNA incubated with human cell extracts. Rather than scoring only for recircularized DNA molecules, the total DNA from the joining reaction was recovered and analyzed *(14,48)*. In contrast to the previous studies, recircularization of the plasmid DNA was observed rarely, if at all. The vast majority of the recovered DNA was joined in head-to-head (H:H) or tail-to-tail (T:T) orientations. In addition, sequence analysis of the junctions demonstrated that in most cases, a uniform number of nucleotides was lost depending on the enzyme used for the original digestion, consistent with a joining mechanism dependent on short homologous sequences near the DNA ends (Fig. 1A,B). Partial purification of the activity yielded a single-strand ligase activity independent of the known DNA ligases (NHR-ligase) and a homologous pairing activity (HPP-1) *(14)*.

In conclusion, experiments in intact amphibian and mammalian cells appear to be consistent with each other. Thus, these studies may reflect true DSBR. However, the results from different experiments utilizing cell extracts were inconsistent in several important ways. Some studies detected plasmid recircularization, whereas others demonstrated mostly concatamer formation with H:H and T:T joining. In addition, experiments in cell extracts demonstrated nucleotide loss very rarely, even though deletions of up to 20 bp were observed routinely in experiments in intact cells. These discrepancies may be the result of differences in substrate or assay conditions, but suggest that the end joining activity observed in cell extracts might not reflect the DSBR that occurs in intact cells.

4.3. V(D)J Recombination

The germline loci of the immunoglobulin and T cell receptor genes contain multiple V (variable), J (junctional), and in some cases D (diversity) elements, arranged in tandem array. During the development of B- and T-cells, these elements generate immunological diversity by undergoing V(D)J recombination, which involves the rearrangement of V, D, and J elements to form functional genes *(39)*. Rearrangements are targeted to recombination signal sequences, which consist of conserved heptamer and nonamer sequences separated by a nonconserved spacer of 12 or 23 bp. Cleavage occurs at the junction of a recombination signal sequences and a V, D, or J coding element. Independent coding elements are then fused at coding joints to form the rearranged genes capable of encoding functional proteins.

V(D)J recombination is initiated by the combined action of the recombination activating genes, *RAG1* and *RAG2*, which are expressed specifically in lymphoid cells. Cotransfection of *RAG1* and *RAG2* confers V(D)J recombination activity to non-lymphoid cells *(49,65)*. Thus, once V(D)J recombination is initiated by RAG1 and RAG2, general factors present in all cells complete the recombination reaction.

The coding ends are processed extensively prior to ligation to generate additional diversity. Individual nucleotides can be added to the coding end in a nontemplated manner by terminal deoxynucleotidyl transferase (TdT), an enzyme expressed specifically in lymphoid cells *(74)*. Oligonucleotides may be added at the joint site even in the absence of TdT, perhaps by oligonucleotide capture *(22)*. A few nucleotides palindromic to one or both of the coding ends may be added by a process called P-nucleotide addition *(36)*. Nucleotides can also be lost from the coding ends.

The joining reaction in V(D)J recombination does not require extensive homology in the recombining DNA. Nevertheless, coding joints are often formed by utilization of microhomology near the DNA termini. As in the case of general DSBR, coding joint formation preferentially utilizes short stretches of homology if they are present *(22,34)*. Coding joint formation can also occur independently of microhomology, since extrachromosomal substrates with homopolymeric coding sequences incompatible with homologous pairing will still undergo V(D)J recombination *(8)*.

In contrast to coding joint formation, recombination signal sequence elements are fused precisely to form signal joints with exact conservation of the DNA ends. Signal joint formation apparently occurs by a simple blunt end joining reaction, and is thus completely independent of homology.

5. X-RAY-SENSITIVE MAMMALIAN CELLS

5.1. *Ataxia Telangiectasia (AT)*

AT is an autosomal-recessive disorder affecting up to 1 in 40,000 live births (Chapter 19). AT is characterized by cerebellar degeneration, abnormal proliferation of small blood vessels, and severe predisposition to cancers, particularly lymphomas. Cells derived from AT patients are hypersensitive to ionizing radiation. AT cells are not defective in DSBR, and their hypersensitivity is the result of a defect in cell cycle arrest after exposure to ionizing radiation *(32)*. The gene mutated in AT *(ATM)* was recently cloned and found to share homology to the phosphatidylinositol 3-kinase superfamily *(64)*. Interestingly, despite their sensitivity to ionizing radiation, AT cells display increased intrachromosomal HR *(43)*, consistent with the idea that HR is not critical for radiation resistance in mammalian cells.

5.2. *The SCID Mouse*

The severe combined immune deficient (SCID) mouse lacks mature B- and T-cells and is highly susceptible to the development of T-cell lymphomas *(7)*. The failure to develop a competent immune system is due to a defect in V(D)J recombination. SCID cells form signal joints at a normal rate, but cannot form coding joints during V(D)J recombination *(41)*. The coding ends accumulate in a hairpin conformation, suggesting that hairpins are a DNA intermediate during coding joint formation and that the SCID mutation leads to defective processing of the hairpins. SCID cells were also found to have a general defect in DSBR affecting all tissues and causing hypersensitivity to ionizing radiation *(3,18,29)*. This was the first suggestion that V(D)J recombination and DSBR might utilize common factors present in all tissues.

5.3. X-Ray Sensitive Hamster Mutants

To search for the genetic basis of resistance to ionizing radiation, a number of easily cultured hamster cell lines have been developed by mutagenesis and screening for hypersensitivity to ionizing radiation *(30)*. (*See* Chapter 17 for a more complete discussion.) Cell fusion experiments show that these cell lines fall into at least 10 genetic complementation groups *(13)*. The corresponding genes are designated *XRCC1–XRCC10 (72)*. Three of the X-ray-sensitive complementation groups (*XRCC4*, *XRCC5*, and *XRCC7*) are defective in DSBR. Cell lines from these complementation groups are extremely sensitive to ionizing radiation, antitumor drugs, such as bleomycin, and other sources of DSBs. These cell lines also show increased incidence of chromosomal aberrations like translocations and inversions after exposure to X-rays.

Homologous recombination has been analyzed in *XRCC5* cell lines. Interestingly, the cells appear to be at most mildly defective in HR, if at all *(26)*. However, cell lines from all three groups are strikingly defective for V(D)J recombination *(37,53,69)*. *XRCC4* and *XRCC5* cells are defective in the formation of both coding and signal joints. Joints were formed at a greatly reduced frequency, and the rare residual joints contained large deletions at the site of the joint. *XRCC7* cells (which include SCID mouse cells) are defective only in the formation of coding joints. Again, the overall frequency was reduced, and large deletions were found in the remaining joints. Signal joint formation in *XRCC7* cells occurred at a normal frequency, but with reduced precision.

Based on the evidence that X-ray-sensitive cells defective in the repair of DSBs are also defective in V(D)J recombination, the pathways of DSBR and V(D)J appear to overlap extensively and utilize common factors present in all cell types. Furthermore, the defects in DSBR do not lead to a significant defect in HR, but do lead to a defect in the NHR pathway of V(D)J recombination. Thus, DSBR in mammalian cells appears to depend critically on NHR and not HR.

6. DSBR PROTEINS IN MAMMALIAN CELLS

6.1. RAD52 *Epistasis Group Homologs*

The genes of the yeast *RAD52* epistasis group are essential for processes involving HR, including DSBR. The mechanisms of action of *RAD52* and *RAD51* are detailed in Chapter 16, vol. 1. Recently, human *RAD52* and *RAD51* genes were isolated by virtue of their homology to the corresponding yeast genes *(47,66)*. When human *RAD52* was overexpressed in wild type monkey cells, it conferred increased resistance to ionizing radiation and increased levels of homologous recombination *(52)*. Human *RAD51* shares sequence homology with *Escherichia coli recA* (Chapter 8, vol. 1) and also has similar DNA binding and helicase activities *(2)*. These human genes may be involved in a pathway for homologous recombination that contributes to the repair of DNA DSBs.

6.2. The Ku Autoantigen

Ku is the target antigen in patients with several autoimmune diseases, including scleroderma-polymyositis overlap syndrome, systemic lupus erythematosus, Graves' disease, and Sjögren's syndrome *(59)*. Ku is a highly stable heterodimer of 70- and 86-kDa polypeptides (Ku70 and Ku86) *(44)*. The cDNAs for both subunits have been

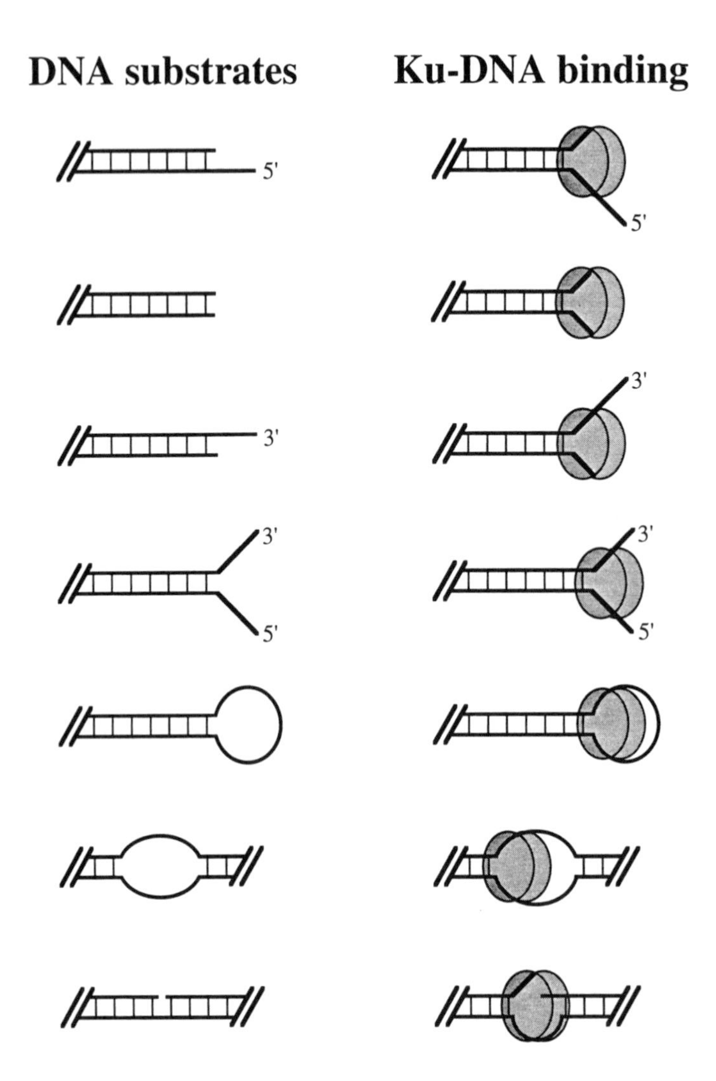

Fig. 2. DNA substrates for Ku binding. Ku is capable of binding to several different DNA substrates, including DNA ends with blunt ends and 5'- or 3'-protruding ends. It will also bind to DNA ending in unpaired strands, DNA ending in stem-loop structures, DNA containing unpaired bubbles, and DNA containing nicks or gaps. Ku will not bind to obligate single-stranded DNA ends, such as those present in poly(dA) or poly(dT). The substrate specificity of Ku is consistent with a model in which Ku interacts with DNA via a transition fork between duplex DNA and two single strands.

cloned *(45,60,75)*. Ku is localized to the nucleus and is moderately abundant, with about 200,000–400,000 molecules/cell.

Ku has an interesting specificity for its DNA substrates. It does not bind to single-stranded DNA ends, but binds tightly to double-stranded ends, having equal affinity for 5'-protruding, 3'-protruding, and blunt ends. Ku also binds to DNA nicks, gaps, bubbles, and DNA ending in stem-loop structures *(4,16)*. This spectrum of DNA binding activity can be explained by a model in which Ku recognizes the transition between a region of double-stranded DNA and two nonannealed single strands (Fig. 2). Once

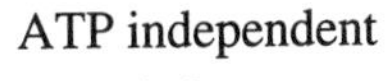

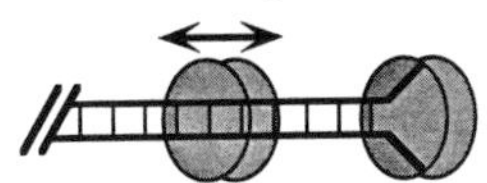
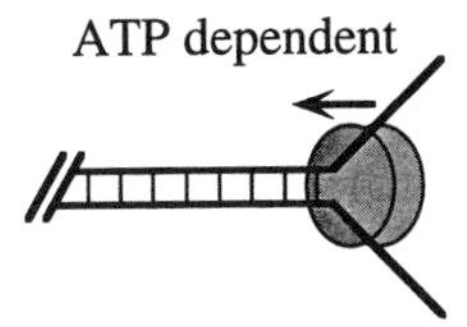

Fig. 3. Ku movement on DNA. Ku is able to move on DNA by an ATP-independent translocation mechanism that requires Ku to be loaded onto the DNA molecule via one of the DNA binding substrates of Ku, such as a DNA end, as shown in the figure. The translocation is ATP-independent. Alternatively, Ku has an ATP-dependent helicase activity, presumably activated by DNA-PK-mediated autophosphorylation.

Ku binds to DNA ends, it is capable of translocating along the DNA, so that three or more molecules of Ku can be bound to a single DNA fragment *(51)* (Fig. 3).

In an independent line of investigation, a DNA end-binding (DEB) factor was detected in cells by an electrophoretic mobility shift assay (EMSA) *(56)*. DEB factor was present in cells from yeast, and from a number of tissues from humans and rodents (*see* Table 1). It was expressed normally in cell lines from X-ray-sensitive complementation groups 1, 4, and 7, in AT cells, in Nijmegan breakage syndrome cells, and in a mutant X-ray sensitive human lymphoblast cell line. DEB activity was notably absent in three cell lines from complementation group 5, XR-V15B, XR-V9B, and xrs5 *(56)*. Azacytidine will induce xrs5 cells to revert to X-ray resistance at a frequency of about 1%. Twenty-three independent clones selected for X-ray resistance following azacytidine treatment all showed restored DEB activity, further supporting a role for DEB factor in X-ray resistance.

DEB factor and Ku share many similarities, including nuclear localization and abundance in the cell *(57)*. DEB factor and Ku also bind to the same DNA substrates, including double-stranded DNA ends with 5'-, 3'-, or blunt ends, and M13 single-stranded circular virion DNA (presumably via binding to stem-loop and bubble structures potentially formed at the M13 origin). Neither binds to the single-stranded DNA ends in poly(dA) or poly(dT). Three or more Ku molecules can load and bind to a single DNA fragment, producing a ladder of mobility shifts similar to that seen with DEB factor.

DEB factor and Ku are also antigenically similar *(20,57)*. The DEB protein–DNA complex observed by EMSA was recognized by Ku antisera, so that antibodies bound to the protein-DNA complex to produce a larger supershifted complex. When these same Ku antisera were used to probe immunoblots, a 70-kDa peptide in hamster extracts cofractionated with DEB activity on heparin agarose and thus was identified as hamster Ku70 *(57)*. Hamster Ku70 was absent or only barely detectable in *XRCC5* hamster cell lines, but present in *XRCC4* and *XRCC7* cell lines, and in wild-type cell lines *(57)*.

Despite these observations, *Ku70* does not define the genetic defect of *XRCC5* cells. Chromosome mapping studies assigned the *Ku70* gene to human chromosome 22q13 and the *Ku86* gene to chromosome 2q33-35 *(9)*. X-ray resistance in *XRCC5* cells was partially restored by human chromosome region 2q34 *(12,25)*. This raised the possibility that the primary defect in *XRCC5* cells resides in the *Ku86* gene.

Table 1
DEB Activity in Cell Lines from Yeast, Human, Mouse, and Hamster[a]

Cell line	Cell type	X-ray resistance	Group	DEB activity
Yeast				
DBY746	*S. cerevisiae*	+	wt	2.6
hdf1	*S. cerevisiae*	+	*Ku70*	<0.02
rad52	*S. cerevisiae*	–	*RAD52*	2.6
Human				
HeLa	Cervical carcinoma	+	wt	2.2
IMR-90	Lung fibroblast	+	wt	1.7
Jurkat	T-cell lymphoma	+	wt	4.4
SJC3	Mutated lymphoblast	–	?	4.3
AT5BI	AT	–	*ATM*	2.4
ARO	AT	–	*ATM*	1.0
BMA	AT	–	*ATM*	0.9
NBS	Nijmegen breakage syndrome	–	?	0.7
LEE	Navajo wild type	?	wt	2.0
HAR	Navajo immunodeficiency	?	?	1.8
NEW	Navajo immunodeficiency	?	?	1.6
Mouse				
NIH3T3	Fibroblast	+	wt	1.9
1255/2-4	Melanoma	?	?	1.4
C.B-17	Fibroblast	+	wt	2.1
SCID	Fibroblast	–	*XRCC7*	1.2
Hamster				
V79	Chinese hamster fibroblast	+	wt	1.0
V79B	Chinese hamster fibroblast	+	wt	0.5
AA8	CHO	+	wt	1.0
EM9	CHO, EMS-sensitive	–	*XRCC1*	2.0
BL-10	CHO, bleomycin sensitive	+	?	1.8
BL-14	CHO, bleomycin sensitive	–	?	2.2
XR-1	CHO	–	*XRCC4*	1.3
V-3	CHO	–	*XRCC7*	1.0
XR-V15B	Chinese hamster fibroblast	–	*XRCC5*	≤0.02
XR-V9B	Chinese hamster fibroblast	–	*XRCC5*	<0.02
xrs5	CHO	–	*XRCC5*	<0.02
xrs5 (rev)	CHO, revertant pool	+	wt	1.0

[a]Shown are data from yeast *(17)* (Cheng, S. and G. Chu, unpublished results), human *(56)* (Rathmell, W. K. and G. Chu, unpublished results), mouse, and hamster *(56)*. The cells have either wild-type resistance (+) or increased sensitivity (–) to X-rays. For the sensitive cells, the X-ray complementation group is indicated when known. The xrs5(rev) cells are xrs5 cells reverted to X-ray resistance by azacytidine. DEB activity was measured by the EMSA, quantitated by scanning densitometry, and expressed as units of binding activity per microgram of cell extract. One unit was defined by the shift of 33% of the DNA probe. Errors in DEB activity were ±30%. Abbreviations: CHO, Chinese hamster ovary; EMS, ethyl methanesulfonate.

Direct evidence that the *Ku86* gene and the XRCC5 gene are identical was provided by transfection experiments. The transfection of a human *Ku86* cDNA gene into hamster *XRCC5* mutant cells partially restored DEB activity, X-ray resistance, and V(D)J recombination *(67,68)*. Full restoration was observed by the transfection of hamster *Ku86* cDNA *(15)*. In addition, Ku70 protein levels were restored to wild-type levels, suggesting that the Ku70 polypeptide is stabilized by the presence of normal Ku86 polypeptide. Finally, *Ku86* cDNA from XR-V15B and XR-V9B cells contained in-frame deletions of 46 and 84 amino acids, respectively, proving that mutation of the *Ku86* gene was responsible for the phenotype of *XRCC5* cells *(15)*.

XRCC6 has been reserved for the human *Ku70* gene. Presumably the cells will prove to be X-ray-sensitive, defective in V(D)J recombination, and lack DEB activity. However, the corresponding cell lines have not yet been identified.

Interestingly, yeast cells disrupted for the *Ku70* gene have undetectable levels of DEB activity, but have wild-type resistance to X-rays (*see* Table 1), consistent with the idea that DSBR in yeast occurs primarily by HR, whereas DSBR in humans occurs primarily by the NHR pathway that involves Ku.

6.3. DNA-Activated Protein Kinase

The DNA-PK is a serine/threonine protein kinase that is activated on binding to DNA ends, nicks, and gaps *(23)*. The regulatory subunit of the kinase is the Ku autoantigen. The catalytic subunit of DNA-PK is a very large (465-kDa) nuclear protein designated DNA-PK_{cs} *(28)*. Sequence analysis has shown that the kinase domain of DNA-PK_{cs} shares strong homology to the phosphatidylinositol 3-kinase superfamily.

XRCC7 mutant cells (V-3 hamster cells and SCID mouse cells) are deficient in DNA-PK activity *(5)*, but have intact Ku DEB activity *(56)*. DNA-PK activity is restored when *XRCC7* mutant cell extracts are supplemented with purified DNA-PK_{cs}. Furthermore, transfection of *XRCC7* mutant cells with yeast artificial chromosomes carrying the DNA-PK_{cs} gene restored kinase activity, X-ray resistance, and V(D)J recombination to wild-type levels *(5,33)*.

DNA-PK phosphorylates a number of in vitro substrates including histones, topoisomerases, various transcription factors, the tumor suppressor p53, RP-A, and both subunits of the Ku autoantigen *(1)*. However, the in vivo substrates of DNA-PK remain to be defined. On phosphorylation by DNA-PK, Ku acquires an ATPase activity *(10)*. Furthermore, Ku has an ATP-dependent helicase activity *(73)* (Fig. 3), suggesting that autophosphorylation by DNA-PK confers helicase activity to Ku.

6.4. XRCC4 *Gene Product*

XR-1 is the only mutant cell line so far assigned to X-ray-sensitive complementation group 4. Like *XRCC5* mutants, XR-1 is defective for DSBR, and for both coding and signal joint formation during V(D)J recombination *(69)*. However, XR-1 cell extracts contain normal levels of DEB activity *(56)*. The X-ray sensitivity of XR-1 varies dramatically during the cell cycle, with extreme sensitivity during G1 and near-normal resistance during late S and G2 *(21)*. The *XRCC4* gene was recently cloned by complementation of the V(D)J recombination deficiency *(40)*. The encoded protein is not homologous to other known proteins, and the function of this protein has yet to be identified.

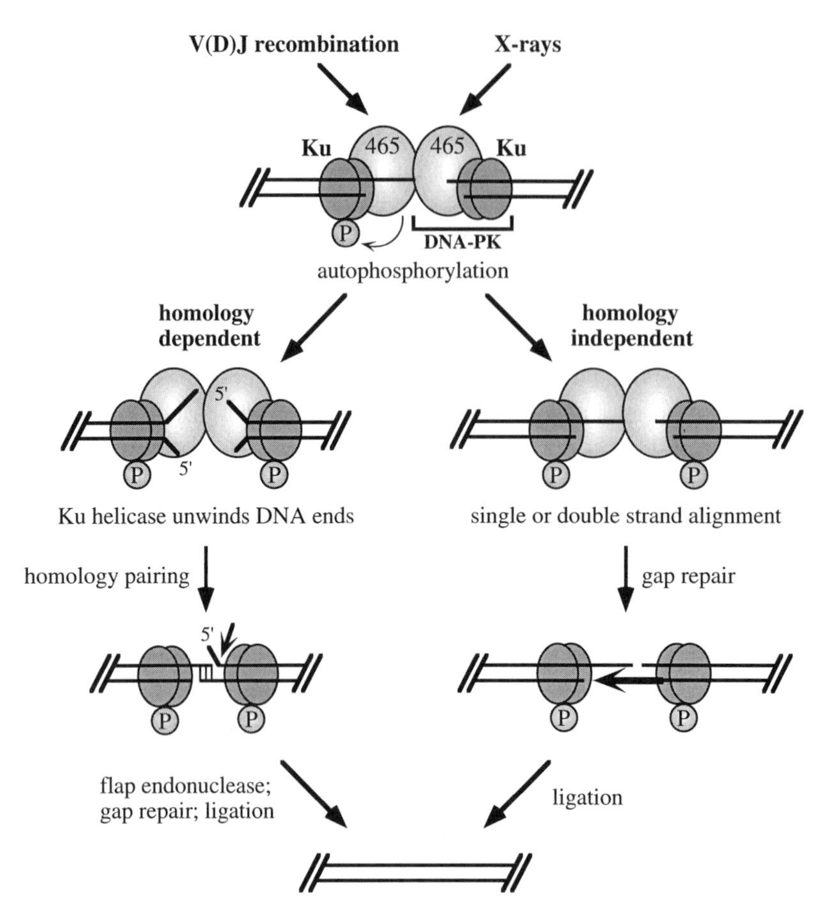

Fig. 4. Role for Ku in DSBR. DSBs produced either by ionizing radiation or during V(D)J recombination are recognized by the binding of Ku to the DNA ends. Ku recruits DNA-PK_{cs} to form an activated DNA-PK, which then autophosphorylates Ku. The repair process can then proceed by homology-dependent or independent pathways. In the homology-dependent pathway, we hypothesize that Ku helicase activity unwinds the DNA ends to allow homology pairing in regions of microhomology from 1–5 bp. Unpaired DNA ends would then be processed by exonuclease or flap endonuclease activity, and repair of the DSB would be completed by gap filling and ligation. In the homology-independent pathway, we hypothesize that the DNA ends are aligned along either both strands or one pair of single strands. DNA polymerase would then fill in gaps present across aligned single strands, and repair would be completed by ligation of the resulting nicked DNA. Presumably, the homology-independent pathway would not require autophosphorylation and activation of Ku helicase activity.

7. MODEL FOR NONHOMOLOGOUS DSBR IN MAMMALIAN CELLS

To explain the DNA intermediates of DSBR in terms of the biochemical properties of Ku and DNA-PK, we propose a model for nonhomologous DSBR in mammalian cells (Fig. 4). In this model, autophosphorylation of Ku by DNA-PK confers helicase activity to Ku, which then mediates microhomology pairing and resolution for DSBR. The individual steps proceed as follows.

1. On the generation of a DSB, Ku binds to the free DNA ends and prevents their degradation by exonucleases.
2. Ku recruits DNA-PK_{cs} to the DNA ends to form activated DNA-PK.
3. DNA-PK autophosphorylates Ku and activates its ATP-dependent helicase activity. Ku unwinds the DNA ends to allow a search for microhomology. Duplex DNA may also be processed by 5'- or 3'-exonuclease activity to produce single-stranded ends available for microhomology pairing.
4. Microhomology pairing is presumably stimulated by a protein factor, since 1–5 bp may not be long enough for efficient spontaneous pairing.
5. Following microhomology pairing, unpaired DNA ends extending beyond the region of paired microhomology are processed, either by exonucleases or by a flap endonuclease *(27)*.
6. As an alternative to steps 3 and 4, for some end joining events, Ku facilitates the joining reaction without unwinding or utilization of microhomology.
7. DNA synthesis occurs across any gaps, and DNA ligase seals the nick.

This model is incomplete, since roles for several factors remain undefined. The *XRCC4* gene product must play a role that is somehow related to the cell-cycle-dependent X-ray sensitivity of XR-1 cells. The kinase activity of DNA-PK may be important for DSBR by acting on other substrates in addition to Ku. DNA-PK_{cs} is a very large protein and may have activities in addition to its kinase activity. Several aspects of the current model will require experimental verification, including: DNA-PK autophosphorylation of Ku in intact cells; requirement of the helicase activity associated with Ku protein; identification of participating exonucleases; involvement of the flap endonuclease *(27)*; and identification of potential factors involved in microhomology pairing.

8. CONCLUSIONS

8.1. Summary of Recent Progress

Recent studies of DSBR in eukaryotic cells have resulted in several significant advancements in the field. Experiments in intact cells and cell extracts have shown that eukaryotic cells are able to repair DSBs by both HR and NHR mechanisms. In mammalian cells, NHR mechanisms appear to be more important. Three complementation groups (*XRCC4*, *XRCC5*, and *XRCC7*) of X-ray-sensitive mutants are defective in both DSBR and V(D)J recombination. The genes that rescue these mutants encode the XRCC4 protein, Ku86, and DNA-PK_{cs}, respectively. Therefore, DNA ends generated by ionizing radiation and cleavage of recombination signal sequences during V(D)J recombination are resolved by a common mechanism. The end joining mechanism may utilize microhomology or proceed independently of microhomology.

8.2. Questions for the Future

Can the Ku-dependent DSBR pathway be reconstituted in a cell-free system? As discussed above, several attempts have been made to study DSBR in vitro, but the results have been inconsistent. It is unclear whether different investigators are studying different parts of one DSBR pathway or completely different pathways. These concerns can be addressed with the mutant *XRCC4*, *XRCC5*, and *XRCC7* rodent cell lines. Unfortunately, at least one cell-free DSBR system does not measure the DSBR pathway defined by DNA-PK, since in that system, extracts from SCID cells supported normal levels of end joining *(48)*. To reflect the dominant pathway for DSBR in mammalian cells, cell-free systems should demonstrate a requirement for Ku, DNA-PK_{cs}, and the XR-1 gene product. It is possible that conditions inside the nucleus can be re-created only with great difficulty in a cell-free system for DSBR. For example, the accessibility of chromatin structure (Chapter 13), the stability of the nuclear matrix, or the compartmentalization of repair factors may be essential features of DSBR not easily studied with soluble factors.

Are other factors essential for vertebrate DSBR? We have alluded to the possible need for factors involved in the alignment of DNA ends. These factors may be involved in simply maintaining the free DNA ends in close proximity or in pairing DNA ends in regions of microhomology. Exonucleases or flap endonucleases are also required to resolve the DNA intermediates formed by microhomology pairing.

Is DSBR coupled to control of the cell cycle? In vitro studies have shown that DNA-PK can phosphorylate the G1 checkpoint protein p53 *(38)*. We have analyzed p53 accumulation and cell-cycle arrest in primary SCID mouse cells in response to ionizing radiation, and have observed no defect compared to wild-type cells *(58)*. AT cells do display a delay in p53 induction in response to ionizing radiation *(32,42)*. The *ATM* gene has a kinase domain homologous to the kinase domain of DNA-PK_{cs}, and is therefore a candidate for phosphorylating p53 and signaling for cell-cycle arrest after ionizing radiation (*see* Chapter 21). It is possible that Ku is the regulatory subunit for the ATM kinase as well as for DNA-PK_{cs}.

Clearly many questions remain to be addressed to define DSBR further in vertebrates. The discovery that Ku and DNA-PK_{cs} are involved in this pathway has been an important step forward in our understanding of DSBR.

ACKNOWLEDGMENTS

Work in the authors' laboratory was supported in part by a grant from the Department of the Army (DAMD17-94-J-4350) to G. C. and by a Howard Hughes Medical Institute predoctoral fellowship to W. K. R. The authors thank Vaughn Smider, Sara Cheng, and Sharon Hays for reading the manuscript.

REFERENCES

1. Anderson, C. 1993. DNA damage and the DNA-activated protein kinase. *Trends Biochem. Sci.* **18:** 433–437.
2. Benson, F. E., A. Stasiak, and S. C. West. 1994. Purification and characterization of the human Rad51 protein, an analogue of RecA. *EMBO J.* **13:** 5764–5771.
3. Biedermann, K., J. Sun, A. Giaccia, L. Tosto, and J. Brown. 1991. Scid mutation in mice confers hypersensitivity to ionizing radiation and a deficiency in DNA double-strand break repair. *Proc. Natl. Acad. Sci. USA* **88:** 1394–1397.
4. Blier, P. R., A. J. Griffith, J. Craft, and J. A. Hardin. 1992. Binding of Ku protein to DNA: measurement of affinity for ends and demonstration of binding to nicks. *J. Biol. Chem.* **268:** 7594–7601.
5. Blunt, T., N. Finnie, G. Taccioli, G. Smith, J. Demengeot, T. Gottlieb, R. Mizuta, A. Varghese, F. Alt, P. Jeggo, and S. Jackson. 1995. Defective DNA-dependent protein kinase activity is linked to V(D)J recombination and DNA repair defects associated with the murine scid mutation. *Cell* **80:** 813–823.

6. Bollag, R. J., A. S. Waldman, and R. M. Liskay. 1989. Homologous recombination in mammalian cells. *Annu. Rev. Genet.* **23:** 199–225.
7. Bosma, M. and A. Carroll. 1991. The SCID mouse mutant: definition, characterization, and potential uses. *Annu. Rev. Immunol.* **9:** 323–350.
8. Boubnov, N., Z. Wills, and D. Weaver. 1993. V(D)J recombination coding junction formation without DNA homology: processing of coding termini. *Mol. Cell. Biol.* **13:** 6957–6968.
9. Cai, Q.-Q., A. Plet, J. Imbert, M. Lafage-Pochitaloff, C. Cerdan, and J.-M. Blanchard. 1994. Chromosomal location and expression of the genes coding for Ku p70 and p80 in human cell lines and normal tissues. *Cytogenet. Cell Genet.* **65:** 221–227.
10. Cao, Q., S. Pitt, J. Leszyk, and E. Baril. 1994. DNA-dependent ATPase from HeLa cells is related to human Ku autoantigen. *Biochemistry* **33:** 8548–8557.
11. Carroll, D., S. H. Wright, R. K. Wolff, E. Grzesiuk, and E. B. Maryon. 1986. Efficient homologous recombination of linear DNA substrates after injection into Xenopus laevis oocytes. *Mol. Cell. Biol.* **6:** 2053–2061.
12. Chen, D. J., B. L. Marrone, T. Nguyen, M. Stackhouse, Y. Zhao, and M. J. Siciliano. 1994. Regional assignment of a human DNA repair gene (XRCC5) to 2q35 by X-ray hybrid mapping. *Genomics* **21:** 423–427.
13. Collins, A. R. 1993. Mutant rodent cell lines sensitive to ultraviolet light, ionizing radiation and cross-linking agents: a comprehensive survey of genetic and biochemical characteristics. *Mutat. Res.* **293:** 99–118.
14. Derbyshire, M. K., L. H. Epstein, C. S. H. Young, P. L. Munz, and R. Fishel. 1994. Nonhomologous recombination in human cells. *Mol. Cell. Biol.* **14:** 156–169.
15. Errami, A., V. Smider, W. Rathmell, D. He, E. Hendrickson, M. Zdzienicka, and G. Chu. 1996. Ku86 defines the genetic defect and restores X-ray resistance and V(D)J recombination to complementation group 5 hamster cells. *Mol. Cell. Biol.* **16:** 1519–1526.
16. Falzon, M., J. W. Fewell, and E. L. Kuff. 1993. EBP–80, a transcription factor closely resembling the human autoantigen Ku, recognizes single- to double-strand transitions in DNA. *J. Biol. Chem.* **268:** 10,546–10,552.
17. Feldmann, H. and E. Winnacker. 1993. A putative homologue of the human autoantigen Ku from *Saccharomyces cerevisiae*. *J. Biol. Chem.* **268:** 12,895–12,900.
18. Fulop, G. and R. Phillips. 1990. The scid mutation in mice causes a general defect in DNA repair. *Nature* **347:** 479–482.
19. Ganesh, A., P. North, and J. Thacker. 1993. Repair and misrepair of site-specific DNA double-strand breaks by human cell extracts. *Mutat. Res.* **299:** 251–259.
20. Getts, R. and T. Stamato. 1994. Absence of a Ku-like DNA end binding activity in the xrs double-strand DNA repair-deficient mutant. *J. Biol. Chem.* **269:** 15,981–15,984.
21. Giaccia, A., R. Weinstein, J. Hu, and T. D. Stamato. 1985. Cell-cycle-dependent repair of double-strand breaks in a gamma-ray-sensitive Chinese hamster cell. *Somatic Cell Mol. Genet.* **11:** 485–491.
22. Gilfillan, S., A. Dierich, M. Lemeur, C. Benoist, and D. Mathis. 1993. Mice lacking TdT: mature animals with an immature lymphocyte repertoire. *Science* **261:** 1175–1178.
23. Gottlieb, T. and S. Jackson. 1993. The DNA-dependent protein kinase: requirement for DNA ends and association with Ku antigen. *Cell* **72:** 131–142.
24. Grzesiuk, E. and D. Carroll. 1987. Recombination of DNAs in Xenopus oocytes based on short homologous overlaps. *Nucleic Acids Res.* **15:** 971–985.
25. Hafezparast, M., G. Kaur, M. Zdzienicka, R. Athwal, A. Lehmann, and P. Jeggo. 1993. Subchromosomal localization of a gene (XRCC5) involved in double-strand break repair to the region 2q34–36. *Somatic Cell Mol. Genet.* **19:** 413–421.
26. Hamilton, A. A. and J. Thacker. 1987. Gene recombination in X-ray sensitive hamster cells. *Mol. Cell. Biol.* **7:** 1409–1414.
27. Harrington, J. J. and M. R. Lieber. 1994. The characterization of a mammalian DNA structure-specific endonuclease. *EMBO J.* **13:** 1235–1246.

28. Hartley, K. O., D. Gell, G. C. M. Smith, H. Zhang, N. Divecha, M. A. Connelly, A. Admon, S. P. Lees-Miller, C. W. Anderson, and S. P. Jackson. 1995. DNA-dependent protein kinase catalytic subunit: a relative of phosphatidylinositol 3-kinase and the ataxia telangiectasia gene product. *Cell* **82:** 849–856.
29. Hendrickson, E., X. Q. Qin, E. Bump, D. Schatz, M. Oettinger, and D. Weaver. 1991. A link between double-strand break related repair and V(D)J recombination: the scid mutation. *Proc. Natl. Acad. Sci. USA* **88:** 4061–4065.
30. Jeggo, A. P. 1990. Studies on mammalian mutants defective in rejoining double-strand breaks in DNA. *Mutat. Res.* **239:** 1–16.
31. Jeong-yu, S. and D. Carroll. 1992. Test of the double strand break repair model of recombination in *Xenopus laevis* oocytes. *Mol. Cell. Biol.* **12:** 112–119.
32. Kastan, M., Q. Zhan, F. El-Deiry, T. Jacks, W. Walsh, B. Plunkett, B. Vogelstein, and A. Fornace. 1992. A mammalian cell cycle checkpoint pathway utilizing p53 and GADD45 is defective in ataxia-telangiectasia. *Cell* **71:** 587–597.
33. Kirschgessner, C., C. Patil, J. Evans, C. Cuomo, L. Fried, T. Carter, M. Oettinger, and J. M. Brown. 1995. DNA-dependent kinase (p350) as a candidate gene for the murine SCID defect. *Science* **267:** 1178–1185.
34. Komori, T., A. Okada, V. Stewart, and F. Alt. 1993. Lack of N regions in antigen receptor variable region genes of TdT-deficient lymphocytes. *Science* **261:** 1171–1175.
35. Kramer, K., J. Brock, K. Bloom, J. Moore, and J. Haber. 1994. Two different types of double-strand breaks in *Saccharomyces cerevisiae* are repaired by similar *RAD52*-independent, nonhomologous recombination events. *Mol. Cell. Biol.* **14:** 1293–1301.
36. Lafaille, J., A. DeCloux, M. Bonneville, Y. Takagaki, and S. Tonegawa. 1989. Junctional sequences of T cell receptor γδ genes: implications for γδ T cell lineages and for a novel intermediate of V-(D)-J joining. *Cell* **59:** 859–870.
37. Lee, S. E., C. Pulaski, D. M. He, D. M. Benjamin, J. M. Voss, J. Y. Um, and E. Hendrickson. 1995. Isolation of mammalian mutants that are X-ray sensitive, impaired in DNA double-strand break repair and defective for V(D)J recombination. *Mutat. Res.* **336:** 279–291.
38. Lees-Miller, S., K. Sakaguchi, S. Ullrich, E. Appela, and C. Anderson. 1992. Human DNA-activated protein kinase phosphorylates serines 15 and 37 in the amino-terminal transactivation domain of human p53. *Mol. Cell. Biol.* **12:** 5041–5049.
39. Lewis, S. 1994. The mechanism of V(D)J joining: lessons from molecular, immunological, and comparative analyses. *Adv. Immunol.* **56:** 27–150.
40. Li, Z., T. Otevrel, Y. Gao, H.-L. Cheng, B. Seed, T. D. Stamato, G. E. Taccioli, and F. W. Alt. 1995. The XRCC4 gene encodes a novel protein involved in DNA double-strand break repair and V(D)J recombination. *Cell* **83:** 1079–1089.
41. Lieber, M. R., J. E. Hesse, S. Lewis, G. C. Bosma, N. Rosenberg, K. Mizuuchi, M. J. Bosma, and M. Gellert. 1988. The defect in murine severe combined immune deficiency: joining of signal sequences but not coding segments in V(D)J recombination. *Cell* **55:** 7–16.
42. Lu, X. and D. Lane. 1993. Differential induction of transcriptionally active p53 following UV or ionizing radiation: defects in chromosome instability syndromes? *Cell* **75:** 765–778.
43. Meyn, M. S. 1993. High spontaneous intrachromosomal recombination rates in ataxia-telangiectasia. *Science* **260:** 1327–1330.
44. Mimori, T. and J. A. Hardin. 1986. Mechanism of interaction between Ku protein and DNA. *J. Biol. Chem.* **261:** 10,375–10,379.
45. Mimori, T., Y. Ohosone, N. Hama, M. Akizuki, M. Homma, A. J. Griffith, and J. A. Hardin. 1990. Isolation and characterization of cDNA encoding the 80-kDa subunit protein of the human autoantigen Ku (p70/p80) recognized by autoantibodies from patients with scleroderma-polymyositis overlap syndrome. *Proc. Natl. Acad. Sci. USA* **87:** 1777–1781.
46. Moore, P. D., K. Y. Song, L. Cheruki, L. Wallace, and R. Kucherlapati. 1986. Homologous recombination in a Chinese hamster X-ray-sensitive mutant. *Mutat. Res.* **160:** 149–155.

47. Muris, D. F. R., O. Bezzubova, J. Buerstedde, K. Vreeken, A. S. Balajee, C. J. Osgood, C. Troelstra, J. H. J. Hoeijmakers, K. Ostermann, H. Schmidt, A. T. Natarajan, J. C. J. Eeken, P. H. M. Lohman, and A. Pastink. 1994. Cloning of human and mouse genes homologous to RAD52, a yeast gene involved in DNA repair and recombination. *Mutat. Res.* **315:** 295–305.
48. Nicolas, A. L. and C. S. H. Young. 1994. Characterization of DNA end joining in a mammalian cell nuclear extract: junction formation is accompanied by nucleotide loss, which is limited and uniform, but not site specific. *Mol. Cell. Biol.* **14:** 170–180.
49. Oettinger, M., D. Schatz, C. Gorka, and D. Baltimore. 1990. RAG–1 and RAG–2, adjacent genes that synergistically activate V(D)J recombination. *Science* **248:** 1517–1523.
50. Orr-Weaver, T. L. and J. W. Szostack. 1985. Fungal recombination. *Microbiol. Rev.* **49:** 33–58.
51. Paillard, S. and F. Strauss. 1991. Analysis of the mechanism of interaction of simian Ku protein with DNA. *Nucleic Acids Res.* **19:** 5619–5624.
52. Park, M. 1995. Expression of human RAD52 confers resistance to ionizing radiation in mammalian cells. *J. Biol. Chem.* **270:** 15,467–15,470.
53. Pergola, F., M. Z. Zdzienicka, and M. R. Lieber. 1993. V(D)J recombination in mammalian cell mutants defective in DNA double-strand break repair. *Mol. Cell. Biol.* **13:** 3464–3471.
54. Pfeiffer, P., S. Thode, J. Hancke, and W. Vielmetter. 1994. Mechanisms of overlap formation in nonhomologous DNA end joining. *Mol. Cell. Biol.* **14:** 888–895.
55. Pfeiffer, P. and W. Vielmetter. 1988. Joining of nonhomologous DNA double strand breaks in vitro. *Nucleic Acids Res.* **16:** 907–924.
56. Rathmell, W. K. and G. Chu. 1994. A DNA end-binding factor involved in double-strand break repair and V(D)J recombination. *Mol. Cell. Biol.* **14:** 4741–4748.
57. Rathmell, W. K. and G. Chu. 1994. Involvement of the Ku autoantigen in the cellular response to DNA double-strand breaks. *Proc. Natl. Acad. Sci. USA* **91:** 7623–7627.
58. Rathmell, W. K., W. K. Kaufmann, J. C. Hurt, L. L. Byrd, and G. Chu. 1997. DNA-dependent protein kinase is not required for accumulation of p53 or cell cycle arrest after DNA damage. *Cancer Res.* **57:** 68–74.
59. Reeves, W. 1992. Antibodies to the p70/p80 (Ku) antigens in systemic lupus erythematosus. *Rheumatic Dis. Clin. North Am.* **18:** 391–415.
60. Reeves, W. H., and Z. M. Sthoeger. 1989. Molecular cloning of cDNA encoding the p70 (Ku) lupus autoantigen. *J. Biol. Chem.* **264:** 5047–5052.
61. Roth, D. B., X. Chang, and J. H. Wilson. 1989. Comparison of filler DNA at immune, nonimmune, and oncogenic rearrangements suggests multiple mechanisms of formation. *Mol. Cell. Biol.* **9:** 3049–3057.
62. Roth, D. B., T. N. Porter, and J. H. Wilson. 1985. Mechanisms of nonhomologous recombination in mammalian cells. *Mol. Cell. Biol.* **5:** 2599–2607.
63. Roth, D. B. and J. H. Wilson. 1986. Nonhomologous recombination in mammalian cells: role for short sequence homologies in the joining reaction. *Mol. Cell. Biol.* **6:** 4295–4304.
64. Savitsky, K., A. Bar-Shira, S. Gilad, G. Rotman, Y. Ziv, L. Vanagaite, D. Tagle, S. Smith, T. Uziel, S. Sfez, M. Ashkenazi, I. Pecker, M. Frydman, R. Harnik, S. Patanjali, A. Simmons, G. Clines, A. Sartiel, R. Gatti, L. Chessa, O. Sanal, M. Lavin, N. Jaspers, M. Taylor, C. Arlett, T. Miki, S. Weissman, M. Lovett, F. Collins, and Y. Shiloh. 1995. A single ataxia-telangiectasia gene with product similar to PI–3 kinase. *Science* **268:** 1749–1753.
65. Schatz, D. G., M. A. Oettinger, and D. Baltimore. 1989. The V(D)J recombination activating gene, RAG–1. *Cell* **59:** 1035–1048.
66. Shinohara, A., H. Ogawa, Y. Matsuda, N. Ushio, K. Ikeo, and T. Ogawa. 1993. Cloning of human, mouse and fission yeast recombination genes homologous to RAD51 and recA. *Nature Genet.* **4:** 239–243.
67. Smider, V., W. K. Rathmell, M. Lieber, and G. Chu. 1994. Restoration of X-ray resistance and V(D)J recombination in mutant cells by Ku cDNA. *Science* **266:** 288–291.

68. Taccioli, G., T. Gottlieb, T. Blunt, A. Priestly, J. Demengeot, R. Mizuta, A. Lehmann, F. Alt, S. Jackson, and P. Jeggo. 1994. Ku80: product of the XRCC5 gene and its role in DNA repair and V(D)J recombination. *Science* **265:** 1442–1445.
69. Taccioli, G., G. Rathbun, E. Oltz, T. Stamato, P. Jeggo, and F. Alt. 1993. Impairment of V(D)J recombination in double-strand break repair mutants. *Science* **260:** 207–210.
70. Thacker, J., J. Chalk, A. Ganesh, and P. North. 1992. A mechanism for deletion formation in DNA by human cell extracts: the involvement of short sequence repeats. *Nucleic Acids Res.* **20:** 6183–6188.
71. Thode, S., A. Schafer, P. Pfeiffer, and W. Vielmetter. 1990. A novel pathway of DNA end-to-end joining. *Cell* **60:** 921–928.
72. Thompson, L. and P. A. Jeggo. 1995. Nomenclature of human genes involved in ionizing radiation sensitivity. *Mutat. Res.* **337:** 131–134.
73. Tuteja, N., R. Tuteja, A. Ochem, P. Taneja, N. Huang, A. Simoncsits, S. Susic, K. Rahman, L. Marusic, J. Chen, J. Zhang, S. Wang, S. Pongor, and A. Falaschi. 1994. Human DNA helicase II: a novel DNA unwinding enzyme identified as the Ku autoantigen. *EMBO J.* **13:** 4991–5001.
74. Weaver, D. T. 1995. V(D)J recombination and double-strand break repair. *Adv. Immunol.* **58:** 29–85.
75. Yaneva, M., J. Wen, A. Ayala, and R. Cook. 1989. cDNA-derived amino acid sequence of the 86-kDa subunit of the Ku antigen. *J. Biol. Chem.* **264:** 13,407–13,411.

17

Mutant Rodent Cells Defective in DNA Double-Strand Break Repair

P. A. Jeggo

1. INTRODUCTION

Ionizing radiation (IR) induces an array of lesions in DNA, including base damage, single- and double-strand breaks (DSBs), and damage to the phosphodiester backbone. It might therefore be anticipated that a range of DNA repair pathways will be utilized to restore the integrity of the DNA following exposure to IR. In addition, cells respond to radiation damage by activating signaling mechanisms to arrest the cells at various checkpoints to prevent inappropriate progression through the cell cycle. Yeast cells defective in checkpoint pathways are sensitive to DNA-damaging agents, including IR *(1,91)*. Additionally, it has recently been shown that cellular changes can result in a different rate of induction of DNA DSBs, indicating that altered radiation responses can arise for reasons not directly associated with defects in DNA repair or damage-recognition checkpoints *(69)*. It is therefore likely that defects in a large number of genes will render cells sensitive to IR. In keeping with this prediction, rodent mutants in at least 11 complementation groups have been shown to exhibit sensitivity to IR and display differing phenotypes, with some showing only minor radiosensitivity *(46,85)*. However, mutants in four complementation groups (IR groups 4–7) display exquisite radiosensitivity and similar, though not identical, phenotypes, including a defect in DSB rejoining and an inability to carry out V(D)J recombination effectively *(80,81,86)*. This overlap in defects indicates that the gene products defined by these complementation groups may operate in a single pathway involved in the repair of DSBs. In line with this, the genes defective in some of these complementation groups have recently been identified and shown to encode components of the DNA-dependent protein kinase (DNA-PK) complex *(5,7,50,64,75,83)*. This chapter reviews the phenotypes of the mutants defective in these four complementation groups, and this information is discussed in light of the recently identified genetic defects. It should be mentioned, however, that additional pathways of DSB rejoining are likely to exist in mammalian cells and that the mutants isolated to date are not likely to represent all the possible complementation groups.

From: DNA Damage and Repair, Vol. 2: DNA Repair in Higher Eukaryotes
Edited by: J. A. Nickoloff and M. F. Hoekstra © Humana Press Inc., Totowa, NJ

2. THE MUTANTS REPRESENTING COMPLEMENTATION GROUPS 4–7

The mutants representing the four complementation groups defective in DSB rejoining are listed in Table 1. Consistent with the nomenclature utilized for mutants and genes defective in excision repair, the genes defective in these complementation groups have been designated *XRCC4–7* (X-ray crosscomplementing) *(86)*. Considering the defects of the *xrs* and *scid* V-3 cell lines in the Ku80 and DNA-PK_{cs} components of DNA-PK *(see below)*, it was anticipated that mutants defective in Ku70 would have a similar phenotype, and on this basis, complementation group 6 was allocated for mutants with this defect. Based on preliminary data, *sxi-1* was initially thought to be a member of this group, but more recent evidence suggests that it is another group 5 mutant *(53,54,94)*. The majority of the cell lines listed in Table 1 were isolated from mutagenized cultures of established rodent cell lines on the basis of their sensitivity to IR. *scid* cell lines, in contrast, were derived from the *scid* mouse, identified by its severe combined immunodeficiency phenotype *(6)*. Cell lines established from the *scid* mouse were subsequently shown to be radiosensitive, and to belong to the same complementation group as the radiosensitive hamster V-3 cells *(24,80)*.

3. THE GENES DEFINED BY THESE COMPLEMENTATION GROUPS

A number of approaches have been adopted in attempts to clone the genes defective in these mutants, and information gleaned from these studies recently converged and culminated in the identification of the genes defined by these groups. Several recent publications have reviewed the data leading to the identification of these genes, and only an outline will be presented here *(38,39,87)*. Two quite different approaches led to the identification of the first of these genes, namely *XRCC5*, which was found to encode the 80-kDa subunit of the Ku protein. Biochemical studies on the group 5 mutants identified a defect in a DNA end binding activity with properties similar to the previously characterized Ku protein *(26,66)*. In parallel, a positional cloning strategy localized a complementing human gene to a 3-Mbp fragment mapping to 2q33-35, and when the gene encoding the Ku80 subunit of Ku was localized to the same chromosomal locus, *Ku80* became a strong candidate gene *(9,30)*. Finally, Ku80 cDNA was shown to complement the radiosensitivity and V(D)J recombination defects of group 5 mutants *(7,75,83)*.

The Ku protein was identified in 1981 as an antigen present in the sera of certain autoimmune patients, and was subsequently shown to be a heterodimer comprising subunits of 70 and 80 kDa *(59,60)*. It is a highly abundant nuclear protein in human cells and has the characteristic property of binding to double-strand DNA ends. Despite extensive study, its function had before this elusively escaped detection, although a role in DNA repair had been suggested. Recently, Ku was identified as the DNA-binding component of DNA-PK, a complex whose catalytic component (DNA-PK_{cs}) is a large protein of approx 460 kDa *(19,29)*. The discovery that Ku80 is the product of IR group 5 therefore implicated an involvement of Ku70 and DNA-PK_{cs} in DNA repair, and suggested these were potential candidate proteins defective in the remaining complementation groups. Recently, a number of lines of evidence have shown that DNA-PK_{cs} is defective in the two group 7 mutants. First, although end binding activity

Table 1
Mutants Belonging to IR Complementation Groups 4–7[a]

Complementation group	Mutant	Gene	Chromsomal localization	DNA end binding activity	DNA-PK activity	References
4	XR-1	*XRCC4*	5q13-14	Wt	Wt	*(79)*
5	*xrs*1-6 XRV-15B XRV9-B sxi-3	*XRCC5* *Ku80*	2q34-36	Absent	Absent	*(44,54,94)*
6	Not identified	*XRCC6* *Ku70*	22q13	Unknown (anticipated to be absent)	Unknown (anticipated to be absent)	
7	V-3 *scid*	*XRCC7* *SCID* *DNA-PK$_{cs}$*	8p11-q11	Wt	Absent	*(24,92)*

[a]References given are for the identification of the mutants only. References for the chromosomal localization and gene identification are given in *(86)* and in the text.

of Ku is normal, both V-3 and *scid* cells are missing DNA-PK activity, and contain little or no protein crossreacting with DNA-PK_{cs} antibody. Second, YACs encoding DNA-PK_{cs} are able to complement hamster V-3 mutants, and finally, the human DNA-PK_{cs} gene maps to a region of chromosome 8, which complements *scid* cells, and is syntenic to a segment of mouse chromosome 16 close to the locus to which the *SCID* gene was mapped by mouse genetic crosses *(5,50,51,64,74)*. To complete the cloning of the genes from these groups, a cDNA capable of complementing the group 4 mutant, XR-1, has also recently been identified *(56)*.

Ku80 has a 2.2-kb open reading frame (ORF) encompassing a leucine zipper motif, but no other obvious domains, such as DNA binding or helicase motifs *(61)*. It is ubiquitously expressed, but higher levels are found in testis, spleen and brain. *Ku70* has a slightly smaller ORF (1.8 kb), but is similar to *Ku80* in having only a leucine zipper motif *(67)*. The mutants from group 5 all have decreased Ku80 protein levels, and for some, mutations in the *Ku80* gene have been identified *(20,73)*. DNA-PK_{cs} has one of the largest cDNAs identified (14 kb), which includes a kinase domain at its carboxy-terminus. It is a member of the phosphatidylinositol-(PI) 3-kinase superfamily, but in vitro, it is able to phosphorylate proteins, but not lipids *(32)*. It has some sequence similarity to the recently identified *ATM* gene, which is mutated in ataxia telangiectasia (AT) patients, to the *Saccharomyces pombe rad3*, *Saccharomyces cerevisiae MEC1*, and *Drosophila melanogaster mei-41* genes, all of which are involved in cell-cycle control. No information is yet available concerning the molecular nature of the defects in V-3 and *scid* cells.

4. BIOCHEMICAL PROPERTIES OF DNA-PK AND ITS DEFECTIVE MUTANTS

The biochemistry of DNA-PK and its defective mutants is discussed in Chapter 16, and only a summary is presented here. The Ku protein can exist in the cell without being complexed to DNA-PK_{cs}, and can bind to double-stranded DNA ends in the absence of the large DNA-PK subunit. DNA-bound Ku, however, can bind to DNA-PK_{cs} with the concomitant activation of its kinase function *(19,29)*. DNA-PK is therefore a complex specifically resulting in the activation of a kinase following the introduction of DNA DSBs. DNA-PK is able to phosphorylate many DNA binding proteins in vitro, including the transcription factors Sp-1, c-Jun, and p53 *(3)*. Phosphorylation is most effective when these substrates are bound to the same DNA molecule as DNA-PK. However, the substrates of DNA-PK in vivo are unknown although one report has shown that RPA is not phosphorylated in *scid* cells *(8)*. All group 5 mutants examined to date lack Ku-specific DNA end binding activity, whereas members of the other complementation groups (XR-1, V-3, and *scid* cells) possess parental levels *(26,65)*.

The group 5 mutants also lack DNA-dependent protein kinase activity, as predicted from in vitro data showing that Ku is a necessary component of DNA-PK *(21)*. This result is significant in showing that Ku is the major and, possibly only, component activating DNA-PK_{cs} in vivo as well as in vitro. As mentioned above, V-3 and *scid* cells also lack DNA-PK activity *(5,50)*. XR-1 cells, in contrast, contain parental levels of this activity. Consistent with the differing genotypes of *xrs* compared to V-3 and *scid* cells, biochemical complementation of the kinase activity is obtained by mixing

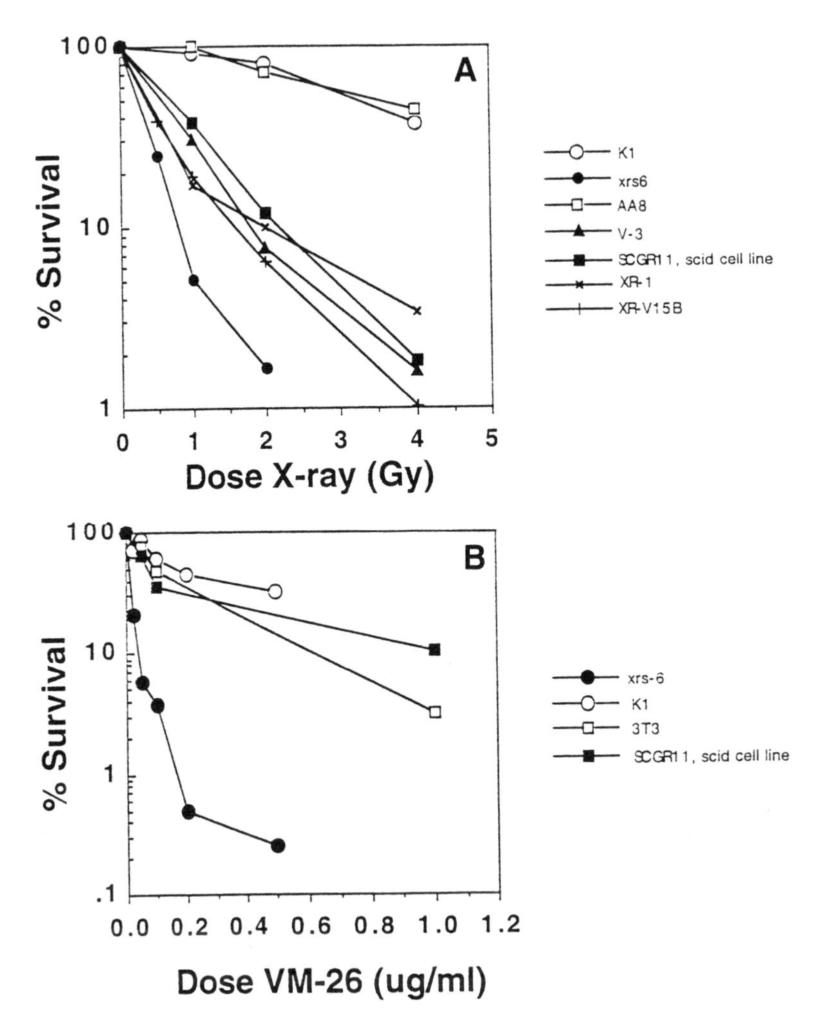

Fig. 1. (A) Sensitivities to IR of group 4, 5, and 7 mutants. **(B)** Sensitivities of group 4 and 5 mutants to VP-16, VM-26, and mAMSA.

extracts of *xrs* and V-3 or *scid* cells, but not by mixing extracts of V-3 and *scid* cells *(5)*. No biochemical defect has yet been identified in XR-1 cells.

5. SENSITIVITY TO DNA-DAMAGING AGENTS

5.1. Sensitivity to Agents Inducing DNA DSBs

Members of these complementation groups are exquisitely sensitive to IR. The sensitivities of group 4, 5, and 7 mutants observed in our laboratory are shown in Fig. 1A. Two members of group 5, *xrs-6*, and to a lesser extent, XR-V15B are more sensitive than the two group 7 mutants, V-3 and *scid*. However, conclusions concerning any significance of this cannot be made until the mutational changes in both group 5 and 7 mutants have been identified. XR-1 displays a biphasic survival curve, which varies with the proportion of S-phase cells in the population *(79)*. The initial slope of the survival curve, which represents the survival of G1 cells, appears to be similar to that of the group 5 mutants.

These mutants are also sensitive to DSBs generated by restriction endonucleases introduced into cells by permeabilization, and to bleomycin, a radiomimetic agent, but they show little cross-sensitivity to UV radiation or alkylating agents *(12,14,28,44)*. There is also some sensitivity to DNA crosslinking agents *(11)*. These results are consistent with a sensitivity to agents introducing DNA DSBs, either directly as in the case of IR and bleomycin, or indirectly, for example, during the repair of other DNA damage.

Group 4 and 5 mutants also display extreme sensitivity to a range of topoisomerase II inhibiting drugs, including the anticancer drugs VP-16, VM-26, and mAMSA (Fig. 1) *(42)*. These drugs inhibit the ligation step of topisomerase II, thereby resulting in the accumulation of protein-bridged DSBs in DNA. Such lesions have been called "cleavable complexes," since they are only revealed as DSBs following treatment with protein denaturants. Such lesions are reversible, and removal of the drug results in the parallel loss of cleavable complexes and DSBs, and consequently enhanced survival. Although cleavable complexes are lost at the same rate and to the same extent in normal and group 5 mutants after drug removal, a fraction of the breaks remain unrejoined in the mutant cells *(10)*. This suggests that Ku80 is required for efficient repair of these lesions, or at least a subset of them. One possibility is that a subset of cleavable complexes are converted to open DSBs not associated with topoisomerase II. An intriguing and surprising observation is that neither of the group 7 mutants (V-3 or *scid*) is appreciably sensitive to VP-16 or VM-26 (Fig. 1). (The sensitivity to VM-26 for *scid* and *xrs-6* cells is shown in Fig. 1B, and the sensitivity to VP-16 for V-3 and *xrs-6* cells is shown in reference *[42]*). If sensitivity to these drugs caused by a percentage of cleavable complexes being handled as frank, nonprotein-associated breaks, then the expectation would be that the group 7 mutants would also show sensitivity. Signal join formation, which occurs during V(D)J recombination, is another situation in which group 5 mutants show a major defect, whereas group 7 mutants show little or no defect *(see below)*. Possible explanations for these results will be considered in Section 11.

5.2. Cell-Cycle Sensitivity

Another characteristic of these mutants is a marked cell stage-specific alteration in radiosensitivity. Before assessing the change in sensitivity of the mutants at various stages in the cell cycle, it is necessary to consider the response of parental cells. Several studies have shown that both human and rodent cell lines show maximum radioresistance in S phase, and maximum radiosensitivity in G2 *(41,92)*. Since these repair-proficient cells possess a shoulder to their survival curve, it is necessary to analyze survival at a range of doses for each cell-cycle stage, which reveals that these cell-cycle changes involve alterations in both the shoulder and the slope of the survival curves. The shape of these curves must be considered when comparing the relative sensitivities of mutant cells and any comparative increase in survival.

XR-1, *xrs*, and V-3 cells have all been examined for sensitivity through the cell cycle, and consistently show enhanced resistance at S (or S/G2) *(37,41,77,79,92)*. XR-1 shows the most dramatic effect, changing from 11-fold more sensitive compared to parental cells in G1 to only 2.6-fold more sensitive in S phase. The enhanced resistance of S phase cells of XR-1 is also evident from the survival of asynchronous cultures, which exhibit a characteristic resistant tail that can be removed by exposure to ^{3}H-TdR, which selectively kills S-phase cells. V-3 and the *xrs* mutants (*xrs-5* and *xrs-7*) also gain radioresistance as they progress through the cell cycle, but in contrast to V-3 and XR-1, the *xrs* mutants achieve maximum radioresistance at a slightly later stage in late S/G2 phase. There is no apparent explanation for this cell-cycle difference of *xrs* cells, which has nonetheless been observed in two laboratories *(41,92)*. It has been suggested that resistance of S-phase *xrs* cells could result from transient hypomethylation of a silent hypermethylated copy of the *XRCC5* gene following replication, an explanation suggested by the induced reversion of *xrs* cells following azacytidine treatment *(17,43)*. Such an explanation cannot, however, explain the transient radioresistance of V-3 and XR-1 cells, which are not revertible by azacytidine. Where examined, these changes in radiosensitivity correlate with changes in ability to rejoin DSBs, suggesting that they do reflect cell-cycle alterations in repair capacity *(27,58)*. Taken together, these data show that the DNA-PK-dependent pathway of DSB repair exerts its greatest contribution during G1 phase of the cell cycle. Several explanations are possible for the enhanced radioresistance seen in S or G2 cells of which one is the existence of a second, DNA-PK-independent pathway of DSB repair which operates during this specific phase of the cell cycle. A likely candidate for such a mechanism is homologous recombination.

6. DSB REJOINING IN THE MUTANTS

One of the earliest features characterized in the *xrs* mutants was a defect in ability to rejoin DSBs measured using the neutral elution technique *(49,90)*. Subsequently, this defect has been characterized using a number of other techniques, including pulse-field gel electrophoresis (PFGE), nucleoid sedimentation, DNA unwinding, and the comet assay *(13,35,68)*. The defect in all the mutants represented by these complementation groups is manifest as a higher percentage of unrejoined DSBs, and the magnitude of the defect increases with dose *(36)*. The defect is not large (two- to fourfold), but is of a similar magnitude for all the mutants examined *(4,27,33)*. Thus, all the mutants show residual DSB rejoining, but there is some discrepancy regarding whether or not this residual rejoining occurs at a slower rate *(13,15)*. One study has argued that there is a fraction of irreparable breaks observable in wild-type cells, and that the *xrs* mutants have a higher fraction of irreparable breaks, but that the residual rejoining occurs at the same rate as that observed for parental cells *(15)*. The residual DSB rejoining probably indicates an alternative mechanism for handling DSBs since it is unlikely that all these mutants (*scid*, V-3, XR-1, and *xrs*) would have the same level of leakiness. Moreover, when wild-type cells are held under nondividing conditions following irradiation, a recovery process takes place that has been called potentially lethal damage repair. None of these mutants (*xrs*, XR-1, and *scid* cells) show any evidence of potentially lethal damage repair *(4,37,77,84)*, which suggests that the mutants do not merely repair their breaks slowly, and supports the notion that they are not leaky. There are several possible explanations to reconcile these apparently contradictory results, one of which is that the breaks in group 5 and 7 mutants are rejoined by an alternative mechanism, which is nonconservative and does not enhance survival.

7. DEFECTS IN V(D)J RECOMBINATION

In 1983, four littermates having a severe combined immunodeficiency *(scid)* phenotype were noted by Bosma and coworkers *(6)*, and have become the forebears of *scid* mice. This is a somewhat confusing terminology, since defects in many genes are now known to cause a SCID phenotype, and the defect most commonly found in human *scid* patients is quite distinct from the defect now identified in the *scid* mouse. It is therefore important to be aware of the distinction between a SCID phenotype and the *scid* mouse genotype. As outlined above, the mouse *SCID* gene, *XRCC7*, and the *DNA-PK*$_{cs}$ gene are now known to be one and the same. The immune defects of *scid* mice have been described in several reviews and only an outline relevant to the present context will be presented here *(2,25,55)*. V(D)J rearrangement is a process which occurs during development of T- and B-cells, in which the four ends generated by two site-specific DSBs are rejoined recombinogenically in a very defined way. The DSBs are introduced between partially conserved sequences known as recombination signal sequences (RSSs), which probably serve to direct the site of the break, and their adjacent sequences, known as coding sequences, which are potentially destined to become part of the coding region of immunoglobulin and TCR genes. The ends of the RSSs are accurately and precisely joined together, whereas nucleotide loss and insertion are frequently seen at the junctions of the coding sequences. This process generates a large repertoire of different coding sequences and plays a major role in enhancing the diversity of the immune system. *scid* mice exhibit an intriguing phenotype; RSS junctions are formed at close to normal frequencies and with normal accuracy, whereas coding junctions occur at a frequency approximately three orders of magnitude lower than normal, and display differences in their junctional sequences compared to normal cells. Although V(D)J recombination is normally restricted to take place in lymphoblastoid cells, an assay has recently enabled the process to be examined in somatic cells *(34,81,82)*. Cell lines derived from *scid* mice show essentially the same phenotype in this assay to that observed in vivo *(81)*. Members of complementation groups 4 and 5 are also defective in ability to undergo V(D)J recombination when examined by this assay, but differ in exhibiting a major defect in both signal and coding junction formation *(63,81)*. In contrast, V-3 cells also display a similar phenotype to *scid* cells *(63,80)*.

8. DEFECTS IN OTHER RECOMBINATION PROCESSES

At least three categories of recombination have been identified in yeast cells, the organism most amenable to genetic analysis. The most well-characterized process is that involving large regions of homology, and it is clear that this encompasses several mechanisms, including gene conversion and crossing over, and is dependent on *RAD50–57 (23)*. When first characterized, the mammalian mutants defective in DSB repair appeared similar to *rad52* mutants of yeast, and my prediction, and that of others, was that the mutants would be defective in homologous recombination. Two studies have examined homologous recombination between extrachromosomal plasmid sequences in *xrs* and parental cells *(31,62)*. One study found a fourfold defect, comparable to the defect in DSB rejoining, whereas the other study found no defect. These studies are complicated, however, in that they require integration of the recombined sequences for expression, an event itself decreased in the mutant cells *(see below) (31,45)*. Nevertheless, the results show that there is not a major defect in homologous recombination comparable to that seen in *rad52* mutants. However, these experiments have been carried out on extrachromosomal sequences, which are subject to exceptionally high levels of recombination and may not be representative of events taking place on chromosomal sequences. Further experiments are therefore needed to measure homologous recombination using plasmids integrated into genomic DNA.

A second class of recombination identified in yeast cells is the single-strand annealing pathway, which occurs by the pairing of short direct repeat sequences exposed in single-stranded tails *(22)*. A third class, which may, like homologous recombination, represent more that one mechanism, occurs in the absence of any homology or utilizes just 1 or 2 bp of homology *(52)*. This mechanism can involve deletions or insertions of a few base pairs. A similar mechanism of direct end-joining has been characterized in mammalian cells, and has been proposed to be the dominant mechanism by which mammalian cells integrate foreign DNA. Significantly, V(D)J recombination does not rely on any significant homology, and involves a process of direct end joining. In line with this relationship, *xrs* cells appear to have a decreased ability to integrate foreign DNA compared to parental cells *(31,45)*. With current knowledge, the most likely prediction is therefore that this mechanism might be a process dependent on DNA-PK.

9. CELL-CYCLE ARREST AFTER DNA DAMAGE

The ability of Ku and DNA-PK_{cs} to activate a protein kinase specifically following the introduction of DSBs has fueled speculation that one function of DNA-PK is to participate in a signaling pathway alerting cells to the presence of DSBs. The fact that DNA-PK_{cs} is a member of the "PI-3 kinase" superfamily, which includes the *ATM* and *S. pombe rad3* genes, both of which are involved in checkpoint pathways, has added further spark to these ideas. Studies with yeast cells have highlighted the importance of cell-cycle arrest following DNA damage, and show that defects in signaling pathways can result in sensitivity to DNA-damaging agents *(1,91)*. Following irradiation, mammalian cells show pronounced arrests at G1/S and at the G2/M boundary, as well as a reduced rate of DNA synthesis. AT cells are defective in all of these checkpoints following radiation. The *xrs* cells show a markedly different phenotype to AT cells, and exhibit a more pronounced delay in both the S-phase and G2 checkpoints *(40,90)*. Indeed the available data suggest that the arrest in G2 could be terminal in *xrs* cells and may be the ultimate cause of lethality. These observations are compatible with a model in which an unrepaired DSB signals an arrest that cannot be overcome until the break has been repaired. Since *xrs* cells have an enhanced number of unrejoined DSBs, their delay is greater and possibly permanent. The inability of *xrs* cells to induce DNA-PK activity suggests that DNA-PK is not involved in this signaling pathway. The arrest of mammalian cells in G1 following irradiation is dependent on a functional p53 protein, although lack of radiosensitivity of p53-defective cell lines indicates that this "checkpoint" is of little significance to the final viability of normal cells *(47)*. Cultured rodent cell lines show a less pronounced G1 arrest following irradiation compared to human cells, making it difficult to evaluate the defective mutants. Although this may raise questions regarding the significance of this arrest, a role of DNA-PK cannot, at present, be excluded. Since V(D)J recombination occurs during G1, it is possible that an arrest at this stage could be of some functional significance *(57,71)*.

10. CHROMOSOME DAMAGE

Following exposure to ionizing radiation, damage to DNA and its consequences can be observed in the cell by examining metaphase chromosomes. Aberrant chromosome structures observed microscopically can be divided into two classes: chromosome aberrations in which both chromatids have been damaged, and chromatid aberrations in which only one of the pair of chromatids is involved. Chromosome aberrations include such structures as chromosome breaks, ring, and dicentric chromosomes. Examples of chromatid aberrations are chromatid gaps and breaks, and exchange aberrations, such as triradials *(70)*. Cells irradiated at different stages of the cell cycle yield a different pattern of chromosome or chromatid aberrations, and the *xrs* cells, which have been the most extensively studied for aberration formation, yield an enhanced frequency of aberrations as well as a different pattern of chromosome vs chromatid aberration formation compared to parental cells *(16,48)*. These differences include the formation of chromatid as well as chromosome aberrations following irradiation in G1 phase in *xrs* cells, whereas only chromosome aberrations arise in G1 irradiated parental cells. A significant observation emerging from the analysis of metaphase chromosomes is that *xrs* and *scid* cells, in addition to showing enhanced frequencies of chromosome and chromatid gaps and breaks, indicative of unrejoined DSBs, additionally show enhanced dicentrics, ring chromosomes, and exchange type aberrations *(16,18,48)*. Thus, although defective in DSB rejoining by physical assays (PFGE, neutral elution, and so forth), these cells show elevated aberrant rejoining of broken chromosomes. This suggests the existence of a DNA-PK-independent mechanism of rejoining the ends of broken chromosomes. Surprisingly, the patterns of aberration formation observed in *xrs* and AT cells are remarkably similar, despite the fundamental difference in the basis of their radiosensitivity *(16)*. It thus appears that an elevated level of DSBs in mitosis, which presumably can arise either through defective repair or defective cell cycle arrest, can lead to enhanced chromosome exchanges.

11. SPECULATIONS AND FUTURE PERSPECTIVES

The identification of components of DNA-PK as the gene products defined by IR complementation groups 5–7 has fulfilled many of the predictions based on the properties of the defective mutants and the biochemical characterization of the enzyme complex. The combined defects of all of these mutants in DSB rejoining and in V(D)J recombination suggested that the gene products might operate in a single pathway and possibly within the same complex. The specificity of DNA-PK for double-stranded ends correlates with the specific and pronounced sensitivity of the group 5–7 mutants to agents that induce DSBs. Finally, the more severe defect in V(D)J recombination in the group 5 compared to group 7 mutants rendered the prediction that *XRCC5* operates upstream of *XRCC7*, a speculation now fulfilled by the known binding of Ku to DNA ends prior to binding of DNA-PK_{cs}. There are, however, many loose ends to the parcel, and the precise mechanisms by which DNA-PK participates in DSB rejoining is far from understood. A consideration of the biological properties of the defective mutants together with the biochemical properties of DNA-PK may help to tidy the ragged packaging.

Taken together, the data are consistent with a model in which the Ku protein, without necessarily being preassociated with DNA-PK_{cs}, binds to double-stranded DNA ends, and protects them from nucleolytic degradation *(39,83)*. DNA-bound Ku then binds DNA-PK_{cs}, inducing its kinase activity, and facilitates, in some way, the rejoining of the DNA ends (Fig. 2). The function of the protein kinase activity is currently

Step 1. Ku protein binds to ds DNA ends

Step 2. DNA-PKcs binds to DNA bound Ku and activates DNA-PK activity

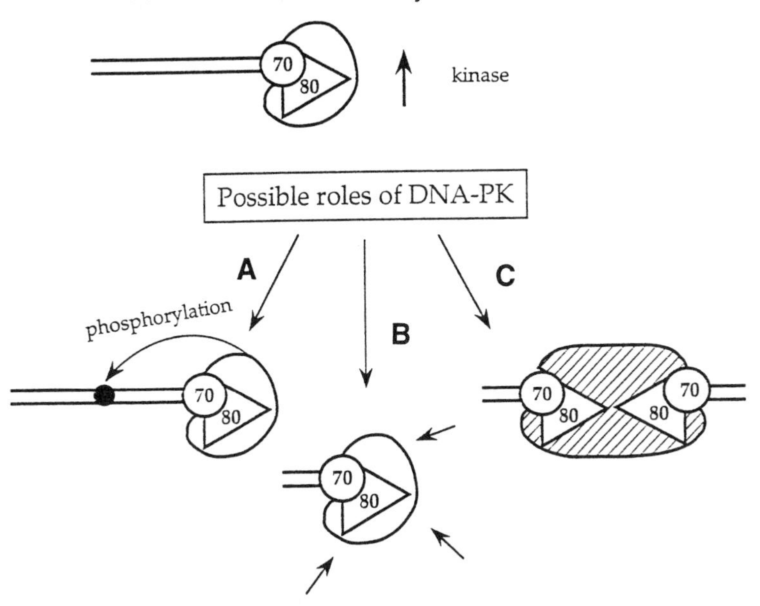

Fig. 2. Model for role of DNA-PK in DSB repair. Possible roles of DNA-PK are: **(A)** function as a kinase, e.g. arrest of the transcriptional apparatus in the vicinity of a DSB by phosphorylation, **(B)** framework around which other enzymes involved in DNA repair bind, and **(C)** Holding broken ends within the chromatin structure to prevent degradation and/or aberrant rejoining. There are no data regarding the stoichiometry of binding of DNA-PK to DNA ends to favor any particular mechanism for this, and just one possibility is shown in Fig. 2.

unclear. Despite the many proteins DNA-PK can phosphorylate in vitro, which includes transcription factors such as Sp-1, *c-jun*, and p53, its in vivo target remains to be identified *(3)*. Due consideration should, however, be given to the observation that DNA-PK is most efficient in phosphorylating proteins bound to the same DNA molecule. This may suggest a function in affecting metabolic events taking place in the vicinity of the DSB. The lack of full complementation by human DNA-PK_{cs} in a hamster background, despite efficient induction of protein kinase activity, may indicate that DNA-PK_{cs} plays a role in addition to its function as a kinase *(5)*. The large size of the protein raises the possibility that it could play a structural role in the rejoining process.

The group 4 and 5 mutants differ from the group 7 mutants in two ways. First, the group 7 mutants are insensitive to topoisomerase II-inhibiting drugs, and second, they are able to carry out signal join formation during V(D)J recombination *(2,25,42,55)*. It is tempting to seek a clue to the mechanism of the process from these observations. These two events (rejoining of topoisomerase II-associated breaks and signal join formation) can clearly take place in the absence of DNA-PK activity, but since the mutation has not yet been identified in either *scid* or V-3 cells, an additional role for DNA-PK, such as a structural role, cannot yet be excluded. An intriguing possibility is that both of these examples could involve protein-associated breaks if the V(D)J site-specific endonuclease, now known to be RAG1/RAG2 *(88)*, or an associated complex, remains attached to the break. One possibility is that Ku is able to bind to such protein-associated ends and serves a function in protecting them, but that the large DNA-PK_{cs} component is precluded from binding. In the case of topoisomerase II-associated breaks, the final rejoining step may be provided by topoisomerase II itself, and in the case of signal join formation, an associated V(D)J complex may facilitate the ligation. Both models imply a role for Ku independently of its role within the DNA-PK complex.

Another intriguing phenotype of the mutants is that all of them show only a two- to fourfold defect in DSB rejoining compared to a major sensitivity to ionizing radiation, and a major defect in V(D)J recombination *(4,27,33,49,63,80,82,90)*. Indeed V(D)J coding junctions have not been recovered from *xrs* cells. All the mutants show residual DSB rejoining, yet are devoid of recovery when held under nongrowing conditions in which a process enhancing survival (potentially lethal damage repair) is known to take place in wild-type cells. An additional feature is that analysis of chromosome damage following irradiation shows not only an increase in gaps and breaks indicative of unrejoined DSBs, but also an increase in rearrangements indicative of abnormal rejoining. One possible explanation is that the cell possesses several mechanisms for DSB rejoining, of which the DNA-PK-dependent process handles a defined subset, and is the only mechanism operating on the site-specific DSBs induced during V(D)J recombination. An additional possibility is that these mutants are not defective in rejoining *per se*, and indeed, they have normal levels of ligase activity *(12,78)*, but rather that the absence of DNA-PK either renders the ends irreparable, or causes them to be channeled into a mechanism of rejoining that does not enhance survival. Such an explanation would be compatible with the enhanced frequency of rearrangements as well as broken chromosomes seen in metaphase preparations of *xrs* cells, and with the observation that *xrs* cells possess a greater number of "irreparable" breaks. The absence of Ku in *xrs* cells, as suggested previously, may result in enhanced nucleolytic degradation of DNA ends, release them from the normal constraints of chromatin structure, and allow aberrant rejoining to take place. However, it should be noted that *scid* cells also show enhanced rearranged metaphase chromosomes, so that these observations are not merely owing to the lack of "end protection" resulting from an absence of Ku, but may require the intact DNA-PK complex *(18)*.

A consideration of the mechanisms of DSB rejoining indicates that the DNA-PK-dependent pathway may represent a process of direct end joining with limited or no dependence on DNA homology. This process appears to play a major role in handling breaks in higher organisms, but its role in lower eukaryotes appears to be of less significance *(93)*. Thus, in contrast to mammalian cells, the majority of DSBs in yeast appear to be handled by a *RAD52*-dependent mechanism involving homologous recombination. Moreover, a yeast mutant defective in an apparent Ku70 homolog shows normal radioresistance *(8a,72)*. Nevertheless, recent studies have shown that yeast cells do possess a direct end joining mechanism for rejoining DSBs *(52)*. This mechanism appears frequently to involve small insertions and deletions, and it is possible that these may be of greater significance to a small organism containing few introns than to a larger organism containing copious amounts of redundant DNA. The extent of utilization of the repair process may therefore have changed to suit the needs of the organism.

Reviewed here are the phenotypes of the defective mutants in light of their recently identified genetic defects. However, many unanswered questions remain. It is intriguing that DNA-PK is a highly abundant protein in human cells, whereas rodent cells contain 50-fold lower levels. Rodent and human cells display very similar radiosensitivities and abilities to carry out V(D)J recombination, and the rodent DNA-PK-defective mutants (including the *scid* mouse) grow normally in the absence of DNA damage. The construction of knockout mice should provide further information concerning any other roles played by components of DNA-PK. Additionally, the features of the mutants provide few clues to the function of the kinase activity induced specifically following the introduction of DSBs. Additional radiosensitive mutants, some with defects in DSB rejoining and some proficient in DSB repair, have been described and could provide important tools to identify further enzymes and pathways of DSB repair *(76,89,94)*. The mutants described here are undergoing extensive analysis, which should provide exciting insights into the mechanism by which DNA-PK participates in handling DSBs.

ACKNOWLEDGMENTS

I wish to thank G. Taccioli, S. P. Jackson, and members of the MRC CMU, including A. R. Lehmann in particular, for their input into this work. I apologize to the authors of the many studies that I have been unable to reference because of lack of space.

REFERENCES

1. Al-Khodairy, F. and A. M. Carr. 1992. DNA repair mutants defining G2 checkpoint pathways in Schizosaccharomyces pombe. *EMBO J.* **11:** 1343–1350.
2. Alt, F. W., E. M. Oltz, F. Young, J. Gorman, G. Taccioli, and J. Chen. 1992. (VDJ) recombination. *Immunol. Today* **13:** 306–314.
3. Anderson, C. W. 1993. DNA damage and the DNA-activated protein kinase. *TIBS* **18:** 433–437.
4. Biedermann, K. A., J. Sun, A. J. Giaccia, L. M. Tosto, and J. M. Brown. 1991. *scid* mutation in mice confers hypersensitivity to ionizing radiation and a deficiency in DNA double-strand break repair. *Proc. Natl. Acad. Sci. USA* **88:** 1394–1397.
5. Blunt, T., N. J. Finnie, G. E. Taccioli, G. C. M. Smith, J. Demengeot, T. M. Gottlieb, R. Mizuta, A. J. Varghese, F. W. Alt, P. A. Jeggo, and S. P. Jackson. 1995. Defective DNA-dependent protein kinase activity is linked to V(D)J recombination and DNA repair defects associated with the murine *scid* mutation. *Cell* **80:** 813–823.
6. Bosma, G. C., R. P. Custer, and M. J. Bosma. 1983. A severe combined immunodeficiency mutation in the mouse. *Nature* **301:** 527–530.
7. Boubnov, N. V., K. T. Hall, Z. Wills, E. L. Sang, M. H. Dong, D. M. Benjamin, C. R. Pulaski, H. Band, W. Reeves, E. A. Hendrickson, and D. T. Weaver. 1995. Complementation of the ionizing radiation sensitivity, DNA end binding, and V(D)J recombination defects of double-strand break repair mutants by the p86 Ku autoantigen. *Proc. Natl. Acad. Sci. USA* **92:** 890–894.
8. Boubnov, N. V. and D. T. Weaver. 1995. *scid* cells are deficient in Ku and replication protein A phosphorylation by the DNA-dependent protein kinase. *Mol. Cell. Biol.* **15:** 5700–5706.

8a. Boulton, S. J. and S. P. Jackson. 1996. *Saccharomyces cerevisiae* Ku70 potentiates illegitimate DNA double-strand break repair and serves as a barrier to error-prone DNA repair pathways. *EMBO J.* **15:** 5093–5103.
9. Cai, Q.-Q., A. Plet, J. Imbert, M. Lafage-Pochitaloff, C. Cerdan, and J.-M. Blanchard. 1994. Chromosomal location and expression of the genes coding for Ku p70 and p80 in human cell lines and normal tissues. *Cytogenet. Cell Genet.* **65:** 221–227.
10. Caldecott, K., G. Banks, and P. A. Jeggo. 1990. DNA double-strand break repair pathways and cellular tolerance to inhibitors of topoisomerase II. *Cancer Res.* **50:** 5778–5783.
11. Caldecott, K. and P. Jeggo. 1991. Cross-sensitivity of gamma-ray-sensitive hamster mutants to cross-linking agents. *Mutat. Res.* **255:** 111–121.
12. Chang, C., K. A. Biedermann, M. Mezzina, and J. M. Brown. 1993. Characterization of the DNA double strand break repair defect in *scid* mice. *Cancer Res.* **53:** 1244–1248.
13. Costa, N. D. and P. E. Brayant. 1988. Repair of DNA single-strand and double-strand breaks in the Chinese hamster xrs-5 mutant cell line as determined by DNA-unwinding. *Mutat. Res.* **194:** 93–99.
14. Costa, N. D. and P. E. Bryant. 1991. Elevated levels of DNA double-strand breaks (dsb) in restriction enfidonuclease-treated *xrs5* cells correlate with the reduced capacity to repair dsb. *Mutat. Res.* **255:** 219–226.
15. Dahm-Daphi, J., E. Dikomey, C. Pyttlik, and P. A. Jeggo. 1993. Reparable and non-reparable DNA strand breaks induced by X-irradiation in CHO K1 cells and the radiosensitive mutants xrs1 and xrs5. *Radiat. Biol.* **64:** 19–26.
16. Darroudi, F. and A. T. Natarajan. 1987. Cytogenetical characterization of Chinese hamster ovary X-ray-sensitive mutant cells xrs 5 and xrs 6: I. Induction of chromosomal aberrations by X-irradiation and its modulation by 3-aminobenzamide and caffeine. *Mutat. Res.* **177:** 133–148.
17. Denekamp, J., G. F. Whitmore, and P. A. Jeggo. 1989. Biphasic survival curves for xrs radiosensitive cells: Subpopulations or transient expression of repair competence? *Int. J. Radiat. Biol.* **55:** 605–617.
18. Disney, J. E., A. L. Barth, and L. D. Schultz. 1992. Defective repair of radiation-induced chromosomal damage in *scid/scid* mice. *Cytogenet. Cell Genet.* **59:** 39–44.
19. Dvir, A., S. R. Peterson, M. W. Knuth, H. Lu, and W. S. Dynan. 1992. Ku autoantigen is the regulatory component of a template-associated protein kinase that phosphorylates RNA polymerase II. *Proc. Natl. Acad. Sci. USA* **89:** 11,920–11,924.
20. Errami, A., V. Smider, W. K. Rathmeli, D. M. He, E. A. Hendrickson, M. Z. Zdzienicka, and G. Chu. 1996. Ku86 defines the genetic defect and restores X-ray resistance and V(D)J recombination to complementation group 5 hamster cell mutants. *Mol. Cell. Biol.* **16:** 1519–1526.
21. Finnie, N. J., T. M. Gottlieb, T. Blunt, P. A. Jeggo, and S. P. Jackson. 1995. DNA-dependent protein-kinase activity is absent in xrs-6 cells—implications for site-specific recombination and DNA double-strand break repair. *Proc. Natl. Acad. Sci. USA* **92:** 320–324.
22. Fishman-Lobell, J., N. Rudin, and J. Haber. 1992. Two alternative pathways of double-strand beak repair that are kinetically separable and independently modulated. *Mol. Cell. Biol.* **12:** 1292–1303.
23. Friedberg, E. C. 1991. Yeast genes involved in DNA-repair processes: new looks on old faces. *Mol. Microbiol.* **5:** 2303–2310.
24. Fulop, G. M. and R. A. Phillips. 1990. The *scid* mutation in mice causes a general defect in DNA repair. *Nature* **374:** 479–482.
25. Gellert, M. 1992. Molecular analysis of V(D)J recombination. *Annu. Rev. Genet.* **26:** 425–446.
26. Getts, R. C., and T. D. Stamato. 1994. Absence of a Ku-like DNA end binding-activity in the xrs double-strand DNA repair-deficient mutant. *J. Biol. Chem.* **269:** 15,981–15,984.

27. Giaccia, A., R. Weinstein, J. Hu, and T. D. Stamato. 1985. Cell cycle-dependent repair of double-strand DNA breaks in a gamma-ray sensitive Chinese hamster cell. *Somatic Cell Molec. Gen.* **11:** 485–491.
28. Giaccia, A. J., R. A. MacLaren, N. Denko, D. Nicolaou, and T. D. Stamato. 1990. Increased sensitivity to killing by restriction enzymes in the XR-1 DNA double-strand break repair-deficient mutant. *Mutat. Res.* **236:** 67–76.
29. Gottlieb, T. M. and S. P. Jackson. 1993. The DNA-dependent protein kinase: requirement of DNA ends and association with Ku Antigen. *Cell* **72:** 131–142.
30. Hafezparast, M., G. P. Kaur, M. Zdzienicka, R. S. Athwal, A. R. Lehmann, and P. A. Jeggo. 1993. Sub-chromosomal localisation of a gene *(XRCC5)* involved in double strand break repair to the region 2q34-36. *Somatic Cell Mol. Genet.* **19:** 413–421.
31. Hamilton, A. A., and J. Thacker. 1987. Gene recombination in X-ray sensitive hamster cells. *Mol. Cell. Biol.* **7:** 1409–1414.
32. Hartley, K. O., D. Gell, G. C. M. Smith, H. Zhang, N. Divecha, M. A. Connelly, A. Admon, S. P. Lees-Miller, C. W. Anderson, and S. P. Jackson. 1995. DNA-dependent protein kinase catalytic subunit: a relative of phosphatidylinositol 3-kinase and the ataxia telangiectasia gene product. *Cell* **82:** 849–856.
33. Hendrickson, E. A., X.-Q. Qin, E. A. Bump, D. G. Schatz, M. Oettinger, and D. T. Weaver. 1991. A link between double-strand break-related repair and V(D)J recombination: the *scid* mutation. *Proc. Natl. Acad. Sci. USA* **88:** 4061–4065.
34. Hesse, J. E., M. R. Lieber, M. Gellert, and K. Mizuuchi. 1987. Extrachromosomal DNA substrates in pre-B cells undergo inversion or deletion at immunoglobulin V-(D)-J joining signals. *Cell* **49:** 775–783.
35. Iliakis, G. 1991. The role of DNA double strand breaks in ionizing radiation-induced killing of eukaryotic cells. *BioEssays* **13:** 641–648.
36. Iliakis, G., R. Mehta, and M. Jackson. 1992. Level of DNA double-strand break rejoining in Chinese hamster *xrs-5* cells is dose-dependent: implications for the mechanism of radiosensitivity. *Int. J. Radiat. Biol.* **61:** 315–321.
37. Iliakis, G. and R. Okayasu. 1990. Radiosensitivity throughout the cell cycle and repair of potentially lethal damage and DNA double-strand breaks in an X-ray-sensitive CHO mutant. *Int. J. Radiat. Biol.* **57:** 1195–1211.
38. Jackson, S. P. and P. A. Jeggo. 1995. DNA double-strand break repair and V(D)J recombination: involvement of DNA-PK. *TIBS* **20:** 412–415.
39. Jeggo, A. P., G. E. Taccioli, and S. P. Jackson. 1995. Menage à trois: double strand break repair, V(D)J recombination and DNA-PK. *Bioessays* **17:** 949–957.
40. Jeggo, P. A. 1985. X-ray sensitive mutants of Chinese hamster ovary cell line: Radiosensitivity of DNA synthesis. *Mutat. Res.* **145:** 171–176.
41. Jeggo, P. A. 1990. Studies on mammalian mutants defective in rejoining double-strand breaks in DNA. *Mutat. Res.* **239:** 1–16.
42. Jeggo, P. A., K. Caldecott, S. Pidsley, and G. R. Banks. 1989. Sensitivity of chinese hamster ovary mutants defective in DNA double strand break repair to topoisomerase II inhibitors. *Cancer Res.* **49:** 7057–7063.
43. Jeggo, P. A., and R. Holliday. 1986. Azacytidine induced reactivation of a DNA repair gene in Chinese hamster ovary cells. *Mol. Cell. Biol.* **6:** 2944–2949.
44. Jeggo, P. A., and L. M. Kemp. 1983. X-ray sensitive mutants of Chinese hamster ovary cell line. Isolation and cross-sensitvity to other DNA damaging agents. *Mutat. Res.* **112:** 313–327.
45. Jeggo, P. A. and J. Smith-Ravin. 1989. Decreased stable transfection frequences of six X-ray-sensitive CHO strains, all members of the xrs complementation group. *Mutat. Res.* **218:** 75–86.
46. Jeggo, P. A., J. Tesmer, and D. J. Chen. 1991. Genetic analysis of ionising radiation sensitive mutants of cultured mammalian cell lines. *Mutat. Res.* **254:** 125–133.

47. Kastan, M. B., Q. Zhan, W. S. El-Deiry, F. Carrier, T. Jacks, W. V. Walsh, B. S. Plunkett, B. Vogelstein, and A. J. Fornace. 1992. A mammalian cell cycle checkpoint pathway utilizing p53 and GADD45 is defective in ataxia-telangiectasia. *Cell* **71:** 587–597.
48. Kemp, L. M. and P. A. Jeggo. 1986. Radiation induced chromosome damage in X-ray sensitive mutants (xrs) of the chinese hamster ovary cell line. *Mutat. Res.* **166:** 255–263.
49. Kemp, L. M., S. G. Sedgwick, and P. A. Jeggo. 1984. X-ray sensitive mutants of Chinese hamster ovary cells defective in double-strand break rejoining. *Mutat. Res.* **132:** 189–196.
50. Kirchgessner, C. U., C. K. Patil, J. W. Evans, C. A. Cuomo, L. M. Fried, T. Carter, M. A. Oettinger, J. M. Brown, G. Iliakis, R. Mehta, and M. Jackson. 1995. DNA-dependent kinase (p350) as a candidate gene for murine SCID defect. *Science* **267:** 1178–1183.
51. Kirchgessner, C. U., L. M. Tosto, K. A. Biedermann, M. Kovacs, D. Araujo, E. J. Stanbridge, and J. M. Brown. 1993. Complementation of the radiosensitive phenotype in severe combined immunodeficient mice by human chromosome 8. *Cancer Res.* **53:** 6011–6016.
52. Kramer, K. M., J. A. Brock, K. Bloom, J. K. Moore, and J. E. Haber. 1994. Two different types of double-strand breaks in *Saccharomyces cerevisiae* are repaired by similar *RAD52*-independent, nonhomologous recombination events. *Mol. Cell. Biol.* **14:** 1293–1301.
53. Lee, S. E., D. M. He, and E. A. Hendrickson. 1996 Characterization of Chinese hamster cell lines that are X-ray-sensitive, impaired in DNA double-strand break repair and defective for V(D)J recombination, in *Current Topics in Microbiology and Immunology, Volume 217 on Molecular Analysis of DNA Rearrangements in the Immune System* (Jessberger R. and M. R. Lieber, eds.), Springer-Verlag, Heidelberg, pp. 133–142.
54. Lee, S. E., C. R. Pulaski, D. M. He, D. M. Benjamin, M. Voss, J. Um, and E. A. Hendrickson. 1995. Isolation of mammalian cell mutants that are X-ray sensitive, impaired in DNA double-strand break repair and defective for V(D)J recombination. *Mutat. Res.* **336:** 279–291.
55. Lewis, S. M. 1994. The mechanism of V(D)J joining: Lessons from molecular, immunological, and comparative analyses. *Adv. Immunol.* **56:** 27–150.
56. Li, Z., T. Otevrel, Y. Gao, H.-L. Cheng, B. Seed, T. D. Stamato, G. E. Taccioli, and F. W. Alt. 1995. The *XRCC4* gene encodes a novel protein involved in DNA double-strand break repair and V(D)J recombination. *Cell* **83:** 1079–1089.
57. Lin, W. C. and S. Desiderio 1994. Cell cycle regulation of V(D)J recombination-activating protein RAG-2. *Proc. Natl. Acad. Sci. USA* **91:** 2733–2737.
58. Mateos, S., P. Slijepcevic, R. A. F. MacLeod, and P. E. Bryant. 1994. DNA double-strand break rejoining in *xrs5* cells is more rapid in the G2 than in the G1 phase of the cell cycle. *Mutat. Res.* **315:** 181–187.
59. Mimori, T., M. Akizuki, H. Yamagata, S. Inada, S. Yoshida, and M. Homma. 1981. Characterization of a high molecular weight acidic nuclear protein recognized by autoantibodies in sera from patients with polymyositis-scleroderma overlap syndrome. *J. Clin. Invest.* **68:** 611–620.
60. Mimori, T., J. A. Hardin, and J. A. Steitz. 1986. Characterization of the DNA-binding protein antigen Ku recognized by autoantibodies from patients with rheumatic disorders. *J. Biol. Chem.* **261:** 2274–2278.
61. Mimori, T., Y. Ohosone, N. Hama, A. Suwa, M. Akizuki, M. Homma, A. J. Griffith, and J. A. Hardin. 1990. Isolation and characterization of cDNA encoding the 80-kDa subunit protein of the human autoantigen Ku (p70/p80) recognized by autoantibodies from patients with scleroderma-polymyositis overlap syndrome. *Proc. Natl. Acad. Sci. USA* **87:** 1777–1781.
62. Moore, P. D., K.-Y. Song, L. Chekuri, L. Wallace, and R. S. Kucherlapati. 1986. Homologous recombination in a Chinese hamster X-ray sensitive mutant. *Mutat. Res.* **160:** 149–155.
63. Pergola, F., M. Z. Zdzienicka, and M. R. Lieber. 1993. V(D)J recombination in mammalian cell mutants defective in DNA double strand break repair. *Mol. Cell. Biol.* **13:** 3464–3471.

64. Peterson, S. R., A. Kurimasa, M. Oshimura, W. S. Dynan, E. M. Bradbury, and D. J. Chen. 1995. Loss of the catalytic subunit of the DNA-dependent protein kinase in DNA double-strand-break-repair mutant mammalian cells. *Proc. Natl. Acad. Sci. USA* **92:** 3171–3174.
65. Rathmell, W. K. and G. Chu. 1994. A DNA end-binding factor involved in double-strand break repair and V(D)J recombination. *Mol. Cell. Biol.* **14:** 4741–4748.
66. Rathmell, W. K., and G. Chu. 1994. Involvement of the Ku autoantigen in the cellular response to DNA double-strand breaks. *Proc. Natl. Acad. Sci. USA* **91:** 7623–7627.
67. Reeves, W. H. and Z. M. Sthoeger. 1989. Molecular cloning of cDNA encoding the p70 (Ku) lupus autoantigen. *J. Biol. Chem.* **264:** 5047–5052.
68. Ross, G. M., J. J. Eady, N. P. Mithal, C. Bush, G. G. Steel, and P. A. Jeggo. 1995. DNA strand break rejoining defect in *xrs-6* is complemented by transfection with the human Ku80 gene. *Cancer Res.* **55:** 1235–1238.
69. Ruiz de Almodovar, J. M., M. I. Nunez, T. J. McMillan, N. Olea, C. Mort, M. Villalobos, V. Pedraza, and G. G. Steel. 1994. Initial radiation-induced DNA damage in human tumour cell lines: a correlation with intrinsic cellular radiosensitivity. *Br. J. Cancer* **69:** 457–462.
70. Savage, J. R. K. 1975. Classification and relationships of induced chromosomal structural changes. *J. Med. Genet.* **12:** 103–122.
71. Schlissel, M., A. Constantinescu, T. Morrow, M. Baxter, and A. Peng. 1993. Double-strand signal sequence breaks in V(D)J recombination are blunt, 5'-phosphorylated, RAG-dependent, and cell cycle regulated. *Genes and Development* **7:** 2520–2532.
72. Siede, W., I. Dianova, A. S. Friedberg, A. Friedl, F. Eckardt-Schupp, and E. C. Friedberg. 1995. Characterization of G1 checkpoint arrest in *S. cerevisiae* following treatment with DNA-damaging agents. *J. Biochem.* **S21A:** 344.
73. Singleton, B. K., A. Priestley, D. Gell, T. Blunt, S. P. Jackson, A. R. Lehmann, and P. A. Jeggo. 1997. Molecular and biochemical characterisation of mutants defective in Ku80. *Mol. Cell Biol.* **17**, in press.
74. Sipley, J. D., J. C. Menninger, K. O. Hartley, D. C. Ward, S. P. Jackson, and C. W. Anderson. 1995. Gene for the catalytic subunit of the human DNA-activated protein kinase maps to the site of the *XRCC7* gene on chromosome 8. *Proc. Natl. Acad. Sci. USA* **92:** 7515–7519.
75. Smider, V., W. K. Rathmell, M. R. Lieber, and G. Chu. 1994. Restoration of X-ray resistance and V(D)J recombination in mutant-cells by Ku cDNA. *Science* **266:** 288–291.
76. Stackhouse, M. A. and J. S. Bedford. 1993. An ionizing radiation-sensitive mutant of CHO cells: irs–20. I. Isolation and initial characterization. *Radiat. Res.* **136:** 241–249.
77. Stamato, T. D., A. Dipatri, and A. Giaccia. 1988. Cell-cycle-dependent repair of potentially lethal damage in the XR–1 gamma-ray-sensitive Chinese hamster ovary cell. *Radiat. Res.* **115:** 325–333.
78. Stamato, T. D. and J. Hu. 1987. Normal DNA ligase activity in a gamma-ray-sensitive Chinese hamster mutant. *Mutat. Res.* **183:** 61–67.
79. Stamato, T. D., R. Weinstein, A. Giaccia, and L. Mackenzie. 1983. Isolation of cell cycle-dependent gamma ray-sensitive Chinese hamster ovary cell. *J. Somatic Cell Genet.* **9:** 165–173.
80. Taccioli, G. E., H.-L. Cheng, A. J. Varghese, G. Whitmore, and F. W. Alt. 1994. A DNA repair defect in Chinese hamster ovary cells affects V(D)J recombination similarly to the murine *scid* mutation. *J. Biol. Chem.* **269:** 7439–7442.
81. Taccioli, G. E., G. Rathbun, E. Oltz, T. Stamato, P. A. Jeggo, and F. W. Alt. 1993. Impairment of V(D)J recombination in double-strand break repair mutants. *Science* **260:** 207–210.
82. Taccioli, G. E., G. Rathbun, Y. Shinkai, E. M. Oltz, H. Cheng, G. Whitmore, T. Stamato, P. Jeggo, and F. W. Alt. 1992. Activities involved in V(D)J recombination. *Curr. Topics Microbiol. Immunol.* **182:** 108–114.
83. Taccioli, T. G., T. M. Gottlieb, T. Blunt, A. Priestley, J. Demengeot, R. Mizuta, A. R. Lehmann, F. W. Alt, S. P. Jackson, and P. A. Jeggo. 1994. Ku80: product of the XRCC5 gene. Role in DNA repair and V(D)J recombination. *Science* **265:** 1442–1445.

84. Thacker, J. and A. Stretch. 1985. Responses of 4 X-ray-sensitive CHO cell mutants to different radiation and to irrradiation conditions promoting cellular recovery. *Mutat. Res.* **146:** 99–108.
85. Thacker, J. and R. E. Wilkinson. 1991. The genetic basis of resistance to ionising radiation damage in cultured mammalian cells. *Mutat. Res.* **254:** 135–142.
86. Thompson, L. H. and P. A. Jeggo. 1995. Nomenclature of human genes involved in ionizing radiation sensitivity. *Mutat. Res.* **337:** 131–134.
87. Troelstra, C. and N. G. J. Jaspers. 1994. Ku starts at the end. *Curr. Biol.* **4:** 1149–1151.
88. van Gent, D. C., J. F. McBlane, D. A. Ramsden, M. J. Sadofsky, J. E. Hesse, and M. Gellert. 1995. Initiation of V(D)J recombination in a cell-free system. *Cell* **81:** 925–934.
89. Warters, R. L., L. R. Barrows, and D. J. Chen. 1995. DNA double-strand break repair in two radiation-sensitive mouse mammary carcinoma cell lines. *Mutat. Res.* **336:** 1–7.
90. Weibezahn, K. F., H. Lohrer, and P. Herrlich. 1985. Double strand break repair and G2 block in Chinese hamster ovary cells and their radiosensitive mutants. *Mutat. Res.* **145:** 177–183.
91. Weinert, T. A. and L. H. Hartwell. 1988. The RAD9 gene controls the cell cycle response to DNA damage in *Saccharomyces cerevisiae*. *Science* **241:** 317–322.
92. Whitmore, G. F., A. J. Varghese, and S. Gulyas. 1989. Cell cycle responses of two X-ray sensitive mutants defective in DNA repair. *Int. J. Radiat. Biol.* **56:** 657–665.
93. Wilson, J. H., P. B. Berget, and J. M. Pipas. 1982. Somatic cells efficiently join unrelated DNA segments end-to-end. *Mol. Cell. Biol.* **2:** 1258–1269.
94. Zdzienicka, M. Z. 1995. Mammalian mutants defective in the response to ionizing radiation-induced DNA damage. *Mutat. Res.* **336:** 203–213.

18

Nucleotide Excision Repair

Its Relation to Human Disease

Larry H. Thompson

1. INTRODUCTION

1.1. Biological Importance of Nucleotide Excision Repair (NER)

Historically, the major contribution of excision repair processes to recovery of cells from DNA damage was evident from the greatly increased sensitivity of mutant cells that were defective in this repair process *(121)*. Ultraviolet (UV) (usually referring to UV-C, 240–290 nm) and ionizing radiations are often used as prototype DNA-damaging agents because of their ease of delivery to cells and their relative selectivity for damaging nucleic acids vs other cellular macromolecules. The spectra of DNA damage produced by these two types of radiation differ greatly and consequently, so do the repair pathways that act on these damages. Repair of UV damage produces a much longer "repair patch" than damage from ionizing radiation, i.e., ~30 nucleotides *(124,143)* vs one or several nucleotides, respectively *(76,97,298)*. Although ionizing radiation results in a complex mixture of base damage and strand breaks (*see* Chapter 4), the biologically relevant UV-induced photoproducts are predominantly cyclobutane pyrimidine dimers (CPDs) and (6-4) photoproducts ([6-4]PDs), with the latter being 15–25% of the total lesions *(89,102,252)*. In mammalian cells these major photoproducts are repaired exclusively by the NER pathway (recently reviewed in *[135,136,198,218a,317,319,401]*).

The biological consequences of each of these types of lesions will depend on DNA sequence context, the relative efficiency of repair, and the particular end point (e.g., cell survival, mutation induction, or inhibition of DNA replication or transcription). The relative biological importance of the (6-4)PDs in mutagenesis and carcinogenesis has recently become much more appreciated *(99,252)*. Indeed, in many contexts, such as in vivo repair synthesis measured immediately after UV damage, or in vitro repair synthesis in human cell extracts, (6-4)PDs are responsible for most of the observed effect *(102,249,450)*. The same NER process that protects skin cells by excising UV photoproducts *(251,313,427)* also acts on a multitude of bulky chemical adducts *(37)* and may protect internal tissues against cancer *(180)*. Examples of such chemicals are the polycyclic aromatic hydrocarbons (e.g., benzo[*a*]pyrene) associated with the combustion of fossil fuels, the heterocyclic amines (e.g., 2-amino-1-methyl-6-

From: DNA Damage and Repair, Vol. 2: DNA Repair in Higher Eukaryotes
Edited by: J. A. Nickoloff and M. F. Hoekstra © Humana Press Inc., Totowa, NJ

phenylimidazo[4,5-*b*]pyridine, PhIP) produced during the cooking of food *(399)*, the synthetic carcinogen *N*-acetoxy-2-acetylaminofluorene *(391)*, and many agents used in cancer chemotherapy (e.g., mitomycin C, cisplatin, melphalan). The interstrand crosslinks produced by these latter agents are probably repaired by a recombinational mechanism that requires certain components of the NER pathway and other, unidentified gene products *(156,393)*.

The base damage produced by ionizing radiation, and by oxidizing and methylating agents is repaired predominately by the base excision repair pathway, which is distinct from NER (*99*; *see* Chapter 3, vol. 1 and Chapters 3 and 4, this vol.). Although some fraction of lesions that are repairable by base excision repair may also be dealt with by NER *(141)*, there is no backup system for NER. This conclusion stems from the total absence of UV photoproduct removal in many mutant cell lines derived from xeroderma pigmentosum (XP) patients or Chinese hamster cells. Bulky lesions that go unrepaired by the time DNA is replicated are potentially mutagenic and may be dealt with through other damage-tolerance mechanisms known as post-replication repair (*169,266,428*; *see* Chapter 12, vol. 1).

1.2. Methods for Detecting NER

A variety of in vivo methods are used for detecting the lesions themselves, strand breaks resulting from incision, or repair synthesis. Radioimmunoassays that detect UV photoproducts or bulky adducts represent a sensitive means of quantifying damage and repair *(249)*. Under conditions of drug inhibition of repair synthesis, breaks resulting from incision accumulate and can be detected by alkaline elution of DNA from filters *(179)* or by alkaline unwinding and hydroxylapatite chromatography *(6,364)*. Measurement of repair synthesis employs isopycnic sedimentation and fractionation of radiolabeled DNA by density *(357)*. Autoradiographic detection of incorporated thymidine readily allows visualizing repair synthesis in individual cells in non-S phases of the cell cycle, often referred to as unscheduled DNA synthesis (UDS) *(177,427)*.

The amount of repair in a specific genomic sequence can be measured by Southern blotting *(99,363)*. The band of interest is diminished in intensity when breaks are introduced at sites of damage within the restriction fragment, as accomplished by prior digestion with a CPD-specific endonuclease or the bacterial UvrABC excision nuclease. Repair capacity can also be assessed by transfecting cells with damaged reporter plasmids and measuring reporter gene function (e.g., chloramphenicol acetyltransferase) *(183,449)*. Reliable in vitro methods are now available for measuring repair capacity in cell extracts by monitoring the excised fragment from an adducted oligonucleotide *(260)* or repair synthesis *(451)* in a plasmid. More detailed discussion of methods is presented in Chapter 24.

1.3. Intragenomic Heterogeneity of Repair

1.3.1. Transcription-Dependent Repair

An anomaly arose early in repair studies with the observation that human and rodent cells generally have very similar survival responses to UV radiation, whereas the rate and extent of CPD removal are much lower in rodent cells *(245,418,464)*. This dilemma became even more intriguing with the discovery that repair-deficient mutants of rodent cells were quite sensitive to killing by UV radiation, i.e. three- to fourfold more sensitive than wild-type for mouse lymphoma cells *(348)* and six- to sevenfold for hamster CHO cells *(42,396,397)*, although not as sensitive as human cells from XP complementation group A (≥10-fold). A plausible explanation for this rodent repair paradox came with the finding that in both rodent cells *(25,26,222,430)* and human cells *(192,242)*, the removal of CPDs in active genes is generally more efficient than in inactive genes or in bulk DNA. This difference in repair between active and inactive sequences is much larger in rodent cells than in human cells. Preferential repair in active genes has been confirmed in mouse epidermis *(313)*.

Within active genes, the kinetics of repair of both CPDs and (6-4)PDs are similar in rodent and human cells *(25,222,430)*. In this regard, cell survival correlates with the efficiency of repair in active genes *(217)*. This inference is strengthened by comparing the relatively UV-sensitive XP group C cells with wild-type CHO cells under conditions where little repair occurs in the active *DHFR* gene in the XP-C cells *(25)*. Preferential repair of active genes has been extended to (6-4)PDs *(214,392)* and agents that produce bulky adducts, i.e., nitrogen mustard *(189)* and cisplatin *(165,189)*. However, there are notable exceptions in the case of 4-nitro-quinoline 1-oxide (4NQO) *(361)*, *N*-acetoxy-2-acetylaminofluorene *(388)*, psoralen monoadducts *(149)*, and cisplatin at high dose *(165)*. In some instances, transcription-dependent repair is dose-dependent *(165,189)*. Moreover, some nontranscribed genes are repaired relatively well *(188,419,443)*. In cultured cells induced to undergo differentiation in culture, repair efficiency is usually *(134,150)*, but not always *(23)*, correlated with transcriptional activity.

The concept of active gene repair is supported by the finding of transcription-coupled repair (TCR), but there appear to be other factors, relating to the more open structure of chromatin in active genes, that enhance the efficiency of repair *(356,419)*. In monkey cells treated with furocoumarins or aflatoxin B1, the highly repetitive α DNA sequences are poorly repaired compared with bulk DNA, but with UV damage, there is no difference *(193,468)*. These results further suggest that differences in chromatin structure can modify access of repair enzymes and do so in a lesion-specific manner.

1.3.2. Preferential Repair of the Transcribed Strand: TCR

Analysis of gene-specific repair using strand-specific RNA probes gave new insights into the mechanism of transcription-dependent repair. In both human and hamster cells, CPDs in the transcribed (noncoding) strand of the *DHFR* gene are repaired more efficiently than in the nontranscribed strand (differences of ~twofold and ~10-fold, respectively), which shows little or no repair in hamster cells *(243)*. These findings can be generalized to other genes *(51,263,430)*, including *p53* in both cultured cells and mouse epidermis *(88,95,314)*, and to other types of bulky adducts *(44,51)*. Preferential repair also occurs throughout the cell cycle *(216,294)*. The predicted strand bias in favor of mutation induction in the nontranscribed strand in repair-proficient cells was confirmed *(43,49,240,244,315,432)*. This bias is not owing to preferential damage to the transcribed strand since repair-deficient cells from XP group A do not usually show such bias *(50,78,240)*. Repair-deficient hamster and human cells actually show a reverse strand bias for UV-induced mutations in the *hprt* gene, with almost all mutations occurring in the transcribed strand *(240,244,431,433)*, and can be explained in terms of differential efficiency of mutation fixation by different polymerases during bypass replication of lesions in the transcribed and nontranscribed strands *(431)*. Alternatively, it could be owing to selection for nonfunctional HPRT protein *(240)*.

It is well established that the enhanced repair of transcribed strands is the result of an efficient coupling of repair with the transcription machinery, which helps ensure that transcription can proceed in the face of adduct damage (*122*; Chapter 10). An inhibitor of RNA polymerase II, α-amanitin, inhibits repair of the transcribed strand in human and hamster cells *(52,191)*. Genes transcribed by RNA polymerase I do not show strand-specific repair for CPDs or most bulky chemical adducts *(53,371,429)*. These findings are consistent with the idea that specific factors associated with RNA polymerase II are involved in coupling transcription and repair. The concept of a TCR pathway that is separate from the pathway operating on bulk DNA is further supported by the finding that XP group C cells retain repair only in the transcribed strands of active class II genes *(54,413,422)*. Thus, XP-C cells have chromatin domains showing selective repair *(166,167)*. Recent reviews discuss the mechanism of gene-specific and strand-specific repair *(24,80,122,123)*.

2. THE NER SYNDROMES

2.1. *Xeroderma pigmentosum*

XP, indicating dry or parchment skin with pigmentation abnormalities, is a prototype of a multigenic human disorder in which mutations in several genes cause similar phenotypes. Seven complementation groups (XP-A–XP-G) were defined using restoration of UDS in heterokaryons arising from fused cells *(160,177,425)*. Thus far, these groups can be explained most simply on the basis of single-locus mutations rather than a corecessive inheritance model *(187)*. The severity of the repair deficiency varies from <10% of normal usually for groups A, B, and G up to ~50% in groups D and E *(182)*. Groups A, C, D are relatively common (ranging from ~160 to ~50 reported cases), while the other groups are much rarer *(182)*. In addition, a variant form (XP-V) having normal NER occurs at a frequency similar to that of XP-A *(182)*.

2.1.1. *Clinical Features*

XP is the classical DNA repair disorder in which cutaneous abnormalities, including cancer, are associated (usually) with an overall deficiency in NER (reviewed in *13,63,99,182,186*). This autosomal-recessive disease occurs at a frequency of ~10^{-6} in the United States. A 2000-fold increased risk of skin cancer in XP individuals *(180)* is associated with a median age of onset of 8 yr, representing a dramatic ~50 yr reduction in age compared with the general population *(63,180)*. A variety of complications are associated with the sun-exposed areas of the skin (Table 1 and Fig. 1A,B). The eyes and the tip of the tongue are often affected. The onset of both early dermatological symptoms of XP and tumor growth is correlated with the extent of DNA repair deficiency measured by cell survival curves (e.g., D_o) and incision activity *(391)*. XP patients also appear at increased risk for some internal neoplasms *(180)*.

Neurological abnormalities are present in ~20% of XP patients (XP-A > XP-D = XP-G) and are progressive *(63,180)*. They include hyporeflexia, sensorineural deafness, mental retardation, and microcephaly, which are associated with neuronal degeneration from loss of neurons, especially in the cerebellum and cerebral cortex *(63)*. These abnormalities are sometimes associated with slow growth and delayed secondary sexual characteristics *(181)*. The neurological dysfunction seems to correlate with the degree of repair deficiency *(9,308,309)* and is thought to arise from unrepaired

Table 1
Comparison of Clinical and Cellular Features of XP, TTD, and CS

Characteristics	XP	TTD	CS
Clinical features			
Cutaneous photosensitivity	++	– or +	+
Pigmentation disturbances	+[a]	–	–
Ocular abnormalities (sunlight)	++[a]	–	–
Oral abnormalities	+	–	–
Actinic keratoses	+	–	–
Basal and squamous cell carcinomas, melanomas	++	–	–
Neurological abnormalities[b]	– or +	+	++
Slow rate of growth	– or +	+	++
Impaired sexual development	– or +	+	+
Sensorineural deafness	– or +	+	++
Mental retardation	– or +	+	++
Ichthyosis (fish–like scaly skin)	–	++	–
Brittle hair (sulfur-deficient)	–	++	–
Peculiar facies	–	+[c]	+[c]
Cataracts/pigmentary retinopathy	–	+	++
Dental caries	–	+	++
Cellular features			
Intrinsic chromosomal instability	–	–	–
Hypermutability	+	– or +	+
Excision defect in bulk DNA[d]	+	– or +	–
Excision defect in transcription–dependent repair[d]	+	– or +	+

[a]Traits that are hallmarks of the disease are indicated by (++), and traits that are sometimes associated with the disorder are indicated by (+); unassociated traits are shown by (–).

[b]These features are present in about 20% of XP patients. The molecular basis of the neurological dysfunctions in XP differs from that of CS and TTD. Demyelination of neurons is seen in TTD and CS *(293,325)*. XP involves primary neuronal degeneration thought to be related to the severity of repair deficiency and metabolic damage to DNA *(308)*.

[c]The facial abnormalities of TTD and CS overlap in term of protruding ears. TTD often has receding chin and small thin or beaked nose, whereas CS tends to have large nose and projecting jaw.

[d]XP-C is exceptional in being deficient in repairing bulk DNA, but proficient in TCR.

oxidative damage in the brain *(59)*. XP-A cells are known to be repair-deficient for certain types of oxidative damage *(328)*.

2.1.2. Cellular Features

Hypersensitivity of XP cells to killing by UV radiation or chemicals that produce bulky adducts is a hallmark feature of the disease as shown by numerous studies *(9,14,223,391)*. Hypersensitivity to UV radiation varies widely, both within and among complementation groups, ranging from >10-fold for cells from group A to ~1.5-fold for cells from group E *(63,391)*. Hypersensitivity can extend to small alkyl groups, such as the ethyl groups produced by *N*-ethyl-*N*-nitrosourea *(32,355)*. The repair of

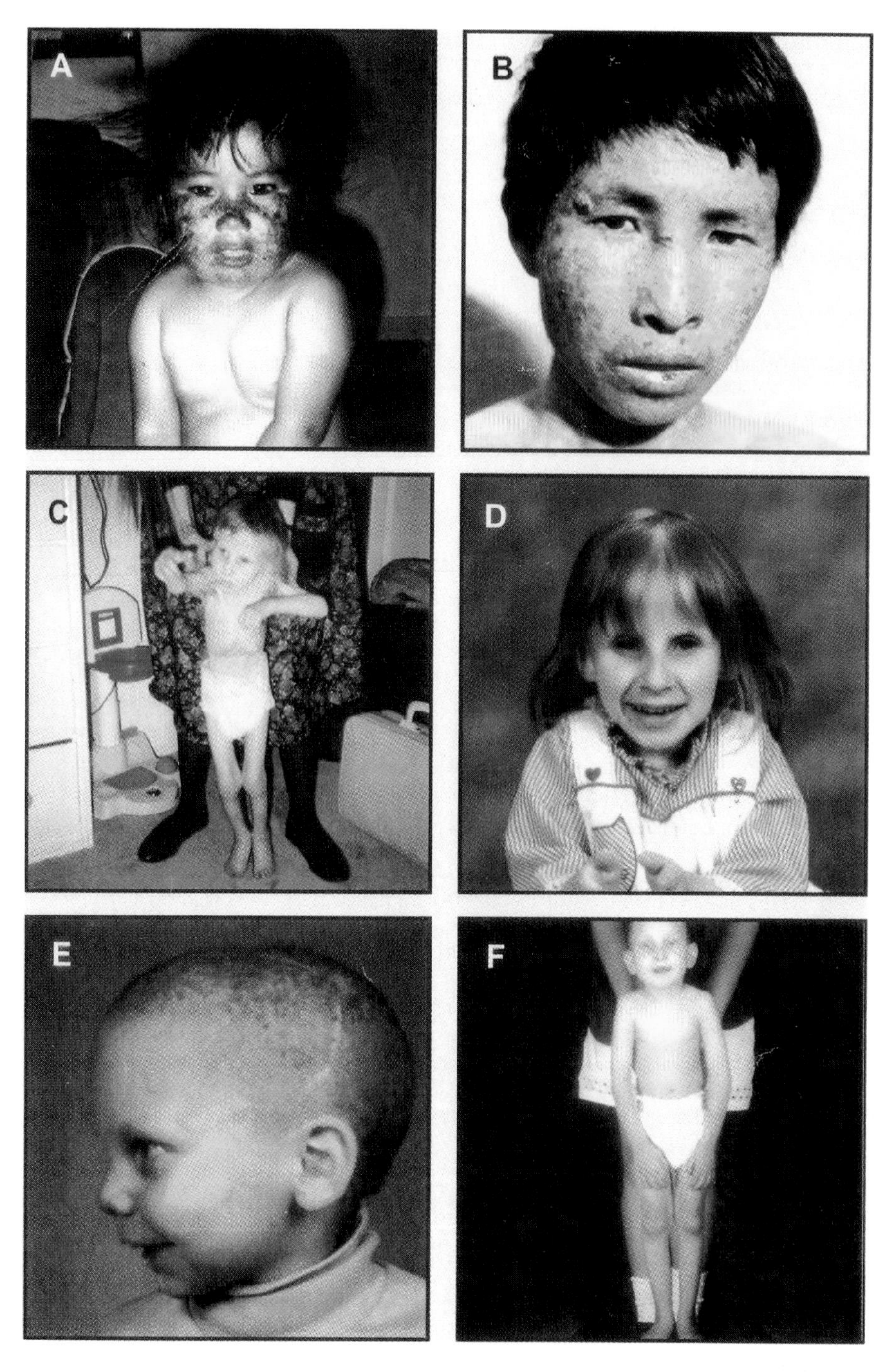

Fig. 1. (A,B) Two XP patients of undefined complementation group; the patient in panel A subsequently showed decline of the central nervous system. **(C,D)** A CS patient at ages 4 and almost 7 yr, respectively. **(E,F)** two TTD patients. Pictures were kindly provided by J. Cleaver, University of California, San Francisco (A and B), T. Wall, Cockayne Syndrome Network, P. O. Box 552, Stanleytown, VA (C and D), and C. Weber, Lawrence Livermore National Laboratory, Livermore, CA (E and F).

O^6-ethylguanine appears to require NER in cooperation with O^6-alkylguanine-DNA alkyltransferase *(31,224)*. Even the repair of methyl groups in XP may be somewhat diminished in XP *(391)*, although the cytotoxicity and mutagenicity of *N*-methyl-*N'*-nitro-*N*-nitrosoguanidine is apparently independent of NER *(224)*. Hypersensitivity to mutation induction generally parallels that of cytotoxicity *(106,114,225,389)*. Transfection of XP cells with shuttle vectors facilitated an analysis of the spectrum of induced mutations, which is more restricted with fewer transversions and fewer multiple mutations than in normal cells *(30,340,455)*. The frequency of UV-induced neoplastic transformation to anchorage independence is higher in XP cells than in normal cells *(228)*.

A DNA repair deficiency in XP cells was first demonstrated in 1968 by Cleaver *(58)* using repair synthesis and autoradiography, and confirmed by Setlow and coworkers *(345)* using velocity sedimentation to detect incision-generated breaks. The phenotypic heterogeneity seen with XP fibroblasts from different patients, as illustrated by different residual levels of UDS *(63)*, prompted genetic complementation analysis. The finding of seven NER groups implied that many gene products were required to perform the incision reaction, which is deficient in XP cells *(73,96,364,384)*. XP lines from all seven groups have deficiencies in the repair of UV-endonuclease-susceptible sites *(463)*.

The degree of reduction in UDS, which is often used as a measure of repair capacity, sometimes does not correlate with degree of sensitivity as measured by cell survival or the extent of removal of endonuclease-sensitive sites *(63,391)*. The UDS assay, which is normally performed at early times after treatment, detects mainly the rapid removal of all (6-4)PDs and CPDs in transcribed strands *(34,250)* and may not be indicative of overall repair. The defects in XP lines from different complementation groups differentially affect the repair of the two major classes of UV photoproducts *(102)*. This observation was used to argue that CPDs and (6-4)PDs are repaired by different NER subpathways. Moreover, the residual repair in XP-C cells is clustered in a small fraction of the genome, now known to contain active genes *(166,167)*. Some XP-G cell lines show sensitivity to ionizing radiation *(10)*, which appears to be owing to a defect in repairing oxidative base damage, such as thymine glycols *(68)*.

In XP patients, there are elevated levels of *hprt* mutations in circulating T lymphocytes *(66)*. The mutagenic consequences of failing to repair UV damage can be detected in the skin tumors of XP patients. These tumors show high levels of mutations in the *p53* tumor suppressor gene *(85,326,327)* and in *ras* proto-oncogenes *(72,326,372)* with a spectrum characteristic of that produced by UV radiation *(327)* and similar to those seen in skin tumors from normal individuals *(467)*.

The variant form of XP (XP-V), although clinically indistinguishable from other XP groups, is normal for excision repair *(60)* and forms a single complementation group *(158)*. XP-V has a reduced ability to elongate nascent DNA strands when replicating past UV damage *(29,113,205)*. XP-V cells are slightly UV-hypersensitive to killing and severalfold hypersensitive to both mutation induction *(226)* and transformation to anchorage independence by UV *(238)*. Caffeine specifically potentiates both cytotoxicity and mutagenicity of XP-V cells *(227)*, an effect that should provide adequate selection conditions for cloning the XP-V gene by functional complementation. At the molecular level, XP-V cells show error-prone DNA replication in the presence of DNA damage *(439)* and are less likely than normal cells to incorporate adenine opposite sites of pyrimidine dimers *(440)*. Bypass replication in psoralen-treated XP-V cells is defective in both rRNA genes *(248)* and in a transfected *supF* shuttle vector, the latter showing enhanced mutagenesis *(301)*.

2.2. *Trichothiodistrophy*

2.2.1. *Clinical Features*

TTD is the term that has been given to an autosomal-recessive disorder characterized by sulfur-deficient brittle hair, scaly skin, and physical and mental retardation *(296,299)*. The hair shafts have trichoschisis (transverse fractures) and alternating light and dark bands under polarization microscopy, defects related to decreased sulfur-rich proteins *(105,151)*. Delayed or arrested physical maturation often occurs *(151,299)*. Patients may have an unusual facial appearance (Fig. 1E,F) and altered body proportions (Fig. 1F). TTD actually embraces a complex, heterogeneous set of rare sulfur-deficient brittle-hair disorders *(151)*. The acronym PIBIDS has been used to refer to the photosensitive form of TTD (*p*hotosensitivity, *i*chthyosis, *b*rittle hair, *i*mpaired intelligence, *d*ecreased fertility, and *s*hort stature) *(151,306)*. About 50% of TTD patients have severe photosensitivity linked to deficient excision repair like that seen in XP cell lines, particularly group D *(306,369,457)*. Nevertheless, these patients do not show the same kinds of skin abnormalities seen in XP (cf Table 1), and there is a notable absence of skin cancer associated with these TTD patients *(197,200,368)*. Although both TTD and XP include symptoms of neurological abnormalities and impaired growth (e.g., low intelligence, areflexia, microcephaly), the molecular basis of the neurological dysfunction differs. Demyelination is the predominant cause in TTD patients *(293)*. The early age of death of TTD patients has prompted prenatal diagnosis based on defective DNA repair in families known to carry mutant alleles *(324)*.

2.2.2. *Repair Deficiency and Genetic Heterogeneity*

The photosensitivity of some TTD patients prompted complementation analysis based on UDS, with the unexpected finding that most UV-sensitive TTD lines fall into the XP-D complementation group *(368)*. A demonstration that the cloned *XPD* gene could correct the repair deficiency in these TTD lines strongly suggested that defects in the *XPD* gene were responsible for the repair deficiency in TTD cell lines assigned to the XP-D complementation group *(246)*. These findings remained enigmatic until the discovery that the XPD protein plays an essential role in both transcription and repair (*see* Section 3.3.2.)

Detailed studies of DNA repair parameters in cell lines from repair-defective TTD patients show broad heterogeneity, overlapping the spectrum seen in XP-D cells *(34,89,161,201,381)*. In one study, TTD cell lines showed either reduced repair of only CPDs or reduced repair of both CPDs and (6-4)PDs *(89)*. In an earlier study, TTD lines showed either normal repair of CPDs and reduced repair of (6-4)PDs, or much reduced repair of both classes of damage *(34)*. Some cell lines common to the two studies gave conflicting results.

Recently, two additional complementation groups of TTD were identified *(368,370)*. One of these (TTD6VI) corresponds to the XP-B group as shown using UDS measurements in fused cells or microinjection of the *XPB* cDNA *(426)*. The other TTD line (TTD1BR) represents a new gene that encodes one of the proteins (TTD-A protein) of the basal transcription factor TFIIH (*see* Section 3.3.2.). This conclusion was based on immunodepletion experiments in which HeLa extracts were rendered deficient in their ability to complement the repair defect in TTD-A cells in a cell-free repair assay after depletion of the TFIIH complex. Confirmatory results came from microinjection of TTD-A cells with purified TFIIH complex and correction of the repair deficiency *(426)*.

2.3. *Cockayne's Syndrome*

2.3.1. *Clinical Features*

In 1936 Cockayne described a disorder in which postnatal growth failure and progressive neurological dysfunction are cardinal features *(65)*, and in which there is frequent involvement of cutaneous photosensitivity, deafness, cataracts, pigmentary retinopathy, and dental caries (Table 1) *(206,271)*. In addition to a dwarfed appearance, disproportionately long arms and sunken eyes are characteristic (*see* Fig. 1C,D). The disease has the appearance of premature aging with a median life-span of about 12 yr. Despite the sun sensitivity, CS patients show neither a predisposition to skin cancer *(197,271)* nor the associated skin changes seen in XP cells (cf Table 1). As in the case of TTD, the neurological abnormalities in CS are associated with demyelination in nerve tissue *(271,325)*. In a few rare individuals, features of both CS and XP are found (designated here as XP-CS). These complex syndromes are almost certainly the result of single-locus mutations in view of the rarity of the individual XP and CS disorders. The association with CS has been reported for XP group B *(339,424)*, D *(36,425)*, and G *(423)*.

2.3.2. *UV Sensitivity and Repair Characteristics*

CS cells in culture show pronounced hypersensitivity to killing by UV radiation and bulky adduct mutagens *(8,230,414,434)*. The magnitude of sensitivity to UV is approximately four- to eightfold, which is less than the most sensitive XP cell lines *(434)*. The response of CS cells to ionizing radiation was at first thought to be in the normal range *(11)*, but recent studies point to modest, significant sensitivity of CS cells *(68,190)*. Repair synthesis and dimer removal in bulk DNA are normal in CS cells after UV irradiation *(5,8,434)* despite their high UV sensitivity and hypermutability *(12)*. The recovery of DNA synthesis seen in normal cells after irradiation fails to occur in CS cells *(203,385)*. CS cells also fail to recover RNA synthesis *(206,237,414)*, which led to the prescient suggestion of a defect in repairing damage in transcribing regions of DNA *(206)*. Based on the end points of recovery of RNA or DNA synthesis after UV irradiation, cell fusion experiments with >30 CS cell lines identified two complementation groups (CS-A and CS-B) that are distinct from XP *(199,367,385)*. Most patients belong to the CS-B group *(367)*.

Moreover, examples are arising in which the clinical and cellular features of CS do not fit the initial paradigm. One study described two siblings who had an XP-like clinical presentation with neurological involvement *(111)*. The cells from these patients were two- to threefold hypersensitive to UV killing and appeared normal for excision repair. However, they showed the characteristic CS-like defect in the recovery of RNA synthesis and were found to belong to the CS-B complementation group *(152)*.

CS-A and CS-B cell lines are defective in the repair of active genes after UV damage *(420)*, specifically with a lack of repair of the transcribed strand *(412)*. There is also a somewhat diminished repair of the nontranscribed strand *(54,412)*. In rRNA genes, which are transcribed by RNA polymerase I, there is diminished repair in both CS-A and CS-B cells *(54)*. Although it was suggested that the CS gene products may act as transcription repair coupling factors *(320)*, these results point to a more generalized role of the CS proteins, which appear to promote access to damage by repair proteins. Here it is of interest that in both transfected reporter genes and a shuttle vector system, CS cells did not repair CPDs, but did repair the nonphotoreactivatable biologically important UV photoproducts (mainly [6-4]PDs) *(20,291)*. Since (6-4)PDs are probably more structurally distorting than CPDs *(401)*, these findings are consistent with the idea of the defects in CS cells causing restricted access to damage. Recent studies of the repair of aminofluorene adducts, which did not show strand-specific repair in normal cells, led to a model in which the CSA and CSB proteins play an active role in initiating transcription after damage is repaired by routing the transcription/repair complex TFIIH (which is associated with RNA polymerase II transcription; *see* Section 3.3.2.) from repair to transcription *(414)*. With this model, restricted access to damage in class II genes of CS cells could be a consequence of reduced transcription, but the reduced repair seen in class I genes is difficult to explain.

XP cells, except for XP-C, are generally deficient in both global and transcription-dependent repair pathways. However, one XP-F cell line, which retains partial repair capacity, resembles CS in being relatively deficient in the repair of CPDs in active genes *(87)*.

2.4. Uncharacterized UV-Sensitive Disorders

Additional UV-sensitive cell lines from patients have been identified that do not fit into any of the categories described above for XP, TTD, CS, and XP-CS. Three patients from two nonconsanguinous families show clinical features of mild XP, but cellular features like CS cells, i.e., failure of recovery of RNA synthesis after UV irradiation *(100,153,154)*. Itoh and coworkers assigned the cell lines from these patients to a new complementation group *(153,154)* that differs from the eight groups of XP (including XP-V), the two groups of CS, and the rodent *ERCC1* complementation group, for which no human syndrome has been identified. The cellular phenotype of this new group shows considerable UV hypersensitivity to cell killing (three- to fourfold for nonimmortalized cells), while having normal UDS. There is a clear defect in the recovery of RNA synthesis, which can be overcome by microinjecting the cells with T4 endonuclease V, implying that the cellular defect involves a failure to remove CPDs in active genes *(154)*.

3. GENETIC AND BIOCHEMICAL DISSECTION OF NER

3.1. Genetic and Biochemical Complementation Analysis

3.1.1. Biochemical Complementation

Complementation studies that led to the identification of XP, CS, and TTD groups were based largely on UDS, as discussed above. With the availability of cloned genes, complementation tests on new lines can be greatly simplified if done by gene transfection using reporter genes *(45)*.

Assays employing cell extracts to measure NER have demonstrated defective repair in XP cells and provided a biochemical means of determining complementation groups in several instances *(305,452)*. These assays were useful in defining the overlap between XP cell lines and phenotypically similar rodent repair mutants *(22,284)*. However, some combinations of mutant extracts, e.g., rodent group 1 vs group 4 or XP-F cells, do not show biochemical complementation *(305,415)*, which can be explained by a multiprotein complex in which ERCC1 and ERCC4 reside *(1,21,287)*. In a tight complex, subunits are not free to exchange and complement in vitro.

3.1.2. Genetic Complementation of Rodent Cell Mutants

The study of NER in mammalian cells was greatly aided by the production and characterization of UV-sensitive rodent cells that display repair-deficient phenotypes analogous to those of the human complementation groups. The first reports of mutant isolation were with 4NQO-sensitive cells from mouse lymphoma L5178Y *(348)* and UV-sensitive clones from Chinese hamster cells (CHO and V79) *(40,396,397,461)*. Several laboratories independently devised various methods for identifying rare (~10^{-3}) UV-sensitive clones of cells. The biochemical and genetic properties of these cell lines have been extensively summarized *(67,136,446)*.

Many subsequent advances came with mutants derived from the CHO lines, which were particularly suitable for genetic studies because of their relatively stable karyotype and partial hemizygosity *(353)*. By far, the most productive mutant isolations involved the large-scale growth of CHO colonies on a soft agar surface under conditions where colony size could be monitored photographically before and after UV irradiation *(41,42)*. Colonies that failed to increase in diameter after irradiation were candidate mutants for further characterization. D. Busch and coworkers obtained >200 mutant clones *(41,42)*, most of which were assigned to the first six genetic complementation groups involving NER deficiency, and one mutant that is not defective in NER was identified *(42a)*. Mutants from other laboratories led to the identification of groups 7–11 *(307,367,398,462)*.

There is considerable overlap between these rodent groups and the XP and CS groups. However, it is noteworthy that some mutants in rodent groups 1 and 4 show extreme sensitivity to DNA crosslinking agents, a feature not identified among human NER mutants. These groups show extreme hypersensitivity (10- to 100-fold) to mitomycin C, cisplatin, melphalan, diepoxybutane, and other agents that produce interstrand crosslinks *(140,362,397,453)*. High sensitivity to crosslinking agents is not associated with XP or CS cells, but rather is one of the hallmarks of cells from Fanconi's anemia, which have normal NER *(79,101,147,258)*. Thus, the rodent mutants in groups 1 and 4 represent a unique class of NER-deficient phenotypes among mammalian cell mutants, which will undoubtedly prove valuable for determining the mechanisms of DNA crosslink repair.

3.2. Gene Cloning and Characterization

3.2.1. Overlap of Genes Identified Using Rodent and Human Cell Mutants

There was skepticism regarding whether NER processes would prove similar in human and rodent cells, since rodent cells remove CPDs inefficiently *(418,464)*. Nevertheless, with the advent of DNA transfection methods, strategies were put forth to clone human repair genes based on complementing the UV (or mitomycin C) sensitive phenotypes of CHO mutants. Using CHO mutants from the first six rodent complementation groups, complementing human genes were isolated and given the designation *ERCC1*, *ERCC2*, and so forth, where ERCC signifies excision repair cross complementing (a human gene correcting across species) and the number refers to the rodent complementation group.

3.2.2. Genes Identified Using Rodent Cell Mutants

3.2.2.1. ERCC1

ERCC1, the first NER-specific gene isolated *(409)*, is a relatively small gene that shows interesting homology with the *RAD10* gene of the yeast *Saccharomyces cerevisiae*, which is required for NER. Table 2 summarizes the properties of the isolated human NER genes and lists their *S. cerevisiae* homologs. (It should be noted that for nomenclature simplification, the definition of rodent groups 1 and 2 was interchanged subsequent to the reported isolation of the *ERCC1* gene *[394]*). Somewhat unexpectedly, *ERCC1* did not correct any of the XP complementation groups and thus represents a class of cell mutants unique to rodent cells *(410)*. This observation helped reinforce the earlier notion that the mechanism of NER might differ substantially between rodent and human cells. *ERCC1* maps to chromosome 19q13.2 and lies <250 kbp from the *ERCC2* gene *(256)*.

3.2.2.2. ERCC2/XPD, ERCC3/XPB, and ERCC6/CSB

Isolation and characterization of the *ERCC2 (445)*, *ERCC3 (447)*, and *ERCC6 (405)* genes revealed that their encoded proteins shared a common set of seven DNA helicase motifs that are highly conserved across eukaryotic and prokaryotic proteins *(109)*, and led to the prediction that the cognate proteins would possess DNA unwinding activity. Homologs of *ERCC2* proved to be *RAD3*, a gene in *S. cerevisiae* encoding a DNA helicase that is required for cell viability *(125,132,272,375)*, and *Schizosaccharomyces pombe rad15 (264)*. No *RAD* homologs of *ERCC3* and *ERCC6* had been reported when these human sequences were identified. Subsequently, the human cDNAs were used to isolate the yeast homologs from *S. cerevisiae (290,411)*. *RAD25 (SSL2)*, the homolog of *ERCC3*, was found to be essential for cell viability, like *RAD3 (290)*. *SSL2* was independently identified as a gene affecting translational initiation whose product could suppress the phenotype associated with an artificial stem-loop structure in the 5'-untranslated region of *HIS4* mRNA *(115)*. Mutants of *rad26* (*RAD26* is homologous to *ERCC6*) surprisingly showed no UV sensitivity *(411)* and explained the previous lack of *rad26* mutants. *ERCC3* maps to chromosome 2q21 *(448)*, and *ERCC6* maps to 10q11-21 *(404)*.

At this point, since all three of the *ERCC* genes having helicase motifs were involved in the human repair disorders, it became clear that rodent and human cell excision repair processes are much more similar than dissimilar. *ERCC2* was found to correct XP-D cells *(93,382)*, *ERCC3* was defective in XP-B cells *(447)*, and *ERCC6* showed causal mutations in Cockayne syndrome group B *(405)*. Once an *ERCC* gene is implicated in a specific disorder and mutation analysis of the locus from mutant cells confirms the involvement, the gene is named after the disorder *(202)*. Thus, *ERCC2* = *XPD*, *ERCC3* = *XPB*, *ERCC6* = *CSB*, etc. Depending on the context, the original designation is still sometimes used for clarity.

3.2.2.3. *ERCC4/XPF*

The functional *ERCC4* genomic sequence was isolated as a cosmid after transfection experiments using a chromosome-specific library *(395)*. The complete cDNA sequence was determined *(33)* and found to have homology with *RAD1* of *S. cerevisiae*, *rad16* of *S. pombe*, and *mei-9* of *Drosophila melanogaster (341)*, a gene conferring UV resistance and required for meiotic recombination. Recent biochemical studies have shown that both purified native and recombinant ERCC4 can correct the repair deficiency in extracts of XP-F cells when complexed with *ERCC1 (20a,288)*, and mutations have been identified in the *ERCC4* gene in XP-F cells *(354)*. Previous results indicating apparent complementation between rodent group 4 and XP-F cells *(215)* can now be explained by the residual repair in the XP-F cells *(101)*. *ERCC4* also corrects the CHO mutant UVS1 *(33)*, which was assigned to group 11 *(127,307)*, indicating the equivalence of NER groups 11 and 4. *ERCC4* maps to 16p13.13-p13.2 *(215)*.

3.2.2.4. *ERCC5/XPG*

ERCC5 was first isolated in CHO UV135 cells by intercosmid complementation *(261)* and cosmid-cDNA complementation *(221)* and later isolated using mouse cells *(347)*. *ERCC5* has limited homology with *RAD2* in *S. cerevisiae (221,336)*. The isolation of a *Xenopus* cDNA led independently to the chance discovery of *ERCC5* and its identification as the *XPG* gene *(284,336)*. The conserved regions of *XPG* are restricted to two domains that are shared with four genes in *S. cerevisiae* and *S. pombe (221)*. *XPG* maps to chromosome 13q33 *(316)*.

No complementing human genes are identified for rodent mutants in groups 7, 9, and 10 *(366,398,462)*. The modest UV sensitivity of these mutants (approximately two-fold) makes it technically very difficult to devise effective selection procedures for gene isolation. The unidentified rodent groups could be equivalent to some of the genes in Table 2 that have not been tested (e.g., *XPA*, *XPC*, and so on). However, cell extracts of the mutants in these three remaining groups show no defect in the in vitro oligonucleotide excision assay *(304)*, which argues that the defects in these mutants are most likely in genes controlling cellular functions not directly involved in the general excision pathway.

3.2.3. *Genes Identified Using Human Cell Mutants*

3.2.3.1. *XPA*

Many early attempts to transfect and correct human XP or CS cells with genomic DNA were unsuccessful because of the low efficiency of stable integration of gene-size DNA sequences *(137,236)*. A frequent pitfall in experiments with immortalized XP-A cells was the presence of revertant cells that could be confused with bona fide transformants *(312,338)*. These limitations notwithstanding, Tanaka et al. isolated the partially correcting mouse *XPA* gene *(387)*, which was used to obtain the human counterpart *(387)*. The translated cDNA sequence maps to chromosome 9q34.1 and contains two zinc-finger motifs (C_4 and C_2H_2 classes), suggesting that the protein interacts directly with DNA *(386)*. Another putative XP-A correcting gene was reported, but was later thought to reflect phenotypic reversion *(162)*.

Table 2
Cloned Nucleotide Excision Repair Genes and Their Encoded Proteins[a]

Gene[b]	Human disease involved	Gene, kb	ORF, aa	Chrom. location	Protein, kDa	Homologous protein in *S. cerevisiae*	Specific protein function and properties
ERCC1	?	15	297	19q13.2	33	Rad10p	Endonucleolytic incision on the 5'-side of damage; complexed with XPF; interaction with XPA; putative recombination function(s)
XPA	XP-A	25	273	9q34.1	31	Rad14p	Damage recognition/DNA binding; zinc finger motif; binding to RPA and ERCC1-XPF complex
XPB/ERCC3	XPB-CS TTD	45	782	2q21	89	Rad25p/ Ssl2p	3' → 5' DNA helicase; member of TFIIH complex; essential for transcription initiation; formation of preincision complex
XPC	XP-C	24	940	3p25.1	106	Rad4p	Specific to global repair pathway; complexed with HHR23B; tight binding to ssDNA; stabilization of preincision complex
XPD/ERCC2	XP-D, TTD XPD-CS	19	760	19q13.2	87	Rad3p	5' → 3' DNA helicase; member of TFIIH complex; essential for transcription initiation; formation of preincision complex
DDB1	XP-E?	?	1140	11q12-13	127	Unknown	Recognition/DNA binding of UV photoproducts and some bulky lesions; absent in some XP-E lines
XPF/ERCC4	XP-F	30	916	16p13.13	104	Rad1p	Endonucleolytic incision on the 5'-side of damage; complexed with ERCC1; putative recombination function(s)
XPG/ERCC5	XP-G XPG-CS	32	1186	13q33	133	Rad2p	Endonucleolytic incision on the 3'-side of damage; junction-specific nuclease; interaction with XPC?
CSA/ERCC8	CS-A	?	396	5	44	Snf2p	Preferential repair of active genes; repair of transcribed strand; transcriptional regulator?
CSB/ERCC6	CS-B	~85	1493	10q11.2	168	Rad26p	Preferential repair of active genes; repair of transcribed strand; transcriptional regulator?

HHR23A	?	?	363	19p13.2	40	Rad23p	Ubiquitin-like N-terminus; identified by homology with *HHR23B*, a homolog of *RAD23*
HHR23B	?	?	409	3p25.1	43	Rad23p	Specific to global repair pathway; complexed with XPC; ubiquitin-like N-terminus
p62TFIIH	?	?	548	11p14	62	Tfb1p	Component of TFIIH; essential for viability
p44TFIIH	?	?	395	5q1.3	44	Ssl1p	Component of TFIIH; essential for viability; TFIIIA-like zinc finger motif
LIG1	46BR	53	919	19q13.2	102	Cdc9p	Nick closure in the final step of repair

[a]In addition to the proteins listed, the complete excision repair reaction requires trimeric RPA, additional components of the TFIIH complex, polymerase acccessory factors RFC, PCNA, and polymerase δ/ε. References for information given are given in the text.

[b]By agreement *(202)*, the name *"XPB"* has replaced *XPBC* and *ERCC3*, and so forth, but *ERCC* names can be preferable in certain contexts.

3.2.3.2. *XPC*

Theoretically, it should be straightforward to correct cellular repair deficiency by transfecting a cDNA expression library, assuming the existence of near full-length sequences in the library. Legerski and Peterson *(196)* obtained transformants of XP-C cells using an Epstein-Barr virus vector (pEBS7), which replicates episomally, and were able to rescue the complementing *XPC* cDNA. Although the cloned sequence gave excellent correction for cell survival, it was later discovered that the clone was slightly truncated at the 5'-end of the open reading frame *(231)*. Interestingly, the mouse XPC protein is only 74% identical to the human and is truncated by 40 aa at the N-terminus *(210a)*. *XPC* maps to chromosome 3p25.1 *(195,408)*, which is also the location of another gene, *HHR23B*, which encodes a homolog of yeast Rad23p. Surprisingly, HHR23B forms a tight complex in vivo with the XPC protein *(231)*. *HHR23A,* a second human homolog of *RAD23*, was also identified and mapped to 19p13.2. Rad23p and both human homologs carry at the N-terminus a ubiquitin-like domain *(231)*.

3.2.3.3. *CSA/ERCC8*

Using the episomally replicating pEBS7 cDNA libraries developed by Legerski, functional *CSA* cDNA was transferred into SV40-immortalized CS-A fibroblasts and then rescued *(131)*. The *CSA* sequence maps to chromosome 5 and encodes a protein that belongs to the WD repeat family of proteins, which are associated with diverse aspects of metabolism and appear to be regulatory, rather than enzymatic, in nature *(273)*. These results are consistent with the CSA protein being involved in transcription or its control *(131)*. The rodent *ERCC8* group, which is represented by a mouse lymphoma mutant (US31), corresponds to *CSA (155)*.

3.3. Biochemical Mechanism of NER

3.3.1. Role of Individual Proteins in Damage Excision

The current state of knowledge of NER mechanism (recently reviewed in *3,135,218a,266,319–321,449a*) is based on in vitro studies involving both recombinant proteins and proteins purified by conventional chromatographic methods from cells or tissue. An outline of the steps is presented in terms of the proteins known to be essential for the process as it occurs in global DNA (Fig. 2). TCR requires most of these components and likely additional proteins not yet clearly identified. Compared with mammalian systems, the yeast *S. cerevisiae* NER genes and proteins are relatively well conserved in terms of structure and function, as indicated in Table 2, and they have been well studied at the biochemical level *(99,297)*. However, interspecies mammalian-yeast genetic complementation rarely occurs *(117,185,373)*. In *S. pombe*, a collection of mutants representing homologous genes has also been developed *(99,204)*.

While the NER reactions are described here as a series of individual steps, these enzymatic reactions may be concerted and form a continuum without the presence of stable intermediates. The term excinuclease has been coined to refer to the excision nuclease activity that cuts on both sides of the damaged site *(317)*.

3.3.1.1. Damage Recognition

A typical mammalian cell contains $\geq 1.2 \times 10^{10}$ nucleotides in its nuclear DNA. A critical role for the XPA protein in damage recognition is suggested by the severity

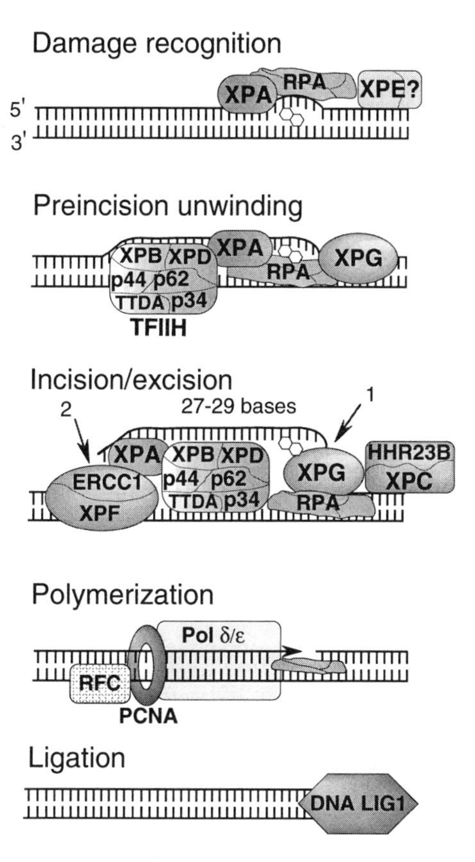

Fig. 2. Model of NER in the global DNA repair pathway. In particular, the roles of XPE as a putative damage-recognition factor and the XPC–HHR23B complex are presently unclear. Proteins are shown to be overlapping in cases where protein interactions have been reported; most such interactions appear to be present in vivo. TFIIH is not drawn to imply specific protein contacts *(see* ref. *155a)*.

of the disease in most XP-A patients and the extreme UV sensitivity of XP-A cell lines *(9,63)*. Purified XPA protein has a higher affinity for UV-irradiated DNA (and DNA damaged by cisplatin or OsO_4) than undamaged DNA *(15,163)*. Oxygen radical-induced damage is also recognized *(328)*. Binding of XPA to (6-4)PDs appears to be much stronger than to CPDs *(163)*. XPA also binds preferentially to undamaged single-stranded vs undamaged double-stranded DNA *(86,163)*. Binding at damaged sites may be mediated by their partially single-stranded character or destabilized double-helical conformation, rather than the structure of the photoproduct or covalent adduct *(115a)*. The sequence of the nuclear *(254)* XPA protein reveals a zinc finger motif, a glutamic acid cluster, and other motifs that are required for its biological activity *(210,255)*. A truncated XPA polypeptide of 122 amino acids (M98 to F219) was identified as the minimal polypeptide necessary for damage recognition *(184)*. In cell-extract assays, no repair incision is observed in the absence of XPA *(259,349)*. Mutations in the XPA protein may have subtle effects on damage recognition. One XPA reversion mutation (R207G) showed a differential ability to remove (6-4)PDs vs CPDs in bulk

DNA *(62,217)*, suggesting altered substrate specificity associated with the missense mutation *(239)*.

Replication protein A (RPA) is also implicated in damage recognition *(39,64,130,210)*. The RPA heterotrimeric complex (p70, p32, p14 subunits; also known as HSSB) *(107)*, which is required for SV40 replication in vitro, is necessary for the incision step of the repair reaction *(69,259)*. RPA binds to single-stranded DNA through the p70 subunit *(107)* and to XPA through the p70 and p32 subunits *(210,232)*. RPA binds tightly to XPA in vitro and shows a cooperative interaction with XPA in binding to damaged DNA *(130,210,232)*. Thus, it is likely that an XPA–RPA complex serves to enhance the specificity of damage recognition. In yeast, RPA is required for replication and recombination, as well as repair *(218)*.

XPA also associates specifically with ERCC1 *(209,289)*, and regions of XPA necessary for this interaction have been identified *(211)*. This interaction may also enhance damage recognition *(269)*. However, another study did not detect an association between XPA and ERCC1–ERCC4 complex *(416)*. As discussed below, XPA also interacts with the TFIIH helicase complex. These multiple interactions argue that XPA plays a central role in recruiting other proteins that are essential for the incision process.

Another protein, initially known as the DNA damage binding (DDB) protein or XPE binding factor (XPE-BF), which binds preferentially to damaged DNA in gel-retardation assays, is implicated in the damage-recognition step, but its role remains elusive. DDB was identified as a UV-damage-inducible protein in monkey cells *(133)*, and a similar protein identified in human cells was absent in a subset of XP-E cell lines *(56,168,176)*. The human protein, which has been purified to near homogeneity *(4,146,173)*, has a high affinity for UV-damaged DNA owing to strong binding to (6-4)PDs and some CPDs *(146,173,303,403,407)*. Examination of the spectrum of lesions recognized by DDB showed binding to alterations resulting from nitrogen mustard, *N*-methyl-*N'*-nitro-*N*-nitrosoguanidine, depurination, and denaturation, but undetectable binding to DNA adducts from trans-diamminedichloroplatinum(II), 4-nitroquinoline-*N*-oxide, and 8-methoxypsoralen *(146,292)*. Some cisplatin-resistant cells show increased levels of DDB and crossresistance to UV *(48,55)*.

The 127-kDa subunit of DDB (DDB1) has been isolated as a heterodimer with a 48-kDa DDB2 *(173)*. In the XP-E lines that lack binding activity (DDB$^-$), microinjection of purified complex efficiently restored the repair synthesis, suggesting that DDB is an XPE factor *(174)*. However, in cell extracts, purified DDB did not complement the overt excision defect seen in both DDB$^-$ and DDB$^+$ cells *(170)*. Even more surprisingly, RPA corrected these defects, but no mutations in any of the three subunits of RPA were found in XPE cells *(170)*. In reconstituted systems, DDB is not an essential component; addition of DDB to the excision nuclease assay gave a slight inhibition *(170)*, whereas addition to the repair synthesis assay gave a slight stimulation *(1)*.

The *DDB1* and *DDB2* genes were isolated *(84,146a,380)*, and an analysis for causative mutations in both DDB$^+$ and DDB$^-$ diploid XP-E fibroblasts showed the presence of substitution mutations in the *DDB2* gene of three DDB$^-$ cell strains *(276a)*. No mutations were found in either gene in three DDB$^+$ cell strains, suggesting that DDB$^+$ cells may be altered in a third gene whose product could interact with DDB. Thus, the relationship of *DDB1* and *DDB2* to the phenotypes of XP-E cells remains unclear, and at present it is difficult to propose any model that accommodates these often conflicting observations.

Other modes of damage recognition are possible. In the case of yeast Rad3p, it has been proposed that DNA helicase activity may be involved in scanning DNA and recognizing damage *(269,270)*. However, ATP-dependent helicase scanning of DNA over long distances would likely be very energy-consuming.

3.3.1.2. Preincision Conformation Changes

Before incision at the damaged site can occur, a region of DNA containing the lesion is likely converted to single-stranded form. This conformation change is thought to be brought about by the activities contained in the XPB and/or XPD helicases *(219,311,335,373)*, which are highly conserved *(290,445)* and essential for cell viability as well as repair *(41,373a,417)*. XPD helicase activity unwinds DNA in the 5' to 3' direction with respect to the strand to which it is bound *(373)*, and XPB unwinds in the opposite direction *(81,335)*. Both proteins are subunits of the basal transcription factor TFIIH, which is required to initiate basal transcription of most RNA polymerase II class genes (*81,334,335*; Chapter 10). XPB and XPD participate in repair as members of the TFIIH complex, which contains five to nine proteins, depending on the purification scheme *(81,82,334)*. An interaction between TFIIH and XPA *(288)* suggests that XPA might be responsible for recruiting TFIIH to the damaged site, although another study presented differing results *(416)*. RPA may also assist the loading of TFIIH helicase activity and stabilize the open complex until incision occurs.

In the XPB protein, K346R substitution within the nucleoside triphosphate binding domain (GAGKS), a change that is expected to abolish the ATPase and helicase activities of the protein *(220,373)*, abolishes the correcting activity of the protein, and constitutes a dominant negative mutation that abrogates repair and transcription *(220,417)*. All seven helicase domains of XPB, which include a DNA binding domain, appear necessary for the repair function, but the putative nuclear localization signal (NLS) does not *(220)*. In contrast, in the XPD protein, K48R substitution in the nucleotide binding domain (GTGKT) results only in a partial loss of repair synthesis (~50% residual UDS; Weeda, G., W. Vermeulen, and J. H. J. Hoeijmakers, personal communication), suggesting that the helicase activity of XPD is partly dispensible for its repair function and that the protein may have an important structural role. In *S. cerevisiae*, the equivalent mutation in *RAD3* (homolog of XPD) inactivated helicase and repair functions, but not viability *(374)*, whereas a similar mutation in *RAD25* (*SSL2*, homolog of XPB) was defective in transcription and inviable *(118)*.

3.3.1.3. Dual Asymmetrical Incisions

A preincision, open complex is thought to form *(317)* as shown in Fig. 2 (although the results of one study suggested that TFIIH performs its helicase function following the catalysis of the 3' nick *[3]*). The mechanism of lesion removal in mammalian cells and in *S. cerevisiae* involves endonucleolytic incision on both sides of the damaged site as in *Escherichia coli*, but with a much larger distance between the lesion and the incision on the 5'-side of damage. A. Sancar and coworkers demonstrated that the excised oligonucleotides containing the lesion are predominantly 27–29-mers that correspond to one incision at the third to fifth phosphodiester bond on the 3'-side of the lesion and a second incision at the 22nd to 25th phosphodiester bond on the 5'-side *(143,234,380)*. This dual incision requires two distinct nuclease activities.

XPG is a single-strand specific endonuclease *(120,283)* that acts on the 3'-side of damage *(234,282)*, based on substrate specificity and antibody inhibition experiments. This step precedes incision on the 5'-side of damage, as shown by comparing in vitro kinetics of appearance of 3'-uncoupled incision with the excision product in a reconstituted system *(259)*. XPG has homology with human FEN1 (flap endonuclease), which cleaves the 5'-single-stranded end at a single-strand to double-strand junction *(126)*. An exonucleolytic activity of XPG protein (and its Rad2 homolog) was also demonstrated *(119)*. Targeting of XPG to the lesion may be mediated by its interaction with RPA bound to single-stranded sequence *(130)*. RPA is required in vitro for both the 3'- and 5'-incisions *(259)*.

Incision on the 5'-side of damage involves a protein complex composed of ERCC1 and ERCC4 (XPF) *(234,287)*. In vitro, the ERCC1–ERCC4 complex was necessary and sufficient for 5' incision *(20a,259)*. A purified ERCC1–ERCC4 endonuclease complex had a preference for single-stranded DNA and the single-stranded portion of a DNA duplex containing a bubble region *(287)*. Anti-ERCC1 antibodies specifically inhibited the 5'-incision in a defined system, resulting in uncoupled 3'-incision *(234)*. The ERCC1–ERCC4 complex is analogous to the homologous Rad1p–Rad10p endonuclease complex from *S. cerevisiae (376,400)*, which cuts specifically at the junction region of Y-shaped DNA and only in the strand with the 3'-single-stranded terminus *(18)*. Targeting of the ERCC1–ERCC4 complex to the site of the lesion likely occurs through other proteins, particularly XPA. RPA may confer structure-specific incision activity to both the ERCC1–ERCC4 and XPG endonucleases *(235)*.

In vitro, incision on the 3'-side of damage appears obligatory for 5'-incision *(259)*, which follows within seconds in vitro *(259)*. Because pyrimidine dimer endonucleases such as T4 denV, which provide 5'-incision, can restore repair capacity in XP cell lines *(73,384)*, the phenotype of total repair deficiency in CHO *ERCC5/XPG* mutants that are defective in 3'-incision also supports the idea that 3'-incision must precede 5'-incision.

The exact role(s) of the XPC–HHR23B complex in repair remains unclear. Both the complex and XPC alone bind with high affinity to single- and double-stranded DNA *(231,302)*. Although XPC is normally complexed with HHR23B in vivo, in the highly purified, reconstituted excision nuclease assay, there was no requirement for HHR23B *(302)*. In another study, HHR23B stimulated repair in cell extracts *(372a)*. XPC is required for in vitro incision of substrates containing thymine dimers, but not for certain synthetic oligonucleotide substrates *(259)* or for TCR in vivo. XPC–HHR23B complex may bind to the damaged strand of the unwound preincision complex, perhaps on the 3'-side of the damage *(350)*. The complex may help stabilize the bubble structure and help target the incision nuclease activities to ensure their appropriate specificity *(259,317,350)*. The ubiquitin-like domain of HHR23B suggests that it may have chaperone characteristics that could promote assembly of the incision complex *(103,302)*. It might also assist in nucleosome disassembly, which could help explain the requirement for XPC in global repair.

3.3.1.4. Excision of the Damaged Segment

Following dual incision, the oligonucleotide containing the lesion is released, a step that may require the continued presence of RPA. In vitro analysis indicates that the incised oligonucleotides (still bound to protein) are released from the DNA duplex by the incision nuclease machinery without assistance from the repair synthesis proteins *(259)*. In *S. cerevisiae*, release of the damaged oligonucleotide may involve Rad3p helicase activity, which is needed for a postincision event *(374)*.

3.3.1.5. Synthesis and Ligation

The DNA polymerase accessory factor proliferating cell nuclear antigen (PCNA) is required in vitro for the repair synthesis step, but not the incision step *(275,349)* (Fig. 2). PCNA increases excision efficiency by promoting the catalytic turnover of the excinuclease complex *(275)*. PCNA also appears necessary in vivo based on immunofluorescence studies showing a close association of PCNA with repair synthesis *(2,46,157,253,402)*. These observations implicated DNA polymerase δ (pol δ) or pol ε in the polymerization step of repair, since these polymerases are stimulated by PCNA in vitro *(144)*. In cell-free repair assays, both pol δ and pol ε can perform the polymerization step on DNA repair intermediates that have already been incised *(352)*. In the presence of DNA ligases, pol ε is more efficient in producing closed, rather than nicked, circles from the repair intermediate. However, in a similar system, antibody-inhibition studies implicated pol δ as contributing most of the repair synthesis activity *(465)*. In permeabilized cells, pol ε appears to be the enzyme that is responsible for repair synthesis *(175,276,379)*. In vivo, it has been difficult to obtain data that implicate a specific polymerase, since the effects of aphidicolin and other polymerase inhibitors are difficult to interpret *(99)*. It may be that mammalian cells can utilize more than one polymerase in repair, depending on cell type, physiological status, and other factors. In *S. cerevisiae*, genetic evidence points toward the involvement of both pol δ and pol ε in repair of UV damage *(38)*.

The polymerase accessory factors PCNA and replication factor C (RFC) may be required to displace the incision proteins from the postincision complex. The trimeric PCNA protein, which acts as a ring-shaped sliding clamp to increase the processivity of polymerases *(178)*, requires another factor, RFC, for its loading onto a DNA template *(295,406)*. RFC is multiprotein complex consisting of five discrete subunits *(194)*.

The final step of repair requires the action of a DNA ligase. Although four distinct ligases are known in mammalian cells, their specific physiological roles are poorly understood (*see* Chapter 12). Any of these may participate in repair. DNA ligase I participates in DNA replication *(300)*. Evidence for the involvement of ligase I in excision repair comes from a unique human mutant cell line (46BR) isolated from a patient manifesting a complex clinical syndrome, including photosensitivity *(390)*. 46BR cells, which have mutations affecting both alleles of the *LIG1* gene *(19)*, are somewhat UV-sensitive and show an abnormally high level of incision-associated breaks after UV treatment *(266,390)*.

3.3.2. *Repair Proteins as Transcription Factors—Dual Roles*

TFIIH (also called BTF2) is one of seven basal or general transcription factors that determine the specificity of RNA polymerase II (recently reviewed in *136a*). TFIIH is unique in terms of having enzymatic activity. In addition to its DNA-dependent ATPase *(311)* and bidirectional helicase activities *(81,311,315)*, TFIIH can be isolated with an associated cyclin-dependent kinase (Cdk)-activating kinase (CAK) involving cyclin H and CDK7/MO15 *(301a,310,343,346)*. It was argued that CDK7 is also involved in DNA repair, since microinjection of CDK7 antibodies into cells inhibits both transcription and repair *(310)*. However, this inhibition is likely an indirect effect. In vitro reconstitution experiments, using forms of TFIIH containing or lacking CDK7, showed no dependence of incision activity on the presence of this kinase activity *(259)*. In addition to the involvement of XPB and XPD helicases, other components of TFIIH are required for DNA repair. The cloned human p62 *(91)* and p44 subunits *(145)* have homology with the Tfb1p and Ssl1p proteins in yeast (cf Table 2). The properties of temperature-sensitive mutations in these yeast homologs suggest the necessity of the mammalian counterparts for both transcription and NER *(233,443,458)*. It might be expected that proteins known to associate with TFIIH could also be involved in repair. Indeed, XPC is one such protein *(81)*, and TFIIE, which interacts directly with TFIIH and can influence its enzymatic activities *(82)*, could also be a candidate "repair protein" complex. Potential connections between transcription factors and repair are discussed in Chapter 10 and elsewhere *(47,83,98b,136a,344)*.

During transcription initiation of class II promoters, TFIIH is the last factor to assemble in the initiation complex and is required for promoter opening *(139)*, i.e., the creation of a 10-nt single-stranded region, which precedes further opening and formation of the first nucleotide bond. An earlier study had suggested that TFIIH was involved in promoter clearance *(108)*. Once the transcript reaches ~30 nt, TFIIH is released and is not associated with the pol II ternary elongation complex *(460)*. Curiously, in *S. cerevisiae*, only one of these two helicase activities of TFIIH is essential for transcription, although both proteins must be present (*see* Section 3.3.1.2.). The CDK7 kinase of TFIIH is responsible for hyperphosphorylating the carboxy-terminal domain of the largest subunit of RNA pol II. This kinase activity appears necessary for promoter clearance or elongation *(7)*.

The precise relationships between forms of TFIIH involved in transcription vs those involved in repair remain to be clarified. From HeLa cells, a core form designated TFIIH*, was purified and found to lack XPD-CAK, which could be isolated as a separate 7.4S complex *(301a)*. TFIIH* supported in vitro pol II-catalyzed transcription at a low level that was stimulated by XPD-CAK, and TFIIH* and XPD-CAK complexes were active in repair as shown by complementing mutant cell extracts for excision nuclease activity *(301a)*. In yeast, TFIIH (also called transcription factor b) is required for both repair and transcription *(90,442)*, but distinctly different forms of TFIIH may be involved *(377)*. For repair, a complex of TFIIH lacking the kinase activity, but containing Rad3p and Rad25p, and consisting of all other essential repair proteins (Rad1p, Rad2p, Rad4p, Rad10p, and Rad14p) was described and referred to as the "repairosome" *(377)*. The interchange between holo-TFIIH and the repairosome was suggested as a means of coupling transcription with repair.

3.3.3. *Reconstitution of NER*

By fractionating cell extracts and purifying components, it has been possible to reconstitute the reactions using highly purified native proteins and, in some cases, recombinant proteins. Sancar's laboratory has characterized the requirements for reactions leading to dual incision *(20a,259,260)*. A highly purified reconstituted system containing RPA, XPA, XPC-HHR23B, XPG, ERCC1–ERCC4, and TFIIH effects dual incision and excision in linear model substrates containing thymine dimers or a cholesterol moiety *(20a,259)*. These findings suggest that all proteins essential for excision are now identified, although other proteins may well influence the efficiency of the process. In the reconstituted excision system, the incised, released oligonucleotides carry bound proteins. The minimum size substrate for human excinuclease is ~100 bp in length *(142)*.

Wood and coworkers used a plasmid repair synthesis assay to define the essential proteins *(1)*. For the incision step, an additional, unidentified protein (IF7) appeared necessary to obtain specificity of repair synthesis. To accomplish repair synthesis and ligation of plasmids into closed circular forms, the incised products were purified and incubated with RPA, PCNA, RFC, pol ε, and DNA ligase I. The overall incision and repair synthesis reaction was also accomplished in a single incubation. More detailed mechanistic studies of the various enzymatic steps will be aided by the larger quantities of proteins that can now be obtained from cDNA overexpression (e.g., the ERCC1–ERCC4 complex *[20a]*).

Reconstitution of the incision step of the reaction was also achieved using highly purified proteins from *S. cerevisiae*, where a very similar absolute requirement for the following homologous yeast proteins was found: RPA, Rad14p for damage recognition, Rad2 endonuclease, Rad1p–Rad10p endonuclease complex, Rad4p–Rad23p complex, and TFIIH consisting of Rad3p, Rad25p, Tfb1p, Ssl1p, p55, and p38 subunits *(116)*.

3.3.4. Unique Requirements of TCR vs. Global Genome Repair

TCR provides a means of promoting the resumption of transcription complexes that are blocked by DNA lesions. In the case of *E. coli*, RNA polymerase stalled at a UV photoproduct interacts with the Mfd protein (or transcription repair coupling factor [TRCF]), which releases the stalled polymerase and recruits the UvrA damage recognition protein, thereby targeting repair to the transcribed strand *(342)*. Mammalian cells may have an analogous, but more complex mechanism. The CSA and CSB proteins were suggested to be analogs of the bacterial TRCF *(412)*, but more recent studies suggest a different role (*414*; *see* Section 2.3.2.). Although the precise roles of the CSA and CSB proteins in transcription and repair remain to be determined *(98a)*, clues are provided by the recent finding that the CSA protein interacts with both CSB and TFIIH *(131)*. Although the CSB protein has helicase motifs *(447)*, it lacks helicase activity while showing DNA-stimulated ATPase activity *(341a)*. Moreover, unlike Mfd, CSB did not dissociate stalled RNA pol II in a ternary complex *(341a)*. Since CSB binds to XPA, TFIIH, and TFIIE *(341a)*, it may serve to recruit proteins to a transcription-blocking lesion and/or assist transcript shortening by upstream translocation of stalled pol II *(77)* to facilitate repair.

The components of the global repair process (Fig. 2) are also required for TCR, with one notable exception. XP-C cells, although grossly defective in global repair of the genome, are proficient in repairing the transcribed strand of active genes *(413,421,422)*. Thus, XPC protein, and possibly the associated HHR23B, are required for processing damage in inactive parts of the genome as well as the nontranscribed strands of active genes. During TCR, the requirement for XPC is absent and its putative role in the 3'-side incision step appears to be alleviated by the nature of the DNA structure associated with the stalled RNA polymerase complex *(350)*. With certain in vitro model substrates, such those containing a cholesterol adduct, the requirement for XPC is also alleviated *(259)*.

3.3.5. Links among DNA Damage, p53, Repair, and Cell-Cycle Regulation

The biological effectiveness of NER will be governed by the specific cell type and the growth/differentiation status of the cell. In response to DNA damage, cells actively modulate their cell-cycle progression in a coordinated manner, by arresting progression or undergoing apoptosis. The p53 tumor suppressor protein plays a key regulatory role in both responses *(71,285,360)*. By its action as a transcriptional activator, p53 mediates a delay in the start of S phase, thereby providing a G1 checkpoint *(71)*. The importance of p53 in maintaining genomic stability is evident from the commonness of mutations in the *p53* gene in human tumors *(110,138)*.

UV irradiation results in an immediate and prolonged increase in the activity *(466)* and level of p53 through posttranslational stabilization *(229)*. For a variety of DNA-damaging agents, p53 increase correlates with the presence of strand breaks *(274)*, but the precise nature of the DNA intermediate(s) that acts as a signal for p53 stabilization is unclear. p53 shows multiple interactions with DNA, including inhibition of DNA replication through a mechanism that is independent of transcription *(70)*. p53 catalyzes renaturation of complementary single strands of DNA or RNA *(17,281)*, exhibits 3'-to-5' exonuclease activity *(263)*, binds to single-stranded DNA ends *(17,281)*, binds to internal segments of single-stranded molecules *(16)* and to short oligonucleotides of the size range that results from the excision reaction *(159)*. Therefore, short oligonucleotides are a reasonable candidate for the mediator that leads to increased affinity of p53 for its target sequences in DNA.

Attempts to link cellular p53 to repair and recovery from DNA damage have explored several possibilities. First, there could a direct physical connection between p53 and repair proteins. p53 indeed interacts in vitro with each of the XPB, XPD, and p62 components of TFIIH *(207,436,438,454)* and inhibits XPB and XPD helicase activities *(207,436,438)*. This helicase inhibition does not appear to involve the strand-annealing activity of p53 *(207,281)*. However, p53 did not affect NER in vitro *(207)*. It was suggested that p53 might promote repair by acting as a sensor of DNA damage (*see* Chapter 15) and recruiting the TFIIH complex to sites where it is needed in repair *(438)*. A comparison of UV-induced p53 accumulation at low-UV fluence found an increased sensitivity in XP-A and CS cell lines vs normal and XP-C cells. It was suggested that the inducing signal might be the stalled RNA polymerase blocked by a UV lesion in an active gene *(456)*. From these and other observations, a model was suggested in which TFIIH interacts with p53 to effect its activation and stabilization by phosphorylation, thereby promoting G1/S arrest *(164)*. Recent studies also implicate the XPB and XPD helicases of TFIIH in the p53-dependent apoptosis pathway, and suggest that this pathway is mediated through direct interaction between p53 and these two helicases *(437)*.

A second mode by which p53 might regulate repair is in terms of transactivation of genes involved in repair. Gadd45, a protein that is upregulated by p53 levels, interacts with PCNA *(358)*, which is required for DNA repair *(275,349)* as well as replication. Repair synthesis in cell extracts was reported to be stimulated by adding Gadd45 *(321,358)*, but other investigators using both excision and repair synthesis assays did not find this effect *(171,172)*. Another key gene induced by p53 is $p21^{CIP1/WAF1}$. p21 interacts with PCNA and inhibits PCNA-dependent DNA replication *(435)*. p21 was reported to inhibit DNA replication selectively and not repair synthesis *(212,351)*, suggesting a means of enhancing cell survival via a p53 response. However, these differential effects of p21 could not be confirmed *(286)*.

Finally, examining the effects of p53 status on DNA repair, mutagenesis, and cell survival after damage have given rather complex results. In Li-Fraumeni human diploid fibroblasts (p53 +/–) and their immortalized p53 (–/–) derivatives, mutational inactivation of p53 correlated with reduced UV-stimulated NER *(94,247)*, and this reduction affected global repair, but not TCR *(94)*. The increased UV resistance associated with reduced repair was attributed to reduced p53-dependent apoptosis *(94)*. In contrast, human diploid fibroblasts having viral E6 protein abrogation of p53 function showed increased cytotoxicity from cisplatin and other DNA crosslinking agents *(129)*. Moreover, in diploid embryonic mouse fibroblasts, p53 knockout deficiency was associated with enhanced resistance to killing (approximately twofold), but not enhanced mutagenesis in a *lacI* transgene after exposure to UV-mimetic damage (e.g., 4NQO) *(323)*. In human RKO colon carcinoma cells having abrogation of p53 by viral E6 protein or by mutation, cell survival after UV, reactivation of irradiated plasmid, and repair synthesis in cell extracts were all decreased *(128,359)*. In transformed mouse cells expressing a temperature-sensitive mutant p53, conditions permissive for p53 function conferred increased survival and decreased mutagenesis in a *supF* reporter gene after UV exposure *(459)*. However, another study with mouse cells found no difference in cell survival or UV photoproduct removal as a function of p53 status (+/+, +/–, –/–) *(148)*. It should be emphasized that the magnitude of greater sensitivity or resistance for UV survival afforded by a normal p53 status in these instances is quite modest (less than or equal to twofold) when compared with that of the intact repair system itself (~6- to 15-fold enhancement). Some of the above apparent contradictions may be owing to differences between the behavior of diploid and transformed cells, and possibly differences between human and rodent cells. A clearer understanding of the role of p53 in suppressing damage-induced carcinogenesis will require a consideration of broader mutational end points and the contribution of apoptosis pathways to cell killing in the context of specific cell types.

4. RELATIONSHIPS BETWEEN MOLECULAR AND CLINICAL PHENOTYPES

4.1. Genes Involved Only in Repair: XPA *and* XPC

Mutations characterized in the *XPA* gene involve base substitutions producing abnormal splicing, missense or nonsense codons, or small deletions that result in premature termination of the protein *(61,278,329–333)*. In Japan almost all cases of XP-A are caused by one or two of three different mutations: one mutation affecting mRNA splicing and leading to premature termination and inactive protein, and two mutations causing nonsense codons *(277)*. Eighty-six percent of Japanese XP-A patients carry one or two copies of the allele affecting splicing, and the severity of disease correlates with genotype. The predominant splicing mutation, which truncates the XPA protein or deletes an exon, is associated with early skin manifestations and progressive neurologic abnormalities *(277)*. These results support the idea that the neurological deterioration seen in some XP patients is caused by severe repair deficiency combined with oxidative damage in nerve tissue, leading to premature death of neurons *(59,308)*. However, the developmental abnormalities (e.g. microcephaly, slow growth) seen in some patients remain unexplained. These symptoms may derive from abnormal interactions of mutant XPA proteins with the TFIIH complex *(288)*, leading to subtle perturbations in transcription that give rise to abnormal gene expression during development (*see* Section 4.2.1.2.).

Several mutations in the *XPC* genes from five patients have been identified *(208)*. Curiously, four of the lines were homo- or hemizygous for the detected mutation. Levels of *XPC* mRNA are reduced substantially in all cell lines and correlate roughly with the degree of cellular UV sensitivity. Instability of mRNA appears to be associated with nonsense mutations.

4.2. Genes Involved in Both Repair and Transcription

4.2.1. Genes for TFIIH Repair/Transcription Proteins: XPD, XPB

4.2.1.1. Spectra of Mutations Underlying the Genetic Disorders

As members of TFIIH, XPD, XPB, and TTDA are required for both repair and transcription. Mutations in the *XPD* and *XPB* genes result in clinical patterns that are more varied and complex than those resulting from mutations in *XPA* or *XPC*, which are only involved in DNA repair. *XPD/ERCC2* is unique in terms of the known phenotypic heterogeneity both with respect to clinical manifestations (XPD, XPD-CS, and TTD) and alterations in DNA repair. Most TTD patients who show repair deficiency are assigned to the *XPD* complementation group *(368)*.

DNA sequencing analysis shows a variety of mutations in the *ERCC2* gene in patients of XPD, TTD, or XPD-CS origin. Most mutations (summarized in Fig. 3) are in the C-terminal half of the protein, which includes four of the seven helicase domains and the putative NLS. Mutations in XPD patients do not appear to map into a region of the protein that is distinct from the locations of mutations found in TTD patients—although it is conceivable that there are discrete clusters of sites in the three-dimensional folded protein. Usually, mutant cells express two different mutant alleles. Mutations in XP-D cell lines include amino acid substitutions, deletions owing to splicing alterations, defects in expression, and a nonsense substitution that should truncate the protein by 34 amino acids *(98,382)*. At least four independent XP-D lines have an R683T substitution within the putative NLS motif *(382)*. Since this mutation was not seen among 15 TTD lines or two XPD-CS lines, it is correlated with the XP clinical presentation *(35,36,381–383)*. This mutation was not seen in a study in which mutations were identified in only one of two alleles in each of two additional XPD cell lines *(98)*. Two patients who combine the features of XP and CS were analyzed and found to have distinct *ERCC2* mutations not seen in other XPD or TTD patients *(36,382)*.

Also shown in Fig. 3 (lower panel, boxes) are the mutations present in two highly UV-sensitive mutant lines of CHO cells that are hemizygous for the hamster *ERCC2* locus *(444)*. These mutations abolish repair capacity without substantially impairing the growth rate of the cell lines. Thus, the presence of a single allele is sufficient to support the TFIIH transcription function of ERCC2 in cultured cells.

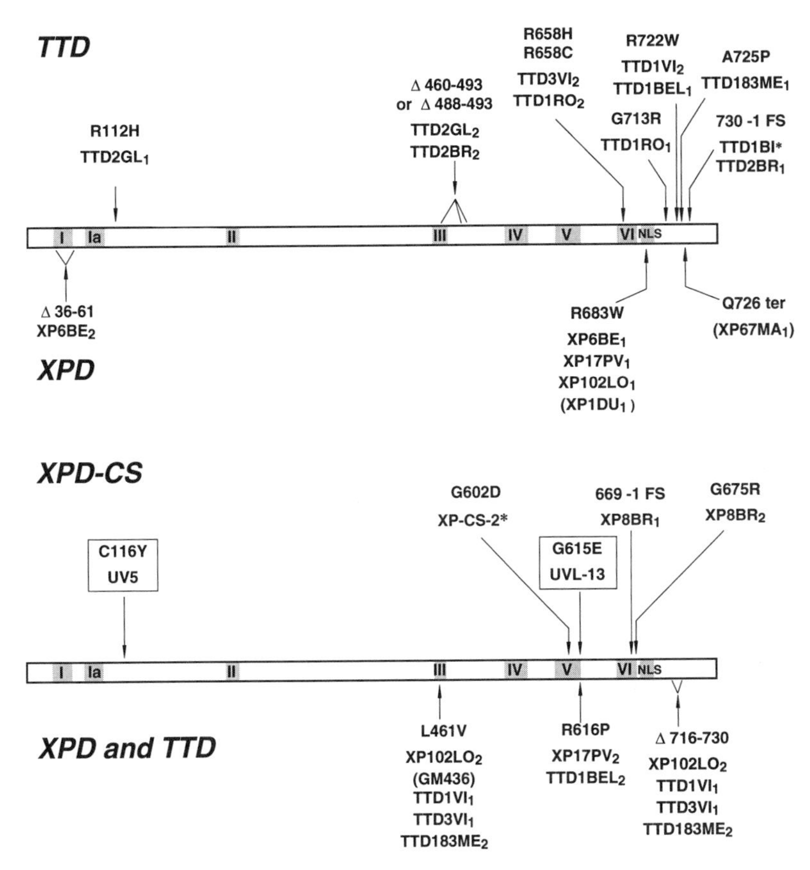

Fig. 3. Mutations affecting the ERCC2 (XPD) protein. A schematic of the protein showing the seven conserved helicase domains *(445)* and the region containing the putative nuclear localization signal (NLS) is shown in each panel. The amino acid changes are shown above the cell line(s) having the specific change. Cell lines in which only one mutation was identified, but two alleles are expressed, are shown in parentheses. Cell lines in which only one allele is expressed are indicated by (*). All other cell lines are given a suffix to designate the two alleles. **(Top)** top, TTD patients *(381,383)*, bottom, XPD *(382)*. **(Bottom)** top, two XPD-CS patients *(36,382)*, bottom, alleles that are present in both XPD and TTD patients, as well as mutations identified in CHO mutants *(444)* shown in boxes. XP line GM436 was reported *(98)* to express only one allelic form, but the identified L461V mutation argues that it carries the common XP102LO2 allele, which has the additional Δ716–730. Since this allele is likely nonviable because of the deletion, a second, expressed allele is likely to be responsible for the viability of GM436 cells.

TTD cell lines are usually compound heterozygotes with changes qualitatively similar to those in XP-D lines *(35,381,383)*. One TTD cell line (TTD183ME) with a moderate repair deficiency (UV-sensitive only at doses >4 J/m^2 and ~65% of normal repair after 6 h) clearly shows critical mutations in the *XPD* gene *(381)*. It is particularly interesting that in some instances, XP-D and TTD individuals share a common allele, which suggests that the clinical phenotype is determined by the second allele. One example is an allele that is present in at least one XP-D cell line and three TTD lines, including TTD183ME. This allele has compound mutations: a L461V substitution in helicase domain III and a 15 amino acid deletion beginning at position 716 in the 760 amino acid protein *(381–383)*. Another example is the R616P substitution seen in lines XP17PV and TTD1BEL *(35,383)*.

In the rare XP-B complementation group (three patients in two families), XP has occurred only in combination with CS symptoms. *XPB* has been sequenced in patients from each family, and in each case one allele was not detectably expressed. The second allele in one cell line (XPCS1BA) had an F99S substitution at this highly conserved position *(424)*. In spite of a severe NER defect in this cell line, the patient and his or her sibling both showed late onset of neurological abnormalities, mild cutaneous symptoms, and a dramatic lack of skin cancer at age 40. The other patient (XP11BE, with combined XP-CS) had an intron-derived 4-bp insertion that resulted in an inactivating frame shift in the C-terminus of the XPB protein *(447)*. This mutation in the maternal allele inactivated the repair function of XPB, and the paternal allele was not significantly expressed *(447)*. Characterization of the purified mutant TFIIH complex isolated from XP11BE lymphoblasts showed that its activity in an in vitro transcription assay was impaired by ~30% and that XPB helicase activity was reduced compared with the maternal control TFIIH *(146b,447)*. In one instance (TTD6VI), a repair-deficient TTD line falls into the XP-B group as shown by measuring UDS after either cell fusion or microinjection of the *XPB* cDNA *(426)*. The *XPB* mutation in TTD6VI is a T119P substitution, which partially inactivates repair *(422a)*.

4.2.1.2. Concept of Transcription Syndromes

XPD and XPB proteins reside in the TFIIH repair/transcription complex, and the mutations described above affect these proteins in XP, TTD, and XP-CS individuals, who have very different clinical presentations. It appears that these distinct diseases can result from different mutations in a single protein. Although many TTD patients have normal repair function, mutations in *ERCC2* in the cells from such patients have not yet been analyzed. Unlike XP-CS, no patient who has combined XP and TTD has been reported. Therefore, XP and TTD appear to be mutually exclusive.

To explain these observations, the concept of transcription syndromes was proposed *(27,28,135,426)*, suggesting that mutations in *ERCC2* or *ERCC3* can affect only the repair function (XP), only the transcription function (repair-proficient TTD), or both functions (repair-deficient TTD and XP-CS). Since XPD and XPB proteins are essential for normal transcription, subtle mutations may interfere with transcription in ways that are detrimental to physiological development and tissue integrity, either pre- or postnatally. Certain genes that are normally transcribed at high efficiency and/or at critical times may be preferentially affected. During hair shaft development in TTD, the synthesis of sulfur-rich proteins may be a limiting factor. Other features of TTD (e.g., delayed growth, mental retardation) could also be owing to inadequate or inappropriate gene expression. According to the transcription syndrome model, XP-CS would then represent the additive effects of faulty gene expression and repair deficiency/cancer proneness. As indicated above, XPB-CS cells (XP11BE) provide the first example in which a mutation in TFIIH is associated with a transcription deficiency in vitro *(146b)*. These cells show a reduced growth rate in culture, which may reflect a generalized in vivo impairment of class II gene transcription *(146b)*.

Additional evidence supporting the idea of defective transcription in TTD alleles comes from studies in *S. cerevisiae* expressing mutant alleles of *XPD*. In *rad3* null mutants, wild-type XPD protein (partially) rescued the lethality, but at least two different TTD mutant alleles did not *(117)*. These results indicate that TTD mutations, which may be causing a subtle effect on transcription in these patients, lead to an overt and lethal transcription defect in yeast, as might be expected in view of the partial conservation of transcriptional function.

A eukaryotic model for the transcription syndrome concept is provided by mutations in *D. melanogaster* in the *haywire* gene, the homolog of *ERCC3 (257)*. Most *haywire* mutant alleles are lethal, but some are associated with UV sensitivity, sterility, reduced life-span, and central nervous system defects, i.e., abnormalities similar to those in TTD and CS. In summary, defective transcription is a plausible explanation for the complex symptoms seen in NER disorders *(135)*, but it does not solve the dilemma of how repair-deficient TTD patients avoid skin cancer and other associated cutaneous effects.

4.2.2. Genes Affecting Transcription-Dependent Repair: CSA, CSB, *and* XPG

CS and TTD patients show a number of broad similarities in terms of clinical features: neurological abnormalities, mental retardation, and slow rate of growth (Table 1), prompting the question of whether CS might also be interpreted as a transcription syndrome. The *CSA* and *CSB* genes are needed for efficient repair of UV damage in transcribed sequences. These genes are apparently not essential for cell viability, because the yeast homolog of *CSB* is not essential *(411)* and because identified mutations in *CSA* and *CSB* involve large deletions *(131)* and a frameshift *(405)*, respectively, which would likely inactivate protein functions. Although the biochemical roles of the CSA and CSB proteins are not understood, both show interaction with TFIIH *(131,343a)*. Therefore, both proteins are likely involved in transcription or its control. Recently, reduced levels of transcription from the adenovirus major late promoter were demonstrated in vitro in both CS-A and CS-B cell extracts, and this deficiency is corrected in *CSA* and *CSB* cDNA transformants *(75)*. These results support the transcription syndrome hypothesis for CS.

CS can occur in combination with XP owing to mutations in the *XPB*, *XPD*, and *XPG* genes. Unlike *CSA* and *CSB*, *XPG* is not selectively associated with defective transcription-dependent repair of UV damage, nor is XPG an essential protein or a component of TFIIH, as are XPB and XPD. However, CSA, CSB, and XPG proteins do seem to share a role in the transcription-dependent repair of certain oxidative lesions (e.g., thymine glycols) produced by ionizing radiation *(68,190)*. Cell lines from XPG-CS patients, but not regular XP-G patients, show defective repair of this class of damage *(68)*. The role of XPG protein in transcription-dependent repair appears to involve a function that is distinct from its endonucleolytic activity in NER (Section 3.3.1.3.), likely a function involving protein interactions. This idea is supported by mutation analysis. In three XPG-CS patients, the mutations in *XPG* substantially truncate the protein *(280)*, whereas, in two non-CS XP-G siblings, one allele has a less disruptive point mutation in the putative active site of the endonuclease *(279)*. The suggestion was made that the defect causing progressive neurological deterioration in CS may lie in the failure to repair certain classes of oxidative damage in active genes *(68,122)*, akin to the rationale used to explain the neurological effects in some XP patients. Neuronal cell death during early development might cause the characteristic demyelination of neurons seen in CS.

Another speculative hypothesis to explain the clinical manifestations of CS (and TTD) proposed that the CSA, CSB, XPG, and TFIIH proteins are involved in a pathway that normally promotes gene expression by excising methylated cytosine *(57)*. A failure to demethylate could result in the critical loss of gene expression needed, for example, for the synthesis of myelin or sulfur-rich hair proteins during development. In this context, a notable association of XPG protein with TFIIH was observed *(155a,260)*, and coimmunoprecipitation of XPG and CSB proteins was reported *(155a)*.

4.2.3. Dilemma of Inconsistent Cancer Proneness vs Repair Deficiency

None of the three genetic diseases discussed here has a high spontaneous chromosomal instability as is found in cancer-prone repair disorders not involving NER, namely Fanconi's anemia, Bloom syndrome, and ataxia telangiectasia (*104*; *see* Chapter 19). The cancer associated with XP is specifically owing to NER deficiency, which is also a skin cancer risk factor for non-XP individuals *(449)*. Exceptions are seen in certain XP cases, such as the 38- and 41-yr-old brothers who showed very low repair capacity (5%), but no evidence of tumors in spite of severe sun sensitivity *(339)*. Conversely, in cells from another patient (group D XP1PO), a high level of repair (~60% incision and UDS) is associated with severe neurological abnormalities and serious developmental aberrations *(161)*. To understand these exceptions, more detailed repair studies are needed. UDS or incision levels may not always reflect completed repair.

Repair-deficient TTD presents a major dilemma, since the patients have none of the skin abnormalities seen in most XP cases. One possibility is that qualitative differences in repair deficiency are responsible for the differing clinical outcome in the two disorders. A recent study comparing the efficiency of removal of (6-4)PDs vs CPDs found relatively efficient repair of the former lesions, but not the latter *(89)*. The suggestion was made that (6-4)PDs may be the critical class of damage for skin carcinogenesis. The relatively proficient repair of (6-4)PDs in TTD might prevent the important class of mutations that initiates the progression to cancer. However, the results of another study are in conflict *(34)*. A second possibility is that there might be a difference between TTD and XP in terms of the role the immune system plays in skin tumor promotion following mutation *(198)*. Compared with XPD, TTD may be resistant to the immunosuppressive effects of UV-B radiation that are part of the mechanism of photocarcinogenesis *(112)*. Finally, the putative transcriptional defect in TTD could impair the high expression of certain proto-oncogenes (e.g. c-*ras*, c-*myc*) that may be involved in the progression to malignancy.

5. TRANSGENIC MODELS

Mice homozygous for *ERCC1* disruption, which has no counterpart among the human disorders, showed unexpected pathologies, including runtedness at birth, early death, liver aneuploidy, and elevated p53 levels in liver, brain, and kidney *(241)*, possibly because of the lack of repair of spontaneous oxidative damage to DNA *(213)*. The alterations in p53 were considered to support the idea that p53 serves as a monitor of DNA damage. Another laboratory confirmed many of these results and found that the survival of the *ERCC1*$^{-}$ animals varies with the mouse genetic background (Weeda, G., personal communication).

Mice carrying disruption mutations (–/–) in the *XPA* and *XPC* loci were found to mimic, although incompletely, XP patients in their pathology. Greatly increased susceptibility to UV radiation *(74,270,322)* and dimethylbenz[*a*]anthracene-induced skin tumors *(74,270)* was observed, along with other skin and eye lesions consistent with the human disease. *XPA* (–/–) mutant mice appeared physiologically normal at birth and did not develop (by 18 mo) the nervous system pathology seen in XP-A patients *(74,270)*. However, in one study, ~50% of the (–/–) embryos died in the midfetal period with signs of anemia *(74)*, a phenomenon not seen in the other study. This difference between studies might be attributable to mouse line differences as seen with ERCC1-deficient mice. The less severe effect of *XPA* mutations compared with *ERCC1* mutations in the mice suggests an involvement of the *ERCC1* gene in some critical process during development other than NER, perhaps a role in recombination. The UV sensitivity of *XPC*⁻ mice indicates that the global repair pathway is important in rodents even though they have little capacity to remove CPDs.

A transgenic mouse with a homozygous mutation resembling one of the alleles of a CS-B patient was prepared. The repair phenotype matches that of CS, but the clinical abnormalities are only slightly apparent *(409a)*.

6. CONCLUDING REMARKS

In cancer biology, the study of NER has moved from the background toward center stage. With the realization of several NER proteins as participants in transcription and transcription-dependent repair, apoptosis, DNA replication, and recombination, it is clear that repair proteins often have multiple roles in DNA and cellular metabolism. A likely involvement of certain NER proteins (i.e., ERCC1, ERCC4) in recombination and recombinational repair can be anticipated from the properties of CHO mutants and the homologous proteins in yeasts (*92,337*; discussed in *393*). DNA interstrand crosslinks are a class of damage whose repair is not understood and likely involves a recombinational pathway *(156)*.

From the recent reconstitution experiments, an outline of the sequence of biochemical events in NER is now apparent. However, there are puzzling and unresolved issues that include:

1. Whether the helicase activities of both the XPB and XPD proteins are required for normal incision;
2. The identity of the *XPE* gene product(s) and its relationship of to the DDB1, DDB2, and RPA;
3. The role of XPC and its associated HHR23B; and
4. The identity of the gene underlying the new human complementation group described by Itoh *(153,154)*.

Overexpressed, purified proteins will certainly aid in mechanistic studies and allow each of the steps of NER to be examined in greater detail. The availability of the CSA and CSB proteins should facilitate an examination of their relationship to transcription and TCR and whether they play a role analogous to the Mfd protein in *E. coli* or some other role *(414)*. Success in devising an in vitro system that links repair with transcription would undoubtedly accelerate progress in this area.

Results so far with knockout mutations in transgenic mice suggest that there are significant interspecies differences in terms of the roles of NER genes (e.g., *XPA*, *CSB*) in functions that are tied to normal development in humans. An absence of anticipated developmental defects is also seen with the first mouse model for Fanconi's anemia using the *FAC* gene *(48a)*. Answers to these puzzles, as well as many surprises, are expected to come from the study of future transgenic animals, including those that carry subtle mutations mimicking the ones found in the putative transcription syndromes. DNA repair studies offer the intellectual challenge and excitement of explaining complex human diseases in terms of specific molecular pathologies.

To understand the relationships between repair deficiency vs cancer proneness (or lack thereof) and developmental defects, additional systematic, comprehensive data on repair kinetics are needed for the cell lines from patients. Contradictions in the repair parameters of TTD lines need to be resolved *(34,89)* in terms of whether they differ from XP-D lines. Radioimmunoassays *(249)* and chromatographic detection of photoproducts in excision products *(102)* can provide sensitive, independent measures of repair capacity.

Finally, there is the hope that our knowledge of DNA repair will aid in cancer treatment and cancer prevention. Understanding the interplay among DNA repair, replication, recombination, and cell-cycle regulation should help identify the critical genomic events that cause the progressive genetic instability resulting in cancer. The absence of skin tumors in TTD patients reminds us that mutagenesis is an insufficient prerequisite for carcinogenesis, which requires multiple genetic perturbations and alterations in gene expression.

ACKNOWLEDGMENTS

I thank Christine Weber and Christopher Parris for their helpful comments on the manuscript, and Aziz Sancar and James Cleaver for valuable discussions and preprints of publications. I also acknowledge Errol Friedberg, Stuart Linn, Leon Mullenders, and Geert Weeda for allowing me to cite their unpublished studies. This work was done under the auspices of the US Department of Energy by Lawrence Livermore National Laboratory under contract No. W-7405-ENG-48.

REFERENCES

1. Aboussekhra, A., M. Biggerstaff, M. K. K. Shivji, J. A. Vilpo, V. Moncollin, V. N. Podust, M. Protic, U. Hubscher, J. M. Egly, and R. D. Wood. 1995. Mammalian DNA nucleotide excision repair reconstituted with purified protein components. *Cell* **80:** 859–868.
2. Aboussekhra, A. and R. D. Wood. 1995. Detection of nucleotide excision repair incisions in human fibroblasts by immunostaining for PCNA. *Exp. Cell Res.* **221:** 326–332.
3. Aboussekhra, A. and R. D. Wood. 1994. Repair of UV-damaged DNA by mammalian cells and *Saccharomyces cerevisiae*. *Curr. Opinion Genet. Dev.* **4:** 212–220.
4. Abramic, M., A. S. Levine, and M. Protic. 1991. Purification of an ultraviolet-inducible, damage-specific DNA-binding protein from primate cells. *J. Biol. Chem.* **266:** 22,493–22,500.
5. Ahmed, F. E. and R. B. Setlow. 1978. Excision repair in ataxia telangiectasia, Fanconi's anemia, Cockayne syndrome, and Bloom's syndrome after treatment with ultraviolet radiation and *N*-acetoxy–2-acetylaminofluorene. *Biochim. Biophys. Acta* **521:** 805–817.
6. Ahnstrom, G. and K. Erixon. 1981. Measurement of strand breaks by alkaline denaturation and hydroxylapatite chromatography, in *DNA Repair—A Laboratory Manual of Research Procedures*, vol. 1, part B (Friedberg, E. C. and P. C. Hanawalt, eds.), Marcel Dekker, New York, pp. 403–418.

7. Akoulitchev, S., T. P. Mälelä, R. A. Weinberg, and D. Reinberg. 1995. Requirement for TFIIH kinase activity in transcription by RNA polymerase II. *Nature* **377:** 557–560.
8. Andrews, A. D., S. F. Barrett, F. W. Yoder, and J. H. Robbins. 1978. Cockayne's syndrome fibroblasts have increased sensitivity to ultraviolet light but normal rates of unscheduled DNA synthesis. *J. Invest. Dermatol.* **70:** 237–239.
9. Andrews, A. D., S. F. Barrett, and J. H. Robbins. 1978. Xeroderma pigmentosum neurological abnormalities correlate with colony-forming ability after ultraviolet radiation. *Proc. Natl. Acad. Sci. USA* **75:** 1984–1988.
10. Arlett, C. F., S. A. Harcourt, A. R. Lehmann, S. Stevens, M. A. Ferguson-Smith, and W. N. Morley. 1980. Studies on a new case of xeroderma pigmentosum (XP3BR) from complementation group G with cellular sensitivity to ionizing radiation. *Carcinogenesis* **1:** 745–751.
11. Arlett, C. F., and S. A. Harcourt. 1980. Survey of radiosensitivity in a variety of human cell strains. *Cancer Res.* **40:** 926–932.
12. Arlett, C. F., and S. A. Harcourt. 1983. Variation in response to mutagens amongst normal and repair-defective human cells, in I*nduced Mutagenesis. Molecular Mechanisms and Their Implications for Environmental Protection* (C. W. Lawrence, ed.), Plenum, New York, pp. 249–266.
13. Arlett, C. F. and A. R. Lehmann. 1995. Xeroderma pigmentosum, Cockayne syndrome, and trichothiodystrophy: sun sensitivity, DNA repair defects and skin cancer, in *Genetic Predisposition to Cancer* (Eeles, R., B. Ponder, D. Easton, and, A. Horwich, eds.), Chapman & Hall, London, pp. 185–206.
14. Arlett, C. F., J. E. Lowe, S. A. Harcourt, A. P. Waugh, J. Cole, L. Roza, B. L. Diffey, T. Mori, O. Nikaido, and M. H. Green. 1993. Hypersensitivity of human lymphocytes to UV-B and solar irradiation. *Cancer Res.* **53:** 609–614.
15. Asahina, H., I. Kuraoka, M. Shirakawa, E. H. Morita, N. Miura, I. Miyamoto, E. Ohtsuka, Y. Okada, and K. Tanaka. 1994. The XPA protein is a zinc metalloprotein with an ability to recognize various kinds of DNA damage. *Mutat. Res.* **315:** 229–237.
16. Bakalkin, G., G. Selivanova, T. Yakovleva, E. Kiseleva, E. Kashuba, K. P. Magnusson, L. Szekely, G. Klein, L. Terenius, and K. G. Wiman. 1995. p53 binds single-stranded DNA ends through the C-terminal domain and internal DNA segments via the middle domain. *Nucleic Acids Res.* **23:** 362–369.
17. Bakalkin, G., T. Yakovleva, G. Selivanova, K. P. Magnusson, L. Szekely, E. Kiseleva, G. Klein, L. Terenius, and K. G. Wiman. 1994. p53 binds single-stranded DNA ends and catalyzes DNA renaturation and strand transfer. *Proc. Natl. Acad. Sci. USA* **91:** 413–417.
18. Bardwell, A. J., L. Bardwell, A. E. Tomkinson, and E. C. Friedberg. 1994. Specific cleavage of model recombination and repair intermediates by the yeast Rad1-Rad10 endonuclease. *Science* **265:** 2082–2085.
19. Barnes, D. E., A. E. Tomkinson, A. R. Lehmann, A. D. Webster, and T. Lindahl. 1992. Mutations in the DNA ligase I gene of an individual with immunodeficiencies and cellular hypersensitivity to DNA-damaging agents. *Cell* **69:** 495–503.
20. Barrett, S. F., J. H. Robbins, R. E. Tarone, and K. H. Kraemer. 1991. Evidence for defective repair of cyclobutane pyrimidine dimers with normal repair of other DNA photoproducts in a transcriptionally active gene transfected into Cockayne syndrome cells. *Mutat. Res.* **255:** 281–291.

20a. Bessho, T., Sacar, A., Thompson, L. H., and Thelen, M. P. 1997. Reconstitution of human excision nuclease with recombinant XPF-ERCC1 complex. *J. Biol. Chem.* **272:** 3833–3837.

21. Biggerstaff, M., D. E. Szymkowski, and R. D. Wood. 1993. Co-correction of the ERCC1, ERCC4 and xeroderma pigmentosum group F DNA repair defects in vitro. *EMBO J.* **12:** 3685–3692.

22. Biggerstaff, M. and R. D. Wood. 1992. Requirement for *ERCC–1* and *ERCC–3* gene products in DNA excision repair in vitro. Complementation using rodent and human cell extracts. *J. Biol. Chem.* **267:** 6879–6885.
23. Bill, C. A., B. M. Grochan, R. E. Meyn, V. A. Bohr, and P. J. Tofilon. 1991. Loss of intragenomic DNA repair heterogeneity with cellular differentiation. *J. Biol. Chem.* **266:** 21,821–21,826.
24. Bohr, V. A. 1995. DNA repair fine structure and its relations to genomic stability. *Carcinogenesis* **16:** 2885–2892.
25. Bohr, V. A., D. S. Okumoto, and P. C. Hanawalt. 1986. Survival of UV-irradiated mammalian cells correlates with efficient DNA repair in an essential gene. *Proc. Natl. Acad. Sci. USA* **83:** 3830–3833.
26. Bohr, V. A., C. A. Smith, D. S. Okumoto, and P. C. Hanawalt. 1985. DNA repair in an active gene: removal of pyrimidine dimers from the DHFR gene of CHO cells is much more efficient than in the genome overall. *Cell* **40:** 359–369.
27. Bootsma, D. and J. H. Hoeijmakers. 1993. DNA repair. Engagement with transcription. *Nature* **363:** 114,115.
28. Bootsma, D. and J. H. Hoeijmakers. 1994. The molecular basis of nucleotide excision repair syndromes. *Mutat. Res.* **307:** 15–23.
29. Boyer, J. C., W. K. Kaufmann, B. P. Brylawski, and M. Cordeiro-Stone. 1990. Defective postreplication repair in xeroderma pigmentosum variant fibroblasts. *Cancer Res.* **50:** 2593–2598.
30. Bredberg, A., K. H. Kraemer, and M. M. Seidman. 1986. Restricted ultraviolet mutational spectrum in a shuttle vector propagated in xeroderma pigmentosum cells. *Proc. Natl. Acad. Sci. USA* **83:** 8273–8277.
31. Bronstein, S. M., J. E. Cochrane, T. R. Craft, J. A. Swenberg, and T. R. Skopek. 1991. Toxicity, mutagenicity, and mutational spectra of *N*-ethyl-*N*-nitrosourea in human cell lines with different DNA repair phenotypes. *Cancer Res.* **51:** 5188–5197.
32. Bronstein, S. M., T. R. Skopek, and J. A. Swenberg. 1992. Efficient repair of O^6-ethylguanine, but not O^4-ethylthymine or O^2-ethylthymine, is dependent upon O^6-alkylguanine-DNA alkyltransferase and nucleotide excision repair activities in human cells. *Cancer Res.* **52:** 2008–2011.
33. Brookman, K. W., J. E. Lamerdin, M. P. Thelen, M. Hwang, J. T. Reardon, A. Sancar, Z. Q. Zhou, C. A. Walter, C. N. Parris, and L. H. Thompson. 1996. *ERCC4 (XPF)* encodes a human nucleotide excision repair protein with eukaryotic recombination homologs. *Mol. Cell. Biol.* **16:** 6553–6562.
34. Broughton, B. C., A. R. Lehmann, S. A. Harcourt, C. F. Arlett, A. Sasrasin, W. J. Kleijer, F. A. Beemer, R. Nairn, and D. L. Mitchell. 1990. Relationship between pyrimidine dimers, 6-4 photoproducts, repair synthesis and cell survival: studies using cells from patients with trichothiodystrophy. *Mutat. Res.* **235:** 33–40.
35. Broughton, B. C., H. Steingrimsdottir, C. A. Weber, and A. R. Lehmann. 1994. Mutations in the xeroderma pigmentosum group D DNA repair/transcription gene in patients with trichothiodystrophy. *Nature Genet.* **7:** 189–194.
36. Broughton, B. C., A. F. Thompson, S. A. Harcourt, W. Vermeulen, J. H. J. Hoeijmakers, E. Botta, M. Stefanini, M. King, C. Weber, J. Cole, C. F. Arlett, and A. R. Lehmann. 1995. Molecular and cellular analysis of the DNA repair defect in a patient in xeroderma pigmentosum group D with the clinical features of xeroderma pigmentosum and Cockayne syndrome. *Am. J. Hum. Genet.* **56:** 167–174.
37. Brown, A. J., T. H. Fickel, J. E. Cleaver, P. H. M. Lohman, M. H. Wade, and R. Waters. 1979. Overlapping pathways for repair of damage from ultraviolet light and chemical carcinogens in human fibroblasts. *Cancer Res.* **39:** 2522–2527.
38. Budd, M. E. and J. L. Campbell. 1995. DNA polymerases required for repair of UV-induced damage in *Saccharomyces cerevisiae*. *Mol. Cell. Biol.* **15:** 2173–2179.

39. Burns, J. L., S. N. Guzder, P. Sung, S. Prakash, and L. Prakash. 1996. An affinity of human replication protein A for ultraviolet-damaged DNA. Implications for damage recognition in nucleotide excision repair. *J. Biol. Chem.* **271:** 11,607–11,610.
40. Busch, D. B., J. E. Cleaver, and D. A. Glaser. 1980. Large-scale isolation of UV-sensitive clones of CHO cells. *Somatic Cell. Genet.* **6:** 407–418.
41. Busch, D., C. Greiner, K. L. Rosenfeld, R. Ford, J. de Wit, J. H. J. Hoeijmakers, and L. H. Thompson. 1994. Complementation group assignments of moderately UV-sensitive CHO mutants isolated by large-scale screening (FAECB). *Mutagenesis* **9:** 301–306.
42. Busch, D., C. Greiner, K. Lewis, R. Ford, G. Adair, and L. Thompson. 1989. Summary of complementation groups of UV-sensitive CHO cell mutants isolated by large-scale screening. *Mutagenesis* **4:** 349–354.
42a. Busch, D. B., M. Z. Zdzienicka, A. T. Natarajan, N. J. Jones, W. I. J. Overkamp, A. Collins, D. L. Mitchell, M. Stefanini, E. Botta, R. B. Albert, N. Liu, D. A. White, A. J. van Gool, and L. H. Thompson. 1996. A CHO mutant, UV40, that is sensitive to diverse mutagens and represents a new complementation group of mitomycin C sensitivity. *Mutat. Res.* **363:** 209–221.
43. Carothers, A. M., J. Mucha, and D. Grunberger. 1991. DNA strand-specific mutations induced by (+/-)–3 alpha,4 beta-dihydroxy- 1 alpha,2 alpha-epoxy–1,2,3,4-tetrahydrobenzo[c]phenanthrene in the dihydrofolate reductase gene. *Proc. Natl. Acad. Sci. USA* **88:** 5749–5753.
44. Carothers, A. M., W. Zhen, J. Mucha, Y. J. Zhang, R. M. Santella, D. Grunberger, and V. A. Bohr. 1992. DNA strand-specific repair of (+-)–3 alpha,4 beta-dihydroxy–1 alpha,2 alpha-epoxy–1,2,3,4-tetrahydrobenzo[c]phenanthrene adducts in the hamster dihydrofolate reductase gene. *Proc. Natl. Acad. Sci. USA* **89:** 11,925–11,929.
45. Carreau, M., E. Eveno, X. Quilliet, O. Chevalier-Lagente, A. Benoit, B. Tanganelli, et al. 1995. Development of a new easy complementation assay for DNA repair deficient human syndromes using cloned repair genes. *Carcinogenesis* **16:** 1003–1009.
46. Celis, J. E. and P. Madsen. 1986. Increased nuclear cyclin/PCNA antigen staining of non S-phase transformed human amnion cells engaged in nucleotide excision DNA repair. *FEBS Lett.* **209:** 277–283.
47. Chalut, C., V. Moncollin, and J. M. Egly. 1994. Transcription by RNA polymerase II: a process linked to DNA repair. *Bioessays* **16:** 651–655.
48. Chao, C. C., S. L. Huang, H. M. Huang, and S. Lin-Chao. 1991. Cross-resistance to UV radiation of a cisplatin-resistant human cell line: overexpression of cellular factors that recognize UV-modified DNA. *Mol. Cell. Biol.* **11:** 2075–2080.
48a. Chen, M., D. J. Tomkins, W. Auerbach, C. McKerlie, H. Youssoufian, L. Liu, O. Gan, M. Carreau, A. Auerbach, T. Groves, C. J. Guidos, M. H. Freedman, J. Cross, D. H. Percy, J. E. Dick, A. L. Joyner, and M. Buchwald. 1996. Inactivation of *Fac* in mice produces inducible chromosomal instability and reduced fertility reminiscent of Fanconi anemia. *Nature Genet.* **12:** 448–451.
49. Chen, R. H., V. M. Maher, and J. J. McCormick. 1990. Effect of excision repair by diploid human fibroblasts on the kinds and locations of mutations induced by (+/-)–7 beta,8 alpha-dihydroxy–9 alpha,10 alpha-epoxy–7,8,9,10- tetrahydrobenzo[a]pyrene in the coding region of the HPRT gene. *Proc. Natl. Acad. Sci. USA* **87:** 8680–8684.
50. Chen, R. H., V. M. Maher, and J. J. McCormick. 1991. Lack of a cell cycle-dependent strand bias for mutations induced in the HPRT gene by (+/-)–7 beta,8 alpha-dihydroxy–9 alpha,10 alpha-epoxy–7,8,9,10-tetrahydrobenzo(a)pyrene in excision repair-deficient human cells. *Cancer Res.* **51:** 2587–2592.
51. Chen, R. H., V. M. Maher, J. Brouwer, P. van de Putte, and J. J. McCormick. 1992. Preferential repair and strand-specific repair of benzo[*a*]pyrene diol epoxide adducts in the HPRT gene of diploid human fibroblasts. *Proc. Natl. Acad. Sci. USA* **89:** 5413–5417.

52. Christians, F. C. and P. C. Hanawalt. 1992. Inhibition of transcription and strand-specific DNA repair by alpha-amanitin in Chinese hamster ovary cells. *Mutat. Res.* **274:** 93–101.
53. Christians, F. C. and P. C. Hanawalt. 1993. Lack of transcription-coupled repair in mammalian ribosomal RNA genes. *Biochemistry* **32:** 10,512–10,518.
54. Christians, F. C. and P. C. Hanawalt. 1994. Repair in ribosomal RNA genes is deficient in xeroderma pigmentosum group C and in Cockayne's syndrome cells. *Mutat. Res.* **323:** 179–187.
55. Chu, G. and E. Chang. 1990. Cisplatin-resistant cells express increased levels of a factor that recognizes damaged DNA. *Proc. Natl. Acad. Sci. USA* **87:** 3324–2227.
56. Chu, G. and E. Chang. 1988. Xeroderma pigmentosum group E cells lack a nuclear factor that binds to damaged DNA. *Science* **242:** 564–567.
57. Chu, G. and L. Mayne. 1996. Xeroderma pigmentosum, Cockayne syndrome and trichothiodystrophy: do the genes explain the diseases? *Trends Genet.* **12:** 187–192.
58. Cleaver, J. E. 1968. Defective repair replication of DNA in xeroderma pigmentosum. *Nature* **218:** 652–656.
59. Cleaver, J. E. 1990. Do we know the cause of xeroderma pigmentosum? *Carcinogenesis* **11:** 875–882.
60. Cleaver, J. E. 1972. Xeroderma pigmentosum: variants with normal DNA repair and normal sensitivity fo ultraviolet light. *J. Invest. Dermatol.* **58:** 124–128.
61. Cleaver, J. E., W. C. Charles, G. H. Thomas, and M. L. McDowell. 1995. A deletion and an insertion in the alleles for the xeroderma pigmentosum (XPA) DNA-binding protein in mildly affected patients. *Hum. Mol. Genet.* **4:** 1685–1687.
62. Cleaver, J. E., F. Cortes, L. H. Lutze, W. F. Morgan, A. N. Player, and D. L. Mitchell. 1987. Unique DNA repair properties of a xeroderma pigmentosum revertant. *Mol. Cell. Biol.* **7:** 3353–3357.
63. Cleaver, J. E. and K. H. Kraemer. 1995. Xeroderma pigmentosum and Cockayne syndrome, in *The Metabolic and Molecular Bases of Inherited Disease*, 7th ed., vol. III (Scriver, C. R., A. L. Beaudet, W. S. Sly, D. Valle, eds.), McGraw-Hill Book Co., New York, pp. 4393–4419.
64. Clugston, C. K., K. McLaughlin, M. K. Kenny, and R. Brown. 1992. Binding of human single-stranded DNA binding protein to DNA damaged by the anticancer drug cis-diamminedichloroplatinum (II). *Cancer Res.* **52:** 6375–6379.
65. Cockayne, E. A. 1936. Dwarfism with retinal atrophy and deafness. *Arch. Dis. Child.* **11:** 1–8.
66. Cole, J., C. F. Arlett, P. G. Norris, G. Stephens, A. P. Waugh, D. M. Beare, and M. H. Green. 1992. Elevated *hprt* mutant frequency in circulating T-lymphocytes of xeroderma pigmentosum patients. *Mutat. Res.* **273:** 171–178.
67. Collins, A. R. 1993. Mutant rodent cell lines sensitive to ultraviolet light, ionizing radiation and cross-linking agents: a comprehensive survey of genetic and biochemical characteristics. *Mutat. Res.* **293:** 99–118.
68. Cooper, P. K., T. Nouspikel, S. G. Clarkson, and S. A. Leadon. 1997. Defective transcription-coupled repair of oxidative base damage in Cockayne syndrome patients from XP group G. *Science* **275:** 990–993.
69. Coverley, D., M. K. Kenny, D. P. Lane, and R. D. Wood. 1992. A role for the human single-stranded DNA binding protein HSSB/RPA in an early stage of nucleotide excision repair. *Nucleic Acids Res.* **20:** 3873–3880.
70. Cox, L. S., T. Hupp, C. A. Midgley, and D. P. Lane. 1995. A direct effect of activated human p53 on nuclear DNA replication. *EMBO J.* **14:** 2099–2105.
71. Cox, L. S. and D. P. Lane. 1995. Tumour suppressors, kinases and clamps: how p53 regulates the cell cycle in response to DNA damage. *Bioessays* **17:** 501–508.
72. Daya-Grosjean, L., C. Robert, C. Drougard, H. Suarez, and A. Sarasin. 1993. High mutation frequency in *ras* genes of skin tumors isolated from DNA repair deficient xeroderma pigmentosum patients. *Cancer Res.* **53:** 1625–1629.

73. de Jonge, A. J., W. Vermeulen, W. Keijzer, J. H. Hoeijmakers, and D. Bootsma. 1985. Microinjection of *Micrococcus luteus* UV-endonuclease restores UV-induced unscheduled DNA synthesis in cells of 9 xeroderma pigmentosum complementation groups. *Mutat. Res.* **150:** 99–105.
74. de Vries, A., C. T. M. van Oostrom, F. M. A. Hofhuis, P. M. Dortant, R. J. W. Berg, F. R. de Gruijl, P. W. Wester, C. F. van Kreijl, P. J. A. Capel, H. van Steeg, and S. J. Verbeek. 1995. Increased susceptibility to ultraviolet-B and carcinogens of mice lacking the DNA excision repair gene *XPA*. *Nature* **377:** 169–173.
75. Dianov, G., J. F. Houle, N. Iyer, V. A. Bohr, and E. C. Friedberg. 1996. Unpublished results.
76. Dianov, G., A. Price, and T. Lindahl. 1992. Generation of single-nucleotide repair patches following excision of uracil residues from DNA. *Mol. Cell. Biol.* **12:** 1605–1612.
77. Donahue, B. A., S. Yin, J. S. Taylor, D. Reines, and P. C. Hanawalt. 1994. Transcript cleavage by RNA polymerase II arrested by a cyclobutane pyrimidine dimer in the DNA template. *Proc. Natl. Acad. Sci. USA* **91:** 8502–8506.
78. Dorado, G., H. Steingrimsdottir, C. F. Arlett, and A. R. Lehmann. 1991. Molecular analysis of ultraviolet-induced mutations in a xeroderma pigmentosum cell line. *J. Mol. Biol.* **217:** 217–222.
79. dos Santos, C. C., H. Gavish, and M. Buchwald. 1994. Fanconi anemia revisited: old ideas and new advances. *Stem Cells* **12:** 142–153.
80. Downes, C. S., A. J. Ryan, and R. T. Johnson. 1993. Fine tuning of DNA repair in transcribed genes: mechanisms, prevalence and consequences. *Bioessays* **15:** 209–216.
81. Drapkin, R., J. Reardon, A. Ansari, J. C. Huang, L. Zawel, K. Ahn, A. Sancar, and D. Reinberg. 1994. TFIIH, a link between RNA polymerase II transcription and DNA excision repair. *Nature* **368:** 769–772.
82. Drapkin, R. and D. Reinberg. 1994. The multifunctional TFIIH complex and transcriptional control. *Trends Biochem. Sci.* **19:** 504–508.
83. Drapkin, R., A. Sancar, and D. Reinberg. 1994. Where transcription meets repair. *Cell* **77:** 9–12.
84. Dualan, R., T. Brody, S. Keeney, A. F. Nichols, A. Admon, and S. Linn. 1995. Chromosomal localization and cDNA cloning of the gene (*DDB1* and *DDB2*) for the p127 and p48 subunits of a human damage-specific DNA binding protein. *Genomics* **29:** 62–69.
85. Dumaz, N., C. Drougard, A. Sarasin, and L. Daya-Grosjean. 1993. Specific UV-induced mutation spectrum in the p53 gene of skin tumors from DNA-repair-deficient xeroderma pigmentosum patients. *Proc. Natl. Acad. Sci. USA* **90:** 10,529–10,533.
86. Eker, A. P., W. Vermeulen, N. Miura, K. Tanaka, N. G. Jaspers, J. H. Hoeijmakers, and D. Bootsma. 1992. Xeroderma pigmentosum group A correcting protein from calf thymus. *Mutat. Res.* **274:** 211–224.
87. Evans, M. K., J. H. Robbins, M. B. Ganges, R. E. Tarone, R. S. Nairn, and V. A. Bohr. 1993. Gene-specific DNA repair in xeroderma pigmentosum complementation groups A, C, D, and F. Relation to cellular survival and clinical features. *J. Biol. Chem.* **268:** 4839–4847.
88. Evans, M. K., B. G. Taffe, C. C. Harris, and V. A. Bohr. 1993. DNA strand bias in the repair of the p53 gene in normal human and xeroderma pigmentosum group C fibroblasts. *Cancer Res.* **53:** 5377–5381.
89. Eveno, E., F. Bourre, X. Quilliet, O. Chevallier-Lagente, L. Roza, and A. P. M. Eker. 1995. Different removal of ultraviolet photoproducts in genetically-related xeroderma pigmentosum and trichothiodistrophy diseases. *Cancer Res.* **55:** 4325–4332.
90. Feaver, W. J., J. Q. Svejstrup, L. Bardwell, A. J. Bardwell, S. Buratowski, K. D. Gulyas, T. F. Donahue, E. C. Friedberg, and R. D. Kornberg. 1993. Dual roles of a multiprotein complex from *S. cerevisiae* in transcription and DNA repair. *Cell* **75:** 1379–1387.
91. Fischer, L., M. Gerard, C. Chalut, Y. Lutz, S. Humbert, M. Kanno, P. Chambon, and J. M. Egly. 1992. Cloning of the 62-kilodalton component of basic transcription factor BTF2. *Science* **257:** 1392–1395.

92. Fishman-Lobell, J. and J. E. Haber. 1992. Removal of nonhomologous DNA ends in double-strand break recombination: the role of the yeast ultraviolet repair gene *RAD1*. *Science* **258:** 480–484.
93. Flejter, W. L., L. D. McDaniel, D. Johns, E. C. Friedberg, and R. A. Schultz. 1992. Correction of xeroderma pigmentosum complementation group D mutant cell phenotypes by chromosome and gene transfer: involvement of the human *ERCC2* DNA repair gene. *Proc. Natl. Acad. Sci. USA* **89:** 261–265.
94. Ford, J. M. and P. C. Hanawalt. 1995. Li-Fraumeni syndrome fibroblasts homozygous for p53 mutations are deficient in global DNA repair but exhibit normal transcription-coupled repair and enhanced UV resistance. *Proc. Natl. Acad. Sci. USA* **92:** 8876–8880.
95. Ford, J. M., L. Lommel, and P. C. Hanawalt. 1994. Preferential repair of ultraviolet light-induced DNA damage in the transcribed strand of the human p53 gene. *Mol. Carcinog.* **10:** 105–109.
96. Fornace Jr., A. J., K. W. Kohn, and H. E. Kann, Jr. 1976. DNA single-strand breaks during repair of UV damage in human fibroblasts and abnormalities in xeroderma pigmentosum. *Proc. Natl. Acad. Sci. USA* **73:** 39–43.
97. Francis, A. A., R. D. Snyder, W. C. Dunn, and J. D. Regan. 1981. Classification of chemical agents as to their ability to induce long- or short-patch DNA repair in human cells. *Mutat. Res.* **83:** 159–169.
98. Frederick, G. D., R. H. Amirkhan, R. A. Schultz, and E. C. Friedberg. 1994. Structural and mutational analysis of the xeroderma pigmentosum group D *(XPD)* gene. *Hum. Mol. Genet.* **3:** 1783–1788.
98a. Friedberg, E. C. 1996. Cockayne syndrome—a primary defect in DNA repair, transcription, both or neither? *Bioessays* **18:** 731–738.
98b. Friedberg, E. C. 1996 Relationships between DNA repair and transcription. *Ann. Rev. Biochem.* **65:** 15–42.
99. Friedberg, E. C., G. C. Walker, and W. Siede. 1995. *DNA Repair and Mutagenesis*, American Soc. Microbiol., Washington, DC.
100. Fujiwara, Y., M. Ichihashi, Y. Kano, K. Goto, and K. Shimizu. 1981. A new human photosensitive subject with a defect in the recovery of DNA synthesis after ultraviolet-light irradiation. *J. Invest Dermatol.* **77:** 256–263.
101. Fujiwara, Y., M. Tatsumi, and M. S. Sasaki. 1977. Cross-link repair in human cells and its possible defect in Fanconi's anemia cells. *J. Mol. Biol.* **113:** 635–649.
102. Galloway, A. M., M. Liuzzi, and M. C. Paterson. 1994. Metabolic processing of cyclobutyl pyrimidine dimers and (6-4) photoproducts in UV-treated human cells. Evidence for distinct excision-repair pathways. *J. Biol. Chem.* **269:** 974–980.
103. Garrett, K. P., T. Aso, J. N. Bradsher, S. I. Foundling, W. S. Lane, R. C. Conaway, and J. W. Conaway. 1995. Positive regulation of general transcription factor SIII by a tailed ubiquitin homolog. *Proc. Natl. Acad. Sci. USA* **92:** 7172–7176.
104. German, J. 1990. The chromosome-breakage syndromes: rare disorders that provide models for studying somatic mutation, in *Detection of Cancer Predisposition: Laboratory Approaches* (Spatz, L., A. D. Bloom, and N. W. Paul, eds.), March of Dimes Birth Defects Foundation, White Plains, NY, pp. 85–111.
105. Gillespie, J. M. and R. C. Marshall. 1983. A comparison of the proteins of normal and trichothiodystrophic human hair. *J. Invest. Dermatol.* **80:** 195–202.
106. Glover, T. W., C. C. Chang, J. E. Trosko, and S. S. Li. 1979. Ultraviolet light induction of diptheria toxin-resistant mutants of normal and xeroderma pigmentosum human fibroblasts. *Proc. Natl. Acad. Sci. USA* **76:** 3982–3986.
107. Gomes, X. V. and M. S. Wold. 1995. Structural analysis of human replication protein A. Mapping functional domains of the 70-kDa subunit. *J. Biol. Chem.* **270:** 4534–4543.
108. Goodrich, J. A. and R. Tjian. 1994. Transcription factors IIE and IIH and ATP hydrolysis direct promoter clearance by RNA polymerase II. *Cell* **77:** 145–156.

109. Gorbalenya, A. E., E. V. Koonin, A. P. Donchenko, and V. M. Blinov. 1989. Two related superfamilies of putative helicases involved in replication, recombination, repair and expression of DNA and RNA genomes. *Nucleic Acids Res.* **17:** 4713–4730.
110. Greenblatt, M. S., W. P. Bennett, M. Hollstein, and C. C. Harris. 1994. Mutations in the p53 tumor suppressor gene: clues to cancer etiology and molecular pathogenesis. *Cancer Res.* **54:** 4855–4878.
111. Greenhaw, G. A., A. Hebert, M. E. Duke-Woodside, I. J. Butler, J. T. Hecht, J. E. Cleaver, G. H. Thomas, and W. A. Horton. 1992. Xeroderma pigmentosum and Cockayne syndrome: overlapping clinical and biochemical phenotypes. *Am. J. Hum. Genet.* **50:** 677–689.
112. Grewe, M., C. Ahrens, S. Grether-Beck, K. Gyufko, S. A. Harcourt, A. R. Lehmann, C. F. Arlett, and J. Krutmann. 1995. Evidence for the importance of UVB radiation (UVBR)-induced immunosuppression in human photocarcinogenesis. *J. Invest. Dermatol.* **104:** 574 (abstract).
113. Griffiths, T. D. and S. Y. Ling. 1991. Effect of UV light on DNA chain growth and replicon initiation in xeroderma pigmentosum variant cells. *Mutagenesis* **6:** 247–251.
114. Grosovsky, A. J. and J. B. Little. 1983. Mutagenesis and lethality following S phase irradiation of xeroderma pigmentosum and normal human diploid fibroblasts with ultraviolet light. *Carcinogenesis* **4:** 1389–1393.
115. Gulyas, K. D. and T. F. Donahue. 1992. *SSL2*, a suppressor of a stem-loop mutation in the *His4* leader encodes the yeast homolog of human *ERCC–3*. *Cell* **69:** 1031–1042.
115a. Gunz, D., M. T. Hess, and H. Naegeli. 1996. Recognition of DNA adducts by human nucleotide excision repair. *J. Biol. Chem.* **271:** 25,089–25,098.
116. Guzder, S. N., Y. Habraken, P. Sung, L. Prakash, and S. Prakash. 1995. Reconstitution of yeast nucleotide excision repair with purified Rad proteins, replication protein A, and transcription factor TFIIH. *J. Biol. Chem.* **270:** 12,973–12,976.
117. Guzder, S. N., P. Sung, S. Prakash, and L. Prakash. 1995. Lethality in yeast of trichothiodistrophy (TTD) mutations in the human xeroderma pigmentosum group D gene. *J. Biol. Chem.* **270:** 17,660–17,663.
118. Guzder, S. N., P. Sung, V. Bailly, L. Prakash, and S. Prakash. 1994. RAD25 is a DNA helicase required for DNA repair and RNA polymerase II transcription. *Nature* **369:** 578–581.
119. Habraken, Y., P. Sung, L. Prakash, and S. Prakash. 1994. A conserved 5' to 3' exonuclease activity in the yeast and human nucleotide excision repair proteins RAD2 and XPG. *J. Biol. Chem.* **269:** 31,342–31,345.
120. Habraken, Y., P. Sung, L. Prakash, and S. Prakash. 1994. Human xeroderma pigmentosum group G gene encodes a DNA endonuclease. *Nucleic Acids Res.* **22:** 3312–3316.
121. Hanawalt, P. C. 1995. DNA repair comes of age. *Mutat. Res.* **336:** 101–113.
122. Hanawalt, P. C. 1994. Transcription-coupled repair and human disease. *Science* **266:** 1957,1958.
123. Hanawalt, P. C., B. A. Donahue, and K. S. Sweder. 1994. Repair and transcription. Collision or collusion? *Curr. Biol.* **4:** 518–521.
124. Hansson, J., M. Munn, W. D. Rupp, R. Kahn, and R. D. Wood. 1989. Localization of DNA repair synthesis by human cell extracts to a short region at the site of a lesion [published erratum appears in *J. Biol. Chem.* 1990. **265(7):** 4172]. *J. Biol. Chem.* **264:** 21,788–21,792.
125. Harosh, I., L. Naumovski, and E. C. Friedberg. 1989. Purification and characterization of Rad3 ATPase/DNA helicase from *Saccharomyces cerevisiae*. *J. Biol. Chem.* **264:** 20,532–20,539.
126. Harrington, J. J. and M. R. Lieber. 1994. The characterization of a mammalian DNA structure-specific endonuclease. *EMBO J.* **13:** 1235–1246.
127. Hata, H., M. Numata, H. Tohda, A. Yasui, and A. Oikawa. 1991. Isolation of two chloroethylnitrosourea-sensitive Chinese hamster cell lines. *Cancer Res.* **51:** 195–198.

128. Havre, P. A., J. Yuan, L. Hedrick, K. R. Cho, and P. M. Glazer. 1995. p53 inactivation by HPV16 E6 results in increased mutagenesis in human cells. *Cancer Res.* **55:** 4420–4424.
129. Hawkins, D. S., G. W. Demers, and D. A. Galloway. 1996. Inactivation of p53 enhances sensitivity to multiple chemotherapeutic agents. *Cancer Res.* **56:** 892–898.
130. He, Z., L. A. Henricksen, M. S. Wold, and C. J. Ingles. 1995. RPA involvement in the damage-recognition and incision steps of nucleotide excision repair. *Nature* **374:** 566–569.
131. Henning, K. A., L. Li, N. Iyer, L. D. McDaniel, M. S. Reagan, R. Legerski, R. A. Schultz, M. Stefanini, A. R. Lehmann, L. V. Mayne, and E. C. Friedberg. 1995. The Cockayne syndrome group A gene encodes a WD repeat protein that interacts with CSB protein and a subunit of RNA polymerase II TFIIH. *Cell* **82:** 555–564.
132. Higgins, D. R., S. Prakash, P. Reynolds, R. Polakowski, S. Weber, and L. Prakash. 1983. Isolation and characterization of the *RAD3* gene of *Saccharomyces cerevisiae* and inviability of *rad3* deletion mutants. *Proc. Natl. Acad. Sci. USA* **80:** 5680–5684.
133. Hirschfeld, S., A. S. Levine, K. Ozato, and M. Protic. 1990. A constitutive damage-specific DNA-binding protein is synthesized at higher levels in UV-irradiated primate cells. *Mol. Cell. Biol.* **10:** 2041–2048.
134. Ho, L., and P. C. Hanawalt. 1991. Gene-specific DNA repair in terminally differentiating rat myoblasts. *Mutat. Res.* **255:** 123–141.
135. Hoeijmakers, J. H. J. 1994. Human nucleotide excision repair syndromes: molecular clues to unexpected intricacies. *Eur. J. Cancer* **30A:** 1912–1924.
136. Hoeijmakers, J. H. 1993. Nucleotide excision repair. II: From yeast to mammals. *Trends Genet.* **9:** 211–217.
136a. Hoeijmakers, J. H. J., J. M. Egly, and W. Vermeulen. 1996. TFIIH: a key component in multiple DNA transactions. *Current Opin. Genet. Dev.* **6:** 26–33.
137. Hoeijmakers, J. H. J., H. Odijk, and A. Westerveld. 1987. Differences between rodent and human cell lines in the amount of integrated DNA after transfection. *Exp. Cell Res.* **169:** 111–119.
138. Hollstein, M., D. Sidransky, B. Vogelstein, and C. C. Harris. 1991. p53 mutations in human cancers. *Science* **253:** 49–53.
139. Holstege, F. C. P., P. C. van der Vliet, and H. T. M. Timmers. 1996. Opening of an RNA polymerase II promoter occurs in two distinct steps and requires the basal transcription factors IIE and IIH. *EMBO J.* **15:** 1666–1677.
140. Hoy, C. A., L. H. Thompson, C. L. Mooney, and E. P. Salazar. 1985. Defective DNA cross-link removal in Chinese hamster cell mutants hypersensitive to bifunctional alkylating agents. *Cancer Res.* **45:** 1737–1743.
141. Huang, J. C., D. S. Hsu, A. Kazantsev, and A. Sancar. 1994. Substrate spectrum of human excinuclease: repair of abasic sites, methylated bases, mismatches, and bulky adducts. *Proc. Natl. Acad. Sci. USA* **91:** 12,213–12,217.
142. Huang, J. C. and A. Sancar. 1994. Determination of minimum substrate size for human excinuclease. *J. Biol. Chem.* **269:** 19,034–19,040.
143. Huang, J. C., D. L. Svoboda, J. T. Reardon, and A. Sancar. 1992. Human nucleotide excision nuclease removes thymine dimers from DNA by incising the 22nd phosphodiester bond 5' and the 6th phosphodiester bond 3' to the photodimer. *Proc. Natl. Acad. Sci. USA* **89:** 3664–3668.
144. Hubscher, U. and S. Spadari. 1994. DNA replication and chemotherapy. Physiol. Rev. **74:** 259–304.
145. Humbert, S., H. van Vuuren, Y. Lutz, J. H. Hoeijmakers, J. M. Egly, and V. Moncollin. 1994. p44 and p34 subunits of the BTF2/TFIIH transcription factor have homologies with SSL1, a yeast protein involved in DNA repair. *EMBO J.* **13:** 2393–2398.
146. Hwang, B. J., and G. Chu. 1993. Purification and characterization of a human protein that binds to damaged DNA. *Biochemistry* **32:** 1657–1666.

146a. Hwang, B. J., J. C. Liao, and G. Chu. 1996. Isolation of a cDNA encoding a UV-damaged DNA binding factor defective in xeroderma pigmentosum group E cells. *Mutat. Res.* **362:** 105–117.

146b. Hwang, J. R., V. Moncollin, W. Vermeulen, T. Seroz, H. van Vuuren, J. H. J. Hoeijmakers, and J. M. Egly. 1996. A 3' → 5' XPB helicase defect in repair-transcription factor TFIIH of xeroderma pigmentosum group B affects both DNA repair and transcription. *J. Biol. Chem.* **271:** 15,898–15,904.

147. Ishida, R., and M. Buchwald. 1982. Susceptibility of Fanconi's anemia lymphoblasts to DNA-cross-linking and alkylating agents. *Cancer Res.* **42:** 4000–4006.

148. Ishizaki, K., Y. Ejima, T. Matsunaga, R. Hara, A. Sakamoto, M. Ikenaga, Y. Ikawa, and S. Aizawa. 1994. Increased UV-induced SCEs but normal repair of DNA damage in p53-deficient mouse cells. *Int. J. Cancer* **58:** 254–257.

149. Islas, A. L., F. J. Baker, and P. C. Hanawalt. 1994. Transcription-coupled repair of psoralen cross-links but not monoadducts in Chinese hamster ovary cells. *Biochemistry* **33:** 10,794–10,799.

150. Islas, A. L. and P. C. Hanawalt. 1995. DNA repair in the MYC and FMS proto-oncogenes in ultraviolet light-irradiated human HL60 promyelocytic cells during differentiation. *Cancer Res.* **55:** 336–341.

151. Itin, P. H. and M. R. Pittelkow. 1990. Trichothiodystrophy: review of sulfur-deficient brittle hair syndromes and association with the ectodermal dysplasias. *J. Am. Acad. Dermatol.* **22:** 705–717.

152. Itoh, T., J. E. Cleaver, and M. Yamaizumi. 1996. Cockayne syndrome complementation group B associated with xeroderma pigmentosum phenotype. *Hum. Genet.* **97:** 176–179.

153. Itoh, T., Y. Fujiwara, T. Ono, and M. Yamaizumi. 1995. UVs syndrome, a new general category of photosensitive disorder with defective DNA repair, is distinct from xeroderma pigmentosum variant and rodent complementation group 1. *Am. J. Hum. Genet.* **56:** 1267–1276.

154. Itoh, T., T. Ono, and M. Yamaizumi. 1994. A new UV-sensitive syndrome not belonging to any complementation groups of xeroderma pigmentosum or Cockayne syndrome: siblings showing biochemical characteristics of Cockayne syndrome without typical clinical manifestations. *Mutat. Res.* **314:** 233–248.

155. Itoh, T., T. Shiomi, N. Shiomi, Y. Harada, M. Wakasugi, T. Matsunaga, O. Nikaido, E. C. Friedberg, and M. Yamaizumi. 1996. Rodent complementation group 8 (ERCC8) corresponds to Cockayne syndrome complementation group A. *Mutat. Res.* **362:** 167–174.

155a. Iyer, N., M. S. Reagan, K. J. Wu, B. Canagarajah, and E. C. Friedberg. 1996. Interactions involving the human RNA polymerase II transcription/nucleotide excision repair complex TFIIH, the nucleotide excision repair protein XPG, and Cockayne syndrome group B (CSB) protein. Biochemistry **35:** 2157–2167.

156. Jachymczyk, W. J., R. C. von Borstel, M. R. Mowat, and P. J. Hastings. 1981. Repair of interstrand cross-links in DNA of *Saccharomyces cerevisiae* requires two systems for DNA repair: the *RAD3* system and the *RAD51* system. *Mol. Gen. Genet.* **182:** 196–205.

157. Jackson, D. A., A. B. Hassan, R. J. Errington, and P. R. Cook. 1994. Sites in human nuclei where damage induced by ultraviolet light is repaired: localization relative to transcription sites and concentrations of proliferating cell nuclear antigen and the tumour suppressor protein, p53. *J. Cell Sci.* **107:** 1753–1760.

158. Jaspers, N. G., V. D. Jansen, G. Kuilen, and D. Bootsma. 1981. Complementation analysis of xeroderma pigmentosum variants. *Exp. Cell Res.* **136:** 81–90.

159. Jayaraman, J. and C. Prives. 1995. Activation of p53 sequence-specific DNA binding by short single strands of DNA requires the p53 C-terminus. *Cell* **81:** 1021–1029.

160. Johnson, R. T., G. C. Elliott, S. Squires, and V. C. Joysey. 1989. Lack of complementation between xeroderma pigmentosum complementation groups D and H. *Hum. Genet.* **81:** 203–210.

161. Johnson, R. T. and S. Squires. 1992. The XPD complementation group. Insights into xeroderma pigmentosum, Cockayne's syndrome and trichothiodystrophy. *Mutat. Res.* **273:** 97–118.
162. Jones, C. J., R. S. Lloyd, and R. D. Wood. 1994. Analysis of cells harboring a putative DNA repair gene reveals a lack of evidence for a second independent xeroderma pigmentosum group A correcting gene. *Mutat. Res.* **324:** 159–164.
163. Jones, C. J. and R. D. Wood. 1993. Preferential binding of the xeroderma pigmentosum group A complementing protein to damaged DNA. *Biochemistry* **32:** 12,096–12,116.
164. Jones, C. J. and D. Wynford-Thomas. 1995. Is TFIIH an activator of the p53-mediated G1/S checkpoint? *Trends Genet.* **11:** 165,166.
165. Jones, J. C., W. P. Zhen, E. Reed, R. J. Parker, A. Sancar, and V. A. Bohr. 1991. Gene-specific formation and repair of cisplatin intrastrand adducts and interstrand cross-links in Chinese hamster ovary cells. *J. Biol. Chem.* **266:** 7101–7107.
166. Kantor, G. J., L. S. Barsalou, and P. C. Hanawalt. 1990. Selective repair of specific chromatin domains in UV-irradiated cells from xeroderma pigmentosum complementation group C. *Mutat. Res.* **235:** 171–180.
167. Karentz, D. and J. E. Cleaver. 1986. Excision repair in xeroderma pigmentosum group C but not group D is clustered in a small fraction of the total genome. *Mutat. Res.* **165:** 165–174.
168. Kataoka, H. and Y. Fujiwara. 1991. UV damage-specific DNA-binding protein in xeroderma pigmentosum complementation group E. *Biochem. Biophys. Res. Commun.* **175:** 1139–1143.
169. Kaufmann, W. K. 1989. Pathways of human cell post-replication repair. *Carcinogenesis* **10:** 1–11.
170. Kazantsev, A., D. Mu, A. F. Nichols, X. Xhao, S. Sinn, and A. Sancar. 1996. Functional complementation of xeroderma pigmentosum group E by replication protein A (RPA) in an *in vitro* system. *Proc. Natl. Acad. Sci. USA* **93:** 5014–5018.
171. Kazantsev, A. and A. Sancar. 1995. Does the p53 up-regulated gadd45 protein have a role in excision repair (Technical Comments). *Science* **270:** 1003,1004.
172. Kearsey, J. M., M. K. K. Shivji, P. A. Hall, and R. D. Wood. 1995. Does the p53 up-regulated gadd45 protein have a role in excision repair (Technical Comments). *Science* **270:** 1004,1005.
173. Keeney, S., G. J. Chang, and S. Linn. 1993. Characterization of a human DNA damage binding protein implicated in xeroderma pigmentosum E. *J. Biol. Chem.* **268:** 21,293–21,300.
174. Keeney, S., A. P. Eker, T. Brody, W. Vermeulen, D. Bootsma, J. H. Hoeijmakers, and S. Linn. 1994. Correction of the DNA repair defect in xeroderma pigmentosum group E by injection of a DNA damage-binding protein. *Proc. Natl. Acad. Sci. USA* **91:** 4053–4056.
175. Keeney, S. and S. Linn. 1990. A critical review of permeabilized cell systems for studying mammalian DNA repair. *Mutat. Res.* **236:** 239–252.
176. Keeney, S., H. Wein, and S. Linn. 1992. Biochemical heterogeneity in xeroderma pigmentosum complementation group E. *Mutat. Res.* **273:** 49–56.
177. Keijzer, W., N. G. J. Jaspers, P. J. Abrahams, A. M. R. Taylor, C. F. Arlett, B. Zelle, H. Takebe, P. D. S. Kinmont, and D. Bootsma. 1979. A seventh complementation group in excision-deficient xeroderma pigmentosum. *Mutat. Res.* **62:** 183–190.
178. Kelman, Z. and M. O'Donnell. 1995. DNA polymerase III holoenzyme: structure and function of a chromosomal replicating machine. *Annu. Rev. Biochem.* **64:** 171–200.
179. Kohn, K. W., R. A. G. Ewig, L. C. Erickson, and L. A. Zwelling. 1981. Measurement of strand break and cross-links by alkaline elution, in *DNA Repair, a Laboratory Manual of Research Procedures*, vol. 1, part B (Friedberg, E. C. and P. C. Hanawalt, eds.), Marcel Dekker, New York, pp. 379–401.
180. Kraemer, K. H., M. M. Lee, and J. Scotto. 1984. DNA repair protects against cutaneous and internal neoplasia: evidence from xeroderma pigmentosum. *Carcinogenesis* **5:** 511–514.

181. Kraemer, K. H., M. M. Lee, and J. Scotto. 1987. Xeroderma pigmentosum. *Arch. Dermatol.* **123:** 241–250.
182. Kraemer, K. H., D. D. Levy, C. N. Parris, E. M. Gozukara, S. Moriwaki, S. Adelberg, and M. M. Seidman. 1994. Xeroderma pigmentosum and related disorders: examining the linkage between defective DNA repair and cancer. *J. Invest. Dermatol.* **103:** 96S–101S.
183. Kraemer, K. H., M. Protic-Sabljic, A. Bredberg, and M. M. Seidman. 1987. Plasmid vectors for study of DNA repair and mutagenesis. *Curr. Probl. Derm.* **17:** 166–181.
184. Kuraoka, I., E. H. Morita, M. Saijo, T. Matsuda, K. Morikawa, M. Shirakawa, and K. Tanaka. 1996. Identification of a damaged-DNA binding domain of the XPA protein. *Mutat. Res.* **362:** 87–95.
185. Lambert, C., L. B. Couto, W. A. Weiss, R. A. Schultz, L. H. Thompson, and E. C. Friedberg. 1988. A yeast DNA repair gene partially complements defective excision repair in mammalian cells. *EMBO J.* **7:** 3245–3253.
186. Lambert, W. C., H. R. Kuo, and M. W. Lambert. 1995. Xeroderma pigmentosum. *Dermatol. Clin.* **13:** 169–209.
187. Lambert, W. C. and M. W. Lambert. 1992. Co-recessive inheritance: a model for DNA repair and other surveillance genes in higher eukaryotes. *Mutat. Res.* **273:** 179–192.
188. Larminat, F., E. J. Beecham, C. J. J. Link, A. May, and V. A. Bohr. 1995. DNA repair in the endogenous and episomal amplified c-*myc* oncogene loci in human tumor cells. *Oncogene* **10:** 1639–1645.
189. Larminat, F., W. Zhen, and V. A. Bohr. 1993. Gene-specific DNA repair of interstrand cross-links induced by chemotherapeutic agents can be preferential. *J. Biol. Chem.* **268:** 2649–2654.
190. Leadon, S. A. and P. K. Cooper. 1993. Preferential repair of ionizing radiation-induced damage in the transcribed strand of an active human gene is defective in Cockayne syndrome. *Proc. Natl. Acad. Sci. USA* **90:** 10,499–10,513.
191. Leadon, S. A. and D. A. Lawrence. 1991. Preferential repair of DNA damage on the transcribed strand of the human metallothionein genes requires RNA polymerase II. *Mutat. Res.* **255:** 67–78.
192. Leadon, S. A. and M. M. Snowden. 1988. Differential repair of DNA damage in the human metallothionein gene family. *Mol. Cell. Biol.* **8:** 5331–5338.
193. Leadon, S. A., M. E. Zolan, and P. C. Hanawalt. 1983. Restricted repair of aflatoxin B1 induced damage in alpha DNA of monkey cells. *Nucleic Acids Res.* **11:** 5675–5689.
194. Lee, S. H., A. D. Kwong, Z. Q. Pan, and J. Hurwitz. 1991. Studies on the activator 1 protein complex, an accessory factor for proliferating cell nuclear antigen-dependent DNA polymerase delta. *J. Biol. Chem.* **266:** 594–602.
195. Legerski, R. J., P. Liu, L. Li, C. A. Peterson, Y. Zhao, R. J. Leach, S. L. Naylor, and M. J. Siciliano. 1994. Assignment of xeroderma pigmentosum group C *(XPC)* gene to chromosome 3p25. *Genomics* **21:** 266–269.
196. Legerski, R. and C. Peterson. 1992. Expression cloning of a human DNA repair gene involved in xeroderma pigmentosum group C. *Nature* **359:** 70–73.
197. Lehmann, A. R. 1987. Cockayne's syndrome and trichothiodystrophy: Defective repair without cancer. *Cancer Rev.* **7:** 82–103.
198. Lehmann, A. R. 1995. Nucleotide excision repair and the link with transcription. *TIBS* **20:** 402–405.
199. Lehmann, A. R. 1982. Three complementation groups in Cockayne syndrome. *Mutat. Res.* **106:** 347–356.
200. Lehmann, A. R. 1989. Trichothiodystrophy and the relationship between DNA repair and cancer. *Bioessays* **11:** 168–170.
201. Lehmann, A. R., C. F. Arlett, B. C. Broughton, S. A. Harcourt, H. Steingrimsdottir, M. Stefanini, A. Malcolm, R. Taylor, A. T. Natarajan, S. Green, M. D. King, R. M. MacKie, J. B. P. Stephenson, and J. L. Tolmie. 1988. Trichothiodystrophy, a human DNA repair

disorder with heterogeneity in the cellular response to ultraviolet light. *Cancer Res.* **48:** 6090–6096.

202. Lehmann, A. R., D. Bootsma, S. G. Clarkson, J. E. Cleaver, P. J. McAlpine, K. Tanaka, L. H. Thompson, and R. D. Wood. 1994. Nomenclature of human DNA repair genes. *Mutat. Res.* **315:** 41,42.
203. Lehmann, A. R., A. M. Carr, F. Z. Watts, and J. M. Murray. 1991. DNA repair in the fission yeast, *Schizosaccharomyces pombe. Mutat. Res.* **250:** 205–210.
204. Lehmann, A. R., S. Kirk-Bell, and L. Mayne. 1979. Abnormal kinetics of DNA synthesis in ultraviolet light-irradiated cells from patients with Cockayne's syndrome. *Cancer Res.* **39:** 4237–4241.
205. Lehmann, A. R., S. Kirk-Bell, C. F. Arlett, M. C. Paterson, P. M. H. Lohman, E. A. De Weerd-Kastelein, and D. Bootsma. 1975. Xeroderma pigmentosum cells with normal levels of excision repair have a defect in DNA synthesis after UV-irradiation. *Proc. Natl. Acad. Sci. USA* **72:** 219–223.
206. Lehmann, A. R., A. F. Thompson, S. A. Harcourt, M. Stefanini, and P. G. Norris. 1993. Cockayne's syndrome: correlation of clinical features with cellular sensitivity of RNA synthesis to UV irradiation. *J. Med. Genet.* **30:** 679–682.
207. Léveillard, T., L. Andera, N. Bissonnette, L. Schaeffer, L. Bracco, J. M. Egly, and Wasylyk B. 1996. Functional interactions between p53 and the TFIIH complex are affected by tumour-associated mutations. *EMBO J.* **15:** 1615–1624.
208. Li, L., E. S. Bales, C. A. Peterson, and R. J. Legerski. 1993. Characterization of molecular defects in xeroderma pigmentosum group C. *Nature Genet.* **5:** 413–417.
209. Li, L., S. J. Elledge, C. A. Peterson, E. S. Bales, and R. J. Legerski. 1994. Specific association between the human DNA repair proteins XPA and ERCC1. *Proc. Natl. Acad. Sci. USA* **91:** 5012–5016.
210. Li, L., X. Lu, C. A. Peterson, and R. J. Legerski. 1995. An interaction between the DNA repair factor XPA and replication protein A appears essential for nucleotide excision repair. *Mol. Cell. Biol.* **15:** 5396–5402.

210a. Li, L., C. Peterson, and R. Legerski. 1996. Sequence of the mouse XPC cDNA and genomic structure of the human XPC gene. *Nucleic Acids Res.* **24:** 1026–1028.

211. Li, L., C. A. Peterson, X. Lu, and R. J. Legerski. 1995. Mutations in XPA that prevent association with ERCC1 are defective in nucleotide excision repair. *Mol. Cell. Biol.* **15:** 1993–1998.
212. Li, R., S. Waga, G. J. Hannon, D. Beach, and B. Stillman. 1994. Differential effects by the p21 CDK inhibitor on PCNA-dependent DNA replication and repair. *Nature* **371:** 534–537.
213. Lindahl, T. 1993. Instability and decay of the primary structure of DNA. *Nature* **362:** 709–715.
214. Link, C. J. J., D. L. Mitchell, R. S. Nairn, and V. A. Bohr. 1992. Preferential and strand-specific DNA repair of (6–4) photoproducts detected by a photochemical method in the hamster DHFR gene. *Carcinogenesis* **13:** 1975–1980.
215. Liu, P., J. Siciliano, B. White, R. Legerksi, D. Callen, S. Reeders, M. J. Siciliano, and L. H. Thompson. 1992. Regional mapping of human DNA excision repair gene *ERCC4 to* chromosome 16p13.13-p13.2. *Mutagenesis* **8:** 199–205.
216. Lommel, L., C. Carswell-Crumpton, and P. C. Hanawalt. 1995. Preferential repair of the transcribed DNA strand in the dihydrofolate reductase gene throughout the cell cycle in UV-irradiated human cells. *Mutat. Res.* **336:** 181–192.
217. Lommel, L. and P. C. Hanawalt. 1993. Increased UV resistance of a xeroderma pigmentosum revertant cell line is correlated with selective repair of the transcribed strand of an expressed gene. *Mol. Cell. Biol.* **13:** 970–976.
218. Longhese, M. P., P. Plevani, and G. Lucchini. 1994. Replication factor A is required in vivo for DNA replication, repair, and recombination. *Mol. Cell. Biol.* **14:** 7884–7890.

218a. Ma, L., J. H. Hoeijmakers, and A. J. van der Eb. 1995. Mammalian nucleotide excision repair. *Biochim. Biophys. Acta* **1242:** 137–163.
219. Ma, L., E. D. Siemssen, H. M. Noteborn, and A. J. van der Eb. 1994. The xeroderma pigmentosum group B protein ERCC3 produced in the baculovirus system exhibits DNA helicase activity. *Nucleic Acids Res.* **22:** 4095–4102.
220. Ma, L., A. Westbroek, A. G. Jochemsen, G. Weeda, A. Bosch, D. Bootsma, J. H. Hoeijmakers, and A. J. van der Eb. 1994. Mutational analysis of ERCC3, which is involved in DNA repair and transcription initiation: identification of domains essential for the DNA repair function. *Mol. Cell. Biol.* **14:** 4126–4134.
221. MacInnes, M. A., J. A. Dickson, R. R. Hernandez, D. Learmonth, G. Y. Lin, J. S. Mudgett, M. S. Park, S. Schauer, R. J. Reynolds, G. F. Strniste, and J. Y. Yu. 1993. Human *ERCC5* cDNA-cosmid complementation for excision repair and bipartite amino acid domains conserved with RAD proteins of *Saccharomyces cerevisiae* and *Schizosaccharomyces pombe*. *Mol. Cell. Biol.* **13:** 6393–6402.
222. Madhani, H. D., V. A. Bohr, and P. C. Hanawalt. 1986. Differential DNA repair in transcriptionally active and inactive proto-oncogenes: *c-abl* and *c-mos*. *Cell* **45:** 417–423.
223. Maher, V. M., N. Birch, R. J. Otto, and J. J. McCormick. 1975. Cytotoxicity of carcinogenic aromatic amides in normal and xeroderma pigmentosum fibroblasts with different DNA repair characteristics. *J. Natl. Cancer Inst.* **54:** 1287–1294.
224. Maher, V. M., J. Domoradzki, N. P. Bhattacharyya, T. Tsujimura, R. C. Corner, and J. J. McCormick. 1990. Alkylation damage, DNA repair and mutagenesis in human cells. *Mutat. Res.* **233:** 235–245.
225. Maher, V. M., D. J. Dorney, A. L. Mendrala, B. Konze-Thomas, and J. J. McCormick. 1979. DNA excision-repair processes in human cells can eliminate the cytotoxic and mutagenic consequences of ultraviolet irriadiation. *Mutat. Res.* **62:** 311–323.
226. Maher, V. M., L. M. Ouellette, R. D. Curren, and J. J. McCormick. 1976. Frequency of ultraviolet light-induced mutation is higher in xeroderma pigmentosum variant cells than in normal human cells. *Nature* **261:** 277–284.
227. Maher, V. M., L. M. Oullette, R. D. Curren, and J. J. McCormick. 1976. Caffeine enhancement of the cytotoxic and mutagenic effect of ultraviolet irradiation in a xeroderma pigmentosum variant strain of human cells. *Biochem. Biophys. Acta* **71:** 228–234.
228. Maher, V. M., L. A. Rowan, K. C. Silinskas, S. A. Kateley, and J. J. McCormick. 1982. Frequency of UV-induced neoplastic transformation of diploid human fibroblasts is higher in xeroderma pigmentosum cells than in normal cells. *Proc. Natl. Acad. Sci. USA* **79:** 2613–2617.
229. Maltzman, W. and L. Czyzyk. 1984. UV irradiation stimulates levels of p53 cellular tumor antigen in nontransformed mouse cells. *Mol. Cell. Biol.* **4:** 1689–1694.
230. Marshall, R. R., C. F. Arlett, S. A. Harcourt, and B. A. Broughton. 1980. Increased sensitivity of cell strains from Cockayne's syndrome to sister-chromatid-exchange induction and cell killing by UV light. *Mutat. Res.* **69:** 107–112.
231. Masutani, C., K. Sugasawa, J. Yanagisawa, T. Sonoyama, M. Ui, T. Enomoto, K. Takio, K. Tanaka, P. J. van der Spek, D. Bootsma, J. H. J. Hoeijmakers, and F. Hanaoka. 1994. Purification and cloning of a nucleotide excision repair complex involving the xeroderma pigmentosum group C protein and a human homologue of yeast RAD23. *EMBO J.* **13:** 1831–1843.
232. Matsuda, T., M. Saijo, I. Kuraoka, T. Kobayashi, Y. Nakatsu, A. Nagai, T. Enjoji, C. Masutani, K. Sugasawa, F. Hanaoka, A. Yasui, and K. Tanaka. 1995. DNA repair protein XPA binds replication protein A (RPA). *J. Biol. Chem.* **270:** 4152–4157.
233. Matsui, P., J. De Paulo, and S. Buratowski. 1995. An interaction between the Tfb1 and Ssl1 subunits of yeast TFIIH correlates with DNA repair activity. *Nucleic Acids Res.* **23:** 767–772.

234. Matsunaga, T., D. Mu, C. H. Park, J. T. Reardon, and A. Sancar. 1995. Human DNA repair excision nuclease. *J. Biol. Chem.* **270:** 20,862–20,869.
235. Matsunaga, T., C. H. Park, T. Bessho, D. Mu, and A. Sancar. 1996. Replication protein A confers structure-specific endonuclease activities to the XPF-ERCC1 and XPG subunits of human DNA repair excision nuclease. *J. Biol. Chem.* **271:** 11,047–11,050.
236. Mayne, L. V., T. Jones, S. W. Dean, S. A. Harcourt, J. E. Lowe, A. Priestley, H. Steingrimsdottir, H. Sykes, M. H. Green, and A. R. Lehmann. 1988. SV40-transformed normal and DNA-repair-deficient human fibroblasts can be transfected with high frequency but retain only limited amounts of integrated DNA [published erratum in *Gene* 1989. Nov 30; **83(2)**: 395]. *Gene* **66:** 65–76.
237. Mayne, L. V. and A. R. Lehmann. 1982. Failure of RNA synthesis to recover after UV irradiation: an early defect in cells from individuals with Cockayne's syndrome and xeroderma pigmentosum. *Cancer Res.* **42:** 1473–1478.
238. McCormick, J. J., S. Kateley-Kohler, M. Watanabe, and V. M. Maher. 1986. Abnormal sensitivity of human fibroblasts from xeroderma pigmentosum variants to transformation to anchorage independence by ultraviolet radiation. *Cancer Res.* **46:** 489–492.
239. McDowell, M. L., T. Nguyen, and J. E. Cleaver. 1993. A single-site mutation in the *XPAC* gene alters photoproduct recognition. *Mutagenesis* **8:** 155–161.
240. McGregor, W. G., R. H. Chen, L. Lukash, V. M. Maher, and J. J. McCormick. 1991. Cell cycle-dependent strand bias for UV-induced mutations in the transcribed strand of excision repair-proficient human fibroblasts but not in repair-deficient cells. *Mol. Cell. Biol.* **11:** 1927–1934.
241. McWhir, J., J. Selfridge, D. J. Harrison, S. Squires, and D. W. Melton. 1993. Mice with DNA repair gene *(ERCC–1)* deficiency have elevated levels of p53, liver nuclear abnormalities and die before weaning. *Nature Genetics* **5:** 217–224.
242. Mellon, I., V. A. Bohr, C. A. Smith, and P. C. Hanawalt. 1986. Preferential DNA repair of an active gene in human cells. *Proc. Natl. Acad. Sci. USA* **83:** 8878–8882.
243. Mellon, I., G. Spivak, and P. C. Hanawalt. 1987. Selective removal of transcription-blocking DNA damage from the transcribed strand of the mammalian DHFR gene. *Cell* **51:** 241–249.
244. Menichini, P., H. Vrieling, and A. A. van Zeeland. 1991. Strand-specific mutation spectra in repair-proficient and repair-deficient hamster cells. *Mutat. Res.* **251:** 143–155.
245. Meyn, R. E., D. L. Vizard, R. R. Hewitt, and R. M. Humphrey. 1974. The fate of pyrimidine dimers in the DNA of ultraviolet-irradiated Chinese hamster cells. *Photochem. Photobiol.* **20:** 221–226.
246. Mezzina, M., E. Eveno, O. Chevallier-Lagente, A. Benoit, M. Carreau, W. Vermeulen, J. H. Hoeijmakers, M. Stefanini, A. R. Lehmann, C. A. Weber, and A. Sarasin. 1994. Correction by the *ERCC2* gene of UV sensitivity and repair deficiency phenotype in a subset of trichothiodystrophy cells. *Carcinogenesis* **15:** 1493–1498.
247. Mirzayans, R., L. Enns, K. Dietrich, R. D. C. Barley, and M. C. Paterson. 1966. Faulty DNA polymerase α/ε-mediated excision repair in response to γ radiation or ultraviolet light in p53-deficient fibroblast strains from affected members of a cancer-prone family with Li-Fraumeni syndrome. *Carcinogenesis* **17:** 691–698.
248. Misra, R. R. and J. M. Vos. 1993. Defective replication of psoralen adducts detected at the gene-specific level in xeroderma pigmentosum variant cells. *Mol. Cell. Biol.* **13:** 1002–1012.
249. Mitchell, D. L. 1996. Radioimmunoassay of DNA damaged by ultraviolet light, in *Technologies for Detection of DNA Damage and Mutations* (Pfeifer, G., ed.), Plenum Publishing Corp., NY, pp. 73–85.
250. Mitchell, D. L., D. E. Brash, and R. S. Nairn. 1990. Rapid repair kinetics of pyrimidine(6-4)pyrimidone photoproducts in human cells are due to excision rather than conformational change. *Nucleic Acids Res.* **18:** 963–971.

251. Mitchell, D. L., J. E. Cleaver, and J. H. Epstein. 1990. Repair of pyrimidine(6-4) pyrimidone photoproducts in mouse skin. *J. Invest. Dermatol.* **95:** 55–59.
252. Mitchell, D. L. and R. S. Nairn. 1989. The biology of the (6-4) photoproduct. *Photochem. Photobiol.* **49:** 805–819.
253. Miura, M., M. Domon, T. Sasaki, S. Kondo, and Y. Takasaki. 1992. Restoration of proliferating cell nuclear antigen (PCNA) complex formation in xeroderma pigmentosum group A cells following cis-diamminedichloroplatinum (II)-treatment by cell fusion with normal cells. *J. Cell Physiol.* **152:** 639–645.
254. Miura, N., I. Miyamoto, H. Asahina, I. Satokata, K. Tanaka, and Y. Okada. 1991. Identification and characterization of xpac protein, the gene product of the human XPAC (xeroderma pigmentosum group A complementing) gene. *J. Biol. Chem.* **266:** 19,786–19,789.
255. Miyamoto, I., N. Miura, H. Niwa, J. Miyazaki, and K. Tanaka. 1992. Mutational analysis of the structure and function of the xeroderma pigmentosum group A complementing protein. Identification of essential domains for nuclear localization and DNA excision repair. *J. Biol. Chem.* **267:** 12,182–12,187.
256. Mohrenweiser, H. W., A. V. Carrano, A. Fertitta, B. Perry, L. H. Thompson, J. D. Tucker, and C. A. Weber. 1989. Refined mapping of the three DNA repair genes, *ERCC1*, *ERCC2*, and *XRCC1*, on human chromosome 19. *Cytogenet. Cell Genet.* **52:** 11–14.
257. Mounkes, L. C., R. S. Jones, B. C. Liang, W. Gelbart, and M. T. Fuller. 1992. A Drosophila model for xeroderma pigmentosum and Cockayne's syndrome: *haywire* encodes the fly homolog of *ERCC3*, a human excision repair gene. *Cell* **71:** 925–937.
258. Moustacchi, E., D. Papadopoulo, C. Diatloff-Zito, and M. Buchwald. 1987. Two complementation groups of Fanconi's anemia differ in their phenotypic response to a DNA-crosslinking treatment. *Hum. Genet.* **75:** 45–47.
259. Mu, D., D. S. Hsu, and A. Sancar. 1996. Reaction mechanism of human DNA repair excision nuclease. *J. Biol. Chem.* **271:** 8285–8294.
260. Mu, D., C. H. Park, T. Matsunaga, D. S. Hsu, J. T. Rearson, and A. Sancar. 1995. Reconstitution of human DNA repair excision nuclease in a highly defined system. *J. Biol. Chem.* **270:** 2415–2418.
261. Mudgett, J. S. and M. A. MacInnes. 1990. Isolation of the functional human DNA excision repair gene *ERCC5* by intercosmid recombination. *Genomics* **8:** 623–633.
262. Mummenbrauer, T., F. Janus, B. Müller, L. Wiesmüller, W. Deppert, and F. Grosse. 1996. p53 protein exhibits 3'-to-5' exonuclease activity. *Cell* **85:** 1089–1099.
263. Murad, A. O., J. de Cock, D. Brown, and M. J. Smerdon. 1995. Variations in transcription-repair coupling in mouse cells. *J. Biol. Chem.* **270:** 3949–3957.
264. Murray, J. M., C. L. Doe, P. Schenk, A. M. Carr, A. R. Lehmann, and F. Z. Watts. 1992. Cloning and characterisation of the *S. pombe rad15* gene, a homologue to the *S. cerevisiae RAD3* and human *ERCC2* genes. *Nucleic Acids Res.* **20:** 2673–2678.
265. Naegeli, H. 1995. Mechanisms of DNA damage recongnition in mammalian nucleotide excision repair. *FASEB J.* **9:** 1043–1050.
266. Naegeli, H. 1994. Roadblocks and detours during DNA replication: mechanisms of mutagenesis in mammalian cells. *Bioessays* **16:** 557–564.
267. Naegeli, H., L. Bardwell, and E. C. Friedberg. 1992. The DNA helicase and adenosine triphosphatase activities of yeast Rad3 protein are inhibited by DNA damage. A potential mechanism for damage-specific recognition. *J. Biol. Chem.* **267:** 392–398.
268. Naegeli, H., P. Modrich, and E. C. Friedberg. 1993. The DNA helicase activities of Rad3 protein of *Saccharomyces cerevisiae* and helicase II of *Escherichia coli* are differentially inhibited by covalent and noncovalent DNA modifications. *J. Biol. Chem.* **268:** 10,386–10,392.
269. Nagai, A., M. Saijo, I. Kuraoka, T. Matsuda, N. Kodo, Y. Nakatsu, T. Mimaki, M. Mino, M. Biggerstaff, R. D. Wood, A. Sijbers, J. H. J. Hoeijmakers, and K. Tanaka. 1995.

Enhancement of damage-specific DNA binding of XPA by interaction with the ERCC1 DNA repair protein. *Biochem. Biophys. Res. Commun.* **211:** 960–966.

270. Nakane, H., S. Takeuchi, S. Yuba, M. Saijo, Y. Nakatsu, H. Murai, Y. Nakatsuru, T. Ishikawa, S. Hirota, Y. Kitamura, Y. Kato, Y. Tsunoda, H. Miyauchi, T. Horio, T. Tokunaga, T. Matsunaga, O. Nikaido, Y. Nishimune, Y. Okada, and K. Tanaka. 1995. High incidence of ultraviolet-B- or chemical-carcinogen-induced skin tumours in mice lacking the xeroderma pigmentosum group A gene. *Nature* **377:** 165–168.
271. Nance, M. A. and S. A. Berry. 1992. Cockayne syndrome: review of 140 cases. *Am. J. Med. Genet.* **42:** 68–84.
272. Naumovski, L., and E. C. Friedberg. 1986. Analysis of the essential and excision repair functions of the *RAD3* gene of *Saccharomyces cerevisiae* by mutagenesis. *Mol. Cell. Biol.* **6:** 1218–1227.
273. Neer, E. J., C. J. Schmidt, R. Nambudripad, and T. F. Smith. 1994. The ancient regulatory-protein family of WD-repeat proteins. *Nature* **371:** 297–300.
274. Nelson, W. G. and M. B. Kastan. 1994. DNA strand breaks: the DNA template alterations that trigger p53-dependent DNA damage response pathways. *Mol. Cell. Biol.* **14:** 1815–1823.
275. Nichols, A. F. and A. Sancar. 1992. Purification of PCNA as a nucleotide excision repair protein [corrected and republished with original paging, article originally printed in Nucleic Acids Res 1992. **20(10):** 2441–2446]. *Nucleic Acids Res.* **20:** 2441–2446.
275a. Nichols, A. F., P. Ong, and S. Linn. 1996. Mutations specific to the xeroderma pigmentosum group E Ddb$^-$ phenotype. *J. Biol. Chem.* **271:** 24,317–24,320.
276. Nishida, C., P. Reinhard, and S. Linn. 1988. DNA repair synthesis in human fibroblasts requires DNA polymerase delta. *J. Biol. Chem.* **263:** 501–510.
277. Nishigori, C., S. Moriwaki, H. Takebe, T. Tanaka, and S. Imamura. 1994. Gene alterations and clinical characteristics of xeroderma pigmentosum group A patients in Japan. *Arch. Dermatol.* **130:** 191–197.
278. Nishigori, C., M. Zghal, T. Yagi, S. Imamura, M. R. Komoun, and H. Takebe. 1993. High prevalence of the point mutation in exon 6 of the xeroderma pigmentosum group A-complementing *(XPAC)* gene in xeroderma pigmentosum group A patients in Tunisia. *Am. J. Hum. Genet.* **53:** 1001–1006.
279. Nouspikel, T., P. Lalle, S. A. Leadon, P. K. Cooper, and S. G. Clarkson. 1997. A common mutational pattern in XP-G/Cockayne syndrome patients: implications for a second XPG function. *Proc. Natl. Acad. Sci. USA*, in press.
280. Nouspikel, T., P. K. Cooper, and S. G. Clarkson. 1996. A mutational pattern in Cockayne syndrome patients belonging to xeroderma pigmentosum group G. Submitted.
281. Oberosler, P., P. Hloch, U. Ramsperger, and H. Stahl. 1993. p53-catalyzed annealing of complementary single-stranded nucleic acids. *EMBO J.* **12:** 2389–2396.
282. O'Donovan, A., A. A. Davies, J. G. Moggs, S. C. West, and R. D. Wood. 1994. XPG endonuclease makes the 3' incision in human DNA nucleotide excision repair. *Nature* **371:** 432–435.
283. O'Donovan, A., D. Scherly, S. G. Clarkson, and R. G. Wood. 1994. Isolation of active recombinant XPG protein, a human DNA repair endonuclease. *J. Biol. Chem.* **269:** 15,965–15,968.
284. O'Donovan, A. and R. D. Wood. 1993. Identical defects in DNA repair in xeroderma pigmentosum group G and rodent *ERCC* group 5. *Nature* **363:** 185–188.
285. Oren, M. 1992. p53: the ultimate tumor suppressor gene? *FASEB J.* **6:** 3169–3176.
286. Pan, Z. Q., J. T. Reardon, L. Li, H. Fores-Rozas, R. Legerski, A. Sancar, and J. Hurwitz. 1995. Inhibition of nucleotide excision repair by the cyclin-dependent kinase inhibitor p21. *J. Biol. Chem.* **270:** 22,008–22,016.
287. Park, C. H., T. Bessho, T. Matsunaga, and A. Sancar. 1995. Purification and characterization of the XPF-ERCC1 complex of human DNA repair excision nuclease. *J. Biol. Chem.* **270:** 22,657–22,660.

288. Park, C. H., D. Mu, J. T. Reardon, and A. Sancar. 1995. The general transcription-repair factor TFIIH is recruited to the excision repair complex by the XPA protein independent of the TFIIE transcription factor. *J. Biol. Chem.* **270:** 4896–4902.
289. Park, C. H. and A. Sancar. 1994. Formation of a ternary complex between human XPA, ERCC1, and ERCC4(XPF) excision repair proteins. *Proc. Natl. Acad. Sci. USA* **91:** 5017–5021.
290. Park, E., S. N. Guzder, M. H. Koken, I. Jaspers-Dekker, G. Weeda, J. H. Hoeijmakers, S. Prakash, and L. Prakash. 1992. *RAD25 (SSL2)*, the yeast homolog of the human xeroderma pigmentosum group B DNA repair gene, is essential for viability. *Proc. Natl. Acad. Sci. USA* **89:** 11,416–11,420.
291. Parris, C. N. and K. H. Kraemer. 1993. Ultraviolet-induced mutations in Cockayne syndrome cells are primarily caused by cyclobutane dimer photoproducts while repair of other photoproducts is normal. *Proc. Natl. Acad. Sci. USA* **90:** 7260–7264.
292. Payne, A. and G. Chu. 1994. Xeroderma pigmentosum group E binding factor recognizes a broad spectrum of DNA damage. *Mutat. Res.* **310:** 89–102.
293. Peserico, A., P. A. Battistella, and P. Bertoli. 1992. MRI of a very rare hereditary ectodermal dysplasia: PIBI(D)S. *Neuroradiology* **34:** 316,317.
294. Petersen, L. N., D. K. Orren, and V. A. Bohr. 1995. Gene-specific and strand-specific DNA repair in the G1 and G2 phases of the cell cycle. *Mol. Cell. Biol.* **15:** 3731–3737.
295. Podust, V. N., A. Georgaki, B. Strack, and U. Hubscher. 1992. Calf thymus RF-C as an essential component for DNA polymerase delta and epsilon holoenzymes function. *Nucleic Acids Res.* **20:** 4159–4165.
296. Pollitt, R. J., F. A. Fenner, and M. Davies. 1968. Sibs with mental and physical retardation and trichorrhexis nodosa with abnormal amino acid composition of the hair. *Arch. Dis. Child.* **43:** 211–216.
297. Prakash, S., P. Sung, and L. Prakash. 1993. DNA repair genes and proteins of *Saccharomyces cerevisiae*. *Ann. Rev. Genet.* **27:** 33–70.
298. Price, A. 1993. The repair of ionising radiation-induced damage to DNA. *Semin. Cancer Biol.* **4:** 61–71.
299. Price, V. H., R. B. Odom, W. H. Ward, and F. T. Jones. 1980. Trichothiodystrophy: sulfur-deficient brittle hair as a marker for a neuroectodermal symptom complex. *Arch. Dermatol.* **116:** 1375–1384.
300. Prigent, C., M. S. Satoh, G. Daly, D. E. Barnes, and T. Lindahl. 1994. Aberrant DNA repair and DNA replication due to an inherited enzymatic defect in human DNA ligase I. *Mol. Cell. Biol.* **14:** 310–317.
301. Raha, M., G. Wang, M. M. Seidman, and P. M. Glazer. 1996. *Mutagenesis* by third-strand-directed psoralen adducts in repair-deficient human cells: high frequency and altered spectrum in xeroderma pigmentosum variant. *Proc. Natl. Acad. Sci. USA* **93:** 2941–2946.

301a. Reardon, J. T., H. Ge, E. Gibbs, A. Sancar, J. Hurwitz, and Z. Q. Pan. 1996. Isolation and characterization of two human transcription factor IIH (TFIIH)-related complexes: ERCC2/CAK and TFIIH*. *Proc. Natl. Acad. Sci. USA* **93:** 6482–6487.

302. Reardon, J. T., D. Mu, and A. Sancar. 1996. Overproduction, purification, and characterization of the XPC subunit of the human DNA repair excision nuclease. *J. Biol. Chem.* **271:** 19,451–19,456.
303. Reardon, J. T., A. F. Nichols, S. Keeney, C. A. Smith, J. S. Taylor, S. Linn, and A. Sancar. 1993. Comparative analysis of binding of human damaged DNA-binding protein (XPE) and *Escherichia coli* damage recognition protein (UvrA) to the major ultraviolet photoproducts: T[c,s]T, T[t,s]T, T[6–4]T, and T[Dewar]T*. *J. Biol. Chem.* **268:** 21,301–21,308.
304. Reardon, J. T., L. H. Thompson, and A. Sancar. 1993. Excision repair in man and the molecular basis of xeroderma pigmentosum syndrome. *Cold Spring Harbor Symp. Quant. Biol.* **58:** 605–617.

305. Reardon, J. T., L. H. Thompson, and A. Sancar. 1997. Rodent UV sensitive mutant cell lines in complementation groups 6 through 10 have normal general excision repair activity. *Nucleic Acids Res.* **25:** 1015–1021.
306. Rebora, A. and F. Crovato. 1988. Trichothiodystrophy, xeroderma pigmentosum and PIBI(D)S syndrome. *Hum. Genet.* **78:** 106–108.
307. Riboni, R., E. Botta, M. Stefanini, M. Numata, and A. Yasui. 1992. Identification of the eleventh complemenation group of UV-sensitive excision repair-defective rodent mutants. *Cancer Res.* **52:** 6690,6691.
308. Robbins, J. H. 1988. Xeroderma pigmentosum. Defective DNA repair causes skin cancer and neurodegeneration. *JAMA* **260:** 384–388.
309. Robbins, J. H., R. A. Brumback, M. Mendiones, S. F. Barrett, J. R. Carl, S. Cho, M. B. Denckla, M. B. Ganges, L. H. Gerber, R. A. Guthrie, and et al. 1991. Neurological disease in xeroderma pigmentosum. Documentation of a late onset type of the juvenile onset form. *Brain* **114:** 1335–1361.
310. Roy, R., J. P. Adamczewski, T. Seroz, W. Vermeulen, J. P. Tassan, L. Schaeffer, E. A. Nigg, J. H. Hoeijmakers, and J. M. Egly. 1994. The MO15 cell cycle kinase is associated with the TFIIH transcription-DNA repair factor. *Cell* **79:** 1093–1101.
311. Roy, R., L. Schaeffer, S. Humbert, W. Vermeulen, G. Weeda, and J. M. Egly. 1994. The DNA-dependent ATPase activity associated with the class II basic transcription factor BTF2/TFIIH. *J. Biol. Chem.* **269:** 9826–9832.
312. Royer-Pokora, B. and W. A. Haseltine. 1984. Isolation of UV-resistant revertants from a xeroderma pigmentosum complementation group A cell line. *Nature* **311:** 390–392.
313. Ruven, H. J., R. J. Berg, C. M. Seelen, J. A. Dekkers, P. H. Lohman, L. H. Mullenders, and A. A. van Zeeland. 1993. Ultraviolet-induced cyclobutane pyrimidine dimers are selectively removed from transcriptionally active genes in the epidermis of the hairless mouse. *Cancer Res.* **53:** 1642–1645.
314. Ruven, H. J., C. M. Seelen, P. H. Lohman, H. van Kranen, A. A. van Zeeland, and L. H. Mullenders. 1994. Strand-specific removal of cyclobutane pyrimidine dimers from the p53 gene in the epidermis of UVB-irradiated hairless mice. *Oncogene* **9:** 3427–3432.
315. Sage, E., E. A. Drobetsky, and E. Moustacchi. 1993. 8-Methoxypsoralen induced mutations are highly targeted at crosslinkable sites of photoaddition on the non-transcribed strand of a mammalian chromosomal gene. *EMBO J.* **12:** 397–402.
316. Samec, S., T. A. Jones, J. Corlet, D. Scherly, D. Sheer, R. D. Wood, and S. G. Clarkson. 1994. The human gene for xeroderma pigmentosum complementation group G *(XPG)* maps to 13q33 by fluorescence in situ hybridization. *Genomics* **21:** 283–285.
317. Sancar, A. 1996. DNA excision repair. *Ann. Rev. Biochem.* **65:** 43–81.
318. Sancar, A. 1995. Excision repair in mammalian cells. *J. Biol. Chem.* **270:** 15,915–15,918.
319. Sancar, A. 1994. Mechanisms of DNA excision repair. *Science* **266:** 1954–1956.
320. Sancar, A. and M. S. Tang. 1993. Nucleotide excision repair. *Photochem. Photobiol.* **57:** 905–921.
321. Sanchez, Y. and S. J. Elledge. 1995. Stopped for repairs. *Bioessays* **17:** 545–548.
322. Sands, A. T., A. Abuin, A. Sanchez, C. J. Conti, and A. Bradley. 1995. High susceptibility to ultraviolet-induced carcinogenesis in mice lacking XPC. *Nature* **377:** 162–165.
323. Sands, A. T., M. B. Suraokar, A. Sanchez, J. E. Marth, L. A. Donehower, and A. Bradley. 1995. p53 deficiency does not affect the accumulation of point mutations in a transgene target. *Proc. Natl. Acad. Sci. USA* **92:** 8517–8521.
324. Sarasin, A., C. Blanchet-Bardon, G. Renault, A. Lehmann, C. Arlett, and Y. Dumez. 1992. Prenatal diagnosis in a subset of trichothiodistrophy patients defective in DNA repair. *Br. J. Dermatol.* **127:** 485–491.
325. Sasaki, K., N. Tachi, M. Shinoda, N. Satoh, R. Minami, and A. Ohnishi. 1992. Demyelinating peripheral neuropathy in Cockayne syndrome: a histopathologic and morphometric study. *Brain Dev.* **14:** 114–117.

326. Sato, M., C. Nishigori, Y. Lu, M. Zghal, T. Yagi, and H. Takebe. 1994. Far less frequent mutations in *ras* genes than in the *p53* gene in skin tumors of xeroderma pigmentosum patients. *Mol. Carcinog.* **11:** 98–105.
327. Sato, M., C. Nishigori, M. Zghal, T. Yagi, and H. Takebe. 1993. Ultraviolet-specific mutations in *p53* gene in skin tumors in xeroderma pigmentosum patients. *Cancer Res.* **53:** 2944–2946.
328. Satoh, M. S., C. J. Jones, R. D. Wood, and T. Lindahl. 1993. DNA excision-repair defect of xeroderma pigmentosum prevents removal of a class of oxygen free radical-induced base lesions. *Proc. Natl. Acad. Sci. USA* **90:** 6335–6339.
329. Satokata, I., K. Tanaka, N. Miura, I. Miyamoto, Y. Satoh, S. Kondo, and Y. Okada. 1990. Characterization of a splicing mutation in group A xeroderma pigmentosum. *Proc. Natl. Acad. Sci. USA* **87:** 9908–9912.
330. Satokata, I., K. Tanaka, S. Yuba, and Y. Okada. 1992. Identification of splicing mutations of the last nucleotides of exons, a nonsense mutation, and a missense mutation of the XPAC gene as causes of group A xeroderma pigmentosum. *Mutat. Res.* **273:** 203–212.
331. Satokata, I., K. Tanaka, and Y. Okada. 1992. Molecular basis of group A xeroderma pigmentosum: a missense mutation and two deletions located in a zinc finger consensus sequence of the XPAC gene. *Hum. Genet.* **88:** 603–607.
332. Satokata, I., K. Tanaka, N. Miura, M. Narita, T. Mimaki, Y. Satoh, S. Kondo, and Y. Okada. 1992. Three nonsense mutations responsible for group A xeroderma pigmentosum. *Mutat. Res.* **273:** 193–202.
333. Satokata, I., M. Uchiyam, and K. Tanaka. 1995. Two novel splicing mutations in the XPA gene in patients with group A xeroderma pigmentosum. *Hum. Mol. Genet.* **4:** 1993,1994.
334. Schaeffer, L., V. Moncollin, R. Roy, A. Staub, M. Mezzina, A. Sarasin, G. Weeda, J. H. Hoeijmakers, and J. M. Egly. 1994. The ERCC2/DNA repair protein is associated with the class II BTF2/TFIIH transcription factor. *EMBO J.* **13:** 2388–2392.
335. Schaeffer, L., R. Roy, S. Humbert, V. Moncollin, W. Vermeulen, J. H. Hoeijmakers, P. Chambon, and J. M. Egly. 1993. DNA repair helicase: a component of BTF2 (TFIIH) basic transcription factor [*see* comments]. *Science* **260:** 58–63.
336. Scherly, D., T. Nouspikel, J. Corlet, C. Ucla, A. Bairoch, and S. G. Clarkson. 1993. Complementation of the DNA repair defect in xeroderma pigmentosum group G cells by a human cDNA related to yeast *RAD2*. *Nature* **363:** 182–185.
337. Schiestl, R. H. and S. Prakash. 1990. *RAD10*, an excision repair gene of *Saccharomyces cerevisiae*, is involved in the *RAD1* pathway of mitotic recombination. *Mol. Cell. Biol.* **10:** 2485–2491.
338. Schultz, R. A., D. P. Barbis, and E. C. Friedberg. 1985. Studies on gene transfer and reversion to UV resistance in xeroderma pigmentosum cells. *Somat. Cell Mol. Genet.* **11:** 617–624.
339. Scott, R. J., P. Itin, W. J. Kleijer, K. Kolb, C. Arlett, and H. Muller. 1993. Xeroderma pigmentosum-Cockayne syndrome complex in two patients: absence of skin tumors despite severe deficiency of DNA excision repair. *J. Am. Acad. Dermatol.* **29:** 883–889.
340. Seetharam, S., M. Protic-Sabljic, M. M. Seidman, and K. H. Kraemer. 1987. Abnormal ultraviolet mutagenic spectrum in plasmid DNA replicated in cultured fibroblasts from a patient with the skin cancer-prone disease, xeroderma pigmentosum. *J. Clin. Invest.* **80:** 1613–1617.
341. Sekelsky, J. J., K. S. McKim, G. M. Chin, and R. S. Hawley. 1995. The Drosophila meiotic recombination gene *mei–9* encodes a homologue of the yeast excision repair protein Rad1. *Genetics* **141:** 619–627.
341a. Selby, C. P. and A. Sancar. 1997. Human transcription-repair coupling factor CSB/ERCC6 is a DNA-stimulated ATPase but is not a helicase and does not disrupt the ternary transcription complex of stalled RNA polymerase II. *J. Biol. Chem.* **272:** 1885–1890.

342. Selby, C. P. and A. Sancar. 1993. Molecular mechanism of transcription-repair coupling. *Science* **260:** 53–58.
343. Serizawa, H., T. P. Makela, J. W. Conaway, R. C. Conaway, R. A. Weinberg, and R. A. Young. 1995. Association of Cdk-activating kinase subunits with transcription factor TFIIH. *Nature* **374:** 280–282.
344. Seroz, T., J. R. Hwang, V. Moncollin, and J. M. Egly. 1995. TFIIH: a link between transcription, DNA repair and cell cycle regulation. *Curr. Opin. Genet. Dev.* **5:** 217–221.
345. Setlow, R. B., J. D. Regan, J. German, and W. L. Carrier. 1969. Evidence that xeroderma pigmentosum cells do not perform the first step in the repair of ultraviolet damage to their DNA. *Proc. Natl. Acad. Sci. USA* **64:** 1035–1041.
346. Shiekhattar, R., F. Mermelstein, R. P. Fisher, R. Drapkin, B. Dynlacht, H. C. Wessling, D. O. Morgan, and D. Reinberg. 1995. Cdk-activating kinase complex is a component of human transcription factor TFIIH. *Nature* **374:** 283–287.
347. Shiomi, T., Y. Harada, T. Saito, N. Shiomi, Y. Okuno, and M. Yamaizumi. 1994. An *ERCC5* gene with homology to yeast *RAD2* is involved in group G xeroderma pigmentosum. *Mutat. Res.* **314:** 167–175.
348. Shiomi, T., N. Hieda-Shiomi, and K. Sato. 1982. Isolation of UV-sensitive mutants of mouse L5178Y cells by a suspension spotting method. *Somatic Cell. Genet.* **8:** 329–345.
349. Shivji, K. K., M. K. Kenny, and R. D. Wood. 1992. Proliferating cell nuclear antigen is required for DNA excision repair. *Cell* **69:** 367–374.
350. Shivji, M. K., A. P. Eker, and R. D. Wood. 1994. DNA repair defect in xeroderma pigmentosum group C and complementing factor from HeLa cells. *J. Biol. Chem.* **269:** 22,749–22,757.
351. Shivji, M. K., S. J. Grey, U. P. Strausfeld, R. D. Wood, and J. J. Blow. 1994. Cip1 inhibits DNA replication but not PCNA-dependent nucleotide excision-repair. *Curr. Biol.* **4:** 1062–1068.
352. Shivji, M. K., V. N. Podust, U. Hubscher, and R. D. Wood. 1995. Nucleotide excision repair DNA synthesis by DNA polymerase epsilon in the presence of PCNA, RFC, and RPA. *Biochemistry* **34:** 5011–5017.
353. Siciliano, M. J., R. L. Stallings, G. M. Adair, R. M. Humphrey, and J. Siciliano. 1983. Provisional assignment of *TPI*, *GPI*, and *PEPD* to Chinese hamster autosomes 8 and 9: a cytogenetic basis for functional haploidy of an autosomal linkage group in CHO cells. *Cytogenet. Cell Genet.* **35:** 15–20.
354. Sijbers, A. M., W. L. de Laat, R. R. Ariza, M. Biggerstaff, Y. F. Wei, J. G. Moggs, K. C. Carter, B. K. Shell, E. Evans, M. C. de Jong, S. Rademakers, j. de Rooij, N. G. J. Jaspers, J. H. J. Hoeijmakers, and R. D. Wood. 1996. Xeroderma pigmentosum group F caused by a defect in a structure-specific DNA repair endonuclease. *Cell* **86:** 811–822.
355. Simon, L., R. M. Hazard, V. M. Maher, and J. J. McCormick. 1981. Enhanced cell killing and mutagenesis by ethylnitrosourea in xeroderma pigmentosum cells. *Carcinogenesis* **2:** 567–570.
356. Smerdon, M. J., and F. Thoma. 1990. Site-specific DNA repair at the nucleosomal level in a yeast minichromosome. *Cell* **61:** 675–684.
357. Smith, C. A., P. K. Cooper, and P. C. Hanawalt. 1981. Measurement of repair replication by equilibrium sedimentation in *DNA Repair—A Laboratory Manual of Research Procedures*, vol. 1, part B (Friedberg, E. C. and P. C. Hanawalt, eds.), Marcel Dekker, New York, pp. 289–305.
358. Smith, M. L., I. T. Chen, Q. Zhan, I. Bae, C. Y. Chen, T. M. Gilmer, M. B. Kastan, P. M. O'Connor, and A. J. J. Fornace. 1994. Interaction of the p53-regulated protein Gadd45 with proliferating cell nuclear antigen. *Science* **266:** 1376–1380.
359. Smith, M. L., I. T. Chen, Q. Zhan, P. M. O'Connor, and A. J. J. Fornace. 1995. Involvement of the p53 tumor suppressor in repair of U.V.-type DNA damage. *Oncogene* **10:** 1053–1059.

360. Smith, M. L. and A. J. J. Fornace. 1995. Genomic instability and the role of p53 mutations in cancer cells. *Curr. Opinion Oncol.* **7:** 69–75.
361. Snyderwine, E. G., and V. A. Bohr. 1992. Gene- and strand-specific damage and repair in Chinese hamster ovary cells treated with 4-nitroquinoline 1-oxide. *Cancer Res.* **52:** 4183–4189.
362. Sorenson, C. M. and E. Eastman. 1988. Influence of *cis*-diamminedichloroplatinum(II) on DNA synthesis and cell cycle progession in excision repair proficient and deficient Chinese hamster ovary cells. *Cancer Res.* **48:** 6703–6707.
363. Spivak, G. and P. C. Hanawalt. 1995. Determination of damage and repair in specific DNA sequences. *Methods: A companion to Methods in Enzymol.* **7:** 147–161.
364. Squires, S. and R. T. Johnson. 1988. Kinetic analysis of UV-induced incision discriminates between fibroblasts from different xeroderma pigmentosum complementation groups, XPA heterozygotes and normal individuals. *Mutat. Res.* **193:** 181–192.
365. Squires, S. and R. T. Johnson. 1983. U.V. induces long-lived DNA breaks in Cockayne's syndrome and cells from an immunodeficient individual (46BR): defects and disturbance in post incision steps of excision repair. *Carcinogenesis* **4:** 565–572.
366. Stefanini, M., A. R. Collins, R. Riboni, M. Klaude, E. Botta, D. L. Mitchell, and F. Nuzzo. 1991. Novel Chinese hamster ultraviolet-sensitive mutants for excision repair form complementation groups 9 and 10. *Cancer Res.* **51:** 3965–3971.
367. Stefanini, M., H. Fawcett, E. Botta, T. Nardo, and A. R. Lehmann. 1996. Genetic analysis of twenty-two patients with Cockayne syndrome. *Hum. Genet.* **97:** 418–423.
368. Stefanini, M., P. Lagomarsini, S. Giliani, T. Nardo, E. Botta, A. Peserico, W. J. Kleijer, A. R. Lehmann, and A. Sarasin. 1993. Genetic heterogeneity of the excision repair defect associated with trichothiodystrophy. *Carcinogenesis* **14:** 1101–1105.
369. Stefanini, M., P. Lagomarsini, C. F. Arlett, S. Marinoni, C. Borrone, F. Crovato, G. Trevisan, G. Cordone, and F. Nuzzo. 1986. Xeroderma pigmentosum (complementation group D) mutation is present in patients affected by trichothiodystrophy with photosensitivity. *Hum. Genet.* **74:** 107–112.
370. Stefanini, M., W. Vermeulen, G. Weeda, S. Giliani, T. Nardo, M. Mezzina, A. Sarasin, J. I. Harper, C. F. Arlett, J. H. Hoeijmakers, and A. R. Lehmann. 1993. A new nucleotide-excision-repair gene associated with the disorder trichothiodystrophy. *Am. J. Hum. Genet.* **53:** 817–821.
371. Stevnsner, T., A. May, L. N. Petersen, F. Larminat, M. Pirsel, and V. A. Bohr. 1993. Repair of ribosomal RNA genes in hamster cells after UV irradiation, or treatment with cisplatin or alkylating agents. *Carcinogenesis* **14:** 1591–1596.
372. Suarez, H. G., L. Daya-Grosjean, D. Schlaifer, P. Nardeux, G. Renault, J. L. Bos, and A. Sarasin. 1989. Activated oncogenes in human skin tumors from a repair-deficient syndrome, xeroderma pigmentosum. *Cancer Res.* **49:** 1223–1228.
372a. Sugasawa, K., C. Masutani, A. Uchida, T. Maekawa, P. J. van der Spek, D. Bootsma, J. H. J. Hoeijmakers, and F. Hanaoka. 1996. HHR23B, a human Rad23 homolog, stimulates XPD protein in nucleotide excision repair in vitro. *Mol. Cell. Biol.* **16:** 4852–4861.
373. Sung, P., V. Bailly, C. Weber, L. H. Thompson, L. Prakash, and S. Prakash. 1993. Human xeroderma pigmentosum group D gene encodes a DNA helicase. *Nature* **365:** 852–855.
373a. Sung, P., S. N. Guzder, L. Prakash, and S. Prakash. 1996. Reconstitution of TFIIH and requirement of its DNA helicase subunits, Rad3 and Rad25, in the incision step of nucleotide excision repair. *J. Biol. Chem.* **271:** 10,821–10,826.
374. Sung, P., D. Higgins, L. Prakash, and S. Prakash. 1988. Mutation of lysine–48 to arginine in the yeast RAD3 protein abolishes its ATPase and DNA helicase activities but not the ability to bind ATP. *EMBO J.* **7:** 3263–3269.
375. Sung, P., L. Prakash, S. W. Matson, and S. Prakash. 1987. RAD3 protein of *Saccharomyces cerevisiae* is a DNA helicase. *Proc. Natl. Acad. Sci. USA* **84:** 8951–8955.

376. Sung, P., P. Reynolds, L. Prakash, and S. Prakash. 1993. Purification and characterization of the *Saccharomyces cerevisiae* RAD1/RAD10 endonuclease. *J. Biol. Chem.* **268:** 26,391–26,399.
377. Svejstrup, J. Q., Z. Wang, W. J. Feaver, X. Wu, D. A. Bushnell, T. F. Donahue, E. C. Friedberg, and R. D. Kornberg. 1995. Different forms of TFIIH for transcription and DNA repair: holo-TFIIH and a nucleotide excision repairosome. *Cell* **80:** 21–28.
378. Svoboda, D. L., J. S. Taylor, J. E. Hearst, and A. Sancar. 1993. DNA repair by eukaryotic nucleotide excision nuclease. Removal of thymine dimer and psoralen monoadduct by HeLa cell-free extract and of thymine dimer by *Xenopus laevis* oocytes. *J. Biol. Chem.* **268:** 1931–1936.
379. Syvaoja, J., S. Suomensaari, C. Nishida, J. S. Goldsmith, G. S. Chui, S. Jain, and S. Linn. 1990. DNA polymerases alpha, delta, and epsilon: three distinct enzymes from HeLa cells. *Proc. Natl. Acad. Sci. USA* **87:** 6664–6668.
380. Takao, M., M. Abramic, M. J. Moos, V. R. Otrin, J. C. Wootton, M. McLenigan, A. S. Levine, and M. Protic. 1993. A 127 kDa component of a UV-damaged DNA-binding complex, which is defective in some xeroderma pigmentosum group E patients, is homologous to a slime mold protein. *Nucleic Acids Res.* **21:** 4111–4118.
381. Takayama, K., D. M. Danks, E. P. Salazar, J. E. Cleaver, and C. A. Weber. 1996. DNA repair characteristics and mutations in the *ERCC2* DNA repair and transcription gene in a trichothiodistrophy patient. *Hum. Mutat.*
382. Takayama, K., E. P. Salazar, A. R. Lehmann, M. Stefanini, L. H. Thompson, and C. A. Weber. 1995. Defects in the DNA repair and transcription gene *ERCC2* in the cancer-prone disorder xeroderma pigmentosum group D. *Cancer Res.* **55:** 5656–5663.
383. Takayama, K., E. P. Salazar, B. C. Broughton, A. R. Lehmann, A. Sarasin, L. H. Thompson, and C. A. Weber. 1996. Defects in the DNA repair and transcription gene *ERCC2 (XPD)* in trichothiodistrophy. *Am. J. Hum. Genet.* **58:** 263–270.
384. Tanaka, K., H. Hayakawa, M. Sekiguchi, and Y. Okada. 1977. Specific action of T4 endonuclease V on damaged DNA in xeroderma pigmentosum cells *in vivo*. *Proc. Natl. Acad. Sci. USA* **74:** 2958–2962.
385. Tanaka, K., K. Y. Kawai, Y. Kumahara, and M. Ikenaga. 1981. Genetic complementation groups in Cockayne syndrome. *Somatic Cell. Genet.* **7:** 445–456.
386. Tanaka, K., N. Miura, I. Satokata, I. Miyamoto, M. C. Yoshida, Y. Satoh, S. Kondo, A. Yasui, H. Okayama, and Y. Okada. 1990. Analysis of a human DNA excision repair gene involved in group A xeroderma pigmentosum and containing a zinc-finger domain. *Nature* **348:** 73–76.
387. Tanaka, K., I. Satokata, Z. Ogita, T. Uchida, and Y. Okada. 1989. Molecular cloning of a mouse DNA repair gene that complements the defect in group-A xeroderma pigmentosum. *Proc. Natl. Acad. Sci. USA* **86:** 5512–5516.
388. Tang, M. S., V. A. Bohr, X. S. Zhang, J. Pierce, and P. C. Hanawalt. 1989. Quantification of aminofluorene adduct formation and repair in defined DNA sequences in mammalian cells using the UVRABC nuclease. *J. Biol. Chem.* **264:** 14,455–14,462.
389. Tatsumi, K., M. Toyoda, T. Hashimoto, J. I. Furuyama, T. Kurihara, M. Inoue, and H. Takebe. 1987. Differential hypersensitivity of xeroderma pigmentosum lymphoblastoid cell lines to ultraviolet light mutagenesis. *Carcinogenesis* **8:** 53–57.
390. Teo, I. A., C. F. Arlett, S. A. Harcourt, A. Priestley, and B. C. Broughton. 1983. Multiple hypersensitivity to mutagens in a cell strain (46BR) derived from a patient with immuno-deficiencies. *Mutat. Res.* **107:** 371–386.
391. Thielmann, H. W., O. Popanda, L. Edler, and E. G. Jung. 1991. Clinical symptoms and DNA repair characteristics of xeroderma pigmentosum patients from Germany. *Cancer Res.* **51:** 3456–3470.
392. Thomas, D. C., D. S. Okumoto, A. Sancar, and V. A. Bohr. 1989. Preferential DNA repair of (6-4) photoproducts in the dihydrofolate reductase gene of Chinese hamster ovary cells. *J. Biol. Chem.* **264:** 18,005–18,010.

393. Thompson, L. H. 1996. Evidence that mammalian cells possess homologous recombinational repair pathways. *Mutat. Res.* **363:** 77–88.
394. Thompson, L. and D. Bootsma. 1988. Designation of mammalian complementation groups and repair genes, in *Mechanisms and Consequences of DNA Damage Processing* (Friedberg, E.C. and P.C. Hanawalt, eds.), Liss, New York, p. 279.
395. Thompson, L. H., K. W. Brookman, C. A. Weber, E. P. Salazar, J. T. Reardon, A. Sancar, Z. Deng, and M. J. Siciliano. 1994. Molecular cloning of the human nucleotide-excision-repair gene *ERCC4*. *Proc. Natl. Acad. Sci. USA* **91:** 6855–6859.
396. Thompson, L. H., D. B. Busch, K. W. Brookman, C. L. Mooney, and D. A. Glaser. 1981. Genetic diversity of UV-sensitive DNA repair mutants of Chinese hamster ovary cells. *Proc. Natl. Acad. Sci. USA* **78:** 3734–3737.
397. Thompson, L. H., J. S. Rubin, J. E. Cleaver, G. F. Whitmore, and K. Brookman. 1980. A screening method for isolating DNA repair-deficient mutants of CHO cells. *Somatic Cell. Genet.* **6:** 391–405.
398. Thompson, L. H., T. Shiomi, E. P. Salazar, and S. A. Stewart. 1988. An eighth complementation group of rodent cells hypersensitive to ultraviolet radiation. *Somatic Cell. Mol. Genet.* **14:** 605–612.
399. Thompson, L. H., R. W. Wu, and J. S. Felton. 1995. Genetically modified CHO cells for studying the genotoxicity of heterocyclic amines from cooked foods. *Toxicol. Lett.* **82/83:** 883–889.
400. Tomkinson, A. E., A. J. Bardwell, L. Bardwell, N. J. Tappe, and E. C. Friedberg. 1993. Yeast DNA repair and recombination proteins Rad1 and Rad10 constitute a single-stranded-DNA endonuclease. *Nature* **362:** 860–862.
401. Tornaletti, S. and G. P. Pfeifer. 1996. UV damage and repair mechanisms in mammalian cells. *Bioessays* **18:** 221–228.
402. Toschi, L., and R. Bravo. 1988. Changes in cyclin/proliferating cell nuclear antigen distribution during DNA repair synthesis. *J. Cell Biol.* **107:** 1623–1628.
403. Treiber, D. K., Z. Chen, and J. M. Essigmann. 1992. An ultraviolet light-damaged DNA recognition protein absent in xeroderma pigmentosum group E cells binds selectively to pyrimidine (6-4) pyrimidone photoproducts. *Nucleic Acids Res.* **20:** 5805–5810.
404. Troelstra, C., R. M. Landsvater, J. Wiegant, M. van der Ploeg, G. Viel, C. H. Buys, and J. H. Hoeijmakers. 1992. Localization of the nucleotide excision repair gene *ERCC6* to human chromosome 10q11-q21. *Genomics* **12:** 745–749.
405. Troelstra, C., A. van Gool, J. de Wit, W. Vermeulen, D. Bootsma, and J. H. Hoeijmakers. 1992. *ERCC6*, a member of a subfamily of putative helicases, is involved in Cockayne's syndrome and preferential repair of active genes. *Cell* **71:** 939–953.
406. Tsurimoto, T. and B. Stillman. 1991. Replication factors required for SV40 DNA replication in vitro. I. DNA structure-specific recognition of a primer-template junction by eukaryotic DNA polymerases and their accessory proteins. *J. Biol. Chem.* **266:** 1950–1960.
407. van Assendelft, G. B., E. M. Rigney, and I. D. Hickson. 1993. Purification of a HeLa cell nuclear protein that binds selectively to DNA irradiated with ultra-violet light. *Nucleic Acids Res.* **21:** 3399–3407.
407a. van der Horst, G. T. J., A. J. van Gool, J. de Wit, H. van Steeg, C. F. van Kreijl, R. J. W. Berg, F. R. de Gruijl, G. Weeda, D. Bootsma, and J. H. J. Hoeijmakers. 1997. Defective transcription-coupled repair in Cockayne syndrome B mice is associated with skin cancer predisposition. *Cell*, in press.
408. van der Spek, P. J., E. M. Smit, H. B. Beverloo, K. Sugasawa, C. Masutani, F. Hanaoka, J. H. Hoeijmakers, and A. Hagemeijer. 1994. Chromosomal localization of three repair genes: the xeroderma pigmentosum group C gene and two human homologs of yeast *RAD23*. *Genomics* **23:** 651–658.
409. van Duin, M., J. de Wit, H. Odijk, A. Westerveld, A. Yasui, M. H. M. Koken, J. Hoeijmakers, and D. Bootsma. 1986. Molecular characterization of the human excision

repair gene *ERCC–1*: cDNA cloning and amino acid homology with the yeast DNA repair gene *RAD10*. *Cell* **44:** 913–923.

410. van Duin, M., G. Vredeveldt, L. V. Mayne, H. Odijk, W. Vermeulen, B. Klein, G. Weeda, J. H. Hoeijmakers, D. Bootsma, and A. Westerveld. 1989. The cloned human DNA excision repair gene *ERCC–1* fails to correct xeroderma pigmentosum complementation groups A through I. *Mutat. Res.* **217:** 83–92.
411. van Gool, A. J., R. Verhage, S. M. Swagemakers, P. van de Putte, J. Brouwer, C. Troelstra, D. Bootsma, and J. H. Hoeijmakers. 1994. *RAD26*, the functional *S. cerevisiae* homolog of the Cockayne syndrome B gene *ERCC6*. *EMBO J.* **13:** 5361–5369.
412. van Hoffen, A., A. T. Natarajan, L. V. Mayne, A. A. van Zeeland, L. H. Mullenders, and J. Venema. 1993. Deficient repair of the transcribed strand of active genes in Cockayne's syndrome cells. *Nucleic Acids Res.* **21:** 5890–5895.
413. van Hoffen, A., J. Venema, R. Meschini, A. A. van Zeeland, and L. H. Mullenders. 1995. Transcription-coupled repair removes both cyclobutane pyrimidine dimers and 6-4 photoproducts with equal efficiency and in a sequential way from transcribed DNA in xeroderma pigmentosum group C fibroblasts. *EMBO J.* **14:** 360–367.
414. van Oosterwijk, M. F., A. Versteegh, R. Filon, A. A. van Zeeland, and L. H. F. Mullenders. 1996. The sensitivity of Cockayne's syndrome cells to DNA-damaging agents is not due to defective transcription-coupled repair of active genes. *Mol. Cell. Biol.* **16:** 4436–4444.
415. van Vuuren, A. J., E. Appeldoorn, H. Odijk, A. Yasui, N. G. J. Jaspers, D. Bootsma, and J. H. J. Hoeijmakers. 1993. Evidence for a repair enzyme complex involving ERCC1 and complementing activities of ERCC4, ERCC11 and xeroderma pigmentosum group F. *EMBO J.* **12:** 3693–3701.
416. van Vuuren, A. J., E. Appeldoorn, H. Odijk, S. Humbert, V. Moncollin, A. P. Eker, N. G. Jaspers, J. M. Egly, and J. H. Hoeijmakers. 1995. Partial characterization of the DNA repair protein complex, containing the ERCC1, ERCC4, ERCC11 and XPF correcting activities. *Mutat. Res.* **337:** 25–39.
417. van Vuuren, A. J., W. Vermeulen, L. Ma, G. Weeda, E. Appeldoorn, N. G. Jaspers, A. J. van der Eb, D. Bootsma, J. H. Hoeijmakers, S. Humbert, L. Schaeffer, and J. M. Egly. 1994. Correction of xeroderma pigmentosum repair defect by basal transcription factor BTF2 (TFIIH). *EMBO J.* **13:** 1645–1653.
418. van Zeeland, A. A., C. A. Smith, and P. C. Hanawalt. 1981. Sensitive determination of pyrimidine dimers in DNA of UV-irradiated mammalian cells. Introduction of T4 endonuclease V into frozen and thawed cells. *Mutat. Res.* **82:** 173–189.
419. Venema, J., Z. Bartosova, A. T. Natarajan, A. A. van Zeeland, and L. H. Mullenders. 1992. Transcription affects the rate but not the extent of repair of cyclobutane pyrimidine dimers in the human adenosine deaminase gene. *J. Biol. Chem.* **267:** 8852–8856.
420. Venema, J., L. H. Mullenders, A. T. Natarajan, A. A. van Zeeland, and L. V. Mayne. 1990. The genetic defect in Cockayne syndrome is associated with a defect in repair of UV-induced DNA damage in transcriptionally active DNA. *Proc. Natl. Acad. Sci. USA* **87:** 4707–4711.
421. Venema, J., A. van Hoffen, A. T. Natarajan, A. A. van Zeeland, and L. H. Mullenders. 1990. The residual repair capacity of xeroderma pigmentosum complementation group C fibroblasts is highly specific for transcriptionally active DNA. *Nucleic Acids Res.* **18:** 443–448.
422. Venema, J., A. van Hoffen, V. Karcagi, A. T. Natarajan, A. A. van Zeeland, and L. H. Mullenders. 1991. Xeroderma pigmentosum complementation group C cells remove pyrimidine dimers selectively from the transcribed strand of active genes. *Mol. Cell. Biol.* **11:** 4128–4134.

422a. Vermeulen, W. 1995. DNA repair and transcription deficiency syndromes. Ph.D. dissertation, Offsetdrukkerij Ridderprint B.V., Ridderkerk.

423. Vermeulen, W., J. Jaeken, N. G. Jaspers, D. Bootsma, and J. H. Hoeijmakers. 1993. Xeroderma pigmentosum complementation group G associated with Cockayne syndrome. *Am. J. Hum. Genet.* **53:** 185–192.
424. Vermeulen, W., R. J. Scott, S. Rodgers, H. J. Muller, J. Cole, C. F. Arlett, W. J. Kleijer, D. Bootsma, J. H. Hoeijmakers, and G. Weeda. 1994. Clinical heterogeneity within xeroderma pigmentosum associated with mutations in the DNA repair and transcription gene *ERCC3*. *Am. J. Hum. Genet.* **54:** 191–200.
425. Vermeulen, W., M. Stefanini, S. Giliani, J. H. Hoeijmakers, and D. Bootsma. 1991. Xeroderma pigmentosum complementation group H falls into complementation group D. *Mutat. Res.* **255:** 201–208.
426. Vermeulen, W., A. J. van Vuuren, M. Chipoulet, L. Schaeffer, E. Appeldoorn, G. Weeda, N. G. J. Jaspers, A. Priestley, C. F. Arlett, A. R. Lehmann, M. Stefanini, M. Mezzina, A. Sarasin, D. Bootsma, J. M. Egly, and J. H. J. Hoeijmakers. 1994. Three unusual repair deficiencies associated with transcription factor BTF2(TFIIH). Evidence for the existence of a transcription syndrome. *Cold Spring Harbor Symp.* **59:** 317–329.
427. Vink, A. A., R. J. Berg, F. R. de Gruijl, P. H. Lohman, L. Roza, and R. A. Baan. 1993. Detection of thymine dimers in suprabasal and basal cells of chronically UV-B exposed hairless mice. *J. Invest. Dermatol.* **100:** 795–799.
428. Vos, J. M. H. 1995. Replication of genotoxic lesions: mechanisms and medical impact, in *DNA Repair Mechanisms: Impact on Human Diseases and Cancer* (Vos, J.-M. H., ed.), R. C. Landes, Austin, pp. 187–218.
429. Vos, J. M. and E. L. Wauthier. 1991. Differential introduction of DNA damage and repair in mammalian genes transcribed by RNA polymerases I and II. *Mol. Cell. Biol.* **11:** 2245–2252.
430. Vreeswijk, M. P., A. van Hoffen, B. E. Westland, H. Vrieling, A. A. van Zeeland, and L. H. Mullenders. 1994. Analysis of repair of cyclobutane pyrimidine dimers and pyrimidine 6-4 pyrimidone photoproducts in transcriptionally active and inactive genes in Chinese hamster cells. *J. Biol. Chem.* **269:** 31,858–31,863.
431. Vrieling, H., M. L. Van Rooijen, N. A. Groen, M. Z. Zdzienicka, J. W. Simons, P. H. Lohman, and A. A. van Zeeland. 1989. DNA strand specificity for UV-induced mutations in mammalian cells. *Mol. Cell. Biol.* **9:** 1277–1283.
432. Vrieling, H., J. Venema, M. L. van Rooyen, A. van Hoffen, P. Menichini, M. Z. Zdzienicka, J. W. Simons, L. H. Mullenders, and A. A. van Zeeland. 1991. Strand specificity for UV-induced DNA repair and mutations in the Chinese hamster HPRT gene. *Nucleic Acids Res.* **19:** 2411–2415.
433. Vrieling, H., L. H. Zhang, A. A. van Zeeland, and M. Z. Zdzienicka. 1992. UV-induced hprt mutations in a UV-sensitive hamster cell line from complementation group 3 are biased towards the transcribed strand. *Mutat. Res.* **274:** 147–155.
434. Wade, M. H. and E. H. Chu. 1979. Effects of DNA damaging agents on cultured fibroblasts derived from patients with Cockayne syndrome. *Mutat. Res.* **59:** 49–60.
435. Waga, S., G. J. Hannon, D. Beach, and B. Stillman. 1994. The p21 inhibitor of cyclin-dependent kinases controls DNA replication by interaction with PCNA. *Nature* **369:** 574–578.
436. Wang, X. W., K. Forrester, H. Yeh, M. A. Feitelson, J. R. Gu, and C. C. Harris. 1994. Hepatitis B virus X protein inhibits p53 sequence-specific DNA binding, transcriptional activity, and association with transcription factor ERCC3. *Proc. Natl. Acad. Sci. USA* **91:** 2230–2234.
437. Wang, X. W., W. Vermeulen, J. D. Coursen, M. Gibson, S. E. Lupold, K. Forrester, G. Xu, L. Elmore, H. Yeh, J. H. J. Hoeijmakers, and C. C. Harris. 1996. The XPB and XPD DNA helicases are components of the p53-mediated apoptosis pathway. *Genes Dev.* **10:** 1219–1232.
438. Wang, X. W., H. Yeh, L. Schaeffer, R. Roy, V. Moncollin, J. M. Egly, Z. Wang, E. C. Friedberg, M. K. Evans, B. G. Taffe, V. A. Bohr, G. Weeda, J. H. J. Hoeijmakers, K.

Forrester, and C. C. Harris. 1995. p53 modulation of TFIIH-associated nucleotide excision repair activity. *Nature Genet.* **10:** 188–195.
439. Wang, Y. C., V. M. Maher, D. L. Mitchell, and J. J. McCormick. 1993. Evidence from mutation spectra that the UV hypermutability of xeroderma pigmentosum variant cells reflects abnormal, error-prone replication on a template containing photoproducts. *Mol. Cell. Biol.* **13:** 4276–4283.
440. Wang, Y. C., V. M. Maher, and J. J. McCormick. 1991. Xeroderma pigmentosum variant cells are less likely than normal cells to incorporate dAMP opposite photoproducts during replication of UV-irradiated plasmids. *Proc. Natl. Acad. Sci. USA* **88:** 7810–7814.
441. Wang, Z., S. Buratowski, J. Q. Svejstrup, W. J. Feaver, X. Wu, R. D. Kornberg, T. F. Donahue, and E. C. Friedberg. 1995. The yeast *TFB1* and *SSL1* genes, which encode subunits of transcription factor IIH, are required for nucleotide excision repair and RNA polymerase II transcription. *Mol. Cell. Biol.* **15:** 2288–2293.
442. Wang, Z., J. Q. Svejstrup, W. J. Feaver, X. Wu, R. D. Kornberg, and E. C. Friedberg. 1994. Transcription factor b (TFIIH) is required during nucleotide-excision repair in yeast. *Nature* **368:** 74–76.
443. Wassermann, K., P. M. O'Connor, J. Jackman, A. May, and V. A. Bohr. 1994. Transcription-independent repair of nitrogen mustard-induced N-alkylpurines in the c-myc gene in Burkitt's lymphoma CA46 cells. *Carcinogenesis* **15:** 1779–1783.
444. Weber, C. A., K. M. Kirchner, E. P. Salazar, and K. Takayama. 1994. Molecular analysis of *CXPD* mutations in the repair-deficient hamster mutants UV5 and UVL–13. *Mutat. Res.* **324:** 147–152.
445. Weber, C. A., E. P. Salazar, S. A. Stewart, and L. H. Thompson. 1990. *ERCC2*: cDNA cloning and molecular characterization of a human nucleotide excision repair gene with high homology to yeast RAD3. *EMBO J.* **9:** 1437–1447.
446. Weeda, G., J. H. Hoeijmakers, and D. Bootsma. 1993. Genes controlling nucleotide excision repair in eukaryotic cells. *Bioessays* **15:** 249–258.
447. Weeda, G., R. C. A. van Ham, W. Vermeulen, D. Bootsma, A. J. van der Eb, and J. H. J. Hoeijmakers. 1990. A presumed DNA helicase encoded by *ERCC–3* is involved in the human repair disorders xeroderma pigmentosum and Cockayne's syndrome. *Cell* **62:** 777–791.
448. Weeda, G., J. Wiegant, M. van der Ploeg, A. H. Geurts van Kessel, A. J. van der Eb, and J. H. Hoeijmakers. 1991. Localization of the xeroderma pigmentosum group B-correcting gene *ERCC3* to human chromosome 2q21. *Genomics* **10:** 1035–1040.
449. Wei, Q., G. M. Matanoski, E. R. Farmer, M. A. Hedayati, and L. Grossman. 1993. DNA repair and aging in basal cell carcinoma: a molecular epidemiology study. *Proc. Natl. Acad. Sci. USA* **90:** 1614–1618.
449a. Wood, R. D. 1996. DNA repair in eukaryotes. *Annu. Rev. Biochem.* **65:** 135–167.
450. Wood, R. D. 1989. Repair of pyrimidine dimer ultraviolet light photoproducts by human cell extracts. *Biochemistry* **28:** 8287–8292.
451. Wood, R. D., M. Biggerstaff, and M. K. K. Shivji. 1995. Detection and measurement of nucleotide excision repair synthesis by mammalian cell extracts in vitro. Methods: *A Companion to Methods in Enzymol.* **7:** 163–175.
452. Wood, R. D., P. Robins, and T. Lindahl. 1988. Complementation of the xeroderma pigmentosum DNA repair defect in cell-free extracts. *Cell* **53:** 97–106.
453. Wu, Z. N., C. L. Chan, A. Eastman, and E. Bresnick. 1992. Expression of human O^6-methylguanine-DNA methyltransferase in a DNA excision repair-deficient Chinese hamster ovary cell line and its response to certain alkylating agents. *Cancer Res.* **52:** 32–35.
454. Xiao, H., A. Pearson, B. Coulombe, R. Truant, S. Zhang, J. L. Regier, et al. 1994. Binding of basal transcription factor TFIIH to the acidic activation domains of VP16 and p53. *Mol. Cell. Biol.* **14:** 7013–7024.

455. Yagi, T., J. Tatsumi-Miyajima, M. Sato, K. H. Kraemer, and H. Takebe. 1991. Analysis of point mutations in an ultraviolet-irradiated shuttle vector plasmid propagated in cells from Japanese xeroderma pigmentosum patients in complementation groups A and F. *Cancer Res.* **51:** 3177–3182.
456. Yamaizumi, M. and T. Sugano. 1994. U.V.-induced nuclear accumulation of p53 is evoked through DNA damage of actively transcribed genes independent of the cell cycle. *Oncogene* **9:** 2775–2784.
457. Yong, S. L., J. E. Cleaver, G. D. Tullis, and M. M. Johnston. 1984. Is trichothiodystrophy part of the xeroderma pigmentosum spectrum? *Clin. Genet.* **36:** 82S.
458. Yoon, H., S. P. Miller, E. K. Pabich, and T. F. Donahue. 1992. SSL1, a suppressor of a HIS4 5'-UTR stem-loop mutation, is essential for translation initiation and affects UV resistance in yeast. *Genes Dev.* **6:** 2463–2477.
459. Yuan, J., T. M. Yeasky, P. A. Havre, and P. M. Glazer. 1995. Induction of p53 in mouse cells decreases mutagenesis by UV radiation. *Carcinogenesis* **16:** 2295–2300.
460. Zawel, L., K. P. Kumar, and D. Reinberg. 1995. Recycling of the general transcription factors during RNA polymerase II transcription. *Genes Dev.* **9:** 1479–1490.
461. Zdzienicka, M. Z., and J. W. I. Simons. 1987. Mutagen-sensitive cell lines are obtained with a high frequency in V79 Chinese hamster cells. *Mutat. Res.* **178:** 235–244.
462. Zdzienicka, M. Z., G. P. van der Schans, and J. W. I. Simons. 1988. Identification of a new seventh complementation group of UV-sensitive mutants in Chinese hamster cells. *Mutat. Res.* **194:** 165–170.
463. Zelle, B. and P. H. Lohman. 1979. Repair of UV-endonuclease-susceptible sites in the 7 complementation groups of xeroderma pigmentosum A through G. *Mutat. Res.* **62:** 363–368.
464. Zelle, B., R. J. Reynolds, M. J. Kottenhagen, A. Schuite, and P. H. Lohman. 1980. The influence of the wavelength of ultraviolet radiation on survival, mutation induction and DNA repair in irradiated Chinese hamster cells. *Mutat. Res.* **72:** 491–509.
465. Zeng, X. R., Y. Jiang, S. J. Zhang, H. Hao, and M. Y. W. Lee. 1994. DNA polymerase δ is involved in the cellular response to UV damage in human cells. *J. Biol. Chem.* **269:** 13,748–13,751.
466. Zhan, Q., F. Carrier, and A. J. J. Fornace. 1993. Induction of cellular p53 activity by DNA-damaging agents and growth arrest [published erratum in *Mol. Cell. Biol.* 1993. **9:** 5928]. *Mol. Cell. Biol.* **13:** 4242–4250.
467. Ziegler, A., D. J. Leffell, S. Kunala, H. W. Sharma, M. Gailani, J. A. Simon, et al. 1993. Mutation hotspots due to sunlight in the *p53* gene of nonmelanoma skin cancers. *Proc. Natl. Acad. Sci. USA* **90:** 4216–4220.
468. Zolan, M. E., G. A. Cortopassi, C. A. Smith, and P. C. Hanawalt. 1982. Deficient repair of chemical adducts in alpha DNA of monkey cells. *Cell* **28:** 613–619.

19

Cellular Responses to DNA Damage and Human Chromosome Instability Syndromes

KumKum Khanna, Richard Gatti, Patrick Concannon, Corry M. R. Weemaes, Merl F. Hoekstra, Martin Lavin, and Alan D'Andrea

1. INTRODUCTION

Ataxia telangiectasia (AT), Fanconi's anemia (FA), Werner's syndrome (WS), Bloom's syndrome (BS), and Nijmegen breakage syndrome (NBS) are pleiotropic disorders that have occasionally been referred to as premature aging diseases. In part, this designation refers to the clinical observation that patients with these severe diseases show a variety of disorders that tend to accompany aging: skin changes, endocrine abnormalities, and higher relative risk and incidence of cancer. These disease have also been grouped together because cells from patients with AT, FA, WS, BS, and NBS have defects in maintaining genomic stability and integrity, even though patients with these syndromes have not been shown to carry germ-line p53 mutations like Li-Fraumeni patients (*see* Chapters 14 and 15). Further, cells from patients with any of these five syndromes have in common certain unique sensitivities to DNA damaging agents. However, the patients and their cells can be contrasted to xeroderma pigmentosum (XP), trichothiodistrophy, and Cockayne's syndrome (CS) patients (Chapter 18) in that AT, FA, WS, BS, and NBS patients tend to show little, if any, sensitivity to UV radiation. In this chapter, we summarize some of the clinical presentations of the human chromosome instability diseases, and discuss the molecular properties of the genes encoding AT, FA, WS, BS, and NBS.

2. MOLECULAR AND CELLULAR BIOLOGY OF ATAXIA TELANGIECTASIA

A high frequency of chromosome breakage was first described in AT over 30 years ago *(122,178)*. These observations together with other data led to the classification of AT as a chromosomal instability syndrome together with BS and FA *(55,113,120,125)*. Clinical radiosensitivity in AT became evident about the same time *(73,108,180)*, and an adverse response was also demonstrated in AT cells in vitro where it was demonstrated that AT fibroblasts and lymphoblastoid cells were three to four times more sensitive to ionizing radiation than controls *(46,233)*. This increased sensitivity to ion-

From: DNA Damage and Repair, Vol. 2: DNA Repair in Higher Eukaryotes
Edited by: J. A. Nickoloff and M. F. Hoekstra © *Humana Press Inc., Totowa, NJ*

izing radiation was not associated with a defect in repair of DNA single-strand breaks *(78,162,245)*.

Although early attempts failed to reveal any significant defect in double-strand break (DSB) rejoining at short times after irradiation, Cornforth and Bedford *(58)* demonstrated the existence of residual breaks in DNA from AT cells utilizing the premature chromosome condensation assay. Further support for a defect in DSB rejoining in AT was provided by Coquerelle et al. *(57)*, who reported that rejoining of γ-radiation-induced DNA DSBs was slower in AT fibroblasts than in controls. Determination of the extent of repair over a 24 h period or after low dose-rate irradiation also revealed that more residual DNA damage existed in AT cells *(15,77)*. Approximately 10% of the total induced breaks were present in AT cells at longer times after radiation exposure. Pandita and Hittleman *(190)* provided evidence for an intrinsic abnormality in chromatin structure in AT that appears to cause more proficient translation of DNA damage into chromosome damage throughout the cell cycle. It seems likely that residual breaks in DNA in AT cells are caused by a defect in detecting this damage in DNA, and as a consequence its translation, into chromosome breaks that remain unrepaired.

2.1. Clinical Features of AT

AT is characterized by an early-onset neurological degeneration that always involves progressive cerebellar ataxia and disabling mutations in the *ATM* gene. The syndrome also includes ocular apraxia, elevated alphafetoprotein, conjunctival and cutaneous telangiectasiae, immunodeficiency with recurrent sinopulmonary infections, cancer susceptibility, and radiosensitivity (both clinical and in vitro). The disorder is transmitted in an autosomal recessive pattern. The incidence has been estimated at 1:40,000 to 1:100,000 live births; the carrier frequency is 0.5–1.0%. Despite the rarity of this disorder, it is the most common progressive ataxia of childhood. The patients appear quite normal until 1–2 yr of age, at which time they become clumsy and frequently lose their balance. By 10 yr of age, repeated falling and slow reflexes necessitate use of a wheelchair. Although muscle strength is normal at first, disuse of certain muscles creates contractures later in life unless aggressive physical therapy is administered early. The lack of physical activity and repeated sinopulmonary infections also leads to loss of pulmonary function. There is no known treatment for the progressive ataxia. The immunodeficiency is sometimes responsive to supportive therapy; however, only one-third of AT patients have severe enough immune impairment to warrant this. Life expectancy is 20–40 yr. Most patients die either from pulmonary failure or malignancy. For further clinical details and references, *see* ref. *88*.

One-third of AT patients develop a malignancy during their lifetimes. Young patients usually develop either a lymphoma or leukemia. Older patients also develop non-lymphoid cancers, such as breast or stomach, and occasionally T-cell prolymphocytic leukemia *(120–122)*. Because AT patients are about 30% more sensitive to ionizing radiation than normals, conventional doses of radiation therapy for cancer are life-threatening and contraindicated *(73,108,180)*. Thus, a dangerous clinical situation arises whenever a malignancy presents before the diagnosis of AT is suspected. This can be avoided if all young patients in need of radiation therapy are first checked for neurological symptoms. Epidemiological studies suggest that AT heterozygotes are also at an increased risk for cancer, perhaps comprising as much as 5% of all cancer patients. These studies further suggest that the mothers of AT patients are at a fivefold increased risk of breast cancer. Despite this, several large cohorts of breast cancer patients and breast cancer families have been screened for *ATM* mutations over the past few years and none have revealed a convincing increase in mutations.

Karyotyping of AT patients shows an increase in the frequency of reciprocal translocations on chromosomes 2, 7, 14, and 22. The breakpoints for these translocations coincide with the sites of T and B cell receptor gene complexes; presently, these are the only known sites of normal gene rearrangement in the genome. These nonrandom changes in AT lymphocytes are not seen in AT fibroblasts, which instead have an increased frequency of random chromosomal aberrations. Telomeric fusions are also commonly observed in AT karyotypes. Although *atm* knockout mice are infertile secondary to improper chromosome synapsing during gamete formation, as discussed below, most AT patients have normal sexual development. However, a small subset of patients do not enter puberty and at postmortem examination may have gonadal streaks in place of gonads, similar to the knockout mice.

2.2. *The* ATM *Gene*

Gatti et al. *(87)* mapped the AT gene to chromosome 11q 22–23 and over a period of 7 years several groups carried out physical mapping to narrow the region of interest to approx 200 kbp *(161,238,240)*. This heralded the cloning of the gene by Savitsky et al. *(206)*, who identified a 5.9-kbp partial cDNA for *ATM* (ataxia telangiectasia mutated), which was mutated in all patient samples analyzed. Subsequent analysis has revealed that the majority of AT mutations are predicted to give rise to truncated proteins *(41,103,235)*. *ATM* extends over 150 kbp of genomic DNA, codes for a 13-kb mRNA, and is composed of 66 exons *(239)*. The *ATM* ORF codes for a protein related to a family of yeast, *Drosophila*, and mammalian proteins containing a phosphatidylinositol 3-kinase-like (PIK) protein kinase domain *(131,164,270)*. Figure 1 shows a schematic alignment of many family members and Fig. 2 shows an amino acid alignment of the protein kinase domain. These model system orthologous proteins are involved in cell cycle control and/or the detection of DNA damage and include *TEL1* and *MEC1/ESR1* in *Saccharomyces cerevisiae, rad3* in *Schizosaccharomyces pombe*, *MEI-41* in *Drosophila melanogaster*, and, in mammals, the catalytic subunit of DNA-dependent protein kinase (DNA-PK_{cs}), ataxia telangiectasia and $rad3^+$-related (ATR/FRP1 [FRAP-related protein kinase]) and FRAP1 (FKBP12 and rapamycin associated protein) *(53,116,117,143)*.

The prototype PI3-kinase is a heterodimeric protein composed of 85 (regulatory) and 110 kDa (catalytic) subunits with dual specificity for the inositol ring of phosphatidylinositol and Ser-608 in the p85 subunit *(42)*. No lipid kinase activity has been reported for DNA-PK_{cs} *(117)*, mTOR/FRAP *(32)*, ATR/FRP, or ATM *(139,143)*. Although these enzymes have been shown to have protein kinase activity with a range of substrates, it remains undetermined whether some or all of these are physiological substrates. Two other sequences have been identified in the *ATM* ORF that code for potential domains in the protein, including a leucine zipper (nucleotides 3651–3714), which would enable ATM to interact with other leucine zipper-containing proteins, and a sequence coding for a proline-rich region (nucleotides, 4113–4152) provides additional evidence for a role for ATM in signaling pathways. This proline-rich region is a potential binding site for SH3 groups of proteins involved in signal transduction pathways *(193)*.

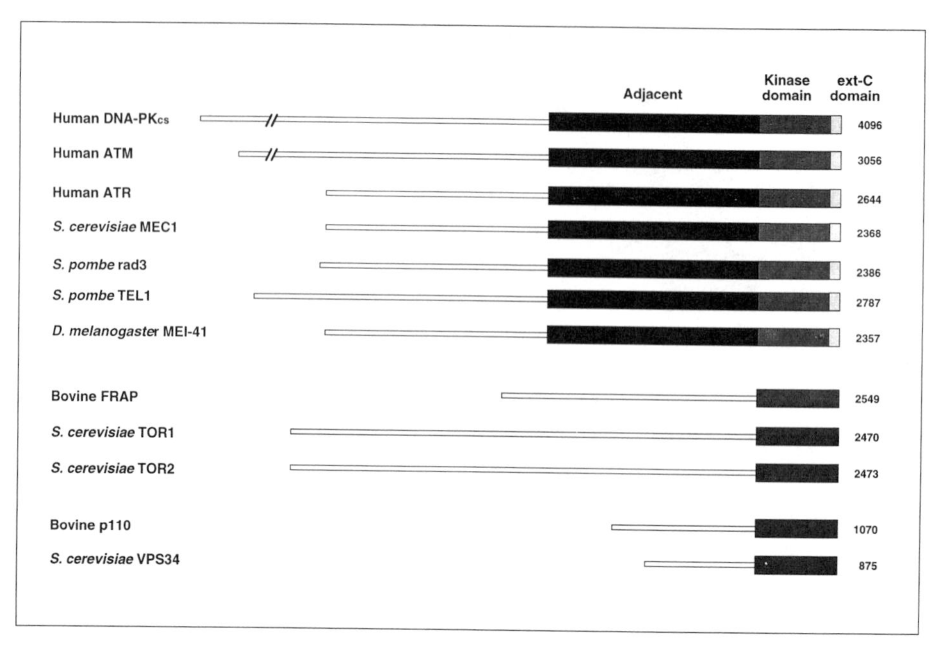

Fig. 1. Phosphatidylinositol 3-kinase family members. The members of this family are involved in signal transduction, DNA damage recognition, cell cycle checkpoint control, meiosis and chromosome segregation. DNA-PK_{cs}, DNA-dependent protein kinase catalytic subunit; *ATM*, ataxia telangiectasia mutated; *ATR*, ataxia telangiectasia and *rad3*$^{+}$ related; *MEC1*, cell cycle arrest in S and G2; *rad3*, G2 checkpoint control; *TEL1*, telomerase; *MEI41*, *Drosophila* mutant defective in meiotic recombination; FRAP, FKBP-rapamycin associated protein; TOR1/2, targets for rapamycin; p110, PI-3 kinase catalytic subunit.

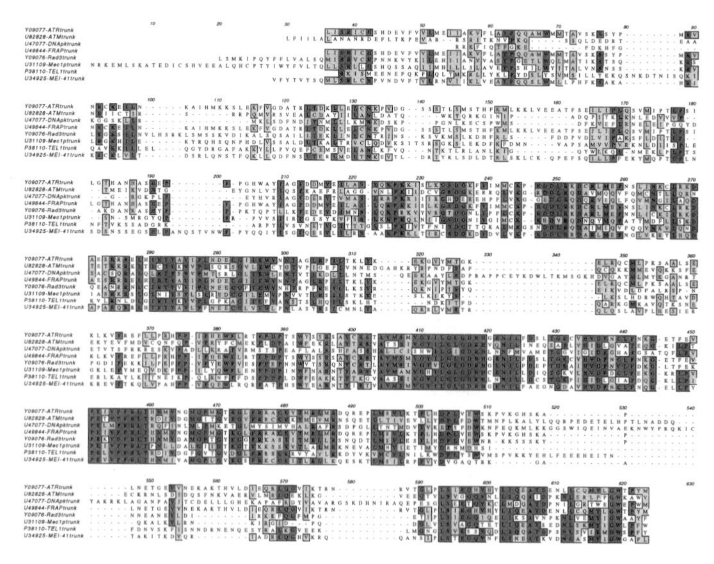

Fig. 2. Amino acid alignment of the protein kinase members of the PIK kinase family. Shown is a Clustal W alignment of ATR, ATM, DNA-PK, rad3p, Mec1p, Tel1p, and MEI41. The accession numbers are given next to the protein name. Identical amino acids are boxed in dark gray shading and similar amino acids have light gray shading.

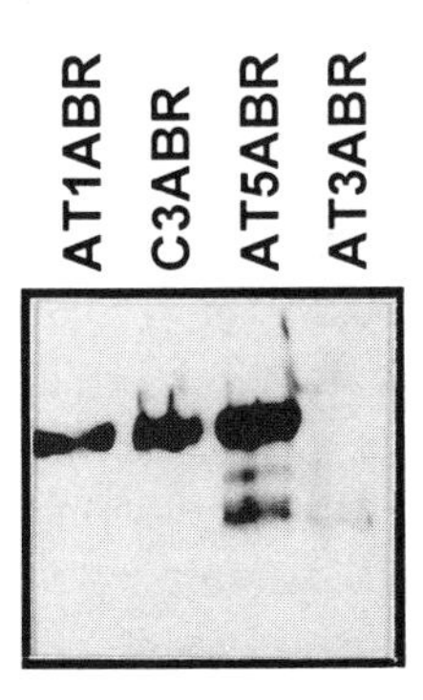

Fig. 3. Detection of ATM in control and AT cells by immunoblotting. The protein was detected using ATM-3BA antibody, which was raised in rabbits against an ATM peptide (amino acids 2581–2599). The mutations in the AT cells were AT1ABR, 9 nucleotide in-frame deletion (7638–7646), AT3ABR (A 8266T), and AT5ABR (6412 del AG).

2.3. *Analysis of* ATM *in AT Patients*

Based on an ORF of 9.168 kb, the predicted size for the ATM protein (3056 amino acids) is 350.6 kDa *(207)*. Antibodies first reported by Keegan et al. *(143)* and subsequently produced in several laboratories against ATM peptides and recombinant proteins detected a protein of approx 350 kDa in size *(35,48,139,159,249)*. In one study the presence of the 350-kDa protein was detected in 12 of 23 AT cell lines *(249)*. As expected, near full-length protein was detected in a sibling pair that was shown to be homozygous for an inframe deletion of 9 nucleotides (Fig. 3 and ref. *206*). Another cell line with a compound heterozygote genotype, had a mutation predicted to give rise to a truncated protein from one allele that was expressed at a higher than normal level (Fig. 3). The change in the second allele is suggested to be caused by a missense mutation that stabilizes the protein. It is of interest that truncated proteins were not detected in any of the studies examining AT mutations, implying that these incomplete forms of the protein are unstable and rapidly degraded.

Heterogeneity is well described in AT, with patients showing considerable variability in clinical and cellular/molecular features *(47,121,132)*. One of the original (but now defunct) complementation groups reported in AT, group VI, includes patients with NBS (*see* Section 7.), was characterized by radiosensitivity, microcephaly, immunodeficiency, and cancer predisposition but without ataxia or telangiectasia *(132,133)*. Only recently has it been possible to relate these types of heterogeneity to the type of mutation in *ATM* or the amount of ATM protein present. For example, AT patients with a 137-nucleotide insertion in *ATM* at nucleotide 5762 have a later onset of symptoms, a slower rate of neurological deterioration, intermediate radiosensitivity, absence of tumors at an early age, and intermediate levels of stabilization of p53 postirradiation. Preliminary data is also available on AT variants, those cases that reveal a milder form of the disease or are lacking in some of the hallmarks of AT. In general these variants have a later stage of onset of the major symptoms, the clinical features are milder, infections are absent, cancer predisposition is not apparent, and intermediate radiosensitivity is observed in some cases *(50,74,234)*.

One of the variants, AT9RM, characterized by intermediate radiosensitivity and radioresistant DNA synthesis *(50)*, is homozygous for a truncating mutation in codon

3047 (C-T), which gives rise to a protein with 9 amino acids missing from the C-terminus *(102)*. Immunoblotting detected 17% of the normal level of ATM protein in AT9RM cells. It seems likely that the level of ATM protein is important in determining the extent of radiosensitivity. Clearly in AT cells, where truncation mutations give rise to unstable, shortened proteins, there is effectively no ATM present and the cells show classical radiosensitivity. In the AT9RM variant, and presumably other variants, protein levels of the order of 20% are associated with reduced radiosensitivity. Transfection of normal control cells with full-length ATM cDNA in the reverse orientation leads to a 60–70% reduction in ATM protein by 6 h postinduction with $CdCl_2$ *(274)*. Under these conditions radiosensitivity was enhanced approximately threefold, approaching that observed in AT cells. Here again it is evident that a significant reduction in ATM protein sensitizes cells to ionizing radiation. These observations have some bearing on previous results demonstrating intermediate radiosensitivity in AT heterozygotes *(47,163)*. This suggests that in AT carriers reduced ATM (50% of normal) leads to increased radiosensitivity but not to other features of AT.

2.4. *Function of the ATM Protein*

As pointed out above, ATM is a member of the PIK family and is likely to function as a protein kinase *(143,164,272)*. In this respect its mechanism of action would be expected to be similar to that of DNA-PK, recognizing and phosphorylating specific substrates in response to DNA damage caused by ionizing radiation *(131)*. The well-described defect in G1/S checkpoint control *(24)* and the reduced p53 expression in response to DNA damage *(141,145)* in AT cells postirradiation suggests that p53 may be downstream of ATM. Immunoprecipitation of cell lysates with anti-p53 antibody followed by immunoblotting with anti-ATM antibody revealed that ATM was present in these p53 precipitates *(249)*. The coprecipitation between p53 and ATM was either absent in AT lysates, where truncated protein was predicted, or occurred at low affinity in AT cells producing mutated but near full-length ATM. It seems likely that the p53-ATM association is direct since an ATM protein kinase domain encoded by a 5.9-kbp cDNA interacted with p53 in the yeast two-hybrid system *(249)*. Binding of p53 to ATM was constitutive and was not influenced by exposure of cells to ionizing radiation. The observations that the p53 response to radiation damage is defective in AT cells, and that ATM binds p53, suggest that ATM plays an important role in stabilizing p53 postirradiation. This might be achieved by phosphorylating p53 directly or by activating another serine/threonine kinase that utilizes p53 as a substrate. Since p53 is stabilized to a normal extent in AT cells after UV exposure *(145)*, it is evident that more than one signaling pathway exists, pointing to the involvement of other kinases independent of ATM. The identification of DNA-PK as one such candidate was recently described *(217)*. In that report it was demonstrated that p53 is phosphorylated at serine 15, which leads to reduced interaction of p53 with its negative regulator MDM2. Purified DNA-PK phosphorylates p53 on serines 15 and 37 and also impairs the ability of MDM2 to inhibit p53-dependent transactivation. Recent data from the Lavin laboratory demonstrate that p53 may be phosphorylated by ATM (Khanna et al., unpublished data). However, neither Jung et al. *(139)* nor Keegan et al. *(143)* could demonstrate in vitro phosphorylation of various p53 substrates by immunoprecipitated ATM or ATR. Hence, these data must be interpreted with caution since a multitude of substrates have been identified for DNA-PK by this approach, many of which are probably not of biological significance *(6)*. Also, potential substrates should not be eliminated based on an inability to demonstrate activity by an immunoprecipitated kinase.

Overexpression of the nonreceptor tyrosine kinase c-Abl induces the expression of *WAF1*, downregulates cyclin-dependent kinase cdk2, and blocks cells in G1 phase *(269)*. Cells lacking c-Abl are impaired in their ability to downregulate cdk2 activity or undergo G1 arrest in response to ionizing radiation exposure *(269)*. Since c-Abl is activated by ionizing radiation and since it appears to play a role in cell cycle control, the possibility that ATM and c-Abl might interact was explored. Constitutive binding of c-Abl to ATM, which was not further enhanced by ionizing radiation exposure, was observed in control cells but not in AT cells *(216)*. Constitutive interaction of c-Abl with DNA-PK has also been demonstrated but unlike that with ATM, this interaction is stimulated by exposure of cells to ionizing radiation *(148)*. Inability to bind ATM in AT cells resulted in diminished levels of c-Abl tyrosine kinase activation in irradiated cells. Baskaran et al. *(21)* demonstrated phosphorylation of c-Abl using a recombinant protein corresponding to the PI3-kinase domain in irradiated AT cells, further supporting a role for ATM in direct activation of c-Abl kinase. In addition, c-Abl-deficient cells failed to activate Jun kinase (JNK/SAPK), providing evidence that c-Abl is upstream of SAPK in a stress response pathway *(147)*; this kinase is not activated in response to ionizing radiation in AT cells *(146,216)*. In AT cells the defect in SAPK activation appears to be confined to the damaging agent to which these cells are hypersensitive (ionizing radiation) since SAPK activity is comparable in AT and control cells after UV treatment or on exposure to the protein synthesis inhibitor anisomycin *(216)*.

The involvement of ATM with c-Abl may be in a pathway separate to the p53 pathway operating through SAPK and leading to the induction of stress response genes via c-jun. Alternatively, there may be some overlap between these molecules in a single pathway involving ATM. Yuan et al. *(269)* have demonstrated that c-Abl functions in the cellular response to DNA damage through a p53-dependent mechanism. Furthermore c-Abl-mediated transactivation of p53 is kinase independent and the downregulation of cyclin-dependent kinase activity for growth arrest is independent of WAF1 *(269)*. In essence, ATM may function in regulating checkpoint control not only through p53 and its effector gene WAF1 to inhibit cyclin kinase activity, and thus cell cycle progress, but also through c-Abl to induce genes mediated by c-jun and also to contribute to growth arrest via p53 in a *WAF1*-independent pathway. These pathways are summarized in Fig. 4.

2.5. *The* ATM *DNA Damage Response*

In its most simplistic form, response to DNA damage would involve the recognition and repair of lesions in DNA as part of a single process. However, it is evident that DNA damage response is more complex than this, involving sensors for different types of lesions, transducer molecules, enzymes for DNA repair and a variety of effector molecules. The response is directed not only to repairing DNA damage in isolation but also to regulating the passage of cells through the cycle and activating transcriptional events pertinent to the overall process. In relation to the sensing mechanism the best characterized enzyme is DNA-PK_{cs}, which is recruited to DSBs in DNA by the Ku heterodimer where it presumably phosphorylates proteins involved in DNA repair,

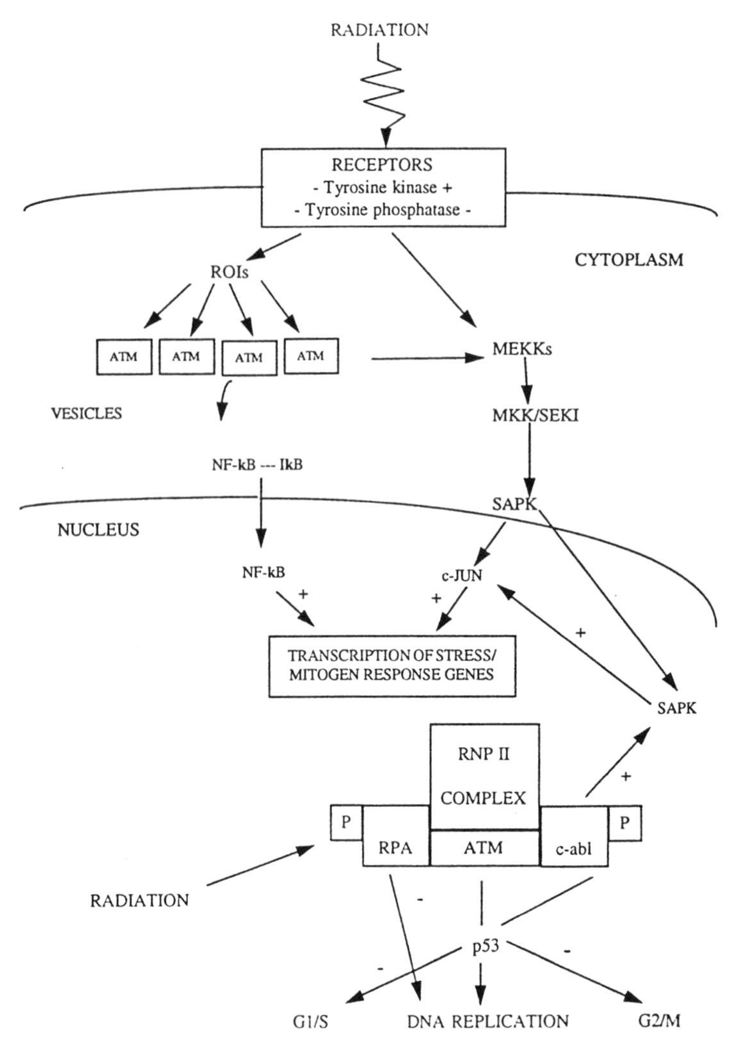

Fig. 4. Model for signal transduction mediated through ATM. In the model ATM is represented both inside the nucleus and in vesicles in the cytoplasm. In proliferating cells (fibroblasts, lymphoblastoid cells) the bulk of ATM is present in the nucleus. Radiation damage to DNA activates ATM by an unknown mechanism possibly involving p53 and other adapter proteins. This in turn activates checkpoint control for the G1/S phase transition and possibly G2/M and S phase. By activating c-Abl in response to radiation damage ATM may add to the complexity of cell cycle control and/or it may activate SAPK and in turn c-jun for transcriptional regulation. The presence of ATM in cytoplasmic vesicles is more difficult to understand. ATM in this location may respond to reactive oxygen intermediates (ROIs) similar to redox regulated proteins to activate transcription factors, such as NF-kB. + and – refer to activation and inhibition, respectively.

activating them for their role in the process of excision and repair of damage (*109*; Chapters 15–17). Since this DNA-PK_{cs}-dependent process appears to be coupled to transcriptional events it is likely that DNA-PK may also phosphorylate proteins in the transcriptional complex to inhibit RNA synthesis and allow repair of the lesion to occur prior to transcription *(131)*. Considerably less is known about the DNA damage sensor activity of ATM. It is evident that damage induced by ionizing radiation or radiomimetic agents, such as bleomycin and neocarcinostatin, is recognized directly or indirectly by ATM *(165)*. However, the nature of the specific lesion remains unknown. Single-strand breaks are repaired with normal kinetics in AT cells *(78,162,245)* but there is evidence that a portion of DSBs remain unrepaired in these cells *(57,58)*. It is unlikely that the defect in AT cells is in the repair of radiation-induced DSBs in DNA since it might be expected that DNA-PK would cope with these. More likely, in the absence of functional ATM protein, an appropriate response to specific lesions in DNA is not initiated or is inefficient and as a consequence they remain unrepaired and eventually are converted to DSBs not recognized by Ku heterodimer.

The appearance of chromosome breaks and rearrangements in DNA represents one consequence of mutated ATM. Failure to effectively activate multiple cell-cycle checkpoints *(24)* is another such defect. In the case of the G1/S checkpoint this is caused by an inefficient p53 response. Since p53 coimmunoprecipitates with ATM and since this interaction appears to be direct as assayed in the yeast two-hybrid system *(249)*, one model suggests that ATM responds to DNA damage by stabilizing and activating p53 by phosphorylation and thus activating the G1/S checkpoint. However, this may be only one of several pathways to activate p53 since DNA-PK is also implicated in modifying p53 in response to DNA damage *(217)*. Purification of ATM will provide an approach to determining whether ATM phosphorylates p53 and at what sites. The physiological significance of these results can be tested by comparison of in vivo data in control and AT cells. The interaction of c-Abl with ATM adds another dimension to the involvement of ATM in cell-cycle control. A role for c-Abl in the G1/S checkpoint remains somewhat controversial in that data with $abl^{-/-}$ cells both supports and provides evidence to the contrary for involvement of c-Abl at this level of control *(166,269)*. Clearly the binding of c-Abl to ATM, its activation by ionizing radiation as a kinase only in cells with functional ATM *(146)*, and its phosphorylation by the kinase domain of ATM *(21)* strongly suggest that c-Abl is a downstream effector of ATM. As referred to in Section 2.3., the ATM/c-Abl pathway may play a role in cell-cycle control or perhaps more likely this pathway is responsible for transcriptional control. Indeed, there is evidence that c-Abl is upstream of SAPK in a radiation signal transduction pathway *(147)* and activation of SAPK by ionizing radiation is defective in AT cells *(146,216)*.

It is possible that ATM is part of a complex that includes p53, c-Abl, RPA, and other factors that control transcription, replication, and cell cycle progression (Fig. 4). In this context it detects damage in DNA and the appropriate proteins are activated to mediate a variety of aspects of radiation signal transduction. Identification of other partner proteins for ATM and purification of the protein will assist in understanding its multifaceted role.

3. RESPONSES TO DNA DAMAGE AND REGULATION OF CELL-CYCLE CHECKPOINTS BY THE ATM PROTEIN KINASE FAMILY

In mammalian cells, four protein kinases form the PIK kinase superfamily (Fig. 1). FRAP, DNA-PK, ATM, and ATR are distinguished by their large size (all are >2500 aa), their common primary sequence relatedness through the C-terminal protein kinase domain (Fig. 2), and their sequence similarity to the p110 lipid kinase subunit of PI3-

Table 1
Mammalian PIK Kinase and Model Organism Homologs

Protein kinase	Mammalian synonyms	Model system homologs	Comment
FRAP	RAPT mTOR RAFT	Tor1p (Sc)[a] Tor2p (Sc)	Protein kinase
DNA-PKcs	Scid XRS-6	—	Catalytic subunit of protein kinase
Ku70	—	Hdf1p/Ku70 (Sc) IRBP (Dm)	DNA binding cofactor for DNA-PK Dimerizes with Ku86
KU 86	—	Ku80 (Sc)	DNA binding cofactor for DNA-PK. Dimerizes with Ku70
ATM	—	Tel1p (Sc) MEI41 (Dm)	Protein kinase
ATR	FRP MCCS1 MRK1	rad3p (Sp) Mec1p/Esr1p (Sc) MEI41 (Dm)	Protein kinase

[a]Sc = *Saccharomyces cerevisiae*; Sp = *Schizosaccharomyces pombe*; Dm = *Drosophila melanogaster*. Refer to the text for appropriate references.

kinase (Table 1). FRAP participates in mitogenic and growth factor responses in G1 and may regulate specific mRNA translation signals. DNA-PK, ATM, and ATR participate in responses to nuclear cues that activate DNA rearrangements or cell-cycle arrest. Furthermore, ATM and ATR may participate in a meiotic surveillance mechanism that may regulate proper chromosome transmission.

A cell's primary response to DNA damage can follow one of several routes. UV-induced lesions, for example, can be repaired by direct reversal through photoreactivation or by excision repair pathways *(43)*. DNA strand lesions are recombinogenic and can be repaired by direct religation or by homologous or nonhomologous recombination pathways. DNA strand breaks can effect cell-cycle checkpoint surveillance mechanisms, and one of the primary and conserved mechanisms for detecting damage and activating cell-cycle checkpoint signal transduction is through the PIK kinase family members *(144)*.

DNA damage checkpoint surveillance is a regulatory pathway and a signal transduction response *(66)*. Checkpoints effect cell-cycle arrest in response to internal errors and external agents. Checkpoints are genetic and biochemical pathways. In *S. cerevisiae*, for example, two genes form the upstream sensor *(MEC1)* and subsequent amplifier *(RAD53)* for checkpoint signal transduction *(1,204)*. *MEC1* is a PIK kinase superfamily member and is the homolog of *S. pombe rad3* and mammalian ATR *(25)*. Rad53p is a protein kinase that is phosphorylated and activated in response to DNA damage. Rad53p phosphorylation depends on *MEC1*, placing Rad53p downstream in a protein kinase cascade. In *S. pombe*, the chk1p protein kinase, which links radiation responses to cell cycle, is apparently downstream of the *MEC1* homolog *rad3 (205)*. Clearly, protein phosphorylation cascades effect signal transduction and the notion that protein kinases act as sensors and amplifiers of checkpoint responses demonstrates the concept of surveillance as a genetic and biochemical pathway.

3.1. *FRAP*

Rapamycin inhibits cell growth by interfering with the function of the FRAP (mTOR/RAPT/RAFT) protein kinase *(32,51,202,203)*. Rapamycin is an inhibitor of G1 progression and can lead to effects on translation of specific mRNAs. The inhibitory affect of rapamycin on cell-cycle progression results from effects on signal transduction that inputs into mRNA translation *(34,38)*. Inhibition of cell-cycle progression depends on formation of a FKBP12-rapamycin complex that provides high-affinity inhibition of FRAP protein kinase activity. The intrinsic protein kinase activity of FRAP is capable of autophosphorylating the enzyme and directly phosphorylating at two substrate proteins, PHAS-1 and $p70^{56}$ kinase *(38)*. Further, FRAP regulates $p70^{s6k}$ activation in vivo and this activation relies on a small portion of the N-terminus of FRAP as well as an active FRAP protein kinase domain *(33)*.

In the presence of rapamycin, IL-2-stimulated T cells accumulate in G1. Drug-treated cells accumulate the cdk inhibitor protein, $p27^{kip1}$ *(187)*. The inhibitory effect of rapamycin delineates a G1 signaling cascade from FRAP through to $p70^{s6k}$ and cyclin-dependent protein kinase. It is unclear if FRAP action on these activities is direct or indirect.

A further response that is affected by rapamycin is mitogenic protein synthesis. The rate-limiting step in translation of certain mRNAs with extensive 5'-UTR secondary structure is the binding of eIF-4E to the mRNA cap. In quiescent cells, the ability of eIF-4E to interact and form an active initiation complex is suppressed by PHAS-1. Growth factor treatment leads to PHAS-1 phosphorylation and relief of PHAS-1 inhibition of eIF-4E *(26,106,187)*.

3.2. *DNA-PK*

Significant interest in DNA-PK has arisen from an array of experiments. First, p53 is an in vitro substrate for DNA-PK *(160)*. The region phosphorylated is at the N-terminus and DNA-damaging agents stimulate N-terminal p53 phosphorylation. Unfortunately, the p53 nuclear kinase that is stimulated by DNA damage is unlikely to be solely DNA-PK and suggestive evidence implicates casein kinase I as the damage-induced p53 N-terminus protein kinase *(29)*. Second, scid mice, which have immunologic defects resulting from an inability to correctly undergo V(D)J and T cell receptor rearrangements, have an underlying sensitivity to γ-irradiation *(27)*. The sensitivity of scid cells was akin to that observed in CHO mutant lines that were identified in screens for radiation sensitivity (*see* Chapters 16 and 17). The common feature and underlying defect in the CHO cell lines and scid mouse was a defect in DNA-PK activity *(29)*. The clear connection between a defined genetic defect associated with radiation sensitivity and a protein kinase activity that is stimulated by broken DNA has led to a number of studies that have examined specific substrates for radiation-induced phosphorylation.

The DNA-PK enzymic component DNA-PK resides in a 460 kDa catalytic subunit ($DNA\text{-}PK_{cs}$) *(117)*. Optimal DNA-PK activity requires cofactors: Ku70 and Ku86 subunits and DNA ends. Ku70 and Ku86 were originally identified using autoantibodies from lupus patients. However, Ku-like genes have been reported in *S. cerevisiae* and *Drosophila melanogaster (20,30,75)*. Knock-out mice that lack Ku70 and Ku86 have been developed and, like scid mice, these mutants show severe immunodeficiencies *(189)*.

Perhaps the most critical question with regard to understanding DNA-PK is how DNA-PK participates in DNA damage signal transduction. Given that scid mice lack DNA-PK activity, the connection between protein kinase activity and DNA strand break repair needs to be understood. Several studies have investigated the possible interrelationship among cell cycle arrest, V(D)J recombination, scid defects, and p53 *(114,182)*.

p53 contributes to genomic stability through its involvement in cell-cycle checkpoints (*see* Chapter 15). G1 checkpoints are activated by DNA damage and ribonucleotide pool depletion. A metaphase/anaphase checkpoint is activated by incomplete spindle formation (*see* Chapter 21). The response to DNA damage is mediated by posttranslational modifications that alter p53 conformation, stability, and transactivation potential. Because DNA-PK phosphorylates the p53 amino terminus in a DNA-dependent fashion, several experiments have addressed whether DNA-PK deficiency alters p53-dependent checkpoint activity. In a direct examination of p53 induction following ionizing radiation, p53 levels showed normal induction kinetics in scid cells *(79,182)*, indicating that DNA-PK activity is unlikely to be important for p53 stabilization. Likewise, intact p53-dependent arrest pathways were observed in G1 and metaphase/anaphase scid cells following γ-irradiation, ribonucleotide depletion, or colcemid treatment *(128)*, indicating that DNA-PK does not regulate p53 checkpoint properties. Finally, the most direct test of p53 regulation by DNA-PK came from an analysis of scid p53 double mutant mice *(114,182)*. In genetic terms, if two genes are in a linear pathway then a double mutant should demonstrate only the most severe single mutant defect. If two genes are in different pathways then an additive or synergistic effect is predicted for the double mutant. If DNA-PK regulates p53 function then a scid p53 double mutant should be scid-like or p53-like. The scid p53 double mutants confirmed the prediction that DNA-PK is in a separate pathway from the tumor suppressor. The double mutants showed cooperativity in tumorigenesis onset and the scid mutation did not interfere with p53-dependent thymocyte apoptosis. Taken together, the induction of p53 in scid cell lines, the maintenance of cell cycle arrest, and the cooperativity in the double mutant all indicate that p53 is unlikely to be a relevant in vivo target for DNA-PK and its regulatory capacity.

3.3. *ATM and ATR*

ATM is a nuclear phosphoprotein with associated protein kinase activity *(48,139,143)*. Even though extensive patient analysis has resulted in collection of more than 100 *ATM* mutations, a direct correlation between AT and ATM protein kinase activity has not been reported.

The cell-cycle defects following DNA-damaging agent treatment reveals the varying roles ATM plays in humans and mice. Cellular defects in AT patient cells and $ATM^{-/-}$ mouse cells reveal hypersensitivity to γ-irradiation and defects in cell-cycle arrest after irradiation *(17,141,262,263)*. Both human and mouse cells carrying ATM defects are defective in upregulation of p53, suggesting that p53 is downstream of ATM in a signaling pathway. Furthermore, $ATM^{-/-}$ mice are infertile because of meiotic failure. Meiosis is arrested at zygotene/pachytene in prophase I as a result of abnormal chromosome synapsis leading to chromosome fragmentation *(262)*. Consistent with a role in meiosis I, the ATM protein directly associates with synapsed regions of zygotene/pachytene stage chromosomes *(143)*. The mouse results extend our molecular understanding of AT patient gonadal defects.

The notion that ATM signals p53 has been pioneered by work from Kastan *(141,181)*. In part because AT homozygotes lack the p53-mediated G1/S DNA damage checkpoint, and in part because the kinetics of p53-, p21-, and GADD45-induction by radiation are abnormal in AT cells, p53 is thought to function downstream of ATM in a damage-response pathway. A number of experiments summarized below have been designed and performed to address this proposal. However, given the paradox that AT patients and their cells are radiosensitive whereas p53 cells are radioresistant relative to control cells, the biological relevance of reduced p53 induction kinetics in AT cells is unclear.

Two different types of experiments have been performed to address an ATM-p53 connection. First, the types of DNA-damaging agents that induce p53 have been surveyed and tested in AT cells *(145)*. Both UV- and γ-ray treatment induce p53 in many cell types. Where directly examined, the p53 induction response to UV and γ-rays is dampened in AT, even though patients and their cells show minimal survival defects following UV treatment whereas γ-rays have a significant effect. Clearly, a linear, direct connection between ATM and p53 is not apparent from these experiments. Second, like DNA-PK/scid and p53, double mutant mice have been constructed by a number of groups *(256,257)*. In these studies, ATM and p53 mutations cooperate in that tumorigenesis is significantly advanced relative to either single mutant background. Furthermore, $atm^{-/-}$ mice and $atm^{-/-}$ $p53^{-/-}$ mice show similar radiation toxicity and viability profiles, suggesting that ATM cannot act in a direct linear fashion upstream of p53. Likewise, in $atm^{-/-}$ $p53^{-/-}$ mice, differential activation of p53 appears to be ATM-independent, suggesting a nonlinear ATM-p53 interaction.

The ATR protein kinase is also known as FRP (FRAP Related Protein, *7*). *ATR* is most closely related to *S. pombe rad3* and *S. cerevisiae ESR1/MEC1 (25)*. Indeed, *ATR* can complement *esr1-1/esr1-1* UV sensitivity. The protein kinase activity of the *ATR/rad3/ESR1(MEC1)* homologues appears to be essential for mediating checkpoint regulation *(25)*. *S. pombe rad3* mutants are defective in checkpoint responses to X-ray, UV, and HU treatment. Changes in conserved protein kinase subdomain residues in rad3p are unable to complement any measured checkpoint responses and these subdomain changes lead to reduced protein kinase activity.

S. pombe rad3p homomultimerizes *(25)*; this property can account for its original identification as a truncated cDNA that complemented *rad3* defects *(25,211)*. Intragenic complementation through protein multimerization could account for complementation by a truncated cDNA of the expressed point mutant protein. Further, the notion that *ATR* is the functional homolog of *rad3/ESR1(MEC1)* can be extended by multimerization experiments. Human ATR can heteromultimerize with *S. pombe* rad3p.

Like ATM, ATR is a nuclear protein kinase *(143)*. ATR directly associates with meiosis I chromatin. Indirect immunofluorescence experiments reveal that ATR shows a complementary pattern to ATM in meiosis I: ATR associates with asynapsed regions and ATM associates with synapsed regions in zygotene and pachytene nuclei.

3.4. A Model for Meiotic Regulation and Cell-Cycle Regulation by PIK Kinases

A variety of observations suggest a model for the role of ATR and ATM in mammalian meiosis (Fig. 5), including the localization of ATR on asynapsed zygotene meiotic chromosomes, the interaction of ATM with pachytene meiosis I chromosomes, the

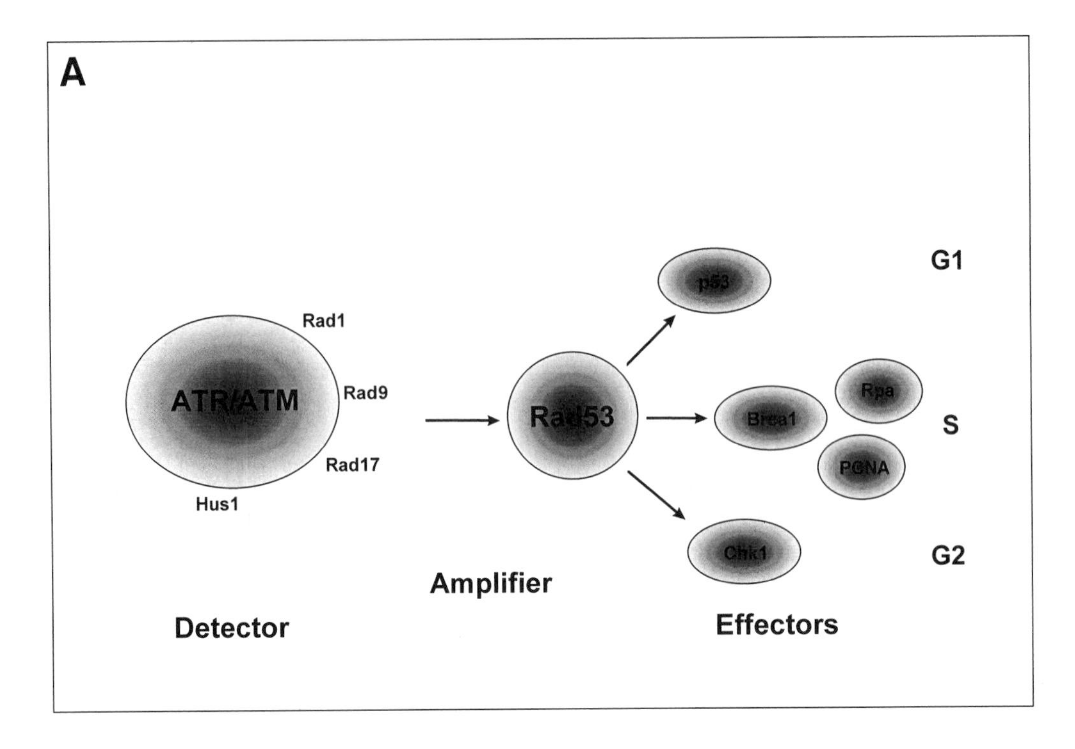

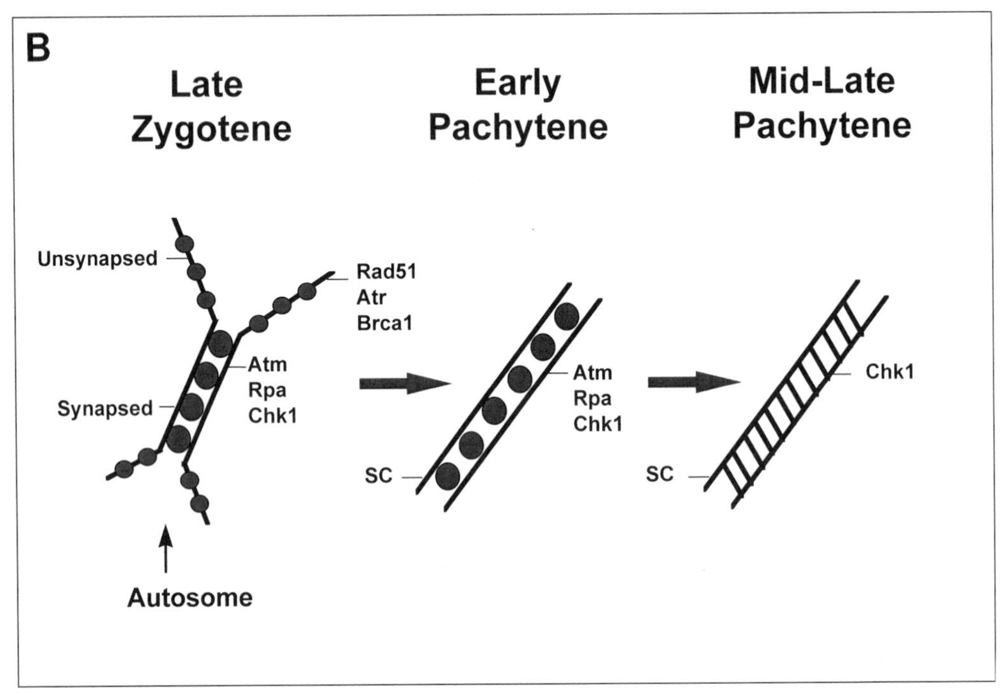

Fig. 5. DNA damage signaling pathways in which ATM family members participate. **(A)** Signaling during the mitotic cycle. ATM and ATR are suggested to be upstream detectors for DNA damage detection. In model organisms, signals are amplified through Rad53 homologs. The effectors in such a cascade could be p53 for G1/s, BRCA1 for S phase *(211)*, and Chk1 for G2/M *(76,85,205)*. **(B)** In meiosis, the temporal signaling program can be determined for autosomes by indirect immunofluorescence *(143)* and by chromosome fragmentation in atm$^{-/-}$ knockout mice *(262)*.

meiosis I failure of atm$^{-/-}$ mice, the meiotic and sporulation defect present during DSB repair in *S. cerevisiae esr1-1/esr1-1* strains *(142)*, the multimerization of rad3p and ATR, the observation that the *D. melanogaster* ATR/ATM homolog *mei41 (116)* has mutant defects on the frequency of meiotic recombination, the similar localization of ATM and RPA on meiotic chromosomes, and the association of Chk1 with meiotic chromosomes *(194)*.

In meiotic DNA-damage response and signaling pathways, the association of ATR with asynapsed regions could represent the primary response to early lesions, such as direct DSBs, that might initiate meiotic recombination and synapsis. As for yeast meiotic DSB repair, DSBs are nucleolytically processed during recombination. If such processing occurs in mammalian cells, the loss of ATR on synapsed chromosomes and the gain of ATM might occur through a multimerization dynamic change. Since ATR can heteromultimerize with *S. pombe* rad3p, DSBs in meiosis I may be processed and an ATR multimer would lose affinity and allow for heteromultimeric recruitment of ATM to ATR at a strand-break site. As processing of a strand lesion continues during meiosis I, ATR would disappear and the heralded ATM could form its own homomultimer at the synapsed region.

This model predicts a protein complex at ATR-detected sites of strand interruption. To date, two additional proteins have been shown to interact with asynapsed regions on meiosis I chromosomes: RAD51 and BRCA1 *(211)*. The RAD51 interaction is not limited to asynapsed chromosomes because RAD51 is also observed on synapsed chromosomes. The breast cancer susceptibility gene product, BRCA1, is found solely on asynapsed chromosomes in a pattern identical to ATR. It is tempting to suggest that ATR might interact with, and directly phosphorylate, BRCA1 in vivo, which would implicate *ATR* itself as a cancer susceptibility gene.

RPA interacts with meiotic chromosomes in a fashion like ATM *(194)*, implicating this single-strand DNA binding protein complex as an effector or substrate for ATM. Further, mammalian Chk1, a protein kinase that phosphorylates Cdc25c *(76,84,205)*, also associates with synapsed chromosomes in meiosis I. The association of Chk1 is temporally distinct and dependent on ATM: atm$^{-/-}$ mice show little Chk1 in their testes *(76)*. These data implicate Chk1 in mammalian cells as a temporally late effector for ATM and ATR in meiosis and in the mammalian cell cycle.

4. FANCONI ANEMIA

4.1. Clinical Presentation of FA Patients

FA is a rare autosomal recessive disease characterized by multiple congenital abnormalities, bone marrow failure, and cancer susceptibility. The mean age of anemia onset is 8 yr, and the mean survival is 16 yr. Death usually results from complications of bone marrow failure. The congenital abnormalities and clinical course of FA have been extensively reviewed *(2,5,10,12,65,97,98,266)*. In general, patients with FA have growth retardation and abnormalities of the skin (generalized hyperpigmentation and/or cafe-au-lait spots), upper extremities (frequently with defects in the thumbs or forearms), kidneys, and gastrointestinal system. Male FA patients have underdeveloped gonads and defective spermatogenesis. The large range of organ systems affected in FA implicates the FA genes in a general developmental process required during

Fig. 6. Clinical abnormalities associated with FA. This child with diagnosed FA has several common anomalies of the syndrome, including the typical facial features and the abnormalities of the arm and thumb. Of note, the most common abnormalities observed in FA are skin pigment abnormalities (75%), thumb abnormalities (40%), and kidney abnormalities (32%). Approximately 33% of FA patients have no detectable congenital abnormalities.

normal human embryogenesis. A child demonstrating many of the clinical features of FA is shown in Fig. 6.

The hematological complications of FA have also been extensively reviewed *(4,5)*. FA patients develop bone marrow failure typically during the first decade of life. Deficiencies in platelets or red cells usually precede white-blood-cell abnormalities. The patients have "fetal-like" erythropoiesis with increased hemoglobin F, and generally have high serum erythropoietin levels. The progression to bone marrow failure is highly varied among FA patients.

At least 20% of patients with FA develop cancers *(5)*. Since many FA patients die of other causes before they might have developed cancer, the actuarial risk of cancer is even higher *(40)*. FA patients primarily develop acute myeloblastic leukemia; however, cancers of several organ systems, including skin, gastrointestinal, and gynecological systems, have been described. The skin and gastrointestinal tumors are usually squamous cell carcinomas *(185,231)*. In addition, FA patients receiving androgen therapy for bone marrow failure are prone to liver tumors. Cancer tends to be a disease of older FA patients. The average age of patients who develop cancers is 15 yr for leukemia, 16 yr for liver tumors, and 23 yr for other tumors. There is a slightly higher risk of cancer for female FA patients, irrespective of the increased risk of gynecological cancers *(3)*.

The diagnosis of FA exploits the sensitivity of FA cells to bifunctional alkylating agents. FA cells have increased spontaneous chromosomal breakage that is amplified by the addition of the crosslinking agents, diepoxybutane (DEB) or mitomycin C

(MMC). Similar spontaneous, but not DEB-induced, chromosomal changes are observed in BS and AT *(127,177)*. The DEB test is a highly sensitive and specific test for FA *(9)*. Although useful in the diagnosis of FA patients, the DEB test does not distinguish FA carriers from the general population. Also, the DEB test can give false negative results, particularly in patients with cellular mosaicism *(64)*. The DEB test has been used successfully in prenatal diagnosis of FA *(11,218)*.

The diagnosis of FA is complicated by the wide variability in FA patient phenotype *(2)*. The differential diagnosis of FA includes other genetic syndromes, including neurofibromatosis, VACTERL association, and TAR (thrombocytopenia with absence of radii). Although the DEB test is highly effective in discriminating FA from these syndromes, the test remains underutilized *(9)*. Further confounding diagnosis, approx 33% of patients with FA have no obvious congenital abnormalities *(98)*. These FA patients may first present with bone marrow failure or cancer *(13)*. Despite this range in phenotypic variation, there exists no obvious correlation among the severity of the disease, the extent of cellular sensitivity to DEB, or the FA complementation group *(see* Section 4.2.*)*.

The treatment of FA is similar, but not identical, to the treatment of other forms of acquired aplastic anemia. Patients are treated with supportive care (i.e., blood transfusions) for their bone marrow failure. FA patients respond transiently to therapy with androgens and the cytokines, GM-CSF *(115)* and G-CSF *(198)*. The treatment of choice for FA is allogenic bone marrow transplant with a histocompatible sibling donor. In one recent study *(149)*, 18 patients with FA had allogenic bone marrow transplants from matched sibling donors (MSD). Seventeen patients had sustained engraftment and transfusion independence. Still, FA patients have severe toxicity from graft-versus-host (GVH) disease, perhaps resulting from an increased cellular sensitivity to endogenous cytokines released during GVH. T cell depletion has reduced GVH disease for bone marrow transplantation from unrelated donors *(62)*. Umbilical cord blood also offers a potential source of hematopoietic stem cells for FA patients without sibling matches *(36,37,156,157)*.

4.2. Molecular and Cellular Properties of FA

Because of the cellular sensitivity to crosslinking agents, FA is often compared to other syndromes of drug sensitivity and genomic instability, including AT, XP, CS *(80,81)*, BS, and hereditary nonpolyposis colon carcinoma *(71)*. The eight genes for XP encode proteins that comprise a DNA excision repair complex (*see* Chapters 10 and 18 and refs. *81,224*). *ATM* encodes a protein with homology to protein- and PI-3 kinases (*see* Sections 1. and 2.), suggesting that ATM plays a role in sensing DNA damage. More recent evidence demonstrates that the ATM protein product associates with meiotically pairing chromosomes *(118,143)*. The gene for BS encodes a protein related to a known yeast protein that regulates the cell cycle (*see* Section 6.). The BS protein may therefore regulate a G2/M checkpoint in the normal mammalian cell cycle. The genomic instability of these syndromes may result from a cellular defect in one of several processes, including DNA repair, cell cycle regulation, or DNA replication.

In addition to sensitivity to crosslinking agents, FA cells have several other phenotypic abnormalities (reviewed in ref. *61*). These abnormalities include increased sensitivity to oxygen radicals and increased apoptosis in response to interferon *(260)*. FA patients also appear to have a selective defect in stem cell proliferation, as demon-

Table 2
Complementation Groups of Fanconi Anemia

Subtype	Estimated percentage of FA patients	Chromosome location	Protein product
A	66	16q24.3	163 kDa (predicted nuclear localization)
B	4	?	?
C	12	9q22.3	63 kDa (cytoplasmic localization)
D	4	3p22–26	?
E	12	?	?
F	Rare	?	?
G	Rare	?	?
H	Rare	?	?

strated by decreased hematopoietic colony growth in vitro *(4)* and decreased gonadal stem cell survival *(260)*. In addition, FA cells appear to have an intrinsic defect in cell-cycle progression, demonstrating a baseline prolongation in G2 phase *(140,152)*. Many of these abnormalities have been identified in FA cells from multiple complementation groups. Accordingly, it remains unclear whether these cellular abnormalities correspond to all FA complementation groups or to only a subset. Most of the abnormalities described for FA cells are probably epiphenomena and do not relate directly to the primary cellular defect in each FA complementation group.

The complementation analysis of FA cells, using somatic cell fusion studies, has allowed the identification of at least eight complementation groups *(39,135–137,224)* (Table 2). At least three of the eight groups (A, C, and D) map to discrete chromosomal loci *(196,225,229)*.

4.3. *Biochemistry and Molecular Biology of* FAC

FAC was cloned by functional complementation of an EBV-immortalized type C FA cell line, HSC536 *(226)*. As predicted by the complementation test, the *FAC* cDNA corrects the MMC sensitivity and DEB sensitivity of FA(C) cell lines, but does not correct the MMC sensitivity of FA cells derived from patients from other complementation groups. Cells derived from patients with FA(C) have mutations in both alleles of *FAC*, consistent with the autosomal recessive inheritance of the FA syndrome. The *FAC* cDNA encodes a 558 amino-acid polypeptide (63 kDa) with no homology to other proteins in GenBank. *FAC* is composed of 14 exons *(99)*, spans approx 80 kbp, and maps to human chromosome 9q22.3 *(99)*. Murine *FAC* has been isolated, but no *FAC* homologs from other species have been reported. Murine *FAC* is only 66% identical to the human protein, but is able to functionally complement human cells from FA(C) patients *(258)*.

Mutational analysis of *FAC* has revealed a relatively small number of characteristic mutations, represented in specific ancestral backgrounds. The IVS4+4 A → T mutation is found in patients of Ashkenazi-Jewish ancestry and accounts for >80% of FA in this population *(261)*. Patients homozygous for this mutation have more severe FA, with multiple congenital abnormalities and early onset of hematological disease *(262)*. The 322delG mutation is found in patients of Northern European ancestry, particularly from Holland. Patients homozygous for this mutation have a comparatively mild FA, with fewer congenital abnormalities and later onset of hematological disease. Additional mutant alleles of *FAC* have been identified with mutations in exon 1, exon 6 *(100)*, and exon 14 *(101,226)*. A total of seven pathogenic mutations have now been identified, of which three are located in exon 14. Most of these mutations result in either truncation or internal deletion of the FAC protein. Only one pathogenic missense mutation (L554P) has been described.

Two murine models of FA(C) were developed, using targeted recombination in embryonic stem cells *(49,260)*. In one knockout mouse, exon 8 of *FAC* was deleted *(49)* and replaced by a neomycin resistance gene; in the other, exon 9 was deleted *(260)*. These mutations are likely to represent null alleles. The phenotype of homozygous *FAC* mutant animals was similar in the two strains. *FAC* mutants are viable and show no obvious birth defects of the skeletal system or urinary system. Cells derived from these animals show the classic hypersensitivity to bifunctional crosslinking agents. Similar chromosomal abnormalities and G2 cell-cycle abnormalities were also observed. Nonetheless, pancytopenia did not develop during the first year of life. Similarly, no leukemia or increased cancer susceptibility was observed during this time span. Although no peripheral blood abnormalities were detectable, mice with deleted exon 9 had an age-dependent decrease in BFU-E and CFU-GM progenitor assays. In addition, hematopoietic progenitor cells showed a distinct hypersensitivity to interferon-γ (IFN-γ). Other mitotic inhibitors had no differential effect. Increased cell susceptibility to IFN-γ is mediated by fas-induced apoptosis *(124,260)*. Cells derived from the *FAC* knockout mouse demonstrated increased levels of fas expression at lower levels of IFN-γ. It is possible that IFN-γ hypersensitivity is the major pathogenic mechanism in the development of progressive anemia in FA patients. The relationship between this phenotype and the cellular response to DNA crosslinks is currently unclear.

4.4. Biochemistry and Molecular Biology of FAA

The cloning of the major FA gene, *FAA*, was reported more recently *(7,162)*, and less is known about the FAA protein. The *FAA* mRNA transcript is 5.5 kb long and found in low abundance in all adult tissues examined. In addition to this major mRNA species, multiple transcripts, both larger and smaller, can be seen on Northern blot, suggesting a complex pattern of mRNA splicing of the gene. The mRNA contains an open reading frame of 4368 nt, predicting a 1455 amino-acid protein with a relative molecular mass of 163 kDa. Functional complementation has been achieved with this protein, but the 5' end of the mRNA has not yet been definitively mapped. Since no stop codons have been found 5' to the putative translation start site in the cDNA, additional 5' protein coding sequence may exist. Similar to FAC, the predicted FAA protein has no significant homologies to other proteins in sequence databases. Also, no common motifs are shared between the two FA proteins. The predicted FAA protein has two overlapping bipartite nuclear localization signals, located at amino acids 18–34 and 19–35. In addition, a partial leucine zipper consensus sequence is found at position 1069–1090. These homologies are consistent with a function of FAA in the nucleus, perhaps as a DNA binding protein, a transcriptional activator, or a DNA repair enzyme. The chromosomal location of *FAA* at 16q23.4, between markers D16S3121 and D16S303 *(196)*, was confirmed by direct methods. The gene spans approx 80 kbp and consists of at least 43 exons *(129)*.

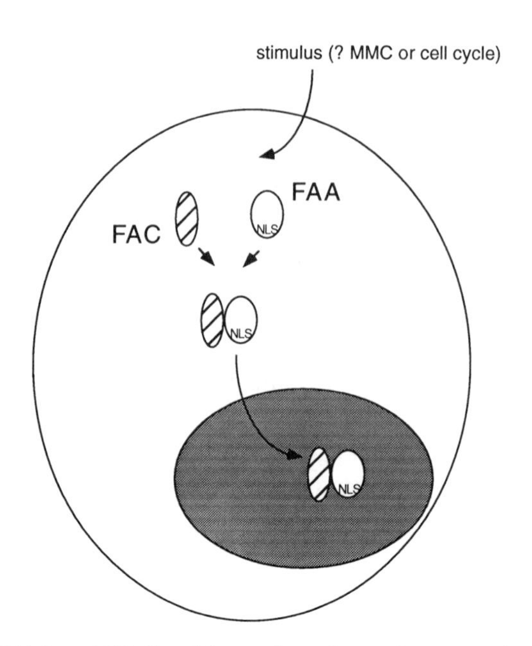

Fig. 7. Interaction of FAA and FAC and formation of a nuclear complex. The FAA and FAC proteins form a protein complex, which is localized to both the cytoplasm and nucleus of normal cells. The FAA protein has a functional NLS. Transport of the FAC protein to the nucleus requires functional FAA protein. The regulatory signals controlling nuclear translocation as well as the nuclear function of the protein complex remain unknown.

Several *FAA* gene mutations have been described, but no predominant alleles are apparent *(7,162)*. Approximately 40 variants of *FAA* were identified with probable pathogenic mutations. Seventeen of these 40 variants were microdeletions or microinsertions, suggesting a mechanism of slipped-strand mispairing during DNA replication. Two mutations (1115–1118del and 3788–3790del) accounted for 2 and 5% of *FAA* mutant alleles, respectively. The absence of a common *FAA* mutant allele makes the subtyping of FA(A) more tedious than for FA(C), where a few mutant alleles comprise the majority of patients.

Less is known about the structure and function of the FAA protein. In normal human lymphoblasts, the FAA protein is 163 kDa, as predicted by the cDNA sequence. Many FA(A) cell lines express no detectable FAA protein, as demonstrated by anti-FAA immunoblots *(155)*. Interestingly, a mutant FAA protein lacking a single phenylalanine residue at amino acid 1263 (FAA delF1263) is encoded by the mutant *FAA* allele, 3788–3790del. The FAA delF1263 protein is stable but not functional, suggesting that the mutation falls in a relevant functional domain of the protein.

4.5. Interaction of FAA and FAC Proteins in a Pathway

Based on the striking clinical similarities among FA patients from all complementation groups, the proteins encoded by the FA genes presumably interact in a protein complex or cellular pathway. Consistent with this hypothesis, recent data demonstrate that FAA and FAC bind in a protein complex *(155)* (Fig. 7). FAA and FAC are expressed as monomeric proteins in the cytoplasm of normal cells. In addition, FAA and FAC bind and form a protein complex that can be immunoprecipitated with antisera either to FAA or FAC. Mutant FAA, expressed in patient-derived FA(A) cell lines, fails to bind to wild-type FAC. For instance, FAA delF1263 fails to bind to FAC (A. D'Andrea, unpublished observation). Likewise, mutant FAC protein, expressed in patient-derived FA(C) cell lines, fails to bind to wild-type FAA. For instance, FAC L554P fails to bind to FAA. Whether wild-type FAA and FAC bind directly or indirectly through the interaction of adapter proteins remains unknown. It is possible that other proteins in the FA protein complex are encoded by other FA genes.

FAA contains a functional nuclear localization signal (NLS) at its amino terminus. The FAA/FAC complex is observed in relatively equal abundance in both the cytoplasm and the nucleus. The regulation of nuclear transport of the FAA/FAC complex is not known, but may result from stimulation with various crosslinking agents, such as MMC, or may result from various cell cycle cues. The nuclear FAA/FAC complex presumably plays some critical nuclear function, such as DNA repair, transcriptional regulation, or RNA splicing.

4.6. *Future Directions in FA Research*

Further understanding of the FA pathway will require the identification and cloning of additional FA genes. Multiple independent approaches are being utilized to accomplish this task. First, FA genes may be cloned by expression cloning approaches, similar to those utilized for *FAA* and *FAC (162,226)*. Second, additional FA genes may be cloned by positional cloning approaches. For instance, *FAD* has been mapped by microcell fusion to human chromosome 3p *(226)*. Third, additional FA genes may be cloned by identifying and cloning proteins that are bound in the FA protein complex. The cellular function of the FA genes (i.e., DNA repair, cell cycle regulation, transcriptional regulation) may be evident from the sequence homologies and presumed biochemical functions of any newly cloned FA proteins.

While these cloning approaches are ongoing, the precise cellular phenotype of FA cells will require further delineation. Since the underlying mechanism of FA results in genomic instability, it is possible that FA gene products will interact with other proteins involved in genomic instability, such as p53, ATM, or XP gene products. Mouse models, with knockouts of FA genes, will allow detailed studies of synergy with these other known cancer susceptibility genes.

5. WERNER'S SYNDROME

WS (MIM 277700) is a rare autosomal recessive disorder characterized by premature development of age-related diseases and features and may represent a useful model of some aspects of the normal human aging process *(70,107,171)*. Cells from WS patients display a limited replicative capacity in culture and have a "mutator" phenotype characterized by frequent chromosomal translocations and intergenic deletions *(83,126,208,209)*. However, unlike other disorders discussed in this chapter, neither WS patients nor their cells in culture have been demonstrated to have any increased sensitivity to DNA-damaging agents *(89,222)*. This is despite the fact that the product of the gene mutated in WS has helicase activity *(110)*, as do the genes responsible for XP and BS—disorders that are characterized by hypersensitivity to certain DNA-dam-

Fig. 8. Schematic representation of the members of the RecQ-like DNA helicase family. The name of the gene product and the organism is shown to the left and the number of amino acids in each protein is indicated on the right. The conserved helicase domain is shown as lighter shading with bars denoting the seven helicase motifs. Regions of extended homology are shown as black boxes; acidic domains are shown as darker shading and regions with no sequence homology are empty boxes.

aging agents. Indeed, WS gene, *WRN*, and BS genes are related to DNA helicases that are conserved in *E. coli*, yeast, and mammalian cells (Figs. 8 and 9).

5.1. Clinical Features of WS

The first clinical manifestation of WS to appear is typically in the second decade of life when patients fail to undergo the growth increase associated with adolescence *(70)*. As a result, their height remains significantly reduced compared to unaffected family members for the remainder of their adult life. By the third decade, graying and loss of hair is apparent and bilateral cataracts often develop *(199)*. Premature occurrence of age-related disorders, such as arteriosclerosis, osteoporosis, type 2 diabetes, and/or malignancies, is often a problem. The median lifespan is 47 yr, with death resulting primarily from cardiovascular pathology *(70)*.

Although many of the clinical features of WS are consistent with an accelerated aging process, there are some clear differences between the progression of WS and the process of normal human aging. For example, WS patients display an unusual calcification of soft tissues and ulcerations around the ankles. Moreover, the distribution of lesions in patients developing osteoporosis or arteriosclerosis is atypical. There are endocrine abnormalities, including gonadal atrophy *(70,107)*. Finally, there is a strong bias toward the development of malignancies of nonepithelial origins, including some very rare cancers, in WS patients *(105)*.

5.2. Cellular Findings for WS

Fibroblast and lymphoblastoid cell lines from WS patients display an attenuated lifespan in culture as well as an increased frequency of chromosomal abnormalities. Primary fibroblast cultures from WS patients usually undergo only about 20 doublings as compared to 50–60 doublings for normal cells *(172)*. Fibroblasts from WS patients also display reduced mitogenic responses to various growth factors in culture despite normal receptor levels *(22)*. The S phase of the cell cycle is prolonged in WS cells, and there is a tendency for the cells to arrest in S phase *(195)*. WS cells also display a propensity for mutation in culture. Lymphocytes resistant to 6-thioguanine are isolated at an elevated level from WS patients *(84)*. An examination of the *HPRT* gene in these cells reveals large intergenic deletions that appear to arise by nonhomologous recombination *(83,179)*.

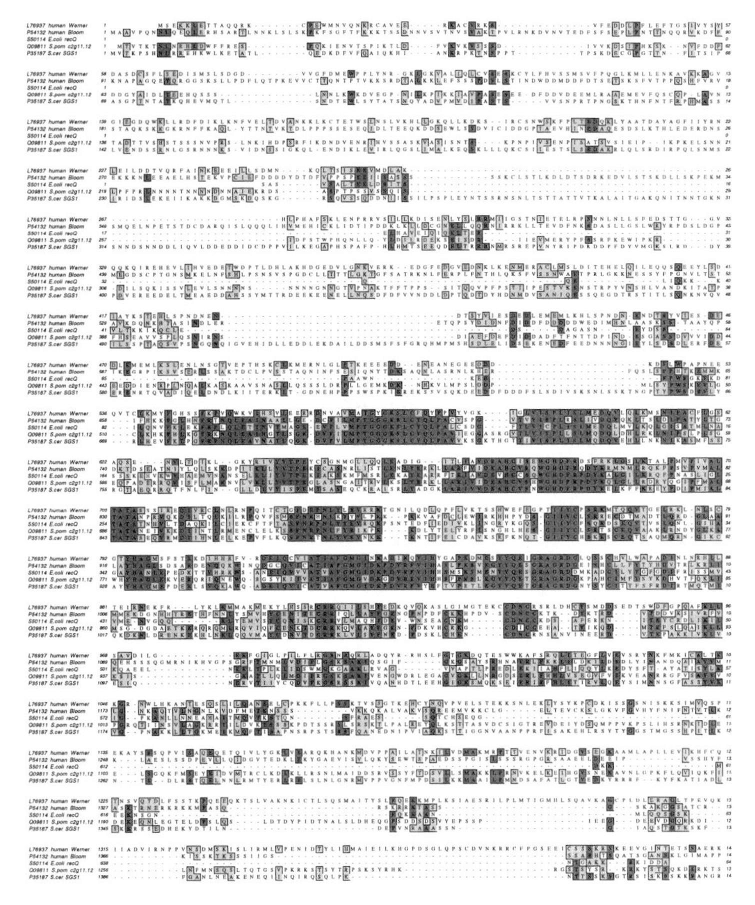

Fig. 9. Amino acid alignment of RecQ-like helicases. The Clustal W alignment is as described in Fig. 2.

5.3. Genetics of WS

Linkage studies in Japanese WS families initially localized *WRN* to 8p12–p11 *(106)*. This localization was confirmed and further refined in subsequent linkage studies that took advantage of the consanguinity present in many WS families by mapping for homozygosity in affected individuals *(184,210,236)*. These linkage studies placed *WRN* within a minimal 8.3 c*M* region near the marker D8S339. This region was further narrowed by the detection of linkage disequilibrium between *WRN* and alleles of flanking markers *(265,269)*. A physical map of the *WRN* region was developed and a minimal region of 1.2–1.4 Mbp was established based on recombinants. *WRN* was isolated from this region by genomic sequencing and database searching and its identity confirmed by the detection of 4 inactivating mutations in WS patients *(268)*.

WRN encodes a protein of 1432 amino acids with significant sequence similarity to DNA helicases (Fig. 8). Across the 7 consensus helicase domains present in the WRN protein there is >60% similarity to the human RecQL and *E. coli* RecQ molecules as well as to helicases from *C. elegans* and *S. cerevisiae (268)*. Recombinant WRN protein, affinity-purified from a baculovirus system, catalyzes ATP-dependent strand displacement *(110)*. An amino-acid substitution in WRN that, by analogy to other known helicases, is located within a conserved, active site of the molecule, eliminates this activity. Thus, the WRN protein is not only similar to other known helicases, but has demonstrable helicase activity in vitro.

All known *WRN* mutations in WS patients are predicted to lead to the production of a truncated protein. There are 19 such mutations reported that include nonsense mutations, insertions, deletions, and frameshifts *(104,189,268,270)*. These mutations are distributed across the gene and include mutations that are located far enough to the 3' end of the gene that they would not be predicted to disrupt the helicase domain. Indeed, at least one of these mutant forms of WRN has in vitro helicase activity *(110)*. However, WS patients with different mutations, including this one, have similar clinical phenotypes, suggesting a common mechanism for the effects of these mutations. One possibility for such a mechanism is an effect on protein localization *(175)*. Native WRN protein, when expressed in HeLa cells with an N-terminal green fluorescent protein tag, exhibits nuclear localization. C-terminal truncated forms, such as the mutant proteins from WS patients that have more than the last 69 amino acids deleted, do not localize to the nucleus. Further, the last 69 amino acids alone, when expressed with a fluorescent tag, are sufficient to direct nuclear localization. Thus, the various mutated forms of *WRN* detected in WS patients may have the common effect of inhibiting correct nuclear localization.

5.4. *Concluding Remarks on WS*

Much of the impetus for the cloning of *WRN* came from the supposition that it might provide a window into the mechanisms involved in normal human aging. The question of whether lessons learned from the study of the biochemistry and cell biology of *WRN* will be specific to WS or can be generalized to the normal aging process is very much unresolved. One interesting connection that has arisen from the gene cloning is the similarity of the WRN protein to a helicase in *S. cerevisiae*, Sgs1p *(220)*. Recently, mutant *sgs1* yeast cells have been demonstrated to have a reduced lifespan (approx 40% of wild-type). The *sgs1* mutants also appear to age prematurely, based on the early appearance of age-related phenotypes, such as sterility and nucleolar enlargement and fragmentation. These results suggest the intriguing possibility that there are, at the cellular level, some common mechanisms underlying the process of aging in eukaryotes.

6. BLOOM'S SYNDROME

Over 40 years ago Bloom *(28)* described congenital telangiectatic erythema resembling a skin disorder associated with severe growth retardation. Eleven years later German et al. *(93)* reported that the disease was associated with chromosomal breakage and cancer and was later shown to be the result of autosomal recessive inheritance *(91)*. The major clinical and laboratory abnormalities in BS are summarized below.

6.1. *Clinical Abnormalities*

Growth retardation, which involves severe prenatal and postnatal growth deficiency, is one of the major and constant clinical manifestations in BS homozygotes, whose average size and weight at term birth are at the 50th percentile for a 32-wk premature infant *(93)*. Slow growth is also a well known attribute of BS cells grown in culture.

Most BS homozygotes have generalized immunodeficiency documented as low levels of circulating immunoglobulins (Igs) and most individuals have defective delayed hypersensitivity *(160,250)*. BS homozygotes are susceptible to persistent bacterial sinopulmonary infections as a result of this generalized immunodeficiency, making chronic respiratory failure the second leading cause of death in BS *(93)*.

BS was the first chromosomal instability syndrome in which a link was recognized between genetic instability and cancer risk *(90)*. BS patients have a high risk of cancer—tumors occur early in life and multiple primary neoplasms are frequent. Lymphoreticular malignancies are the most common form of cancer in BS homozygotes, with leukemias and lymphoma occurring in roughly equal proportion. Together, they account for 45% of the malignant tumors reported in the Bloom's Syndrome Registry established in the early 1960s by James German *(92)*.

A variety of common solid tumors account for the remaining malignancies in BS homozygotes *(92)*, the most frequent carcinomas being colon, skin, breast, and cervix. Nonsurgical treatment of cancer in BS is made difficult by the apparent sensitivity of some, but not all, BS homozygotes to radiation therapy and/or chemotherapy. Unlike AT homozygotes, however, an in vivo sensitivity to DNA-damaging agents does not appear to be universal in BS since many BS homozygotes have tolerated normal therapeutic regimes without incident *(191)*.

BS homozygotes express a characteristic sun-sensitive erythematous facial rash that usually develops in the first summer of life *(28)*. Like AT, gonadal abnormalities are a feature of the syndrome. Diabetes has been reported in older BS homozygotes. BS homozygotes typically express near-average intelligence, although learning disabilities are frequent and occasional BS homozygotes have moderate mental retardation *(93)*.

6.2. Spontaneous Genetic Instability and DNA Damage Responses

Genetic instability is a hallmark of the BS phenotype. Beginning with German's initial observation of an increased frequency of chromosomal rearrangements *(90,94)*, a variety of spontaneous chromosomal aberrations have been observed in cells from BS homozygotes, including breaks in the chromosomes that are visible in metaphase spreads by light microscopy, a spontaneous 10-fold increase in rate of exchanges between sister-chromatids present in all affected individuals, and a high incidence of symmetrical, quadriradial chromosomes (presumed to be homologous chromosomes caught in the act of recombination), which are indicative of an elevated frequency of crossing-over between homologous chromosomes *(54,154)*. Multiple cell lineages express chromosomal instability in BS, including lymphocytes, lymphoblasts, keratinocytes, and fibroblasts *(54)*. Increased frequencies of spontaneous mitotic chromosomal recombination have been documented in BS T lymphocytes in vivo *(112)* and in B lymphoblasts and fibroblasts in culture *(31)*. Finally it should be noted that the *BLM* locus itself is a target for interchromosomal recombination in BS cells, with intralocus rearrangements restoring a wild-type gene and normal genetic stability *(67,68)*. An important clinical consequence of the excessive genomic instability of BS is an enormous predisposition to the generality of human cancers.

Many BS cells are sensitive to X-rays *(8,253)* and UV light *(130)*, although the degree of sensitivity is variable, with some BS cells expressing essentially normal resistance to these agents and to DSBs induced by restriction enzyme treatment *(59)*. BS cells do not have major functional defects in the several DNA repair pathways that have been examined to date *(54,72)*. Damage-induced cell cycle checkpoints have not been extensively studied in BS. However, the G1/S checkpoint is triggered normally after UV irradiation in BS fibroblasts *(243)*, despite impaired induction of p53 following irradiation in some BS cell lines *(169)*. The p53-independent, X-ray-induced S phase checkpoint also appears to be intact *(265)*.

6.3. *The* BLM *Gene and Structural Homologues*

McDonnel and Schults *(173)* used microcell-mediated chromosome transfer to demonstrate that chromosome 15 complements the high spontaneous sister-chromatid exchange (SCE) phenotype of BS cells in culture, suggesting a single locus for BS on chromosome 15. This observation facilitated the work of German's group, who subsequently linked the BS locus to chromosome 15q 26.1 markers *(96)*. They subsequently isolated a gene from 15q 26.1 that was mutated in multiple BS homozygotes *(67)*. *BLM* was identified using a powerful new approach termed Somatic Crossover Point (SCP) mapping *(67)*. One of the perplexing observations in BS has been that a minor population of blood lymphocytes shows a normal SCE rate (low-SCE) rather than the typical high SCE rate *(95)*. These exceptional cells are made wild-type by an intragenic recombination event, provided the cell carries mutations at two different sites within the gene. Ellis et al. *(68)* examined lymphoblastoid cell lines (LCLs) from 11 patients with low SCE. In LCLs from five patients polymorphic DNA markers distal to D155S127 had become homozygous, whereas loci proximal to D15S1108 remained heterozygous in all low-SCE LCLs. The region between these markers spanning about 250 kbp had to contain *BLM*, according to the principle of SCP mapping.

Ellis et al. *(67)* isolated and sequenced a 4437 bp cDNA in this interval containing a long open reading frame that encodes a 1417-amino acid peptide. The presence of chain terminating mutations in the candidate gene in individuals with BS proved that it was *BLM*. Four of the mutations introduced premature nonsense codons, and three produced amino acid substitutions. Interestingly, four patients of Ashkenazi Jewish ancestry had the same frameshift mutation, a 6-bp deletion/7-bp insertion at nucleotide 2281. This must be a founder mutation, since BS is known to occur very frequently among Askhenazi Jews compared to other populations *(95)*. All mutations detected thus far lead to loss of function, which is consistent with the autosomal recessive inheritance of BS.

BLM is predicted to encode a 159-kDa protein with homology to a family of DNA and RNA helicases, the prototype of which is the RecQ gene in *E. coli (183)*. This family also includes *S. cerevisiae* Sgs1p *(86,248)*, *S. pombe* rqh1p *(223)*, the human RecQ-L protein *(197,214)*, and *WRN (266)*. The predicted gene product contains seven helicase motifs (Figs. 8 and 9) that are present in members of the RecQ subfamily of DExH box-containing DNA helicases (D for aspartic acid, E for glutamic acid, H for histidine, x for any amino acid). Although all six members of the RecQ helicase family share sequence homology within the core helicase domain, BLM appears to belong to a subfamily that also includes Sgs1p, rqh1p, and WRN (Fig. 8). These four proteins are approximately the same length, but considerably larger than the *E. coli* prototype. They also show an extended region of sequence homology and have a similar positioning of the helicase domain (Fig. 8). In addition, although the N- and C-termini of BLM, Sgs1p, rqh1p, and WRN show little sequence similarity, they are all rich in charged and polar amino acids, especially serines, and they have patches of acidic residues *(67,268)*.

6.4. Phenotypic Homologs of BLM and Implications for BLM Function

The *E. coli* RecQ is a constituent of the RecF genetic recombination pathway *(255)*. Although some similarities exist between *recQ* and *BLM* mutant cells, it is clear the RecQ protein is required for efficient genetic recombination and DNA repair, whereas the phenotype of BS cell lines is one of genetic hyperrecombination coupled with an apparent proficiency in DNA repair. Mutations in *SGS1*, the gene most closely related *BLM*, share phenotypic similarities with BS homozygotes. The phenotype of *sgs1* null mutants includes slow growth, high frequencies of both inter- and intrachromosomal homologous mitotic recombination, and a high level of genomic instability *(86,248)*. Sgs1p normally interacts in vivo with the Topo II and Topo III, suggesting that Sgs1p functions with these topoisomerases in a complex. Although Sgs1p has been shown to be a helicase biochemically, an *sgs1* mutant lacking helicase activity is wild-type with regard to its interaction with topo III mutants *(168)*. It is possible that helicase activity is required for some but not all the in vivo functions of these proteins. Moreover, we still await the demonstration that BLM is an active helicase. BLM may interact with the human homologs to the yeast Topo II and Topo III enzymes *(14,82)*, consistent with the finding that Topo II activity in BS cells is abnormally low following exposure to BrdUr *(119)*. The sequence and phenotypic similarities between Sgs1p and BLM suggest that these two proteins perform similar functions.

6.5. How Defects in BLM Might Promote Genetic Instability and Cancer

The function of most DNA helicases in cells has not been established *(167)*. Mutations in RecQ and RecQ-like DNA helicases results in genomic instability in a variety of organisms. With the cloning of *BLM*, insight has been gained into the molecular basis of the genomic instability that is hallmark of BS phenotype. In general, BLM has been implicated in DNA replication. Defects in BLM could lead to genomic instability via one of several different routes.

1. Inability of mutant cells to resolve specific DNA structures generated during DNA replication may lead to DNA breaks, which are largely recombinogenic.
2. The absence of a functional BLM helicase may allow more recombinational events to proceed to completion, resulting in an increased frequency of SCE and other homologous recombination events.
3. The predicted helicase function of BLM suggests that wild-type BLM may unwind DNA to facilitate access to damaged DNA by repair complexes. Loss of BLM function in BS homozygotes would limit the access of repair complexes to sites of DNA damage and the resultant increase in unrepaired lesions would contribute to genetic instability.
4. In general, BLM has not been implicated in DNA repair although some proteins in the DExH family have been implicated in DNA repair and mutations in three of them, XPB, XPD, and ERCC6, have been identified in the human disease phenotypes XP and CS *(170,230)*.
5. Absence of BLM may destabilize other enzymes, including DNA ligase I *(45)*, Topo II in BrdUr-treated BS cells *(119)*, thymidylate synthetase *(219)*, and superoxide dismutase

Fig. 10. NBS patient.

(186), that participate in DNA replication and repair, perhaps through direct interactions or through general responses to DNA damage. Alternatively, direct interaction of BLM with other proteins may be required for their correct function.

Although many pieces of the puzzle remain to be fitted into place regarding the molecular mechanism of pleiotropic cytogenetic and biochemical abnormalities in BS, it is hoped that further studies aimed at elucidation of the enzymatic activities of BLM, the factors with which it interacts, and the substrates on which it operates will help to reveal the exact mechanisms by which this family of helicases maintain the integrity of the eukaryotic genome and consequently suppress cancer. The purification of active BLM should allow direct tests of the generation of genomic instability at the cytogenetic and molecular levels.

7. NIJMEGEN BREAKAGE SYNDROME

NBS is a rare autosomal recessive disorder (MIM 261260) first recognized in 1981 (Fig. 10; *251*). A number of separate reports of clinically similar syndromes, most notably Berlin Breakage Syndrome (BBS) (MIM 600885), appear to refer to this same disorder *(56,213,221,254)*. NBS is one of the family of inherited human disorders including BS, FA, and AT that are characterized by specific cellular defects in response to DNA-damaging agents. NBS shares a number of cellular features with AT, in particular a specific sensitivity to ionizing radiation and radiomimetic chemicals. As a result, NBS has often been described as a variant of AT. However, NBS patients lack many of the clinical hallmarks of AT, such as progressive cerebellar ataxia, oculocutaneous telangiectasia, and elevated serum alpha-fetoprotein levels. There is now abundant evidence from genetic and cell biology studies that NBS is a separate disorder from AT arising from mutations in a distinct, but as yet unidentified, gene.

7.1. Clinical Features of NBS

Clinically, NBS is characterized by microcephaly, a distinctive facial appearance that is frequently described as "bird-like," growth retardation, immunodeficiency, frequent sino-pulmonary infections, chromosomal instability, radiation hypersensitivity, and an increased incidence of malignancies, particularly lymphomas *(16,52,56, 111,192,213,221,232,237,241,242,251,252,254)*. An NBS patient registry maintained in Nijmegen contains entries for a total of 53 patients, the oldest of whom is 28 yr of age at this writing. The majority of NBS patients are of eastern European ancestry with the largest concentration of patients ascertained in Poland *(52)*. As noted in Section 7.3., this probably reflects a founder effect of a single mutation causing this disorder. The clinical features of NBS are briefly summarized here. Readers seeking more detailed clinical information are referred to other recent reviews *(52,242)*.

Most children affected with NBS are microcephalic at birth. In a few cases, children are born with normal head circumference but become microcephalic in the first 2–3 mo of life. Growth retardation is apparent from the first developmental stages. Height is typically below the 10th percentile in all patients, and weight is proportionally reduced. NBS patients have a distinguishing facial appearance that becomes more pronounced with age. This face is characterized by a receding forehead and mandible, prominent midface, large ears, epicanthic folds, and upward slant of the palpebral fissures.

Developmental milestones in NBS patients are generally reached at normal times during the first year of life. However, mild to moderate retardation is reported in about half of older NBS patients. A survey of patients in the NBS registry reported normal mental development in 40%, borderline retardation in 14%, mild retardation in 36%, and moderate retardation in the remaining 10%. Severe mental retardation has not been reported.

Recurrent infections of the respiratory tract, including pneumonia, bronchitis, otitis media, sinusitis, and mastoiditis, are reported in most but not all *(16,111)* NBS patients. These infections are not typically caused by opportunistic pathogens, a situation similar to that which is observed in AT patients. Serious complications occur in some cases. Four of the patients in the NBS registry have died as a result of bronchopneumonia *(242)*. Recurrent infections of the urinary tract are also reported in a smaller percentage (15%) of cases.

Immunodeficiency, usually with defects in both cellular and humoral compartments, is also a consistent finding in NBS patients. Agammaglobulinemia has been found in 35% and IgA deficiency in 20% of patients. Deficiencies in IgG2 and IgG4 are frequent, even when the IgG level is normal. The most commonly reported defects in cellular immunity are reduced percentages of total CD3+ T cells and CD4+ cells. The frequency of CD8+ cells usually does not appear to be affected and NK cells are increased. T cells from NBS patients are markedly impaired in responses to mitogenic stimuli, such as phytohemagglutinin, or crosslinking with anti-CD3 or anti-T cell receptor antibodies. Cytokine production is also reduced in response to these stimuli. However, addition of exogenous lymphokines, such as IL-2, to cultured T cells does not overcome the defect in proliferation *(241)*.

Constitutional karyotypes of NBS patients are generally normal. However, cytogenetic aberrations are observed in 10–45% of metaphase chromosome spreads from PHA-cultured T cells. Because of the poor proliferative capacity of T cells from these patients, these cytogenetic analyses can be difficult. As is observed in AT, the vast majority of chromosome aberrations in cultured T cells from NBS patients involve chromosomes 7 and 14 and are typically inversions and translocations with breakpoints at the sites of immunoglobulin or T cell receptor genes. Inv(7) is among the most frequently detected aberration in NBS cells as it is in AT cells. However, large clones of cells with rearrangements of chromosome 14 are rare in NBS patients. Both the tendency to express rearrangements of chromosomes 7 and 14 and the tendency to develop a malignancy is much higher in NBS than in AT *(252)*. Chromosomal rearrangements typically are not increased in fibroblasts from NBS patients and, in cases where increased rearrangements in fibroblasts are reported, no bias in the sites of rearrangement analogous to that in lymphocytes is observed *(56)*.

Malignancies occur at an elevated frequency and at an early age in NBS patients. Of the 53 patients recorded in the NBS registry, 17 have had malignancies, of which 14 were lymphomas. The majority of these are B cell lymphomas. Among patients ascertained in Poland, all malignancies were lymphomas arising under the age of 15 yr and 8 of 11 of these were B-cell lymphomas *(52)*. Because of the hypersensitivity of these patients to ionizing radiation and to radiomimetic drugs, these agents should be avoided in treatment or used at reduced dosages.

Only limited autopsy reports are available for NBS patients. An extensive postmortem examination has been reported for only one 4-yr-old child *(241)*. Key findings included a thymus that was small, dysplastic, and relatively devoid of lymphoid cells. This observation is consistent with the reported defects in cellular immunity in these patients and similar to postmortem findings in AT patients. There was no evidence of cerebellar degeneration in the NBS patient, unlike what would be expected in AT *(88)*.

7.2. Cellular Findings for NBS

Primary cells (fibroblasts and lymphocytes) cultured from NBS patients typically display poor growth in culture, a feature they share with cells from AT patients. Similarly, primary cells and transformed cell lines from NBS patients have aberrant responses to ionizing radiation. In colony-forming assays, NBS cells are three- to fivefold more sensitive to ionizing radiation or radiomimetic drugs than normal cells *(134,138,173,192,229,237,254)*. This is comparable to the range of sensitivity displayed by AT cells in such assays *(158)*. Radiation-induced chromosome aberrations occur with increased frequency in NBS cells *(228)*. NBS cells also display radioresistant DNA synthesis (RDS), the inability to halt or slow S phase progression after exposure to high doses of X-rays, another key feature of AT cells *(102,134)*. Some reports suggest that RDS in NBS cells may be intermediate between normal and AT cells *(237)*.

In complementation studies, fibroblasts from nine different NBS patients and 41 AT patients were fused and assayed for RDS *(133,134,254)*. These studies defined a total of six complementation groups. All NBS cell lines were grouped into two complementation groups termed Variant 1 (V1) and V2. The V1 group consisted of cell lines from the original NBS patients reported by Weemaes et al. *(251)* and Seemanova et al. *(213)* and one cell line from a patient with AT-Fresno. AT-Fresno is an extremely rare condition in which patients share the clinical features of both NBS and AT *(60)*. The V2 group included both NBS and BBS patients strongly suggesting homogeneity between these two syndromes *(254)*. NBS cells were able to complement all AT cells assayed indicating that AT and NBS arise from distinct mutations *(133,254)*. NBS is still frequently referred to as an AT variant, implying that the disorders are allelic, a conclusion that is not excluded based on the complementation data. However, as summarized below, a preponderance of genetic and cellular data now clearly indicate that NBS and AT result from defects in distinct genes. Stumm et al. *(228)* have re-examined the complementation approach and have assayed radiation-induced chromosome aberrations in fusions between AT and NBS cells. They report noncomplementation in these experiments and suggest that although AT and NBS result from defects in distinct genes, the products of these genes may interact, perhaps in a multisubunit complex or in a signaling cascade.

Biochemical studies of NBS cells reveal defects in cell-cycle checkpoint control similar to those reported in AT cells. Analyses of DNA content by fluorescence-activated cell sorting (FACS) in irradiated and unirradiated NBS cells suggest that they are unable to activate the G1/S checkpoint in response to radiation *(16,138,229)*. After exposure to X-ray doses in the range of 5 Gy, NBS cells fail to halt at the G1/S checkpoint, and at later times postirradiation (24–72 h) a significant accumulation of cells in G2 is observed *(138)*. In normal cells, exposure to ionizing radiation results in the accumulation of the p53 protein via a posttranslational stabilization mechanism. Since p53 regulates the transcription of p21, this increase in p53 protein results in a concomitant rise in the steady-state levels of p21 mRNA. In NBS and AT cell lines (lymphoblastoid or fibroblast), the radiation-induced increase in p53 protein is absent or significantly delayed and the level of mRNA for p21 remains constant *(138)*. The absence of a p53-dependent G1/S checkpoint in NBS cells cannot be easily attributed to a transcriptional effect on *ATM*, the gene defective in AT, because *ATM* transcript levels in both irradiated and unirradiated NBS cells are normal *(138)*. However, effects of the NBS gene product on the ATM protein or vice-versa at the level of protein translation or modification remain a distinct possibility.

7.3. Genetics of NBS

The increasing numbers of NBS families that have been ascertained over the last few years combined with the higher efficiency of PCR-based genome scanning has made it possible to apply genetic linkage approaches to the localization and positional cloning of the gene defective in NBS. The isolation of *ATM* provided significant insight into the nature of the pathways in humans that regulate cellular responses to DNA-damaging agents. Given the similarity in many aspects of the cellular phenotype between AT and NBS, and the clustering of the clinical features of both diseases in the AT-Fresno syndrome, the first important step in identifying an NBS gene was to evaluate the possibility that mutations in *ATM* might also be responsible for NBS. An examination of haplotypes for markers spanning the *ATM* gene at 11q22–23 excludes linkage for a number of NBS families *(111,227)*. No mutations in *ATM* have been detected in the NBS cell lines that have been screened *(44)*. However, AT-Fresno cell lines do have mutations in *ATM (102)*. Lastly, microcell-mediated transfer of a normal chromosome 11 into NBS cells does not complement the phenotype of radiation sensitivity *(150)*.

Saar et al. *(200)* carried out a whole genome screen in a mixed panel of 14 NBS and BBS families of Polish, Czech, and German origin. They reported significant evidence of linkage (maximum lod score = 6.86) to markers on chromosome 8q21. Based on recombination events observed in the families and the reported map distances between markers, these results localized the gene to a 1 c*M* region flanked by the markers D8S270 and D8S271. Within this region, the marker D8S1811 revealed a highly significant allelic association with disease, being present on 86% of mutant chromosomes in NBS families and only 3% of normal chromosomes. These results confirmed the suspected homogeneity of the NBS and BBS disorders, and suggested a founder effect for many of the mutations in this disorder. Genotyping of additional markers in the 8q21 region in NBS families has now refined the map, and provided evidence for a common haplotype of markers present in families with eastern European origins *(44,201)*. The presence of a common founder haplotype may facilitate further fine mapping of the NBS gene by inferring ancestral recombination events at breakpoints in this haplotype in individual patients. Examination of haplotypes in additional NBS patients reveals that many, but not all patients with classical clinical phenotypes and linkage to chromosome 8q21 share this common haplotype *(44,201)*. Thus, some limited allelic heterogeneity is still expected in NBS, which may help to explain some of the reported clinical variation in phenotype *(63,237)*.

One advantage in positional cloning of genes involved in DNA-damage response disorders is the opportunity to use functional complementation in cell lines from patients as an alternative, or adjunct, to genetic linkage approaches. Using complementation of radiation sensitivity as a phenotype, Matsuura et al. *(173)* were able to independently map an NBS gene to chromosome 8. In these studies, individual chromosomes tagged with selectable markers were transferred by microcell fusion into NBS cell lines. Only chromosome 8 was able to complement the radiation sensitivity of the NBS cells. This was true in cells classified as V1 or V2. Further mapping of reduced chromosome 8 hybrids resulting from the random fragmentation that occurs in the microcell transfer procedure localized the complementing gene to the 8q21–24 region.

7.4. Concluding Remarks on NBS

NBS was originally described as a variant of AT because of the significant overlap in the phenotypes of cells and patients with AT and NBS patients. However, genetic studies and complementation studies of both NBS and AT clearly indicate that NBS is a distinct disorder arising from mutations not in the *ATM* gene at chromosome 11q22–23, but in a separate gene located at chromosome 8q21. The similarity in the cellular phenotypes of these two disorders, and some aspects of the clinical features, suggests that the products of the AT and NBS genes will likely act in a common biochemical pathway that connects the effects of ionizing radiation to apoptosis and/or cell-cycle checkpoint control. The finding of noncomplementation for the phenotype of induced chromosome aberrations in heterodikaryons between AT and NBS cells suggests a direct interaction between ATM and the putative product of the NBS gene. Isolation of the NBS gene should help to clarify this relationship.

The majority of NBS patients have a common marker haplotype in the 8q21 region where the NBS gene has been localized *(44,201)*. This haplotype is present in virtually all patients of eastern European ancestry who constitute the majority of patients identified to date. However, there are some NBS families in which patients appear to be compound heterozygotes, sharing the common founder haplotype and a second unique haplotype as well as at least one patient with a typical NBS phenotype who does not carry even a single copy of this haplotype *(44,60,117)*. Although there is some variability reported in the clinical findings in NBS patients, it is not yet possible to conclude whether these haplotypic differences correspond to specific phenotypic differences.

An additional question is the nature of the genetic defects in patients with AT-Fresno. These patients display all of the clinical and cellular features of both AT and NBS. All of the known AT-Fresno patients have mutations in the *ATM* gene but there is no consistent molecular pattern in the mutations observed in these patients *(44,102)*. Whether they will also carry NBS mutations awaits the cloning of the NBS gene. Thus far, none of the AT-Fresno patients that have been examined share the common NBS founder haplotype *(44)*.

The region of chromosome 8q21 where the gene for NBS has been mapped contains no obvious candidates for a gene whose product might be involved in responses to DNA damage resulting from exposure to ionizing radiation. However, the region appears to be rich in expressed sequence tags (ESTs) mapped by radiation hybrid analysis. The availability of cDNA clones corresponding to these mapped ESTs should facilitate mutation screening as part of positional cloning strategy for identifying the NBS gene once a minimum region containing the gene can be defined. Further, a singly linked YAC contig covers the entire region from the flanking markers D8S270 to D8S271. With appropriate modification of these YACs to allow selection in mammalian hosts, it may be possible to localize the NBS gene by a complementation approach using YAC clones. Thus, the prospects are good for the eventual isolation and identification of a gene for NBS by either genetic or functional approaches. Besides explaining the underlying cause for this intriguing disorder, the availability of the NBS gene should further our understanding of cellular responses to ionizing radiation.

REFERENCES

1. Allen, J. B., Z. Zhou, W. Siede, E. C. Friedberg, and S. J. Elledge. 1994. The SAD1/RAD53 protein kinase controls multiple checkpoints and DNA damage-induced transcription in yeast. *Genes. Dev.* **8:** 2401–2415.
2. Alter, B. P. 1993. Fanconi's anaemia and its variability. *Br. J. Haematol.* **85:** 9.
3. Alter, B. P. 1996. Fanconi's anemia and malignancies. *Am. J. Hematol.* **53:** 99–110.
4. Alter, B. P., M. E. Knobloch, and R. S. Weinberg. 1991. Erythropoiesis in Fanconi's anemia. *Blood* **78:** 602–608.
5. Alter, B. P., and N. S. Young. 1993. The bone marrow failure syndromes, in *Hematology of Infancy and Childhood*, 4th ed., vol. 1. (Nathan, D. G. and F. A. Oski, eds.), Saunders, Philadelphia, PA, pp. 216–316.
6. Anderson, C. W. and T. H. Carter. 1996. The DNA-activated protein-kinase—DNA-PK, in *Current Topics in Microbiology and Immunology, vol. 217: Molecular Analysis of DNA Rearrangements in the Immune System* (Jessberger, R. and M. R. Lieber, eds.), Springer-Verlag, Berlin, pp. 91–111.
7. Apostolou, S., S. A. Whitmore, J. Crawford, et al. 1996. Positional cloning of the Fanconi anaemia group A gene. *Nat. Genet.* **14:** 324–328.
8. Arlett, C. F. and S. A. Harcourt. 1980. Survey of radiosensitivity in a variety of human cell strains. *Cancer Res.* **40:** 926–932.
9. Auerbach, A. D. 1993. Fanconi anemia diagnosis and the diepoxybutane (DEB) test. *Exp. Hemat.* **21:** 731–733.
10. Auerbach, A. D. 1995. Fanconi anemia. *Dermatol. Clin.* **13:** 41.
11. Auerbach, A. D. and B. P. Alter. 1989. Prenatal and postnatal diagnosis of aplastic anemia, in *Methods in Hematology: Perinatal Hematology* (Alter, B. P., ed.), Churchill Livingstone, Edinburgh, p. 225.
12. Auerbach, A. D., A. Rogatko, and T. M. Schroeder-Kurth. 1989. International Fanconi anemia registry: relation of clinical symptoms to diepoxybutane sensitivity. *Blood* **73:** 391.

13. Auerbach, A. D., M. A. Weiner, K. Warburton, L. L. Yeboa, and H. E. Broxmeyer. 1982. Acute myeloid leukemia as the first hematologic manifestation of Fanconi Anemia. *Am. J. Hemat.* **12:** 289–300.
14. Austin, C. A., J. H. Sang, S. Patel, and L. M. Fisher. 1993. Novel HeLa topoisomerase II is the II beta isoform: complete coding sequence and homology with other type II topoisomerases. *Biochim. Biophys. Acta* **1172:** 283–291.
15. Badie, C., G. Iliakis, N. Foray, G. Alsbeih, G. E. Pantellias, R. Okayasu, N. Cheong, N. S. Russell, A. C. Begg, C. F. Arlett, and E. P. Malaise. 1995. Defective repair of DNA double-strand breaks and chromosome damage in fibroblasts from a radiosensitive leukemia patient. *Cancer Res.* **55:** 1232–1234.
16. Barbi, G., J. M. J. C. Scheres, D. Schindler, R. D. F. M. Taalman, K. Rodens, K. Mehnert, M. Müller, and H. Seyschab. 1991. Chromosome instability and X-ray hypersensitivity in a microcephalic and growth-retarded child. *Am. J. Med. Genet.* **40:** 44–50.
17. Barlow, C., S. Hirotsume, R. Paylor, M. Liyanage, M. Eckhaus, F. Collins, Y. Shiloh, J. N. Crowley, T. Ried, D. Tagle, and A. Wynshaw-Boris. 1996. ATM-deficient mice: a paradigm of ataxia telangiectasia. *Cell* **86:** 159–171.
18. Barlow, C., K. D. Brown, C.-X. Deng, D. A. Tagle, and A. Wynshaw-Boris. 1997. Atm selectively regulates distinct p53-dependent cell-cycle checkpoint and apoptotic pathways. *Nat. Genet.* **17:** 453–456.
19. Barlow, C, M. Liyanage, P. B. Moens, C.-X. Deng, T. Ried, and A. Wynshaw-Boris. 1997. Partial rescue of the prophase I defects of *Atm*-deficient mice by *p53* and *p21* null alleles. *Nat. Genet.* **17:** 462–466.
20. Barnes, G. and D. Rio. 1997. DNA double-strand-break sensitivity, DNA replication, and cell cycle arrest phenotypes of Ku-deficient *Saccharomyces cerevisiae. Proc. Natl. Acad. Sci. USA* **94:** 867–872.
21. Baskaran, R., L. D. Wood, L. L. Whitaker, C. E. Canman, S. E. Morgan, Y. Zu, C. Barlow, D. Baltimore, A. Wynshaw-Boris, M. B. Kastan, and J. Y. J. Wang. 1997. Ataxia-telangiectasia mutated gene product activates c-Abl tyrosine kinase in response to ionizing radiation. *Nature* **387:** 516–520.
22. Bauer, E. A., N. Silverman, D. F. Busiek, A. Kronberger, and T. F. Deuel. 1986. Diminished response of Werner's syndrome fibroblasts to growth factors PDGF and FGF. *Science* **234:** 1240–1243.
23. Beamish, H. and M. F. Lavin. 1994. Radiosensitivity in ataxia-telangiectasia: anomalies in radiation-induced cell cycle delay. *Int. J. Radiat. Biol.* **65:** 175–184.
24. Beamish, H., R. Williams, P. Chen, and M. F. Lavin. 1996. Defect in multiple cell cycle checkpoints in ataxia-telangiectasia post-irradiation. *J. Biol. Chem.* **271:** 20,486–20,493.
25. Bentley, N. J., D. A. Holtzman, G. Flaggs, K. S. Keegan, A. DeMaggio, J. C. Ford, M. Hoekstra, and A. M. Carr. 1996. The *S. pombe rad3* checkpoint gene. *EMBO J.* **15:** 6641–6651.
26. Beretta, L., A.-C. Gingras, Y. V. Soitkin, M. N. Hall, and N. Sonenburg. 1996. Rapamycin blocks the phosphorylation of 4E-BP1 and inhibits cap-dependent initiation of translation. *EMBO J.* **15:** 658–664.
27. Biedermann, K. A., J. Sun, A. J. Giaccia, T. M. Tosto, and J. M. Brown. 1991. Scid mutation in mice confers hypersensitivity to ionizing radiation and a deficiency in DNA double-strand break repair. *Proc. Natl. Acad. Sci. USA* **88:** 1394–1397.
28. Bloom, D. 1954. Congenital telangiectatic syndrome erythema resembling lupus erythematosus in dwarfs. *Am. J. Dis. Child.* **88:** 754–758.
29. Blunt, T., N. J. Finnie, G. E. Taccioli, G. C. M. Smith, J. Demengeot, T. M. Gottlieb, R. Mizuta, A. J. Varghese, F. W. Alt, P. A. Jeggo, and S. P. Jackson. 1995. Defective DNA-dependent protein kinase activity is linked to V(D)J recombination and DNA repair defects associated with the murine scid mutation. *Cell* **80:** 813–823.

30. Boulton, S. J. and S. P. Jackson. 1996. Identification of a *Saccharomyces cerevisiae* Ku80 homologue: roles in DNA double-strand break rejoining and in telomeric maintenance. *Nucleic Acids Res.* **23:** 4639–4648.
31. Brainard, E., L. B. K. Herzing, J. Bainton, and M. S. Meyn. 1991. A common feature of the chromosome instability syndromes: increased spontaneous intrachromosomal mitotic recombination. *Am. J. Hum. Genet.* 49A: 449.
32. Brown, E. J., M. W. Albers, T. B. Shin, K. Ichikawa, C. T. Keith, W. S. Lane, and S. L. Schreiber. 1994. A mammalian protein targeted by G1-arresting rapamycin-receptor complex. *Nature* **369:** 756–758.
33. Brown, E. J., P. A. Beal, C. T. Keith, J. Chen, T. B. Shin, and S. L. Schreiber. 1995. Control of p70 56 kinase activity of FRAP in vivo. *Nature* **377:** 442–446.
34. Brown, E. J. and S. L. Schreiber. 1996. A signaling pathway to translational control. *Cell.* **86:** 517–520.
35. Brown, K. D., Y. Ziv, S. N. Sadanandan, L. Chessa, F. S. Collins, Y. Shiloh, Y., and D. A. Tagle. 1997. The ataxia-telangiectasia gene product, a constitutively expressed nuclear protein that is not up-regulated following genome damage. *Proc. Natl. Acad. Sci. USA* **94:** 1840–1845.
36. Broxmeyer, H. E., G. W. Douglas, G. Hangoc, S. Cooper, J. Bard, D. English, M. Arny, L. Thomas, and E. A. Boyse. 1989. Human umbilical cord blood as a potential source of transplantable hematopoietic stem/progenitor cells. *Proc. Natl. Acad. Sci. USA* **86:** 3828–3832.
37. Broxmeyer, H. E., J. Kurtzberg, E. Gluckman, et al. 1991. Umbilical cord blood hematopoietic stem and repopulating cells in human clinical transplantation. *Blood* **17:** 313–329.
38. Brunn, G. J., J. Williams, C. Saber, G. Wiederrecht, J. C. Lawrence, Jr., and R. T. Abraham. 1996. Direct inhibition of the signaling functions of the mammalian target of rapamycin by the phosphoinositide 3-kinase inhibitors, wortmannin and LY294002. *EMBO J.* **15:** 5256–5267.
39. Buchwald, M. 1995. Complementation groups: one or more per gene. *Nat. Genet.* **11:** 228–230.
40. Butturini, A., R. P. Gale, P. C. Verlander, B. Adler-Brecher, A. Gillio, and A. D. Auerbach. 1994. Hematologic abnormalities in Fanconi anemia. An international Fanconi anemia registry study. *Blood* **84:** 1650–1655.
41. Byrd, P. J., C. M. McConville, P. Cooper, J. Parkhill, T. Stankovic, G. M. McGuire, J. A. Thick, and A. M. R. Taylor. 1996. Mutations revealed by sequencing the 5' half of the gene for ataxia- telangiectasia. *Hum. Mol. Genet.* **5:** 145–149.
42. Carpenter, C. L. and L. C. Cantley. 1990. Phosphoinositide kinases. *Biochemistry* **29:** 11,147–11,156.
43. Carr, A. M. and M. F. Hoekstra. 1995. The cellular response to DNA damage. *Trends Cell Biol.* **5:** 32–40.
44. Cerosaletti, K. M., E. Lange, H. M. Stringham, C. M. R. Weemaes, D. Smeets, B. Solder, B. H. Belohradsky, A. M. R. Taylor, P. Karnes, A. Elliott, K. Komatsu, R. A. Gatti, M. Boehnke, and P. Concannon. 1998. Fine localization of the Nijmegen Breakage Syndrome gene at 8q21: evidence for a common founder haplotype. *Am. J. Hum. Genet.*, in press.
45. Chan, J. Y. H., F. F. Becker, and J. German. 1987. Altered DNA ligase I activity in Bloom's syndrome cells. *Nature* **325:** 357–359.
46. Chen, P. C., M. F. Lavin, C. Kidson, and D. Moss. 1978. Identification of ataxia telangiectasia heterozygotes, a cancer prone population. *Nature* **274:** 484–486.
47. Chen, P., A. Farrell, K. Hobson, A. Girjes, and M. F. Lavin. 1994. Comparative study of radiation-induced G2 phase delay and chromatid damage in families with ataxia-telangiectasia. *Cancer Genet. Cytogenet.* **76:** 43–46.
48. Chen, G. and E. Y.-H. P. Lee. 1996. The product of the *ATM* gene is a 370-kDa nuclear phosphoprotein. *J. Biol. Chem.* **271:** 33,693–33,697.

49. Chen M., D. J. Tomkins, W. Auerbach, C. McKerlie, H. Youssoufian, L. Liu, O. Gan, M. Carreau, A. Auerbach, T. Groves, C. J. Guidos, M. H. Freeman, J. Cross, D. H. Percy, J. E. Dick, A. L. Joyner, and M. Buchwald. 1996. Inactivation of Fac in mice produces inducible chromosomal instability and reduced fertility reminiscent of Fanconi anaemia. *Nat. Genet.* **12:** 448–451.
50. Chessa, L., P. Petrinelli, P. A. Antonelli, M. Fiorelli, R. Elli, L. Marcucci, A. Federico, and E. Gandini. 1992. Heterogeneity in Ataxia-telangiectasia: classical phenotype associated with intermediate cellular radiosensitivity. *Am. J. Med. Genet.* **42:** 741–746.
51. Chiu, M. I., H. Katz, and V. Berlin. 1994. RAPT, a mammalian homolog of yeast Tor, interacts with the FKBP12/rapamycin complex. *Proc. Natl. Acad. Sci. USA* **91:** 12,574–12,578.
52. Chrzanowska, K. H., W. J. Kleijer, M. Krajewska-Walasek, M. Bialecka, A. Gutkowska, B. Goryluk-Kozakiewicz, J. Michalkiewicz, J. Stachowski, H. Gregorek, G. Lyson-Wojciechowska, W. Janowicz, and S. Jozwiak. 1995. Eleven Polish patients with microcephaly, immunodeficiency and chromosomal instability: the Nijmegen breakage syndrome. *Am. J. Med. Genet.* **57:** 462–471.
53. Cimprich, K. A., T. B. Shin, C. T. Keith, and S. L. Schreiber. 1996. cDNA cloning and gene mapping of a candidate human cell cycle checkpoint protein. *Proc. Natl. Acad. Sci. USA*. **93:** 2850–2855.
54. Cohen, M. M. and H. P. Levy. 1989. Chromosome instability syndromes. Adv. *Hum. Genet.* **18:** 43–149.
55. Cohen, M. M., M. Shaham, J. R. K. Dagan, E. Shmueli, and G. Kohn. 1975. Cytogenetic investigations in families with ataxia-telangiectasia. *Cytogenet. Cell. Genet.* **15:** 338–356.
56. Conley, M. E., M. B. Spinner, B. S. Emanuel, P. C. Nowell, and W. W. Nichols. 1986. A chromosome breakage syndrome with profound immunodeficiency. *Blood* **67:** 1251–1256.
57. Coquerelle, T. M., K. F. Weibezahn, and C. Lucke-Huhle. 1987. Rejoining of double strand breaks in normal human and ataxia-telangiectasia fibroblasts after exposure to ^{60}Co γ-rays, ^{241}Am α-particles or bleomycin. *Int. J. Radiat. Biol.* **51:** 209–211.
58. Cornforth, M. W. and J. S. Bedford. 1985. On the nature of a defect in cells from individuals with ataxia-telangiectasia. *Science* **227:** 1589–1591.
59. Costa, N. D. and J. Thacker. 1993. Response of radiation-sensitive human cells to defined DNA breaks. *Int. J. Radiat. Biol.* **64:** 523–529.
60. Curry, C. J. R., J. Tsai, H. T. Hutchinson, N. G. J. Jaspers, D. Wara, and R. A. Gatti. 1989. AT(Fresno): a phenotype linking ataxia-telangiectasia with the Nijmegen breakage syndrome. *Am. J. Hum. Genet.* **45:** 270–275.
61. D'Andrea, A. D. and M. Grompe. 1997. Molecular biology of Fanconi anemia: implications for diagnosis and therapy. *Blood* **90:** 1725–1736.
62. Davies, S. M., S. Khan, J. E. Wagner, D. C. Arthur, A. D. Auerbach, N. K. C. Ramsay, and D. J. Weisdorf. 1996. Unrelated donor bone marrow transplantation for Fanconi anemia. *Bone Marrow Transplant.* **17:** 43–47.
63. Der Kaloustian, V. M., W. Kleijer, A. Booth, A. D. Auerbach, B. Mazer, A. M. Elliott, S. Abish, R. Usher, G. Watters, M. Vekemans, and P. Eydoux. 1996. Possible new variant of Nijmegen breakage syndrome. *Am. J. Med. Genet.* **65:** 21–26.
64. Dokal, I., A. Chase, N. V. Morgan, S. Coulthard, G. Hall, C. G. Mathew, and I. Roberts. 1996. Positive diepoxybutane test in only one of two brothers found to be compound heterozygotes for Fanconi's anaemia complementation group C mutations. *Br. J. Haematol.* **93:** 813–816.
65. dos Santos, C. C., H. Gavish, and M. Buchwald. 1994. Fanconi anemia revisited: old ideas and new advances. *Stem Cells* **12:** 142.
66. Elledge, S. J. 1996. Cell cycle checkpoints: preventing an identity crisis. *Science* **274:** 1664–1672.

67. Ellis, N. A., J. Groden, T-Z. Ye, J. Straughen, D. J. Lennon, S. Ciocci, M. Proytcheva, and J. German. 1995. Bloom's syndrome gene product is homologous to recQ helicases. *Cell* **83:** 655–666.
68. Ellis, N. A., D. J. Lennon, M. Proytcheva, B. Alhadeff, E. E. Henderson, and J. German. 1995. Somatic intragenic recombination within the mutated locus BLM can correct the high sister-chromatid exchange phenotype of Bloom syndrome cells. *Am. J. Hum. Genet.* **57:** 1019–1027.
69. Elson, A., Y. Wang, C. J. Daugherty, C. C. Morton, F. Zhou, J. Campos-Torres, et al. 1996. Pleiotropic defects in ataxia-telangiectasia protein-deficient mice. *Proc. Natl. Acad. Sci. USA* **93:** 13,084–13,089.
70. Epstein, C. J., G. M. Martin, A. L. Schultz, A. G. Motulsky. 1966. Werner's syndrome: a review of its symptomatology, natural history, pathologic features, genetics and relationship to the natural aging process. *Medicine* **45:** 177–222.
71. Eshlemean, J. and S. Markowitz. 1996. Mismatch repair defects in human carcinogenesis. *Hum. Mol. Genet.* **5:** 489–494.
72. Evans, M. K. and V. A. Bohr. 1994. Gene-specific DNA repair of UV-induced cyclobutane pyrimidine dimers in some cancer-prone and premature-aging human syndromes. *Mutat. Res.* **314:** 221–231.
73. Feigin, R. D., T. J. Vietti, R. G. Wyatt, D. G. Kaufmann, and C. H. Smith, Jr. 1970. Ataxia-telangiectasia with granulocytopenia. *J. Pediatr.* **77:** 431–438.
74. Fiorilli, M., A. Antonelli, G. Russo, M. Crescenzi, M. Carbonary, and P. Petrinelli. 1985. Variant of ataxia-telangiectasia with low level radiosensitivity. *Hum. Genet.* **70:** 274–277.
75. Feldmann, H. and E. L. Winnacker. 1993. A putative homologue of the human autoantigen Ku from *Saccharomyces cerevisiae. J. Biol. Chem.* **268:** 12,885–12,900.
76. Flaggs, G., A. Plug, K. M. Dunks, K. E. Mundt, J. C. Ford, M. R. E. Quiggle, E. M. Taylor, C. H. Westphal, T. Ashley, M. F. Hoekstra, and A. M. Carr. 1997. Atm-dependent interactions of a mammalian Chk1 homolog with meiotic chromosomes. *Curr. Biol.* **7:** 977–986.
77. Foray, N., C. F. Arlett, and E. P. Malaise. 1995. Dose-rate effect on induction and repair rate of radiation-induced DNA double-strand breaks in a normal and an ataxia-telangiectasia human fibroblast cell line. *Biochimie* **77:** 900–905.
78. Fornace, A. J., Jr. and J. B. Little. 1980. Normal repair of DNA single-strand breaks in patients with ataxia-telangiectasia. *Biochim. Biophys. Acta* **607:** 432–437.
79. Fried, L. M., C. Koumeris, S. R. Peterson, S. L. Green, P. Van Zijl, J. Allalunis-Turner, D. J. Chen, R. Fishel, A. J. Giaccia, J. M. Brown, and C. U. Kirchgessner. 1996. The DNA damage response in DNA-dependent protein kinase-deficient SCID mouse cells: replication protein A hyperphosphorylation and p53 induction. *Proc. Natl. Acad. Sci. USA* **93:** 13,825–13,830.
80. Friedberg, E. 1996. Cockayne Syndrome—a primary defect in DNA repair, transcription, both, or neither. *Bioessays* **18(9):** 731–738.
81. Friedberg, E. 1996. Relationships between DNA repair and transcription. *Ann. Rev. Biochem.* **65:** 15–42.
82. Fritz, E., S. H. Elsea, P. I. Patel, and M. S. Meyn. 1997. Overexpression of a truncated human topoisomerase III partially corrects multiple aspects of the ataxia-telangiectasia phenotype. *Proc. Natl. Acad. Sci. USA* **94:** 4538–4542.
83. Fukuchi, K., G. M. Martin, and R. J. Monnat, Jr. 1989. Mutator phenotype of Werner syndrome is characterized by extensive deletions. *Proc. Natl. Acad. Sci. USA* **86:** 5893–5897.
84. Fukuchi, K., K. Tanaka, Y. Kumahara, K. Marumo, M. B. Pride, G. M. Martin, and R. J. Monnat, Jr. 1990. Increased frequency of 6-thioguanine-resistant peripheral blood lymphocytes in Werner syndrome patients. *Hum. Genet.* **84:** 249–252.

85. Furnari, B., N. Rhind, and P. Russell. 1997. Cdc25 mitotic inducer targeted by Chk1 DNA damage checkpoint kinase. *Science* **277:** 1495–1497.
86. Gangloff, S., J. P. McDonald, C. Bendixen, L. Arthur, and R. Rothstein. 1994. The yeast type 1 topoisomerase Top3 interacts with SGS1, a DNA helicase homolog: a potential eukaryotic reverse gyrase. *Mol. Cell. Biol.* **14:** 8391–8398.
87. Gatti, R. A., I. Berkel, E. Boder, G. Braedt, P. Charrmley, P. Concannon, F. Ersoy, T. Foroud, N. G. J. Jaspers, K. Lange, G. M. Lathrop, M. Leppert, Y. Nakamura, P. O'Connel, M. Paterson, W. Salser, O. Sanal, J. Silver, R. S. Sparkes, E. Susi, D. E. Weeks, S. Wei, R. White, and F. Yoder. 1988. Localization of an ataxia-telangiectasia gene to chromosome 11q22–23. *Nature* **336:** 577–580.
88. Gatti, R. A. 1998. Ataxia-telangiectasia, in *The Genetic Basis of Human Cancer* (Vogelstein, B. and K. W. Kinzler, eds.), McGraw-Hill, New York, pp. 275–300.
89. Gebhart, E., R. Bauer, U. Raub, M. Schinzel, K. W. Ruprecht, and J. B. Jonas. 1988. Spontaneous and induced chromosomal instability in Werner syndrome. *Hum. Genet.* **80:** 135–139.
90. German, J. 1964. Cytological evidence for crossing-over in vitro in human lymphoid cells. *Science* **144:** 298–301.
91. German, J. 1969. Bloom syndrome. I. Genetic and clinical observations in the first twenty seven patients. *Am. J. Hum. Genet.* **21:** 196–227.
92. German, J. 1993. Bloom syndrome: a Mendelian prototype of somatic mutational disease. *Medicine (Baltimore)* **72:** 393–406.
93. German, J. 1995. Bloom's syndrome. *Dermatol. Clin.* **13:** 7–18.
94. German, J., R. Archibald, and D. Bloom. 1965. Chromosomal breakage in a rare and probably genetically determined syndrome of man. *Science* **148:** 506,507.
95. German, J., S. Schonberg, E. Louie, and R. S. K. Chaganti. 1977. Bloom's syndrome. IV. Sister chromatid exchanges in lymohocytes. *Am. J. Hum. Genet.* **29:** 248–255.
96. German, J., A. M. Roe, M. F. Leppert, and N. A. Ellis. 1994. Bloom syndrome: an analysis of consanguineous families assigns the locus mutated to chromosome band 15q26. 1. *Proc. Natl. Acad. Sci. USA* **91:** 6669–6673.
97. Giampietro, P. F., B. Adler-Brecher, P. C. Verlander, S. G. Pavlakis, J. G. Davis, and A. D. Auerbach. 1993. The need for more accurate and timely diagnosis in Fanconi anemia. A report from the International Fanconi Anemia Registry. *Pediatrics* **91:** 1116–1120.
98. Giampietro, P. F., P. C. Verlander, J. G. Davis, and A. D. Auerbach. 1997. Diagnosis of Fanconi anemia in patients without congenital malformations: an International Fanconi Anemia Registry study. *Am. J. Med. Genet.* **68:** 58–61.
99. Gibson, R. A., M. Buchwald, R. G. Roberts, and C. G. Mathew. 1993. Characterisation of the exon structure of the Fanconi anaemia group C gene by vectorette PCR. *Hum. Mol. Genet.* **2:** 35–38.
100. Gibson, R. A., A. Hajkanpour, M. Murer-Orlando, M. Buchwald, and G. C. Mathew. 1993. A nonsense mutation and exon skipping in the Fanconi anemia group C gene. *Hum. Mol. Genet.* **2:** 797–799.
101. Gibson, R. A., N. V. Morgan, L. H. Goldstein, I. C. Pearson, I. P. Kesterton, M. J. Foot, S. Jansen, C. Havenga, T. Pearson, T. J. de Ravel, R. J. Cohn, I. M. Marques, I. Dokal, I. Roberts, J. Marsh, S. Ball, R. D. Milner, J. C. Llerena, E. Samochatova, S. P. Mohan, P. Vasudevan, F. Birjandi, A. Hajianpour, M. Murer-Orlando, and C. G. Mathew. 1996. Novel mutations and polymorphisms in the Fanconi anaemia group C gene. *Hum. Mutat.* **8:** 140–148.
102. Gilad, S., L. Chessa, R. Khosravi, P. Russell, Y. Galanty, M. Piane, R. A. Gatti, T. J. Jorgensen, Y. Shiloh, and A. Bar-Shira. 1997. Genotype-phenotype relationships in ataxia-telangiectasia (AT) and AT variants. *Am. J. Hum. Genet.* **62:** 551–562.
103. Gilad, S., R. Khosravi, D. Shkedy, T. Uziel, Y. Ziv, K. Savitsky, G. Rotman, S. Smith, L. Chessa, T. J. Jorgensen, R. Harnik, M. Frydman, O. Sanal, S. Portnoi, Z. Goldwicz,

N. G. J. Jaspers, R. A. Gatti, G. Lenoir, M. F. Lavin, K. Tatsuni, R. D. Wegner, Y. Shiloh, and A. Bar-Shira. 1996. Predominance of null mutations in ataxia-telangiectasia. *Hum. Mol. Genet.* **5:** 433–440.
104. Goto, M., O. Imamura, J. Kuromitsu, T. Matsumoto, Y. Yamabe, Y. Tokutake, N. Suzuki, B. Mason, D. Drayna, M. Sugawara, M. Sugimoto, and Y. Furuichi. 1997. Analysis of helicase gene mutations in Japanese Werner's syndrome patients. *Hum. Genet.* **99:** 191–193.
105. Goto M., R. W. Miller, Y. Ishikawa, and H. Sugano. 1996. Excess of rare cancers in Werner syndrome (adult progeria). *Cancer Epidemiol. Biomarkers Prevent.* **5:** 239–246.
106. Goto, M., M. Rubenstein, J. Weber, K. Woods, and D. Drayna. 1992. Genetic linkage of Werner's syndrome to five markers on chromosome 8. *Nature* **355:** 735–738.
107. Goto, M., K. Tanimoto, Y. Horiuchi, and T. Sasazuki. 1981. Family analysis of Werner's syndrome: a survey of 42 Japanese families with a review of the literature. *Clin. Genet.* **19:** 8–15.
108. Gotoff, S. P., E. Amirmokri, and E. J. Liebner. 1967. Ataxia-telangiectasia. Neoplasia, untoward response to X-irradiation, and tuberous sclerosis. *Am. J. Dis. Child.* **114:** 617–625.
109. Gottlieb, T. M. and S. P. Jackson. 1994. Protein kinases and DNA damage. *Trends Biochem. Sci.* **19:** 500–503.
110. Gray, M. D., J.-C. Shen, A. S. Kamath-Loeb, A. Blank, B. L. Sopher, G. M. Martin, J. Oshima, and L. A. Loeb. 1997. The Werner syndrome protein is a DNA helicase. *Nat. Genet.* **17:** 100–103.
111. Green, A. J., J. R. W. Yates, A. M. R. Taylor, P. Biggs, G. M. McGuire, C. M. McConville, C. J. Billing, and N. D. Barnes. 1995. Severe microcephaly with normal intellectual development: the Nijmegen breakage syndrome. *Arch. Dis. Childhood* **73:** 431–434.
112. Groden, J., Y. Nakamura, and J. German. 1990. Molecular evidence that homologous recombination occurs in proliferating human somatic cells. *Proc. Natl. Acad. Sci. USA* **87:** 4315–4319.
113. Gropp, A. and G. Flatz. 1967. Chromsome breakage and blastic transformation of lymphocytes in ataxia-telangiectasia. *Hum. Genet.* **5:** 77–79.
114. Guidos, C. J., C. J. Williams, I. Grandal, G. Knowles, T. F. Manley, L.-C. Huang, and J. S. Danska. 1996. V(D)J recombination activates a p53-dependent DNA damage checkpoint in scid lymphocyte precursors. *Genes Dev.* **10:** 2038–2054.
115. Guinan, E. C., K. D. Lopez, R. D. Huhn, J. M. Felser, and D. G. Nathan. 1994. Evaluation of granulocyte-macrophage colony-stimulating factor for treatment of pancytopenia in children with Fanconi anemia. *J. Pediatr.* **124:** 144.
116. Hari, K. L., A. Santerre, J. J. Sekelsky, K. S. McKim, J. B. Boyd, and R. S. Hawley. 1995. The *mei-4l* gene of Drosophila melanogaster is functionally homologous to the human ataxia telangiectasia gene. *Cell* **82:** 815–821.
117. Hartley, K. O., D. Gell, G. C. M. Smith, H. Zhang, N. Divecha, M. A. Connelly, A. Admon, S. P. Lees-Miller, C. W. Anderson, and S. P. Jackson. 1995. DNA-dependent protein kinase catalytic subunit: a relative of phosphatidyl-inositol 3-kinase and the ataxia telangiectasia gene product. *Cell* **82:** 849–856.
118. Hawley, R. S. and S. H. Friend. 1996. Strange bedfellows in even stranger places: the role of ATM in meiotic cells, lymphocytes, tumors, and its functional links to p53. *Genes Dev.* **10:** 2383–2388.
119. Heartlein, M. W., H. Tsuji, and S. A. Latt. 1987. 5-Bromodeoxyuridine-dependent increase in sister chromatid exchange formation in Bloom's syndrome is associated with reduction in topoisomerase II activity. *Exp. Cell Res.* **169:** 245–254.
120. Hecht, F., B. K. McCaw, and R. Koler. 1973. Ataxia-telangiectasia-clonal growth of translocation lymphocytes. *N. Engl. J. Med.* **289:** 286–291.

121. Hecht, F. and B. Kaiser-McCaw. 1982. Ataxia-telangiectasia: genetics and heterogeneity, in *Ataxia-Telangiectasia* (Bridges, B. A. and D. G. Harnden, eds.), Wiley, New York, pp. 197–201.
122. Hecht, F., R. D. Koler, D. A. Rigas, G. S. Dahnke, M. P. Case, V. Tisdale, and R. W. Miller. 1996. Leukaemia and lymphocytes in ataxia-telangiectasia. *Lancet* **2:** 1193.
123. Heitman, J., N. R. Movva, and M. N. Hall. 1991. Targets for cell cycle arrest by the immunosuppressant rapamycin in yeast. *Science* **253:** 905–909.
124. Heneveld, M., G. Faulkner, G. Royle, M. Heinrich, W. Keeble, R. Kahn, M. K. R. Grompe, and G. C. Bagby. 1996. The pathogenesis of bone marrow failure in Fanconi anemia: the role of IFNγ-induced FAS expression. *Blood* **88:** 1346.
125. Higurashi, M. and P. E. Cohen. 1973. *In vitro* chromosomal radiosensitivity in chromosomal breakage syndromes. *Cancer* **32:** 380–383.
126. Hoehn, H., Bryant, E. M., Au, K., Norwood, T. H., Boman, H., and Martin, G. M. 1975. Variegated translocation mosaicism in human skin fibroblast cultures. *Cytogenet. Cell Genet.* **15:** 282–298.
127. Hojo, E. T., P. C. van Diemen, F. Darroudi, and A. T. Natarajan. 1995. Spontaneous chromosomal aberrations in Fanconi anaemia, ataxia telangiectasia fibroblast and Bloom's syndrome lymphoblastoid cell lines as detected by conventional cytogenetic analysis and fluorescence in situ hybridisation (FISH) technique (published erratum appears in *Mutat. Res.* **334:** 268–270). *Mutat. Res.* **334:** 59–69.
128. Huang, L.-C., K. C. Clarkin, and G. M. Wahl. 1996. p53-dependent cell cycle arrests are preserved in DNA-activated protein kinase-deficient mouse fibroblasts. *Cancer Res.* **56:** 2940–2944.
129. Ianzano, L.,M. D'Apolito, M. Centra, M. Savino, O. Levran, A. D. Auerbach, A.-M. Cleton-Jansen, N. A. Doggett, J. C. Pronk, A. J. Tipping, R. A. Gibso, C. G. Mathew, S. A. Whitmore, S. Apostolou, D. F. Callen, L. Zelante, and A. Savoia. 1997. The genomic organization of the Fanconi Anemia Group A (FAA) gene. *Genomics* **41:** 309–314.
130. Ishizaki, K., T. Yagi, M. Inoue, O. Nikaido, and H. Takebe. 1981. DNA repair in Bloom's syndrome fibroblasts after UV irradiation or treatment with mitomycin C. *Mutat. Res.* **80:** 213–219.
131. Jackson, S. P. 1995. Ataxia-telangiectasia at the crossroads. *Curr. Biol.* **5:** 1210–1212.
132. Jaspers, N. G. and D. Bootsma. 1982. Genetic heterogeneity in ataxia-telangiectasia studied by cell fusion. *Proc. Natl. Acad. Sci. USA* **79:** 2641–2644.
133. Jaspers, N. G., R. D. Taalman, and C. Baan. 1988. Patients with an inherited syndrome characterized by immunodeficiency, microcephaly, and chromosomal instability; genetic relationship to ataxia-telangiectasia. *Am. J. Hum. Genet.* **42:** 66–73.
134. Jaspers, N. G., R. A. Gatti, C. Baan, P. C. Linssen, and D. Bootma. 1988. Genetic complementation analysis of ataxia-telangiectasia and Nijmegen breakage syndrome: a survey of 50 patients. *Cytogenet. Cell Genet.* **49:** 259–263.
135. Joenje, H. 1996. Fanconi anemia complementation groups in Germany and the Netherlands. *Hum. Genet.* **97:** 280–282.
136. Joenje, H., A. B. Oostra, M. Wijker, F. M. di Summa, C. G. M. van Berkel, M. A. Rooimans, W. Ebell, M. van Weel, J. C. Pronk, M. Buchwald, and F. Arwert. 1997. Evidence for at least eight Fanconi anemia genes. *Am. J. Hum. Genet.* **61:** 940–944.
137. Joenje, H., F. L. Ten, A. Oostra, C. V. Berkel, M. Rooimans, S. Schroeder, T. Kurth, R. Wegner, J. Gille, M. Buchwald, and F. Arwert. 1995. Classification of Fanconi anemia patients by complementation analysis: evidence for a fifth genetic subtype. *Blood* **86:** 2156.
138. Jongmans, W., M. Vuillaume, K. Chrzanowska, D. Smeets, K. Sperling, and J. Hall. 1997. Nijmegen breakage syndrome cells fail to induce the p53-mediated DNA damage response following exposure to ionizing radiation. *Mol. Cell. Biol.* **17:** 5016–5022.
139. Jung, M., A. Kondratyev, S. Lee, A. Dimtchev, and A. Dritschilo. 1997. ATM gene product phosphorylates IkB. *Cancer Res.* **57:** 24–27.

140. Kaiser, T. N., A. Lojewski, C. Dougherty, L. Juergens, E. Sahar, and S. A. Latt. 1982. Flow cytometric characterization of the response of Fanconi's Anemia cells to mitomycin C treatment. *Cytometry* **2:** 291–297.
141. Kastan, M. B., O. Zhan, W. S. EL-Deiry, F. Carrier, T. Jacks, W. V. Walsh, B. S. Plunkett, B. Vogelstein, and A. J. Fornace. 1992. A mammalian cell cycle checkpoint pathway utilizing p53 and GADD45 is defective in ataxia-telangiectasia. *Cell* **71:** 587–597.
142. Kato, R. and H. Ogawa. 1994. An essential gene, *ESR1*, is required for mitotic cell growth, DNA repair and meiotic recombination in *Saccharomyces cerevisiae. Nucleic Acids Res.* **22:** 3104–3112.
143. Keegan, K. S., D. A. Holtzman, A. W. Plug, E. R. Christenson, E. E. Brainerd, G. Flaggs, N. J. Bentley, E. M. Taylor, M. S. Meyn, S. B. Moss, A. M. Carr, T. Ashley, and M. F. Hoekstra. 1996. The Atr and Atm protein kinases associate with different sites along meiotically pairing chromosomes. *Genes Dev.* **10:** 2423–2437.
144. Keith, C. T. and S. L. Schreiber. 1995. PIK-related kinases: DNA repair, recombination, and cell cycle checkpoints. *Science* **270:** 50,51.
145. Khanna, K. K. and M. F. Lavin. 1993. Ionizing radiation and UV induction of p53 protein by different pathways in ataxia-telangiectasia cells. *Oncogene* **8:** 3307–3312.
146. Khanna, K. K., T. Shafman, P. Kedar, K. Spring, S. Kozlov, T. Yen, K. Hobson, M. Gatei, N. Zhang, D. Watters, M. Egerton, Y. Shiloh, S. Kharbanda, D. W. Kufe, and M. F. Lavin. 1997. Role of the ATM protein in stress response to DNA damage: evidence for interaction with c-Abl. *Nature* **387:** 520–523.
147. Kharbanda, S., R. Ren, P. Pandey, T. D. Shafman, S. M. Feller, R. R. Weichselbaum, and D. W. Kufe. 1995. Activation of the c-Abl tyrosine kinase in the stress response to DNA-damaging agents. *Nature* **376:** 785–788.
148. Kharbanda, S., P. Pandey, S. Jin, S. Inoue, A. Bharti, Z.-M. Yuan, R. Weichselbaum, D. Weaver, and D. Kufe. 1997. Functional interaction between DNA-PK and c-Abl in response to DNA damage. *Nature* **386:** 732–735.
149. Kohli-Kumar, M., C. Morris, C. Delaat, J. Sambrano, M. Masterson, R. Mueller, N. T. Shahidi, G. Yanik, K. Desantes, D. J. Friedman, A. D. Auerbach, and R. E. Harris. 1994. Bone marrow transplantation in Fanconi Anemia using matched sibling donors. *Blood* **84:** 2050–2054.
150. Komatsu, K., S. Matsuura, H. Tauchi, S. Endo, S. Kodama, D. Smeets, C. Weemaes, and M. Oshimura. 1996. The gene for Nijmegen breakage syndrome (V2) is not located on chromosome 11. *Am. J. Hum. Genet.* **58:** 885–888.
151. Krippschild, U., D. Milne, L. Campbell, and D. Meek. 1996. p53 N-terminus-targeted protein kinase activity is stimulated in response to wild type p53 and DNA damage. *Oncogene* **13:** 1387–1393.
152. Kubbies, M., D. Schindler, H. Hoehn, A. Schinzel, and P. S. Rabinovich. 1985. Endogenous blockage and delay of the chromosome cycle despite normal recruitment and growth phase explain poor proliferation and frequent edomitosis in Fanconi Anemia Cells. *Am. J. Hum. Genet.* **37:** 1022–1030.
153. Kunz, J., R. Henriquez, U. Schneider, M. Deuter-Reinhard, R. Movva, and M. N. Hall. 1993. Target of rapamycin in yeast, TOR2, is an essential phosphatidylinisitol kinase homolog required for G1 progression. *Cell* **73:** 585–596.
154. Kuhn, E. M. and E. Therman. 1986. Cytogenetics of Bloom's syndrome. *Cancer Genet. Cytogenet.* **22:** 1–18.
155. Kupfer, G. M., D. Naf, A. Suliman, M. Pulsipher, and A. D. D'Andrea. 1997. The Fanconi anemia proteins, FAA and FAC, interact to form a nuclear complex. *Nat. Genet.* **17:** 487–490.
156. Kurtzberg, J. 1996. Umbilical cord blood: a novel alternative source of hematopoietic stem cells for bone marrow transplantation (editorial). *J. Hematother.* **5:** 95,96.

157. Kurtzberg, J., M. Laughlin, M. L. Graham, C. Smith, J. F. Olson, E. Halperin, G. Ciocci, et al. 1996. Placental blood as a source of hematopoietic stem cells for transplantation into unrelated recipients. *N. Engl. J. Med.* **335:** 157–166.
158. Kuo, Y. K., Z. Wang, J.-H. Hong, L. Chessa, W. H. McBride, S. L. Perlman, and R. A. Gatti. 1994. Radiosensitivity of ataxia-telangiectasia, X-linked agammaglobulinemia and related syndromes. *Cancer Res.* **54:** 2544–2547.
159. Lakin, N. D., P. Weber, T. Stankovic, S. T. Rottinghaus, A. Malcolm, R. Taylor, and S. P. Jackson. 1996. Analysis of the ATM protein in wild-type and ataxia telangiectasia cells. *Oncogene* **13:** 2707–2716.
160. Landau, J. W., M. S. Sasaki, V. D. Newcomer, and A. Norman. 1966. Bloom's syndrome: the syndrome of telangiectatic erythema and growth retardation. *Arch. Dermatol.* **94:** 687–694.
161. Lange, E., A.-L. Borreson, X. Chen, L. Chessa, S. Chiplunkar, P. Concannon, S. Dandekar, S. Gerken, K. Lange, T. Liang, C. McConville, J. Polakow, O. Porras, G. Rotman, O. Sanal, M. Telatar, S. Sheikhavdndi, Y. Shiloh, E. Sobel, M. Taylor, N. Udar, N. Uhrhammer, L. Vanagaite, Z. Wang, H-M. Yang, L. Yang, Y. Ziv, and R. A. Gatti. 1995. Localization of an ataxia-telangiectasia gene to a 850 kb interval on chromosome 11q23. 1 by linkage analysis of 176 families in an international consortium. *Am. J. Hum. Genet.* **57:** 112–119.
162. Lavin, M. F. and M. Davidson. 1981. Repair of strand breaks in superhelical DNA of ataxia telangiectasia lymphoblastoid cells. *J. Cell Sci.* **48:** 383–391.
163. Lavin, M. F. 1992. Biochemical defects in ataxia-telangiectasia, in *Ataxia-Telangiectasia NATO ISI Series H* (Gatti, R. A. and R. B. Painter, eds.), Springer-Verlag, Berlin, pp. 235–255.
164. Lavin, M. F., K. K. Khanna, H. Beamish, K. Spring, D. Watters, and Y. Shiloh. 1995. Relationship of the ataxia-telangiectasia protein ATM to phosphoinositide 3-kinase. *Trends Biochem. Sci.* **20:** 382,383.
165. Lavin, M. F. and Y. Shiloh. 1997. The genetic defect in ataxia-telangiectasia. *Ann. Rev. Immunol.* **15:** 177–202.
166. Liu, Z.-G., R. Baskaran, E. T. Lea-Chon, L. D. Wood, Y. Chen, M. Karin, and J. Y. J. Wang. 1996. Three distinct signalling responses by murine fibroblasts to genotoxic stress. *Nature* **384:** 273- 276.
167. Lohman, T. M. 1993. Helicases-catalysed DNA unwinding. *J. Biol. Chem.* **268:** 2269–2272.
168. Lu, J., J. R. Mullen, S. J. Brill, S. Kleff, A. M. Romeo, and R. Sternglanz. 1996. Human homologues of yeast helicase. *Nature* **383:** 678,679.
169. Lu, X. and D. P. Lane. 1993. Differential induction of transcriptionally active p53 following UV or ionizing radiation: defects in chromosome instability syndromes? *Cell* **75:** 765–778.
170. Ma. L., A. Westbroek, A. G. Jochemsen, G. Weeda, A. Bosch, D. Boothsma, J. H. J. Hoeijmakers, and A. J. van der Ed. 1994. Mutational analysis of ERCC3, which is involved in DNA repair and transcription initiation: identification of domains essential for the DNA function. *Mol. Cell. Biol.* **14:** 4126–4134.
171. Martin, G. M. 1997. The Werner mutation: does it lead to a 'public' or 'private' mechanism of aging? *Mol. Med.* **3:** 356–358.
172. Martin, G. M., C. A. Sprague, and C. J. Epstein. 1970. Replicative life-span of cultivated human cells. Effects of donor's age, tissue, and genotype. *Lab. Invest.* **23:** 86–92.
173. McDaniel, L. D. and R. A. Schults. 1992. Elevated sister chromatid exchange phenotype of Bloom syndrome cells is complemented by human chromosome 15. *Proc. Natl. Acad. Sci. USA* **89:** 7968–7972.
174. Matsumoto, T., O. Imamura, Y. Yamabe, J. Kuromitsu, Y. Tokutake, A. Shimamoto, N. Suzuki, M. Satoh, S. Kitao, K. Ichikawa, H. Kataoka, K. Sugawara, W. Thomas,

B. Mason, Z. Tsuchihashi, D. Drayna, M. Sugawara, M. Sugimoto, Y. Furuichi, and M. Goto. 1997. Mutation and haplotype analyses of the Werner's syndrome gene based on its genomic structure: genetic epidemiology in the Japanese population. *Hum. Genet.* **100:** 123–130.

175. Matsumoto, T., A. Shimamoto, M. Goto, Y. Furuichi. 1997. Impaired nuclear localization of defective DNA helicases in Werner's syndrome. (Letter) *Nat. Genet.* **16:** 335,336.
176. Matsuura, S., C. Weemaes, D. Smeets, H. Takami, N. Kondo, S. Sakamoto, N. Yano, A. Nakamura, H. Tauchi, S. Endo, M. Oshimura, and K. Komatsu. 1997. Genetic mapping using microcell-mediated chromosome transfer suggests a locus for Nijmegen breakage syndrome at chromosome 8q21–24. *Am. J. Hum. Genet.* **60:** 1487–1494.
177. Meyn, M. S. 1993. High spontaneous intrachromosomal recombination rates in Ataxia-Telangiectasia. *Science* **260:** 1327–1330.
178. Miller, M. E. and J. Chatten. 1967. Ovarian changes in ataxia-telangiectasia. *Acta. Paediat. Scand.* **56:** 559–561.
179. Monnat, R. J., Jr., A. F. M. Hackmann, and T. A. Chiaverotti. 1992. Nucleotide sequence analysis of human hypoxanthine phosphoribosyltransferase (HPRT) gene deletions. *Genomics* **13:** 777–787.
180. Morgan, J. L., T. M. Holcomb, and R. W. Morrissey. 1968. Radiation reaction in ataxia-telangiectasia. *Am. J. Dis. Child.* **116:** 557,558.
181. Morgan, S. E. and M. B. Kastan. 1997. p53 and ATM: cell cycle, cell death, and cancer. *Adv. Cancer Res.* **71:** 1–25.
182. Nacht, M., A. Strassor, Y. R. Chan, A. W. Harris, M. Schlisel, R. T. Bronson, and T. Jacks. 1996. Mutations in the p53 and SCID genes cooperate in tumorigenesis. *Genes Dev.* **10:** 2055–2066.
183. Nakayama, H., K. Nakayama, R. Nakayama, N. Irino, Y. Nakayama, and P. C. 1992. Isolation and genetic characterisation of a thymineless death-resistant mutant of Escherichia coli K12: identification of a new mutation (recQ1) that blocks the RecF recombination. *Ann. Rev. Biochem.* **61:** 474–480.
184. Nakura, J., E. M. Wijsman, T. Miki, K. Kamino, C.-E. Yu, J. Oshima, K. Fukuchi, J. L. Weber, C. Piussan, M. I. Melaragno, C. J. Epstein, S. Scappaticci, M. Fraccaro, T. Matsumura, S. Murano, S. Yoshida, Y. Fujiwara, T. Saida, T. Ogihara, G. M. Martin, and G. D. Schellenberg. 1994. Homozygosity mapping of the Werner syndrome locus (WRN). *Genomics* **23:** 600–608.
185. Nara, N., T. Miyamoto, et al. 1980. Two siblings with Fanconi's anemia developing squamous cell carcinomas. *Rinsho Ketsueki* **21:** 1944.
186. Nicotera, T. M., J. Notaro, S. Notaro, J. Schumer, and A. A. Sandberg. 1989. Elevated superoxide dismutase in Bloom's syndrome: a genetic condition of oxidative stress. *Cancer Res.* **49:** 5239–5343.
187. Nourse, J., E. Fripo, W. M. Flanagan, S. Coats, K. Polyak, M.-H. Lee, J. Massague, G. R. Crabtree, and J. M. Roberts. 1994. Interleukin-2-mediated elimination of the $p27^{kip1}$ cyclin-dependent kinase inhibitor prevented by rapamycin. *Nature* **372:** 570–573.
188. Nussenzweig, A., C. Chen, V. Da Costa Soares, M. Sanchez, K. Sokol, M. C. Nussenzweig, and G. C. Li. 1996. Requirement for Ku80 in growth and immunoglobulin V(D)J recombination. *Nature* **382:** 551–555.
189. Oshima, J., C.-E. Yu, C. Piussan, G. Klein, J. Jabkowski, S. Balci, T. Miki, J. Nakura, T. Ogihara, J. Ells, M. A. C. Smith, M. I. Melaragno, M. Fraccaro, S. Scappaticci, J. Matthews, S. Ouais, A. Jarzebowicz, G. D. Schellenberg, and G. M. Martin. 1996. Homozygous and compound heterozygous mutations at the Werner syndrome locus. *Hum. Mol. Genet.* **5:** 1909–1913.
190. Pandita, T. K. and W. N. Hittleman. 1992. Initial chromosome damage but not DNA damage is greater in ataxia telangiectasia cells. *Radiat. Res.* **130:** 94–103.
191. Passarge, E. 1991. Bloom's syndrome: the German experience. *Ann. Genet.* **34:** 179–197.

192. Perez-Véra, P., A. G. Angel, B. Molina, L. Gómez, S. Frías, R. A. Gatti, and A. Carnevale. 1997. Chromosome instability with bleomycin and X-ray hypersensitivity in a boy with Nijmegen breakage syndrome. *Am. J. Med. Genet.* **70:** 24–27.
193. Pleiman, C. M., W. M. Hertz, and J. C. Cambier. 1994. Activation of phosphatidylinositol-3 kinase by Src-family kinase SH_3 binding to the p85 subunit. *Science* **263:** 1609–1612.
194. Plug, A. W., A. H. F. M. Peters, Y. Xu, K. S. Keegan, M. F. Hoekstra, D. Baltimore, P. De Boer, and T. Ashley. 1997. ATM and RPA in meiotic chromosome synapsis and recombination. *Nat. Genet.* **17:** 457–461.
195. Poot, M., H. Hoehn, T. M. Runger, and G. M. Martin. 1992. Impaired S-phase transit of Werner syndrome cells expressed in lymphoblastoid cell lines. *Exper. Cell Res.* **202:** 267–273.
196. Pronk, J. C., et al. 1995. Localization of the Fanconi Anemia complementation group A gene to chromosome 16q24. 3. *Nat. Genet.* **11:** 338–340.
197. Puranam, K. L. and P. J. Blackshear. 1994. Cloning and characterisation of RECQL, a potential human homologue of the Escherichia coli DNA helicase RecQ. *J. Biol. Chem.* **269:** 29,838–29,845.
198. Rackoff, W. R., A. Orazi, C. A. Robinson, R. J. Cooper, B. P. Alter, M. H. Freedman, R. E. Harris, and D. A. Williams. 1996. Prolonged administration of granulocyte colony-stimulating factor (Filgrastim) to patients with Fanconi anemia: a pilot study. *Blood* **88:** 1588–1593.
199. Ruprecht, K. W. 1989. Ophthalmological aspects in patients with Werner's syndrome. *Arch. Gerontol. Geriat.* **9:** 263–270.
200. Saar, K., K. H. Chrzanowska, M. Stumm, M. Jung, G. Nurnberg, T. F. Wienker, E. Seemanova, R.-D. Wegner, A. Reis, and K. Sperling. 1997. The gene for the ataxia-telangiectasia variant, Nijmegen breakage syndrome, maps to a 1-cM interval on chromosome 8q21. *Am. J. Hum. Genet.* **60:** 605–610.
201. Saar, K., R. Varon, K. H. Chrzanowska, M. Platzer, M. Stumm, M. Jung, G. Nurnberg, T. F. Wienker, E. Seemanova, R. D. Wegner, M. Digweed, A. Rosenthal, K. Sperling, and A. Reis. 1997. Nijmegen and Berlin breakage syndromes are caused by mutations in the same gene mapping to chromosome 8q21. *Seventh Ataxia-Telangiectasia Workshop*, Abstract 25. Baltimore, MD.
202. Sabatini, D. M., H. Erdjument-Bromage, M. Lui, P. Tempst, and S. H. Snyder. 1994. RAFT1: a mammalian protein that binds to FKBP12 in a rapamycin-dependent fashion and is homologous to yeast TORs. *Cell* **78:** 35–43.
203. Sabers, C. J., M. M. Martin, G. J. Brunn, J. M. Williams, F. J. Dumont, G. Wiederrecht, and R. T. Abraham. 1995. Isolation of a protein target of the FKBP12-rapamycin complex in mammalian cells. *J. Biol. Chem.* **270:** 815–822.
204. Sanchez, Y., B. A. Desany, W. J. Jones, W. Liu, B. Wang, and S. J. Elledge. 1996. Regulation of *RAD53* by the ATM-like kinases *MEC1* and *TEL1* in yeast cell cycle checkpoint pathways. *Science* **271:** 357–360.
205. Sanchez, Y., C. Wong, R. S. Thoma, R. Richman, Z. Wu, H. Piwnica-Worms, et al. 1997. Conservation of the Chk1 checkpoint pathway in mammals: linkage of DNA damage to Cdk regulation through Cdc25. *Science* **277:** 1497–1501.
206. Savitsky, K., A. Bar-Shira, S. Gilad, G. Rotman, Y. Ziv, L. Vanagaite, D. A. Tagle, S. Smith, T. Uziel, S. Sfez, M. Ashkenazi, I. Pecker, M. Frydman, R. Harnik, S. R. Patanjali, A. Simmons, G. A. Clines, A. Sartiel, R. A. Gatti, L. Chessa, O. Sanal, M. F. Lavin, N. G. J. Jaspers, A. M. R. Taylor, C. F. Arlett, T. Miki, S. M. Weissman, M. Lovett, F. S. Collins, and Y. Shiloh. 1995. A single ataxia-telangiectasia gene with a product similar to Pl-3 kinase. *Science* **268:** 1749–1753.
207. Savitsky, K., S. Sfez, D. Tagle, Y. Ziv, A. Sartiel, F. S. Collins, Y. Shiloh, and G. Rotman. 1995. The complete sequence of the coding region of the ATM gene reveals similarity to cell cycle regulators in different species. *Hum. Mol. Genet.* **4:** 2025–2032.

208. Scappaticci, S., D. Cerimele, and M. Fraccaro. 1982. Clonal structural chromosomal rearrangements in primary fibroblast cultures and in lymphocytes of patients with Werner's syndrome. *Hum. Genet.* **62:** 16–24.
209. Scappaticci, S., A. Forabosco, G. Borroni, G. Orecchia, and M. Fraccaro. 1990. Clonal structural chromosomal rearrangements in lymphocytes of four patients with Werner's syndrome. *Ann. Genet.* **33:** 5–8.
210. Schellenberg, G. D., G. M. Martin, E. M. Wijsman, J. Nakura, T. Miki, and T. Ogihara. 1992. Homozygosity mapping and Werner's syndrome (letter). *Lancet* **339:** 1002.
211. Scully, R., J. Chen, A. Plug, Y. Ziao, D. Weaver, J. Feunteun, T. Ashley, and D. M. Livingston. 1997. Association of BRCA1 with Rad51 in mitotic and meiotic cells. *Cell* **88:** 265–275.
212. Seaton, B. L., J. Yucel, P. Sunnerhagen, and S. Subramani. 1992. Isolation and characterization of the *Schizosaccharomyces pombe* rad3 gene which is involved in the DNA damage and DNA synthesis checkpoints. *Gene* **119:** 83–89.
213. Seemanova, E., E. Passarge, D. Beneskova, J. Houstek, P. Kasal, and M. Sevcikova. 1985. Familial microcephaly with normal intelligence, immunodeficiency, and risk of lymphoreticular malignancies: a new autosomal recessive disorder. *Am. J. Med. Genet.* **20:** 639–648.
214. Seki, M., H. Miyazawa, S. Tada, J. Yanagisawa, T. Yamaoka, S-I. Hoshino, K. Ozawa, T. Eki, M. Nogami, K. Okumura. et al. 1994. Molecular cloning of cDNA encoding human DNA helicase and localisation of the gene at chromosome 12p12. *Nucleic Acids Res.* **22:** 4566–4573.
215. Shafman, T., K. K. Khanna, P. Kedar, K. Spring, S. Kozlov, and T. Yen. 1997. Interaction between ATM protein and c-Abl in response to DNA damage. *Nature* **387:** 520–523.
216. Shafman, T. D., A. Saleem, J. Kyriakis, R. Weichselbaum, S. Kharbanda, and D. W. Kufe. 1995. Defective induction of stress-activated protein kinase activity in ataxia-telangiectasia cells exposed to ionizing radiation. *Cancer Res.* **55:** 3242–3245.
217. Shieh, S-Y., M. Ikeda, Y. Taya, and C. Prives. 1997. DNA damage-induced phosphorylation of p53 alleviates inhibition of MDM2. *Cell* **91:** 325–334.
218. Shipley, J., C. H. Rodek, et al. 1984. Mitomycin C-induced chromsome damage in fetal blood cultures and prenatal diagnosis of Fanconi's anaemia. *Prenat. Diagn.* **4:** 217–221.
219. Shiraishi, Y., T. Taguchi, M. Ozawa, and R. Bamezai. 1989. Different mutations responsible for the elevated sister-chromatid exchange frequencies in Bloom syndrome and X-irradiated B-lymphoblastoid cell lines originating from acute leukemia. *Mutat. Res.* **211:** 273–278.
220. Sinclair, D. A., K. Mills, and L. Guarente. 1997. Accelerated aging and nucleolar fragmentation in yeast sgs1 mutants. *Science* **277:** 1313–1316.
221. Sperling, K. 1983. Analisi dell' eterogeneita nell'uomo. *Prospective Pediat.* **49:** 53–66.
222. Stefanini, M., S. Scappaticci, P. Lagomarsini, G. Borroni, E. Berardesca, and F. Nuzzo. 1989. Chromosome instability in lymphocytes from a patient with Werner's syndrome is not associated with DNA repair defects. *Mutat. Res.* **219:** 179–185.
223. Stewart, E., C. R. Chapman, F. Al-Khodairy, A. M. Carr, and T. Enoch. 1997. rqh1$^+$, a fusion yeast related to the Bloom's and Werner's Syndrome genes, is required for reversible S phase arrest. *EMBO J.* **16:** 2682–2692.
224. Strathdee, C. A., A. M. V. Duncan, and M. Buchwald. 1992. Evidence for at least four Fanconi Anemia genes including FACC on chromosome 9. *Nat. Genet.* **1:** 196–198.
225. Strathdee, C. A., A. M. Duncan, and M. Buchwald. 1992. *Nat. Genet.* **1:** 196–198.
226. Strathdee, C. A., H. Gavish, W. R. Shannon, and M. Buchwald. 1992. Cloning of cDNAs for Fanconi's anaemia by functional complementation. *Nature* **356:** 763–767.
227. Stumm, M., R. A. Gatti, A. Reis, N. Udar, K. Chrzanowska, E. Seemanova, K. Sperling, and R.-D. Wegner. 1995. The ataxia-telangiectasia-variant genes 1 and 2 are distinct from the ataxia-telangiectasia gene on chromosome 11q23. 1. *Am. J. Hum. Genet.* **57:** 960–962.

228. Stumm, M., K. Sperling, and R.-D. Wegner. 1997. Noncomplementation of radiation-induced chromosome aberrations in ataxia-telangiectasia/ataxia-telangiectasia-variant heterodikaryons. *Am. J. Hum. Genet.* **60:** 1246–1251.
229. Sullivan, K. E., E. Veksler, H. Lederman, and S. P. Lees-Miller. 1997. Cell cycle checkpoints and DNA repair in Nijmegen breakage syndrome. *Clin. Immun. Immunopath.* **82:** 43–48.
230. Sung, P., V. Bailly, C. Weber, L. H. Thompson, L. Prakash, and S. Prakash. 1993. Human xeroderma pigmentosum group D gene encodes a DNA helicase. *Nature* **365:** 852–855.
231. Swift, M, D. Zimmerman, and E. R. McDonough. 1971. Squamous cell carcinomas in Fanconi's anemia. *JAMA* **216:** 325,326.
232. Taalman, R. D. F. M., T. W. J. Hustinx, C. M. R. Weemaes, E. Seemanova, A. Schmidt, E. Passarge, and J. M. J. C. Scheres. 1989. Further delineation of the Nijmegen breakage syndrome. *Am. J. Med. Genet.* **32:** 425–431.
233. Taylor, A. M., D. G. Harnden, C. F. Arlett, S. A. Harcourt, A. R. Lehmann, S. Stevens, and B. A. Bridges. 1975. Ataxia-telangiectasia: a human mutation with abnormal radiation sensitivity. *Nature* **4:** 427–429.
234. Taylor, A. M. R., E. Flude, B. Laher, M. Stacer, E. McKay, J. Watt, S. H. Green, and A. E. Harding. 1987. Variant forms of ataxia-telangiectasia. *J. Med. Genet.* **24:** 669–677.
235. Telatar, M., Z. Wang, W. Udar, T. Liang, P. Concannon, E. Bernatowska-Matuscklewicz, M. F. Lavin, Y. Shiloh, R. A. Good, and R. A. Gatti. 1996. Ataxia-telangiectasia: mutations in cDNA detected by protein truncation screening. *Am. J. Hum. Genet.* **59:** 40–44.
236. Thomas, W., M. Rubenstein, M. Goto, and D. Drayna. 1993. A genetic analysis of the Werner syndrome region on human chromosome 8p. *Genomics* **16:** 685–690.
237. Tupler, R., G. L. Marseglia, M. Stefanini, E. Prosperi, L. Chessa, T. Nardo, A. Marchi, and P. Maraschio. 1997. A variant of the Nijmegen breakage syndrome with unusual cytogenetic features and intermediate cellular radiosensitivity. *J. Med. Genet.* **34:** 196–202.
238. Urhammer, N., E. Lange, O. Porras, A. Naeim, X. Chen, S. Sheikhavandi, S. Chiplunkar, L. Yang, S. Dandekar, T. Liang, T. Lange, N. Patel, S. Teraoka, N. Udar, N. Calvo, P. Concannon, K. Lange, and R. A. Gatti. 1995. Sublocalization of an ataxia-telangiectasia gene distal to D11S384 by ancestral haplotyping in Costa Rican families. *Am. J. Hum. Genet.* **57:** 103–111.
239. Uziel, T., K. Savitsky, M. Platzer, Y. Ziv, T. Helbitz, M. Nehls, T. Boehm, A. Rosenthal, Y. Shiloh, and G. Rotman. 1996. Genomic organization of the ATM gene. *Genomics* **33:** 317–320.
240. Vanagaite, L., M. R. James, G. Rotman, K. Savitsky, A. Bar-Shira, S. Gilad, Y. Ziv, V. Uchenik, A. Sartiel, F. S. Collins, V. C. Sheffield, J. Weissenbach, and Y. Shiloh. 1995. A high-density microsatellite map of the ataxia-telangiectasia locus. *Hum. Genet.* **95:** 451–455.
241. Van de Kaa, C. A., C. M. R. Weemaes, P. Wesseling, H. E. Schaafsma, and A. Haraldsson. 1994. Postmortem findings in the Nijmegen breakage syndrome. *Pediatric Pathol.* **14:** 787–796.
242. van der Burgt, I., K. H. Chrzanowska, D. Smeets, and C. Weemaes. 1996. Nijmegen breakage syndrome. *J. Med. Genet.* **33:** 153–156.
243. van Laar, T., W. T. Steegenga, A. G. Jochemsen, C. Terleth, and A. J. van der Eb. 1994. Bloom's syndrome cells GM1492 lack detectable p53 protein but exhibit normal G_1 cell-cycle arrest after UV irradiation. *Oncogene* **9:** 981–983.
244. van Vuuren, A. J., E. Appeldoorn, H. Odijk, A. Yasui, N. G. J. Jaspers, D. Bootsma, and J. H. J. Hoeijmakers. 1993. Evidence for a repair enzyme complex involving ERCC1 and complementing activities of ERCC4, ERCC11 and xeroderma pigmentosum group F. *EMBO J.* **12:** 3693–3701.
245. Vincent, R. A. Jr., R. B. Sheriden 3rd., and P. C. Huang. 1975. DNA strand breakage repair in ataxia-telangiectasia fibroblast-like cells. *Mutat. Res.* **33:** 357–366.

246. Wagner, J. E., J. Rosenthal, R. Sweetman, X. O. Shu, S. M. Davies, N. K. C. Ramsay, P. B. McGlave, et al. 1996. Successful transplantation of HLA-matched and HLA-mismatched umbilical cord blood from unrelated donors: analysis of engraftment and acute graft-versus-host disease. *Blood* **88:** 795–802.
247. Waghray, M., S. Al-Sedairy, P. T. Ozand, and M. A. Hannan. 1990. Cytogenetic characterization of ataxia-telangiectasia (AT) heterozygotes using lymphoblastoid cell lines and chronic gamma-irradiation. *Hum. Genet.* **84:** 532–534.
248. Watt, P. M., E. J. Louis, R. H. Borts, and I. D. Hickson. 1995. Sgs1: a eukaryotic homolog of *E. coli RecQ* that interacts with topoisomerase II in vivo and is required for faithful chromosome segregation. *Cell* **81:** 253–260.
249. Watters, D., K. K. Khanna, H. Beamish, G. Birrell, K. Spring, P. Kedar, M. Gatei, D. Stenzel, K. Hobson, S. Kozlov, A. Farrell, J. Ramsay, R. Gatti, and M. F. Lavin. 1997. Cellular localisation of the ataxia-telangiectasia (ATM) gene proteins and discrimination between mutated and normal forms. *Oncogene* **14:** 1911–1921.
250. Weemaes, C. M., J. A. Bakkeren, A. Haraldsson, and D. F. Smeets. 1991. Immunological studies in Bloom's syndrome. A follow-up report. *Ann. Genet.* **34:** 201–205.
251. Weemaes, C. M. R., T. W. J. Hustinx, J. M. J. C. Scheres, P. J. J. van Munster, J. A. J. M. Bakkeren, and R. D. F. M. Taalman. 1981. A new chromosomal instability disorder: the Nijmegen breakage syndrome. *Acta Paediat. Scand.* **70:** 557–564.
252. Weemaes, C. M. R., D. F. C. M. Smeets, and C. J. A. M. Burgt. 1994. Nijmegen breakage syndrome: a progress report. *Int. J. Radiat. Biol.* **66:** S185-S188.
253. Weichselbaum, R. R., J. Nove, and J. B. Little. 1980. X-ray sensitivity of fifty-three human diploid fibroblast cell strains from patients with characterized genetic disorders. *Cancer Res.* **40:** 920–925.
254. Wegner, R.-D., M. Metzger, F. Hanefeld, N. G. J. Jaspers, C. Baan, K. Magdorf, J. Kunze, and K. Sperling. 1988. A new chromosomal instability disorder confirmed by complementation studies. *Clin. Genet.* **33:** 20–32.
255. West, S. C. 1992. Enzymes and molecular mechanisms of genetic recombination. *Ann. Rev. Biochem.* **61:** 603–640.
256. Westphal, C. H., S. Rowan, C. Schmaltz, A. Elson, D. E. Fisher, and P. Leder. 1997. ATM and p53 cooperate in apoptosis and suppression of tumorigenesis, but not in resistance to acute radiation toxicity. *Nat. Genet.* **16:** 397–401.
257. Westphal, C. H., C. Schmaltz, S. Rowan, A. Elson, D. E. Fisher, and P. Leder. 1997. Genetic interactions between ATM and p53 influence cellular proliferation and irradiation-induced cell cycle checkpoints. *Cancer Res.* **57:** 1664–1667.
258. Wevrick, R., C. A. Clarke, and M. Buchwald. 1993. Cloning and analysis of the murine Fanconi anemia group C cDNA. *Hum. Mol. Genet.* **2:** 655–662.
259. Whitney, M. et al. 1995. Microcell mediated chromosome transfer maps the Fanconi anemia group D gene to chromosome 3p. *Nat. Genet.* **11:** 341–343.
260. Whitney, M. A., G. Royle, M. J. Low, M. A. Kelly, M. K. Axthelm, C. Reifsteck, S. Olson, R. E. Braun, M. C. Heinrich, R. K. Rathbun, G. C. Bagby, and M. Grompe. 1996. Germ cell defects and hematopoietic hypersensitivity to gamma-interferon in mice with a targeted disruption of the Fanconi anemia C gene. *Blood* **88:** 49–58.
261. Whitney, M. A., H. Saito, P. M. Jakobs, R. A. Gibson, R. E. Moses, and M. Grompe. 1993. A common mutation in the FACC gene causes Fanconi anemia in Ashkenazi Jews. *Nat. Genet.* **4:** 202–205.
262. Yamashita, T., N. Wu, G. Kupfer, C. Corless, H. Joenje, M. Grompe, and A. D. D'Andrea. 1996. The clinical variability of Fanconi Anemia (Type C) results from expression of an amino terminal truncated FAC polypeptide with partial activity. *Blood* **87:** 4424–4432.
263. Ye, L., J. Nakura, N. Mitsuda, Y. Fujioka, K. Kamino, T. Ohta, Y. Jinno, N. Niikawa, T. Miki, and T. Ogihara. 1995. Genetic association between chromosome 8 microsatellite

(MS8–134) and Werner syndrome (WRN): chromosome microdissection and homozygosity mapping. *Genomics* **28:** 566–569.
264. Young, N. S. and B. P. Alter. 1994. *Clinical Features of Fanconi's Anemia*. Saunders, Philadelphia.
265. Young, B. R. and R. B. Painter. 1989. Radioresistant DNA synthesis and human genetic disease. *Hum. Genet.* **82:** 113–117.
266. Yu, C.-E., J. Oshima, Y.-H. Fu, E. M. Wijsman, F. Hisama, R. Alish, S. Matthews, J. Nakura, T. Miki, S. Ouais, G. M. Martin, J. Mulligan, and G. D. Schellenberg. 1996. Positional cloning of the Werner's syndrome gene. *Science* **272:** 258–262.
267. Yu, C.-E., J. Oshima, K. A. B. Goddard, T. Miki, J. Nakura, T. Ogihara, M. Poot, H. Hoehn, M. Fraccaro, C. Piussan, G. M. Martin, G. D. Schellenberg, and E. M. Wijsman. 1994. Linkage disequilibrium and haplotype studies of chromosome 8p11.1–21.1 markers and Werner syndrome. *Am. J. Hum. Genet.* **55:** 356–364.
268. Yu, C.-E., J. Oshima, E. M. Wijsman, J. Nakura, T. Miki, C. Piussan, S. Matthews, Y.-H. Fu, J. Mulligan, G. M. Martin, and G. D. Schellenberg. 1997. Werner's Syndrome Collaborative Group: Mutations in the consensus helicase domains of the Werner syndrome gene. *Am. J. Hum. Genet.* **60:** 330–341.
269. Yuan, A. M., Y. Huang, Y. Whang, C. Sawyers, R. Weichselbaum, S. Kharbanda, and D. Kufe. 1996. Role for c-Abl tyrosine kinase in growth arrest response to DNA damage. *Nature* **382:** 272–274.
270. Zakian, V. A. 1995. ATM-related genes: What do they tell us about functions of the human gene? *Cell* **82:** 685–687.
271. Zdzienicka, M. Z. 1995. Mammalian mutants defective in the response to ionizing radiation-induced DNA damage. *Mutat. Res.* **336:** 203–213.
272. Zhang, N., P. Chen, P. Gatei, S. Scott, K. K. Khana, and M. F. Lavin. 1998. An anti-sense construct of full-length *ATM* cDNA imposes a radiosensitive phenotype on normal cells. *Oncogene,* in press.

20

Genetics of DNA Mismatch Repair, Microsatellite Instability, and Cancer

Tomas A. Prolla, Sean Baker, and R. Michael Liskay

1. INTRODUCTION

In both prokaryotes and eukaryotes mismatched or unpaired bases in a DNA duplex can arise spontaneously through errors in DNA replication, genetic recombination, and deamination of 5-methylcytosine to thymine (T). Errors generated during the process of DNA replication can be thought of as a highly "primitive" type of DNA damage. Systems designed to minimize the production of errors, or to correct mismatches, are likely to have evolved very early in order to assure the fidelity of genetic transmission. Paradoxically, such repair systems cannot result in error-free DNA synthesis, since the constant generation of mutant alleles is a requirement for variation and evolution. In a changing environment, increased spontaneous mutation rates can confer benefits to a bacterial population *(38)*. In fact, the discovery of bacterial mutants that have reduced spontaneous mutation rates (antimutators) suggests that optimum mutation rates have evolved *(31,32)*. Two biological processes, DNA polymerase proofreading and DNA mismatch repair (DMR), are known to affect spontaneous mutation rates dramatically in both prokaryotes and eukaryotes *(83,98)*.

There is now solid evidence for the concept that cancer is the end result of clonal evolution of tumor cell populations *(104)*. Accumulation of a multiple genetic changes, such as loss of tumor suppressor genes and activation of oncogenes, has been observed in a number of tumor types *(30)*. Given that the number of genetic alterations characterized is likely to be only a fraction of those actually required for the process of tumorigenesis, and the fact that spontaneous mutation rates are low, it has been proposed that the acquisition of a mutator phenotype may be a requirement for tumor development *(81,84)*. The recent discovery of DMR defects in hereditary cancer syndromes and in sporadic tumors has provided strong support for the role of spontaneous mutation in cancer development, and has given further impetus for research in the area of DNA repair. This chapter briefly reviews bacterial and yeast DMR as a foundation for discussion of the mammalian DMR system, and its role in both inherited and sporadic cancer. It also includes a discussion of the recent reports of DMR-deficient mice generated by gene-targeting techniques.

From: DNA Damage and Repair, Vol. 2: DNA Repair in Higher Eukaryotes
Edited by: J. A. Nickoloff and M. F. Hoekstra © Humana Press Inc., Totowa, NJ

2. OVERVIEW OF DMR IN *ESCHERICHIA COLI* AND *SACCHAROMYCES CEREVISIAE*

2.1. *Bacterial DMR*

Genetic evidence for the existence of a DMR system in *Escherichia coli* was obtained by the isolation of strains displaying elevated levels of spontaneous mutation ("mutator" strains), followed by the isolation and characterization of the respective genes and their encoded proteins. Several alleles of the *Salmonella typhimurium* and *E. coli mutL* gene were isolated independently after treatment with mutagens *(77,128)*, and shown to be transition and frameshift mutators that display wild-type resistance to UV light *(128)*. *mutS* mutants were also isolated in a genetic screen, and shown to display increased frequencies of transition and frameshift mutations *(127,129)*.

After characterization of several bacterial mutator genes and their products, it became clear that bacteria possess multiple mismatch repair systems that fall into two classes, long patch repair (LPMR) and very short patch mismatch repair (VSPMR). For a review of short-patch repair systems, please refer to Chapter 8. LPMR, also known as methyl-directed repair, appears to be the major replicative DMR system, and is characterized by "long" excision and resynthesis tracts of up to several kilobases in length *(131,143)*. In *E. coli*, LPMR involves the mutator genes *mutS*, *mutL*, *mutH*, *uvrD*, and *dam (43)*. The presence of hemimethylated 5'-GATC-3' sequence, directs discrimination between parental and newly synthesized strands *(71,86,115)*. Since methylation lags behind replication in *E. coli* by several minutes *(55,90)*, LPMR is believed to occur on newly synthesized strands immediately behind the replication fork. *E. coli dam* mutants are deficient in adenine methylation in the 5'-GATC-3' sequence and are therefore unable to discriminate parental and newly synthesized strands *(91)*.

Isolation of the *mutHLS* and *uvrD* genes, and the purification of their encoded proteins have allowed characterization of their biochemical functions: MutS is a 97-kDa DNA-mismatch binding protein that can bind to a variety of mispaired bases and small (1–5 bases) single-stranded loops *(109,136,137)*. UvrD is DNA helicase II, a protein that unwinds DNA in an ATP-dependent manner *(52)*. MutH is a latent endonuclease that incises the transiently unmethylated strands of hemimethylated 5'-GATC-3' sequences, with fully methylated strands being resistant to cleavage *(146)*. No biochemical activity for the 70-kDa MutL protein has been identified, but it may couple mismatch recognition by MutS to MutH incision at 5'-GATC-3' sequences in an ATP-dependent manner *(43)*. Sancar and Hearst have proposed that MutL acts as a "molecular matchmaker," a protein that promotes an effective complex between two or more DNA binding proteins in an ATP-dependent manner, without necessarily being part of the final effector complex *(123)*. Single-strand DNA binding protein (SSB), DNA polymerase II holoenzyme, DNA ligase, exonuclease I and either exonuclease VII or RecJ are also required for in vitro reconstitution of bacterial DNA mismatch repair *(20,68)*. For a detailed review of bacterial DMR please refer to Chapter 11, vol. 1.

2.2. *Genetics of DMR in* S. cerevisiae

Fogel et al. first reported the identification of a yeast mutant, *pms1-1*, displaying elevated levels of post-meiotic segregation (PMS) for several genetic markers, a 50-fold increase in levels of spontaneous forward mutation rate to canavanine resistance, and a several thousandfold increase in the reversion rate of the *hom3-10* allele, a single-base frameshift mutation *(34,147)*. PMS reflects the lack of correction of heteroduplex DNA formed during genetic recombination and is therefore indicative of a defect in DMR. Cloning of the yeast *PMS1* gene revealed that it encodes a protein that displays homology to bacterial MutL *(66)*. Using a degenerate PCR approach, Reenan and Kolodner *(116,117)* cloned two additional *S. cerevisiae* DMR gene homologs, *MSH1* and *MSH2* (MutS homologs 1 and 2). Disruption of the *MSH2* gene affects nuclear mutation rates and PMS levels in a manner similar to Pms1p, whereas Msh1p appears to act in the repair of mitochondrial DNA *(116,117)*. Using a similar PCR approach, Prolla et al. *(113)* isolated a novel *S. cerevisiae* gene, *MLH1*, that encodes a predicted protein product with sequence similarity to bacterial MutL. A detailed genetic study of *mlh1*, *pms1*, and *mlh1 pms1* double-mutant strains showed indistinguishable phenotypes, suggesting that Mlh1p and Pms1p are components of the same DMR pathway *(113)*.

Although it is presumed that the yeasts Mlh1p, Pms1p, and Msh2p are components of a pathway that is mechanistically similar to bacterial LPMR, little is known regarding the biochemistry of yeast DMR. Msh2p binds to mismatched DNA in vitro *(2)* and appears to be responsible for most DNA mismatch binding activities in *S. cerevisiae* nuclear extracts, as determined by gel shift studies *(96)*. A gel-shift assay also suggests that Mlh1p, Pms1p, and Msh2p form a ternary complex in the presence of heteroduplex DNA containing a G-T mismatch *(114)*. Interestingly, Mlh1p and Pms1p appear to form a heterodimer complex as determined by experiments with the yeast two-hybrid system and affinity chromatography *(114)*. Since bacterial MutL exists as a homodimer in solution, it has been suggested *(114)* that the yeast Mlh1p-Pms1p heterodimer is the functional homolog of the bacterial MutL homodimer *(44)*.

Mutations in either *MLH1*, *PMS1*, or *MSH2* lead not only to a general increase in spontaneous mutation rates, but also result in a 100- to 700-fold increase in simple-repeat tract instability *(134)*. A similar, but less pronounced destabilization of simple-sequence repeat DNA has also been previously observed in bacterial DMR mutants *(75)*. Tracts of simple repeat DNA, such as poly(CA) · poly(TG), are widespread in eukaryotic genomes and change length at frequencies that are considerably higher than expected for standard point mutations *(46,132,138)*. This genetic instability is presumed to be owing largely to DNA replication slippage: during replication of the tract, primer and template strand may dissociate and then reassociate in a misaligned configuration. DNA replication will result in either deletion or expansion of the tract, depending on whether the unpaired bases occurred because of slippage of the template or primer strand. The slippage intermediates are likely to be recognized by the DMR system, since both MutS and Msh2p have high affinity for short unpaired heterologies *(3,109)*. Mutations in the proofreading domains of DNA polymerases have small effects on the stability of simple repeat tracts *in vivo (134)*. DNA polymerases appear to be unable to correct small heterologies in newly synthesized DNA, especially as repeat tracts become longer *(67)*.

More recently, two additional *S. cerevisiae* DMR homologs involved in replicative error correction have been identified, *MSH3* and *MSH6 (64,100)*. Disruption of the *MSH3* gene results in a small increase in the reversion rate of the *hom3-10* allele and no detectable increase in forward mutation rates to canavanine resistance *(100)*. In contrast, *MSH3* disruption does have a modest effect on the stability of $(GT)_n$

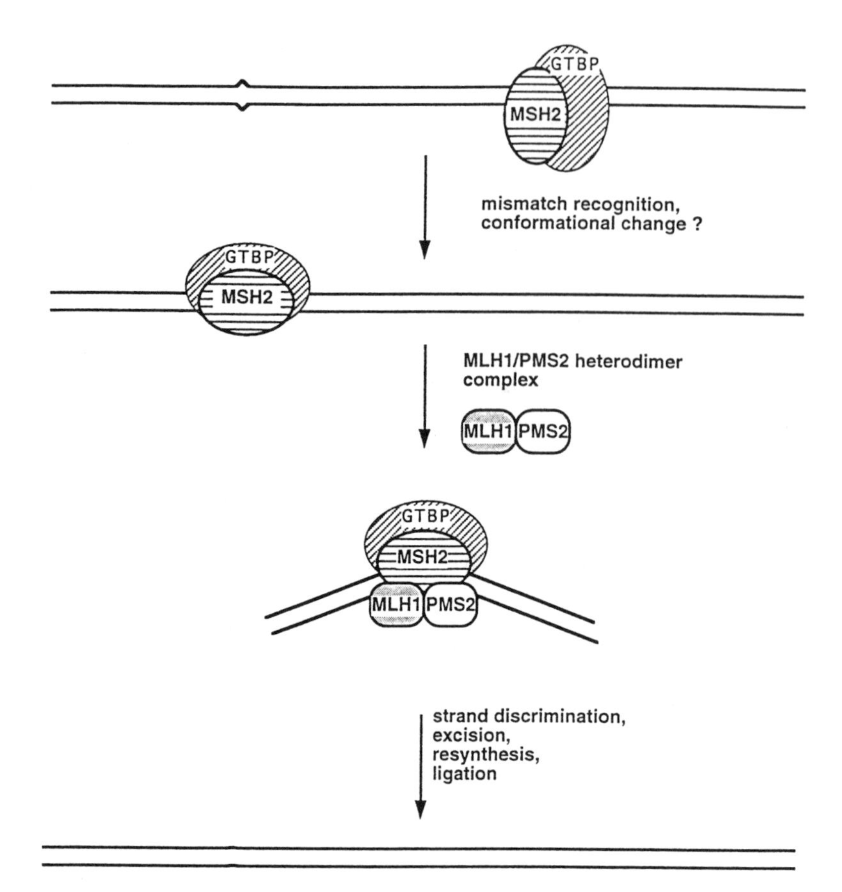

Fig. 1. A model for the initiation of eukaryotic DMR. Based on previous studies with the yeast Mlh1p, Pms1p, and Msh2p, a model was proposed for the initiation of eukaryotic DMR *(114)*. In this model, DNA mismatches or unpaired small heterologies are recognized by Msh2p, followed by binding of an Mlh1–Pms1 heterodimer complex, resulting in the formation of a DNA-bound ternary complex. This stable ternary complex is likely to recruit additional repair factors, such as helicases and exonucleases. This updated "human" DMR model includes a second homolog, GTBP, which has been shown to associate physically with *h*MSH2 *(26)*.

tracts *(133)*. These results suggest that Msh3p function may be limited to correction of very small heterologies. Msh3p may be a factor that increases Msh2p affinity for such heterologies *(133)*. Disruption of *MSH6* leads to a similar weak mutator phenotype in the *hom3-10* reversion assay, but strains deficient for both Msh3p and Msh6p have a much higher reversion rate of the *hom3-10* allele than either single mutant, indicating that the Msh3p and Msh6p have redundant functions in the repair of single-base heterologies *(64)*. The formation of a hetereodimeric complex between the human G-T binding protein (GTBP) and hMSH2, homologs of yeast Msh6p and Msh2p, respectively, has been reported *(26)*. Based on these recent findings, an updated review of our previous model for the initiation of eukaryotic DMR is offered *(114)* (*see* Fig. 1).

Two additional yeast mutS homologs, Msh4p and Msh5p, appear to have no role in mitotic DMR *(54,122)*. *msh4* and *msh5* mutant strains have reduced levels of spore viability owing to a defect in meiotic crossing over *(122)*. It has been proposed that Msh4p plays a role in the resolution of recombination intermediates *(122)*. For a more detailed discussion of the yeast DMR system, please refer to Chapter 19, vol. 1.

3. IDENTIFICATION OF DMR GENES INVOLVED IN HEREDITARY CANCER SYNDROMES

3.1. Hereditary Nonpolyposis Colon Cancer

Evidence for a link between human cancer and DNA replication errors was first obtained when a high frequency of mutations within $(CA)_n$ and other simple repeated sequences (microsatellites) was observed in a subset of sporadic colorectal cancers and in a large percentage of colon cancers from hereditary nonpolyposis colon cancer (HNPCC) families *(1,57,140)*. HNPCC is a common autosomal-dominant cancer predisposition syndrome of high penetrance, characterized mainly by early onset (usually before age 50) of colon carcinomas, as well as cancers of the endometrium, stomach, upper urinary tract, small intestine, and ovary *(88,89,145)*. HNPCC is distinct from the less frequent familial adenomatous polyposis (FAP) syndrome, which is characterized by the presence of multiple colorectal adenomas and associated with germline alterations of the *APC* gene *(45,61,103)*. One clue that the genes affected in HNPCC were involved DMR was provided by the previously mentioned work with *S. cerevisiae* and *E. coli* that together demonstrated that mutations in DMR genes result in destabilization simple-sequence repeats *(75,134)*.

Homology-based approaches were used by several groups to clone human homologs of yeast DNA mismatch repair genes. To date, germline mutations in four homologs of yeast DNA mismatch repair genes have been shown to be associated with HNPCC: *hMSH2*, *hMLH1*, *hPMS1*, and *hPMS2 (15,33,74,102,108)*. Affected family members inherit one defective allele of the gene involved, but tumors usually have lost or suffered a mutation in the allele inherited from the unaffected parent *(51,74,102,108)*, and are therefore DMR-defective. As expected for a defect in a major DMR pathway, a tumor cell line defective in *hMLH1* shows not only microsatellite instability, but also an increased level of base substitution mutations, resulting in a spontaneous "mutator" phenotype *(11)*. Nonneoplastic cells from most affected individuals do not display biochemical evidence of DMR defects or high mutability *(111)*. A notable exception was the observation of mismatch repair deficiency in phenotypically normal cells from two unrelated HNPCC patients that harbor mutations in *hPMS2* and *hMLH1 (110)*. Since these individuals are developmentally normal, it appears that a high mutational load is compatible with development and does not result in very early tumor development.

Mutations in human *hPMS2* and *hPMS1* particularly are relatively rare in HNPCC families *(79)*. Since only one HNPCC family member has been shown to have a germline mutation in *hPMS1* and no tumor analysis for microsatellite instability has been performed in this patient, the role of *hPMS1* in replicative DMR is unclear. As previously mentioned, human *hPMS2* is more closely related to the yeast *PMS1*, and it has been demonstrated that like yeast *PMS1* and *MLH1*, *hPMS2* and *hMLH1* form a heterodimer complex *(76)*. Possibly, *hPMS1* is a component of a secondary DMR pathway functionally homologous to bacterial "short-patch" repair (*see* Chapter 3 in vol. 1 and Chapter 8 in this vol.). Generation of cell lines deficient for *hPMS1* should clarify this issue. The higher frequency of *hMSH2* and *hMLH1* mutations in HNPCC families

Table 1
Mammalian DNA Mismatch Repair Homologs

Mammalian	Yeast functional homolog	Function
MLH1	Mlh1p	Forms a heterodimer with PMS1, binds MSH2 in the presence of mismatches, recruits additional repair factors?, required for proper chromosome pairing in mouse meiosis, tumor suppressor in HNPCC
PMS2	Pms1p	Forms a heterodimer with MLH1, binds to MSH2 in the presence of mismatches, recruits additional repair factors?, required for chiastmata formation in mice, tumor suppressor in HNPCC?
PMS1	?	Function in DMR unknown, tumor suppressor in HNPCC?
MSH2	Msh2p	Recognizes DNA mismatches and small (1–5 bp) insertions, no meiotic role in mice, tumor suppressor in HNPCC
MSH6/GTBP	Msh6p	MSH6/GTBP gene is mutated in some sporadic tumors, MSH6 may increase MSH2 affinity for DNA mismatches
REP3	Msh3p	?

relative to *hPMS2* and *hPMS1* could be explained by a number of factors, such as variable penetrance of mutations and founder effects. Another plausible explanation is that different DMR genes suffer different rates of loss of heterozygosity in somatic tissues, leading to different numbers of cells homozygous for DMR mutations. Interestingly, about a third of HNPCC patients do not have mutations in *hMLH1*, *hPMS1*, *hPMS2*, or *hMSH2 (79)*. The recent discovery of seven novel human genes that are highly homologous to *hPMS2* raises the possibility that additional DMR genes are involved in HNPCC *(101)*. (*See* Table 1 for a list of mammalian DMR homologs.)

Although few studies have addressed the issue of mutations in known tumor suppressor genes and oncogenes in DMR-deficient tumors, one of the original reports on the association between HNPCC and alterations in repeated sequences showed that the incidence of mutations in *KRas*, *p53,* and *APC* was similar between HNPCC and sporadic colorectal cancer *(1)*. Lazar et al. *(73)* have reported accumulation of a surprisingly large number of mutations in tumor suppressor genes during colorectal carcinogenesis in two HNPCC patients. One tumor had accumulated four mutations in *p53*, whereas another tumor had six mutations in the *APC* tumor suppressor gene. Most mutations were either base substitutions or frameshift mutations, a mutation spectrum compatible with loss of DMR *(73)*. The accumulation of such a large number of mutations in a single locus is unlikely to be a random event even in a DMR-deficient genetic background, and may indicate a mechanism of tumor suppressor loss unique to DMR-deficient cells.

3.2. Muir-Torre Syndrome

Muir-Torre syndrome (MTS) encompasses a rare subset of HNPCC families that display sebaceous gland tumors *(19)*. Affected individuals may develop several sebaceous tumors, but these rarely metastasize *(42,124)*. Criteria for establishing diagnosis of MTS are the identification of at least one sebaceous gland tumor and a minimum of one internal malignancy, usually a colorectal carcinoma *(19)*. Even though the majority of MTS patients have multiple internal malignancies, they have a generally good prognosis *(42,124)* with a 12-yr 50% survival rate *(19)*. An analysis of microsatellite instability in MTS suggests that this syndrome may be composed of two subgroups, with one group having widespread microsatellite instability in sebaceous and colorectal tumors *(55)*. Kolodner et al. *(65)* have found frameshift and nonsense mutations in the *hMSH2* gene of affected members of two MTS families. It will be interesting to determine if there is a correlation between MTS and specific alleles of *hMSH2* or other DMR genes. Alternatively, the sebaceous tumors observed in MTS could be owing to the action of other factors, such as modifier genes. A correlation between immune suppression and the appearance of sebaceous tumors in two MTS patients has been observed *(124)*.

3.3. Turcot's Syndrome

Turcot's syndrome (TS) is another rare cancer syndrome, characterized by the presence of colorectal cancers and a primary tumor of the central nervous system. Many TS families meet the clinical criteria for FAP, although TS patients usually have fewer polyps and appear to have an earlier development of colonic carcinoma. Recently, germline *APC* mutations were found in 10 TS families, whereas germline mutations in *hMLH1* and *hPMS2* were found in two families *(47)*. The predominant type of central nervous tumors in the APC-mutated families was medulloblastoma. Gliomas were observed in the two families with mutations in *hMLH1* and *hPMS1*, and also in the families for which no mutations in either *APC* or the known HNPCC genes were observed *(47)*.

Although not mentioned in the Hamilton et al. study *(47)*, the TS patients that harbor mutations in *hMLH1* and *hPMS2* are the same patients previously classified as having HNPCC and shown to have a DMR deficiency in nonneoplastic cells *(110)*. This observation suggests certain mutations (i.e., potential dominant negative mutations) in *hMLH1* and *hPMS2* may lead to a form of HNPCC characterized by a higher risk of gliomas.

4. DNA MISMATCH REPAIR DEFECTS IN SPORADIC CANCER

4.1. Microsatellite Instability in Sporadic Tumors

The PCR-based analysis of genomic dinucleotide loci has facilitated rapid analysis of microsatellite instability in a variety of sporadic tumors. In fact, because of the clonal nature of tumors, microsatellite instability appears to be an excellent molecular marker of loss of DMR function *(82)*. Microsatellite instability is observed in a significant fraction of a variety of common tumors examined, including endometrial *(16,27,120)*, lung *(35,94,112,126)*, breast *(35,48,80,148,150)*, pancreatic *(48)*, gastric *(18,48,78, 112,119,135)*, and prostate *(37)*. These observations suggest a more general role for loss of DMR in cancer development. Comprehensive compilations of microsatellite instability studies in sporadic tumors have been recently published elsewhere *(22,29)*.

The reported incidence of microsatellite instability varies among different tumors, and also varies for different studies addressing the same tumor type *(22,29)*. The incidence of microsatellite instability in breast cancer has been reported to be 20% in one study *(150)* and 0% in another *(85)*. In addition to the obvious problem of the relatively low number of tumors examined in each study, possible explanations for differing results are:

1. The use of different endogenous repeat loci to access microsatellite instability;
2. Different classification criteria for microsatellite instability (i.e., number of repeat loci showing instability/number of loci examined); and
3. Technical problems involved in the commonly used PCR-based assay.

Another often overlooked caveat in the interpretation of microsatellite instability studies of tumors is the fact that it is difficult to establish the "background" microsatellite instability of nonneoplastic cells, since unlike tumors, tissues are not clonal in origin. Genomic instability of dinucleotide repeats in leukocytes is approx 0.1%, whereas hypervariable repeats have a mutation rate as high as 5% *(22)*.

Although many more microsatellite instability studies of sporadic tumors need to be done before firm conclusions can be drawn, some interesting findings are worth mentioning. Obviously, microsatellite instability is not restricted to tumors types observed in HNPCC patients. Microsatellite instability has been consistently observed in a small fraction of lung cancers *(35,94,112,126)*, although this type of cancer appears to be rare in HNPCC families *(145)*. It is also clear that the generally good prognosis of HNPCC *(89,145)*, MTS *(19,42,124)* and TS *(47)* tumors does not necessarily apply to equivalent sporadic tumors reportedly displaying microsatellite instability. One informative example is glioblastoma multiforme, a usually lethal brain tumor that can originate from clonal expansion of astrocytomas *(17)*. In one study, TS with glioblastoma multiforme displaying microsatellite instability had an exceptionally long survival time *(47)*. However, microsatellite instability has been reported to be present in 50% of sporadic glioblastoma multiformes *(22)*. A plausible explanation for this observation is that DMR deficiency occurs as a later event in sporadic glioblastomas compared to the glioblastomas observed in TS. This is supported by the observation that microsatellite instability is absent in low-grade astrocytomas *(22,121)*, but common in high-grade glioblastoma multiforme *(22)*. Perhaps the number of cell divisions after the establishment of a microsatellite instability phenotype is inversely correlated with tumor aggressiveness because of the progressive accumulation of mutations that either impair cell function and/or elicit an immune response.

If DMR-deficient tumor cells display mutated, nonself-antigens on their surface, these cells might elicit a strong host immune response. A role for defects in cellular immunity in the emergence of DMR-deficient tumors is supported by the observation that microsatellite instability is common in Kaposi's sarcoma and lymphomas of HIV-infected patients, but apparently rare in similar tumors of HIV-negative patients *(10)*. Further, microsatellite unstable colorectal cancer-derived cell lines have either mutations in β2 microglobulin or do not express HLA surface proteins *(12,14)*.

Since these proteins are required for cytotoxic T-cell responses, such mutations may be an early step in DMR-associated cancer that allows escape from immune surveillance *(12,14)*. Inherent to this proposal is the idea that a large number of DMR-deficient cells are generated throughout the human life-span and eliminated by immune surveillance, and that only selected variants may be able to establish tumors in immune competent individuals.

4.2. DMR Defects and Mutability of Tumor Derived Cell Lines

A number of mutator human cell lines have been shown to harbor mutations in specific DMR genes. The HCT116 colorectal carcinoma cell line, which is devoid of wild-type *hMLH1 (108)*, has a 200- to 300-fold increase in *HPRT* mutation rate *(11)*. The mismatch repair deficiency of HCT116 nuclear extracts is restored by a HeLa cell activity comprised of a heterodimer of *hMLH1* and *hPMS2 (76)*, thus supporting the idea that *hPMS2* is the true functional homolog of yeast Pms1p. An *h*PMS2-defective cell line (HEC-1-A) displaying microsatellite instability is defective in strand-specific DMR *(120)*. Cell lines derived from PMS2-deficient mice also display a mutator phenotype (*see* Section 5.).

Addition of a heterodimer of *hMSH2* and p160, also known as G-T binding protein (GTBP) *(56)*, can restore strand-specific DMR to extracts of the *hMSH2*-deficient cell line LoVo *(26)*. GTBP was isolated independently by its ability to form a complex with G-T mismatches *(106)*. p160 (GTBP) encodes an MutS homolog *(106)*. Cell lines deficient in GTBP display instability in $(A)_n$ tracts, and a low-level instability in $(CA)_n$ tracts *(107)*. Since GTBP is required for base–base mismatch binding activity in cell extracts *(106)*, the major function of GTBP may be in the recognition of single-base loops and possibly base–base mismatches. The mutation spectrum of yeast cells deficient in Msh6p, an *S. cerevisiae* GTBP homolog, supports this view *(64)*. Given that GTBP germline mutations have not been detected in HNPCC, Papadopoulos et al. has suggested that GTBP deficiency may be either insufficient to induce cancer predisposition or, alternatively, may be incompatible with embryonic development *(107)*. Since mutations in both the *MSH6* and *MSH3* yeast DMR genes are required to induce a strong mutator phenotype in *S. cerevisiae*, the first explanation appears more likely.

Another interesting aspect of DMR-deficient cell lines is resistance to killing by *N*-methyl-*N'*-nitro-*N*-nitrosoguanidine (MNNG), and the guanine base analog 6-thioguanine (6-TG). The initial evidence implicating MMR in MNNG toxicity was obtained with bacteria mutated at the *dam*, *mutL*, and *mutS* genes *(59,60)*. DMR status has no effect on MNNG-induced mutability, suggesting that toxicity and mutations result from different events *(60)*. Cell lines deficient for either mouse *Msh2* or *hMSH2* have been shown to be MNNG-resistant *(23,63)*, although the molecular mechanism is unknown. It has been proposed that MNNG induces DNA lesions that can be recognized by the DMR system, but that lead to abortive attempts of repair *(41,59,60)*.

The *hMLH1*-deficient cell line, HCT116 *(11)*, has been reported to be defective in an MNNG-induced G2 cell-cycle checkpoint *(50)*. It was suggested that in wild-type cells, the presence of lesions than can be sensed by the DMR system leads to signaling between the DMR machinery and proteins involved in cell-cycle arrest *(50)*. An alternative explanation for these observations is that MNNG and 6-TG lead to continuous, futile attempts of DMR, resulting in strand discontinuities that are the actual signal leading to cell-cycle arrest. Treatment with alkylation agents does induce strand breaks in mammalian cells *(36)*, and it is also known that some yeast mutants defective in DNA strand break repair are hypersensitive to alkylating agents *(24,53)*.

5. GENERATION AND ANALYSIS OF MICE DEFICIENT IN DMR

Recent developments in the genetic manipulation of mouse embryonic stem (ES) cells have permitted the production and analysis of mice with predetermined gene mutations *(13)*. To understand better the role of DMR processes in mammals, several groups have used gene targeting with mouse ES cells to generate animals with null mutations in either of the DMR gene homologs, *Pms*2 *(8)* or *Msh*2 *(23,118)* and *Mlh*1 *(7,28)*. As discussed below, these studies have resulted in the production of animals that appear to lack DMR activity in all tissues.

5.1. DMR-Deficient Mice Are Viable

Mice homozygous for inactivating mutations in either *Msh2*, *Pms2*, or *Mlh1* are viable and show no detectable increases in embryonic lethality based on expected Mendelian ratios from heterozygous breedings. The viability of DMR-deficient animals suggests that murine development can accommodate elevated levels of spontaneous mutation. We offer the following factors as relevant to the somewhat unexpected health of mice with elevated spontaneous mutation:

1. Functional diploidy for most of the genome;
2. DMR-deficiency expected to result primarily in point mutations, which in turn, will produce recessive loss of function mutations; and
3. Compensatory mechanisms, such as apoptosis or cell-cycle checkpoint-mediated cell removal systems.

The study of mice with mutations in DMR and other cell processes could shed some light on these issues.

5.2. Mice Deficient in DMR Are Prone to Cancer

Both *Pms2*-deficient and *Msh2*-deficient animals are prone to lymphomas and sarcomas *(8,23,118)*. In two studies, the lymphomas were reported as T-cell in origin *(23,118)*. An earlier age of incidence of lymphomas for one line of *Msh2*-deficient animals has been reported *(23)*, as compared to the other *Msh2*-deficient mice *(118)*. However, differences in genetic background of the mouse strains might account for this apparent difference. Therefore, at least in the first months of life, the predisposition to cancer of *Msh2*- and *Pms2*-deficient animals appears to be similar both tumors in different tissues in timing and tumor type. Animals heterozygous for DMR mutations may develop, although at a later age. Relevant to this point, different tumor types do occur in mice heterozygous vs homozygous for null *p53* mutation *(49)*.

5.3. Msh2-, Pms2-, and Mlh1-Deficient Mice Are General Mutators

Studies indicate that the *Msh2*, *Pms2*, and *Mlh1* knockout mice do in fact display a mutator phenotype in all tissues examined. Frequent mutation in microsatellite sequences was observed in tumors from homozygous *Pms2* mutant mice *(8)*, from

Msh2-deficient ES cells *(23)*, and from single-molecule PCR experiments with tail and spleen DNA from homozygous *Pms2* and *Mlh1* mutant mice *(7,8)*. In addition to microsatellite instability in somatic tissues of homozygous deficient mice, microsatellite instability was also observed in the male germline *(7,8)*. In fact, mutation frequencies for dinucleotide repeats in spermatozoa from *Pms2*- and *Mlh1*-deficient males was 1000-fold greater than that estimated for wild-type animals *(25)*. Preliminary data indicate that *PMS2* is also involved in DMR in the female germline *(9)*.

Cell studies indicate a general role of *Msh2* and *Pms2* and *Mlh1* in mutation avoidance. ES cells deficient in *Msh2* show increased microsatellite mutation and elevated forward mutation at the *HPRT* locus *(23)*. Likewise, fibroblasts derived from *Pms2*-deficient animals show not only increased microsatellite instability, but also elevated rates of resistance to ouabain and *HPRT* mutation *(9)*. The human HCT116 colorectal carcinoma cell line, which is devoid of wild-type *hMLH1 (108)*, has a 200- to 300-fold increase in *HPRT* mutation rate *(11)*. Therefore, *hMSH2*, *hPMS2*, and *hMLH1* appear to be involved in a major DMR pathway in most if not all cell types.

5.4. Role of Msh2 in Regulating Targeted Recombination Between Nonidentical Sequences

Previous studies indicated that efficient gene targeting in wild-type ES cells at the *Rb* locus requires the use of isogenic DNA in the targeting vector, indicating a sensitivity to even small amounts (e.g., ~1%) of mismatched sequences *(139)*. This preference for perfectly matched sequences is absent in *Msh2*-deficient ES cells, indicating that the MSH2 protein functions to prevent recombination between diverged sequences *(23)*. In vitro studies using a RecA-driven strand exchange indicate that bacterial proteins MutS and MutL regulate heteroduplex DNA formation between homologous sequences *(149)*. By analogy with the bacterial DMR system, MSH2 and presumably mammalian MutL homologs have a similar function of monitoring heteroduplex DNA formation in genetic recombination.

5.5. Involvement of Pms2 and Mlh1 in Meiosis

A prominent difference between the *Pms2* and *Msh2* knockouts is that the *Pms2*-deficient males are sterile *(8)*, whereas *Msh2*-deficient males are fertile *(23)*. *Pms2*-deficient males produce only abnormally shaped spermatozoa in reduce numbers and show an abnormal testicular histology. Subsequent analysis of chromosome synapsis (pairing) revealed frequent synaptic aberrations during meiotic prophase I (Fig. 2) *(8)*. Results of other cases of male-specific infertility suggested that the degree of synaptic abnormalities was sufficient to account for meiotic failure and resultant infertility. It was further suggested that the synaptic problems reflected a role for PMS2 in a recombination-driven homology search *(8)*. The recent analysis of *Mlh1*-deficient mice has shown that both males and females are sterile *(8,28)*. *Mlh1*-deficient spermatocytes exhibit high levels of prematurely separated chromosomes and arrest in first division meiosis *(8,28)*. MLH1 appears to localize to sites of crossing over on meiotic chromosomes *(8)*, a finding that suggests that MLH1 plays a role in reciprocal recombination. The different phenotypes of *Mlh1*- and *Pms2*-deficient mice suggest that these proteins can function independently during

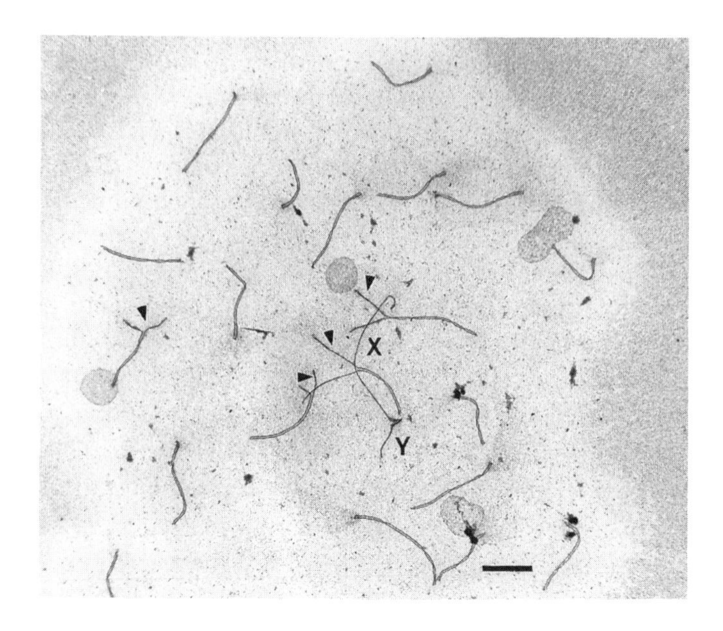

Fig. 2. Chromosome pairing abnormalities in *Pms2*-deficient male mice. Electron micrograph illustrating synapsis (pairing) abnormalities in the male germline from a *Pms2*-deficient male mouse. Shown is an example of a nucleus at late zygotene or early pachytene stage. Although most of the bivalents are fully synapsed, the nucleus contains four bivalents that are not completely synapsed, indicated by arrowheads. Three bivalents are clustered around the XY. Scale bar, 500 nm.

meiosis. Because *Msh2*-deficient males are fertile, another MutS-like protein is likely to be involved in this aspect of meiosis.

6. TUMOR TISSUE SPECIFICITY

Given that DMR deficiency in any dividing tissue should result in a high mutational load, the tumor tissue specificity observed in human HNPCC and DMR-deficient mice is of interest. Affected HNPCC family members typically develop colon cancer, but they also have statistically significant increased frequencies of cancers of the endometrium, stomach, upper urinary tract, small intestine, and ovary *(88,89,145)*. In contrast, mice deficient in either *Pms2* or *Msh2* develop mainly lymphomas and a few sarcomas. de Wind et al. have proposed that rapid cell expansion is required to unveil the effects of DMR deficiency in tumorigenesis, and that the developing immune system in newborn mice provides an opportunity for DMR-induced mutations *(23)*. Reitmar et al. suggested that enhanced recombination between evolutionary divergent endogenous retroviruses, resulting in translocations, may be a factor in the development of hematopoietic tumors in DMR-deficient mice *(118)*. Since HNPCC family members are heterozygous for DMR gene mutations, whereas most current tumor data from DMR-deficient animals come from homozygous animals, it would be premature to draw any conclusions regarding species-related tumor specificity. A relevant observation is that *p53*$^{-/-}$ homozygous mice develop lymphomas, but in heterozygous mice, osteosarcomas and soft tissue sarcomas predominate *(49)*.

Dietary factors are known to play a major role in human carcinogenesis *(39,40,141)*, so the predominance of colonic tumors in HNPCC families may be a result of exposure to dietary tumor promoters that increase colonic cell proliferation. In fact, there is compelling evidence that most carcinogens are mitogens *(4)*. These issues can be addressed by altering diets, including exposing DMR-deficient mice to typical human dietary carcinogens, especially those resulting from food cooking, such as oxidized fats and heterocyclic amines *(5)*.

Important clues regarding the relationship of tumor tissue specificity to genetic or environmental factors can also be obtained through studies of large "cancer-prone" families. A study of HNPCC family "G," originally undertaken by Warthin in 1895 *(144)*, and followed up by Hauser and Weller in 1936, and more recently by Lynch and Krush in 1971 *(87)*, provides interesting information regarding familial cancer. The progenitor of family G had ten progeny, six of which developed histologically verified cancer. These individuals originated family branches A–J, which have been followed up. Although affected family members develop the typical HNPCC tumors, such as adenocarcinomas of the colon and endometrium, they also appear to display an elevated level of chronic lymphocytic leukemia, sarcomas, and brain tumors. These studies suggest that HNPCC is in fact associated with a wider tumor spectrum than currently assumed, and that genetic and/or environment factors may be important determinants of tumor specificity in HNPCC. Surprisingly, no mutations in either *hMSH2*, *hMLH1*, or *hPMS2* have been found in affected members of family G *(142)*. The definition of HNPCC as a syndrome of "autosomal dominantly inherited susceptibility of colorectal cancer with early age of onset" *(89)* is likely to be biasing the identification of families with DMR mutations toward those families where colorectal tumors predominate. Whether families that are susceptible to other types of malignancies have defects in DMR genes remains to be determined.

Recent observations indicate that most colon and gastric cancer cells displaying microsatellite instability lack expression of TGF-β type II receptors *(92)*. Sequence analysis of the TGF-β type II receptor gene (RII) revealed mutations within the coding region in two simple-sequence repeat tracts, $(GT)_3$ and $(A)_{10}$ *(92)*. Since TGF-β may play a role in growth inhibition of epithelial tumors, it has been proposed that RER^+ associated colon cancer usually proceeds through specific mutations in this particular gene *(92)*. Interestingly, the mouse TGF-βRII gene lacks the $(A)_{10}$ sequence found in the human gene *(72)*, and as noted above, DMR-deficient mice do not appear to develop early CRC. Although limited and preliminary, these findings suggest that the occurrence of particular mutational hotspots in oncogenes or tumor suppressor genes may be an important factor in the observed tissue and species specificity of DMR-deficient tumors.

7. THE ROLE OF MUTATION IN CANCER DEVELOPMENT

A causal role for mutation in cancer development is strongly supported by the detection of multiple mutations in human tumors, by an association between carcinogens and mutagens, and more importantly, by the finding that at least one somatic mutation is required for the development of tumors in cancer susceptibility syndromes, such as retinoblastoma (reviewed in refs. *5* and *81*). A more controversial issue is whether the accumulation of mutations *per se* is actually a rate limiting step for tumor development *(110)*. Cancer incidence increases roughly with the sixth power of age, which has been suggested to be owing to the need for multiple events or mutations in the development of most tumors *(6)*. The finding that mice *(8,23)* and possibly humans *(110)* can develop normally, without early tumor development in the absence of DMR, has challenged previous assumptions regarding the role of mutation in cancer. Since DMR mutant cells have spontaneous mutation rates that are at least a hundredfold higher than normal *(11,23)*, one would expect embryonic tumor development in DMR-deficient animals, based on either two-hit *(99)*, or even multple-hit cancer development models *(6)*. Such is clearly not the case for either *Msh2*- or *Pms2*- deficient animals, which typically develop lymphomas several months postdevelopment *(8,23)*. Clearly, a high mutational load is not sufficient to induce embryonic or neonatal tumors in mice.

If a high mutational load is insufficient to induce early tumor growth, then which factors are truly rate-limiting for tumor development? Several lines of evidence indicate that epigenetic factors may be involved in determining cancer development rates, but any such factor must somehow correlate with the marked dependency of cancer incidence on age *(6)*.

One factor that apparently satisfies this criteria is DNA methylation, since cellular aging is characterized by in vivo alterations in methylation status *(62,93,97,105,130)*. Since changes in methylation correlate with alterations in expression levels of a variety of genes, DNA methylation status may have a role in tumor development. This hypothesis, reviewed recently *(21)*, is supported by several observations:

1. Chemically induced reduction of the activity of DNA methyltransferase, dramatically reduces the rate of adenoma formation in Apc^{Min} mice *(69)*;
2. Alterations of methylation patterns have been observed in both cellular oncogenes and tumor suppressor genes during the process of tumor promotion *(70)*; and
3. Methyl-deficient diets lead to spontaneous liver tumors in mice *(95,125)* and may result in higher number of colonic polyps in humans *(39)*.

One theory to explain the role of DNA methylation in tumor development is that DNA methyltransferase affects the rate of tumor development through a mutagenic mechanism that leads to higher levels of deamination of 5-methylcytosine in critical target genes *(21,58,69)*.

8. FUTURE DIRECTIONS

The recent discovery that mutations in DMR genes are involved in HNPCC has prompted many studies aimed at characterizing the eukaryotic DMR system and its involvement in human cancer, both inheritable and sporadic. Important issues that should be addressed in the near future include tumor tissue specificity in HNPCC and DMR-deficient mice, the role of spontaneous mutation rate in tumor development, and the identification of novel "mutator" genes involved in HNPCC and sporadic cancer. Finally, the availability of DMR-deficient mice should facilitate the development of compounds that specifically target DMR-deficient cells, providing a rational approach for HNPCC-related tumors and sporadic tumors displaying microsatellite instability.

REFERENCES

1. Aaltonen, L. A., P. Peltomaki, F. Leach, P. Sistonen, S. M. Pylkkanen, J.-P. Mecklin, H. Jarvinen, S. Powell, J. Jen, S. R. Hamilton, G. M. Petersen, K. W. Kinzler, B. Vogelstein, and A. de la Chapelle. 1993. Clues to the pathogenesis of familial colorectal cancer. *Science* **260:** 812–816.
2. Alani, E., N.-W. Chi, and R. Kolodner. 1995. The *Saccharomyces cerevisiae* Msh2 protein specifically binds to duplex oligonucleotides containing mismatched DNA base pairs and insertions. *Genes Dev.* **9:** 234–247.
3. Alani, E., N.-W. Chi, and R. D. Kolodner. 1995. The *Saccharomyces cerevisiae* MSh2 protein specifically binds to duplex oligonucleotides containing mismatched DNA base pairs and insertions. *Genes Dev.* **9**:234–247.
4. Ames, B. N. and L. S. Gold. 1990. Too many rodent carcinogens: mitogenesis increases mutagenesis. *Science* **249:** 970,971.
5. Ames, B. N., R. Magaw, and L. S. Gold. 1987. Ranking possible carcinogenic hazards. *Science* **236:** 271–279.
6. Armitage, P. and R. Doll. 1954. The age distribution of cancer and a multi-stage theory of carcinogenesis. *Br. J. Cancer* **8:** 1–12.
7. Baker, S., A. Plug, T. Prolla, C. Bronner, A. Harris, X. Yao, D.-M. Christie, C. Monell, N. Arnheim, A. Bradley, T. Ashley, and R. Liskay. 1996. Involvement of *Mlh1* in DNA mismatch repair and meiotic crossing over. *Nature Genet.* **13:** 336–342.
8. Baker, S. M., C. E. Bronner, L. Zhang, A.W. Plug, M. Robatzek, G. Warren, E. A. Elliot, J. Wu, T. Ashley, N. Arnheim, and R. M. Liskay. 1995. Male mice defective in the DNA mismatch repair gene PMS2 exhibit abnormal chromosome synapsis and meiosis. *Cell* **82:** 309–319.
9. Baker, S. M., C. E. Bronner, and R. M. Liskay. Unpublished observations.
10. Bedi, G. C., W. H. Westra, H. Farazdegan, P. M. Pitha, and D. Sidransky. 1995. Microsatellite instability in primary neoplasms from HIV+ patients. *Nature Med.* **1:** 65–68.
11. Bhattacharyya, N. P., A. Skandalis, A. Ganesh, J. Groden, and M. Meuth. 1994. Mutator phenotype in human colorectal carcinoma cell lines. *Proc. Natl. Acad. Sci. USA* **91:** 6319–6323.
12. Bicknell, D. C., A. Rowan, and W. F. Bodmer. 1994. β2M-microglobulin gene mutations: a study of established colorectal cell lines and fresh tumor. *Proc. Natl. Acad. Sci. USA* **91:** 4751–4755.
13. Bradley, A., P. Hasty, A. Davis, and R. Ramirez-Solis. 1992. Modifying the mouse: design and desire. *BioTechnology* **10:** 534–539.
14. Branch, B., D. C. Bicknell, A. Rowan, W. F. Bodmer, and P. Karran. 1995. Immune surveilance in colorectal carcinoma. *Nature Genet.* **9:** 231,232.
15. Bronner, C. E., S. M. Baker, P. T. Morrison, G. Warren, L. G. Smith, M. K. Lescoe, M. Kane, C. Earabino, J. Lipford, A. Lindblom, P. Tannergard, R. J. Bollag, A. R. Godwin, D. C. Ward, M. Nordenskjold, R. Fishel, R. Kolodner, and R. M. Liskay. 1994. Mutation in the DNA mismatch repair gene homologue *hMLH1* is associated with hereditary nonpolyposis colon cancer. *Nature* **368:** 258–261.
16. Burks, R. T., T. D. Kessis, K. R. Cho, and L. Hedrick. 1994. Microsatellite instability in endometrial carcinoma. *Oncogene* **9:** 1163–1166.
17. Cavenee, W. B., H. J. Scrable, and C. D. James. 1991. Molecular genetics of human cancer predisposition and progression. Mutation Research **247:** 199–202.
18. Chong, J.-M., M. Fukayama, Y. Hayashi, T. Takizawa, M. Koike, M. Konishi, R. Kikuchi-Yanoshita, and M. Miyaki. 1994. Microsatellite instability in the progression of gastric carcinoma. *Cancer Res.* **54:** 4595–4597.
19. Cohen, P. R. and S. R. Kohn. 1991. Association of sebaceous gland tumors and internal malignancy: the Muir-Torre syndrome. *Am. J. Med.* **90:** 606–613.

20. Cooper, D. L., R. S. Lahue, and P. Modrich. 1993. Methyl-directed Mismatch Repair Is Bidirectional. *J. Biol. Chem.* **268:** 11823–11829.
21. Counts, J. L. and J. I. Goodman. 1995. Alterations in DNA methylation may play a variety of roles in carcinogenesis. *Cell* **83:** 13–15.
22. Dams, E., E. J. Van de Kelft, J. J. Martin, J. Verlooy, and P. J. Willems. 1995. Instability of microsatellites in human gliomas. *Cancer Res.* **55:** 1547–1549.
23. de Wind, N., M. Dekker, A. Bern, M. Radman, and H. te Reile. 1995. Inactivation of the mouse *Msh2* gene results in mismatch repair deficiency, methylation tolerance, hyper-recombination, and predisposition to cancer. *Cell* **82:** 321–330.
24. Dhillon, N. and M. F. Hoekstra. 1994. Characterization of two protein kinases from *Schizosaccaromyces pombe* involved in the regulation of DNA repair. *EMBO J.* 2777–2788.
25. Dietrich, W., H. Katz, S. E. Lincoln, H.-S. Shin, J. Friedman, N. Dracopoli, and E. S. Lander. 1992. A genetic map of the mouse suitable for typing intraspecific crosses. *Genetics* **131:** 423–447.
26. Drummond, J. T., G.-M. Li, M. J. Longley, and P. Modrich. 1995. Isolation of an hMSH2-p160 heterodimer that restores DNA mismatch repair to tumor cells. *Science* **268:** 1909–1912.
27. Duggan, B. D., J. C. Felix, L. I. Muderspach, D. Tourgeman, J. Zheng, and D. Shibata. 1994. Microsatellite instability in sporadic endometrial carcinoma. *J. Natl. Cancer. Inst.* **86:** 1216–1221.
28. Edelman, W., P. Cohen, M. Kane, K. Lau, B. Morrow, S. Bennet, A. Umar, T. Kunkel, G. Cattoretti, R. Chaganti, J. Pollard, R. Kolodner, and R. Kucherlapati. 1996. Meiotic pachytene arrest in MLH1-deficient mice. *Cell* **85:** 1125–1134.
29. Eshleman, J. R. and S. D. Markowitz. 1995. Microsatellite instability in inherited and sporadic neoplasms. *Curr. Opinion Oncol.* **7:** 83–89.
30. Fearon, E. R. and B. Vogelstein. 1990. A genetic model for colorectal tumorigenesis. *Cell* **61:** 759–767.
31. Fijalkowska, I. J., R. L. Dunn, and R. M. Schaaper. 1993. Antimutator mutations in the alpha subunit of *Escherichia coli* DNA polymerase III: identification of the responsible mutations and alignment with other DNA polymerases. *Genetics* **134:** 1039–1044.
32. Fijalkowska, I. J., R. L. Dunn, and R. M. Schaaper. 1993. Mutants in *Escherichia coli* with increased fidelity of DNA replication. *Genetics* **134:** 23–30.
33. Fishel, R. A., M. K. Lescoe, M. R. S. Rao, N. Copeland, N. Jenkins, J. Garber, M. Kane, and R. Kolodner. 1993. The human mutator gene homolog MSH2 and its association with hereditary nonpolyposis colon cancer. *Cell* **75:** 1027–1038.
34. Fogel, S., R. K. Mortimer, and K. Lusnak. 1981. Mechanisms of meiotic gene conversion, or "wanderings on a foreign strand," in *The Molecular Biology of the Yeast* Saccharomyces: *Life Cycle and Inheritance* (Strathern, J. N., ed.), Cold Spring Harbor Laboratory, Cold Spring Harbor, NY, p. 289.
35. Fong, K. W., P. V. Zimmerman, and P. J. Smith. 1995. Microsatellite instability and other molecular abnormatilies in non-small cell lung cancer. *Cancer Res.* **55:** 28–30.
36. Fujiwara, Y. 1975. Postreplication repair of alkylation damage to DNA of mammalian cells in culture. *Cancer Res.* **35:** 2780–2789.
37. Gao, X., N. Wu, D. Grignon, A. Zacharek, A. Liu, A. Salkowski, G. Li, W. Sakr, F. Sarkar, A. T. Porter, Y. Q. Chen, and K. V. Honn. 1994. High frequency of mutator phenotype in human prostatic adenocarcinoma. *Oncogene* **9:** 2999–3003.
38. Gibson, T. C., M. L. Scheppe, and E. C. Cox. 1970. Fitness of an *Escherichia coli* mutator gene. *Science* **169:** 686–688.
39. Giovannucci, E., M. J. Stampfer, G. A. Colditz, E. B. Rimm, D. Trichopoulos, B. A. Rosner, F. E. Speizer, and W. C. Willet. 1993. Folate, methionine, and alcohol intake and risk of colorectal adenoma. *J. Natl. Cancer Inst.* **85:** 875–884.
40. Giovannuci, E. and W. C. Willet. 1994. Dietary factors and risk of colon cancer. *Ann. Rev. Med.* **26:** 443–452.

41. Goldmacher, V. S., R. A. Cuzick, and W. G. Thilly. 1986. Isolation and partial characterization of human cell mutants differing in sensitivity to killing and mutation by methylnitrosourea and *N*-methyl-*N*'-nitro-*N*-nitrosoguanidine. *J. Biol. Chem.* **261:** 12,462–12,469.
42. Graham, R., P. McKee, D. McGibbon, and E. Heyderman. 1985. Torre-Muir syndrome. *Cancer* **55:** 2868–2873.
43. Grilley, M., J. Holmes, B. Yashar, and P. Modrich. 1990. Mechanisms of DNA-mismatch correction. *Mutat. Res.* **236:** 253–267.
44. Grilley, M., K. M. Welsh, S.-S. Su, and P. Modrich. 1989. Isolation and characterization of the *Escherichia coli mutL* gene product. *J. Biol. Chem.* **264:** 1000–1004.
45. Groden, J., A. Thliveris, W. Samowitz, M. Carlson, L. Gelbert, H. Albertsen, G. Joslyn, J. Stevens, L. Spirio, M. Robertson, L. Sargeant, K. Krapcho, E. Wolff, R. Burt, J. P. Hughes, J. Warrington, J. McPherson, J. Wasmuth, D. Le Paslier, H. Abderrahim, D. Cohen, M. Leppert, and R. White. 1991. Identification and characterization of the Familial Adenomatous Polyposis Coli gene. *Cell* **66:** 589–600.
46. Hamada, H. and T. Kakunaga. 1982. Potential Z-DNA forming sequences are highly dispersed in the human genome. *Nature* 396–398.
47. Hamilton, S. R., B. Liu, R. E. Parsons, N. Papadopoulos, J. Jen, S. M. Powell, A. J. Krush, T. Berk, Z. Cohen, B. Tetu, P. C. Burger, P. A. Wood, F. Taqi, S. V. Booker, G. M. Petersen, J. A. Offerhaus, A. C. Tersmette, F. M. Giardiello, B. Vogelstein, and K. Kinzler. 1995. The molecular basis of Turcot's syndrome. *N. Engl. J. Med.* **332:** 839–847.
48. Han, H., A. Yanagisawa, Y. Kato, J. Park, and Y. Nakamura. 1993. Genetic instability in pancreatic cancer and poorly differentiated type of gastric cancer. *Cancer Res.* **53:** 5087–5089.
49. Harvey, M., M. J. McArthur, C. A. Montgomery, Jr., J. S. Butel, A. Bradley, and L. A. Donehower. 1993. Spontaneous and carcinogen-induced tumorigenesis in p53-deficient mice. *Nature Genet.* **5:** 225–229.
50. Hawn, M. T., A. Umar, J. M. Carethers, G. Marra, T. A. Kunkel, C. R. Boland, and M. Koi. 1995. Evidence for a connection between the mismatch repair system and the G2 cell cycle checkpoint. *Cancer Res.* **55:** 3721–3725.
51. Hemminki, A., P. Peltomaki, J.-P. Mecklin, H. Jarvinen, R. Salovaara, M. Nystrom-Lahti, A. de la Chapelle, and L. A. Aaltonen. 1994. Loss of the wild type *MLH1* gene is a feature of hereditary nonpolyposis colorectal cancer. *Nature Genet.* **8:** 405–410.
52. Hickson, I. D., H. M. Arthur, D. Bramhill, and P. T. Emmerson. 1983. The *E. coli uvrD* gene product is DNA helicase II. *Mol. Gen. Genet.* **190:** 265–270.
53. Hoekstra, M. F., R. M. Liskay, A. C. Ou, A. J. DeMaggio, D. G. Burbee, and F. Hiffron. 1991. *HRR25*, a putative protein kinase from budding yeast: association with repair of damaged DNA. *Science* **253:** 1031–1034.
54. Hollingsworth, N. M., L. Ponte, and C. Halsey. 1995. MSH5, a novel MutS homolog, facilitates meiotic reciprocal recombination between homologs in *Saccharomyces cerevisiae* but not mismatch repair. *Genes Dev.* **9:** 1728–1739.
55. Honchel, R., K. C. Halling, D. J. Schaid, M. Pittelkow, and S. N. Thibodeau. 1994. Microsatellite instability in Muir-Torre syndrome. *Cancer Res.* **54:** 1159–1163.
56. Hughes, M. J. and J. Jiricny. 1992. The purification of a human mismatch-binding protein and identification of its associated ATPase and helicase activities. *J. Biol. Chem.* **267:** 23,876–23,882.
57. Ionov, Y., M. A. Peinado, S. Malkbosyan, D. Shibata, and M. Perucho. 1993. Ubiquitous somatic mutations in simple repeated sequences reveal a new mechanism for colonic carcinogenesis. *Nature* **260:** 558–561.
58. Jiang-Cheng, W. M. Rideout, III, and P. A. Jones. 1992. High frequency mutagenesis by a DNA methyltransferase. *Cell* **71:** 1073–1080.

59. Jones, M. and R. Wagner. 1981. N-methyl-N'-nitro-N-nitrosoguanidine sensitivity of *E. coli* mutants deficient in DNA methylation and mismatch repair. *Mol. Gen. Genet.* **184:** 562,563.
60. Karran, P. and M. G. Marinus. 1982. Mismatch correction of O^6-methylguanine residues in *E. coli* DNA. *Nature* **296:** 868,869.
61. Kinzler, K. W., M. C. Nilbert, L.-K. Su, B. Vogelstein, T. M. Bryan, D. B. Levy, K. J. Smith, A. C. Preisinger, P. Hedge, D. McKechnie, R. Finniear, A. Markham, J. Groffen, M. S. Boguski, S. F. Altschul, A. Horii, H. Ando, Y. Miyoshi, Y. Miki, I. Nishisho, and Y. Nakamura. 1991. Identification of FAP locus genes from chromosome 5q21. *Science* **253:** 661–665.
62. Kitraki, E., E. Bozas, H. Philippidis, and F. Stylianopoulou. 1993. Aging-related changes in IGF-II and c-*fos* gene expression in the rat brain. *Intl. J. Dev. Neurosci.* **11:** 1–9.
63. Koi, M., A. Umar, D. P. Chauhan, S. P. Cherian, J. M. Carethers, T. A. Kunkel, and C. R. Boland. 1994. Human chromosome 3 corrects mismatch repair deficiency and microsatellite instability and reduces *N*-methyl-*N'*-nitro-*N*-nitrosoguanidine tolerance in colon tumor cells with homozygous *hMLH1* mutation. *Cancer Res.* **54:** 4308–4312.
64. Mersischky, G. T., M. F. Kane, and R. Kolodner.1996. Redundancy of *Saccharomyces cerevisiae* MSH6 and MSH3 in MSH2-dependent mismatch repair. *Genes Dev.* **10**: 407–420.
65. Kolodner, R. D., N. R. Hall, J. Lipford, M. F. Kane, M. R. S. Rao, P. Morrison, L. Wirth, P. J. Finan, J. Burn, P. Chapman, C. Earabino, E. Merchant, and D. T. Bishop. 1994. Structure of the human MSH2 locus and analysis of two Muir-Torre kindreds for msh2 mutations. *Genomics* **24:** 516–526.
66. Kramer, W., B. Kramer, M. S. Williamson, and S. Fogel. 1989. Cloning and nucleotide sequence of DNA mismatch repair gene *PMS1* from *Saccharomyces cerevisiae*: homology of PMS1 to procaryotic MutL and HexB. *J. Bacteriol.* **171:** 5339–5346.
67. Kunkel, T. A., S. S. Patel, and K. A. Johnson. 1994. Error-prone replication of repeated DNA sequences by T7 DNA polymerase in the absence of its processivity subunit. *Proc. Natl. Acad. Sci. USA* **91:** 6830–6834.
68. Lahue, R. S., K. G. Au, and P. Modrich. 1989. DNA mismatch correction in a defined system. *Science* **245:** 160–164.
69. Laird, P. W., L. Jackson-Grusby, A. Fazeli, S. L. Dickinson, W. E. Jung, E. Li, R. A. Weinberg, and R. Jaenisch. 1995. Suppression of intestinal neoplasia by DNA hypomethylation. *Cell* **81:** 197–205.
70. Laird, P. W. and R. Jaenisch. 1994. DNA methylation and cancer. *Hum. Mol. Genet.* **3:** 1487–1495.
71. Längle-Rouault, F., M. G. Maenhaut, and M. Radman. 1987. GATC sequences, DNA nicks and the MutH function in *Escherichia coli* mismatch repair. *EMBO J.* **6:** 1121–1127.
72. Lawler, S., A. F. Candia, R. Ebner, L. Shum, A. R. Lopez, H. L. Moses, C. V. E. Wright, and R. Derynck. 1994. The murine type II TGF-β receptor has a coincident embryonic expression and binding preference for TGF-β1. *Development* **120:** 165–175.
73. Lazar, V., S. Grandjouan, C. Bognel, D. Couturier, P. Rougier, D. Bellet, and B. B. Pallerets. 1994. Accumulation of multiple mutations in tumor suppressor genes during colorectal tumorigenesis in HNPCC patients. *Hum. Mol. Genet.* **3:** 2257–2260.
74. Leach, F. S., N. C. Nicolaides, N. Papadopoulos, B. Liu, J. Jen, R. Parsons, P. Peltomaki, P. Sistonen, L. A. Aaltonen, M. Nystrom-Lahti, X.-Y. Guan, J. Zhang, P. S. Meltzer, J.-W. Yu, F.-T. Kao, D. J. Chen, K. M. Cerosaletti, R. E. K. Fournier, S. Todd, T. Lewis, R. J. Leach, S. L. Naylor, J. Weissenbach, J.-P. Mecklin, H. Jarvinen, G. M. Petersen, S. R. Hamilton, J. Green, J. Jass, P. Watson, H. T. Lynch, J. M. Trent, A. de la Chapelle, K. W. Kinzler, and B. Vogelstein. 1993. Mutations of a mutS homolog in hereditary nonpolyposis colorectal cancer. *Cell* **75:** 1215–1225.
75. Levinson, G. and G. A. Gutman. 1987. High frequencies of short frameshifts in poly-CA/TG tandem repeats borne by bacteriophage M13 in *Escherichia coli* K–12. *Nucleic Acids Res.* **15:** 5313–5338.

76. Li, G.-M. and P. Modrich. 1995. Restoration of mismatch repair to nuclear extracts of H6 colorectal tumor cells by a heterodimer of human MutL homologs. *Proc. Natl. Acad. Sci. USA* **92:** 1950–1954.
77. Liberfarb, R. M. and V. Bryson. 1970. Isolation, characterization and genetic analysis of mutator genes in Escherichia coli B and K12. *J. Bacteriol.* **104:** 363–375.
78. Lin, J.-T., M.-S. Wu, C.-T. Shun, W.-J. Lee, J.-C. Sheu, and T.-H. Wang. 1995. Ocurrence of microsatellite instability in gastric carcinoma is associated with enhanced expression of erbB–2 oncoprotein. *Cancer Res.*1428–1430.
79. Liu, B., N. C. Nicolaides, S. Markowitz, J. K. V. Willson, R. E. Parsons, J. Jen, N. Papdopolous, P. Peltomaki, A. de la Chapelle, S. R. Hamilton, K. W. Kinzler, and B. Vogelstein. 1995. Mismatch repair gene defects in sporadic colorectal cancers with microsatellite instability. *Nature Genet.* **9:** 48–55.
80. Loathe, R. A. 1993. Genomic instability in colorectal cancer: Relationship to clinicopathological variables and family history. *Cancer Res.* **53:** 5849–5852.
81. Loeb, L. A. 1991. Mutator phenotype may be required for multistage carcinogenesis. *Cancer Res.* **51:** 3075–3079.
82. Loeb, L. A. 1994. Microsatellite instability: marker of a mutator phenotype in cancer. *Cancer Res.* **54:** 5059–5063.
83. Loeb, L. A. and K. C. Cheng. 1990. Errors in DNA synthesis: a source of spontaneous mutations. *Mutat. Res.* **238:** 297–304.
84. Loeb, L. A., C. F. Springgate, and N. Battula. 1974. Errors in DNA replication as a basis of malignant changes. *Cancer Res.* **34:** 2311–2321.
85. Lothe, R. A., P. Peltomaki, G. I. Meling, L. A. Aaltonen, M. Nystrom-Lahti, L. Pylkkanen, K. Heimdal, T. I. Andersen, P. Moller, T. O. Rognum, S. D. Fossa, T. Haldorsen, F. Langmark, A. Brogger, A. de la Chapelle, and A.-L. Borresen. 1993. Genomic instability in colorectal cancer: relationship to clinicopathological variables and family history. *Cancer Res.* **53:** 5849–5852.
86. Lu, A.-L., S. Clark, and P. Modrich. 1983. Methyl-directed repair of DNA base pair mismatches in vitro. *Proc. Natl. Acad. Sci. USA* **80:** 4639–4643.
87. Lynch, H. T. and A. J. Krush. 1971. Cancer family "G" revisited: 1895–1970. *Cancer* **27:** 1505–1511.
88. Lynch, H. T., T. Smyrk, P. Watson, S. J. Lanspa, B. M. Boman, P. M. Lynch, J. F. Lynch, and J. Cavalieri. 1991. Hereditary colorectal cancer. *Semin. Oncol.* **18:** 337–366.
89. Lynch, H. T., T. C. Smyrk, P. Watson, S. J. Lanspa, J. F. Lynch, P. M. Lynch, R. J. Cavalieri, and C. R. Boland. 1993. Genetics, natural history, tumor spectrum, and pathology of hereditary nonpolyposis colorectal cancer: an updated review. *Gastroenterology* **104:** 1535–1549.
90. Lyons, S. M. and P. F. Schendel. 1984. Kinetics of methylation in *Escherichia coli* K–12. *J. Bacteriol.* **159:** 421–423.
91. Marinus, M. G. and N. R. Morris. 1975. Pleiotropic effects of a DNA adenine methylation mutation *(dam–3)* in *Escherichia coli* K12. *Mutat. Res.* **28:** 15–26.
92. Markowitz, S., J. Wang, L. Myeroff, R. Parsons, L. Sun, J. Lutterbaugh, R. S. Fan, E. Zborowska, K. W. Kinzler, V. B., M. Brattain, and J. K. V. Willson. 1995. Inactivation of the type II TGF-β receptor in colon cancer cells with microsatellite instability. *Science* **268:** 1336–1338.
93. Mays-Hoopes, L. L. 1985. DNA methylation: A possible correlation between aging and cancer. p. 49–65. in R. S. Sohal, L. S. Birnbaum and R. G. Cutler (ed.), *Molecular Biology of Aging*, vol. 29. Raven, New York.
94. Merlo, A., M. Mabry, E. Gabrielson, R. Vollmer, S. B. Baylin, and D. Sidransky. 1994. Frequent microsatellite instability in primary small cell lung cancer. *Cancer Res.* **54:** 2098–2101.
95. Mikol, Y. B., K. L. Hoover, D. Creasia, and L. A. Porier. 1983. Hepatocarcinogenesis in rats fed methyl-deficient, amino acid-defined diets. *Carcinogenesis* **4:** 1619–1629.

96. Miret, J. J., M. Milla, and R. S. Lahue. 1993. Characterization of a DNA mismatch binding activity in yeast extracts. *J. Biol. Chem.* **268:** 3507–3513.
97. Miyamura, Y., H. Ohtsu, O. Niwa, A. Kurishita, H. Watanabe, T. Sado, and T. Ono. 1993. Comparison of the age- and tumor-associated changes in the c-myc gene methylation in mouse liver. *Gerontology* **39(Suppl.) 1:** 3–10.
98. Modrich, P. 1991. Mechanisms and biological effects of mismatch repair. *Ann. Rev. Genet.* **25:** 229–253.
99. Moolgavakar, S. H. and A. G. Knudson. 1981. Mutation and cancer: a model for human carcinogenesis. *J. Natl. Cancer Inst.* **66:** 1037–1051.
100. New, L., K. Liu, and G. F. Crouse. 1993. The yeast gene *MSH3* defines a new class of eukaryotic MutS homologues. *Mol. Gen. Genet.* **239:** 97–108.
101. Nicolaides, N. C., K. C. Carter, B. K. Shell, N. Papadopoulos, B. Vogelstein, and K. W. Kinzler. 1995. Genomic organization of the hPMS2 gene family. *Genomics* **30:** 195–206.
102. Nicolaides, N. C., N. Papadopoulos, B. Liu, Y.-F. Wei, K. C. Carter, S. M. Ruben, C. A. Rosen, W. A. Haseltine, R. D. Fleischmann, C. M. Fraser, M. D. Adams, J. C. Venter, M. G. Dunlop, S. R. Hamilton, G. M. Petersen, A. de la Chapelle, B. Vogelstein, and K. W. Kinzler. 1994. Mutations of two *PMS* homologues in hereditary nonpolyposis colon cancer. *Nature* **371:** 75–80.
103. Nishisho, I., Y. Nakamura, Y. Miyoshi, Y. Miki, H. Ando, and A. Horii. 1991. Mutations of chromosome 5q21 genes in FAP and and colorectal cancer patients. *Science* **253:** 665–669.
104. Nowell, P. C. 1976. The clonal evolution of tumor cell populations. *Science* **194:** 23.
105. Ono, T., S. Yamamoto, A. Kurishita, K. Yamomoto, Y. Yamamoto, Y. Ujeno, K. Sagisaka, Y. Fukui, M. Miyamoto, and R. E. A. Tawa. 1990. Comparison of age associated changes of c-*myc* gene methylation in liver between man and mouse. *Mutat. Res.* **237:** 239–246.
106. Palombo, F., P. Gallinari, I. Iaccarino, T. Lettier, M. Hughes, A. D'Arrigo, O. Truong, J. J. Hsuan, and J. Jiricny. 1995. GTBP, a 160-Kilodalton protein essential for mismatch-binding activity in human cells. *Science* **268:** 1912–1914.
107. Papadopoulos, N., N. C. Nicolaides, B. liu, R. Parsons, C. Lengauer, F. Palombo, A. D'Arrigo, S. Markowitz, J. K. V. Willson, K. W. Kinzler, J. Jiricny, and B. Vogelstein. 1995. Mutations of GTBP in genetically unstable cells. *Science* 1915–1917.
108. Papadopoulos, N., N. C. Nicolaides, Y.-F. Wei, S. M. Ruben, K. C. Carter, C. A. Rosen, W. A. Haseltine, R. D. Fleischmann, C. M. Fraser, M. D. Adams, J. C. Venter, S. R. Hamilton, G. M. Petersen, P. Watson, H. T. Lynch, P. Peltomaki, J.-P. Mecklin, A. de la Chapelle, K. W. Kinzler, and B. Vogelstein. 1994. Mutation of a mutL homolog in hereditary colon cancer. *Science* **263:** 1625–1629.
109. Parker, B. O. and M. G. Marinus. 1992. Repair of DNA heteroduplexes containing small heterologous sequences in *Escherichia coli*. *Proc. Natl. Acad. Sci. USA* **89:** 1730–1734.
110. Parsons, R., G.-M. Li, M. Longley, P. Modrich, B. Liu, T. Berk, S. R. Hamilton, K. W. Kinzler, and B. Vogelstein. 1995. Mismatch repair deficiency in phenotypically normal human cells. *Science* **268:** 738–740.
111. Parsons, R., G.-M. Li, M. J. Longley, W.-H. Fang, N. Papadopoulos, J. Jen, A. de la Chapelle, K. W. Kinzler, B. Vogelstein, and P. Modrich. 1993. Hypermutability and mismatch repair deficiency in RER+ tumor cells. *Cell* **75:** 1227–1236.
112. Peltomaki, P., R. A. Lothe, L. A. Aaltonen, L. Pylkkanen, M. Nystrom-Lahti, R. Seruca, L. David, R. Holm, D. Ryberg, A. Haugen, A. Brogger, A.-L. Borresen, and A. de la Chapelle. 1993. Microsatellite instability is associated with tumors that characterize the hereditary non-polyposis colorectal carcinoma syndrome. *Cancer Res.* **53:** 5853–5855.
113. Prolla, T., D.-M. Christie, and R. M. Liskay. 1994. Dual requirement in yeast DNA mismatch repair for MLH1 and PMS1, two homologs of the bacterial mutL gene. *Mol. Cell. Biol.* **14:** 407–415.

114. Prolla, T. A., Q. Pang, E. Alani, R. D. Kolodner, and R. M. Liskay. 1994. Interactions between the MSH2, MLH1 and PMS1 proteins during the initiation of DNA mismatch repair. *Science* **265:** 1091–1093.
115. Pukkila, P. J., J. Peterson, G. Herman, P. Modrich, and M. Meselson. 1983. Effects of high levels of DNA adenine methylation on methyl-directed mismatch repair in *Escherichia coli. Genetics* **104:** 571–582.
116. Reenan, R. A. G. and R. D. Kolodner. 1992. Characterization of insertion mutations in the *Saccharomyces cerevisiae MSH1* and *MSH2* genes: evidence for separate mitochondrial and nuclear functions. *Genetics* **132:** 975–985.
117. Reenan, R. A. G. and R. D. Kolodner. 1992. Isolation and characterization of two *Saccharomyces cerevisiae* genes encoding homologs of the bacterial HexA and MutS mismatch repair proteins. *Genetics* **132:** 963–973.
118. Reitmair, A. H., R. Schmits, A. Ewel, B. Bapat, M. Redston, A. Mitri, P. Waterhouse, H.-W. Mittrucker, A. Wakeham, B. Liu, A. Thomason, H. Griesser, S. Gallinger, W. G. Ballhausen, R. Fishel, and T. W. Mak. 1995. MSH2 deficient mice are viable and susceptible to lymphoid tumors. *Nature Genet.* **11:** 64–70.
119. Rhyu, M. G., W. S. Park, and S. J. Meltzer. 1994. Microsatellite instability occurs frequently in human gastric carcinoma. *Oncogene* **9:** 29–32.
120. Risinger, J. I., A. Berchuck, M. F. Kohler, P. Watson, H. T. Lynch, and J. Boyd. 1993. Genetic instability of microsatellites in endometrial carcinoma. *Cancer Res.* **53:** 5100–5103.
121. Ritland, S. R., V. Ganju, and R. B. Jenkins. 1995. Region-specific loss of heterozigosity on chromosome 19 is related to the morphologic type of human glioma. *Genes, Chromosomes and Cancer* **12:** 277–282.
122. Ross-Macdonald, P. and G. S. Roeder. 1994. Mutation of a meiosis-specific MutS homolog decreases crossing over but not mismatch correction. *Cell* **79:** 1069–1080.
123. Sancar, A. and J. E. Hearst. 1993. Molecular matchmakers. *Science* **259:** 1415–1420.
124. Schwartz, R. A., D. J. Goldberg, F. Mahmood, R. L. DeJager, W. C. Lambert, A. Z. Najem, and P. J. Cohen. 1989. The Muir-Torre syndrome: a disease of sebaceous and colonic neoplasms. *Dermatologica* **178:** 23–28.
125. Shivapurkar, N., M. J. Wilson, K. L. Hoover, Y. B. Mikol, D. Creasia, and L. A. Poirier. 1986. Hepatic DNA methylation and liver tumor formation in male C3H mice fed methionine- and choline-deficient diets. *J. Natl. Cancer Inst.* **77:** 213–217.
126. Shridhar, V., J. Siegfried, J. Hunt, M. del Mar Alonso, and D. I. Smith. 1994. Genetic instability of microsatellite sequences in many non-small cell lung carcinomas. *Cancer Res.* **54:** 2084–2087.
127. Siegel, E. C. and V. Bryson. 1964. Selection of resistant strains of *Escherichia coli* by antibiotics and antibacterial agents: role of normal and mutator strains. *Antimicrobial Agents and Chemother.* 629–634.
128. Siegel, E. C. and J. J. Ivers. 1975. mut–25, a mutation to mutator linked to purA in *Escherichia coli. J. Bacteriol.* **121:** 524–530.
129. Siegel, E. C. and F. Kamel. 1974. Reversion of frameshift mutations by mutator genes in *Escherichia coli. J. Bacteriol.* **117.**
130. Singhal, R. P., L. L. Mays-Hoopes, and G. L. Eichhorn. 1987. DNA methylation in aging of mice. *Mech. Ageing and Dev.* **41:** 199–210.
131. Solaro, P. C., K. Birkenkamp, P. Pfeiffer, and B. Kemper. 1993. Endocuclease VII of phage T4 triggers mismatch correction *in vitro. J. Mol. Biol.* **230:** 868–877.
132. Stallings, R. L., A. F. Ford, D. C. Torney, C. E. Hildebrand, and R. K. Moyzis. 1991. Evolution and distribution of (GT)n repetitive sequences in mammalian genomes. *Genomics* **10:** 807–815.
133. Strand, M., M. C. Earley, G. F. Crouse, and T. D. Petes. 1995. Mutations in the *MSH3* gene preferentially lead to deletions within tracts of simple repetitive DNA in *Saccharomyces cerevisiae. Proc Natl. Acad. Sci. USA* **92:** 10,418–10,421.

134. Strand, M., T. A. Prolla, R. M. Liskay, and T. D. Petes. 1993. Destabilization of tracts of simple repetitive DNA in yeast by mutations affecting DNA mismatch repair. *Nature* **365:** 274–276.
135. Strickler, J. G., J. Zheng, Q. Shu, L. J. Burgart, S. R. Albert, and D. Shibata. 1994. p53 mutations and microsatellite instability in sporadic gastric cancer: when guardians fail. *Cancer Res.* **54:** 4750–4755.
136. Su, S.-S., R. S. Lahue, K. G. Au, and P. Modrich. 1988. Mispair specificity of methyl-directed DNA mismatch correction *in vitro*. *J. Biol. Chem.* **263:** 6829–6835.
137. Su, S.-S. and P. Modrich. 1986. *Escherichia coli mutS*-encoded protein binds to mismatched DNA base pairs. *Proc. Natl. Acad. Sci. USA* **83:** 5057–5061.
138. Tautz, D. and M. Renz. 1984. simple sequences are ubiquitous components of eukaryotic genomes. *Nucleic Acids Res.* **12:** 4127–4136.
139. te Riele, H., E. R. Maandag, and A. Berns. 1992. Highly efficient gene targeting in embryonic stem cells through homologous recombination with isogenic DNA constructs. *Proc. Natl. Acad. Sci. USA* **89:** 5128–5132.
140. Thibodeau, S. N., G. Bren, and D. Schaid. 1993. Microsatellite instability in cancer of the proximal colon. *Science* **260:** 816–819.
141. Vargas, P. A. and D. S. Alberts. 1992. Primary prevention of colorectal cancer through dietary modification. *Cancer* **70:** 1229–1235.
142. Vogelstein, B. Personal communication.
143. Wagner, R. and M. Meselson. 1976. Repair tracts in mismatched DNA heteroduplexes. *Proc. Natl. Acad. Sci. USA* **73:** 4135–4139.
144. Warthin, A. S. 1913. Heredity with reference to carcinoma as shown by studies of the cases examined in the pathological laboratory of the University of Michigan, 1895–1913. *Arch. Intern. Med.* **12:** 546–555.
145. Watson, P. and H. T. Lynch. 1993. Extracolonic cancer in hereditary nonpolyposis colorectal cancer. *Cancer* **71:** 677–685.
146. Welsh, K. M., A.-L. Lu, S. Clark, and P. Modrich. 1987. Isolation and characterization of the *Escherichia coli mutH* gene product. *J. Biol. Chem.* **262:** 15,624–15,629.
147. Williamson, M. S., J. C. Game, and S. Fogel. 1985. Meiotic gene conversion mutants in *Saccharomyces cerevisiae*. I. Isolation and characterization of *pms1–1* and *pms1–2*. *Genetics* **110:** 609–646.
148. Wooster, R., J. A. Cleton, N. Collins, J. Mangion, R. S. Cornelis, C. S. Cooper, B. A. Gusterson, B. A. Ponder, D. A. von, O. D. Wiestler, et al. 1994. Instability of short tandem repeats (microsatellites) in human cancers. *Nature Genet.* **6:** 152–156.
149. Worth, L., Jr., S. Clark, M. Radman, and P. Modrich. 1994. Mismatch repair proteins MutS and MutL inhibit RecA-catalyzed strand transfer between diverged DNAs. *Proc. Natl. Acad. Sci. USA* **91:** 3238–3241.
150. Yee, C. J., N. Roodi, C. S. Verrier, and F. F. Parl. 1994. Microsatellite instability and loss of heterozigozity in breast cancer. *Cancer Res.* **54:** 1641–1644.

21

Mammalian Cell-Cycle Responses to DNA-Damaging Agents

Roy Rowley

1. INTRODUCTION

DNA-damaging agents may shorten *(29,110)* or prolong the cell cycle, depending on the agent used, the dose, and the cell type. This chapter focuses on cell-cycle delays, since they are by far the most frequently observed cell-cycle response to DNA damage.

Cell-cycle delays can be broadly categorized into active and passive responses. Active responses are now generally known as checkpoint control-mediated delays (*140*; *see* Chapter 17, vol. 1) and represent an effort by the cell to improve its chances of surviving DNA damage. For example, radiation-induced G2 delay is presumed to provide time for the cell to repair chromosome damage before attempting chromosome segregation and thereby risking death by loss of acentric chromosome fragments. Passive responses reflect mechanical interference with normal cellular functions *(39)*. It is sometimes difficult to distinguish between the two, but the existence of checkpoint mutants is diagnostic. Prior to the identification of checkpoint-defective mutants, correlations between reduced cell-cycle delays and increased sensitivity for killing, as for example following caffeine treatment of irradiated cells, had led to the same conclusion: that delays allow time for repair of potentially lethal damage (HeLa-S3[*5*]). In the case of mutants showing sensitivity to DNA damage-induced killing or mutation, it may also be difficult to determine whether the primary defect is repair of DNA damage *per se* or loss of the checkpoint-mediated delay, which provides time for such repair. The defect is likely to result from loss of a checkpoint control function if the increased sensitivity for killing or mutation is ameliorated by the artificial imposition of a period of delay. For example, the radiosensitive budding yeast *rad9* checkpoint mutant can be rendered radiation-resistant by temporarily suspending postirradiation cell progress into mitosis with methylbenzimide-2-yl-carbamate, a microtubule polymerization inhibitor *(142)*. Thus, *rad9* is repair-competent, but checkpoint defective.

DNA damage-induced cell-cycle delays are observed in all phases of interphase and in mitosis (Fig. 1). For historical reasons and because subsequent sections will be easier to understand, this chapter on cell-cycle responses to DNA-damaging agents will proceed backwards through the cell cycle.

From: DNA Damage and Repair, Vol. 2: DNA Repair in Higher Eukaryotes
Edited by: J. A. Nickoloff and M. F. Hoekstra © *Humana Press Inc., Totowa, NJ*

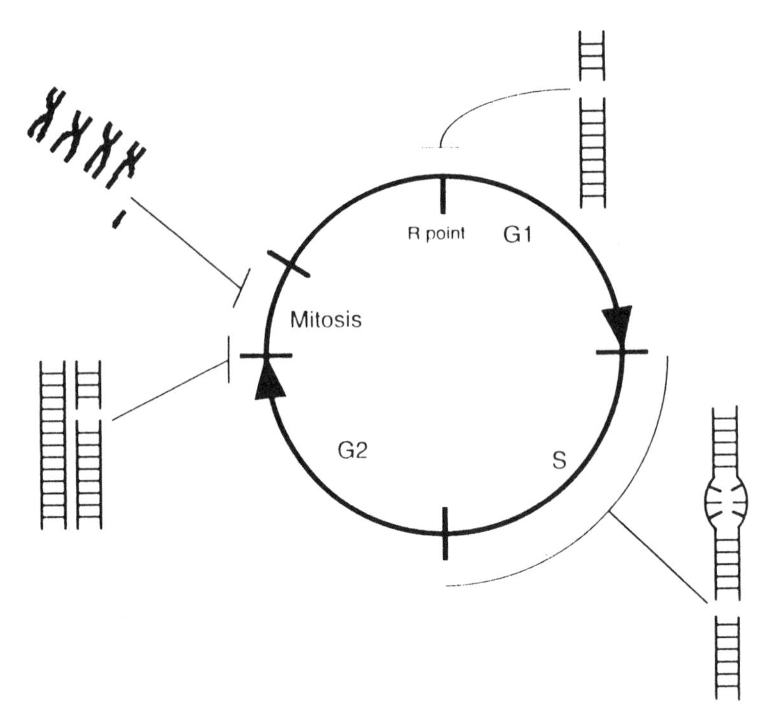

Fig. 1. A representation of the mammalian cell cycle indicating known points at which checkpoint-mediated delays are imposed in response to the presence of a DNA double-strand or chromosome break.

2. G2 DELAY

2.1. Description

Delay in passage through G2 results in a postponement of mitosis and cell division. This cell-cycle response has thus been variously termed mitotic delay, division delay, G2 delay, and G2 arrest. Although formally there is a difference between mitotic and division delays, and between arrest (which implies permanence) and delay, these differences have not been emphasized in the literature and the terms have been used interchangeably. G2 delay has been observed in all eukaryotic organisms examined, including ciliates *(41)*, fungi *(140)*, algae *(99)*, flowering plants *(56)*, slime molds *(109)*, sea urchins *(108)*, insects *(8)*, and mammals *(13)*.

The results of early, descriptive studies of G2 delay in mammalian cells can be summarized as follows: Irradiation of growing cell populations results in a delay in entering mitosis, which depends on the radiation dose (55 min delay/Gy in Chinese hamster ovary [CHO] cells *[53]*), the degree of cellular differentiation (mouse intestinal crypt cells *[10]*) and the phase of the cell cycle (V79 hamster and HeLa-S3 *[121]*). In CHO cells the relationship between delay duration and cell-cycle age is exponential such that, for a given dose, $t = \text{constant} \times \text{age}^{0.11}$ *(60)*. An hour delay per Gy is a reasonable approximation for most asynchronously growing cultured cell lines provided that they are repair-proficient. Once a cell has entered mitosis, it is no longer delayed in that cell cycle (HeLa *[31]*). The precise position in the cell cycle of this transition point was determined by a detailed microscopic analysis of viable CHO cells *(8)*. Cells irradiated prior to middle prophase at the time of exposure with 1.5–4.0 Gy were delayed. At lower doses, cells were only delayed if irradiated prior to early prophase. Note, however, that prophase, as defined by Carlson, appears well back in G2. In fact, prophase cells that had initiated chromosome condensation underwent decondensation on irradiation and then delayed in the decondensed state, a phenomenon termed "reversion." The geography of G2 has also been extensively analyzed using the mitotic shake-off procedure (CHO *[114]*). Mammalian cells usually "round up" on entry to mitosis. Mitotic cells may thus be selectively detached from the substrate on which they are growing by shaking the culture flask. By this means, the number of cells in mitosis can be determined in real time at intervals after any treatment that does not alter cell attachment. In CHO cells, the arrest point in G2, determined by mitotic cell selection, is 49 minutes before completion of cell division, 17 min before metaphase, and 18 min after the cycloheximide transition point (the point in the cell cycle after which this protein synthesis inhibitor can no longer delay entry to mitosis). Errors in this study were approx ±5 min. Similar observations, using less precise techniques, have been made for mouse leukemic cells *(17)*. The X-ray transition point follows the cycloheximide transition point implying that cells are blocked by DNA damage after all proteins required for entry to mitosis have been synthesized.

2.2. Inhibitor Studies

Initial attempts to define a biochemical pathway for the induction of delay relied on the use of inhibitors of more or less well-defined specificity (reviewed in *102*). Cycloheximide, itself a cytostatic agent, blocks both the induction of and recovery from G2 delay, indicating requirements for protein synthesis (CHO *[13]*). Recovery also requires RNA synthesis, demonstrated by the effect of actinomycin-D, but only when the duration of delay exceeds the life-time of a postulated RNA species *(136)*. Considerable attention has focused on the action of caffeine and related xanthines and purines, prompted by the observation that caffeine sensitizes bacteria to UV-induced cell killing *(75)*. Mammalian cells are also radiosensitized by caffeine (Hela-S3 *[5]*) and cells exposed to caffeine immediately postirradiation are not delayed in G2 (CHO *[135]*). To test whether caffeine functions by inhibiting cyclic AMP phosphodiesterase, cells have been exposed to the more potent inhibitor, methyl isobutyl xanthine (MIX). The results are variable. In established cell lines, MIX is without effect (CHO *[54]*), whereas radiation-induced delay is reduced in MIX-treated mouse embryo cells *(47)*. Theophylline, theobromine, 2-aminopurine, and cordycepin also reduce G2 delay. These caffeine-like agents share a number of other activities *(52)*, including inhibition of protein kinases *(23)*, but to date the mechanism of caffeine action remains unknown.

At a cellular level, if caffeine is administered to irradiated cells immediately postirradiation and then removed after an interval of as long as 12 h, a period of delay follows that is identical in duration to that expressed in cells that do not receive caffeine (CHO *[105]*). It appears that caffeine not only blocks the expression of the delay, but also preserves the damage that gives rise to it, allowing the expression of delay on caffeine removal. Caffeine may thus sensitize cells to DNA damage by two mechanisms: indirect inhibition of repair via abolition of checkpoint controls and direct inhibition of repair functions, perhaps by binding sites of damage, as postulated for UV-irradiated bacteria *(116)*.

A very early radiation response is the induction of protein kinase C (PKC) activity *(142)*. Syrian hamster embryo fibroblasts exposed to PKC inhibitors, especially the isoquinoline sulfonamide H7, are made radiosensitive. H7 did not alter DNA strand break repair kinetics of irradiated cells, as measured by filter elusion techniques *(34)*, but delayed the onset of G2 delay (human tumor cells *[33]*). How H7 affects the checkpoint response was not determined, so these results remain correlative. However, the UV-induction of *jun*, usually characterized as the "mammalian UV response" (*see* Section 6. and Chapters 15 and 16), is also blocked by PKC inhibitors.

2.3. Effects of G2 Delay on Survival and Mutagenesis

No mammalian cell mutant exists that is defective for G2 delay alone. Thus, it is not possible to determine, without the use of drug treatments and other manipulations, the consequences of losing the G2 checkpoint. Cells from patients with the inheritable disease ataxia telangiectasia (AT), for example, are purported to show defects in checkpoint controls in all three phases of interphase *(1,147)*. (Note that Beamish and Lavin *[1]* measured G2 delay in cells synchronized to the end of the cell cycle with nocodazole, a microtubule polymerization inhibitor. Since such inhibitors usually block cell progress in mitosis, it seems possible that the synchronous population was actually beyond the X-ray transition point when irradiated and, thus, beyond the point at which G2 delay is normally inducible. When irradiated in S phase, before the X-ray transition point, AT cells suffer abnormally long G2 delays *[83]*, indicating that the G2 checkpoint control is intact. *See* Chapter 19). Most experiments to determine the effects of abolishing G2 delay have utilized caffeine (HeLa *[5]*; V79 *[69]*; CHO *[103]*). When combined with the mitotic cell selection procedure, it is possible to measure survival only in those cells that were treated in G2 and delayed or failed to delay in G2. Cells that received 5 m*M* caffeine immediately and continuously after irradiation (250 kVp X-rays) were radiosensitive for killing and showed the same radiosensitivity as mitotic cells. Cells harbored increased numbers of achromatic gaps, deletions, and exchanges. If an interval was allowed before the addition of caffeine, permitting a period of delay, aberration frequencies declined with the length of the interval *(101)*. An alternative approach, used by Lucke-Huhle *(69)*, was to arrest V79 cells with α particle irradiation, which is more efficient at inducing G2 delay than S-phase delay. α-Irradiation thus caused the entire cell population to arrest in G2. Caffeine treatment of the arrested cell population increased cell killing. According to Beetham and Tolmach *(2)*, HeLa-S3 cells released from G2 arrest in this manner suffer a reproductive death *(21)*, indistinguishable from that resulting from irradiation alone; however, in a more recent study, caffeine-induced abolition of arrest was reported to trigger apoptosis (HeLa-S3 cells *[3]*). Assuming that chromosome aberrations result from DNA double-strand breaks (DSBs; *see* Chapter 25) and because caffeine blocks the repair of chromosome aberrations, then caffeine should reduce the rate of DSB repair. When DSBs are measured at intervals after caffeine treatment of CHO cells, the initial (rapid) phase of repair is unaltered, but the following (slower) phase of repair appears to be inhibited *(104)*, although the statistical significance of that observation was marginal.

2.4. *What Is the Cellular Target for G2 Delay?*

The most sensitive region of the mammalian cell for G2 delay induction is the nuclear periphery. Also, radioactive iodine incorporated into the DNA induces G2 delay far more effectively than radioactive isotopes in the cytoplasm. These observations have been used to argue (i) that the target for delay induction is therefore nuclear and probably DNA or chromatin (CHO *[79,137]*), and (ii) that the target is not DNA, but may be associated with the DNA at some point in the cell cycle (CHO *[115]*). Sandell and Zakian *(111)* have shown that the induction of a single DNA DSB in the genome of the budding yeast is sufficient to induce a cell-cycle delay, strongly arguing that DNA damage initiates G2 delay and that the delay provides time for repair of DNA damage, prior to chromosome condensation and segregation. Consistent with that result, hamster cell DSB repair mutants suffer abnormally long delays in G2 *(83,138)*.

2.5. *Possible Mechanisms of G2 Delay Induction*

The biochemical pathway that carries a signal from damaged DNA to the G2/M cell-cycle progression apparatus is still unknown. Since transition from G2 into mitosis is mediated by a cyclin-dependent kinase, attention has focused on pathways regulating cdc2 activity, such as changes in $p34^{cdc2}$ phosphorylation, availability/activity of the requisite B-type cyclin, and although not demonstrated for cells in G2, association with a cyclin kinase inhibitor (CKI) *(78)*. In brief, for a cell to enter mitosis, $p34^{cdc2}$, the "mitotic kinase," must be phosphorylated at threonine-161, dephosphorylated at threonine-14 and tyrosine-15, associated with a B-type cyclin, and presumably not blocked by association with a CKI. Cyclin B protein levels increase from mid- to late S-phase onward.

Cells arrested in G2 after clastogen treatment contain $p34^{cdc2}$ in the inactive state (CHO *[67]*; human lymphoma cells *[88]*; HeLa *[131]*). Activity of the mitotic kinase complex ($p34^{cdc2}$ in association with cyclin B) is suppressed for the duration of delay and $p34^{cdc2}$ remains phosphorylated at tyrosine-15 (1.75 Gy, HeLa S3 cells *[74]*). This result appears specific for Cdc2/Cdkl (compared with Cdk2) in lymphoma cells *(87)*. Consistent with these results, irradiation was also found to block the cdc2-mediated hyperphosphorylation and activation of Cdc25C, the phosphatase required for dephosphorylation/activation of $p34^{cdc2}$ at tyrosine-15 *(86)*. Although these results are compatible with the hypothesis that G2 delay is induced by phosphorylation and inactivation of the cyclin-dependent kinase (CDK) required for transition from G2 to mitosis, they remain only a correlation and do not distinguish between cause and effect. Cells prior to mitosis, arrested or not, surely would retain the mitotic kinase in an inactive state.

Of particular interest is a proposal by Uckun *(132)* that irradiation activates the *lyn* protein tyrosine kinase, which phosphorylates $p34^{cdc2}$ at tyrosine-15, thereby inactivating it and blocking progress to mitosis. The *lyn* tyrosine kinase is an *src*-like nonreceptor protooncogene. The *src*-like kinases are activated directly, by association with the cytosolic domain of some transmembrane receptors, or indirectly, possibly by G-proteins or PKC. Substrates of the *src* family of kinases include phosphatidylinositol-3-kinase, phosphatidylinositol phospholipase c-γ, *ras*-GAP, and $p74^{c\text{-}raf}$ *(7)*. The *lyn* kinase in particular is activated following exposure to ionizing radiation *(50)* mitomycin C (MMC) *(51)* and 1-β-D-arabinofuranosylcytosine *(146)*. Significantly, kinase activation is also accompanied by association of p53/56 *lyn* with $p34^{cdc2}$, as demonstrated by coimmunoprecipitation with anti-34^{cdc2} antibodies. The mitotic kinase was in fact phosphorylated by p53/56 *lyn* at tyrosine-15 in vitro. Although it is at first counterintuitive that a plasma-membrane-associated kinase might mediate a signal from damaged DNA to the mitotic control system, induction of the mammalian cell "UV response" (*see* Section 6.) also appears to initiate at the membrane and terminates with the induction of the nuclear proto-oncogenes *fos* and *jun*. Interestingly, the UV response is also activated by the action of *src* protein kinases *(14)*, perhaps suggesting a common signal origin.

Irradiation may also affect cyclin B mRNA and protein levels. At low radiation doses (1.75 Gy), cyclin B1 protein accumulates during delay (HeLa-S3 *[76]*). Similar observations were made in human lymphoma cell lines exposed to nitrogen mustard; cells exhibit a progressive accumulation of cyclin B during G2 delay (the dose used was 0.5 μ*M*, compared with 1 μ*M* required for 50% population growth inhibition *[88]*). At higher doses (6 Gy), irradiation of G2 HeLa cells delays the normal accumulation of cyclin B1 protein, providing a correlation between progression and cyclin levels. However, irradiation during S-phase, before the increase in cyclin protein and before cells have arrested in G2, delays the accumulation of cyclin mRNA *(80,81)*. Delayed message synthesis suggests that a period of arrest in G2 is anticipated, indicative of an active (checkpoint-mediated) response to DNA damage. Further, cyclin A mRNA and protein levels are not suppressed by irradiation, indicating specificity for cyclin B1 *(84)*. A reduction of the mRNA half-life is a potential mechanism by which cyclin synthesis may be regulated *(71)*. In HeLa cells, the cyclin B1 mRNA half-life increases through the cell cycle: 1–2 h in G1, 8 h in S phase, and 12 h in G2. Following irradiation, the half-life of cyclin B mRNA measured in S phase cells after irradiation in S-phase was 5 h and 3 h for cyclin B mRNA measured during the G2 block after cells were irradiated in the preceding S phase. Although it is sometimes difficult to distinguish cause from effect, in this instance, the reduction in half-life cannot be explained by cell-cycle redistribution. As yet there is no published report of a CKI that might block cdc2 activation and progress into mitosis after DNA damage.

2.6. Future Directions

Identification of a *lyn* kinase checkpoint-defective mutant would be powerful confirmation that *lyn* kinase inactivates the mitotic kinase in response to DNA damage and thereby arrests progress into mitosis. Alternatively, a CKI might mediate G2 delay, and tests of this possibility are underway. The identification of mammalian homologs of yeast checkpoint genes is also receiving much attention. Of particular note, the genetic defect responsible for the human condition AT has recently been identified. The defective gene *(ATM)* resembles *MEC1* and *rad3,* checkpoint genes from *Saccharomyces cerevisiae* and *Schizosaccharomyces pombe*, respectively, that function in G2 *(112)* and to mammalian DNA-PK *(37)*. Elaboration of the pathway which incorporates this gene may bear on the mechanism of DNA damage-induced G2 delay *(70)*.

3. S-PHASE DELAY

3.1. Description

When cultured cells in S phase are exposed to ionizing radiation doses of a few to tens of Gy, incorporation of DNA precursors is temporarily inhibited (HeLa-S3 cells

[129]). The degree and duration of incorporation inhibition are dose-dependent. If the log of the minimum observed level of incorporation, expressed as a fraction of that measured in an unirradiated control, is plotted against radiation dose a biphasic curve results; doses of 10 Gy and below inhibit uptake of tritiated thymidine more effectively than doses above 10 Gy (human diploid fibroblasts *[93]*). In that study, the size of newly synthesized DNA was measured by sedimentation through alkaline sucrose gradients in cells 30 min after irradiation (5 and 20 Gy) and in unirradiated cells. Incorporation into replicon-sized DNA fragments was suppressed disproportionately, presumably reflecting preferential inhibition of DNA synthesis initiation. The biphasic shape of the radiation dose–response curve may therefore result from the differential radiosensitivity of initiation vs elongation of the nascent DNA molecule. A target size of 10^9 Dalton was calculated from the initial portion of the radiation dose–response, corresponding to the size of a replicon cluster and to the molecular weight of the DNA in the chromosome structural subunit identified by Cook and Brazel *(11)*.

Does the suppression of DNA synthesis initiation, elongation, or both reflect activation of a checkpoint? A mutation in *ATM,* the gene mutated in AT patients *(112)*, alters the shape of the radiation dose–response curve for inhibition of DNA synthesis *(145)*. The curve is log-linear and parallels the terminal portion of the wild-type dose–response curve. Analysis of the nascent DNA size after irradiation indicates that the normal suppression of DNA replicon initiation does not occur *(91)*. The existence of a mutant for radiation-induced inhibition of replicon initiation argues that this response is indeed checkpoint-mediated. A similar conclusion was reached by Lamb et al. *(58)*. These authors irradiated a human cell line containing an extrachromosomal EBV-based plasmid, which was replicated in tight coordination with the genome. The doses chosen were sufficient to damage the cellular genome (≥30 Gy), but insufficient to damage the plasmid (assessed by the degree of reduction of plasmid supercoiling, consequent to the induction of single- or double-stranded DNA breaks). Nonetheless irradiation suppressed plasmid replication. Replication was therefore blocked actively in the manner of a checkpoint control, presumably by transmission of a signal from the damaged genome to the plasmid replication machinery. The stage at which replication was inhibited was not determined.

3.2. *Inhibitor Studies*

As with G2 delay, S-phase or replicon initiation delay is abolished by continuous caffeine treatment (HeLa, V79 *[90]*) and postponed by finite treatments (HeLa *[130]*). Incorporation of DNA precursors into replicon-sized DNA thus continues after irradiation, as observed for AT cells *(145)*. Caffeine studies of replicon initiation delay have not elucidated a mechanism. This checkpoint function is not blocked by 3-aminobenzamide *(92)*, indicating that synthesis of poly(adenosine-diphosphoribose) is not a requirement for replicon initiation delay.

3.3. *Effects of S-Phase Delay on Survival and Mutagenesis*

No mammalian cell mutant exists that is defective for the replicon initiation checkpoint alone, and nothing is known of the consequences of this checkpoint defect. We may speculate, however, that as with loss of the p53-dependent checkpoint control, the role of replicon initiation delay is to allow repair of DNA lesions before their replication. Such repair is relatively error-free, compared with postreplication repair of chromatid breaks, when the fusion of sister chromatids is possible as well as consequent formation of bridges at mitosis, nondisjunction and gene amplification (*see* Section 4.).

3.4. What Causes Replicon Initiation Delay?

The nature of the lesion causing replicon initiation delay remains speculative. Based on a derivation of target size from the dose–response data for normal human fibroblasts, Painter *(93)* has suggested that a single DNA DSB in a replicon cluster may result in a deformation of chromatin structure in that domain and this acts to signal DNA synthesis delay.

3.5. Possible Mechanisms of S-Phase Delay Induction

In mammalian cells, only inactivation of the *ATM* gene is known to eliminate the S phase checkpoint control. How *ATM* regulates DNA replicon initiation is entirely unknown. It is known, however, that although induction of the tumor suppressor protein, p53, is suboptimal in AT cells *(6)*, loss of p53 function does not result in the replicon initiation checkpoint defect characteristic of AT cells *(59)*. Other than this single piece of negative evidence, nothing is known of how the S-phase checkpoint control operates.

3.6. Future Directions

In all systems studied, DNA synthesis regulation takes place at the level of replicon initiation. The first step is the sequence-specific recognition of the origins of replication, which in yeast are termed the autonomously replicating sequence (ARS) elements. Subdomains of ARS elements have been identified as essential for replication. Protein binding to ARS elements varies with the cell cycle, such that pre- and postreplicative states of binding can be discerned *(15)*. There are thus compositional and/or conformational and/or posttranslational changes in the proteins binding ARS elements, which permit or inhibit replication of the DNA. It is reasonable to hypothesize that such changes may be involved in checkpoint-mediated regulation of mammalian cell replicon initiation. However, even in the absence of a defined mechanism, a "simple" analysis of the consequences for cell survival and mutation of a failure to delay replicon initiation would be of interest. Such experiments should be possible in manipulatable systems, such as yeasts.

4. G1 DELAY

4.1. Description

Little *(64)* reported that modest doses of X-rays caused a marked delay in the entry of primary human amnion cells into S phase. For example, a delay of 2–3 d was induced for a dose of 1 Gy. Given that delays to DNA replicon initiation are only detectable after doses of 510 times higher, it seems likely that a delay in traversing G1 was being observed, more recently shown to depend on the function of p53. Kastan et al. *(48)* measured the effects of γ-irradiation on cell progression into S phase in cells with normal *p53* (normal human myeloid progenitors and human myeloblastic leukemia, ML-1, cells) and cells lacking p53 or overexpressing a mutant form of p53. Normal cells suffered delays in both G1 and G2; the duration of the delay in G1 approaching 2 days for a 1-Gy exposure. Cells with mutant or no p53 delayed in G2 only. Radiation-induced G1 arrest is thus an active cellular response.

It seems likely that the point in G1 at which cells are delayed by the p53-dependent checkpoint control coincides with the Restriction point *(94)*, that is, the point in the cell cycle beyond which the cell commits to reproduction and loses the potential to differentiate, mate (in lower eukaryotes), or enter a state of quiescence. This conclusion is based on the observations that the activities of Cdk4 and Cdk6 can be modulated by *p53* (*see* Section 4.5.), that both Cdks associate with D-type cyclins *(96)* and that these cyclins appear to define the restriction point, as assessed by the influence of their expression on the duration of G1 and the cellular dependence on mitogens (reviewed in *119*). Consistent with this notion, tyrosine phosphorylation of Cdk4 is required to block cells in G1 after UV irradiation *(128)* and normal diploid human fibroblasts can be arrested by irradiation in early or mid G1, but not in late G1, presumably after the Restriction point *(16)*.

4.2. *Inhibitor Studies*

Kastan et al. *(48)* reported that 2 and 4 m*M* of caffeine inhibited the induction of *p53* by irradiation and concomitantly eliminated the period of G1 delay in human myeloblastic leukemia ML-1 cells. Powell et al. *(97)* showed no effect of 2 m*M* caffeine on G1 delay in rat embryo fibroblasts, but an apparent elimination of the G1 delay in mouse embryo fibroblasts. In A549 human lung adenocarcinoma cells, Russell et al. *(107)* showed no effect of caffeine on G1 delay. A safe conclusion is that *p53*-mediated G1 delay following DNA damage may be abolished by caffeine, but the effective concentration is cell-line-dependent. This being so, it suggests the existence of a caffeine-susceptible checkpoint-control component that mediates delays in all three phases of interphase. A review of the agents that block checkpoint-control functions further suggests that such a component may be a protein kinase *(102)*. Cycloheximide, an inhibitor of translation, blocked the induction of *p53* by irradiation and partially reduced the G1 delay *(48)*.

4.3. *Effects of G1 Delay on Survival and Mutagenesis*

When HeLa cell progression is arrested by treatment with cycloheximide, before cells are irradiated then released, cell survival increases with increasing time to release from the cycloheximide-induced arrest *(95)*. The interpretation is that, with additional time before progress into S phase, the cell is capable of repairing potentially lethal damage—that is, damage that may be rendered lethal if present when the cell replicates its DNA or attempts chromosome segregation. Surprisingly then, loss of the G1 checkpoint control (i.e., failure to delay in G1 after irradiation) does not sensitize cells to radiation-induced killing *(46,123,125)*. This paradox remains to be resolved, but a possible explanation is that cells that fail to delay in G1 go on to delay longer at subsequent checkpoints in S phase or G2, allowing compensatory repair. Fan et al. *(22)*, for example, show that a higher proportion of a *p53*-defective MCF-7, cell population is arrested in G2 at 16 h after 6.3 Gy than the *p53* wild-type cell population.

It was recently demonstrated that cells that lack a functional *p53*-dependent checkpoint control, and thus fail to arrest in G1 in response to DNA damage, are sensitized to irradiation or cisplatin by caffeine (or similar) treatment to a greater extent than cells that arrest in G1 *(22,97,108,128)*. With respect to irradiation, the difference in effect of caffeine on survival may have resulted from a difference in effect on G2 delay: *p53*-deficient cells could be released from G2 arrest by caffeine at lower concentrations than *p53* wild-type cells *(97)*. This observation suggests that caffeine, or a caffeine-like agent suitable for human use, might be used specifically to enhance the radiosensitivity of *p53*-defective human tumors.

Failure to delay in G1 increases the frequency of gene amplification in irradiated cells, as measured by CAD gene-mediated resistance to *N*-(phosphoanacetyl)-L-(aspartate) (PALA). *(66,141,144)*. The CAD gene encodes a single protein with three enzyme activities: carbamyl phosphate synthetase, aspartate transcarbamylase and dihydrooratase. PALA is a specific inhibitor of *de novo* uridine biosynthesis and is thus toxic. The degree of resistance depends on the number of CAD gene copies, so that depending on the PALA dose, amplification can be monitored by colony formation. Gene amplification occurred at a rate of 10^{-5}–10^{-3} in cells without a functional *p53*-dependent checkpoint control and at an undetectable rate ($<10^{-9}$) in normal diploid fibroblasts *(66)*. A likely explanation for the amplification defect was suggested by Hartwell *(38)*. Failure to delay in G1 after clastogen treatment leaves insufficient time.for repair of chromosome breaks. Such breaks may then be replicated on passage through S phase, making possible the union of sister chromatids, with consequent bridge formation in mitosis, nondisjunction and the unequal segregation of specific genes. Repetition of this break-fusion-bridge process then leads to gene amplification. Repair of a chromosomal break in G1, when only two fragments require rejoining, is thus relatively error-free compared with repair during G2, when the break has been replicated. Mice deleted for both copies of *p53* develop relatively normally, but by 6 mo of age are prone to develop sarcomas and lymphomas, confirming the predicted relationship between increased genomic instability and predisposal to malignancy *(18)*.

4.4. What Causes G1 Delay?

G1 delay is observed after irradiation (ionizing *[48]*; UV *[128]*), treatment with topoisomerase II inhibitors *(84)*, a DNA restriction enzyme *(68)*, alkylating agents, depleted medium *(149)*, heat, or hypoxia *(30)*. After irradiation, the duration of G1 delay is increased if the cell is first allowed to incorporate a halogenated pyrimidine into the DNA *(148)*. G1 delay may thus be induced by a variety of factors, including DNA damage. Nelson and Kastan *(84)* noted that p53 in particular is induced by agents that either induce DNA damage directly (e.g., ionizing radiation) or initiate processes such as repair or replication, which lead to the formation of DNA strand breaks. A model for induction of cell-cycle delays in budding yeast *(70)*, suggests that various types of DNA lesions must be processed to a form that is recognizable by the checkpoint apparatus. Given the emerging homologies between yeast and mammalian systems, the same principle may apply in mammalian cells. There are no reports that G1 delays are prolonged in DNA repair mutants.

4.5. Possible Mechanisms of G1 Delay Induction

A checkpoint control is likely to consist of at least two major components: a detector, of DNA damage in this instance, and an effector of progression arrest. Intermediary signal transduction components may also exist. Most observations support the view that the effector for induction of G1 cell-cycle progression delay in response to DNA damage is one or more Cdks. Several pathways determine Cdk activity *(78)*:

1. Phosphorylation of the Cdk itself;
2. Association of the Cdk with a specific cyclin and the state of posttranslational modification of that cyclin;
3. Association with one of a variety of CKIs; and
4. Association with other proteins, specific to certain cell-cycle functions (e.g., proliferating cell nuclear antigen; PCNA).

The detector remains unknown, but candidate proteins are now being put forward, such as the Ku protein and the associated double-stranded DNA-dependent protein kinase (*DNA-PK*; *see* Chapters 16 and 17). Signal transduction elements may include the tumor suppressor p53 and, by analogy with the budding yeast *(70)*, the product of the *ATM* gene. The p53 protein is not always essential for G1 delay induction. Evidence for these suggestions and for their interrelationships is as follows.

Kastan et al. *(48)* and Kuerbitz et al. *(57)* have shown that p53 levels increase in response to DNA damage in human and other cells. *p53* mRNA levels do not change after irradiation *(48)*, but the protein half-life increases. Protein half-life values for mouse keratinocytes are 30 min before and 200 min after UV-B irradiation *(65)*; for mouse 3T3 cells, they are 35 min before and 150 min after UV *(73)* and 15 min before and 60 min after X-irradiation *(98)*. Degradation of p53 is ubiquitin-dependent, a fact that has been exploited by the oncogenic tumor viruses. The E6 protein from human papilloma type 16 and type 18 viruses, for example, binds with and directs p53 for ubiquitin-mediated degradation *(113)*. DNA damage-induced stabilization may thus be mediated by blocking a ubiquitin-dependent degradation pathway.

p53 regulation is defective in AT cells *(6)*, suggesting that the ATM protein kinase, regulates p53 levels. The p53 protein is phosphorylated at multiple sites by a variety of kinases *(28)*, including *DNA-PK (61)*; however, *DNA-PK* and the associated DNA binding protein Ku are unlikely to modulate p53 stability, since *p53* induction is normal in SCID mouse fibroblasts (J. M. Brown, personal communication). The fact that mutation of the *DNA-PK* phosphorylation site (serine-15) of human p53 reduces the ability of *p53* to block G1 cell progress *(26)* may suggest then that the DNA-PK phosphorylation site is actually an ATM-dependent phosphorylation site. A total of 10 phosphorylation sites have been identified in rat p53 *(28)*, phosphorylated variously by *MAP* kinase, casein kinase I, casein kinase II *(CKII)*, and $p34^{cdc2}$. Mutation of the *CKII* site also results in loss of growth suppression by *p53 (76)*, and both *DNA-PK* and *CKII* can promote tetramerization of *p53,* a conformation required for DNA binding *(44)*. Phosphorylation of serine-12 by $p34^{cdc2}$ modulates the nuclear localization signal of *p53 (77)*.

Downstream effectors of p53 function are better characterized. Using subtractive hybridization, El-Diery et al. *(22)* identified a transcript encoding a 21-kDa protein that is induced in response to elevated p53 levels generated by induction from a dexamethasone-inducible promoter. *WAF 1* (also called cip 1 *[36]*, CAP20 *[32]*, *p21 [143]*, and SDI1 *[85]*) is a universal inhibitor of cyclin kinases *(32,143)*. *WAF1* is not induced in cells lacking p53 or in cells expressing a mutant form of p53. The same protein was simultaneously shown, by means of a yeast two-hybrid system, to complex with a variety of Cdk proteins, blocking protein kinase activity *(36)*. This same inhibitory activity is presumably also required for WAF1's role in coordinating S phase and mitosis. In the absence of *WAF1*, HCT116 human colon cancer cells arrested in G2 by irradiation undergo additional rounds of DNA replication without intervening mitoses *(134)*.

The retinoblastoma tumor suppressor protein (Rb) is a substrate for the cyclin D-dependent protein kinase Cdk4. Rb phosphorylation results in release of transcription factors (E2Fs) and transcription of a variety of genes, including those required for entry to S phase: dihydrofolate reductase, thymidylate synthetase, ribonucleotide reductase, thymidine kinase, and c-*fos*/*AP-1* (*100*; reviewed in *139*). Inhibition of Cdk4 kinase activity may thus block entry to S phase (more accurately, transit past the Restriction point in G1) by Rb-mediated down regulation of the transcription of "S-phase genes." In confirmation of this hypothesis, Cdk4 is found in the tyrosine-phosphorylated form in cells blocked in G1 by UV irradiation *(128)* and cells expressing the viral oncoprotein E7, which binds and inactivates Rb, and other Rb-like proteins, do not arrest in G1 *(122)*.

If delays provide time for repair of DNA damage, then a system to coordinate repair and progression seems likely to exist. In normal mammalian cells, the CKI *WAF1* is found complexed with a cyclin, a Cdk, and of relevance here, PCNA *(133)*. PCNA is an essential component of the replication machinery, acting as a processivity factor for polymerase δ, the principal replicative DNA polymerase *(55)*. *PCNA* thus usually localizes to sites of replication, but may relocalize to putative sites of DNA damage repair following irradiation *(45)*. Using an SV40-based in vitro DNA synthesis assay, Waga et al. showed that *WAF1* directly blocked PCNA-dependent DNA synthesis and activation of polymerase δ. Competitive inhibition studies, in the absence of the cyclin and Cdk components, indicated a probable direct interaction of WAF1 and *PCNA*. Li et al. *(63)* then demonstrated that although *WAF1* blocks replicative DNA synthesis, it does not inhibit the function of PCNA in repair synthesis required to fill the gap left after damage excision: short patch synthesis was more refractory to WAF1 inhibition than long (replicative) synthesis. Inhibition of Cdk activity by WAF1 required the amino-terminal, whereas inhibition of PCNA function required the carboxy-terminal residues of WAF1 *(9)*. Concomitant with relocalization of PCNA after high-dose (75 J/m^2) UV irradiation, G1 human lung fibroblasts lose cyclin D1 *(89)*. This cyclin is found in association with Cdk4 and Cdk6, both of which are required for G1/S-phase progression. Cyclin D1 is also found in association with PCNA, in the absence of Cdk4. When cyclin D1 was moderately overexpressed, both replicative and repair synthesis of DNA were suppressed, whereas expression of cyclin D1 antisense RNA as well as coexpression of *PCNA* with cyclin D1 (sense) restored repair synthesis. Pagano et al. *(89)* speculate that cyclin D1 acts in a checkpoint control to block G1 progression by keeping PCNA in an inactive form. This, however, appears to be at odds with both the observation that cyclin D1 is lost after irradiation and the assertion that decreased levels of cyclin D1 permit DNA repair synthesis. Under normal (unirradiated/stress-free) conditions, D1 cyclin in conjunction with Cdk4 promotes progression past the Restriction point. D1 cyclin then reverses roles and blocks progress past the G1/S boundary until degraded. It may thus require analysis of synchronously growing cell populations to clarify the role of D1 cyclin in repair.

Like WAF1, *GADD45* is induced by ionizing radiation in a *p53*-dependent manner *(49)* and associates with PCNA. When *GADD45* was removed from an in vitro excision repair reaction by immunodepletion, the rate of repair was decreased *(124)*.

Although the precise role of *GADD45* is not known, a reasonable hypothesis is that it interacts with *(35)* and inactivates a form of PCNA required for DNA replication, thus blocking entry to S phase while simultaneously stimulating repair. *GADD45* may also be induced by UV or MMS, but not ionizing radiation, in the absence of a functional p53 protein *(100)*.

The process of excision repair provides another example of how repair may be coupled to cell-cycle progression, as well as to transcription. The following summary is based in part on observations by Svestrup et al. *(126)* and reviews by Bootsma and Hoejimakers *(4)* and Drapkin et al. *(19)*. Repair of UV-induced damage in particular occurs preferentially on the transcribed DNA strand and simultaneously, at a lower rate, on the entire genome. In brief, transcription by RNA polymerase II (RNA pol II) is initiated by formation of a complex between the polymerase and seven accessory proteins, collectively termed general transcription factors. One such factor, TFIIH, is itself comprised of 8–10 *(19)* or 5–9 *(118)* polypeptides (*see* Chapter 10). The p80 and p89 TFIIH subunits are the helicases *XP-D* and *XP-B*, respectively, mutations in which result in the UV-sensitive human genetic disorder, xeroderma pigmentosum (*see* Chapter 18). The p34 and p40 subunits are cyclin H and the accompanying cyclin-dependent kinase *MO15*, also termed CAK *(24,106,117)*. Substrates of *MO15* include RNA pol II (the carboxy-terminal domain [CTD], phosphorylation of which allows the polymerase to dissociate from the TATA binding protein and thus clear the promoter) and the related cyclin-dependent kinases $p34^{cdc2}$ and $p33^{cdk2}$. *MO15* activates these Cdks by phosphorylation at threonine-161 *(25,72)*. A more detailed description of the multiple roles of TFIIH is available for yeast *(118)*, to which the mammalian system bears many similarities. TFIIH can be isolated in two forms: The form active in transcription is comprised of a five-subunit core, the *SSL2* gene product (human *XP-BC)* and a complex of 47-, 45-, and 33-kDa polypeptides. The entire complex is termed holo-TFIIH; the latter polypeptides contain the Cdk/CTD kinase activity. The form active in nucleotide excision repair, on the other hand, lacks the CTD-kinase activity, but is associated with a number of genes required for excision repair, i.e., it is a complex of the five-subunit core, the *SSL2* product, and the products of the genes *RAD1, RAD2 (XP-G), RAD4 (XP-C), RAD10*, and *RAD14 (XP-A)*. The latter complex has been termed a repairosome *(118)*. Precisely what role *MO15* plays in modulating the progression of clastogen-treated cells remains to be seen.

4.6. Future Directions

It is reasonable to speculate that ubiquitin-mediated protein degradation plays a significant role in the induction of cell-cycle responses to DNA damage. Indeed the checkpoint protein p53, the CKI p27, and cyclins are all regulated to some extent by this pathway. However, in addition to these major fields of investigation, it is worthwhile considering a somewhat neglected area. Almost all mammalian cell radiobiology has been established using cells that lacked p53-dependent checkpoint controls. Since we are now aware that p53 checkpoint-mediated delays are extremely long, comparing dose for dose with G2 delay, it seems likely that the radiobiology of p53-defective cells will be different from normal cells. How, for example, does normal cell sensitivity for killing vary with cell cycle age? What is the age response for sensitivity to transformation (if measurable), replicon-initiation delay, G2 delay, and metaphase delay? Do repair deficient cells suffer long G1 delays or do they apoptose? Are DNA strand break repair kinetics different in G1-arrested cells? Is the spectrum of chromosome aberrations normally observed after irradiating asynchronous cell populations altered by the action of the G1 checkpoint? Judging by the volume of literature on related subjects, we should not have to wait long for the answers to these questions.

5. METAPHASE DELAY

D'Hooge et al. *(14)* used time-lapse cinematography to show that EMT6 mouse tumor cells in vitro are delayed in completing mitosis by exposure to γ-rays. The delay was dose- and cell-cycle-age-dependent, such that no marked delay was observed until the cell had passed the X-ray transition point in G2 (i.e., the last point in G2 at which a cell may be delayed by irradiation). Once past that point, the delay increased linearly, so that the duration of mitosis was doubled by a dose of 10 Gy. The delay was greater in cells destined to produce inviable daughters and was almost entirely owing to a prolongation of metaphase. The authors thus suggested that delay resulted from a mechanical interference with the separation of sister chromatids by the radiation-induced formation of chromosome bridges. However, since budding yeast possess checkpoint controls to block chromatid separation in the presence of an incompletely assembled spindle *(43,62)*, it is conceivable that the observed mitotic delay was the equivalent mammalian response to the presence of chromosome damage.

6. THE MAMMALIAN UV RESPONSE

There is considerable overlap between the mammalian "UV response" (reviewed in *40* and *42*. *See also* Chapters 15 and 16) and the radiation-induced molecular events described above. Within 5 minutes of exposure to UV, the rate of *fos* and *jun* transcription is increased. Together Fos and Jun form the heterodimeric transcription factor AP-1, which triggers transcription of a number of genes, including metalothionein, collagenase, and *jun* itself. The pathway involved in AP-1 activation is just now being characterized. The initial signal appears to be generated at the plasma membrane, since agents that block growth factor receptor activation also block the UV response. Activation of AP-1 requires prior activation of Ras and Raf, probably by Src-dependent phosphorylation *(12)*. Both Jun and Fos show alterations in phosphorylation. Jun is phosphorylated on serines 63 and 73 by the Jun-NH_2-terminal kinase *(JNK1)*, which itself is phosphorylated and partially activated by Ha-Ras, accounting for Ras involvement. *JNK1* is related to the MAP kinases and to the *spc1* gene product recently identified in the fission yeast, which is involved in cell cycle regulation in response to environmental stress *(120)*. Some AP-1-responsive transcription is required to suppress the induction of oxidative damage, and a number of genes induced by UV in mammalian cells are also induced by ionizing radiation *(27)*. *fos* and *jun* induction is normal in AT cells, suggesting that the UV response is a separate pathway from that involving *p53*, *WAF1*, *GADD45*, and *mdm2* *(6)*. Conversely, *fos* can be downregulated by an Rb-dependent mechanism *(100)*, which, according to the widely accepted scheme for Rb regulation presented in Section 4.5., would place AP-1 downstream of p53 in the UV-response pathway. In short, and not surprisingly, there are clearly myriad interconnections between the stress-activated pathways, so that a "nudge" to any corner of the system has repercussions throughout the network.

REFERENCES

1. Beamish, H. and M. F. Lavin. 1994. Radiosensitivity in ataxia-telangiectasia: anomalies in radiation-induced cell cycle delay. *Int. J. Radiat. Biol.* **65:** 175–184.
2. Beetham, K. L. and L. J. Tolmach. 1984. The action of caffeine on X-irradiated HeLa cells VII Evidence that caffeine enhances expression of potentially lethal damage. *Radiat. Res.* **100:** 585–593.
3. Bernhard, E. J., R. J. Muschel, V. J. Bakanauskas, and W. G. McKenna. 1996. Reducing the radiation-induced G_2 delay causes HeLa cells to undergo apoptosis instead of mitotic death. *Int. J. Radiat. Biol.* **69:** 575–584.
4. Bootsma, D. and J. H. J. Hoejimakers. 1993. Engagement with transcription. *Nature* **363:** 114,115.
5. Busse, P. M., S. K. Bose, R. W. Jones, and L. J. Tolmach. 1978. The acffon of caffeine on X-irradiated HeLa cells: III. Enhancement of X-ray-induced killing during G_2 arrest. *Radiat. Res.* **76:** 292–307.
6. Canman, C. E., A. C. Wolff, C.-Y. Chen, A. J. Fornace, and M. B. Kastan. 1994. The *p53*dependent G_1 cell cycle checkpoint pathway and *ataxia telangiectasia. Cancer Res.* **54:** 5054–5058.
7. Cantley, L. C., K. R. Auger, C. Carpenter, B. Duckworth, A. Graziani, R. Kapeller, and S. Soltoff. 1991. *Oncogene*s and signal transduction. *Cell* **64:** 281–302.
8. Carlson, J. G. 1989. Chinese hamster ovary cell mitosis and its response to ionizing radiation: amorphological analysis. *Radiat. Res.* **118:** 311–323.
9. Chen, J., P. K. Jackson, M. W. Kirschner, and A. Duffa. 1995. Separate domains of *p21* involved in the inhibition of cdk kinase and PCNA. *Nature* **374:** 386–388.
10. Chwalinski, S. and C. S. Potten. 1986. Radiation-induced mitotic delay: Duration, dose and cell position dependence in the crypts of the small intestine of the mouse. *Int. J. Radiat. Biol.* **49:** 809–819.
11. Cook, P. R. and I. A. Brazel. 1975. Supercoils in human DNA. *J. Cell Sci.* **19:** 261–279.
12. Devary, Y., R. A. Gofflieb, T. Smeal, and M. Karin. 1992. The mammalian ultraviolet response is triggered by activation of *src* tyrosine kinases. *Cell* **71:** 1081–1091.
13. Dewey, W. C. and D. P. Highfield. 1976. G_2 block in Chinese hamster cells induced by X-irradiation, hyperthermia, cycloheximide, or actinomycin-D. *Radiat. Res.* **65:** 511–528.
14. d'Hooge, M C., D. Hemon, A. J. Valleron, and E. P. Malaise. 1980. Comparative effects of ionizing radiations on cycle time and mitotic duration. A time-lapse cinematography study. *Radiat. Res.* **81:** 383–392.
15. Diffley, J. F. X., J. H. Cocker, S. J. Dowell, and A. Rowley. 1994. Two steps in the assembly of complexes at yeast replication origins *in vivo. Cell* **78:** 303–316.
16. Di-Leonardo, A., S. P. Linke, K. Clarkin, and G. M. Wahl. 1994. DNA damage triggers a prolonged *p53*-dependent G. arrest and long term induction of cipl in normal human fibroblasts. *Genes Dev.* **8:** 2540–2551.
17. Doida, Y. and S. Okada. 1969. Radiation-induced mitotic delay in cultured mammalian cells (L5178Y). *Radiat. Res.* **38:** 513–529.
18. Donehower, L. A., M. Harvey, B. L. Slagle, M. J. McArthur, C. A. Montgomery, J. S. Butel, and A. Bradley. 1992. Mice deficient for *p53* are developmentally normal but susceptible to spontaneous tumors. *Nature* **356:** 215–221.
19. Drapkin, R. A. Sancar, and D. Reinberg. 1994. Where transcription meets repair. *Cell* **77:** 9–12.
20. El-Diery, W. S., T. Tokino, V. E. Velculescu, D. B. Levy, R. Parsens, J. M. Trent, D. Lin, W. E. Merceer, K. W. Kinzler, and B. Vogelstein. 1993. WAF1, a potential mediator of *p53* tumor suppresion. *Cell* **75:** 817–825.
21. Elkind, M. M., A. Han, and K. W. Volz. 1963. Radiation response of mammalian cells grown in culture. IV. Dose dependence of division delay and postirradiation growth of surviving and non-surviving Chinese hamster cells. *J. Natl. Cancer Inst.* **30:** 705–772.

22. Fan, S., M. L. Smith, D. J. Rivet, Q. Zhan, K. W. Kohn, A. J. Fornace, and P. M. O'Connor. 1995. Disruption of *p53* sensitizes breast cancer MCF–7 cells to Cisplatin and pentoxyfylline. *Cancer Res.* **55:** 1649–1654.
23. Farrell, P. J., K. Balkow, T. Hunt, and R. J. Jackson. 1977. Phosphorylation of initiation factor eIF–2 and the control of reticulocyte protein synthesis. *Cell* **11:** 187–200.
24. Feaver, WJ., J. Q. Svestrup, N. L. Henry, and R. D. Kornberg. 1994. Relationship between CDK-activating kinase and RNA polymerase II CTD kinase TFIIH/TFIIK. *Cell* **79:** 1103–1109.
25. Fesquet, D., J.-C. Labbe, J. Derancourt, J.-P. Capony, S. Galas, F. Girard, T. Lorca, J. Shuttleworth, M. Doree, and J.-C. Cavadore. 1993. The MO15 gene encodes the catalytic subunit of a protein kinase that activates cdc2 and other cyclin-dependent kinases (CDKs) through phosphorylation of Thrl61 and its homologues. *EMBO J.* **12:** 3111–3121.
26. Fiscella, M., S. J. Ullrich, N. Zambrano, M. T. Shields, D. Lin, S. P. Lees-Miller, C. W. Anderson, W. Mercer, and E. Appella. 1993. Mutation of the serine 15 phosphorylation site of human *p53* reduces the ability of *p53* to inhibit cell cycle progress. *Oncogene* **8:** 1519–1528.
27. Fornace, A. J. 1992. Mammalian genes induced by radiation: Activation of genes associated with growth control. *Ann. Rev. Genet.* **26:** 507–526.
28. Fuchs, B., D. O'Connor, L. Fallis, K. H. Scheidtmann, and X. Lu. 1995. *p53* phosphorylation mutants retain transcription activity. *Oncogene* **10:** 789–793.
29. Giese, A. C. 1947. Radiations and cell division. *Q. Rev. Biol.* **22:** 253–264.
30. Graeber, T. G., J. F. Peterson, M. Tsai, K. Monica, A. J. Fornace, and A. J. Giaccia. 1994. Hypoxia induces accumulation of *p53* protein, but activation of a G_1-phase checkpoint by low oxygen is independent of *p53* status. *Mol. Cell. Biol.* **14:** 6264–6277.
31. Griffith, T. D. and L. J. Tolmach. 1976. Lethal response of HeLa cells to X-irradiation in the latter part of the generation cycle. Biophys. J. **16:** 303–317.
32. Gu,Y., C. W. Turek, and D. O. Morgan. 1993. Inhibition of CDK2 activity *in vivo* by an associated 20K regulatory subunit. *Nature* **366:** 707–710.
33. Halahan, D. E., S. Virudachalam, D. Grdina, and R. R. Weichselbaum. 1992. The isoquinoline sulfonamide H7 attenuates radiation-mediated protein kinase C activation and delays the onset of X-ray-induced G_2 arrest. *Int. J. Radiat. Oncol. Biol. Phys.* **24:** 687–692.
34. Halahan, D. E., S. Virudachalam, J. L. Schwartz, N. Panje, R. Mustafi, and R. R. Weichselbaum. 1992. Inhibition of protein kinases sensitizes human tumor cells to ionizing radiation. *Radiat. Res.* **129:** 345–350.
35. Hall, P. A., J. M. Kearsey, P. J. Coates, D. G. Norman, E. Warbrick, and L. S. Cox. 1995. Characterization of the interaction between PCNA and gadd45. *Oncogene* **10:** 2427–2433.
36. Harper, J. W., G. R. Adami, N. Wei, K. Keyomarsi, and S. J. Elledge. 1993. The p21 cdk interacting protein cipl is a potent inhibitor of G_1 cyclin dependent kinases. *Cell* **75:** 805–816.
37. Hartley, K. O., D. Gell, G. C. Smith, H. Zhang, N. Divecha, M. A. Conelly, A. Admon, S. P. Lees-Miller, C. W. Anderson, and S. P. Jackson. 1995. DNA-dependent protein kinase catalytic subunit: A relative of phosphatidylinositol 3-kinase and the *ataxia telangiectasia* gene product. *Cell* **82:** 849–856.
38. Hartwell, L. H. 1992. Defects in cell cycle checkpoints may be responsible for the genomic instability of cancer cells. *Cell* **71:** 543–546.
39. Hartwell, L. H. and T. A. Weinert. 1989. Checkpoints: controls that ensure the order of cell cycle events. *Science* **246:** 629–634.
40. Herrlich, P., C. Sachsenmaier, A. Radler-Pohl, S. Gebel, C. Blattner, and H. J. Rahmsdorf. 1994. The mammalian UV response: mechanism of DNA damage induced gene expression. *Adv. Enzyme Regul.* **34:** 381–395.

41. Hedge, F. A. and D. S. Nachtwey. 1972. X-ray-induced delay of cell division in synchronized *Tetrahymena pyriformis*. *Radiat. Res.* **52:** 603–617.
42. Holbrook, N. J. and A. J. Fornace. 1991. Response to adversity: Molecular control of gene activation following genotoxic stress. *New Biol.* **3:** 825–833.
43. Hoyt, M. A., L. Totis, and T. Roberts. 1991. S. *cerevisiae* genes required for cell cycle arrest in response to loss of microtubule function. *Cell* **66:** 507–517.
44. Hupp, T. R. and D. P. Lane. 1994. Allosteric activation of latent *p53* tetramers. *Curr. Biol.* **4:** 865–875.
45. Jackson, D. A., A. B. Hassan, R. J. Errington, and P. R. Cook. 1994. Sites in human nuclei where damage induced by ultraviolet light is repaired: Localization relative to transcription sites and concentrations of proliferating cell nuclear antigen and the tumor suppressor protein, *p53*. *J. Cell Sci.* **107:** 1753–1760.
46. Jung, M., V. Notario, and A. Dritschilo. 1992. Mutations in the *p53* gene in radiation-sensitive and -resistant human squamous carcinoma cells. *Cancer Res.* **52:** 6390–6393.
47. Jung, T. and C. Streffer. 1992. Effects of caffeine on protein phosphorylation and cell cycle progression in X-irradiated two-cell mouse embryos. *Int. J. Radiat. Biol.* **62:** 161–168.
48. Kastan, M. B., O. Onyekwere, D. Sidransky, B. Vogelstein, and R. Craig. 1991. Participation of *p53* protein in the cellular response to DNA damage. *Cancer Res.* **51:** 6304–6311.
49. Kastan, M. B., Q. Zhan, W. S. El-Deiry, F. Carrier, T. Jacks, W. V. Walsh, B. S. Plunkett, B. Vogelstein, and A. J. Fornace. 1992. A mammalian cell cycle checkpoint pathway utilizing *p53* and GADD45 is defective in ataxia-telangiectasia. *Cell* **71:** 587–597.
50. Kharbanda, S., Z.-M. Yuan, E. Rubin, R. R. Weichselbaum, and D. Kufe. 1994. Activation of src-like *p56/p53 (lyn)* tyrosine kinase by ionizing radiation. *J. Biol. Chem.* **269:** 20,739–20,743.
51. Kharbanda, S., Z.-M. Yuan, N. Taneja, R. R. Weichselbaum, and D. Kufe. 1994. *p56/p53 lyn* tyrosine kinase activation in mammalian cells treated with mitomycin C. *Oncogene* **9:** 3005–3011.
52. Kihlman, B. A. 1977. *Caffeine and Chromosomes*. Elsevier Scientific Publishing Co. Amsterdam.
53. Kimler, B. F., D. B. Leeper, and M. H. Schneiderman. 1981. Radiation-induced division delay in Chinese hamster ovary fibroblast and carcinoma cells: dose effect and ploidy. *Radiat. Res.* **74:** 430–438.
54. Kimler, B. F., D. B. Leeper, M. H. Snyder, R. Rowley, and M. H. Schneiderman. 1982. Modification of radiation-induced division delay by caffeine analogs and dibutyryl cyclic AMP. *Int. J. Radiat. Biol.* **41:** 47–58.
55. Kornberg A. and T. A. Baker. 1991. *DNA Replication*, 2nd ed. W. H. Freeman, New York.
56. Kovacs, C. J. and J. Van't Hof. 1971. Mitotic delay and regulating events in plant cell proliferation: DNA replication by a G_1/S population. *Radiat. Res.* **48:** 95–106.
57. Kuerbitz, S. J., B. S. Plunkett, W. V. Wabh, and M. B. Kastan. 1992. Wild-type *p53* is a cell cycle checkpoint determinant following irradiation. *Proc. Natl. Acad. Sci. USA* **89:** 7491–7495.
58. Lamb, J. R., B. C. Petit-Frere, B. C. Broughton, A. R Lehmann, and M. H. Green. 1989. Inhibition of DNA replication by ionizing radiation is mediated by a *trans*-acting factor. *Int. J. Radiat. Biol.* **56:** 125–130.
59. Larner, J. M., H. Lee, and J. L. Hamlin. 1994. Radiation effects on DNA synthesis in a defined chromosomal replicon. *Mol. Cell. Biol.* **14:** 1901–1908.
60. Leeper, D. B., M. H. Schneiderman, and W. C. Dewey. 1972. Radiation-induced division delay in synchronized Chinese hamster ovary cells in monolayer culture. *Radiat. Res.* **50:** 401–417.

61. Lees-Miller, S. P., K. Sakaguchi, S. J. Ullrich, E. Appela, and C. W. Anderson. 1992. Human DNA-activated protein kinase phosphorylates serines 15 and 37 in the amino-terminal transactivation domain of human *p53*. *Mol. Cell. Biol.* **12:** 5041–5049.
62. Li, R. and A. W. Murray. 1991. Feedback controls of mitosis in budding yeast. *Cell* **66:** 519–531.
63. Li, R., S. Waga, G. J. Hannon, D. Beach, and B. Stillman. 1994. Differential effects by the p21 CDK inhibitor on PCNA-dependent DNA replication and repair. *Nature* **371:** 534–537.
64. Little, J. B. 1968. Delayed initiation of DNA synthesis in irradiated human diploid cells. *Nature* **218:** 1064,1065.
65. Liu, M., K. R. Dhanwada, D. F. Birt, S. Hecht, and J. C. Pelling. 1994. Increase in *p53* protein half-life in mouse keratinocytes following UV-B irradiation. *Carcinogenesis* **15:** 1089–1092.
66. Livingstone, L. R., A. White, J. Sprouse, E. L. Livanos, T. Jacks, and D. T. Tlsty. 1992. Altered cell cycle arrest and gene amplification potential accompany loss of wild-type *p53*. *Cell* **70:** 923–935.
67. Lock, R. B. and W. E. Ross. 1990. Inhibition of $p34^{cdc2}$ kinase activity by etoposide or irradiation as a mechanism of G_2 arrest in Chinese hamster ovary cells. *Cancer Res.* **50:** 3761–3766.
68. Lu, X. and D. P. Lane. 1993. Differential induction of transcriptionally active *p53* following UV or ionizing radiation: Defects in chromosome instability syndromes? *Cell* **75:** 765–778.
69. Lucke-Huhle, C. 1982. Alpha-irradiation-induced G_2 delay: a period of cell recovery. *Radiat. Res.* **89:** 298–308.
70. Lydall, D. and T. A. Weinert. 1995. Yeast checkpoint genes in DNA damage processing: Implications for repair and arrest. *Science* **270:** 1488–1491.
71. Maity, A., W. G. McKenna, and R. J. Muschel. 1994. The molecular basis for cell cycle delays following ionizing radiation: A review. *Radiother. Oncol.* **31:** 1–13.
72. Makela, T. P., J.-P. Tassan, E. A. Nig, S. Frutiger, G. J. Hughes, and R. A. Weinberg. 1994. A cyclin associated with the CDK-activating kinase MO15. *Nature* **371:** 254–257.
73. Maltzman, W. and L. Czyzyk. 1984. UV irradiation stimulates levels of *p53* cellular tumor antigen in nontransformed mouse cells. *Mol. Cell. Biol.* **4:** 1689–1694.
74. Metting, N. and J. B. Little. 1995. Transient failure to dephosphorylate the cdc2-cyclin B1 complex accompanies radiation-induced G_2-phase arrest in HeLa cells. *Radiat. Res.* **143:** 286–292.
75. Metzger, K. 1964. On the dark reactivation mechanism in ultraviolet irradiated bacteria. *Biochem. Biophys. Res. Commun.* **15:** 101–109.
76. Milne, D. M., R. H. Palmer, D. G. Campbell, and D. W. Meek. 1992. Phosphorylation of the *p53* tumor suppressor protein at three N-terminal sites by a novel casein kinase I-like enzyme. *Oncogene* **7:** 1361–1369.
77. Milner, J., A. Cook, and J. Mason. 1990. *p53* is associated with $p34^{cdc2}$ in transformed cells. *EMBO J.* **9:** 2885–2889.
78. Morgan, D. O. 1995. Principles of CDK regulation. *Nature* **374:** 131–134.
79. Munro, T. R. 1970. The site of the target region for radiation-induced mitotic delay in cultured mammalian cells. *Radiat. Res.* **44:** 748–757.
80. Muschel, R. J., H. B. Zhang, G. Iliakis, and W. G. McKenna. 1991. Cyclin B expression in HeLa cells during the G_2 block induced by ionizing radiation. *Cancer Res.* **51:** 5113–5117.
81. Muschel, R. J., H. B. Zhang, G. Iliakis, and W. G. McKenna. 1992. Effects of ionizing radiation on cyclin expression in HeLa cells. *Radiat. Res.* **132:** 153–157.
82. Muschel, R. J., H. B. Zhang, and W. G. McKenna. 1993. Differential effects of ionizing radiation on the expression of cyclin A and cyclin B in HeLa cells. *Cancer Res.* **53:** 1128–1135.

83. Nagasawa, H., P. Keng, R. Harley, W. Dahlberg, and J. B. Little. 1994. Relationship between gamma-ray-induced G_2 delay and cellular radiosensitivity. *Int. J. Radiat. Biol.* **66:** 373–379.
84. Nelson, W. G. and M. B. Kastan. 1994. DNA strand breaks: The DNA template alterations that trigger *p53*-dependent DNA damage response pathways. *Mol. Cell. Biol.* **14:** 1815–1823.
85. Noda, A., Y. Ning, S. F. Venable, O. M. Periera-Smith, and J. R. Smith. 1994. Cloning of senescent cell-derived inhibitors of DNA synthesis using an expression screen. *Exp. Cell Res.* **211:** 90–98.
86. O'Connor, P. M., D. K. Ferris, I. Hoffman, J. Jackman, G. Draetta, and K. W. Kohn. 1994. Role of the cdc25C phosphatase in G_2 arrest induced by nitrogen mustard. *Proc. Natl. Acad. Sci. USA* **91:** 9480–9484.
87. O'Connor, P. M., D. K Ferris, M. Pagano, G. Draetta, J. Pines, T. Hunter, D. L. Longo, and K.W. Kohn. 1993. G_2 delay induced by nitrogen mustard in human cells affects cyclin A/cdk2 and cyclin B/cdc2-kinase complexes differentially. *J. Biol. Chem.* **268:** 8298–8308.
88. O'Connor, P. M., D. K Ferris, G. A. White, J. Pines, T. Hunter, D. L. Longo, and K. W. Kohn. 1992. Relationship between cdc2 kinase, DNA crosslinking, and cell cycle perturbations induced by nitrogen mustard. *Cell Growth Differ.* **3:** 43–52.
89. Pagano, M., A. M. Theodoras, S. W. Tam, and G. F. Draetta. 1994. Cyclin D1-mediated inhibition of repair and replicative DNA synthesis in human fibroblasts. *Genes Dev.* **8:** 1627–1639.
90. Painter, R. B. 1980. Effect of caffeine on DNA synthesis in irradiated and unirradiated mammalian cells. *J. Mol. Biol.* **143:** 289–301.
91. Painter, R. B. 1983. Are lesions produced by ionizing radiation direct blocks to DNA chain elongation? *Radiat. Res.* **95:** 421–426.
92. Painter, R. B. 1985. 3-aminobenzamide does not affect radiation-induced inhibition of DNA synthesis in human cells. *Mutat. Res.* **143:** 113–115.
93. Painter, R. B. 1986. Inhibition of mammalian cell DNA synthesis by ionizing radiation. *Int. J. Radiat. Biol.* **49:** 771–781.
94. Pardee, A. B. 1974. A restriction point for control of normal animal cell proliferation. *Proc. Natl. Acad. Sci. USA* **71:** 1286–1290.
95. Phillips, R. A. and L. J. Tolmach. 1966. Repair of potentially lethal damage in X-irradiated HeLa cells. *Radiat. Res.* **29:** 413–432.
96. Pines, J. 1994. The cell cycle kinases. *Semin. Cancer Biol.* **5:** 305–313.
97. Powell, S. N., J. S. DeFrank, P. Connell, P., M. Eogan, F. Pfeffer, D. Dombkowski, W. Tang, and S. Friend. 1995. Differential sensitivity of *p53*(–) and *p53*(+) cells to caffeine-induced radiosensitization and override of G_2 delay. *Cancer Res.* **55:** 1643–1648.
98. Price, B. D. and S. K. Calderwood. 1993. Increased sequence-specific *p53*-DNA binding activity after DNA damage is attenutated by phorbol esters. *Oncogene* **8:** 3055–3062.
99. Pujara, C. M., R. J. Horsley, and S. N. Banerjee. 1970. Radiation-induced division and mitotic delay studies in green alga *(Oedogonium cardiacum)*. *Radiat. Res.* **44:** 413–420.
100. Robbing, P. D., J. M. Horowitz, and R. C. Mulligan. 1990. Negative regulation of human *c-fos* expression by the retinoblastoma gene product. *Nature* **346:** 668–671.
101. Rowley, R. 1990. Repair of radiation-induced chromatid aberrations: relationship to radiation-induced G_2-arrest. *Int. J. Radiat. Biol.* **58:** 489–498.
102. Rowley, R. 1992. Reduction of radiation-induced G_2 arrest by caffeine. *Radiat. Res.* **129:** 224–227.
103. Rowley, R. and M. J. Egger. 1988. Method for probing cells in radiation-induced G_2-arrest: demonstration of potentially lethal damage repair. *Cell Tissue Kinet.* **21:** 395–403.
104. Rowley, R. and L. Kort. 1988. Effect of modulators of radiation-induced G_2-arrest on the repair of radiation-induced DNA damage detectable by neutral filter elusion. *Int. J. Radiat. Biol.* **54:** 749–759.

105. Rowley, R., M. Zorch, and D. B. Leeper. 1984. Effect of caffeine on radiation-induced mitotic delay: delayed expression of G_2 arrest. *Radiat. Res.* **97:** 178–185.
106. Roy, R. J. P. Adamczewski, T. Seroz, W. Vermeulen, J.-P. Tassan, L. Schaeffer, E. A. Nigg, J. H. J. Hoeijmakers, and J.-M. Egly. 1994. The MO15 cell cycle kinase is associated with the TFIIH transcription-DNA repair factor. *Cell* **79:** 1093–1101.
107. Russell, K. J., L. W. Wiens, G. W. Demers, D. A. Galloway, S. E. Plon, and M. Groudine. 1995. Abrogation of the G_2 checkpoint results in differential radiosensitization of G_1 checkpoint deficient and G_1 checkpoint-competent cells. *Cancer Res.* **55:** 1639–1642.
108. Rustad, R. C. 1970. Variations in the sensitivity of X-ray-induced mitotic delay during the cell division cycle of the sea urchin egg. *Radiat. Res.* **42:** 498–512.
109. Sachsenmaier, W., E. Bohnert, B. Clausinizer, and O. Nygnard. 1970. Cycle dependent variation of X-ray effects on synchronous mitosis and thymidine kinase induction in *Physarum polycephalum. FEBS Lett.* **10:** 185–189.
110. Sachsenmaier, W., K. H. Donges, H. Rupff, and G. Czihak. 1970. Advanced initiation of synchronous mitoses in *Physarum polycephalum* following UV irradiation. *Zeitschrift fur Naturforschung* **25:** 866–871.
111. Sandell, L. L. and V. A. Zakian. 1993. Loss of yeast telomere: arrest, recovery and chromosome loss. *Cell* **75:** 729–739.
112. Savitsky, K., A. Bar-Shira, S. Gilad, G. Rotman, Y. Ziv, L. Vanagaite, D. A. Tagle, S. Smith, T. Uziel, S. Sfez, M. Ashkenazi, I. Pecker, M. Frydman, R. Harnik, S. R. Patanjali, A. Simmons, G. A. Clines, A. Sartiel, R. A. Gaffi, L. Chessa, O. Sanal, M. F. Lavin, N. G. J. Jaspers, A. M. R. Taylor, C. F. Arlett, T. Miki, S. M. Weissman, M. Lovett, F. S. Collins, and Y. Shiloh. 1995. A single *ataxia telangiectasia* gene with a product similar to PI–3 kinase. *Science* **268:** 1749–1753.
113. Scheffner, M., B. A. Werness, J. M. Huibregtse, A. J. Levine, and P. M. Howley. 1990. The E6 oncoprotein encoded by human papilloma virus types 16 and 18 promotes the degradation of *p53. Cell* **63:** 1129–1136.
114. Schneiderman, M. H., W. C. Dewey, D. B. Leeper, and H. Nagasawa. 1972. Use of the mitotic cell selection procedure for cell cycle analysis. *Exp. Cell Res.* **74:** 430–438.
115. Schneiderman, M. H. and K. G. Hofer. 1980. The target for radiation-induced division delay. *Radiat. Res.* **84:** 462–476.
116. Selby, C. P. and A. Sancar. 1990. Molecular mechanisms of DNA repair inhibition by caffeine. *Proc. Natl. Acad. Sci. USA* **87:** 3522–3525.
117. Serizawa, H., T. P. Makela, J. W. Conaway, R. C. Conaway, R. A. Weinberg, and R. A. Young. 1995. Association of cdk-activating kinase subunits with transcription factor TFIIH. *Nature* **374:** 280–282.
118. Seroz, T., J. R. Hwang, V. Moncollin, and J. M. Egly. 1995. TFIIH: a link between transcription, DNA repair and cell cycle regulation. *Curr. Biol.* **5:** 217–221.
119. Sherr, C. J. 1994. G_1 phase progression: Cycling on cue. *Cell* **79:** 551–555.
120. Shiozaki, K. and P. Russell. 1995. Cell-cycle control linked to extracellular environment by MAP kinase pathway in fission yeast. *Nature* **378:** 739–743.
121. Sinclair, W. K. 1968. Cyclic X-ray responses in mammalian cells *in vitro. Radiat. Res.* **33:** 620–643.
122. Slebos, R. J., M. H. Lee, B. S. Plunkett, T. D. Kessis, B. O. Williams, T. Jacks, L. Hedrick, M. B. Kastan, and K. R. Cho. 1994. *p53*-dependent G_1 arrest involves pRB-related proteins and is disrupted by human papillomavirus 16 E7 oncoprotein. *Proc. Natl. Acad. Sci. USA* **91:** 3520–5324.
123. Slichenmyer, W. J., W. G., R. J. Slebos, and M. B. Kastan. 1993. Loss of *a p53* associated checkpoint does not decrease cell survival following DNA damage. *Cancer Res.* **53:** 4146–4168.
124. Smith, M. L., I-T., Chen, Q. Zhan, I. Bae, C-Y. Chen, T. M. Gilmer, M. B. Kastan, P. M. O'Connor, and A. J. Fornace. 1994. Interaction of *p53*-regulated protein *GADD54* with proliferating cell nuclear antigen. *Science* **266:** 1376–1380.

125. Su, L. N. and J. B. Little. 1992. Transformation and radiosensitivity of human diploid skin fibroblasts transfected with SV40 T-antigen mutants defective in Rb and *p53* binding domains. *Int. J. Radiat. Biol.* **62:** 461–468.
126. Svestrup, J. Q., Z. Wang, W. J. Feaver, X. Wu, D. A. Bushnell, T. F. Donahue, E. C. Friedberg, and R. D. Kornberg. 199S. Different forms of TFIIH for transcription and DNA repair: HoloTFIIH and a nucleotide excision repairosome. *Cell* **80:** 21–28.
127. Taylor, Y. C., A. J. Pasian, and P. G. Duncan. 1993. Differential post-irradiation caffeine response in normal diploid versus SV40 transformed human fibroblasts: potential role of nuclear organization. *Int. J. Radiat. Biol.* **64:** 57–70.
128. Terada, Y., M. Tatsuka, S. Jinno, and H. Okayama. 1995. Requirement for tyrosine phosphorylation of Cdk4 in G_1 arrest induced by ultraviolet irradiation. *Nature* **376:** 358–362.
129. Tolmach, L. J., R. B. Hawkins, and P. M. Busse. 1980. The relationship between depressed synthesis of DNA and killing in X-irradiated HeLa cells, in *Radiation Biology in Cancer Research* (Meyn, R. E. and H. R. Withers, eds.), Raven, New York, pp. 125–142.
130. Tolmach, L. J., R. W. Jones, and P. M. Busse. 1977. The action of caffeine on X-irradiated HeLa cells. I. Delayed inhibition of DNA synthesis. *Radiat. Res.* **71:** 653–665.
131. Tsao, Y-P., P. d'Arpa, and L. F. Liu. 1992. The involvement of active DNA synthesis in camptothecin-induced G_2 arrest: altered regulation of $p34^{cdc2}$/cyclin B. *Cancer Res.* **52:** 1823–1829.
132. Uckun, F. M., L. Tuel-Ahlgren, K. G. Waddick, Jun Xiao, Jin Jizhong, D. E. Myers, B. R. Rowley, A. L. Burkhardt, and J. B. Bolen. 1996. Physical and functional interactions between Lyn and $p34^{cdc2}$ kinases in irradiated human B-cell precursors. *J. Biol. Chem.* **271:** 6389–6397.
133. Waga, S., G. J. Hannon, D. Beach, and B. Stillman. 1994. The p21 inhibitor of cyclin-dependent kinases controls DNA replication by interaction with PCNA. *Nature* **369:** 574–578.
134. Waldman, T., C. Lengauer, K. W. Kinzler, and B. Vogelstein. 1996. Uncoupling of S phase and mitosis induced by anticancer agents in cells lacking *p21*. *Nature* **381:** 713–716.
135. Walters, R. A., L. R. Gurley, and R. A. Tobey. 1974. Effects of caffeine on radiation-induced phenomena associated with cell cycle traverse of mammalian cells. *Biophys. J.* **14:** 99–118.
136. Walters, R. A. and D. F. Petersen. 1968. Radiosensitivity of mammalian cells: II. Radiation effects on macromolecular synthesis. *Biophys. J.* **8:** 1487–1504.
137. Warters, R. L., K. G. Hofer, and R. Harris. 1977. Radionuclide toxicity in cultured mammalian cells: Elucidation of the primary site of radiation damage. *Curr. Top. Radiat. Res. Q.* **12:** 389–407.
138. Weibezahn, K. F., H. Lohrer, and P. Herrlich. 1985. Double-strand break repair and G_2 block in Chinese hamster ovary cells and their radiosensitive mutants. *Mutat. Res.* **145:** 177–183.
139. Weinberg, R. A. 1995. The retinoblastoma protein and cell cycle control. *Cell* **81:** 323–330.
140. Weinert, T. A. and L. H. Hartwell. 1988. The *RAD9* gene controls the cell cycle response to DNA damage in *Saccharomyces cerevisiae*. *Science* **241:** 317–322.
141. White, A. E., E. M. Livanos, and T. D. Tlsty. 1994. Differential disruption of genomic integrity and cell cycle regulation in normal human fibroblasts by the HPV oncoproteins. *Genes Dev.* **8:** 666–677.
142. Woloschak, G. E., C. M. Ghang-Liu, and P. Shearin-Jones. 1990. Regulation of protein kinase C by ionizing radiation. *Cancer Res.* **50:** 3963–3967.
143. Xiong, Y., G. J. Hannon, H. Zhang, D. Casso, R. Kobyashi, and D. Beach. 1993. *p21* is a universal inhibitor of cyclin kinases. *Nature* **366:** 701–704.
144. Yin, Y., M. A. Tainsly, F. Z. Bischoff, L. C. Strong, and G. M. Wahl. 1992. Wild-type *p53* restores cell cycle control and inhibits gene amplification in cells with mutant *p53* alleles. *Cell* **70:** 937–948.

145. Young, B. R. and R. B. Painter. 1989. Radioresistant DNA synthesis and human genetic diseases. *Hum. Genet.* **82:** 113–117.
146. Yuan, Z.-M., S. Kharbanda, and D. Kufe. 1995. 1-β-D-arabinofuranosylcytosine activates tyrosine phosphorylation of $p34^{cdc2}$ and its association with the *src*-like p56/*p53 lyn* tyrosine kinase in human myeloid leukemia cells. *Biochemistry* **34:** 1058–1063.
147. Zampetti-Boseller, F. and D. Scott. 1981. Cell death, chromosome damage and mitotic delay in normal human, *ataxia telangiectasia* and retinoblastoma fibroblasts after X-irradiation. *Int. J. Radiat. Biol.* **39:** 547–558.
148. Zhan, Q., I. Bae, M. B. Kastan, and A. J. Fornace. 1994. The *p53*-dependent gamma-ray response of gadd45. *Cancer Res.* **54:** 2755–2760.
149. Zhan, Q., F. Carrier, and A. J. Fornace. 1993. Induction of cellular *p53* activity by DNA-damaging agents and growth arrest. *Mol. Cell. Biol.* **13:** 4242–4250.

22

Poly(ADP-Ribose) Polymerase in Response to DNA Damage

Satadal Chatterjee and Nathan A. Berger

1. INTRODUCTION

1.1. Discovery

The first observation that paved the way for the discovery of poly(ADP-ribose) and poly(ADP-ribose) polymerase (PARP) was reported by Chambon, et al. in 1963 *(30)*. They found that a particulate fraction from nuclei of chicken liver catalyzed the nicotinamide mononucleotide (NMN)-dependent incorporation of ^{14}C-adenine-labeled ATP into trichloroaceticacid-insoluble material and that the reaction was markedly inhibited by deoxyribonuclease, but not by ribonuclease. It was later found that the DNase-dependent inhibition of the catalytic activity could be restored by DNA, establishing a dependency on DNA. Furthermore, incorporation of ^{14}C-adenine-ATP in the presence of NMN was drastically decreased by addition of unlabeled NAD, suggesting that NAD was an obligatory intermediate.

It was initially proposed that the reaction product was polyadenylic acid. However, the correct identity was soon established as described below *(29)*. Similar observations were also reported by Sugimura's group and Hayaishi's group *(57,58,103)*. However, it was later found that NAD rather than ATP was the immediate substrate for the reaction, and that ATP was not incorporated until it was converted to NAD by NAD pyrophosphorylase, which was also present in the particulate fraction from nuclei. The acid-insoluble reaction product was found to be resistant to mild alkaline hydrolysis, pancreatic DNase, RNase, *Neurospora* NADase, and various proteinases. On treatment with snake venom phosphodiesterase, more than 95% of the acid-insoluble reaction product became acid-soluble. The major hydrolysis product was found to contain a ratio of one molecule of adenine, two molecules of ribose, and two molecules of phosphate, and was deduced to be 2'-(5"-phosphoribosyl)-5'AMP, which was named "phosphoribosyl AMP (PR-AMP)" *(57,58,103)*. Thus, the reaction product was thought to be a macromolecule composed of repeating ADP-ribose units. In summary, these findings showed that NAD was formed from NMN and ATP by NAD pyrophosphorylase in the nuclear extract, and that the ADP-ribose moiety of NAD was subsequently converted enzymatically to poly(ADP-ribose) with concomitant release of nicotinamide. The enzyme responsible for this reaction was later purified to homoge-

From: DNA Damage and Repair, Vol. 2: DNA Repair in Higher Eukaryotes
Edited by: J. A. Nickoloff and M. F. Hoekstra © Humana Press Inc., Totowa, NJ

neity from various sources and named PARP (EC 2.4.2.30). The enzyme is also known as ADP-ribosyltransferase or poly(ADP-ribose) synthetase.

1.2. Cellular and Subcellular Distribution

PARP has been found in nearly all eukaryotes though not in the yeasts *Saccharomyces* and *Torulae*. Several terminally differentiated cells, such as mature granulocytes, epidermal cells, intestinal epithelial cells, and mature erythrocytes, lack the enzyme activity. However, the enzyme is present in their precursor cells, and is lost or inactivated as part of the differentiation process *(2)*. More than 95% of the enzyme is localized in the nucleus, and a preferential nucleolar distribution has been observed. Some studies show association of PARP with nuclear membrane and nuclear matrix *(27,80,88)*. The estimated number of enzyme molecules present in the cell varies from 2×10^5 molecules in HeLa cells to 2×10^6 molecules in the human CEM lymphoblastic cell line *(91,144)*. Considering 3×10^9 bp of DNA/cell this would represent about 1 PARP molecule for each 1500–15,000 bp of DNA, or one molecule for every 7–70 nucleosomes.

The enzyme is activated by DNA strand breaks to catalyze the formation of ADP-ribose polymers primarily linked to itself and possibly to other acceptor proteins by using NAD^+ as a substrate and releasing nicotinamide as a byproduct *(136)*. This poly(ADP-ribosyl)ation reaction will be described in Section 2.3. in detail. Although a number of studies have been done to elucidate the role of this posttranslational modification in various cellular processes, such as DNA replication and repair, differentiation, gene expression, recombination, and transformation, its major function appears to be in response to DNA damage *(2,13,24,28,44,49,50,88,92,122,128,129,136)*. This chapter focuses on studies aimed at elucidating PARP involvement in DNA repair.

2. ENZYMOLOGY AND MOLECULAR BIOLOGY

2.1. Structure and Function of the Enzyme

Human PARP is a polypeptide that consists of 1014 amino acid residues with a calculated mol wt of 113 kDa *(85)*. The complete primary structure of PARP in various organisms was deduced from analysis of cDNAs *(85)*. As illustrated in Fig. 1, three distinct domains of PARP have been identified following partial proteolysis with papain and/or α-chymotrypsin. These domains include:

1. A 42-kDa DNA binding fragment located at the amino-terminal region;
2. A central 16-kDa fragment containing the automodification sites; and
3. A carboxy-terminal fragment of 55 kDa bearing the NAD binding domain *(76,77,85,102)*.

Although the DNA binding and automodification domains are rich in basic lysine residues (net charge +15 and +7, respectively), the NAD binding domain is neutral *(85)*.

The N-terminal region of the human PARP DNA binding domain contains a repeated amino acid sequence (residues 2–97 and 106–207) in which 35 amino acid residues are highly conserved. Interestingly, two zinc-finger-like motifs of the form $C\text{-}X_2\text{-}CX_{28,30}\text{-}H\text{-}X_2\text{-}C$ are contained in these sequences, suggesting that the enzyme belongs to the "zinc finger" DNA binding protein family. The first finger (F1) that is closer to the amino-terminus is essential for the enzyme activity in response to either nicked DNA or double-strand DNA breaks (DSBs), whereas the second (F2) is responsible for the

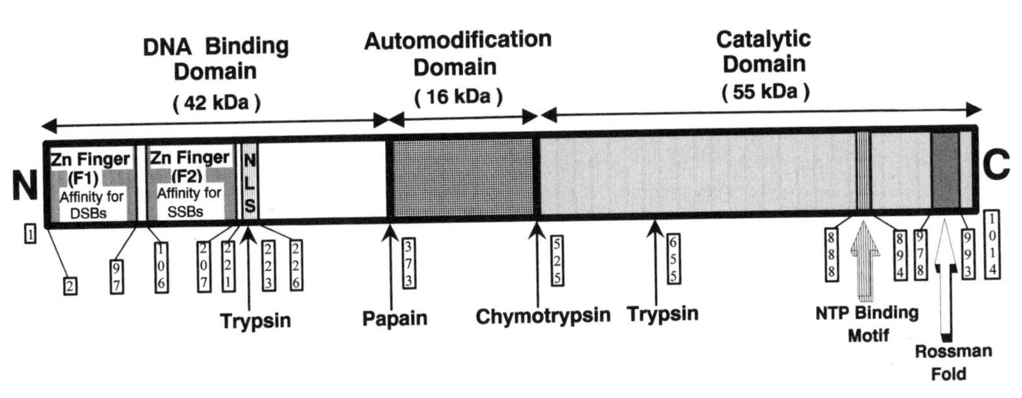

Fig. 1. Schematic representation of the different functional domains of PARP obtained after digestion with papain and chymotrypsin. Sites specific for trypsin digestion are also shown. Boxed numbers represent positions of amino acid residues. Two zinc fingers, F1 and F2, with affinity for DSBs and SSBs, respectively, are located in the DNA binding domain. Nuclear localization signal and nucleotide triphosphate binding domains are indicated by NLS and NTP, respectively. The NTP binding motif, found in numerous NTP hydrolytic proteins, is known to be involved in the binding of the pyrophosphate moiety of NTP. Several NAD-dependent dehydrogenases have a well-known structure for NAD binding known as Rossman fold. This consists of a β-fold-α-helix-β-fold structure containing the consensus sequence G/AXXXGKT/S/G found in the NTP binding motif.

activity corresponding to nicked DNA but not DSBs *(59,67)*. PARP is a metalloenzyme that requires zinc for activity. Recently, using the energy dispersive X-ray fluoresence technique, de Murcia's group has determined that each PARP molecule contains two zinc ions *(49)*. Furthermore, this DNA-binding domain binds to nicked DNA in a zinc-dependent manner. In addition to the zinc fingers, the DNA binding domain also contains the nuclear localization signal (NLS). Amino acid residues 207–226 constitute a bipartite NLS of the form KKK-X_{11}-KKKSKK that is conserved from *Xenopus* to mammals.

The automodification domain contains 15 highly conserved glutamate residues. These glutamate residues are the potential auto-poly(ADP-ribosyl)ation sites *(132)*. Interestingly, this domain in the *Drosophila melanogaster* gene also contains a leucine zipper motif *(132)* thought to play an important role in the homodimerization of PARP as well as heterodimerization of PARP with other chromatin proteins. In this dimerization process, protein–protein interactions may help stabilize $(PARP)_2$–DNA complexes and thus facilitate the characteristic intermolecular auto-poly(ADP-ribosyl)ation reaction *(93)*.

In the carboxy-terminal catalytic domain, PARPs from various species contain a highly conserved dinucleotide binding motif, GXXXGKG (corresponding to amino acids 888–894 in the human PARP sequence) *(133)*, which is very similar to the NTP binding motif or Rossman fold (G/AXXXGKT/S/G), typically associated with proteins, such as NAD-dependent dehydrogenases, that recognize and tightly bind to phosphoanhydride bonds of mono- and dinucleotides, e.g., NAD *(116)*.

Recently, PARP cDNA sequences from a variety of species, including mammals, chicken, *Xenopus laevis*, and *D. melanogaster*, have been determined. The mean conservation of the entire amino acid sequence among different species is 72%. Cys, His, and basic residues in the zinc finger consensus regions are conserved. The carboxyl-terminal region corresponding to the NAD binding domain shows an identity of 83%. The dinucleotide binding consensus sequence, β1-αA-β2 Rossman fold structure, and β-sheet structures are completely conserved from insects to mammals *(133)*.

2.2. *PARP Gene Structure and Regulation*

The PARP gene is localized on chromosome 1q41-42 in humans. The gene is 43 kbp in length and contains 23 exons ranging in size from 62–553 bp. Introns vary from 0.2–8 kbp in length, and the intron–exon boundaries show the canonical consensus splicing sequence. Although the DNA binding and automodification domains are encoded by seven and four exons, respectively, the NAD binding domain contains 12 exons *(7)*. Two pseudogenes have been found on chromosomes 13 and 14 *(41)*. Recent studies on the characterization of human PARP promoter have revealed multiple transcription initiation sites *(105)*. However, the major initiation site is located ~160 bp upstream from the translational start site. Interestingly, the PARP promoter structure contains CCAAT and TATA boxes, SP1 binding sites, and two sets of inverted repeats. However, none of the CCAAT/TATA boxes are located sufficiently close to an SP1 binding site to allow formation of an active polymerase II transcription complex. Although the SP1 site resides very close to the translational initiation site, the closest CCAAT/TATA box is located 580 bp upstream from the SP1 site. The first inverted repeat is positioned between the SP1 site and the CCAAT/TATA box closest to the transcriptional initiation site. The two sets of inverted repeats allow the formation of two potential DNA cruciform structures. The formation of one cruciform would not greatly reduce the gap between the SP1 site and CCAAT/TATA box. However, the other (the first inverted repeat) would considerably shorten the distance between the most proximal SP1 site and the nearest upstream CCAAT/TATA box, thereby allowing the formation of an active promoter structure. Recently, it has been suggested that PARP may regulate its own promoter activity *(105)*.

Human PARP mRNA is 3.7 kb by northern blot analysis *(41,61,85,134)*. The 3'-untranslated region of the message contains 459 bases and has a polyadenylation sequence. The 5'-untranslated region shows sequences flanking the first ATG that match the sequences identified as initiation regions *(41)*. The PARP mRNA has an open reading frame of 3042 nucleotides and codes for a protein of 113 kDa consisting of 1014 amino acids.

2.3. *Poly(ADP-Ribosyl)ation Reaction*

As indicated earlier, NAD is the sole substrate of PARP. The enzyme hydrolyzes NAD following activation by DNA strand breaks and catalyzes the formation of poly(ADP-ribose) in three chemically distinct enzymatic steps, as described in Fig. 2 *(2,4,5,13,50,85)*. The initiation reaction involves transfer of the ADP-ribose portion of NAD to a glutamic acid residue of the acceptor protein to form mono-ADP-ribosyl protein. The second reaction, termed the ADP-ribose chain elongation reaction, involves the transfer of the ADP-ribose portion of NAD to mono-ADP-ribosyl protein. In this reaction, the second ADP-ribose is linked by a (1"–2') glycosidic

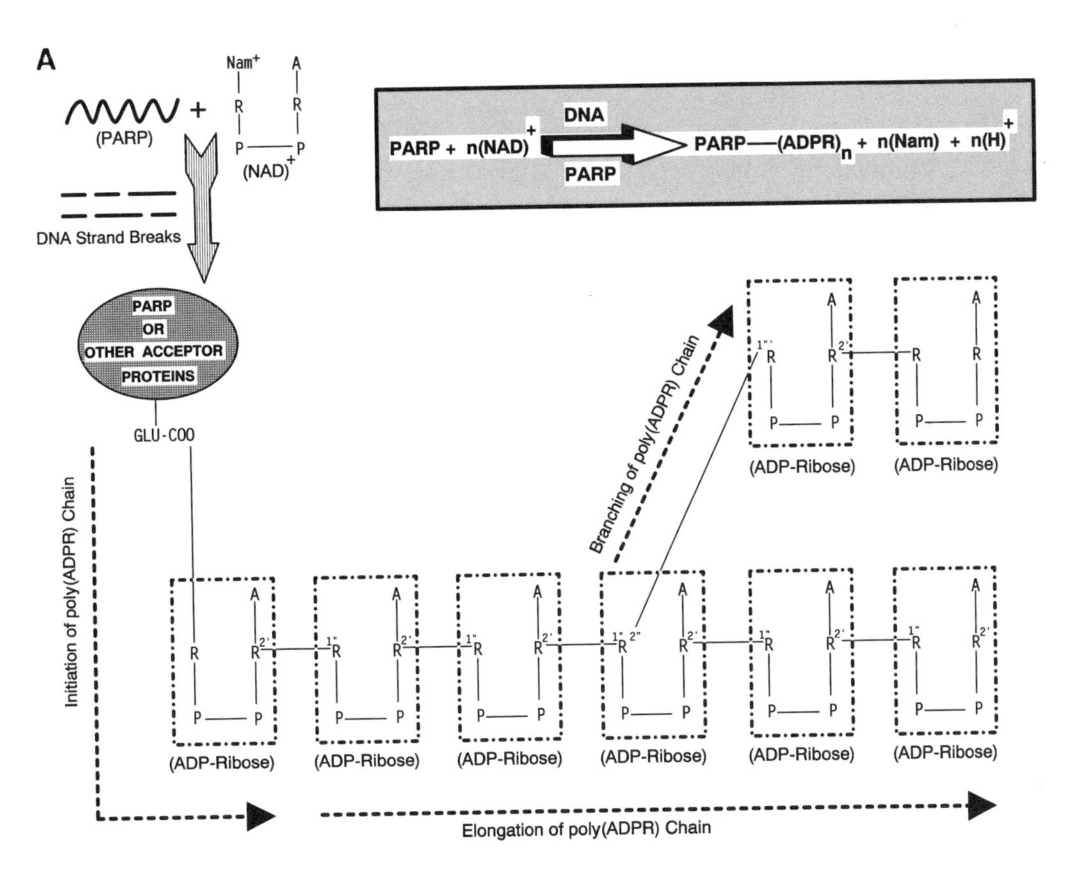

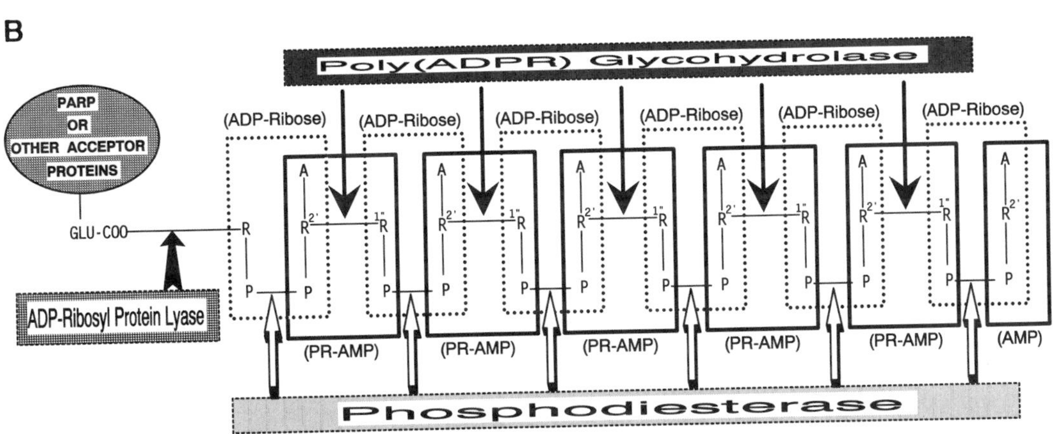

Fig. 2. Synthesis and degradation of poly(ADP-ribose). **(A)** Synthesis includes (1) transfer of the first ADP-ribose moiety to a glutamate residue of PARP or other acceptor proteins (initiation); (2) further addition of ADP-ribose units through a 1"–2' glycosidic bond to the "adenine" ribose of the preceding ADP-ribose (elongation); and (3) transfer of ADP-ribose to the "nicotinamide" ribose of a poly(ADP-ribose) chain through a 2"–1' ribose–ribose bonding (branching). The inset shows that on activation by DNA breaks, PARP utilizes "*n*" molecules of cellular NAD to synthesize ADP-ribose polymers of "*n*" subunits attached to acceptor proteins. In this reaction, "*n*" molecules of nicotinamide are liberated, which can be recycled by the cells to synthesize NAD. **(B)** Degradation involves the hydrolysis of poly(ADP-ribose) chains (1) by poly(ADP-ribose) glycohydrolase to produce ADP-ribose, (2) by phosphodiesterase to produce phosphoribosyl-AMP, (3) by phosphodiesterase to produce 5'-AMP from the terminus distal to the protein attachment site, and (4) by ADP-ribosyl protein lyase to cleave the ester bond between the terminal ribose of the last ADP-ribose unit and the glutamic acid residue of PARP or other acceptor proteins.

bond to the "adenine" ribose of the preceding ADP-ribose. On repetition of this second reaction a linear homopolymer of ADP-ribose bound to the acceptor protein is formed. The third reaction is the branching reaction that involves the transfer of the ADP-ribose portion of NAD to the "nicotinamide" ribose of a poly(ADP-ribose) chain to form a (2"–1"') ribose-ribose bonding between ADP-ribose units. Thus, these three reactions can generate either mono-ADP-ribosylated protein or oligo- or poly-ADP-ribosylated protein in which the polymer may be linear or branched. It has been estimated that for every initiation step, PARP can catalyze 5–7 branching reactions and over 200 elongation events *(5)*. Target proteins that undergo poly(ADP-ribosyl)ation in vitro include PARP, histones, and many DNA-metabolizing enzymes; the major target in vivo is PARP itself *(88)*. Recent studies employing an in vitro assay system have suggested that PARP performs the automodification reaction as a catalytic dimer *(5,93)*.

2.4. Catabolism of Poly(ADP-Ribose)

Poly(ADP-ribose) is hydrolyzed "endonucleolytically" by snake venom phosphodiesterase. This enzyme splits the pyrophosphate bond of poly(ADP-ribose) yielding oligomers/monomers of PR-AMP. Consequently, one molecule of 5'-AMP is released from the free end of the poly(ADP-ribose) chain. Thus, the chain length of the polymer can be calculated from the molar ratio of PR-AMP and 5'-AMP *(128,130)*. In contrast, rat liver phosphodiesterase cleaves poly(ADP-ribose) "exonucleolytically" at pyrophosphate bonds. The direction of hydrolysis in this case is from the open end AMP terminus to the closed end terminus of the polymer where poly(ADP-ribosyl)ation is initiated *(128,130)*. The most important enzyme for the in vivo catabolism of polymers is the chromatin-bound poly(ADP-ribose) glycohydrolase. Although some controversy remains regarding whether this enzyme acts endoglycosidically or exoglycosidically, the general consensus is that poly(ADP-ribose) is degraded exoglycosidically to the last ADP-ribose on the acceptor protein *(128,130)*. Thus, poly(ADP-ribose) glycohydrolase cleaves ribose–ribose bonds of the polymer to yield monomers of ADP-ribose (Fig. 2b). In vivo-polymer synthesis and degradation occur rapidly, and at high levels of DNA damage, the catabolic half-life of individual polymer molecules may be <30 s. The rapid turnover contrasts with slower catabolism of a constitutive polymer fraction exhibiting a half-life of 7.7 h in undamaged cells *(1,3,5)*. The rate of polymer shortening by poly(ADP-ribose) glycohydrolase varies as a function of the length of the polymer *(86)*. Following glycohydrolase action, one molecule of ADP-ribose remains bound to the protein. This last monomeric residue is removed from the protein by ADP-ribosyl protein lyase *(108)*. Based on these observations, polymer catabolism emerges as a highly organized process, in which the order and kinetics of degradation are largely determined by the size and perhaps structural complexities of the substrate.

2.5. Shuttle Effect

As outlined above, PARP binds to DNA single- and double-strand breaks that arise directly as a consequence of DNA damage (e.g., ionizing radiation), or indirectly following enzymatic incision of DNA during DNA repair. Consequently, the enzyme becomes activated and undergoes the auto-poly(ADP-ribosyl)ation reaction. The polymers attached to the DNA-bound enzyme are more acidic than DNA (two negative charges per monomer unit compared to only one per nucleotide of DNA, and thus, electrostatic repulsion dissociates the automodified PARP from the DNA. This dissociated polymer-bound PARP is inactive. However, rapid action of poly(ADP-ribose) glycohydrolase releases PARP from the bound polymers, thus enabling the enzyme to bind to DNA breaks and regain activity. Thus, with the cooperation of poly(ADP-ribose) glycohydrolase, PARP cycles between its DNA-bound active state and automodified inactive state *(55,146)*. A cyclical relationship between automodified PARP and poly(ADP-ribose) glycohydrolase may eliminate the need for new synthesis of PARP after DNA damage.

2.6. Inhibitors, Activators, and Cofactors

Most of our knowledge about the physiological role of PARP (discussed in Section 3.) evolved from studies employing enzyme inhibitors. Since nicotinamide is a critical component of NAD, essential in the enzyme substrate interaction, and an effective inhibitor, it was used extensively in early studies. Other inhibitors include thymidine, theophylline, and 5-methylnicotinamide. Following the discovery in 1980 that substituted benzamides, such as 3-aminobenzamide (3-AB) and 3-methoxybenzamide (3-MB), are potent and specific inhibitors of PARP, these compounds, especially 3-AB, have been used frequently for studies aimed at defining the biological roles of poly(ADP-ribosyl)ation *(112,126)*. Both the nicotinamide and benzamide analogs are competitive inhibitors of PARP with respect to its substrate NAD^+. The systematic analysis of the structure-activity relationship of various inhibitors *(126)* has shown that an aromatic ring system and an intact carboxamide group are important for effective inhibition and that a ring nitrogen is not always necessary. Many strong inhibitors have been identified, although their specificity and usefulness in cellular or animal systems has not yet been fully documented *(8,9,113)*. These studies have contributed to a better understanding of the structure–activity relationship of various other inhibitors.

PARP is dependent on DNA for its activity. Short double-stranded DNA (dsDNA) fragments, such as an octamer d(GGAATTCC), potently activate the enzyme *(16)*. Although a hexamer, such as d(GTTAAC), alone is unable to stimulate enzyme activity, it is highly stimulatory in the presence of histone H_1, suggesting that the minimum length of a dsDNA necessary for enzyme activation is between 6 and 8 bp. Free DNA strand ends, either blunt, 5'-, or 3'-protruding, are necessary for activation of the enzyme. Dephosphorylated ends are more effective than phosphorylated ends and covalently closed dsDNA has no activating effect *(12,106)*.

Mg^{2+} stimulates enzyme activity in the presence of both DNA and histones. However, in the presence of a large excess of histones over DNA, Mg^{2+} is inhibitory *(131,145)*. Although auto(ADP-ribosyl)ation of PARP is minimal in the absence of Mg^{2+}, exogenously added acceptor proteins are still modified. Mg^{2+} can be replaced by Mn^{2+} or Ca^{2+} to stimulate PARP activity in a purified enzyme preparation *(26,101)*. Other metal ions, such as Cu^{2+}, Hg^{2+}, and Cd^{2+}, are highly inhibitory *(68)*. Polyamines, such as spermine, spermidine, and putrescine, are stimulatory in isolated nuclei or purified enzyme systems; spermine is the most effective, followed by spermidine; putrescine has minimal effect *(9)*.

3. BIOCHEMISTRY OF POLY(ADP-RIBOSE) POLYMERASE

3.1. Response to DNA Damage

Involvement of PARP in DNA repair was originally suggested from inhibitor studies *(122)*. Further studies utilizing genetic and molecular systems are described in this section. Three types of DNA-damaging agents have been used, including alkylating agents, UV radiation, and direct break-inducing agents, such as ionizing radiation, neocarzinostatin, and bleomycin. Lesions generated by monofunctional alkylating agents are removed by base excision repair which involves the sequential action of several enzymes (*see* Chapter 3, vol. 1 and Chapter 3, this vol.). DNA glycosylases excise damaged bases thereby generating abasic sites that are substrates for apurinic/apyrimidinic endonucleases, which induce single-strand breaks, followed by limited exonuclease action, polymerization, and ligation. UV produces mainly pyrimidine dimers that are recognized and eliminated by the nucleotide excision repair pathway (*see* Chapters 2 and 15, vol. 1 and Chapter 18, this vol.). This pathway involves lesion recognition and incision of DNA on either side of the damaged site by an endonuclease, followed by excision of the damaged oligonucleotide, gap filling DNA synthesis, and ligation. Large carcinogenic adducts are also repaired by this pathway. Direct DNA strand breaks produced by X-rays and bleomycin are possibly repaired by some enzymatic modification to the DNA termini to render them suitable for subsequent polymerization and ligation (*see* Chapters 16 and 17).

3.1.1. Low to Moderate Levels of DNA Damage and PARP Activity

The relationship between DNA strand breaks and PARP activity that results in increased synthesis of poly(ADP-ribose) has been established using both in vitro and in vivo systems. Using in vitro systems consisting of radiolabeled NAD^+, purified PARP, and SV40 minichromosomes, it was found that closed circular SV40 minichromosomes do not activate PARP, whereas linear minichromosomes activate the enzyme markedly *(46)*. To confirm further that poly(ADP-ribose) synthesis occurs in response to strand breaks, as opposed to other types of DNA damage that do not directly produce strand breaks, the effect of UV irradiation on the ability of the minichromosomes to stimulate PARP activity was examined. UV doses ranging from 50–1000 J/m^2 have little or no effect on the ability of the minichromosomes to activate PARP. However, treatment of UV-irradiated minichromosomes with *Micrococcus luteus* UV endonuclease to create DNA strand incisions at the sites of UV damage results in a considerable increase in their ability to activate PARP *(45)*, clearly indicating that DNA strand breaks rather than damaged or modified bases are responsible for stimulating poly(ADP-ribose) synthesis.

Studies with cells from patients with xeroderma pigmentosum (XP) were used to determine whether PARP responds to DNA strand breaks or to base damage. XP cells are deficient in UV damage repair because of their inability to form the initial incision at UV lesion sites. However, they can efficiently repair MNNG-induced damage. Treatment of XP cells with MNNG results in an increase in poly(ADP-ribose) synthesis, thereby demonstrating that these cells have PARP and that it responds to DNA strand breaks resulting from repair of MNNG-induced DNA damage. However, treatment of XP cells with UV irradiation does not increase poly(ADP-ribose) synthesis thereby implying that PARP is not activated in response to unrepaired UV-induced DNA damage. When XP cells are UV-irradiated and subsequently permeabilized and treated with *M. luteus* UV endonuclease to produce DNA strand incisions at the site of UV damage, a marked augmentation of poly(ADP-ribose) synthesis is observed, further indicating that PARP is activated only by DNA strand breaks, but not by any other kind of DNA damage *(18,20)*. Other studies have reached similar conclusions using different approaches *(13,44,122)*.

NAD levels drop subsequent to the treatment of cells with DNA-damaging agents, and this can be prevented or delayed by PARP inhibitors *(13,122)*. Clearly, these studies suggest that PARP activation is responsible for the loss of NAD and that the increase in poly(ADP-ribose) synthesis results from the concomitant loss of NAD. This concept was proven directly by Jacobsons' group who developed a robust technique to measure the levels of poly(ADP-ribose) generated in vivo *(74,82)*. Treatment of SV40-transformed 3T3 mouse cells with MNNG produced a dramatic increase in the intracellular levels of poly(ADP-ribose) concomitantly with a decrease in intracellular NAD levels. Furthermore, neither NAD glycohydrolase activity nor the rate of NAD biosynthesis was affected by such an increase in poly(ADP-ribose) synthesis, thus clearly suggesting that PARP is the only enzyme that responds to DNA breaks by consuming NAD and producing ADP-ribose homopolymers *(72,75)*. Other studies have reached the same conclusion *(13,17,19,122,125,126)*.

Thus, PARP is a damage-responsive enzyme, activated specifically by DNA strand breaks. However, PARP is different from the class of damage-inducible gene products; PARP is a constitutive enzyme whose gene transcription and mRNA levels remain constant after DNA damage, whereas transcription of damage-inducible genes is increased dramatically following DNA damage (*22*; *see* Chapter 14).

3.1.2. High Levels of DNA Damage and Role of PARP in Suicide Response

Low to moderate levels of DNA damage are associated with an increase in PARP activity. In addition, inhibition of PARP activity after low to moderate levels of DNA damage is associated with an increase in repair replication, DNA strand breaks, sister chromatid exchange (SCE) formation, and cytotoxicity, suggesting that PARP activity may be necessary to rescue cells after low to moderate levels of DNA damage. In contrast, when cells sustain massive levels of DNA damage and DNA strand breaks, the consequent activation of PARP may in itself lead to a crisis situation. A dose-dependent decrease in NAD levels occurs following treatment of cells with increasing concentrations of MNNG *(13,122)*. For example, 1–2 μg/mL MNNG causes nearly 50% depletion in NAD level, whereas 10–20 μg/mL MNNG almost completely depletes the cells of their NAD in <60 min. This MNNG-induced, dose-dependent decrease in NAD is also accompanied by a dose-dependent decrease in ATP. Concurrent presence of PARP inhibitors with MNNG prevents the depletion of both NAD and ATP. This observation suggests that the depletion of both NAD and ATP is secondary to activation of PARP *(13,14,21)*.

The dose-dependent depletion of NAD and ATP may contribute to the decrease in DNA repair that occurs with increasing concentrations of MNNG. It has indeed been found that treatment of cells with increasing doses of MNNG results in a progressive decrease in DNA replicative synthesis. However, DNA repair synthesis increases as a function of MNNG dose and then decreases as higher doses of MNNG are used *(13,21)*.

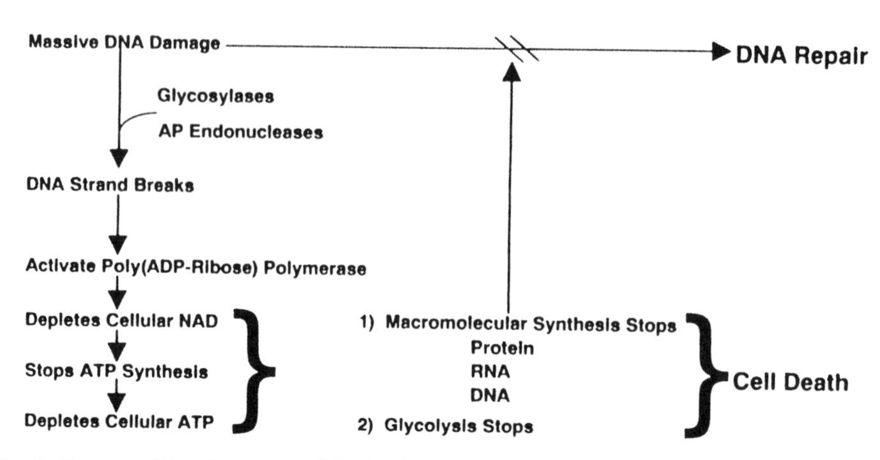

Fig. 3. Proposed involvement of PARP in the suicide response to massive DNA damage.

This is unexpected, since DNA repair synthesis should continue to increase in proportion to the amount of damage until the DNA repair mechanism is saturated and then remain at a constant level. The fact that repair synthesis reaches a peak and decreases suggests that the system becomes depleted of some essential components. Consistent with this proposal is the finding that treatment of cells with high doses of MNNG results in drastic reductions in energy-dependent processes, including synthesis of DNA, RNA, and protein. The use of inhibitors of PARP to block the NAD and ATP depletion results in partial restoration of these energy-dependent processes. Based on these observations, the suicide response hypothesis was proposed *(13–15,21)* as outlined below and illustrated in Fig. 3.

In this view, activation of PARP by massive DNA damage is actually a suicide response, since it causes rapid NAD and ATP depletion and leads to cell death before there is an opportunity to repair the DNA damage. Such a suicide mechanism limits the possibility that massively damaged cells will try to repair and survive with a high mutation rate.

3.2. Consequence of PARP Inhibition on Repair of DNA Strand Breaks and Clonogenic Cytotoxicity

3.2.1. Studies with Enzyme Inhibitors

As outlined above, DNA strand breaks are created either directly by X-rays and radiomimetic agents, or indirectly, during excision of damaged DNA produced by various DNA-damaging agents. The amount of breaks detected at any particular time following exposure to these agents represents the result of the dynamic production and repair of DNA breaks over time. Although PARP inhibitors by themselves do not cause any DNA strand breaks, their presence in association with alkylating agents or ionizing radiations significantly increases the strand break frequency compared to that produced by DNA-damaging agents alone. In addition, in most cases, a delay in rejoining of strand breaks is observed in the presence of these inhibitors *(13,85)*. It is possible that inhibitors prevent poly(ADP-ribosyl)ation, thereby delaying the release of PARP from the DNA strand ends. Thus, these strand ends would remain inaccessible to DNA repair enzymes for longer times.

Although inhibitors of PARP are usually not cytotoxic by themselves, at higher concentrations, they can cause cycle delay. PARP inhibitors potentiate the X-ray-induced cell killing to varying degrees in different cell types that are cycling *(10)*. Controversies still exist concerning whether PARP inhibitors can increase UV-induced cell killing *(19,43,48,124,126)*. Data suggest that the potentiating effect of the inhibitors on UV-inflicted cytotoxicity is possibly cell-type-specific, e.g., lymphocytes are sensitive, but HeLa cells are not. PARP inhibitors are most effective in potentiating the cytotoxicity of alkylating agents in dividing cells. The degree of potentiation depends on the type of alkylating agents and inhibitors, but in all cases, a substantial increase in cell killing has been observed *(13,44,122)*. It is pertinent to mention a study by Boorstein and Pardee *(23)*, who have shown that the lethality of 3-AB to a population of asynchronously cycling cells treated with the alkylating agent, methyl methane sulfonate (MMS), is maximum at S phase, and this lethality can be prevented by treating the cells with the DNA synthesis inhibitor aphidicolin. This study concluded that DNA synthesis is required to reach a step in DNA repair at which 3-AB causes lethality. Thus, these studies suggest that PARP inhibitors may not be effective in potentiating cytotoxicity of alkylating agents in growth-arrested confluent cells that are mostly in G1 phase of the cell cycle.

Growth-arrested confluent cells, held in growth-inhibited conditions following DNA damage, exhibit a mode of repair designated potentially lethal damage repair (PLDR) *(138)*. This implies that DNA-damaged, growth-arrested cells have the capacity to repair damage that is potentially lethal, i.e., damage that ordinarily causes cell death, but whose expression can be prevented by appropriate postdamage treatment, such as by holding cells at high density and preventing them from proliferating for a period of time when damage is presumably repaired. Although the mechanism of PLDR is not clearly understood, it has been reported that treatment of cells with PARP inhibitors reduces X-ray-induced PLDR *(11,65,83)*. Jacobsons' group has reported that although 3-MB increases DNA strand breaks following MNNG in both quiescent and dividing cells, it has no effect on MNNG-induced PLDR in terms of cell survival, but is very cocytotoxic with MNNG in dividing cells *(70,71)*. These studies suggest that the substantial increase in alkylating agent induced DNA strand breaks observed in the presence of PARP inhibitors may not directly contribute to the increased cytotoxicity. Furthermore, although PARP inhibitors effectively potentiate MNNG-induced cytotoxicity in exponentially growing cells, they fail to produce such effect in growth-arrested confluent cells, suggesting a role of PARP in DNA repair in logarithmically growing cells, but not in growth-arrested cells.

3.2.2. Reducing PARP Activity by Expression of an Antisense RNA Construct

To understand the consequences of inhibition of PARP on the repair of MMS-induced DNA strand breaks, Smulson's group used HeLa cells (PADPRP-as[7]) that are 90% depleted of PARP because of the expression of an antisense RNA construct *(51,52)*. Following treatment with 2 m*M* MMS for 1 h, both control and PARP-depleted cells show similar levels of SSBs as monitored by alkaline elution. In the first 20 min following removal of the drug, control cells show 60% disappearance of SSBs as opposed to no repair in PARP-depleted cells. In contrast, in the next 70 min, SSBs are removed at a much faster rate (5.7 SSB/min) in PARP-depleted cells compared to a rate of 2 SSB/min in control cells. Thus, although the number of SSBs remaining in control or PARP-depleted cells are the same at 90 min after DNA damage, the rate of repair of MMS-induced SSB appears to be different *(51)*. Using the same system, DNA damage and repair were examined in the dihydrofolate reductase *(DHFR)* gene after exposure of the cells to either UV or nitrogen mustard *(127)*. Both control and PARP-depleted cells show similar rates of repair of pyrimidine dimers over a period of 24 h after UV exposure. However, PARP-deficient cells are deficient in the gene-specific repair of nitrogen mustard-induced lesions. Furthermore, based on clonogenic survival assays, this cell line exhibits increased susceptibility to nitrogen mustard and MMS relative to the control cells. These studies indicate that in HeLa cells, PARP appears to participate in gene-specific repair of alkylation damage, but not in the repair of UV-induced pyrimidine dimers.

3.2.3. Reducing PARP Activity by Overexpression of the PARP DNA Binding Domain

Overexpression of the DNA binding domain (DBD) of PARP in HeLa cells (PARP-DBD) *(84)*, diminished PARP activity because of a competition for DNA breaks between overexpressed DBDs and resident PARP molecules. Overexpression of DBD blocks unscheduled DNA synthesis (UDS) induced by MNNG treatment, but not that induced by UV irradiation *(94)*, indicating that PARP is involved in base excision repair, but not in nucleotide excision repair. Further studies with exponentially growing PARP-DBD cell lines have shown that following exposure to low doses of MNNG, these cells exhibit increased doubling time, G2 + M accumulation, and a marked reduction in the clonogenic cell survival compared to the parental HeLa cells, thus suggesting a role of PARP in repair of MNNG-inflicted damage in HeLa cells. However, these responses were not observed with UV-C radiation *(120)*.

3.2.4. Mutant Cell Lines Deficient in PARP Activity

The role of PARP in rejoining of DNA strand breaks was studied using V79 Chinese hamster-derived cell lines, ADPRT54 and ADPRT351, that are deficient in PARP activity *(38,39)*. Under conditions of maximum DNA damage, these cells have only 5–10% PARP activity compared to the parental V79 cells. The level of DNA strand breaks induced by X-rays and MNNG and their rates of repair are similar in control and PARP-depleted cell lines, thus suggesting that the absence of PARP does not contribute to any alterations in either the production or repair of DNA strand breaks *(32)*. The results obtained with PARP-deficient cells should be contrasted with inhibitor-treated cells where a higher levels of DNA breaks as well as a delay in rejoining are observed. In the case of inhibitor-treated cells, PARP molecules are present and bound to the DNA ends. However, they cannot undergo automodification and subsequently dissociate from DNA, thus possibly prolonging the life of the breaks and interfering with their rejoining. In the absence of PARP, free DNA ends are readily accessible by repair enzymes, and therefore, no significant delay in rejoining of DNA strand breaks is observed.

ADPRT54 and ADPRT351 cell lines were also used to elucidate the effect of PARP deficiency on the clonogenic cell survival following exposure of these cells to a variety of DNA-damaging agents *(33,35,37,40)*. These experiments have been conducted with logarithmically growing cells. These cells are hypersensitive to a variety of antitumor agents, including melphalan, BCNU, mitomycin, and bleomycin, compared to their parental V79 cells. They are also hypersensitive to UV and X-rays, the UV-mimetic agent, 4-nitroquinoline-1-oxide, the topoisomerase I inhibitor camptothecin, and to a series of alkylating agents that includes alkylsulfonates, alkylnitrosoureas, and nitrosoguanidine. Similar observations have been made using cell lines that are deficient in PARP activity because of a deficiency in substrate NAD. Thus, these findings suggest that PARP may be involved in the repair of damaged DNA in replicating cells. However, these cell lines are significantly resistant to topoisomerase II inhibitors, such as etoposide, m-AMSA, and adriamycin *(37,40)*. An explanation for this observation is given in Section 4.2.1.

PARP-deficient cell lines that are hypersensitive to alkylating agents and X-rays during exponential growth become more resistant to X-rays and MNNG under growth-arrested conditions and show essentially the same sensitivity as the parental V79 cells *(32)*. In addition, PARP-deficient cell lines and the parental V79 cells exhibit a similar level of proficiency in performing X-ray-induced PLDR. Similar results were obtained by Jacobsons' group *(70,71)*, who showed that the inhibition of PARP did not result in the potentiation of alkylating agent-induced cytotoxicity in growth-arrested cells. These results suggest that although the absence of PARP does not interfere with the production or repair of DNA strand breaks, its presence definitely protects the replicating cells from undergoing faulty repair resulting in cell death, since following DNA damage, replicating cells deficient in PARP activity exhibit a significant reduction in cell survival compared to the cells having normal PARP activity. In contrast, X-ray or MNNG-induced cytotoxicity in growth-arrested confluent cells is independent of PARP activity suggesting a specific protective role of PARP following DNA damage only in replicating cells but not in cells that are unable to replicate. These results are consistent with the hypothesis that the normal function of PARP is to protect DNA ends from recombinational events, and that when the system is stressed by too many DNA strand ends resulting from DNA damage or by too little enzyme activity, SCEs, homologous, and nonhomologous recombination increase, with the latter leading to toxicity and cell death. Growth-arrested cells do not undergo recombination leading to an increase in SCE and therefore are protected from such a recombination-mediated mode of cell death accentuated by PARP deficiency. The role of PARP in preventing DNA recombination will be outlined in detail in Section 3.3., and this hypothesis is elaborated in Section 3.4.

3.2.5. Cell Extracts Lacking PARP Activity by Depletion of PARP or NAD

Lindahl's group used an in vitro assay system to elucidate the role of PARP in DNA repair *(117–119)*. The assay system consists of human cell extracts, and plasmid DNA damaged with UV, γ-rays, bleomycin, neocarzinostatin, MNNG, or MMS. Repair of damaged DNA has been examined in cell extracts under the following conditions:

1. Cell extracts depleted of PARP, but containing NAD;
2. Cell extracts containing PARP, but no NAD; and
3. Cell extracts containing PARP and NAD.

Rejoining of DNA SSBs from γ-rays, bleomycin, or neocarzinostatin is catalyzed inefficiently in cell extracts containing PARP, but not NAD. However, addition of

NAD greatly increases the efficiency of end rejoining. This efficient DNA rejoining is blocked by 3-AB. Furthermore, selective depletion of PARP from cell extracts improves the repair of DNA exposed to a variety of DNA-damaging agents by removing the NAD-dependent mode of repair *(117–119)*. Similar phenomena have been observed with alkylated DNA. However, no repair replication has been observed with UV-irradiated DNA that contains only CPDs. Thus, NAD is required for facilitating efficient rejoining of DNA breaks in the presence of PARP. However, in the absence of PARP, DNA breaks are rejoined efficiently. These observations suggest that the association of PARP with NAD may play an important role in the initiation of end rejoining in cells with normal PARP activity. Furthermore, these observations agree with the observation that PARP-deficient cells with normal NAD levels are equally efficient in rejoining DNA strand breaks when compared with cells that are normal in PARP-NAD metabolism *(32)*.

It has been concluded that in the absence of NAD, PARP interferes with base excision repair processes, because the bound enzyme molecules block DNA strand interruptions. However, in the presence of NAD, PARP can undergo automodification and dissociate from DNA, thereby allowing the repair enzymes to act on the interrupted DNA strands. In the absence of PARP no enzyme is bound to the DNA ends, and therefore, DNA rejoining is independent of NAD *(117–119)*.

3.3. Role of PARP in Recombination

3.3.1. Effect of PARP Inhibition on Recombination in Undamaged Cells

A variety of cytogenetic and molecular genetic studies suggest a role of PARP in preventing DNA recombination. A stably transformed cell line obtained by transfection of NIH 3T3 cells with human c-Ha-Ras converted to morphologically normal flat cells following 15 d of incubation with benzamide. Further investigation has revealed that benzamide led to deletion of the transfected oncogene resulting in reversion of the transformed phenotype to flat cell phenotype. However, endogenous mouse c-Ha-Ras is not lost during the culture in presence of benzamide, suggesting that PARP may be involved in DNA recombination and that its inhibition results in the rejection of randomly integrated DNA *(99)*. Shall's group has demonstrated that the integration of transfected DNA into the host cell genome requires PARP activity, because inhibition of poly(ADP-ribosyl)ation results in the reduction of random integration of transfected DNA molecules into the genome *(54)*. Also, treatment of mouse fibroblasts with 3-methoxybenzamide inhibits random integration of transfected DNA, but such treatment does not affect the rate of extrachromosomal homologous recombination among the transfected molecules *(140)*. These observations suggest that these two recombination pathways in mammalian cells are biochemically distinct. Growth of mouse fibroblasts with 2 m*M* 3-MB increases the rate of intrachromosomal homologous recombination by nearly fourfold *(139)*. Both gene conversions as well as single crossovers are increased to similar extents, suggesting a mechanistic association between these two types of intrachromosomal rearrangements in mammalian cells and their regulation by PARP activity.

Involvement of PARP in the DNA recombination process is also supported by cytogenetic studies aimed at measuring SCE frequency *(63,96,98,100,107,137)*. The precise mechanisms of formation of SCE is not known. SCE likely reflects homologous recombination and is thus not cytotoxic. Cell lines treated with various inhibitors of PARP show a dose-dependent increase in SCE frequency. For example, 3-AB, which is usually not cytotoxic and does not cause any chromosomal aberrations and mutations, is a potent inducer of SCE *(96)*. Cell lines deficient in PARP also show a nearly 10-fold increase in SCE frequency *(38,39)*. In addition, cell lines capable of growing indefinitely in nicotinamide-free medium are deficient in substrate NAD, and under this growth condition, they exhibit nearly a 10-fold increase in SCE. However, on growth in regular nicotinamide-containing medium, NAD and SCE levels return to normal *(38)*. Furthermore, PARP-DBD cells that are functionally deficient in PARP activity also exhibit an increase in spontaneous SCE frequency compared to that of control cells *(120)*. Taken together, the results indicate that the inhibition of PARP activity causes a substantial increase in baseline SCE frequency, thus suggesting that PARP activity prevents homologous recombination. Despite the numerous conflicting reports on the role of PARP in various cellular processes, one constant phenotypic effect of interfering with PARP activity is a substantial increase in SCE frequency.

3.3.2. Effect of PARP Inhibition on Recombination and Chromosomal Aberrations in Damaged Cells

Cytogenetic effects of DNA-damaging agents can be measured in metaphase chromosomes as chromosomal aberrations and/or SCE. Chromosomal aberrations may result from unrejoined or aberrantly rejoined DSBs and generally result in cell death. Inhibition of PARP by an inhibitor, such as 3-AB, usually does not produce any chromosomal aberrations. However, in cells treated with X-rays or alkylating agents, 3-AB greatly enhances chromosomal aberrations *(44,85)*. It has been shown that 3-AB has dramatically different effects on X-ray-induced chromosomal aberrations in human lymphocytes depending on the stage of the cell cycle in which cells are irradiated *(60,110,142)*. These studies indicate that 3-AB potentiates X-ray-induced chromosomal aberrations maximally at S phase, whereas no potentiation is observed when cells are irradiated in G0 *(60,110,142)*. The electroporation of restriction enzymes into mammalian cells results in DSBs that can lead to chromosome aberrations. Morgan's group has shown that 3-AB dramatically increases the yield of both exchange-type and deletion-type aberrations and causes a transient increase in the yield of DSBs induced by restriction enzymes *(42)*.

Treatment of cells with a PARP inhibitor and alkylating agent synergistically increases SCE frequencies. The degree of synergism is dependent on the concentrations of both the inhibitor and alkylating agents *(97,111,121)*. PARP-DBD cells, deficient in PARP activity, exhibit chromosomal instability as demonstrated by higher frequencies of MNNG-induced SCE compared to the parental HeLa cells *(120)*.

3.4. Role of PARP Following DNA Damage

Studies outlined in the previous three sections can be summarized as follows.

1. PARP is a damage-responsive enzyme activated specifically by DNA strand breaks;
2. NAD is required for efficient rejoining of DNA strand breaks in the presence of PARP, however, in the absence of PARP, rejoining occurs in an NAD-independent mode;
3. PARP inhibition increases SCE frequency and exposure of cells to DNA-damaging agents with PARP inhibition results in an synergistic increase in SCE frequency;

4. Inhibition of PARP potentiates cytotoxicity induced by DNA-damaging agents;
5. The substantial increase in alkylating agent-induced DNA strand breaks observed in the presence of PARP inhibitors is not associated with the observed increased cytotoxicity;
6. Under exponentially growing conditions, PARP mutant cell lines are hypersensitive to X-irradiation and MNNG, suggesting that PARP may be involved in the repair of DNA damage in replicating cells; however, it is not the difference in strand break formation or their rate of repair that contributes to the enhanced cell killing in exponentially growing PARP-deficient cell lines; and
7. Under growth-arrested conditions, wild-type and mutant cell lines become similarly sensitive to both X-irradiation and MNNG, suggesting that PARP is not involved in the repair of DNA damage in growth-arrested cells.

The challenge is to explain how an enzyme can function in the DNA repair in only certain growth phases, when it is functionally active in the different growth phases. One hypothesis is that PARP does not have a direct role in DNA repair, but does function normally to prevent DNA recombination and facilitate DNA ligation (*32*; Fig. 4). Under normal growth conditions, semiconservative DNA replication consists of DNA strand separation, discontinuous DNA synthesis, and ligation of newly synthesized DNA fragments. In addition, cells undergo some recombination during the normal DNA replication process, as demonstrated by the low, but constant level of background SCE in untreated cells. These recombination events may function as a bypass mechanism for spontaneously damaged DNA and/or they may function to close gaps between fragments that are not acted on by the normal ligation process.

Under normal circumstances, recombination occurs infrequently despite the multitude of strand ends created during discontinuous DNA synthesis. We propose that the normal function of PARP is to bind transiently to these DNA strand ends, and protect them from the action of nucleases or from entering into recombination processes. Binding of PARP to DNA strand ends activates synthesis of poly(ADP-ribose), and subsequent dissociation of the enzyme from DNA allows ligation to take place. However, interruption or interference with any of the steps involved in the normal function of PARP will facilitate the recombination process and nuclease action leading to the appearance of SCE and genomic deletions.

Assuming that replicating cells contain an excess of strand ends relative to ligase, then many ends will undergo recombination unless they are shielded from this process until ligase becomes available. This may explain why cells from Bloom syndrome patients exhibit an increase in spontaneous SCE *(31)*. In this view, PARP protects the DNA strand ends, cycles on and off of these ends in a fashion that controls their rate of availability, and favors their utilization in the ligase reaction relative to the recombination process. However, in the absence of normal PARP activity, the DNA strand ends will not be protected, and this will predispose to recombination and an increase in SCE.

Exposure of cells to DNA-damaging agents leads to:

1. Creation of DNA strand breaks;
2. Excision of damaged DNA;
3. Synthesis of repair patches; and
4. Ligation of strand breaks and/or repair patches.

Persistent DNA damage or errors introduced during DNA repair can lead to mutations and/or cell death. However, when replicating cells are exposed to DNA damaging

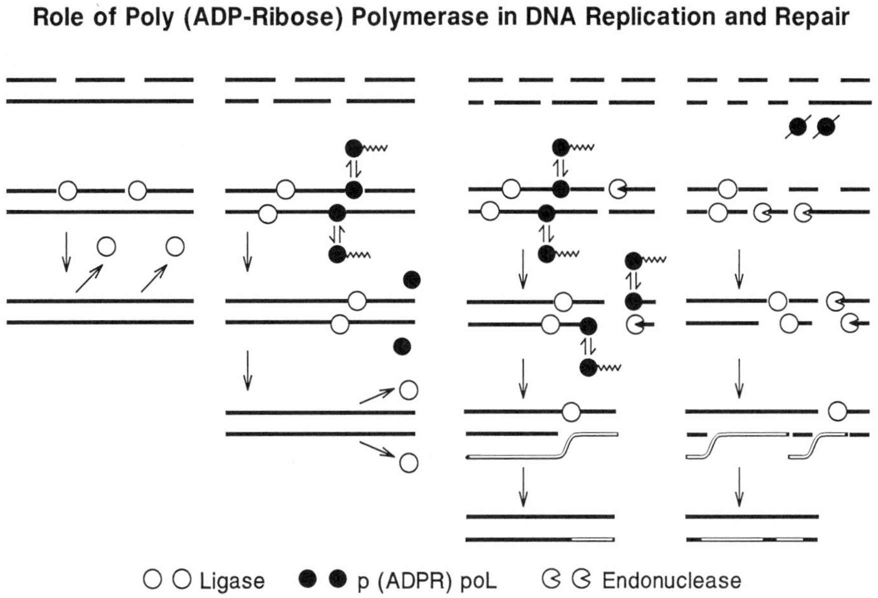

Role of Poly (ADP-Ribose) Polymerase in DNA Replication and Repair

○ ○ Ligase ● ● p (ADPR) poL Endonuclease

Fig. 4. Model of PARP role in DNA replication and repair. Open circles indicate ligase, closed circles represent PARP, jagged lines represent poly(ADP-ribose), and notched circles represent endonuclease. This model assumes a limited number of molecules of DNA ligase and PARP. Although two molecules of each enzyme are indicated, this does not imply that there are equal numbers of each enzyme present. The panel on the left shows DNA with a limited number of strand ends and there are sufficient ligase molecules to seal all DNA strand disruptions directly and restore DNA integrity.

In the second column, DNA strand ends exceed the available number of ligase molecules and DNA strand ends that exceed the ligase capacity are protected by PARP, which is activated by binding DNA strand ends to synthesize poly(ADP-ribose). Polymer synthesis causes the enzyme to dissociate from the DNA. Subsequent degradation of poly(ADP-ribose) allows the enzyme to return, bind another DNA strand end, and reinitiate polymer synthesis. PARP cycles on and off of DNA strand ends until ligase is available.

In the third column the number of DNA strand ends exceeds the number of ligase and poly(ADP-ribose) molecules, i.e., in replicating cells exposed to DNA damaging agents. Under these conditions, endonucleases act on some of the free ends, and recombination repairs other sections resulting in deletions, SCEs, and other recombinational events.

The fourth column illustrates the situation with increased DNA strand breaks and inactive PARP. In this situation, the number of free strand ends far exceeds the capacity to be joined by the existing number of ligase molecules, but PARP is not available to protect the excess free ends. As a result, DNA strand ends are subject to more extensive endonuclease action, resulting in DNA deletions. DNA strand ends and gaps are ultimately rejoined by ligation and recombination, resulting in chromosomal aberrations, SCE, increased nonhomologous recombination, gene deletion, and increased cytotoxicity.

agents, the situation becomes more complicated. In addition to the normal replication/repair process described above, there is an increase in recombination processes, as demonstrated by the dose-dependent increase in SCE formation that occurs in response to treatment with DNA damaging agents. According to the hypothesis, this increase in SCE occurs because the number of DNA strand ends generated by the combination of semiconservative replication and the excision repair process exceeds the capacity of PARP to protect against recombination events completely.

In general, SCE represent homologous recombination events and thus are not cytotoxic. However, as the susceptibility of ends to nucleases and frequency of recombination increase, it is likely that there is a proportional increase in imperfect (nonhomologous) recombination. The latter produce DNA rearrangements and deletions that are potentially mutagenic and cytotoxic. Thus, when replicating cells are subjected to DNA damage, two different processes contribute to cell death: Errors during replication repair, and nonhomologus DNA recombination, whereas when growth-arrested cells are subjected to DNA damage, only the former contributes to cell death.

Replicating cells with impaired PARP activity show a 10- to 20-fold increase in SCE compared to normal, untreated cells. This can be explained by an insufficient amount of PARP relative to the number of DNA strand ends produced by discontinuous DNA synthesis. When these cells are further stressed by increasing doses of DNA-damaging agents, the number of recombinational events is further increased and the chance of nonhomologus recombination also increases. Thus, cells with impaired PARP activity will be more susceptible to the induction of recombination and will undergo both homologous and nonhomologous recombination at much lower doses of DNA-damaging agents compared to cells with normal PARP activity. Consequently, treatment of replicating cells with a given dose of DNA-damaging agent will result in enhanced cell killing in poly(ADP-ribose)-deficient cells relative to the cells with normal poly(ADP-ribose) synthetic activity. If only replicating cells can undergo recombination, impaired PARP activity will result in increased cytotoxicity in replicating cells treated with DNA-damaging agents, but not in growth arrested cells.

This hypothesis fits well with a wide variety of studies examining the interaction between PARP inhibitors and DNA-damaging agents. It is in agreement with the suggestion by Creissen and Shall *(47)* that poly(ADP-ribose) synthesis facilitates the DNA ligation process. However, it does not support direct ADP-ribosylation and subsequent activation of ligase *(32)*. This hypothesis provides an explanation for the observations by Jacobson et al. *(70,71)*, that PARP inhibitors potentiate the effect of MNNG in proliferating cells, but not in quiescent cells. It also provides an explanation for the demonstration by Boorstein and Pardee *(23)* that MMS-induced cytotoxicity can be maximally potentiated by 3-AB during S phase. It also explains the observation by Wiencke and Morgan *(142)* that inhibitors of PARP in combination with DNA-damaging agents produce chromatid aberrations in a cell-cycle-dependent fashion with maximal aberrations occurring in cells treated during S phase.

4. INTERACTION OF PARP WITH OTHER CELLULAR PROCESSES AND SYSTEMS

4.1. Role in Apoptosis

Apoptosis is an active mode of cell death that is encountered among normal as well as tumor cells in physiological and pathological situations. This mode of cell death is distinctly different from passive cell death apoptotic necrosis *(6)*. In vivo apoptotic cells are phagocytosed by macrophages, thus disappearing without activating the immune system *(25)*. In a tissue-culture system, however, apoptotic cells are not phagocytosed. These cells show chromatin condensation and margination in a crescent shape, nuclear fragmentation in several membrane-bound vesicles, and DNA cleavage at internucleosomal linker regions *(143)*. Since activity of PARP is dependent on fragmented DNA and DNA fragmentation is associated with apoptosis, several studies have investigated the role of PARP in apoptosis. Conflicting results have been obtained regarding a direct role for PARP activation in apoptosis. However, there is general agreement that a specific PARP cleavage process occurs during apoptosis that may be a marker of cellular commitment to programmed cell death.

4.1.1. PARP Inhibition and Apoptosis

Following treatment with 1 m*M* hydrogen peroxide and hyperthermia at 43°C for 1 h, NAD levels in U937 human myeloid leukemia cells drop by 50–70% within 1 h of treatment. Subsequently, NAD levels recover with a concomitant appearance of apoptotic cells at 8 h and then drop slowly. The presence of 3-AB during and after the treatment preserves the NAD pool, reduces the extent of apoptosis, and enhances survival, suggesting that the presence of PARP is required for the induction of apoptosis *(104)*. The death of target cells by cytotoxic effector cells occurs through an apoptotic program, e.g., K562 target cells are killed by human effector natural killer cells (NK) or by lymphokine-activated killer cells (LAK). It has been shown that PARP inhibitors, including 3-AB, nicotinamide, 4-AB, and luminol, strongly inhibit K562 target cell killing by NK cells, but they are much less effective in preventing target cell killing by LAK cells *(95)*. These studies indicate that although PARP is involved in the apoptotic death of target cells, the differential sensitivity of NK and LAK activities to PARP inhibitors suggests that the molecular mechanisms underlying these two types of cytotoxicity are at least partially different. Although PARP-deficient mutant cells are susceptible to cytotoxic T-lymphocyte (CTL)-induced DNA fragmentation, they are resistant to CTL-induced cell lysis *(114)*. CTL-triggered genome fragmentation, and subsequent cell lysis correlates with activation of PARP in many SV40-transformed NIH 3T3 clones *(135)*. However, one clone was resistant to genome fragmentation, but susceptible to CTL-induced cell lysis, suggesting that neither fragmentation nor its consequences, such as activation of PARP, are essential for cell lysis resulting from CTL attack. Another study has shown that exposure of mouse thymocytes to 1 μ*M* dexamethasone (Dex) results in a substantial increase in DNA strand breaks within 4–6 h of incubation that is accompanied by a marked decrease in the number of total and viable cells. DNA fragmentation is also apparent by this time. 3-AB protects mouse thymocytes from Dex-mediated apoptotic cell death and cytolysis, although it has no effect on Dex-mediated DNA strand breakage *(64)*. Another study has shown that Ca^{2+}-activated DNA fragmentation in rat liver nuclei can be inhibited by 3-AB *(73)*. These results have been interpreted as indicating that activation of PARP contributes to apoptosis. Kun's group has shown that two C-nitroso compounds (6-nitroso-1,2-benzopyrone and 3-nitrosobenzamide) inactivate PARP by oxidizing one of the zinc fingers of PARP *(115)*. Although these result in the ejection of a zinc ion, DNA binding by PARP is not abrogated. These compounds have been shown to be cytostatic and apoptotic in leukemic and other malignant human cells. However, at tumoricidal concentrations, they are relatively harmless to bone marrow progenitor cells. The apoptotic effect of these compounds has been traced to the derepression of a calcium/magnesium-dependent endonuclease because of the inhibition of poly(ADP-ribosyl)ation.

Although the compounds used in this study are noncompetitive inhibitors of the enzyme and can potentially have other consequences, the results suggest that inhibition of PARP induces apoptosis, a result in disparity with the earlier conclusions. Additionally, PARP- or NAD-deficient cells are resistant to VP-16-induced apoptosis *(141)*. However, as discussed in Section 4.2.2., inhibition of PARP activity may not be the immediate cause for the observed resistance to apoptosis.

4.1.2. Apoptosis and PARP Cleavage

Treatment of various cell lines with a number of different chemotherapeutic agents, such as etoposide (VP-16), methotrexate, cisplatin, and γ-irradiation, results in apoptotic death *(81)*. The apoptotic changes are accompanied by quantitative proteolytic cleavage of PARP to a 25-kDa fragment containing the amino-terminal DNA-binding domain of PARP and a 85 kDa fragment containing the automodification and catalytic domains *(81)*. The 85-kDa fragment retains basal PARP activity, but is not stimulated by exogenous nicked DNA. This apoptosis-associated cleavage of PARP is distinctly different from the fragmentation that occurs during the course of normal protein turnover. During normal turnover, PARP is cleaved toward its carboxyl-terminus to yield an enzymatically inactive 80-kDa fragment containing the DNA binding and automodification domains *(78)*. These studies suggest that proteolytic cleavage of PARP is an early marker of apoptosis and that this may represent one of the several proteolytic events that commonly occur during apoptosis. Recent studies have identified a novel protease resembling interleukin 1-β-converting enzyme to be responsible for PARP cleavage *(89)*.

4.2. Relation to Stress Protein Network

4.2.1. Relation to Glucose Regulated Stress Protein

GRP78 is located in the endoplasmic reticulum of virtually all mammalian cells where it binds to mutated and malfolded proteins to prevent their secretion *(90)*. GRP78 levels can be upregulated by several inducers of cellular stress, including calcium ionophore, 2-deoxyglucose, anoxia, or glucose deprivation *(90)*. Studies indicate that:

1. GRP78 is overexpressed in PARP-deficient cell lines;
2. GRP78 is also overexpressed in cell lines capable of growing persistently in absence of nicotinamide or any of its analogs in the growth medium, thus having an NAD level of 1–3% compared to those growing in medium containing nicotinamide;
3. This overexpression of GRP78 is regulated at the transcriptional level; and
4. Restoration of normal NAD levels is followed by restoration of GRP78 to normal levels *(36)*.

These studies suggest that GRP78 is associated with the functional decrease in poly(ADP-ribose) synthesis. These studies were extended to identify the specific component of poly(ADP-ribose) synthesis system that is responsible for GRP78 overexpression. Decreased NAD^+ levels, rather than inhibition of PARP, may play a more direct role in the upregulation of GRP78 *(34)*. Furthermore, the refractory nature of PARP/NAD-deficient cell lines to topoisomerase II inhibitors, including VP-16, adriamycin, and m-AMSA, can be explained by association of GRP78 overexpression with the development of resistance to topoisomerase II inhibitors, since previous studies have documented that treatment of cells with agents or conditions capable of inducing GRP78 confers resistance to topoisomerase II-targeted drugs *(66,123)*.

4.2.2. Relation to p53

p53, the product of a tumor suppressor gene, is a cell-cycle regulatory protein *(109)*. Like PARP, p53 activity is known to increase in response to genotoxic insults and contribute to the ability of the cell to recover from DNA damage (*79,87*; *see* Chapter 15). Overactivation of PARP and overexpression of p53 may contribute to cell death and/or apoptosis *(13,87,109)*. Moreover, PARP activity is increased within seconds to minutes after induction of DNA strand breaks, whereas p53, which is usually maintained at a constant level by a balance between protein synthesis and degradation, is increased within hours of DNA damage by stabilization of protein degradation *(56,79,109)*. Thus, the time sequence of PARP activation and p53 upregulation suggests the possibility that PARP activity may be involved in the regulation of p53 expression. NAD/PARP deficient cell lines exhibit a significant reduction in both baseline p53 expression and its activity compared to parental V79 cells *(141)*. Furthermore, VP-16, which causes an increase in p53 expression and subsequent apoptosis in V79 cells, fails to produce any significant increase in p53 expression or apoptotic DNA fragmentation in PARP/NAD-deficient cell lines. Thus, these studies suggest that NAD/poly(ADP-ribose) synthesis may be involved in the regulation of p53 and its dependent pathways and that PARP/NAD may modulate an interactive network of stress-responsive proteins. Whereas the mechanism of these interactions and modulations remains to be determined, the demonstration that NAD/PARP metabolism may have a role in regulation of p53 levels has several important implications. Failure of normal p53 response is associated with an increased level of tumor occurrence *(53,62)*. Thus, p53 knockout mice develop a high incidence of spontaneous tumors *(53)*. Inactivation of p53 through mutation and/or loss of heterozygosity at the p53 locus is frequently associated with human tumors *(62)*. Recent studies suggest that niacin nutrition status may still be a prevalent problem, leading to low cellular NAD levels *(69)*. As discussed above, low NAD levels contribute to a significant reduction in baseline p53 and abrogate the p53 response to DNA damage in Chinese hamster V79 cells. These observations may provide an important link between micronutrient metabolism and susceptibility to environmental carcinogens and oxidative stress, resulting in an increased incidence of cancer.

ACKNOWLEDGMENT

Our recent studies presented in this chapter were supported in part by National Cancer Institute Grants CA65920 to S. Chatterjee and CA48735, CA51183, and CA43703 to N. A. Berger.

REFERENCES

1. Althaus, F. R., L. Hofferer, H. E. Kleczkowska, M. Malanga, H. Naegeli, P. L. Panzeter, and C. A. Realini. 1994. Histone shuttling by poly ADP-ribosylation. *Mol. Cell. Biochem.* **138:** 53–59.
2. Althaus, F. R. and C. Richter. 1987. ADP-ribosylation of proteins. Enzymology and biological significance. *Mol. Biol. Biochem. Biophys.* **37:** 1–237.
3. Alvarez Gonzalez, R. and F. R. Althaus. 1989. Poly(ADP-ribose) catabolism in mammalian cells exposed to DNA-damaging agents. *Mutat. Res.* **218:** 67–74.
4. Alvarez Gonzalez, R. and M. K. Jacobson. 1987. Characterization of polymers of adenosine diphosphate ribose generated in vitro and in vivo. *Biochemistry* **26:** 3218–3224.

5. Alvarez Gonzalez, R., G. Pacheco Rodriguez, and H. Mendoza Alvarez. 1994. Enzymology of ADP-ribose polymer synthesis. *Mol. Cell Biochem.* **138:** 33–37.
6. Arends, M. J. and A. H. Wyllie. 1991. Apoptosis: mechanisms and roles in pathology. *Int. Rev. Exp. Pathol.* **32:** 223–254.
7. Auer, B., U. Nagl, H. Herzog, R. Schneider, and M. Schweiger. 1989. Human nuclear NAD^+ ADP-ribosyltransferase(polymerizing): organization of the gene. *DNA* **8:** 575–580.
8. Banasik, M., H. Komura, M. Shimoyama, and K. Ueda. 1992. Specific inhibitors of poly(ADP-ribose) synthetase and mono(ADP-ribosyl)transferase. *J. Biol. Chem.* **267:** 1569–1575.
9. Banasik, M. and K. Ueda. 1994. Inhibitors and activators of ADP-ribosylation reactions. *Mol. Cell Biochem.* **138:** 185–197.
10. Ben Hur, E. 1984. Involvement of poly (ADP-ribose) in the radiation response of mammalian cells. *Int. J. Radiat. Biol.* **46:** 659–671.
11. Ben Hur, E. and M. M. Elkind. 1984. Poly(ADP-ribose) metabolism in X-irradiated Chinese hamster cells: its relation to repair of potentially lethal damage. *Int. J. Radiat. Biol.* **45:** 515–523.
12. Benjamin, R. C. and D. M. Gill. 1980. Poly(ADP-ribose) synthesis in vitro programmed by damaged DNA. A comparison of DNA molecules containing different types of strand breaks. *J. Biol. Chem.* **255:** 10,502–10,508.
13. Berger, N. A. 1985. Poly(ADP-ribose) in the cellular response to DNA damage. *Radiat. Res.* **101:** 4–15.
14. Berger, N. A. and S. J. Berger. 1986. Metabolic consequences of DNA damage: the role of poly (ADP-ribose) polymerase as mediator of the suicide response. *Basic Life Sci.* **38:** 357–363.
15. Berger, N. A., S. J. Berger, and S. L. Gerson. 1987. DNA repair, ADP-ribosylation and pyridine nucleotide metabolism as targets for cancer chemotherapy. *Anticancer Drug Des.* **2:** 203–209.
16. Berger, N. A. and S. J. Petzold. 1985. Identification of minimal size requirements of DNA for activation of poly(ADP-ribose) polymerase. *Biochemistry* **24:** 4352–4355.
17. Berger, N. A. and G. W. Sikorski. 1980. Nicotinamide stimulates repair of DNA damage in human lymphocytes. *Biochem. Biophys. Res. Commun.* **95:** 67–72.
18. Berger, N. A. and G. W. Sikorski. 1981. Poly(adenosine diphosphoribose) synthesis in ultraviolet-irradiated xeroderma pigmentosum cells reconstituted with Micrococcus luteus UV endonuclease. *Biochemistry* **20:** 3610–3614.
19. Berger, N. A., G. W. Sikorski, S. J. Petzold, and K. K. Kurohara. 1979. Association of poly(adenosine diphosphoribose) synthesis with DNA damage and repair in normal human lymphocytes. *J. Clin. Invest.* **63:** 1164–1171.
20. Berger, N. A., G. W. Sikorski, S. J. Petzold, and K. K. Kurohara. 1980. Defective poly(adenosine diphosphoribose) synthesis in xeroderma pigmentosum. *Biochemistry* **19:** 289–293.
21. Berger, N. A., J. L. Sims, D. M. Catino, and S. J. Berger. 1983. Poly(ADP-ribose) polymerase mediates the suicide response to massive DNA damage: studies in normal and DNA-repair defective cells. *Princess Takamatsu Symp.* **13:** 219–226.
22. Bhatia, K., Y. Pommier, C. Giri, A. J. Fornace, M. Imaizumi, T. R. Breitman, B. W. Cherney, and M. E. Smulson. 1990. Expression of the poly(ADP-ribose) polymerase gene following natural and induced DNA strand breakage and effect of hyperexpression on DNA repair. *Carcinogenesis* **11:** 123–128.
23. Boorstein, R. J. and A. B. Pardee. 1984. 3-Aminobenzamide is lethal to MMS-damaged human fibroblasts primarily during S phase. *J. Cell Physiol.* **120:** 345–353.
24. Boulikas, T. 1993. Poly(ADP-ribosyl)ation, DNA strand breaks, chromatin and cancer. *Toxicol. Lett.* **67:** 129–150.

25. Bursch, W., S. Paffe, B. Putz, G. Barthel, and R. Schulte Hermann. 1990. Determination of the length of the histological stages of apoptosis in normal liver and in altered hepatic foci of rats. *Carcinogenesis* **11:** 847–853.
26. Caplan, A. I., C. Niedergang, H. Okazaki, and P. Mandel. 1979. Poly ADP-ribose polymerase: self-ADP-ribosylation, the stimulation by DNA, and the effects on nucleosome formation and stability. *Arch. Biochem. Biophys.* **198:** 60–69.
27. Cardenas Corona, M. E., E. L. Jacobson, and M. K. Jacobson. 1987. Endogenous polymers of ADP-ribose are associated with the nuclear matrix. *J. Biol. Chem.* **262:** 14,863–14,866.
28. Cerutti, P. A. 1985. Prooxidant states and tumor promotion. *Science* **227:** 375–381.
29. Chambon, P., J. D. Weil, J. Doly, M. T. Strosser, and P. Mandel. 1966. On the formation of a novel adenylic compound by enzymatic extracts of liver nuclei. *Biochem. Biophys. Res. Commun.* **25:** 638–643.
30. Chambon, P., J. D. Weil, and P. Mandel. 1963. Nicotinamide mononucleotide activation of a new DNA-dependent polyadenylic acid synthesizing nuclear enzyme. *Biochem. Biophys. Res. Commun.* **11:** 39–43.
31. Chan, J. Y., F. F. Becker, J. German, and J. H. Ray. 1987. Altered DNA ligase I activity in Bloom's syndrome cells. *Nature* **325:** 357–359.
32. Chatterjee, S. and N. A. Berger. 1994. Growth-phase-dependent response to DNA damage in poly(ADP-ribose) polymerase deficient cell lines: basis for a new hypothesis describing the role of poly(ADP-ribose) polymerase in DNA replication and repair. *Mol. Cell Biochem.* **138:** 61–69.
33. Chatterjee, S., M. F. Cheng, and N. A. Berger. 1990. Hypersensitivity to clinically useful alkylating agents and radiation in poly(ADP-ribose) polymerase-deficient cell lines. *Cancer Commun.* **2:** 401–407.
34. Chatterjee, S., M. F. Cheng, R. B. Berger, S. J. Berger, and N. A. Berger. 1995. Effect of inhibitors of poly(ADP-ribose) polymerase on the induction of GRP78 and subsequent development of resistance to etoposide. *Cancer Res.* **55:** 868–873.
35. Chatterjee, S., M. F. Cheng, S. J. Berger, and N. A. Berger. 1991. Alkylating agent hypersensitivity in poly(adenosine diphosphate-ribose) polymerase deficient cell lines. *Cancer Commun.* **3:** 71–75.
36. Chatterjee, S., M. F. Cheng, S. J. Berger, and N. A. Berger. 1994. Induction of M(r) 78,000 glucose-regulated stress protein in poly(adenosine diphosphate-ribose) polymerase- and nicotinamide adenine dinucleotide-deficient V79 cell lines and its relation to resistance to the topoisomerase II inhibitor etoposide. *Cancer Res.* **54:** 4405–4411.
37. Chatterjee, S., M. F. Cheng, D. Trivedi, S. J. Petzold, and N. A. Berger. 1989. Camptothecin hypersensitivity in poly(adenosine diphosphate-ribose) polymerase-deficient cell lines. *Cancer Commun.* **1:** 389–394.
38. Chatterjee, S., N. V. Hirschler, S. J. Petzold, S. J. Berger, and N. A. Berger. 1989. Mutant cells defective in poly(ADP-ribose) synthesis due to stable alterations in enzyme activity or substrate availability. *Exp. Cell Res.* **184:** 1–15.
39. Chatterjee, S., S. J. Petzold, S. J. Berger, and N. A. Berger. 1987. Strategy for selection of cell variants deficient in poly(ADP-ribose) polymerase. *Exp. Cell Res.* **172:** 245–257.
40. Chatterjee, S., D. Trivedi, S. J. Petzold, and N. A. Berger. 1990. Mechanism of epipodophyllotoxin-induced cell death in poly(adenosine diphosphate-ribose) synthesis-deficient V79 Chinese hamster cell lines. *Cancer Res.* **50:** 2713–2718.
41. Cherney, B. W., O. W. McBride, D. F. Chen, H. Alkhatib, K. Bhatia, P. Hensley, and M. E. Smulson. 1987. cDNA sequence, protein structure, and chromosomal location of the human gene for poly(ADP-ribose) polymerase. *Proc. Natl. Acad. Sci. USA* **84:** 8370–8374.
42. Chung, H. W., J. W. Phillips, R. A. Winegar, R. J. Preston, and W. F. Morgan. 1991. Modulation of restriction enzyme-induced damage by chemicals that interfere with cellu-

lar responses to DNA damage: a cytogenetic and pulsed-field gel analysis. *Radiat. Res.* **125:** 107–113.
43. Cleaver, J. E., W. J. Bodell, W. F. Morgan, and B. Zelle. 1983. Differences in the regulation by poly(ADP-ribose) of repair of DNA damage from alkylating agents and ultraviolet light according to cell type. *J. Biol. Chem.* **258:** 9059–9068.
44. Cleaver, J. E. and W. F. Morgan. 1991. Poly(ADP-ribose)polymerase: a perplexing participant in cellular responses to DNA breakage. *Mutat. Res.* **257:** 1–18.
45. Cohen, J. J. and N. A. Berger. 1981. Activation of poly(adenosine diphosphate ribose) polymerase with UV irradiated and UV endonuclease treated SV 40 minichromosome. *Biochem. Biophys. Res. Commun.* **98:** 268–274.
46. Cohen, J. J., D. M. Catino, S. J. Petzold, and N. A. Berger. 1982. Activation of poly(adenosine diphosphate ribose) polymerase by SV 40 minichromosomes: effects of deoxyribonucleic acid damage and histone H1. *Biochemistry* **21:** 4931–4940.
47. Creissen, D. and S. Shall. 1982. Regulation of DNA ligase activity by poly(ADP-ribose). *Nature* **296:** 271,272.
48. de Murcia G. and J. Menissier de Murcia. 1994. Poly(ADP-ribose) polymerase: a molecular nick-sensor [published erratum appears in *Trends Biochem. Sci.* 1994. **19(6):** 250] *Trends Biochem. Sci.* **19:** 172–176.
49. de Murcia, G., J. Menissier de Murcia, and V. Schreiber. 1991. Poly(ADP-ribose) polymerase: molecular biological aspects. *Bioessays* **13:** 455–462.
50. de Murcia, G., V. Schreiber, M. Molinete, B. Saulier, O. Poch, M. Masson, C. Niedergang, and J. Menissier de Murcia. 1994. Structure and function of poly(ADP-ribose) polymerase. *Mol. Cell Biochem.* **138:** 15–24.
51. Ding, R., Y. Pommier, V. H. Kang, and M. Smulson. 1992. Depletion of poly(ADP-ribose) polymerase by antisense RNA expression results in a delay in DNA strand break rejoining. *J. Biol. Chem.* **267:** 12,804–12,812.
52. Ding, R. and M. Smulson. 1994. Depletion of nuclear poly(ADP-ribose) polymerase by antisense RNA expression: influences on genomic stability, chromatin organization, and carcinogen cytotoxicity. *Cancer Res.* **54:** 4627–4634.
53. Donehower, L. A., M. Harvey, B. L. Slagle, M. J. McArthur, C. A. Montgomery, Jr., J. S. Butel, and A. Bradley. 1992. Mice deficient for p53 are developmentally normal but susceptible to spontaneous tumours. *Nature* **356:** 215–221.
54. Farzaneh, F., G. N. Panayotou, L. D. Bowler, B. D. Hardas, T. Broom, C. Walther, and S. Shall. 1988. ADP-ribosylation is involved in the integration of foreign DNA into the mammalian cell genome. *Nucleic Acids Res.* **16:** 11,319–11,326.
55. Ferro, A. M. and B. M. Olivera. 1982. Poly(ADP-ribosylation) in vitro. Reaction parameters and enzyme mechanism. *J. Biol. Chem.* **257:** 7808–7813.
56. Fritsche, M., C. Haessler, and G. Brandner. 1993. Induction of nuclear accumulation of the tumor-suppressor protein p53 by DNA-damaging agents [published erratum appears in *Oncogene* 1993. **8(9):** 2605]. *Oncogene* **8:** 307–318.
57. Fujimura, S., S. Hasegawa, Y. Shimizu, and T. Sugimura. 1967. Polymerization of the adenosine 5'-diphosphate-ribose moiety of nicotinamide-adenine dinucleotide by nuclear enzyme. I. Enzymatic reactions. *Biochim. Biophys. Acta* **145:** 247–259.
58. Fujimura, S., S. Hasegawa, and T. Sugimura. 1967. Nicotinamide mononucleotide-dependent incorporation of ATP into acid-insoluble material in rat liver nuclei preparation. *Biochim. Biophys. Acta* **134:** 496–499.
59. Gradwohl, G., J. M. Menissier de Murcia, M. Molinete, F. Simonin, M. Koken, J. H. Hoeijmakers, and G. de Murcia. 1990. The second zinc-finger domain of poly(ADP-ribose) polymerase determines specificity for single-stranded breaks in DNA. *Proc. Natl. Acad. Sci. USA* **87:** 2990–2994.
60. Heartlein, M. W. and R. J. Preston. 1985. The effect of 3-aminobenzamide on the frequency of X-ray- or neutron-induced chromosome aberrations in cycling or non-cycling human lymphocytes. *Mutat. Res.* **148:** 91–97.

61. Herzog, H., B. U. Zabel, R. Schneider, B. Auer, M. Hirsch Kauffmann, and M. Schweiger. 1989. Human nuclear NAD+ ADP-ribosyltransferase: localization of the gene on chromosome 1q41-q42 and expression of an active human enzyme in Escherichia coli. *Proc. Natl. Acad. Sci. USA* **86:** 3514–3518.
62. Hollstein, M., D. Sidransky, B. Vogelstein, and C. C. Harris. 1991. p53 mutations in human cancers. Science **253:** 49–53.
63. Hori, T. 1981. High incidence of sister chromatid exchanges and chromatid interchanges in the conditions of lowered activity of poly(ADP-ribose)polymerase. *Biochem. Biophys. Res. Commun.* **102:** 38–45.
64. Hoshino, J., G. Beckmann, and H. Kroger. 1993. 3-aminobenzamide protects the mouse thymocytes in vitro from dexamethasone-mediated apoptotic cell death and cytolysis without changing DNA strand breakage. *J. Steroid Biochem. Mol. Biol.* **44:** 113–119.
65. Huet, J. and F. Laval. 1985. Influence of poly(ADP-ribose) synthesis inhibitors on the repair of sublethal and potentially lethal damage in gamma-irradiated mammalian cells. *Int. J. Radiat. Biol. Relat. Stud. Phys. Chem. Med.* **47:** 655–662.
66. Hughes, C. S., J. W. Shen, and J. R. Subjeck. 1989. Resistance to etoposide induced by three glucose-regulated stresses in Chinese hamster ovary cells. *Cancer Res.* **49:** 4452–4454.
67. Ikejima, M., S. Noguchi, R. Yamashita, T. Ogura, T. Sugimura, D. M. Gill, and M. Miwa. 1990. The zinc fingers of human poly(ADP-ribose) polymerase are differentially required for the recognition of DNA breaks and nicks and the consequent enzyme activation. Other structures recognize intact DNA. *J. Biol. Chem.* **265:** 21,907–21,913.
68. Ito, S., Y. Shizuta, and O. Hayaishi. 1979. Purification and characterization of poly(ADP-ribose) synthetase from calf thymus. *J. Biol. Chem.* **254:** 3647–3651.
69. Jacobson, E. L. 1993. Niacin deficiency and cancer in women. *J. Am. Coll. Nutr.* **12:** 412–416.
70. Jacobson, E. L., R. Meadows, and J. Measel. 1985. Cell cycle perturbations following DNA damage in the presence of ADP-ribosylation inhibitors. *Carcinogenesis* **6:** 711–714.
71. Jacobson, E. L., J. Y. Smith, K. Wielckens, H. Hilz, and M. K. Jacobson. 1985. Cellular recovery of dividing and confluent C3H10T1/2 cells from *N*-methyl-*N'*-nitro-*N*-nitrosoguanidine in the presence of ADP-ribosylation inhibitors. *Carcinogenesis* **6:** 715–718.
72. Jacobson, M. K., V. Levi, H. Juarez Salinas, R. A. Barton, and E. L. Jacobson. 1980. Effect of carcinogenic N-alkyl-N-nitroso compounds on nicotinamide adenine dinucleotide metabolism. *Cancer Res.* **40:** 1797–1802.
73. Jones, D. P., D. J. McConkey, P. Nicotera, and S. Orrenius. 1989. Calcium-activated DNA fragmentation in rat liver nuclei. *J. Biol. Chem.* **264:** 6398–6403.
74. Juarez Salinas, H., H. Mendoza Alvarez, V. Levi, M. K. Jacobson, and E. L. Jacobson. 1983. Simultaneous determination of linear and branched residues in poly(ADP-ribose). *Anal. Biochem.* **131:** 410–418.
75. Juarez Salinas, H., J. L. Sims, and M. K. Jacobson. 1979. Poly(ADP-ribose) levels in carcinogen-treated cells. *Nature* **282:** 740,741.
76. Kameshita, I., M. Matsuda, M. Nishikimi, H. Ushiro, and Y. Shizuta. 1986. Reconstitution and poly(ADP-ribosyl)ation of proteolytically fragmented poly(ADP-ribose) synthetase. *J. Biol. Chem.* **261:** 3863–3868.
77. Kameshita, I., Z. Matsuda, T. Taniguchi, and Y. Shizuta. 1984. Poly (ADP-Ribose) synthetase. Separation and identification of three proteolytic fragments as the substrate-binding domain, the DNA-binding domain, and the automodification domain. *J. Biol. Chem.* **259:** 4770–4776.
78. Kameshita, I., Y. Mitsuuchi, M. Matsuda, and Y. Shizuta. 1989. Biosynthesis and degradation of poly(ADP-ribose) synthetase, in *ADP-ribose Transfer Reactions. Mechanism and Biological Significance* (Jacobson, M. K. and E. L. Jacobson, eds.), Springer-Verlag, New York, pp. 71–75.
79. Kastan, M. B., O. Onyekwere, D. Sidransky, B. Vogelstein, and R. W. Craig. 1991. Participation of p53 protein in the cellular response to DNA damage. *Cancer Res.* **51:** 6304–6311.

80. Kaufmann, S. H., G. Brunet, B. Talbot, D. Lamarr, C. Dumas, J. H. Shaper, and G. Poirier. 1991. Association of poly(ADP-ribose) polymerase with the nuclear matrix: the role of intermolecular disulfide bond formation, RNA retention, and cell type. *Exp. Cell Res.* **192:** 524–535.
81. Kaufmann, S. H., S. Desnoyers, Y. Ottaviano, N. E. Davidson, and G. G. Poirier. 1993. Specific proteolytic cleavage of poly(ADP-ribose) polymerase: an early marker of chemotherapy-induced apoptosis. *Cancer Res.* **53:** 3976–3985.
82. Kiehlbauch, C. C., N. Aboul Ela, E. L. Jacobson, D. P. Ringer, and M. K. Jacobson. 1993. High resolution fractionation and characterization of ADP-ribose polymers. *Anal. Biochem.* **208:** 26–34.
83. Kumar, A., J. Kiefer, E. Schneider, and N. E. Crompton. 1985. Enhanced cell killing, inhibition of recovery from potentially lethal damage and increased mutation frequency by 3-aminobenzamide in Chinese hamster V79 cells exposed to X-rays. *Int. J. Radiat. Biol. Relat. Stud. Phys. Chem. Med.* **47:** 103–112.
84. Kupper, J. H., G. de Murcia, and A. Burkle. 1990. Inhibition of poly(ADP-ribosyl)ation by overexpressing the poly(ADP-ribose) polymerase DNA-binding domain in mammalian cells. *J. Biol. Chem.* **265:** 18,721–18,724.
85. Kurosaki, T., H. Ushiro, Y. Mitsuuchi, S. Suzuki, M. Matsuda, Y. Matsuda, et al. 1987. Primary structure of human poly(ADP-ribose) synthetase as deduced from cDNA sequence. *J. Biol. Chem.* **262:** 15,990–15,997.
86. Lagueux, J., G. M. Shah, L. Menard, H. Thomassin, C. Duchaine, C. Hengartner, and G. G. Poirier. 1994. Poly(ADP-ribose) catabolism in mammalian cells. *Mol. Cell Biochem.* **138:** 45–52.
87. Lane, D. P. 1992. p53, guardian of the genome. *Nature* **358:** 15,16.
88. Lautier, D., J. Lagueux, J. Thibodeau, L. Menard, and G. G. Poirier. 1993. Molecular and biochemical features of poly (ADP-ribose) metabolism. *Mol. Cell. Biochem.* **122:** 171–193.
89. Lazebnik, Y. A., S. H. Kaufmann, S. Desnoyers, G. G. Poirier, and W. C. Earnshaw. 1994. Cleavage of poly(ADP-ribose) polymerase by a proteinase with properties like ICE. *Nature* **371:** 346,347.
90. Lee, A. S. 1987. Coordinated regulation of a set of genes by glucose and calcium ionophores in mammalian cells. *Trends Biochem. Sci.* **12:** 20–23.
91. Ludwig, A., B. Behnke, J. Holtlund, and H. Hilz. 1988. Immunoquantitation and size determination of intrinsic poly(ADP-ribose) polymerase from acid precipitates. An analysis of the in vivo status in mammalian species and in lower eukaryotes. *J. Biol. Chem.* **263:** 6993–6999.
92. Mandel, P., H. Okazaki, and C. Niedergang. 1982. Poly(adenosine diphosphate ribose). *Prog. Nucleic Acid Res. Mol. Biol.* **27:** 1–51.
93. Mendoza Alvarez, H. and R. Alvarez Gonzalez. 1993. Poly(ADP-ribose) polymerase is a catalytic dimer and the automodification reaction is intermolecular. *J. Biol. Chem.* **268:** 22,575–22,580.
94. Molinete, M., W. Vermeulen, A. Burkle, J. Menissier de Murcia, J. H. Kupper, J. H. Hoeijmakers, and G. de Murcia. 1993. Overproduction of the poly(ADP-ribose) polymerase DNA-binding domain blocks alkylation-induced DNA repair synthesis in mammalian cells. *EMBO J.* **12:** 2109–2117.
95. Monti, D., A. Cossarizza, S. Salvioli, C. Franceschi, G. Rainaldi, E. Straface, R. Rivabene, and W. Malorni. 1994. Cell death protection by 3-aminobenzamide and other poly(ADP-ribose)polymerase inhibitors: different effects on human natural killer and lymphokine activated killer cell activities. *Biochem. Biophys. Res. Commun.* **199:** 525–530.
96. Morgan, W. F., J. Bodycote, Y. Doida, M. L. Fero, P. Hahn, and L. N. Kapp. 1986. Spontaneous and 3-aminobenzamide-induced sister-chromatid exchange frequencies estimated by ring chromosome analysis [published erratum appears in *Mutagenesis* 1988. **3(3):** 286]. *Mutagenesis* **1:** 453–459.

97. Morgan, W. F. and J. E. Cleaver. 1982. 3-Aminobenzamide synergistically increases sister-chromatid exchanges in cells exposed to methyl methanesulfonate but not to ultraviolet light. *Mutat. Res.* **104:** 361–366.
98. Morgan, W. F. and S. Wolff. 1984. Induction of sister chromatid exchange by 3-aminobenzamide is independent of bromodeoxyuridine. *Cytogenet. Cell Genet.* **38:** 34–38.
99. Nakayasu, M., H. Shima, S. Aonuma, H. Nakagama, M. Nagao, and T. Sugimura. 1988. Deletion of transfected oncogenes from NIH 3T3 transformants by inhibitors of poly(ADP-ribose) polymerase. *Proc. Natl. Acad. Sci. USA* **85:** 9066–9070.
100. Natarajan, A. T., I. Csukas, and A. A. van Zeeland. 1981. Contribution of incorporated 5-bromodeoxyuridine in DNA to the frequencies of sister-chromatid exchanges induced by inhibitors of poly-(ADP-ribose)-polymerase. *Mutat. Res.* **84:** 125–132.
101. Niedergang, C., H. Okazaki, and P. Mandel. 1979. Properties of purified calf thymus poly(adenosine diphosphate ribose) polymerase. Comparison of the DNA-independent and the DNA-dependent enzyme. *Eur. J. Biochem.* **102:** 43–57.
102. Nishikimi, M., K. Ogasawara, I. Kameshita, T. Taniguchi, and Y. Shizuta. 1982. Poly(ADP-ribose) synthetase. The DNA binding domain and the automodification domain. *J. Biol. Chem.* **257:** 6102–6105.
103. Nishizuka, Y., K. Ueda, K. Nakazawa, and O. Hayaishi. 1967. Studies on the polymer of adenosine diphosphate ribose. I. Enzymic formation from nicotinamide adenine dinuclotide in mammalian nuclei. *J. Biol. Chem.* **242:** 3164–3171.
104. Nosseri, C., S. Coppola, and L. Ghibelli. 1994. Possible involvement of poly(ADP-ribosyl) polymerase in triggering stress-induced apoptosis. *Exp. Cell Res.* **212:** 367–373.
105. Oei, S. L., H. Herzog, M. Hirsch Kauffmann, R. Schneider, B. Auer, and M. Schweiger. 1994. Transcriptional regulation and autoregulation of the human gene for ADP-ribosyltransferase. *Mol. Cell Biochem.* **138:** 99–104.
106. Ohgushi, H., K. Yoshihara, and T. Kamiya. 1980. Bovine thymus poly(adenosine diphosphate ribose) polymerase. Physical properties and binding to DNA. *J. Biol. Chem.* **255:** 6205–6211.
107. Oikawa, A., H. Tohda, M. Kanai, M. Miwa, and T. Sugimura. 1980. Inhibitors of poly(adenosine diphosphate ribose) polymerase induce sister chromatid exchanges. *Biochem. Biophys. Res. Commun.* **97:** 1311–1316.
108. Okayama, H., M. Honda, and O. Hayaishi. 1978. Novel enzyme from rat liver that cleaves an ADP-ribosyl histone linkage. *Proc. Natl. Acad. Sci. USA* **75:** 2254–2257.
109. Oren, M. 1992. p53: the ultimate tumor suppressor gene? *FASEB J.* **6:** 3169–3176.
110. Pantelias, G. E., G. Politis, C. D. Sabani, J. K. Wiencke, and W. F. Morgan. 1986. 3-Aminobenzamide does not affect X-ray-induced cytogenetic damage in G0 human lymphocytes. *Mutat. Res.* **174:** 121–124.
111. Park, S. D., C. G. Kim, and M. G. Kim. 1983. Inhibitors of poly(ADP-ribose) polymerase enhance DNA strand breaks, excision repair, and sister chromatid exchanges induced by alkylating agents. *Environ. Mutagen.* **5:** 515–525.
112. Purnell, M. R., and W. J. Whish. 1980. Novel inhibitors of poly(ADP-ribose) synthetase. *Biochem. J.* **185:** 775–777.
113. Rankin, P. W., E. L. Jacobson, R. C. Benjamin, J. Moss, and M. K. Jacobson. 1989. Quantitative studies of inhibitors of ADP-ribosylation in vitro and in vivo. *J. Biol. Chem.* **264:** 4312–4317.
114. Redegeld, F. A., S. Chatterjee, N. A. Berger, and M. V. Sitkovsky. 1992. Poly-(ADP-ribose) polymerase partially contributes to target cell death triggered by cytolytic T lymphocytes. *J. Immunol.* **149:** 3509–3516.
115. Rice, W. G., C. D. Hillyer, B. Harten, C. A. Schaeffer, M. Dorminy, D. A. d. Lackey, E. Kirsten, J. Mendeleyev, K. G. Buki, A. Hakam, et al. 1992. Induction of endonuclease-mediated apoptosis in tumor cells by C-nitroso-substituted ligands of poly(ADP-ribose) polymerase. *Proc. Natl. Acad. Sci. USA* **89:** 7703–7707.

116. Rossmann, M. G., D. Moras, and K. W. Olsen. 1974. Chemical and biological evolution of nucleotide-binding protein. *Nature* **250:** 194–199.
117. Satoh, M. S., and T. Lindahl. 1992. Role of poly(ADP-ribose) formation in DNA repair. *Nature* **356:** 356–358.
118. Satoh, M. S., G. G. Poirier, and T. Lindahl. 1994. Dual function for poly(ADP-ribose) synthesis in response to DNA strand breakage. *Biochemistry* **33:** 7099–7106.
119. Satoh, M. S., G. G. Poirier, and T. Lindahl. 1993. NAD(+)-dependent repair of damaged DNA by human cell extracts. *J. Biol. Chem.* **268:** 5480–5487.
120. Schreiber, V., D. Hunting, C. Trucco, B. Gowans, D. Grunwald, G. de Murcia, and J. M. De Murcia. 1995. A dominant-negative mutant of human poly(ADP-ribose) polymerase affects cell recovery, apoptosis, and sister chromatid exchange following DNA damage. *Proc. Natl. Acad. Sci. USA* **92:** 4753–4757.
121. Schwartz, J. L., W. F. Morgan, and R. R. Weichselbaum. 1985. Different efficiencies of interaction between 3-aminobenzamide and various monofunctional alkylating agents in the induction of sister chromatid exchanges. *Carcinogenesis* **6:** 699–704.
122. Shall, S. 1984. ADP-ribose in DNA repair: a new component of DNA excision repair. *Adv. Radiat. Biol.* **11:** 1–69.
123. Shen, J., C. Hughes, C. Chao, J. Cai, C. Bartels, T. Gessner, and J. Subjeck. 1987. Coinduction of glucose-regulated proteins and doxorubicin resistance in Chinese hamster cells. *Proc. Natl. Acad. Sci. USA* **84:** 3278–3282.
124. Sims, J. L., S. J. Berger, and N. A. Berger. 1981. Effects of nicotinamide on NAD and poly(ADP-ribose) metabolism in DNA-damaged human lymphocytes. *J. Supramol. Struct. Cell Biochem.* **16:** 281–288.
125. Sims, J. L., S. J. Berger, and N. A. Berger. 1983. Poly(ADP-ribose) Polymerase inhibitors preserve nicotinamide adenine dinucleotide and adenosine 5'-triphosphate pools in DNA-damaged cells: mechanism of stimulation of unscheduled DNA synthesis. *Biochemistry* **22:** 5188–5194.
126. Sims, J. L., G. W. Sikorski, D. M. Catino, S. J. Berger, and N. A. Berger. 1982. Poly(adenosine diphosphoribose) polymerase inhibitors stimulate unscheduled deoxyribonucleic acid synthesis in normal human lymphocytes. *Biochemistry* **21:** 1813–1821.
127. Stevnsner, T., R. Ding, M. Smulson, and V. A. Bohr. 1994. Inhibition of gene-specific repair of alkylation damage in cells depleted of poly(ADP-ribose) polymerase. *Nucleic Acids Res.* **22:** 4620–4624.
128. Sugimura, T. 1973. Poly(adenosine diphosphate ribose). *Prog. Nucleic Acid Res. Mol. Biol.* **13:** 127–151.
129. Sugimura, T. and M. Miwa. 1983. Poly(ADP-ribose) and cancer research. *Carcinogenesis* **4:** 1503–1506.
130. Sugimura, T. and M. Miwa. 1994. Poly(ADP-ribose): historical perspective. *Mol. Cell Biochem.* **138:** 5–12.
131. Tanaka, Y., T. Hashida, H. Yoshihara, and K. Yoshihara. 1979. Bovine thymus poly(ADP-ribose) polymerase histone-dependent and Mg^{2+}-dependent reaction. *J. Biol. Chem.* **254:** 12,433–12,438.
132. Uchida, K., S. Hanai, K. Ishikawa, Y. Ozawa, M. Uchida, T. Sugimura, and M. Miwa. 1993. Cloning of cDNA encoding Drosophila poly(ADP-ribose) polymerase: leucine zipper in the auto-modification domain. *Proc. Natl. Acad. Sci. USA* **90:** 3481–3485.
133. Uchida, K. and M. Miwa. 1994. Poly(ADP-ribose) polymerase: structural conservation among different classes of animals and its implications. *Mol. Cell Biochem.* **138:** 25–32.
134. Uchida, K., T. Morita, T. Sato, T. Ogura, R. Yamashita, S. Noguchi, H. Suzuki, H. Nyunoya, M. Miwa, and T. Sugimura. 1987. Nucleotide sequence of a full-length cDNA for human fibroblast poly(ADP-ribose) polymerase. *Biochem. Biophys. Res. Commun.* **148:** 617–622.

135. Ucker, D. S., P. S. Obermiller, W. Eckhart, J. R. Apgar, N. A. Berger, and J. Meyers. 1992. Genome digestion is a dispensable consequence of physiological cell death mediated by cytotoxic T lymphocytes. *Mol. Cell Biol.* **12:** 3060–3069.
136. Ueda, K. and O. Hayaishi. 1985. ADP-ribosylation. *Annu. Rev. Biochem.* **54:** 73–100.
137. Utakoji, T., K. Hosoda, K. Umezawa, M. Sawamura, T. Matsushima, M. Miwa, and T. Sugimura. 1979. Induction of sister chromatid exchanges by nicotinamide in Chinese hamster lung fibroblasts and human lymphoblastoid cells. *Biochem. Biophys. Res. Commun.* **90:** 1147–1152.
138. Utsumi, H. and M. M. Elkind. 1979. Potentially lethal damage versus sublethal damage: independent repair processes in actively growing Chinese hamster cells. *Radiat. Res.* **77:** 346–360.
139. Waldman, A. S. and B. C. Waldman. 1991. Stimulation of intrachromosomal homologous recombination in mammalian cells by an inhibitor of poly(ADP-ribosylation). *Nucleic Acids Res.* **19:** 5943–5947.
140. Waldman, B. C. and A. S. Waldman. 1990. Illegitimate and homologous recombination in mammalian cells: differential sensitivity to an inhibitor of poly(ADP-ribosylation). *Nucleic Acids Res.* **18:** 5981–5988.
141. Whitacre, C. M., H. Hashimoto, M. L. Tsai, S. Chatterjee, S. J. Berger, and N. A. Berger. 1995. Involvement of NAD-poly(ADP-ribose) metabolism in p53 regulation and its consequences. *Cancer Res.* **55:** 3697–3701.
142. Wiencke, J. K. and W. F. Morgan. 1987. Cell cycle-dependent potentiation of X-ray-induced chromosomal aberrations by 3-aminobenzamide. *Biochem. Biophys. Res. Commun.* **143:** 372–376.
143. Wyllie, A. H. 1980. Glucocorticoid-induced thymocyte apoptosis is associated with endogenous endonuclease activation. *Nature* **284:** 555,556.
144. Yamanaka, H., C. A. Penning, E. H. Willis, D. B. Wasson, and D. A. Carson. 1988. Characterization of human poly(ADP-ribose) polymerase with autoantibodies. *J. Biol. Chem.* **263:** 3879–3883.
145. Yoshihara, K., Y. Tanaka, H. Yoshihara, T. Hashida, H. Ohgushi, R. Arai, and T. Kamiya. 1980. Effect of histone, DNA, and Mg2+ on purified poly(ADP-ribose) polymerase reaction, in *Novel ADP-Ribosylations of Regulatory Enzymes and Proteins* (M. E. Smulson and T. Sugimura, eds.), Elsevier, North Holland, pp. 33–44.
146. Zahradka, P. and K. Ebisuzaki. 1982. A shuttle mechanism for DNA-protein interactions. The regulation of poly(ADP-ribose) polymerase. *Eur. J. Biochem.* **127:** 579–585.

23

DNA Topoisomerases in DNA Repair and DNA Damage Tolerance

John L. Nitiss

1. INTRODUCTION

DNA topoisomerases catalyze the interconversion of topological isomers of DNA. Topological changes are required for a wide variety of cellular processes, including transcription, replication and chromosome segregation. The centrality of topoisomerases in DNA metabolism has frequently led to the suggestion that topoisomerases might play important or essential roles in DNA repair and DNA damage tolerance (for an earlier review, *see 18*). This chapter provides a detailed examination of that thesis.

In addition to their biological functions, DNA topoisomerases have become an important class of targets for both antibacterial antibiotics as well as for antineoplastic therapy (reviewed in *22,52,65*). They also have potential roles in the treatment of parasitic and fungal diseases *(3,28,76)*. Information relating to the roles of topoisomerases in repair will be of use in understanding the roles and limitations of antitopoisomerase therapeutic agents.

2. BIOCHEMICAL REACTIONS AND REACTION MECHANISMS

The DNA topoisomerase reaction consists of breaking one or both DNA strands, passing another DNA strand(s) through the break, and then resealing the break *(66)*. The strand breakage or cleavage reaction is a transesterification reaction, in which the phosphodiester bond of DNA is broken with the formation of a phosphotyrosine bond between the enzyme and DNA. Strand passage occurs while the enzyme is covalently bound to DNA. The religation step reverses the cleavage reaction. Therefore a high-energy cofactor is not required for the religation reaction.

Two different classes of topoisomerases have been identified, in both prokaryotes and eukaryotes. Type I enzymes make a single-strand break (SSB), and therefore change the linking number of DNA by steps of one, whereas the type II enzymes make a double-strand break (DSB), and pass both DNA strands at the same time, thereby changing the linking number by steps of two. Some prokaryotic type II enzymes can introduce negative supercoils into DNA and hence are termed DNA gyrases, whereas no eukaryotic enzyme has been identified that is able to supercoil DNA. In addition to the relaxation of supercoiling, topoisomerases carry out several related reactions,

From: DNA Damage and Repair, Vol. 2: DNA Repair in Higher Eukaryotes
Edited by: J. A. Nickoloff and M. F. Hoekstra © Humana Press Inc., Totowa, NJ

including the catenation/decatenation and knotting/unknotting of circular DNA molecules (reviewed in refs. *66,90*).

2.1. Biochemical Reactions and Biological Roles

DNA topoisomerases play a critical role in DNA replication in both prokaryotic and eukaryotic cells. It was the topological problems that arise from semiconservative replication of a circular DNA molecule that motivated the studies that led to the discovery of DNA topoisomerases (reviewed in *58*). In eukaryotic cells, either topoisomerase I or topoisomerase II can unlink the parental strands of DNA during elongation of DNA replication. However, at the end of DNA replication, a type II topoisomerase is absolutely required to separate the replicated daughter molecules *(17,33,37,85)*. In mitotic yeast cells, this appears to be the only reason why topoisomerase II is essential for viability, since topoisomerase II mutants lose viability specifically at mitosis. The inviability arises from attempting to segregate DNA molecules that remain intercatenated.

Topoisomerases also play several important roles in transcription. In some instances, an enzyme able to remove superhelical turns is needed for transcription as well as for replication. As is the case for elongation of replication, either topoisomerase I or topoisomerase II can serve this purpose. Topoisomerases are also good candidates for regulators of gene expression. The data are most clear in bacterial systems, where several genes, including the structural genes for topoisomerase I *(topA)* and DNA gyrase (*gyrA* and *gyrB*), are regulated by DNA supercoiling *(54)*. The opposite action of these two enzymes, with bacterial topoisomerase I only able to relax negative supercoils and the negative supercoiling activity of DNA gyrase, results in the maintenance of an appropriate level of supercoiling. Other bacterial genes that are regulated by supercoiling have also been described (reviewed in *91*). This last point has some relevance to understanding the effects of bacterial topoisomerase mutations on DNA repair. A mutation in a topoisomerase gene may alter the expression of other genes that are required for DNA repair. This appears to be the case in *Salmonella topA* mutants (*see* Section 5.).

Topoisomerases also play other critical roles in eukaryotic cells. A key reaction is the condensation of chromosomes prior to mitosis and the decondensation immediately following mitosis. The chromosome condensation reaction absolutely requires topoisomerase II activity both in yeast cells and in *Xenopus* oocytes *(57,85)*. Topoisomerase II may play other roles in chromosome structure as well (for review, see *48*). Although it is possible that these other roles of topoisomerase II may be specifically important when cells are exposed to DNA-damaging agents, this has not been examined directly.

2.2. Genome Stability

An issue closely related to DNA repair and DNA damage tolerance is genome stability. Recent studies have suggested that DNA topoisomerases contribute to genome stability and that a lack of specific topoisomerases can lead to various genomic alterations. The instability is seen even when cells lack a topoisomerase that is not essential for viability (like topoisomerase I or III in yeast, or topoisomerase III in *Escherichia coli*). The first demonstration that a lack of topoisomerase activity could lead to genomic instability was by Christman et al. *(11)*, who found that null mutations in the yeast *top1* gene or temperature-sensitive mutations in the *top2* gene grown at semi-permissive temperatures resulted in a hyperrecombination phenotype in the tandemly repeated ribosomal RNA genes (rDNA). No elevation of recombination in genes that are present in a single copy was observed, nor was a higher level of recombination observed in the *CUP1* gene cluster, which like rDNA is a tandemly repeated array of identical units. There are several unique aspects to rDNA genes, including their high copy number (approx 200/cell), very high transcription levels by RNA polymerase I, and their organization as part of a discrete nuclear structure, the nucleolus. Any of these features could be responsible for the observed hyperrecombination phenotype.

An analogous reaction in rDNA has been described that requires mutations in both topoisomerase I and topoisomerase II. *Δtop1 top2*^{ts} double mutants excise single repeat units from rDNA, and the excised units are maintained as extrachromosomal rings *(45)*. If a wild-type topoisomerase activity is restored to the cell, the units reintegrate into the array on chromosome XII. It is not known whether this reaction is mechanistically the same as the elevated recombination reaction described by Christman et al., but it is tempting to guess that the lack of both topoisomerase I and topoisomerase II activity influences the ultimate product of a similar reaction. The fact that the rings reintegrate also suggests that the process is dynamic, with the total topoisomerase level influencing the position of the equilibrium state.

Another eukaryotic topoisomerase also appears to play a role in genome stability *(89)*. The yeast *top3* gene was originally identified as a mutation termed *edr1* that conferred an enhanced level of δ-δ recombination in yeast (δ elements are repeat elements found at the end of the yeast retro-transposon Ty1). Several regions of the yeast genome are rich in "solo δs"; single copies of the long terminal repeat (LTR) of Ty1, and in yeast *top3* mutants, an elevated level of deletions in these regions are observed. Since yeast Top3p has limited activity in relaxing negative supercoiling (and like all topoisomerases that are homologous to bacterial topoisomerase I, no activity against positive supercoils), it has been suggested that the role in suppressing recombination may relate to another activity, perhaps its decatenation activity *(90)*, rather than its relaxation of supercoiling.

3. INHIBITORS OF DNA TOPOISOMERASES

Perhaps no class of enzymes has such a diverse range of inhibitors as DNA topoisomerases. Agents targeting DNA gyrase played a major role in the identification of the genes encoding the two subunits of this enzyme *(70)*. Such agents have been used both as tools to study the biological roles of DNA topoisomerases and clinically as antibacterial antibiotics and anticancer agents. Since topoisomerase inhibitors have been extensively used to study DNA repair, and since their use has several potential pitfalls, a description of their action is in order. DNA topoisomerase inhibitors have also been the subject of several recent reviews *(29,52,65)*.

Two classes of agents have been described that act against prokaryotic topoisomerases. Novobiocin and coumarin antibiotics, such as coumermycin A1, inhibit the ATPase activity of topoisomerase II, and thereby block enzyme turnover. A second class of agents, the quinolones, block an intermediate in the enzyme reaction termed the cleavable complex or the cleavage complex *(65,70)*. These drugs lead to an elevated level of the intermediate of the topoisomerase reaction in which DNA strands are broken and the enzyme is covalently attached to DNA. This can occur either by increasing the rate of formation of the cleavage complex, or by inhibition of the resealing reaction by the enzyme *(29,66)*. The cleavage complex has properties that are important for understanding drug action. If drug is removed, the ionic strength of the enzyme reaction is drastically altered, or the reaction is heated, little if any cleaved DNA is observed. Thus, the cleavage complex is freely reversible.

Quinolone antibiotics are potent cytotoxic agents in bacteria, even though the lesion provoked by them is freely reversible. An explanation for the cytotoxicity is that quinolones convert the enzyme to a cellular poison that creates a type of DNA damage. The cleavage complex interferes with cellular nucleic acid metabolism, particularly DNA replication. The notion of drugs acting to convert a topoisomerase into a cellular poison was first proposed by Kreuzer and Cozzarelli based on the different effects of quinolones vs temperature-sensitive gyrase mutations on *E. coli* cells *(47)*. Quinolones can result in much higher levels of cytotoxicity than conditional gyrase mutations, and in general, have much greater inhibitory effects on transcription and DNA replication than gyrase mutations.

3.1. How DNA Replication Results in Cell Killing in the Presence of Cleavage-Stabilizing Agents

The mechanism of action of quinolone antibiotics has important implications for studies of DNA repair. For the most part, quinolones do not kill cells by depriving the cells of topoisomerase activity. Concentrations that kill a majority of the cells do not completely block topoisomerase II reactions, but they do block DNA replication and transcription, and induce DSBs. DSB induction when a replication fork collides with a covalent complex of topoisomerase II and DNA is illustrated in Fig. 1. The latent DSB of the covalent complex can no longer be resealed after a collision with the replication fork. Not all of the DSBs in quinolone treated cells are protein linked, lending support to the fork collision model.

Recent in vitro experiments suggest that DNA helicases may also be able to generate DSBs when they collide with a covalent complex of topoisomerase II *(38)*, whereas RNA polymerase does not seem to be able to generate a DSB in vitro *(92)*. This scenario is illustrated in Fig. 2, which shows the consequences of a tracking protein colliding with a topoisomerase II cleavage complex. All of the DSBs generated by a tracking protein would be expected to be protein linked. Note that if the covalent complex involves topoisomerase I rather than topoisomerase II, only SSBs (protein linked) will be generated. If DSBs are the main lethal lesion, these models predict that drugs targeting topoisomerase I should be lethal only during ongoing DNA replication, a result borne out in eukaryotic cells *(58,63)*.

The considerations above predict that quinolones acting on topoisomerase II will generate high levels of DNA damage. The DNA damage so generated would be extensive and could saturate repair systems. Since DSBs are one of the major lesions generated, recombinational repair would be most affected. Since quinolones generate DNA damage, they are probably unreliable as probes for assessing DNA damage. By generating DNA damage and inhibiting DNA replication, quinolone antibiotics efficiently induce SOS responses, and this induction would affect repair and survival to other types of DNA damage (*69*; *see also* Section 5.).

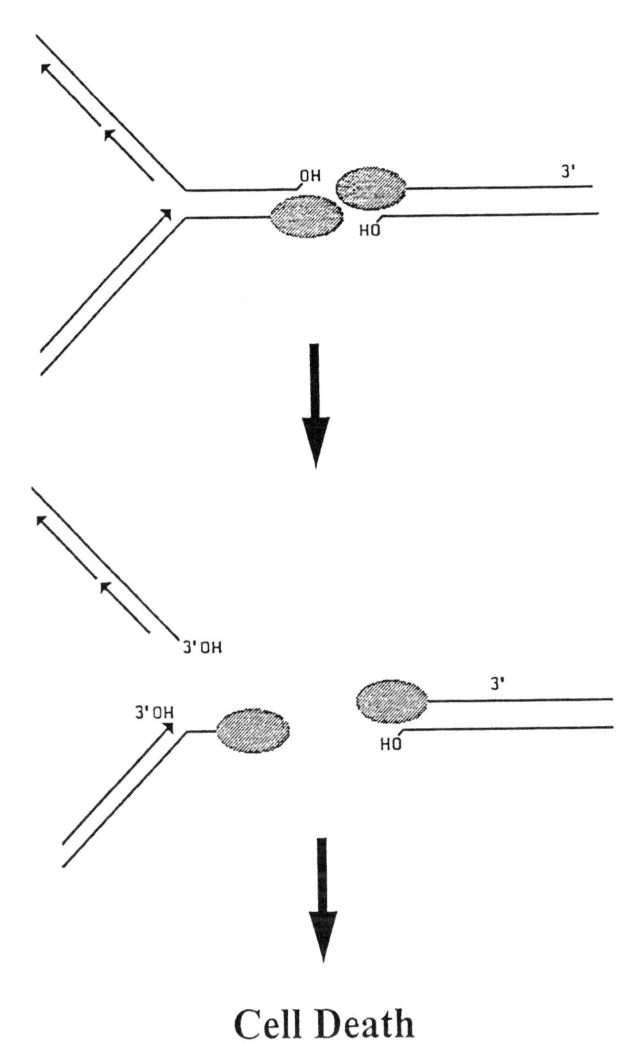

Fig. 1. Generation of irreversible DSBs by replication fork collision with topoisomerase II covalent complexes. The covalent complex of topoisomerase II, DNA, and drug is a reversible complex, and drug removal results in rapid reversal of the DNA strand breaks. Collision of a replication fork with the complex results in DSBs. Note that ends with protein covalently linked, as well as ends lacking protein, are generated. If the covalent complex involves topoisomerase I instead of topoisomerase II, a DSB in DNA is also generated by collision with a replication fork.

3.2. Inhibitors of Eukaryotic Topoisomerases

Most of the inhibitors of prokaryotic DNA topoisomerase II have little activity against eukaryotic DNA topoisomerases, even though DNA gyrase has considerable homology to eukaryotic topoisomerase II. Most quinolones are at least 100 times less active against eukaryotic topoisomerase II, although some quinolones have recently been described that are very active against the eukaryotic enzyme *(24,72)*. Novobiocin has frequently been used as a eukaryotic topoisomerase inhibitor, but this compound requires millimolar concentrations to inhibit the eukaryotic enzyme. At these concen-

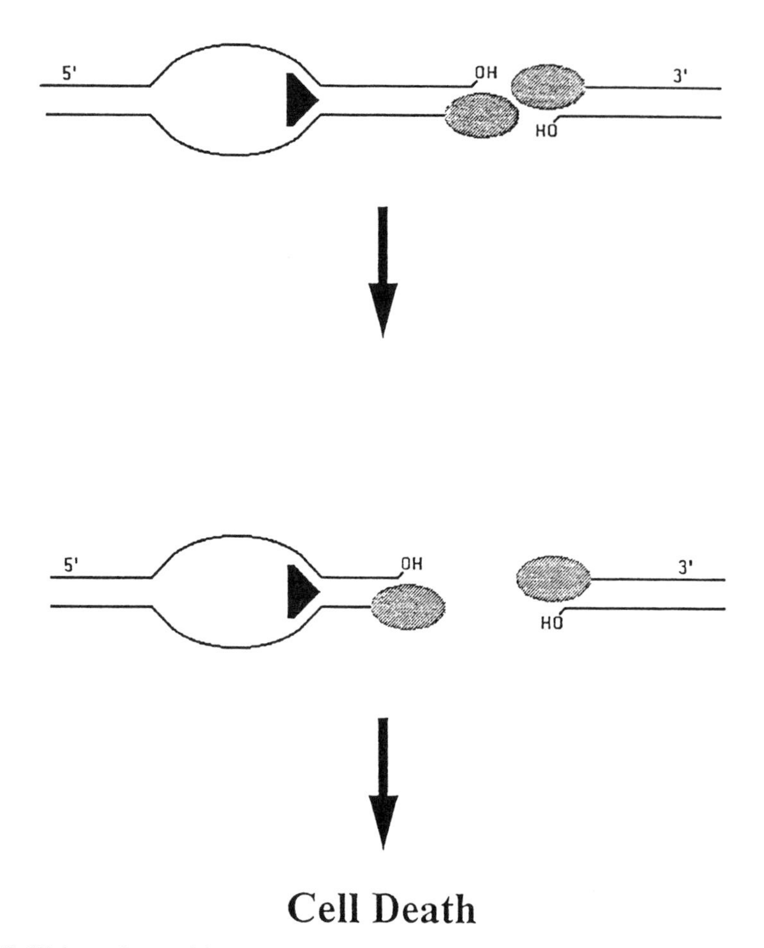

Fig. 2. Collision of a tracking protein with a covalent topoisomerase II complex. Proteins tracking along DNA, such as helicases or RNA polymerase, also have the potential to generate a DSB when the protein collides with a topoisomerase II covalent complex. Results discussed in Section 3.1. suggest that replicative helicases are effective at breaking the covalent complex, whereas RNA polymerase is relatively ineffective. Note also that a DSB would not be generated with a topoisomerase I covalent complex.

trations, novobiocin inhibits a wide range of eukaryotic enzymes, including DNA polymerases. It also inhibits mitochondrial function *(20)* and can precipitate histones *(12)*. Recently, several catalytic inhibitors of eukaryotic topoisomerases with greater specificity have been described. Most noteworthy are the bisdioxopiperazine compounds ICRF159 and ICRF193. These agents appear to be very specific for topoisomerase II *(40)*. They do not appear to inhibit the ATPase activity of the enzyme. Instead, they lock the enzyme into a form that is unable to cleave DNA and is therefore completely inactive *(73)*. Other agents have recently been suggested to act specifically as catalytic inhibitors of eukaryotic topoisomerase II, e.g., merbarone *(21)*, fostriecin *(16)*, and aclarubicin *(41)*. Of the catalytic inhibitors, bisdioxopiperazine compounds and merbarone appear to be most promising as probes for studying topoisomerases. The other agents appear to have significant activities against other intracellular targets, e.g., fostriecin inhibits protein phosphatases *(71)*, whereas aclarubicin is a potent intercalating agent that is likely to interfere with many other DNA binding proteins.

The inhibitors that have been most widely used in studying the biological functions of eukaryotic topoisomerases are agents that stabilize the cleavage complex. These agents, which have been actively studied as anticancer agents, include the topoisomerase I inhibitor camptothecin, the non-intercalating topoisomerase II inhibitors epipodophyllotoxins, teniposide, and etoposide, and a wide range of DNA intercalators, including doxorubicin, amsacrine, and mitoxantrone. Camptothecin and its derivatives have been shown to be very specific for topoisomerase I *(25,61)*. The epipodophyllotoxins, mitoxantrone, and amsacrine all appear to kill cells through action on topoisomerase II, whereas doxorubicin also affects other cellular components *(59)*. The basic mechanisms of cell killing by antitopoisomerase agents appear to be similar to that described for quinolones and bacterial cells. The (reversible) lesion generated by an antitopoisomerase poison is converted into an irreversible lesion by the DNA replication and other cellular processes. Cell killing by the antitopoisomerase I agents requires active DNA synthesis *(63)*. In mammalian cells, processes other than replication convert the cleavage complex generated by topoisomerase II into an irreversible, lethal lesion, whereas in yeast, DNA replication appears most important *(52)*. For both antitopoisomerase I and antitopoisomerase II agents, recent evidence suggests that apoptosis is an important mechanism of cell killing, suggesting that the irreversible lesion (a DSB?) commits cells to programmed cell death *(56)*.

The problems of using quinolones to study bacterial DNA repair also apply to eukaryotic complex-stabilizing drugs. Most of the inhibitors of eukaryotic topoisomerase II are potent intercalating agents, and the intercalation *per se* might be expected to influence DNA structure and DNA repair. An exception is the epipodophyllotoxins, etoposide and teniposide, which do not appear to interact strongly with DNA. Although epipodophyllotoxins are highly specific for topoisomerase II *(59)*, like quinolones, they introduce DNA damage into cells. Therefore, these agents would seem to be most useful for measuring specific steps in DNA repair, such as incision following UV damage, especially in vitro. The results of such studies are discussed in Section 6.

4. MODELS FOR TOPOISOMERASE ACTION IN DNA REPAIR

Although many investigators have suggested that topoisomerases might play a role in DNA repair, there have been relatively few models for how topoisomerases might influence repair. There are three general roles that topoisomerases could play:

1. Topoisomerases could directly participate in DNA repair. This could include more efficient damage recognition, more efficient binding of repair enzymes to sites of DNA damage due to altered DNA topology *(81)*, or a direct participation in the repair reaction.
2. Topoisomerases could regulate the expression of genes important for DNA repair (e.g., SOS response genes).
3. Topoisomerases could influence survival following exposure to DNA damage by indirect mechanisms. Topoisomerase-mediated changes in DNA supercoiling might alter the reactivity of DNA to particular DNA-damaging agents. For example, DNA intercalating agents bind more avidly to negatively supercoiled DNA than to relaxed DNA *(53)*. A wide variety of other indirect mechanisms are also possible.

The first possibility is clearly testable, particularly by use of in vitro repair systems. At present, there is no clear evidence that in vitro nucleotide excision repair, mismatch repair, or repair DNA synthesis requires any topoisomerase. A lack of specific requirement for topoisomerases in repair reactions is also generally supported by other types of studies described in Sections 5. and 6., particularly in the case of nucleotide excision repair.

One direct role for topoisomerases in repair that has not been tested in detail relates to DNA damage surveillance. Some types of DNA damage inhibit DNA topoisomerases *(13,68)*. In particular, UV photoproducts have been shown to be potent inhibitors of prokaryotic topoisomerase I and eukaryotic topoisomerase II. The inhibition of topoisomerases by DNA damage could alter the level of supercoiling in a chromosomal domain. The higher level of supercoiling might lead to the recruitment of other enzymes that directly act on DNA damage sites. For example, Sung et al. *(81)* have shown that yeast Rad3p preferentially binds to negatively supercoiled DNA. By contrast, DNA strand breaks immediately release any supercoiling of a domain and might also lead to the recruitment of a different set of enzymes required for strand break repair.

The other two possible modes of action of topoisomerases in repair do not lend themselves to clear models. Experiments relating to SOS and other DNA damage-inducible systems in prokaryotes are described in Section 5. Here direct tests of SOS induction and mutagenesis mediated by the SOS system are possible. For eukaryotic systems, the experiments have been rather indirect, and rely either on inhibitors or on examining how topoisomerase levels change in response to DNA damage, as described in Section 6.

5. ROLES OF TOPOISOMERASES IN PROKARYOTIC DNA REPAIR

Prokaryotic systems have distinct advantages for assessing whether topoisomerases participate in DNA repair. Conditional mutants have been constructed in *E. coli* or *Salmonella* for the topoisomerases that are essential for viability, and null mutations have been constructed for nonessential topoisomerases. However, there are some specific complications in bacteria.

E. coli has four known DNA topoisomerases, *topA* and *topC* encode type I enzymes, whereas DNA gyrase and topoisomerase IV *(44)* are type II enzymes. Both type II enzymes are essential, but topoisomerase III, the *topC* gene product, is nonessential (reviewed in *58,90*). DNA gyrase has many important roles in replication, transcription, and chromosome segregation, so it is difficult to determine how mutations or inhibition of this enzyme influence DNA repair. As discussed above, *topA* and gyrase act in opposition to maintain the negative superhelical density of the *E. coli* genome, and alterations in these two enzymes can change the level of supercoiling in vivo. Such alterations may change the sensitivity of DNA to chemical mutagens and may also alter the activities of other DNA repair enzymes. Therefore, alterations in DNA damage sensitivity may be due to indirect effects.

These considerations may explain the paradoxical results that have been obtained when examining the effects of DNA damage. For example, Chao and Tillman *(9)* observed decreased killing and mutagenesis in *E. coli gyrB* strains by nitrosoguanidine. Similarly, Bouayadi et al. *(6)* found enhanced resistance to cisplatin in a nalidixic acid-resistant mutant. In this case, although the resistance to nalidixic acid was owing to a gyrase mutation, the resistance to nalidixic acid and to cisplatin was probably the result distinct genetic lesions. von Wright and Bridges *(88)* observed a more complicated pattern in that *gyrB* mutants showed increased efficiency of excision repair, but a decreased efficiency in postreplication repair. All *gyrB* mutants had reduced levels of supercoiling in chromosomal DNA compared to wild-type cells. These results are consistent with two possibilities: either more relaxed DNA is a better substrate for excision repair and a poorer substrate for postreplication repair, or relaxed DNA expresses higher levels of proteins required for excision repair and lower levels of proteins required for postreplication repair.

Further evidence implicating DNA gyrase in postreplication DNA repair was reported by Hays and Boehmer *(36)*. They used a λ transfection assay to assess the effects of gyrase inhibitors on DNA repair and recombination under conditions where transcription and replication of the phage are completely blocked. The phage also was defective in phage-specific recombination functions. The phage carried a duplication that rendered phage growth EDTA-sensitive; homologous recombination deleting the duplicated segment resulted in EDTA-resistant phage growth. Repair was estimated by determining the infectivity of the phage. In this system, phage recombination frequencies are low and are stimulated by UV irradiation. Recovery of infective phage, as well as EDTA-resistant phage, was greatly reduced by either oxolinic acid (a quinolone) or coumermycin. This result is consistent with a requirement for gyrase in both recombination and removal of UV damage from DNA. It is unlikely that the recombination observed in this system is gyrase-mediated illegitimate recombination, since it depends on *recA*, whereas illegitimate recombination catalyzed by DNA gyrase is *recA*-independent, nor are the effects likely to be the result of changes in the expression of recombination functions *(55,87)*. Urios et al. *(86)* showed that *topA* mutations reduce *recA* expression. Since *topA* mutations usually produce elevated supercoiling, a reduction of gyrase activity would be expected to increase *recA* expression, ruling out straightforward models that involve alterations in the expression of DNA repair genes. Taken together with the results of von Wright and Bridges *(88)*, these results support a role for DNA gyrase in postreplication DNA repair.

Mutations in *E. coli topA*, the structural gene for DNA topoisomerase I, result in a decreased frequency of transposition by *Tn*5, but not by *Tn*3 *(78)*. Fishel and Kolodner *(27)* observed that plasmid recombination was decreased by more than 1000-fold in *E. coli topA* strains. However, an unusual temperature effect was seen in those experiments. At 42°C the *topA* mutant had wild-type recombination frequencies (the mutant allele was a null allele). Although it seems likely that the effects on Tn5 transposition are owing to effects on the supercoiling of the substrate DNA, further studies are needed in order to determine what role if any *E. coli* topoisomerase I plays in homologous recombination and recombinational repair.

E. coli has a poor tolerance for *topA* mutations, and in the absence of compensatory mutations in gyrase or other genes, *topA* null mutations are lethal. This is not true for *Salmonella*, which tolerates *topA* deletions in the absence of compensatory mutations. Strains of *Salmonella typhimurium* that are completely deficient in *topA* activity are viable, but they are UV-sensitive and hypomutable *(67)*. Recently, Smith et al. *(77)* suggested that *topA* mutants have DNA repair defects in *Salmonella* because they are unable to induce damage-inducible genes. Using *lac*-operon fusions to DNA damage-inducible *(din)* loci, they found that *topA* mutations abolished the induction of *din* genes. They also found that the lack of inducibility of *din* genes could also occur in hyperosmotic medium, which like *topA* mutations, leads to increased negative supercoiling. These studies suggest that *topA* mutations mainly affect DNA repair by affecting the expression of genes important for repair, rather than by a more direct participation in a repair reaction.

The results described in this section demonstrate an indirect role for topoisomerases in DNA repair in bacteria. Changes in either gyrase activity or topoisomerase I activity alter supercoiling and thereby alter the expression of DNA repair genes. DNA gyrase may also play a more direct role in postreplication repair. A possible role for gyrase could be to control the topology of the substrate molecules. The biochemical details of this remain to be elucidated.

6. EUKARYOTIC TOPOISOMERASES IN DNA REPAIR

Powerful genetic systems are available in lower eukaryotic systems (*Saccharomyces cerevisiae* and *Schizosaccharomyces pombe*) to assess the importance of topoisomerases on DNA repair, but such systems are not available in higher eukaryotes. The most frequent alternative has been the use of topoisomerase inhibitors. Early studies with novobiocin and nalidixic acid strongly suggested a role for DNA topoisomerases in eukaryotic DNA repair. However, as described in Section 3.2., neither agent is specific for eukaryotic topoisomerase II; in fact, nalidixic acid has little inhibitory activity against eukaryotic topoisomerase II *(32,34)*. Novobiocin clearly affects DNA repair, but probably because of its action on mitochondrial function *(20)* or through its action on other ATP-requiring enzymes *(18)*. Nonetheless, many subsequent studies have utilized either nalidixic acid or, more commonly, novobiocin (e.g., *75*), although these studies are not likely to be informative concerning the roles of topoisomerases in DNA repair.

6.1. Studies with Specific Eukaryotic Topoisomerase Inhibitors

Downes et al. *(19)* used etoposide to study various steps of excision repair in human cells. They found that etoposide did not affect incision of UV damage, DNA resynthesis, or ligation of strand breaks, nor did it alter cellular sensitivity to UV damage. Legerski et al. *(50)* used novobiocin, etoposide, and camptothecin alone and in combination in an in vitro *Xenopus* oocyte system. Although novobiocin inhibited repair, neither camptothecin nor etoposide had any effect *(50)*.

Contrasting results were obtained by Jones et al. *(42)*. They separately examined the repair of UV-induced damage in CHO cells in two contexts: repair in a transcribed gene, dihydrofolate reductase *(DHFR)*, and genome wide repair. *DHFR* repair was unaffected when cells were treated either with a combination of camptothecin and a topoisomerase II inhibitor, a complex-stabilizing inhibitor (etoposide), or a catalytic inhibitor (merbarone). Genome wide repair was diminished by all three inhibitors. These workers subsequently reported a minimal effect on genome wide repair by topoisomerase inhibitors, but repair of UV damage in the transcribed strand of *DHFR* was inhibited by camptothecin in combination with the topoisomerase II inhibitors merbarone or etoposide *(79)*. No significant inhibition of repair was noted when only a single inhibitor was used. A similar inhibition was observed when merbarone alone was used in a cell line with reduced topoisomerase I *(79)*. These data suggest that either topoisomerase I or topoisomerase II is required for gene-specific repair, and that inhibition of either enzyme alone is insufficient to inhibit repair. Inhibition of both enzymes appears to affect repair of UV damage significantly. The same authors found that no combination of topoisomerase inhibitors significantly affected repair of cisplatin or nitrogen mustard adducts from DNA *(79)*.

Thielmann et al. *(84)* also showed that merbarone was ineffective at inhibiting incision at UV lesions, but it inhibited DNA repair synthesis at concentrations similar to that required to inhibit the enzyme in vivo. Provided that merbarone is a specific topoisomerase II inhibitor, these results suggest that topoisomerase II has some role in cell survival following DNA damage, though it is not specifically indicative of a role in repair.

Although all these experiments with specific topoisomerase inhibitors are a clear improvement from earlier experiments using nonspecific topoisomerase inhibitors, the results are still equivocal, especially since many of the results depend on the use of inhibitors that stabilize covalent complexes. A similar criticism can be applied to studies that have shown synergistic cell killing with camptothecins and other DNA-damaging agents, such as ionizing radiation *(5)*. Synergistic cell killing may result from saturation of a repair pathway. The antitopoisomerase agent, because it leads to more DNA damage, can cause a large increase in cell killing, even though topoisomerases do not participate in repair. Nonetheless, the effects of merbarone alone on repair suggest that topoisomerase II may influence DNA repair following DNA damage.

6.2. Resistance to Alkylating Agents Correlates with Increased Topoisomerase II Levels

An alternate approach to understanding the roles of specific enzymes in repair is to study mammalian cell lines selected for enhanced resistance to DNA-damaging agents. If topoisomerases play important roles in repair, then cell lines that are resistant to DNA-damaging agents might have elevated topoisomerase levels. Tan et al. *(83)* found that a nitrogen mustard-resistant Burkitt's lymphoma cell line had decreased nitrogen mustard-induced interstrand crosslinks compared to wild-type cells and a threefold elevation in topoisomerase II activity. This cell line is hypersensitive to topoisomerase II inhibitors, such as etoposide, a result that would be expected if topoisomerase II were overexpressed. The authors also found the cell line to be hypersensitive to novobiocin. If novobiocin were a selective catalytic inhibitor of topoisomerase II, then overexpression of topoisomerase II should lead to novobiocin resistance. Clearly, this cell line contains other important alterations.

Recently, Barret et al. *(2)* observed a similar increase in topoisomerase II activity in a cell line selected for resistance to cisplatin. In contrast, Bramson et al. *(7)* measured topoisomerase activity in lymphocytes from B-cell chronic lymphocytic leukemia patients and found no difference in topoisomerase activity from the lymphocytes of patients who failed to respond to nitrogen mustard, even though the lymphocytes had a three- to fivefold elevated resistance to nitrogen mustard.

To test directly whether topoisomerase II overexpression can confer resistance to alkylating agents, Eder et al. *(23)* ectopically expressed Chinese hamster topoisomerase

II α in a mouse mammary tumor cell line, and observed enhanced resistance to nitrogen mustards and cisplatin. Although these results seem to support a role for topoisomerase II in providing resistance to alkylating agents, the authors could only obtain topoisomerase II overexpression in serum-starved cells. It has proven to be very difficult to achieve overexpression of topoisomerase II in proliferating mammalian cells. It is not clear why enhanced expression of topoisomerase II could lead to the observed resistance to DNA-damaging agents. It would be desirable to examine the effect of overexpression of topoisomerase II on DNA repair using a more tightly regulated expression system.

6.3. Direct Assays for DNA Damage Sensitivity in Eukaryotic Topoisomerase Mutants

Yeast topoisomerase mutants are an excellent eukaryotic system for assessing potential roles of topoisomerases in DNA repair. Neither *top1* nor *top2*^{ts} mutants have enhanced sensitivity to MMS, UV, or ionizing radiation *(74)*. However, since topoisomerase II is essential for viability, it was not possible to assay cell survival following DNA damage in the complete absence of topoisomerase II activity.

To overcome this potential difficulty, two approaches were taken. First, the results described above with mammalian cells suggested that higher levels of topoisomerase II should lead to enhanced survival following DNA damage. When topoisomerase II was overexpressed about 10- to 20-fold in yeast, no difference in sensitivity to MMS, UV or ionizing radiation was observed *(60,74)*. Second, the induction of point mutations was examined in *top2*^{ts} mutant cells held at the nonpermissive temperature during and after exposure to MMS. *top2*^{ts} mutant cells are mutable by MMS, but the induced mutation frequency does not differ from isogenic wild-type cells *(74)*. Taken together, these results indicate that topoisomerases do not play an essential role in the repair of MMS-induced DNA damage in yeast *(60)*.

6.4. Repair of DNA Damage Owing to the Presence of Antitopoisomerase Agents

Although drugs that stimulate cleavage by DNA topoisomerases may be of uncertain value in determining the roles of topoisomerases in DNA repair and DNA damage tolerance, antitopoisomerase agents are fascinating reagents for the introduction of lesions into DNA. Cytotoxicity by these agents requires cellular processes, such as replication or transcription. The genomic insults produced are DNA damage at sites of ongoing DNA metabolism, e.g., in front of a replication fork or a transcribing RNA polymerase.

Early studies in *E. coli* had suggested that SOS induction by antitopoisomerase agents played an important role in cell killing *(43)*. Subsequent work suggested that this SOS dependence was owing to the use of *E. coli* strains that lack the *lon* protease. The absence of the *lon* protease prevents the deactivation of the *sulA* protein, a component of the SOS response that induces filamentation following DNA damage (*14,39, see* Chapter 7, vol. 1). Studies with sets of defined mutations have made clear that recombinational repair is critical for cell survival following exposure to quinolones. *recA* mutants with defective recombinational repair are hypersensitive to quinolones, whereas mutations in *recA* or other proteins that abolish SOS induction without affecting recombination have wild-type sensitivities to quinolones *(69)*. Although SOS induction is therefore not the key factor in cell killing by fluoroquinolones, the induction of point mutations by either nalidixic acid or more potent quinolones, such as ciprofloxacin and norfloxacin, appears to depend on SOS induction. For example, reversion of His$^-$ mutations by quinolones depends on a functional SOS system *(51)*.

Eukaryotic cells appear to require similar DNA repair pathways for survival following exposure to antitopoisomerase agents. Yeast *rad52*, *rad50*, or *rad54* mutants are all hypersensitive to both camptothecin and complex stabilizing inhibitors of topoisomerase II *(25,61,62)*. By contrast, *rad52* mutants are only modestly sensitive to catalytic inhibitors of topoisomerase II *(40)*. Similarly, CHO cells defective in the repair of damage owing to ionizing radiation are hypersensitive to topoisomerase inhibitors *(15)*, as are cells derived from xeroderma pigmentosum patients *(64)*.

Other DNA repair pathways appear to play less of a role in cell survival following exposure to antitopoisomerase agents. Yeast mutants lacking *rad2* or *rad6* do not show increased sensitivity to either camptothecin or antitopoisomerase II agents *(62)*. These results suggest that the most critical form of DNA damage induced by antitopoisomerase agents is a DSB, such as the secondary DSBs induced by replication. At present, it is not known whether the DNA/topoisomerase covalent complex is also a substrate for repair. If it is, such repair pathways would need to deal with a unique type of lesion: a protein covalently attached to DNA. Additional work on mutants that are resistant or hypersensitive to antitopoisomerase agents should be informative on this point.

6.5. DNA Topoisomerases and DNA Damage Tolerance

If topoisomerases are not required for DNA repair, do they have any influence on cell survival following DNA damage? A possible answer to this question comes from studies on the enzymatic mechanisms of topoisomerases. DNA topoisomerase I is able to cleave single-stranded DNA or a double-stranded DNA molecule containing a single-stranded region *(1,35)*. When topoisomerase I cleaves single-stranded DNA, a covalent complex on the 3'-phosphate of the cleaved DNA is formed, and the enzyme does not adhere tightly to the other strand (carrying a 5'-OH). If the enzyme cleaves near a single-stranded region of a double-stranded DNA molecule, a DSB can be produced (Fig. 3). Westergaard and colleagues have demonstrated this using oligonucleotides that are partially double-stranded and that carry a strong topoisomerase I recognition site (reviewed in *1*). This molecule can be efficiently cleaved by topoisomerase I, producing a defined DSB. The DNA molecule carrying topoisomerase I is competent for carrying out a strand transfer reaction to a free 5'-OH *(1,10,35)*. Thus, the action of topoisomerase I on single-stranded gaps of DNA could produce DSBs, as well as a substrate for DNA rearrangements *(8)*. A similar reaction has been demonstrated for topoisomerase II *(30,31)*, and may also be competent for mediating DNA rearrangements.

To examine whether topoisomerases might exacerbate DNA damage, the effects of topoisomerase I overexpression in yeast cells exposed to DNA damage were examined. Cells that are expressing elevated level of *TOP1* from the yeast *GAL1* promoter are considerably more sensitive to MMS than wild-type cells (Fig. 4). Similar results were observed with cells exposed to ionizing radiation or UV light. These results sug-

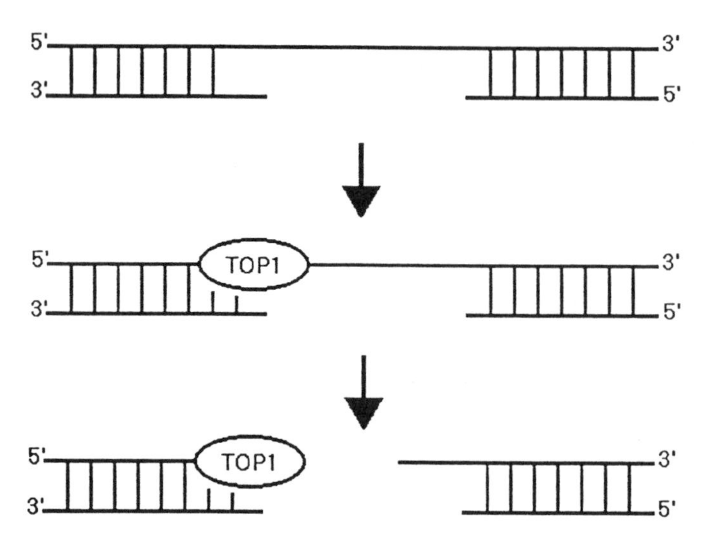

Fig. 3. A model for the generation of DSBs by topoisomerase I. DNA damage and its repair generate nicks and gaps in DNA. The action of topoisomerase I at the junction of single- and double-stranded DNA can result in a DSB *(1)*. This reaction is analogous to model reactions with partially double-stranded oligonucleotides *(82)*. The DNA molecule with topoisomerase I covalently attached to DNA may represent a difficult lesion for the cell to repair. Alternately, topoisomerase I can catalyze illegitimate recombination by religating to any fragment with a free 3'-OH *(35)*.

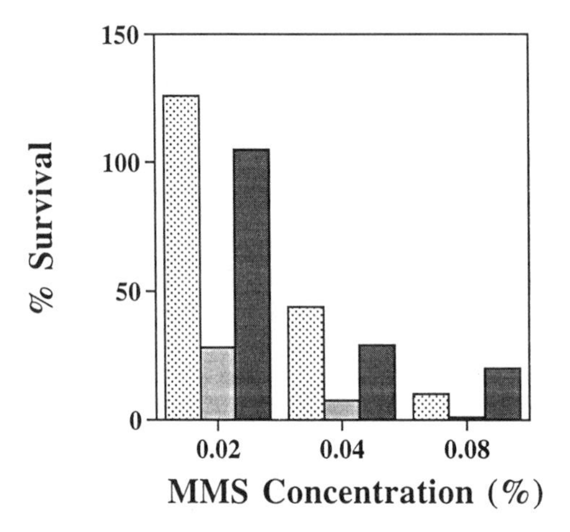

Fig. 4. Overexpression of topoisomerase I sensitizes cells to MMS. Yeast cells were treated with MMS at the indicated concentrations for 3 h in growth media. The cells expressed either wild-type levels of topoisomerases, or overexpressed topoisomerase I (from the yeast *GAL1* promoter) or overexpressed topoisomerase II (from the yeast *DED1* promoter) After MMS exposure, cells were diluted and plated to determine viability. Survival is shown relative to the beginning of exposure. Cells overexpressing topoisomerase I are hypersensitive to MMS compared to cells overexpressing topoisomerase II or cells expressing wild-type levels of topoisomerases. Similar results were obtained when the cells were exposed for <3 h (not shown). □ wild-type; ■ *TOP1* overexpression; ■ *TOP2* overexpression.

gest that yeast mutants that lack topoisomerase I should be more resistant to DNA damage than isogenic *TOP1* mutants. This is indeed the case *(74)*.

A similar sensitization to DNA damage was not observed when topoisomerase II was overexpressed from the yeast *DED1* promoter. There are two possible reasons for this observation. First, the *DED1* promoter is considerably weaker than the *GAL1* promoter. In fact, overexpression of yeast *TOP2* from the *GAL1* promoter is lethal in the absence of DNA damage. A more interesting possibility is that high levels of topoisomerase II in the presence of DNA damage may be less deleterious than high levels of topoisomerase I.

There also may be alterations in topoisomerase activity at sites of DNA damage. Osheroff and colleagues have observed that topoisomerase II is strongly inhibited by photoproducts in DNA *(13)*, and a similar observation with *E. coli* topoisomerase I was made by Pedrini and Ciarrocchi *(68)*. This inhibition of activity would seem to require enhanced levels of topoisomerase II, in agreement with the studies cited above. However, Kingma et al. *(46)* have recently observed that topoisomerase II cleavage is strongly enhanced at abasic sites in DNA. This result suggests that abasic sites can mimic the action of antitopoisomerase agents, and that topoisomerase II may also be deleterious in the presence of DNA damage.

Lanza and colleagues have recently observed that topoisomerase I can also form stable covalent complexes at sites of DNA damage *(49)*. UV-irradiated DNA accumulates higher levels of topoisomerase I covalent complexes than unirradiated DNA, and the location of the complexes corresponds to the hotspots of UV photoproducts *(49)*. Muller and colleagues have made a similar observation by showing that UV-irradiation leads to a higher level of topoisomerase I covalent complexes in vivo in UV irradiated cells *(80)*. They did not observe increased levels of covalent complexes with topoisomerase II in irradiated cells. These results clearly suggest that DNA damage can lead to the formation of covalent complexes with either topoisomerase I or topoisomerase II. The enzyme involved may depend on the specific type of DNA damage.

Whether topoisomerases interfere with DNA repair or act directly at lesions, cells must be able to deal efficiently with topoisomerases in the presence of DNA damage. A plausible mechanism to prevent exacerbation of DNA damage by topoisomerases is posttranslational inactivation of the enzyme(s) by either modification or targeted degradation. In mammalian cells, poly(ADP-ribose) polymerase (PARP) is activated by DNA strand breaks (*see* Chapter 22), and a major substrate of the enzyme is topoisomerase I. Addition of poly(ADP-ribose) to topoisomerase I presumably inactivates the enzyme *(26)*, and if this modification occurs rapidly, inactivation of topoisomerase I could prevent the generation of lethal lesions. This type of model has been suggested for mammalian cells following exposure to DNA damage, and results by Boothman and colleagues suggest that topoisomerase I activity is reduced following DNA damage *(4)*. Yeast cells lack PARP, so if topoisomerase I is inactivated in the presence of DNA damage, another mechanism must be operating. Topoisomerase I activity is indeed rapidly lost when cells are exposed to MMS, suggesting that yeast cells have an efficient mechanism for inactivation of topoisomerase I following DNA damage *(74)*. Details of this mechanism await further studies.

7. PERSPECTIVES—TOPOISOMERASES AFFECT CELL KILLING BY DNA-DAMAGING AGENTS

There is little direct evidence that DNA topoisomerases are required for any DNA repair reaction. Nonetheless, alterations of the levels of topoisomerases are able to alter cellular sensitivity to DNA-damaging agents. In bacterial systems, it is clear that topoisomerases play important roles in regulating gene expression, so it is not surprising that genes required for DNA repair are also affected by topoisomerase activity. Still, even in prokaryotic systems, additional studies to assess possible direct roles of topoisomerases in DNA repair are needed.

There is an even less clear picture regarding topoisomerases in eukaryotic DNA repair. The results described above suggest that topoisomerases may be undesirable enzymes when cells are exposed to DNA damage. If these enzymes are deleterious when cells are exposed to DNA damage, then their activity must be attenuated to prevent the generation of lethal damage. A scenario that has some experimental support is that cells cope with DNA damage by inactivating and perhaps degrading topoisomerase I *(4,74)*. Since a DNA topoisomerase is required for other transactions, such as replication and transcription *(8)*, topoisomerase II may need to be induced in response to DNA damage. This would explain the observed higher levels of expression of topoisomerase II in cell lines resistant to alkylating agents. However, the results of Kingma and colleagues would suggest that topoisomerase II activity is also deleterious when DNA damage is present *(46)*. Perhaps topoisomerase II is the lesser of two evils, in cells with damaged DNA.

An examination of sites of illegitimate recombination, in both yeast and mammalian cells, has found topoisomerase I consensus cleavage sites at the recombination junction. Illegitimate recombination may represent one consequence of DNA damage in the presence of topoisomerase activity. Given the importance of translocations in the activation of cellular oncogenes, studies of the roles of topoisomerases in DNA damage tolerance may yield valuable insights into this mode of carcinogenesis.

ACKNOWLEDGMENTS

This chapter is dedicated to James C. Wang on the occasion of his 60th birthday. His influence in the field has made topoisomerase studies an enjoyable as well as a rewarding pursuit. I gratefully acknowledge the contributions of Angela Rose and Karin Sykes to the work described in this chapter. Research in my laboratory has been supported by grant CA52814 from the National Cancer Institute and the American Lebanese Syrian Associated Charities (ALSAC).

REFERENCES

1. Andersen A. H., J. Q. Svejstrup, and O. Westergaard. 1994. The DNA binding, cleavage, and religation reactions of eukaryotic topoisomerases I and II. *Adv. Pharmacol.* **29:** 83–101.
2. Barret, J. M., P. Calsou, A. K. Larsen, and B. Salles. 1994. A cisplatin-resistant murine leukemia cell line exhibits increased topoisomerase II activity. *Mol. Pharmacol.* **46:** 431–436.
3. Bodley, A. L. and T. A. Shapiro. 1995. Molecular and cytotoxic effects of camptothecin, a topoisomerase I inhibitor, on trypanosomes and Leishmania. *Proc. Natl. Acad. Sci. USA* **92:** 3726–3730.

4. Boothman D. A., N. Fukunaga, and M. Wang. 1994. Down-regulation of topoisomerase I in mammalian cells following ionizing radiation. *Cancer Res.* **54:** 4618–4626.
5. Boothman D. A., M. Wang, R. A. Schea, H. L. Burrows, S. Strickfaden, and J. K. Owens. 1992. Posttreatment exposure to camptothecin enhances the lethal effects of x-rays on radioresistant human malignant melanoma cells. *Int. J. Radiat. Oncol. Biol. Phys.* **24:** 939–48.
6. Bouayadi, K., G. Villani, and B. Salles. 1993. Resistance to cisplatin in an *E. coli* B/r Nal^R mutant. *Mutat. Res.* **294:** 77–87.
7. Bramson J., A. McQuillan, and L. C. Panasci. 1995. DNA repair enzyme expression in chronic lymphocytic leukemia vis-a-vis nitrogen mustard drug resistance. *Cancer Lett.* **90:** 139–148.
8. Brill, S. J., S. DiNardo, K. Voelkel-Meiman, and R. Sternglanz. 1987. Need for DNA topoisomerase activity as a swivel for DNA replication and for transcription of ribosomal RNA. *Nature* **326:** 414–416.
9. Chao L. and D. M. Tillman. 1982. Enhanced resistance to nitrosoguanidine killing and mutagenesis in a DNA gyrase mutant of *Escherichia coli. J. Bacteriol.* **151:** 764–770.
10. Christiansen K, A. B. Svejstrup, A. H. Andersen, and O. Westergaard. 1993. Eukaryotic topoisomerase I-mediated cleavage requires bipartite DNA interaction. Cleavage of DNA substrates containing strand interruptions implicates a role for topoisomerase I in illegitimate recombination. *J. Biol. Chem.* **268:** 9690–9701.
11. Christman, M. F., F. S. Dietrich, and G. R. Fink. 1988. Mitotic recombination in the rDNA of *S. cerevisiae* is suppressed by the combined action of DNA topoisomerase I and II. *Cell* **55:** 413–425.
12. Cotten, M., D. Bresnahan, S. Thompton, L. Sealy, and R. Chalkly. 1986. Novobiocin precipitates histones at concentrations normally used to inhibit eukaryotic type II topoisomerase. *Nucleic Acids Res.* **14:** 3671–3686.
13. Corbett A. H., E. L. Zechiedrich, R. S. Lloyd, and N. Osheroff. 1991. Inhibition of eukaryotic topoisomerase II by ultraviolet-induced cyclobutane pyrimidine dimers. *J. Biol. Chem.* **266:** 19,666–19,671.
14. D'Ari R. and O. Huisman. 1983. Novel mechanism of cell division inhibition associated with the SOS response in *Escherichia coli. J. Bacteriol.* **156:** 243–250.
15. Davies, S. M., S. L. Davies, A. L. Harris, and I. D. Hickson. 1989. Isolation of two Chinese hamster ovary cell mutants hypersensitive to topoisomerase II inhibitors and cross-resistant to peroxides. *Cancer Res.* **49:** 4526–4530.
16. de Jong, S., J. G. Zijlstra, N. H. Mulder, and E. G. de Vries. 1991. Lack of cross-resistance to fostriecin in a human small-cell lung carcinoma cell line showing topoisomerase II-related drug resistance. *Cancer Chemother. Pharmacol.* **28:** 461–464.
17. DiNardo, S., K. Voelkel, and R. Sternglanz. 1984. DNA topoisomerase II mutant of *Saccharomyces cerevisiae* topoisomerase II is required for the segregation of daughter molecules at the termination of DNA replication. *Proc. Natl. Acad. Sci. USA* **81:** 2616–2620.
18. Downes, C. S. and R. T. Johnson. 1988. DNA Topoisomerases and DNA repair. *Bioessays* **8:** 179–184.
19. Downes, C. S., A. M. Mullinger and R. T. Johnson. 1987. Action of etoposide (VP-16-123) on human cells: no evidence for topoisomerase II involvement in excision repair of u. v.-induced DNA damage, nor for mitochondrial hypersensitivity in ataxia telangiectasia. *Carcinogenesis* **8:** 1613–1618.
20. Downes C. S., M. J. Ord, A. M. Mullinger, A. R. Collins, and R. T. Johnson. 1985. Novobiocin inhibition of DNA excision repair may occur through effects on mitochondrial structure and ATP metabolism, not on repair topoisomerases. *Carcinogenesis* **6:** 1343–1352.
21. Drake, F. H., G. H. Hofmann, S.-H. Mong, J. O. Bartus, R. P. Hertzberg, R. K. Johnson, M. R. Mattern, and C. K. Mirabelli. 1989. *In vitro* and intracellular inhibition of topoisomerase II by the antitumor agent merbarone. *Cancer Res.***49:** 2578–2583.

22. Drlica, K. and R. J. Franco. 1988 Inhibitors of DNA topoisomerases. *Biochemistry* **27:** 2253–2259.
23. Eder, J. P., V. T.-W. Chan, S.-W. NG, N. A. Rizvi, S. Zacharoulis, B. A. Teicher, and L. E. Schipper. 1995. DNA topoisomerase II α expression is associated with alkylating agent resistance. *Cancer Res.* **55:** 6109–6116.
24. Elsea, S. H., N. Osheroff, and J. L. Nitiss. 1992. Cytotoxicity of quinolone toward eukaryotic cells. *J. Biol. Chem.* **267:** 13,150–13,153.
25. Eng, W.-K., Faucette, L., Johnson, R. K., and Sternglanz, R. 1988. Evidence that DNA topoisomerase I is necessary for the cytotoxic effects of camptothecin. *Mol. Pharmacol.* **34:** 755–760.
26. Ferro, A. M., M. C. McElwain, and B. M. Olivera. 1984. Poly(ADP-ribosylation) of DNA topoisomerase I: a nuclear response to DNA-strand interruptions. *Cold Spring Harb. Symp. Quant. Biol.* **49:** 683–690.
27. Fishel R. A. and R. Kolodner. 1984. Escherichia coli strains containing mutations in the structural gene for topoisomerase I are recombination deficient. *J. Bacteriol.* **160:** 1168–1170.
28. Fostel, J. and D. Montgomery. 1995. Identification of the aminocatechol A–3253 as an in vitro poison of DNA topoisomerase I from *Candida albicans*. *Antimicrob. Agents Chemother.* **39:** 586–592.
29. Froelich-Ammon, S. J. and N. Osheroff. 1995. Topoisomerase poisons: harnessing the dark side of enzyme mechanism. *J. Biol. Chem.* **270:** 21,429–21,432.
30. Gale K. C. and N. Osheroff. 1990. Uncoupling the DNA cleavage and religation activities of topoisomerase II with a single-stranded nucleic acid substrate evidence for an active enzyme-cleaved DNA intermediate. *Biochemistry* **29:** 9538–9545.
31. Gale K. C. and N. Osheroff. 1992. Intrinsic intermolecular DNA ligation activity of eukaryotic topoisomerase II. Potential roles in recombination. *J. Biol. Chem.* **267:** 12,090–12,097.
32. Goto, T., P. Laipis, and J. C. Wang. 1982. The purification and characterization of DNA topoisomerases I and II of the yeast *Saccharomyces cerevisiae*. *J. Biol. Chem.* **259:** 10,422–10,429.
33. Goto, T. and J. C. Wang. 1984. Yeast DNA topoisomerase II is encoded by a single-copy, essential gene. *Cell* **36:** 1073–1080.
34. Goto, T. and J. C. Wang. 1985. Cloning of yeast *TOP1*, the gene DNA encoding DNA topoisomerase I, and construction of mutants defective in both DNA topoisomerase I and topoisomerase II. *Proc. Natl. Acad. Sci. USA* **82:** 7178–7182.
35. Halligan B. D., J. L. Davis, K. A. Edwards, and L. F. Liu. 1992. Intra- and intermolecular strand transfer by HeLa DNA topoisomerase I. *J. Biol. Chem.* **257:** 3995–4000.
36. Hays J. B. and S. Boehmer. 1978. Antagonists of DNA gyrase inhibit repair and recombination of UV-irradiated phage lambda. *Proc. Natl. Acad. Sci. USA* **75:** 4125–4129.
37. Holm, C., T. Goto, J. C. Wang, and D. Botstein. 1985. DNA topoisomerase II is required at the time of mitosis in yeast. *Cell* **41:** 553–563.
38. Howard, M. T., S. H. Neece, S. W. Matson, and K. N. Kreuzer. 1994. Disruption of a topoisomerase-DNA cleavage complex by a DNA helicase. *Proc. Natl. Acad. Sci. USA* **91:** 12,031–12,035.
39. Huisman, O., R. D'Ari, and S. Gottesman. 1984. Cell-division control in *Escherichia coli*: specific induction of the SOS function SfiA protein is sufficient to block septation. *Proc. Natl. Acad. Sci. USA* **81:** 4490–4494.
40. Ishida, R., M. Hamatake, R. Wasserman, J. L. Nitiss, J. C. Wang, and T. Andoh. 1995. DNA topoisomerase II as the molecular target of bisdioxopiperazine derivatives ICRF-159 and ICRF-193 in budding yeast *Saccharomyces cerevisiae*, *Cancer Res.* **55:** 2299–2303.
41. Jensen, P. B., B. S. Sorensen, M. Sehested, E. J. Demant, E. Kjeldsen, E. Friche, and H. H. Hansen. 1993. Different modes of anthracycline interaction with topoisomerase II. Sepa-

rate structures critical for DNA-cleavage and for overcoming topoisomerase II-related drug resistance. *Biochem. Pharmacol.* **45:** 2025–2035.

42. Jones, J. C., T. Stevnsner, M. R. Mattern, and V. A. Bohr. 1991. Effect of specific enzyme inhibitors on replication, total genome DNA repair and on gene-specific DNA repair after UV irradiation in CHO cells. *Mutat. Res.* **255:** 155–162.
43. Kantor, G. J. and R. A. Deering. 1968. Effect of nalidixic acid and hydroxyurea on division ability of *Escherichia coli* fil^+ and lon^- strains. *J. Bacteriol.* **95:** 520–530.
44. Kato, J., Y. Nishimura, R. Imamura, H. Niki, S. Hiraga, and H. Suzuki. 1990. New topoisomerase essential for chromosome segregation in *E. coli*. *Cell* **63:** 393–404.
45. Kim R. A. and J. C. Wang. 1989. A subthreshold level of DNA topoisomerases leads to the excision of yeast rDNA as extra-chromosomal rings. *Cell* **57:** 975–985.
46. Kingma, P. S., A. H. Corbett, P. C. Burcham, L. J. Marnett, and N. Osheroff. 1995. Abasic sites stimulate double-stranded DNA cleavage mediated by topoisomerase II. DNA lesions as endogenous topoisomerase II poisons. *J. Biol. Chem.* **270:** 21,441–21,444.
47. Kreuzer K. and N. Cozzarelli. 1979. *Escherichia coli* mutants thermosensitive for DNA gyrase subunit A: effects on DNA replication, transcription and bacteriophage growth. *J. Bacteriol.* **140:** 425–430.
48. Laemmli, U. K., E. Kas, L. Poljak, and Y. Adachi. 1992. Scaffold-associated regions: cis-acting determinants of chromatin structural loops and functional domains. *Curr. Opin. Genet. Dev.* **2:** 275–285.
49. Lanza, A., S. Tornaletti, C. Rodolfi, M. Scanavini, and A. Pedrini. 1996. Human DNA topoisomerase I-mediated cleavages stimulated by ultraviolet light-induced DNA damage. *J. Biol. Chem.* **271:** 6978–6986.
50. Legerski, R. J., J. E. Penkala, C. A. Peterson, and D. A. Wright. 1987. Repair of UV-induced lesions in *Xenopus laevis* oocytes. *Mol. Cell. Biol.* **7:** 4317–4323.
51. Lewin, C. S., B. M. Howard, N. T. Ratcliffe, and J. T. Smith. 1989. 4-quinolones and the SOS response. *J. Med. Microbiol.* **29:** 139–144.
52. Liu, L. F. 1989. DNA topoisomerase poisons as anti-tumor drugs. *Ann. Rev. Biochem.* **58:** 351–375.
53. Menichini, P., G. Fronza, S. Tornaletti, S. Galiegue-Zouitina, B. Bailleul, M. H. Loucheux-Lefebvre, A. Abbondandolo, and A. M. Pedrini. 1989. In vitro DNA modification by the ultimate carcinogen of 4-nitroquinoline–1-oxide: influence of superhelicity. *Carcinogenesis* **10:** 1589–1593.
54. Menzel, R. and M. Gellert. 1983. Regulation of the genes for *E. coli* DNA gyrase: homeostatic control of DNA supercoiling. *Cell* **34:** 105–113.
55. Miura-Masuda, A. and H. Ikeda. 1990. Spontaneous deletion by *recA*-independent recombination in vivo. *Mol. Gen. Genet.* **220:** 345–352.
56. Nelson, W. G. and M. B. Kastan. 1994. DNA strand breaks: the DNA template alterations that trigger p53-dependent DNA damage response pathways. *Mol. Cell. Biol.* **14:** 1815–1823.
57. Newport, J. and T. Spann. 1987 Disassembly of the nucleus in mitotic extracts: membrane vesicularization, lamin disassembly, and chromosome condensation are independent processes. *Cell* **48:** 219–230.
58. Nitiss, J. L. 1994. Roles of DNA topoisomerases in chromosomal replication and segregation *Adv. Pharmacol.* **29A:** 103–134.
59. Nitiss, J. L., Y. X. Liu, and Y. Hsiung. 1993. A temperature sensitive topoisomerase II allele confers temperature dependent drug resistance to amsacrine and etoposide: a genetic system for determining the targets of topoisomerase II inhibitors. *Cancer Res.* **53:** 89–93.
60. Nitiss, J. L., A. Rose, K. C. Sykes, J. Harris, and J. Zhou. 1996. Using yeast to understand drugs that target topoisomerases. *Ann. NY Acad. Sci.* **803:** 32–43.
61. Nitiss, J. and J. C. Wang. 1988. DNA topoisomerase-targeting drugs can be studied in yeast. *Proc. Natl. Acad. Sci. USA* **85:** 7501–7505.

62. Nitiss, J. and J. C. Wang. 1991. Yeast as a genetic system in the dissection of the mechanism of cell killing by topoisomerase-targeting anti-cancer drugs, in *DNA Topoisomerases and Cancer* (Potmesil, M. and K. Kohn, eds.), Oxford University Press, London, pp. 77–91.
63. Nitiss, J. L. and J. C. Wang. 1996. On the mechanism of cell killing by anti-topoisomerase agents. *Mol. Pharmacol.* **50:** 1095–1102,
64. Ohta S., M. Shimada, S. Matsukawa, T. Taga, and S. Yamazaki. 1990. Flow-cytometric analysis of DNA pattern of cells derived from xeroderma pigmentosum A-hypersensitivity to vincristine, etoposide and methotrexate. *Acta Paediatr. Jpn.* **32:** 262–268.
65. Osheroff, N. and A. Corbett. 1993. When good enzymes go bad: conversion of topoisomerase II to a cellular toxin by anti-neoplastic drugs. *Chem. Res. Toxicol.* **6:** 585–597.
66. Osheroff, N., E. L. Zechiedrich, and K. C. Gale. 1991. Catalytic function of DNA topoisomerase II. *Bioessays* **13:** 269–273.
67. Overbye, K. M. and P. Margolin. 1981. Role of the supX gene in ultraviolet light-induced mutagenesis in *Salmonella typhimurium. J. Bacteriol.* **146:** 170–178.
68. Pedrini, A. M. and G. Ciarrocchi. 1983. Inhibition of *Micrococcus luteus* DNA topoisomerase I by UV photoproducts. *Proc. Natl. Acad. Sci. USA* **80:** 1787–1791.
69. Piddock, L. J. and R. N. Walters. 1992. Bactericidal activities of five quinolones for *Escherichia coli* strains with mutations in genes encoding the SOS response or cell division. *Antimicrob. Agents Chemother.* **36:** 819–825.
70. Reece, R. J. and A. Maxwell. 1991. DNA gyrase: structure and function. *Crit. Rev. Biochem. Mol. Biol.* **26:** 335–375.
71. Roberge, M., C. Tudan, S. M. Hung, K. W. Harder, F. R. Jirik, and H. Anderson. 1994. Antitumor drug fostriecin inhibits the mitotic entry checkpoint and protein phosphatases 1 and 2A. *Cancer Res.***54:** 6115–6121.
72. Robinson, M. J., B. A. Martin, T. A. Gootz, P. R. McGuirk, and N. Osheroff. 1992. Effects of novel fluoroquinolones on the catalytic activities of eukaryotic topoisomerase II: influence of the C–8 fluorine group. *Antimicrob. Agents Chemother.* **36:** 751–756.
73. Roca, J., R. Ishida, J. M. Berger, T. Andoh, and J. C. Wang. 1994. Antitumor bisdioxopiperazines inhibit yeast DNA topoisomerase II by trapping the enzyme in the form of a closed protein clamp. *Proc. Natl. Acad. Sci. USA* **91:** 1781–1785.
74. Rose, A., K. C. Sykes, and J. L. Nitiss. Unpublished observations.
75. Roy, M., R. Ghosh, S. K. Dey, and S. B. Bhattacharjee. 1991. Response of V79 cells to *N*-methyl-*N*′-nitro-*N*-nitrosoguanidine (MNNG) treatment: inhibition of poly(ADP-ribose) and topoisomerase activity. *Mutat. Res.* **249:** 195–199.
76. Shen, L. L., J. Baranowski, J. Fostel, D. A. Montgomery, and P. A. Lartey. 1992. DNA topoisomerases from pathogenic fungi: targets for the discovery of antifungal drugs. *Antimicrob. Agents Chemother.* **36:** 2778–2784.
77. Smith, C. M., Z. Arany, C. Orrego, and E. Eisenstadt. 1992. Mutations in topA interfere with the inducible expression of DNA damage response loci in *Salmonella typhimurium. Environ. Mol. Mutagen.* **19:** 185–194.
78. Sternglanz R., S. DiNardo, K. A. Voelkel, Y. Nishimura, Y. Hirota, K. Becherer, L. Zumstein, and J. C. Wang. 1981 Mutations in the gene coding for *Escherichia coli* DNA topoisomerase I affect transcription and transposition. *Proc. Natl. Acad. Sci. USA* **78:** 2747–2751.
79. Stevnsner, T. and V. A. Bohr. 1993. Studies on the role of topoisomerases in general, gene- and strand-specific DNA repair. *Carcinogenesis* **14:** 1841–1850.
80. Subramanian, D., B. Rosenstein, and M. T. Muller. 1995. Solar UV-irradiation stimulates formation of endogenous topoisomerase I/DNA covalent complexes in human cells *Proc. Annu. Meet. Am. Assoc. Cancer Res.* **36:** A2696.
81. Sung, P., J. F. Watkins, L. Prakash, and S. Prakash. 1994. Negative superhelicity promotes ATP-dependent binding of yeast RAD3 protein to ultraviolet-damaged DNA. *J. Biol. Chem.* **269:** 8303–8308.

82. Svejstrup, J. Q., K. Christiansen, I. I. Gromova, A. H. Andersen, and O. Westergaard. 1991. New technique for uncoupling the cleavage and religation reactions of eukaryotic topoisomerase I. The mode of action of camptothecin at a specific recognition site. *J. Mol. Biol.* **222:** 669–678.
83. Tan K. B., M. R. Mattern, R. A. Boyce, and P. S. Schein. 1987. Elevated DNA topoisomerase II activity in nitrogen mustard-resistant human cells. *Proc. Natl. Acad. Sci. USA* **84:** 7668–7671.
84. Thielmann, H. W., O. Popanda, H. Gersbach, and F. Gilberg. 1993. Various inhibitors of DNA topoisomerases diminish repair-specific DNA incision in UV-irradiated human fibroblasts. *Carcinogenesis* **14:** 2341–2351.
85. Uemura, T., H. Ohkura, Y. Adachi, K. Morino, K. Shiozaki, and M. Yanagida. 1987. DNA topoisomerase II is required for condensation and separation of mitotic chromosomes in *S. pombe*. *Cell* **50:** 917–925.
86. Urios, A., G. Herrera, V. Aleixandre, and M. Blanco. 1991. Influence of *recA* mutations on *gyrA* dependent quinolone resistance. *Biochimie* **73:** 519–521.
87. Urios, A., G. Herrera, V. Aleixandre, and M. Blanco. 1991. Expression of the *recA* gene is reduced in *Escherichia coli* topoisomerase I mutants. *Mutat. Res.* **243:** 267–272.
88. von Wright A. and Bridges B. A. 1981. Effect of gyrB-mediated changes in chromosome structure on killing of *Escherichia coli* by ultraviolet light: experiments with strains differing in deoxyribonucleic acid repair capacity. *J. Bacteriol.* **146:** 18–23.
89. Wallis, J. W., G. Chrebet, G. Brodsky, M. Rolfe, and R. Rothstein. 1989. A hyper-recombination mutation in *Saccharomyces cerevisiae* identifies a novel eukaryotic topoisomerase. *Cell* **58:** 409–419.
90. Wang, J. C. 1991. DNA topoisomerases: Why so many? *J. Biol. Chem.* **266:** 6659–6662.
91. Wang, J. C. and A. S. Lynch. 1993. Transcription and DNA supercoiling. *Curr. Opin. Genet. Dev.* **3:** 764–768.
92. Willmott, C. J., S. E. Critchlow, I. C. Eperon, and A. Maxwell. 1994. The complex of DNA gyrase and quinolone drugs with DNA forms a barrier to transcription by RNA polymerase. *J. Mol. Biol.* **242:** 351–363.

24

Molecular Approaches for Detecting DNA Damage

Peggy L. Olive

1. INTRODUCTION

In 1966, McGrath and Williams *(51)* introduced alkaline sucrose gradient sedimentation as a method to measure DNA single-strand breaks (SSBs) in irradiated cells. This technique was regarded as an important innovation, and was soon in widespread use for examining DNA damage caused by a variety of agents, as well as for measurement of the kinetics of strand break rejoining as an indicator of DNA "repair." Alkaline sucrose gradient sedimentation remains somewhat unique in that the physical principle behind the measurement of DNA damage is well understood. Unfortunately, the original method proved insensitive to detection of strand breaks produced by "biologically relevant" doses in mammalian cells, largely because conditions required for "ideal" sedimentation required that DNA already contain a significant number of strand breaks. Ability to analyze only a few samples at a time and the long time required for the assay were also viewed as serious limitations.

These limitations provided the impetus for the development of a variety of novel assays for detecting DNA strand breaks. Methods in current use are 100–1000 times more sensitive than the original McGrath and Williams method for measuring SSBs, and speed and sample handling capacity have improved considerably. Many of the methods used to detect SSBs have been adapted for measurement of the less abundant DNA double-strand breaks (DSBs), often considered more relevant than SSBs for chromosome aberrations and cell killing.

The ability to detect SSBs with excellent sensitivity led to modifications of the original methods to allow indirect detection of a variety of other DNA damages, including DNA–protein crosslinks, interstrand crosslinks, and base damage. When damage assays were combined with methods to detect specific genes, preferential gene repair became appreciated (*see* ref. *12* for review), and was later associated with a specific DNA repair disorder *(55,110)*. Methods have recently been developed to quantify SSBs or DSBs at the level of the individual cell so that heterogeneity in response within a population can now be measured (*see* ref. *23* for review). Several authors have reviewed specific DNA damage assays in more detail than is possible here *(2,30,35,43,76,79)*.

Table 1 summarizes the large variety of factors believed to contribute to the response of a cell to a DNA-damaging agent. Any single method for measuring DNA damage is generally limited to examining only one of these factors. Moreover, the relative contri-

From: DNA Damage and Repair, Vol. 2: DNA Repair in Higher Eukaryotes
Edited by: J. A. Nickoloff and M. F. Hoekstra © *Humana Press Inc., Totowa, NJ*

Table 1
Factors Important in Cell Sensitivity to DNA-Damaging Agents

Initial amount, location, and distribution of DNA damage (chromatin structure, transcriptional status, metabolic status)
Initial types of DNA damage (strand breaks, crosslinks, base damage, abasic sites)
Rate and extent of repair of lesions (lesion accessibility, recognition, tolerance)
Accuracy of repair of lesions (type of repair process, efficiency of repair enzymes)
Global cell response to DNA damage (apoptosis, gene induction, cell-cycle block, cell–cell interactions, genetic instability)

bution of each of these factors can vary depending on cell type, DNA-damaging agent, or cell environment. For these reasons, determining the significance of any one factor in the response of a cell to a DNA-damaging agent can be a considerable challenge. Mutant cell lines deficient in one aspect of repair have been valuable in determining the importance of specific genes and pathways involved in cell recovery. Molecular methods are also evolving in an effort to keep pace with the rapidly expanding knowledge in areas of lesion distribution, influence of chromatin organization, and fidelity of DNA repair. Each of the various methods for measuring DNA damage will be reviewed with particular emphasis on the measurement of damage following exposure of mammalian cells to ionizing radiation.

2. DNA STRAND BREAKS

2.1. DNA SSBs

SSBs are produced directly by a large variety of genotoxic agents, including ionizing radiation, oxidizing agents, topoisomerase I inhibitors, alkylating agents, and anthracyclines. Many techniques for measuring SSBs have evolved, as summarized in Table 2. To detect SSBs, most assays require denaturation of the duplex DNA, which is easily accomplished by using alkali. The size or number of DNA pieces can be detected directly or indirectly, and the chemical nature of the end groups is often not important. An exception is the nick translation method where broken ends of DNA molecules may serve as substrates for DNA polymerase *(25,57)*. Incorporation of radiolabeled bases, or biotin-dUTP followed by streptavidin-conjugated to fluorescein can subsequently be measured by autoradiography or flow cytometry.

DNA will begin to denature at about pH 11.6, but complete denaturation of undamaged DNA generally requires exposure to NaOH at a pH >12.3. Although loss of hydrogen bonding is effectively instantaneous at this pH, unwinding and separation of duplex DNA are a relatively slow process. Ahnstrom and Erixon *(3)* estimate that 20 min of alkali treatment are required to unwind DNA containing 1 break every 2×10^9 Da. The time required for complete lysis for alkaline sucrose gradient sedimentation is therefore quite long (8–24 h). Alkali elution requires a shorter lysis

Table 2
Methods Used to Detect DNA Strand Breaks in Mammalian Cells

Assay	Sensitivity[a] (breaks per cell)	Approx time required	Expense	Effort[b]	Refs.
Assays used to detect SSB					
Zonal rotor gradient sedimentation	50–1000	8 h	Moderate	High	*106*
Alkaline elution	100–5000	24 h	Moderate	High	*41*
Alkaline unwinding	50–50,000	4 h	Low	Moderate	*3,88*
Alkaline DNA precipitation	500–25,000	2 h	Low	Low	*62*
Nucleoid sedimentation	100–20,000	2 h	Low	Moderate	*19*
Halo assay	200–15,000	2 h	Moderate	Moderate	*85,103*
Alkaline comet assay	50–15,000	6 h	Moderate	Moderate	*23,67*
Nick translation	250–20,000	2 h	Low	Moderate	*25,57,97*
Assays used to detect DSB					
Neutral filter elution	125–2500	24 h	Moderate	High	*41*
Neutral sucrose gradient sedimentation	125–5000	96 h	Moderate	High	*8*
Neutral gel electrophoresis	125–2500	24–96 h	Low	Low	*1,10,99*
Neutral comet assay	125–7500	24 h	Moderate	Moderate	*72*

[a]Estimates of range of sensitivity are based on response to ionizing radiation, with 1 Gy producing 1000 SSBs and 25 DSBs/diploid cell.

[b]Effort reflects technical proficiency required and amount of time required per sample.

period, but the flow of elution fluid through the filter effectively increases the rate of strand denaturation. The alkali unwinding assay, in which each SSB becomes a swivel for DNA unwinding, makes use of the kinetics of unwinding as the basis for detecting SSBs; lysis times are subsequently reduced to minutes in this assay. The alkali DNA precipitation assay relies on detergent solubilization of cell membranes and proteins followed by salt precipitation of high-mol-wt DNA *(62)*. This technique also requires only minutes for lysis, but repeated heating and precipitation steps appear to accelerate DNA strand separation. For most strand break assays, DNA has generally been detected by radiolabeling DNA. However, fluorescent detection methods (e.g., *7*) and immunochemical detection of single-stranded regions *(107)* have also been used. Detection of DNA damage in nonproliferating cells can also be accomplished by hybridizing recovered DNA with radiolabeled sequence-specific probes *(14)*.

When using strong alkali to separate the two strands of DNA, some types of base damage and abasic sites can also be converted to SSBs (so-called alkali-labile lesions). The "nucleoid" assays avoid this problem by measuring relaxation of DNA supercoiling by SSBs examined under neutral conditions. Cells are placed in a solution containing an nonionic detergent (generally 0.5% Triton X-100), 2*M* NaCl to remove histones, and a fluorescent DNA binding stain like propidium iodide. The DNA "halos" that are observed using a fluorescence microscope consist of 50–100 kbp DNA loops attached to a nuclear protein matrix. The change in the diameter of these structures in response to DNA damage is dependent on the number of strand breaks, and on the concentration of propidium iodide: low concentrations relax DNA supercoils, and high concentrations induce rewinding by creating negative supercoils. This effect can be observed microscopically *(18,86)*, by examining sedimentation of nucleoids in neutral sucrose gradients *(19)*, or by analysis of forward light scatter of nucleoids using a flow cytometer *(52)*. Damage detected by these assays is also influenced by higher-order DNA structure (*see* Section 5.1.).

2.2. DNA DSBs

Direct double-strand breaking agents, which are relatively uncommon among genotoxic agents, include ionizing radiation, bleomycin, topoisomerase II inhibitors, and carcinostatin. Damage can also be mimicked by restriction enzymes (blunt and staggered cutting). Since DSBs are also correlated more closely than SSBs to chromosome damage and cell killing by ionizing radiation, a major focus in recent years has been to compare initial numbers of DSBs, their kinetics of rejoining, and residual numbers of DSBs with cell response to ionizing radiation measured using clonogenic assays. The expectation is that measures of DNA damage or repair would be simpler and more rapid to perform than clonogenic assays for predicting the intrinsic radioresistance of tumors and normal tissues. Results from a number of studies have been somewhat contradictory (e.g., *71*), although there seems to be a developing consensus that strand break assays lack the sensitivity for detecting small differences in radiosensitivity between human tumor and normal cells, and that a measure of residual damage could be a more useful end point. In general, the same intensity of effort has not been attempted for monitoring the DNA-damaging effects of chemotherapy agents.

A variety of methods have been developed to detect DSBs (Table 2). Although the sensitivity for detecting damage is similar among these assays, they differ significantly in ease of performance. The advent of pulsed-field gel electrophoresis (PFGE) as a method to examine larger-mol-wt DNA *(91)* led to the application of this technique for measuring DSBs in mammalian cells *(1,9,99)*. Constant field gel electrophoresis fails to separate DNA pieces on the basis of size once they exceed about 50 kb. The theory behind PFGE is that by alternating the direction of the electric field at regular intervals, DNA is forced to reorient in the new field direction, and smaller DNA pieces will reorient faster than larger ones because of the physical resistance of the agarose matrix. The result is a far superior separation of large-mol-wt DNA according to size. This method has many technical advantages over other methods used to measure DSB, including good reproducibility and large sample handling capacity. To perform the assay, about 10^6 cells are embedded in low-gelling-temperature agarose in plugs (100 μL), which are then immersed in a lysing solution generally containing the detergent SDS and proteinase K, and often incubated at high temperatures (50°C) for several hours. Individual colonies of cells have also been analyzed in this fashion, which should facilitate clonal analysis of DNA damage *(13)*. The plugs are then sealed into the wells of a preformed gel, and subjected to PFGE for various times. The method of embedding intact cells in agarose prior to lysis considerably reduces the background level of strand breaks, thus increasing the sensitivity of the method *(99)*. Fluorescence detection of DNA, or sectioning of gels containing radiolabeled DNA can then be performed *(82,100)*. In theory, PFGE provides information on DNA fragment size. In practice, the amount of DNA that migrates out of the well and into the lane is often used as a sensi-

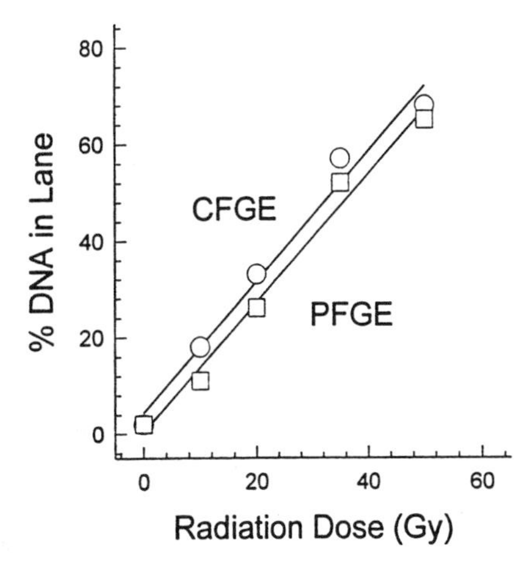

Fig. 1. Comparison between constant-field (CFGE) and pulsed-field (PFGE) gel electrophoresis for detecting DNA DSBs in CHO cells exposed to X-rays. For both methods, agarose plugs containing 3×10^6 cells in 0.75% agarose were lysed in detergent and proteinase K and then inserted into the wells of a 0.75% agarose gel. CFGE was performed at 0.6 V/cm for 20 h, whereas PFGE was performed using a CHEF DRII at 1.7 V/cm with a 60-s pulse time and 30-h run time. (Redrawn from ref. *112*.)

tive indicator of DSBs; doses as low as 1–2 Gy can produce detectable damage. A variety of hardware systems with different electrode configurations have been used, resulting in a number of confusing acronyms. Orthogonal field agarose gel electrophoresis (OFAGE), transverse alternating field electrophoresis (TAFE), clamped homogeneous electric field (CHEF) and field inversion gel electrophoresis (FIGE) appear to produce similar dose–response relationships for DNA damage when the amount of DNA able to migrate out of the well is used as the end point, although these systems can differ in terms of speed of separation, resolution within a particular size range, and "straightness" of lanes. However, for this simple end point (amount of DNA smaller than a defined exclusion size), conventional constant-field gel electrophoresis is equally sensitive (Fig. 1). In all cases, it would appear that DNA larger than about 10 Mbp remains within the plug.

More sophisticated analysis of fragment size and accurate measurement of DNA-rejoining rates require separation and analysis using PFGE *(17,44,87)*. Although the theory behind PFGE is well developed, it is not possible to measure fragment sizes over the complete DNA size range. Each combination of voltage, pulse time, and duration of electrophoresis is generally optimum for detection of a specific range of fragment sizes, and larger fragments of different sizes can still comigrate as "compression bands." Moreover, a problem of size inversion can occur where large fragments migrate faster than smaller ones *(31)*, so that rates of DSB induction are generally calculated over well defined "working" ranges for limited fragment sizes defined using yeast chromosome size markers. Another consideration using PFGE methods is the influence of DNA replicative status on DNA migration. DNA from cells in S phase migrates three to four times more slowly than DNA from cells in other phases *(36,65)*. This appears to be a limitation for all DSB assays, although when DNA is detected by radiolabeling methods, a chase time for several hours in isotope-free medium following radiolabeling can minimize the problem.

Calibration of DSB assays is possible by using ^{125}IdUrd incorporated into DNA and by making the assumption that each decay of this isotope produces a DSB with near unit probability *(42)*. Increased sensitivity for the detection of SSBs or DSBs often comes at the expense of a clear understanding of the physical basis for detection. As stated by Lange et al. *(43)*, methods that calibrate a measurement parameter against a known number of DSBs do not measure DSBs—they measure something that under some conditions is directly or indirectly proportional to DSBs. Neutral filter elution, like alkali filter elution, was originally thought to detect DNA pieces of different sizes that elute from polycarbonate filters: small pieces elute early, whereas large pieces elute after a longer period. However, analysis of the sizes of DNA fragments indicates that the pieces that elute from the filter are of similar size, regardless of time of elution *(11,113)*. Similar-sized fragments probably result from the shear forces of the eluting solution: it is the number of pieces available for elution at any point in time, not their size, that determines the amount of DNA eluted. The method of viscoelastometry measures relaxation recoil of large polymers like DNA once the sheer-stressing torque is released *(43)*. This method has the potential of detecting very low numbers of breaks and is also based on a well-developed theory. However, in mammalian cells, analysis becomes quite complex owing to higher-order chromatin structure.

2.3. Use of Strand Break Assays to Detect Apoptotic Cells

It should not be surprising that assays useful for detecting DSBs or SSBs would also prove useful in identifying apoptotic cells that contain highly fragmented DNA. The possible emergence of cells dying by apoptosis or necrosis has always been of concern in measurements of strand-break rejoining during prolonged repair periods, and in some studies, apoptotic cells were clearly present (e.g., *40*). In applying methods that detect overall DNA damage, it is necessary to consider the possibility that limited rejoining is a result of the presence of a small population of heavily damaged cells rather than the inability of all of the cells to repair the damage. The application of the comet assay (*see* Section 5.3.) can resolve this question nicely and also provides more information on the process of apoptosis than can be obtained by examining fragment patterns on gels *(70)*. Application of a crosslinking agent, mechlorethamine, to apoptotic cells can also be used in conjunction with the comet assay to size DNA fragments in individual apoptotic cells *(66)*.

A number of flow cytometry methods have been developed to detect DNA strand breaks in apoptotic cells *(34,104,114)*. An early method simply involved fixing cells in alcohol and then allowing sufficient time for very small fragments of DNA to diffuse out of the cells. When stained with a DNA binding dye like Hoechst 33342, the apoptotic cells form a population with a DNA content smaller than the diploid G1 DNA content. This method was not particularly sensitive to the presence of small populations of apoptotic cells, nor could it distinguish apoptosis in cells where the DNA was cleaved to much larger fragment sizes (50 or 300 kb). More recent methods *(20,34)* have applied two DNA binding dyes, one of which does not penetrate the membrane of an intact cell and the other whose binding can apparently be reduced by chromatin compaction during apoptosis. Apoptotic cells are identified as those that exclude the first dye and stain less intensely with the second, a distinction that is not always clear, and that can be cell-type-dependent. The use of nick-translation and end-labeling methods to detect apoptotic cells has perhaps been most successful. Here the permeabilized cells are incubated with terminal deoxynucleotidyl transferase or DNA polymerase and biotin-labeled d-UTP. Subsequent incorporation at nicks produced by the apoptotic endonuclease(s) is detected by flow cytometry using fluoresceinated avidin. Alternatively, a monoclonal antibody (MAb) specific for single-stranded DNA has also been used to identify apoptotic cells *(28)*. With all of these methods, the distinction between apoptotic and necrotic cells can be difficult. Unfortunately, although these flow cytometry methods have advantages of speed and large sample handling capacity, they currently lack sensitivity for detecting very low numbers of SSBs or DSBs.

3. BASE DAMAGE

Base damage predominates for many genotoxic chemicals and for UV radiation, and although a frequent event for ionizing radiation, the large variety of base changes induced by oxidizing agents makes it difficult to assess their individual contributions to cell response. Specific base damage in vivo has been typically measured using gas chromatography–mass spectrometry after exposure to doses from 40 up to 500 Gy or more *(54,56)*, still above the relevant biological range. Techniques such as ^{32}P-postlabeling significantly improve sensitivity for detecting some types of damage, and the typical sensitivity of 1 adduct/10^7 nucleotides can be increased by orders of magnitude using various enrichment methods *(16,75)*. In the basic postlabeling method, damaged DNA is enzymatically digested producing nucleoside monophosphates that are subsequently end-labeled with ^{32}P-ATP; normal and modified bases are then separated using two-dimensional thin-layer chromatography or high-performance liquid chromatography. In tumors, the ability to quantify drug or radiation-induced base damages is complicated further by high "background" base modifications, which can be many-fold greater than present in normal tissues *(50)*. Once a specific lesion is identified, the application of site-directed mutagenesis using chemically modified nucleotides allows further insight into the mutagenic potential of specific base lesions (e.g., *4*). MAb can now be used to detect specific DNA adducts in various tissues in immunoassays sensitive at the femtomole to subfemtomole range, or to provide information at the level of the individual cell *(6,22,94,107,109)*. Antibodies against thymine glycol and pyrimidine dimers have also been used to examine repair in DNA in transcribed vs nontranscribed strands *(12,45)*.

3.1. Endonuclease Sensitive Sites

A less specific approach that has been used for many years to quantify base damage is to apply enzymes that cleave DNA at or near sites of base damage *(74)*. Once the enzyme cleaves the DNA backbone, sensitive strand break assays can be used to measure base damage after doses as low as a few Gy *(27,108)*. A similar approach has been applied to the analysis of damage by chemicals and peroxide (e.g., *18*), and can provide a global picture of base damage when many different damages are present. Although it has been common to use fairly crude enzyme preparations (e.g., extracts from the bacterium *Micrococcus luteus*), the use of purified endonucleases improves specificity of base damage detection (e.g., *53*).

3.2. *Inhibition of DNA Ligation*

Base (and other) damage can also be detected indirectly using such drugs as cytosine arabinoside, which have multiple effects, one of which is to inhibit religation of DNA strands during repair *(117)*. SSBs produced during the process of DNA repair will then accumulate in the presence of ligase inhibitors and can be used as an indirect indicator of the presence of base damage.

3.3. *Detection of Repair Patches*

"Unscheduled" DNA synthesis occurs in response to DNA damage and can be detected in nondividing cells by measuring incorporation of ^{3}H-thymidine using autoradiography *(48)*. A technical advance is to use flow cytometry to analyze binding of antibodies to 5'-bromodeoxyuridine (BrdU) incorporated into repair patches *(5,95)*. Antibody binding to bromouracil in repair patches of cells exposed to UV or carcinogens was also used as the basis of a method to compare repair in specific sequences of the genome with overall genome repair *(45,46)*. Photolysis of BrdU incorporated into repair patches using 313 nm light *(80)* will also produce SSBs which can be detected with conventional assays *(83)*. More recently, Kalle et al. *(39)* have used streptavidin-coated magnetic beads and biotinylated antibodies against BrdU to investigate induction and repair of DNA damage in specific sequences.

4. DNA–PROTEIN AND DNA–DNA CROSSLINKS

DNA crosslinks are often detected using the alkali elution method of Kohn, which is performed with proteinase K included or omitted from the elution solution (for review, *see* ref. *41*). DNA crosslinks can be readily observed as a loss of ability to detect DNA SSBs following exposure to radiation. The kinetics of induction and loss of crosslinks can be quite complex. For the nitrogen mustard, mechlorethamine, crosslinking is essentially instantaneous, but for other drugs, like chloroethylnitrosourea, 12 h or more may be required after treatment for maximum crosslink production *(21)*. Analysis of DNA–protein crosslinks continues to be studied using a simple membrane filter binding assay first described by Strniste and Rall *(101)*.

DNA interstrand crosslinks can be detected indirectly using any of the SSB assays. The simplest approach is to treat the cell with the crosslinking agent and then expose cells (on ice to inhibit repair) to ionizing radiation. Crosslinks between broken strands effectively increase the molecular weight and size of the DNA pieces, and decrease the ability to detect the radiation-induced SSBs. However, this method does not give the exact number of crosslinks present in a cell. Interstrand crosslinks are also measured using a method based on ethidium bromide fluorescence of heat-denatured DNA (e.g., *98*).

5. METHODS TO MEASURE HETEROGENEITY IN DNA DAMAGE

5.1. *Effect of Higher-Order Chromatin Organization on Detection of DNA Damage*

A small number of strand break assays appear to discriminate between certain radiosensitive and radioresistant cells based only on measurement of initial damage (*see* ref. *62* for review). An appealing explanation is that there is a fundamental, but as yet undefined, property of chromatin organization that influences fidelity of DNA repair and that can be detected indirectly using some assays, such as neutral filter elution or nucleoid sedimentation *(78,93,103)*, but not using others, like PFGE *(26,71)*. Perhaps, as suggested by a number of authors, a more stable association between chromatin and the nuclear matrix influences the detection of DNA damage, and might also influence the efficiency of DNA strand break rejoining *(63)*. Alternatively, the spatial orientation of lesions might influence radioresponse by determining whether multiple lesions are likely to be more difficult to repair *(38)*.

One of the most dramatic changes in chromatin structure is the transition from interphase chromatin to metaphase chromosomes, yet this transition is apparently not accompanied by a change in the number of DSBs induced by ionizing radiation. A model proposed by Ronne et al. *(81)* offers a solution to this dilemma by suggesting that this transition may not involve dramatic changes in local chromatin structure because DNA loop-anchoring complexes are simply drawn together at mitosis to increase the degree of chromatin packing. Direct methods available for probing the role of chromatin organization in the response of cells to DNA-damaging agents are fairly limited. Gross observation of condensed chromosomes can sometimes be informative *(92)*, as can electron microscopic analysis of nuclear appearance *(115)*. Relative rates of digestion of DNA with enzymes may also provide information on accessibility of damaged sites to attack *(116)*, although it is interesting that DNA damage itself may release torsional stress in transcriptionally active chromatin domains that can make them resistant to digestion by DNase *(111)*. However, without a clear indication of which specific aspect(s) of organization influences DNA response to insult, it is difficult to develop effective methods to probe this aspect of early DNA damage and repair.

5.2. Heterogeneity in DNA Damage at the Genome Level

The DNA strand break methods described above do not address the question of the placement of the lesion within the chromatin, but instead assume that strand breaks are induced randomly, and that all cells within the population respond in an identical fashion. The "microheterogeneity" of DNA damage by ionizing radiation, as reviewed by Oleinick et al. *(60,61)*, clearly indicates that not all chromatin sustains equal amounts of damage even by an agent, such as ionizing radiation, which has been assumed to induce damage randomly. (Here the distinction between random induction of DNA damage and the presence of sensitive sites distributed randomly throughout the genome becomes important.) "The production of radiation-induced DNA damage is influenced by the chromatin environment, features of which include association of DNA with chromosomal proteins, attachment to the nuclear matrix, the site-specific binding of metal ions and the compaction of chromatin structure" *(59)*. Other agents (e.g., UV, cisplatin) can produce DNA damage which is clearly sequence-specific, and initial adduct formation by some drugs (nitrogen mustard and psoralen) is dependent on gene activity *(12)*. The combination of alkali unwinding with Southern hybridization provides a way to detect DNA damage within specific chromatin regions *(15)*. Direct visualization of the localization of biotin-dUTP-labeled DNA repair patches relative to transcription sites as well as other interesting proteins (p53, PCNA) is also possible *(37)*.

Recently, Rydberg *(89)* has examined the concept of randomness of lesions using a gel electrophoresis method to size DNA fragments following exposure to X-rays or nitrogen ions. Small DNA fragments were produced linearly with dose and in amounts dependent on linear energy transfer (LET). Rydberg found an unexpected number of small DNA fragments following irradiation, which could be explained as a consequence of chromatin organization, with the 30-nm chromatin fiber being the target *(33)*. The proposed concept of "regionally multiply damaged sites" would explain why high LET radiations produce relatively low yields of SSBs, why deletions may be common with ionizing radiation, and why total rejoining rates are less efficient after exposure of cells to high LET radiation *(89)*.

5.3. Resolution of Subpopulations Using Single-Cell Assays

Several methods are available that can be used to measure DNA damage and repair in individual mammalian cells. Table 3 lists these assays along with a brief indication of specific advantages and disadvantages and a representative reference. Chromosome aberrations and micronuclei (*see* Chapter 25), although providing an important measure of misrepair, do not generally produce a sufficient number of events per cell to provide information on the possible presence of subpopulations. For assays like the nick translation method, each cell provides a response, but the variation (in relatively homogeneous populations) is often so large that subpopulations cannot be easily resolved *(25)*. The comet assay (so-called because of the appearance of individual nuclei in this method) offers a powerful new approach, because it can be used to identify subpopulations differing in damage by as little as factor as 2.

The comet assay evolved from the method of Ostling and Johanson *(73)*. Single cells embedded in agarose are lysed to remove proteins and then exposed to an electric field. The amount of DNA that migrates is proportional to the number of strand breaks (SSBs using alkali lysis and electrophoresis; DSBs using neutral lysis and electrophoresis). Once DNA is stained using a fluorescent DNA binding drug, such as propidium iodide, damage can be quantified using an image analysis system *(23)*. The ability to quantify damage to individual cells provides a valuable indication of heterogeneity in response within the population. For example, the response of cells to the topoisomerase II poison, etoposide, can be used as a sensitive indicator of their proliferative status, since noncycling cells show a negligible number of strand breaks after treatment with this drug *(64)*. Irradiation of solid tumors and analysis using the alkaline comet assay can be used as an indicator of the presence of hypoxic cells *(68,69)*, since the latter show threefold fewer DNA strand breaks after irradiation than aerobic cells. The ability to examine individual cells provides a way to ensure that slow rejoining kinetics are not the result of a small population of degrading or apoptotic cells *(58,67)*.

In the results shown in Fig. 2, cells grown as monolayers or multicell spheroids were incubated with three DNA-damaging agents. Spheroids, a three-dimensional tissue-culture model for a solid tumor, contain cells growing in a variety of microenvironments *(102)*. The internal cells are poorly nourished, often lacking oxygen. Conversely the external cells are accessible to drugs and nutrients, and are actively proliferating. RSU-1069 is a bifunctional alkylating agent that produces SSBs in well-oxygenated cells and interstrand crosslinks in hypoxic cells (Fig. 2A). The topoisomerase II poison, etoposide, produces SSBs and DSBs selectively in the external proliferating cells of the spheroid (Fig. 2C). Rapid metabolism of 4-nitroquinoline-*N*-oxide (4NQO) is responsible for poor penetration into the center of spheroids, so that DNA damage is

Table 3
Assays Useful in Measuring DNA Damage at the Level of the Individual Cell

Assay	Advantages	Disadvantages	Refs.
Micronucleus	Provides a measure of fidelity of repair	Only dividing cells evaluable Difficult to detect subpopulations Sample time critical	*96*
Chromosome damage (premature chromosome condensation)	Provides a measure of fidelity of repair	Only some cells and some cell-cycle phases evaluable Difficult to detect subpopulations	*32*
Nick translation	Applicable to any single cell	Also detects necrosis or apoptosis Only 3'-OH end groups labeled Difficult to detect subpopulations	*25*
Halo	Applicable to any single cell	Samples must be analyzed immediately Detects only SSBs Lysis methods must be optimized for different cell types	*85*
Comet	Applicable to any single cell/nucleus Detects subpopulations	Image analysis system required	*23*

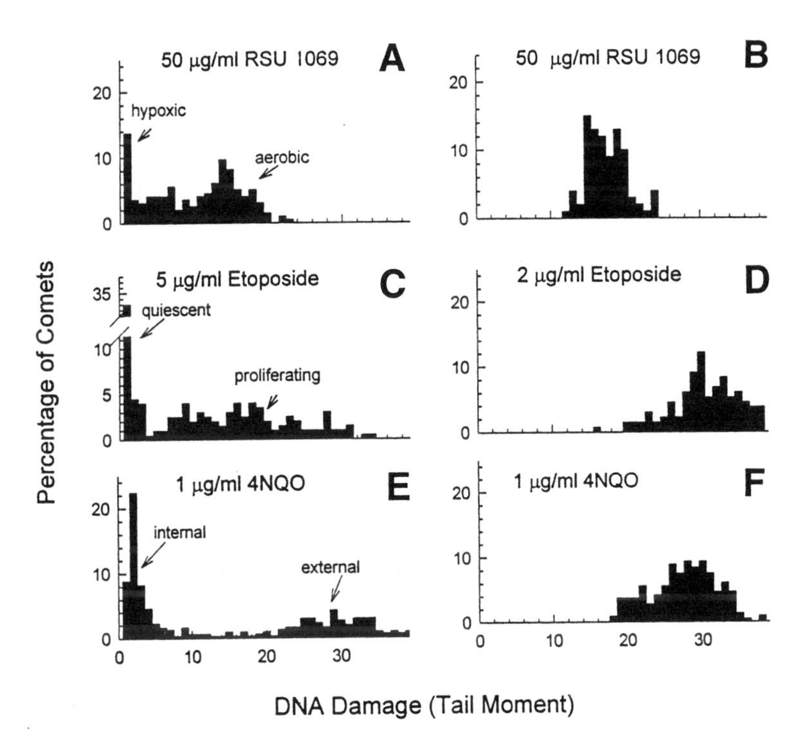

Fig. 2. Heterogeneity in response to DNA-damaging agents. Chinese hamster V79 multicell spheroids **(A,C,E)** or monolayers **(B,D,F)** were exposed for 1–2 h to various agents prior to dissaggregation and analysis of individual cells for DNA damage using the alkaline comet assay. Tail moment is directly proportional to the number of SSB.

concentrated in the external cells (Fig. 2E). The response of single cells grown as monolayers is quite different; all of the cells of the population show a similar amount of DNA damage since all of the cells are proliferating, accessible to drugs, and well-oxygenated (Fig. 2B,D,F). Although the average amount of DNA damage would adequately describe the single-cell response, it is clear that measuring the average amount of DNA damage produced in cells of spheroids (or tumors) can be of limited value in estimating their sensitivity to treatment.

In addition to applications in cancer biology, the comet assay/single-cell gel electrophoresis method is in widespread use in genetic toxicology. Cells or nuclei from virtually any tissue of the body are amenable to measurement of DNA damage using this method, and only a small sample size is required (equivalent to the number of white blood cells present in a drop of blood).

6. MEASUREMENT OF FIDELITY OF REJOINING

A number of molecular approaches can provide information on the accuracy of removal of lesions. Thacker *(105)* developed an integrating DNA vector, encoding a selectable marker gene that could be broken in defined sites in the coding sequence by restriction endonucleases. Plasmids were created with two selectable markers, a *gpt* gene, which is cut with restriction enzymes (*Kpn*I or *Eco*RV representing staggered and blunt cutters), and a *neo* gene as a marker of transfection. The fidelity of the rejoining process could then be followed by measuring survival of vector-transformed cells in medium that would only support growth of cells containing the functional *gpt* gene. Alternatively, cell extracts have been used to repair plasmids containing a break in the *lacZ* gene, and subsequent transfection of bacteria (or Southern analysis) is then used to determine mis-rejoining frequency *(29)*. A similar plasmid-based approach has been applied to understanding the mechanisms of DNA end rejoining *(84)* or to probing the nature of enzymes involved in rejoining *(24)*. Powell and McMillan *(77)* have also used the Thacker method as a way to probe differential cell repair capacity in a series of human tumor cell lines varying in intrinsic radiosensitivity.

Lobrich et al. *(49)* have recently developed a method to examine DSB production in specific genes. DNA from irradiated cells embedded in agarose was exposed to a rare cutting restriction endonuclease, *Not*I. After PFGE, DNA is transferred to nitrocellulose and probed by Southern analysis using a human lactoferrin cDNA probe (about 0.8 kbp). The absolute yield of breaks can be obtained easily using this method, and preferential damage and repair can be examined using different DNA probes. Most importantly, since inaccurate rejoining can lead to changes in fragment size, information on the fidelity of rejoining can also be obtained. On the negative side, small deletions (<2 Mbp) would not be detectable, and the method lacks sensitivity owing to the relatively small target being analyzed.

7. SUMMARY

Several new methods and approaches have been developed in recent years to measure DNA damage to mammalian cells. Although each has advantages and disadvantages, the method of choice ultimately depends on the question being addressed. At present, the ability to detect damage in specific genes comes at the price of decreased sensitivity. Population-based methods for detecting damage are adequate, providing heterogeneity in DNA damage is not a consideration. Measurement of DNA damage rejoining rates and residual damage can be useful in identifying some, but not all repair-deficient cell types, and assays that measure fidelity of repair are therefore receiving more attention. At present, limits of sensitivity appear to be about one break in 10^8 nucleotides and perhaps one specific adduct in 10^5–10^7 nucleotides, so that information in the biologically relevant dose range is possible. The development of methods for measuring the heterogeneity in lesion placement within the genome, the recognition of the importance of higher-order chromatin structure, and the ability to evaluate heterogeneity of DNA damage within a population are providing new insights into the processes of DNA damage and repair in mammalian cells.

ACKNOWLEDGMENTS

This work was supported by the National Cancer Institute of Canada with funds provided by the Canadian Cancer Society. The advice of R. Durand and P. Johnston, and the able assistance of J. P. Banath and W. Cottingham are gratefully acknowledged.

REFERENCES

1. Ager, D. D. and W. C Dewey. 1990. Calibration of pulsed field gel electrophoresis for measurement of DNA double-strand breaks. *Int. J. Radiat. Biol.* **58:** 249–259.
2. Ahnstrom, G. 1988. Techniques to measure DNA single-strand breaks in cells: a review. *Int. J. Radiat. Biol.* **54:** 695–707.
3. Ahnstrom, G. and K. Erixon. 1973. Radiation induced strand breakage in DNA from mammalian cells. Strand separation in alkaline solution. *Int. J. Radiat. Biol.* **23:** 285–289.
4. Basu, A. K., E. L. Loechler, S. A. Leadon, and J. M. Essigmann. 1989. Genetic effects of thymine glycol: site-directed mutagenesis and molecular modeling studies. *Proc. Natl. Acad. Sci. USA* **86:** 7677–7681.
5. Beisker, W. and W. N. Hittelman. 1988. Measurement of the kinetics of DNA repair synthesis after UV irradiation using immunochemical staining of incorporated 5-bromo–2'-deoxyuridine and flow cytometry. *Exp. Cell Res.* **174:** 156–167.
6. Bianchini, F. and C. P. Wild. 1994. 7-Methyldeoxyguanosine as a marker of exposure to environmental methylating agents. *Toxicol. Lett.* **72:** 175–184.
7. Birnboim, H. C. 1990. Fluorometric analysis of DNA unwinding to study strand breaks and repair in mammalian cells. *Methods Enzymol.* **186:** 550–555.
8. Blocher, D. 1982. DNA double-strand breaks in Ehrlich ascites tumour cells at low doses of x-rays. I. Determination of induced breaks by centrifugation at reduced speed. *Int. J. Radiat. Biol.* **42:** 317–328.
9. Blocher, D. 1990. In CHEF electrophoresis a linear induction of dsb corresponds to a nonlinear fraction of extracted DNA with dose. *Int. J. Radiat. Biol.* **57:** 7–12.
10. Blocher, D., M. Einspenner, and J. Zalackowski. 1989. CHEF electrophoresis, a sensitive technique for the determination of DNA double-strand breaks. *Int. J. Radiat. Biol.* **56:** 437–448.
11. Blocher, D. and G. Iliakis. 1991. Size distribution of DNA molecules recovered from non-denaturing filter elution. *Int. J. Radiat. Biol.* **59:** 919–926.
12. Bohr, V. A. 1991. Gene specific DNA repair. *Carcinogenesis* **12:** 1983–1992.
13. Boultwood, J., M. Thompson, C. Fidler, S. A. Lorimore, M. S. Lewis, J. S. Wainscoat, and E. Wright. 1993. Pulsed field gel electrophoresis on single murine hemopoietic colonies. *Leukemia* **7:** 1635–1636.
14. Buatti, J. M., L. R. Rivero, and T. J. Jorgensen. 1992. Radiation-induced DNA single-strand breaks in freshly isolated human leukocytes. *Radiat. Res.* **132:** 200–206.

15. Bunch, R. T., D. A. Gewirtz, and L. F. Povirk. 1992. A combined alkaline unwinding/ Southern blotting assay for measuring low levels of cellular DNA breakage within specific genomic regions. *Oncol. Res.* **4:** 7–15.
16. Cadet, J., O. Francette, J. Mouret, M. Polverelli, A. Audie, P. Giacomoni, A. Favier, and M. Richard. 1992. Chemical and biochemical post-labeling methods for singling out specific oxidative lesions. *Mutat. Res.* **275:** 343–354.
17. Cedervall, B., P. Kallman, and W. C. Dewey. 1995. Repair of double-strand breaks: errors encountered in the determination of half-life times in pulsed field gel electrophoresis and neutral filter elution. *Radiat. Res.* **142:** 23–28.
18. Collins, A. R., S. J. Duthie, and V. L. Dobson. 1993. Direct enzymic detection of endogenous oxidative base damage in human lymphocyte DNA. *Carcinogenesis* **14:** 1733–1735.
19. Cook P. R. and I. A. Brazell. 1975. Supercoils in human DNA. *J. Cell Sci.* **19:** 261–279.
20. Dive, C. C. D. Gregory, D. J. Phipps, D. L. Evans, A. E. Milner, and A. H. Wyllie. 1992. Analysis and discrimination of necrosis and apoptosis (programmed cell death) by multiparameter flow cytometry. *Biochem. Biophys. Acta* **1133:** 275–285.
21. Ewig, R. A. G. and K. W. Kohn. 1978. DNA-protein cross-linking and DNA interstrand cross-linking by haloethylnitrosoureas in L1210 cells. *Cancer Res.* **38:** 3197–3203.
22. Fadlallah, S., M. Lachapelle, K. Krzystyniak, S. Cooper, F. Denizeau, F. Guertin, and M. Fournier. 1994. O^6-methlguanine-DNA adducts in rat lymphocytes after in vivo exposure to N-nitrosodimethylamine (NDMA). *Int. J. Immunopharmacol.* **16:** 583–591.
23. Fairbairn, D. W., P. L. Olive, and K. L. O'Neill. 1993. The comet assay: A comprehensive review. *Mutat. Res.* **339:** 37–59.
24. Fairman, M. P., A. P. Johnson, and J. Thacker. 1992. Multiple components are involved in the efficient joining of double-stranded DNA breaks in human cell extracts. *Nucleic Acids Res.* **20:** 4145–4152.
25. Fertil, B., S. Modak, N. Chavaudra, H. Debry, F. Meyer, and E. P. Malaise. 1984. Detection in situ of gamma-ray-induced DNA strand breaks in single cells: enzymatic labeling of free 3'OH ends. *Int. J. Radiat. Biol.* **46:** 529–540.
26. Flentje, M., B. Asadpour, D. Latz, and K. J. Weber. 1993. Sensitivity of neutral filter elution but not PFGE can be modified by non-dsb chromatin damage. *Int. J. Radiat. Biol.* **63:** 715–724.
27. Fohe, C. and E. Dikomey. 1994. Induction and repair of DNA base damage studies in X-irradiated CHO cells using the M. luteus extract. *Int. J. Radiat. Biol.* **66:** 697–704.
28. Frankfurt O. S. 1994. Detection of apoptosis in leukemic and breast cancer cells with monoclonal antibody to single-stranded DNA. *Anticancer Res.* **14:** 1861–1869.
29. Ganesh, A., P. North, and J. Thacker. 1993. Repair and misrepair of site-specific DNA double-strand breaks by human cell extracts. *Mutat. Res.* **299:** 251–259.
30. George, A. M. and W. A. Cramp. 1987. The effects of ionizing radiation on structure and function of DNA. *Prog. Biophys. Mol. Biol.* **50:** 121–169.
31. Gunderson, K. and G. Chu. 1991. Pulsed-field electrophoresis of megabase-sized DNA. *Mol. Cell. Biol.* **11:** 3348–3354.
32. Hittelman, W. N. 1986. The technique of premature chromosome condensation to study the leukemic process: review and speculations. *Crit. Rev. Oncol. Hematol.* **6:** 147–221.
33. Holley, W. R. and A. Chatterjee. 1996. Clusters of DNA damage induced by ionizing radiation: formation of short DNA fragments. I. Theoretical modeling. *Radiat. Res.* **145:** 188–199.
34. Hotz, M. A., J. Gong, F. Traganos, and Z. Darzynkiewicz. 1994. Flow cytometric detection of apoptosis: Comparison of the assays of in situ DNA degradation and chromatin changes. *Cytometry* **15:** 237–244.
35. Iliakis, G. 1991. The role of DNA double-strand breaks in ionizing radiation-induced killing of eukaryotic cells. *Bioessays* **13:** 641–648.

36. Iliakis, G., O. Cicilioni, and L. Metzger. 1991. Measurement of DNA double-strand breaks in CHO cells at various stages of the cell cycle using pulsed field gel electrophoresis: calibration by means of 125I decay. *Int. J. Radiat. Biol.* **59:** 343–358.
37. Jackson D. A., A. B. Hassan, R. J. Errington, and P. R. Cook. 1994. Sites in human nuclei where damage induced by ultraviolet light is repaired: localization relative to transcription sties and concentrations of proliferating cell nuclear antigen and the tumour suppressor protein, p53. *J. Cell Sci.* **107:** 1753–1760.
38. Johnston, P. J. and P. E. Bryant. 1994. A component of DNA double strand break repair is dependent on the spatial orientation of the lesions with the higher order structures of chromatin. *Int. J. Radiat. Biol.* **66:** 531–536.
39. Kalle, W. H., A. M. Hazekamp-van Kokkum, P. H. Lohman, A. T. Natarajan, A. A. van Zeeland, and L. H. Mullenders. 1993. The use of streptavidin-coated magnetic beads and biotinylated antibodies to investigate induction and repair of DNA damage: analysis of repair patches in specific sequences of uv-irradiated human fibroblasts. *Anal. Biochem.* **208:** 228–236.
40. Kantor, P. M. and H. S. Schwartz 1980. Post-repair DNA damage in X-irradiated cultured human tumour cells. *Int. J. Radiat. Biol.* **38:** 483–493.
41. Kohn, K. W. 1991. Principles and practice of DNA filter elution. *Pharmacol. Ther.* **49:** 55–77.
42. Krish, R. E., E. Krasin, and C. J. Sauri. 1976. DNA breakage, repair and lethality after ^{125}I decay in rec^{+} and recA strains of Escherichia coli. *Int. J. Radiat. Biol.* **29:** 37–50.
43. Lange, C. S., A. Cole, and J. Y. Ostashevsky. 1993. Radiation-induced damage in chromosomal DNA molecules: deduction of chromosomal DNA organization from the hydrodynamic data used to measure DNA double-strand breaks and from stereo electron microscopic observations. *Adv. Radiat. Biol.* **17:** 261–421.
44. Lawrence, T. S., D. P. Normolle, M. A. Davis, and J. Maybaum. 1993. The use of biphasic linear ramped pulsed field gel electrophoresis to quantify DNA damage based on fragment size distribution. *Int. J. Radiat. Oncol. Biol. Phys.* **27:** 659–663.
45. Leadon, S. A. and P. C. Hanawalt. 1983. Monoclonal antibody to DNA containing thymidine glycol. *Mutat. Res.* **112:** 191–200.
46. Leadon, S. A. 1986. Differential repair of DNA damage in specific nucleotide sequences in monkey cells. *Nucleic Acids Res.* **14:** 8979–8995.
47. Leadon, S. A. and D. A. Lawrence. 1992. Strand-selective repair of DNA damage in the yeast GAL7 gene requires RNA polymerase II. *J. Biol. Chem.* **267:** 23,175–23,182.
48. Lieberman, M. W., R. N. Baney, R. E. Lee, S. Sell, and E. Farber. 1971. Studies on DNA repair in human lymphocytes treated with proximate carcinogens and alkylating agents. *Cancer Res.* **31:** 1297–1306.
49. Lobrich, M., S. Ikpeme, and J. Kiefer. 1994. DNA double-strand break measurement in mammalian cells by pulsed field gel electrophoresis: an approach using restriction enzymes and gene probing. *Int. J. Radiat. Biol.* **65:** 623–630.
50. Malins, D. C. and R. Haimanot. 1991. Major alterations in the nucleotide structure of DNA in cancer of the female breast. *Cancer Res.* **51:** 5430–5432.
51. McGrath, R. A. and R. W. Williams. 1966. Reconstruction in vivo of irradiated Escherichia coli deoxyribonucleoic acid: the rejoining of broken pieces. *Nature* **212:** 534–535.
52. Milner, A. E., A. T. M. Vaughan, and I. P. Clark. 1987. Measurement of DNA damage in mammalian cells using flow cytometry. *Radiat. Res.* **110:** 108–117.
53. Moran, M. F. and K. Ebisuzaki. 1987. Base excision repair of DNA in irradiated human cells. *Carcinogenesis* **8:** 607–609.
54. Mori, T., Y. Hori, and M. Dizdaroglu. 1993. DNA base damage generated in vivo in hepatic chromatin of mice upon whole body gamma irradiation. *Int. J. Radiat. Biol.* **64:** 645–650.
55. Mullenders, L. H., H. Vrieling, J. Venema, and A. A. Van Zeeland. 1991. Hierarchies of DNA repair in mammalian cells: biological consequences. *Mutat. Res.* **250:** 223–228.

56. Nackerdien, A., R. Olinski, and M. Dizdaroglu. 1992. DNA base damage in chromatin of gamma irradiated cultured human cells. *Free Radical Res. Commun.* **16:** 259–273.
57. Nose K. and H. Okamoto. 1983. Detection of carcinogen-induced DNA breaks by nick translation in permeable cells. *Biochem. Biophys. Res. Commun.* **111:** 383–389.
58. Nygren, J., B. Cedervall, S. Eriksson, M. Dusinska, and A. Kolman. 1994. Induction of DNA strand breaks by ethylene oxide in human diploid fibroblasts. *Environ. Mol. Mutagen.* **24:** 161–167.
59. Oleinick, N. L. 1995. Higher order DNA structure and radiation damage, in Radiation Damage in DNA. *Structure/Function Relationships at Early Times* (Fuciarelli, A. F. and J. D. Zimbrick, eds.), Batelle, Richland, WA, pp. 395–408.
60. Oleinick, N. L., U. Balasubramaniam, L. Xue, and S. Chiu. 1994. Nuclear structure and the microdistribution of radiation damage in DNA. *Int. J. Radiat. Biol.* **66:** 523–529.
61. Oleinick, N. L. and S. M. Chiu. 1994. Nuclear and chromatin structures and their influence on the radiosensitivity of DNA. *Radiat. Protec. Dosimetry* **52:** 353–358.
62. Olive, P. L. 1988. The DNA precipitation assay: a rapid and simple method for detecting DNA damage in mammalian cells. *Environ. Mol. Mutagen.* **11:** 487–495.
63. Olive, P. L. 1992. DNA organization affects cellular radiosensitivity and detection of initial DNA strand breaks. *Int. J. Radiat. Biol.* **62:** 389–396.
64. Olive, P. L. and J. P. Banáth. 1992. Tumour growth fraction measured using the comet assay. *Cell Proliferation* **25:** 447–457.
65. Olive, P. L. and J. P. Banáth 1993. Detection of DNA double-strand breaks through the cell cycle after exposure to X-rays, bleomycin, etoposide and ^{125}IdUrd. *Int. J. Radiat. Biol.* **64:** 349–358.
66. Olive, P. L. and J. P. Banáth. 1995. Sizing highly fragmented DNA in individual apoptotic cells using the comet assay and a DNA crosslinking agent. *Exp. Cell Res.* **221:** 19–26.
67. Olive, P. L., J. P. Banáth, and R. E. Durand. 1990. Heterogeneity in radiation-induced DNA damage and repair in tumor and normal cells measured using the "comet" assay. *Radiat. Res.* **122:** 69–72.
68. Olive, P. L. and R. E. Durand. 1992. Detecting hypoxic cells in a murine tumor using the comet assay. *J. Natl. Cancer Inst.* **85:** 707–711.
69. Olive, P. L., R. E. Durand, J. LeRiche, I. Olivotto, and S. M. Jackson. 1993. . Gel electrophoresis of individual cells to quantify hypoxic fraction in human breast cancers. *Cancer Res.* **53:** 733–736.
70. Olive, P. L., G. Frazer, and J. P. Banáth. 1993. Radiation-induced apoptosis measured in TK6 human B lymphoblast cells using the comet assay. *Radiat. Res.* **136:** 130–136.
71. Olive, P. L, S. H. MacPhail, and J. P. Banáth. 1994. Lack of correlation between DNA double-strand break induction/rejoining and radiosensitivity in six human tumor cell lines. *Cancer Res.* **54:** 3939–3946.
72. Olive, P. L., D. Wlodek, and J. P. Banáth. 1991. DNA double-strand breaks measured in individual cells subjected to gel electrophoresis. *Cancer Res.* **51:** 4671–4676.
73. Ostling, O. and K. J. Johanson. 1984. Microelectrophoretic study of radiation-induced DNA damages in individual mammalian cells. *Biochem. Biophys. Res. Commun.* **123:** 291–298.
74. Paterson, M. C. 1978. Use of purified lesion-recognizing enzymes to monitor DNA repair in vivo, in *Advances in Radiation Biol.*, vol. 7 (Lett, J. T. and H. Adler, eds.), Academic, New York, NY, pp. 1–53.
75. Phillips, D. A., M. Castegnaro, and H. Bartsch. 1993. *Postlabelling methods for detection of DNA adducts.* IARC, Lyon.
76. Powell, S. and T. J. McMillan. 1991. DNA damage and repair following treatment with ionizing radiation. *Radiother. Oncol.* **19:** 95–108.
77. Powell, S. N. and T. J. McMillan. 1994. The repair fidelity of restriction enzyme-induced double strand breaks in plasmid DNA correlates with radioresistance in human tumor cell lines. *Int. J. Radiat. Oncol. Biol. Phys.* **29:** 1035–1040.

78. Radford, I. R. 1985. The level of induced DNA double-strand breakage correlates with cell killing after X-irradiation. *Int. J. Radiat. Biol.* **48:** 45–54.
79. Radford, I. R. 1988. The dose-response for low-LET radiation-induced DNA double-strand breakage: methods of measurement and implications for radiation action models. *Int. J. Radiat. Biol.* **54:** 1–11.
80. Regan, J. D., R. B. Setlow, and R. D. Ley. 1971. Normal and defective repair of damaged DNA in human cells: a sensitivity assay utilizing the photolysis of bromodeoxyuridine. *Proc. Natl. Acad. Sci. USA* **68:** 708–712.
81. Ronne, M., A. O. Glydenholm, and C. O. Storm. 1995 Minibands and chromosome structure. A theory. *Anticancer Res.* **15:** 249–254.
82. Rosemann, M., B. Kanon, A. W. Konings, and H. H. Kampinga. 1993. An image analysis technique for detection of radiation-induced DNA fragmentation after CHEF electrophoresis. *Int. J. Radiat. Biol.* **64:** 245–249.
83. Rosenstein, B. S., J. T. Murphy, and J. M. Ducore. 1985. Use of a highly sensitive assay to analyze the excision repair of dimer and nondimer DNA damages induced in human skin fibroblasts by 254 nm and solar ultraviolet radiation. *Cancer Res.* **45:** 5536–5531.
84. Roth, D. B. and J. H. Wilson. 1986. Nonhomologous recombination in mammalian cells: role for short sequence homologies in the joining reaction. *Mol. Cell. Biol.* **6:** 4295–4304.
85. Roti Roti, J. L. and W. D. Wright. 1987. Visualization of DNA loops in nucleoids from HeLa cells: Assays for DNA damage and repair. *Cytometry* **8:** 461–467.
86. Roti Roti, J. L., W. D. Wright, and Y. C. Taylor. 1993. DNA loop structure and radiation response. *Adv. Radiat. Biol.* **17:** 227–259.
87. Ruiz de Almodovar, J. M., G. G. Steel, S. J. Whitaker, and T. J. McMillan. 1994. A comparison of methods for calculating DNA double-strand break induction frequency in mammalian cells by pulsed field gel electrophoresis. *Int. J. Radiat. Biol.* **65:** 641–649.
88. Rydberg, B. 1980. Detection of induced DNA strand breaks with improved sensitivity in human cells. *Radiat. Res.* **81:** 492–495.
89. Rydberg, B. 1996. Clusters of DNA damage induced by ionizing radiation: formation of short DNA fragments. II. Experimental detection. *Radiat. Res.* **145:** 200–209.
90. Schwartz, D. C. and C. R. Cantor. 1984. Separation of yeast chromosome-sized DNA molecules by pulsed field gradient gel electrophoresis. *Cell* **37:** 67–75.
91. Schwartz, J. L., J. Shadley, D. R. Jaffe, J. Whitlock, J. Rotmensch, J. M. Cowan, D. J. Gordon, and A. T. M. Vaughan. 1990. Association between radiation sensitivity, DNA repair and chromosome organization in the Chinese hamster ovary cell line xrs–5, in *Mutation and the Environment. Part A: Basic Mechanisms* (Mendelsohn, M. L. and R. J. Albertini, eds.),Wiley-Liss, New York, pp. 255–264.
92. Schwartz, J. L. and A. T. M. Vaughan. 1989. Association among DNA chromosome break rejoining rates, chromatin structure alterations and radiation sensitivity in human tumor cell lines. *Cancer Res.* **49:** 5054–5057.
93. Seiler, F., U. Kirstein, G. Eberle, K. Hochleitner, and M. R. Rajewsky. 1993. Quantification of specific DNA *O*-alkylation products in individual cells by monoclonal antibodies and digital imaging of intensified nuclear fluorescence. *Carcinogenesis* **14:** 1907–1913.
94. Selden, J. R., F. Dolbeare, J. H. Clair, W. W. Nichols, J. E. Miller, R. M. Kleemeyer, R. J. Hyland, and J. G. DeLuca. 1993. Statistical confirmation that immunofluorescent detection of DNA repair in human fibroblasts by measurement of bromodeoxyuridine incorporation is stoichiometric and sensitive. *Cytometry* **14:** 154–167.
95. Shibamoto, Y., C. Streffer, C. Fuhrmann, and V. Budach. 1991. Tumor radiosensitivity prediction by the cytokinesis-block micronucleus assay. *Radiat. Res.* **128:** 293–300.
96. Snyder, R. D. and D. W. Matheson. 1985. Nick translation—a new assay for monitoring DNA damage and repair in cultured human fibroblasts. *Environ. Mutagen.* **7:** 267–279.

97. Spengler, S. J. and B. Singer. 1988. Formation of interstrand cross-links in chloroacetaldehyde-treated DNA demonstrated by ethidium bromide fluorescence. *Cancer Res.* **48:** 4804–4806.
98. Stamato, T. and N. Denko. 1990. Asymmetric field inversion gel electrophoresis: A new method for detecting DNA double-strand breaks in mammalian cells. *Radiat. Res.* **121:** 196–205.
99. Story, M. D., E. A. Mendoza, R. E. Meyn, and P. J. Tofilon. 1994. Pulsed field gel electrophoretic analysis of DNA double-strand breaks in mammalian cells using photostimulable storage phosphor imaging. *Int. J. Radiat. Biol.* **65:** 523–528.
100. Strniste, G. F. and R. C. Rall. 1976. Induction of stable protein-deoxyribonucleic acid adducts in Chinese hamster cell chromatin by ultraviolet light. *Biochemistry* **15:** 1712–1719.
101. Sutherland, R. M. 1988. Cell and environment interaction in tumor microregions: the multicell spheroid model. *Science* **240:** 177–184.
102. Taylor, Y. C., P. G. Duncan, X. Zhang, and W. D. Wright. 1991. Differences in the DNA supercoiling response of irradiated cell lines from ataxia-telangiectasia versus unaffected individuals. *Int. J. Radiat. Biol.* **59:** 359–371.
103. Telford, W. G., L. E. King, and P. J. Fraker. 1994. Rapid quantitation of apoptosis in pure and heterogeneous cell populations using flow cytometry. *J. Immunol. Methods* **172:** 1–16.
104. Thacker, J. 1989. The use of integrating DNA vectors to analyze the molecular defects in ionizing radiation-sensitive mutants of mammalian cells including ataxia telangiectasia. *Mutat. Res.* **145:** 177–183.
105. Ueno, A. M., E. M. Goldin, A. B. Cox, and J. T. Lett. 1979. Deficient repair and degradation of DNA in X-irradiated L5178Y S/S cells: cell-cycle and temperature dependence. *Radiat. Res.* **79:** 377–389.
106. van Delft, J. H. M., M. J. M. van Winden, A. Luiten-Schuite, L. R. Ribeiro, and R. A. Baan. 1994. Comparison of various immunochemical assays for the detection of ethylene oxide-DNA adducts with monoclonal antibodies against imidazole ring-opened N7-(2-hydroxylethyl) guanosine: application in a biological monitoring study. *Carcinogenesis* **15:** 1867–1873.
107. van Loon, A. A. W. M., R. H. Groenendijk, G. P. Van der Schans, P. H. M. Lohman, and R. A. Baan. 1991. Detection of base damage in DNA in human blood exposed to ionizing radiation at biologically relevant doses. *Int. J. Radiat. Biol.* **59:** 651–660.
108. van Loon, A. A., P. J. Den Boer, G. P. Van der Schans, P. Mackenbach, J. A. Grootegoed, R. A. Baan, and P. H. Lohman. 1991. Immunochemical detection of DNA damage induction and repair at different cellular stages of spermatogenesis of the hamster after in vitro or in vivo exposure to ionizing radiation. *Exp. Cell Res.* **193:** 303–309.
109. Venema, J., L. H. F. Mullenders, A. T. Natarajan, A. A. van Zeeland, and L. F. Mayne. 1990. The genetic defect in Cockayne syndrome is associated with a defect in repair of UV-induced DNA damage in transcriptionally active DNA. *Proc. Natl. Acad. Sci. USA* **87:** 4707–4711.
110. Villeponteau, B. and H. G. Martinson. 1987. Gamma rays and bleomycin nick DNA and reverse the DNase I sensitivity of β-globin gene chromatin in vivo. *Mol. Cell. Biol.* **7:** 1917–1924.
111. Wlodek, D., J. P. Banáth, and P. L. Olive. 1991. Comparison between pulsed-field and constant-field gel electrophoresis for measurement of DNA double-strand breaks in irradiated CHO cells. *Int. J. Radiat. Biol.* **60:** 779–790.
112. Wlodek, D. and P. L. Olive. 1990. Basis for detecting DNA double-strand breaks using neutral filter elution. *Radiat. Res.* **124:** 326–333.
113. Yang, C. S., C. Wang, M. D. Minden, and E. A. McCulloch. 1994. Fluorescence-labeling of nicks in DNA from leukemic blast cells as a measure of damage following

cytosine arabinoside. Application to the study of regulated drug sensitivity. *Leukemia* **8:** 2052–2059.

114. Yasui, L. S., T. J. Fink, and A. M. Enrique. 1994. Nuclear scaffold organization in the X-ray sensitive Chinese hamster mutant cell line, xrs–5. *Int. J. Radiat. Biol.* **65:** 185–192.
115. Yasui, L. S., S. Ling-Indeck, B. Johnson-Wint, T. J. Fink, and D. Molsen. 1991. Changes in the nuclear structure in the radiation-sensitive CHO mutant cell, xrs–5. *Radiat. Res.* **127:** 269–277.
116. Zittoun, J., J. Marquet, and J. C. David. 1991. Mechanism of inhibition of DNA ligase in Ara-C treated cells. *Leukemia* Res. **15:** 157–164.

25

Radiation-Induced Damage and the Formation of Chromosomal Aberrations

Michael N. Cornforth

1. INTRODUCTION

Studies involving cytogenetic endpoints following DNA damage have been conducted for half a century, providing a great deal of information relevant to DNA damage repair. More recently developed techniques and approaches have expanded considerably the scope and power of cytogenetic analysis, yielding new insights into the mechanisms of chromosome aberration formation.

Contributions of classical cytogenetics are briefly discussed in a historical framework in order to highlight a few long-standing issues, but the main objective of this chapter is to examine more recent contributions to the field of radiation cytogenetics as they relate to the repair of chromosome damage. The focus is further confined to the discussion of lesions produced by ionizing radiation (IR) in mammalian cells and their postirradiation processing leading to the formation of structural changes in chromosomes, principally those found at the first mitosis following exposure.

1.1. Significance of Aberrations Produced by IRs

Virtually all agents capable of producing structural aberrations (clastogens) are both mutagenic and carcinogenic; conversely, the vast majority of carcinogens are also clastogens. Chromosomal rearrangements are a hallmark of cancerous cells, and a number of cancers are associated with rather specific chromosome anomalies *(71,112)*. The ongoing formation of aberrations reflects genomic instability associated with neoplastic development *(89,112,124)*, and has been postulated as a mechanism to explain some types of gene amplification *(73,135)*. Finally, certain aberration types kill cells with remarkable efficiency *(25,31,93)*, a property of fundamental relevance for cancer therapy, but that is also relevant to mechanisms of mutagenesis.

As a model damage-inducing agent, IR is unique. Sparsely IRs, such as X- and γ-rays, deposit their energy in cellular structures through discrete ionization events that are essentially randomly distributed in space. From the standpoint of the experimentalist, there are a number of advantages that derive from (or are associated with) this behavior. These include the ability to define the concept of absorbed dose, as well as the ability to estimate the dimensions of critical cellular sites (i.e., target size). IR is typically highly penetrating and produces damage that is not affected by the sorts of

From: DNA Damage and Repair, Vol. 2: DNA Repair in Higher Eukaryotes
Edited by: J. A. Nickoloff and M. F. Hoekstra © Humana Press Inc., Totowa, NJ

subcellular structures that influence the diffusion of chemical agents. The physics and subsequent chemistry associated with photon absorption, and the ionizations that occur along fast electron tracks thus produced, are complete within a few microseconds. Therefore, in the context of cellular repair processes, initial damage from radiation can be considered essentially instantaneous; this gives an "on/off" character to the delivery of damage. Finally, the spacing of ionizations along charged particle tracks (i.e., the lineal energy transfer or LET) can be manipulated over several orders of magnitude, allowing IR to be used as a "biophysical probe" of cell structure.

1.2. For Noncytogeneticists

The microscopic analysis of chromosomes proper is possible only during mitosis, cytogenetic analysis being typically confined to cells in the first postirradiation mitosis following exposure to a potentially clastogenic agent. With rare exception, IR does not cause chromosome aberrations in cells that are both irradiated and analyzed in the same mitosis. Aberrations seen at metaphase, therefore, represent the manifestation of damage to the cell that was registered during the preceding interphase. With the possible of exception of aberrations found in cells that are exposed during G2, the time interval between exposure and analysis at metaphase will be substantial compared to the kinetics of DNA damage repair. As such, the analysis of metaphase chromosomes is essentially a measure of residual damage that remains in spite (or because) of repair mechanisms that operate to remove initial lesions.

When normal cells are exposed to IR in G1, first postirradiation mitoses yield exclusively chromosome-type aberrations. These are characterized as lesions affecting both sister chromatids at the same site. A similar exposure to S or G2 cells, however, produces aberrations of a different class. Termed chromatid-types, these display changes affecting only one chromatid of a chromosome at a particular site. IR shares this property with only a handful of chemical agents. In contrast, UV radiation, along with the vast majority chemical clastogens, produce chromatid-type aberrations in cells exposed during G1 *(8)*.

The term "restitution" describes the situation whereby the broken ends of a chromosome (*see* Section 2.) reunite with each other; in this case, the break is said to be rejoined or repaired, and no structural aberration is formed. Restitution, of course, does not guarantee fidelity of repair at the nucleotide level. Nevertheless, because it serves to re-establish the original linkage relationships, restitution (at least at the cytogenetic level) may be equated with repair. It is crucial to recognize that the bulk of aberrations take the form of exchanges that represent illegitimate recombination between different chromosomes or between different sites on the same chromosome. Unrepaired chromosome breaks, resulting in terminal deletions, make up a relatively minor fraction of total aberrations. The term "misrepair" is used here to describe exchange events. Thus, to a first approximation, the residual lesions seen as aberrations at metaphase reflect misrepaired rather than unrepaired damage.

Recombination leading to exchanges is thought by most investigators to occur between two or more damaged sites, which explains the characteristic dose–response relationships observed with radiation of different ionization densities. For the sparsely ionizing X- and γ-rays, moderate- to high-dose rates produce a response that is curvilinear upward, whereas densely IRs, such as α-particles, produce aberrations that increase linearly and more steeply with dose. In addition, split-dose or reduced-dose rate exposures cause a pronounced reduction in the yield of aberrations for sparsely, but not densely IR. The "classical" explanation for these phenomena is that the damaged sites (e.g., on different chromosomes) need to be close to one another in both time and space for pairwise interaction (recombination) to occur. The spatial component of this interaction relates to the track structure of charged particles initiating the damage, whereas the temporal component is affected by the removal (repair) of damaged sites before they can participate in an illegitimate recombination. This topic was recently reviewed by Cornforth and Bedford *(26)*.

The above viewpoint is consistent with the vast majority of radiobiological data, including experiments specifically designed to test its validity *(20,21,28,39)*. However, on the basis of biophysical considerations, a few investigators have suggested that chromosome exchanges may result from a damaged site on one chromosome interacting with an undamaged site on another chromosome *(44,45,62,121)*. Additional support for such a "one-hit" mechanism of exchange formation derives from molecular analysis of recombination (e.g., *59,123*) or recombinational models of DNA repair *(91)* in mammalian cells. Thus, even at this most fundamental level, there is disagreement among investigators.

2. THE INITIAL LESION

Early investigators *(61,104,105)* developed the concept that radiation produced a frank "chromosome break" as the initial lesion, the broken ends of which were free to unite with those produced by another break when the two were close in both time and space. In essence, the most widely subscribed view regarding the identity of this initial "precursor" lesion has not changed since that time. The current dogma is that chromosome breaks are produced by DNA double-strand breaks (DSBs) and that each DSB is produced by the passage of a single charged particle track. Although considerable indirect evidence supports this viewpoint, it falls somewhat short of being conclusive.

2.1. *The Case for DSBs*

With some exceptions noted *(60)*, a uninemic model of mammalian chromosome structure is universally accepted. Consequently, a chromosome break must involve the equivalent of at least one DSB. The proposition that radiation-induced DSBs represent initial uncommitted precursor lesions to aberration formation has obvious appeal to advocates of classical cytogenetic theory. Supporting data are briefly summarized below.

As previously mentioned, few chemical agents are capable of eliciting the classical "nondelayed" clastogenic response (i.e., chromosome-type aberrations after G1 exposure; chromatid-type aberrations following exposure to S and G2 cells). They include the bleomycins and neocarzinostatin, which produce damage to DNA via attack by radical species that mimic the indirect effects of IR *(54,129)*. These same agents are specifically known for their ability to produce copious quantities of DNA DSBs.

When mammalian cells are exposed to hydrogen peroxide under conditions of reduced temperature, large amounts of DNA single-strand breakage (SSB) as well as a wide spectrum of base damage products are produced. However, under these same conditions (e.g., the equivalent of 100 Gy of IR for yields of SSB and base damage), few if any DNA DSBs are produced, and no significant cell killing is detected *(134)*.

This observation is important, in light of the previously mentioned fact that chromosome aberrations account for the vast majority of radiation-induced cell killing in most cell types *(25,31,93)*.

Restriction enzymes, when directly introduced into living cells, are potent inducers of chromosome aberrations *(2,16,72,79)* and also elicit the classical nondelayed response for aberration formation *(136)*. These enzymes apparently cleave DNA in vivo as they do in vitro (*86*; *see also* Section 4.2.). Restriction enzymes also cause frank chromosome breaks when introduced into human cells, as detected by the technique of premature chromosome condensation (PCC) (Loucas and Bedford, personal communication; *see also* Section 3.). Folle and Obe *(35)* reported that restriction enzymes electroporated into CHO cells produced exchanges that were preferentially located in the G-light band regions of chromosomes. This observation is potentially significant in light of previous studies showing that radiation-induced exchanges also tended to be preferentially expressed in a subset of G-light bands, which, interestingly, tend to colocalize with actively transcribing genes *(50)*. Thus, agents that presumably produce only DNA DSBs mimic the effects of IR, both in terms of the types of aberrations produced with respect to the cell cycle, and their distribution with respect to chromosomal substructure. These results confirm the conclusions of earlier work by Natarajan and colleagues *(80,81)*, who treated X-irradiated mammalian cells with *Neurospora* endonuclease and observed an increase in chromosome aberrations compared to cells exposed to X-rays alone. They argued this increase was owing to the conversion of X-ray-induced DNA single-strand damage into DSBs by the endonuclease. Qualitatively similar conclusions regarding the importance of DSBs in aberration formation were reached by Sen and Hittelman *(108)*, who found that *Neurospora* endonuclease increased aberrations in cells previously treated with nitrogen mustard.

2.2. *The Exchange Hypothesis*

An alternative view to the idea that exchanges form from frank radiation-induced chromosome "breaks" is the Exchange Hypothesis, first introduced by Revell *(92)*. This hypothesis holds that the primary lesions destined for pairwise interaction of an exchange are not true breaks in the chromosome, but some other form of damage that does not immediately compromise the integrity of the chromosome. Only after such primary lesions attempt the process of exchange is gross chromosomal structure disrupted so as to produce the aberrations visible at mitosis. An important consequence of this notion is that most terminal deletions are thought to result from failed exchange events, not unrejoined breaks. Although formally this hypothesis relates to the formation of chromatid-type exchanges, its underlying principles have often been extended to include chromosome-types aberrations as well *(13)*. The Exchange Hypothesis is couched in the abstract, in the sense that it does not address the molecular nature of the primary lesions produced by IR. Nevertheless, it gives a credible biophysical underpinning to models that consider radiation damage other than DSBs to represent the precursor lesion. For example, Preston *(87)* argued that the increase in radiation-induced aberrations seen following exposure to β-Ara C was owing to the accumulation of DNA single-strand gaps caused by incomplete repair of radiation-induced base damage.

Base damage is produced in great excess compared to DSBs for all types of IRs, and its role in UV- and chemically induced aberrations is well established. Despite such considerations, most investigators have not found the base damage argument compelling. Garnering support for a central role of base damage in radiation-induced aberrations will be difficult, partly because of the plethora of biochemical modifications that fall into this category, and because not all types of base damage would be expected to have the same biological impact. In any event, if some form of radiation-induced base damage (or cluster of base damage) were responsible for aberration production, it must be a unique type that, to date, has not been shown to be produced by any known chemical agent. Investigators may need to find ways to introduce candidate lesions selectively into the chromosomal DNA of mammalian cells before the issue is finally settled.

3. LESION PROCESSING

Important information regarding the postirradiation processing of damage can be deduced by investigating the influence of dose rate and dose fractionation on the yield of chromosome aberrations. Such studies indicated that the broken ends of most radiation-induced breaks restitute (rejoin correctly) and that breaks destined for illegitimate rejoining remain "open" (capable of recombination) for a limited length of time. More recent data indicate that lesions responsible for X-ray-induced asymmetrical exchanges in human cells are repaired with first-order kinetics, with a half-time of 1.25 *(106)* to 1.7 h *(6)*.

3.1. Chromosome Breaks and Their Rejoining During Interphase

Many of the limitations that metaphase analysis presents for the study of chromosome repair kinetics were circumvented when Johnson and Rao *(55)* discovered that the chromatin of interphase cells became visibly condensed as "interphase chromosomes" when such a cell was fused with another mitotic cell. The prematurely condensed chromosomes (PCC) technique allows direct visualization of breaks in interphase chromatin within minutes following exposure to IR, providing information about the extent of initial chromosome breakage and the characteristics of break repair as a function of time following exposure *(47,26,90,132)*.

Dose–response curves for the induction of initial PCC breaks (i.e., immediately after exposure) are linear for virtually all IRs, with slopes that increase with increasing ionization density. For example, for human fibroblasts exposed to X- or γ-rays, the yield of initial PCC breaks is 5–6 Gy^{-1} *(23,24,85)*, whereas for α-particles, the yield is twice as high *(7,27)*. Clearly, these yields account for only a fraction (e.g., 10–15%) of the total DNA DSBs initially produced in such cells. In most, but perhaps not all cases, the kinetics of PCC break rejoining can be adequately described as a first-order process, with half times ranging from about 0.5 *(51)* to 2 h *(24)*, depending on cell type. Notable exceptions include the additional appearance of what has been termed the "fast component" of PCC break repair ($T_{1/2}$ of 2–5 min) for some cell types, especially following hypertonic treatment of irradiated cells *(52)*. Okayasu and Iliakis *(84)* argued that the repair defect associated with the radiation-sensitive *xrs-5* hamster cell line—mutant Ku protein (*115*; *see* Chapters 16 and 17)—abolishes the fast repair component. (*See* Cornforth and Bedford *[26]*, for an alternative explanation of the hypertonic phenomena). In either case, PCC break rejoining kinetics can explain the decrease in aberrations seen at metaphase (and associated increase in cell survival) that accompanies the recovery from potentially lethal (delayed plating experiments) *(25)* and sublethal

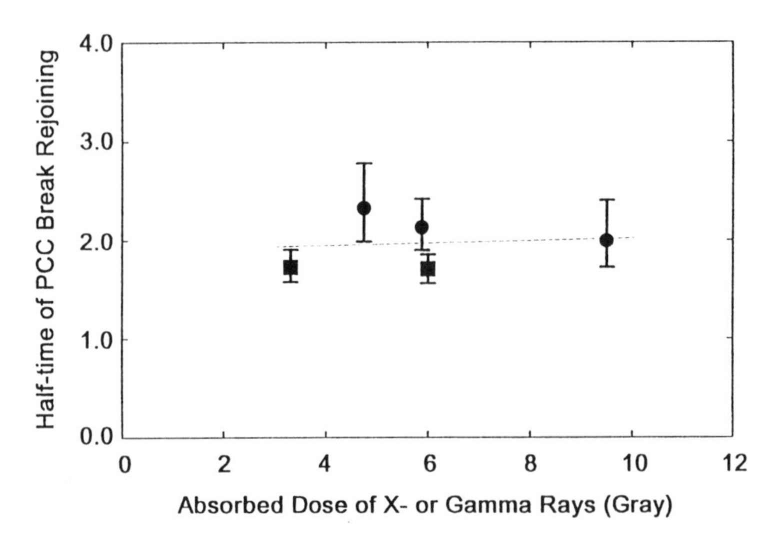

Fig. 1. Half-time (in hours) for the rejoining of PCC breaks in normal human fibroblasts, as a function of radiation dose *(22)*. Solid squares and circles represent exposures to X- or γ-rays, respectively. There is no evidence of a systematic change in break rejoining kinetics over the dose range examined. Each data point represents the fit of an individual rejoining profile to a first-order kinetic model containing a residual term *(23,24)*. Bars indicate standard errors.

(fractionated exposure) damage *(6)*. PCC break rejoining is also in reasonable agreement with the repair kinetics of radiation-induced DSBs, as measured by neutral velocity sedimentation *(9)*. On balance, PCC data are consistent with the idea that a subset of DSBs are effectively converted to chromosome breaks, and that the latter represent the pool of uncommitted precursor lesions underlying aberration formation.

More recent PCC studies have uncovered additional information pertinent to aberration formation. The rate constants for both PCC break rejoining (*see* Fig. 1) and DSB repair *(83)* appear to be independent of dose. That is, over the range of doses considered biologically relevant for mammalian cells, there is no evidence that repair systems become "saturated" as the level of damage increases, as might be expected for biophysical models of radiation action that are at odds with the notion of lesion interaction (*41,98*; *see* Section 1.2.), as well as related "one-hit" cytogenetic models, which argue that exchanges result from a single damaged site *(44,45,62,118,121)*. These same models explain the increased effectiveness of high LET radiations (compared to X- or γ-rays) in producing chromosome aberrations on the basis of high LET radiation causing a more "severe" type of damage *(42,88)*. However, in contrast to the expectations of these models, the rate constants for PCC break rejoining in primary human fibroblasts are not systematically altered for radiations whose ionization densities range from about 1 (X-rays) to 200 keV/μm (e.g., accelerated helium ions; *65*).

3.2. *Analysis of Misrepair by PCC*

By the time an irradiated cell reaches mitosis, most exchanges are complete, meaning that all four free ends produced by a pair of chromosome breaks find illegitimate partners and rejoin. Occasionally, however, the exchange is incomplete, leaving broken ends, which then appear as terminal deletions. By using whole-chromosome painting to study the rejoining of X-ray-induced PCC breaks in chromosome 4 of normal human fibroblasts, Brown et al. *(15)* observed, at any point in time postirradiation, complete as well as incomplete exchanges. The peak frequency of incomplete exchanges preceded that for complete exchanges, as though each of the two free ends of a chromosome break destined for exchange was capable of entering into an illegitimate recombination independently of the other. Because both partners of a complete exchange did not tend to form together, at the same time, these data may be taken as evidence against the Exchange Hypothesis (Section 2.2.) as it relates to chromosome-type aberrations *(26)*.

Evans and coworkers *(33)* studied PCC break rejoining in a SCID mouse cell line, which has a reduced ability to form active DNA-dependent protein kinase complexes (*10*; *see* Chapter 17). Compared to wild-type control cells, PCC break rejoining was slower; half-times being about 1.5 vs 2 h, respectively. Use of a pancentromeric fluorescence *in situ* hybridization (FISH) probe allowed study of the kinetics of dicentric formation postirradiation. Despite the typically large errors associated with the parameters of resulting curve fits, the kinetics of radiation-induced dicentric formation for both cell types could be adequately described as a first-order process in which no dicentrics were presumed to occur immediately following exposure. This latter result contrasts those of Greinert et al. *(43)*, who employed PCC in conjunction with C-banding to study X-ray-induced breaks and the kinetics of their repair in unstimulated human lymphocytes. They observed that the yields of initial breaks, as well as the rate constants for their first-order repair were virtually identical to those previously published *(24,85)*. However, as opposed to the conclusions of these earlier studies and those of Evans et al. *(33)*, radiation-induced dicentric formation did not follow similar kinetics to PCC break repair. Instead, roughly half of the dicentrics destined to form following a 4-Gy exposure were present at the earliest times postirradiation. Furthermore, the frequency of these exchanges remained relatively unchanged for about 2 h postirradiation, gradually increasing over the next 3–5 h. The resulting sigmoidal shape to the curve was interpreted as evidence for a "two-step" process of exchange formation operating on lesions produced by independent charged particle tracks. The first step was proposed to be a bimolecular reaction between DNA DSBs on separate chromosomes leading to exchanges of "intermediate" character that are not yet stable enough to survive the subsequent stress of (premature) chromosome condensation. Intermediate exchanges are then proposed to become transformed to a more stable state by a slower process, having a time constant of about 3 h.

The observation that a significant fraction of total radiation-induced exchanges are formed within a few minutes postirradiation is reminiscent of earlier work by Hittelman and Rao *(47)*, who studied chromatid-type exchange aberration formation in X-irradiated CHO cells, and by Vyas et al. *(131)*, who studied dicentric formation in human lymphocytes. It is interesting to speculate whether differences in rates of PCC break rejoining and the formation of exchange aberrations seen in some cell types reflect basic differences between repair and misrepair events at the molecular level. A plausible alternative explanation is that such differences merely reflect the spatial distributions of break pairs produced by single versus multiple particle tracks, since, on average, the latter will be separated by larger distances inside the cell *(11,43)*.

3.3. *Inter- vs Intrachromosomal Exchange*

Essentially all of the potential "break pairs" produced by dense IRs (e.g., α-particles) are thought to be produced by the passage of single charged particle tracks, whereas for sparse IRs (e.g., X-rays), many such break pairs are produced by different (independent) tracks. By considering the traditional cytogenetic view that the interaction of broken chromosome ends is effectively confined to break pairs within a few hundred nanometers of one another, it has been proposed that the spectra for certain types of aberrations seen at metaphase will change as a function of ionization density *(48,97)*, specifically that sparsely IRs favor the production of chromosomal interchanges, i.e., exchanges that involve two or more different chromosomes. Densely IRs, by comparison, should allow for the increased probability of intrachange formation, exchanges whose breakpoints are both within the same chromosome. Thus, the ratio of interchanges to intrachanges, termed the "F-ratio," should be LET-dependent, providing a cytogenetic "fingerprint" that might be used to estimate the quality of prior radiation exposure to human populations when physical methods of dosimetry are not appropriate. Preliminary tests of the F-ratio hypothesis are encouraging. For example, the ratio of dicentrics to its intrachromosomal alternative (centric rings) is reportedly about 15 for X- or γ-rays, whereas this ratio for α-particle-induced aberrations is about 6 *(12,97)*. However, since cells harboring dicentrics are quickly eliminated from a dividing population, it will probably become necessary to examine the F-ratio for stably inherited aberration types (perhaps translocations-to-inversions) before such an approach would be appropriate as a "biodosimeter" of past exposures in human populations *(12)*.

3.4. *Symmetrical vs Asymmetrical Exchange*

Misrepair appears in two forms at the cytogenetic level—asymmetrical exchanges such as dicentrics and rings, which always result in acentric fragment formation, and symmetrical exchanges (i.e., translocations and inversions), which do not. A long-standing question in radiation cytogenetics is whether symmetrical exchanges and their respective asymmetrical forms represent "alternative" outcomes of misrepair *(102)*, as expected if the two aberration types shared common recombinational mechanism(s) during their formation. If so, it would be expected, for example, that the average yields of radiation-induced reciprocal translocations would be equal to that for dicentrics. The question is also important from the standpoint of radiation risk assessment because the vast majority of cytogenetic data used to derive quantitative dose–response relationships have focused exclusively on the scoring of (largely lethal) asymmetrical exchanges, such as dicentrics, under the tacit assumption that their yields followed those of their nonlethal symmetrical counterparts, i.e., reciprocal translocations. Older literature, based on banding analysis of X- or γ-irradiated cells, is decidedly split regarding the relative contributions of these two aberration classes, various investigators having concluded that symmetrical types are formed with equal (e.g., *49,102*), greater *(17)*, or lower *(99)* frequency than asymmetrical types.

Because it is more robust in detecting symmetrical interchanges than previous approaches based on banding, whole chromosome painting ostensibly allows a more rigorous examination of this issue. Unfortunately, here too a consensus is lacking. Several studies have indicated that X- or γ-induced reciprocal translocations were produced in significant excess over dicentrics *(30,68,82,107,127)*, whereas others found the frequencies of both aberration types to be equal or nearly so *(3,40,67)*.

It would appear that part of this disparity stems from the fact that dicentrics can sometimes be misscored as translocations. When more stringent measures of centromere identification were used in conjunction with painting, previously observed excesses in the symmetrical class disappeared, and translocation to dicentric ratios of essentially 1:1 were restored *(77,113)*. Ratios of unity were also subsequently reported by Finnon et al. *(34)* for X-irradiated human lymphocytes, using a combination of painting probes for chromosomes 2, 3, and 5, together with a pancentromeric probe to mark all centromeres.

The potential misscoring of dicentrics, however, fails to explain the data of Bauchinger et al. *(5)*, who exposed human lymphocytes to γ-rays, and then painted chromosomes 1, 4, and 12, together with a pancentromeric probe. Following graded doses from 0.1–6.0 Gy, translocation/dicentric ratios ranged from 1.8–1.2, respectively. Further analysis involving four additional chromosome groupings indicated that three of the five groupings examined had an excess of translocations compared to dicentrics *(58)*. They also evaluated exchanges for individual chromosomes, and found a similar translocation bias in chromosomes 1, 4, 6, 7, 8, and X. Comparable results were later reported by Griffin et al. *(44)*, for chromosome 4 of primary human fibroblasts following exposure to 120 keV/μm ^{238}Pu α-particles, although, oddly, this effect was not found for X-rays. The following section explains how the issue is further complicated by the fact that the terms "symmetrical" and "asymmetrical" no longer have a clear-cut meaning for a sizable proportion of exchanges produced by IR.

3.5. FISH Reveals Additional Complexity

One of the more striking and unexpected results from whole-chromosome painting relates to the detection of "complex exchanges" that result from three (or more) illegitimate breakpoints among two (or more) chromosomes. An example of a complex aberration type involving three chromosomes that is frequently observed in irradiated cells is shown in Fig. 2F–J. In reality, the "three-way exchange" pictured is among the most straightforward of an extensive array of possible complex exchanges that exist *(103)*. Note that the ability to detect complex exchanges depends on which chromosome(s) of the three is painted in a given exchange. When a single chromosome is painted, the same exchange can appear in different forms, and the same painted form may arise from different types of exchanges *(103)*. Thus, it is not possible to deduce the extent of any rearrangement from a single painted chromosome. It follows that many exchanges (e.g., Fig. 2I) will be classified as simple, when in fact, they involve multiple chromosomes and breakpoints of exchange.

Although the potential for complex exchanges had been previously considered (e.g., *36*) and directly observed using conventional banding analysis *(96)*, their significance was not fully appreciated until whole-chromosome painting techniques demonstrated their prevalence. Lucas and Sachs *(66)* used a two-color painting scheme to identify definitively exchanges simultaneously involving chromosomes 1 and 2 in human lymphocytes following a 2.3-Gy dose of ^{137}Cs γ-rays. They reported that about 7% of the exchanges involving both painted chromosomes were actually "three-way" complex exchanges that also involved anonymous remaining (counterstained) chromosomes.

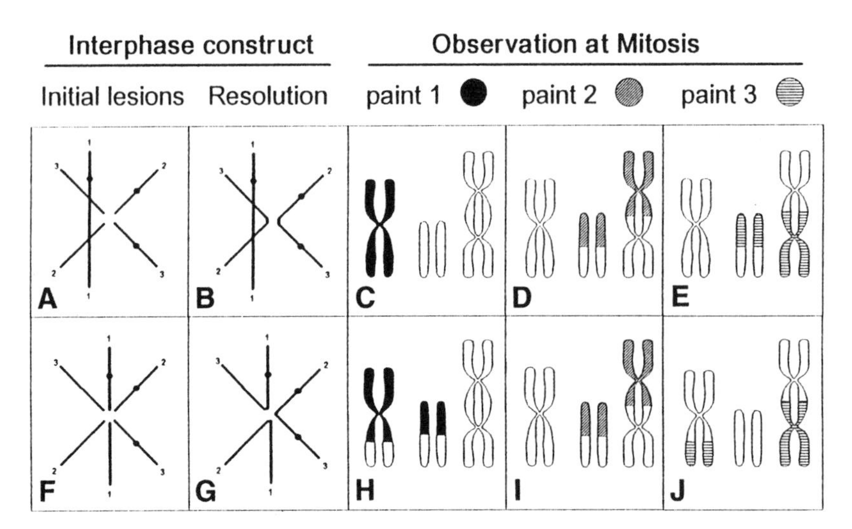

Fig. 2. Resolution of initial radiation-induced chromosome breakage, leading to either a simple asymmetrical exchange **(A–E)** or a complex exchange, involving three or more chromosomes **(F–J)**. Depending on the particular painting probe used, identical rearrangements may appear to be different (e.g., compare **H** with **I**). Conversely, different rearrangements may look identical (e.g., compare **D** with **I**). Reproduced, with permission, from Tucker et al. *(126)*.

Other studies have used a single-color painting scheme with qualitatively similar results. After 6 Gy of X-rays, Brown and Kovacs *(14)* observed that 26% of the complete exchanges involving chromosome 1, 4, or 8 of primary human fibroblasts were visibly complex (here referred to as nonreciprocal exchanges), but estimated that up to half of the total exchanges formed may be of this type. They also predicted that the relative frequency of complex exchanges should increase with dose. These results were later confirmed and extended for exchanges among chromosomes 1, 5, and 7 of human fibroblasts that were exposed to either 4 Gy or 6 Gy of X-rays *(109,110)*. About 23% and 33%, respectively, of exchanges involving these chromosomes were visibly complex. Employing a break-rejoining model that describes interactions between breaks of a complex exchange as independent events, 35% of total exchanges induced by the 4-Gy dose were calculated to be truly complex and 54% at 6 Gy. Equally striking was the conclusion that the majority of these complex exchanges involved five breaks in four chromosomes! Griffin et al. *(44)* later reported that a 2-Gy dose of X-rays delivered to the same cells resulted in no visibly complex aberrations in chromosome 1 or 4, indicating that the formation of complex aberrations resulting from low LET radiations is highly dose-dependent.

Interestingly, a 0.4 Gy dose of high LET ^{238}Pu α-particles—which produces the same level of cell kill (and, by inference, roughly the same overall yield of chromosome aberrations) as the aforementioned 2-Gy X-ray dose—resulted in about 40% of the exchanges being visibly complex *(44)*. Moreover, complex exchange formation was dose-independent from 0.4–1.0 Gy. At 1.0 Gy, 48% of the total exchanges were visibly complex, but since no correction for the fraction of complex aberrations appearing as simple exchanges (i.e., "psuedosimples") was attempted, this value clearly underestimates the actual contribution of complex exchanges. There is also evidence that complex aberration spectra differ substantially for α-particles and X-rays *(103)*. Whereas complex aberrations from low LET exposure tend to involve multiple chromosomes, the most abundant classes of (visible) complexes produced by α-particles were insertion and ring types ostensibly involving a pair of chromosomes, and visible complexes involving four chromosomes were not seen at all. Since, for the moderate doses of α-particles used, interaction of damage caused by independent charged particle tracks is extremely unlikely, all chromosome breaks leading to exchange must have resulted from a single particle track. Griffin et al. *(44)* estimate that, at the doses used, the trajectory of a single α-particle passes through an average of two to three interphase chromosome domains. These considerations are consistent with the notion that high LET tracks from α-particles tend to produce breaks within chromosomes, which, on average, are much closer to one another than breaks produced by low LET radiations. This would explain the increased yield of high LET-induced exchanges (including complexes) and the reduced complexity of the exchanges observed, but only if it is assumed that broken chromosome ends can interact over large (submicron) distances. Biophysical considerations have led some to conclude that such interaction distances are limited to a few nanometers (e.g., *121*). Under this latter assumption, Griffin et al. *(44,45)* have considered the possibility that exchanges may be formed between a radiation-induced DSB at one chromosomal site and an enzymatically induced DSB (via recombinational repair) at another site (i.e., "one-hit" exchanges; Section 1.2.).

Certain complex exchange types are relevant to the previous discussion of the Exchange Hypothesis (Section 2.2.), namely those formed when each free end of a chromosome break can freely exchange with any other contemporary break, as implied by Fig. 2F-J. When this situation exists, the two ends of each break can rejoin to different chromosomes. Exchanges that result from this "musical chairs" scenario have been referred to as "obligate complexes" *(103)*, and also as nonreciprocal exchanges involving multiple mutual sites (NEMMS; *126*). In fact, the complex exchanges described by Lucas and Sachs *(66)* in their dual-painting experiments, as well as most of the "nonreciprocal exchanges" detected by single-color painting (Brown and Kovacs, *[14]*) were precisely of this type. The appearance of these complex types has been considered damaging to the Exchange Hypothesis as it relates to chromosome-type exchanges *(14,66)*, because the hypothesis assumes the interaction of unspecified primary lesions (i.e., not chromosome breaks) prior to the initiation of exchange. This, in turn, implies that sites for such interaction for obligate complex exchanges must simultaneously accommodate several chromosomes at a single site. Although conceivable, our current view of interphase chromosome structural relationships makes such an assumption tenuous. In either case, it seems to represent an unnecessary complication that can be avoided altogether by more classical views of aberration formation that identify precursor lesions as frank chromosome breaks *(26)*.

For complex exchanges, the common terminology used to describe aberrations in routine Giemsa-stained preparations (e.g., *100*) is useless, and the ISCN nomenclature *(53)* applied to banded material has a number of shortcomings as well *(101,126)*. Two alternative systems of nomenclature have been used to describe complex aberrations. The first is the CAB system of Savage and Simpson *(103)*, which classifies exchanges on the basis of the number of chromosomes (C), chromosome arms (A), and breaks (B) involved. Complex exchanges are categorized into families and subfamilies from recognizable patterns produced by single-color whole-chromosome painting. The second is the PAINT system developed by Tucker and colleagues *(126)*, which, unlike the CAB system, is a purely descriptive terminology with no connection to underlying mechanism implied. Rather than attempting to classify the entire exchange event, PAINT describes each visible breakpoint of an exchange separately, as though they were independent events. PAINT terminology can be applied to mutlicolor chromosome painting.

It is not obvious how complex aberrations should be handled in order to construct meaningful dose–response relationships. For the purposes of "biodosimetry," the total number of "color junctions" (visible exchange breakpoints) can be plotted against dose *(125)*, and compared against "standard curves" generated for each radiation and cell type. However, this leads to data that are dose-dependently (and perhaps unpredictably) overdispersed, making it difficult to specify uncertainties associated with curve fits *(34)*. For radiation risk assessment, this empirical approach has additional limitations, particularly as it relates to models of radiation action whose parameters themselves suggest underlying biophysical processes, such as the popular "one-track/two-track" interpretation of the characteristic linear quadratic response observed for low LET radiations *(56,61,64)*. Alternatively, expectancy tables may be constructed for any exchange and analyzed to establish relative frequencies of visible complexes. Various empirical "mixes" can then be used to model the relative frequencies actually observed *(110)*. One problem is that tables have not yet been constructed that consider all complex aberrations arising from more than four breaks. Another problem, as pointed out by Savage and Simpson *(103)*, is that each particular combination of chromosomes participating in a complex exchange will be part of a family that will require its own contingency table.

4. MOLECULAR INSIGHTS AND IMPLICATIONS

Molecular analysis is beginning to yield insights into the fundamental mechanisms of radiation-induced aberration formation. DNA sequence information associated with the breakpoints of chromosomal exchanges is an important step in this direction, since it is here where underlying recombinational processes leave their signature.

The very properties that make IR an attractive model genotoxic agent also present certain problems for the molecular analysis of cellular changes they produce. For example, exposure to sparsely IR produces initial lesions that are effectively randomly distributed in space. For biologically relevant doses, lesion processing yields few visible aberrations whose recombinational junctions are scattered throughout the genome. Consequently, a molecular search for the generic radiation-induced exchange is difficult, and little information on the junctions themselves is available. Most molecular data relevant to aberration formation come indirectly, through the analysis of mutations occurring at selectable genetic loci.

Abrahamson and Wolff *(1)* argued on theoretical grounds that most radiation-induced germline mutations in the mouse were owing to "large scale" genetic alterations, including gross structural chromosomal changes. This suspicion has since been substantiated for several mammalian genetic loci for IRs of various qualities *(118)*. Large-scale deletions and rearrangements often dominate radiation-induced mutational spectra of mammalian cells, and for systems designed to allow for the viable recovery of such events, multilocus deletions and rearrangements overwhelm the contributions of point mutations and small-scale intragenic changes *(32,133)*. Thus, although IR is capable of producing in mammalian cells *(46)* the small-scale genetic changes commonly recovered in bacteria (e.g., nucleotide base transition, transversion, and frameshift mutations), a hallmark feature of its effects on most loci of higher eukaryotes is to cause large-scale genetic alterations, most notably deletions. These often involve the excision of the entire locus under study and frequently are cytogenetically visible *(37,119,128,130)*.

Simpson and colleagues *(111)* performed a high-resolution cytogenetic analysis of seven independent X-ray-induced *hprt* mutants of primary human fibroblasts: five mutants were previously known to have total gene deletions by Southern and PCR analysis; two were partial deletions having one breakpoint contained within the remaining *HPRT* sequence. No visible cytogenetic changes were found in either partial deletion, but strikingly, all five of the total deletion mutants examined revealed cytogenetic changes that were consistent with the size and scope of the underlying molecular events. Molecular analysis demonstrates that a significant fraction of radiation-induced mutants are cytologically visible because they contain megabase deletions *(75)*, suggesting that mechanisms producing deletions detectable by molecular and cytogenetic approaches are similar, if not identical. More specifically, an identity is implied between molecular deletions recovered at selectable genetic loci and asymmetric intrachanges (interstitial deletions and acentric rings) observed in individual cells at their first postirradiation mitosis. With the following caveats in mind, this assumption is reasonable.

The most informative sequencing will probably continue to come from intragenic deletions whose breakpoints are both contained within target loci, since these are the most amenable to sequence analysis. In human material, the largest radiation-induced deletion sequenced to date is 4.2 kbp *(74)*, a sizable event by molecular standards. However, this is two orders of magnitude smaller than the smallest interstitial deletion observable by routine microscopy. It is unclear whether mechanisms underlying the formation of kilo- and megabase deletions are sufficiently similar, given the influence of higher-order chromatin structure and nuclear architecture on aberration formation (for review *see 26*). Moreover, the isolation of mutant DNA requires the clonal expansion of a single progenitor cell. It is worth considering the influence of radiation-induced genomic instability under these conditions, specifically whether certain rearrangements in mutant cell populations several generations postirradiation represent subsequent rather than initial recombinational events. Finally, there is also the issue of whether the analysis of asymmetrical events, such as deletions, is directly applicable to other types of aberrations (*see* Section 3.4.).

4.1. Asymmetrical Events

Strobel et al. *(114)* sequenced a multilocus germline deletion in the offspring of C57BL male mice whose spermatogonia received 1.5 Gy of neutrons. Later both Nakatsu et al. *(78)* and Thomas et al. *(122)* reported DNA sequences of other X-ray-induced heritable deletion breakpoints in mice. The results of all three studies are remarkably similar on three counts. First, junctions were "flush," indicating that no extraneous nucleotides of unknown origin (orphan, filler DNA) had been inserted. This result differs from insertion-type junctions seen in DNA transfection experiments, immune system rearrangements, and certain spontaneous deletions *(94,95)*. Second, short 1–2 bp regions of homology were found at the junction, similar to that previously reported for somatic mammalian cells following exposure to IR *(46)*. Third, no extensive sequence homologies were found between the proximal and distal sites of any of the deletions that might indicate a homologous recombinational mode of formation. Although an 80% homology to long interspersed elements (LINES) of the mouse was found associated with one of the deletions studied *(122)*, it occurred only on one side. Thus, contrary to the implications of numerous studies involving spontaneous, UV, and chemically induced deletions *(69,70)*, radiation-induced germline deletions in mice do not appear to involve recombination between repetitive elements.

These conclusions are strengthened by the work of Morris and Thacker *(74)*, who sequenced breakpoints in two X-ray-induced *hprt* deletion mutants of primary human fibroblasts. In both cases, *Alu* consensus sequences overlapped the upstream, but not the downstream breakpoint. One of the new junctions contained a single adenine residue common to both breakpoints involved, reminiscent of the short sequence homologies shared by the germline deletions described above. The other involved a 4-bp insertion of filler DNA that, in the context of surrounding sequences, formed a direct 10 bp repeat with residues close to the downstream breakpoint. The authors explained these phenomena as reflecting single stranded regions of DNA "looping out" or exhibiting "strand slippage." In the case of simple (flush) deletions, such loops may be stabilized through imperfect palindromes formed by base pairing between short inverted repeats. Such a structure would serve to juxtapose sequences destined for ligation (the new junction), as intervening sequences were deleted. An interesting prediction of this model is that all sequences destined for deletion would be contained within the quasipalidromic sequence. Although this is not difficult to envision for very small deletions, the prospect of such structures large enough to accommodate megabase sized deletions is another matter. To explain the formation of the "insertion-duplication" deletion observed, a second mechanism based on double-strand gap repair was invoked *(95)*. In this scenario, looping out and strand slippage allow imperfect pairing to occur between adjacent strands via short repeat sequence homologies; subsequent gap filling produces an intact template for synthesis of the remaining strand. It is worth noting that no evidence for the involvement of V(D)J-type recombination was found, although such a mechanism would appear to play an important role in the genesis of spontaneous *hprt* mutations during human fetal development *(38)*. Topoisomerase I and II consensus recognition sites were found in the regions analyzed, but, as pointed out, this finding is of questionable significance given the abundance and degeneracy of these sequences. Regardless of the model used to explain these findings, it is clear that recombination junctions at radiation-induced deletions can involve processes that are more complex than the simple end-joining type of reactions implied by breakage-first models of aberration formation *(61,104)*. Moreover, we are now forced to consider than such junctions are formed through more than one mechanism.

Both of these generalizations are supported by the work of Phillips and Morgan *(86)*, who analyzed CHO *aprt* mutants induced by electroporation of restriction enzymes producing blunt ends, each having a single recognition sequence within this gene. A wide variety of sequence alterations were found in mutants that lost the target restriction site, which comprised 70% of all mutants recovered. These consisted of small (1–36 bp) deletions, insertions of various sizes, and combinations thereof. Most of the small deletions involved complementary 1–4 bp overlaps at the recombination junction, similar to those found in SV40 and plasmid rejoining studies *(57,94,120)*. Small insertions that were direct repeats of sequences located near the enzyme cleavage site were also found, and these appear to be identical to the "insertion-duplication" events mentioned above *(74,95)*. Because these studies focus on the processing of single lesions within a target sequence, they may not directly relate to mechanisms of (exchange) aberration formation that are predicated on the notion of interaction between damaged sites. They are nevertheless very informative because of their implications concerning the fidelity of repair at the cytogenetic level. For example, although it seems likely that the vast majority of initial enzyme-induced DSBs in chromosomes are repaired with fidelity, these studies clearly show this is not always the case. Since IR is likely to produce at least some types of DSBs less amenable to repair than those produced by simple phosphodiester enzymatic cleavage, it follows that a portion of what radiation cytogeneticists call "restitution" cannot represent faithful repair. Moreover, some of the sequence modifications found in these studies appear to occur for much larger deletions as well. Since larger deletions may well result from the pairwise interaction (recombination) of distant DSBs, it could be argued that molecular evidence is now in hand to support the long-held view that restitution and exchange really are alternative forms of damage processing that utilize similar cellular machinery *(61)*.

4.2. Symmetrical Events

It may be instructive to consider that inversions and reciprocal translocations together can occur with sufficient frequency to explain most radiation-induced mutations. Taking primary human fibroblasts as an example, an X-ray dose of 3 Gy typically produces an average of 1 asymmetrical aberration/cell and, by inference, roughly 1 symmetrical aberration/cell. A single symmetrical exchange (i.e., inversion or translocation) yields two breakpoints, each with a finite probability of occurring within a target gene. The 50-kbp *HPRT* gene represents 8×10^{-6} of the 6×10^{9} bp contained in a human diploid cell. Therefore, a dose of 3 Gy might be expected to produce an exchange in which one of the breakpoints occurs within this gene with a frequency of about 1.6×10^{-5}. This estimate may be conservative, since it does not consider the possible contributions of position effect (discussed below), but it is consistent with radiation-induced *hprt* mutation frequencies (e.g., *19,75,118,137)*, suggesting that symmetrical exchanges could contribute significantly to radiation-induced mutation. In fact, it is somewhat surprising that mutants harboring such changes are not recovered with greater abundance.

Unlike asymmetrical exchanges, inversions and reciprocal translocations do not usually produce acentric fragments. As such, they are not associated with gross loss of genetic information that often leads to cell death. The breakpoint junctions of acentric fragments associated with interstitial deletions are not recoverable, but sequence analysis of inversions and translocations can provide information about both junctions of the exchange that would be useful in determining the extent of postirradiation lesion processing. Although symmetrical and asymmetrical exchanges likely occur via similar pathways, comparative molecular sequence analysis will be needed to substantiate this assumption. Since both inversions and translocations have been repeatedly implicated in carcinogenesis *(71,138)*, it is unfortunate that no sequence information is yet available for radiation-induced, microscopically visible inversions or translocations. Nevertheless, there are a number of studies that ostensibly provide insight into their formation.

Because radiation-induced inversions and interstitial deletions are found with similar frequencies at the cytogenetic level *(76)*, it is not surprising that inversions are also significant contributors to radiation-induced mutation. Urlaub et al. *(128)* examined 11 γ-ray induced *dhfr* mutants in CHO cells; all had major structural changes. Three were disruptions of *DHFR*, without any detectable sequence loss (by Southern blotting). At least two were accompanied by inversions that were visible at the cytogenetic level, indicating that inversion was second only to deletion as the most common type of mutation formed by low LET radiation.

Thacker *(116)* described an *hprt* mutant of hamster V-79 cells produced following exposure to α-particles that contained a large X-chromosome inversion at or near *HPRT*, yet produced normal Southern hybridization patterns in several different restriction digests *(117)*. Cattanach et al. *(18)* showed that nearly 80% of specific locus mutations in the progeny of X-irradiated mice involved gross rearrangements that were cytologically visible, including deletions comprising 2.5–30% the length of target chromosomes; of these, about 15% involved either inversions or translocations. Inversion mutants in these studies may have been caused by a breakpoint(s) that physically disrupted gene sequences *(116)*. However, both groups speculated that the inversion may have led to mutation through classical "position effect," whereby DNA near heterochromatin becomes transcriptionally silenced (*4*; *see also* ref. *29*). This explanation is supported for the *hprt* mutant in question, since the inversion placed the gene near a large region of heterochromatin. If this interpretation turns out to be correct, it would have far ranging implications for radiation mutagenesis. For example, it would radically alter our perception of what is considered the "target size" for virtually all large-scale mutational events.

Translocations are also significant contributors to radiation mutagenesis. Cox and Mason *(29)* found that some 40% of IR-induced *hrpt* mutations of primary human fibroblasts were associated with cytogenetic changes in the long arm of the X chromosome. About half of these mapped to Xq26 (the location of *HPRT*), most of which were translocations; the remainder were deletions. These authors did not specifically consider the possibility that simple translocations may have directly disrupted target gene sequences. Instead, they postulated that loss of *HPRT* function was either a consequence of position effect *(18,116)* or that submicroscopic *hprt* deletions may have accompanied the translocation. The latter interpretation may prove to be prophetic, in light of more recent data discussed below. Many studies have implicated translocations in mutagenesis. Translocation breakpoints at (or very near) target gene sequences have been demonstrated cytogenetically, but their involvement has also been inferred from molecular analysis *(70)*.

Only one radiation-induced "translocation mutation" has been identified cytogenetically and characterized at the molecular level *(111)*, from a set of seven primary human mutants previously known to contain complete or partial *hprt* deletions. Interestingly,

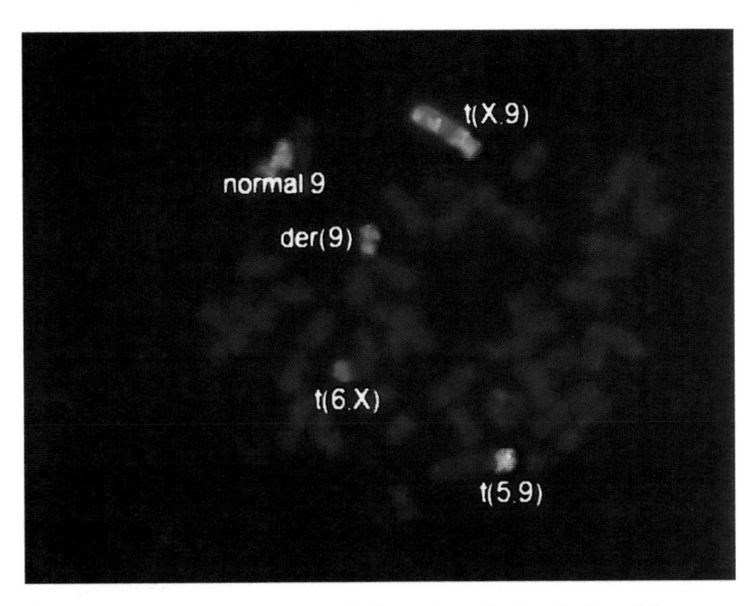

Fig. 3. Multicolor whole-chromosome painting of a radiation-induced *hprt* mutant. Only the X (green, FITC) and chromosome 9 (red, rhodamine) are probed in the photomicrograph; the remaining chromosomes have been counterstained with DAPI (blue). A portion of chromosome 9 is translocated to the terminal q-arm portion of the X at or near Xq26. Note, however, that the resulting acentric X-fragment of the exchange is not reciprocally translocated to chromosome 9, but instead has joined with chromosome 6 (not probed). The derivative 9 is also involved in a second exchange with chromosome 2 (not probed). In all, there are five chromosomes involved in this complex rearrangement.

this mutant was shown (by high-resolution G-banding and whole-chromosome painting) to be associated with a translocation between the terminal short arm of chromosome 11 and the distal long arm of X. As is often the case *(101)*, G-banding was not successful in unequivocally resolving chromosomal breakpoints of the exchange. Nevertheless, it would appear that the translocation must have involved an Xq26 breakpoint. Of significance is the fact that the translocation was associated with a >1-Mbp total gene deletion. In addition, two other total deletion mutants in this study were proposed to harbor "microtranslocations" (presumably insertions) involving DNA from the Xq26 region.

Two putative γ-ray-induced *hprt* translocation mutants of human epithelial origin have been shown to harbor X-autosomal chromosome translocations that involve Xq26 *(22)*. In agreement with the results of Simpson et al. *(111)*, one of these suffered a total deletion of *HPRT*. Multicolor chromosome painting indicates that the translocation is part of a complex rearrangement, involving four additional autosomes (Fig. 3). In contrast, the other mutant contained a simple reciprocal translocation between X and chromosome 13, yet produced normal Southern hybridization patterns. The mutation clearly does not involve a large-scale deletion in *HPRT* and probably arose by a simple reciprocal translocation with one breakpoint in the gene.

These results, though limited, indicate that exchange processes can be more complex than classical cytogenetic theory would suggest, with large-scale deletions sometimes associated with translocations (and probably also with inversions). Important unanswered questions include whether codeletion occurs before repair (e.g., as a result of nonspecific degradation at damaged sites) or whether it occurs as an integral part of the recombination process *(111)*. Also, are translocation/codeletion events diagnostic of IR damage, or do other clastogens evoke this response as well? Some translocations, however, do not involve large-scale loss of sequences. At the molecular level, are these essentially reciprocal events?

From the standpoint of symmetrical exchanges, interesting information derives from the analysis of restriction enzyme-induced *aprt* mutants that failed to support PCR amplification across target sequences *(86)*. About half of these appeared to contain the entire *APRT* sequence and gave normal Southern hybridization patterns when digested with the enzyme used to produce the mutation. These mutants, however, yielded abnormal Southern patterns when restricted with other enzymes, consistent with large inversion or translocation events—bona fide exchange events that conceivably involve rearrangements detectable at the microscopic level. If this is true, it would imply that a simple blunt-ended rejoining mechanism can account for a large fraction of radiation-induced exchanges. It would also serve to validate earlier assumptions concerning the ability of restriction enzymes to cut chromosomal DNA in vivo with defined specificity (e.g., without significant star activity). Finally, and perhaps most importantly, it provides strong molecular evidence that illegitimate chromosomal exchange occurs through the pairwise interaction of separate lesions. In the context of IR, this supports the view that exchanges are truly "two-hit" in nature, a notion considered axiomatic by most radiation cytogeneticists *(21,26)*, but that, as discussed in Section 1.2., has been challenged on the basis of biophysical *(44,45,62,121)* and molecular arguments *(118)*.

5. CONCLUDING REMARKS

Insights into the mechanisms of radiation-induced aberration formation derive from a variety of approaches, biophysical, biochemical, cellular, and molecular. Conclusions drawn from these sources will be by way of general consensus, which is by no means universal. Nevertheless, there are some firm conclusions that can be made, as well a number of others that can be offered *pro tem*.

Although indirect, the bulk of data supports the view that the most important initial radiogenic lesions producing aberrations are DSBs, each of which are produced by the passage of a single charged particle track. Clearly, DSBs are necessary, if not sufficient, for aberration formation. Explaining the increased clastogenic effect that accompanies increases in ionization density then raises the question of whether densely IRs produce a qualitatively different (e.g., more severe) kind of DSB than sparsely IRs or, alternatively, whether the high LET effect derives from a more effective spatial distribution of DSBs. For now, the latter explanation is preferred for its simplicity. It is clear that LET-dependent differences in track structure can lead to patterns of energy deposition that are qualitatively different with respect to theoretical target volumes, and these volumes may well relate in size to critical cellular components. However, whether such differences translate into classes of lesions that the cell recognizes as being fundamentally different is another matter. Despite the intuitive appeal of the qualitative argument, it does not need to be invoked to explain the vast majority of data. On that basis, the model must be rejected until the additional complexity it brings can be justified; this may require the development of additional biochemical measures of lesion severity.

It is clear that the vast majority of radiation damage manifested at the chromosomal level is in the form of exchange aberrations, but whether these arise from a single DSB or "pairs" of radiation-induced breaks via lesion interaction, remains contested. Previously relegated to abstract biophysical arguments, this issue reappears as perhaps the most important point of controversy regarding radiation-induced damage to chromosomes.

The study of illegitimate recombination (i.e., misrepair resulting in exchange) by PCC analysis has produced data both supporting and contradicting the conventional wisdom. In support are the facts that the kinetics of the removal of lesions giving rise to PCC breaks are not affected by lesion burden (i.e., dose), nor does it appear to be affected by changes in the ionization density of radiations, as might be expected if high LET radiations were capable of producing a more "severe" lesion class than low LET radiations. Also, the initial dose response for PCC breaks is linear, whereas residual damage following full repair becomes curvilinear, as would be expected under the tenets of lesion interaction. Some studies have concluded that the kinetics governing the formation of PCC break repair and those underlying exchange formation are similar. However, there are a growing number of instances where the kinetics of the two end points appear to be dramatically different, and it is no longer possible to dismiss these conclusions as artifact. Whereas alternative models may be constructed to restore consistency with the traditional viewpoint that repair and misrepair are alternative outcomes to a similar molecular process, even traditionalists are now forced at least to consider the possibility that they are not.

Observations that indicate unequal frequencies between symmetrical and asymmetrical exchanges recovered in cells at their first postirradiation mitosis are vexing to all models of aberration formation, and no adequate explanation for the phenomenon exists. Certainly the newly appreciated complexity associated with exchanges tends to cloud the conclusion that symmetrical to asymmetrical ratios do, in fact, deviate from unity, but, to date, no evidence has been offered to suggest that this complexity itself leads to a systematic misclassification of aberration types. It is difficult to imagine, given our current knowledge of interphase chromosome structure, how a cell could recognize the asymmetry of new linkage relationships relative to the placement of centromeres that, in many instances, are several tens of megabases removed from the actual recombinational event. Conceivably, a subset of translocations might exist whose formation is dependent on limited homology between certain abundant (e.g., repetitive) sequences, and that these sequences tend to be oriented in the same 5' $\rightarrow$ 3' direction with respect to centromeres.

The caveats of Section 4. notwithstanding, the available data tell us that exchange formation is driven by recombinational processes that are nonhomologous in nature. Consequently, "one-hit" models of aberration formation based on extensive sequence homologies must also be rejected; this would appear to include interactions proposed to occur via substantial homologies provided by repetitive elements. At least in that sense, molecular studies do little to refute classical cytogenetic models based on lesion interaction. However, it is becoming increasing apparent that if the classical viewpoint is correct, then it is correct only as an approximation; aberration formation is almost certainly more complicated at the "nuts and bolts" level. A case in point is recurrent observations suggesting that recombination may utilize very short sequence homologies, in some instances strongly implicating a template-directed repair/misrepair process. Additionally, it appears as though deletions are sometimes (perhaps even typically) associated with otherwise symmetrical events. Obviously, much work remains to be done in the molecular arena before a meaningful consensus emerges on these and other points.

Whole chromosome painting studies aptly demonstrate that our naiveté has also extended beyond what might have been expected at the sequence level of breakpoint junctures themselves. For example, a significant portion of radiation-induced exchanges are complex; therefore, some of these must involve multiple recombinational events occurring among several chromosomes. It is worth pondering how this might occur in the context of a spatially constrained recombinational process that must account for the fact that chromosomes during interphase occupy discrete domains *(63)* whose boundaries severely limit interchromosomal contact. The answers to these and other questions related to aberration formation may prove elusive until further insight is gained about the architecture of interphase nuclei.

ACKNOWLEDGMENTS

I thank Joel S. Bedford, William F. Morgan, and Bradford D. Loucas for their suggestions on the manuscript. Thanks also given to John Thacker for his input and for making available unpublished results.

REFERENCES

1. Abrahamson, S. and S. Wolff. 1976. Re-analysis of radiation induced specific locus mutations of the mouse. *Nature* **264:** 715–719.
2. Ager, D. D., J. W. Phillips, E. A. Columna, R. A. Winegar, and W. F. Morgan. 1991. Analysis of restriction enzyme-induced DNA double-strand breaks in Chinese hamster ovary cells by pulsed-field gel electrophoresis: implications for chromosome damage. *Radiat. Res.* **128:** 150–156.
3. Bahari, I. B., J. S. Bedford, A. J. Giaccia, and T. D. Stamato. 1990. Measurement of the relative proportion of symmetrical and asymmetrical chromosome-type interchanges induced by γ-irradiation in human-hamster hybrid cells. *Radiat. Res.* **123:** 105–107.
4. Baker, W. K. 1968. Position-effect variegation. *Adv. Genet.* **14:** 133–169.
5. Bauchinger, M., E. Schmid, H. Zitselsberger, H. Braselmann, and U. Nahrstedt. 1993. Radiation-induced chromosome aberrations analysed by two-colour fluorescence *in situ* hybridization with composite whole chromosome-specific DNA probes and a pancentromeric DNA probe. *Int. J. Radiat. Biol.* **64:** 179–184.
6. Bedford, J. S. and M. N Cornforth. 1987. Relationship between the recovery from sublethal X-ray damage and the rejoining of chromosome breaks in normal human fibroblasts. *Radiat. Res.* **111:** 406–423.
7. Bedford, J. S. and D. T. Goodhead. 1989. Breakage of human interphase chromosomes by alpha particles and X-rays. *Int. J. Radiat. Biol.* **55:** 211–216.
8. Bender, M. A, H. G. Griggs, and J. S. Bedford. 1974. Mechanisms of chromosomal aberration production III. Chemicals and ionizing radiation. *Mutat. Res.* **23:** 197–212.
9. Blöcher, D. 1982. DNA double strand breaks in Ehrlich ascites tumor cells at low doses of X-rays. I. Determination of induced breaks by centrifugation at reduced speed. *Int. J. Radiat. Biol.* **42:** 317–328.

10. Blunt, T., N. J. Finnie, G. E. Taccioli, G. C. Smith, J. Demengeot, T. M. Gottlieb, R. Mizuta, A. J. Varghese, F. W. Alt, and P. A. Jeggo. 1995. Defective DNA-dependent protein kinase activity is linked to V(D)J recombination in DNA repair defects associated with the murine scid mutation. *Cell* **80:** 813–823.
11. Brenner, D. J. 1990. Track structure, lesion development, and cell survival. *Radiat. Res.* **124:** S29–S37.
12. Brenner, D. J. and R. K. Sachs. 1994. Chromosomal "fingerprints" of prior exposure to densely ionizing radiation. *Radiat. Res.* **140:** 134–142.
13. Brewen, J. G. and R. D. Brock. 1968. The exchange hypothesis and chromosome-type aberrations. *Mutat. Res.* **6:** 245–255.
14. Brown, J. M. and M. S. Kovacs. 1993. Visualization of non-reciprocal chromosome exchanges in irradiated human fibroblasts by fluorescent *in situ* hybridization. *Radiat. Res.* **136:** 71–76.
15. Brown, J. M., J. W. Evans, and M. S. Kovacs. 1993. Mechanism of chromosome exchange formation in human fibroblasts: insights from chromosome painting. *Environ. Mol. Mutagen.* **22:** 218–224.
16. Bryant, P. E. 1984. Enzymatic restriction of mammalian cell DNA using Pvu II and Bam H1: Evidence for the double-strand break origin of chromosomal aberrations. *Int. J. Radiat. Biol.* **46:** 57–65.
17. Caspersson, T., U. Haglund, B. Lindell, and L. Zech. 1972. Radiation-induced non-random chromosome breakage. *Exp. Cell Res.* **75:** 541–543.
18. Cattanach, B. M., M. D. Burtenshaw, C. Rasberry, and E. P. Evans. 1993. Large deletions and other gross forms of chromosome imbalance compatible with viability and fertility in the mouse. *Nature Genet.* **3:** 56–61.
19. Chen, D. J., G. F. Strniste, and N. Tokita. 1984. The genotoxicity of alpha particles in human embryonic skin fibroblasts. *Radiat. Res.* **100:** 321–327.
20. Cornforth, M. N. 1989. On the nature of interactions leading to radiation-induced chromosomal exchange. *Int. J. Radiat. Biol.* **56:** 635–643.
21. Cornforth, M. N. 1990. Testing the notion of the one-hit exchange. *Radiat. Res.* **121:** 21–27.
22. Cornforth, M. N. 1996. Unpublished results.
23. Cornforth, M. N. and J. S. Bedford. 1983. X-ray-induced breakage and rejoining of human interphase chromosomes. *Science* **222:** 1141–1143.
24. Cornforth, M. N. and J. S. Bedford. 1985. On the nature of a defect in cells from individuals with Ataxia Telangiectasia. *Science* **227:** 1589–1591.
25. Cornforth. M. N. and J. S. Bedford. 1987. A quantitative comparison of potentially lethal damage repair and the rejoining of interphase chromosome breaks in low passage normal human fibroblasts. *Radiat. Res.* **111:** 385–405.
26. Cornforth, M. N. and J. S. Bedford. 1993. Ionizing radiation damage and its early development in chromosomes, in *Advances in Radiation Biology*, vol. 17 (Lett, J. T. and W. K. Sinclair, eds.), Academic, San Diego, pp. 423–496.
27. Cornforth, M. N. and E. H. Goodwin. 1991. The dose-dependent fragmentation of chromatin in human fibroblasts by 3. 5 MeV α particles from 238Pu: Experimental and theoretical considerations pertaining to single-track effects. *Radiat. Res.* **127:** 64–74.
28. Cornforth, M. N., J. Meyne, L. G. Littlefield, S. M. Bailey, and R. K. Moyzis. 1989. Telomere staining of human chromosomes and the mechanism of radiation-induced dicentric formation. *Radiat. Res.* **120:** 205–212.
29. Cox, R. and W. K. Mason. 1978. Do radiation-induced thioguanine resistant mutants of cultured mammalian cells arise by HGPRT gene mutation or X-chromosome rearrangement? *Nature* **276:** 629–630.
30. Cremer, T., S. P. Popp, P. Emmerich, P. Lichter, and C. Cremer. 1990. Rapid metaphase and interphase detection of radiation-induced aberrations in human lymphocytes by chromosomal suppression *in situ* hybridisation. *Cytometry* **11:** 110–118.

31. Dewey, W. C., H. H. Miller, and D. B. Leeper. 1971. Chromosomal aberrations and mortality of X-irradiated mammalian cells: emphasis on repair. *Proc. Natl. Acad. Sci. USA* **68:** 667–671.
32. Evans, H. H., M. Nielsen, J. Mencl, M.-F. Horng, and M. Ricanati. 1990. The effect of dose rate on X-radiation-induced mutant frequency and the nature of DNA lesions in mouse lymphoma cells. *Radiat. Res.* **122:** 316–325.
33. Evans, J. W., X. F. Liu, C. U. Kirchgessner, and J. M. Brown. 1996. Induction and repair of chromosome damage in scid cells measured by premature chromosome condensation. *Radiat. Res.* **145:** 39–46.
34. Finnon, P., D. C. Lloyd, and A. A. Edwards. 1995. Fluorescence *in situ* hybridization detection of chromosomal aberrations in human lymphocytes: applicability to biological dosimetry. *Int. J. Radiat. Biol.* **68:** 429–435.
35. Folle, G. A. and G. Obe. 1995. Localization of chromosome breakpoints induced by AluI and BamHI in Chinese hamster ovary cells treated in the G1 phase of the cell cycle. *Int. J. Radiat. Biol.* **68:** 437–445.
36. Fox, D. P. 1967. The effects of X-rays on the chromosomes of locust embryos. III. The chromatid aberration types. *Chromosoma (Berl.)* **20:** 386–412.
37. Fuscoe, J. C., C. H. Ockey, and M. Fox. 1986. Molecular analysis of X-ray-induced mutants at the HPRT locus in V79 Chinese hamster cells. *Int. J. Radiat. Biol.* **49:** 1011–1020.
38. Fuscoe, J. C., L. J. Zimmerman, K. Harrington-Brock, L. Burnette, M. M. Moore, J. A. Nicklas, J. P. O'Neill, and R. J. Albertini. 1992. V(D)J recombinase-mediated deletion of the *hprt* gene in T-lymphocytes from adult humans. *Mutat. Res.* **283:** 13–20.
39. Geard, C. R. 1985. Charged particle cytogenetics: Effects of LET, fluence, and particle separation on chromosome aberrations. *Radiat. Res.* **104s:** s112–s121.
40. Geard, C. R. and G. Jenkins. 1995. Human chromosome-specific changes in a human-hamster hybrid cell line (A_L) assessed by fluorescence *in situ* hybridization (FISH). *Int. J. Radiat. Oncol. Biol. Phys.* **32:** 113–120.
41. Goodhead, D. T. 1985. Saturable repair models of radiation action in mammalian cells. *Radiat. Res.* **104s:** s58–s67.
42. Goodhead, D. T. and H. Nikjoo. 1989. Track structure analysis of ultrasoft X-rays compared to high and low LET radiations. *Int. J. Radiat. Biol.* **55:** 513–529.
43. Greinert, R., E. Detzler, B. Volkmer, and D. Harder. 1995. Kinetics of the formation of chromosome aberrations in x-irradiated human lymphocytes: analysis by premature chromosome condensation with delayed fusion. *Radiat. Res.* **144:** 190–197.
44. Griffin, C. S., S. J. Marsden, D. L. Stevens, P. Simpson, and J. R. K. Savage. 1995. Frequencies of complex chromosome exchange aberrations induced by ^{238}Pu α-particles and detected by fluorescence *in situ* hybridization using single chromosome-specific probes. *Int. J. Radiat. Biol.* **67:** 431–439.
45. Griffin, C. S., D. L. Stevens, and J. R. K. Savage. 1996. Ultrasoft 1. 5 keV Al K X-rays are efficient producers of complex exchange aberrations as revealed by fluorescence *in situ* hybridization. *Radiat. Res.* **146:** 144–150.
46. Grosovsky, A. J., J. G. de Boer, P. J. de Jong, E. A. Drobetsky, and B. W. Glickman. 1988. Base substitutions, frameshifts, and small deletions constitute ionizing radiation-induced point mutations in mammalian cells. *Proc. Natl. Acad. Sci. USA* **85:** 185–188.
47. Hittelman, W. N. and P. N. Rao. 1974. Premature chromosome condensation. I. Visualization of X-ray induced chromosome damage in interphase cells. *Mutat. Res.* **23:** 251–258.
48. Hlatky, L., R. K. Sachs, and P. Hahnfeldt. 1992. The ratio of dicentrics to centric rings produced in human lymphocytes by acute low-LET radiation. *Radiat. Res.* **129:** 304–308.
49. Holmberg, M. and J. Jonasson. 1973. Preferential location of X-ray induced chromosome breakage in the R-bands of human chromosomes. *Heriditas* **74:** 57–68.
50. Holmquist, G. P. 1992. Chromosome bands, their chromatin flavors, and their functional features. *Am. J. Hum. Genet.* **51:** 17–37.

51. Iliakis, G. E. and G. E. Pantelias. 1990. Production and repair of chromosome damage in an X-ray sensitive CHO mutant visualized and analyzed in interphase using the technique of premature chromosome condensation. *Int. J. Radiat. Biol.* **57:** 1213–1223.
52. Iliakis, G., R. Okayasu, J. Varlotto, C. Shernoff, and Y. Wang. 1994. Hypertonic treatment during premature chromosome condensation allows visualization of interphase chromosome breaks repaired with fast kinetics in irradiated CHO cells. *Radiat. Res.* **135:** 160–170.
53. ISCN. 1985. *An International System for Human Cytogenetic Nomenclature* (Harnden, D. G. and H. P. Kliinger, eds.), S. Karger, Basel/New York, pp. 1–117.
54. Ishida, R. and T. Takahashi. 1975. Increased DNA strand breakage by combined action of bleomycin and superoxide radical. *Biochem. Biophys. Res. Commun.* **66:** 1432–1438.
55. Johnson, R. T. and P. N. Rao. 1970. Mammalian cell fusion. II. Induction of premature chromosome condensation in interphase nuclei. *Nature* **226:** 717–722.
56. Kellerer, A. M. and H. H. Rossi. 1972. The theory of dual radiation action. *Curr. Top. Radiat. Res. Q.* **8:** 85–158.
57. King, J. S., E. R. Valcarcel, J. T. Rufer, J. W. Phillips, and W. F. Morgan. 1993. Noncomplementary DNA double-strand-break rejoining in bacterial and mammalian cells. *Nucleic Acids Res.* **21:** 1055–1059.
58. Knehr, S., H. Zitzelsberger, H. Braselmann, and M. Bauchinger. 1994. Analysis of DNA-proportional distribution of radiation-induced chromosome aberrations in various triple combinations of human chromosomes using fluorescence *in situ* hybridization. *Int. J. Radiat. Biol.* **65:** 683–690.
59. Kucherlapati, R. S., E. M. Eves, D. Song, B. S. Morse, and O. Smithies. 1984. Homologous recombination between plasmids in mammalian cells can be enhanced by treatment of input DNA. *Proc. Natl. Acad. Sci. USA* **81:** 3153–3157.
60. Lange, C. S., A. Cole, and J. Y. Ostashevsky. 1993. Radiation-induced damage in chromosomal DNA molecules: deduction of chromosomal DNA organization from the hydrodynamic data used to measure DNA double-strand breaks and from stereo electron microscopic observations, in *Advances in Radiation Biology*, vol. 17 (Lett, J. T. and W. K. Sinclair, eds.), Academic, San Diego, pp. 261–421.
61. Lea, D. E. 1946. *Actions of Radiations on Living Cells*, 2nd ed. Cambridge University Press 1955, London.
62. Leenhouts, H. P. and K. H. Chadwick. 1978. The crucial role of DNA double strand breaks in cellular radiobiological effects. *Advan. Radiat. Res.* **9:** 55–101.
63. Lichter, P., T. Cremer, J. Borden, L. Manuelidis, and D. C. Ward. 1988. Delineation of individual human chromosomes in metaphase and interphase cells by *in situ* hybridisation using recombinant DNA libraries. *Hum. Genet.* **80:** 224–238.
64. Lloyd, D. C. and A. A. Edwards. 1983. Chromosome aberrations in human lymphocytes: effect of radiation quality, dose, and dose rate, in *Radiation-Induced Chromosome Damage in Man* (Ishihara, T. and M. S. Sasaki, eds.), Liss, New York, pp. 23–49.
65. Loucas, B. D. and C. R. Geard. 1994. Kinetics of chromosome rejoining in normal human fibroblasts after exposure to low- and high-LET radiations. *Radiat. Res.* **138:** 352–360.
66. Lucas, J. N. and R. K. Sachs. 1993. Using three-color chromosome painting to test chromosome aberration models. *Proc. Natl. Acad. Sci. USA* **90:** 1484–1487.
67. Lucas, J. N., A. Awa, T. Straume, M. Poggensee, Y. Kodama, M. Nakano, K. Ohtaki, H.-U. Weier, D. Pinkel, J. Gray, and G. Littlefield. 1992. Rapid translocation frequency analysis in humans decades after exposure to ionizing radiation. *Int. J. Radiat. Biol.* **62:** 53–63.
68. Lucas, J. N., T. Tenjin, T. Straume, D. Pinkel, D. Moore, M. Litt, and J. W. Gray. 1989. Rapid human chromosome aberration analysis using fluorescence *in situ* hybridization. *Int. J. Radiat. Biol.* **56:** 34,35.
69. Meuth, M. 1989. Illegitimate recombination in mammalian cells, in *Mobile DNA* (Berg, D. E. and M. M. Howe, eds.), American Society for Microbiology, Washington, pp. 833–860.

70. Meuth, M. 1990. The structure of mutation in mammmalian cells. *Biochim. Biophys. Acta* **1032:** 1–17.
71. Mitelman, F. 1991. *Catalog of Chromosome Aberrations in Cancer*, 4th ed., Wiley-Liss, New York, 1991.
72. Morgan, W. F. and R. A. Winegar. 1990. The use of restriction endonucleases to study the mechanisms of chromosome damage, in *Chromosomal Aberrations: Basic and Applied Aspects* (Obe, G. and A. T. Natarajan, eds.), Springer-Verlag, Berlin, pp. 70–78.
73. Morgan, W. F., J. P. Day, M. I. Kaplan, E. M. McGhee, and C. L. Limoli. 1996. Genomic instability induced by ionizing radiation. *Radiat. Res.* **146:** 247–258.
74. Morris, T. and J. Thacker. 1993. Formation of large deletions by illegitimate recombination in the HPRT gene of primary human fibroblasts. *Proc. Natl. Acad. Sci. USA* **90:** 1392–1396.
75. Morris, T., W. Masson, B. Singleton, and J. Thacker. 1993. Analysis of large deletions in the HPRT gene of primary human fibroblasts using the polymerase chain reaction. *Somatic Cell Mol. Genet.* **19:** 9–19.
76. Mühlmann-Diaz, M. C., and J. S. Bedford. 1995. Comparison of gamma-ray-induced chromosome ring and inversion frequencies. *Rad. Res.* **143:** 175–180.
77. Nakano, M., E. Nakashima, D. J. Pawel, Y. Kodama, and A. Awa. 1993. Frequency of reciprocal translocations and dicentrics induced in human blood lymphocytes by X-irradiation as determined by fluorescence *in situ* hybridization. *Int. J. Radiat. Biol.* **64:** 565–569.
78. Nakatsu, Y., R. F. Tyndale, T. M. DeLorey, D. Durham-Pierre, J. M. Gardner, H. J. McDanel, Q. Nguyen, J. Wagstaff, M. Lalande, J. M. Sikela, R. W. Olsen, A. J. Tobin, and M. H. Brilliant. 1993. A cluster of three $GABA_A$ receptor subunit genes is deleted in a neurological mutant of the mouse *p* locus. *Nature* **364:** 448–450.
79. Natarajan, A. T. and G. Obe. 1984. Molecular mechanisms involved in the production of chromosomal aberrations. III. Restriction endonucleases. *Chromosoma (Berl.)* **90:** 120–127.
80. Natarajan, A. T. and T. S. B. Zwanenburg. 1982. Mechanisms for chromosomal aberrations in mammalian cells. *Mutat. Res.* **95:** 1–6.
81. Natarajan, A. T., G. Obe, A. A. van Zeeland, F. Palitti, M. Meijers, and E. A. M. Verdegaal-Immerzeel. 1980. Molecular mechanisms involved in the production of chromosomal aberrations, II. Utilization of Neurospora endonculease for the study of aberration production by X-rays in G1 and G2 stages of the cell cycle. *Mutat. Res.* **69:** 293–305.
82. Natarajan, A. T., R. C. Vyas, F. Darroudi, and S. Vermeulen. 1992. Frequencies of X-ray induced chromosome translocation in human peripheral lymphocytes as detected by *in situ* hybridization using chromosome-specific DNA libraries. *Int. J. Radiat. Biol.* **61:** 199–203.
83. Nevaldine, B., J. A. Longo, M. Vilenchik, G. A. King, and P. J. Hahn. 1994. Induction and repair of DNA double-strand breaks in the same dose range as the shoulder of the survival curve. *Radiat. Res.* **140:** 161–165.
84. Okayasu, R. and G. Iliakis. 1994. Evidence that the product of the xrs gene is predominantly involved in the repair of a subset of radiation-induced interphase chromosome breaks rejoining with fast kinetics. *Radiat. Res.* **138:** 34–43.
85. Pantelias, G. E. and H. D. Maillie. 1985. Direct analysis of radiation-induced chromosome fragments and rings in unstimulated human peripheral blood lymphocytes by means of the premature chromosome condensation technique. *Mutat. Res.* **149:** 67–72.
86. Phillips, J. W. and W. F. Morgan. 1994. Illegitimate recombination induced by DNA double-strand breaks in a mammalian chromosome. *Mol. Cell. Biol.* **14:** 5794–5803.
87. Preston, R. J. 1982. The use of inhibitors of DNA repair in the study of the mechanisms of chromosome aberrations. *Cytogenet. Cell Genet.* **33:** 20–26.
88. Prise, K. M., M. Folkard, H. C. Newman, and B. D. Michael. 1994. Effect of radiation quality on lesion complexity in cellular DNA. *Int. J. Radiat. Biol.* **66:** 537–542.
89. Rabbitts, T. H. 1994. Chromosomal translocations in human cancer. *Nature* **372:** 143–149.

90. Rao, P. N., R. T. Johnson, and K. Sperling (eds.). 1982. *Premature Chromosome Condensation: Application in Basic, Clinical, and Mutation Research.* Academic, New York.
91. Resnick, M. A. and P. D. Moore. 1979. Molecular recombination and the repair of DNA double-strand breaks in CHO cells. *Nucleic Acids Res.* **6:** 3145–3160.
92. Revell, S. H. 1959. The accurate estimation of chromatid breakage, and its relevance to a new interpretation of chromatid aberrations induced by ionizing radiations. *Proc. R. Soc. (Lond.) Ser. B* **150:** 563–589.
93. Revell, S. H. 1983. Relationship between chromosome damage and cell death, in *Radiation-Induced Chromosome Damage in Man* (Ishihara, T. and M. S. Sasaki, eds.), Liss, New York, pp. 215–233.
94. Roth, D. B. and J. H. Wilson. 1986. Nonhomologous recombination in mammalian cells: role for short sequence homologies in the joining reaction. *Mol. Cell. Biol.* **6:** 4295–4304.
95. Roth, D. B., T. N. Porter, and J. H. Wilson. 1985. Mechanisms of nonhomologous recombination in mammalian cells. *Mol. Cell. Biol.* **5:** 2599–2607.
96. Sabatier, L., W. Al Achkar, F. Hoffschir, C. Luccioni, and B. Dutrillaux. 1987. Qualitative study of chromosomal lesions induced by neutrons and neon ions in human lymphocytes at G_0 phase. *Mutat. Res.* **178:** 91–97.
97. Sachs, R. K., A. A. Awa, Y. Kodama, M. Nakano, K. Ohtaki, and J. N. Lucas. 1993. Ratios of radiation-produced chromosome aberrations as indicators of large-scale DNA geometry during interphase. *Radiat. Res.* **133:** 345–350.
98. Sanchez-Reyes, A. 1992. A simple model of radiation action in cells based on a repair saturation mechanism. *Radiat. Res.* **130:** 139–147.
99. San Roman, C. and M. Bobrow. 1973. The sites of radiation induced breakage in human lymphocyte chromosomes determined by quinacrine fluorescence. *Mutat. Res.* **18:** 325–331.
100. Savage, J. R. K. 1975. Classification and relationships of induced chromosomal structural changes. J. *Med. Genet.* **12:** 103–122.
101. Savage, J. R. K. 1977. Assignment of aberration break points in banded chromosomes. *Nature* **270:** 513,514.
102. Savage, J. R. K. and D. G. Papworth. 1982. Frequency and distribution of asymmetrical versus symmetrical chromosome aberrations. *Mutat. Res.* **95:** 7–18.
103. Savage, J. R. K., and P. Simpson. 1994. FISH "painting" patterns resulting from complex exchanges. *Mutat. Res.* **312:** 51–60.
104. Sax, K. 1938. Chromosome aberrations induced by X-rays. *Genetics* **23:** 494–516.
105. Sax, K. 1939. Time factor in X-ray production of chromosome aberrations. *Proc. Natl. Acad. Sci. USA* **25:** 225–233.
106. Schmid, E., M. Bauchinger, and W. Mergenthaler. 1976. Analysis of the time relationship for the interaction of X-ray-induced primary breaks in the formation of dicentric chromosomes. *Int. J. Radiat. Biol.* **30:** 339–346.
107. Schmid, E., H. Zitselsberger, H. Braselman, J. W. Gray, and M. Bauchinger. 1992. Radiation-induced chromosome aberrations analysed by fluorescence *in situ* hybridization with a triple combination of composite whole chromosome-specific DNA probes. *Int. J. Radiat. Biol.* **62:** 673–678.
108. Sen, P. and W. N. Hittelman. 1984. Induction of chromosome damage by *Neurospora* endonuclease in repair-inhibited quiescent normal human fibroblasts. *Mutat. Res.* **129:** 359–364.
109. Simpson, P. J., and J. R. K. Savage. 1994. Identification of X-ray-induced complex chromosome exchanges using fluorescence *in situ* hybridization: a comparison at two doses. *Int. J. Radiat. Biol.* **66:** 629–632.
110. Simpson, P. J. and J. R. K. Savage. 1995. Estimating the true frequency of X-ray-induced complex chromosome exchanges using fluorescence *in situ* hybridization. *Int. J. Radiat. Biol.* **67:** 37–45.

111. Simpson, P., T. Morris, J. Savage, and J. Thacker. 1993. High-resolution cytogenetic analysis of X-ray induced mutations of the HPRT gene of primary human fibroblasts. *Cytogenet. Cell Genet.* **64:** 39–45.
112. Solomon, E., J. Borrow, and A. D. Goddard. 1991. Chromosome aberrations and cancer. *Science* **254:** 1153–1160.
113. Straume, T., and J. N. Lucas. 1993. Comparison of the yields of translocations and dicentrics measured using fluorescence *in situ* hybridization. *Int. J. Radiat. Biol.* **64:** 185–187.
114. Strobel, M. C., P. K. Seperack, N. G. Copeland, and N. A. Jenkins. 1990. Molecular analysis of two mouse dilute locus deletion mutations: spontaneous dilute lethal^{20J} and radiation-induced dilute prenatal lethal Aa2 alleles. *Mol. Cell. Biol.* **10:** 501–509.
115. Taccioli, G. E., T. M. Gottlieb, T. Blunt, A. Priestley, J. Demengoet, R. Mizuta, A. R. Lehmann, F. W. Alt, S. P. Jackson, and P. A. Jeggo. 1994. Ku80: product of the *XRCC5* gene and its role in DNA repair and V(D)J recombination. *Science* **265:** 1442–1445.
116. Thacker, J. 1981. The chromosomes of a V79 Chinese hamster line and a mutant subline lacking HPRT activity. *Cytogenet. Cell Genet.* **29:** 16–25.
117. Thacker, J. 1986. The nature of mutants induced by ionising radiation in cultured hamster cells. III. Molecular characterization of HPRT-deficient mutants induced by γ-rays or α particles showing that the majority have deletions of all or part of the *hprt* gene. *Mutat. Res.* **160:** 267–275.
118. Thacker, J. 1992. Radiation-induced mutation in mammalian cells at low doses and dose rates, in *Advances, in Radiation Biology*, vol. 16 (Nygaard, O. F., W. K. Sinclair, and J. T. Lett, eds.), Academic, San Diego, pp. 77–124.
119. Thacker, J. and R. Cox. 1983. The relationship between specific chromosome aberrations and radiation-induced mutations in cultured mammalian cells, in *Radiation-Induced Chromosome Damage in Man* (Ishihara, T. and M. S. Sasaki, ed.), Liss, New York, pp. 235–275.
120. Thacker, J., J. Chalk, A. Ganesh, and P. North. 1992. A mechanism for deletion formation in DNA by human cell extracts: the involvement of short sequence repeats. *Nucleic Acids Res.* **20:** 6183–6188.
121. Thacker, J., R. E. Wilkinson, and D. T. Goodhead. 1986. The induction of chromosome exchange aberrations by carbon ultrasoft x-rays in V79 hamster cells. *Int. J. Radiat. Biol.* **49:** 645–656.
122. Thomas, J. W., B. C. Holdener, and T. Magnuson. 1994. Sequence analysis of a radiation-induced deletion breakpoint fusion in mouse. *Mammalian Genome* **5:** 518,519.
123. Thomas, K. R., K. R. Folger, and M. R. Capecchi. 1986. High frequency targeting of genes to specific sites in the mammalian genome. *Cell* **44:** 419–428.
124. Tlsty, T. D., P. Jonczyk, A. White, M. Sage, I. Hall, D. Schaefer, A. Briot, E. Livanos, H. Roelofs, B. Poulose, and J. Sanchez. 1993. Loss of chromosomal integrity in neoplasia. *Cold Spring Harbor Symp. Quant. Biol.* **58:** 645–654.
125. Tucker, J. D., D. A. Lee, and D. H. Moore. 1995. Validation of chromosome painting. II. A detailed analysis of aberrations following high doses of ionizing radiation in vitro. *Int. J. Radiat. Biol.* **67:** 19–28.
126. Tucker, J. D., W. F. Morgan, A. A. Awa, M. Bauchinger, D. Blakey, M. N. Cornforth, L. G. Littlefield, A. T. Natarajan, and C. Shasserre. 1995. A proposed system for scoring structural aberrations detected by chromosome painting. *Cytogenet. Cell Genet.* **68:** 211–221.
127. Tucker, J. D., M. J. Ramsey, D. A. Lee, and J. L. Minkler. 1993. Validation of chromosome painting as a biodosimeter in human peripheral lymphocytes following acute exposure to ionizing radiation *in vitro*. *Int. J. Radiat. Biol.* **64:** 27–37.
128. Urlaub, G., P. J. Mitchell, E. Kas, L. A. Chasin, V. L. Funanage, T. T. Myoda, and J. Hamlin. 1986. Effect of gamma rays at the dihydrofolate reductase locus: deletions and inversions. *Somat. Cell Mol. Genet.* **12:** 555–566.

129. Vig, B. K. and R. Lewis. 1978. Genetic toxicology of bleomycin. *Mutat. Res.* **55:** 121–145.
130. Vrieling, H., J. W. I. M. Simons, F. Arwert, A. T. Natarajan, and A. A. van Zeeland. 1985. Mutations induced by X-rays at the HPRT locus in cultured Chinese hamster cells are mostly large deletions. *Mutat. Res.* **144:** 281–286.
131. Vyas, R. C., F. Darroudi, and A. T. Natarajan. 1991. Radiation-induced chromosomal breakage and rejoining in interphase-metaphase chromosomes of human lymphocytes. *Mutat. Res.* **249:** 29–35.
132. Waldren, C. A. and R. T. Johnson. 1974. Analysis of interphase chromosome damage by means of premature chromosome condensation after X-ray and ultraviolet irradiation. *Proc. Natl. Acad. Sci. USA* **71:** 1137–1141.
133. Waldren, C. A., L. Correll, M. A. Sognier, and T. T. Puck. 1986. Measurement of low levels of X-ray mutagenesis in relation to human disease. *Proc. Natl. Acad. Sci. USA* **83:** 4839–4843.
134. Ward, J. F., W. F. Blakely, and E. I. Joiner. 1985. Mammalian cells are not killed by DNA single-strand breaks caused by hydroxy radicals from hydrogen peroxide. *Radiat. Res.* **103:** 383–392.
135. Windle, B. E. and G. M. Wahl. 1992. Molecular dissection of mammalian gene amplification: new mechanistic insights revealed by analysis of very early events. *Mutat. Res.* **276:** 199–224.
136. Winegar, R. A. and R. J. Preston. 1988. The induction of chromosome aberrations by restriction endonucleases that produce blunt-end or cohesive-end double-strand breaks. *Mutat. Res.* **197:** 141–149.
137. Whaley, J. M. and J. B. Little. 1990. Molecular characterization of hprt mutants induced by low- and high-LET radiations in human cells. *Mutat. Res.* **243:** 35–45.
138. Zech, L., T. Godal, L. Hammarström, H. Mellstedt, C. I. E. Smith, T. Tötterman, and M. Went. 1986. Specific chromosome markers involved with chronic T-lymphocyte tumors. *Cancer Genet. Cytogenet.* **21:** 67–77.

26

Whole-Organism Responses to DNA Damage

Modulation by Cytokines of Damage Induced by Ionizing Radiation

Ruth Neta and Scott K. Durum

1. INTRODUCTION

This volume is primarily devoted to genotoxin effects at the cell and the molecular biology levels of DNA repair. In this chapter, we leap to the organismic level, which is clearly much more complex. The evidence for the radioprotective and sensitizing effects of cytokines in mice is reviewed in order to address a major question, namely: What molecular mechanisms account for these effects?

In the first half of this century, ionizing radiation was regarded as a promising means of diagnosis and cure for a number of ailments. The carcinogenic and life-threatening effects of radiation were recognized and studied in the second half of the century. Acute death after single doses of radiation to the whole body of mice was shown to occur in three dose-dependent phases:

1. High doses (over 100 Gy) induced death within hours resulting from neurologic and cardiovascular breakdown, a process termed the "cerebrovascular syndrome;"
2. Intermediate doses (over 15 Gy), induced death in a matter of days, resulting from destruction of gastrointestinal (GI) tissue, a process termed the "GI syndrome;" and
3. Lower doses (7–13 Gy depending on the strain of mice) induced death within 1–4 wk, resulting from hematopoietic failure, and is referred to as the "hematopoietic syndrome" *(4)*.

It was found that the hematopoietic syndrome could be influenced by stimulating the reticuloendothelial system (RES), for example, with bacterial lipopolysaccharide (LPS) protecting mice from death and accelerating hematopoietic recovery *(59)*. Despite evidence that protection of hematopoiesis was a prerequisite to host survival, the basis for such protection by RES stimulation was unclear. Then, a decade ago, it was observed that IL-1 and TNF, two cytokines produced by the activated RES, protected mice from the hematopoietic syndrome when injected prior to irradiation *(40,43)*.

Cytokines are hormone-like proteins produced by stimulated cells and tissues that serve as intercellular messengers. In the past two decades, a multitude of cytokines have been cloned that have effects on the hematopoietic and immune system. Many cytokines are pleiotropic, i.e., they act on multiple cells and tissues, and produce a broad range of effects. For example, IL-1, TNF, and IL-6 all affect hemopoiesis, but also elicit powerful effects on many other tissues, including brain, liver, vasculature,

From: DNA Damage and Repair, Vol. 2: DNA Repair in Higher Eukaryotes
Edited by: *J. A. Nickoloff and M. F. Hoekstra © Humana Press Inc., Totowa, NJ*

and muscle *(46)*. Other cytokines are more restricted to hematopoietic effects, such as IL-3, G-CSF, GM-CSF, M-CSF, EPO, TPO, and SCF, all of which affect proliferation and differentiation of hematopoietic cells at various stages of their lineage progression *(39)*. The immune system is influenced by additional cytokines, including IL-2, IL-4, IL-5, IL-7, IL-12, IL-13, and IL-15, affecting growth, differentiation, and function of T- and B-lymphocytes *(53)*. A number of cytokines downregulate proliferation of hematopoietic progenitor cells *(39)*; these include; TGFβ, MIP-1α, and the interferon family: IFNα, IFNβ, and IFNγ. However, each of these latter cytokines, in addition to inhibiting hematopoiesis, also have a wide range of additional target cells. For example, MIP-1α also acts as chemotactic factor on lymphoid cells, TGFβ acts as a differentiation factor on multiple cells, and IFN acts as a differentiation and antiviral factor. Cytokines were shown to act as a first line of defense in infections and promote wound healing. Stress and genotoxic agents, including ionizing radiation, induce production of a number of cytokines. Cytokines, in turn, induce a cascade of additional cytokines, all of which may up- or downregulate the activity of the initial signal. Many cytokines were shown to upregulate or downregulate cellular receptors and adhesion molecules on cell membranes. Such changes alter functions of many cells and their interactions. In that way, cytokines orchestrate cellular and tissue response to damage, promote repair, and participate in elimination of damaged cells. Cytokines can also be destructive. For example, the tissue damage of rheumatoid arthritis is mediated by chronic local production of inflammatory cytokines. Death from septic shock may also be mediated by high blood levels of inflammatory cytokines.

Most cytokines act in the local microenvironment, i.e., as "paracrine" or "autocrine" signals. Systemic or "endocrine"-like effects (such as fever or a drop in blood pressure) can be transiently observed with high levels produced in tissues or by blood cells. Cytokines are rapidly cleared from blood via the kidney because of their small size (generally <20 kDa), and hence, pharmacologically administered cytokines have a short half-life.

Cytokines act via high-affinity membrane receptors, usually consisting of two to three distinct protein chains, and are expressed by a wide range of cells in the body. Many cytokines can interact with at least two distinct specific receptors (e.g., two for IL-1, three for TGFβ). In several families of cytokines that share structural homology, their receptors also share a common chain; for example, IL-6, LIF, CNTF, and OSM all use the common gp130 receptor chain together with a receptor chain specific for the cytokine *(29)*. Cytokine receptors are generally triggered by crosslinking, which brings protein and lipid kinases and substrates together on the intracellular domains of the receptor. This initiates intracellular cascades leading to cellular responses, including gene induction, cell division, cell cycle arrest, motility, and apoptotic death.

This chapter focuses primarily on recent findings relating cytokines with radioprotection and radiosensitization of mice exposed to acute effects of whole body irradiation on the hematopoietic and GI system of mice. These observations are outlined in Fig. 1 illustrating increased radioprotection, i.e. increase in doses of radiation lethal to kill 100% of mice in 30 d ($LD_{100/30}$) in response to IL-1, TNF, SCF, bFGF, or IL-12 (hematopoietic death), and for $LD_{100/6}$ (GI death) in response to IL-11, SCF, or IL-1. Radiosensitization occurs in response to TGFβ, IL-6, or IFN for $LD_{100/30}$ and to IL-12,

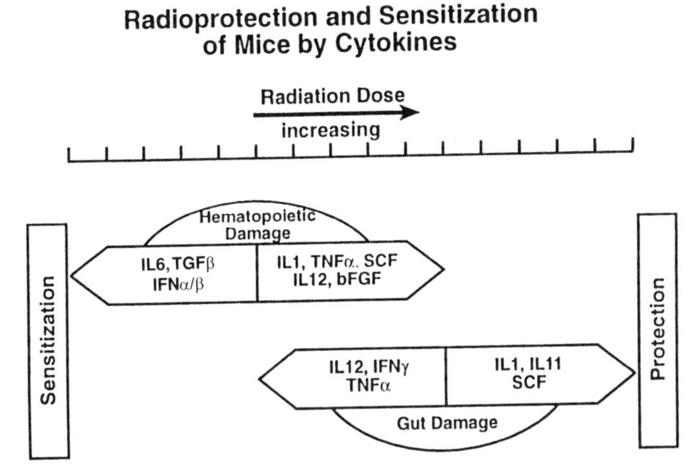

Fig. 1. Radioprotection and sensitization of mice by cytokines.

IFNγ, or TNF for $LD_{100/6}$. The implications of these findings are discussed for radiotherapy in humans, and suggestions for a number of possible mechanisms to explain radioprotection and sensitization by cytokines are presented.

2. MYELOPROTECTION

2.1. Myeloprotection with Cytokines

Early studies using mice demonstrated that LPS, given prior to exposure to ionizing radiation, protected from death resulting from hematopoietic failure *(59)*. Subsequently, the proinflammatory cytokines, IL-1 and TNF, were shown similarly to protect mice from myelotoxicity and death induced by ionizing radiation *(40,43)*. LPS radioprotection is mediated, at least in part, by IL-1 and TNF, since anti-IL-1 and anti-TNF antibodies blocked LPS radioprotection. Therefore, these cytokines can be produced endogenously and serve as radioprotective agents *(45)*.

Of the many cytokines that were tested for their survival-promoting effect (given in the optimal regime as a single ip injection 18–24 h prior to irradiation), only IL-1, TNF, SCF, and IL-12 acted alone to protect mice from $LD_{100/30}$ radiation. The cytokines IL-2, IL-3, IL-4, IL-6, IL-11, GM-CSF, G-CSF, M-CSF, LIF, IFNγ, and TGFβ, given before irradiation, did not promote survival of lethally irradiated mice *(38)*. In fact, administration of IL-6, TGFβ, or IFN 24 h before irradiation resulted in increased lethality of $LD_{50/30}$ irradiated mice *(38)*. This course of experimentation revealed several insights into the bases of protection of mice from the lethal hematopoietic syndrome as discussed in Sections 2.2.–2.5.

2.2. IL-1 and TNF Protect Irradiated Mice from Hematopoietic Syndrome

IL-1 and TNF are two proinflammatory cytokines with biological effects vital to host defenses, but contributing also to host morbidity. Thus, their administration to patients leads to desirable as well as toxic effects *(48)*. The rapidly developing understanding of their structure–function relationship, and of the downstream events of their action in different cells and tissues aids in identifying their role in each of these processes.

Such knowledge in turn should lead to developing means of intervention to harness the beneficial effects while diminishing the undesirable effects of these cytokines.

In most murine strains tested, the dose modifying factor for IL-1 is greater than for TNF. A single dose of 100–300 ng of IL-1 given 20 h prior to radiation had a greater protective action than 5 μg of TNF. The combined action of these cytokines is synergistic, suggesting that the two molecules employ radioprotective pathways that are in part distinct *(43)*. The combined effect of IL-1 and TNF is also more protective than the optimal dose of bacterial LPS. Conversely, blocking the activity of IL-1 and TNF with antibodies in LPS-treated mice revealed a radiosensitizing effect of LPS *(45)*. One explanation for this may be that LPS induces radiosensitizing cytokines, such as TGFβ, IL-6, IFNα, and IFNβ.

Complex interactions of cytokines occur in vivo. For example, despite IL-6-sensitizing mice to radiation lethality when given alone, coadministration of this cytokine with suboptimal doses of IL-1 results in synergistic radioprotection *(44)*. Conversely, blocking IL-6 abolished IL-1- and TNF-induced radioprotection *(47)*. In addition, anti-TNF antibody abolished IL-1-induced protection, and anti-IL-1 antibody similarly blocked TNF-induced radioprotection. Because IL-1 and TNF induce one another and both induce IL-6, it is apparent that the joint action of IL-1, IL-6, and TNF in vivo is required for optimal radioprotection.

Even in mice that have not been treated with exogenous cytokines, injection of antibody to IL-1, IL-6, or TNF resulted in $LD_{50/30}$ doses of radiation becoming lethal to 100% of mice *(45,47)*. This indicates that the endogenous production of each of these cytokines contributes to the animal's ability to survive midlethal doses of radiation. Radiation induces several of these cytokines, as documented in cell and animal studies *(44)*, suggesting that the cytokines serve as natural defenses against radiation. The selective value of producing cytokines in response to radiation is presumably not related to lethal radiation doses that do not occur in nature. Perhaps the benefit lies in responding to any genotoxic agents in the environment that elicit oxygen radicals or oxidative DNA damage; a cell responding to such insults produces cytokines that would then trigger neighboring cells to protect themselves from such oxidative damage.

2.3. The Cellular Target of IL-1 and TNF Radioprotection: Direct Effects on Stem Cells vs Indirect Effects via Stromal Products

The identity of the radioprotected cells includes hematopoietic stem cells with both short-term and long-term repopulating capacity based on experiments using male donor cells in female recipient mice *(67)*. Treatment of donor cells with IL-1 before radiation protects both short- and long-term repopulation. In short-term repopulation (3 mo post-transplantation), the female recipients receiving IL-1-pretreated cells had greatly increased expression of the male marker. Furthermore, in serial transplantation studies, male cells could only be detected in tertiary recipients who received IL-1-pretreated bone marrow cells.

How these stem cells become radioprotected is not clear as yet and could involve indirect effects of IL-1 and TNF on stromal cells, and direct effects on the stem cells. In support of an indirect mechanism, both IL-1 and TNF stimulate production of a cascade of hematopoietic growth factors, including G-CSF, GM-CSF, IL-6, and IL-3, as well as platelet-derived growth factor (PDGF) *(46)*. It is possible that the induction of these myeloproliferative growth factors accounts for the myelorestorative action of high doses of IL-1 and TNF given after irradiation of mice *(42)*. In addition, G-CSF and GM-CSF given prior to lethal irradiation of mice can synergistically cooperate with suboptimal doses of IL-1 to enhance survival *(43)*. Evidence discussed previously shows that in the cascade of cytokines induced by IL-1 and TNF, other cytokines detrimental to radioprotection and their neutralization with the specific antibody may further enhance the radioprotective effects of IL-1 and TNF. Conversely, some cytokines may synergize in reducing the cytotoxic damage to normal tissue.

Thus, circumstantial evidence supports an indirect action of IL-1 and TNF on stem cells, acting through secondary hematopoietic cytokines produced perhaps by bone marrow stromal cells. On the other hand, a few studies have reported that in vitro treatment of marrow preparations impart radioprotection to stem cells *(44)*. However, these studies were not conducted with purified hematopoietic precursors and could reflect a contribution by contaminating stromal cells.

Although radioprotection of hematopoietic stem cells by IL-1 and TNF may be indirect, a number of other cell types have been shown to be directly rendered radioresistant in vitro. Both IL-1 and TNF induce the mitochondrial scavenging enzyme, manganese superoxide dismutase (MnSOD) *(36,64)*. Fibrosarcoma cells transfected with cDNA for MnSOD become more radioresistant than untreated cells and those transfected with antisense cDNA for MnSOD become more radiosensitive *(22)*. TNF-treated mononuclear cells show enhanced resistance to radiation *(64)*. As in the case of TNF, bone marrow cells obtained from mice 6 h after IL-1 treatment display a dose-dependent increase of mRNA for MnSOD *(13)*. Likewise, progenitor cell-enriched bone marrow cells from 5-fluorouracil (5-FU)-treated mice had elevated levels of mRNA for MnSOD following IL-1 treatment in vitro. Irradiation of in vitro IL-1-pre-treated cells also protected long-term culture-initiating cells. Together, these experiments provide evidence that MnSOD induced by IL-1 or TNF may serve as an important secondary mediator in myeloprotection. Other scavenging proteins induced by these cytokines include metallothionin and ceruloplasmin *(38)*, both shown to be protective against radiation lethality.

In contrast to TNF radioprotection of primary cells, there are a few examples of TNF radiosensitization of cell lines. For example, in the case of several tumor cell lines producing TNF in response to radiation, TNF and radiation were synergistic in lethal effect *(15,16)*. This interaction was demonstrated for human soft tissue and bone sarcomas, squamous cell carcinomas, and myeloid leukemias. These cell lines produce oxygen radicals in response to TNF (as do many cell types), but do not produce MnSOD. Therefore, this type of synergy between radiation and TNF may result from their combined radical production, exacerbated by a failure to scavenge radicals. Support for this oxygen radical mechanism was shown by reduced killing in a low oxygen atmosphere or in the presence of radical scavengers *(64)*.

2.4. Radioprotection with Stem Cell Factor (SCF)

SCF (also known as Steel factor or Kit Ligand) and its receptor c-Kit are important for normal development of hematopoietic cells. Strains of mice that are heterozygous for mutations in the *steel* (*Sl*) and *white spotting* (*W*) loci coding for SCF and c-Kit, respectively, exhibit hematopoietic abnormalities and increased sensitivity to ionizing radiation *(2,56)*. SCF treatment prior to radiation protects normal mice from radiation lethality *(66)*. Conversely, antibody to its receptor, c-Kit, given to $LD_{50/30}$-irradiated mice, leads to 100% lethality *(49)*. Likewise, treatment with this antibody blocks entirely LPS and IL-1-induced radioprotection. These experiments document that like IL-1, IL-6, and TNF, normal expression of SCF and its receptor is required for maximal protection from ionizing radiation.

The coadministration of IL-1 and SCF to mice before lethal irradiation confers synergistic protection from death *(50)*. This protection is associated with a profound increase in the numbers of hematopoietic progenitor cells recovered within 1–4 d after irradiation, indicating that these cells probably survive radiation insult. Anti-SCF antibody blocks IL-1-induced radioprotection, whereas anti-IL-1 antibody greatly reduces SCF radioprotection, indicating that endogenous production and interaction in a network of these cytokines are required for optimal protection.

SCF, unlike IL-1, does not induce hematopoietic growth factors, such as CSF or IL-6, nor does it induce MnSOD *(50)*; this radioprotection by SCF cannot be explained by secondary hematopoietic cytokines, as invoked for IL-1 or by radical scavenging. However, SCF injection results in cycling of hematopoietic progenitors *(3)*, suggesting that their cycling may be the basis of the radioprotective effect of SCF—this cell cycle hypothesis is expanded below as a general mechanism for stem cell radioprotection. However, SCF apparently has an additional mechanism of radioprotection in mast cell lines that is independent of cell cycle. Radioprotection in this cell line by SCF is not attributable to Bcl-2 induction, in contrast to radioprotection by IL-3 *(65)*.

2.5. Hypothesis: IL-1 Induces Radioprotection by Driving Stem Cells into S Phase of the Cell Cycle

In addition to scavenging-promoting effects, the radioprotective effects of IL-1 on hematopoiesis may also be based on driving the stem cell to late S phase of the cell cycle, which is relatively radioresistant. This hypothesis was suggested by comparing the radioprotection with IL-1 that induces cycling of hematopoietic cells, and radiosensitization with TGFβ that inhibits cycling of such cells. This hypothesis was further supported by experiments in which the radioprotective effect of IL-1 depended on the schedule of its administration, with no significant protection at 4 h, but optimal protection at 18–24 h after IL-1.

Early findings indicated that in vivo IL-1 induced an increase in bone marrow progenitors reaching a maximum at 48 h after administration *(38)*. This suggested that an expansion of progenitor cells may be the basis for the myeloprotective action of IL-1. However, mice receiving irradiation 48 h following IL-1 treatment are not protected from death *(52)*, indicating that the mere increase in the numbers of progenitor cells is not sufficient for IL-1 radioprotection.

A profound difference was demonstrated in radiosensitivity of various phases cell cycle. Whereas the transition of G1 to S and G2/M phases are highly radiosensitive, radioresistance is greatly increased in late S phase *(9,58,63)*. The kinetics of the radioprotective effect of IL-1 suggested that it may be attributed to the radioresistant phase of IL-1-induced progenitor's cycling. Indeed, administration of IL-1 18–24 h prior to irradiation coincided with increased sensitivity of progenitor cells of various lineages to hydroxyurea (HU), selectively toxic for cells in S phase *(41,57)*. These results implied that IL-1 induced the relevant progenitor cells to progress to S phase at the same time they became radioresistant.

The cytotoxic drug 5-FU normally spares early progenitor cells owing to their quiescence, but is highly toxic to cycling cells at all stages of cycle *(31)*. Pretreatment with a single dose of IL-1 resulted within 10–14 d in death of 5-FU-treated mice *(52)*. The death occurred only when IL-1 was given 18 h, but not at 4 or 48 h prior to administration of sublethal doses of 5-FU. The death was owing to hematopoietic failure as evidenced by only a minute fraction (2%) of primitive hematopoietic progenitor cells surviving in IL-1/5-FU compared to 5-FU treatment alone. Apparently, IL-1 induced the remaining 98% of these normally 5-FU-resistant, slow proliferating, or resting cells to cycle, perhaps synchronously reaching S phase at 18–24 h. On the other hand, after 48 h, IL-1 no longer sensitized mice to sublethal doses of 5-FU and did not affect significantly the numbers of surviving progenitor cells, suggesting that the IL-1-induced cycling was transient.

Importantly, in contrast to radioprotection by pretreatment with a single dose of IL-1 at 18–24 h, two injections of IL-1 48 h apart, at 72 and 24 h before irradiation, abrogated radioprotection. Likewise, the killing of progenitor cells by 5-FU was substantially reduced when two injections of IL-1, 48 h apart, were administered prior to 5-FU. These findings suggest that the 48-h pretreatment with IL-1 induced one cell cycle, followed by resistance to a second, IL-1-induced cycling. This effect could be owing to induction of TGFβ, which inhibits hematopoiesis *(26)* and is upregulated 36–48 h following IL-1 treatment *(52)*.

TGFβ radiosensitizes mice *(45)*, and this effect may also be attributable to its being a potent inhibitor of cycling of the hematopoietic stem cells (*see* Section 4.). TGFβ, in addition to preventing S-phase entry, has additional antiproliferative activities. It reduces expression of multiple growth factor receptors, including c-Kit *(12,19,21,23,26)*, the receptor for SCF expressed on early hematopoietic progenitors. TGFβ also inhibits production of SCF by stromal cells *(19)*. TGFβ inhibited the in vitro growth of primitive progenitor cells (HPP-CFC), whereas the growth of more differentiated granulocyte progenitors (G-CFU) was stimulated *(27)*. Interestingly, five daily injections of IL-1 resulted in highly granulocytic BM (Neta, unpublished results), perhaps via induction of TGFβ.

In addition to TGFβ, other factors induced by IL-1 and inhibitory to cycling and proliferation of hematopoietic progenitors may be involved. For example, prostaglandin and TNF, both induced by IL-1, were reported to inhibit the growth of progenitor cells through yet undefined pathways *(39)*. More recently, IL-1 was shown first to upregulate and then downregulate GM-CSF production in human fibroblasts *(54)*. The downregulation was associated with the production of prostaglandins. Thus, a cascade of cytokines and other mediators induced by IL-1 includes positive as well as negative regulators of cycling of primitive progenitors.

Together, these findings suggest that the myeloprotective effects of IL-1 against ionizing radiation and cytoablative chemotherapeutic drugs are probably mediated by a complexity of intermediary products. Nevertheless, the outcome may commonly be determined by the cell-cycle stage of hematopoietic stem cells at the time of radiation.

3. GI EFFECTS

3.1. *Effects of Cytokines on GI Tissue*

Experimental evidence in animal models indicates that IL-11, IL-1, and SCF protect the gut from radiation damage. In mice receiving radiochemotherapy (5-FU and irradiation) resulting in death caused by severe damage to the small intestinal mucosa, IL-11 increased survival and led to a rapid recovery of intestinal mucosa *(11)*. The recovery was associated with an increase in the mitotic index of crypt cells. Similarly, daily treatment with IL-11 of mice receiving radiation only resulted in increased survival of intestinal clonogenic crypt cells *(55)*. The D_0 for these cells in mice given IL-11 for 2 d prior to radiation was 2.0 Gy compared to 1.8 Gy in control mice. The D_0 was substantially increased to 2.3 Gy when the treatment was continued for three additional days after irradiation. The level of protection afforded by posttreatment alone was minimal. Of particular interest was the observation that the extrapolation number *(N)*, which is the measure of the shoulder size of the survival curve, was significantly reduced in the post- vs pretreated groups. Such a reduction is thought to represent a reduction in repair capacity. The results of this study suggest that IL-11 treatment protects the most resistant populations of reserve clonogenic cells from ionizing radiation.

IL-1 given to mice 20 h prior to whole body irradiation modestly protected duodenal crypt cells *(17)*. However, when given within 4–8 h before irradiation, IL-1 sensitized crypt cells to radiation. Thus, once again, depending on the time of treatment, IL-1 may have opposing effects on tissue damage by radiation. IL-1 exposure did not substantially alter the slope of the survival curve, but affected the shoulder. When given 20 h before irradiation, IL-1 expanded the shoulder by 1 Gy, whereas given 4 h before irradiation, it shortened the shoulder by 1.28 Gy. The protective effect may therefore be based on the repair of sublethal injury.

SCF treatment was also shown to enhance the survival of mouse duodenal crypt stem cells *(30)*. The dose modification factor for $LD_{50/6}$ was 1.28 and was increased from 14.9–19.0 Gy. The time of treatment to achieve the optimal protection was from 2–24 h before treatment. Although the exact mechanism for this protection remains to be established, the previously discussed antiapoptotic effects of SCF may underlie such protection *(65)*.

3.2. *Contrasting Effects of IL-12 on Hematopoietic and GI Tissue*

IL-12 has potent antitumor and antimetastatic activity in a number of murine tumor models *(61)*. IL-12 is also a potent costimulator of the early hematopoietic progenitor cells *(24)*. Administration of IL-12 within 24 h prior to lethal irradiation protects a significant fraction of mice from lethal hematopoietic syndrome *(51)*. The radioprotection is associated with a significant increase in the numbers of hematopoietic progenitor cells surviving a supralethal dose of 12 Gy. However, mice receiving IL-12 and 12 Gy die within 4–6 d of irradiation from the GI syndrome, as shown by gross necroscopy and histologic evaluation. Induction of a similar syndrome in mice not treated with IL-12 requires doses <16 Gy. Thus, at doses of radiation at which IL-12 still protects bone marrow cells, it sensitizes the intestinal tract to radiation damage. Whereas protection of hematopoietic cells can be abrogated with anti-IL-1R and anti-SCF antibody, the sensitization of the intestinal tract can be prevented with anti-IFNγ and anti-TNF antibodies. Therefore, cytokines interacting with IL-12 to protect the bone marrow differ from those sensitizing the gut.

Sensitization of the gut epithelial cells by IL-12-induced IFNγ may involve upregulation of Fas antigen, ligation of which induces apoptosis in lymphocytes and hepatocytes. Although gut epithelial cells have not been reported to express Fas, other cell types respond to IFNγ and TNF (both induced by IL-12) by increasing Fas expression and thereby increasing their susceptibility to Fas ligand-induced apoptosis *(34)*. Radiation killing conceivably could synergize with Fas signaling. An alternative explanation for IFNγ-induced apoptosis may be based on its ability to induce inappropriate entry into S phase in some cells.

4. THE CELLULAR BASIS OF CYTOKINE PROTECTION AND SENSITIZATION: APOPTOSIS AND CELL CYCLE

Understanding how cytokines are radioprotective at the cellular level will only become clear when we know how the irradiated cells die. Apoptotic death is presumed for hematopoietic stem cells and gut epithelial cells. There is little direct information about the radiation-induced apoptotic process in these cells, because they are rare cells in the host tissues, and to culture them outside the host environment requires conditions that are probably not physiological.

A likely pathway may include p53, which detects radiation damage and can induce either cell-cycle arrest or apoptosis *(5)*. Studies in p53 knockout mice *(33)* clearly showed that thymocytes, which undergo apoptosis in response to very low doses of radiation, require p53 for this response. There is a limited amount of evidence that the stem cells of bone marrow and gut epithelium also respond in this way, but at higher radiation doses. Bone marrow myeloid colony-forming cells from p53 knockout mice were much more resistant to irradiation than wild-type, showing a fivefold difference in the numbers of colonies surviving 3 Gy *(32)*. In the small intestine, histological studies showed that 30- to 40-fold fewer presumed stem cells underwent apoptosis following 8 Gy in p53 knockout mice than in wild-type mice *(37)*. Therefore, if p53 mediates radiation-induced apoptosis of the hematopoietic and gut stem cells, then cytokines presumably modify events upstream or downstream of p53. Other p53-independent pathways could also be involved. Rb can mediate cell death in a cell line following irradiation *(10)*. IRF-1 induces cell death in mature T-cells following irradiation *(60)*.

The understanding of the downstream events in cytokine-induced signaling pathways that lead to either growth arrest, apoptotic death, or progression in cell cycle has been developing rapidly. However, many cytokine effects depend on the cell type and other converging signals. Therefore, the specific cells in question, i.e., hematopoietic stem cells and intestinal crypt cells, may not conform to the many cell lines that are studied.

As previously suggested, the radiosensitizing effect of TGFβ may be related to this cytokine arresting early hematopoietic progenitors in G1 phase of cell cycle, rendering them more susceptible to radiation *(7,14)*. Although not completely understood, the growth-arresting effect of TGFβ may be exerted on several molecular targets *(1)*. Among the suggested targets are downregulation of c-*myc* RNA and protein levels, downregulation of transcriptional induction of cyclin A and cyclin E, and upregulation of inhibitors for Cdk–cyclin complexes, p15, p16, p21, and p27. Evidence accumulating from different cell types suggests that there may be substantial differences between different cells in the mechanisms by which TGFβ is able to inhibit their proliferation *(1)*. For example, in HaCaT cells TGFβ uniquely upregulated p15 mRNA *(18)*.

In contrast to its apparent detrimental effects through reducing survival of hematopoietic progenitors from radiation, TGFβ was recently recognized as an important cancer growth inhibitor. Mutations in TGFβ type II receptor were identified in several types of colorectal and gastric cancer and in Sezary leukemia *(6,35)*.

Several cytokines may either promote or prevent radiation-induced apoptotic death. For example, TNFα and IFNγ, by promoting upregulation of Fas antigen on $CD34^+$ cells *(34)*, may render them more susceptible to apoptosis. An additional mechanism was suggested in cell lines, in which TNF and irradiation induce apoptotic death involving the activation of hydrolysis of sphingomyelin to ceramide. In HL-60 and U-937, two myeloid leukemic cell lines treatment with either radiation, TNF, or ceramide resulted in downregulation of Bcl-2 mRNA expression, suggesting that modulation of Bcl-2 gene expression may be a target for ceramide-induced apoptosis characteristic of radiation and TNF *(8)*.

On the other hand, IL-3 through the Ras signaling pathway, upregulates Bcl-2 leading to the rescue of hematopoietic cells from radiation-induced apoptosis *(28)*. Although the mechanism of prevention of apoptotic death by SCF is not entirely clear, experimental evidence indicates that this cytokine protects mast cells from radiation-induced apoptotic death *(65)*. Another cytokine, a vascular endothelial growth factor (VEGF), identified as a peptide growth factor specific for vascular endothelial cells, prevented apoptosis of hematopoietic progenitor colonies, BFU-E, CFU-GEMM, and CFU-GM *(25)* all of which express the two VEGF receptors, KDR and FLT-1. Addition of VEGF before irradiation prevented apoptosis, but did not induce proliferation. It is of interest to note that the ability to prevent apoptotic death was independent of cell-cycle distribution.

Several growth-stimulatory cytokines induce cell-cycle progression by induction of proto-oncogene c-*myc* in a variety of cell types. The mechanisms underlying the positive cell-cycle regulation by c-*myc* are currently not well understood. However, experimental evidence indicates that c-*myc* upregulation and downregulation may be critical events in determining cell-cycle progression or arrest. Downregulation of c-*myc* is a hallmark of growth-inhibitory cytokines, including TGFβ, IFNγ, IFNα/β, IL-6, and LIF. In contrast, upregulation of c-*myc* by growth factors abrogates p53-induced cell-cycle arrest. Despite high expression of p53, activation of c-*myc* results in cell-cycle progression, indicating that c-*myc* can override a p53-induced cell-cycle arrest *(62)*. Several mechanisms were suggested to explain this effect. Some of these include activation of Cdk leading to hyperphosphorylation of pRB. Most recent evidence suggests that c-*myc* induces a heat-labile inhibitor for p21, a Cdk–cyclin complex inhibitor *(20)*.

5. CONCLUSIONS

The response to radiation is modified by cytokines given systemically to mice. Moreover, endogenously produced cytokines serve as a natural defense against ionizing radiation, as shown by blocking endogenous cytokines with antibodies in irradiated mice. Radioprotective and sensitizing effects of cytokines on animals and tissues appear to be based on a cascade of events involving the release of additional cytokines, and the ensuing receptor-mediated downstream signals that converge on the most sensitive and critical target cells: the hematopoietic stem cells and the epithelial gut cells. The cellular mechanisms by which cytokines protect these cells against radiation or enhance their killing by radiation remain somewhat conjectural at this point. Information about responses of these cells to cytokines that have effects in irradiated mice will hopefully reveal important intracellular pathways that protect against radiation-induced cell death.

Several possible mechanisms of protection have emerged:

1. Reduction of oxidative damage through induction of mitochondrial enzymes, such as MnSOD, and other scavenging proteins, as demonstrated for IL-1, TNF, and IL-6;
2. Reduction of apoptotic response through induction of Bcl-2 (by IL-3) and via some, as yet unknown pathways for SCF; and
3. Induction of normally quiescent early progenitor cells to enter cell cycle and possibly reach the relatively radioresistant late S phase, as suggested by the kinetics of IL-1 and SCF radioprotection and the susceptibility that these cells develop to cycle dependent drugs, HU and 5-FU.

In contrast, the sensitizing mechanisms may include:

1. Increased oxidative damage, that may occur in the absence of scavenger induction, as observed for TNF for many tumor cell lines;
2. Enhanced apoptosis by upregulation of Fas antigen, as reported for TNF and IFNγ; and
3. Arrest of cells in G1 phase of the cycle at the time of exposure to radiation, which may promote apoptosis.

The effect of a given cytokine may vary depending on the target cells and tissues. For example, IL-12, which protects the bone marrow, sensitizes the gut. TNF may protect cells capable of upregulation of MnSOD, but may be sensitizing cells lacking MnSOD by inducing oxidative damage.

The study of cytokines has had an important impact on biological science. Our current concepts of receptor function, gene induction, signal transduction, induction of cell cycling, cycle arrest, and apoptosis have all been illuminated through the studies of cytokines. It is possible that cytokine-mediated pathways that affect radiation-induced DNA damage and repair will be important in understanding the mechanisms protecting against radiation.

REFERENCES

1. Aleksandrow, M. G. and H. L. Moses. 1995. Transforming growth factor β and cell cycle regulation. *Cancer Res.* **55:** 1452–1457.
2. Bernstein, S. E. 1962. Acute radiosensitivity in mice of differing W genotype. *Science* **137:** 428.
3. Bodine, D. M., N. E. Seidel, K. M. Zsebo, and D. Orlic. 1993. In vivo administration of stem cell factor to mice increases the absolute number of pluripotent hematopoietic stem cells. *Blood* **82:** 445–455.
4. Bond, V. P., T. M. Fliedner, and J. O. Archambeau. 1965. *Mamalian Radiation Lethality; a Disturbance in Cellular Kinetics.* Academic, New York.
5. Canman, C. E. and M. B. Kastan. 1995. Induction of apoptosis by tumor suppressr genes and oncogenes. *Semin. Cancer Biol.* **5:** 17–25.
6. Capocasale, R. J., R. J. Lamb, E. C. Vonderheid, F. E. Fox, A. H. Rook, P. C. Nowell, and J. S. Moore. 1995. Reduced surface expression of transforming growth factor β receptor type II in mitogen-activated T-cells from Sezary patients. *Proc. Natl. Acad. Sci. USA* **92:** 5501–5505.

7. Cashman, J. D., A. C. Eaves, E. W. Raines, R. Ross, and C. J. Eaves. 1990. Mechanisms that regulate the cell cycle status of very primitive hematopoietic cells in long-term human marrow cultures. I. stimulatory role of a variety of mesenchymal cell activators and inhibitory role of TGF?. *Blood* **72:** 96–103.
8. Chen, M., J. Quintans, Z. Fuks, C. Thompson, D. Kufe, and R. R. Weichselbaum. 1995. Suppression of Bcl-2 messenger RNA production may mediate apoptosis after ionizing radiation, tumor necrosis factor α, and ceramide. *Cancer Res.* **55:** 991–994.
9. Denekamp, J. 1986. Cell kinetics and radiation biology. *Int. J. Radiat. Biol.* **49:** 357–380.
10. Dou, Q. P., B. An, and P. L. Will. 1995. Induction of a retinoblastoma phosphatase-activity by anticancer drugs accompanies p53-independent G(1) arrest and apoptosis. *Proc. Natl. Acad. Sci. USA* **92:** 9019–9023.
11. Du, X. X., C. M. Doerschuk, A. Orazi, and D. A. Williams. 1994. A bone marrow stromal-derived growth factor, interleukin-11, stimulates recovery of small intestinal mucosal cells after cytoablative surgery. *Blood* **83:** 33–37.
12. Dubois, C. M., F. W. Ruscetti, J. Stankova, and J. R. Keller. 1994. Transforming growth factor-β regulates c-kit message stability and cell-surface protein expression in hematopoietic progenitors. *Blood* **83:** 3138–3145.
13. Eastgate, J., J. Moreb, H. S. Nick, K. Suzuki, N. Taniguchi, and J. R. Zucali. 1993. A role for manganese superoxide dismutase in radioprotection of hematopoietic stem cells by interleukin-1. *Blood* **81:** 639–646.
14. Geng, Y. and R. A. Weinberg. 1993. Transforming growth factor β effects on expression of G1 cyclins and cyclin-dependent protein kinases. *Proc. Natl. Acad. Sci. USA* **90:** 10,315–10,319.
15. Hallahan, D. E., D. R. Spriggs, M. A. Beckett, D. W. Kufe, and R. R. Weichelbaum. 1989. Increase tumor necrosis factor β mRNA after cellular exposure to ionizing radiation. *Proc. Natl. Acad. Sci. USA* **86:** 10,104–10,107.
16. Hallahan, D. E., M. A. Beckett, D. Kufe, and R. R. Weichelbaum. 1990. The interaction between recombinant human tumor necrosis factor and radiation in 13 human tumor cell lines. *Int. Radiat. Oncol. Biol. Phys.* **19:** 69–74.
17. Hancock, S. L., R. T. Chung, R. S. Cox, and Rÿ. F. Kallman. 1991. Interleukin-1β initially sensitizes and subsequently protects murine intestinal stem cells exposed to photon radiation. *Cancer Res.* **51:** 2280–2285.
18. Hannon, G. J., and D. Beach. 1994. p15 INK4B, is a potential effector of TGF-β-induced cell cyclearrest. *Nature* **371:** 257–261.
19. Heinrich, M. C., D. C. Dooley, and W. W. Keeble. 1995. Transforming growth factor β1 inhibits expression of the gene products for Steel factor and its receptor (c-kit). *Blood* **85:** 1769–1780.
20. Hermeking, H., J. O. Funk, M. Reichert, J. W. Ellwart, and D. Eick. 1995. Abrogation of p53-induced cell cycle arrest by c-Myc: evidence for an inhibition of p21WAF1/CIP1/SDI1. *Oncogene* **11:** 1409–1415.
21. Hestdal, K., E. W. Jacobsen, F. W. Ruscetti, C. M. Dubois, D. L. Longo, R. Chizzonite, J. J. Oppenheim, and J. R. Keller. 1992. In vivo effect of Interlukin-1α on hematopoiesis; role of colony stimulating factor receptor modulation. *Blood* **80:** 2486–2494.
22. Hirose K., D. L. Longo, J. J. Oppenheim, and K. Matsushima. 1991. Overexpression of mitochondrial manganese superoxide dismutase confers resistance on tumor cells to interlukin-1, tumor necrosis factor, selected anti-cancer drugs and ionizing radiation. *FASEB J.* **7:** 361–368.
23. Jacobsen, S. E. W., S. W. Ruscetti, C. M. Dubois, J. Lee, T. C. Boone, and J. R. Keller. 1991. Transforming growth factor-β trans-modulates the expression of colony stimulating factor receptors on murine hematopoietic progenitor cell lines. *Blood* **77:** 1706–1716.

24. Jacobsen, S. E. W., O. P. Veiby, and E. B. Smeland. 1993. Cytotoxic lymphocyte maturation factor (interleukin 12) is a synergistic growth factor for hematopoietic stem cells. *J. Exp. Med.* **178:** 413–423.
25. Katoh, O., H. Tauchi, K. Kawaishi, A. Kimura, and Y. Satow. 1995. Expression of the vascular endothelial growth factor (VEGF) receptor gene KDR, in hematopoietic cells and inhibitory effect of VEGF on apoptotic cell death caused by ionizing radiation. *Cancer Res.* **55:** 5687–5692.
26. Keller, J. R., S. E. W. Jacobsen, C. M. Dubois, K. Hestdal, and F. W. Ruscetti. 1991. Transforming growth factor β: A bidirectional regulator of hematopoietic cell growth. *Int. J. Cell Cloning* **10:** 2–11.
27. Keller, J. R., S. E. W. Jacobsen, K. T. Sill, L. R. Ellingsworth, and F. W. Ruscetti. 1991. Stimulation of granulopoiesis by transforming growth factor β: synergy with granulocy-macrophage colony stimulating factor. *Proc. Natl. Acad. Sci. USA* **88:** 7190–7194.
28. Kinoshita, T., T. Yokota, K. I. Arai, and A. Miyajima. 1995. Regulation of Bcl-2 expression by oncogenic Ras protein in hematopoietic cells. *Oncogene* **10:** 2207–2212.
29. Kishimoto, K., S. Akira, M. Narazaki, and T. Taga. 1995. Interleukin-6 family of cytokines and gp130. *Blood* **86:** 1243–1254.
30. Leigh, B. R., W. Khan, S. L. Hancock, and S. J. Knox. 1995. Stem cell factor enhances the survival of murine intestinal stem cells after photon irradiation. *Radiat. Res.* **142:** 12–15.
31. Lerner, C. and D. E. Harrison. 1990. 5-Fluorouracil spares hematopoietic stem cells responsible for long term repopulation. *Exp. Hematol.* **18:** 114–120.
32. Lotem, J. and L. Sachs. 1993. Hematopoietic cells from mice deficient in wild-type p53 are more resistant to induction of apoptosis by some agents. *Blood* **82:** 1092–1096.
33. Lowe, S. W. E. M. Schmitt, S. W. Smith, B. A. Osborne, and T. Jacks. 1993. p53 is required for radiation-induced apoptosis in mouse thymocytes. *Nature* **362:** 847–849.
34. Maciejewski, J., C. Selleri, S. Anderson, and N. S. Young. 1995. Fas antigen expression on $CD34^+$ human marrow cells is induced by interferon γ and tumor necrosis factor α and potentiates cytokine-mediated hematopoietic suppression in vitro. *Blood* **85:** 3183–3190.
35. Markowitz, S., J. Wang, L. Myeroff, R. Parsons, L. Sun, J. Lutterbaugh, R. S. Fan, E. Zborowska, K. W. Kinzler, B. Vogelstein, M. Brattain, and J. K. V. Willson. 1995. Inactivation of the type II TGF-β receptor in colon cancer cells with microsatellite instability. *Science* **268:** 1336–1338.
36. Masuda, A., D. L. Longo, Y. Kobayashi, J. J. Oppenheim, and K. Matsushima. 1988. Induction of mitochondrial manganese superoxide dismutase by interleukin-1. *FASEB J.* **2:** 3087–3091.
37. Merritt, A. J., C. S. Potten, C. J. Kemp, J. A. Hickman, A. Balmain, D. P. Lane, and P. A. Hall. 1994. The role of p53 in spontaneous and radiation-induced apoptosis in the gastrointestinal tract of normal and p53-deficient mice. *Cancer Res.* **54:** 614–617.
38. Miller, L. L. and R. Neta. 1993. Therapeutic utility of cytokines in counteracting the bone marrow suppression of radio- and chemo-therapy, in *Clinical Applications of Cytokines: Role in Pathogenesis, Diagnosis and Therapy* (Gearing, A., J. Rossio, and J. J. Oppenheim, eds.), Oxford University Press, Oxford, pp. 225–236.
39. Moore, M. A. S. 1991. Clinical implications of positive and negative hematopoietic stem cell regulators. *Blood* **78:** 1–19.
40. Neta, R., S. D. Douches, and J. J. Oppenheim. 1986. Interleukin-1 is a radioprotector. *J. Immunol.* **136:** 2483–2485.
41. Neta, R., M. B. Sztein, J. J. Oppenheim, S. Gillis, and S. D. Douches. 1987. In vivo effects of IL-1. I. Bone marrow cells are induced to cycle following administration of IL-1. *J. Immunol.* **139:** 1861–1866.
42. Neta, R. and J. J. Oppenheim. 1988. Cytokines in therapy of radiation injury. *Blood* **72:** 1093–1095.

43. Neta, R., J. J. Oppenheim, and S. D. Douches. 1988. Interdependence of the radioprotective effects of human recombinant IL-1, TNF, G-CSF, and murine recombinant G-CSF. *J. Immunol.* **140:** 108–111.
44. Neta, R. and J. J. Oppenheim. 1991. Radioprotection with cytokines. Learning from nature to cope with radiation damage. *Cancer Cell.* **3:** 391–396.
45. Neta, R., J. J. Oppenheim, R. D. Schreiber, R. Chizzonite, G. D. Ledney, and T. J. MacVittie. 1991. Role of cytokines (interleukin 1, tumor necrosis factor, and transforming growth factor b) in natural and lipopolysaccharide-enhanced radioresistance. *J. Exp. Med.* **173:** 1177–1182.
46. Neta, R., T. Sayers, and J. J. Oppenheim. 1991. Relationship of tumor necrosis factor to interleukins, in *Tumor Necrosis Factor: Structure, Function and Mechanism of Action* (Vilcek, J., and Aggarwal, B. B., eds.), Marcel Dekker, New York, pp. 499–566.
47. Neta, R., R. Perlstein, S. N. Vogel, G. D. Ledney, and J. Abrams. 1992. Role of IL 6 in protection from lethal irradiation and in endocrine responses to IL 1 and TNF. *J. Exp. Med.* **175:** 689–694.
48. Neta, R. and J. J. Oppenheim. 1992. IL-1: Can we exploit Jekyll and subjugate Hyde? in *Biologic Therapy of Cancer Updates*, vol. 2 (DeVita, V. and S. Rosenberg, eds.), pp. 1–11.
49. Neta, R., D. Williams, F. Selzer, and J. Abrams. 1993. Inhibition of c-kit ligand/steel factor by antibodies reduces survival of lethally-irradiated mice. *Blood* **81:** 324–327.
50. Neta, R., J. J. Oppenheim, J. M. Wang, C. M. Snapper, M. A. Moorman, and C. M. Dubois. 1994. Synergy of IL-1 and c-kit Ligand (KL) in radioprotection of mice correlates with IL-1 upregulation of mRNA and protein expression for c-kit on bone marrow cells. *J. Immunol.* **153:** 1536–1543.
51. Neta, R., S. M. Stiefel, F. Finkelman, S. Herrmann, and N. Ali. 1994. Interleukin-12 protects bone marrow from and sensitizes intestinal tract to ionizing radiation. *J. Immunol.* **153:** 4230–4237.
52. Neta, R., J. K. Keller, N. Ali, F. Blanchette and C. M. Dubois. 1996. Contrasting mechanisms of myeloprotective effects of IL-1 against ionizing radiation and cytoablative 5-fluorouracil (5-FU). *Radiat. Res.* **145:** 624–631.
53. Paul, W. E. and R. A. Seder. 1994. Lymphocyte responses and cytokines. *Cell* **76:** 241–251.
54. Patil, R. R. and R. F. Borch. 1995. Granulocyte-macrophage colony-stimulating factor expression by human fibroblasts is both upregulated and subsequently downregulated by interleukin-1. *Blood* **85:** 80–86.
55. Potten, C. S. 1995. Interleukin-11 protects the clonogenic stem cells in murine small-intestinal crypts from impairment of their reproductive capacity by radiation. *Int. J. Cancer* **62:** 356–361.
56. Russell, E. S., S. E. Bernstein, E. C. McFarland, and W. R. Modeen. 1963. The cellular basis of differential radiosensitivity of normal and genetically anemic mice. *Radiat. Res.* **20:** 677.
57. Schwartz, G. N., T. J. MacVittie, R. M. Vigneulle, M. L. Patchen, S. D. Douches, J. J. Oppenheim, and R. Neta. 1987. Enhanced hematopoietic recovery in irradiated mice pretreated with interleukin-1 (IL-1). *Immunopharmacol. Immunotoxicol.* **9:** 371–389.
58. Sinclair, W. K. and R. A. Morton. 1966. X-ray sensitivity during the cell generation cycle of cultured Chinese hamster cells. *Radiat. Res.* **29:** 450–474.
59. Smith, W. W., I. M. Alderman, and R. I. Gillespie. 1957. Increased survival in irradiated animalstreated with bacterial endotoxins. *Am. J. Physiol.* **191:** 124.
60. Tamura, T., M. Ishihara, M. S. Lamphier, N. Tanaka, I. Oishi, S. Aizawa, T. Matsuyama, T. W. Mak, S. Taki, and T. Taniguchi. 1995. An IRF-1-dependent pathway of DNA damage-induced apoptosis in mitogen-activated T lymphocytes. *Nature* **376:** 596–599.
61. Trincheri, G. 1995. Interleukin 12: a proinflammatory cytokine with immunoregulatory functions that bridge innate resistance and antigen specific adaptive immunity. *Annu. Rev. Immunol.* **13:** 251–276.

62. Vairo, G., P. K. Vadiveloo, A. K. Royston, S. P. Rockman, C. O. Rock, S. Jackowski, and J. A. Hamilton. 1995. Deregulated c-myc expression overrides IFN-γ-induced macrophage growth arrest. *Oncogene* **10:** 1969–1976.
63. Withers, H. R., K. Mason, B. O. Reid, N. Dubraysky, H. T. Barkley, B. W. Brown, and J. B. Smathers. 1974. Response of mouse intestine to neutrons and gamma rays in relation to dose fraction and cell cycle. *Cancer* **34:** 39–47.
64. Wong, G. H. W. and D. V. Goeddel. 1988. Induction of manganous superoxide dismutase by tumor necrosis factor: Possible protective mechanism. *Science* **242:** 941–944.
65. Yee, N. S., I. Paek, and P. Besmer. 1994. Role of kit-Ligand in proliferation and suppression of apoptosis in mast cells: basis for radiosensitivity of White Spotting and Steel mutant mice. *J. Exp. Med.* **179:** 1777–1787.
66. Zsebo, K. M., K. A. Smith, C. A. Hartley, M. Greenblatt, K. Cooke, W. Rich, and I. K. McNiece. 1992. Radioprotection of mice by recombinant rat stem cell factor. *Proc. Natl. Acad. Sci. USA* **89:** 9464.
67. Zucali, J. R., J. Moreb, W. Gibbons, J. Alderman, A. Suresh, Y. Zhang, and B. Shelby. 1994. Radioprotection of hematopoietic stem cells by interleukin-1. *Exp. Hematol.* **22:** 130–135.

27

DNA Damage and Repair in the Clinic

David B. Mansur and Ralph R. Weichselbaum

1. INTRODUCTION

In this chapter, some of the basic radiobiological principles of DNA damage and repair and their potential application to clinical radiation oncology are reviewed. In addition, newer technologies to modify tumor cell killing that may have applications to radiotherapy are discussed.

2. THE THERAPEUTIC RATIO

A central problem in radiation oncology is to eradicate the tumor while maintaining damage to normal tissue at an acceptable level. The relationship between desired and undesired effects of therapy is referred to as the therapeutic ratio. As shown in Fig. 1, both the tumor control probability and the complication probability are a sigmoidal function of dose. An improved therapeutic ratio is represented by a greater distance between the two curves and allows a high probability of control with low risk of complications. For example, a local control rate of 80% at 5 yr in patients with organ-confined prostate cancer is obtained with external beam radiation therapy with an expected rate of radiation proctitis of approx 2%. With conventional therapy, further gains in local control by increasing dose are not pursued owing to the increased risk of severe complications. Acute reactions during therapy are common and can be managed with attentive supportive care. In contrast, the late complications of radiotherapy are often severe and are dose-limiting. Many strategies have been investigated in an attempt to increase the therapeutic ratio. These include altered fractionation schemes, radiation sensitizers or protectors, use of cytokines, and the concomitant use of chemotherapeutic agents with radiotherapy.

3. CLASSICAL RADIOBIOLOGY

3.1. *The Mechanism of Cell Killing*

The interaction of ionizing radiation with DNA has classically been described as resulting from a direct or an indirect interaction (*see* Chapter 5). In the direct interaction, a photon is absorbed by the medium resulting in a secondary fast moving electron. The indirect interaction occurs following the generation of a hydroxyl radical, which interacts with the DNA molecule. Experiments using free radical scavengers have led

From: DNA Damage and Repair, Vol. 2: DNA Repair in Higher Eukaryotes
Edited by: J. A. Nickoloff and M. F. Hoekstra © Humana Press Inc., Totowa, NJ

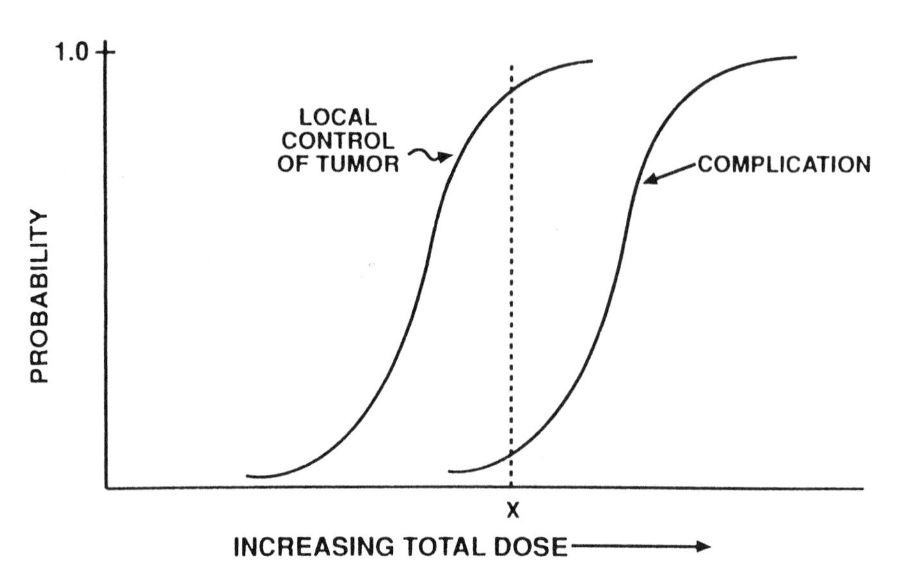

Fig. 1. The therapeutic ratio. A total dose of *X* will yield a high probability of tumor control with acceptable risk of complications.

to the estimation that two-thirds of DNA damage from photon irradiation are indirect. The most clinically relevant definition of cell death is the loss of reproductive integrity of the cell. The cell may continue to function, but attempts to divide will prove lethal. Therefore, the cell is no longer clonogenic.

Many lines of evidence support the view that DNA damage is the principal target affecting cell reproductive death. For example, using a polonium-tipped microneedle, Munro compared the effect of preferentially irradiating the cytoplasm verses the nucleus in Chinese hamster fibroblasts *(53)*. Doses in excess of 250 Gy to the cytoplasm did not affect proliferation, but very small doses to the nucleus were lethal. Further evidence suggests it is the DNA molecule that is the target in radiation-induced cell lethality. For instance, the incorporation of the short-range β-emitter tritiated thymidine into DNA is lethal, and incorporation of halogenated pyrimidines into DNA results in increased radiosensitivity, which is directly proportional to the amount incorporated into DNA *(32)*. Historically, radiobiology has concerned itself with mitotic death. As will be discussed in more detail later, the more recent significance attributed to programmed cell death or apoptosis, which was previously referred to as interphase death, has added another dimension to classical radiobiological concepts.

3.2. Cell Survival Curves

The radiosensitivity of many cell lines has been determined experimentally. A known number of cells in culture are exposed to various doses of photon irradiation and incubated. The surviving number of clonogens is determined by counting colonies. The plating efficiency is considered, and the surviving fraction is calculated. Using the surviving fractions at various radiation doses, cell survival curves are usually expressed with the surviving fraction on a logarithmic scale on the abscissa and radiation dose on a linear scale on the ordinate. Puck and Marcus published the first mammalian cell

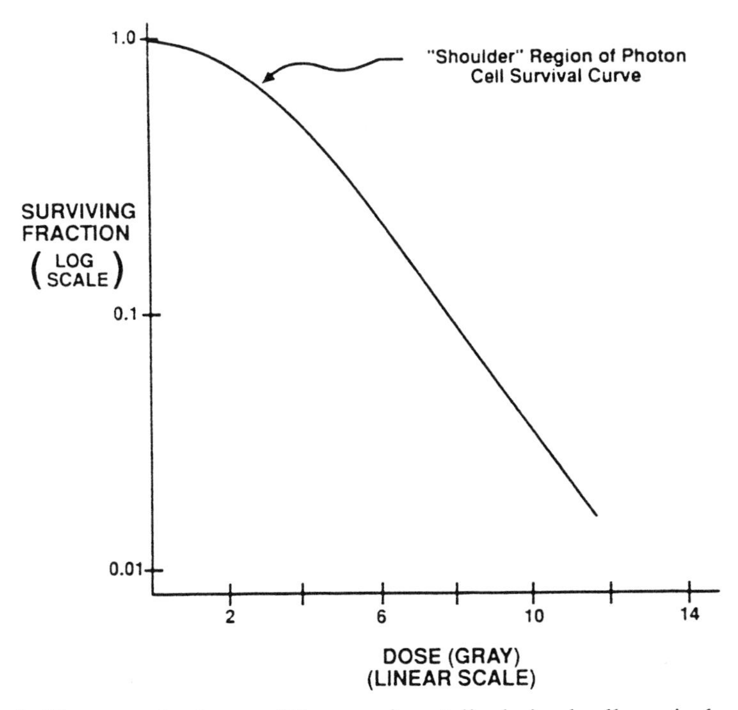

Fig. 2. The general scheme of the experimentally derived cell survival curve.

survival curve in 1956 *(65)*. An idealized representation of the curve is shown in Fig. 2. The general shape of the cell survival curve with photon irradiation includes an initial linear region representing exponential cell killing as a function of dose at lower dose ranges. With increasing doses, however, the curve bends, showing an increased rate of cell kill over a finite dose range. Finally, at higher doses, the curve again approaches a straight line. The intermediate curved region is often referred to as the "shoulder on the curve" *(85)*.

3.3. Radiobiological Models Utilized in Radiotherapy

Two mathematical models are commonly used to describe mammalian cell survival curves, the linear-quadratic and the two component. The linear quadratic model is ascribed to Read, who, building on the work of Lea and Catcheside *(46)*, quantified growth delay and chromosome aberrations in terms of dose coefficients in the irradiated bean root *(66)*. He showed that the effect of irradiation was proportional to the sum of two factors:

$$\text{Effect} \propto \alpha d + \beta d^2 \tag{1}$$

where d = dose, α = linear dose coefficient, and β = quadratic dose coefficient. α and β are constants, unique to tissue type, describing the relative impact of high dose (β) vs low dose (α) on an effect. The curve resulting from the product of these two factors represents the survival curve (Fig. 3). The ratio of α/β is that dose where low-dose effect and high-dose effect contribute equally to cell killing *(32)*. A higher α/β ratio (approx 10 Gy) is characteristic of acute responding tissues such as mucosa, skin, and

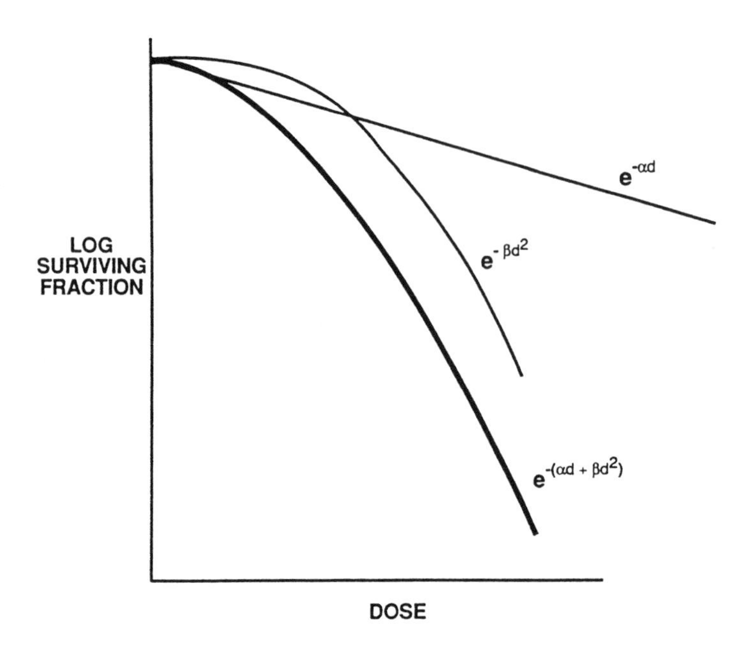

Fig. 3. The linear-quadratic model of the cell survival curve. *See text.*

tumors, and results in a curve with a steep initial slope and gentle shoulder. A lower α/β ratio (approx 2 Gy) is characteristic of late responding tissues, like brain, spinal cord, kidney, and lung, and results in a shallow initial slope, but a sharply curving shoulder (Fig. 4). Thus, effects on late responding tissues, are predicted to be more dependent on fraction size (the daily dose delivered during treatment) than effects on early responding tissues including tumors. By decreasing fraction size, acute responding tissues are affected little, but significant sparing occurs in late responding tissues *(61)*. This has important clinical implications to be discussed later.

An alternative mathematical representation is the two-component model. In contrast to the linear-quadratic, which has a continuously bending terminal portion, the two-component model has an initial linear region followed by a shoulder and then a return to a terminal linear portion (Fig. 5). D_0 is defined as the dose that reduces the survival by one natural logarithm. This curve is described by the slope of the initial linear region ($1/D_1$), the slope of the terminal linear region ($1/D_0$), with the extrapolation number *(n)* and D_q value giving additional information regarding the size of the shoulder *(85)*. Although the terminal region of experimentally derived cell survival curve is better described by the terminal portion of the two component model, the shoulder region is better approximated by the linear-quadratic model, as described by the α/β ratio. A distinct advantage to the linear-quadratic model is that α/β ratios of different cell lines can be determined experimentally, which can help to predict their response to fractionation *(61)*.

The biological mechanism resulting in the characteristic shape of the cell survival curve is unknown. Lethality presumably results from chromosome deletions and from exchange-type chromosome aberrations (rings and dicentrics), which require the interaction of two double-strand breaks. The initial linear region is thought to represent lethality from DNA double-strand breaks from single-hit phenomena (both breaks

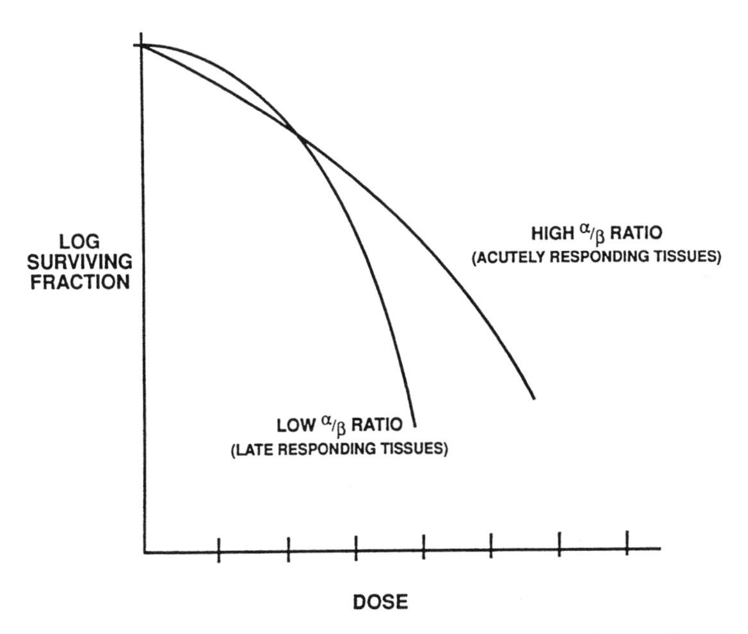

Fig. 4. A comparison of linear-quadratic curves for high vs low α/β values. Acutely responding tissues, such as skin, mucosa, and most tumors, have characteristically high ratios. Late-responding tissues such as kidney, spinal cord, and brain, have lower ratios.

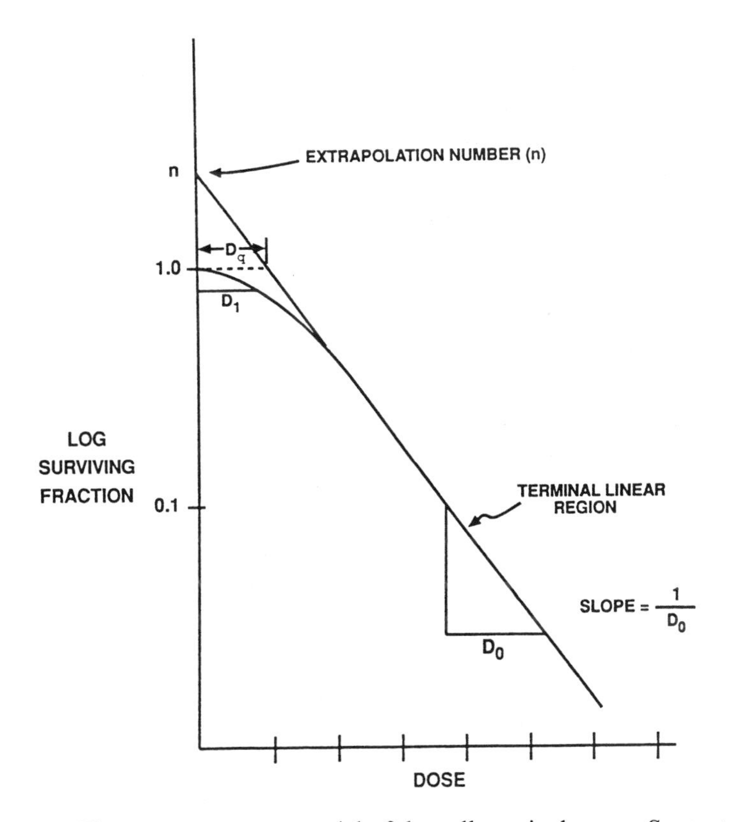

Fig. 5. The two-component model of the cell survival curve. *See text.*

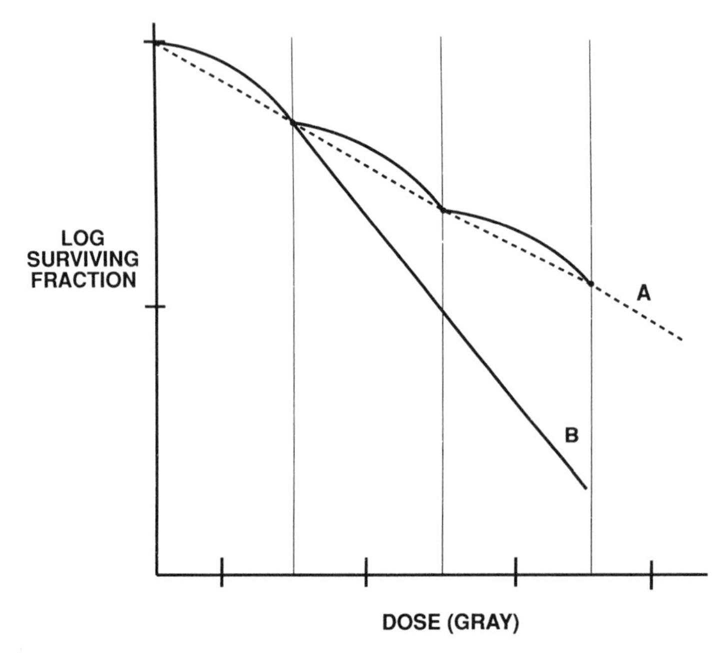

Fig. 6. An idealized comparison of the multifraction survival curve **(A)** with the single-dose survival curve **(B)**. The multifraction curve is obtained experimentally by three 4-Gy fractions with a few hours between each dose.

caused by the same electron). Over this region of the curve, the probability of an exchange-type chromosome aberration is proportional to dose (D). The steeper region beyond the shoulder represents lethality from double-strand breaks from multiple-hit phenomena. In this range, the probability of a chromosome aberration is proportional to the dose squared (D^2) *(32)*. In the two-component model (Fig. 5), the cell has a greater capacity to repair DNA damage at the lower dose ranges associated with the shoulder region. A broader shoulder, quantified by the extrapolation number and/or the value of D_q, signifies greater ability of the cell to repair sublethal damage. Sublethal damage repair (SLDR) is defined operationally as a process that results in a larger surviving fraction after a given amount of radiation if the dose is split into two equal fractions and separated by a time interval. The mechanisms are not known, but it is hypothesized that SLDR results from repair of double-strand DNA breaks, given a sufficient time interval and normal metabolic activity, thereby preventing formation of exchange-type chromosome aberrations. This process has been demonstrated in split dose experiments *(22)* and also in multifraction survival curves. With fractionated doses, the shoulder is recapitulated with each fraction, as depicted in Fig. 6. Table 1 demonstrates that the ultimate surviving fraction can be predicted mathematically by the survival of the individual fraction raised to an exponent equal to the total number of fractions *(38,80)*. The probability of ultimate survival, or in clinical terms lack of tumor control, is therefore profoundly influenced by the size of the shoulder region. Interestingly, in split dose experiments of different established cell lines from human tumors with considerable variation in clinical radiocurability, little variation among SLDR capacity has been demonstrated *(83)*.

Table 1
Calculated Cumulative Survival Fraction[a]

Cumulative survival fraction	X^{32} X=	X^{20} X=
10^{-11}	0.45	0.28
10^{-10}	0.49	0.32
10^{-9}	0.52	0.35
10^{-8}	0.56	0.40
10^{-7}	0.60	0.45
10^{-6}	0.65	0.50
10^{-5}	0.70	0.56

[a]Calculated cumulative survival for either 32 or 20 equal fractions when the survival after each fraction *(X)* is varied *(38)*.

When irradiated cell populations are maintained in suboptimal conditions and thereby prevented from proceeding through the cell cycle, an increase in the surviving fraction is observed. This phenomenon has been described as resulting from repair of potentially lethal DNA damage. Since human tumors contain a majority of noncycling cells, Little and Hahn suggested that density inhibited plateau phase cultures might be a better in vitro model of clinical conditions *(31,47)*. When irradiated cells are maintained in a density-inhibited state for longer time intervals, the surviving fraction, when eventually plated in low-density conditions, is increased. This observation, analogous to liquid holding recovery in bacteria and yeast *(27,60)*, is an example of potentially lethal damage repair (PLDR).

Nine established human cell lines from tumors with varied radiocurability were studied with respect to their capacity for PLDR *(83)*. The cell lines from tumors generally considered radiation-resistant, such as osteosarcoma and melanoma, displayed a greater capacity for PLDR than those of more radiocurable histologies, such as breast and neuroblastoma. Since the D_0 and extrapolation number *(n)* were similar in the nine cell lines, it was postulated that the difference in PLDR represents at least one factor influencing radiocurability of human cancer *(83)*.

The amount of PLDR has been shown to be influenced by the dose in some tumors. The relationship between PLDR and dose in melanoma was studied in a single fraction experiment *(82)*. Although the absolute cell kill was greater, PLDR was increased with higher single doses (*see* Fig. 7). This has implications for clinical radiation oncology, since often large fraction sizes are used to treat supposedly radioresistant histologies. The in vitro evidence, however, indicates ultimate control might be improved by limiting PLDR using smaller fraction sizes *(82)*.

In keeping with the in vitro data on PLDR, when permanent local tumor control and not regression or palliation is used as the end point, clinical experience favors smaller fraction sizes. In a clinical study, Eichorn *(21)* compared the tumor control in patients with lung cancer treated with various fractionation schemes. Control was evaluated in thoracotomy specimens in patients given preoperative radiation therapy and also at

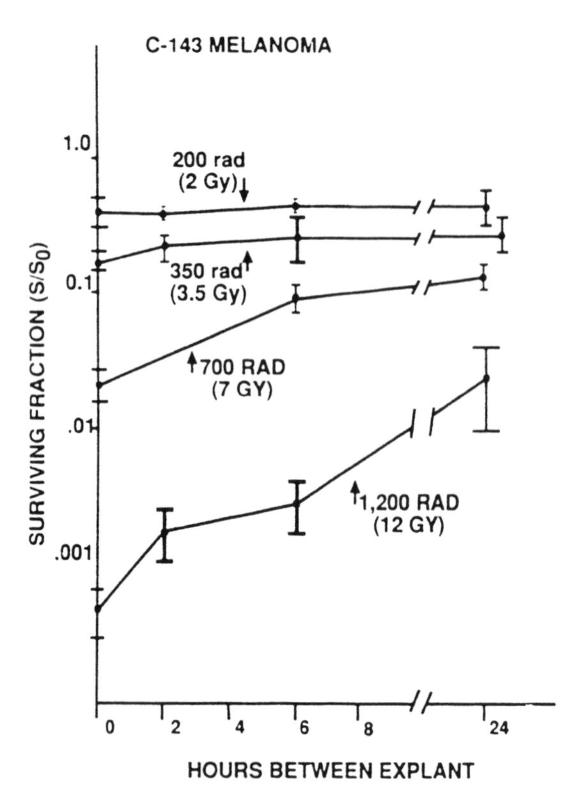

Fig. 7. Enhancement in survival following subculture as a function of dose in human melanoma line C-143. Twenty-four-hour survival levels may be more analogous to the in vivo situation in clinical radiotherapy than are survival levels obtained from exponentially growing cells or 0 h subculture survival points. Note increase in repair with increasing X-ray doses *(80)*.

autopsy in patients treated for unresectable cancers. The fractionation varied from 2.5 Gy × 22 to 10 Gy × 2. Though the total effective dose (nominal standard dose) was similar in all patients, the percentage of complete responders (those with no tumor detectable) was significantly higher in patients treated with multiple smaller fractions. Not only is tumor control improved with fractionation, but as predicted by the linear-quadratic model and supported by many clinical observations, late-responding dose-limiting normal tissues (low α/β ratio) are more sensitive to the effects of fractionation which results in relatively more sparing of these tissues than those of acutely responding tissues (high α/β ratio, including tumors) (*see* Fig. 8). Owing to the relative shapes of the curves in the low dose range, a change to a smaller fraction size results in the preferential sparing of late responding tissues.

It must be emphasized that although the two-component and linear-quadratic models of cell killing are based on calculation and biological observation, neither theory has a firm basis in modern molecular biology.

3.4. Fractionation in Clinical Radiotherapy

For almost a century, fractionation has been central to clinical radiotherapy resulting in a workable therapeutic ratio. The concept of fractionation was applied early in this

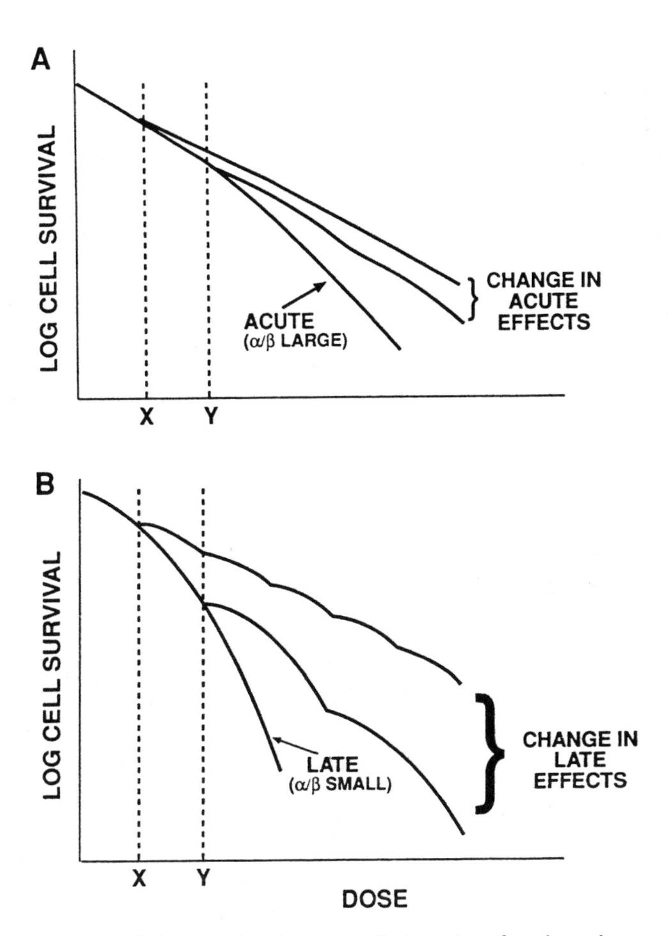

Fig. 8. A comparison of the relative impact of changing fraction size on late- or acutely responding tissues. Reducing fraction size from *Y* to *X* has more sparing of late effects.

century after two decades of clinical observation. In 1932 Coutard published his results of fractionated external beam radiotherapy in 212 patients treated for carcinomas of the head and neck *(13)*. Daily treatments of 2 Gy were given over several weeks. After healing of severe skin reactions inflicted by the low-energy X-rays used, 20% of patients were cured of their malignancy. The theoretical basis for the success of fractionation has been attributed by radiobiologists to three main phenomena: repair (SLDR and PLDR), redistribution within the cell cycle, and reoxygenation.

Generally speaking, tumor cells are asynchronously distributed throughout the cell cycle. Cells have different radiosensitivities depending on their position within the cell cycle, with G2 and M being most sensitive and late S phase being most resistant. After one fraction, the initially asychronous population becomes partially synchronized in resistant phases. Redistribution within the cell cycle in theory allows, given a sufficient time interval, a population of cells to redistribute so that more cells will be in sensitive phases for the next fraction.

Experimental rodent tumors have a substantial proportion of relatively hypoxic cells, which confers radioresistance. After a first fraction, killing of the well-oxygenated radiosensitive cells can result in improved oxygenation and therefore sensitization of the remaining, previously hypoxic cells. Since normal and late-responding tissues do not ordinarily have large hypoxic regions, this resistance to radiation killing, which is in part overcome by fractionation, is limited to the tumor. There is much debate regarding the role of hypoxic cells (and thus reoxygenation) in human radiotherapy. Much indirect evidence points to the presence of hypoxic cells in human tumors, and some direct measurements of hypoxia in human tumors have recently been made *(37)*.

3.5. Clinical Trials with Altered Fractionation

According to the linear-quadratic model, late-responding tissues would be preferentially spared by hyperfractionation, which is defined as the use of smaller than conventional fraction sizes, allowing an increase in overall dose to a tumor with no increase in late complications. This hypothesis that the total dose can be increased safely by decreasing the fraction size was tested in two prospective trials by the Radiation Therapy Oncology Group (RTOG) *(14,15)*. One thousand eighty-five patients with advanced or unresectable cancers of the head and neck or lung were treated with 1.2 Gy twice a day with a 4-h interfraction interval. The patients were randomly allocated to different dose levels ranging from 60–81.6 Gy. The late complication rate in the higher-dose patients was only slightly increased in spite of a 20–30% increase in total dose. Thus, there is emerging clinical evidence that hyperfractionation may allow significant increase in total dose without significantly increasing complication rates.

There have been several prospective randomized clinical trials directly comparing conventional fractionation to hyperfractionation. The European Organization for Research on the Treatment of Cancer (EORTC) randomized 356 patients with oropharyngeal carcinoma of intermediate stage to receive either conventional radiation therapy consisting of 2 Gy daily to a total dose of 70 Gy or hyperfractionated radiation therapy consisting of 1.15 Gy twice daily to a total of 80.5 Gy *(41)*. A significant improvement in local tumor control for the larger tumors receiving hyperfractionation was observed (42 vs 17%), but late complications were similar in both groups. The potential benefit of hyperfractionation for certain subsets of patients was demonstrated in this study. Edsmyr et al. *(20)* randomized 168 patients with transitional cell carcinoma of the bladder to receive either the conventional 2.0 Gy daily to a total of 64 Gy or the hyperfractionated 1.0 Gy 3X/d to a total of 84 Gy. Local control was improved in the hyperfractionation arm (65 vs 36%). However, small bowel complications were increased twofold. This study has been criticized for excessive total dose in the hyperfractionated arm resulting in increased normal tissue complications *(61)*. Datta et al. *(16)* conducted a randomized trial of 212 patients with intermediate-stage carcinomas of the head and neck region comparing 2 Gy once daily to a total of 66 Gy to 1.2 Gy twice daily to a total of 79.2 Gy. Complications were similar in the two groups, but disease-free survival was improved from 34 to 63% with hyperfractionation. Pinto et al. *(63)* demonstrated an improved local control at the primary site of 84 vs 64% in patients treated with 1.1 Gy twice daily to a total of 70.4 Gy compared to those treated with conventional fractionation. Late toxicity was similar in both groups.

Clinical data to date suggest that in select patients, hyperfractionated external beam radiotherapy will result in an improved therapeutic ratio. Further studies are under way comparing hyperfractionation to other innovative techniques, including concomitant chemoradiotherapy.

Hyperfractionation should be distinguished from another form of altered fractionation, accelerated fractionation. Accelerated fractionation is defined as the use of an overall shorter treatment time to lessen the likelihood of tumor repopulation. As tumors sustain damage from therapy, individual clonogens can show a dramatic increase in proliferation ("repopulation") even as the gross tumor decreases in volume *(85)*. The phenomenon of repopulation is thought to contribute to the poor radiotherapy control rates of epidermoid carcinomas seen clinically if overall treatment time is protracted. Treatment time can be reduced in a number of ways, such as increasing the daily fraction size (which requires a decrease in overall dose owing to increased risk of adverse effects to normal late-responding tissue), giving twice daily conventional size fractions, or eliminating any conventional treatment breaks including weekends.

4. APOPTOSIS

4.1. History

At the turn of the century, embryologists observed pyknotic nuclei occurring at various locations in developing systems and interpreted them as "mitotic metabolites," suggesting their presence was a byproduct of growing regions rather than evidence of concurrent cell death. In 1951, Glucksmann temporally and spatially detailed the occurrence of "cell deaths" in vertebrate embryology *(28)*. He described nuclear disruption with chromatin clumping, cell shrinkage, "chromatolysis," and phagocytosis by neighboring cells. He noted the absence of mitochondrial changes observed prior to cell death from injurious agents. This process was integral in the phenomena of invagination, segmentation, lumen formation, and other embryological processes. He concluded that this process is an integral and deliberate part of development.

In 1972 Kerr et al. *(43)* proposed the term "apoptosis" (a Greek term describing the falling or dropping off of leaves or petals) for the active and controlled cell death now recognized to occur not only during embryogenesis, but in many facets of biology, including cell renewal systems and hormone-related atrophy. Apoptosis emerged as integral in all of development and homeostasis, playing an equally important, but opposite role to that of mitosis. Modern ultrastructural descriptions of apoptosis *(88)* echo the earlier observations of Glucksmann and others. There are four important elements that define apoptosis. The first is a rapid volume reduction of approximately one-third that occurs within minutes. Second, chromatin condensation results from nuclear endonuclease activity at internucleosomal sites. Third, recognition of new cell membrane structures by phagocytic cells occurs. Finally, active protein synthesis is required as evidenced by the blocking of apoptosis in thymocytes treated with glucocorticoids or ionizing radiation in the presence of protein synthesis inhibitors *(88)*.

4.2. Control of Apoptosis

There exists a dynamic transition between susceptibility and resistance to apoptosis, which is controlled by regulatory genes *(89)*. In a wide range of mammalian cells, the *bcl*-2 gene family functions to suppress apoptosis. Other genes with little homology to *bcl*-2 also function to suppress apoptosis including adenovirus E1b, activated T24-*ras*, v-abl, p35, and *iap*. In opposition to the above genes are important inducers of apoptosis, including p53 and c-*myc* (*84*). In addition to these genetic controls, many inducers and inhibitors of apoptosis have been identified. These include physiological factors, such as hormones, growth factors, and neurotransmitters, as well as pharmacological agents and toxins. Many chemotherapeutic agents as well as ionizing radiation and free radicals induce apoptosis. A summary of inducers and inhibitors of apoptosis can be found in recent reviews *(76,26)*. The classic radiobiological models assume cell lethality results from damage to DNA prior to attempted division. Apoptosis results in a linear in vitro cell survival curve. Thus, it is proportional to dose (D) rather than D^2, and in a population of cells will contribute to the linear portion of the curve. Although DNA damage may lead to cell death, it may do so, at least in part, through an apoptotic mechanism. Thus, cells may have inherent apoptosis checkpoints, such that irradiation beyond a threshold dose triggers a programmed cell death *(26)*.

4.3. Apoptosis in the Clinic

4.3.1. Tumorigenesis

Historically, neoplasia has been viewed in broad terms as an increasing rate and tenaciousness of proliferation. With the realization that apoptosis is an integral component of development and homeostasis, however, an alternative conceptual framework has become popular, namely that cancer can result from either increased cell proliferation or decreased rate of cell death within an otherwise normally homeostatic system *(76)*. Thus, a cell's loss of apoptotic tendencies could disrupt homeostasis and result in malignancy.

Using a myeloid leukemic cell line that lacks p53 protein and mRNA, Yonish-Rouach et al. *(90)* studied the effect of introducing p53 activity, which is associated with apoptosis. By transfecting the cells with a temperature-sensitive p53 mutant, normal p53 was expressed when the temperature was decreased to 32.5°C. This resulted not only in loss of viability of this cell line, but cell death via apoptosis was induced. This is consistent with the loss of normal p53 activity being involved in tumorigenesis by disruption of a regulatory control involving apoptosis. Indeed, p53 has been shown to be mutated in many common human cancers. *(4,6,42,44,77,91)*.

4.3.2. Cancer Therapy

Apoptosis has implications not only in tumorigenesis, but also in cancer therapy. Changes in the balance regulating apoptosis, if selectively applied to the cancer cell, could significantly improve the therapeutic ratio. The prospect of selectively using the efficient destructive power of programmed cell death against cancer is quite exciting.

The apoptotic response to irradiation has been documented for a number of human cancer cell lines, and the amount of apoptosis occurring is often directly proportional to the radiosensitivity of the cell line. Of note is the heterogeneity of the apoptotic radiation response both between cell lines and between cells of the same cell line. Some cell lines undergo no apoptosis even in response to large doses of radiation. Within a radiosensitive population, the proportion of cells observed to undergo apoptosis peaks at approx 35% in spite of further increases in dose *(52)*. In the radiation oncology clinic, the only sizable tumors with significant cure rates are those, such as lymphomas and germ cell tumors, which are known to respond to radiation predominantly by apoptosis.

The biochemical mechanisms responsible for the presence or absence of the apoptotic response to therapeutic doses of radiation can be influenced by various cytokines, oncogenes, and growth factors, which appear to exert their effect by way of calcium metabolism *(50,52,72)*, transcription and protein synthesis *(52,72)*, signal transduction *(49,51,52)*, and polyamine metabolism *(7,52)*. For example, the human lymphoma LY-TH cell line has been shown in vivo to undergo radiation-induced apoptosis readily. However, when intracellular calcium is chelated, apoptosis is blocked. Surprisingly, when actinomycin D and cyclohexamide were used to inhibit protein synthesis (a technique that often is used to inhibit apoptosis), these agents induced apoptosis spontaneously. The influence of signal transduction was shown by the opposing influences of protein kinase A (PKA) and protein kinase C (PKC) in affecting radiation-induced apoptosis. Both activation of the PKC pathway and inhibition of the PKA pathway inhibited radiation induced apoptosis. Likewise, the addition of the polyamine spermine to media containing LY-TH cell cultures inhibited radiation induced apoptosis *(52)*.

The preferential induction of apoptosis within tumor cells resulting in apparent improved therapeutic ratio in vivo has been accomplished by combining gene therapy and external beam radiation therapy *(34)*. The radiation-inducible promotor of the Egr-1 gene was ligated to tumor necrosis factor-α (TNF-α) cDNA. This construct, carried by a virus vector, was injected into radioresistant human epithelial tumors growing in nude mice. Irradiation increased TNF-α expression, increased apoptosis as well as necrosis, and improved tumor control in the treated tumors compared with those serving as controls (either untreated, treated with construct but no radiation, treated with null virus and irradiated, or treated with radiation alone). Acute toxicity was comparably mild in the various groups. Thus, the expression of genes and their tumoricidal effects can be spatially and temporally controlled, which may result in increased therapeutic ratio and have far-reaching implications for cancer therapy (*see* Section 6.).

5. USE OF CONCOMITANT CHEMORADIOTHERAPY

The simultaneous administration of radiation therapy and chemotherapeutic agents has become increasingly used as evidence of efficacy. Chemotherapy reduces the risk of distant micrometastatic disease in many clinical settings. When administered concomitantly with radiation therapy, improvements in local tumor control can also result. In general terms, the interaction within the irradiated tissue can be either additive with each mode of therapy acting on different tumor cell populations or by a separate mechanism, or the interaction can be synergistic with resultant radiosensitization or potentiation. There exists considerable debate regarding the mechanism of interaction between chemotherapeutic agents and ionizing radiation. Though many clinical observations attest to the efficacy of concomitant chemoradiotherapy, whether these interactions fit within the classic radiobiological models and how they affect apoptosis are unclear.

The chemotherapy agents commonly used concomitantly with radiation therapy include hydroxyurea (HU), cisplatin, 5-fluorouracil (5-FU), and mitomycin C. HU is an S-phase-specific agent that inhibits ribonucleotide reductase and has efficacy when used concomitantly with radiotherapy. HU is proposed to increase killing of cells in the relatively radioresistant S phase. Randomized clinical trials in patients with cervical cancer treated with radiation and concomitant HU have shown improved tumor control rates compared to control groups *(64,71)*. Cisplatin is cytotoxic by the formation of intrastrand and interstrand crosslinks within DNA and RNA *(12)*. Cisplatin has been shown to enhance cell killing from radiation in mammalian cells in vitro *(18,10)*. The antimetabolite, 5-FU, interferes with DNA synthesis by the inhibition of thymidylate synthase *(87)*. Many in vitro studies have shown improved cell killing with combinations of 5-FU and radiation compared to these agents alone *(78)*. The RTOG compared the efficacy of concomitant cisplatin, 5-FU, and radiation therapy (50 Gy) to radiation therapy alone (64 Gy) in a prospective randomized trial of patients with carcinoma of the esophagus *(39)*. After accruing 121 patients, the trial was prematurely stopped after a statistically significant improvement in survival was detected in the group receiving combination therapy. The combined modality group had improved local tumor control as well as less distant metastatic disease. Mitomycin C is another chemotherapeutic agent used clinically with concomitant radiotherapy, which is activated to an alkylating agent within the cell. Thus, mitomycin C is cytotoxic by the production of DNA strand crosslinks. The intracellular activation of mitomycin C may occur preferentially in hypoxic cells, which provides a rationale for additive effects with radiotherapy *(78)*. Mitomycin C has been shown in vitro to kill radioresistant populations *(67)*. In the clinical setting, dramatic responses in patients with carcinoma of the anus have been observed. Nigro et al. *(57)* utilized concomitant mitomycin C, 5-FU, and radiation therapy (30 Gy) prior to abdominoperineal resection in anal cancer patients. Long considered the standard of care, the abdominoperineal resection inflicts significant morbidity in that the anal sphincter is removed, leaving the patient with a permanent colostomy. Of 44 patients initially entered into this study, 40 had a clinical complete response (complete disappearance of all disease) to preoperative concomitant chemoradiotherapy, and the surgical specimens from these revealed no microscopic evidence of tumor in 39 patients. The abdominoperineal resection has subsequently been performed only on those few patients with less than a complete response. Many similar trials in anal cancer have followed using definitive concomitant therapy with resection reserved for only those patients with a noncomplete response. Results yield excellent tumor control rates with a high percentage of patients retaining their sphincters. Though the mechanisms remain unclear, the successful use of concomitant chemoradiotherapy has been accomplished with improved patient survival, improved local tumor control, and often improved quality of life.

6. RADIATION-INDUCIBLE CYTOKINES WITH POTENTIAL APPLICATION TO CLINICAL RADIOTHERAPY

The observation that radiation-inducible cytokines may offer killing or protective effects in addition to those of direct radiation effect on DNA introduces yet another dimension into the classic radiobiological models. Also, the dose-limiting late effects of radiotherapy may result from induction of various cytokines (*see* Chapter 26). Therefore, the ability to exploit or inhibit these effects preferentially in the clinic offers a potential means to improve the therapeutic ratio.

6.1. *TNF-α*

Early this century, certain patients with advanced cancers and simultaneous bacterial infections were observed to have regression of their malignancy. Coley and others demonstrated hemorrhagic necrosis of mouse tumors with the administration of killed bacteria mixtures, so-called Coley's toxins. Subsequently, the active agent in these tumor regressions was shown to be lipopolysaccharide (LPS), a major component of Gram-negative bacterial cell walls. In the 1970s, the necrotizing and cytotoxic actions elicited by LPS were shown to be mediated directly by a protein termed tumor necrosis factor (TNF) *(58)*. TNF-α is a peptide of 157 amino acids produced mainly by monocytes, but also by a number of other cell types, including lymphocytes, neutrophils, and glial cells. In addition, half of all tumor cell lines tested express TNF mRNA. A closely related cytokine, lymphotoxin, has also been named TNF-β *(19)*.

There are multiple biological effects of TNF. Theses cytokines produce fibrin formation on endothelial cell surfaces resulting in thrombosis and hemorrhage. TNF induces expression of IL-2 receptors, stimulates T- and B-cell proliferation, and activates neutrophils and macrophages. A number of secondary cytokines are induced by TNF, including IL-1 and IL-6. Clinically, administration of TNF produces a syndrome similar to Gram-negative shock including fever, hypotension, and disseminated intravascular coagulation *(19)*.

The excitement generated by sustained responses or even cures in mice with experimental tumors eventually led to human trials using exogenous TNF. Many phase I (toxicity) trials have been conducted with disappointing response rates *(70)*. These trials have been characterized by dose-limiting hypotension. Old and Spriggs *(59)* have estimated the highest phase I dose level used in human trials is 5–25 times lower than that required for a hemorrhagic response in the mouse. Phase II trials involving patients with metastatic melanoma *(25)* and breast cancer *(8)* had similarly disappointing results with virtually no responses.

The ability of ionizing radiation to stimulate production of TNF-α has been demonstrated. Hallahan et al. *(35)* showed that several sarcoma cell lines increased TNF-α production fivefold following exposure to radiation compared to nonirradiated controls. Radiation also induces TNF gene transcription in human promyelocytic and monocytic leukemia cells *(69)*. In addition to the in vitro data, clinical studies also demonstrate increased TNF production after therapeutic irradiation. Patients receiving total body irradiation (TBI) prior to bone marrow transplantation have increased serum TNF *(5,40)*, and monocytes cultured from irradiated breast cancer patients show elevated TNF production *(79)*. In contrast, many other human tumor cell lines show no increased TNF production following irradiation *(35)*.

It is noteworthy that TNF has radioprotective effects in some tissues, yet potentiates radiation sequelae in others. The majority of tissues are sensitized to irradiation by TNF. For instance, the serum concentration of TNF is correlated with severity of hepatic dysfunction, pneumonitis, and renal insufficiency following TBI *(5)*. In other clinical settings, TNF may potentiate demyelination and the resultant central nervous system complications *(33)*. Although sensitizing many tissues, TNF functions to protect the bone marrow from effects of irradiation *(1,55,56,68)*.

The differing effects of TNF on irradiated tissues have been partially explained by the discovery of two distinct TNF receptors, R1 and R2, with separate signaling pathways. *(54,75)*. The effects of TNF, either radioprotection or radiosensitization, could therefore be explained by the relative concentrations R1 and R2 on different tissues *(33)*.

There are two main mechanisms in TNF-related cytotoxicity. First, TNF is directly cytotoxic to some tumor cell lines. This action is blocked by anaerobic conditions as well as antioxidants suggesting free radical DNA damage as a mechanism for cell killing *(45,48,58,73,92)*. The radiosensitization effect of TNF is maximal when TNF is administered 4 h prior to irradiation. It is therefore suggested that TNF depletes the bioreductive capability of sensitive cells resulting in enhanced effect of irradiation *(33)*.

6.2. Clinical Trials with TNF and Radiation Therapy

The potential radiosensitizing effect of TNF in humans was studied in a phase I trial of recombinant human TNF-α and concomitant external beam radiotherapy *(36)*. Thirty-one patients with locally advanced or metastatic tumors were given iv human recombinant TNF-α 4 h prior to radiotherapy. Similar to previous trials, systemic toxicity led to seven patients being removed from the study. There was no increase in normal tissue toxicity within the radiation field. In the 20 patients evaluated for response to treatment within the irradiated field, there were four complete responders, four partial responders, four patients with stable disease, and four patients with progressive disease. This trial is significant in a number of ways. TNF-α produced no toxic reactions in the irradiated fields, and showed significant response rates at relatively low radiotherapy doses for tumors historically thought to be radioresistant. This picture is consistent with that of TNF-α selectively sensitizing some tumors while sparing normal tissue *(36)*.

As described in Section 4.3.2., Weichselbaum et al. *(81)* have employed gene therapy in an attempt to avoid systemic toxicity while increasing TNF concentration within tumors. By using a radiation-inducible promotor attached to the TNF-α gene, TNF-α expression can be limited to the region that is irradiated. Human trials are planned.

6.3. Other Radiation-Inducible Cytokines with Potential Application to Clinical Radiotherapy

The secretion of both basic fibroblast growth factor (bFGF) and platelet-derived growth factor (PDGF) from bovine aortic endothelial cells (BAEC) is stimulated in vitro by ionizing radiation *(86)*. Doses as low as 1.25 Gy result in a twofold increase in bFGF. Doses of 2.5 Gy (near those of standard fraction sizes in clinical radiotherapy) result in maximal bFGF mRNA levels experimentally *(33)*. Also, Fuks et al. *(33)* showed that the radiosensitivity of BAEC was dependent on its in vitro microenvironment. When cultured on an autologous BAEC basement membrane-like extracellular matrix, these cells demonstrated improved repair of radiation damage compared to cells cultured on uncoated culture dishes or on unrelated matrix components. These observations have led investigators *(33)* to postulate that in BAEC, bFGF synthesis results as a stress response to DNA damage, and that secreted bFGF initiates a DNA repair pathway via an autocrine loop. The late effects of radiotherapy on normal tissue are the major factors limiting radiotherapy dose in the clinic. Specifically, endothelial damage at the microvascular level is a major mechanism inducing late normal tissue radiation induced necrosis *(24)*. However, in other regions of the circulatory system, such as the heart and larger vessels, the endothelium is less affected by irradiation. Immunohistochemistry studies show a relative paucity of bFGF activity in capillaries compared to intermediate or larger vessels, thereby correlating bFGF activity with protected regions of the circulation. The logical clinical extension of this, proposed by Hallahan et al. *(33)*, would involve the iv administration of bFGF during radiotherapy to increase normal tissue tolerance by preventing microvascular damage, thereby improving the therapeutic ratio.

Based on immunohistochemistry analyses of nontumor-bearing regions of colon, Carney and Dean *(9)* demonstrated transforming growth factor-β (TGF-β) production in four of six patients treated with preoperative radiation therapy, and this was associated with the expected histological lesions of endothelial proliferation and destruction of normal epithelium. No TGF-β was detected in patients resected without preoperative radiation. Others have shown that in the irradiated liver, the amount of TGF-β-positive hepatocytes correlated with the resulting extent of fibrosis, suggesting that the production of TGF-β in response to radiation injury may play a role in the pathogenesis of radiation hepatitis *(2)*. These findings have led observers to consider TGF-β a radiation-inducible cytokine associated with late responding tissue damage, and possible clinical applications rest on inhibiting the effect of this cytokine in patients receiving radiation therapy.

Therefore, the radiation-inducible cytokines have emerged as important in the radiation response in normal tissues as well as tumors. Great potential exists in selectively exploiting or inhibiting these factors in clinical radiotherapy.

7. RADIOSENSITIZATION WITH HALOGENATED PYRIMIDINES

Another approach to improving the therapeutic ratio is the use of radiosensitizers and radioprotectors. These agents have been reviewed previously *(11,32)*, and this section focuses on halogenated pyrimidines, which function as nonhypoxic sensitizers by direct alteration of DNA radiosensitivity.

The initial observation that human cells in vitro were more sensitive to radiation when 5-bromodeoxyuridine (BUDR) or 5-iododeoxyuridine (IUDR) was incorporated into DNA was published in 1960 *(17)*. The structures of the halogenated analogs thymidine are shown in Fig. 9.

Halogenated pyrimidines are agents capable of selectively sensitizing the DNA of cells to the effects of ionizing radiation. Though the initial generalization that tumors proliferate at a faster rate than normal tissue is now accepted as false, certain clinical settings exist where malignant cells proliferate more rapidly than adjacent normal tissue, as in high-grade gliomas surrounded by normal, nonproliferating brain cells. Only in rare clinical settings such as this are pyrimidine analogs incorporated preferentially into tumor cells. Since cells containing these analogs in their DNA have been shown to be several times more radiosensitive than controls, a larger therapeutic ratio could theoretically be obtainable *(74)*. The degree of radiosensitization is proportional to the amount of halogenated analog that is incorporated. Though the mechanism of radiosensitization is not clear, it most likely involves direct effects on the substituted DNA molecule, rather than other cellular effects of these analogs, such as enzyme inhibition *(44)*. Other evidence that the biochemistry of substituted DNA is altered includes observed increases in both temperature- and pH-sensitivity with incorporation of halogenated pyrimidines *(30)*.

The experimental work with halogenated pyrimidines throughout the 1960s (reviewed by Szybalski; *74*) led to initial clinical trials in the US. A small prospective

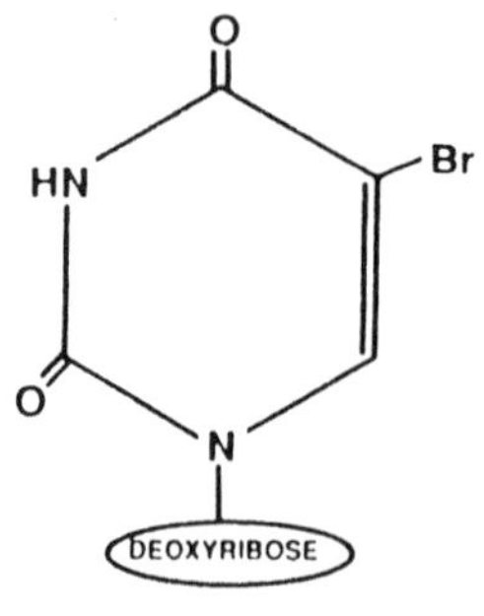

5 - bromodeoxyuridine

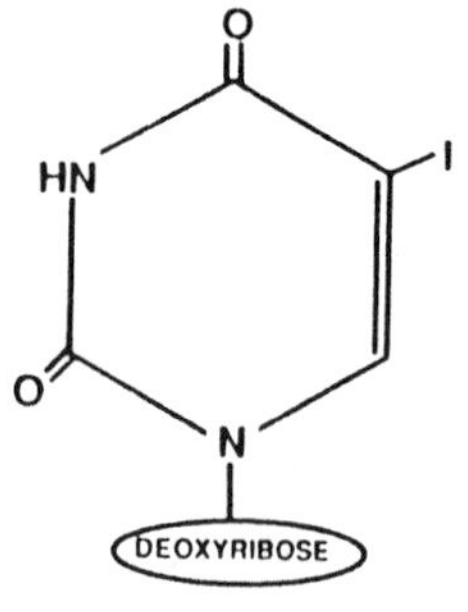

5 - iododeoxyuridine

Fig. 9. The halogenated pyrimidines used clinically as radiosensitizers.

randomized trial was done in the late 1960s involving BUDR in patients with advanced head and neck cancers *(3)*. No improvement in local control was obtained. However, significant increased acute mucositis within irradiated tissues perfused with analog was seen. These results are not surprising, since normal mucosal epithelium is a tissue with significant proliferation rate. These disappointing initial results diminished clinical enthusiasm for these agents *(44)*.

A number of advancements have been made since this initial clinical study. The use of IUDR, which is less toxic owing to less photosensitization, became popular. Monoclonal antibody (MAb) against the halogenated pyrimidines has allowed the quantitation of analog uptake into tumors from biopsy specimens taken after administration and has resulted in the ability to correlate clinical outcome with degree of uptake *(23)*. In addition, since continuous iv infusion was shown to yield similar serum plasma levels as intra-arterial infusion, the route of administration became considerably less cumbersome, and is achievable on an outpatient basis *(44)*. These factors led to a renewed interest in IUDR and BUDR as radiosensitizers, and modern clinical trials have followed.

High-grade brain tumors, which are proliferating, but surrounded by normal nonproliferating brain, form the correct paradigm in which to test the ability of the halogenated pyrimidines to improve the therapeutic ratio in clinical radiotherapy. In a phase II study at the National Cancer Institute, 45 patients with glioblastoma multiforme (GBM) were treated with iv IUDR and hyperfractionated radiation therapy *(29)*. The 2-yr actuarial survival was only 9%, and the median survival was 11 mo, representing no improvement over historic controls treated with radiation alone. Acute reactions were not increased over that seen in historic controls. Two patients had tumor biopsies after IUDR infusion, and peak thymidine replacement by the halogenated analog was only 4%.

In a separate phase II trial, 173 patients with GBM received BUDR and radiation therapy as well as adjuvant chemotherapy consisting of procarbazine, CCNU, and vincristine *(62)*. Median survival was 14 mo, and interestingly, the patients who received a higher BUDR dose had significantly longer time to failure and improved median survival than those receiving lower doses. However, the influence of BUDR dose is clouded by the many other variables present in this study. Based on multivariate analysis, no significance was seen with regard to BUDR dose. No postinfusion biopsy data were presented.

The feasibility of safely administering the halogenated pyrimidines with radiation therapy has been established, but adequate incorporation of analog remains to be achieved. Until improved incorporation can be achieved with acceptable toxicity, the benefit to overall therapy remains to be demonstrated.

8. CONCLUSION

There have been many advances in cancer treatment over the past few decades. The radiation oncologist has seen an explosion of knowledge that has resulted in an alteration of traditional concepts of cell killing and new possibilities for improving the therapeutic ratio. Although the true victories in clinical oncology have been few, the benefits of altered fractionation schemes, the molecular mechanisms of programmed cell death, the effects of radiation-inducible cytokines, the feasibility of radiosensitizers, as well as the combined therapeutic effects of radiation and chemotherapy all provide hope for future successes in cancer therapy.

REFERENCES

1. Ainsworth, E. J. and H. B. Chase. 1959. Effect of microbial antigens on irradiation mortality in mice. *Proc. Soc. Exp. Biol. Med.* **102:** 483.
2. Anscher, M., I. Crocker, and R. Jirtle. 1990. Transforming growth factor beta 1 expression in irradiated liver. *Radiat. Res.* **122:** 77–85.
3. Bagshaw, M. A., S. Doggett, K. C. Smith, H. S. Kaplan, and T. S. Nelsen. 1967. Intra-arterial 5-bromodeoxyuridine and x-ray therapy. *Am. J. Roentgenology* **99:** 886–894.
4. Baker, S. J., E. R. Fearon, J. M. Nigro, S. R. Hamilton, A. C. Preisinger, J. M. Jessup, P. vanTuinen, D. H. Ledbetter, D. F. Barker, Y. Nakamura, R. White, and B. Vogelstein. 1989. Chromosome 17 deletions and p53 gene mutations in colorectal carcinomas. *Science* **244:** 217–221.
5. Bianco, J., F. Applebaum, J. Nemunaitis, J. Almgren, F. Andrews, P. Kettner, et al. 1991. Phase I-II trial of pentoxifylline for the prevention of transplant-related toxicities following bone marrow transplantation. *Blood* **78:** 1205–1211.
6. Blount, P. L., S. Ramel, W. H. Raskind, R. C. Haggitt, C. A. Sanchez, P. J. Dean, P. S. Rabinovitch, and B. J. Reid. 1991. 17p Allelic deletions and p53 protein overexpression in Barrett's adenocarcinoma. *Cancer Res.* **51:** 5482–5486.
7. Brune, B., P. Hartzell, P. Nicotera, and S. Orrenius. 1991. Spermine prevents endonuclease activation and apoptosis in thymocytes. *Exp. Cell Res.* **195:** 323–329.
8. Budd, G. T., S. Green, L. H. Baker, E. P. Hersh, J. K. Weick, and C. K. Osborne. 1991. A southwest oncology group phase II trial of recombinant tumor necrosis factor in metastatic breast cancer. *Cancer* **68:** 1694,1695.
9. Carney, P. and S. Dean. 1990. Transforming growth factor beta: A promotor of late connective tissue injury following radiotherapy? *Br. J. Radiol.* **63:** 620–623.
10. Carde, P. ,and F. Laval. 1981. Effects of cis-dichlorodiammine platinum II and X-rays on mammalian cell survival. *Int. J. Radiat. Oncol. Biol. Phys.* **7:** 929–933.
11. Coleman, N. C., D. J. Glover, and A. T. Turrisi. 1990. Radiation and chemotherapy sensitizers and protectors, in *Cancer Chemotherapy: Principles and Practice* (Chabner, B. A. and J. M. Collins, eds.), Lippincott, Philadelphia, PA, pp. 424–448.
12. Coughlin, C. T. and R. C. Richmond. 1989. Biologic and clinical developments of cisplatin combined with radiation: concepts, utility, projections for new trials, and the emergence of carboplatin. *Semin. Oncol.* **16:** 31–43.

13. Coutard, H. 1932. Roentgen therapy of epitheliomas of the tonsillar region, hypopharynx and larynx from 1920 to 1926. *Am. J. Roentgenology* **28:** 313–331.
14. Cox, J. D., N. Azarnia, R. W. Byhardt, K. H. Shin, B. Emami, and T. F. Pajak. 1990. A Randomized phase I/II trial of Hyperfractionated radiation therapy with total doses of 60. 0 Gy to 79. 2 Gy: possible survival benefit with ≥69.6 Gy in favorable patients with Radiation Therapy Oncology Group stage III non-small cell lung carcinoma: Report of Radiation Therapy Oncology Group 83-11. *J. Clin. Oncol.* **8:** 1543–1555.
15. Cox, J. D., T. F. Pajak, V. A. Marcial, G. E. Hanks, M. Mohiuddin, K. K. Fu, R. W. Byhardt, and P. Rubin. 1990. Dose-response for local control with hyperfractionated radiation therapy in advanced carcinomas of the upper aerodigestive tracts: Preliminary report of Radiation Therapy Oncology Group protocol 83-13. *Int. J. Rad. Oncol. Biol. Phys.* **18:** 515–521.
16. Datta, N. R., A. D. Choudhry, S . Gupta, and A. K. Bose. 1989. Twice a day versus once a day radiation therapy in head and neck cancer. *Int. J. Rad. Oncol. Biol. Phys.* **17(Suppl. 1):** 132,133.
17. Djordjevic, B. and W. Szybalski. 1960. Genetics of cell lines III. Incorporation of 5-bromo and 5-iododeoxyuridine into the deoxyribonucleic acid of humancells and its effect of radiation sensitivity. *J. Exp. Med.* **112:** 509–531.
18. Douple, E. B. and R. C. Richmond. 1978. Platinum complexes as radiosensitizers of hypoxic mammalian cells. *Br. J. Cancer* **37:** 98–102.
19. Economou, J. S. 1993. Tumor necrosis factor, in *Cancer Medicine* (Holland, J., F., E. Frei, R. C. Bast, D. W. Kufe, D. L. Morton, and R. R. Weichselbaum, eds.), Lea and Febiger, Philadelphia, pp. 937–940.
20. Edsmyr, F., L. Andersson, P. L. Esposti, B. Littbrand, and B. Nilsson. 1985. Irradiation with multiple small fractions per day in urinary bladder cancer. *Radiother. Oncol.* **4:** 197–203.
21. Eichorn, H. J. 1981. Different fractionation schemes tested by histological examination of autopsy specimens from lung cancer patients. *Br. J. Radiol.* **54:** 132–135.
22. Elkind, M. M., G. H. Sutton, W. B. Moses, T. Alescio, and R. W. Swain. 1965. Radiation response of mammalian cells grown in culture. *Radiation Res.* **25:** 359–376.
23. Epstein, A. H., J. A. Cook, T. Goffman, and E. Glatstein. 1992. Tumour radiosensitization with the halogenated pyrimidines 5-bromo- and 5-iododeoxyuridine. *Br. J. Radiol.* **Suppl. 24:** 209–214.
24. Fajardo, L. F. and M. Berthrong. 1988. Vascular lesions following radiation. *Pathol. Ann.* **23:** 297–330.
25. Feldman, E. R., E. T. Creagan, D. J. Schaid, and D. L. Ahmann. 1992. Phase II trial of recombinant tumor necrosis factor in disseminated malignant melanoma. *Am. J. Clin. Oncol.* **15:** 256–259.
26. Fisher, D. E. 1994. Apoptosis in cancer therapy: crossing the threshold. *Cell* **78:** 539–542.
27. Ganesan, A. K. and K. C. Smith. 1968. Dark recovery processes in *Escherichia coli* irradiated with ultraviolet light; Effect of rec$^-$ mutations on liquid-holding recovery. *J. Bacteriol.* **96:** 365–373.
28. Glucksmann, A. 1951. Cell deaths in normal vertebrate ontogeny. *Biol. Rev.* **26:** 59.
29. Goffman, T. E., L. J. Dachowski, H. Bobo, E. H. Oldfield, M. Steinberg, J. Cook, et al. 1992. Long-term follow-up on national cancer institute phase I/II study of glioblastoma multiforme treated with iododeoxyuridine and hyperfractionated irradiation. *J. Clin. Oncol.* **10:** 264–268.
30. Goz, B. 1978. The effects of incorporation of 5-halogenated deoxyuridines into the DNA of eukaryotic cells. *Pharmacol. Rev.* **29:** 249–272.
31. Hahn, G. M. and J. B. Little. 1972. Plateau phase cultures of mammalian cells: An *in vitro* model for human cancer. *Curr. Top. Radiation Res.* **8:** 39–71.
32. Hall, E. J., ed. 1994. Cell survival curves, in *Radiobiology for the Radiologist.* Lippincott, Philadelphia, PA, pp. 29–43.

33. Hallahan, D. E., A. Friedman-Haimovitz, D. W. Kufe, Z. Fuks, and R. R. Weichselbaum. 1993. The role of cytokines in radiation oncology, in *Important Advances in Oncology* (DeVita, V. T., Hellman, S., and Rosenberg, S. A., eds.), Lippincott, Philadelphia, PA, pp. 71–80.
34. Hallahan, D. E., H. Mauceri, L. P. Seung, E. J. Dunphy, J. D. Wayne, N. N. Hanna, A. Toledano, S. Hellman, D. W. Kufe, and R. R. Weichselbaum. 1995. Spatial and temporal control of gene therapy using ionizing radiation. *Nature Med.* **1:** 786–791.
35. Hallahan, D. E., D. R. Spriggs, M. A. Beckett, D. W. Kufe, and R. R. Weichselbaum. 1989. Increased tumor necrosis factor α mRNA after cellular exposure to ionizing radiation. *Proc. Natl. Acad. Sci. USA* **86:** 10,104–10,107.
36. Hallahan, D. E., E. E. Vokes, S. J. Rubin, S. O'Brien, B. Samuels, S. Vijayakumar, et al. 1995. Phase I dose-escalation study of tumor necrosis factor-alpha and concomitant radiation therapy. *Cancer J. Sci. Am.* **1:** 204–209.
37. Halpern, H., C. Yu, M. Peric, E. Barth, D. J. Grdina, and B. A. Teicher. 1996. Oxymetry deep in tissues with low frequency electron paramagnetic resonance. *Proc. Natl. Acad. Sci. USA* **91:** 13,047–13,051.
38. Hellman, S. 1975. Cell kinetics, models, and cancer treatment—Some principles for the radiation oncologist. *Radiology* **114:** 219–223.
39. Herskovic, A., K. Martz, M. A. Al-Sarraf, L. Leichman, J. Brindle, V. Vaitkevicius, J. Cooper, R. Byhardt, L. Davis, and B. Emami. 1992. Combined chemotherapy and radiotherapy compared with radiotherapy alone in patients with cancer of the esophagus. *New Engl. J. Med.* **326:** 1593–1598.
40. Holler, E., H. Kolb, A. Moller, J. Kempeni, S. Liesenfeld, H. Pechumer, et al. 1990. Increased serum levels of TNF precede major complications of bone marrow transplantation. *Blood* **75:** 1011–1016.
41. Horiot, J. C., R. Le Fur, T. N'Guyen, C. Chenal, S. Schraub, S. Alfonsi, G. Gardani, W. Van Den Bogaert, S. Danczak, M. Bolla, M. Van Glabbeke, and M. De Pauw. 1993. Hyperfractionated versus conventional fractionation in oropharyngeal carcinoma: final analysis of a randomized trial of the EORTC cooperative group of radiotherapy. *Radiother. Oncol.* **21:** 231–241.
42. Iggo, R., K. Gatter, J. Barter, D. Lane, and A. Harris. 1990. Increased expression of mutant forms of p53 oncogene in primary lung cancer. *Lancet* **335:** 675–679.
43. Kerr, J. F. R., A. H. Wyllie, and A. R. Currie. 1972. Apoptosis: a basic biological phenomenon with wide-ranging implications in tissue kinetics. *Br. J. Cancer* **26:** 239–257.
44. Kinsella, T. J., J. B. Mitchell, A. Russo, G. Morstyn, and E. Glatstein. 1984. The use of halogenated thymidine analogs as clinical radiosensitizers: Rationale, current status, and future prospects: non-hypoxic cell sensitizers. *Int. J. Radiation Oncol. Biol. Phys.* **10:** 1399–1406.
45. Larrick, J. W. 1990. Cytotoxic mechanisms of tumor necrosis factor-alpha. *FASEB J.* **4:** 3215–3223.
46. Lea, D. E., and D. G. Catcheside. 1945. The mechanism of induction by radiation of chromosome aberrations in *Tradescantia*. *J. Genet.* **44:** 216–245.
47. Little, J. B. 1969. Repair of sub-lethal and potentially lethal radiation damage in plateau phase cultures of human cells. *Nature* **224:** 804–806.
48. Matthews, N., M. Neale, S. Jackson, J. M. Stark. 1987. Tumor cell killing by TNF inhibited by anaerobic conditions, free radical scavengers and inhibitors of arachidonate metabolism. *Immunology* **62:** 153–155.
49. McConkey, D. J., M. Hartzell, M. Jondal, and S. Orrenius. 1989. Inhibition of DNA fragmentation in thymocytes and isolated thymocyte nuclei by agents that stumulate protein kinase C. *J. Biol. Chem.* **264:** 13,399–13,402.
50. McConkey, D. J., P. Nicotera, P. Hartzell, G. Bellomo, A. H. Wyllie, and S. Orrenius. 1989. Glucocorticoids activate a suicide process in thymocytes through an elevation of cytosolic Ca^{++} concentration. *Arch. Biochem. Biophys.* **269:** 365–370.

51. McConkey, D. J., and S. Orrenius. 1991. Cellular signaling in thymocyte apoptosis, in *Apoptosis: The Molecular Basis of Cell Death* (Tomei, L. D. and F. O. Cope, eds.), Cold Spring Harbor Laboratory, Cold Spring Harbor, NY, pp. 227–246.
52. Meyn, R. E., L. C. Stephens, D. W. Voehringer, M. D. Story, N. Mirkovic, and L. Milas. 1993. Biochemical modulation of radiation-induced apoptosis in murine lymphoma cells. *Radiation Res.* **136:** 327–334.
53. Munro, T. R. 1970. The relative radiosensitivity of the nucleus and cytoplasm of the Chinese hamster fibroblasts. *Radiat. Res.* **42:** 451–470.
54. Naume, B., R. Shalaby, W. Lesslauer, and T. Espevik. 1991. Involvement of the 55- and 75-kDa tumor necrosis factor receptors in the generation of lymphokine-activated killer cell activity and proliferation of natural killer cells. *J. Immunol.* **146:** 3045–3048.
55. Neta, R. 1988. The role of cytokines in radiation protection. *Pharm. Ther.* **39:** 261–266.
56. Neta, R., J. J. Oppenheim, R. D. Schreiber, R. Chizzonite, G. D. Ledney, and T. J. MacVittie. 1991. Role of cytokines (interleukin 1, tumor necrosis factor, and transforming growth factor beta) in natural and lipopolysaccharide-enhanced radioresistance. *J. Exp. Med.* **173:** 1177–1182.
57. Nigro, N. D., V. K. Vaitkevicius, and J. B. Considine. 1989. Dynamic management of squamous cell cancer of the anal canal. *Invest. New Drugs* **7:** 83–89.
58. Old, L. J. 1985. Tumor necrosis factor. *Science* **230:** 630–632.
59. Old, L. J. and D. R. Spriggs. 1990. Clinical trials with tumor necrosis factor, in *Tumor Necrosis Factor: Structure, Mechanism of Action, Role in Disease and Therapy* (Bonavida, B. B. and G. Granger, eds.), Karger, Basel, pp. 1–30.
60. Patrick, M. A., R. H. Haynes, and R. B. Uretz. 1964. Dark recovery phenomena in yeast. *Radiation Res.* **21:** 144–163.
61. Peters, L. J., K. K. Ang, and H. D. Thames. 1992. Altered fractionation schedules, in *Principles and Practice of Radiation Oncology* (Perez, C. A. and L. W. Brady, eds.), Lippincott, Philadelphia, pp. 97–113.
62. Phillips, T. L., V. A. Levin, D. K. Ahn, P. H. Gutin, R. L. Davis, C. B. Wilson, et al. 1991. Evaluation of bromodeoxyuridine in glioblastoma multiforme: A northern California cancer center phase II study. *Int. J. Radiation Oncol. Biol. Phys.* **21:** 709–714.
63. Pinto, L. H. J., P. C. V. Canary, C. M. M. Araujo, S. C. Bacelar, and L. Souhami. 1991. Prospective randomized trial comparing hyperfractionated versus conventional radiotherapy in stages III and IV oropharyngeal carcinoma. *Int. J. Rad. Oncol. Biol. Phys.* **21:** 557–562.
64. Piver, M. S., M. Khalil, and L. J. Emrich. 1989. Hydroxyurea plus pelvic irradiation versus placebo plus pelvic irradiation in nonsurgically staged stage IIIB cervical cancer. *J. Surg. Oncol.* **42:** 120–125.
65. Puck, T. T. and P. I. Marcus. 1956. Action of x-rays on mammalian cells. *J. Exp. Med.* **103:** 653–666.
66. Read, J. 1959. *Biology of Vica Faba in Relation to the General Problem.* Blackwell Scientific Publications, Oxford.
67. Rockwell, S. 1983. Effects of mitomycin C alone and in combination with x-rays on EMT6 mouse mammary tumors *in vivo*. *J. Natl. Cancer Inst.* **71:** 765–771.
68. Sersa, G. V. W. and L. Milas. 1988. Anti-tumor effects of tumor necrosis factor alone or combined with radiotherapy. *Int. J. Cancer* **42:** 129–134.
69. Sherman, M. L., R. Datta, D. E. Hallahan, R. R. Weichselbaum, and D. W. Kufe. 1991. Tumor necrosis factor gene expression is transcriptionally and posttranscriptionally regulated by ionizing radiation in human myeloid leukemia cells and peripheral blood monocytes. *J. Clin. Invest.* **87:** 1794–1797.
70. Spriggs, D. R., M. L. Sherman, E. Frei, and D. W. Kufe. 1990. Clinical trials with tumor necrosis factor, in *Tumor Necrosis Factor: Structure, Mechanism of Action, Role in Disease and Therapy* (Bonavida, B. B., and G. Granger, eds.), Karger, Basel, pp. 233–239.
71. Stehman, F. B., B. N. Bundy, G. Thomas, H. M. Keys, G. d'Ablaing, W. C. Fowler, R. Mortel, and W. T. Creasman. 1993. Hydroxyurea versus misonidazole with radiation in

cervical carcinoma: Long-term follow-up of a gynecological oncology group trial. *J. Clin. Oncol.* **11:** 1523–1528.
72. Story, D. J., S. P. Stephens, S. P. Tomasovic, and R. E. Meyn. 1992. A role for calcium in regulating apoptosis in rat thymocytes irradiated *in vitro*. 1992. *Int. J. Radiat. Biol.* **61:** 243–251.
73. Sugarman, B. J., B. B. Aggarwal, R. E. Hass, I. S. Figari, M. A. Palladino, and H. M. Shepard. 1985. Recombinant tumor necrosis factor-alpha: effects on proliferation of normal and transformed cells *in vitro*. *Science* **230:** 943–945.
74. Szybalski, W. 1974. X-ray sensitization by halopyrimidines. *Cancer Chemother. Rep.* **58:** 539–557.
75. Tartaglia, L. A., R. F. Weber, I. S. Figari, C. Reynolds, M. A. Palladino, and D. V. Goeddel. 1991. The two different receptors for tumor necrosis factor mediate distinct cellular responses. *Proc. Natl. Acad. Sci. USA* **88:** 9292–9296.
76. Thompson, C. B. 1995. Apoptosis in the pathogenesis and treatment of disease. *Science* **267:** 1456–1462.
77. Varley, J. M., W. J. Brammar, D. P. Lane, J. E. Swallow, C. Dolan, and R. A. Walker. 1991. Loss of chromosome 17p13 sequences and mutation of p53 in human breast carcinomas. *Oncogene* **6:** 413–417.
78. Vokes, E., E. and R. R. Weichselbaum. 1990. Concomitant chemoradiotherapy: rationale and clinical experience in patients with solid tumors. *J. Clin. Oncol.* **8:** 911–934.
79. Wasserman, J., B. Petrini, G. Wolk, I. Vedin, U. Glas, H. Blomgren, et al. 1991. Cytokine release from mononuclear cells in patients irradiated for breast cancer. *Anticancer Res.* **11:** 461–464.
80. Weichselbaum, R. R. 1984. The role of DNA repair processes in the response of human tumors to fractionated radiotherapy. *Int. J. Rad. Oncol. Biol. Phys.* **10:** 1127–1134.
81. Weichselbaum, R. R., D. E. Hallahan, M. A. Beckett, H. J. Mauceri, H. Lee, V. P. Sukhatme, et al. 1994. Gene Therapy treated by radiation preferentially radiosensitizes tumor cells. *Cancer Res.* **54:** 4266–4269.
82. Weichselbaum, R. R., A. Malcolm, and J. B. Little. 1982. Fraction size and the repair of potentially lethal damage in a human melanoma cell line. *Radiology* **142:** 225–227.
83. Weichselbaum, R. R., A. Schmit, and J. B. Little. 1982. Cellular factors influencing radiocurability of human malignant tumors. *Br. J. Cancer* **45:** 10–16.
84. Williams, G. T. and C. A. Smith. 1993. Molecular regulation of apoptosis: genetic controls on cell death. *Cell* **74:** 777–779.
85. Withers, H. R. 1992. Biological basis of radiation therapy, in *Principles and Practice of Radiation Oncology* (Perez, C. A. and L. W. Brady, eds.), Lippincott, Philadelphia, PA, pp. 64–96.
86. Witte, L., Z. Fuks, A. Haimovits-Friedman, I. Vlodarsky, D. S. Goodman, and A. Eldor. 1989. Effects of radiation on the release of growth factors from cultured bovine, porcine and human endothelial cells. *Cancer Res.* **49:** 5066–5072.
87. Wittes, R. E. and S. M. Hubbard. 1992. Chemotherapy: the properties and uses of single agents, in *Manual of Oncologic Therapeutics* (Wittes, R., E., ed.), Lippincott, Philadelphia, PA, pp. 66–121.
88. Wyllie, A. H. 1987. Apoptosis: cell death in tissue regulation. *J. Pathol.* **153:** 313–316.
89. Wyllie, A. H. 1993. Apoptosis (The 1992 Frank Rose memorial lecture). *Br. J. Cancer* **67:** 205–208.
90. Yonish-Rouach, E., D. Resnitzky, J. Lotem, L. Sachs, A. Kimichi, and M. Oren. 1991. Wild-type p53 induces apoptosis of myeloid leukaemic cells that is inhibited by interleukin-6. *Nature* **352:** 345–347.
91. Younes, M., R. M. Lebovitz, L. V. Lechago, and J. Lechago. 1993. p53 Protein accumulation in Barrett's metaplasia, dysplasia, and carcinoma: a follow-up study. *Gastroenterology* **105:** 1637–1642.
92. Zimmerman, R. J., A. Chan, and S. A. Leadon. 1989. Oxidative damage in murine tumor cells treated *in vitro* by recombinant human tumor necrosis factor. *Cancer Res.* **49:** 1644–1648.

Index

Most mammalian genes and proteins (but not *p53* and *p21*$^{WAF1/CIP1}$) are listed under "Human/mammalian" – some species-specific genes are listed under that species (e.g., mouse). Genes and their cognate gene products are listed either by gene or protein names, but not both